Structural
Engineering
Handbook

OTHER McGRAW-HILL HANDBOOKS OF INTEREST

Baumeister · Marks' Standard Handbook for Mechanical Engineers
Brady and Clauser · Materials Handbook
Brater · Handbook of Hydraulics
Callender · Time-Saver Standards for Architectural Design Data
Carrier Air Conditioning Company · Handbook of Air Conditioning System Design
Conover · Grounds Maintenance Handbook
Considine · Energy Technology Handbook
Crocker and King · Piping Handbook
Croft, Carr, and Watt · American Electricians' Handbook
Davis and Sorensen · Handbook of Applied Hydraulics
Emerick · Handbook of Mechanical Specifications for Buildings and Plants
Emerick · Heating Handbook
Emerick · Troubleshooters' Handbook for Mechanical Systems
Fink and Beaty · Standard Handbook for Electrical Engineers
Foster · Handbook of Municipal Administration and Engineering
Gieck · Engineering Formulas
Harris · Handbook of Noise Control
Harris and Crede · Shock and Vibration Handbook
Havers and Stubbs · Handbook of Heavy Construction
Heyel The Foreman's Handbook
Hicks · Standard Handbook of Engineering Calculations
Higgins and Morrow · Maintenance Engineering Handbook
King and Brater · Handbook of Hydraulics
La Londe and Janes · Concrete Engineering Handbook
Leonards · Foundation Engineering
Lund · Industrial Pollution Control Handbook
Manas · National Plumbing Code Handbook
Mantell · Engineering Materials Handbook
Merritt · Building Construction Handbook
Merritt · Standard Handbook for Civil Engineers
Merritt · Structural Steel Designers' Handbook
Myers · Handbook of Ocean and Underwater Engineering
O'Brien · Contractor's Management Handbook
Peckner and Bernstein Handbook of Stainless Steels
Perry · Engineering Manual
Rossnagel · Handbook of Rigging
Smeaton · Switchgear and Control Handbook
Stanair · Plant Engineering Handbook
Streeter · Handbook of Fluid Dynamics
Timber Engineering Co. Timber Design and Construction Handbook
Tuma · Engineering Mathematics Handbook
Tuma · Handbook of Physical Calculations
Tuma · Technology Mathematics Handbook
Urquhart · Civil Engineering Handbook
Waddell · Concrete Construction Handbook
Woods · Highway Engineering Handbook

Structural Engineering Handbook

Edited by

Edwin H. Gaylord, Jr.
**Professor of Civil Engineering, Emeritus
University of Illinois, Urbana**

Charles N. Gaylord
**Professor of Civil Engineering, Emeritus
University of Virginia**

Second Edition

McGRAW-HILL BOOK COMPANY

New York St. Louis San Francisco Auckland Bogotá
Düsseldorf Johannesburg London Madrid
Mexico Montreal New Delhi Panama
Paris São Paulo Singapore
Sydney Tokyo Toronto

Library of Congress Cataloging in Publication Data

Gaylord, Edwin Henry.
    Structural engineering handbook.

    Includes bibliographies and index.
    1.   Structural engineering—Handbooks, manuals,
etc.   I.   Gaylord, Charles N.   II.   Title.
TA635.G3   1979   624'.1   78-25705

1234567890   KPKP   7865432109

The editors for this book were Harold B. Crawford and Joseph Williams,
the designer was Naomi Auerbach, and the production supervisor
was Thomas G. Kowalczyk. It was set in Caledonia by University Graphics, Inc.

Printed and bound by The Kingsport Press.

Contents

Preface   xiii
Contributors   xv

Section 1   STRUCTURAL ANALYSIS **1-1**

Part 1.   Elastic Systems   *J. Graham and W. G. Godden*   1-1

Degrees of Freedom—Principles of Analysis—Equilibrium of Forces; **Energy Theorems:** Classification—The Principle of Virtual Work—The Principle of Minimum Potential Energy—The Minimum-Complementary-Potential-Energy Theorem—Castigliano's Theorems—The Reciprocal Theorem—Inelastic Analysis; **Classification of Structural Types:** External Forces and Reactions—Independent Force Components—Pin-jointed Plane and Space Frames—Stiffjointed Frameworks; **Methods of Analysis: Forces in Statically Determinate Structures:** Pin-jointed Lattice Frameworks—Beam Systems; **Deflections:** Williot-Mohr Diagram—Numerical-Integration Method for Beams; **Deflections by Unit-load Method:** General Form of Solution—Linearly Elastic Pin-jointed Frameworks—Linearly Elastic Beam Deflections—Linearly Elastic Beam-type Structures—Relative Deflections—Linearly Elastic Structures: General Case—Nonlinearly Elastic and Plastic Structures—Deflection Due to Self-straining—The Complementary-Energy Method and Castigliano's Theorem; **Influence Lines:** Statically Determinate Systems—Statically Indeterminate Systems; **Statically Indeterminate Structures:** Choice of Method—Flexibility Coefficients—The Unit-load Method—Castigliano's Theorem—The Three-Moment Equation—The Column Analogy—Stiffness Coefficients—The Unit-Displacement Method—Slope-Deflection Equations—Moment-Distribution Method—The Minimum-Complementary-Energy Method—The Minimum-Potential-Energy Method; **Matrix Methods:** Deflections—Statically Determinate Structures—Statically Indeterminate Structures—Choice of Method—Matrix Force Method—Choice of Redundancies—Self-straining—Stiffness Matrix—Direct Calculation of Stiffness Matrix—The Matrix-Displacement Method—Condensed Stiffness Matrix—Self-straining—Nonlinear Structures; **Elements of Matrix Algebra:** Definitions—Matrix Operations

Part 2.   The Finite-Element Method   *W. C. Schnobrich and J. Graham*   1-96

Discretization of the Structure—Guidelines for Selection of Grid—Element Models—Plane-Stress and Plain-Strain Elements—Plane Stress Analysis—Beam and Plate Bending—Shells and Combined Direct Stress and Bending—Three Dimensional Problems—Element Stiffness Matrix and Nodal Force Vector—Equilibrium Equations for the Assemblage—Solution for Displacements—Solution for Element Strains and Stresses

Section 2   COMPUTER APPLICATIONS IN STRUCTURAL ENGINEERING   *Steven J. Fenves* 2-1

Basic Concepts—The Structural Design Process—The Program-Development Process; **Program-Development Tools:** Structured Programming—Programming Languages—Flowcharting—Decision Tables—Tools for Organizing Programs—Tools for Organizing Data—Computer Hardware; **Interaction with Computers:** Modes of Interaction—Media of Interaction—Level of Communication; **Program Types and Sources:** Program Types—Sources of Programs—Other Services; **Typical Applications:** Analysis—Proportioning—Detailing—Preparation of Final Documents

Section 3   EARTHQUAKE-RESISTANT DESIGN   *N. M. Newmark and W. J. Hall* 3-1

Response of Simple Structures to Earthquake Motions—Earthquake Motions; **Response Spectra:** Elastic Systems—Design Response Spectra—Inelastic Systems—Multi-Degree-of-Freedom Systems, Use of Response Spectra, Use of Time History—Motions in Several Directions; **Computation of Period of Vibration:** Fundamental Mode—Higher Modes—Modal Participation Factors—Spring Constant for Equivalent Shear Beam; **Design:** General Considerations—Effects of Design on Behavior—Design Lateral Forces—Seismic Forces for Overturning Moments and Shear Distribution—Damping—Gravity Loads—Vertical and Horizontal Excitation—Unsymmetrical Structures in Torsion—Curtain-Wall Buildings—Core Walls—Parts of Buildings—Effects of Soil Conditions—Detailing and Quality Control—Cost

Section 4   FATIGUE AND BRITTLE FRACTURE   *W. H. Munse* 4-1

Fatigue of Structural Steel: Significance of Fatigue—Fatigue of Structural Steels—Factors Affecting Fatigue Strength—Structural Members—Fatigue Resistance of Beams and Girders—Riveted Connections—Bolted Connections—Welded Connections—Design for Fatigue—Type of Detail for Fatigue Design—Basic Allowable Fatigue Design Stress—Protecting Against Fatigue; **Brittle Fracture of Structural Steel:** Sigificance of Brittle Fractures—Examples of Brittle Fracture—Initiation, Propagation, and Arrest—Factors Affecting Brittle Fracture—Susceptibility of Structural Steels to Strain Aging—Designing against Brittle Fracture—Fabrication—Inspection

Section 5   SOIL MECHANICS AND FOUNDATIONS 5-1

Part 1.   Soil Mechanics   *H. G. Larew* 5-1

Nature and Causes of Soil Deformation: Settlement—Frost Movements—Shrinkage—Subsidence—Soil Deformation—Time-dependent Deformation; **Strength Properties of Soils:** Compressive Strength—Effect of Confining Pressures—Transient and Repeated Loads; **Factors Affecting Bearing Pressure:** Allowable Bearing Pressure—Permissible Settlement—Elastic-plastic Deformation—Shear Failures—Consolidation—Time Rate of Settlement; **Cuts and Embankments:** Slope Stability; **Compaction and Permeability:** Compaction—Permeability; **Appendix**

Part 2.   Soil Exploration   *Thomas H. Thornburn* 5-23

General Foundation Conditions: Sources of Information—Glacial Materials—Wind-laid Materials—Water-laid Materials—Organic Soils—Residual Soils—Frozen Soils; **Exploratory Procedures:** Test Pits—Borings—Sampling—Borehole Cameras—Boring Reports—Standard Penetration Tests—In Situ Strength Tests—Correlations with Surficial Geology—Geophysical Surveys—Load

Tests—Special Observations—Preliminary Site Inspection—Fact-finding Survey—Borings

Part 3.   Retaining Structures and Foundations   *Herbert O. Ireland*   5-43

Earth Pressures: Stresses in Earth Mass—Rankine's Theory—Coulomb's Theory—Equivalent-Fluid Method—Trial-Wedge Method; **Retaining Walls:** Types and Behavior—Determining Earth Pressure—Bases on Piles or Piers—Bearing Capacity—Stability—Drainage—Other Considerations—Gravity Walls—Semigravity Walls—Cantilever Walls—Counterfort Walls—Joints; **Abutments; Bulkheads:** Forces on Bulkheads—Penetration of Piles—Anchorage; **Footing Foundations:** Footings on Clay—Footings on Sand—Footings on Silt and Loess; **Raft Foundations:** Raft on Clay—Rafts on Sand; **Pier Foundations:** Open Excavations—Drilled Piers—Piers on Clay—Piers on Sand—Caisson Foundations—Foundation Requirements; **Pile Foundations:** Pile-driving Equipment—Pile-driving Formulas—Pile Tests—Piles in sand—Piles in Clay—Settlement of Pile Foundations—Laterally Loaded Piles—Batter Piles—Lateral Stability of Poles—Guy Anchors—Foundations Subjected to Uplift—Improvement of Subsoil—Construction Problems

Section 6   DESIGN OF STEEL STRUCTURAL MEMBERS   *William J. LeMessurier, Hans William Hagen, and Lee C. Lim* **6-1**

Design Procedures—Types of Steel—Shapes; **Tension Members:** Concentrically Loaded Tension Members—Threaded Members—Member Types and Selection—Truss Members; **Compression Members:** Column Strength—Concentrically Loaded Columns—Effective Length—Amplification Factors and Frame Stability—Proportioning—Local Buckling—Lacing and Perforated Cover Plates—Tapered Columns—Slender Compression Elements; **Beams:** Allowable Stresses—Biaxial Bending—Shear—Deflection—Combined Bending and Compression; **Plate Girders:** Web—Flanges—Lengths of Flange Plates—Lateral Buckling—Requirements for Stiffeners—Combined Bending and Shear; **Welded Connections:** Welding Processes—Weld Classification—Weldability—Electrodes—Inspection—Fillet-welded Joints—Groove-welded Joints—Concentrically Loaded Connections—Beam Seat Connections—Stiffened Beam-Seat Connections—Framed Beam Connections—Moment-resistant Beam Connections; **Riveted and Bolted Connections:** Rivets—High-Strength Bolts—Installation of High-Strength Bolts—Inspection of High-Strength Bolts—Unfinished Bolts—Turned Bolts—Ribbed Bolts—Bearing Bolts—High-Strength Tension Control and Tension Set Bolts—Friction-type Connections—Bearing-type Connections—Behavior of Riveted and Bolted Connections—Allowable Stresses—Eccentrically Loaded Connections, Fasteners in Shear—Eccentrically Loaded Connections, Fasteners in Tension—Flexible Beam-Seat Connections—Stiffened Beam-Seat Connections—Framed Beam Connections—Moment-resistant Beam Connections—Pinned Connections; **Bearing Plates and Splices:** Beam Bearing Plates—Column Bases—Compression-Member Splices

Section 7   PLASTIC DESIGN OF STEEL FRAMES   *Lynn S. Beedle and T. V. Galambos* **7-1**

Inelastic Bending—Indeterminate Structures; **Analysis:** Theorems—Statical Method—Mechanism Method—Moment Check—Instantaneous Center—Distributed Loads—Moment Balancing; **Deflection Analysis:** Deflection at Ultimate Load—Deflection at Working Load; **Design Requirements:** Specifications—Loads and Forces—General Design Procedure—Preliminary Design—Analysis; **Secondary Design Considerations:** Axial Force—Lateral Bracing—Local Buckling—Shear—Frame Instability; **Connections:** Corner Connections—Interior Beam-to-Column Connections

Section 8   FABRICATION AND ERECTION OF STRUCTURAL STEEL   8-1

Part 1.   Fabrication   *C. F. Harris*   8-1

Drawing-Room Operations—Material Orders—Erection Diagrams—Shop Drawings—Shop Operations—Fitting and Assembling—Suggestions for Design Engineers

Part 2.   Erection   *D. B. Rees*   8-6

Drawings—Specifications—Consultation with Erector—Budget Cost Estimates—Bidding—Schedules; **Design Considerations:** Economy in Design—Lateral Stability—Girder Bridges—Splices—Geometric and Cambered Shapes—Stress Participation—Expansion and Construction Joints—Subpunching and Reaming—Bolting, Welding, and Other Fasteners—Weighing Reactions—Strain-gage Measurements—Erection Stresses—Maintenance—Storage and Shipment; **Equipment and Methods:** Equipment—Responsibility of Engineer and Contractor—Erection of Buildings—Erection of Bridges—Falsework—Erector's Responsibility

Section 9   DESIGN OF COLD-FORMED STEEL STRUCTURAL MEMBERS   *George Winter*   9-1

Thicknesses and Weights of Uncoated Steel Sheets—Materials—Shapes and Uses—Shear Diaphragms—Shear Strengths of Diaphragms—Folded-plate and Shell Roofs; **Design:** Basic Design Stresses—Section Properties—Thin Compression Elements—Compression Members—Flexural Members—Stiffened Compression Flanges—Beam Webs—Lateral Buckling of Beams—Combined Bending and Axial Compression—Bracing of Channels and Z's—Connections—Effects of Cold Forming on Steel Properties—Test Determination of Structural Performance

Section 10   DESIGN OF ALUMINUM STRUCTURAL MEMBERS   *John W. Clark*   10-1

Materials and Specifications: Shapes—Codes and Specifications—Characteristics of Aluminum Alloys; **Design of Tension Members:** Yielding and Fracture—Welded Tension Members; **Design of Compression Members:** Column Formulas—Lacing—Local Buckling of Plates, Legs, and Flanges in Edge Compression—Stiffeners for Flat Plates in Edge Compression—Local Buckling of Round Tubes in End Compression—Welded Compression Members; **Beams:** Yielding and Ultimate Strength—Lateral-Torsional Buckling—Local Buckling of Beams—Shear Strength of Beam Webs; **Plate Girders:** Lateral Buckling—Design of Web—Vertical Stiffeners—Longitudinal Stiffeners—Bearing Stiffeners; **Combined Loading:** Combined Bending and Axial Load—Plate Buckling under Combined Stress; **Connections:** Riveted Connections—Bolted Connections—Welded Connections; **Design for Repeated Loads**

Section 11   DESIGN OF REINFORCED CONCRETE STRUCTURAL MEMBERS   *Raymond C. Reese and Phil M. Ferguson*   11-1

Concrete—Reinforcement—Specifications, Codes, and Standards—Strength Design and Working Stress Design—ACI Load and Reduction Factors—Precision—Rectangular Beams—Continuity—Doubly Reinforced Beams—Tee Beams—Special Beam Shapes—Shear and Diagonal Tension—Development and Anchorage of Reinforcement—Splices—Bar Cutoffs and Bend Points—Deflection—Column Design—Combined Compression and Bending—Column Splices—Column With Biaxial Bending—Stairs—Wall Footings—Column Footings—Walls—Slabs—Structural Framing Systems

Section 12   DESIGN OF PRESTRESSED CONCRETE STRUCTURAL MEMBERS *T. Y. Lin and Paul Zia* **12-1**

Notation; Materials: Concrete—Steel—Grouting; Methods and Systems of Prestressing: Tensioning Methods—Pretensioning—Posttensioning Systems; Loss of Prestress: Elastic Shortening of Concrete—Creep—Shrinkage—Relaxation in Steel—Slippage of Tendons during Anchoring—Friction—Effective Prestress—Elongation of Tendons; Analysis for Flexure: Basic Concepts—Stress in Steel—Cracking Moment—Ultimate Moment—Composite Sections; Design for Flexure: Preliminary Design—Elastic Design—Ultimate Design—Balanced-load Design—Deflections; Shear, Bond, and Bearing: Principal Tension—Web Reinforcement—Prestress Transfer Bond—Anchorage; Typical Sections: Beam Sections—Span-Depth Ratios—Cable Layouts—Tendon Protection and Spacing—Partial Prestress—Combination of Prestressed and Reinforced Concrete; Continuous Beams: Continuous-beam C Lines—Load-balancing Method—Ultimate Strength of Continuous Beams

Section 13   CONCRETE CONSTRUCTION METHODS *Francis A. Vitolo* **13-1**

General Considerations—Formwork—Reinforcing Steel—Concrete—Embedded Items—Special Designs—Tolerances—Shop Drawings—Material Samples; Inspection: The Resident Engineer; Contract Documents: Preparation—Specifications—Intent—Scope of Work—Drawings

Section 14   COMPOSITE CONSTRUCTION *W. H. Fleischer, D. C. Frederickson, W. C. Hansell, and I. M. Viest* **14-1**

Definitions—Elastic Properties of Cross Section—Plastic Strength of Cross Section—Shear Connectors—Unsymmetrical Steel Sections—Negative-moment Sections—Deflections and Vibrations; Building Design: Assumptions—Design of Composite Beams—Steel Member Selection—Design of Shear Connectors—Design Example; Bridge Design: Assumptions—Design of Composite Beams—Steel Member Selection—Design of Shear Connectors—Design Example

Section 15   MASONRY CONSTRUCTION *Walter L. Dickey* **15-1**

Materials: Burned-Clay Units—Brick—Structural Clay Tile—Concrete Units—Mortar; Reinforced Masonry: Materials—Design—Allowable Stresses—Beams—Walls—Columns—Diaphragms; Tests and Inspection: Compressive Strength of Masonry; Detailing and Construction: Detailing—Concrete Foundations—Workmanship

Section 16   TIMBER STRUCTURES *Kenneth P. Milbradt* **16-1**

Structural Properties of Wood: Anisotropic Nature of Wood—Elastic Constants—Directional Strength Properties—Factors Affecting Strength—Working Stresses for Sawn Lumber—Glued-laminated Lumber—Plywood; Fasteners: Bolts—Split Rings and Shear Plates—Truss Plates; Beams: Flexure—Shear—Bearing—Deflections—Lateral Stability—Continuous Spans—Pitched and Tapered Beams; Columns: Solid Columns—Box Columns—Spaced Columns—Beam Columns; Trusses: Proportions—Design of Members—Deflections—Camber—Bracing—Trussed Joists; Arches: Three-hinged Tudor Arch—Two-hinged Arches; Shell Structures: Domes—Barrel Vaults—Hyperbolic Paraboloids

Section 17   ARCHES AND RIGID FRAMES   *Thomas C. Kavanagh (deceased) and Robert C. Y. Young* **17-1**

Nomenclature and Classification; **Analysis:** Assumptions—Kern Relationships—Arches and Closed Rings—Rigid Frames; **Design of Arches:** General Procedure—Preliminary Selection of Shape—Approximation for Special Shapes—Intermediate Design—Approximations of Whitney Data—Final Design—Unsymmetrical Arches—Ultimate Design of Concrete Arches; **Design of Frames:** Steel Frames—Concrete Rigid-frame Bridges—Design—Arched Bents, Continuous Arches on Elastic Piers; **Special Topics:** Second-order Theory—Interaction of Arch and Deck—Buckling of Arches—Laterally Loaded Arches and Frames—Skewed Barrel Arches and Rigid-frame Slabs; **Construction and Details:** Concrete Arches and Frames—Steel Arches and Rigid Frames—Economics

Section 18   BRIDGES **18-1**

Part 1.   Steel and Concrete Bridges   *Arthur L. Elliott* 18-1

Loads—Maximum Moments and Shears in Simple Spans—Positive Moments in Continuous Spans—Negative Moments in Continuous Spans—Shears in Continuous Spans—Impact—Wind—Other Loads—Grouping of Loads; Floor Systems—Concrete Floors—Steel Floors—Floor Beams; **Bearing and Expansion Details:** End Bearings—Expansion Hangers—Deck Expansion Joints; **Beam and Plate-Girder Bridges:** Beam Bridges—Plate-Girder Bridges—Composite Beam Bridges—Continuous Spans—Spacing—Lateral Systems—Deflection—Welded Plate Girders—Field Splices; **Truss Bridges:** Proportions—Loads and Stresses—Secondary Stresses—Truss Members—Lateral Forces; **Concrete Bridges:** Camber, Plastic Flow, and Shrinkage; **Slab Bridges:** Simple Spans—Continuous Spans—Design of Bents—Typical Details, Continuous Slabs; **T-beam Bridges:** Economics—Design of a T-beam Bridge—Design of Substructure—Typical Details; **Box-Girder Bridges:** Economics—Proportions—Design—Substructure; **Prestressed-Concrete Bridges:** Standard Sections—Stresses—Path of Prestressing Force—Friction Loss—Ultimate Load—Web Reinforcement—Uplift—Live-Load Deflection; **Bridge Railings:** Railing Design—Curbs and Sidewalks—Pedestrian Railings

Part 2.   Steel-plate-deck Bridges   *Roman Wolchuk* 18-103

Applications—Economic Considerations—Structural Behavior—Deck Plate—Rib Criteria—Design of Ribs—Design of Floor Beams—Stresses in Ribs and Floor Beams—Design of Closed Ribs; **Box Girders:** Analysis of Box Girders—Design in accordance with Linear Elastic Theory; **Nonlinear Analysis of Box Girders:** Effect of Imperfections on Behavior of Steel Plating—Unstiffened Plate Panel under Axial Compression—Stiffened Plate Panel under Axial Compression—Plate Panel in Shear—Load Bearing Diaphragm; **Construction Details:** Fabrication and Erection of Decks—Erection of Box Girders; **Wearing Surfaces:** Seal and Tack Coating—Surface Courses; **Railroad Bridges**

Section 19   BUILDINGS **19-1**

Part 1.   General Design Considerations   *Stephen J. Y. Tang and S. G. Haider* .. 19-1

Planning Building Structures: Selection of Structural Scheme—Spatial Requirements—Wind Systems—Deflection—Structural Materials—Fire Resistance—Deterioration—Provision for Environmental Control Systems—Limitations of Various Systems; **Loads:** Dead Load—Live Load—Snow Load—Wind Loads; **Floor and Roof Construction:** Floor and Roof Systems—Floor Finish—Roofing; **Wall Construction:** Types of Walls—Nonbearing Walls—Bearing Walls—Windows; **Stairs:** Planning—Types—Framing—Steel Stairs—Concrete Stairs—Escalators; **Miscellaneous Considerations:** Openings and Voids—Thermal Movement

Part 2.   Industrial Buildings *E. Alfred Picardi* 19-50

Design Philosophy—Planning—Framing Systems—Wall Systems—Bracing Systems for Lateral Loads—Materials Handling

Part 3.   Tall Buildings *Morton H. Eligator and Anthony F. Nassetta* 19-63

Framing: Bay Sizes—Columns—Elevator Shafts—Moving Stairs—Stairwells—Transfer Girders and Trusses; **Wind Bracing:** Medium-Rise Buildings (20 to 60 Stories)—Braced Bents, Rigid Frames, and Shear Walls—High-Rise (Above 60 Stories)—Tubular Frames, Tube Within a Tube, and Combinations—Fixed and Partially Fixed Joints in Steel Structures—Wind Load Determination—Wind Deflection—Wind-Shear Dissipation—Approximate Methods of Analysis—Computer Methods

Section 20   THIN-SHELL CONCRETE STRUCTURES *David P. Billington* **20-1**

Thin-Shell Concrete Roofs—Behavior of Roof Structures—Thin-Shell Curtain Walls; **Structural Analysis:** Thin-Shell Theory—Stability—Dynamic Behavior—Behavior of Domes—Membrane Theory; **Shell Walls:** Cylindrical Tanks—Hyperboloids; **Barrel Shells:** Long Barrels—Short Barrels—General Procedure for Shallow Shells—Shell with Edge Beams—Transverse Frames—Barrel-Shell Reinforcement; **Folded Plates:** Analysis of Folded Plate—Continuous Folded Plates—Prestressed Folded Plates—Membrane Theory—Elliptic Paraboloids—Hyperbolic Paraboloids; **Dimensioning; Construction**

Section 21   SUSPENSION ROOFS *Lev Zetlin and I. Paul Lew* **21-1**

Examples of Suspension Structures; **Design of Suspension Systems:** Anchorage Forces—Dynamic Behavior—Single Cable under Uniformly Distributed Load—Configuration and Shapes of Suspension Structures; **Double Layer of Prestressed Cables:** Damped Suspension Systems—Structural Relationships—Notation—Preliminary Design of Double Layer Cable System—Analysis of Double Layer Cable System—Behavior of Pair-set of Cables—Application to Preliminary Design of Cable Grids—Load Combinations for Selection of Cables—Types of Cables—Fittings—Membranes

Section 22   REINFORCED-CONCRETE BUNKERS AND SILOS *German Gurfinkel* **22-1**

Introduction—Bin Pressures—Emptying Pressures in Funnel-Flow Silos—Emptying Pressures in Funnel-Flow Silos—ACI 313—Shock Effects from Collapse of Domes—Pressures Induced by Dustlike Materials—Earthquake Forces; **Wall Forces:** Circular Silos—Rectangular and Polygonal Silos—Thermal Effects; **Design of Walls:** Minimum Thickness of Circular Walls—Maximum Crack Width—Walls in Tension—Walls in Tension and Flexure—Walls in Compression—Walls in Compression and Flexure—In-Place Bending of Walls—Walls Subjected to Thermal Stresses—Vertical Reinforcement—Details and Placement of Reinforcement; **Design of Bottoms:** Bottom Pressure—Plane Bottoms—Conical Hoppers—Pyramidal Hoppers—Hopper Supporting Beams—Columns—Roofs—Failures—Dust Explosions in Grain Elevators and Flour Mills; **Examples**

Section 23   STEEL TANKS   *Robert S. Wozniak* 23-1

Reservoirs: Capacity—Shell Design—Bottom Plates—Concrete Ringwall—
Roofs; **Standpipes:** Design—Anchorage—Foundations; **Elevated Tanks:**
Roofs—Suspended Bottoms—Balcony or Ring Girder—Columns—Founda-
tions; **Accessories; Bins:** Forces—Circular Bins—Shallow Bins—Miscellaneous
Details; **Materials**

Section 24   TOWERS AND TRANSMISSION POLE
STRUCTURES   *Max Zar and Joseph R. Arena* 24-1

Types of Towers—Materials—Height Limitations—Loads—Candelabra; **Free-
standing Towers:** Stresses—Foundations; **Guyed Towers:** Wind—Design of
Guys—Ice Loading—Guy Tensioning—Guy Vibration—Design of Mast—
Foundations—Erection; **Transmission Towers:** Types—Loads—Vibration—
Stress Analysis—Steel Tension Members—Aluminum Tension Members—
Steel Compression Members—Aluminum Compression Members—Limiting
Slenderness Ratios—Tower Tests; **Pole Structures:** Design—Material—Pole
Splices—Foundations

Section 25   BURIED CONDUITS   *Raymond J. Krizek* 25-1

Types of Conduits—Analysis and Design; **Loads on Conduits:** Loads on Ditch
Conduits—Loads on Projecting Conduits—Conduits in Wide Ditches—Loads
on Negative Projecting and Imperfect Ditch Conduits—Surface Loads; **Rigid
Conduits:** Supporting Strength—Bedding Classes for Trench Conduits—Bedding Classes for Embankment Installation—Monolithic Conduits; **Flexible Con-
duits:** Ring Compression—Deflection—Pipe Arches—Arches on Rigid Foundations; **Pressure Conduits:** Flexible Pressure Conduits—Rigid Pressure
Conduits; **Modern Design Methodology:** Elasticity Solution—Finite-Element
Solution—CANDE (Culvert ANalysis and DEsign); **Additional Design Considerations:** Handling Criteria—Durability—Camber—Wrappings and Coatings;
Construction Considerations: Site Preparation—Bedding—Fill Construction—
Compaction Procedures—Strutting—Joints—Backpacking; Long-Span Corru-
gated Metal Conduits

Section 26   CHIMNEYS   *Max Zar and Shih-Lung Chu* 26-1

Materials—Diameter and Height; **Design Loads:** Dead Loads—Wind Loads—
Earthquake Forces—Pressure Differentials—Temperature Differentials—Natu-
ral Frequency of Vibration; **Steel Stacks:** Allowable Stresses—Cone-to-Cylinder
Junction—Circumferential Stiffeners—Anchor Bolts—Base Ring for Anchor
Bolts—Guyed Stacks—Braced Stacks—Resonant Vibrations; **Reinforced Con-
crete Chimneys:** ACI Standard—Vibration Due to Wind; Linings; Foundations

Appendix A-1

Torsional Properties of Solid Cross Sections; Torsional Properties of Closed
Thin-walled Cross Sections; Torsional Properties of Open Cross Sections; Effec-
tive Length Coefficients for Columns; Buckling of Plates under Edge Stress;
Stiffened Beam Webs

Index follows the Appendix.

Preface

The second edition of "Structural Engineering Handbook," like the first, provides engineers, architects, and students of civil engineering and architecture with an authoritative reference work on structural engineering by assembling in one volume a concise, up-to-date treatment of the planning, design, and construction of a variety of engineered structures. Every section has been updated, and many have been revised extensively. Among topics not covered in the first edition which have been added are finite-element analysis, reinforced-concrete silos and bunkers, design for the P-Δ effect in tall-building columns, steel poles for transmission lines, and buried-conduit design by recently developed soil-structure interaction methods.

Among the structures covered are industrial buildings, tall buildings, bridges, thin-shell structures, arches, suspension roofs, tanks for liquid storage, bins and silos for granular materials, retaining walls, bulkheads, steel transmission towers and poles, chimneys, and buried conduits. Design in reinforced concrete, prestressed concrete, steel, composite construction, wood, aluminum, and masonry are covered. Sections on soil mechanics and foundations, construction methods, fabrication and erection, and a comprehensive treatment of structural analysis, including matrix methods, the finite-element method, and computer applications, give the designer the information likely to be needed for these phases of design. Finally, earthquake-resistant design and design against fatigue and brittle fracture are treated.

The 26 sections have been written by 47 authors chosen for their eminence and wide experience in specific areas of analysis, design, and construction. They have presented their material in ready-to-use form wherever possible. To this end, derivations of formulas are omitted in all but a few instances and many worked-out examples are given. Background information, descriptive matter, and explanatory material have been condensed and omitted. Because each section treats a subject which is broad enough to fill a book in itself, the authors have had to select that material which in their judgment is likely to be most useful to the greatest number of users. However, sources of additional material are noted for most of the topics which could not be treated in sufficient detail.

Each section was edited to minimize duplication, to arrange the contents of the book in a logical order, and to see that important topics were not overlooked. The editors gratefully acknowledge the authors' painstaking efforts, their cooperation in our editing of their work, and their patience during the time it has taken to bring the second edition to completion.

Edwin H. Gaylord, Jr.
Charles N. Gaylord

Contributors

Joseph R. Arena *Transmission Line Consultant, Chicago, Ill. (Towers and Transmission Pole Structures)*

Lynn S. Beedle *Director, Fritz Engineering Laboratory, Lehigh University (Plastic Design of Steel Frames)*

David P. Billington *Professor of Civil Engineering, Princeton University (Thin-Shell Concrete Structures)*

John W. Clark *Manager, Engineering Properties and Design Division, Alcoa Technical Center, Alcoa Center, Pa. (Design of Aluminum Structural Members)*

Walter L. Dickey *Consulting Civil and Structural Engineer, Los Angeles, Cal. (Masonry Construction)*

Morton H. Eligator *Partner, Weiskopf and Pickworth, Consulting Engineers, New York, N.Y. (Tall Buildings)*

Arthur L. Elliott *Bridge Engineer, Sacramento, Cal. (Bridges)*

Steven J. Fenves *University Professor of Civil Engineering, Carnegie-Mellon University (Computer Applications in Structural Engineering)*

Phil M. Ferguson *Professor Emeritus of Civil Engineering, University of Texas, Austin (Design of Reinforced-Concrete Structural Members)*

W. H. Fleischer *Structural Consultant, Sales Engineering Division, Bethlehem Steel Corporation, Bethlehem, Pa. (Design of Composite Beams and Girders)*

D. C. Frederickson *Sales Engineer, Sales Engineering Division, Bethlehem Steel Corporation, Bethlehem, Pa. (Design of Composite Beams and Girders)*

T. V. Galambos *H. D. Jolley Professor of Civil Engineering, Washington University (Plastic Design of Steel Frames)*

W. G. Godden *Professor of Civil Engineering, University of California, Berkeley (Structural Analysis—Elastic Systems)*

J. Graham *Lecturer in Aeronautical Engineering, Queen's University, Belfast (Structural Analysis—Elastic Systems and Finite-Element Method)*

German Gurfinkel *Professor of Civil Engineering, University of Illinois, Urbana (Reinforced-Concrete Bunkers and Silos)*

Hans William Hagen *Partner, LeMessurier Associates/SCI, Consulting Engineers, Cambridge, Mass. (Design of Steel Structural Members)*

S. G. Haider *Professor of Architecture and Engineering, Carleton University (Buildings—General Design Considerations)*

William J. Hall *Professor of Civil Engineering, University of Illinois, Urbana (Earthquake-Resistant Design)*

W. C. Hansell *Consultant, Wiss, Janney, Elstner and Associates, Northbrook, Ill. (Design of Composite Beams and Girders)*

C. F. Harris *Chief Engineer—Drawing Rooms, United States Steel Corporation, Pittsburgh, Pa. (Fabrication of Structural Steel)*

Herbert O. Ireland *Professor of Civil Engineering, University of Illinois, Urbana (Retaining Structures and Foundations)*

Thomas C. Kavanagh (deceased) *Former Vice-President, Iffland Kavanagh Waterbury, New York, N.Y. (Arches and Rigid Frames)*

Raymond J. Krizek *Professor of Civil Engineering, Northwestern University (Buried Conduits)*

H. G. Larew *Professor of Civil Engineering, University of Virginia (Soil Mechanics)*

William J. LeMessurier *Senior Partner, LeMessurier Associates/SCI, Cambridge, Mass. (Design of Steel Structural Members)*

I. Paul Lew *Partner, Lev Zetlin Associates, Inc., Consulting Engineers, New York, N.Y. (Suspension Roofs)*

Lee C. Lim *Associate, LeMessurier Associates/SCI, Consulting Engineers, Cambridge, Mass. (Design of Steel Structural Members)*

T. Y. Lin *Professor Emeritus of Civil Engineering, University of California, Berkeley (Design of Prestressed Concrete Structural Members)*

Kenneth P. Milbradt *Associate Professor of Civil Engineering, Illinois Institute of Technology (Timber Structures)*

W. H. Munse *Professor of Civil Engineering, University of Illinois, Urbana (Fatigue and Brittle Fracture)*

Anthony F. Nassetta *Partner, Weiskopf and Pickworth, Consulting Engineers, New York, N.Y. (Tall Buildings)*

N. M. Newmark *Professor Emeritus of Civil Engineering, University of Illinois, Urbana (Earthquake-Resistant Design)*

E. Alfred Picardi *Consulting Engineer, Richmond, Va. (Industrial Buildings)*

D. B. Rees *Director—Construction Procedures, American Bridge Division, United States Steel Corporation, Pittsburgh, Pa. (Erection of Structural Steel)*

Raymond C. Reese *Consulting Engineer, Toledo, Ohio (Design of Reinforced-Concrete Structural Members)*

W. C. Schnobrich *Professor of Civil Engineering, University of Illinois, Urbana (Structural Analysis—Finite Element Method)*

Shih-Lung Chu *Associate and Head, Structural Analysis Division, Sargent and Lundy Engineers, Chicago, Ill. (Chimneys)*

Stephen J. Y. Tang *Professor of Architecture, University of Oregon (Buildings—General Design Considerations)*

Thomas H. Thornburn *Consulting Soils Engineer, Las Vegas, Nev. (Soil Exploration)*

Ivan M. Viest *Assistant Manager, Sales Engineering Division, Bethlehem Steel Corporation, Bethlehem, Pa. (Design of Composite Beams and Girders)*

Francis A. Vitolo *President, Corbetta Construction Company, Inc., White Plains, N.Y. (Concrete Construction Methods)*

George Winter *Professor of Engineering, emeritus, Cornell University (Design of Cold-Formed Steel Structural Members)*

Roman Wolchuk *Partner, Wolchuk and Mayrburl, Consulting Engineers, New York, N.Y. (Steel-Plate-Deck Bridges)*

Robert S. Wozniak *Senior Engineer, Chicago Bridge and Iron Company, Oak Brook, Ill. (Steel Tanks)*

Robert C. Y. Young *Vice President, URS/Madigan-Praeger, Inc., New York, N.Y. (Rigid Frames and Arches)*

Max Zar *Partner and Manager of Structural Department, Sargent and Lundy Engineers, Chicago, Ill. (Towers and Transmission Pole Structures; Chimneys)*

Lev Zetlin *Zetlin-Argo Liaison and Guidance Corporation, New York, N.Y. (Suspension Roofs)*

Paul Zia *Professor of Civil Engineering, North Carolina State University (Design of Prestressed Concrete Structural Members)*

Structural Analysis

Part 1. Elastic Systems

J. GRAHAM
Lecturer in Aeronautical Engineering, Queen's University, Belfast

W. G. GODDEN
Professor of Civil Engineering, University of California, Berkeley

1. Degrees of Freedom The concept of degrees of freedom provides a basis for the study of *structural equilibrium* and *redundancy,* and for the formulation of the *problem coordinates* used in analysis. The number of degrees of freedom of a system is the number of independent coordinates required to locate it in space; alternatively, it can be considered as the number of independent displacement configurations that can be given to the system. In a structural assembly the forces and displacements at the *joints* are studied, with the degrees of freedom related solely to the joints of the assembly. These joints can be either physical connections between individual members (for example, the pins in a pin-jointed lattice frame, or the beam-column connections in a building frame) or joints (sometimes called *nodes*) which are arbitrarily introduced within a member when for purposes of integration it is subdivided into a series of elements.

The total number of degrees of freedom of a system of joints in space is the sum of the degrees of freedom of all the joints. Any one joint may have up to six degrees of freedom in a space frame. In practice, the freedom of the system is related to a datum; this datum can be chosen quite arbitrarily, but in most practical cases it is obviously taken as the earth. Joints on this datum have their freedom reduced according to the nature of the restraint, and relative to the other (free) joints of the system these restraints are called *external restraints.* The freedom of the joints relative to this datum is the *structural freedom,* which equals the number of degrees of freedom of the joint system minus the number of independent external restraints.

There is one equilibrium equation corresponding to each degree of freedom, and from this the following can be deduced:

1. Equilibrium. A joint system in equilibrium has not less than one independent restraint (corresponding to an internal force or external reaction) for every degree of freedom.

2. Statically determinate. When the number of independent restraints equals the total number of degrees of freedom of the joint system and the restraints are properly arranged (Art. 13) the system is in equilibrium and is statically determinate.

3. Statically indeterminate. When the number of independent restraints is greater than the degrees of freedom, the system is in equilibrium but statically indeterminate.

The solution to any structural problem can be expressed in terms of problem coordinates. These can be chosen as one of the following:

1. The redundant forces. In this case the solution appears as a set of simultaneous equations equal in number to the number of redundant forces, and the method is called the "force method."

2. The displacements in the structural degrees of freedom. In this case the solution appears as a set of simultaneous equations equal in number to the structural degrees of freedom, and the method is called the "displacement method."

2. Principles of Analysis   The solution to every structural-analysis problem must satisfy two conditions:

1. Equilibrium of forces. The structure considered as a whole must be in equilibrium under the action of the applied loads and the reactions: this is the external-force equilibrium condition. Each member of the structure considered in isolation as a free body must be in equilibrium under the action of its own applied loads and boundary forces: this is the internal-force equilibrium condition.

2. Compatibility of displacements. For any loading, the displacements of all the members of the structure due to their respective stress-strain relationships must be consistent with respect to each other: that is, the continuity of the structure must be preserved.

In statically determinate structures condition 2 is automatically satisfied, the internal-force distribution being completely determined from condition 1. In the case of statically indeterminate structures both conditions must be applied, the internal-force distribution being a function of the applied loading, the structural geometry, and the elastic properties of the members.

3. Equilibrium of Forces   The necessary and sufficient conditions for the equilibrium of a system of forces are:

1. The resultant of the forces must be zero.

2. The resultant moment of the forces about any axis must be zero.

It is usually convenient to refer the force system to rectangular coordinates (Fig. 1). The conditions of equilibrium are then expressed by the six equations:

$$\Sigma P_x = 0 \qquad \Sigma P_y = 0 \qquad \Sigma P_z = 0 \tag{1}$$

$$\begin{aligned} \Sigma(yP_z - zP_y) &= 0 \\ \Sigma(zP_x - xP_z) &= 0 \\ \Sigma(xP_y - yP_x) &= 0 \end{aligned} \tag{2}$$

In most problems the applied forces are due to static loads. In a dynamic problem the forces applied to the body at any instant of time, together with appropriate inertia forces, satisfy the equations of equilibrium (d'Alembert's principle).

ENERGY THEOREMS

4. Classification   All energy theorems can be classified into one of two general forms. The first form enables the conditions of equilibrium to be expressed in terms of displacements; the second form enables the conditions for the compatibility of displacements to be expressed in terms of forces. These lead respectively to the displacement and the force methods of analysis. In the displacement method the compatibility conditions alone are first imposed, and hence an infinite number of compatible displacement patterns of the structure is possible. The energy theorems are then applied to select the true displacement pattern, that is, the unique pattern which is in equilibrium with the applied loads. In the force method the equilibrium conditions alone are first imposed, and hence an infinite number of sets of forces in equilibrium with the applied loads is possible. The energy theorems are then applied to select the true equilibrium state, that is, the unique set of forces which satisfies the compatibility condition.

Each of the energy theorems can be derived from the principle of virtual work (Art. 5), which is a restatement of the general principle of conservation of energy; consequently all methods of structural analysis of elastic systems can be reduced to this basic principle. The principal energy theorems used in structural analysis involve the following concepts and definitions. *Force* is used in a generalized sense; that is, "force" can be a point force, a

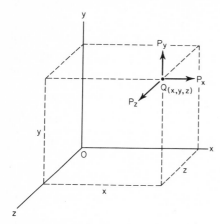

Fig. 1 Forces referred to rectangular coordinates.

moment or couple, or a system of forces. Similarly, *deflection* can be a translation, a rotation, or a deflection configuration. The deflection of a force at its point of application and along its line of action is called a *corresponding deflection*. The corresponding deflection of a point force is a translation, and the corresponding deflection of a moment or couple is a rotation. *Work* is defined as the product of a force and its corresponding displacement. *Energy* is defined as the capacity to do work. A body may possess energy because of its position, called *potential energy*, or by virtue of its motion, called *kinetic energy*. It is assumed that the loads are applied in such a manner that the kinetic energy is zero. A force system is said to be *conservative* if the net work done by a loading and unloading cycle is zero. The work done by a conservative system of forces depends only on the initial and final positions of the points of application of the forces and is "recoverable."

An *elastic body* is one in which the deformations vanish when the loads causing the deformations are removed. Hence the external forces and reactions applied to an elastic body form a conservative system of forces, and the work done by this system as it moves through a set of displacements is stored as *strain energy* in the body and is recoverable. The potential energy V of an elastic body is defined as

$$V = U + V_e \qquad (3)$$

where U = strain energy
$V_e = C - \Sigma W_j \Delta_j$ = potential energy of the external forces, referred to a datum
C = a constant depending on the datum

In Fig. 2, *a* and *b* show *linear* and *nonlinear elastic* relationships in which loading and unloading occur along the same line *OA*. In Fig. 2*c* loading occurs along the line *OA* and unloading along another line *AB*. This is a nonconservative system, and during the loading and unloading process work is done on the body. The body suffers a permanent distortion or "set" Δ_{res}. The work done on the body is not recoverable and is equal to the area *OAB*.

An elastic structure can exhibit *linear* or *nonlinear* characteristics. To possess linear characteristics it is necessary that two conditions be satisfied: the material must have a linear stress-strain relationship, and the structural geometry must result in overall linear behavior. If either of these two conditions is not satisfied, nonlinear behavior will occur. Figure 3 shows some examples of structural geometry which result in nonlinear behavior, even though the material from which the structural elements are made is itself linear.

Work and *complementary work*. Figure 4a shows a nonlinear elastic relationship between an applied load W_j and its corresponding deflection Δ_j.

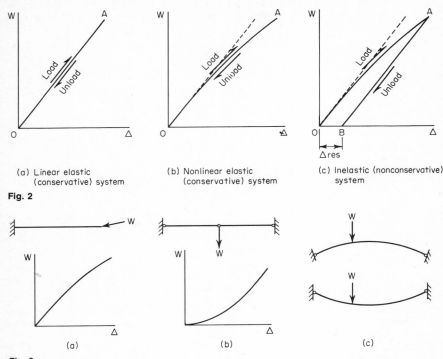

(a) Linear elastic   (b) Nonlinear elastic   (c) Inelastic (nonconservative)
    (conservative) system   (conservative) system   system

Fig. 2

(a)   (b)   (c)

Fig. 3

Strain energy and *complementary energy*. Figure 4b shows the internal force (stress) and the corresponding deformation (strain) relationship produced in a volume element dV of the body by the force system W.

Since the system is conservative, the total strain energy is equal to the work done by the applied loads W, that is,

$$\int_v \left(\int_0^{\epsilon_1} \sigma\, d\epsilon \right) dV = \sum_1^m \left(\int_0^{\Delta_1} W_j\, d\Delta_j \right) \tag{4}$$

Also, the total complementary energy is equal to the complementary work of the applied loads W, that is,

$$\int_v \left(\int_0^{\sigma_1} \epsilon\, d\sigma \right) dV = \sum_1^m \left(\int_0^{W_1} \Delta_j\, dW_j \right) \tag{5}$$

Complementary energy possesses no physical meaning but is a mathematical quantity from which important results can be derived. In the case of a linear load-deflection relationship the strain energy is numerically equal to the complementary energy, and the real work is numerically equal to the complementary work.

5. The Principle of Virtual Work If an elastic body which is in equilibrium under the action of any externally applied loading system is given an infinitesimal deformation from the equilibrium configuration, then the work done by the external loads during this displacement (the virtual work dW) is equal to the change in the internal strain energy (dU).

The body of Fig. 5 is in equilibrium under the applied loads $W_1, W_2, \ldots, W_j, \ldots, W_m$, and the reactions $R_1, R_2, \ldots, R_h, \ldots, R_r$. If U represents the strain energy in the equilibrium position, and $\Delta_1, \Delta_2, \ldots, \Delta_j, \ldots, \Delta_m$ and $d_1, d_2, \ldots, d_h, \ldots, d_r$ the corresponding displacements of the applied forces and reactions respectively, then

$$dU = \sum_1^m W_j \, d\Delta_j + \sum_1^r R_h \, d(d_h) \tag{6}$$

In the case of rigid supports the second term is zero. For nonrigid supports d_h is negative.

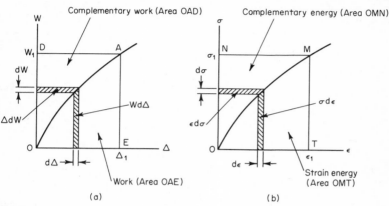

Fig. 4

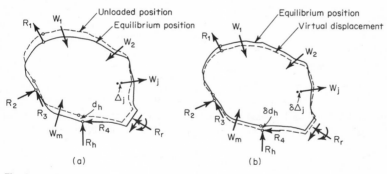

Fig. 5

Equation (6) relates to infinitesimal deformations from the equilibrium configuration, but provided the geometry of the system is not violated, finite virtual displacements may be considered, in which case it becomes

$$\delta U = \sum_1^m W_j \, \delta\Delta_j + \sum_1^r R_h \, \delta d_h \tag{7}$$

In Eqs. (6) and (7) the virtual work is caused by real forces acting through virtual displacements, and for this reason this form of the virtual-work principle is sometimes called the *principle of virtual displacements*. Compatibility of displacements is preserved during the process and the equilibrium requirement is imposed. Hence this form provides the basis for the displacement method of analysis.

Alternatively, a set of self-balancing virtual forces may be applied to the body, equilibrium being preserved. In this case the virtual work is caused by virtual forces acting through real displacements, and this form is called the *principle of virtual forces* or *complementary virtual work*. This states that the increment of complementary strain energy is equal to the increment of the complementary work done by the virtual forces acting through the real displacements. This form imposes the compatibility requirement and is the basis for the force method of analysis.

These principles are entirely general and so are applicable to nonlinear elastic systems. The following dual versions provide a convenient basis for the application of the virtual-work principle to structural analysis:

The Unit-Displacement Theorem.   In this method a unit displacement is applied at the point of application and in the direction of an unknown force. A system of displacements, compatible with the applied unit displacement, is chosen. The unit-displacement theorem is used in the following two ways:

1. Unknown externally applied loads can be expressed in terms of known internal forces. This is the basis for the *displacement method* of analysis of redundant structures.

$$1 \cdot W = \Sigma e_1 P \tag{8}$$

where the summation extends over all the elements of the structure. This equation can be stated as

$$\begin{matrix} \text{Applied} \\ \text{unit} \\ \text{displacement} \end{matrix} \times \begin{matrix} \text{unknown} \\ \text{external} \\ \text{force} \end{matrix} = \begin{matrix} \text{known defor-} \\ \text{mations of elements} \\ \text{compatible with the} \\ \text{unit displacement} \end{matrix} \times \begin{matrix} \text{known} \\ \text{internal} \\ \text{forces} \end{matrix}$$

2. An unknown internal force can be expressed in terms of a known external force. This gives the Müller-Breslau principle for finding influence lines in linearly elastic redundant structures.

The Unit-Load Theorem.   In this method a unit load is applied at the point and in the direction of an unknown displacement. A system of internal forces, statically equivalent to the unit load and its reactions, is chosen. The virtual-work principle gives

$$1 \cdot \Delta = \Sigma pe \tag{9}$$

where the summation extends over all the elements of the structure. This equation can be stated as

$$\begin{matrix} \text{Applied} \\ \text{unit} \\ \text{load} \end{matrix} \times \begin{matrix} \text{unknown displacement} \\ \text{of structure} \\ \text{in direction} \\ \text{of unit load} \end{matrix} = \begin{matrix} \text{known} \\ \text{statically} \\ \text{equivalent} \\ \text{force system} \end{matrix} \times \begin{matrix} \text{known} \\ \text{deformations} \\ \text{of} \\ \text{elements} \end{matrix}$$

6. The Principle of Minimum Potential Energy This theorem states that the potential energy of an elastic structure is a minimum when the structure is in equilibrium. That is,

$$\delta(U + V_e) = 0 \tag{10}$$

The principle is essentially a restatement of the principle of virtual work in terms of variational techniques. It is widely used as a means of deriving approximate solutions to structural stability problems but is also applicable to the solution of structures with a large number of redundancies in which the displacement can be expressed by a few terms of a simple series (Art. 41).

The minimum-potential-energy method is a "displacement method," and the strain energy must be given in terms of displacements because variations of potential energy are expressed in terms of the displacements.

7. The Minimum-Complementary-Potential-Energy Theorem This theorem states that of all the states of stress in an elastic structure which satisfy equilibrium and the specified boundary conditions, the state which satisfies compatibility conditions (that is, the true state of stress) makes the complementary potential energy of the system a minimum. The complementary potential energy is the sum of the complementary energy U' and the work done by the reactions ΣRd. That is,

$$\delta(U' + \Sigma Rd) = 0 \tag{11}$$

This theorem deals with a variation in the state of stress. In application it is a "force method" in which the complementary energy must be expressed in terms of the stresses or forces.

When the supports are fixed, Eq. (11) becomes

$$\delta U' = 0 \tag{12}$$

that is, the complementary energy is a minimum.

The concept of complementary energy enables the deflection of an elastic structure to be calculated. Figure 4 shows that

$$\frac{\partial U'}{\partial W_j} = \Delta_j \tag{13}$$

8. Castigliano's Theorems   When a system is linearly elastic the strain energy is numerically equal to the complementary energy. In this case Eqs. (12) and (13) can be stated in terms of strain energy. In this form they are known as Castigliano's theorems. This particular case of Eq. (12) may be restated as follows:

Of all the possible states of stress in a linearly elastic structure which satisfy equilibrium and the given boundary conditions, the state of stress which satisfies compatibility requirements makes the strain energy a minimum. This theorem is commonly called the "least-work" principle, which is conveniently applied by imposing the condition that the redundant forces make the strain energy a minimum:

$$\delta U = 0 \tag{14}$$

The particular case of Eq. (13) for linear structures is

$$\frac{\partial U}{\partial W_j} = \Delta_j \tag{15}$$

9. The Reciprocal Theorem   This theorem gives a relationship between two equilibrium states of a linearly elastic body. The first state consists of a load system $W_1, W_2, \ldots, W_j,$ $\ldots, W_m$, and the corresponding displacements $\Delta_1, \Delta_2, \ldots, \Delta_j, \ldots, \Delta_m$. A "corresponding displacement" Δ_j is the displacement of W_j in its line of action. The second state consists of a load system $W'_1, W'_2, \ldots, W'_j, \ldots, W'_m$ acting at the same points and in the same directions as those of the first system, and the corresponding displacements $\Delta'_1, \Delta'_2, \ldots, \Delta'_j,$ $\ldots, \Delta'_m$. The reciprocal theorem states that the work which would be done by the loads of the first state (W_j) acting through the displacements of the second state (Δ'_j) is equal to the work which would be done by the loads of the second state (W'_j) acting through the displacements of the first state (Δ_j). That is,

$$\sum_{j=1}^{m} W_j \Delta'_j = \sum_{j=1}^{m} W'_j \Delta_j \tag{16}$$

The particular case of Eq. (16) which gives the relationship between the deflections corresponding to two loads is known as *Maxwell's theorem of reciprocal displacements*. Figure 6a shows the displacements corresponding to the loads W_1 and W_2, due to the load W_1 applied alone. Similarly Fig. 6b shows the displacements due to the load W_2 applied alone. Maxwell's theorem states

$$W_1 \Delta_{12} = W_2 \Delta_{21} \tag{17}$$

It is convenient to state the theorem in terms of unit loads; that is, the deflection of a linearly elastic structure at point 1 due to a unit load applied at point 2 is equal to the deflection at point 2 due to a unit load applied at point 1. Equation (17) becomes

$$1 \times \delta_{12} = 1 \times \delta_{21} \tag{18}$$

The terms "force" and "displacement" are used in a generalized sense, so that

1. The rotation of point 1 due to a unit couple applied at point 2 is equal to the rotation of point 2 due to a unit couple applied at point 1.

2. The deflection (translation) of point 1 due to a unit couple applied at point 2 is numerically equal to the rotation at point 2 due to a unit load applied at point 1.

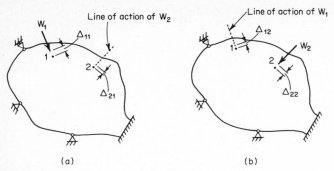

Fig. 6

The main application of the theorem is in the construction of influence lines for linearly elastic systems.

10. Inelastic Analysis The theorems of virtual work, minimum potential energy, and minimum complementary potential energy are applicable to general conservative systems, that is, nonlinearly elastic systems. They can, however, be applied to the analysis of structures possessing elastoplastic characteristics (Fig. 2c) provided unloading does not occur.

CLASSIFICATION OF STRUCTURAL TYPES

11. External Forces and Reactions A plane rigid body has three degrees of freedom, namely, translations in the x and y directions and a rotation about the z axis (Fig. 7a), and the applied loads W and reactions R must satisfy three equilibrium equations:

$$\Sigma V = 0 \qquad \Sigma H = 0 \qquad \Sigma M = 0$$

It follows that three *independent* external restraints are required to maintain the position of the body under any external load system W in its plane. The independence of the three reactions is of fundamental importance.

In Fig. 7 support systems b and c are identical. In each case the external loading W is reacted by support forces R_1, whose direction is fixed, and R_2, whose direction can vary and hence can be resolved into two independent components, for example, vertical and

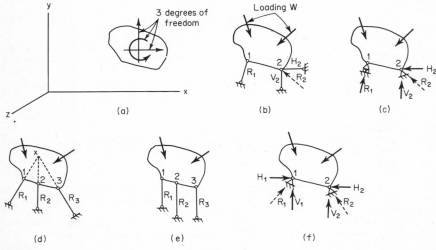

Fig. 7

horizontal components V_2 and H_2. These systems are stable and statically determinate. System d, however, is unstable, because R_1, R_2, and R_3 meet in a fixed point X and so are equivalent to two independent force components at X. System d can only support a loading whose resultant passes through X; for any other loading it is a mechanism. A support system approaching this geometry can give rise to large forces and excessive displacements, and should be avoided. It should be noted that a particular case of this arrangement is a system where the three reactions are parallel as in e. A necessary and sufficient condition is that the three reactions must not be concurrent.

System f has four independent reactions because R_1 and R_2 can each be resolved vertically and horizontally, giving components V_1, H_1, and V_2, H_2. But as the body has only three degrees of freedom, the system is externally redundant to the first degree.

A plane structure with an internal hinge has one further degree of freedom, that is, the rotation at the hinge; thus four properly located external reactions are required (Fig. 8).

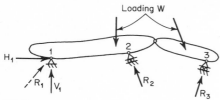

Fig. 8

A rigid body in space has six degrees of freedom; namely, translations along the x, y, and z axes, and rotations about each of these axes. Hence six independent external restraints are required to maintain the position of the body against all possible conditions of loading. No simple geometrical rules are possible to ensure the adequacy of the arrangement of the six restraints; in doubtful cases an application of the equations of equilibrium will always show an unsatisfactory arrangement (in the case of a mechanism the equations are indeterminate, and in cases approaching this some of the reactions become very large). If an additional degree of freedom is introduced into the body (for example, by the introduction of a single-freedom hinge or the omission of one internal member of a pin-jointed frame), one additional and properly located reaction is required to maintain overall equilibrium.

12. Independent Force Components A structure can be regarded for purposes of analysis as an assemblage of elements connected together at a number of "free" joints or nodes and supported at a number of "fixed" joints. The number of independent force components in any individual element depends on its end conditions. Figure 9 shows the components induced in plane frame elements. An element (a) with a stiff joint at each end has three internal forces, axial force P, moment M, shear S; the three independent forces can more conveniently be taken (as shown) as P, M_1, M_2, where M_1 and M_2 are the end moments. A pin at one end (b) reduces the number to two; a pin and roller at one end (c) reduce it to one. An element pinned at each end (d) has one force component P; if, in addition, one end is on rollers (e), there are no force components. Similarly, if one end is free (f), there are no force components.

The total number of independent force components is used in studying the state of equilibrium or the degree of redundancy of a structural assembly.

The operations of structural analysis are applied to a mathematical model which is called the *idealized structure*. This idealization transforms the actual structure, which, strictly speaking, has an infinite number of degrees of freedom, into a system with a finite number. For example, a pin-jointed or a stiff-jointed frame may be idealized as an assemblage of simple discrete elements connected at idealized joints (frictionless pins or rigid joints). The idealization may also relate to variation of cross-sectional properties, as, for example, when a tapered member is idealized as a series of connected elements.

In the idealization process, attention is concentrated on the structural joints and the concept of *equivalent* loading (Art. 37) is introduced to ensure that the displacements in the coordinates of the mathematical model are identical with the corresponding displacements of the real structure due to the actual loading.

13. Pin-Jointed Plane and Space Frames In a plane lattice framework, each element is pinned at each end and so has one force component P. Each joint is located by two rectangular coordinates and thus has two degrees of freedom; consequently an external force applied to a joint is resisted by two independent forces as $\Sigma V = 0$ and $\Sigma H = 0$. If there are j joints in the frame, the total number of degrees of freedom of the joint system is $2j$. If there are e independent external restraints at the fixed points, the number of degrees of freedom of the structure relative to the fixed points (the structural freedom) is $2j - e$.

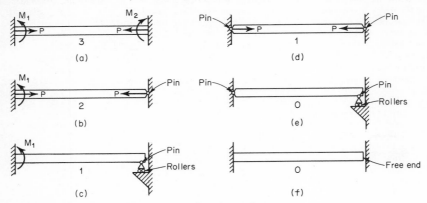

Fig. 9   Independent force components in plane frame elements.

This is the number of independent equilibrium equations associated with the free joints, and hence a necessary condition for a statically determinate frame is that the number of force components (in this case the number of members m) is

$$m = 2j - e \tag{19}$$

Whereas this gives the number of members required for equilibrium and determinacy, it gives no guidance as to the correct arrangement.

In the 7-joint frame of Fig. 10 there are 3 external restraints; hence $2 \times 7 - 3 = 11$ members are required for equilibrium. In most practical cases these 11 members will be

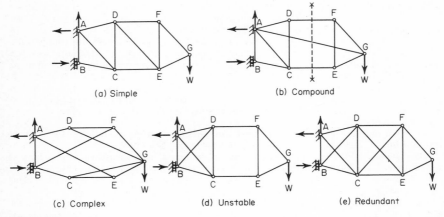

Fig. 10

arranged in a straightforward manner to give an assembly that is statically determinate and where all the internal forces can be found by using either the method of joints or the method of sections. The geometrical arrangement, however, influences both the equilibrium and the analysis. In frame a the geometry is at its simplest; it can be considered as a base member AB from which all other joints are fixed by triangles in a step-by-step manner. Joint C is fixed from A and B, joint D from A and C, and so on in alphabetical order to G. The importance of this geometry lies in the fact that the equilibrium equations when applied to the joints can be solved in a similar step-by-step manner progressing in the reverse order. Starting at G, $\Sigma V = 0$ and $\Sigma H = 0$ give all the information required to solve for P_{FG} and P_{EG}. Then the forces at joint F can be found, and so on to joint A. Mathematically, the simultaneous equations appear in groups of two which can be solved directly when taken in the correct order. This arrangement is called a *simple* frame.

The influence of the arrangement of the 11 members on both equilibrium and analysis can be seen in frames a, b, c, and d. Frame d has an incorrect arrangement which results in an unstable panel $DFEC$. There is no simple rule to guard against this sort of error, and the danger of its occurring is greater the more complicated the frame. The error will, however, always show up in the analysis. Frame b is stable and determinate but cannot be solved in the step-by-step manner used for frame a because three members meet at G. A partial solution can be found by applying the method of sections across XX ($\Sigma V = 0$ gives P_{AG}). From this, all other forces can be found by the method of joints. This arrangement of members is sometimes called a *compound* frame.

Frame c is stable and determinate, but none of the forces can be found by either the method of joints or the method of sections (a vertical section cuts four members). Here the forces have to be found by a formal solution of the $2j - 3$ equilibrium equations set up by the method of joints. For those who prefer the commoner step-by-step analysis, Henneberg's method may be used.[1] Frame c is sometimes called a *complex* frame.

Frame e has 13 members and is thus internally redundant to the second degree. It can be considered, for example, as frame a plus two redundant members BD and CF; in this case P_{BD} and P_{CF} are considered as two redundant self-equilibrating force pairs acting on the *reduced structure* of frame a.

Identical considerations apply to space frames (Fig 11). A pin-jointed space frame with j joints has $3j$ degrees of freedom. Each joint requires three rectangular coordinates to

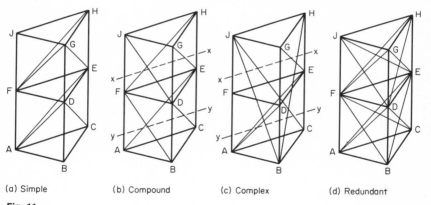

(a) Simple   (b) Compound   (c) Complex   (d) Redundant

Fig. 11

locate it in space; hence an external force applied to a joint is resisted by three independent forces. By "independent" is implied that the three members must not be coplanar. If overall equilibrium is maintained by e independent external restraints at the fixed points (the minimum possible value of e being 6), the structural freedom is $3j - e$. There are $3j - e$ independent equilibrium equations associated with the free joints; thus if there are m members, the equilibrium requirement is

$$m = 3j - e \qquad (20)$$

When these are properly arranged, the frame is in equilibrium and statically determinate.

In Fig. 11a there are 9 joints and 6 independent external restraints; hence $3 \times 9 - 6 = 21$ frame members are required for equilibrium. Cases a, b, and c satisfy this requirement, and further, the members are arranged to maintain equilibrium. However, the interconnection of the bracing on the three vertical faces differs in the three cases, and this causes a difference in the grouping or interconnection of the simultaneous equations derived by considering the three conditions of equilibrium at each joint.

Frame a is the simplest, as it is built up geometrically in a step-by-step manner. The method of joints applied first to J, then to H, and so on in reverse alphabetical order gives the internal forces directly. This is a *simple* space frame.

In frame b, FG and AD are replaced by JD and FB, and a step-by-step solution is no longer possible. A section xx or yy cuts only six members, and as the free body above this plane has six degrees of freedom (and thus six independent equilibrium equations) the forces at the section can be found as a partial solution. If P_{JD} and P_{FB} are found by the method of sections, a step-by-step solution for the remaining members is possible by the method of joints. This is a *compound* space frame.

Frame c is a complex space frame. A step-by-step solution by the method of joints is not possible. Also, sections xx and yy cut more than six members, and the method of sections does not give a direct solution. A direct solution can be effected only by Henneberg's method,[1] but this can be tedious if more than two members have to be rearranged.

If computation is done manually, a is simplest, b next, and c the most difficult. If machine computation is used, these comments do not apply as the grouping of equations presents no problems to a computer.

Frame d has 27 internal members and is thus internally redundant to the sixth degree. It can be considered as the *reduced structure* of frame a plus one redundant diagonal member in each of the six vertical panels.

14. Stiff-Jointed Frameworks In a stiff-jointed framework an element may be taken as a beam or column jointed at both ends or having one end free (Fig. 9). In a plane stiff-jointed framework each joint has three degrees of freedom and is located by three coordinates (two translation coordinates and one rotation); an external force applied to a joint is resisted by three independent forces as $\Sigma V = 0$, $\Sigma H = 0$, $\Sigma M = 0$. If there are j joints the joint system has $3j$ degrees of freedom. If there are e independent external

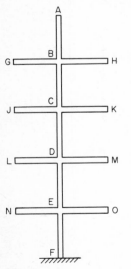

reactions (the minimum possible value of e being 3) the structural freedom is $3j - e$. The number of force components in an element depends on its end conditions; a member stiff-jointed at both ends has three, and if there are m such members there is a total of $3m$ force components. Hence a necessary condition for equilibrium and determinacy in a stiff-jointed plane frame is

$$3m = 3j - e \qquad (21)$$

In Fig. 12 there are 5 joints which give $3 \times 5 = 15$ degrees of freedom in the joint system. There are 9 members of the type shown in Fig. 9f, with no force components, and 4 members of the type shown in Fig. 9a, with 3 force components. Hence the total number of force components is $0 \times 9 + 3 \times 4 = 12$. Considered as a rigid body the structure has 3 external reactions; $12 + 3 = 15$; hence the structure is statically determinate.

In Fig. 13a there are $4 \times 3 = 12$ force components, the joint system has $5 \times 3 = 15$ degrees of freedom, and there are 6 external reactions. Since $12 - 15 + 6 = 3$, the structure is redundant to the third degree. The system can be reduced to a statically determinate structure by changing the number of degrees of freedom of the joint system, or by changing the number of force components, or by changing the number of external restraints. This reduction can also be effected by a suitable combination of these changes. The structure can be taken as b where $3 \times 3 = 4 \times 3 - 3$; frame a is thus considered as a determinate cantilever frame with three unknown external forces V_E, H_E, M_E, which restore displacement compatibility at E. Alternatively the reduced structure can be taken as the determinate three-pinned frame c. In this case all features have been changed; there are 4 members each with 2 force components, 2 joints with 3 degrees of freedom, 3 joints with 2 degrees of freedom, and 4 external reactions; $4 \times 2 = 2 \times 3 + 3 \times 2 - 4$; hence the system is determinate. Frame a can thus be

Fig. 12

considered as frame c with three unknown moments M_A, M_C, M_E which restore continuity of slope at A, C, and E.

METHODS OF ANALYSIS

The principal methods of analysis in common use are summarized in Tables 1, 2, and 3, together with comments indicating their sphere of usefulness and application. The reference in the third column indicates those which are discussed in this section. Methods for finding forces in statically determinate structures are given in Table 1. Methods for computing deflections are given in Table 2. The methods summarized in Table 3 combine those of Tables 1 and 2 for solving statically indeterminate structures.

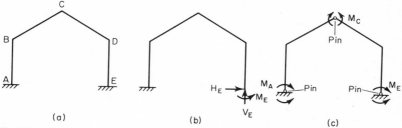

Fig. 13

TABLE 1   Forces in Statically Determinate Structures

Method	Remarks	Article
Pin-jointed frameworks:		
1. Method of joints...........	Eq. (1) is applied to each joint in turn to give forces throughout the frame. In general suitable only for plane frames	15
2. Method of sections.........	Eqs. (1) and (2) are applied to a part of the structure to give the forces across a section. Useful when forces in particular members are required	15
3. Polygon of forces and funicular polygon..............	Graphical solution to 1	15
4. Henneberg's method........	A member-transposition technique for solving complex frames (e.g., Figs. 10c and 11c)	Ref. 1
5. Direction cosines...........	5 and 6 are systematic solutions to 1. 6 embodies 5 in a more useful general form. Area of application: complicated frames, especially space frames	
6. Tension coefficients.........	See 5	15
Beams:		
7. Mathematical integration...	Application of calculus to obtain shear and moment from the loading equation. Macaulay's notation convenient for treatment of discontinuities. In most practical cases 10 is more suitable	
8. Method of sections.........	2 applied to beams. Useful when shears and moments are required at particular points	16
9. Polygon of forces and funicular polygon	Graphical solution for shears and moments along a beam. Unsuited to distributed loading. Useful for moving loads	16
10. Numerical integration......	In most cases preferable to 9; faster, more accurate, and deals effectively with distributed loading	16

FORCES IN STATICALLY DETERMINATE STRUCTURES

15. Pin-Jointed Lattice Frameworks In any particular case the method chosen to solve the equilibrium equations depends on the type and complexity of the problem, and also on the arrangement of the members. When the geometry of the frame is simple (e.g., Fig. 14a), the *method of joints* or *method of sections* gives a quick solution. When the geometry is more complicated (e.g., Fig. 15), the forces in a plane frame are often obtained

TABLE 2   Computation of Deflections

Method	Remarks	Article
Kinematic methods. Vectorial addition of element deformation vectors. Suitable only for structures possessing simple geometrical form. In practice, almost entirely restricted to plane pin-jointed frameworks and beam elements:		
1. Williot-Mohr method.......	Graphical vector-addition solution for plane pin-jointed frameworks	17
2. Mathematical integration of beam equation	Generally suitable only for simple problems. Macaulay's notation convenient for treatment of discontinuities. Not suited to cases of variable moment of inertia and arbitrary loading	
3. Numerical integration of beam equations	Suitable for any variation of moment of inertia and loading	18
4. Moment area, conjugate beam, elastic weights	These are specialized techniques designed to expedite the solution to particular problems. The moment-area and conjugate-beam methods are devices for solving the beam equation and in this text are replaced on grounds of usefulness by 3. The elastic-weights method (sometimes called elastic loads, angle loads, or angle weights) is a device for applying the conjugate-beam technique to plane pin-jointed frameworks	Ref. 1
Energy methods. These are general methods applicable to any structure and any loading:		
5. Unit-load method..........	The most generally useful method for calculating the deflection of any structure subjected to any loading action. Also called Maxwell-Mohr method, virtual work, virtual velocities, virtual loads, dummy loads, unit loads, auxiliary loads, etc. It is the basis for the unit-load method of Art. 32 and the matrix force method of Art. 45	19
6. Complementary energy and Castigliano's theorem	The deflection of a load in its line of action is obtained by differentiating the complementary energy with respect to the load. Castigliano's method is the particular case for linearity and in practical application is usually identical with the unit-load method	27
7. Matrix method............	Systematic method, based on 5, of calculating structural flexibility. Requires only simple input information. Particularly suitable for problems involving a large number of different loading actions, especially if highly redundant	42

more quickly by the graphical method. For *complex* frames *Henneberg's method*[1] is suitable provided not more than one or two member transformations have to be made, but for more difficult cases a formal solution of the equilibrium equations is more satisfactory.

None of these methods is very suitable for the analysis of space frames. For space frames the *method of tension coefficients* is the most useful procedure. This method is also useful for the analysis of plane frameworks subjected to a large number of different loading arrangements.

Method of Joints. The forces at each joint in a plane framework must satisfy two equilibrium equations, $\Sigma V = 0$ and $\Sigma H = 0$. Components referred to any set of rectangular coordinates are equally suitable. When applied to each joint in turn all the forces

TABLE 3   Forces in Redundant Structures

Method	Remarks	Article
Force methods. The redundant forces are the unknowns:		
1. Unit-load method..........	Based on method 5 of Table 2. Also known as Maxwell-Mohr method, δ_{ik} method, flexibility-coefficient method	32
2. Minimum complementary energy and Castigliano's theorem	The complementary energy is minimized with respect to the redundant forces. Castigliano's method is a particular case for linearity	33, 40
3. Three-moment equation....	Equation relating moments at three adjacent support points in a continuous beam	34
4. Elastic center.............	A technique for choosing coordinates to simplify the formulation of the equations for closed rings, frames, and bents	Sec. 17, Art. 4
5. Column analogy...........	Analogy between an eccentrically loaded short column and the redundant bending moments in closed rings, frames, and bents	35
6. Matrix method...........	Systematic method based on 1. Requires only simple input information. Particularly suitable for highly redundant structures	45
Displacement methods. The displacements in the degrees of freedom are the unknowns:		
7. Unit-displacement method..	A unit displacement is applied to each degree of freedom, producing internal forces. These are equated to the applied loads, giving a set of equations for the displacements in the degrees of freedom	37
8. Slope-deflection method....	Equations in terms of slope and deflection for stiff-jointed frames composed of straight elements. Useful only in cases of a small number of degrees of freedom	38
9. Moment-distribution method	A technique for solving the slope-deflection equations of 8 by successive approximations. Useful for stiff-jointed frames with a large number of members, provided there is no joint translation	39
10. Minimum potential energy..	Used to obtain approximate solutions for highly redundant structures where the displacement can be expressed by a few terms of a series	41
11. Matrix method...........	A systematic method based on 7. Requires only simple input information. Particularly suitable for highly redundant structures when the number of degrees of freedom is less than the number of force redundancies	51, 52

throughout the frame are obtained in a step-by-step manner, provided the geometry of the frame is suitable (Art. 13). For example, in Fig. 14a, when the external reactions have been found at A, the method of joints can be started at A and then applied to the other joints in alphabetical order. Assuming all members to be in tension for the purpose of writing the equilibrium equations, we get

At A:
$$\Sigma V = 12 + P_{AB} = 0 \qquad P_{AB} = -12$$
$$\Sigma H = P_{AC} = 0$$

At B:
$$\Sigma V = P_{AB} + P_{BC} \sin 45° = 0 \qquad P_{BC} = 12\sqrt{2}$$
$$\Sigma H = P_{BC} \cos 45° + P_{BD} = 0 \qquad P_{BD} = -12$$

As in all step-by-step solutions, errors are carried forward. In practice the method is useful only when the geometry of the frame is simple.

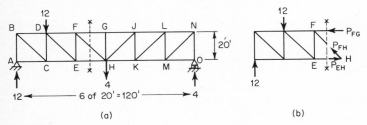

Fig. 14

Method of Sections. If a section is taken through a frame (all the applied loads and reactions being known) and if the section cuts not more than three unknown internal forces, the three equilibrium equations give sufficient information to find these forces. Taking section xx in Fig. 14a, the part of the structure to the left (or right) can be

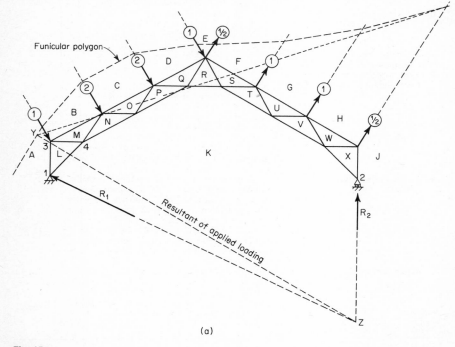

(a)

Fig. 15

considered as a rigid body in equilibrium under the action of the two known external forces and the three unknown member forces P_{FH}, P_{EH}, P_{FG} (Fig. 14b).

$$\Sigma V = 12 - 12 + P_{FH} \sin 45° = 0 \qquad P_{FH} = 0$$
$$\Sigma M_H = 12 \times 60 - 12 \times 40 - 20P_{FG} = 0 \qquad P_{FG} = 12$$
$$\Sigma H = P_{FG} - P_{EH} = 0 \qquad P_{EH} = 12$$

This method is also applicable to space frames. Forces can be found across a section that cuts not more than six members.

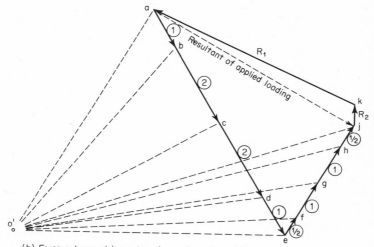

(b) Force polygon giving external reactions R_1 and R_2

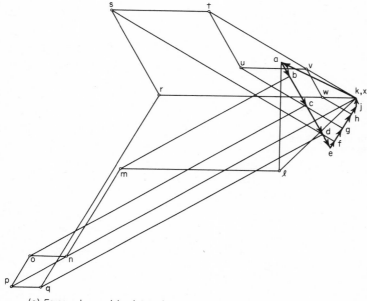

(c) Force polygon giving internal member forces

Fig. 15 (*continued*)

Graphical Method. In this method the reactions are first found by applying the funicular polygon to the complete load system. Once the reactions are found, the polygon of forces is then applied to each joint in turn to give the internal forces. Bow's notation assigns to each space in the frame diagram a reference letter (A, B, C, etc.); then vector *ab* in the force polygon is the force which separates spaces A and B.

In Fig. 15 all forces can be found as follows:

1. Construct the funicular polygon (Fig. 15a) for the applied loading in order to find the external reactions R_1 and R_2. In this construction the loading is drawn as a set of force vectors *ab* to *hj* (Fig. 15b), the resultant in magnitude and direction being *aj*. A pole *o'* is arbitrarily chosen and lines are drawn from *o'* to every point on the vector diagram. In the space diagram (Fig. 15a) a dotted line is drawn in space A parallel to *o'a* and located in any suitable position. In space B a line is drawn parallel to *o'b* joining the previous line, and this procedure is continued to space J. The lines in spaces A and J are produced to intersect at Y, and this locates the position of the resultant *aj*. The position and direction of R_2 are known, and this force intersects the resultant *aj* at Z. The direction of R_1 is formed by joining Z to joint 1. The directions of R_1 and R_2 transferred to the force polygon give the values of these forces. All external forces on the frame are now known.

2. Construct the polygon of forces (Fig. 15c) for the forces in the frame members. This can be superimposed on force polygon *b* but is here drawn separately for clarity. The polygon of forces is drawn for joint 1 to obtain the two unknown member forces. This is repeated at joints 3, 4, etc., in a step-by-step manner across the frame.

Bow's notation is useful in determining the kind of stress in a member. For example, reading clockwise around joint 4 in Fig. 15a (corresponding to the order in which the external forces were drawn in the force polygon), bar 34 must be read as *LM*. The corresponding vector *lm* in Fig. 15c is directed away from joint 4, indicating that bar 34 is in tension.

This method is perhaps the simplest approach to a problem of the type shown, especially if only one or two loading arrangements have to be analyzed.

Tension Coefficients. In this method the equations of equilibrium are referred to a system of rectangular coordinates. The essential geometrical data (the projected lengths of the members on these axes) are the dimensions given in the three projections. The tension coefficient *t* of a member is defined as the ratio of the force *P* in the member to its true length *L*.

At each joint in the framework an equation of equilibrium is written for the components of force along each axis, as follows:

1. All members are assumed to be in tension, which is taken as positive.
2. The origin of the coordinate system is placed at the joint under consideration. In writing the equilibrium equations due regard is paid to the signs of the projected lengths of the members, and of the components of the applied loads, with respect to the positive directions of the coordinate axes.
3. For each axis the equation of equilibrium is then given as

Σ (projected length of member) $\times$ tension coefficient of member
$$+ \, \Sigma \text{ (applied load component)} = 0$$

The formation of the equations of equilibrium by the use of tension coefficients is illustrated by the simple tripod structure of Fig. 16. The resulting equations for joint D are

$$-8t_{AD} - 9t_{BD} - 10t_{CD} - W_x = 0$$
$$t_{AD} + 5t_{BD} - 3t_{CD} - W_y = 0$$
$$4t_{AD} - 4t_{BD} - 2t_{CD} - W_z = 0$$

These equations are solved for the tension coefficients; signs are taken care of automatically, a negative result indicating that the member concerned is in compression. The force in a member is then found by multiplying the tension coefficient from this solution by the true length of the member (which is always positive). In practice the calculations are most conveniently organized in a tabular arrangement of the type shown in Fig. 17.

Three equations of equilibrium exist for each joint. In the case of a simple frame, these can be solved completely at each joint without reference to the equations at the other joints, provided the joints are taken in the correct order (Art. 13). In the case of compound and complex frames, the equations are interrelated to those at other joints. Hence more than three equations have to be solved simultaneously.

If the support points are included in the equations of equilibrium, the total number of equations is sufficient to find all the internal member forces and all the external reactions. If the e external reactions have previously been found from rigid-body considerations, there will be e surplus equations in the tension-coefficient solution. These surplus equations provide a necessary and sufficient check on the accuracy of the solution.

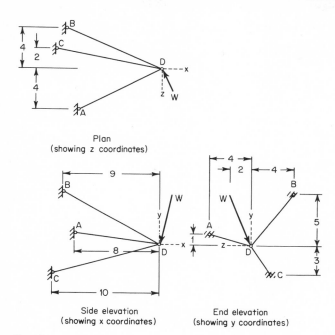

Plan
(showing z coordinates)

Side elevation
(showing x coordinates)

End elevation
(showing y coordinates)

Fig. 16

16. Beam Systems When the loading is simple, for example, consisting of a small number of point forces, shear-force and bending-moment diagrams can usually be drawn by inspection. Shear and bending-moment diagrams due to a moving set of point forces of fixed spacing across a beam can be found quickly by the graphical funicular-polygon method. In the case of distributed loading, especially when it has a complicated form, the simplest solution is that of numerical integration, a method ideally suited to the desk calculator.

Method of Sections. In the case of a horizontal beam with vertical applied loading and reactions, the two equilibrium conditions $\Sigma V = 0$ and $\Sigma M = 0$ applied to the part of a beam to the left (or right) of a section give shear force and bending moment, respectively, at that section. For example, taking a section just to the left of B in Fig. 18,

$$\Sigma V = 0 \qquad S_{BA} = R_A = 6 \text{ kips}$$
$$\Sigma M = 0 \qquad M_B = 10R_A = 60 \text{ kip-ft}$$

By determining shear and moment values at A, B, C, D the complete shear and moment diagrams can be plotted.

Graphical Method. The effect of a moving set of point loads of fixed spacing can be found by the funicular polygon. The polygon is drawn once for the loading, and the span is then placed on it in any position to give the appropriate shear and moment diagrams. For example, the funicular polygon for the vehicle of Fig. 19a can be drawn by the method described in Art. 15. The bending-moment diagram, due to the action of this loading in any position on any span XY, is the area between the funicular polygon and the base XY drawn on it (Fig. 19c). The values of R_X and R_Y are found by drawing a line

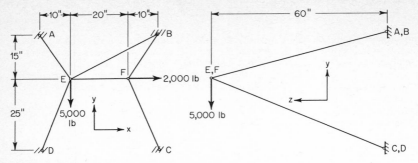

NOTE: The 2,000-lb load is parallel to the x axis

Joint	Coordinate	Equilibrium equation	Member	Tension coefficient t, lb/in.	Length L, in.	Force P, lb
F	x	$10t_{FB}+10t_{FC}-20t_{FE}+2{,}000=0$	FB	0	62.7	0
	y	$15t_{FB}-25t_{FC}=0$	FC	0	65.9	0
	z	$-60t_{FB}-60t_{FC}=0$	FE	100	20.0	2,000
E	x	$-10t_{EA}+30t_{EB}-10t_{ED}+20t_{EF}=0$	EA	175	62.7	10,950
	y	$15t_{EA}+15t_{EB}-25t_{ED}-5{,}000=0$	EB	-50	68.8	-3,440
	z	$-60t_{EA}-60t_{EB}-60t_{ED}=0$	ED	-125	65.9	-8,250

Fig. 17

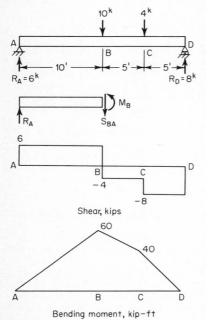

Fig. 18

through pole O' parallel to this base. This line also locates the position of the base of the shear-force diagram (Fig. 19*b*).

Numerical Integration. The integration for shear and moment can often be difficult in the case of distributed loading. One of the simplest ways of dealing with this general problem is by numerical integration. The tabular layout of the technique given here is due to Newmark.[2] The procedure is to reduce the distributed loading to *equivalent concentrated loading* at selected *node* points and, by forward integration from node to node, to find values of *chord* (element) shear and nodal moment.

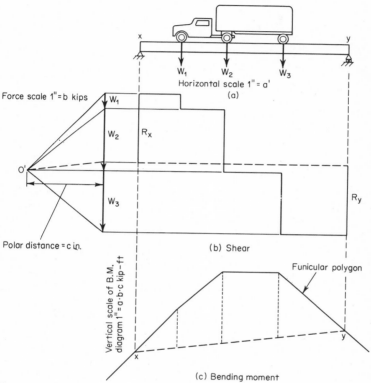

Fig. 19

Concentration formulas for standard cases are given in Fig. 20. When the loading is either *linear* or *parabolic*, these formulas will lead to exact results. If the loading is more complex, intelligent use of the same formulas will lead to good approximations. Point loading, if present, is then added.

The procedure is illustrated in the example of Fig. 21. The nodal concentrations are found from Fig. 20 as follows:

Node B:
$$\frac{h}{6}(a + 4b + c) = \frac{w_0 h}{6}(4 + 8 + 0) = \frac{24 w_0 h}{12}$$

Node C:
$$\frac{h}{6}(a + 2b) = \frac{w_0 h}{6}(2 + 0) = \frac{4 w_0 h}{12}$$

Node C:
$$\frac{h}{24}(7a + 6b - c) = \frac{w_0 h}{24}(7 + 12 - 5) = \frac{7 w_0 h}{12}$$

Node D:
$$\frac{h}{12}(a + 10b + c) = \frac{w_0 h}{12}(1 + 20 + 5) = \frac{26 w_0 h}{12}$$

If there is a point of known shear, chord shears are found directly by adding the W values from left to right. In Fig. 21 there is no point of known shear, and a guessed value of the shear in chord AB, $S_{AB} = 0$, is chosen as a starting value for a set of trial chord shears. The summation is carried out in the direction of the arrows. The change in moment over a chord length is given by

$$\text{Moment increment} = \text{chord shear} \times \text{chord length}$$

Moment increments are tabulated, unless, as in this problem, the chord length is constant, when it becomes a common factor. Starting from a known moment value, $M_A =$

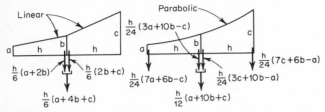

Fig. 20 Nodal concentration formulas for trapezoidal and parabolic distributed loading. *(Trans. ASCE, vol. 108, p. 1161, 1943.)*

0, trial nodal moments are found by summing the moment increments as indicated by the arrows. The value of M_E is incorrect (due to the error in S_{AB}); a correction is applied to make it zero, and the correction to moment at any node is proportional to the distance of the node from A.

As the computation of deflection from a given M/EI curve is similar to that of moment

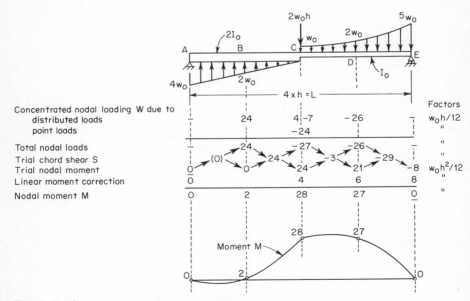

Fig. 21 Bending moment computed by numerical integration.

from a given loading curve, this forward-integration technique is ideal for computing beam deflections (Art. 18) and influence lines (Art. 29).

Table 4 gives some standard beam formulas, showing in most cases maximum values of shear, moment, slope, and deflection.

TABLE 4 Formulas for Beams

Structure	Shear	Moment	Slope	Deflection
		Simply supported beam		
	$S_A = -\dfrac{M_o}{L}$	M_o	$\theta_A = \dfrac{M_o L}{3EI}$ $\theta_B = -\dfrac{M_o L}{6EI}$	$y_{max} = 0.062\,\dfrac{M_o L^2}{EI}$ at $x = 0.422L$
	$S_A = \dfrac{W}{2}$	$M_C = \dfrac{WL}{4}$	$\theta_A = -\theta_B = \dfrac{WL^2}{16EI}$	$y_C = \dfrac{WL^3}{48EI}$
	$S_A = \dfrac{Wb}{L}$ $S_B = -\dfrac{Wa}{L}$	$M_a = \dfrac{Wab}{L}$	$\theta_A = \dfrac{Wab}{6EIL}(L+b)$ $\theta_B = -\dfrac{Wab}{6EIL}(L+a)$	$y_a = \dfrac{Wa^2 b^2}{3EIL}$
	$S_A = \dfrac{wL}{2}$	$M_C = \dfrac{wL^2}{8}$	$\theta_A = -\theta_B = \dfrac{wL^3}{24EI}$	$y_C = \dfrac{5wL^4}{384EI}$
	$S_A = \dfrac{wL}{6}$ $S_B = -\dfrac{wL}{3}$	$M_{max} = 0.064\,wL^2$ at $x = 0.577L$	$\theta_A = \dfrac{7wL^3}{360EI}$ $\theta_B = -\dfrac{8wL^3}{360EI}$	$y_{max} = 0.00652\,\dfrac{wL^4}{EI}$ at $x = 0.519L$

TABLE 4   Formulas for Beams (Continued)

Beam	Shear / Reaction	Moment	Slope	Deflection
	$S_A = \dfrac{wL}{4}$	$M_C = \dfrac{wL^2}{12}$	$\theta_A = -\theta_B = \dfrac{5wL^3}{192EI}$	$y_C = \dfrac{wL^4}{120EI}$
Fixed beam				
	$S_A = \dfrac{W}{2}$	$M_C = \dfrac{WL}{8}$	$\theta_A = \theta_B = 0$	$y_C = \dfrac{WL^3}{192EI}$
	$S_A = \dfrac{Wb^2}{L^3}(3a+b)$ $S_B = -\dfrac{Wa^2}{L^3}(3b+a)$	$M_A = -\dfrac{Wab^2}{L^2}$ $M_B = -\dfrac{Wba^2}{L^2}$	$\theta_A = \theta_B = 0$	$y_a = \dfrac{Wa^3 b^3}{3EIL^3}$
	$S_A = \dfrac{wL}{2}$	$M_A = M_B = -\dfrac{wL^2}{12}$	$\theta_A = \theta_B = 0$	$y_C = \dfrac{wL^4}{384EI}$
	$S_A = \dfrac{3wL}{20}$ $S_B = \dfrac{7wL}{20}$	$M_A = -\dfrac{wL^2}{30}$ $M_B = -\dfrac{wL^2}{20}$	$\theta_A = \theta_B = 0$	$y_{max} = 0.00131\,\dfrac{wL^4}{EI}$ at $x = 0.525L$
	$S_A = \dfrac{wL}{4}$	$M_A = M_B = -\dfrac{5wL^2}{96}$	$\theta_A = \theta_B = 0$	$y_C = \dfrac{0.7wL^4}{384EI}$

TABLE 4 Formulas for Beams (Continued)

Structure	Shear	Moment	Slope	Deflection
Cantilever beam				
	O	M_0	$\theta_A = \dfrac{M_0 L}{EI}$	$y_A = -\dfrac{M_0 L^2}{2EI}$
	W	$M_B = -WL$	$\theta_A = -\dfrac{WL^2}{2EI}$	$y_A = \dfrac{WL^3}{3EI}$
	$S_B = -wL$	$M_B = -\dfrac{wL^2}{2}$	$\theta_A = -\dfrac{wL^3}{6EI}$	$y_A = \dfrac{wL^4}{8EI}$
	$S_B = -\dfrac{wL}{2}$	$M_B = -\dfrac{wL^2}{6}$	$\theta_A = -\dfrac{wL^3}{24EI}$	$y_A = \dfrac{wL^4}{30EI}$
	$S_B = -\dfrac{wL}{2}$	$M_B = -\dfrac{wL^2}{2}$	$\theta_A = -\dfrac{wL^3}{8EI}$	$y_A = \dfrac{11wL^4}{120EI}$

TABLE 4 Formulas for Beams (Continued)

Propped cantilever

Beam	Reaction / Shear	Moment	θ_A	y_{max}
M_0 at A	$S_A = -\dfrac{3M_0}{2L}$	$M_B = -\dfrac{M_0}{2}$	$\theta_A = \dfrac{M_0 L}{4EI}$	$y_{max} = \dfrac{M_0 L^2}{27EI}$ at $x = \dfrac{L}{3}$
W at C	$S_A = \dfrac{5W}{16}$	$M_B = -\dfrac{3WL}{16}$ $M_C = \dfrac{5WL}{32}$	$\theta_A = \dfrac{WL^2}{32EI}$	$y_{max} = 0.00932\,\dfrac{WL^3}{EI}$ at $x = 0.447L$
W at a, b	$S_A = \dfrac{Wb^2}{2L^3}(a+2L)$ $S_B = -\dfrac{Wa}{2L^3}(3L^2-a^2)$	$M_B = -\dfrac{Wab}{L^2}\left(a+\dfrac{b}{2}\right)$	$\theta_A = \dfrac{Wab^2}{4EIL}$	$y_a = \dfrac{Wa^2b^3}{12EIL^3}(3L+a)$
w uniform	$S_A = \dfrac{3wL}{8}$	$M_B = -\dfrac{wL^2}{8}$	$\theta_A = \dfrac{wL^3}{48EI}$	$y_{max} = 0.0054\,\dfrac{wL^4}{EI}$ at $x = 0.422L$
w	$S_A = \dfrac{wL}{10}$	$M_{max} = 0.03wL^2$ at $x = 0.447L$ $M_B = -\dfrac{wL^2}{15}$	$\theta_A = \dfrac{wL^3}{120EI}$	$y_{max} = 0.00239\,\dfrac{wL^4}{EI}$ at $x = 0.447L$
w	$S_A = \dfrac{11wL}{40}$	$M_{max} = 0.0423wL^2$ at $x = 0.329L$ $M_B = -\dfrac{7wL^2}{120}$	$\theta_A = \dfrac{wL^3}{80EI}$	$y_{max} = 0.00305\,\dfrac{wL^4}{EI}$ at $x = 0.402L$

DEFLECTIONS

The calculation of deflections is fundamentally a problem of kinematics which can be solved either directly, by kinematic methods, or indirectly by energy methods (Table 2).

17. Williot-Mohr Diagram The method of graphical vector addition relates the displacement of each joint of a simple plane pin-jointed frame to the preceding two joints; hence the solution progresses in the same direction as the geometrical buildup of the frame. If it is possible to start from a fixed point and a fixed direction, the graphical construction (Williot diagram) is straightforward. When there is no fixed direction in the system a single trial-and-error solution is required (Williot-Mohr diagram).

Figure 22a is a simple frame in which A is a fixed point and AB is fixed in direction. In the displacement diagram A is taken as the fixed point and all other displacements are related to it (Fig. 22b). Vector AB is the zero displacement of B relative to A in both magnitude and direction. Ac_1 is the extension of AC drawn in the direction of AC, and Bc_2 the shortening of BC, drawn in the direction of CB. Perpendiculars at c_1 and c_2 intersect at C, which gives the displacement of C relative to A. In the same way D is fixed from A and C, and so on to F, giving the displacement of each joint in relation to A and in relation to each other.

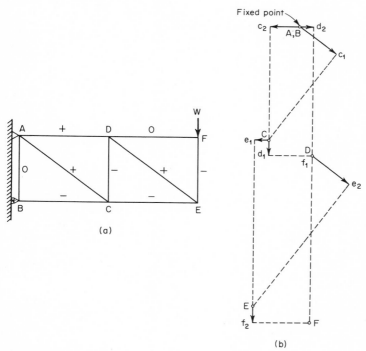

(a)

(b)

Fig. 22

The technique applies only to *simple* frames (Art. 13) and in its direct form is not applicable to compound or complex frames.

The frame of Fig. 23a has a fixed point A, but the fixed direction AK is not directly useful as it is not a single member. In the solution shown in b, E (or F) is assumed fixed and EF is assumed to remain vertical. The vector AK in the resulting diagram is not horizontal, as it should be. This condition is corrected by rotating the frame about A. The rotation vectors are given by the Mohr correction, which is a scale diagram of the frame turned through 90°. Its scale is found by inspection, as $K'K$ is the known displacement of K. Then, for example, $E'E$ is the true displacement of E.

The point and direction chosen as the datum for the Williot diagram is arbitrary. Any internal joint can be used as the solution gives relative displacements. For example, if W is at joint E in Fig. 23a the system is symmetrical; in this case, as EF is vertical, the Williot diagram for half the frame can be started at E, and this gives all the required information without the need for a Mohr correction.

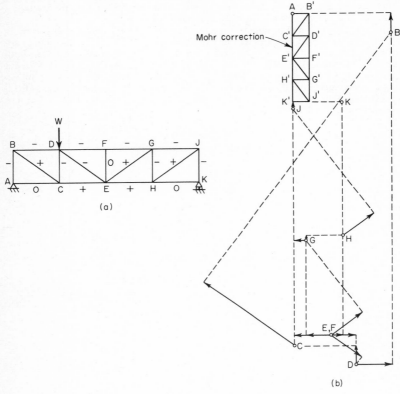

(a)

(b)

Fig. 23

18. Numerical-Integration Method for Beams   The procedure for computing beam deflections from a given curvature (M/EI) diagram is similar to that for computing bending moments from a distributed loading. The numerical-integration technique for the bending-moment problem (Art. 16) is directly applicable to the deflection problem. It is especially suited to computing overall deflections, and for dealing with problems where the M/EI curve is complicated and contains discontinuities. The steps in the solution are as follows:

1. The general shapes of the moment and the M/EI diagrams are constructed from the nodal values of moment. All the important features of the M/EI diagram must be known, including the extent of linearity, curvature, and the location of points of discontinuity. The M/EI graph shows the distribution of curvature y'' over the length of the beam.

2. The M/EI curve is concentrated in Y'' values at the node points, which is equivalent to considering the beam to be constructed of rigid chords connected by elastic hinges at the node points. Y'' is the angle between the rigid chords at the node points. These concentrations are determined by the formulas given in Fig. 20.

3. Starting at a point of known slope, the Y'' values are added from left to right, giving the chord slopes y'. If there is no point of known slope, chord slopes must be based on a trial value.

4. The deflection increment over any chord is given by

$$\text{Deflection increment} = \text{chord slope} \times \text{chord length}$$

This is tabulated for each chord unless the chord length h is constant, when h becomes a common factor.

5. Starting at a point of known deflection, the deflection increments are added from left to right to give the nodal deflections. An error in the trial value of chord slope in step 3 produces a linear error in the deflection line, so that a linear correction must be applied to the trial solution.

This type of solution is shown in Fig. 24 for the beam of Fig. 21. The point load at C in Fig. 21 causes a slope discontinuity in the M diagram at C, but the diagram is curved and without discontinuity over length AC and CE. The sudden change in I at node C causes a discontinuity in M/EI at C, the diagram over lengths AC and CE remaining continuously curved (Fig. 24).

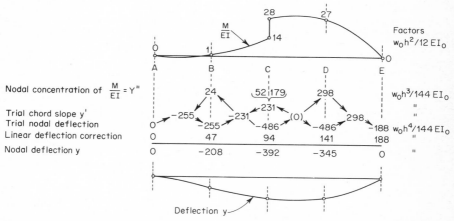

Fig. 24 Computation of beam deflection by numerical integration (see Fig. 21 for moment computation).

The nodal concentrations of curvature are (Fig. 20)

Node B: $\dfrac{h}{12}(a + 10b + c) = \dfrac{h}{12} \times \dfrac{w_0 h^2}{12EI_0}(0 + 10 + 14) = \dfrac{24 w_0 h^3}{144 EI_0}$

Node C: $\dfrac{h}{24}(7c + 6b - a) = \dfrac{h}{24} \times \dfrac{w_0 h^2}{12EI_0}(98 + 6 - 0) = \dfrac{52 w_0 h^3}{144 EI_0}$

Node C: $\dfrac{h}{24}(7a + 6b - c) = \dfrac{h}{24} \times \dfrac{w_0 h^2}{12EI_0}(196 + 162 - 0) = \dfrac{179 w_0 h^3}{144 EI_0}$

Node D: $\dfrac{h}{12}(a + 10b + c) = \dfrac{h}{12} \times \dfrac{w_0 h^2}{12EI_0}(28 + 270 + 0) = \dfrac{298 w_0 h^3}{144 EI_0}$

On account of the lack of symmetry of the system there is no point of known slope, and the chord slopes are based on a trial value $y'_{CD} = 0$. The other y' values are found by summation (in the direction of the arrows). Starting at one of the known deflection values, $y_A = 0$, trial nodal deflections are found by adding the y' values from left to right in the direction of the arrows. This gives a value $y_E = -188$, and as the true value of y_E is known to be zero, the linear correction to deflection must be zero at A and $+188$ at E. The correction at any other node is proportional to the distance of the node from A. The true nodal deflections are the trial plus correction values in each case.

DEFLECTIONS BY UNIT-LOAD METHOD

19. General Form of Solution The structure of Fig. 25a is loaded by the system of externally applied loads $W_1, W_2, \ldots, W_j, \ldots, W_m$, and also by a self-straining system. Equilibrium of the structure is maintained by the reactions $R_1, R_2, \ldots, R_h, \ldots, R_r$. It is assumed that the strains ϵ at every point of the structure, due to the applied loading and self-straining, are known or can be calculated (see note 2 below). It is required to calculate the deflection $\Delta_{Q\theta}$, in a specified direction θ, of a general point Q with reference to some defined datum.

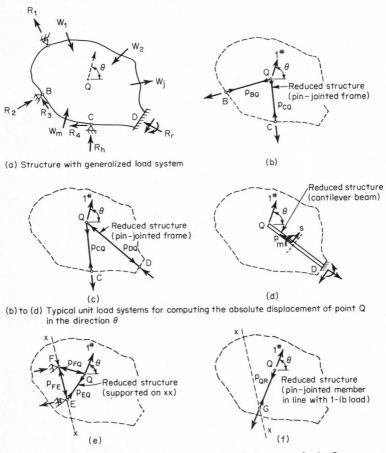

(a) Structure with generalized load system

(b)

(c)

(d)

(b) to (d) Typical unit load systems for computing the absolute displacement of point Q in the direction θ

(e)

(f)

(e) and (f) Typical unit load systems for computing the displacement of point Q in the direction θ relative to datum x–x

Fig. 25 Typical unit-load systems.

To calculate $\Delta_{Q\theta}$, a unit load is applied at Q in the specified direction and reacted at appropriate points on the reference datum by the structure. These are called the "unit-load reactions." The unit load is transmitted to these chosen reactions by a selected system of internal forces within the structure. This force system is called a "statically equivalent force system" and the corresponding structure a "reduced structure." The term "unit-load system" is used to describe the unit load, the unit-load reactions, and the associated statically equivalent force system. The only requirements to be satisfied in the selection of the unit-load system are those of equilibrium: it is not required that the stresses due to the unit-load system should satisfy the compatibility condition for the actual structure (see note 3 below). The required deflection $\Delta_{Q\theta}$ is then given by

$$\Delta_{Q\theta} = \int_V \bar{\sigma} \epsilon \, dV \qquad (22)$$

where $\bar{\sigma}$ is the internal stress at a general point in the reduced structure due to the application of the unit load, and the integral extends over the volume of the actual structure.

Equation (22) gives the unit-load method in its general form. The application of this equation requires two sets of data, the true strains ϵ and the stresses $\bar{\sigma}$ in the reduced structure.

1. The unit-load reactions and relative deflections. The deflection given by Eq. (22) is measured with respect to a datum defined by the position of the unit-load reactions. In calculating the deflection of a redundant structure with respect to the *structural datum* (the support points), the computation is simplified by selecting unit-load reactions which are statically determinate. Three examples, together with corresponding statically determinate reduced structures, are shown in Fig. 25b, c, and d. It will be noted that the unit-load reactions can be selected only from a suitable combination of the actual reactions; it is not permissible to introduce arbitrary reactions on the structural datum. In most problems in structural analysis the reference datum is obvious as it is usually the support points. In cases of *relative deflection* where the datum has to be defined explicitly, an appropriate superscript is added to the deflection symbol (Art. 23). Relative deflections are obtained from Eq. (22) by applying the unit-load reactions at the defined reference datum (Fig. 25e and f).

To calculate a relative rotation due to the true strains ϵ, a unit couple is applied at the point or line concerned and reacted at the reference datum (Fig. 26). The required rotation is obtained by using Eq. (22) in conjunction with an appropriate statically equivalent system due to the unit couple and its reactions.

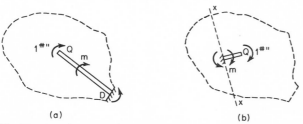

(a)   (b)

Fig. 26 Typical unit-load systems: (a) For computing the rotation of point Q. (b) For computing the rotation of point Q relative to datum x-x.

2. The true strains ϵ in the actual structure. These strains may originate from the external loading or from self-straining, and can be linearly or nonlinearly elastic or plastic; the only conditions to be satisfied are that the continuity of the structure is preserved (the compatibility condition) and that gross distortion does not occur. Gross distortion is defined as a geometrical change sufficiently large to cause significant changes in the statically equivalent forces of the unit-load system.

3. The statically equivalent stress system $\bar{\sigma}$. These are the stresses resulting from the forces in the statically equivalent system. These forces are denoted by p, m, s, and t for axial force, bending moment, shear force, and torsion, respectively. The unit-load system has to satisfy only the equilibrium conditions. When the actual structure is statically determinate, it is defined completely by equilibrium conditions, and hence in this case the actual structure and the reduced structure are identical. In the case of a redundant structure, the reduced structure can be selected from a number of possibilities, which includes a number of statically determinate reduced structures. Also, $\bar{\sigma}$ is zero at all points

of the actual structure except those occupied by the reduced structure. Hence the deflection is obtained by integrating the product (statically equivalent stress) × (corresponding true strain) over the volume of the reduced structure only, and so the computation involved in Eq. (22) can be minimized by a suitable choice of reduced structure.

It will also be observed that a deflection can be found when the true strains are known only over an appropriate reduced structure. For example, in the structure of Fig. 27 there are seven choices of statically determinate reduced structure available to calculate the vertical component of deflection of the point A; five are shown. The choice of b, c, or f results in a minimum amount of computation as it is then necessary to integrate only over the single member AC. Also, if the actual structure is subjected to an arbitrary system of strains which satisfy the general conditions of note 2, and only the strain in the member AC is known, the vertical deflection of the point A can be calculated from the known strain in member AC and the unit-load system of Fig. 27b. Figure 27 shows the principle in its simplest form. It is, however, equally applicable to more complex cases where strain data over the complete structure are not known but sufficient data are available to permit the selection of a reduced structure from which the required deflection can be computed.

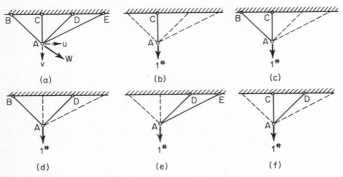

Fig. 27 Choices for statically determinate reduced structure.

4. Superposition of deflection components due to axial force P, bending moment M, shear force S, and torsion T. Each of these forces does work only during its own deformation; hence if an element is simultaneously subjected to two or more types of deformation, the method of superposition can be used and Eq. (22) becomes

$$\Delta_{Q\theta} = \int_V \bar{\sigma}_p \epsilon_P \, dV + \int_V \bar{\sigma}_m \epsilon_M \, dV + \int_V \bar{\sigma}_s \epsilon_S \, dV + \int_V \bar{\sigma}_t \epsilon_T \, dV \tag{23}$$

where ϵ_P, ϵ_M, ϵ_S, and ϵ_T are the true strains due to P, M, S, and T, respectively.

5. Linearly elastic structures. In the case of a linearly elastic frame structure the element is taken as an infinitesimal length dx of the member with section properties A = cross-sectional area, I = moment of inertia, A' = effective area in shear (i.e., $A' = \frac{2}{3}A$ for a rectangular cross section), and J = torsion constant. The unit-load forces are p, m, s, and t, and the corresponding displacements of the element are $P \, dx/AE$, $M \, dx/EI$, $S \, dx/A'G$, and $T \, dx/JG$. Hence Eq. (23) becomes

$$\Delta_{Q\theta} = \sum_M \left\{ \int_0^L p \frac{P}{AE} \, dx + \int_0^L m \frac{M}{EI} \, dx + \int_0^L s \frac{S}{A'E} \, dx + \int_0^L t \frac{T}{JG} \, dx \right\} \tag{24}$$

where the integrals are summed over the $\overline{M}$ elements comprising the selected reduced structure. In many practical cases Eq. (24) reduces to a simple form; for example, in a pin-jointed framework each member of which has a constant area along its length,

$$\Delta_{Q\theta} = \sum_{\overline{M}} p \frac{PL}{AE} \tag{25}$$

6. Note regarding signs. In applying Eqs. (22) through (25), any sign convention can be used provided consistency is maintained between each pair of quantities that appear in a given term. For example, if m and M subject a given fiber of an element to the same sign of stress, their product is positive; similarly for the other corresponding pairs of quantities.

20. Linearly Elastic Pin-Jointed Frameworks   Figure 28a shows a statically determinate, linearly elastic, plane, pin-jointed framework in which the area of each member is constant along its length. To find $\Delta_{C\theta}$, the deflection of point C in the direction θ:

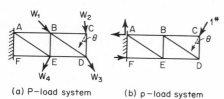

(a) P–load system   (b) p–load system

Fig. 28

1. Apply the loads W_1, W_2, W_3, and W_4 and calculate the force in each member; this gives the P system.

2. The structure is statically determinate; hence the reduced structure is the actual structure. Apply a unit load at point C in the direction θ and calculate the force in each member; this gives the p system (Fig. 28b).

3. The structure is linearly elastic, and the area of each member is constant along its length; hence Eq. (25) applies. The procedure is illustrated by the example of Fig. 29.

In the case of a redundant structure (Fig. 30a) a suitable unit-load system is selected having regard to the computation of Eq. (25). For example, if the vertical deflection of the point K is required, a suitable unit-load system is that shown in Fig. 30b. Two other possible unit-load systems are shown in c and d.

The deflection analysis for a space frame is obtained by the same procedure, but the calculation of the true strain and the unit-load systems (steps 1 and 2) involves the use of the tension-coefficient method (Art. 15).

21. Linearly Elastic Beam Deflections   In most practical problems of beam deflection, shear deformations are negligible. If, in addition, the beam is of constant section along its length, Eq. (24) states that the product EI times the deflection is obtained by an integration of the product mM over the length L. When m and M are each linear functions of the spanwise coordinate x (for example, in many cantilever and simply supported beam problems), the integral of this product can be obtained by applying Eq. (26):

$$\int_0^L mM \, dx = \frac{L}{6} \left[m_0(2M_0 + M_1) + m_1(2M_1 + M_0) \right] \qquad (26)$$

The terms in this equation are defined in Table 5, which also gives the values of the product integrals for some commonly occurring cases of linear diagrams.

Equation (26) can be used only over those parts of the beam where both the m and M diagrams are single straight lines. When discontinuities occur in either diagram the formulas must be applied separately to each part and the partial integrals added. Values of the product integral $\int_0^L mM \, dx$ for a number of common cases in which the M diagram is parabolic are also given in Table 5.

Examples of the application of Table 5 are given in Fig. 31. In Fig. 32 the midspan deflection of the propped cantilever is determined, using the reduced structure shown in c.

Table 5 is directly applicable to other forms of loading which produce the same shapes of loading curves; it is only necessary to substitute AE, $A'G$, or JG for EI; P, S, or T for M; and p, s, or t for m, where appropriate [Eq. (24)].

In the analysis of beams of constant section, the actual bending-moment diagram M may be of arbitrary shape and the m diagram will always consist of not more than two segments of linear variation. The product integral for each segment can always be found

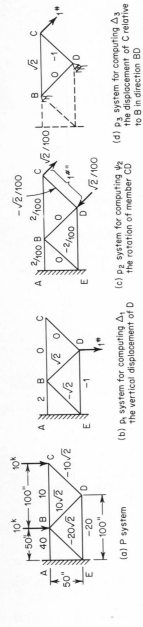

(a) P system

(b) p₁ system for computing Δ_1 the vertical displacement of D

(c) p₂ system for computing ψ_2 the rotation of member CD

(d) p₃ system for computing Δ_3 the displacement of C relative to B in direction BD

Member	Length L, in.	Area A, in.²	Force P, kips	$\frac{PL}{A}$	p_1, lb	$p_1\left(\frac{PL}{A}\right)$	p_2, in-lb	$p_2\left(\frac{PL}{A}\right)$	p_3, lb	$p_3\left(\frac{PL}{A}\right)$
AB	0.5 (100)	1	40	2(10³)	2	4(10³)	2(10⁻²)	4(10)	0	0
BC	1	1	10	1	0	0	2	2	√2	√2(10³)
EB	0.5√2	1	-20√2	-2	-√2	2√2	0	0	0	0
BD	0.5√2	1	10√2	1	√2	√2	0	0	0	0
DC	0.5√2	1	-10√2	-1	0	0	-√2	√2	-1	1
ED	1	1	-20	-2	-1	2	-2	4	0	0
						10.24(10³)		11.41(10)		2.41(10³)

(1) Vertical displacement of D $\quad 1 \times \Delta_1 = \Sigma\, p_1\left(\frac{PL}{AE}\right) = \frac{10.24 \times 10^3}{30 \times 10^3} = 0.34\ \text{in.}$

(2) Rotation of member CD $\quad 1 \times \psi_2 = \Sigma\, p_2\left(\frac{PL}{AE}\right) = \frac{11.41 \times 10}{30 \times 10^3} = 0.0038\ \text{rad}$

(3) Displacement of C relative to B in direction BD $\quad 1 \times \Delta_3 = \Sigma\, p_3\left(\frac{PL}{AE}\right) = \frac{2.41 \times 10^3}{30 \times 10^3} = 0.08\ \text{in.}$

Fig. 29

by a simple numerical integration (for example, by Simpson's rule). Alternatively, and more conveniently, Eq. (27) can be applied:

$$\int_0^L mM \, dx = A_M m_G \tag{27}$$

where A_M and m_G are defined in Fig. 33.

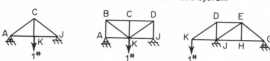

(a) Redundant structure with P-load system

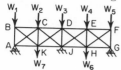

(b) to (d) Some statically determinate reduced structures for the p-load system

Fig. 30

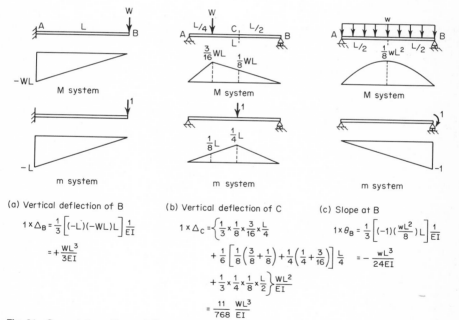

M system   M system   M system

m system   m system   m system

(a) Vertical deflection of B

$$1 \times \Delta_B = \frac{1}{3}\Big[(-L)(-WL)L\Big]\frac{1}{EI}$$

$$= +\frac{WL^3}{3EI}$$

(b) Vertical deflection of C

$$1 \times \Delta_C = \Big\{\frac{1}{3} \times \frac{1}{8} \times \frac{3}{16} \times \frac{L}{4}$$

$$+ \frac{1}{6}\Big[\frac{1}{8}\Big(\frac{3}{8}+\frac{1}{8}\Big) + \frac{1}{4}\Big(\frac{1}{4}+\frac{3}{16}\Big)\Big]\frac{L}{4}$$

$$+ \frac{1}{3} \times \frac{1}{4} \times \frac{1}{8} \times \frac{L}{2}\Big\}\frac{WL^2}{EI}$$

$$= \frac{11}{768}\frac{WL^3}{EI}$$

(c) Slope at B

$$1 \times \theta_B = \frac{1}{3}\Big[(-1)(\frac{wL^2}{8})L\Big]\frac{1}{EI}$$

$$= -\frac{wL^3}{24EI}$$

Fig. 31 Computation of beam deflections (see Table 5).

In the case of beams of variable section, EI must be included under the integral sign. A numerical integration could be applied to the integral $\int_0^L m(M/EI)dx$. Alternatively, the value of the integral can be found by applying Eq. (28), which is an extension of Eq. (27):

$$\int_0^L m\frac{M}{EI} \, dx = A_M^* m_G^* \tag{28}$$

where A_M^* is the area under the M/EI curve and m_G^* is the ordinate of the m curve corresponding to the centroid of the area of the M/EI curve.

TABLE 5 Values of $\int_0^L mM\,dx$

m diagram \ M diagram	Linear M diagrams				Parabolic M diagrams		
	M (rect.)	M_0 (triangle)	M_1 (triangle)	M_0,M_1 (trapezoid)	M_1 ($L/2,L/2$)	M_1 (origin)	M_0,M_1 (origin)
m (rectangle)	mML	$\dfrac{1}{2}mM_0L$	$\dfrac{1}{2}mM_1L$	$\dfrac{1}{2}mL(M_0+M_1)$	$\dfrac{2}{3}mM_1L$	$\dfrac{1}{3}mM_1L$	$\dfrac{1}{3}mL(2M_0-M_1)$
m_0 (triangle)	$\dfrac{1}{2}m_0ML$	$\dfrac{1}{3}m_0M_0L$	$\dfrac{1}{6}m_0M_1L$	$\dfrac{1}{6}m_0L(2M_0+M_1)$	$\dfrac{1}{3}m_0M_1L$	$\dfrac{1}{12}m_0M_1L$	$\dfrac{1}{12}m_0L(5M_0-M_1)$
m_1 (triangle)	$\dfrac{1}{2}m_1ML$	$\dfrac{1}{6}m_1M_0L$	$\dfrac{1}{3}m_1M_1L$	$\dfrac{1}{6}m_1L(2M_1+M_0)$	$\dfrac{1}{3}m_1M_1L$	$\dfrac{1}{4}m_1M_1L$	$\dfrac{1}{4}m_1L(M_0-M_1)$
m_0,m_1 (trapezoid)	$\dfrac{1}{2}ML(m_0+m_1)$	$\dfrac{1}{6}M_0L(2m_0+m_1)$	$\dfrac{1}{6}M_1L(m_0+2m_1)$	$\dfrac{L}{6}\left[m_0(2M_0+M_1)+m_1(2M_1+M_0)\right]$	$\dfrac{1}{3}M_1L(m_0+m_1)$	$\dfrac{1}{12}M_1L(m_0+3m_1)$	$\dfrac{L}{12}\left[m_0(5M_0-M_1)+3m_1(M_0-M_1)\right]$

22. Linearly Elastic Beam-Type Structures The procedure for calculating the deflection of statically determinate and redundant beam-type structures is summarized in Fig. 34, where the horizontal deflection of B is to be determined. In the case of the redundant structure any suitable statically determinate unit-load system can be chosen (Art. 19), and by an appropriate choice, the integral $\int_0^L m(M/EI)dx$ can be simplified. The advantage of the first of two possible m systems shown in b is self-evident.

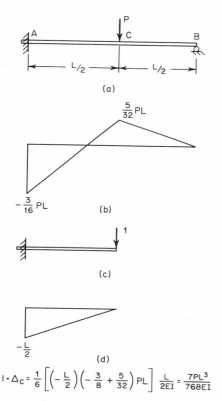

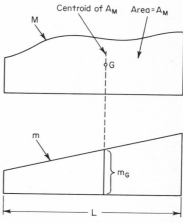

$$1 \cdot \Delta_C = \frac{1}{6}\left[\left(-\frac{L}{2}\right)\left(-\frac{3}{8}+\frac{5}{32}\right)PL\right]\frac{L}{2EI} = \frac{7PL^3}{768EI}$$

Fig. 32 Deflection of propped cantilever (see Table 5).

Fig. 33 Evaluation of $\int_0^L m\mathrm{M}\ dx$ for general M diagram.

23. Relative Deflections of Linearly Elastic Structures In the truss of Fig. 28, to calculate the displacement of joint C relative to joint B, in the direction θ relative to line BE ($\Delta_{C\theta}^{BE}$, where the superscript BE denotes the datum, point B, direction BE), the unit load is applied at C in the direction θ and reacted at the line BE with the fixed point of the reduced structure at B (Fig. 35). The statically equivalent force system is calculated and the relative displacement $\Delta_{C\theta}^{BE}$ is then obtained from Eq. (25).

To calculate a relative rotation, a unit couple is applied at the point or the line concerned and reacted at the reference datum. For example, to calculate the rotation of the member EG of the redundant truss of Fig. 30a relative to the line CK (in its deflected position) the unit-load system can be chosen as in Fig. 36.

To calculate the deflection of the point F relative to the line joining the points C and D of the redundant frame $ABCD$ of Fig. 34b, the statically equivalent force system of Fig. 37a may be used. To calculate the slope at point C relative to the line joining C and D, the procedure is similar but the unit-load system is a unit couple applied at C and reacted at C and D (Fig. 37b).

24. Linearly Elastic Structures: General Case A complete description of the deflected configuration of a three-dimensional structure involves the computation of six displacement components at each joint (three translations and three rotations) for each applied

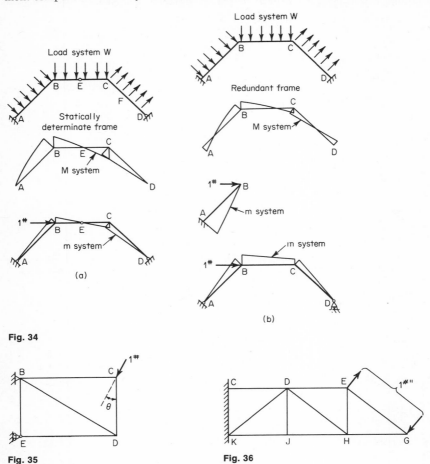

Fig. 34

Fig. 35   **Fig. 36**

loading or specified strain system, in the case of a plane structure three displacement components per joint (two translations and one rotation). The deflected configuration of the whole structure is unlikely to be required in practice, the more usual requirement being the calculation of the displacement components at one or two points within the structure.

Figure 38b shows a reduced structure and the corresponding unit-load system which could be used to calculate the horizontal deflection in the direction JK of joint J of the structure shown in a. Rotation of joint J in planes $AJKB$ and $JKLM$ could be determined using, respectively, the reduced structures and unit-load systems shown in c and d.

25. Nonlinearly Elastic and Plastic Structures If any part of a structure has nonlinear characteristics or is loaded into the plastic range, the procedure described in the preceding sections remains applicable, provided the true strains are known. It is only necessary to substitute the true strain (which may be nonlinearly elastic or plastic) for the elastic strain terms (Art. 19, note 2).

26. Deflection Due to Self-Straining Equation (22) is directly applicable to the problem of calculating deflection due to self-straining. The only new feature in this case is the calculation of the true strains ϵ due to self-straining. When the structure is redundant, the true strains are obtained by the methods of Art. 31. In the case of a statically determinate structure, a change in shape of any element of the structure does not cause self-straining (that is, internal forces are not induced, but the structural geometry is altered). The strains ϵ in this case are the specified deformations of the individual elements of the structure, and the unit-load method permits the calculation of the deformed shape.

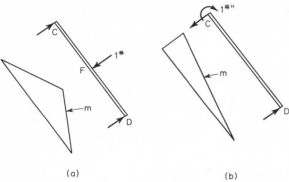

(a)   (b)

Fig. 37

27. The Complementary-Energy Method and Castigliano's Theorem The deflections at a general point in an elastic structure can be obtained by applying the complementary-energy method (Art. 7) in the form given by Eq. (13). When the system is linearly elastic the strain energy is numerically equal to the complementary energy and Eq. (15) of Art. 8 is applicable. These equations enable the deflection Δ_j corresponding to the load W_j to be

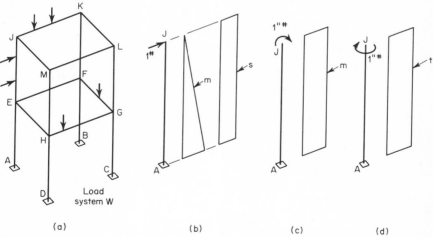

(a)   (b)   (c)   (d)

Fig. 38

calculated. They can also be applied to calculate the deflection at an unloaded point on a structure. In this case a fictitious load is applied at the point and in the direction of the required deflection, and after differentiation the load is made equal to zero.

In applying Castigliano's theorem the load W_j, for which the corresponding deflection is sought, is an independent parameter in the energy integral. For example, in an elastic structure containing members loaded axially and members loaded in bending, the strain energy can be expressed as

$$U = \sum \frac{P^2 L}{2AE} + \int \frac{M^2 dx}{2EI} \tag{29}$$

Differentiating U with respect to the load W_j before the integration is carried out gives

$$\Delta_j = \frac{\partial U}{\partial W_j} = \sum \frac{P(\partial P/\partial W_j)L}{AE} + \int \frac{M(\partial M/\partial W_j)}{EI} dx \tag{30}$$

The practical application of Eq. (30) is thus identical with the basic unit-load equation [Eq. (24)].

INFLUENCE LINES

An *influence line* is a graph showing the variation of a single (generalized) force or deflection, with respect to the position on the structure of a unit load. The ordinates of an influence line are called *influence coefficients* and are denoted by a_{ij}. The influence line for P_i is constructed by erecting at j the coefficient a_{ij}, the value of P_i due to a unit load at j.

For example, in the continuous beam ABC of Fig. 39, the influence line for R_B is the graph showing the value of R_B for any position of a point load on AC. This graph enables the total effect on R_B of a set of point forces or of a distributed force on AC to be obtained as follows:

1. If $W_1, W_2, \ldots, W_j, \ldots, W_m$ is a system of vertical point loads acting on AC, and if $a_{B1}, a_{B2}, \ldots, a_{Bj}, \ldots, a_{Bm}$ are the corresponding coefficients on the influence line (Fig. 39b), the total value of R_B is

$$R_B = W_1 a_{B1} + W_2 a_{B2} + \cdots + W_j a_{Bj} + \cdots + W_m a_{Bm} = \sum_{j=1}^{m} W_j a_{Bj}$$

2. If a distributed load w acts from 1 to 2 (Fig. 39c), the value of R_B is $w \times$ (*influence area* from 1 to 2):

$$R_B = w \int_{x1}^{x2} a_{Bx} \, dx$$

3. If w is not uniform, but varies with x,

$$R_B = \int_{x1}^{x2} w_x a_{Bx} \, dx$$

The influence line for any quantity can be found directly by computing the influence coefficient for each chosen position of the unit-point load. Alternatively, and generally more conveniently, it can be found indirectly by a virtual-displacement diagram.

The influence line for any independent force component is identical with the virtual-displacement diagram of the *load line* (the line at which the loading is applied to the structure) when the force component is given a unit virtual displacement. The technique applies to determinate structures, whether linear or nonlinear. In its application to indeterminate structures the method is usually called the Müller-Breslau principle. In this case the method makes use of the Maxwell reciprocal theorem *and hence is restricted to linear systems.* This technique is useful in three ways:

1. It enables the general shape of an influence line to be visualized from the displacement of the structure.

2. It enables the precise shape of an influence line and its ordinates to be found from a single deflection calculation. Any method which gives the deflected configuration of a structure due to a single relative displacement is especially useful (for example, the Williot diagram for some plane-frame problems, and numerical integration for some beam problems).

3. It enables influence lines to be found experimentally by the use of deformers in model analysis.

28. Statically Determinate Systems   In a statically determinate system, a unit displacement applied to any force component does not cause self-straining; the structure displaces as a set of connected rigid bodies, and hence the influence line consists only of straight lines.

Pin-Jointed Frameworks. In a statically determinate pin-jointed framework, the virtual-displacement diagram (and hence the influence line) can usually be sketched by inspection. For example, if the influence line for P_{XY}, the force in XY (Fig. 40a), is required, it can be determined by giving P_{XY} a corresponding displacement. This is effected by "cutting" the member and producing a unit gap. A positive sign for P_{XY} indicates tension. This applied deformation alters the shape of triangle XYZ, and the structure is displaced as two rigid bodies pivoting about Z. The influence line for P_{XY} is thus a triangle with apex at Z, and this is completely defined by one ordinate (Fig. 40b). This ordinate could be found from the geometrical properties of virtual-displacement diagrams but is more conveniently found by solving for P_{XY} due to a unit point load at Z.

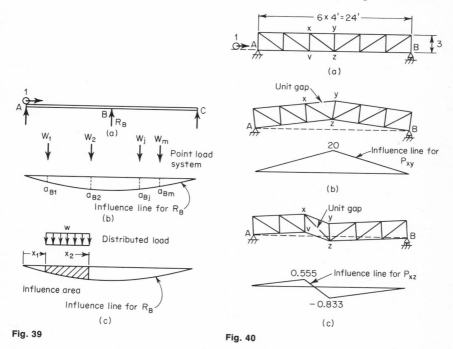

Fig. 39   **Fig. 40**

The influence line for P_{XZ} is found in a similar manner by giving a unit deformation to XZ. This distorts panel $VXYZ$, and the structure displaces as two rigid bodies joined by the deformed panel. This influence line is fully defined by two ordinates, those at V and Z. In this case, AV remains parallel to ZB, so that it is sufficient to calculate one ordinate. Thus, the ordinate 0.555 is determined by solving for P_{XZ} due to a unit point load at V.

Beam Systems. In beam systems, influence lines may be required for three quantities: reaction, shear, and moment. The unit-displacement method can be used in all cases, as illustrated in the structure of Fig. 41.

The influence line for R_B is found by applying a unit downward displacement to support B (Fig. 41b).

The influence line for S_X is found by applying a unit displacement to S_X. For this purpose the beam can be considered cut at X, and the two cut ends given a relative vertical displacement while being kept parallel. The influence line is the displaced position of the beam, and as lengths AX and XC remain parallel the influence line is completely defined by one ordinate. For example, when a unit load is at C, $R_A = 0.5$ and thus $S_X = 0.5$, giving the influence coefficient at C. All other coefficients can be found from the geometry of the system.

The influence line for M_X is found by applying a unit rotation to M_X. For this purpose the beam can again be cut at X, and the two cut ends given a unit relative rotation. The influence line is the displaced position of the beam, and again this is fully defined by one ordinate. For example, due to a unit load at C, $R_A = 0.5$, and hence $M_X = 0.5 \times 30 = 15$.

29. Statically Indeterminate Systems In a statically indeterminate system, a unit displacement applied to any of the redundant forces causes self-straining. The forces due to

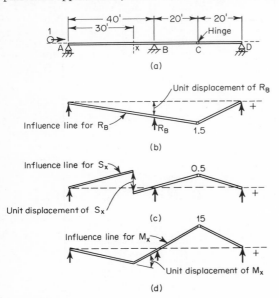

Fig. 41

this self-straining must be computed before the displacement of the structure can be determined. The influence line is defined by the deflected position of the load line.

Pin-Jointed Frameworks. The influence line for a force in a redundant plane pin-jointed framework with one (or perhaps two) redundancies can easily be found by the Williot-Mohr diagram (Art. 17). For example, in the two-pinned frame of Fig. 42a, influence lines for H can be found by giving H a unit displacement and computing the equivalent displacements of all other joints. The reduced structure for this calculation is the frame with the horizontal restraint at A removed, and a point force H is applied at A. Because of symmetry, only half the frame need be considered. From the resulting deflection of the frame, any influence line for H can be found by making $\delta_H = -1$ and scaling all other deflections in proportion. For example, due to normal loading applied to the outer chord of the frame, the influence line for H is as shown in Fig. 42c. The influence coefficient at a point is the component of deflection at that point corresponding to the applied loading; thus the influence coefficient at joint X due to a unit load applied normal to the chord at X is the component of deflection of X in the direction of this load.

The simplicity of this solution derives from the fact that a force analysis due to self-straining involves one less redundancy than exists in the original structure, and hence in the case of a singly redundant structure the force computation is applied to a statically determinate system. In the case of a multiredundant structure there is no such simplicity. The self-straining produces a set of indeterminate forces throughout the frame, which have to be calculated. From these forces and the associated element deformations, the deflected shape can be found by the methods discussed in Art. 19 applied to a suitably chosen reduced structure.

Beams. In the case of a singly redundant beam, the computation of an influence line is a statically determinate problem. For example, the influence line for R_B in Fig. 43 is found by giving R_B a unit displacement. This is effected by applying a point force at B to beam AC supported only at A and C. The deflected shape can be found most conveniently by

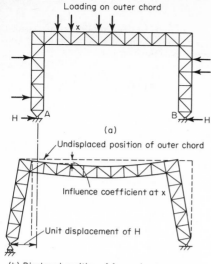

Loading on outer chord

H → A   B ← H

(a)

Undisplaced position of outer chord

Influence coefficient at x

Unit displacement of H

(b) Displaced position of frame due to
unit displacement of H

Fig. 42

−1

(c) Influence line for H due to normal loading
on outer chord of frame

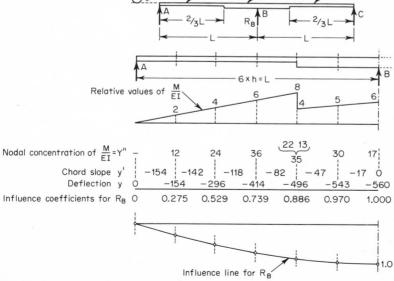

Relative values of $\frac{M}{EI}$

Nodal concentration of $\frac{M}{EI} = Y''$

Chord slope y'
Deflection y

Influence coefficients for R_B

Influence line for R_B

Fig. 43 Computation of influence-line ordinates for center reaction.

numerical integration (Art. 18). As this beam is symmetrical, the influence line is also symmetrical, and only one span need by considered. The solution to this problem is given in Fig. 43. It will be noted that only relative deflections of the beam are required, and thus only relative M/EI values need be used. The deflection values are scaled to make $y_B = 1$.

In a multispan problem the unit-displacement method demonstrates the general shape of an influence line as shown in Fig. 44. Such an influence line usually provides sufficient information for positioning the applied loading to give the maximum value of the force under consideration. If the coefficients of the influence line are required, the redundant bending moments (for example, the moments at the supports) due to self-straining must be found. The reduced structure is a set of simply supported spans, and the deflected position of each span can be found separately by, for example, numerical integration.

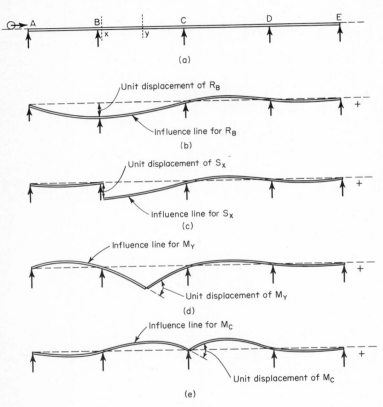

Fig. 44

STATICALLY INDETERMINATE STRUCTURES

30. Choice of Method It is suggested that a basis for the choice of method be considered under, and in the order of, the following headings.

1. Structural behavior. The most generally useful methods for the analysis of linearly elastic redundant structures are the unit-load method (Art. 32) and the unit-displacement method (Art. 37). In the case of nonlinearly elastic systems there is no choice; the correct solution can be obtained only by the direct use of the principle of virtual work in one or another of its forms (Art. 5). Care must be exercised to ensure that techniques which are derived from particular cases of the virtual-work principle—for example, the so-called "least-work" principle—which are restricted to linearly elastic systems, are not applied to nonlinear problems.

The minimum-complementary-potential-energy theorem and the minimum-potential-energy theorem are restatements of the virtual-work principle. The minimum-complementary-energy method provides the basic force method of analysis of nonlinearly elastic structures (Art. 40). This method includes the particular case of linearly elastic behavior (Castigliano's theorem, Art. 33). The basic variational form of the displacement method is provided by the minimum-potential-energy method (Art. 41). This method is of general application but is most commonly used for obtaining approximate solutions to linearly elastic structures.

2. The choice between the *force method* and the *displacement method*. In force methods the problem coordinates are the redundant, self-balancing forces. The number of redundant forces is determined, and appropriate forces designated as redundant. In displacement methods the problem coordinates are the displacements in the degrees of freedom of the structure. Thus the analysis deals directly with the actual complete structure, and the concepts of determinacy and redundancy are irrelevant. For this reason, displacement methods tend to be favored as the number of redundancies increases, because in such cases the selection of the simplest reduced structure for use in the force method becomes increasingly difficult. Further, since the displacement method deals directly with the complete structure, engineering intuition is not required to prepare it for a computer, and so it is more easily adapted to standardized computing techniques.

In the force method, the number of equations to be solved is equal to the number of redundant forces; in the displacement method the number of equations equals the number of degrees of freedom. This is a fundamental distinction between the force and the displacement methods, and since it provides a basis of comparison which can be evaluated quantitatively, it exerts a strong influence on the choice of method. The comparison between the two from this standpoint is illustrated by the structure of Fig. 45. This structure has six redundant forces; hence if the force method is used, there are six unknowns. However, the structure has only two degrees of freedom; hence the solution by the displacement method contains two unknowns.

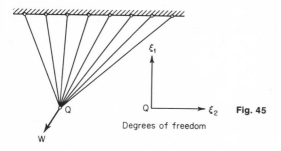

Fig. 45

Degrees of freedom

In practice the configuration of pin-jointed frameworks is usually such that the number of redundant forces is much less than the number of degrees of freedom. The truss of Fig. 30a has four redundant members and one redundant reaction component, but there are 16 degrees of freedom, two at each internal joint and one each at joints A and G. Hence it can be stated that, in general, the force method is preferable to the displacement method for the usual configurations of pin-jointed frameworks.

In the case of beam-type structures, broad conclusions on the choice of method, from the standpoint of the number of unknowns involved, can be stated in terms of the manner of assembly of the elements of the structure. For example, a structure consisting of beam elements in series tends to have fewer unknowns by the force method than by the displacement method. Thus, the structure of Fig. 46 has four degrees of freedom ξ but only two redundant forces. Conversely, a structure consisting of an assemblage of beam elements in parallel (for example, stiff-jointed frameworks of the portal type) tends to have fewer unknowns by the displacement method than by the force method. In the structure of Fig. 47 there are 18 redundant forces but only 10 degrees of freedom.

Other factors may operate to influence the final choice. For example, (1) consideration has to be given to the ease of setting up the equations; (2) the equations in the displacement method for stiff-jointed frameworks are almost invariably well-conditioned, whereas the force-method equations in some cases tend to be ill-conditioned; and (3) the primary objective of the analysis is important, i.e., whether it is the calculation of forces or displacements.

3. The choice of technique. The principle techniques in common use are shown in Table 3. Various other specialized techniques and devices have each been designed to expedite a solution for a particular class of problem. Generally speaking, if such a device is used outside the area for which it has been designed, it results in a laborious solution.

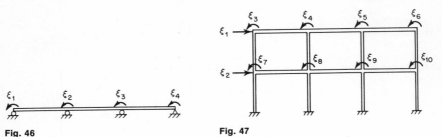

Fig. 46   **Fig. 47**

Linearly Elastic Structures: Force Methods

31. Flexibility Coefficients   The term "flexibility coefficient" is used to define the deflection caused by a unit load or by a load system. In either case, the deflection can relate to an element of the structure, the complete structure, or to selected parts of the complete structure, as, for example, the reduced structure used in the analysis of a redundant structure.

The deflections in the degrees of freedom of a structure (Art. 1) define the deflected shape of the structure. A system of loads $W_1, W_2, \ldots, W_j, \ldots, W_m$ applied to a linearly elastic structure produces corresponding deflections $\Delta_1, \Delta_2, \ldots, \Delta_j, \ldots, \Delta_m$ in the degrees of freedom $\xi_1, \xi_2, \ldots, \xi_j, \ldots, \xi_m$. The deflections can be expressed in terms of flexibility coefficients as follows:

$$
\begin{aligned}
\Delta_1 &= f_{11}W_1 + f_{12}W_2 + \cdots + f_{1j}W_j + \cdots + f_{1m}W_m \\
\Delta_2 &= f_{21}W_1 + f_{22}W_2 + \cdots + f_{2j}W_j + \cdots + f_{2m}W_m \\
&\cdots\cdots\cdots\cdots\cdots\cdots\cdots\cdots\cdots\cdots\cdots\cdots\cdots\cdots\cdots\cdots \\
\Delta_j &= f_{j1}W_1 + f_{j2}W_2 + \cdots + f_{jj}W_j + \cdots + f_{jm}W_m \\
&\cdots\cdots\cdots\cdots\cdots\cdots\cdots\cdots\cdots\cdots\cdots\cdots\cdots\cdots\cdots\cdots \\
\Delta_m &= f_{m1}W_1 + f_{m2}W_2 + \cdots + f_{mj}W_j + \cdots + f_{mm}W_m
\end{aligned}
\tag{31}
$$

That is, the deflection of the typical coordinate j can be written

$$
\Delta_j = \sum_{i=1}^{m} f_{ji}W_i
\tag{32}
$$

where the element f_{ji} is the *flexibility coefficient,* which in this case is defined as the deflection in the degree of freedom (coordinate)* j due to the application of a unit load applied in the degree of freedom (coordinate) i. This definition is the one most commonly used, but the term flexibility coefficient is also used in a more general sense, as, for example, for the δ_{i0} coefficient discussed later in this article. The coefficients f_{ii} and f_{ij} are called direct- and cross-flexibility coefficients, respectively.

Simple examples of flexibility coefficients for structures of one and two degrees of freedom are given in Fig. 48. Flexibility coefficients originate from Maxwell's theorem (Art. 9), and the generalizations relating to "force" and "deflection" noted there also apply to flexibility coefficients [Eq. (18)]. For example, a unit load applied in the coordinate i may cause a rotation in the coordinate j, and a unit couple applied at i may cause a translation in j (Fig. 49).

It is sometimes necessary to define specifically the datum for the deflections, and in

---

*For conciseness in dealing with flexibility and stiffness coefficients, the word "coordinate" is used in a generalized sense, that is, synonymously with the term "degree of freedom."

these cases a superscript is added to the flexibility coefficient. Hence f_{ij}^r is defined as the deflection in the coordinate i with respect to the datum r due to the application of a unit load applied at j and reacted at the coordinates defining the datum r. A flexibility coefficient can be calculated by applying the unit-load theorem of Art. 5 [Eq. (9)]. In the general case, the flexibility coefficient of an element is given by

$$f_{ij} = \int_0^L \frac{p_i p_j}{AE}\,dz + \int_0^L \frac{m_{ix}m_{jx}}{EI_x}\,dz + \int_0^L \frac{m_{iy}m_{jy}}{EI_y}\,dz$$
$$+ \int_0^L \frac{s_{ix}s_{jx}}{A_x'G}\,dz + \int_0^L \frac{s_{iy}s_{jy}}{A_y'G}\,dz + \int_0^L \frac{t_i t_j}{JG}\,dz \quad (33)$$

where x and y are the principal axes of the cross section.

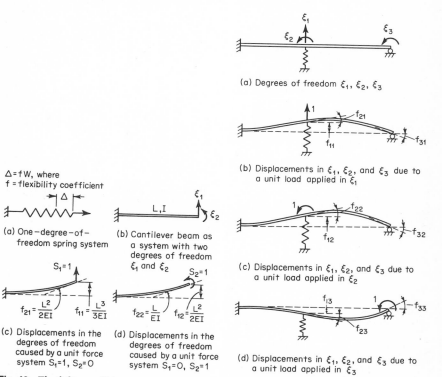

$\Delta = fW$, where
f = flexibility coefficient

(a) One–degree–of–freedom spring system

(b) Cantilever beam as a system with two degrees of freedom ξ_1 and ξ_2

(c) Displacements in the degrees of freedom caused by a unit force system $S_1=1$, $S_2=0$

(d) Displacements in the degrees of freedom caused by a unit force system $S_1=0$, $S_2=1$

Fig. 48 Flexibility coefficients.

(a) Degrees of freedom ξ_1, ξ_2, ξ_3

(b) Displacements in ξ_1, ξ_2, and ξ_3 due to a unit load applied in ξ_1

(c) Displacements in ξ_1, ξ_2, and ξ_3 due to a unit load applied in ξ_2

(d) Displacements in ξ_1, ξ_2, and ξ_3 due to a unit load applied in ξ_3

Fig. 49

Flexibility coefficients of the simple elements which occur commonly in structural analysis are presented in Table 6. The coefficients given for prismatic beam elements can also be deduced from standard beam formulas (Table 4). Values of α and β for the linearly tapered members are shown in Fig. 50. The flexibility matrix given in Table 6 is discussed in Art. 42.

In the analysis of a redundant structure by the unit-load method (Art. 32), the condition for compatibility of displacements is expressed in terms of flexibility coefficients δ_{i0} and δ_{ik}. These coefficients are relative deflections of the reduced structure and are defined as follows:

1. δ_{i0} is the relative deflection of the ends of the "cut" ith redundancy due to the action on the reduced structure of all loads and effects except the redundant forces; that is, δ_{i0} is the relative deflection due to the externally applied loads and any self-straining actions which may be present (Fig. 51b).

2. δ_{ik} is the relative deflection of the ends of the cut ith redundancy due to the action on the reduced structure of a unit value of the kth redundant force (Fig. 51c).

Both δ_{io} and δ_{ik} are taken positive when the relative deflections are in the assumed positive direction of the ith redundant force. Each suffix i and k ranges over all the redundancies of the structure and hence, for a structure with n redundancies, n coeffi-

TABLE 6   Flexibility Coefficients for Beam Elements

Element	Loading action on element	Independent force	Flexibility coefficient* f_{ij}	Flexibility matrix*† f_g
Linear taper bar axial force		S_1	$f_{11} = \dfrac{L}{A_B E}\, \alpha_{22}$	$\dfrac{L}{A_B E}\, \alpha_{22}$
Linear taper cantilever beam		S_1	$f_{11} = \dfrac{L^3}{3EI_B}\, \alpha_{11}$ $f_{21} = \dfrac{L^2}{2EI_B}\, \alpha_{21}$	$\begin{bmatrix} \dfrac{L^3}{3EI_B}\, \alpha_{11} & \dfrac{L^2}{2EI_B}\, \alpha_{12} \\[2mm] \dfrac{L^2}{2EI_B}\, \alpha_{21} & \dfrac{L}{EI_B}\, \alpha_{22} \end{bmatrix}$
		S_2	$f_{12} = \dfrac{L^2}{2EI_B}\, \alpha_{12}$ $f_{22} = \dfrac{L}{EI_B}\, \alpha_{22}$	
Linear taper simply supported beam		S_1	$f_{11} = \dfrac{L}{3EI_B}\, \beta_{11}$ $f_{21} = \dfrac{L}{6EI_B}\, \beta_{21}$	$\begin{bmatrix} \dfrac{L}{3EI_B}\, \beta_{11} & \dfrac{L}{6EI_B}\, \beta_{12} \\[2mm] \dfrac{L}{6EI_B}\, \beta_{21} & \dfrac{L}{3EI_B}\, \beta_{22} \end{bmatrix}$
		S_2	$f_{12} = \dfrac{L}{6EI_B}\, \beta_{12}$ $f_{22} = \dfrac{L}{3EI_B}\, \beta_{22}$	

* Coefficients α and β (Fig. 50) are based on linear taper of element. For prismatic elements $(A_A = A_B, I_A = I_B)$ $\alpha = \beta = 1$

† See Art. 42

cients δ_{io} and n^2 coefficients δ_{ik} are required [see Eqs. (37)]. The calculation of δ_{io} involves only the members of the reduced structure. The calculation of δ_{ik} involves the members of the reduced structure plus the ith member; the ith member is loaded by a unit force and hence, by virtue of its own flexibility, contributes to δ_{ik}. It is convenient in all cases to calculate δ_{io} and δ_{ik} in a single table which lists all members of the structure (see Fig. 54).

These coefficients are most conveniently calculated by applying the methods of Art. 19. Hence, in the case when the structure is loaded only by external loads and not by any self-straining action, Eq. (24) is directly applicable and δ_{io} and δ_{ik} are given by

$$\delta_{i0} = \sum_{\overline{M}} \left[\int_0^L p_i \left(\frac{P}{AE} \right)_0 dz + \int_0^L m_{ix} \left(\frac{M_x}{EI_x} \right)_0 dz + \int_0^L m_{iy} \left(\frac{M_y}{EI_y} \right)_0 dz \right.$$

$$\left. + \int_0^L s_{ix} \left(\frac{S_x}{A_x'G} \right)_0 dz + \int_0^L s_{iy} \left(\frac{S_y}{A_y'G} \right)_0 dz + \int_0^L t_i \left(\frac{T}{GJ} \right)_0 dz \right] \quad (34a)$$

$$\delta_{ik} = \sum_{\overline{M}} \left[\int_0^L p_i \left(\frac{P}{AE} \right)_k dz + \int_0^L m_{ix} \left(\frac{M_x}{EI_x} \right)_k dz + \int_0^L m_{iy} \left(\frac{M_y}{EI_y} \right)_k dz \right.$$

$$\left. + \int_0^L s_{ix} \left(\frac{S_x}{A_x'G} \right)_k dz + \int_0^L s_{iy} \left(\frac{S_y}{A_y'G} \right)_0 dz + \int_0^L t_i \left(\frac{T}{GJ} \right)_k dz \right] \quad (34b)$$

In Eqs. (34a) and (34b) the suffix i denotes the force in an element of the structure due to a unit load applied at the point of required deflection (the point i); the terms $(\ )_0$

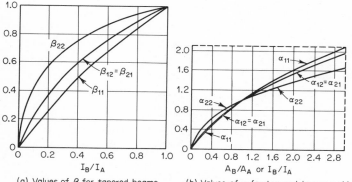

(a) Values of β for tapered beams
(b) Values of α for tapered beams and bars

Fig. 50 Coefficients α and β for Table 6. (From L. B. Wehle and W. Lansing, A Method for Reducing the Analysis of Complex Redundant Structures to a Routine Procedure, J. Aeron. Sci., vol. 19, no. 10, October 1952.)

denote the true strains in the elements of the reduced structure due to the externally applied loads, and the terms $(\ )_k$ denote the true strains in the elements of the reduced structure due to the application of a unit value of the kth redundant force. The terms of Eq. (34a) are summed over the $\overline{M}$ members of the reduced structure. The terms of Eq. (34b) are also summed over the $\overline{M}$ members of the reduced structure, but in the case of direct flexibility coefficients δ_{ii} the terms are summed over the $\overline{M}$ members *plus the member i*.

Self-Straining. If, in addition to the externally applied loading system, the structure is also subjected to self-straining actions, it is necessary to calculate the contribution to δ_{i0} from self-straining [the δ_{ik} coefficients given by Eq. (34b) are not affected by self-straining]. The contribution to δ_{i0} is, in all cases, a change in the geometry of the reduced structure. Whatever the origin of the self-straining, its effect on δ_{i0} is, by definition, the relative deflection of the reduced structure corresponding to the ith redundancy (Fig. 52). This relative deflection is the cumulative effect from all the members comprising the reduced structure and its supports, and may, for example, be a change in length (corresponding to an axial-force redundancy) or a change in slope (corresponding to a bending-moment redundancy). It is convenient to discuss separately the effect on δ_{i0} of self-straining due to internal causes and due to external causes.

1. Internal causes (Fig. 52a). The effects due to internal causes arise from errors in manufacture (lack of fit) or from differential expansions of the members. The simplest case occurs when the redundant member is itself initially oversize (an oversize λ in the positive X_i direction is positive). In this case the contribution to δ_{i0} is simply the specified amount of member oversize λ. If initial lack of fit is present in some members of the reduced structure, the contribution to δ_{i0} is obtained by the unit-load theorem as follows:

$$\delta_{ios} = \sum_M u_i \lambda \qquad (35)$$

where δ_{ios} is the contribution to δ_{io} due to self-straining and u_i is the load distribution due to a unit load in the redundancy i (u_b in general, includes p_i, m_i, s_i, and t_i).

2. External causes (Fig. 52b). The problem here is to calculate the relative deflection of the reduced structure across the position of the ith redundancy due to a prescribed

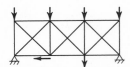

(a) Truss with three redundant members

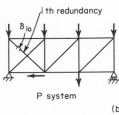 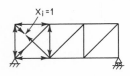

P system p_i system

(b) Calculation of δ_{io}

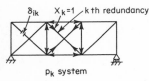 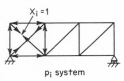

p_k system p_i system

(c) Calculation of δ_{ik}

Fig. 51

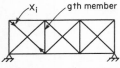 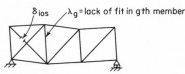

(a) δ_{io} due to initial lack of fit λ_g in gth member

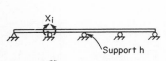 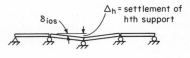

(b) δ_{io} due to settlement Δ_h of hth support

Fig. 52

displacement Δ_h of the support h. Application of the unit-load theorem gives the contribution to δ_{io} as follows:

$$\delta_{ios} = R_h \Delta_h \qquad (36)$$

where R_h is the action of the structure on the support h due to $X_i = 1$.

32. The Unit-Load Method The solution by the unit-load method is obtained by the following eight steps. The procedure is illustrated by Fig. 53.

1. Calculate the number of redundant forces (Arts. 13 and 14).

2. Select the problem coordinates X_i and the associated reduced structure. The problem coordinates are chosen by selecting suitable forces $X_1, X_2, \ldots, X_k, \ldots, X_n$ as redundants (Fig. 53b). The reduced structure c is then obtained by assigning zero values to each of the redundant forces.

3. Calculate the forces in the elements of the reduced structure due to the applied loads (c).

4. Calculate the forces in the elements of the reduced structure due to the action of a unit value of each redundancy in turn (d, e, f).

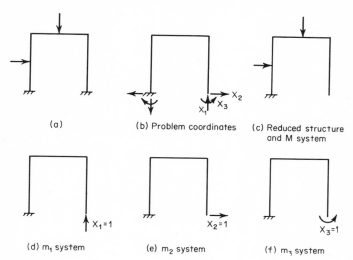

(a)

(b) Problem coordinates

(c) Reduced structure and M system

(d) m_1 system

(e) m_2 system

(f) m_3 system

Fig. 53

5. Apply the compatibility condition. The n equations are expressed in terms of the flexibility coefficients δ_{io} and δ_{ik}:

$$\delta_{11}X_1 + \delta_{12}X_2 + \cdots + \delta_{1i}X_i + \cdots + \delta_{1n}X_n + \delta_{10} = 0$$
$$\cdots\cdots\cdots\cdots\cdots\cdots\cdots\cdots\cdots\cdots\cdots\cdots\cdots\cdots\cdots\cdots\cdots$$
$$\delta_{i1}X_1 + \delta_{i2}X_2 + \cdots + \delta_{ii}X_i + \cdots + \delta_{in}X_n + \delta_{io} = 0 \qquad (37)$$
$$\cdots\cdots\cdots\cdots\cdots\cdots\cdots\cdots\cdots\cdots\cdots\cdots\cdots\cdots\cdots\cdots\cdots$$
$$\delta_{n1}X_1 + \delta_{n2}X_2 + \cdots + \delta_{ni}X_i + \cdots + \delta_{nn}X_n + \delta_{no} = 0$$

If self-straining is present, its contribution to δ_{io} is calculated by Eqs. (35) and (36).

6. Calculate the n coefficients δ_{io} and the n^2 coefficients δ_{ik} (Art. 31).

7. Solve Eqs. (37) for the problem coordinates $X_1, X_2, \ldots, X_k, \ldots, X_n$.

8. Using the results of steps 1 through 4, the internal forces acting on each element of the structure can be expressed in terms of the applied loading system W and the redundant forces X (which, in turn, include the effect of self-straining). In a pin-jointed framework, the force in the general element g can be expressed as

$$P_g + \sum_1^n p_{gi}X_i$$

The above procedure is illustrated in Figs. 54 and 55.

33. Castigliano's Theorem   The procedure for the application of Castigliano's theorem is as follows:

1 through 4. The first four steps are identical with steps 1 through 4 of the unit-load method (Art. 32).

5. The compatibility condition is satisfied by adjusting the redundant forces to make the strain energy a minimum, that is,

$$\frac{\partial U}{\partial X_1} = \frac{\partial U}{\partial X_2} = \cdots = \frac{\partial U}{\partial X_k} = \cdots = \frac{\partial U}{\partial X_n} = 0 \tag{38}$$

These are a set of n linear simultaneous equations in the force redundancies $X_1, X_2, \ldots, X_k, \ldots, X_n$. In most cases it is convenient to differentiate U with respect to the redundancies X_k before the integration is carried out. When this procedure is used, the equations are identical with the compatibility equations of the unit-load method [Eqs. (37)].

Members =, 2 in.², Members −, 1 in.²

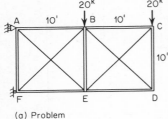

(a) Problem

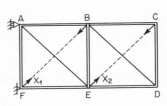

(b) Reduced structure

Member	L, ft	A, in.²	P, kips	p₁	p₂	$\frac{Pp_1L}{A}$	$\frac{Pp_2L}{A}$	$\frac{p_1^2L}{A}$	$\frac{p_2^2L}{A}$	$\frac{p_1p_2L}{A}$
AB	10	2	20	$-\frac{1}{\sqrt{2}}$	0	$-\frac{100}{\sqrt{2}}$	0	$5/2$	0	0
BC	10	2	0	0	$-\frac{1}{\sqrt{2}}$	0	0	0	$5/2$	0
DE	10	2	−20	0	$-\frac{1}{\sqrt{2}}$	0	$\frac{100}{\sqrt{2}}$	0	$5/2$	0
EF	10	2	−60	$-\frac{1}{\sqrt{2}}$	0	$\frac{300}{\sqrt{2}}$	0	$5/2$	0	0
AF	10	2	−40	$-\frac{1}{\sqrt{2}}$	0	$\frac{200}{\sqrt{2}}$	0	$5/2$	0	0
BE	10	2	−40	$-\frac{1}{\sqrt{2}}$	$-\frac{1}{\sqrt{2}}$	$\frac{200}{\sqrt{2}}$	$\frac{200}{\sqrt{2}}$	$5/2$	$5/2$	$5/2$
CD	10	2	−20	0	$-\frac{1}{\sqrt{2}}$	0	$\frac{100}{\sqrt{2}}$	0	$5/2$	0
AE	$10\sqrt{2}$	1	$40\sqrt{2}$	1	0	800	0	$10\sqrt{2}$	0	0
BD	$10\sqrt{2}$	1	$20\sqrt{2}$	0	1	0	400	0	$10\sqrt{2}$	0
BF	$10\sqrt{2}$	1	0	1	0	0	0	$10\sqrt{2}$	0	0
CE	$10\sqrt{2}$	1	0	0	1	0	0	0	$10\sqrt{2}$	0
					Σ	1224.3	682.9	38.28	38.28	2.5
						$=\delta_{10}E$	$=\delta_{20}E$	$=\delta_{11}E$	$=\delta_{22}E$	$=\delta_{12}E$

$$\delta_{10} + \delta_{11}X_1 + \delta_{12}X_2 = 0$$
$$\delta_{20} + \delta_{21}X_1 + \delta_{22}X_2 = 0$$
$$X_1 = -30.95^k, \quad X_2 = -15.82^k$$

Fig. 54

6. Solve Eqs. (38) for the problem coordinates $X_1, X_2, \ldots, X_k, \ldots, X_n$.
7. The internal forces on each element of the structure can now be found in an identical manner to that of step 8 of the unit-load method.

34. The Three-Moment Equation The three-moment equation in its simple form applies to continuous beams of constant section, the supports either being unyielding or settling by known amounts λ. It gives a relationship between the moments at three adjacent supports, in terms of the loading on the two associated spans:

$$M_A \frac{L_1}{I_1} + 2M_B \left(\frac{L_1}{I_1} + \frac{L_2}{I_2} \right) + M_C \frac{L_2}{I_2} = -6 \left(\frac{A_1 \bar{x}_1}{I_1 L_1} + \frac{A_2 \bar{x}_2}{I_2 L_2} \right) - 6E \left(\frac{\lambda_{BA}}{L_1} + \frac{\lambda_{BC}}{L_2} \right) = 0 \quad (39)$$

where A_1, A_2, $\bar{x}_1$, and $\bar{x}_2$ are defined in Fig. 56b.
$\quad \lambda_{BA}$ = downward settlement of B relative to A
$\quad \lambda_{BC}$ = downward settlement of B relative to C

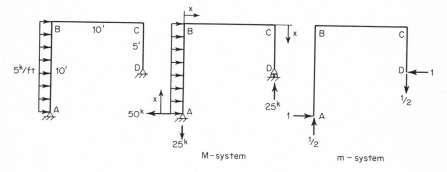

M-system   m − system

Member	M	m
AB	$50x - \frac{5}{2}x^2$	$-x$
BC	$250 - 25x$	$-10 + \frac{x}{2}$
CD	0	$-5 + x$

$$\delta_{io} = \Sigma \int \frac{Mm}{EI} dx$$

$$= \frac{1}{EI} \left\{ \left[-\frac{50}{3}x^3 + \frac{5}{8}x^4 \right]_0^{10} + \left[-2{,}500x + \frac{375}{2}x^2 - \frac{12.5}{3}x^3 \right]_0^{10} \right\}$$

$$= -\frac{10^3}{EI} \times 20.833$$

$$\delta_{ii} = \Sigma \int \frac{m^2}{EI} dx$$

$$= \frac{1}{EI} \left\{ \left[\frac{x^3}{3} \right]_0^{10} + \left[100x - \frac{10}{2}x^2 + \frac{x^3}{12} \right]_0^{10} + \left[25x - \frac{10}{2}x^2 + \frac{x^3}{3} \right]_0^5 \right\}$$

$$= \frac{10^3}{EI} \times 0.9583$$

$$X = -\frac{\delta_{io}}{\delta_{ii}} = 21.73^k$$

Fig. 55

The procedure is illustrated in Fig. 57. In the more general multispan problem, the equation is applied to every pair of adjacent spans. This gives n simultaneous equations for the n redundant moments, a maximum of three redundancies appearing in each equation.

35. The Column Analogy In the case of a fixed beam, a fixed arch or portal, or a continuous ring, the compatibility equations for the reduced structure give an analogy between the moments due to the redundancies, and the stresses in a short column. The technique for finding indeterminate moments by this analogy is illustrated in Fig. 58.

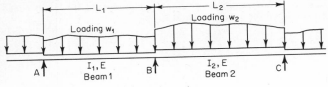

Loading w_1

Loading w_2

I_1, E
Beam 1

I_2, E
Beam 2

A B C

(a) Continuous beam

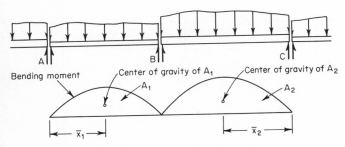

A B C

Bending moment

Center of gravity of A_1

Center of gravity of A_2

A_1

A_2

$\bar{x}_1$

$\bar{x}_2$

(b) Applied loading w on the reduced structure

Fig. 56

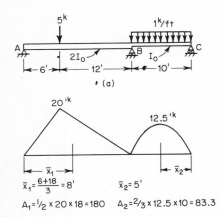

5^k

$1^k/ft$

A $2I_0$ B I_0 C

6' 12' 10'

(a)

$20^{'k}$

$12.5^{'k}$

$\bar{x}_1$ $\bar{x}_2$

$\bar{x}_1 = \dfrac{6+18}{3} = 8'$ $\bar{x}_2 = 5'$

$A_1 = \frac{1}{2} \times 20 \times 18 = 180$ $A_2 = \frac{2}{3} \times 12.5 \times 10 = 83.3$

(b) Moment due to applied loading on
reduced structure

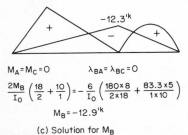

$-12.3^{'k}$

$+$ $-$ $+$

$M_A = M_C = 0$ $\lambda_{BA} = \lambda_{BC} = 0$

$$\frac{2M_B}{I_0}\left(\frac{18}{2}+\frac{10}{1}\right) = -\frac{6}{I_0}\left(\frac{180\times8}{2\times18}+\frac{83.3\times5}{1\times10}\right)$$

$$M_B = -12.9^{'k}$$

(c) Solution for M_B

Fig. 57

M_0 is the moment at any point in the reduced structure due to the applied loading. Any reduced structure can be used; in Fig. 58b it is taken as pinned at A and supported on rollers at D. M_i is the unknown moment at any point in the reduced structure due to the redundancies. The total moment is $M_0 + M_i$. The elevation of the centerline of the frame is considered as the section of a column, the thickness of the walls being taken as

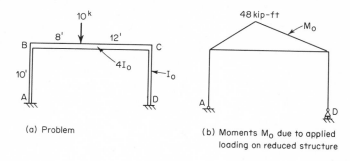

(a) Problem

(b) Moments M_0 due to applied loading on reduced structure

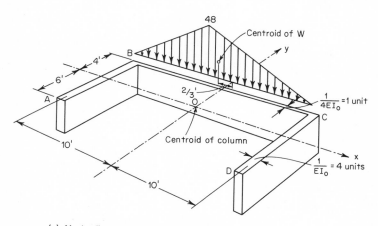

(c) M_0 loading applied to analogous column

Fig. 58

proportional to $1/EI$. The following section properties of the column are computed: the cross-sectional area A, the position of its centroid O, and I_x and I_y, the moments of inertia of A about axes x and y, respectively:

$$A = 10 \times 4 + 10 \times 4 + 20 \times 1 = 100$$
$$\bar{y} = 2 \times 4 \times 10 \times {}^5\!/_{100} = 4$$
$$I_x = 2 \times 4 \times 10^3/12 + 2 \times 4 \times 10 \times 1^2 + 1 \times 20 \times 4^2 = 1067$$
$$I_y = 1 \times 20^3/12 + 2 \times 4 \times 10 \times 10^2 = 8667$$

The analogous column is loaded by the M_0/EI diagram (of area W), and the direct stresses due to this loading computed from

$$\frac{W}{A} \pm \frac{M_y x}{I_y} \pm \frac{M_x y}{I_x} \tag{40}$$

where M_x and M_y are the moments of W about axes x and y, respectively. These direct stresses are the moments M_i.

$$W = 48 \times 1 \times {}^{20}\!\!/_{2} = 480$$
$$M_x = 480 \times 4 = 1920$$
$$M_y = 480 \times {}^{2}\!\!/_{3} = 320$$

Hence the stresses M_i in the analogous column are

$$\frac{480}{100} \pm \frac{320 \times 10}{8667} - \frac{1920 \times 6}{1067} = 4.8 + 0.37 - 10.8 = -5.6 \text{ kip-ft at } A$$

$$= 4.8 - 0.37 - 10.8 = -6.4 \text{ kip-ft at } D$$

and $$\frac{480}{100} \pm \frac{320 \times 10}{8667} + \frac{1920 \times 4}{1067} = 4.8 + 0.37 + 7.2 = 12.4 \text{ kip-ft at } B$$

$$= 4.8 - 0.37 + 7.2 = 11.6 \text{ kip-ft at } C$$

These are the values of M_i. As M_0 at these points is zero, they also represent the total moments.

Care must be taken when applying this technique to unsymmetrical frames or rings. In such cases the *principal axes* of the analogous column must be used in computing the moments of the M_0/EI diagram and the moments of inertia of the column.

Linearly Elastic Structures: Displacement Methods

36. Stiffness Coefficients   The term "stiffness coefficient" is used to define each of the forces which are required to maintain a structure, or a structural element, in a specified deflection mode. This deflection mode is defined by a unit displacement in one specified coordinate,* subject to the condition that the displacements in the other coordinates are zero; that is, every degree of freedom is constrained except the coordinate under consideration. The stiffness coefficients are the values of the forces *resolved in the directions of the coordinates of the structure.*

Equations (31) express deflections in terms of loads and flexibility coefficients referred to unit values of the applied loads. Conversely, the forces can be expressed as linear functions of the deflections as follows:

$$W_1 = k_{11}\Delta_1 + k_{12}\Delta_2 + \cdots + k_{1j}\Delta_j + \cdots + k_{1m}\Delta_m$$
$$W_2 = k_{21}\Delta_1 + k_{22}\Delta_2 + \cdots + k_{2j}\Delta_j + \cdots + k_{2m}\Delta_m$$

$$\vdots \qquad \vdots \qquad \qquad \vdots \qquad \qquad \vdots$$

$$W_j = k_{j1}\Delta_1 + k_{j2}\Delta_2 + \cdots + k_{jj}\Delta_j + \cdots + k_{jm}\Delta_m \qquad (41)$$

$$\vdots \qquad \vdots \qquad \qquad \vdots \qquad \qquad \vdots$$

$$W_m = k_{m1}\Delta_1 + k_{m2}\Delta_2 + \cdots + k_{mj}\Delta_j + \cdots + k_{mm}\Delta_m$$

where the constants k_{ij} are stiffness coefficients.

Some simple illustrations of stiffness coefficients are given in Fig. 59. In the case of a single coordinate system (Fig. 59a), k_{ij} is the spring constant k. In the case of the system containing two coordinates (Fig. 59b and c), the forces W_1 and W_2 corresponding to the displacements Δ_1 and Δ_2 in the coordinates ξ_1 and ξ_2 are

$$W_1 = \frac{12EI}{L^3} \Delta_1 - \frac{6EI}{L^2} \Delta_2$$
$$W_2 = -\frac{6EI}{L^2} \Delta_1 + \frac{4EI}{L} \Delta_2 \qquad (42)$$

Stiffness coefficients originate from Maxwell's theorem (Art. 9), and the generalizations relating to "force" and "deflection" noted there also apply to the stiffness coefficients [Eq. (18)]. For example, the unit translation applied in the coordinate i may require a couple to be applied in the coordinate j, and a unit rotation applied in i may require a point force in j (Fig. 60).

The coefficients k_{ii} and k_{ij} are called direct- and cross-stiffness coefficients, respec-

---

*For conciseness in dealing with flexibility and stiffness coefficients, the word "coordinate" is used in a generalized sense, that is, synonymously with the term "degree of freedom."

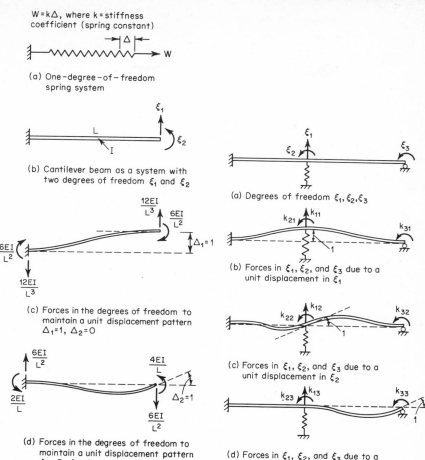

W = kΔ, where k = stiffness
coefficient (spring constant)

→ Δ ←
W

(a) One-degree-of-freedom
    spring system

ξ_1

L

I

ξ_2

(b) Cantilever beam as a system with
    two degrees of freedom ξ_1 and ξ_2

$\dfrac{12EI}{L^3}$ $\dfrac{6EI}{L^2}$

$\dfrac{6EI}{L^2}$ Δ$_1$=1

$\dfrac{12EI}{L^3}$

(c) Forces in the degrees of freedom to
    maintain a unit displacement pattern
    Δ$_1$=1, Δ$_2$=0

$\dfrac{6EI}{L^2}$ $\dfrac{4EI}{L}$

$\dfrac{2EI}{L}$ Δ$_2$=1

$\dfrac{6EI}{L^2}$

(d) Forces in the degrees of freedom to
    maintain a unit displacement pattern
    Δ$_1$=0, Δ$_2$=1

Fig. 59 Stiffness coefficients.

ξ_1
ξ_2 ξ_3

(a) Degrees of freedom ξ_1, ξ_2, ξ_3

k_{21} k_{11} k_{31}

1

(b) Forces in ξ_1, ξ_2, and ξ_3 due to a
    unit displacement in ξ_1

k_{22} k_{12} k_{32}

1

(c) Forces in ξ_1, ξ_2, and ξ_3 due to a
    unit displacement in ξ_2

k_{23} k_{13} k_{33}

1

(d) Forces in ξ_1, ξ_2, and ξ_3 due to a
    unit displacement in ξ_3

Fig. 60

tively. Stiffness coefficients of the simple elements which occur commonly are presented in Table 7. The stiffness matrix in this table is discussed in Art. 49.

37. The Unit-Displacement Method The solution by the unit-displacement method is obtained by the following five steps. The procedure is illustrated by Figs. 61 and 62, which refer respectively to pin-jointed and stiff-jointed frames.

1. Identify the degrees of freedom ξ. The problem coordinates $\Delta_1, \Delta_2, \ldots, \Delta_j, \ldots,$ Δ_m are the displacements in the degrees of freedom.

2. Calculate the equivalent loading in the problem coordinates (Figs. 61a and 62e). The equivalent loading is the sum of the actual loading in the problem coordinates plus (where applicable) the fixed-end forces and moments due to loading applied between the joints (Fig. 62e). The concept of equivalent loading ensures that the displacements in the degrees of freedom of the analytical model are identical with the corresponding displacements of the real structure due to the actual loading. In the case of pin-jointed frameworks, it is necessary for linearity that the load be applied only at the joints; hence the equivalent loading is identical with the actual loading. In most practical cases of stiff-jointed frameworks, deformations due to shear and axial force are negligible. Hence the degrees of

TABLE 7   Stiffness Coefficients for Beam Elements

Element	Displacements on element	Independent displacement	Stiffness coefficient k_{ij}	Stiffness matrix* k_g
Bar with axial force	$v_1 = 1$ k_{11}	v_1	$k_{11} = \dfrac{AE}{L}$	$\dfrac{AE}{L}$
Cantilever beam	k_{11} k_{21} $v_1 = 1$	v_1	$k_{11} = \dfrac{12EI}{L^3}$ $k_{21} = -\dfrac{6EI}{L^2}$	$\begin{bmatrix} \dfrac{12EI}{L^3} & -\dfrac{6EI}{L^2} \\ -\dfrac{6EI}{L^2} & \dfrac{4EI}{L} \end{bmatrix}$
	k_{12} $v_2 = 1$ k_{22}	v_2	$k_{12} = -\dfrac{6EI}{L^2}$ $k_{22} = \dfrac{4EI}{L}$	
Simply supported beam	k_{11} k_{21} $v_1 = 1$	v_1	$k_{11} = \dfrac{4EI}{L}$ $k_{21} = \dfrac{2EI}{L}$	$\begin{bmatrix} \dfrac{4EI}{L} & \dfrac{2EI}{L} \\ \dfrac{2EI}{L} & \dfrac{4EI}{L} \end{bmatrix}$
	k_{12} k_{22} $v_2 = 1$	v_2	$k_{12} = \dfrac{2EI}{L}$ $k_{22} = \dfrac{4EI}{L}$	

* See Art. 49
NOTE: Positive rotation is taken counterclockwise at each end of the beam. This convention facilitates assembly of K from k_g.

freedom can be specified entirely as the joint rotations and the lateral (sway) displacements of the frame (Fig. 62e and f). The loading on an individual element which is loaded between joints is found by applying the principle of superposition (step 5).

3. Form the equation of equilibrium for each problem coordinate. This is obtained by applying a unit displacement in the coordinate concerned and restraining completely all the other coordinates (Figs. 61d and 62f). The deflected configuration defined by this process is maintained by the stiffness coefficients of Art. 36. The unit displacement is applied to each coordinate in turn, and by appropriate summation m equations of equilibrium are written in the m problem coordinates [Eqs. (41)]. These equations can be written

$$W_j = \sum_{i=1}^{m} k_{ji}\Delta_i \tag{43}$$

where i and j range over all the degrees of freedom 1 to m and the stiffness coefficients k_{ji} are calculated by the methods of Art. 36 or (where applicable) are taken from Table 7.

4. Solve Eqs. (43) for the problem coordinates $\Delta_1, \Delta_2, \ldots, \Delta_j, \ldots, \Delta_m$.
5. Calculate the forces in each element of the structure using the appropriate element force-deformation relationship given in Table 7.

The above procedure is illustrated in Examples 1, 2, and 3.

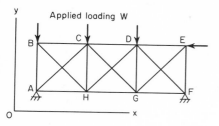

(a) Problem and structural axes

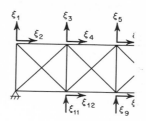

(b) Degrees of freedom

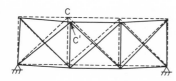

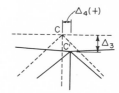

(c) Deflected configuration and typical problem coordinates Δ_3 and Δ_4

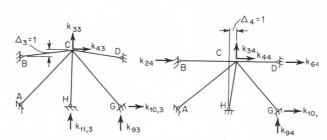

(d) Generation of stiffness coefficients due to unit displacement in typical coordinates Δ_3 and Δ_4

Fig. 61 Unit-displacement method for pin-jointed framework.

Example 1 For the frame of Fig. 63a, the coordinates in the three degrees of freedom are shown in b. The equivalent loading in these coordinates is given in c, where the moments are the fixed-end moments (acting on the joints) for the beam BC. Unit displacements are imposed in each coordinate in turn (d, e, and f), for which the stiffness coefficients follow from Table 7. Thus k_{11} is the sum of the stiffness coefficients k_{11} for AB and CD, k_{21} is the moment at B in AB, etc. Note that, because of the reciprocal relations, it is not necessary to compute k_{12}, k_{13}, etc., once k_{21}, k_{31}, etc., are known.

Substituting the stiffness coefficients into Eqs. (41), with $EI = 1$ and $EI_1 = 2$, results in the equations

$$2.4\Delta_1 + 6\Delta_2 + 6\Delta_3 = 400$$
$$0.6\Delta_1 + 8\Delta_2 + 2\Delta_3 = -400$$
$$0.6\Delta_1 + 2\Delta_2 + 8\Delta_3 = 400$$

from which $\Delta_1 = 239$, $\Delta_2 = -81.0$, and $\Delta_3 = 52.3$. Then

$$M_{AB} = \frac{6EI}{h^2}\Delta_1 + \frac{2EI}{h}\Delta_2 + 0 \times \Delta_3 = 0.06 \times 239 + 0.2(-81)$$
$$= -1.86 \text{ kip-ft}$$

$$M_{BA} = \frac{6EI}{h^2} \Delta_1 + \frac{4EI}{h} \Delta_2 + 0 \times \Delta_3 = 0.06 \times 239 + 0.4(-81)$$
$$= -18.1 \text{ kip-ft}$$
$$M_{BC} = 0 \times \Delta_1 + \frac{4EI_1}{L} \Delta_2 + \frac{2EI_1}{L} \Delta_3 - (-40) = 0.4(-81) + 0.2 \times 52.3 + 40$$
$$= +18.1 \text{ kip-ft}$$

etc.

Example 2 The problem coordinates for the frame of Fig. 64a are shown in b. The distributed load on BC produces the fixed-end forces and moments, acting on the joints, shown in c. The resultant equivalent loading in the coordinate ξ_1 is found from the components of the 5-kip load at B, resolved in

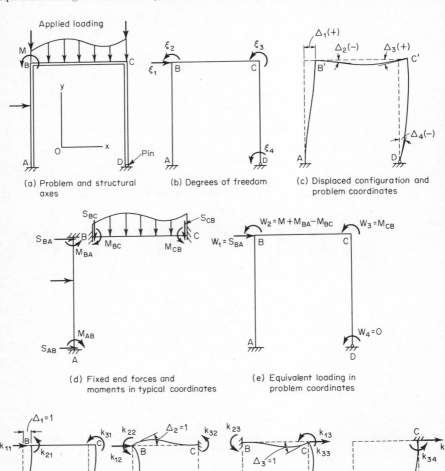

(a) Problem and structural axes

(b) Degrees of freedom

(c) Displaced configuration and problem coordinates

(d) Fixed end forces and moments in typical coordinates

(e) Equivalent loading in problem coordinates

(f) Generation of stiffness coefficients due to unit displacements in problem coordinates

Fig. 62 Unit-displacement method for stiff-jointed framework.

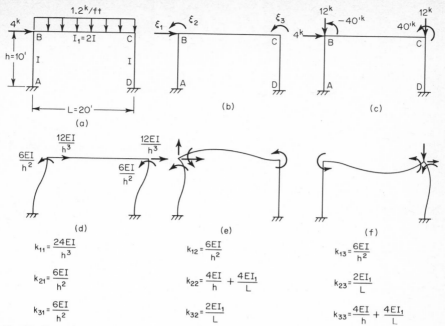

Fig. 63 Example 1.

The figures show:

(a) Frame with 4^k load at B, distributed load $1.2^k/ft$ on BC, $I_1 = 2I$, $h = 10'$, members with I, $L = 20'$, joints A, B, C, D.

(b) ξ_1, ξ_2, ξ_3

(c) 12^k, $-40^{'k}$, 12^k, 4^k, $40^{'k}$

(d) $\dfrac{12EI}{h^3}$, $\dfrac{12EI}{h^3}$, $\dfrac{6EI}{h^2}$, $\dfrac{6EI}{h^2}$

(e)

(f)

$$k_{11} = \frac{24EI}{h^3} \qquad k_{12} = \frac{6EI}{h^2} \qquad k_{13} = \frac{6EI}{h^2}$$

$$k_{21} = \frac{6EI}{h^2} \qquad k_{22} = \frac{4EI}{h} + \frac{4EI_1}{L} \qquad k_{23} = \frac{2EI_1}{L}$$

$$k_{31} = \frac{6EI}{h^2} \qquad k_{32} = \frac{2EI_1}{L} \qquad k_{33} = \frac{4EI}{h} + \frac{4EI_1}{L}$$

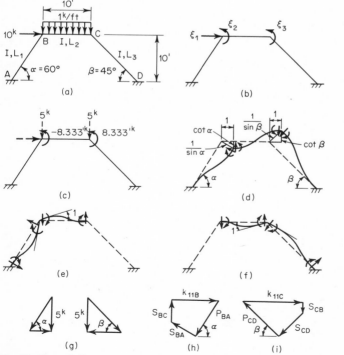

Fig. 64 Example 2.

(a) 10^k, $1^k/ft$, $10'$, I, L_2, B, C, I, L_1, I, L_3, $\alpha = 60°$, $\beta = 45°$, $10'$, A, D

(b) ξ_1, ξ_2, ξ_3

(c) 5^k, 5^k, $-8.333^{'k}$, $8.333^{'k}$

(d) $\cot \alpha$, $\dfrac{1}{\sin \beta}$, $\dfrac{1}{\sin \alpha}$, $\cot \beta$, α, β

(e)

(f)

(g) 5^k, α, 5^k, β

(h) k_{11B}, S_{BC}, P_{BA}, S_{BA}, α

(i) k_{11C}, P_{CD}, S_{CB}, S_{CD}, β

the directions ξ_1 and AB, and from the components of the 5-kip load at C resolved in the directions ξ_1 and CD (Fig. 64g), which are added to the 10-kip load:

$$\text{Equivalent load in } \xi_1 = 10 + 5 \cot \alpha - 5 \cot \beta = 7.887 \text{ kips}$$

A unit displacement in ξ_1 results in the member displacements shown in Fig. 64d. The corresponding moments and shears acting on joints B and C shown in the figure are determined from Table 7. Thus,

$$S_{BA} = \frac{12EI}{L_1^3 \sin \alpha} \qquad S_{BC} = \frac{12EI}{L_2^3} (\cot \alpha + \cot \beta) \qquad M_{BA} = \frac{6EI}{L_1^2 \sin \alpha}$$

etc.

The stiffness coefficient k_{11} can be found from equilibrium of forces at joints B and C (Fig. 64h and i). In these force polygons, P_{BA} and P_{CD} are the (unknown) axial forces in AB and CD. Then

$$k_{11} = k_{11(B)} + k_{11(C)} = \frac{S_{BA}}{\sin \alpha} + S_{BC} \cot \alpha + S_{CB} \cot \beta + \frac{S_{CD}}{\sin \beta}$$

Substitution of the values of the shears into this equation gives k_{11}. The stiffness coefficient $k_{12} = M_{BA} + M_{BC}$, while $k_{13} = M_{CB} + M_{CD}$. The displacement $\xi_2 = 1$ (Fig. 64e) determines k_{22} and k_{23}, and the displacement $\xi_3 = 1$ gives k_{33} (Fig. 64f). The complete set of coeficients is (L in feet)

$$k_{11} = 12EI \left[\frac{1}{L_1^3 \sin^2 \alpha} + \frac{1}{L_3^3 \sin^2 \beta} + \frac{1}{L_2^3} (\cot \alpha + \cot \beta)^2 \right] = 0.04873EI$$

$$k_{22} = 4EI \left(\frac{1}{L_1} + \frac{1}{L_2} \right) = 0.7464EI$$

$$k_{33} = 4EI \left(\frac{1}{L_2} + \frac{1}{L_3} \right) = 0.6828EI$$

$$k_{12} = k_{21} = 6EI \left(\frac{1}{L_1^2 \sin \alpha} - \frac{\cot \alpha + \cot \beta}{L_2^2} \right) = -0.04268EI$$

$$k_{13} = k_{31} = 6EI \left(\frac{1}{L_3^2 \sin \beta} - \frac{\cot \alpha + \cot \beta}{L_2^2} \right) = -0.05222EI$$

$$k_{23} = k_{32} = \frac{2EI}{L_2} = 0.2EI$$

The equations of equilibrium in the problem coordinates are

$$k_{11}\Delta_1 + k_{12}\Delta_2 + k_{13}\Delta_3 = W_1 = 7.887$$
$$k_{21}\Delta_1 + k_{22}\Delta_2 + k_{23}\Delta_3 = W_2 = -8.333$$
$$k_{31}\Delta_1 + k_{32}\Delta_2 + k_{33}\Delta_3 = W_3 = 8.333$$

from which

$$\Delta_1 = 1.8546 \frac{10^5}{EI} \qquad \Delta_2 = -0.0827 \frac{10^5}{EI} \qquad \Delta_3 = 0.2881 \frac{10^5}{EI}$$

The bending moments are given by

$$M_{AB} = \frac{6EI}{L_1^2 \sin \alpha} \Delta_1 + \frac{2EI}{L_1} \Delta_2 + 0 \times \Delta_3 = 8.21 \text{ kip-ft}$$

$$M_{BA} = \frac{6EI}{L_1^2 \sin \alpha} \Delta_1 + \frac{4EI}{L_1} \Delta_2 + 0 \times \Delta_3 = 6.77 \text{ kip-ft}$$

$$M_{BC} = -\frac{6EI}{L_2^2} (\cot \alpha + \cot \beta)\Delta_1 + \frac{4EI}{L_2} \Delta_2 + \frac{2EI}{L_2} \Delta_3 - (-8.333) = -6.77 \text{ kip-ft}$$

$$M_{CB} = -\frac{6EI}{L_2^2} (\cot \alpha + \cot \beta)\Delta_1 + \frac{2EI}{L_2} \Delta_2 + \frac{4EI}{L_2} \Delta_3 - 8.333 = -16.02 \text{ kip-ft}$$

$$M_{CD} = \frac{6EI}{L_3^2 \sin \beta} \Delta_1 + 0 \times \Delta_2 + \frac{4EI}{L_3} \Delta_3 = 16.02 \text{ kip-ft}$$

$$M_{DC} = \frac{6EI}{L_3^2 \sin \beta} \Delta_1 + 0 \times \Delta_2 + \frac{2EI}{L_3} \Delta_3 = 11.94 \text{ kip-ft}$$

Example 3 The truss of Fig. 65a has five degrees of freedom (Fig. 65b). The unit displacement in ξ_1, shown in c, results in the forces $2A_0E/L$ and $(A_0E/L\sqrt{2})(1/\sqrt{2}) = A_0E/2L$ in bars 4 and 5, respectively. Equilibrium of the joint containing k_{11} gives

$$k_{11} = \frac{2A_0E}{L} + \frac{A_0E}{2L}\frac{1}{\sqrt{2}} = 2.3535\frac{A_0E}{L}$$

Similarly,

$$k_{51} = -k_{41} = \frac{A_0E}{2L}\frac{1}{\sqrt{2}} = 0.3535\frac{A_0E}{L}$$

$$k_{21} = k_{31} = 0$$

The remaining coefficients are found by imposing unit displacements in the coordinates ξ_2, ξ_3, etc. The complete set is

$$k_{11} = k_{22} = k_{33} = k_{44} = k_{55} = \frac{2.3535A_0E}{L}$$

$$k_{12} = k_{21} = k_{13} = k_{31} = k_{25} = k_{52} = k_{34} = k_{43} = k_{35} = k_{53} = 0$$

$$k_{14} = k_{41} = k_{45} = k_{54} = -\frac{0.3535A_0E}{L}$$

$$k_{15} = k_{51} = k_{23} = k_{32} = \frac{0.3535A_0E}{L}$$

$$k_{24} = k_{42} = -\frac{2A_0E}{L}$$

The five equations of equilibrium, from Eqs. (41), are

$$2.3535\Delta_1 \qquad\qquad -0.3535\Delta_4 + 0.3535\delta_5 = 0$$

$$2.3535\Delta_2 + 0.3535\Delta_3 \qquad -2\Delta_4 \qquad\qquad = \frac{20L}{A_0E}$$

$$0.3535\Delta_2 + 2.3535\Delta_3 \qquad\qquad = 0$$

$$-0.3535\Delta_1 \quad -2\Delta_2 \qquad\qquad +2.3535\Delta_4 - 0.3535\Delta_5 = 0$$

$$0.3535\Delta_1 \qquad\qquad -0.3535\Delta_4 + 2.3535\Delta_5 = -\frac{10L}{A_0E}$$

The solution of these equations gives

$$\Delta_1 = \frac{4.675L}{A_0E}$$

$$\Delta_2 = \frac{35.469L}{A_0E}$$

$$\Delta_3 = -\frac{5.328L}{A_0E}$$

$$\Delta_4 = \frac{30.797L}{A_0E}$$

$$\Delta_5 = -\frac{0.328L}{A_0E}$$

The forces in the bars are given by

	Δ_1	Δ_2	Δ_3	Δ_4	Δ_5		
$S_1 = 0$		0	$+\dfrac{2A_0E}{L}$	0	0	$=$	-10.66 kips
$S_2 = 0$	$+\dfrac{2A_0E}{L}$		0	$-\dfrac{2A_0E}{L}$	0	$=$	9.34 kips
$S_3 = 0$		0	0	0	$\dfrac{2A_0E}{L}$	$=$	-0.66 kip
$S_4 = \dfrac{2A_0E}{L}$		0	0	0	0	$=$	9.34 kips
$S_5 = \dfrac{A_0E}{2L}$		0	0	$-\dfrac{A_0E}{2L}$	$+\dfrac{A_0E}{2L}$	$=$	-13.23 kips
$S_6 = 0$		$\dfrac{A_0E}{2L}$	$\dfrac{A_0E}{2L}$	0	0	$=$	15.07 kips

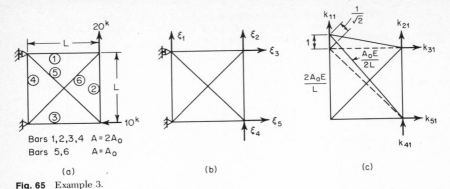

Bars 1,2,3,4   A = 2A_0
Bars 5,6   A = A_0

(a)   (b)   (c)

Fig. 65   Example 3.

38. Slope-Deflection Equations   The end moments in the members of a stiff-jointed frame composed of straight elements are given in terms of joint rotations and relative joint displacements by the slope-deflection equations (Fig. 66)

$$M_{AB} = \frac{2EI}{L}(2\theta_A + \theta_B - 3\phi) + M_{FAB}$$

$$M_{BA} = \frac{2EI}{L}(2\theta_B + \theta_A - 3\phi) + M_{FBA}$$

(44)

where M_{AB}, M_{BA} = end moments in AB, positive clockwise on AB
$\quad \theta_A, \theta_B$ = rotation of joints A and B, positive clockwise
$\quad \phi = \Delta/L$ = rotation of AB, positive clockwise
$\quad \Delta$ = relative displacement of joints A and B, normal to AB
$\quad M_{FAB}, M_{FBA}$ = fixed-end moments resulting from loading between A and B, positive clockwise on AB

Equations sufficient in number to determine the unknown rotations and displacements can be established from consideration of equilibrium. If there are no joint displacements, equilibrium of moments at the joints furnishes the necessary equations. Where joint displacements exist, the joint equations must be supplemented by additional equations of equilibrium; usually these are based on equilibrium of shearing forces on various sections through the frame.

Example 4   For the frame of Fig. 63a (Example 1, Art. 37), taking $EI = 1$, Eqs. (44) give

$$M_{AB} = \frac{2}{10}(\theta_B - 3\phi)$$

$$M_{BA} = \frac{2}{10}(2\theta_B - 3\phi)$$

$$M_{BC} = \frac{2 \times 2}{20}(2\theta_B + \theta_C) - 40$$

$$M_{CB} = \frac{2 \times 2}{20}(2\theta_C + \theta_B) + 40$$

(45)

$$M_{CD} = \frac{2}{10}(2\theta_C - 3\phi)$$

$$M_{DC} = \frac{2}{10}(\theta_C - 3\phi)$$

Two relations among θ_B, θ_C, and ϕ are given by the joint equilibrium equations $M_{BA} + M_{BC} = 0$ and $M_{CB} + M_{CD} = 0$. The third equation is found from $S_{AB} + S_{CD} + 4 = 0$, where $S_{AB} = (M_{AB} + M_{BA})/10$ and $S_{CD} = (M_{CD} + M_{DC})/10$. There results

$$0.8\theta_B + 0.2\theta_C - 0.6\phi = 40$$
$$0.2\theta_B + 0.8\theta_C - 0.6\phi = -40$$
$$0.6\theta_B + 0.6\theta_C - 2.4\phi = -40$$

from which $\theta_B = 81.0$, $\theta_C = -52.3$, and $\phi = 23.9$. Substitution of these values into Eqs. (45) yields the moments. It will be observed that the procedure by the slope-deflection equations is essentially the same as that based on stiffness coefficients.

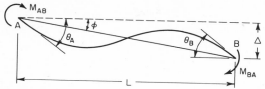

Fig. 66

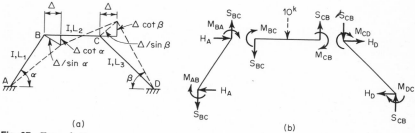

(a)

(b)

Fig. 67   Example 5.

Example 5   Slope-deflection equations for the frame of Fig. 64 (Example 2, Art. 37) can be written by considering the displacement Δ shown in Fig. 67a. The member rotations are as follows:

$$\phi_{AB} = \frac{1}{L_1} \frac{\Delta}{\sin \alpha} = 0.1\Delta$$

$$\phi_{BC} = -\frac{1}{L_2} \Delta(\cot \alpha + \cot \beta) = -0.1577\Delta$$

$$\phi_{CD} = \frac{1}{L_3} \frac{\Delta}{\sin \beta} = 0.1\Delta$$

Therefore,

$$M_{AB} = \frac{2EI}{L_1}(\theta_B - 0.3\Delta)$$

$$M_{BA} = \frac{2EI}{L_1}(2\theta_B - 0.3\Delta)$$

$$M_{BC} = \frac{2EI}{L_2}(2\theta_B + \theta_C + 0.473\Delta) - 8.33$$

$$M_{CB} = \frac{2EI}{L_2}(2\theta_C + \theta_B + 0.473\Delta) + 8.33$$

$$M_{CD} = \frac{2EI}{L_3}(2\theta_C - 0.3\Delta)$$

$$M_{DC} = \frac{2EI}{L_3}(\theta_C - 0.3\Delta)$$

Two relations among the three unknowns are given by joint equilibrium requirements:

$$M_{BA} + M_{BC} = 0 \qquad M_{CB} + M_{CD} = 0$$

The third equation can be found from equilibrium of horizontal forces $H_A + H_D = 10$ (Fig. 67b):

$$\frac{-M_{AB} - M_{BA} + S_{BC}L_1 \cos \alpha}{L_1 \sin \alpha} + \frac{-M_{CD} - M_{DC} + S_{CB}L_3 \cos \beta}{L_3 \sin \beta} = 10$$

where S_{BC} and S_{CB} are found from moment equilibrium of member BC.

Alternatively, the third equation could be obtained by considering equilibrium of forces at joints B and C, as in Fig. 64h and i. It can also be found by the principle of virtual displacements. With the displacement Δ in Fig. 67a considered as a virtual displacement δ, the member rotations are as determined above: $\phi_{AB} = 0.1\delta$, $\phi_{BC} = -0.1577\delta$, and $\phi_{CD} = 0.1\delta$. The work of the internal forces during this displacement is

$$+0.1\delta(M_{AB} + M_{BA}) - 0.1577\delta(M_{BC} + M_{CB}) + 0.1\delta(M_{CD} + M_{DC})$$

The work done by the external distributed load on member BC is

$$\tfrac{1}{2}wL_2\delta(\cot\alpha + \cot\beta) - wL_2\delta\cot\beta = \tfrac{1}{2}wL_2\delta(\cot\alpha - \cot\beta)$$

and that by the horizontal 10-kip load is 10δ. Equating to zero the sum of internal work and external work gives the required equation.

39. Moment-Distribution Method The moment-distribution method is particularly useful for computing the moments in complicated frames when there is rotation of the joints without translation, or with specified translations. Hence the procedure is simple for symmetrical frames with symmetrical loading, or cases of settlement where there is no indeterminate joint translation. Lack of symmetry causes sidesway, and in such cases the procedure has to be adapted to account for joint translation. The complexity of this extra computation depends on the effective number of degrees of freedom of the frame due to joint translation.

In the procedure as applied to frames without translation, all joints are first considered as "locked" against rotation, and the moments due to applied loading computed. The joints are then released and relocked in sequence until all are in equilibrium. The entire procedure is based on the following simple concepts and factors:

1. Fixed-end moment FM. This is the moment acting on the end of a beam when both ends are locked against rotation. This moment is taken as positive if it acts clockwise on the end of the beam. Values for a number of cases are given in Table 4.

2. Stiffness coefficient k. This is the rotational stiffness at a simply supported end of a beam. If the far end is fixed (Fig. 68a)

$$k_{AB} = \frac{M}{\theta} = \frac{4EI}{L} \tag{46}$$

If the far end is pinned (Fig. 68b)

$$k_{AB} = \frac{M}{\theta} = \frac{3EI}{L} \tag{47}$$

3. Distribution factor D. This denotes the proportion of the total moment applied to a

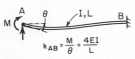

$$k_{AB} = \frac{M}{\theta} = \frac{4EI}{L}$$

(a) Stiffness coefficient for fixed-ended beam

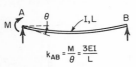

$$k_{AB} = \frac{M}{\theta} = \frac{3EI}{L}$$

(b) Stiffness coefficient for pin-ended beam

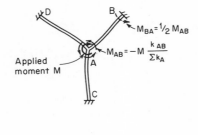

$$M_{BA} = \tfrac{1}{2} M_{AB}$$

$$M_{AB} = -M\frac{k_{AB}}{\Sigma k_A}$$

Applied moment M

(c) Distribution factor

Fig. 68 Constants for moment distribution.

joint which is induced in any particular member. In Fig. 68c a moment M is applied to joint A, and this induces in member AB at A the moment

$$M_{AB} = -D_{AB}M = -M\frac{k_{AB}}{\Sigma k_A} \tag{48}$$

4. Carryover factor C. This gives the moment induced at the far end of a beam if that end is fixed, when a moment is applied at the near end. In Fig. 68c

$$M_{BA} = C_{AB}M_{AB} = \tfrac{1}{2}M_{AB}$$

If the far end is pinned, no moment is induced ($C = 0$).

An example is given in Fig. 69. Relative moments of inertia are given in Fig. 69a, together with the stiffness k determined from Eq. (46) for AB, BC, CF, and FE and from Eq. (47) for CD and FG. Initially joints B, C, and F are locked and the fixed-end moments calculated. Table 4 facilitates these computations. These fixed-end moments and the distribution factors D, determined from Eq. (48), are recorded in Fig. 69b.

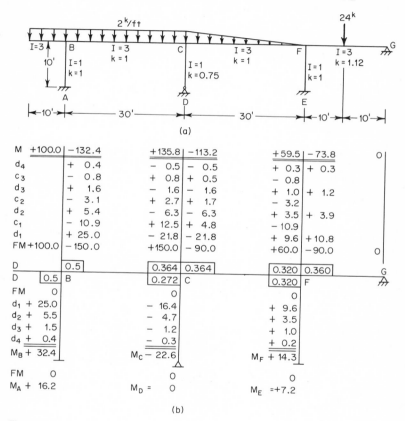

(a)

(b)

Fig. 69

The unbalance at joint B is -50; this joint is released (joint C being held fixed); the consequent moment change in BA at B is $+0.5 \times 50 = 25$ and at B in BC, $0.5 \times 50 = 25$. Similar operations at joints C and D produce the moment changes recorded in line d_1. The moment change in each member is carried over to the far end of the member if the far end is fixed, all carryover factors being $\tfrac{1}{2}$ (line c_1). Moments carried over to A and E are not distributed as these are structurally fixed joints (the carryovers to these joints can be made in one step after M_{BA} and M_{FE} are known). The moment at B in the 10-ft cantilever remains 100 ft-kips. The carryover moments are new unbalances that have to be liquidated by further stages of distribution. When the residuals are liquidated to the required accuracy, the total moment at each joint is found by summing all the moments in the appropriate computation column.

A second example is shown in Fig. 70. Here a line is drawn after each distribution to indicate a temporary balance. The type of solution in these two examples is routine and converges rapidly. It applies directly to all cases where there is no joint translation.

When there is a prescribed amount of joint translation, the procedure is the same, except that the fixed-end moments are due to joint translation and not to applied loading.

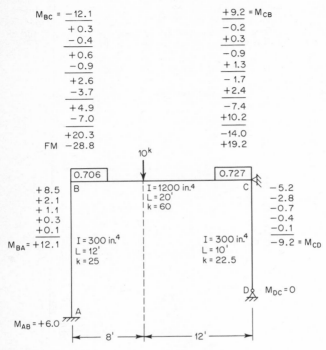

Fig. 70

Fixed-end moments are given in Fig. 71. As an example, the moments due to a prescribed horizontal (1-in) translation of the beam BC of the frame of Fig. 70 will be determined. In determining fixed-end moments, this translation is applied to the structure, joints B and C being locked against rotation. The moments are

$$FM_{AB} = FM_{BA} = \frac{6EI_{AB}\Delta}{L_{AB}^2} = \frac{6 \times 30,000 \times 300 \times 1}{(12 \times 12)^2} = 260.4 \text{ ft-kips}$$

$$FM_{CD} = \frac{3EI_{CD}\Delta}{L_{CD}^2} = \frac{3 \times 30,000 \times 300 \times 1}{(10 \times 12)^2} = 187.5 \text{ ft-kips}$$

To complete the solution, these moments would be distributed as in the example of Fig. 70. This gives the moments throughout the frame due to the applied translation. This procedure is directly applicable to problems of differential settlement in frames when there is no indeterminate sidesway.

When sidesway has to be taken into account, as, for example, in the problem of Fig. 70 if the horizontal reaction at C is absent, the solution can be computed in two steps. The "no-sidesway" solution is first computed (Fig. 70). The unbalanced horizontal shear in panel $ABCD$ is

$$\frac{M_{AB} + M_{BA}}{L_{AB}} + \frac{M_{CD}}{L_{CD}} = \frac{6.0 + 12.1}{12} + \frac{-9.2}{10} = 0.59 \text{ kip}$$

In the second step, the beam BC is translated horizontally, joints B and C being locked.

Since the magnitude of the sidesway is immaterial, only relative values of fixed-end moments are required. Thus (Fig. 71)

$$M_{AB} = M_{BA} = \frac{6EI_{AB}\Delta}{L_{AB}^2}$$

$$M_{CD} = \frac{3EI_{CD}\Delta}{L_{CD}^2}$$

and

$$M_{AB} = M_{BA} = 2\,\frac{I_{AB}}{I_{CD}}\frac{L_{CD}^2}{L_{AB}^2}\,M_{CD} = 1.39 M_{CD}$$

Figure 72 shows the solution for $M_{CD} = 100$ ft-kips. The unbalanced horizontal shear is

$$\frac{M_{AB} + M_{BA}}{L_{AB}} + \frac{M_{CD}}{L_{CD}} = \frac{122 + 104}{12} + \frac{83}{10} = 27.1 \text{ kips}$$

The unbalanced shear to be liquidated by sidesway in the frame of Fig. 70 is 0.59 kip.

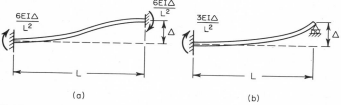

(a)   (b)

Fig. 71 Fixed-end moments due to joint translation.

Hence the proportion of the solution of Fig. 72 to be added to the no-sidesway solution is $-0.59/27.1 = -0.0218$. The final moments are

$$M_{AB} = 6.0 - 0.0218 \times 122 = +3.3 \text{ ft-kips}$$
$$M_{BA} = 12.1 - 0.0218 \times 104 = +9.8 \text{ ft-kips}$$
$$M_{CD} = -9.2 - 0.0218 \times 83 = -11.0 \text{ ft-kips}$$

The comparative simplicity of this solution is due to the fact that there is only one indeterminate translation in sidesway, namely, the horizontal movement of beam BC.

Several factors which complicate the moment-distribution procedure are nonprismatic members, multistory frames with sway, inclined members, and the effect of large axial forces in the members. When nonprismatic members are present, fixed-end moments, stiffness factors, and carryover factors have to be found as the standard values do not apply. Some tables are available.[3] In any nonstandard case the factors have to be computed. Suitable methods are the numerical-integration procedure of Arts. 16 and 18 and the column analogy.

An n-story frame with sidesway has n indeterminate horizontal joint translations. The condition of horizontal shear equilibrium must be satisfied separately in each story, so that n sway corrections have to be solved. These can be solved either simultaneously, or by a technique of successive sway corrections as proposed by Grinter.[4] Alternatively, sway can be allowed for automatically in the initial computation. An approximate solution can be found for certain types of frame by assuming an overall sidesway configuration. This reduces the sidesway problem to one of a single degree of freedom, and hence sway is accounted for in a single correction.[5]

When inclined members are present, care must be taken in computing both the fixed-end moments due to translation and the shear forces to be balanced by sway. This is most easily done by using the virtual-work principle.[6]

The effect of an axial force on fixed-end moments, stiffness factors, and carryover factors depends on its value relative to the Euler buckling load of the member. Stiffness and carryover factors for members carrying axial compression are given in Fig. 73; for more accurate values tables are available.[14] When the factors have been determined, the

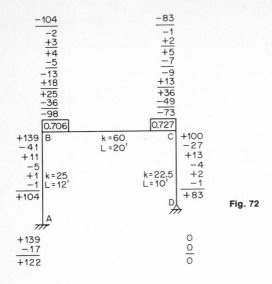

Fig. 72

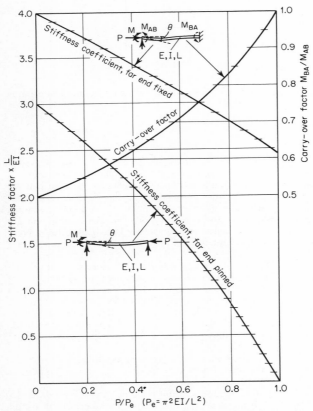

Fig. 73 Stiffness coefficients and carryover factors for members with axial compression (constructed from tabular values in Ref. 14).

moment-distribution procedure is unaltered, within the limits of the assumption that the axial forces do not change the deflection of the structure.

Nonlinear Structures

In practical application nonlinear problems occur in the following areas:

1. Nonlinearly elastic (conservative) systems (Fig. 2b). These characteristics occur in structures made from materials possessing nonlinear elasticity but can also occur because the geometry produces nonlinear behavior, even though the material from which the structural elements are made is itself linear (Fig. 3).

2. Inelastic (nonconservative) systems (Fig. 2c). This behavior occurs in structures made from elastoplastic materials. The methods discussed in Arts. 40 and 41 are applicable to the analysis of structures made from elastoplastic materials provided no unloading occurs.

40. The Minimum-Complementary-Energy Method   This is the general variational method for structures with rigid supports; the redundant forces are the problem coordinates. Castigliano's theorem (Art. 33) is the particular case of this method when the system is linearly elastic. The procedure is as follows:

1 through 4. The first four steps are identical to steps 1 through 4 of the unit-load method (Art. 32).

5. Apply the compatibility condition [Art. 7, Eq. (12)]. In this form, this condition states that the redundant forces minimize the complementary energy, that is,

$$\frac{\partial U'}{\partial X_1} = \frac{\partial U'}{\partial X_2} = \cdots = \frac{\partial U'}{\partial X_k} = \cdots = \frac{\partial U'}{\partial X_n} = 0 \tag{49}$$

These are a set of n nonlinear equations in the force redundancies $X_1, X_2, \ldots, X_k, \ldots, X_n$.

6. Solve Eqs. (49) for the redundancies $X_1, X_2, \ldots, X_k, \ldots, X_n$.

7. The internal force on each element is calculated in an identical manner to that of step 8 of the unit-load method.

41. The Minimum-Potential-Energy Method   Although this method is of general application it is most commonly applied to linearly elastic problems. In practice, the theorem (Art. 6) is usually applied by assuming that the deflection of the system can be expressed in terms of a small number of unknown coefficients (for example, the coefficients of the leading terms of an infinite series). The total energy is expressed in terms of these coefficients and is then differentiated with respect to each coefficient in turn. This yields a set of simultaneous equations in terms of the coefficients. This method is suitable for the approximate solution of systems with a large number of degrees of freedom. The displacement function chosen must satisfy the geometric (prescribed displacement) boundary conditions but need not satisfy the static (prescribed force) boundary conditions. In practice, it is desirable that the function should satisfy as many boundary conditions as possible, since in this case a smaller number of terms is required for an engineering approximation. This application is known as the *Rayleigh-Ritz procedure*.

MATRIX METHODS

42. Deflections   Equations (31) can be written in matrix form as*

$$\begin{bmatrix} \Delta_1 \\ \Delta_2 \\ \cdot \\ \cdot \\ \cdot \\ \Delta_m \end{bmatrix} = \begin{bmatrix} f_{11} & f_{12} & \cdots & f_{1m} \\ f_{21} & f_{22} & \cdots & f_{2m} \\ \cdot \\ \cdot \\ \cdot \\ f_{m1} & f_{m2} & \cdots & f_{mm} \end{bmatrix} \begin{bmatrix} W_1 \\ W_2 \\ \cdot \\ \cdot \\ \cdot \\ W_m \end{bmatrix} \tag{50}$$

or in the abbreviated form

$$\Delta = FW \tag{51}$$

*Matrix operations are discussed briefly in Art. 56.

where Δ = column vector of the m deflections
  F = flexibility matrix of the structure
  W = column vector of the m applied loads
The flexibility matrix F is a square symmetric matrix of the flexibility coefficients f_{ij} (Art. 31). The following relationships can be used to simplify and check the flexibility coefficients, whether determined by calculation or by experiment.

  1. Symmetry. Maxwell's reciprocal theorem (Art. 9) states that $f_{ij} = f_{ji}$. It follows that flexibility matrices are symmetrical about the principal diagonal. This property can be used to reduce the number of calculations or to average experimental results.

  2. Positive definiteness. A structure absorbs energy as a consequence of loading. It follows that the deflection due to, and corresponding to, an applied load cannot be opposite in sense to the applied load. Hence the terms in the principal diagonal of an influence matrix cannot be negative. That is,

$$f_{ii} \geqslant 0 \tag{52}$$

Another relationship derivable from this property is that the product of two adjacent elements on the principal diagonal cannot be less than the square of the corresponding off-diagonal terms. That is,

$$f_{ii}f_{jj} \geqslant f_{ij}^2 \tag{53}$$

The properties defined by Eqs. (52) and (53) provide a means of checking the accuracy of influence coefficients.

  3. Addition of flexibility coefficients. The general expression for a flexibility coefficient [Eq. (33)] is analogous to a set of simple springs in series. It follows that the flexibility matrix F can be written as

$$F = F_P + F_M + F_S + F_T \tag{54}$$

where the subscripts P, M, S, and T refer to axial force, bending moment, shear force, and torsion, respectively.

  The series-addition property of F is used to obtain approximate solutions by neglecting certain flexibilities; for example, the common approximation in beam-type structures of ignoring all the energies except bending energy is equivalent to assuming $1/AE = 1/A'G = 1/JG = 0$. In a similar manner, the process of calculating F by Eq. (67) is equivalent to analyzing the overall structure in a series of steps, in which, at each single step, all the element flexibilities except one are zero. The addition of these results gives the solution for the overall structure (Fig. 74).

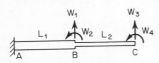

(a) Structure and loading

(b) Elements in series and element loading

Fig. 74

  The relationship between the internal forces chosen to define the loading on a structural element (member) and the corresponding deformations is given by

$$v_g = f_g S_g \tag{55}$$

where S_g = force on element g
  v_g = deformation of element g corresponding to S_g
  f_g = flexibility matrix of element g
The loading on an element of a pin-jointed framework is specified completely by a single quantity, the axial force P (case 1 of Table 6). The corresponding deformation is the elongation of the bar (L/AE for a prismatic member), in which case the matrix f_g is a scalar

quantity. In the general case, however, the force on the element g is defined by the force matrix

$$S_g = \{S_1, S_2, \ldots, S_r\} \tag{56}$$

where the forces $\{S_1, S_2, \ldots, S_r\}$ are generalized forces in the coordinates, and hence may be axial force, moment, shear, or torsion. The corresponding deformed shape is defined by the deformation matrix

$$v_g = \{v_1, v_2, \ldots, v_r\} \tag{57}$$

Thus, in the general case, r independent quantities are required to define the loading on an element and, similarly, r quantities for the corresponding deformations. Hence the flexibility matrix f_g is a square matrix of order r.

A beam-type element is, in general, subjected to four end forces S_1, S_2, S_3, S_4 (Fig. 75a). The four forces are related by means of two equilibrium equations, and hence the internal

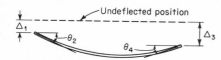

(a) End forces $\{S_1, S_2, S_3, S_4\}$ on beam element

(b) Deflected position of beam element

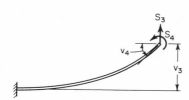

(c) Cantilever element loaded by independent forces $\{S_3, S_4\}$: corresponding deformations $\{v_3, v_4\}$

Fig. 75

(d) Simply supported element loaded by independent forces $\{S_2, S_4\}$: corresponding deformations $\{v_2, v_4\}$

loading is specified completely by a suitable choice of two independent forces. Similarly, the deformation of the element is defined completely in terms of the deformations corresponding to the chosen independent forces (Fig. 75b). In Fig. 75c the element is taken as a cantilever. The corresponding flexibility matrix is

$$\begin{bmatrix} f_{33} & f_{34} \\ f_{43} & f_{44} \end{bmatrix} = \begin{bmatrix} \dfrac{L^3}{3EI} & \dfrac{L^2}{2EI} \\ \dfrac{L^2}{2EI} & \dfrac{L}{EI} \end{bmatrix}$$

In Fig. 75*d* the element is taken as simply supported at each end. The corresponding flexibility matrix is

$$\begin{bmatrix} f_{22} & f_{24} \\ f_{42} & f_{44} \end{bmatrix} = \begin{bmatrix} \dfrac{L}{3EI} & \dfrac{L}{6EI} \\ \dfrac{L}{6EI} & \dfrac{L}{3EI} \end{bmatrix}$$

These correspond to cases 2 and 3, respectively, of Table 6.

The internal forces S throughout the structure, and the corresponding deformations v, are described by assembling the column vectors S_g and v_g of the various elements. Thus, for a structure consisting of elements $a, b, \ldots, g, \ldots, s$,

$$S = \{S_a, S_b, \ldots, S_g, \ldots, S_s\} \tag{58}$$

and

$$v = \{v_a, v_b, \ldots, v_g, \ldots, v_s\} \tag{59}$$

To calculate the deflection at joint j, a unit load is applied at j in the required direction, and a statically equivalent force system $\overline{S}_j$ calculated, where

$$\overline{S}_j = \{\overline{S}_{aj}, \overline{S}_{bj}, \ldots, \overline{S}_{gj}, \ldots, \overline{S}_{sj}\} \tag{60}$$

The deflection Δ_j at j is given by

$$\Delta_j = \Sigma \overline{S}_{gj} v_g = \overline{S}_j' v \tag{61}$$

where $\overline{S}_j'$ is the transpose of $\overline{S}_j$ (Art. 56).

Equation (61) is the unit-load theorem [Eq. (9)] written in matrix form for application to a structure consisting of an assemblage of discrete elements. Notes 1 through 4 qualifying Eq. (22) (Art. 19) are also applicable to Eq. (61) and, in particular, note 2 regarding the origin of the strains ϵ (deformations v).

Repeated application of Eq. (61) gives Δ, the deflection at the points $1, 2, \ldots, j, \ldots, m$. It can be written in the abbreviated form

$$\Delta = b_0' v \tag{62}$$

in which b_0' is the transpose of the matrix

$$b_0 = \{\overline{S}_1, \overline{S}_2, \ldots, \overline{S}_j, \ldots, \overline{S}_m\} \tag{63}$$

The matrix b_0 is called a *statically equivalent unit-load matrix*. It is a unit-load matrix for the *reduced* structure. It is generated by applying a unit value of each external load in turn; the corresponding column vector gives the forces in the members of the reduced structure due to this unit load. In the case of a statically determinate structure, the reduced structure is identical with the true structure. Thus, for the truss of Fig. 76*a* the first column vector of the b_0 matrix consists of the forces in bars a, b, c, etc., due to $W_1 = 1$ (Fig. 76*b*). In Fig. 77*b* the first column vector of the b_0 matrix consists of the values of S due to $W_1 = 1$, i.e.,

$$S_1 = 1 \qquad S_2 = S_3 = S_4 = 0$$

For the second column $W_2 = 1$ and

$$S_1 = 0 \qquad S_2 = 1 \qquad S_3 = S_4 = 0$$

etc.

In the case of a linearly elastic structure loaded only by a force system W, S and v are related through the flexibility matrix f:

$$v = fS \tag{64}$$

In this equation, f is the diagonal matrix of the flexibilities f_g of all the individual elements in the structure and is called the *flexibility matrix of the unassembled structural elements*. Each element flexibility f_g of f is expressed in terms of the unit values of the forces chosen to define the loading on the element. In the case of a pin-jointed truss each element of f is the scalar number L/AE. Thus, for the truss of Fig. 76, the flexibility f_a

of bar a is $L/AE = 50/(1 \times E)$. Similarly, $f_b = 100/(1 \times E)$, etc. Taking $50/E$ as a common factor, the resulting matrix is as shown in the figure.

In the case of loading which varies linearly along the length of an element of constant section, e.g., a linearly varying bending moment, the element f_g of the flexibility matrix f is itself a 2×2 matrix. In the example of Fig. 77, flexibility matrices f can be determined

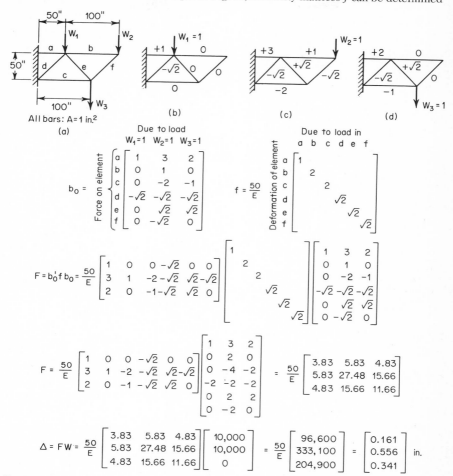

Fig. 76

by using cantilever elements (b), for which the flexibilities f_a and f_b are given in case 2 of Table 6, or by using simply supported elements (c), for which the flexibilities are given in case 3 of Table 6.

The relation between S of Eq. (58) and W of Eq. (51) is given by

$$S = bW \qquad (65)$$

In this equation, b is the *true unit-load matrix*. It is generated by applying a unit value of each external load in turn, and the corresponding column vector gives the forces in the members of the structure due to this unit load. Thus, for the statically determinate structure, the b matrix is identical with the b_0 matrix, since the true structure and the reduced structure are identical. The calculation of b for a redundant structure is discussed in Art. 45.

The flexibility matrix F of the structure can now be expressed in terms of the statically equivalent unit-load matrix b_0 [Eq. (63)], the flexibility matrix f of the unassembled

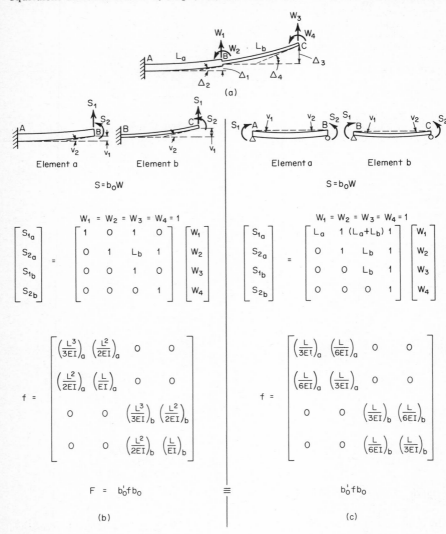

Fig. 77

elements [Eq. (64)], and the true unit-load matrix b [Eq. (65)]. From Eqs. (62), (64), and (65),

$$\Delta = b_0'v = b_0'fS = b_0'fbW \tag{66}$$

Comparing this result with Eq. (51), we find

$$F = b_0'fb \tag{67}$$

43. Statically Determinate Structures Calculation of the deflections of the truss of Fig. 29 are shown in Fig. 76. The truss and loading are shown in *a*. The first step is the determination of the internal force systems for unit values of each external load W (Fig. 76*b, c, d*). The unit-load matrix b_0 and the flexibility matrix f of the unassembled elements are then constructed as explained in Art. 42. After the flexibility matrix $F = b_0'fb_0$ is computed, deflections due to any system of vertical loads can be determined. In this example, the loading system is taken to be that of Fig. 29: $W_1 = W_2 = 10$ kips, $W_3 = 0$. If the deflection of joint 3 were not of interest, and if $W_3 = 0$ for all loading systems to be investigated, the loading system in Fig. 76*a* would consist of only W_1 and W_2, in which case the b_0 matrix would be reduced to 6 × 2, with a corresponding reduction in the computations. The step-by-step matrix multiplication shown in this example is for illustrative purposes only; ordinarily, this is a computer operation.

The elements of the statically determinate frame of Fig. 78 are taken as simply supported beams. The notations accompanying the matrices explain their construction. The 2 × 2 element matrices which make up the flexibility matrix f are from Table 6, case 3. The resulting flexibility matrix F enables the calculation of the displacements $\Delta_1, \Delta_2, \Delta_3$ due to any loading system W_1, W_2, W_3 [Eq. (51)].

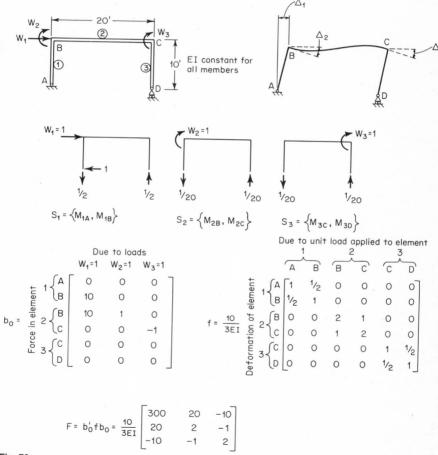

Fig. 78

Statically Indeterminate Structures

The systematic and concise notation of matrix algebra is ideally suited for programming a digital computer to carry out the numerical work involved in the analysis of highly redundant structures. Matrix methods are classified as *matrix force* or *matrix displacement* methods, when forces and displacements are respectively the problem coordinates.

44. Choice of Method   The comments of Art. 30 are generally applicable to matrix methods, but a different emphasis applies to some factors affecting the choice of method. Thus, when digital computation is used, the factors favoring the choice of the displacement method have greater emphasis, for the following reasons:

1. The displacement method does not require a reduced structure to be chosen, but operates directly with the actual complete structure.

2. In the force method it is important to select the reduced structure from the viewpoint of having well-conditioned equations, the best selection of redundancies being those which only affect the structure locally (Art. 47). In the displacement method, however, by the definition of stiffness coefficients (a unit displacement is applied to each coordinate in turn while all the others are held zero), the effect is localized and the displacement-method equations tend to be better conditioned than those in the force method.

3. In vibration problems it is more convenient to work in terms of the stiffness matrix K (Art. 49) rather than the flexibility matrix F, and K is determined directly in the displacement method.

45. Matrix Force Method   The analysis for the commoner case in which the structure is subjected only to the loading system W (no self-straining) is treated first. Self-straining is considered in Art. 48.

The internal forces S are expressed in terms of the applied loads W and the redundancies X by the equilibrium condition

$$S = b_0 W + b_1 X \qquad (68)$$

where b_0 = unit-load matrix of loads W
b_1 = unit-load matrix of redundancies X

The matrix b_0 is the statically equivalent unit-load matrix discussed in Art. 42. The matrix b_1 is generated by applying a unit value of each redundancy in turn to the reduced structure; the corresponding column vector gives the forces in the members of the reduced structure due to this unit value. It is assumed here that the reduced structure is statically determinate, so that the matrices b_0 and b_1 are obtained by statics. The theory is applicable to the general case in which the basic system is itself redundant. In this case b_0 and b_1 would require redundant analyses or would be obtained from standard data.

The conditions for compatibility of displacements are expressed in matrix form through Eq. (62), where b_0 in that equation is replaced by b_1 since the displacements are associated with the redundancies

$$\Delta = b_1' v = 0 \qquad (69)$$

By using Eqs. (69), (64), and (68), specific formulas are obtained which express X and S in terms of the applied loads W and the three basic matrices of the force method: the unit-load matrices b_0 and b_1 and the flexibility matrix f of the unassembled elements (Art. 42). Thus

$$b_1' v = b_1' f S = b_1' f (b_0 W + b_1 X) = 0$$

With the notation

$$D_0 = b_1' f b_0 \qquad (70)$$
$$D = b_1' f b_1 \qquad (71)$$

there results

$$DX + D_0 W = 0 \qquad (72)$$
$$X = -D^{-1} D_0 W \qquad (73)$$

where D^{-1} is the inverse of D (Art. 56).

With the redundant forces X known, the internal forces S can be determined from Eq. (68), which can be put in the form of Eq. (65) as follows:

$$S = b_0 W + b_1 X = b_0 W - b_1 D^{-1} D_0 W$$

from which the true unit-load matrix b is

$$b = b_0 - b_1 D^{-1} D_0 \qquad (74)$$

Thus, the internal forces in a statically indeterminate structure can be found by using Eq. (74) to determine the true unit-load matrix, following which all forces S (including the redundants X) are computed from Eq. (65). Alternatively, the redundants X can be determined by solving the simultaneous equations established by Eq. (72), after which the forces S are given by Eq. (68). This eliminates the inversion of the matrix D, which often results in saving in computer time. The two procedures are illustrated in Art. 46.

The flexibility matrix F of the structure is given by Eq. (67). An alternative form is found by using Eq. (74):

$$F = b_0' f b = b_0' f b_0 - b_0' f b_1 D^{-1} D_0$$

But, by the rule for transposition of matrix products (Art. 56), $b_0' f b_1$ is the transpose of D_0. Therefore,

$$F = F_0 - D_0' D^{-1} D_0 \qquad (75)$$

where the flexibility F_0 of the reduced structure is

$$F_0 = b_0' f b_0 \qquad (76)$$

46. Example Members c and d of the pin-jointed frame of Fig. 79 are chosen redundant. The forces

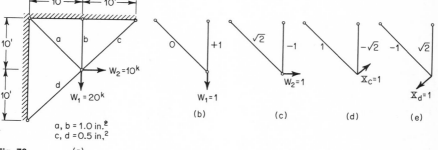

a, b = 1.0 in.^{2}
c, d = 0.5 in.^{2}

Fig. 79 (a)

$\quad$ (b) $\qquad$ (c) $\qquad$ (d) $\qquad$ (e)

due to unit values of W are shown in b and c, those due to unit values of the redundants X_c and X_d in d and e. The three basic matrices b_0, b_1, and f are

$$
\begin{array}{c}
\text{Bar} \\ a \\ b \\ c \\ d
\end{array}
\quad b_0 =
\begin{array}{c}
W_1 \quad W_2 \\
\begin{bmatrix}
0 & \sqrt{2} \\
1 & -1 \\
0 & 0 \\
0 & 0
\end{bmatrix}
\end{array}
\qquad
b_1 =
\begin{array}{c}
X_c \quad X_d \\
\begin{bmatrix}
1 & -1 \\
-\sqrt{2} & \sqrt{2} \\
1 & 0 \\
0 & 1
\end{bmatrix}
\end{array}
\qquad
f = \frac{(10\sqrt{2})}{E}
\begin{array}{c}
a \quad\ b \quad\ c \quad\ d \\
\begin{bmatrix}
1 & 0 & 0 & 0 \\
0 & 1/\sqrt{2} & 0 & 0 \\
0 & 0 & 2 & 0 \\
0 & 0 & 0 & 2
\end{bmatrix}
\end{array}
$$

From Eq. (70)

$$
D_0 = b_1' f b_0 =
\begin{bmatrix}
1 & -\sqrt{2} & 1 & 0 \\
-1 & \sqrt{2} & 0 & 1
\end{bmatrix}
\begin{bmatrix}
1 & 0 & 0 & 0 \\
0 & 1/\sqrt{2} & 0 & 0 \\
0 & 0 & 2 & 0 \\
0 & 0 & 0 & 2
\end{bmatrix}
\begin{bmatrix}
0 & \sqrt{2} \\
1 & -1 \\
0 & 0 \\
0 & 0
\end{bmatrix}
\frac{(10\sqrt{2})}{E}
=
\begin{bmatrix}
-1 & 2.414 \\
1 & -2.414
\end{bmatrix}
\frac{(10\sqrt{2})}{E}
$$

The matrix D is found similarly:

$$
D = b_1' f b_1 =
\begin{bmatrix}
4.414 & -2.414 \\
-2.414 & 4.414
\end{bmatrix}
\frac{10\sqrt{2}}{E}
$$

Inversion of D gives

$$D^{-1} = \begin{bmatrix} 0.3232 & 0.1767 \\ 0.1767 & 0.3232 \end{bmatrix} \frac{E}{10\sqrt{2}}$$

The matrix $b_1 D^{-1} D_0$ is found to be

$$b_1 D^{-1} D_0 = \begin{bmatrix} -0.2930 & 0.7073 \\ 0.4144 & -1.0004 \\ -0.1465 & 0.3537 \\ 0.1465 & -0.3537 \end{bmatrix}$$

The true unit-load matrix b is found next from Eq. (74), so that, from Eq. (65),

$$S = bW = \begin{bmatrix} 0.2930 & 0.7069 \\ 0.5856 & 0.0004 \\ 0.1465 & -0.3537 \\ -0.1465 & 0.3537 \end{bmatrix} \begin{bmatrix} 20 \\ 10 \end{bmatrix}$$

Therefore,

$$\begin{aligned} S_a &= 5.860 + 7.069 = 12.93 \text{ kips} \\ S_b &= 11.712 + 0.004 = 11.72 \text{ kips} \\ S_c &= 2.930 - 3.537 = -0.61 \text{ kip} \\ S_d &= -2.930 + 3.537 = 0.61 \text{ kip} \end{aligned}$$

This completes the solution if only the forces S are required. If the displacements are also wanted, the flexibility matrix F is computed [Eq. (67)]:

$$F = b_0' fb = \begin{bmatrix} 0 & 1 & 0 & 0 \\ \sqrt{2} & -1 & 0 & 0 \end{bmatrix} \begin{bmatrix} 1 & 0 & 0 & 0 \\ 0 & 1/\sqrt{2} & 0 & 0 \\ 0 & 0 & 2 & 0 \\ 0 & 0 & 0 & 2 \end{bmatrix} \begin{bmatrix} 0.2930 & 0.7069 \\ 0.5856 & 0.0004 \\ 0.1465 & -0.3537 \\ -0.1465 & 0.3537 \end{bmatrix} \frac{120\sqrt{2}}{E}$$

$$= \begin{bmatrix} 0.4143 & 0.0003 \\ 0.0003 & 0.9994 \end{bmatrix} \frac{120\sqrt{2}}{E}$$

The deflections are given by

$$\Delta = FW = \begin{bmatrix} 0.4143 & 0.0003 \\ 0.0003 & 0.9994 \end{bmatrix} \begin{bmatrix} 20 \\ 10 \end{bmatrix} \frac{120\sqrt{2}}{E}$$

from which $\Delta_1 = 0.47$ in., $\Delta_2 = 0.57$ in., for $E = 30{,}000$ ksi.

Alternatively, the forces S may be found using Eqs. (72) and (68). From Eq. (72)

$$\begin{bmatrix} 4.414 & -2.414 \\ -2.414 & 4.414 \end{bmatrix} \begin{bmatrix} X_c \\ X_d \end{bmatrix} + \begin{bmatrix} -1 & 2.414 \\ 1 & -2.414 \end{bmatrix} \begin{bmatrix} 20 \\ 10 \end{bmatrix} = 0$$

which yields the equations

$$\begin{aligned} 4.414 X_c - 2.414 X_d + 4.14 &= 0 \\ -2.414 X_c + 4.414 X_d - 4.14 &= 0 \end{aligned}$$

from which $X_c = -0.606$, $X_d = +0.606$. Then, from Eq. (68), using only the upper submatrices of b_0 and b_1 since X_c and X_d are already known,

$$S = \begin{bmatrix} 0 & \sqrt{2} \\ 1 & -1 \end{bmatrix} \begin{bmatrix} 20 \\ 10 \end{bmatrix} + \begin{bmatrix} 1 & -1 \\ -\sqrt{2} & \sqrt{2} \end{bmatrix} \begin{bmatrix} -0.606 \\ 0.606 \end{bmatrix}$$

from which

$$\begin{aligned} S_a &= 0 + 14.14 - 0.606 - 0.606 = 12.93 \text{ kips} \\ S_b &= 20 - 10 + 0.857 + 0.857 = 11.71 \text{ kips} \end{aligned}$$

Because this solution does not produce the true unit-load matrix b, the flexibility matrix F cannot be computed. However, the alternative form $\Delta = b_0' fS$ of Eq. (66) may be used to find the deflections:

$$\Delta = \begin{bmatrix} 0 & 1 & 0 & 0 \\ \sqrt{2} & -1 & 0 & 0 \end{bmatrix} \begin{bmatrix} 1 & 0 & 0 & 0 \\ 0 & 1/\sqrt{2} & 0 & 0 \\ 0 & 0 & 2 & 0 \\ 0 & 0 & 0 & 2 \end{bmatrix} \begin{bmatrix} 12.93 \\ 11.71 \\ -0.606 \\ 0.606 \end{bmatrix} \frac{120\sqrt{2}}{E}$$

It was pointed out in Art. 45 that, because it does not require inversion of the D matrix, the alternative solution is often economical of computer time.

47. Choice of Redundancies   The most important step in a force analysis is the selection of a suitable reduced structure. This determines the form of the b_0 and b_1 matrices and influences the conditioning of the final equations. In Fig. 80 the effect of each redundant is confined to a single panel of the frame. This simplifies the computation of the b_1 matrix and minimizes the cross coupling between redundancies. These effects are illustrated by the form of the b_1 matrix which shows:

1. Each redundancy causes loads only in the panel in which it operates.
2. Cross coupling occurs only between adjacent redundancies, through the member which is common to each panel; for example, only member 10 is loaded by X_1 and X_2 and

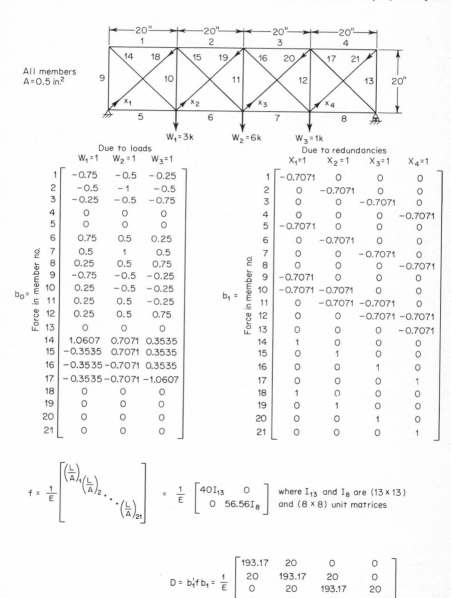

$$f = \frac{1}{E}\begin{bmatrix} \left(\frac{L}{A}\right)_1 & & & \\ & \left(\frac{L}{A}\right)_2 & & \\ & & \ddots & \\ & & & \left(\frac{L}{A}\right)_{21} \end{bmatrix} = \frac{1}{E}\begin{bmatrix} 40I_{13} & 0 \\ 0 & 56.56I_8 \end{bmatrix} \quad \text{where } I_{13} \text{ and } I_8 \text{ are } (13 \times 13) \text{ and } (8 \times 8) \text{ unit matrices}$$

$$D = b_1'f\,b_1 = \frac{1}{E}\begin{bmatrix} 193.17 & 20 & 0 & 0 \\ 20 & 193.17 & 20 & 0 \\ 0 & 20 & 193.17 & 20 \\ 0 & 0 & 20 & 193.17 \end{bmatrix}$$

Fig. 80

there are no members which are loaded by both X_1 and X_3, or X_1 and X_4. This selection of redundancies is the best practical selection for this framework.

Thus, the choice of the redundancies controls the form of the b_1 matrix. This, in turn, through the matrix $D = b_1'fb_1$ [Eq. (71)] is of vital importance to the condition of the equations for final solution [Eq. (73)]. The matrix D has to be inverted to determine X, and if D is ill-conditioned, it is difficult to invert accurately as it is sensitive to small errors.

The example of Fig. 80 produces a well-conditioned matrix D, as shown by the relative strength of the terms in the leading diagonal compared with the other terms. This occurs because the cross coupling between the redundancies is small. In general, the redundant force systems should be selected so that they are as simple as possible and cause only local effects on the structure.

Figure 81 illustrates the effects of a bad selection of redundancies. In this, the computation of b_1 is more involved, and strong cross coupling produces a badly conditioned matrix D.

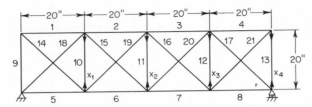

All members
A = 0.5 in.^{2}

Due to redundancies

	$x_1=1$	$x_2=1$	$x_3=1$	$x_4=1$
1	1	−1	1	−1
2	0	1	−1	1
3	0	0	1	−1
4	0	0	0	1
5	1	−1	1	−1
6	0	1	−1	1
7	0	0	1	−1
8	0	0	0	1
9	1	−1	1	−1
10	1	0	0	0
11	0	1	0	0
12	0	0	1	0
13	0	0	0	1
14	−1.4142	1.4142	−1.4142	1.4142
15	0	−1.4142	1.4142	−1.4142
16	0	0	−1.4142	1.4142
17	0	0	0	−1.4142
18	−1.4142	1.4142	−1.4142	1.4142
19	0	−1.4142	1.4142	−1.4142
20	0	0	−1.4142	1.4142
21	0	0	0	−1.4142

$b_1 = $ Force in member no.

$$ f = \frac{1}{E}\begin{bmatrix} \left(\frac{L}{A}\right)_1 & & & \\ & \left(\frac{L}{A}\right)_2 & & \\ & & \ddots & \\ & & & \left(\frac{L}{A}\right)_{21} \end{bmatrix} = \frac{1}{E}\begin{bmatrix} 40I_{13} & 0 \\ 0 & 56.56I_8 \end{bmatrix} $$

where I_{13} and I_8 are (13×13) and (8×8) unit matrices

$$ D = \frac{1}{E}\begin{bmatrix} 386.24 & -346.24 & 346.24 & -346.24 \\ -346.24 & 692.48 & -652.48 & 652.48 \\ 346.24 & -652.48 & 998.72 & -958.72 \\ -346.24 & 652.48 & -958.72 & 1304.96 \end{bmatrix} $$

Fig. 81

In the case of beam-type structures, the extent of the cross coupling between redundancies is indicated by the amount of "overlap" which occurs in the individual unit-load diagrams. For example, Fig. 82 shows a good and a bad choice of redundancies for a five-span continuous beam. In Fig. 82a, each redundancy causes loads in all the structural elements, the computation of b_1 is a comparatively lengthy procedure, and strong cross coupling is present. The opposite effects occur with a good choice of redundancies (Fig.

82b), and as a result the matrix D is much better conditioned. The system of Fig. 82b is, in fact, the redundancy system used for the three-moment equation solution (Art. 34).

48. Self-Straining If the structure is subjected to self-straining, the typical unassembled element g has an initial deformation λ_g. This initial deformation may originate from internal or from external causes (for example, from thermal distortion, error in manufacture, sinking supports), but all cases can be related to an initial deformation of the unassembled structural element [see discussion of δ_{ios}, Eqs. (35) and (36) of Art. 31]. The

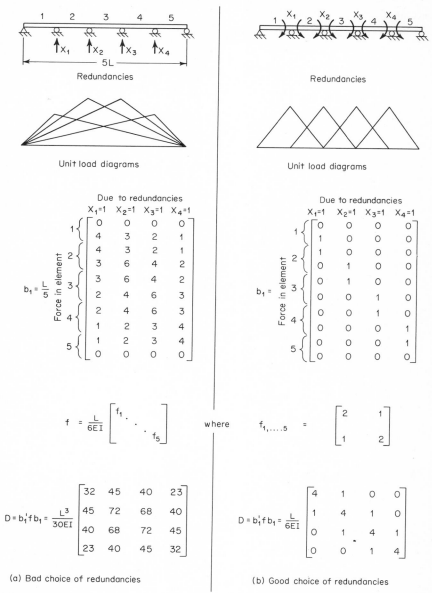

(a) Bad choice of redundancies   (b) Good choice of redundancies

Fig. 82

self-straining of the structure is then defined by the column vector λ, which is dimensionally equal to the deformation matrix v [Eq. (59)].

$$\lambda = \{\lambda_a, \lambda_b, \ldots, \lambda_g, \ldots, \lambda_s\} \tag{77}$$

In this case, the compatibility equation [Eq. (69)] must include the work done by the redundant forces due to the initial deformations λ, and so becomes

$$b_1'v = b_1'(fS + \lambda) = 0 \tag{78}$$

Hence the internal forces S are given by

$$S = (b_0 - b_1 D^{-1} D_0)W - b_1 D^{-1} b_1' \lambda \tag{79}$$

Similarly, the deflections Δ_λ corresponding to the loads W, due to the effect of the initial deformations λ, are given by

$$\Delta_\lambda = (b_0' - b_0' f b_1 D^{-1} b_1') \lambda = b' \lambda \tag{80}$$

where b is the true unit-load matrix corresponding to W. This shows that the deflections due to self-straining can be calculated easily when b is known.

49. Stiffness Matrix   Equations (41) can be written in matrix form as

$$\begin{bmatrix} W_1 \\ W_2 \\ \cdot \\ \cdot \\ \cdot \\ W_m \end{bmatrix} = \begin{bmatrix} k_{11} & k_{12} & \cdots & k_{1m} \\ k_{21} & k_{22} & \cdots & k_{2m} \\ \cdot & & & \cdot \\ \cdot & & & \cdot \\ \cdot & & & \cdot \\ k_{m1} & k_{m2} & \cdots & k_{mm} \end{bmatrix} \begin{bmatrix} \Delta_1 \\ \Delta_2 \\ \cdot \\ \cdot \\ \cdot \\ \Delta_m \end{bmatrix} \tag{81}$$

or in the abbreviated form

$$W = K\Delta \tag{82}$$

where W = column vector of the m applied loads
  K = stiffness matrix of the structure
  Δ = column vector of the m displacements

The stiffness matrix K is a square symmetric matrix of the stiffness coefficients k_{ij} (Art. 36). The relationships discussed in 1 and 2 of Art. 42 apply equally to stiffness coefficients, i.e., $k_{ij} = k_{ji}$, $k_{ii} \geqslant 0$, and $k_{ii}k_{jj} \geqslant k_{ij}^2$. The matrix K is related to the flexibility matrix F [Eq. (51)] by the equation $FK = I$, where I is a unit matrix, i.e., $K = F^{-1}$.

To determine K, the analytical model of the structure is made kinematically determinate by applying a unit displacement to each coordinate in turn, all other coordinates being fixed. In this procedure, compatibility is maintained and equilibrium is enfored by equating the element forces induced by the displacement to the corresponding applied loads. The direct calculation of the stiffness matrix K is therefore an assembly process based on element stiffnesses.

The relationship between the internal forces chosen to define the loading on a structural element (member) and the corresponding deformation is given by

$$S_g = k_g v_g \tag{83}$$

where S_g = force on element g
  v_g = deformation of element g corresponding to S_g
  k_g = stiffness matrix of element g

In the general case, the force on the element g and the corresponding deformed shape are defined by force and deformation matrices, respectively [Eqs. (56) and (57)].

The stiffness matrix k_g and the flexibility matrix f_g [Eq. (55)] are related by the equation

$$f_g k_g = I \tag{84}$$

where I is a unit matrix. Therefore, the stiffness matrix k_g can be obtained by inverting the flexibility matrix f_g. Alternatively, it can be obtained by applying the unit-displacement theorem (Art. 5). In practice, however, it is generally more convenient to select appropriate values from a collection of standard results. Table 7 gives the stiffness matrices for prismatic members for use in the analysis of pin-jointed and stiff-jointed frameworks.

The procedure for obtaining a stiffness matrix by inverting a flexibility matrix is illustrated with reference to the cantilever element of Fig. 83a. The stiffness coefficients for end B are obtained by inverting the appropriate flexibility coefficients, shown in b and c. The forces at end A are obtained by equilibrium considerations (Fig. 83d, e). By symmetry, the complete matrix of stiffness coefficients for the beam regarded as a free body can be constructed from these data (Fig. 84).

Addition of Stiffness Coefficients.   When a number of members meet at a common point, the resultant force necessary to cause a displacement in a given coordinate is the

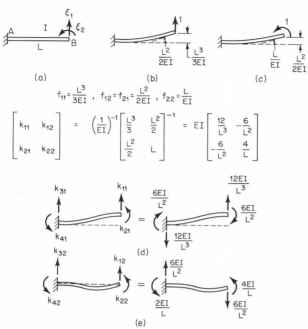

$$f_{11} = \frac{L^3}{3EI} \ , \quad f_{12} = f_{21} = \frac{L^2}{2EI} \ , \quad f_{22} = \frac{L}{EI}$$

$$\begin{bmatrix} k_{11} & k_{12} \\ \\ k_{21} & k_{22} \end{bmatrix} = \left(\frac{1}{EI} \right)^{-1} \begin{bmatrix} \dfrac{L^3}{3} & \dfrac{L^2}{2} \\ \\ \dfrac{L^2}{2} & L \end{bmatrix}^{-1} = EI \begin{bmatrix} \dfrac{12}{L^3} & -\dfrac{6}{L^2} \\ \\ -\dfrac{6}{L^2} & \dfrac{4}{L} \end{bmatrix}$$

Fig. 83   Stiffness matrix by inverting flexibility matrix.

	Unit displacement in coordinate			
	1	2	3	4
1	$\dfrac{12EI}{L^3}$	$-\dfrac{6EI}{L^2}$	$-\dfrac{12EI}{L^3}$	$-\dfrac{6EI}{L^2}$
2	$-\dfrac{6EI}{L^2}$	$\dfrac{4EI}{L}$	$\dfrac{6EI}{L^2}$	$\dfrac{2EI}{L}$
3	$-\dfrac{12EI}{L^3}$	$\dfrac{6EI}{L^2}$	$\dfrac{12EI}{L^3}$	$\dfrac{6EI}{L^2}$
4	$-\dfrac{6EI}{L^2}$	$\dfrac{2EI}{L}$	$\dfrac{6EI}{L^2}$	$\dfrac{4EI}{L}$

Force in coordinate

Fig. 84   Stiffness matrix for beam considered as free body.

sum of the forces necessary to cause the same displacement separately in each element. In other words, the resistance to the displacement of a coordinate is the cumulative effect from all the elements meeting at that point (Fig. 85). Thus, the stiffness coefficients for coupled elements can be obtained by using the appropriate stiffness coefficients for the separate elements and adding terms in common coordinates. This is illustrated by the coupled beam of Fig. 86. In a similar manner, the stiffness matrix K could be assembled by adding element stiffness coefficients having identical subscripts. While this is adequate for simple structures, a simpler and more systematic direct procedure is required to deal with more complicated structures, especially if digital computation is involved.

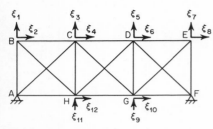
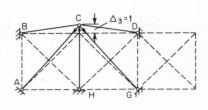

(a) Structure and degrees of freedom   (b) Forces at C resisting a displacement $\Delta_3 = 1$

Fig. 85

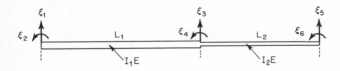

Due to unit displacement in coordinate

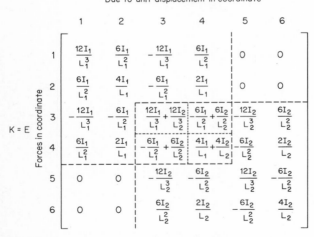

Fig. 86

50. Direct Calculation of Stiffness Matrix In general, the element stiffness matrix k_g is expressed initially with respect to the axes of the element (*local axes*) which differ from the *structural axes* (Fig. 87). The transformation of the element deformations from the local axes into the structural axes is effected by a transformation accordiing to

$$v = a\Delta \tag{85}$$

where a is the *displacement transformation matrix*. To establish a, the system is made kinematically determinate by fixing all the joints; a then expresses the kinematic relationship between the structural coordinate displacements Δ and the structural element deformations v when each coordinate displacement is considered separately. This is illustrated in Fig. 88 for the structure shown in a. A displacement $\Delta_1 = 1$ in the coordinate ξ_1 produces element (member) elongations $v_4 = 1$, $v_5 = 1/\sqrt{2}$, and $v_1 = v_2 = v_3 = v_6 = 0$, which form the first column of the a matrix shown in b. Column 2 is found by imposing a displacement $\Delta_2 = 1$, etc.

The displacement matrix a for the stiff-jointed frame of Fig. 89a is shown in c. Axial deformations of the members are taken into account, so that there are six independent displacements Δ in the six degrees of freedom ξ. Member independent displacements v are shown in Fig. 89b. Displacements v resulting from a displacement $\Delta_1 = 1$, which are shown in Fig. 90a, form the first column of the a matrix. The second column is formed from the v displacements corresponding to $\Delta_2 = 1$ (Fig. 90b), and so on. If axial

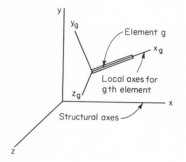

Fig. 87

deformations are neglected, as would usually be the case, the number of degrees of freedom is reduced to three (Art. 37, Example 2, Fig. 64b). The displacements v corresponding to unit displacements in the three coordinates are shown in Fig. 64d, e, and f and the resulting a matrix in Fig. 91.

The element forces S and displacements v are expressed in matrix form by

$$S = kv \tag{86}$$

where k is the *matrix of stiffnesses of the unassembled elements*, which is formed by arranging the stiffnesses of the elements in diagonal form (Fig. 88c). Each element k_g of the k matrix in Fig. 89d is formed from the 1×1 stiffness matrix corresponding to v_1 and the 2×2 beam-element matrix corresponding to v_2 and v_3 (Table 7). If axial deformations are neglected, the element matrix k_g of k is reduced to the beam-element stiffness matrix (Fig. 91).

The unit-displacement theorem (Art. 5) gives $1 \times W = \Sigma vS$, which can be written in matrix form by using the transpose a' of the displacement transformation matrix a,

$$W = a'S \tag{87}$$

Then, from Eqs. (85) and (86),

$$W = a'kv = a'ka\Delta \tag{88}$$

Comparing this result with Eq. (82), the stiffness K of the structure is

$$K = a'ka \tag{89}$$

51. The Matrix-Displacement Method   The matrix-displacement method is a generalization of the unit-displacement method discussed in Art. 37. In this method it is convenient to consider all coordinate displacements as primary unknowns. The analysis for the more general case in which some of the coordinates may not be loaded is discussed in Art. 52. For the case in which each coordinate is subjected to an applied load the analysis is as follows:

1. The stiffness matrix K for the complete structure is computed.
2. The matrix K is inverted to give the flexibility matrix F.
3. The deflections Δ corresponding to the loads W are given by

$$\Delta = FW = K^{-1}W \qquad (90)$$

4. The unit-load matrix b is established through Eqs. (82), (85), and (86):

$$S = kv = ka\Delta = kaFW = kaK^{-1}W \qquad (91)$$

Comparing this result with Eq. (65), the true unit-load matrix is given by

$$b = kaK^{-1} \qquad (92)$$

Equations (89), (90), and (91) give the complete solution to the analysis problem when the two basic matrices a and k have been generated.

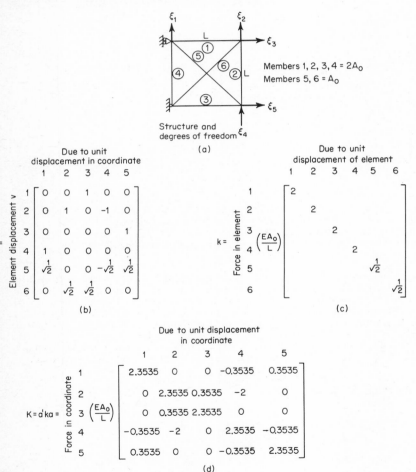

Members 1, 2, 3, 4 = $2A_0$
Members 5, 6 = A_0

Structure and degrees of freedom ξ_4

(a)

Fig. 88

The method is illustrated in Fig. 91. The procedure for obtaining the equivalent loading shown in Fig. 91b is described in Example 2, Art. 37. Formulation of the a and k matrices is discussed in Art. 50. The values of S given by $S = ka\Delta$ are the bending moments in the members due to the equivalent loading. Adding to these the fixed-end moments due to the distributed load gives the bending moments due to the actual loading. Thus,

$$M_{BC} = -15.103 + 8.333 = -6.770$$
$$M_{CB} = -7.682 - 8.333 = -16.015$$

so that the final moments are

$$S = \{8.210, 6.770, -6.770, -16.015, 16.015, 11.940\}$$

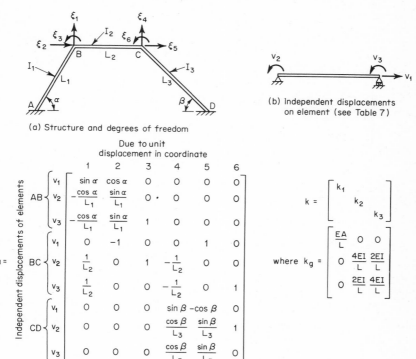

(a) Structure and degrees of freedom

(b) Independent displacements on element (see Table 7)

(c) Matrices a and k for calculation of K, including axial deformations

Fig. 89

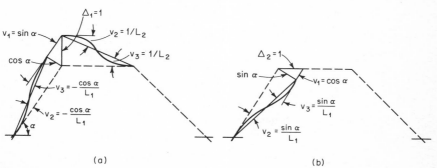

(a)

(b)

Fig. 90

52. Condensed Stiffness Matrix In the more general case where some of the coordinates are not loaded, advantage can be taken of a condensation technique in order to reduce the amount of subsequent computation. In this case, the system coordinate vector $\xi = \{\xi_1, \xi_2, \ldots, \xi_j, \ldots, \xi_m\}$ is partitioned into $\xi = \{\xi_o \mid \xi_1\}$, where ξ_o is the vector of loaded coordinates and ξ_1 the vector of coordinates which are not loaded. The basic load-deflection relation-

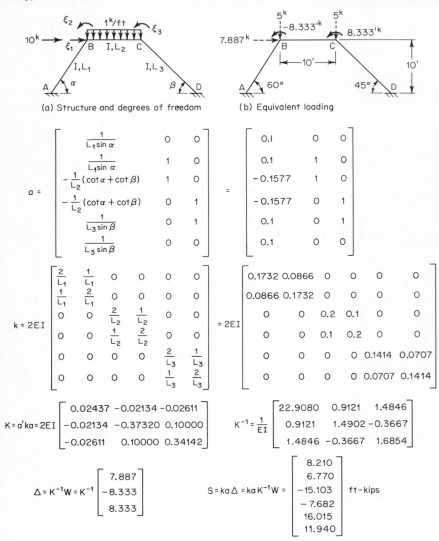

Fig. 91

ships [Eqs. (82) and (51)] and the load transformation matrix a are partitioned in accordance with the partitioning of ξ. Thus,

$$\begin{bmatrix} W_0 \\ \hline 0 \end{bmatrix} = \begin{bmatrix} K_{00} & K_{01} \\ \hline K_{10} & K_{11} \end{bmatrix} \begin{bmatrix} \Delta_0 \\ \hline \Delta_1 \end{bmatrix} \tag{93}$$

$$\begin{bmatrix} \Delta_0 \\ \hline \Delta_1 \end{bmatrix} = \begin{bmatrix} F_{00} & \vdots & F_{01} \\ \hline F_{10} & \vdots & F_{11} \end{bmatrix} \begin{bmatrix} W_0 \\ \hline 0 \end{bmatrix} \tag{94}$$

$$a = [a_0 \vdots a_1] \tag{95}$$

From Eq. (93),

$$W_0 = K_{00}\Delta_0 + K_{01}\Delta_1 \tag{96}$$
$$0 = K_{10}\Delta_0 + K_{11}\Delta_1 \tag{97}$$

Equation (97) gives $\Delta_1 = -K_{11}^{-1}K_{10}\Delta_0$, which upon substitution into Eq. (96) yields

$$W_0 = (K_{00} - K_{01}K_{11}^{-1}K_{10})\Delta_0 = \overline{K}\Delta_0 \tag{98}$$

where $\overline{K}$ is the condensed stiffness matrix, that is,

$$\overline{K} = K_{00} - K_{01}K_{11}^{-1}K_{10} \tag{99}$$

This enables the deflections Δ to be expressed in terms of W_0, $\overline{K}$, and the submatrices of K. Thus, from Eqs. (94) and (98),

$$\Delta_0 = F_{00}W_0 = \overline{K}^{-1}W_0 \tag{100}$$

and from Eqs. (94), (97), and (100),

$$\Delta_1 = F_{10}W_0 = -K_{11}^{-1}K_{10}\overline{K}^{-1}W_0 \tag{101}$$

Equations (100) and (101) determine the complete matrix of deflections Δ and the flexibility matrix F of the structure.

The element deformations v (with respect to the local axes of the elements) are obtained from Eq. (85):

$$v = [a_0 \vdots a_1]\begin{bmatrix} \Delta_0 \\ \hline \Delta_1 \end{bmatrix} = (a_0 - a_1 K_{11}^{-1}K_{10})\overline{K}^{-1}W_0 \tag{102}$$

The element force matrix is obtained from Eq. (86):

$$S = k(a_0 - a_1 K_{11}^{-1}K_{10})\overline{K}^{-1}W_0 \tag{103}$$

which, according to Eq. (65), gives the true unit-load matrix as

$$b = k(a_0 - a_1 K_{11}^{-1}K_{10})\overline{K}^{-1} \tag{104}$$

Equations (99) through (103) are given in terms of the matrices a_0, a_1, k, and the four submatrices of K. Then, from Eq. (89),

$$K = a'ka = \begin{bmatrix} a'_0 \\ \hline a'_1 \end{bmatrix} k[a_0 \vdots a_1]$$

which identifies the submatrices of K in terms of a_0, a_1, and k:

$$K = \begin{bmatrix} K_{00} & \vdots & K_{01} \\ \hline K_{10} & \vdots & K_{11} \end{bmatrix} = \begin{bmatrix} a'_0 k a_0 & \vdots & a'_0 k a_1 \\ \hline a'_1 k a_0 & \vdots & a'_1 k a_1 \end{bmatrix} \tag{105}$$

The complete structural analysis is therefore based fundamentally on the three basic matrices a_0, a_1, and k, which are the counterparts to the three basic matrices b_0, b_1, and f of the matrix force method. The solution by the condensation technique requires the inversion of the matrices K_{11} and $\overline{K}$. Because they are of lower order than K, they are simpler to invert.

The analysis will be illustrated for the truss of Fig. 88 loaded with 20 kips in the coordinate ξ_2 and -10 kips in the coordinate ξ_5, all other coordinates being unloaded (Fig. 65).

The partitioned system coordinate vector is $\{\xi_0 \vdots \xi_1\} = \{\xi_2, \xi_5 \vdots \xi_1, \xi_3, \xi_4\}$. The load transformation matrix a is partitioned accordingly:

Displacement in coordinate

$$
a = [a_0 \mid a_1] = \begin{array}{c} \text{Element} \\ \text{displacement} \end{array} \begin{array}{c} 1 \\ 2 \\ 3 \\ 4 \\ 5 \\ 6 \end{array} \begin{bmatrix} \overset{2}{0} & \overset{5}{0} & \overset{1}{\big|} \overset{1}{0} & \overset{3}{1} & \overset{4}{0} \\ 1 & 0 & 0 & 0 & -1 \\ 0 & 1 & 0 & 0 & 0 \\ 0 & 0 & 1 & 0 & 0 \\ 0 & 1/\sqrt{2} & 1/\sqrt{2} & 0 & -1/\sqrt{2} \\ 1/\sqrt{2} & 0 & 0 & 1/\sqrt{2} & 0 \end{bmatrix}
$$

Thus, the three basic matrices a_0, a_1, and k (Fig. 88) are established. Then, from Eq. (105),

$$
K = \begin{bmatrix} K_{00} & \mid & K_{01} \\ \hline K_{10} & \mid & K_{11} \end{bmatrix} = \begin{bmatrix} a_0' k a_0 & \mid & a_0' k a_1 \\ \hline a_1' k a_0 & \mid & a_1' k a_1 \end{bmatrix}
$$

$$
= \frac{EA_0}{L} \begin{bmatrix} 2.3535 & 0 & \mid & 0 & 0.3535 & -2 \\ 0 & 2.3535 & \mid & 0.3535 & 0 & -0.3535 \\ \hline 0 & 0.3535 & \mid & 2.3535 & 0 & -0.3535 \\ 0.3535 & 0 & \mid & 0 & 2.3535 & 0 \\ -2 & -0.3535 & \mid & -0.3535 & 0 & 2.3535 \end{bmatrix}
$$

The condensed stiffness matrix [Eq. (99)] is

$$
\overline{K} = K_{00} - K_{01}K_{11}^{-1}K_{10} = \frac{EA_0}{L} \begin{bmatrix} 0.5615 & -0.2612 \\ -0.2612 & 2.2612 \end{bmatrix}
$$

and, from Eqs. (100) and (101),

$$
F_{00} = \overline{K}^{-1} = \frac{L}{EA_0} \begin{bmatrix} 1.8821 & 0.2174 \\ 0.2174 & 0.4674 \end{bmatrix}
$$

$$
F_{10} = -K_{11}^{-1}K_{10}\overline{K}^{-1} = \frac{L}{EA_0} \begin{bmatrix} 0.2174 & -0.0327 \\ -0.2827 & -0.0327 \\ 1.6647 & 0.2501 \end{bmatrix}
$$

$$
\Delta = \begin{bmatrix} \Delta_2 \\ \Delta_5 \\ \hline \Delta_1 \\ \Delta_3 \\ \Delta_4 \end{bmatrix} = \frac{L}{EA_0} \begin{bmatrix} 1.8821 & 0.2174 & \dfrac{EA_0}{L} F_{01} \\ 0.2174 & 0.4674 & \\ \hline 0.2174 & -0.0327 & \dfrac{EA_0}{L} F_{11} \\ -0.2827 & -0.0327 & \\ 1.6647 & 0.2501 & \end{bmatrix} \begin{bmatrix} 20 \\ -10 \\ \hline 0 \\ 0 \\ 0 \end{bmatrix} = \frac{L}{EA_0} \begin{bmatrix} 35,468 \\ -326 \\ \hline 4,675 \\ -5,327 \\ 30,793 \end{bmatrix}
$$

The element deformations are obtained from Eq. (102):

$$
v = \{v_1, v_2, v_3, v_4, v_5, v_6\} = [a_0 \mid a_1] \begin{bmatrix} \Delta_0 \\ \Delta_1 \end{bmatrix}
$$

$$
= \frac{L}{EA_0} \{-5327, 4675, -326, 4675, 18,720, 21,300\}
$$

so that, from Eq. (86),

$$
S = \{S_1, S_2, S_3, S_4, S_5, S_6\} = kv = \{-10.654, 9.35, -0.652, 9.35, -13.224, 15.07\} \text{ kips}
$$

53. Self-Straining   The effect of self-straining is treated by forming an initial force vector i. The typical element i_g of this vector is the force system necessary to eliminate the initial deformations λ_g in the unassembled element g. Thus i states the set of internal forces which are necessary to eliminate the initial deformations λ of the unassembled elements:

$$
i = \{i_a, i_b, \ldots, i_g, \ldots, i_s\} \tag{106}
$$

The initial force vector i is related to the initial deformation vector λ through

$$
i = -k\lambda \tag{107}
$$

The forces i are applied to the appropriate elements, assuming all the coordinates of the structure to be held fixed. The coordinates are then released, and as a result, the internal forces are redistributed throughout the structure and deflections occur in the coordinates.

The resulting internal forces (which are in addition to those caused by the applied loads W) are given by

$$S_i = i - ka_1(a_1'ka_1)^{-1}a_i'i \qquad (108)$$

54. Nonlinear Structures In the case of nonlinear structures the solution is given by a set of nonlinear equations and an iterative method is, in general, required for a numerical solution. Matrix techniques for the force and displacement methods have been developed by Argyris[7,8] as an extension of his basic work.[9] Denke has also extended his force method[10] to the nonlinear case.[7,11]

ELEMENTS OF MATRIX ALGEBRA

55. Definitions Matrix methods are useful in preparing problems in structural analysis for the electronic digital computer. This is because many computers can be progammed directly for matrix operations, and because of the simplicity with which systems of equations can be represented in the notation of matrix algebra.

A matrix is a rectangular array of numbers (or symbols) called *elements*. An element is usually denoted by a letter with two subscripts, the first signifying the row and the second the column in which the element is located. The array itself is denoted by a single, bracketed symbol. The brackets are sometimes omitted. Thus,

$$\begin{bmatrix} a_{11} & a_{12} & a_{13} \\ a_{21} & a_{22} & a_{23} \end{bmatrix} = [a_{ij}] = [A] = A$$

The *order* of a matrix signifies the number of rows and columns, the former being specified first. The matrix above is of order 2×3. If there is only one row of elements the matrix is called a *row matrix* (or vector). Similarly, a matrix with only one column is called a column matrix (or vector).

A *square* matrix has the same number of rows as columns. Its diagonal which begins with the element a_{11} is called the *principal* diagonal. A square matrix which is symmetrical about its principal diagonal is a *symmetrical matrix* (Fig. 92).

$$\begin{bmatrix} 3 & 1 & 0 \\ 2 & 4 & -5 \\ 0 & -2 & 1 \end{bmatrix} \qquad \begin{bmatrix} 3 & 1 & 6 \\ 1 & 2 & 4 \\ 6 & 4 & 5 \end{bmatrix} \qquad \begin{bmatrix} 3 & 0 & 0 \\ 0 & -7 & 0 \\ 0 & 0 & 2 \end{bmatrix} \qquad \begin{bmatrix} 1 & 0 & 0 \\ 0 & 1 & 0 \\ 0 & 0 & 1 \end{bmatrix}$$

(a) Square   (b) Symmetrical   (c) Diagonal   (d) Unit
    matrix   matrix   matrix   matrix

Fig. 92

A *diagonal matrix* is a square matrix which has zero elements everywhere except on its principal diagonal. If all the elements of the principal diagonal of a diagonal matrix are unity, the matrix is called a *unit matrix* and is denoted by the symbol I (Fig. 92).

56. Matrix Operations *Addition of matrices* is performed by adding corresponding elements to obtain the elements of the matrix sum. Thus,

$$\begin{bmatrix} 1 & 0 & -2 \\ 4 & -5 & 2 \end{bmatrix} + \begin{bmatrix} 0 & -5 & 4 \\ 1 & 3 & -2 \end{bmatrix} = \begin{bmatrix} 1 & -5 & 2 \\ 5 & -2 & 0 \end{bmatrix}$$

It follows that only matrices of the same order can be added. Subtraction is performed similarly. Two matrices of the same order are said to be *conformable* for addition and subtraction.

Multiplication of a matrix by a scalar is performed by multiplying every element by the scalar:

$$-3 \begin{bmatrix} 2 & 0 & -5 \\ -2 & 1 & 4 \end{bmatrix} = \begin{bmatrix} -6 & 0 & 15 \\ 6 & -3 & -12 \end{bmatrix}$$

Matrix Multiplication.   The product $AB = C$ of the matrices A and B is defined as follows: the element c_{ij} of C is obtained by multiplying, element by element, the ith row of A and the jth column of B and adding the results. Thus,

$$\begin{bmatrix} 3 & -1 & 4 \\ 2 & 0 & -2 \end{bmatrix} \begin{bmatrix} 0 & 2 \\ 1 & -1 \\ -5 & 1 \end{bmatrix} = \begin{bmatrix} -21 & 11 \\ 10 & 2 \end{bmatrix}$$

where the element -21 of the product matrix is given by

$$3 \times 0 + (-1) \times 1 + 4 \times (-5) = -21$$

the element 10 by

$$2 \times 0 + 0 \times 1 + (-2)(-5) = 10$$

the element 11 by

$$3 \times 2 + (-1)(-1) + 4 \times 1 = 11$$

etc.

The reason for defining multiplication in this manner is demonstrated by considering

$$\begin{bmatrix} a_{11} & a_{12} & a_{13} \\ a_{21} & a_{22} & a_{23} \\ a_{31} & a_{32} & a_{33} \end{bmatrix} \begin{bmatrix} x_1 \\ x_2 \\ x_3 \end{bmatrix} = \begin{bmatrix} c_1 \\ c_2 \\ c_3 \end{bmatrix} \tag{109}$$

Applying the rule for multiplication yields

$$\begin{aligned} a_{11}x_1 + a_{12}x_2 + a_{13}x_3 &= c_1 \\ a_{21}x_1 + a_{22}x_2 + a_{23}x_3 &= c_2 \\ a_{31}x_1 + a_{32}x_2 + a_{33}x_3 &= c_3 \end{aligned} \tag{110}$$

Thus, the system of simultaneous equations (110) is represented in matrix notation by (109), which can be written still more concisely as

$$[A][x] = [c] \tag{111}$$

Matrices A and B can be multiplied only if A has the same number of columns as B has rows; that is, the row of A must have the same number of elements as the columns of B. In this case the product AB is defined and A is said to be *conformable* to B for multiplication.

It is important to note that, in general, $AB \neq BA$. That is, if A is conformable to B for multiplication, B is not necessarily conformable to A for multiplication. Therefore, because the sequence is important we *premultiply* B by A (or *postmultiply* A by B) to signify the product AB.

In general, matrix multiplication is not commutative; that is, if $AB = C$, $BA \neq C$. On the other hand, matrix multiplication obeys the associative and distributive laws, so that

$$ABC = (AB)C = A(BC)$$
$$A(B + C) = AB + AC$$

Premultiplication or postmultiplication of a matrix A by a unit matrix I does not alter A. That is,

$$IA = AI = A$$

For example,

$$\begin{bmatrix} 1 & 0 & 0 \\ 0 & 1 & 0 \\ 0 & 0 & 1 \end{bmatrix} \begin{bmatrix} 2 & -1 \\ 1 & 4 \\ 1 & 0 \end{bmatrix} = \begin{bmatrix} 2 & -1 \\ 1 & 4 \\ 1 & 0 \end{bmatrix}$$

Thus, the unit matrix performs the same function in matrix algebra as does unity in ordinary algebra.

Transposition.   A matrix is transposed by converting its rows into columns, in order. The transpose of A is usually denoted by A' or A^t. Thus,

$$A = \begin{bmatrix} 3 & 0 & -1 \\ 2 & 3 & 0 \end{bmatrix} \qquad A' = \begin{bmatrix} 3 & 2 \\ 0 & 3 \\ -1 & 0 \end{bmatrix}$$

Transposition does not alter a symmetrical matrix (Fig. 92b).

The transpose of a matrix product AB is the reversed product of the transposes of the individual matrices, that is,

$$(AB)' = B'A'$$

Furthermore
$$(ABC)' = C'B'A'$$

Inversion. To express the solution of Eq. (111) in matrix notation we write

$$A^{-1}Ax = A^{-1}c$$
$$x = A^{-1}c$$

which requires that A^{-1} and A be reciprocal, that is, $A^{-1}A = I$, a unit matrix. The matrix A^{-1} is called the *inverse* of A; it serves a function analogous to division in ordinary algebra. Only square matrices can be inverted.

There are various procedures for inverting a matrix. However, this is ordinarily a computer operation and is not discussed here. It should be noted that inversion of a large matrix is time-consuming even for a computer, so that solutions which minimize (or eliminate) inversion are sometimes advantageous (Arts. 45, 46).

Partitioning. It is sometimes useful to group the elements of a matrix according to their physical interpretations. This is done by *partitioning*. The groups of elements into which the matrix is partitioned are called *submatrices*. Partitioning is usually indicated by dotted lines, and the submatrices may be denoted by individual symbols. Thus,

$$A = \left[\begin{array}{ccc:cc} 2 & -1 & 4 & 5 & 0 \\ \hdashline 0 & 5 & 2 & 1 & 0 \\ 4 & 2 & 0 & 0 & 1 \end{array}\right] = \left[\begin{array}{c:c} A_{11} & A_{12} \\ \hdashline A_{21} & A_{22} \end{array}\right]$$

Partitioned matrices can be added, subtracted, and multiplied in terms of their submatrices, provided the latter are conformable for these operations. Thus

$$\left[\begin{array}{c:c} A_{11} & A_{12} \\ \hdashline A_{21} & A_{22} \end{array}\right] \left[\begin{array}{c} B_{11} \\ \hdashline B_{21} \end{array}\right] = \left[\begin{array}{c} A_{11}B_{11} + A_{12}B_{21} \\ A_{21}B_{11} + A_{22}B_{21} \end{array}\right]$$

provided that A_{11} has the same number of columns as B_{11} has rows, and similarly for A_{12}, B_{21}, etc.

Part 2. The Finite-Element Method

W. C. SCHNOBRICH

Professor of Civil Engineering, University of Illinois, Urbana

J. GRAHAM

Lecturer in Aeronautical Engineering, Queen's University, Belfast

The basis of the finite-element method is the representation of a structure as a finite number of line, two-dimensional, and/or three-dimensional subdivisions called *finite elements*. These elements are interconnected at joints called *nodes* (Fig. 93b). The external loading is transformed into equivalent forces applied to the nodes, and the behavior of the elements is prescribed by relating their response to that of the nodes.

The method can be considered in three basic steps: discretization of the structure into elements; establishment of the behavior of the individual elements, usually in the form of force-displacement relations; and the assembly and solution of the element array.

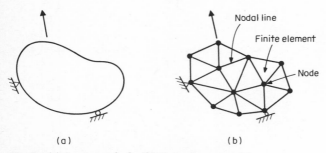

Fig. 93   (*a*) Continuous body. (*b*) Finite-element model.

57. Discretization of the Structure   This step, in which the structure is divided into elements, is critical in establishing the accuracy of the solution. The choice of the number, size, and type of elements is a matter of judgment; some guidelines are given in Art. 58.

The interfaces between adjacent elements are called *nodal lines* in two-dimensional problems (Fig. 93b) and *nodal planes* or *nodal surfaces* in three-dimensional problems. Degrees of freedom, which may be in the form of nodal displacements and/or their derivatives, stresses, or combinations of these, are assigned to the nodes. Displacements are most commonly used; the resulting model is called a *displacement model*. Models based on stresses are called *force models*, while those based on combinations of displacements and stresses are called *mixed models* or *hybrid models* depending upon the variational procedures used and the quantities defined at the nodes.[22] Some advantages of hybrid models for plate and shell problems have been cited, but the displacement model is predominant and only it is discussed here.

Normally the nodes of an element are at the corners, but additional degrees of freedom can be included by using *side nodes,* usually at midside, or *interior (internal)* nodes (Fig. 94). Additional degrees of freedom may also be specified in the form of higher-order derivatives of the nodal displacements.

58. Guidelines for Selection of Grid The following guidelines have evolved from experience with the method but are not subscribed to by all analysts.

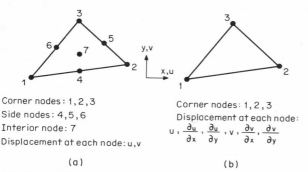

Corner nodes: 1,2,3
Side nodes: 4,5,6
Interior node: 7
Displacement at each node: u,v

Corner nodes: 1,2,3
Displacement at each node:
$u, \dfrac{\partial u}{\partial x}, \dfrac{\partial u}{\partial y}, v, \dfrac{\partial v}{\partial x}, \dfrac{\partial v}{\partial y}$

(a)   (b)

Fig. 94 (*a*) 14 degrees of freedom; (*b*) 18 degrees of freedom.

Element Shape. Quadrilateral elements should be used, except where triangular elements are necessary to accommodate a grid refinement or because of the geometry of the structure. Rectangular elements accommodate sharp stress gradients better than corresponding triangular elements, but if the stress gradient is expected to be reasonably constant over the region this choice is not significant. Most elements respond to uniform stresses, and simple elements such as the constant-strain triangle are effective for these cases.

The quality of the solution is improved by elements which have dimensions of the same order, for example, rectangles nearly square and triangles nearly equilateral. The aspect-ratio limit varies somewhat depending on the element type, but a good general guideline is to keep it below 10 for deformation analysis. This may have to be reduced to 5 for stress analysis. However, elements with aspect ratios near 1 usually produce more accurate solutions.

Geometric Approximations. A curved boundary can be approximated as piecewise linear by the sides of straight-sided elements adjacent to the boundary. Isoparametric elements[26,41] or similar elements capable of accommodating the curved edges are normally more efficient. However, the sides of isoparametric elements should be curved only to represent a curved boundary or to follow the curvature of stress isoclines. The accuracy of a solution is reduced by indiscriminate use of irregularly shaped elements.

Mesh Layout. Gridworks should generally be kept uniform. Of course, this is not possible if a grid refinement within a problem is needed for geometric reasons or because of loading. A refined mesh is necessary where large stress gradients are anticipated.

To obtain more accurate values in regions of stress concentrations, and to examine convergence, the solution process must be repeated. For convergence studies the refined mesh must include the original mesh; otherwise a new approximating sequence is started and with it a new convergence pattern.

Subdivisions at Discontinuities. Nodes and subdivision lines and planes should be located at positions where there are abrupt changes in geometry, loading, and material properties.

Artificial discontinuities may develop in the solution process. Connecting high-order elements to low-order elements can cause irregularities in stress which may require smoothing. If overconforming elements (Art. 59) are used, conflicts between requirements for conformity and the required discontinuities must be watched for.

Element and Node Labels. Elements and nodes must be numbered in a systematic way. The system is optional if a new element is being developed, but a user must adhere to the system once it is developed. The scheme shown in Fig. 95 is in common usage. In this, the nodes are numbered consecutively from left to right with reference to a right-handed coordinate system. An independent system is used to number the elements. Here the system is normally arbitrary and immaterial, unless a data generator is being used.

The numbering system for the nodes can be very important if the program equation solver is bandwidth-dependent, which is the most likely case. The nodes should then be

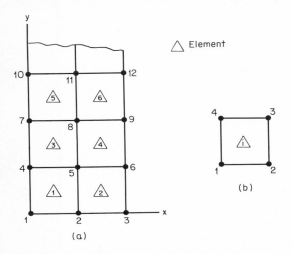

Element index	Global index of element nodes			
	Node 1	Node 2	Node 3	Node 4
1	1	2	5	4
2	2	3	6	5
3	4	5	8	7 etc

(c)

Fig. 95   Element and node numbers. (a) Global nodes; (b) element nodes listed 1–4; (c) incidence table.

numbered in the shortest direction; i.e., the difference in the number of a node and the smallest node number of any adjacent element should be a minimum to achieve the minimum bandwidth. If the structure is cellular, the numbering should be as shown in Fig. 96.

Coordinate Transformations.   In dealing with structures made up of several types of structural members, it will be necessary to transfer some element properties from one coordinate system to another. In a plate-beam system, for example, element properties of the beams must be transferred to the coordinate system of the plate elements. This is accomplished by a simple coordinate transformation similar to the one discussed in Art. 66. Many commercial programs do this through the use of slave nodes, rigid linkages, or some form of constraint equations.

59. Element Models These can be line elements (truss, beam, column, etc.), two-dimensional plane-stress or plane-strain elements, flexural plate and shell elements, axisymmetric elements, general three-dimensional solid elements, etc. Figure 97 shows some elements frequently used. Higher-order elements of similar shapes are also used.

The displacement of any point of an element is approximated by expressing it in terms of the nodal displacements by polynomials whose coefficients are generalized coordinates [Eq. (112)] or by interpolation functions [Eq. (114b)]. The following restrictions must be placed on the form of the polynomials or interpolation functions:

1. Number of terms. The number of terms in the polynomial must equal the number of degrees of freedom associated with the element.

2. Constant-strain and rigid-body modes (completeness). To ensure convergence to the true solution as the number of elements is increased, the displacement function must be capable of representing (1) constant-strain states and (2) rigid-body displacements. To satisfy these conditions, the displacement function must be a complete polynomial to a degree at least equal to the order of the highest derivative which occurs in the strain-

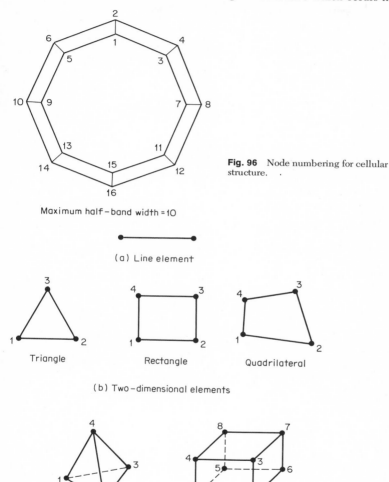

Fig. 96   Node numbering for cellular structure.

Maximum half–band width = 10

(a) Line element

Triangle   Rectangle   Quadrilateral

(b) Two–dimensional elements

Tetrahedron   Rectangular prism

(c) Three–dimensional elements

Fig. 97   Element models.

displacement equations. For a polynomial to be complete, it must contain all possible terms through the order specified. Thus a general, complete polynomial of n degrees requires $n+1$ terms in one dimension, $\frac{1}{2}(n+1)(n+2)$ terms in two dimensions, and $\frac{1}{6}(n+1)(n+2)(n+3)$ terms in three dimensions. For example, a complete quadratic in x and y requires $\frac{1}{2} \times 3 \times 4 = 6$ terms, which gives $\alpha_0 + \alpha_1 x + \alpha_2 y + \alpha_3 x^2 + \alpha_4 xy + \alpha_5 y^2$.

3. Interelement compatibility. Compatibility at the nodes is assured by the assembly procedure (Art. 66), which enforces common displacements at common nodes. If monotonic convergence is to be achieved, the displacement function must maintain compatibility of the displacements and their derivatives along element interfaces up to one order less than the highest derivative in the strain-displacement relations. Elements which satisfy this condition are called *compatible* or *conforming* elements.

Solutions obtained by elements which are conforming and complete provide an upper bound to the stiffness of the structure (more properly, to the energy stored). The solutions converge to the true general displacement configuration as the mesh size is reduced. Convergence is monotonic from below, provided each succeeding discretization contains the grid of the preceding analysis (this is termed a *reducible net*) and provided the form of the element displacement expansions are independent of the orientation of the element (this property is called *geometric isotropy* or *geometric invariance*). A reducible net for a rectangular region is shown in Fig. 98. Monotonic convergence is advantageous because marked reduction in errors can be effected by the use of extrapolation procedures.

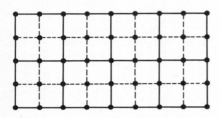

Fig. 98 Reducible net for rectangular region.

——— Nodal lines in former mesh and reduced mesh
— — — Additional nodal lines to form reduced mesh

Although the conformability and completeness requirements are necessary to prove monotonic convergence in the general case, many successful solutions have been obtained by formulations which are complete but nonconforming. A necessary, but not sufficient, test for convergence of such models is the *patch test*.[24] In this test a "patch" of several elements of the structure is isolated and subjected to a uniform-strain state; there must be no residual forces at the interior nodes in this solution.

Because conforming elements are often far too stiff, nonconforming modes (modes with displacements that do not involve the nodes) are sometimes inserted into conforming elements in an effort to accelerate convergence. The use of such elements must be carefully monitored, however, because in irregular grids they may result in convergence problems. Nonconforming elements that converge may not do so monotonically. This is the reason for their success—they oscillate around the true solution, with the error in conformity partly compensating for the overstiffness that results from using a finite, discretized system.

Nonconforming elements have been used most successfully in plate and shell structures, for which the complex displacement expansions required to satisfy the conformability requirement can be replaced by relatively simple expressions if this requirement is relaxed.

Overconforming elements may come about as a result of using higher-order derivatives as nodal unknowns to improve the performance of the element.

60. Plane-Stress and Plane-Strain Elements Many plane-stress and plane-strain elements have been developed; the constant-strain triangle (CST) and the plane-stress rectangle (PSR) were the first.[25] A few of these elements are listed in Table 8. Most general-purpose programs have several such elements in their libraries. The low-order elements (CST, PSR, and general quadrilateral elements) are in most programs. Although a general isoparametric element[26,41] can serve the function of all such elements, it can be expensive to use because of the computational effort involved in computing its stiffness.

61. Plane-Stress Analysis The strain-displacement equations involve only first derivatives. Therefore, to satisfy interelement compatibility, it is only necessary that the displacements u and v be finite and continuous on the element interfaces. Also, to satisfy the constant-strain and rigid-body conditions, the series must contain a complete linear polynomial. These conditions are satisfied by the functions

$$u = \alpha_0 + \alpha_1 x + \alpha_2 y + \text{additional terms}$$
$$v = \alpha_3 + \alpha_4 x + \alpha_5 y + \text{additional terms}$$

The additional terms must be sufficient in number to make the number of terms in the polynomial equal to the number of degrees of freedom of the element. Furthermore, if

TABLE 8   **Plane-Stress Elements**

Element	Nodal parameters	DOF per element	Polynomial degree	Name and reference	Programs
	u,v	6	1	Constant strain triangle (CST) 25	SAP MARC STRUDL NASTRAN ANSYS FINITE
	u,v	12	2	Linear strain triangle (LST) 27	STRUDL FINITE
	u, u_{lx}, u_{ly} v, v_{lx}, v_{ly}	18 plus 2 internal	3	28	ASKA
	u,v	8	Bilinear	Plane stress rectangle (PSR) 25	SAP STRUDL MARC NASTRAN FINITE
	u,v	8	Bilinear in natural coordinates	4-node quadrilateral isoparametric 26	MARC STRUDL NASTRAN ANSYS FINITE
	u,v	16	Quadratic in natural coordinates	8-node quadratic isoparametric 26	MARC SAP STRUDL FINITE
	u,v	8 plus 2 internal	1	Quadrilateral from 4 CST elements 28	SAP STRUDL

interpolation functions are used, they must contain these polynomials. The CST and PSR satisfy these requirements but are much too stiff to accommodate sharp stress gradients, so that they respond with too much shear-strain energy (Fig. 99). This led to the development of the linear-strain triangle (LST) and the isoparametric elements.[26]

Isoparametric elements can be expanded from simple quadrilateral elements to elements of higher order by the addition of side nodes, and have the further advantage of adapting to the geometry of the problem. They are found in some form in most major programs. Computation of the element stiffness requires an integration over the area of the element, but this can be done only approximately, using numerical integration. The general quadrilateral element with only corner nodes has the same overstiffness as the PSR element. For this reason, some programs allow the addition of nonconforming modes and/or the use of reduced integration[41,42,43] (fewer integration points than are needed for an exact integration of the stiffness matrix). When either of these options is used, the patch test should be applied because convergence is no longer guaranteed. Since nonconforming modes will violate some of the boundary conditions along lines of symmetry and support, the user may not want to use them in elements adjacent to these lines.

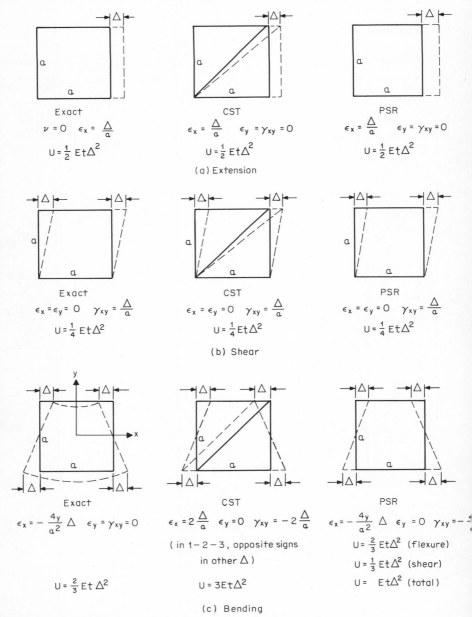

Fig. 99 CST and PSR element strain energies compared with exact solution.

The isoparametric and the LST elements are to be preferred for most plane-stress problems. The eight-node quadratic isoparametric element is very effective. The additional computation time for stiffness generation is offset by less time for equation solving.

62. Beam and Plate Bending   In these problems the strain-displacement equations involve second derivatives. Therefore, to achieve element compatibility, the displacement w together with the derivatives $\partial w/\partial x$ and $\partial w/\partial y$ must be finite and continuous on the element interfaces. Also, to satisfy the constant-strain and rigid-body condition the series must contain a complete quadratic polynomial. These conditions are satisfied by the functions:

One dimension: $w = \alpha_0 + \alpha_1 x + \alpha_2 x^2 +$ additional terms
Two dimensions: $w = \alpha_0 + \alpha_1 x + \alpha_2 y + \alpha_3 x^2 + \alpha_4 xy + \alpha_5 y^2 +$ additional terms

The requirements for additional terms and interpolation functions are the same as for the plane-stress polynomials (Art. 61).

In plate-bending elements the standard plate theories, some with shear deformations, are used to compute middle-surface displacements and bending-moment stress resultants. Some of the numerous elements that have been developed are listed in Table 9. The rectangular elements are generally much better than the triangular ones, since the latter tend to be much stiffer.

The nonconforming rectangular plate element with 12 degrees of freedom (DOF) is a very efficient element. Its major limitation is its shape constraint. The conforming rectangular element with 16 DOF is also very efficient, but it is not available in most programs. Shape constraint limits its applicability and it can cause some problems with such things as step changes in thickness. The corresponding triangular plate elements are more difficult. The nonconforming triangular element with 9 DOF has worked well in many cases but experiences convergence problems with some mesh arrangements. Some arrangements of this element will converge to a result which is slightly in error but normally not of engineering significance. The corresponding conforming element is far too stiff.

Higher orders are normally used to produce efficient, conforming triangular bending elements. The element with a complete fifth-order polynomial has performed well. Felippa's triangular element[28] is in some general programs and has a good record. This element is combined into a general quadrilateral in SAP. Constraints have been applied to the three-dimensional isoparametric element to produce a plate element. It tends to be appreciably too stiff, but reduced integration[41,42,43] can be used to compensate.

63. Shells and Combined Direct Stress and Bending   Many problems involve a combination of plane stress and flexure. Stiffened plates and shells in the form of either curved or prismatic surfaces are examples. Three types of elements can be used for shells:

1. Flat-plate elements that model the shell as a faceted surface
2. Curved "shallow" or "deep" shell elements
3. Three-dimensional elements, either directly or after they have been adapted by constraint equations

Several of the available elements are listed in Table 10. The flat elements are obtained by superimposing plane-stress and plate-bending elements and the curved elements by using the assumed displacement shapes directly in whatever shell theory one desires. A coordinate transfer is needed to convert from element to global coordinates.

Since curved shell elements often violate rigid-body conditions, they should have an opening angle not larger than 10°. Even so, there may be some geometrical differences between the shell and its element array even with grids of small elements, so that some analysts use only flat elements. This restricts the selection for general shells to triangular or quadrilateral elements.

Since the contrained isoparametric elements tend to be much too stiff, reduced integration or nonconforming modes are normally used.

64. Three-Dimensional Problems   Two three-dimensional elements are in common use, the tetrahedron and the isoparametric element. The latter is usually more economical. Three-dimensional problems are quite time-consuming to run. There is a dramatic increase in bandwidth, as well as a significant increase in element-stiffness computation time. A few elements are shown in Table 11.

65. Element Stiffness Matrix and Nodal Force Vector   The element stiffness matrix k and the nodal force vector p are evaluated by the following procedure, expressed here for a

TABLE 9 Plate-Bending Elements

Element	Nodal parameters	DOF per element	Polynomial degree	Com-patible	Refer-ence	Programs
	w_I, w_{Ix}, w_{Iy}	9	Incomplete 3	No	30	ANSYS STRUDL FINITE
	w_I, w_{Ix}, w_{Iy}	9	—	Yes	30	
	w_I, w_{Ix}, w_{Iy}	9	Incomplete 3	Yes	31	SAP EASE NASTRAN
	1. w_I, w_{Ix}, w_{Iy}, 2. w_{In}	12	3	Yes	26	MARC
	1. w_I, w_{Ix}, w_{Iy}, w_{Ixx}, w_{Ixy}, w_{Iyy} 2. w_{In}	21	5	Yes	28,32	ASKA
	w_I, w_{Ix}, w_{Iy}	12	Incomplete 4 (bicubic)	No	33	STRUDL FINITE
	w_I, w_{Ix}, w_{Iy}, w_{Ixy}	16	Incomplete 6 (Hermitians)	Yes	34	FINITE
	w_I, w_{Ix}, w_{Iy}	12	4	No	26	
	w_I, w_{Ix}, w_{Iy}	12 plus 7 internal	—	Yes	35	STRUDL SAP
	w_I, w_{Ix}, w_{Iy}	24	—	Yes	26	

TABLE 10 Shell Elements

Element	Nodal parameters	DOF per element	Flat or curved	Com-patible	Refer-ences	Programs
	u, v, w w_{lx}, w_{ly}	15	Flat triangular	u, v, w, yes w_l, no	36	FINITE ANSYS
	u, v, w α, β	15	Curved triangular		37	MARC
	u, v, w w_x, w_y	20	Flat rectangular	u, v, w, yes w, no	26	FINITE
	u, v, w α, β	20	Curved parallelogram	u, v, w, yes α, β, no	38,39	STRUDL MARC
	u, v, w α, β	20 plus 5	Faceted quadrilateral			SAP
	u, v, w α, β	40	Curved arbitrary		26	FINITE MARC

TABLE 11 Three-Dimensional Elements

Element	Nodal parameters	DOF per element	Polynomial degree	Name and reference	Programs
	u, v, w	12	Linear	Tetrahedron 40	ANSYS
	u, v, w	30	Quadratic in natural coordinates	40	
	u, v, w	24	Bilinear	Hexahedron or 8-node isoparametric 26	SAP STRUDL MARC FINITE
	u, v, w	60	Quadratic in natural coordinates	20-node isoparametric 26	SAP STRUDL MARC NASTRAN FINITE

two-dimensional element with e nodes. The extension to three dimensions, and flexural problems, is obvious.

1. Represent each component of the displacement vector $u(x, y)$ of a point $P(x, y)$ within the element by an assumed polynomial, chosen to satisfy the conditions of Art. 59. Normally, identical polynomial functions apply for each displacement component. Thus

$$u(x, y) = \begin{Bmatrix} u \\ v \end{Bmatrix} = \begin{bmatrix} \phi^t & 0 \\ 0 & \phi^t \end{bmatrix} \{\alpha\} = M\alpha \tag{112}$$

where $\phi^t = [1, x, y, x^2, xy, \ldots$ to e terms$]$
   $\alpha_t = \{\alpha_0, \alpha_1, \alpha_2, \ldots$ to $2e$ terms$\}$
α_t contains the unknown coefficients of the functions ϕ^t.

2. Express the nodal displacements $q = \{q_u \ q_v\}$ in terms of the generalized coordinates α. Substitute the spatial coordinates of the nodes into the matrix M of Eq. (112) to obtain

$$q = A\alpha \tag{113}$$

where $q_u^t = \{u_1, u_2, \ldots, u_e\}$ and $q_v^t = \{v_1, v_2, \ldots, v_e\}$

3. Express the displacements u in terms of the nodal displacements q. Eliminate α between Eqs. (112) and (113) to obtain

$$u = MA^{-1}q = Nq \tag{114}$$

The matrix N is called an *interpolation matrix*. It can be written directly because, by definition, an interpolation function is a function which has unit value at a given node and zero values at all the other nodes. Thus a particular interpolation function can be derived by taking the product of the equations of the lines or surfaces through all nodes other than the one for which the function is sought. For example, to determine the interpolation function for node 1 of the rectangular plane-stress element shown in Fig. 97b, with x axis 1-2, y axis 1-4, side 1-2 of length a, and side 1-4 of length b, the equations $x - a = 0$ of line 2-3 and $y - b = 0$ of line 3-4 are multiplied. This gives $n_1 = c(x - a)(y - b)$, which is zero at nodes 2, 3, and 4. The constant c is set equal to $1/ab$ to give $n_1 = 1$ at node 1.

The expansion of Eq. (114) illustrates the interpolation function expressed in cartesian coordinates. Thus

$$\begin{Bmatrix} u(x, y) \\ v(x, y) \end{Bmatrix} = \begin{bmatrix} N_1^t(x, y) & 0 \\ 0 & N_1^t(x, y) \end{bmatrix} \begin{Bmatrix} q_u \\ q_v \end{Bmatrix}$$

where $N_1^t(x, y) = [n_1, n_2, \ldots, n_j, \ldots, n_e]$ is a row vector containing the shape functions $n_j(x, y), j = 1, 2, \ldots, e$.

Interpolation functions for nonrectangular elements can be expressed most conveniently in terms of *natural coordinates*. Figure 100 shows some systems in common use.

4. Express the element strains $\epsilon(x, y) = \{\epsilon_x, \epsilon_y, \gamma_{xy}\}$ at a point $P(x, y)$ within the element in terms of q. Applying the strain-displacement equations

$$\epsilon = \left\{ \frac{\partial u}{\partial x}, \frac{\partial v}{\partial y}, \frac{\partial u}{\partial x} + \frac{\partial v}{\partial y} \right\}$$

to Eq. (114) gives

$$\epsilon = Bq \tag{115}$$

where the *strain-displacement matrix* B is obtained by appropriate differentiation of the interpolation functions N. If the interpolation functions have been written in a natural coordinate system, a transformation to cartesian coordinates must be made in establishing the strains, since they are defined in an engineering sense in cartesian coordinates.

5. Evaluate the element stresses $\sigma(x, y) = \{\sigma_x, \sigma_y, \tau_{xy}\}$ corresponding to the strains ϵ. Premultiply Eq. (115) by the *stress-strain matrix* D to obtain

$$\sigma = DBq \tag{116}$$

6. Calculate the element stiffness matrix k and the nodal force vector p. The principle of virtual displacements is applied to calculate the element stiffness matrix k,

and to transform the prescribed surface tractions $\overline{T} = \{\overline{T}_x, \overline{T}_y\}$ acting on the part S_1 of the element boundary surface, and the body force intensities $\overline{X} = \{\overline{X}, \overline{Y}\}$ into an equivalent nodal force vector p. The elements of p thus obtained are called *consistent* or *kinematically equivalent* loads. Alternatively, statically equivalent loads may be used.

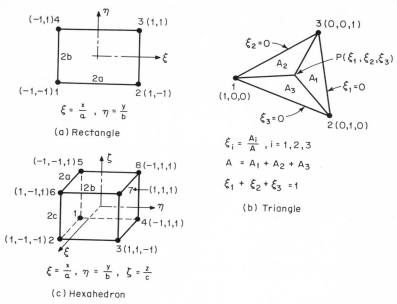

Fig. 100 Natural coordinates.

The internal virtual work δW_i due to the virtual strains $\delta\epsilon = \{\delta\epsilon_x, \delta\epsilon_y, \delta\gamma_{xy},\}$, corresponding to a set of virtual displacements δq, is

$$\delta W_i = \iiint_V (\delta\epsilon)^t\, \sigma\, dV$$

which becomes, on substituting Eqs. (115) and (116),

$$\delta W_i = (\delta q)^t \left(\iiint_V B^t\, DB\, dV \right) q$$

The corresponding external virtual work δW_e is

$$\delta W_e = (\delta q)^t \iiint_V N^t \overline{X}\, dV + \iint_{S_1} (\delta u)^t\, \overline{T}\, dS$$

Equating δW_i to δW_e to obtain the equilibrium equations for the element gives

$$kq = p \tag{117}$$

where

$$k = \iiint_V B^t\, DB\, dV \tag{118}$$

$$p = \iiint_V N^t \overline{X}\, dV + \iint_{S_1} N^t\, \overline{T}\, dS \tag{119}$$

7. Eliminate the displacements of internal nodes. Internal nodes are not connected to adjoining elements; hence they may be solved independently by the condensation

technique of Art. 52. Thus the nodal displacement vector is partitioned into $q = \{q_0 \mid q_1\}$, where q_0 and q_1 are the vectors of the displacements of the external and internal nodes, respectively, and the matrices k and p are partitioned in accordance with the partitioning of q. Thus

$$\left\{ \begin{array}{c} p_0 \\ \hline p_1 \end{array} \right\} = \left[\begin{array}{c|c} k_{00} & k_{01} \\ \hline k_{10} & k_{11} \end{array} \right] \left\{ \begin{array}{c} q_0 \\ \hline q_1 \end{array} \right\}$$

The displacements q_1 are eliminated to give

$$\overline{k} q_0 = \overline{p} \tag{120}$$

where $\overline{k} = k_{00} - k_{01} k_{11}^{-1} k_{10}$ is the *condensed stiffness matrix* and $\overline{p} = p_0 - k_{01} k_{11}^{-1} p_1$ is the *condensed load vector*.

The explicit form of the stiffness matrix for a number of low-order elements is given in Ref. 29. Stiffness matrices for higher-order elements are usually not available and are calculated by computer.

66. Equilibrium Equations for the Assemblage   With the element stiffnesses determined, the equilibrium equations of the nodes can be written. To do this it is necessary to:

1. Transform the element-stiffness matrix k and nodal load vector p for each element into common reference axes, that is, into the global axes of the system. Most large, general-purpose programs do this without the user's being required to initiate it.

2. Eliminate the displacements of all internal nodes by the condensation procedures [Eq. (120)]. This is done automatically in all general-purpose programs that contain elements with internal nodes.

Some matrices k and p are derived in global coordinates, but they are often derived with respect to local axes, which may have directions different from those of the global axes. The required coordinate transformation referred to in item 1 is accomplished by a rotation matrix T to give

$$k_0 = T^t k_l T \tag{121}$$
$$p_0 = T^t p_l \tag{122}$$

where the subscripts 1 and 0 refer to the local and the global axes, respectively. T is a diagonal matrix made up of submatrices t which contain for each node the direction cosines for the local directions with respect to the global directions. T is of the order of the total number of displacements q, while t is of the order of the number of displacements at a node. T is always an orthogonal matrix, that is, $T^{-1} = T^t$.

The element equilibrium equation [Eq. (117)] in global coordinates is

$$k_0 q_0 = p_0 \tag{123}$$

To simplify the notation, the subscripts 0 will be dropped and it is to be understood that all quantities are now referred to the global axes.

The equilibrium equations for the structure are now formed by equating the forces at each node to the sum of the contributions of all the elements meeting at the node. This gives

$$KU = P \tag{124}$$

where K = structure stiffness matrix
U = vector of nodal displacements
P = vector of the $m+r$ nodal forces
m = number of active displacements of structure
r = number of specified support displacements

The matrix K, which is symmetrical and of order $m+r$, is given by

$$K = \sum_{g=1}^{s} k_g \tag{125}$$

where g = subscript denoting a typical element and s = number of elements.

The sum of the corresponding consistent and concentrated nodal forces for each element gives the subvector P_g of the vector P. Thus

$$P = \sum_{g=1}^{s} P_g \qquad (126)$$

Equations (125) and (126) constitute the assembly rules for K and P. The rules are applied by the *direct stiffness method*. In this, the assembly is carried out by use of an *incidence table*, which is an array relating the local and global indices of the element (Fig. 95c). A row of an incidence table is sometimes called the *destination vector* for the element corresponding to that row. Thus, in Fig. 95c the destination vector for element 1 is 1, 2, 5, 4. These vectors are used to insert the submatrices of element stiffness and nodal force into the appropriate locations in the matrices K and P.

67. Solution for the Displacements The structure stiffness matrix K is singular; hence the equilibrium equations [Eq. (124)] cannot be solved until the boundary conditions are enforced. In the displacement method it is not possible to satisfy the force boundary conditions explicitly, but their effect is included in the appropriate equilibrium equations. The geometric boundary conditions are satisfied explicitly, and they must be inserted into the equilibrium equations in order to form the set of nonsingular equations for the assemblage of elements.

The mathematics of inserting the boundary conditions can be thought of as follows: The nodal force vector P is partitioned into $P^t = \{W^t \ R^t\}$, where W is the vector of the m specified forces and R is the vector of the r reactions. The displacement vector q is partitioned into $q^t = \{\Delta^t | d^t\}$, where Δ is the vector of the m unknown displacements and d is the vector of the r specified displacements. The equilibrium equations are partitioned in accordance with the partitioning of P and q. Thus

$$\left[\begin{array}{c|c} K_{11} & K_{12} \\ \hline K_{21} & K_{22} \end{array} \right] \left[\begin{array}{c} \Delta \\ \hline d \end{array} \right] = \left[\begin{array}{c} W \\ \hline R \end{array} \right] \qquad (127)$$

The expansion of Eq. (127) gives

$$K_{11}\Delta = W - K_{12}d$$
$$R = K_{21}\Delta + K_{22}d$$

The geometric boundary conditions can be expressed as $d = \bar{d}$, where $\bar{d}$ is the vector of r prescribed numerical values of the displacements. The structure stiffness matrix and load vector in Eq. (127) can be modified to incorporate these boundary conditions as follows:

$$\left[\begin{array}{cc} K_{11} & 0 \\ 0 & I \end{array} \right] \left\{ \begin{array}{c} \Delta \\ d \end{array} \right\} = \left\{ \begin{array}{c} W_1 - K_{12}\bar{d} \\ \bar{d} \end{array} \right\} \qquad (128)$$

where I is an $r \times r$ unit matrix (Fig. 92d). This procedure maintains the symmetry of the stiffness matrix but sacrifices the direct calculation of the support reactions R.

The equations need not be reordered, as implied by Eq. (127). Instead, for each prescribed displacement the corresponding row and column are set to zero except at their intersection on the diagonal, which is equated to 1. The prescribed value of the displacement is then placed in the corresponding location in the load vector, and the load terms for the remaining equilibrium equations are modified according to Eq. (128).

Some programs (SAP, for example) do not write the equilibrium equations corresponding to a zero displacement. The set of equations $K_{11}\Delta = W$ are then generated directly. The program contains a stiff-spring element to simulate support displacements.

68. Solution for Element Strains and Stresses The solution of the equilibrium equations for the structure gives the displacements at all the interelement nodes. The components of strain and of stress at a point within an element can then be computed using Eqs. (115) and (116). In general, owing to the approximations involved, the predicted stresses for the elements meeting at a node are not compatible, so that the equilibrium conditions for the individual elements are not satisfied. The differences may be quite significant for low-order elements. The disparity generally decreases with increase in the order of the element, so that less interpretation of stresses is needed. This is one of the advantages of higher-order elements. Stresses at midside nodes are normally better behaved; that is, the differences are smaller than at the corner nodes.

It is common practice to average the stresses for all the elements meeting at a node to determine the stresses at the node. The average may be weighted, based on areas or other parameter, or not. Another common practice is to use the stress at the centroid of the element, particularly for low-order elements. For isoparametric elements the stress at the integration points is often used.

Stresses calculated by any of the above procedures should be checked for equilibrium on free bodies cut from the structure. Some form of numerical integration, such as Simpson's rule, must be used to compute the stress resultants. Equilibrium will not be satisfied exactly, since the displacement model in finite-element procedures does not guarantee it. However, if there are errors of more than several percent, the results should be questioned.

In problems involving both in-plane and out-of-plane displacements, such as occur in combinations of beams, plates, shells, etc., the in-plane displacements are normally of lower order. Thus, if elements join at an angle, or if their nodal points are eccentric to one another, etc., not only will the displacement compatibility that exists in the flat be lost, but stress interpretation along the edges between elements becomes more difficult. There will be differences in the variation of the in-plane forces and that of the bending moments along these edges. If beams are eccentric to a plate, the fiber stress in the beam along its connection to the plate will not vary in the same manner as the membrane force in the plate. Averaged nodal stress values may be better in this case. One must know which plane (plate or beam) is the reference for the displacements and stresses; normally the plate reference frame is used. This means that a coordinate transformation may be necessary to put the quantities of interest in a common coordinate system. Some modifications may be necessary in order for the boundary conditions to apply, say, to the bottom of the beam rather than to a point in the plate reference frame. Such special treatments are required in most general-purpose programs.

Few elements include a degree of freedom associated with rotation about a normal to the surface of the element. Therefore, if two elements in combined direct stress and bending intersect orthogonally, or at a large angle, additional constraints should be placed on the rotations about the normals to the line of intersection. This is necessary to take better account of the constraint of the orthogonal plate on the normal displacements of the intersecting plate. There are also difficulties if a beam element intersects a network of plane-stress elements in the same plane. A possible partial solution is to extend the beam one or two elements into the plane-stress grid. This allows a nonzero bending moment at the intersection.

Additional comments on applications, and in particular on stress interpretation, are found in Refs. 26 and 41.

Examples PLATE-GIRDER HAUNCH. Finite-element analysis is useful for investigating stresses in structures, or parts of structures, of irregular configuration. Figure 101 shows some of the results of an analysis of the haunch of a continuous plate girder. The haunch grid is shown in Fig. 101c. The web was modeled with 8-node quadratic isoparametric elements (Table 8), using line elements (truss members) for the flanges and stiffeners. The 8-node element is a good choice for this problem for reasons discussed in Art. 61. The remainder of the girder entered the solution in the form of the stress resultants shown at each end of the haunch section: these were determined by conventional frame analysis. The springs shown at the juncture of the haunch and inclined leg were used to determine the stress resultants at this section.

The floor-beam concentrations (denoted by FB_1, FB_2, and FB_3 in Fig. 101c) were not input directly. FB_1 was accounted for approximately by adjusting the shear and moment at the left end of the grid so as to give the correct shears and moments at FB_1 and FB_2. FB_2 was omitted, which did not affect the shears and moments at the section, but the local effects of the concentration were lost. FB_3 was accounted for correctly by inputting the shear immediately to the left.

FINITE[44] was used for the solution. Bending stresses at two sections and the axial stress along one flange, all for a negative-moment condition, are shown in Figs. 101d, e, and f.

HYPERBOLIC PARABOLOID ROOF. Finite-element analysis is used extensively in shell structures, particularly when the answers sought pertain to shell-beam or shell-stiffener interaction. Problems of this type occur in roof structures, reactors, offshore structures, etc.

Figure 102a shows part of the roof of a water-pollution control structure. The roof consists of 44 hyperbolic paraboloids of the configuration shown in Sec. 20, Fig. 42, and is 4 units wide by 11 units

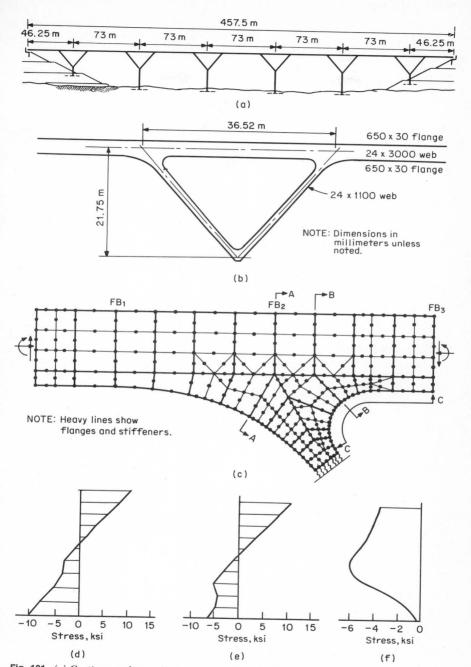

Fig. 101 (a) Continuous plate-girder bridge over Wadi Qaddiyah for Ministry of Communications, Kingdom of Saudi Arabia. Designed jointly by Hansen Engineers, Inc. and Wilson & Company. (b) Typical segment. (c) Finite-element grid for haunch. (d) Stress normal to Section A-A. (e) Stress normal to Section B-B. (f) Flange stress along Section C-D.

long (240 × 660 ft). Each shell has a rise of 9 ft and is 4 in. thick. Ridge beams are 10 in. deep by 18 in. wide, and edge beams are tapered from 10 in. deep at the ridge to 33 in. at the corners. The roof is supported on columns spaced 60 ft in each direction, with tie rods at the column tops.

For economy in construction each shell was cast as a unit. Therefore, since there was little or no continuity of adjacent units, a single unit could be analyzed independently, and because of symmetry one 30 × 30 ft quadrant could be considered. Figure 102c shows a projected view of one quadrant, with an 8 × 8 grid. Because there is no region of sharp stress gradients this structure can be modeled with a uniform grid of square elements. Subdivision of the mesh to investigate a boundary layer effect usually results in a distorted grid with complicated element shapes, and might also make it impossible to use any data-generator capabilities of the program. Furthermore, data preparation can be time-consuming, and mistakes can necessitate a rerun, which could be a serious loss.

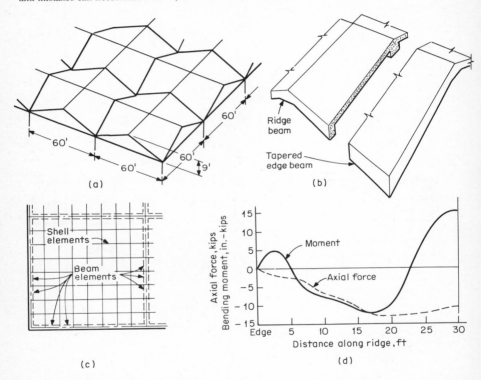

Fig. 102 (a) Portion of hyperbolic paraboloid roof for water pmllution control structure, Oakwood Beach, Staten Island, New York City. Greeley and Hansen, Engineers; Klein & Hoffman, Structural Engineers. (b) View of one quadrant. (c) Finite-element grid for one quadrant. (d) Axial force and bending moment in ridge beam.

The curved parallelogram element of Table 10 was used in this analysis. The flat triangular element in the table could have been used with a shell as shallow as this one, but the resulting accuracy was not expected to be competitive with the curved element.

A 4 × 4 grid will provide adequate preliminary design information for a structure of this type. This results in a problem with 16 shell elements, 16 beam elements, 125 unknown displacements, and a half-bandwidth of 65. Because solution time increases with the square of the bandwidth, it is important to keep it as small as possible. For example, a 6 × 6 grid gives 36 shell elements, 24 beam elements, 245 unknown displacements, and a half-bandwidth of 85, and computation cost will be almost double that of the 4 × 4 grid. The 8 × 8 grid used for the final analysis of this shell has 64 shell elements, 32 beam elements, 405 unknown displacements, and a half-bandwidth of 105. This translates into a computation cost of about three times that of the 4 × 4 grid.

The solution gives the deflected shape, the shell forces, the beam forces, and, if the supports are modeled by stiff springs, the reactions. Shell stresses are usually low, and the primary reason for the analysis is to establish the stresses in the various supporting beams. The axial force and bending moment in the ridge beam are of particular concern, and cannot be determined by an elementary analysis. Because of its interaction with the shell, the ridge beam behaves as a beam on an elastic foundation, which accounts for the wave form of the moment curve in Fig. 102*d*. There have been several failures of hp-roof ridge beams because of inadequate reinforcement due to gross underestimates of the moment, and in at least one case the result was a complete collapse of the roof.

REFERENCES

1. Timoshenko, S. P., and D. H. Young: "Theory of Structures," 2d ed., McGraw-Hill Book Company, New York, 1965.
2. Newmark, N. M.: Numerical Procedure for Computing Deflections, Moments, and Buckling Loads, *Proc. ASCE*, vol. 68, 1942.
3. "Handbook of Frame Constants," Portland Cement Association, Chicago, 1958.
4. Grinter, L. E.: "Theory of Modern Steel Structures," vol. II, 2d ed., The Macmillan Company, New York, 1949.
5. Benjamin, J. R.: "Statically Indeterminate Structures," McGraw-Hill Book Company, New York, 1959.
6. Lightfoot, E.: "Moment Distribution," E. & F. M. Spon Ltd., London, 1961.
7. de Veubeke, F.: "Matrix Methods of Structural Analysis," Pergamon Press, New York, 1964.
8. Argyris, J. H.: "Recent Advances in Matrix Methods of Structural Analysis," Pergamon Press, New York, 1964.
9. Argyris, J. H.: "Energy Theorems and Structural Analysis," Butterworth Scientific Publications, London, 1960.
10. Denke, P. H.: A Matrix Method of Structural Analysis, *Proc. 2d U.S. Congr. Appl. Mech.*, June 1954.
11. Denke, P. H.: A Matrix Solution of Certain Non-Linear Problems in Structural Analysis, *J. Aeronaut. Sci.*, March 1956.
12. Godden, W. G.: "Numerical Analysis of Beam and Column Structures," Prentice-Hall, Inc., Englewood Cliffs, N.J., 1965.
13. Hall, A. S., and R. W. Woodhead: "Frame Analysis," John Wiley & Sons, Inc., New York, 1961.
14. Livesley, R. K., and D. B. Chandler: "Stability Functions for Structural Frameworks," Manchester University Press, Manchester, 1956.
15. Hoff, N. J.: "The Analysis of Structures," John Wiley & Sons, Inc., New York, 1956.
16. Langhaar, H. L.: "Energy Methods in Applied Mechanics," John Wiley & Sons, Inc., New York, 1962.
17. Norris, C. H., and J. B. Wilbur: "Elementary Structural Analysis," 2d ed., McGraw-Hill Book Company, New York, 1960.
18. Rubinstein, M. F.: "Matrix Computer Analysis of Structures," Prentice-Hall, Inc., Englewood Cliffs, N.J., 1966.
19. Martin, H. C.: "Introduction to Matrix Methods of Structural Analysis," McGraw-Hill Book Company, New York, 1966.
20. Asplund, S. O.: "Structural Mechanics: Classical and Matrix Methods," Prentice-Hall, Inc., Englewood Cliffs, N.J., 1966.
21. Wang, P-C: "Numerical and Matrix Methods in Structural Mechanics," John Wiley & Sons, Inc., New York, 1966.
22. Pian, T. H. H., and P. Tong: Basis of Finite Element Methods for Solid Continua, *Int. J. Numer. Methods Eng.*, vol. I, pp. 3–28, 1969.
23. Wilson, E. L.: Solid SAP—A Static Analysis Program for Three Dimensional Solid Structures, *SESM Rept.* 71-19, Department of Civil Engineering, University of California, Berkeley, 1971.
24. Irons, B. M., and A. Razzaque: Experience with the Patch Test for Convergence of Finite Elements, in A. K. Aziz (ed.), "The Mathematical Foundations of the Finite Element Method with Applications for Partial Differential Equations," Academic Press, Inc., New York, 1972.
25. Turner, M. J., et al.: Stiffness and Deflection Analysis of Complex Structures. *J. Aeronaut. Sci.* September 1956.
26. Zienkiewicz, O. C., and I. K. Cheung: "The Finite Element Method in Engineering Science," 2d ed., McGraw-Hill Book Company, New York, 1971.
27. Fraeijs de Veubeke, B.: Displacement and Equilibrium Methods in the Finite Element Method, in O. C. Zienkiewicz and G. S. Holister (eds.), "Stress Analysis," John Wiley & Sons, Inc., New York, 1965.
28. Felippa, C. A.: "Refined Finite Element Analysis of Linear and Nonlinear Two-Dimensional Structures," Structural Engineering Laboratory, University of California, Berkeley, 1966.
29. Przemieniecki, J. S.: "Theory of Matrix Structural Analysis," McGraw-Hill Book Company, New York, 1968.
30. Bazeley, G. P., et al.: Triangular Elements in Plate Bending—Conforming and Nonconforming Solutions, *Proc. 1st Conf. Matrix Methods in Structural Mechanics*, Wright Patterson Air Force Base, 1965.

31. Clough, R. W., and J. L. Tocher: Finite Element Stiffness Matrices for Analysis of Plate Bending, *Proc. 1st Conf. Matrix Methods in Structural Mechanics*, Wright Patterson Air Force Base, 1965.
32. Bell, K.: Triangular Plate Bending Elements, in I. Holand and K. Bell (eds.), "Finite Element Methods in Stress Analysis," Tapir, Trondheim, Norway, 1969.
33. Melosh, R. J.: Basis of Derivation of Matrices for the Direct Stiffness Method, *J. AIAA*, July 1963.
34. Bogner, F. K., R. L. Fox, and L. A. Schmit: The Generation of Interelement Compatible Stiffness and Mass Matrices by Use of Interpolation Formulae, *Proc. 1st Conf. Matrix Methods in Structural Mechanics*, Wright Patterson Air Force Base, 1965.
35. Clough, R. W., and C. A. Felippa: A Refined Quadrilateral Element for the Analysis of Plate Bending, *Proc. 2d Conf. Matrix Methods in Structural Mechanics*, Wright Patterson Air Force Base, 1968.
36. Clough, R. W., and C. P. Johnson: A Finite Element Approximation for the Analysis of Thin Shells, *J. Solids Structures*, vol. 4, 1968.
37. Strickland, G. E., and W. A. Joden: A Doubly Curved Triangular Shell Element, *Proc. 2d Conf. Matrix Methods in Structural Mechanics*, Wright Patterson Air Force Base, 1968.
38. Connor, J., and C. A. Brebbia: Stiffness Matrix for Shallow Rectangular Shell Element, *J. Eng. Mech. Div. ASCE*, October 1967.
39. Pecknold, D. A., and W. C. Schnobrich: Finite-Element Analysis of Skewed Shallow Shells, *J. Struct. Div. ASCE*, April 1969.
40. Argyris, J. H.: Continua and Discontinua, *Proc. 1st Conf. Matrix Methods in Structural Mechanics*, Wright Patterson Air Force Base, 1965.
41. Cook, R. D.: "Concepts and Applications of Finite Element Analysis," John Wiley & Sons, Inc., New York, 1974.
42. Pawsey, S. F., and R. W. Clough: Improved Numerical Integration of Thick Shell Finite Elements, *Int. J. Numer. Methods Eng.*, vol. 3, pp. 575–586, 1971.
43. Zienkiewicz, O. C., R. L. Taylor, and J. M. Too: Reduced Integration Technique in General Analysis of Plates and Shells, *Int. J. Numer. Methods Eng.*, vol. 3, pp. 275–290, 1971.
44. Lopez, L. A., J. Urzua, R. H. Dodds, and D. R. Rehak, Finite: A Polo II Subsystem for Structural Mechanics, Civil Engineering Systems Laboratory, University of Illinois, Urbana, October 1976.

Section **2**

Computer Applications in Structural Engineering

STEVEN J. FENVES
University Professor of Civil Engineering, Carnegie-Mellon
University

1. Basic Concepts Computers used in engineering practice can be defined as general-purpose, stored-program, electronic digital computers. *Digital* means that within the computer numbers are represented by discrete digits in contrast to analog computers, where numbers are represented by continuously varying physical quantities; digital computers can also represent and manipulate symbols other than numbers, such as alphabetic characters or geometric entities. *Electronic* means that the internal operations are performed by electronic circuits. *Stored-program* means that for each application the computer is provided with a sequence of steps or instructions, called the program, which defines the process of solution. *General-purpose* means that the computer is not built specifically for one type of application, so that by using different programs it is capable of solving a wide variety of problems.

2. The Structural Design Process A highly simplified representation of the structural design process is shown in Fig. 1. It begins with information concerning the client's needs and resources, limitations to be imposed on the project (e.g., technical, social, legal), and criteria to be used to evaluate designs. It proceeds through a series of activities of project planning, preliminary design, analysis, and proportioning. Each process generates additional information to be used by the succeeding process. Typically, the information produced by each process is evaluated for consistency, economic and technical feasibility, and the like; if the results turn out to be unsatisfactory, one or more of the preceding processes must be repeated. Certain decision points are explicitly shown in the figure: an iterative analysis-proportion process must converge to some acceptable tolerance before the design can proceed, and the final design must satisfy the original technical and economic criteria (or, ideally, be optimal under the limitations and constraints imposed) before the final information, that is, the design documents, can be produced. Computer programs are used for essentially every phase of each of the design processes shown in this figure.

3. The Program-Development Process The use of the computer requires that problem solving be separated into two phases: development, during which the program is gener-

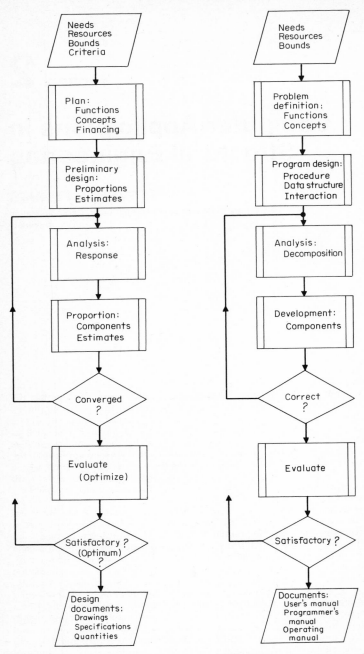

Fig. 1 Schematic diagram of structural design process.

Fig. 2 Schematic diagram of program development process.

ated, and production, involving repeated use of the program. Program development is a major design activity which follows a process similar to the structural design process, as illustrated in Fig. 2. The degree to which the steps given below are followed will naturally depend on the scope and importance of the program, the resources available, and the intended use and distribution of the program. These steps must be followed even if the program is obtained elsewhere and it is only intended to modify or adapt it to the organization's needs, resources, and practices.

Problem Definition. In this phase the needs for a certain computational procedure, the available resources (people, machine), and known bounds are used to develop the major functional and conceptual aspects of the program, including the major input-data types, the scope of the program, and the types of results required. This step includes the review of the procedure currently employed, the evaluation of the methods, assumptions, and limitations to be used; the exploration of possible alternate procedures; and the extrapolation to potential future applications. Since a successful program is likely to change significantly the organization's method of handling future problems of the type being considered, the problem definition must be worked out as a joint effort between the programmers, management and supervisory personnel, and the design personnel who will eventually use the program.

Program Design. In this phase, three sets of important decisions must be made. First, the computational procedures to be used must be determined. They may involve radical departures from manual techniques. Particularly in design application, the alternatives that may present themselves must all be foreseen and the appropriate procedure for each case defined. Provisions for programmed checks on the consistency of the data, as well as for possible expansions, alterations, and changes in specifications must also be considered.

Whenever the mathematical solution selected involves processes which cannot be performed directly by algebraic operations, methods of numerical analysis must be employed. Most common operations are available in the form of subroutines, provided as part of the computer software. In many cases, it is worthwhile to develop alternate formulations which make greater use of existing subroutines.

Second, the organization and structuring of the data used by the program must be carefully developed. This includes the input data to be supplied to the program, the output to be generated, the internal data (such as partial results) directly operated upon by the program, as well as possible data held in long-term storage, such as project files which are used, generated, or modified by the program.

The third aspect deals with the interaction between the program and its intended users. This step includes a careful designation of the input and output data, involving the determination of the range of the variables, the precision required, as well as the layout of the format in which these quantities are to be presented to the computer or generated by the program. This step has a definite bearing on the usefulness of the finished program; a program requiring hours of preparing extraneous data for input and producing pages of unorganized numerical results will be unpopular with users, and therefore its value to the organization will be greatly reduced.

Analysis. The next phase is to carefully analyze the specifications and design decisions made up to this point, in order to determine the best way to decompose the program into manageable components. For all but the smallest programs, such a decomposition is needed, for two reasons. First, at a technical level, it is important to decompose the program into clear-cut functional segments, each performing a specific task related either to the application at hand (e.g., flange or web design for a girder) or to the operating environment within which the program is to function (e.g., read a record of data). Second, from a management standpoint, experience has shown that both initial development and continued updating and modification are vastly improved when the program is developed as an ordered collection of small components.

Development. Work can now proceed by further refinement of the components until they can be completely described in the available programming language, entered on the computer, and tested. Testing is needed to remove coding blunders, programming or logical errors resulting from improper programming, and numerical errors produced by incorrect numerical techniques. Most software systems provide various aids to make program testing (debugging) more efficient. Test solutions obtained by the program must be compared against manual solutions, to verify the correctness and accuracy of the results. In selecting test cases to be run, it is important that all possible alternates provided for in the program be thoroughly investigated.

Typically, the development-testing process will discover errors which need to be corrected by recycling through the analysis process, or even the design and definition phases.

Evaluation. Once the program has been tested and verified to the programmer's satisfaction, it is important to subject it to a careful evaluation by its potential users. Such an evaluation should ascertain that the criteria and objectives have been met, and that the users are satisfied with the program performance and results.

Documentation. As a final phase, if a program is to be used by persons other than the programmer, it is essential that it be properly documented before it is put into production. This documentation (write-up) should consist essentially of four parts: (1) a brief *application description* giving the method of solution, approximations and limitations involved, etc., so that potential users can understand the process and decide whether or not it is applicable to their particular problems; (2) a *user's manual,* describing exactly how the data are to be prepared and options specified, and interpreting the results produced; (3) an *operator's manual* specifying the procedures to be used in running the program on the computer; and (4) a *programmer's manual* describing in detail the methods used and containing flowcharts and program listings.

The above description of the development process is highly idealized. Depending on the importance of the application program and the extent of its intended use, certain phases can be short-cut or eliminated altogether. Conversely, on very large programs, portions of the development may be contracted out to other organizations or performed cooperatively. The important point is that program development is essentially similar to design; it should be noted that the flowcharts in Fig. 1 and 2 are identical in format, the only difference being in the content of the boxes.

Maintenance (updating) is required for several reasons: (1) errors, deficiencies, or other "bugs" can crop up long after initial development and must be rectified; (2) changes in procedures or requirements, such as changes in the design specifications embodied in the program, require modifications to keep the program usable; (3) changes in hardware and interaction modes are almost unavoidable owing to rapidly changing technology, and programs must be modified or "ported" to operate in new environments; and (4) additions, extensions, alternatives to available methods, and integration with other related programs frequently suggest themselves only through continued use, and must be accommodated to increase the usefulness of programs.

Proper maintenance can be conveniently and inexpensively performed only if the program was initially developed according to the guidelines given above. In particular, development should aim at good segmentation of the individual procedures (Art. 9), and the documentation should be as complete as possible.

PROGRAM-DEVELOPMENT TOOLS

Program development from inception to production usage is a complex and demanding process, and a large number of concepts, methods, and tools have been developed under the name of software engineering to organize and expedite this process.

4. Structured Programming   Structural engineers are accustomed to an orderly design process, starting from the "top down" with a basic premise or scheme, which is then repeatedly expanded, analyzed, modified, etc., until the final design emerges (Fig. 1). A similar approach (Fig. 2) is also the best mechanism to develop computer programs. This methodology is frequently referred to as structured programming, top-down programming, or "lead-programmer concept." The latter designation has primarily a management connotation, indicating that a high-level professional performs the decomposition of complex processes, allowing lower-level programming personnel to complete the development of the resulting program components.

Example   A program is to be developed for the computation of loads and selection of members for a floor system consisting of simply supported steel beams (Fig. 3). The program is to accept a definition of the floor geometry, including the location of all beams, the definition of external loads, including area, line, and point loads, and is to produce a list of beam selections obtained from a table of available shapes.

At the highest levels, the program may be structured into three components:

Program::=
1. Input
2. Compute
3. Output

The symbol ":: =" is to be read "is defined as." The second step, "Compute," can now be subdivided or structured as:

2. Compute::=
    2.1 Convert area loads on slabs to line loads on edge beams
    2.2 Carry down beam loads from supported beams to supporting beams
    2.3 Design each beam

In order to carry step 2.1 further, attention must now shift from the process to the representation of the data. Three types of information will be needed as input, and will have to be organized internally for convenient processing:

    a. Geometry, that is, the dimensions involved
    b. Topology, that is, the hierarchical layout of the beams
    c. Information on the slabs, that is, whether one-way or two-way, and if the latter, their orientation

For geometry, a fixed grid, say at 1-ft intervals, may be used, and it may be stipulated that all beams and load locations are assumed to fall on this grid. Such a scheme may be too rigid and inflexible. Alternatively, geometry and topology may be specified by providing as input for each beam:

beam label
beam length
left support: beam label
               location along supporting beam
right support: beam label
               location along supporting beam

as sketched in Fig. 4. The input program segment can generate a variable-dimension grid, which provides a grid line in the X and Y directions only where an actual beam is located, as shown in Fig. 5.

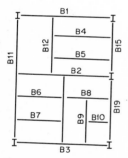

Fig. 3 Typical beam layout.

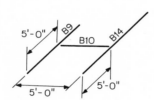

B10;5.00 B9 5.00; B14 5.00

Fig. 4 Input description for a typical beam.

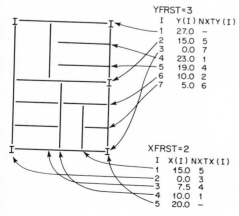

YFRST=3

I	Y(I)	NXTY(I)
1	27.0	–
2	15.0	5
3	0.0	7
4	23.0	1
5	19.0	4
6	10.0	2
7	5.0	6

XFRST=2

I	X(I)	NXTX(I)
1	15.0	5
2	0.0	3
3	7.5	4
4	10.0	1
5	20.0	–

Fig. 5 Internal representation of dimensions. Legend: I = arbitrary sequence number; X(I), Y(I) = dimensions; NXTX(I), NXTY(I) = location of next higher X or Y dimension; XFRST, YERST = location of first (lowest) X or Y dimension.

Such a grid can be readily built as a *list*, with intermediate lines inserted as beam data are read in. With such an array, step 2.1 can be implemented as:

    2.1   Convert area loads::=
        For each area load *do:*
            2.1.1   locate edge beams under area from grid
            2.1.2   determine slab type and direction
            2.1.3   assign line load to edge beams

These steps can be systematically decomposed further. Line and point loads can be similarly handled.

An efficient formulation of step 2.2 can be developed as follows:

    2.2   Carry down loads::=
            2.2.1   *For* each beam *do:*
                increase by 1 the counters for the left and right supports
            2.2.2   *For* each beam *do:*
                if counter = 0 (indicating that this beam does not carry reactions from any supported beam) *then:*
                    2.2.2.1   place beam designation into the next location of a stack
                    2.2.2.2   decrease the counters of the left and right supports by 1
            2.2.3   Repeat 2.2.2 until all beams are in stack

The effect of the above procedure is to determine the sequence in which the beams will be processed: first those carrying only external loads, then those carrying reactions from the beams previously processed, until all beams are processed. The final arrangement of the stack for the sample problem shown is given in Fig. 6. The reverse of this sequence gives the proper order for processing the geometric information: first the beams framing into supports (e.g., B11 and B3), the location of which is known, and then up the stack, using the information from previously located beams and the beam data shown in Fig. 4.

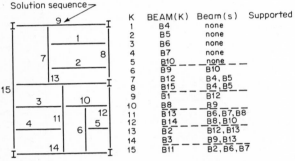

Solution sequence

K	BEAM(K)	Beam(s) Supported
1	B4	none
2	B5	none
3	B6	none
4	B7	none
5	B10	none
6	B9	B10
7	B12	B4, B5
8	B15	B4, B5
9	B1	B12
10	B8	B9
11	B13	B6, B7, B8
12	B14	B8, B10
13	B2	B12, B13
14	B3	B9, B13
15	B11	B2, B6, B7

Fig. 6 Internal representation of solution sequence. Legend: K = sequence number; BEAM(K) = label of beam to be analyzed. Dashed lines represent the status of the stack at each pass through step 2.2.2.

The last step then becomes:

    2.3   Design beams::=
            2.3.1   compute end reactions, shears, and moments
            2.3.2   select a beam from table
            2.3.3   apply the end reactions, with signs reversed, as loads on the supporting beams

Again, each of these steps requires further decomposition. In particular, step 2.3.2 requires the selection of appropriate design criteria to be used, as well as organization of the table of standard sections for efficient searching.

A major advantage of structured programming is that program organization can be reviewed and modified early, before major expenditures of effort have been made. For example, most programs have to deal with multiple loading conditions; a review of step 2.2 shows that it is independent of the actual loading condition and can therefore be put ahead of the other steps and step 2.1 repeated for every load condition. Similarly, it is frequently desirable to make groups of beams identical; therefore the beam-selection step (step 2.3.2) could be removed from the loop shown (with appropriate provisions for estimating the self-weight of the beams), and a selection routine inserted before step 2.3.2 to select the critical member of each group.

The final program incorporating the above observations is:

    Program::=
        1.   General input and preprocessing
            1.1   Input beam and slab data

      1.2 Carry down loads (previously 2.2)

      1.3 Establish geometry

  2. *For* each load condition *do:*

      2.1 Input load data

      2.2 Convert area loads (2.1)

      2.3 Compute beam forces (2.3.1 and 2.3.3)

  3. Beam design

      3.1 *For* each beam *do:* determine critical loading(s)

      3.2 *For* each group *do:*

          3.2.1 select critical member

          3.2.2 select a beam from table

  4. Output

5. Programming Languages As the program statement is refined and expanded, it eventually becomes sufficiently detailed and precise for *coding,* that is, transcription into a language suitable for direct communication with the computer. A computer program exists in two forms: the external form as coded by the programmer, and the internal form as used by the hardware. The internal form is rigidly fixed by the design of the hardware. To execute any program, the external form of the instructions comprising the solution sequence must be converted into the internal, or *machine language,* form and entered into the storage device. However, the software system can provide considerable assistance in this conversion process, and direct machine-language coding is practically never used. The available levels of programming languages are discussed below.

Symbolic Languages. At this level, the programmer still specifies each instruction separately, but uses convenient symbols (mnemonics or abbreviations) for the operations (e.g., ADD, MPY, etc.) and arbitrary symbols for the operands. A program called an *assembler* converts the operation names to their internal representations and assigns storage locations to the operands and then substitutes addresses for the names, thus producing an executable program. Symbolic coding is extremely time-consuming and tedious and should be used only where extremely high efficiency is needed or highly hardware-dependent functions are involved.

Procedural Languages. At the next level, a major increase in programmer efficiency and productivity is achieved by allowing the programmer to specify units of processing in the form of statements. Statements may be algebraic (i.e., SUM = A+B+C), input/output (i.e., READ A,B,C), or control statements implementing *control structures* which affect the sequence in which statements are executed. These structures include: (1) *iteration* statements in the form: *for* a range of variables *do* the following operations repeatedly; (2) *conditional* statements in the form: *if* a condition is true *then* do one operation *else* do another operation (see Art. 4 for examples); and (3) *invocation* statements to *call* a subprocedure and then *return* to the statement following the call (see Art. 7).

The translator which converts the procedural source program into the executable object program is called a *compiler.* On most systems, the compiled object program may be saved, so that on subsequent production runs the compilation can be bypassed. Procedural languages such as FORTRAN, PL/I, ALGOL, and BASIC account for the majority of structural engineering applications in present use. Furthermore, for the more popular languages compilers exist for a number of different computers, so that a given source program can be compiled and executed with little or no reprogramming.

Problem-Oriented Languages. At the topmost level, problem-oriented languages permit users who are not programmers to describe a problem directly in the terminology of their disciplines. Thus, for example, a statement of the form MEMBER 15 PRISMATIC AREA 10. INERTIA 200. is sufficient to describe a member and its properties to a structural analysis program, including the fact that a procedure applicable to prismatic members is to be used. Similarly, statements of the form CHECK BOND or DETAIL CONNECTION, with appropriate data, may be used to execute the appropriate procedures. A problem-oriented language presupposes the existence of a translator which converts the user's source statements into the necessary processing steps, and that the language can be used to specify only those data and procedures for which processing steps have been implemented in the translator.

The availability of problem-oriented languages greatly improves productivity and can eliminate much of the separation of the problem solving described earlier, as the user can go directly from a problem to its specification to the computer by means of the available problem-oriented statements. However, the programming of the procedures incorporated into the software becomes even more crucial, as these procedures will tend to be used more frequently, and under a greater variety of circumstances, than procedures contained in conventional programs.

6. Flowcharting A flowchart is a graphic representation of the flow or logic of the problem-solution procedure, consisting of the operations and decisions involved and of the order in which these must be executed. The common flowcharting symbols are shown in Fig. 7.

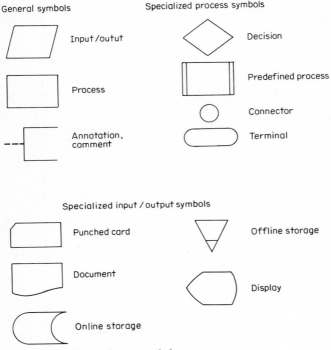

General symbols

Input/outut

Process

Annotation, comment

Specialized process symbols

Decision

Predefined process

Connector

Terminal

Specialized input/output symbols

Punched card

Document

Online storage

Offline storage

Display

Fig. 7 Common flow charting symbols.

Flowcharts are valuable tools both during the development of a program and after it has been turned over for production use. During development, they are the primary communication tool between the individuals working on a common problem. For all but the simplest problems, it is common practice to prepare a one-page block diagram, showing only the major steps. Each of the boxes can then be successively expanded into more detailed charts corresponding to the levels of structured programming (see Art. 4).

Flowcharts are also fundamental communication tools for testing, explaining, reviewing, and modifying programs. Their usefulness is greatly enhanced by liberal use of annotation or comment boxes to provide additional information, and by using problem-oriented rather than procedure-oriented statements in the boxes (e.g., using 'last beam?' rather than 'i = n?' to describe a test).

7. Decision Tables As the complexity of the program logic increases, flowcharts become less suitable to represent all possible paths and to check that all logically possible cases are properly accounted for. The technique of tabular decision logic, or decision tables for short, can be advantageously applied to problems of this type. A decision table is a concise tabulation of logical rules applying to a given problem. The decision table is laid out consisting of four areas as shown below:

Condition stub	Condition entry
Action stub	Action entry

The condition stub is a list of the logical variables involved in the problem. The action stub is a list of the actions involved. The condition entry lists the pertinent combinations of the logical variables in columns, each column specifying a rule. The action entry gives the actions to be taken corresponding to the specified rule.

A decision table is called limited-entry if all logical variables have only two values, so that the elements of the condition entry are limited to yes (Y), no (N), and immaterial (I), and the elements of the action entry are Y or blank, specifying whether a given action is to be taken or not. By contrast, in an extended-entry decision table, the logical variables may be multivalued, the elements of the condition entry may bear any kind of logical relationship (e.g., $=$, $\neq$, $<$, $>$) to the elements of the condition stub, and any number and kind of action may be specified.

A decision table is complete if there are rules for all possible combinations of the logical variables. The presence of an I for immaterial automatically makes the decision table incomplete. Incomplete decision tables often contain one additional rule, called the ELSE rule. The action associated with the ELSE rule is to transfer to an error procedure to handle a combination of parameters for which no specific rule has been provided.

The salient advantages of decision tables are: (1) the decision table displays at a glance all the logical possibilities for which provisions have been made; thus incomplete and inconsistent cases can be located before the conversion to a flowchart is undertaken; (2) the tabular display permits the application of formal methods for checking the consistency of the logic; (3) the simple format of the decision table can be understood, evaluated, and generated by laymen not familiar with computer programming more readily than involved flowcharts; experience has shown that decision tables are also useful in checking, documenting, and updating programs; and (4) decision tables are valuable programming aids for procedural languages such as FORTRAN, as well as being convenient programming languages by themselves.

8. Tools for Organizing Programs   All but the smallest programs consist of several discrete and separate processing steps. The separate steps may be: (1) sequentially related (i.e., input, processing, output); (2) in parallel (e.g., processes for prismatic, tapered, curved, etc., beams); or (3) laid out in the fashion of a tree if many logical possibilities are built into a program.

Subroutines.   The major tool for organizing the separate steps into the overall program is the use of subroutines. A subroutine is a self-contained set of instructions or statements designed to perform a specific task. The subroutine may be *called* from another program; when this occurs, control is transferred from the calling program to the subroutine. When the subroutine is completed, a *return* is made to the calling program and processing resumed in the calling program at the statement following the subroutine call. As far as the programmer is concerned, the composite machine behaves as if the function of the subroutine were available in the actual hardware. The primary reasons for using subroutines are: (1) to combine procedures used repeatedly within a program into a single entity; (2) to incorporate previously developed processes into new programs; (3) to segment large program steps into more manageable parts; and (4) to organize the overall process by implementing structured programming concepts in the actual program.

Modularization.   The organization of large programs into an efficient logical structure is accomplished through modularization and hierarchical arrangement. By modularization is meant that each subroutine is designed to perform a single, clear-cut function. Using such modules, a well-organized program can be laid out much like a pyramid, with the lowest-level subroutines on the bottom and succeedingly higher-level programs as subroutines. Also, the logical interconnection between subroutines should be kept to a minimum.

In a system of subroutines adhering to the two principles above, changes and modifications can be easily implemented by simply replacing the subroutine or subroutines affected. Programs can also be made much more versatile than originally intended with essentially no increase in programming effort by rearranging, under control of input data, the sequence of calling the various subroutines. It is only a small step to convert such a program into a problem-oriented system, by combining it with a control program which reads the user's commands, decodes them, and transfers control to the subroutine(s) corresponding to that command.

9. Tools for Organizing Data   The proper organization of data is as important a consideration in the design of a major program as the organization of procedures. The most serious limitation of many otherwise useful programs is the inflexibility of their data structure, in terms of the size and variety of problems that may be handled. Also, in many cases possible program modifications and extensions are limited by the fact that the data are poorly organized or the data organization is improperly documented. Data organization must be viewed at two levels: (1) at a conceptual level, where the primary emphasis is on *data structures* reflecting the content and organization of data; and (2) at an implementation level, concerned with the actual storage and access of data.

At the conceptual level, *arrays* (vectors, matrices, and tables) are data structures where numerical designations (subscripts) are used to identify specific elements of the array. Other data structures are occasionally very useful. Among these are: (1) *lists*, where each data element designates its own successor(s), thereby allowing compact storage and orderly access to information without the need of sorting; and (2) *stacks*, which allow efficient communication between nonsequential program segments (see Art. 4 for example). More complex structures, such as list structures, data hierarchies, and networks, can be used to provide compact representation or complex relationships among data items.

The tools for the implementation of data organization and management available on most operating systems fall into two categories: (1) those dealing with the management of data in the primary or high-speed storage device; and (2) those dealing with secondary storage devices (disks, drums, etc.). On many computing systems, the distinction between these two largely disappears, as dynamic memory allocation, virtual memory, or paging techniques allow the programmer to deal directly with extremely large virtual volumes of data.

Primary storage can be managed efficiently by placing data needed by several segments into a common area available to all segments. Further economies may be achieved by sharing an area among several mutually exclusive uses.

When the program and its data exceed the available primary storage space, use must be made of secondary storage. Secondary storage must also be used when data are to be saved on a long-term basis, such as in a project file containing all information about a project, which is generated, updated, and modified as the design progresses. Management of secondary storage is based on the recognition that only a small part of the complete file must be brought into primary storage for reference or processing, and can then be written back out into secondary storage. Most operating systems provide subroutines or complete software systems for the convenient management of secondary storage.

10. Computer Hardware *Programmable hand-held calculators* can provide considerable computational facility with great flexibility of use. Typical devices offer 10 numerical registers for data and storage for 100 program steps. Programming is essentially in machine language (Art. 5), each program step corresponding to a keystroke. One of their advantages is the large number of function keys, which perform the equivalent of substantial subroutines on conventional computers. Smaller models require that the program be keyed in every time it is used; larger models permit programs to be saved and reloaded from magnetic strips.

Programmable desk-top computers typically consist of a keyboard terminal for both program and data entry and a cathode-ray-tube (CRT) alphanumeric display. Storage for both data and program steps of the order of 2000 words or more is generally provided. Most have an "algebraic" programming language similar to FORTRAN or BASIC. Optional devices such as additional storage, secondary storage in the form of magnetic cartridges and disks, printers, plotters, card readers, graphic input/output devices, and even multiple terminals can convert these devices into complete data-processing systems. Communications adapters can be added to many desk-top computers, thus converting them into versatile "intelligent terminals."

Minicomputers are available as complete general-purpose computing systems, with up to 64,000 words of storage, FORTRAN and other compilers, a full array of peripherals including terminals, card readers and punches, printers and plotters, secondary storage devices which can access several million words, and operating systems permitting multiple tasks to be executed simultaneously. They are also available as part of small special-purpose systems comprising both hardware and software for specific tasks, such as surveying, drafting, and other applications. They are sometimes purchased separately and incorporated into specialized equipment, such as on-line process control in concrete batch plants. Even smaller machines, called microcomputers, are available and are used to build small dedicated systems; in fact, most pocket and desk-top calculators incorporate a microcomputer as their processing unit. On most microcomputers, a faster and cheaper "read-only" memory is used to store programs, and a separate memory is used for data, which must be stored ("written") as well as retrieved ("read").

Remote-batch terminals provide direct access by means of telecommunication lines to a large computer on a demand basis. They range from simple devices consisting of no more than a medium-speed card reader, a relatively slow printer, and the equipment necessary to interface with the communication system, through "intelligent terminals" allowing some degree of local processing and storage, to sizable minicomputers which can perform communication functions in addition to (and frequently simultaneously with) a full range of local processing operations.

Central computers are available in a range of capabilities from those commensurate with minicomputers to machines with operating speeds of the order of 10^6 arithmetic operations per second and direct-access storage (memory) of the order of 10^6 or more words. The larger machines simultaneously control many input-output devices, including both timesharing and remote-batch terminals, and access a large number and variety of secondary storage devices, having sometimes combined storage capabilities of the order of 10^{10} words. A full range of compilers and other software systems is generally available for tasks such as data management, program development, and text editing.

Computer networks can be assembled by connecting computers through high-speed communication networks. A small "network" may be comprised of a small local computer connected by telecommunication lines to a large central facility. If compatible programs are available on the two machines, it is possible to perform preprocessing (such as editing and checking input data) on the small machine, send the resulting data for processing at the central site, and upon receipt of the basic results, perform postprocessing (such as plotting) again on the local machine. Alternately, the local machine can perform the entire computation for small problems, calling on the central machine only for large problems. More powerful networking capabilities exist through the interconnection of large computers, permitting "distributed computing" through access to remote computers where special programs or particular data bases are available.

INTERACTION WITH COMPUTERS

A prime consideration is the manner in which the user interacts with the computer—the mode of interaction (primarily as it affects the rate at which responses are received from the computer), the medium through which the interaction takes place, and the level at which the engineer and the machine communicate with each other.

11. Modes of Interaction The question of most concern to an engineer is "How soon do I get my results back?" or, alternately, "How soon do I know that there are no errors in my data?" On a pocket or most desk-top calculators the response is immediate, but on larger systems several alternate modes are available.

Batch Processing. Batch processing requires that all input (program and data) be assembled and the entire job run as a unit. The term *batch* means that all jobs are "batched" or queued and executed in sequence. The time from submission of a job to the receipt of results is called *turnaround time,* and may range from minutes to days, depending on the access mechanism (i.e., carrying or mailing the job to the center vs. using a remote-job terminal), the size of the computer, and the priority scheme used by the computer's operating system. Typically, operating systems schedule jobs on the basis of size of memory required, expected duration, and priority levels related to charging rates.

While batch processing is generally the least expensive, from a user's standpoint it has two major disadvantages: (1) when the job is submitted, the user does not know whether he will receive meaningful results, or only an error message which may say: "Missing data on line XX; job terminated."; and (2) users are forced to request voluminous output, so as to avoid resubmitting the job if some additional unanticipated results are needed. The latter fact, of course, makes the job even larger and extends the turnaround time.

Interactive Processing. The alternative to batch processing is to interact directly with the computer by means of a terminal connected to it through a data communication line. Typically, a response is received after each line is input; depending on the program, the response may be an acknowledgment, a message that the line has been checked and errors identified, or a prompting message for the next line of input. A simple terminal will use only a small part of the capacity of the central processing unit (CPU); therefore, the operating system may allow one or more such programs in the *foreground* while batch processing continues in the *background* when the terminals need no servicing, or the computer may *time-share* up to hundreds of interactive programs, allocating to each a small amount of time in turn.

Interactive processing considerably increases the productivity of the man-machine combination but is generally more expensive than the batch mode.

Real-Time Processing. In certain applications, the computer is used as an integral part of a process or experimental setup, the processing being performed in *real time* as the actual process progresses.

Long-Time Interaction. It is becoming increasingly common to store project files and other long-term information on secondary storage devices directly accessible by the computer. The complete design of a project thus becomes a sequence of accesses to the project data base for information retrieval or checking, as source of data for various computations, and for the storage of results. The individual interactions can be performed interactively or in batch mode, with most of the input data to any one program replaced by directives specifying the portions of the project file to be retrieved. Similarly, output from the processing steps can be directed back to the project file for storage and future reference.

12. Media of Interaction *Punched cards* (Hollerith cards) are well suited for batch processing but have two disadvantages: (1) since keypunching is usually done by others than the engineer-users, the punching and verifying can significantly add to the turnaround time between job submission and completion; and (2) the rather inflexible format of cards usually forces rigid formats and conventions on data input.

Alphanumeric Keyboard Entry. For interactive use, the standard medium is a teletypewriter-like keyboard. This permits considerably more flexibility than the punched card, as message length and format become immaterial. Again, as with input mode, hybrid schemes are frequently used. For example, an operator may first enter voluminous tabular data directly into a secondary storage device. Exceptions, corrections, control commands, etc., can then be entered by the engineer by either directly working with the interactive program or by using a *text-editor* program to update the input file.

Many terminals have alphanumeric CRT displays instead of printers. These devices are faster and quieter than printer devices, but they usually require a backup or alternate device for producing *hard-copy* listing or printouts for reference purposes.

Printed listings are by far the most common output medium. Flexible formatting facilities are available to assist in producing tables, text, bar charts, schedules, simple schematic plots, etc., on the printer, either directly connected to the computer or part of a terminal. Considerable attention should be given to produce usable and legible output, paying proper attention to titles, page numbering, headings, units, significant figures, and the like.

Graphic Media. Computer graphics are widely available and cost-effective as an output medium, and occasionally for input as well. Graphics can be produced on either on-line devices, primarily CRT screens, or off-line devices such as plotters. The former is particularly suitable for display of complex information in an interactive design mode (e.g., plots of deflected or buckling shapes), while the latter is generally used for batch production of production drawings, such as steel and concrete detailing information. Graphic input by means of cursors, joy sticks, and light pens is available on interactive CRT devices; their use is not very widespread, primarily because most engineering data require some alphanumeric input (e.g., beam designations, dimensions more accurate than the CRT screen resolution). Hybrid schemes are again used; for example, it is common practice in structural analysis to describe a structure alphanumerically, but prior to analysis produce an isometric plot for checking the consistency of the input data.

13. Level of Communication *Rigid Formats.* Most conventional batch programs require that their input be prepared in a rigid format, with each item of input assigned a specific field within the input medium. Many such programs further require that choices, options, etc., also be numerically coded (e.g., PRINT = 1 will cause printing). Such input is tedious to prepare and provides no ready documentation for subsequent referencing.

Free-Field Input. Most systems, especially interactive ones, permit considerably more freedom in input, allowing alphabetic data mixed with numbers, and requiring only blanks or separators between data items. Input in this form should generally be used, except possibly where extremely large volumes of data are involved.

Prompted Input.   On interactive systems, it is relatively easy to develop prompting programs which lead the user "by the hand," identifying the data to be supplied and performing immediate checks on the data. Such programs tend to be too verbose for experienced users; provisions should be made to allow the prompting to be bypassed and to print only error messages.

Problem-Oriented Languages.   As described in Art. 5, problem-oriented languages eliminate much of the distinction between programming and problem solving, allowing engineers to program a process or solution by commands applicable to their disciplines. As most problem-oriented languages can be entered in free-field form, they can be readily used in both batch and interactive modes.

PROGRAM TYPES AND SOURCES

Whether a given program is appropriate, or even usable, for a particular task depends on a number of factors. Also, since program development is an expensive undertaking, available sources (Art. 15) should be investigated before undertaking the development of a program from scratch.

14. Program Types   No definite scheme exists for completely classifying computer programs in all relevant dimensions. The following is a checklist of the criteria to be used in evaluating available programs or developing new ones.

Scope.   In terms of the scope of available capabilities, programs can be categorized as special-purpose, general-purpose, or software systems, although the distinctions at the interfaces are by no means clear-cut. A special-purpose program addresses itself to a clearly delineated, specific situation, both as to the type of problem handled and the method used. A typical program in this category would be the analysis and proportioning of a two-legged bridge pier. Programs in this class are quite economical to use if the problem at hand exactly fits the scope. By contrast, a general-purpose program deals with broader classes of both problems and methods, as, for example, an analysis program for arbitrary plane frames, or a proportioning program for concrete beams and beam columns allowing the optional use of either the working stress or the strength method. While such programs require more input preparation than special-purpose ones, since the problem type, options to be used, etc., must be specified, they tend to be more economical in the long run because of their broader scope and increased flexibility. Finally, application software systems provide flexible mechanisms to interconnect separate processing steps (each of which may be a substantial program on its own) into automatic sequences.

Intended Usage.   Programs range from "one-shot" programs intended to be discarded after a particular job is finished, through experimental programs concerned primarily with demonstrations of a particular approach or method, to production programs designed for continued use. The level of documentation, the degree of error and consistency checking, the efficiency of the process, and the format and convenience of input and output all depend critically on the intended use of a program.

Language.   Programs written in high-level procedural languages, such as FORTRAN or BASIC, can be transferred to other computers, although not without some difficulty. The style of coding and legibility of the program greatly affect the ease with which a program can be understood, modified for local conditions, and transported to a different computer.

Level of Documentation.   The documentation available with a program is perhaps the most significant factor in determining whether a program is worth adopting or modifying. The requisite four components of program documentation given in Art. 3 should always be insisted upon.

Past Usage.   In acquiring or modifying programs obtained from others, it is useful to know something about the extent of past usage. Programs in extended use tend to be more error-free. Also, especially in design programs, satisfactory usage by several organizations will tend to ensure that the program is sufficiently flexible to accommodate the procedures and assumptions of different organizations.

Hardware-Dependent Aspects.   The principal aspects to consider are the programming language, discussed above, the program size, and its data requirements. The size determines the minimal hardware configuration on which the program can run. Although it is possible to segment large programs and data into smaller units, this process is extremely tedious if the programs have not been designed with such segmentation in mind. The data requirements, that is, the volume, format, and organization of the data on which the program operates, are among the prime factors in determining the difficulty of transferring a program from one machine type to another, inasmuch as the data-management operations tend to be the least standardized among different hardware systems. The data requirements are also generally the limiting factors on the size of problem a given program can handle.

Intended Mode of Access.   It is important to compare the mode (batch or interactive) for which a program was written with the intended mode of use. Small programs can generally be converted from batch to interactive by only rewriting the input-output segments. For large programs this may be prohibitively expensive, and major reprogramming may be needed to make them cost-effective in an interactive mode.

15. Sources of Programs   Computer programs, with various degrees of completeness and reliability, are available from a variety of sources. The most important are described below.

Public Domain.   Many textbooks, technical articles, and research reports contain source listings of programs for various applications in structural engineering. Many such programs are essentially appendixes to technical reports and are therefore to be classified as special-purpose programs, likely to require major modifications before being suitable for production use. Several organizations, such as the National Information Service for Earthquake Engineering (NISEE) maintain catalogs of programs available in the public domain.

Program Libraries.   Most computer manufacturers, some government agencies, and many users' groups maintain program libraries, from which copies of programs and their documentation may be obtained at nominal charges. Most libraries (except for some users' groups discussed below) maintain essentially no control over the scope, reliability, and quantity and quality of documentation of programs accepted into the libraries. Thus testing and verification of such programs may be quite time-consuming and expensive.

Cooperative Users' Groups.   Some users' groups, notably CEPA (Civil Engineering Program Applications) and APEC (Automated Procedures for Engineering Consultants), maintain libraries of programs developed or commissioned by the group's members, with stringent controls on the quality of programs and documentation.

Trade and Service Organizations.   Many trade and service organizations and individual manufacturers provide programs under a variety of access or use mechanisms as essentially an extension of their traditional policy of providing handbooks, design charts, etc., to facilitate the use of the products or services represented by the organizations. Among the organizations providing such services are the Portland Cement Association (PCA) and the American Institute for Steel Construction (AISC).

Proprietary Sources.   Many firms undertake, on a contract basis, the development of new programs to a client's specifications. Most contracts of this type specify a maintenance period, typically one year, during which the supplier is responsible for providing corrections for errors, or bugs, discovered in program use.

16. Other Services   *Machine Access.*   Access to computers, on a charge basis for time actually used, is available from a multiplicity of sources. Time-shared and remote-batch access to computers is available on an even broader basis and, with further reductions in terminal costs and communication charges, will undoubtedly develop into the primary mode of computer use for most engineering organizations, except perhaps a few of the largest ones.

Program Access.   Service bureaus and software firms provide access to their proprietary programs, usually on the basis of a software surcharge multiplier applied to the basic machine rate. Such arrangements are usually accompanied by some form of performance guarantee, frequently limited to reruns at no charge if software malfunctioning can be established. Many engineering firms, even those with small in-house computing facilities, find this type of program access highly advantageous for infrequent use of large-scale or highly specialized programs.

Problem-Solution Services.   Organizations with no in-house capability for computer use may avail themselves of complete problem-solution services offered by many organizations. Under a typical arrangement, the user supplies a diagram of the structure with all the member properties and loads identified, and receives back a complete listing of analysis results. Similar services are also offered in the other application areas described in Secs. 18 through 20.

TYPICAL APPLICATIONS

The range of computer applications in structural engineering covers procedures from preliminary conceptual design to construction control on the site. Most computer applications intended for production use can be classified into the major categories discussed below.

17. Analysis   Analysis comprises the bulk of production use of computers. The reasons for this preponderance are: (1) analysis is a clearly identifiable, time-consuming task in the design process; (2) the increased size and complexity of structures requires modeling and analytical techniques impossible to perform by manual methods; and (3) analysis of a model is based on rational principles of mathematics and behavior and thus is not subject to individual interpretation. It is assumed here that analysis is performed on a fully prescribed mathematical model of the structure, i.e., one for which all relevant member properties (dimensions, stiffnesses, etc.) are assumed to be known.

Methods.   Structural-analysis programs, almost without exception, use the deformation or stiffness method rather than the force or flexibility method commonly used for hand calculations. In the deformation method the assembly of the governing simultaneous equations requires a minimum of input data, follows a simple, logical sequence independent of the choice of redundants, and the resulting equations can be solved quite accurately; while in the flexibility method the governing equations depend on the analyst's choice of redundants, are less suitable for assembly by automation, require considerably more input or precomputation, and are prone to large roundoff errors if an improper primary structure is selected.

Assumptions.   Many of the early frame-analysis programs were based on the slope-deflection and moment-distribution methods, with their built-in assumption of no axial distortions. Later programs incorporate a full-deformation-method formulation, including axial and shearing distortions, for two reasons: (1) in tall, slender buildings, column elongations must be taken into account for realistic modeling; and (2) the efficiency and simplicity of programming the general deformation method far outweighs the penalty of carrying along the additional degrees of freedom associated with the axial deformations.

General finite-element programs have removed the need for modeling structures as assemblages of line elements, allowing plates, shells, and solids to be analyzed with the same convenience.

18. Proportioning   If the element to be proportioned is statically determinate (e.g., a simple-span bridge girder), analysis and proportioning can be combined into a direct design procedure. If, however, the element is a component of an indeterminate structure, analysis and proportioning form an iterative, or trial-and-error, loop, where initial sizes or stiffnesses are assumed, an analysis performed, the elements reproportioned, and the entire cycle repeated until satisfactory results are obtained. The degree to which this cycle is automated, i.e., performed by a program without a designer's intervention, depends on both hardware or cost limitations and the design office's preferences and mode of operation. The iterative design-proportion process may be controlled so as to minimize some desired feature, such as material volume or cost.

Design Assumptions.   Proportioning programs differ in several major respects from analysis programs. Proportioning is governed to a great extent by the applicable design standards or code provisions. Unfortunately, these provisions are not presented in a format suitable for direct conversion to computer programs. The major problem is that the logic of the design codes, such as interactions between various provisions scattered through the text and the specification of the limitations or range of applicability of provisions, is not directly discernible. As a consequence, programming according to any given design code requires a great deal of individual interpretation, and therefore limits the applicability of the program to those users who agree with the interpretations embodied in it.

Design codes by themselves are not a complete design guide, and additional design logic, including assumptions, limitations, shortcuts, search strategies, etc., must be incorporated into any proportioning program. For example, one seldom, if ever, designs every beam of a frame for the most critical combination of load effects acting on it; rather, one chooses a typical or critical beam to be proportioned, and then replicates its design for all similar beams. The definition of what is typical, critical, or similar will generally vary widely among designers or design organizations. It is for reasons such as these that proportioning programs lack the generality of analysis programs, tend to incorporate the assumptions and practices of the originating organizations, and must be carefully reviewed to ascertain whether their assumptions and limitations agree with those in use by the acquiring organizations.

Data-Processing Requirements. From a data-processing standpoint, proportioning programs differ from those for analysis in that they generally involve fewer and simpler calculations, but vastly larger volumes of data. This is particularly true if the programs are integrated into a software system and access all or portions of their data in a project file containing the description of all the important aspects of an entire project.

19. Detailing In concrete design, detailing refers primarily to the determination of the number, size, layout, and location of reinforcement, given the element dimension and the areas of steel required. Frequently, these calculations are combined with printing of schedules and bar bending diagrams, fabrication, inventory, shipping and placement control, and the printing of bundling information and shipping tags. Some programs of this type, developed by reinforcing-steel fabricators, produce directly control tapes for driving numerically controlled cutting and bending machines. Similarly, in steel design, detailing refers to the selection of connections and of the type, number, size, and location of connectors. Again, many of these operations can be combined with fabricating information, shop drawing, production scheduling of fabrication, and control of cutting, drilling, and welding machines.

20. Preparation of Final Documents The most time-consuming and expensive operation in a design office is the preparation of specifications, schedules, and drawings comprising the final documents. Many aspects of final-document preparation are routinely performed by computer programs. Applications contain, by necessity, assumptions and procedures which may not be applicable to every organization. However, in many instances such differences can be handled by differences in the data only, leaving the procedures essentially unchanged. Most applications may be classified into one of the three categories discussed below.

Specifications. Most offices maintain a set of standard or master specifications, and then produce specifications for specific jobs by incorporating additions, modifications, and deletions. There are several computer programs in use which perform exactly the same function: the master specification is stored in computer-readable form, the specification writer produces a set of exceptions keyed to lines or paragraphs of the master, and the program merges the exceptions with the master and produces the text directly on multilith or other medium for direct production. Indexing, cross referencing, etc., of the text may also be performed.

Drawings may be produced directly by plotter devices attached to the computer. Programming problems are quite severe, and successful applications tend to cover only schematic plots, such as erection diagrams, where little detail is to be presented, or highly repetitive structural types, such as footings, beams, and bridge-piers, where the high cost of programming all possible cases can be economically justified.

Schedules. Between the two extremes of pure text and to-scale drawings is the large class of tables, schedules, and schematic diagrams often used in contract documents. The presentation of information in such formats is a natural application for computer-generated data, and much of the output of proportioning and detailing programs is presented in tabular formats ready for direct inclusion into the design documents. It is common practice in many firms to key such tables to not-to-scale generalized sketches and diagrams.

REFERENCES

1. Bashkow, T. R. (ed.): "Engineering Applications of Digital Computers," Academic Press, Inc., New York, 1968.
2. Beaufait, F. W., et al.: "Computer Methods of Structural Analysis," Prentice-Hall, Inc., Englewood Cliffs, N.J., 1970.
3. Bowles, J. E.: "Analytical and Computer Methods in Foundation Engineering," McGraw-Hill Book Company, New York, 1974.
4. Coleman, C. W.: Computer Graphics for Architects and Civil Engineers, *Graphic Sci.*, May 1971.
5. Computer Applications in Concrete Design and Technology, *ACI Spec. Publ.* SP-16, American Concrete Institute, Detroit, 1967.

6. Conference Papers, ASCE Conferences on Electronic Computation (First Conference, 1957; Second Conference, 1960; Third Conference, *J. Struct. Div. ASCE*, August 1963; Fourth Conference, December 1966; Fifth Conference, January 1971; Sixth Conference, April 1975 and (EI2), April 1975.

7. Fenves, S. J.: "Computer Methods in Civil Engineering," Prentice-Hall, Inc., Englewood Cliffs, N.J., 1967.

8. Fenves, S. J., et al. (ed.): "Numerical and Computer Methods in Structural Mechanics," Academic Press, Inc., New York, 1973.

9. Haberman, C. M.: "Use of Digital Computers for Engineering Applications," Charles E. Merrill, Inc., Columbus, Ohio, 1966.

10. Harper, G. N. (ed.): "Computer Applications in Architecture and Engineering," McGraw-Hill Book Company, New York, 1968.

11. Harrison, H. B.: "Computer Methods in Structural Analysis," Prentice-Hall, Inc., Englewood Cliffs, N.J., 1973.

12. Impact of Computers on the Practice of Structural Engineering in Concrete, *ACI Spec. Publ.* SP-33, American Concrete Institute, Detroit, 1972.

13. Litton, E.: "Automatic Computational Techniques in Civil and Structural Engineering," John Wiley & Sons, Inc., New York, 1973.

14. McCuen, R. H.: "FORTRAN Programming for Civil Engineers," Prentice-Hall, Inc., Englewood Cliffs, N.J., 1975.

15. Medearis, K.: "Numerical-Computer Methods for Engineers and Physical Scientists," KMA Research, Denver, Colo., 1974.

16. Pilkey, W., et al., (eds.): "Structural Mechanics Computer Programs," University Press of Virginia, Charlottesville, 1974.

17. Pollock, S. L.: "Decision Tables: Theory and Practice," Wiley-Interscience, New York, 1971.

18. Prager, W.: "Introduction to Basic FORTRAN Programming and Numerical Methods," Blaisdell Publishing Company, New York, 1965.

19. Roos, D.: "ICES System Design," MIT Press, Cambridge, Mass., 1966.

20. Roos, D. (ed.): ICES System: General Description, *Rept.* R67-49, Department of Civil Engineering, MIT, Cambridge, Mass., 1967.

21. Spindell, P.: "Computer Applications in Civil Engineering," Van Nostrand Reinhold, New York, 1971.

22. Ural, O.: "Matrix Operations and Use of Computers in Structural Engineering," International Textbook Company, Scranton, Pa., 1971.

23. Weaver, W., Jr.: "Computer Programs for Structural Analysis," Van Nostrand Reinhold, New York, 1967.

24. Wang, C. K.: "Computer Methods in Advanced Structural Analysis," Intext Educational Publishers, New York, 1973.

Section **3**

Earthquake-Resistant Design

N. M. NEWMARK and W. J. HALL
Professors of Civil Engineering, University of Illinois, Urbana

1. Response of Simple Structures to Earthquake Motions A series of structures of varying size and complexity is shown in Fig. 1, corresponding to a simple, relatively compact machine anchored to a foundation in a, a simple bent or frame in b, a more complex frame in c, multistory buildings of 15 stories in d and of 40 stories in e, an elevated water tank in f, and a suspension bridge responding either laterally or vertically in g. A typical period of vibration T or frequency of vibration f in the fundamental mode of vibration is indicated for each.

Each of the structures shown in Fig. 1 can be represented by a simple oscillator consisting of a single mass supported by a spring and a dashpot (Fig. 2). The relation between the circular frequency of vibration $\omega = 2\pi f$, the natural frequency f, and the period T is given by the following equation in terms of the spring constant k and the mass m:

$$\omega^2 = \frac{k}{m} \tag{1}$$

$$f = \frac{1}{T} = \frac{\omega}{2\pi} = \frac{1}{2\pi}\sqrt{\frac{k}{m}} \tag{2}$$

In general, the effect of the dashpot is to produce damping of free vibrations or to reduce the amplitude of forced vibrations. The damping force is assumed to be equal to a damping coefficient η times the velocity $\dot{u}$ of the mass relative to the ground. That value of η at which the motion loses its vibratory character in free vibration is called the *critical damping coefficient*, i.e., $\eta_{\text{crit}} = 2m\omega$. The amount of damping is most conveniently considered in terms of the proportion β of critical damping,

$$\beta = \frac{\eta}{\eta_{\text{crit}}} = \frac{\eta}{2m\omega} \tag{3}$$

For most practical structures β is relatively small, in the range of 0.5 to 10 or 20 percent, and does not appreciably affect the natural period or frequency of vibration.

2. Earthquake Motions   Strong-motion earthquake acceleration records with respect to time have been obtained for a number of earthquakes. Ground motions from other sources of disturbance, such as quarry blasting and nuclear blasting, are also available and show many of the same characteristics. Among the more intense strong-motion earthquakes recorded so far is that of the El Centro, Calif., earthquake of May 18, 1940. The recorded accelerogram, in the north-south component of horizontal motion, is shown in Fig. 3. On the same figure are shown integration of the ground acceleration a to give the variation of ground velocity v with time, and the integration of velocity to give the variation of ground

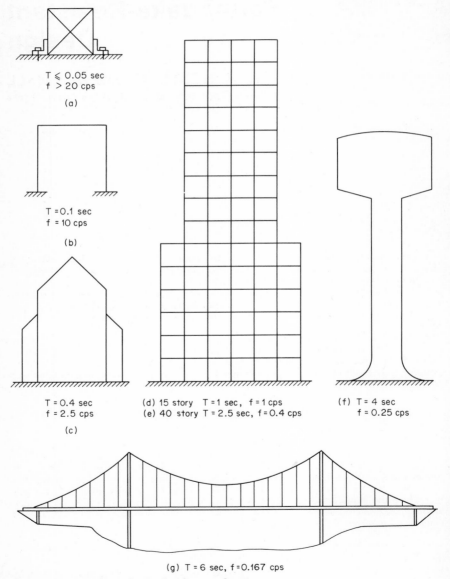

T ≤ 0.05 sec
f > 20 cps

(a)

T = 0.1 sec
f = 10 cps

(b)

T = 0.4 sec
f = 2.5 cps

(c)

(d) 15 story   T = 1 sec, f = 1 cps
(e) 40 story   T = 2.5 sec, f = 0.4 cps

(f) T = 4 sec
f = 0.25 cps

(g) T = 6 sec, f = 0.167 cps

Fig. 1   Structures subjected to earthquake ground motions.

displacement d with time. These integrations require base-line corrections of various sorts, and the magnitude of the maximum displacement may vary depending on how the corrections are made. The maximum acceleration and velocity are relatively insensitive to the corrections, however. For this earthquake, with the integrations shown in Fig. 3, the maximum ground acceleration is 0.32g, the maximum ground velocity 13.7 in./sec, and the maximum ground displacement 8.3 in. These three maximum values are of particular interest because they help to define the response motions of the various structures considered in Fig. 1 most accurately if all three maxima are taken into account.

Spring
k
η
m
Damper
u = x - y

Fig. 2 System considered.

RESPONSE SPECTRA

3. Elastic Systems The response of the simple oscillator shown in Fig. 2 to any type of ground motion can be readily computed as a function of time. The maximum values of the response are of particular interest. These maximums can be stated in terms of the maximum deformation in the spring $u_m = D$, the maximum spring force, the maximum acceleration of the mass (which is related to the maximum spring force directly when there is no damping), or a quantity, having the dimensions of velocity, which gives a measure of the maximum energy absorbed in the spring. This quantity, designated the pseudovelocity V, is defined in such a way that the

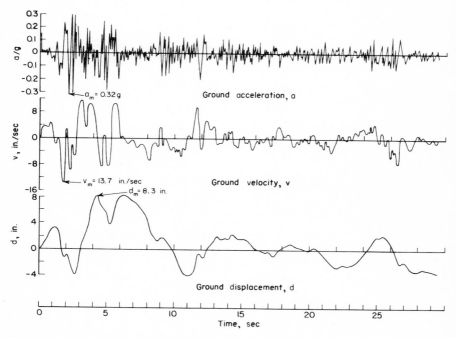

Fig. 3 El Centro, Calif., earthquake of May 18, 1940, north-south component.

energy absorption in the spring is $\frac{1}{2}mV^2$. The relations among the maximum relative displacement of the spring D, the pseudovelocity V, and the pseudoacceleration A, which is a measure of the force in the spring, are as follows:

$$V = \omega D \tag{4}$$
$$A = \omega V = \omega^2 D \tag{5}$$

The pseudovelocity V is nearly equal to the maximum relative velocity for systems with moderate or high frequencies but may differ considerably from the maximum relative velocity for very-low-frequency systems. The pseudoacceleration A is exactly equal to the maximum acceleration for systems with no damping and is not greatly different from the maximum acceleration for systems with moderate amounts of damping, over the whole range of frequencies from very low to very high values.

Typical plots of the response of the system as a function of period or frequency are called response spectra. Plots for acceleration and for relative displacement, for a system with a moderate amount of damping, subjected to an input similar to that of Fig. 3, are shown in Fig. 4. This arithmetic plot of maximum response is simple and convenient to use.

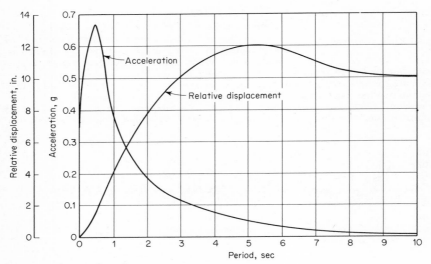

Fig. 4 Arithmetic plots of response.

An equally useful plot, called the *tripartite logarithmic plot*, is shown in Fig. 5. It indicates at one and the same time the response values D, V, and A, and has the additional virtue that it also shows more clearly the extreme or limiting values of the various parameters defining the response. The frequency is plotted on a logarithmic scale. Since the frequency is the reciprocal of the period, the logarithmic scale for period would have exactly the same spacing of the points, or in effect the plot would be turned end for end. The pseudovelocity is plotted on a vertical scale, also logarithmically. Then on diagonal scales along an axis that extends upward from right to left are plotted values of the displacement, and along an axis that extends upward from left to right the pseudoacceleration, in such a way that any one point defines for a given frequency the displacement D, the pseudovelocity V, and the pseudoacceleration A. Points are indicated in Fig. 5 for the seven structures of Fig. 1, plotted at their fundamental frequencies.

A wide variety of motions have been considered in Refs. 1 through 3, ranging from simple pulses of displacement, velocity, or acceleration of the ground, through more complex motions such as those arising from nuclear-blast detonations, and for a variety of earthquakes as taken from available strong-motion records. Response spectra for the El Centro earthquake are shown in Fig. 6. The spectrum for small amounts of damping is much more jagged than indicated by Fig. 5, but for the higher amounts of damping the response curves are relatively smooth. The scales are chosen in this instance to represent the amplifications of the response relative to the ground-motion values of displacement, velocity, or acceleration.

The spectra shown in Fig. 6 are typical of response spectra for nearly all types of ground motion. It is noted that on the extreme left, corresponding to very-low-frequency systems, the response for all degrees of damping approaches an asymptote corresponding to the value of the maximum ground displacement. A low-frequency system corresponds to one having a very heavy mass and a very light spring. When the ground moves relatively rapidly, the mass does not have time to move, and therefore the maximum strain in the

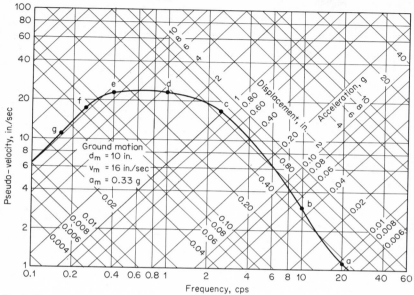

Fig. 5 Response spectrum for typical earthquake.

spring is precisely equal to the maximum displacement of the ground. On the other hand, for a very-high-frequency system, the spring is relatively stiff and the mass very light. Therefore, when the ground moves, the stiff spring forces the mass to move in the same way the ground moves, and the mass therefore must have the same acceleration as the ground at every instant. Hence, the force in the spring is that required to move the mass with the same acceleration as the ground, and the maximum acceleration of the mass is precisely equal to the maximum acceleration of the ground. This is shown by the fact that all the lines on the extreme right-hand side of the figure approach as an asymptote to the maximum ground-acceleration line.

The results of similar calculations for other ground motions are quite consistent with those in Fig. 6, even for simple motions. The general nature of the response spectrum consists of a central region of amplified response and two diminishing regions of response in which, for very-low-frequency systems, the response displacement is equal to the maximum ground displacement, and for high-frequency systems, the response acceleration is equal to the maximum ground acceleration. In general, the amplification factor for displacement is less than that for velocity, which in turn is less than that for acceleration.

4. Design Response Spectra A response spectrum developed to give *design coefficients* is called a *design spectrum*. For any given site, estimates are made of the maximum ground acceleration, maximum ground velocity, and maximum ground displacement, and lines representing these values are drawn on a tripartite logarithmic chart. The lines showing the ground-motion maxima in Fig. 7 are drawn for a maximum ground acceleration a of 1.0g, velocity v of 48 in./sec, and displacement d of 36 in. These data represent motions more intense than those normally considered for a design earthquake. They are, however, approximately in correct proportion for a number of areas of the world where earthquakes occur on firm ground, soft rock, or competent sediments of various kinds. For relatively soft sediments the velocities and displacements might require increases above the values shown. However, it is not likely that maximum ground velocities in excess of 4 to 5 ft/sec are obtainable under any circumstances.

Representative amplification factors for 50 and 84.1 percentile* levels of horizontal response are given in Table 1. Using these factors, and taking points B and A at about 8

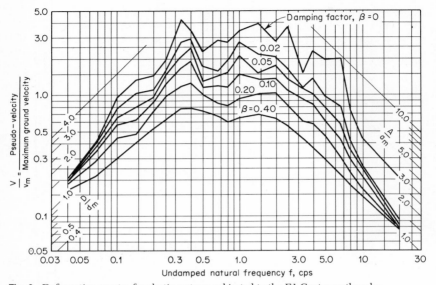

Fig. 6 Deformation spectra for elastic systems subjected to the El Centro earthquake.

and 33 Hz, respectively, for all values of damping, a horizontal response spectrum can be drawn as shown in Fig. 7.

The authors recommend that amplification factors for vertical motion be taken at two-thirds the values given in Table 1.

Further information on construction of elastic-response spectra is given in Refs. 1 to 7.

5. Response Spectra for Inelastic Systems A typical inelastic-spring force-displacement relation is shown in Fig. 8. This can be approximated by an elastoplastic relation as indicated, with an elastic initial region, a plastic ceiling of constant resistance, and an elastic unloading. The approximate relation is drawn so as to have the same area between the origin and u_y, and between u_y and u_m, as the actual curve. The ratio of the maximum permissible or useful displacement u_m to the yield displacement u_y is called the *ductility factor*, denoted by the symbol μ. Unloading is considered to be elastic until yielding is reached in the opposite direction.

For equal yield values in either direction, calculations of the response of the system of Fig. 2 for an elastoplastic resistance function can be made. A variety of such calculations have been made and are reported in Refs. 2 and 3. It is instructive to plot the results of such calculations on a chart similar to the tripartite response-spectrum charts of Figs. 5 and 6. Spectra for the elastic component of the response of elastoplastic systems to the El Centro earthquake, with a damping factor of 2 percent of critical in the elastic range of response, are shown in Fig. 9. The ductility factor μ ranges from 1 (elastic behavior) to 10. The displacement D_y/d_m from this figure must be multiplied by the corresponding ductility factor to obtain the total maximum displacement, but accelerations are correct as read directly from the plot. The curves do not represent the correct value of the maximum pseudo-velocity, which is why it is denoted by V'.

---

*A percentile of 84.1 means that 84.1 percent of the values can be expected to fall at or below that particular amplification.

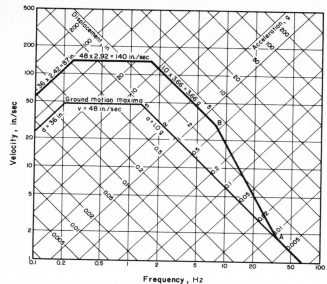

Fig. 7 Basic horizontal-response design spectrum normalized to 1.0g for 2 percent damping, 84.1 percentile level.

TABLE 1 Spectrum Amplification Factors for Horizontal Elastic Response

Damping, % critical	One sigma (84.1%)			Median (50%)		
	A	V	D	A	V	D
0.5	5.10	3.84	3.04	3.68	2.59	2.01
1	4.38	3.38	2.73	3.21	2.31	1.82
2	3.66	2.92	2.42	2.74	2.03	1.63
3	3.24	2.64	2.24	2.46	1.86	1.52
5	2.71	2.30	2.01	2.12	1.65	1.39
7	2.36	2.08	1.85	1.89	1.51	1.29
10	1.99	1.84	1.69	1.64	1.37	1.20
20	1.26	1.37	1.38	1.17	1.08	1.01

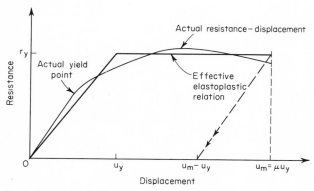

Fig. 8 Resistance-displacement relationship.

A response spectrum for total relative displacement can be drawn for the same conditions as Fig. 9 by multiplying each curve's ordinates by the value of the corresponding ductility factor, as noted above. Such a figure shows that the maximum total displacement is virtually the same for all ductility factors, actually perhaps even decreasing slightly for the larger ductility factors in the low-frequency region (below about 2 Hz). Moreover, it appears from Fig. 9 that the maximum acceleration is nearly the same for all ductility factors for frequencies greater than about 20 or 30 Hz. In between there is a transition. These observations are typical of other earthquake spectra. One can generalize about

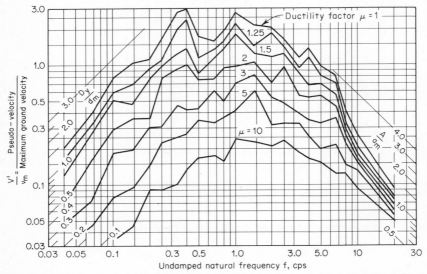

Fig. 9 Deformation spectra for elastoplastic systems with 2 percent critical damping subjected to the El Centro earthquake.

them, for single-degree-of-freedom structures, as follows. For low and intermediate frequencies corresponding to something of the order of about 2 Hz as an upper limit, total relative displacements are preserved and are very nearly the same for all ductility factors. As a matter of fact, inelastic systems have perhaps even a smaller displacement than elastic systems for frequencies below about 0.3 Hz. For frequencies between about 2 and about 8 Hz, the best relationship appears to be to equate the external applied energy with a corresponding resisting energy. There is a transition region between 8 and 30 to 33 Hz, depending on the damping ratio, while above 33 Hz the force or acceleration is nearly the same for all ductility ratios. These observations enable one to derive approximate inelastic-response spectra from an elastic-response spectrum as described in the next paragraph.

To obtain the inelastic-acceleration spectrum from the elastic-response spectrum 0 1 2 3 4 5 shown in Fig. 10, values of D and V are divided by the ductility factor μ to get D' and V' and values of A are divided by $\sqrt{2\mu - 1}$ to get A'. Thus the velocity corresponding to point $1'$ is $1/\mu$ times the velocity for point 1, the acceleration corresponding to point $3'$ is $1/\sqrt{2\mu - 1}$ times the acceleration for point 3, and D', V', and A' are parallel to D, V, and A, respectively. Note that the frequencies corresponding to correspondingly numbered points are the same. The acceleration spectrum is completed by connecting points $3'$ and 4. The amount of the shift from A to A' is such as to give the same energy absorption for the elastoplastic curve as for an elastic curve for the same period of vibration. The curve $0'$ $1'$ $2'$ $3'$ 4 5 also represents the elastic component of displacement as discussed earlier in this article.

The inelastic-displacement spectrum in Fig. 10 is obtained by leaving D and V unchanged, extending V to a point $2''$ vertically above $2'$, and drawing $2''$ $3''$, $3''$ $4''$, and A_0'' parallel to $2'$ $3'$, $3'$ 4, and A_0, respectively, but at a value μ times as great.

Inelastic-response spectra can be used only as an approximation for multi-degree-of-freedom systems but generally give reasonable results for systems with balanced resistances, i.e., systems without major differences in adjacent story stiffnesses.

Additional information on development of inelastic design response spectra may be found in Refs. 1 and 7 to 13.

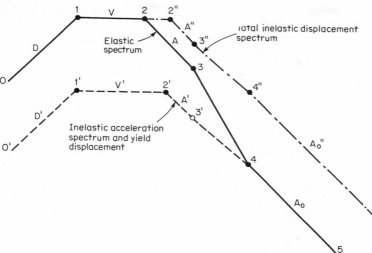

Fig. 10   Inelastic design spectra.

6. Multi-Degree-of-Freedom Systems   A multi-degree-of-freedom system has a number of different modes of vibration. For the shear beam a of Fig. 11 are shown the fundamental mode of lateral oscillation b, the second mode c, and the third mode d. The number of modes equals the number of degrees of freedom, five in this case. In a system that has independent (uncoupled) modes, which condition is usually satisfied for buildings, each

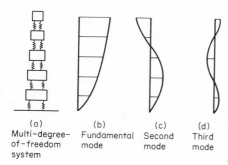

(a)   (b)   (c)   (d)
Multi-degree- Fundamental Second   Third
of-freedom   mode   mode   mode
system

Fig. 11   Modes of vibration of shear beam.

mode responds to the base motion as an independent, single-degree-of-freedom system. Thus the modal responses are nearly independent functions of time. However, the maxima do not necessarily occur at the same time.

Use of Response Spectra.   For multi-degree-of-freedom systems, the concept of the response spectrum can be used for analysis in most cases, although the use of the inelastic-response spectrum is only approximately valid as a design procedure. For a system with a number of masses at nodes in a flexible framework the equation of motion can be written in matrix form as follows:

$$M\ddot{u} + C\dot{u} + Ku = -M(\ddot{y})\{1\}$$

where $\ddot{u}$, $\dot{u}$, and u are vectors, the quantity in brackets represents a unit vector, and $\ddot{y}$ is a scalar quantity. The mass matrix M is usually diagonal, but in all cases both M and the stiffness matrix K are symmetrical. When the damping matrix C satisfies certain conditions, the simplest of which is when it is a linear combination of M and K, the system has normal modes of vibration with modal displacement vectors u_n.

When the modes and frequencies of the system are obtained, the modal responses are determined by using *modal participation* factors γ_n, which may be defined in matrix notation for each mode as follows:

$$\gamma_n = \frac{u_n^T M \{1\}}{u_n^T M u_n}$$

This is restated in scalar form in Eq. (18) for those cases where M is a diagonal matrix.

The particular response quantity desired—say the stress at a particular point, the relative displacement between two reference points, or any other effect—is obtained by combination of the modal values. The procedure is described in Art. 9. A detailed discussion of the design approaches employed for high-rise buildings is given in Refs. 1, 8, and 13.

Use of Time History. Alternatively one may make a calculation of response by considering the motions to be applied and computing the responses by a step-by-step numerical dynamic analysis. This implies a deterministic approach, since a deterministic time history is involved. By use of several time histories independently considered, one can arrive at average or conservative upper bounds of response, but at the expense of a considerably increased amount of calculation. In general, however, there is no real advantage in using a time history as compared with a response-spectrum approach for multi-degree-of-freedom systems.

Motions in Several Directions. Earthquake motions actually occur as random motions in horizontal and vertical directions. In other words, a structure is subjected to components of motion in the vertical direction and in each of two perpendicular horizontal directions. Three components of rotational motion, corresponding to a twist about a vertical axis and rocking motions about the two horizontal axes, also may exist. The linear ground motions are apparently statistically independent. Consequently, if one uses time histories of motion, one must either use actual earthquake records or synthetic records that maintain the same degree of statistical independence as in actual records. Therefore, for time histories that involve inelastic behavior it is an oversimplification to consider each of the components of motion independently since, in general, they all occur at about the same time. However, there is only a small probability that the maximum responses will occur simultaneously, and methods have been derived for handling combined motions in design taking into account the above considerations (Art. 17).

COMPUTATION OF PERIOD OF VIBRATION*

7. Fundamental Mode Procedures are available for the computation of the periods of vibration of undamped† multi-degree-of-freedom systems,[1,8,15,16] and applications to actual buildings are described in several papers.[17,18]

Consider a system with a number of masses "lumped" at particular points, letting m_n represent the nth mass of the system, and assume that the system is vibrating in the jth mode. If the system is vibrating in a steady-state condition, without damping, the displacement u_{nj} of the nth mass can be written in the form

$$u_{nj} \sin \omega_j t \tag{6}$$

The acceleration experienced by the mass during its oscillatory motion is given by the second derivative with respect to time:

$$-\omega_j^2 u_{nj} \sin \omega_j t \tag{7}$$

---

*The following articles are based in part on material from Ref. 14 by permission of the Portland Cement Association.

†Damping of less than 20 percent critical affects the computed periods by less than 2 percent. Hence, nominal damping does not affect the period appreciably.

The negative value of this acceleration, multiplied by the mass m_n, is considered to be a reversed effective force or inertial force applied at the point n. The inertial forces $\overline{Q}_{nj} \sin \omega_j t$ are considered to be applied to the structure at each mass point, where the coefficient of the sine term in the inertial-force expression has the form

$$\overline{Q}_{nj} = m_n \omega_j^2 u_{nj} \tag{8}$$

Since the inertial forces take account of the mass effects, the displacements of the structure due to the forces $\overline{Q}_{nj}$ must be precisely equal to the quantities u_{nj}. Consequently, in order to find the square of the circular frequency for the jth mode ω_j^2, it is necessary merely to find a set of displacements u_{nj} of such magnitudes that forces corresponding to each displacement multiplied by the local mass m_n and by the square of the circular frequency for the jth mode ω_j^2, give rise to the displacements u_{nj}. Any procedure that will establish this condition will give both the modal frequencies and the modal deflection shapes. Multiplying the magnitudes of the modal deflections by a constant does not change the situation, since all the forces, and consequently all the deflections consistent with those forces, will be multiplied by the same constant.

However, it is not possible without other knowledge of the situation to write down directly a correct set of displacements for the jth mode. Therefore, the calculations must make it possible to arrive at these deflections as a result of a systematic method of computation. The most useful procedures, at least for the determination of the fundamental mode, are Rayleigh's method, or modifications thereof, or methods based on a procedure of successive approximations developed originally by Stodola. A description of the successive-approximations procedure follows:

1. Assume a set of deflections at each mass point of magnitude u_{na}. Compute for these deflections an inertial force $\overline{Q}_{na}$ given by

$$\overline{Q}_{na} = m_n u_{na} \omega^2 \tag{9}$$

where the quantity ω is an unknown circular frequency which may be carried in the calculations as an unknown.

2. Apply these forces to the system and compute the corresponding deflections, designated by the symbol u_{nb},

$$u_{nb} = \overline{u}_{nb} \omega^2 \tag{10}$$

3. The problem is to make u_{nb} and u_{na} as nearly equal as possible. To do this, ω may be varied. The value of ω that gives the best fit is a good approximation to the circular frequency for the mode that corresponds to the deflection u_{nb}, which in general will be an approximation to the fundamental mode. In general, u_{nb} will be a better approximation to the fundamental mode shape than was u_{na}.

4. A repetition of the calculations using u_{nb} as the starting point will lead to a new derived deflection that will be an even better approximation.

In most cases, even with a very poor first assumption for the fundamental mode deflection, the process will converge with negligible errors to ω^2 in at most two or three cycles, and one can obtain a good approximation in only one cycle. However, the mode shape will not be so accurately determined unless the calculation is repeated several times.

If the quantity shown in Eq. (11) is made a minimum (in effect minimizing the square of the error between the derived deflection and the assumed deflection) the "best" value of ω^2 consistent with the assumed deflection curve can be determined:

$$\sum_n m_n (u_{nb} - u_{na})^2 = \text{minimum} \tag{11}$$

Substituting Eq. (10) into Eq. (11) and equating to zero the derivative with respect to ω^2 gives

$$\omega^2 = \frac{\sum m_n u_{na} \overline{u}_{nb}}{\sum m_n \overline{u}_{nb}^2} \tag{12}$$

The value of ω^2 given by Eq. (12) exceeds, generally only slightly, the true value for the fundamental mode.

Rayleigh's method is probably the most widely used engineering procedure for computing the period of the fundamental mode. Without modification, however, it does not generally give accurate values of the mode shape. Rayleigh's method for calculating the fundamental frequency of a building frame can be related to the procedure described above by equating u_{na} to unity throughout the structure. In the case of a vertical or horizontal beamlike structure, the derived displacements $\overline{u}_{nb}$ will be proportional to the deflections of the structure due to forces equal to the weight of the structure. Since Rayleigh's procedure using $u_{na} = 1$ gives a quite accurate determination of the fundamental frequency, it is obvious that Eq. (12) will yield a highly accurate value if any more reasonable deflection shape is assumed for the first mode.

The effect of foundation rotation, column shortening, or other contributions to deflection can be readily taken into account in both the successive-approximations procedure and the Rayleigh method.

Example 1 The simplest example of the procedure for computing the period of vibration of a structure is that of the single-degree-of-freedom system. Consider the structure of Fig. 2, which has a single mass m and a spring constant k. For a displacement u_a the inertial force is, from Eq. (9),

$$\overline{Q}_a = mu_a\omega^2 \tag{13}$$

The derived deflection u_b is

$$u_b = \frac{\overline{Q}_a}{k} = \frac{mu_a\omega^2}{k} \tag{14}$$

It is obvious from Eq. (14) that $u_b = u_a$ if $\omega^2 = k/m$, from which the expression for the frequency is

$$f = \frac{\omega}{2\pi} = \frac{1}{2\pi}\sqrt{\frac{k}{m}} \tag{15}$$

and the period is

$$T = \frac{1}{f} = 2\pi\sqrt{\frac{m}{k}} \tag{16}$$

Example 2 Consider the three-degree-of-freedom structure whose relative spring constants and masses are shown in Fig. 12.

For the purposes of this example it is assumed that the structure acts as a shear beam; i.e., the relative deflection of any story is proportional only to the shear in that story. This would be strictly true only if the girders were infinitely stiffer than the columns; but with suitable modifications in column stiffness, as discussed later, the structure can be analyzed quite accurately as a shear beam. The mass m is given in terms of weight divided by the acceleration of gravity. Consequently, the units used in the calculation have to be taken with consistent values. It is apparent from Eq. (16) that, for mass in terms of weight in pounds divided by acceleration in inches per second squared and k in pounds per inch, the period T will be in units of seconds. Another set of consistent units involves displacement in feet, weight in kips, acceleration of gravity in feet per second squared, and spring constant in kips per foot.

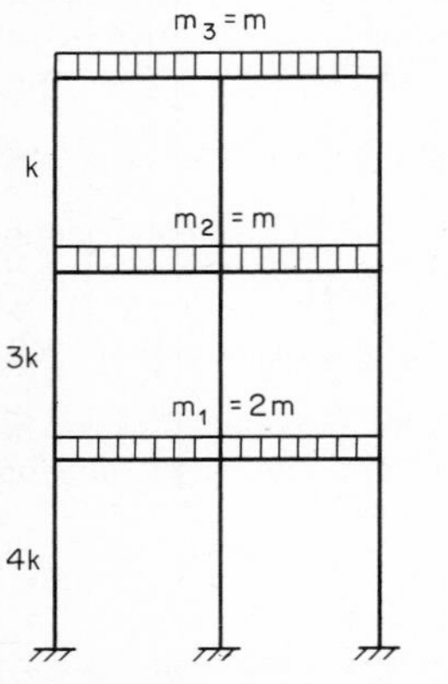

The assumed deflection u_a can have any units whatsoever, and the derived deflection u_b will be in the same units. Consequently, it is convenient to take u_a, or u_{na}, as dimensionless. This does not affect the results and makes it more clear that the modal displacements give the deflection pattern rather than the absolute magnitudes.

Calculations for the fundamental mode are given in Table 2 for an assumed shape of the mode corresponding to deflections at the first, second, and third stories of magnitudes 2, 3, and 3, respectively. The inertial forces $\overline{Q}_{na}$ are computed from Eq. (9) by multiplying the assumed deflections by the corresponding masses. The shears V_n are obtained by summing the inertial forces from the top down, since there is no force at the top. These shears, divided by the spring constants for the corresponding floors, give the increments in story displacement δu_n. From these, by starting with the known zero deflection at the base, one obtains the values of the derived deflection u_{nb}.

The ratios $u_{na}/\overline{u}_{nb}$ give the values of ω^2 at the corresponding mass points for which the derived curve and the assumed curve agree exactly. If ω^2 has the smallest of these values, 0.400, the derived curve lies

Fig. 12 Building frame for Examples 2 and 4.

everywhere inside the assumed curve, while for $\omega^2 = 0.800$ it lies everywhere outside. Any value of ω^2 between 0.400 and 0.800 can make the two curves agree in part. Consequently, the absolute lower and upper limits to the value of ω^2 are 0.400 and 0.800, respectively. (These observations apply only in the case where the assumed curve and the derived curve have no nodal points.)

The pattern of derived deflections 2.5, 4.5, and 7.5 is 1.0, 1.8, and 3.0 relative to a unit deflection at the first mass above the base. If the calculations in Table 2 are repeated for assumed displacements of 1, 2, and 4 the derived displacements are precisely the same. The corresponding (exact) square of the first-mode circular frequency is $\omega_1^2 = 0.5k/m$. Thus, the value $0.517k/m$ determined in Table 2 is seen to be a very good approximation. However, it will be noted that the derived deflection pattern is not nearly so accurate after one cycle, although it is much closer to the first mode pattern than is the assumed pattern.

The calculations for this frame by Rayleigh's method are made in the same way as in Table 2, but for a set of equal (unit) displacements u_{na}. The resulting approximation to the first-mode circular frequency is $\omega_1^2 = 0.532k/m$, which is indeed accurate. However, the absolute lower and upper limits are $0.375k/m$ and $1.0k/m$, which are poorer than those of Table 2, and the derived modal shape, which has relative values of 1, 1.67, and 2.67, is not very good.

8. Higher Modes Several methods are available for computing the frequencies of modes higher than the fundamental mode for a multi-degree-of-freedom system. Two such procedures are described in Ref. 14. However, the high-speed digital computer can determine all the modes and frequencies of even highly complex systems in only a few minutes by use of standard programs, and detailed hand-calculation methods are no longer of interest except for approximating the fundamental frequency or period.

9. Modal Participation Factors Since the mode shapes determined by the methods of Art. 7 are arbitrary, in terms of a uniform multiplier, a *modal participation factor* must be used to determine the actual response of a structure. The general theory is given in Art. 6. In a multi-degree-of-freedom system that has independent or uncoupled modes of deformation (this condition is generally satisfied for buildings), each mode responds to a base excitation as an independent single-degree-of-freedom system. The response of a structure as a function of time can be written in the form

$$\alpha(t) = \sum_j \gamma_j \alpha_j u(t) \tag{17}$$

where $\alpha =$ modal quantity to be determined (e.g., deflection at a particular story, relative story deflection, story shear, acceleration, etc.)
$\alpha_j =$ corresponding modal quantity for jth mode
$\gamma_j =$ modal participation factor defined in Eq. (18)
$u(t) =$ deflection response for a single-degree-of-freedom system subjected to the prescribed base excitation

The modal participation factor is given by

$$\gamma_j = \frac{\displaystyle\sum_n m_n u_{nj}}{\displaystyle\sum_n m_n u_{nj}^2} \tag{18}$$

Because of the complexities in computing the response as a function of time, and because the maximum responses in the individual modes do not necessarily occur at the same time, one is generally interested only in the maximum possible response and the probable response. An upper limit to this maximum is obtained by summing the absolute values of the maximum modal responses:

$$\alpha_{\max} \gtrless \sum_j |\gamma_j \alpha_j D_j| \tag{19}$$

The most probable response (except in the case of closely spaced modes) is given by the square root of the sum of the squares of the appropriate modal quantities:

$$\alpha_{\text{prob}} \simeq \sqrt{\sum_j (\gamma_j \alpha_j D_j)^2} \tag{20}$$

Two equivalent forms of Eq. (19) are convenient to use in certain cases; these give the response of the system in terms of the spectral velocity response V_j or the acceleration response A_j, rather than the displacement response D_j:

$$\alpha_{\max} \gtrless \sum_j \left| \frac{\gamma_j}{\omega_j} \alpha_j V_j \right| \tag{21}$$

TABLE 2 Fundamental Circular Frequency for Frame of Fig. 12

Masses and spring constants	Assumed displacement u_{na}	Inertial force $\frac{\bar{Q}_{an}}{\omega^2} = m_n u_{na}$	Shear $\frac{V_n}{\omega^2} = \Sigma\left(\frac{\bar{Q}_{na}}{\omega^2}\right)$	Increment in story displacement $\frac{\delta u_n}{\omega^2} = \frac{V_n}{\omega^2 k_n}$	Derived displacement $\bar{u}_{nb} = \frac{u_{nb}}{\omega^2} = \Sigma\left(\frac{\delta_{un}}{\omega^2}\right)$	$\dfrac{\text{Assumed displacement}}{\text{Derived displacement}}$ $\dfrac{u_{na}}{\bar{u}_{nb}}$	$m u_{na}\bar{u}_{nb}$	$m\bar{u}_{nb}^2$
m, k ; m, 3k ; 2m, 4k	3	$3m$	$3m$	$3m/k$	$7.5m/k$	$0.400k/m$	$22.5m^2/k$	$56.25m^3/k^2$
	3	$3m$	$6m$	$2m/k$	$4.5m/k$	$0.667k/m$	$13.5m^2/k$	$20.25m^3/k^2$
	2	$4m$	$10m$	$2.5m/k$	$2.5m/k$	$0.800k/m$	$10.0m^2/k$	$12.50m^3/k^2$
						Σ	$46m^2/k$	$89m^3/k^2$

$$\omega_1^2 = \frac{\Sigma m_n u_{na}\bar{u}_{nb}}{\Sigma m_n \bar{u}_{nb}^2} = \frac{46m^2/k}{89m^3/k^2} = 0.517\,\frac{k}{m}$$

$$\alpha_{max} \gtrsim \sum_j \left| \frac{\gamma_j}{\omega_j^2} \alpha_j A_j \right| \tag{22}$$

The equation that is most convenient to use is generally that one in which the spectrum values are most nearly constant for the range of modal frequencies considered. Consequently, Eq. (19) might be used where the spectral displacement is nearly constant, Eq. (21) where the spectral velocity is nearly constant, and Eq. (22) where the spectral acceleration is nearly constant. For example, if the period of the structure in Fig. 12 is very long in all three modes, Eq. (19) might be used, but if the structure is very stiff, so that all three periods are very short, Eq. (22) would be appropriate. In all cases a better estimate of response would be obtained using a square root of the sum of the squares approach as in Eq. (20).

Example 3 The modal participation factors for the frame of Fig. 12 are computed in Table 3.

TABLE 3   Modal Participation Factors, Frame of Fig. 12

Quantity	Mode		
	1	2*	3
ω^2	$0.5k/m$	$2k/m$	$6k/m$
Deflection roof $= u_3$	4	-1	1
Deflection second floor $= u_2$	2	1	-5
Deflection first floor $= u_1$	1	1	3
$\Sigma m_n u_{nj}$	8	2	2
$\Sigma m_n u_{nj}^2$	22	4	44
γ_j	0.364	0.500	0.0454

*Circular frequency and deflection pattern for this mode are computed in Ref. 14.

10. Spring Constant for Equivalent Shear Beam   Many structures can be analyzed as shear beams by suitable modification of the stiffness parameters. Methods of performing the analysis and of modifying the stiffnesses are described here. However, a more accurate analysis, which is perfectly general in applicability, merely requires the determination of the matrix of coefficients relating either (1) the deflections at all mass points due to unit loads at individual mass points or (2) the forces at all mass points required to produce unit displacements at individual mass points. Most accurate analyses must include consideration of shear and flexure, foundation stiffness, and framing interaction effects.

In order to avoid confusing the principles of the earthquake analysis with the details of calculation of structural deflections, only the simple case of the shear beam is discussed here. For a shear beam, the spring constant k for each story is the story stiffness, which is the lateral force required to produce a unit relative lateral displacement of only the story considered.

Frames.   The story stiffness of a moment-resisting frame is the sum of the stiffnesses of the columns in the story. Thus, for a frame with infinitely stiff girders

$$k_{inf} = \sum \frac{12EI}{h^3} \tag{23}$$

The effect of girder flexibility is shown in Fig. 13, where the first-mode deflections of the frame of Fig. 12, with column bases assumed fixed, are shown for rigid girders in *a* and for flexible girders in *b*. Rotation of the joints relieves the column end moments, which reduces the shears and, therefore, the story stiffnesses. The reduced stiffnesses could be determined by computing the column moments for first-mode displacements of the rigid-girder frame and then distributing the moments to account for girder flexibility. The resulting story shears divided by the corresponding relative story deflections give the story stiffnesses.

Third-mode deflections of the frame of Fig. 12 are shown in Fig. 13*c* and *d*. It is evident that the story stiffnesses for this mode differ from those for the first mode if girder flexibility is taken into account. However, it is customary to use first-mode stiffnesses, or approximations to them, for all modes, since the first mode is generally predominant.

The effect of girder flexibility can be determined approximately for a single column by the equation

$$\frac{k}{k_{\text{inf}}} = 1 - \left(\frac{K_{CT}}{\sum K_{JT}} + \frac{K_{CB}}{\sum K_{JB}} \right)$$ (24)

where K_{CT} = stiffness at top of column
  K_{CB} = stiffness at bottom of column
  $\sum K_{JT}$ = sum of all the stiffnesses at joint at top
  $\sum K_{JB}$ = sum of all the stiffnesses at joint at bottom

In deriving Eq. (24) it is assumed that all joint rotations are the same, that the fixed-end moments above and below each joint are equal, and that the fixed-end moments at the top

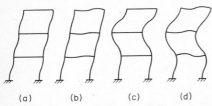

(a)   (b)   (c)   (d)

Fig. 13   Mode shapes of building frame of Fig. 12.

and bottom of each column in a story are equal. In general, this assumption is sufficiently accurate throughout the greater portion of a typical tall building, except in the lower stories and, under certain conditions, in the upper stories. In these regions the story stiffness is sensitive to the relative stiffnesses of columns and girders and generally requires a more accurate determination, particularly if the girder stiffness is appreciably less than that of the columns or the arrangement of beams and columns is irregular. These stiffnesses can be calculated, as outlined previously, by imposing unit displacements on all stories simultaneously and distributing the resulting fixed-end column moments in the stories whose stiffnesses are to be determined.

The contribution of floor systems to stiffness should be taken into account in evaluating girder stiffness.

Shear Walls.   Shearing deformations must be considered in determining the stiffness of a shear panel or wall. Since stiffness is the reciprocal of flexibility, which is defined as the deflection due to a unit force, the stiffness of a wall rotationally restrained at top and bottom is given by

$$\frac{1}{k} = \frac{h^3}{12EI} + \frac{1.2h}{GA}$$ (25)

If only the top (or bottom) is rotationally restrained

$$\frac{1}{k} = \frac{h^3}{3EI} + \frac{1.2h}{GA}$$ (26)

These equations apply only to the uncracked wall. Structures with lateral resistance arising in major part from shear walls cannot be analyzed accurately by the use of the shear-beam concept, but require more accurate treatment.

The above expressions do not reflect the effects of foundation motion, which can be significant in estimating response.

Procedures for estimating the stiffnesses of uncracked shear walls of various shapes, with or without openings, are given in Refs. 19 to 21.

Assemblies.   All vertical resisting elements in a story participate in the story shear provided the upper floor has sufficient stiffness and strength as a diaphragm to distribute the shear among the elements. The elements may be columns, bents, walls, diagonal bracing, etc. The stiffness of such an assembly is the sum of the stiffnesses of the elements.

Example 4 Story shears for the frame of Fig. 12 are determined for an earthquake corresponding approximately to the 1940 El Centro earthquake (Fig. 3) in this example. The elastic-response spectrum shown in Fig. 14 is obtained using the procedures outlined in Art. 4, and the inelastic acceleration and displacement spectra using the procedures outlined in Art. 5. The elastic-response spectrum corresponds to a maximum ground acceleration of 0.3g and damping of 5 percent of critical. The inelastic spectrum is drawn for a ductility factor of 5.

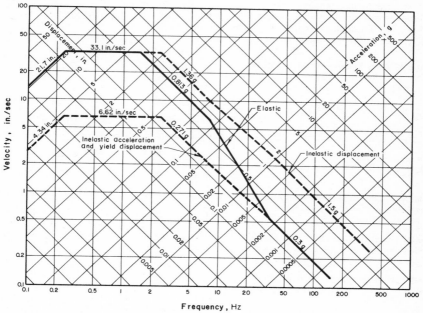

Fig. 14 Response spectrum for Example 4.

The upper-story stiffness k is assumed to be 60 kips/in. and the roof mass to be $m = W/g = 120$ kips/384 in.-sec^2 = 0.313 kip-sec^2/in. Thus, $k/m = 192$ sec^{-2}, and with this value known the three modal frequencies can be determined from the values of ω^2 in Table 3 and the corresponding yield displacement responses read from the elastoplastic spectrum in Fig. 14. The results are shown in Table 4.

TABLE 4 Displacement Response, Frame of Fig. 12

Mode	ω^2, sec^{-2}	ω, sec^{-1}	$f = \omega/2\pi$, Hz	D_j, in.
1	$0.5k/m = 96$	9.8	1.6	0.69
2	$2k/m = 384$	19.6	3.1	0.28
3	$6k/m = 1,152$	33.9	5.4	0.09

Table 5 shows the calculation of the story shears. The modal story deflections are the differences between the modal displacements in Table 3. These are multiplied by the appropriate modal participation factors from Table 3 to obtain the story deflections in inches per inch of spectral displacement, which, when multiplied by the corresponding values of k, give the story shears in kips per inch of modal displacement. These are then multiplied by the appropriate values of D_j from Table 4 to obtain the story shears.

The true response of systems with a small number of degrees of freedom is only slightly less than the sum of the *absolute* values of the modal maxima. These sums are given in Table 5. For comparison, there are also given the square root of the sum of the squares of the values of the modal shears. It is noted that the values computed are consistent with a response spectrum drawn for a ductility factor of 5 and an earthquake intensity corresponding to the El Centro earthquake, for a damping of about 5 percent critical.

DESIGN

11. General Considerations In the design of a building to resist earthquake motions the designer works within certain constraints, such as the architectural configuration of the building, the foundation conditions, the nature and extent of the hazard should failure or collapse occur, the possibility of an earthquake, the possible intensity of earthquakes in

TABLE 5   Story Shears, Frame of Fig. 12

Quantity	Mode	Story		
		2-R	1-2	G-1
Stiffness, kips/in......................	...	60	180	240
Relative modal deflection $u_n - u_{n-1}$.....	1	2	1	1
	2	-2	0	1
	3	6	-8	3
Relative modal deflection, in./in. of spectral displacement $= \gamma_j(u_n - u_{n-1})$	1	0.728	0.364	0.364
	2	-1	0	0.500
	3	0.272	-0.364	0.136
Modal shear, kips/in. of spectral displacement $= k\gamma_j(u_n - u_{n-1})$	1	44	66	87
	2	-60	0	120
	3	16	-66	33
Modal shear for values of D_j in Table 4, kips $= D_j k\gamma_j(u_n - u_{n-1})$	1	30	46	60
	2	-17	0	34
	3	1	-6	3
Max possible story shear, kips..................		48	52	97
Square root of sum of squares, kips..............		34	46	69

the region, the cost or available capital for construction, and similar factors. He must have some basis for the selection of the strength and the proportions of the building and of the various members in it. The required strength depends on factors such as the intensity of earthquake motions to be expected, the flexibility of the structure, and its ductility or reserve strength before damage occurs. Because of the interrelations among flexibility and strength of a structure, and the forces generated in it by earthquake motions, the dynamic design procedure must take these various factors into account. The ideal to be achieved is one involving flexibility and energy-absorbing capacity which will permit the earthquake displacements to take place without unduly large forces being generated. To achieve this end, control of the construction procedures and appropriate inspection practices are necessary. The attainment of the ductility required to resist earthquake motions must be emphasized.

12. Effects of Design on Behavior A structure which is designed for very much larger horizontal forces than are ordinarily prescribed will have a shorter period of vibration because of its greater stiffness. The shorter period results in higher spectral accelerations, so that the stiffer structure may attract more horizontal force. Thus, a structure designed for too large a force will not necessarily be safer than a similar structure based on smaller forces. On the other hand, a design based on too small a force makes the structure more flexible and will increase the relative deflections of the floors.

In general, yielding occurs first in the story that is weakest compared with the magnitudes of the shearing forces to be transmitted. In many cases this will be near the base of the structure. If the system is essentially elastoplastic, the forces transmitted through the yielded story cannot exceed the yield shear for that story. Thus, the shears, accelerations, and relative deflections of the portion of the structure above the yielded floor are reduced compared with those for an elastic structure subjected to the same base motion. Consequently, if a structure is designed for a base shear which is less than the maximum value computed for an elastic system, the lowest story will yield and the shears in the upper stories will be reduced. This means that, with proper provision for energy absorption in the lower stories, a structure will, in general, have adequate strength provided the design shearing forces for the upper stories are consistent with the design base shear. The Uniform Building Code (UBC)[22] recommendations are intended to provide such a consistent set of shears.

A significant inelastic deformation in a structure inhibits the higher modes of oscillation. Therefore, the major deformation is in the mode in which the inelastic deformation predominates, which is usually the fundamental mode. The period of vibration is effectively increased, and in many respects the structure responds almost as a single-degree-of-freedom system corresponding to its entire mass supported by the story which becomes inelastic. Therefore, the base shear can be computed for the modified structure, with its fundamental period defining the modified spectrum on which the design should be based. The fundamental period of the modified structure *generally* will not be materially different from that of the original elastic structure in the case of framed structures. In the case of shear-wall structures it will be longer.

It is partly because of these facts that it is usually appropriate in design recommendations to use the frequency of the fundamental mode, without taking direct account of the higher modes. However, it is desirable to consider a shearing-force distribution which accounts for higher-mode excitations of the portion above the plastic region. This is implied in the UBC and Structural Engineers Association of California (SEAOC)[23] recommendations by the provision for lateral-force coefficients which vary with height. The distribution over the height corresponding to an acceleration varying uniformly from zero at the base to a maximum at the top takes into account the fact that local accelerations at higher levels in the structure are greater than those at lower levels, because of the larger motions at the higher elevations, and accounts quite well for the moments and shears in the structure.

13. Design Lateral Forces   Although the complete response of multi-degree-of-freedom systems subjected to earthquake motions can be calculated, it should not be inferred that it is generally necessary to make such calculations as a routine matter in the design of multistory buildings. There are a great many uncertainties about the input motions and about the structural characteristics that can affect the computations. Moreover, it is not generally necessary or desirable to design tall structures to remain completely elastic under severe earthquake motions, and considerations of inelastic behavior lead to further discrepancies between the results of routine methods of calculation and the actual response of structures.

The UBC and SEAOC recommendations for earthquake lateral forces are, in general, consistent with the forces and displacements determined by more elaborate procedures. A structure designed according to these recommendations will remain elastic, or nearly so, under moderate earthquakes of frequent occurrence but must be able to yield locally without serious consequences if it is to resist an El Centro-type earthquake. Thus, design for the required ductility is an important consideration.

Ductility factors for various types of construction are difficult to characterize briefly. The ductility of the material itself is not a direct indication of the ductility of the structure. Laboratory and field tests and data from operational use of nuclear weapons indicate that structures of practical configurations having frames of ductile materials, or a combination of ductile materials, exhibit ductility factors μ ranging from a minimum of 3 to a maximum of 8.

The ductility factor to be used for design depends on the use of the building, the hazard involved in its failure, the framing or layout of the structure, and above all on the method of construction and the details of fabrication of joints and connectors. Ductility factors commonly used are 2 to 4 for reinforced-concrete structures and 3 to 6 for steel structures, with lower values in both cases if compression behavior controls. A ductility factor of about 4 to 6 iş implicitly assumed for ordinary structures designed to UBC earthquake requirements.

The Applied Technology Council has developed comprehensive seismic design provisions for buildings, which are intended to provide a basis for building codes in the United

States.[24] Provisions of the 1976 UBC Code are given in Sec. 19, Art. 14. The 1975 SEAOC formula for the seismic base shear is the same as the UBC formula. In general, seismic coefficients have been increased in comparison with earlier values, and more factors are considered in arriving at the base shear. The newer values are sometimes 1.5 to 2 times the values in the older codes.

Other parameters and methods of importance in seismic design are given in Refs. 1, 5, 10 to 12, 25, 26, and 29.

14. Seismic Forces for Overturning Moment and Shear Distribution When modal-analysis techniques are not used in a complex structure, or in one having several degrees of freedom, it is generally necessary to define the seismic design forces at each mass point of the structure in order to be able to compute the shears and moments to be used for design. The method described in the SEAOC Code is recommended for this purpose. It is essentially the following:

1. Compute the total base shear corresponding to the seismic coefficient for the structure multiplied by the total weight.

2. Assign a force $F = 0.07TV$, but not more than $0.25V$, to the top of the structure.

3. Assume a linear variation of acceleration in the structure from zero at the base to a maximum at the top.

4. Multiply the acceleration assumed in 3 by the mass at each elevation to find an inertial force acting at each level.

5. Adjust the assumed value of acceleration at the top of the structure in 3 so that the total distributed lateral forces add up to the total base shear computed in 1.

6. Use the resulting seismic forces, assigned to the various masses at each elevation, to compute shears and moments throughout the structure.

7. The resulting overturning moments at each elevation and at the base may cause tension and compression in the columns and walls of the structure. Provision must be made for these forces.

8. The overturning moment at the base should be considered as causing a tilting of the base consistent with the foundation compliance, and also may cause a partial uplift at one edge of the base. The increased foundation compression due to such tilting should be considered in the foundation design.

15. Damping Energy absorption in the linear range of response of structures to dynamic loading is due primarily to damping. For convenience in analysis the damping is generally assumed to be viscous in nature. Damping levels have been determined from observation and measurement, but show a fairly wide spread. Damping values for use in design are generally taken at lower levels than the mean or average estimated values.

Damping is usually taken as a percentage of the critical damping value. Levels of damping summarized from a variety of sources are given in Refs. 4, 25, and 26. Recommended damping values for particular structural types and materials are given in Table 6. The lower levels of the values are considered to be nearly lower bounds and are therefore highly conservative; the upper levels are considered to be average or slightly above average values, and probably are the values that should be used in design when moderately conservative estimates are made of the other parameters entering into the design criteria.

16. Gravity Loads When structures deform laterally by a considerable amount, the effect of gravity loads can be of importance. In accordance with the recommendations of most codes, the effects of gravity loads on member moments (P-Δ effect) as the structure deforms are to be added directly to the primary and earthquake effects. In general, in computing this effect, one must use the actual deflection of the structure, not that corresponding to reduced seismic coefficients.

17. Vertical and Horizontal Excitation Usually the stresses or strains at a point are affected primarily by earthquake motions in only one direction. However, this is not always the case, and certainly not for a simple square building supported on four columns where, in general, a corner column is affected equally by earthquakes in the two horizontal directions and may be affected also by vertical earthquake forces. Since the ground moves in all three directions, and even tilts and rotates, consideration of the combined effects of all these motions must be included in the design. When the response in the various directions may be considered to be uncoupled, the various components of base motion can be considered separately, and individual response spectra can be determined for each component of direction or of transient base displacement. Calculations have been made for the elastic-response spectra in all directions for a number of earthquakes. Studies by the authors indicate that the vertical response spectrum is not more than about two-thirds the horizontal response spectrum for all frequencies, and it is recommended that this ratio be used in design.

For those parts of structures or components which are affected by motions in various directions, the response may generally be computed by either of two methods. The first involves computing the response for each of the directions independently and then taking the square root of the sums of the squares of the resulting stresses (forces, etc.) at a particular point as the combined response. Alternatively, one can combine the seismic forces corresponding to 100 percent of the motion in one direction with 40 percent of each of the motions in the other two orthogonal directions, and add the absolute values of these effects to obtain the maximum resultant forces, strains, etc. This must be done, in some cases, for each of the three principal directions. In general, this method is slightly conservative.

TABLE 6   Recommended Damping Values*

Stress level	Type and condition of structure	Percentage critical damping
Working stress (no more than about ½ yield point)	a. Vital piping	1–2
	b. Welded steel, prestressed concrete, well-reinforced concrete (only slight cracking)	2–3
	c. Reinforced concrete with considerable cracking	3–5
	d. Bolted and/or riveted steel, wood structures with nailed or bolted joints	5–7
At or just below yield point	a. Vital piping	2–3
	b. Welded steel, prestressed concrete (without complete loss in prestress)	5–7
	c. Prestressed concrete with no prestress left	7–10
	d. Reinforced concrete	7–10
	e. Bolted and/or riveted steel, wood structures with bolted joints	10–15
	f. Wood structures with nailed joints	15–20

*Adapted from Refs. 4 and 5.

A related matter that merits attention is the provision for relative motion of parts or elements having multiple supports.

18. Unsymmetrical Structures in Torsion   Consideration should be given to the effects of torsion on unsymmetrical structures and even on symmetrical structures where torsions may arise accidentally for various reasons, including lack of homogeneity of the structure or because of the wave motions developed in earthquakes. Most codes provide values of accidental eccentricity to account for "calculated" and "accidental" torsion. If analyses indicate greater values, the analytical values should be used.

19. Curtain-Wall Buildings*   The strength and rigidity of buildings with noncalculated filler walls, partitions, and stairs are many times those of the frames which were intended to provide the entire structural resistance. This generally accounts for the relatively good seismic and windstorm history of multistory framed buildings of this type. However, the frames cannot function effectively in lateral resistance under severe shocks until the surrounding rigid materials have failed, perhaps with considerable economic loss.

Buildings without any appreciable lateral resistance except in the frame proper may be subject to large story distortions even in moderate earthquakes in spite of meeting present-day seismic codes. Engineers and architects should not inject rigid but brittle elements into otherwise flexible structures without provision for story distortion, since the nonstructural damage may constitute not only a severe financial loss but also a physical danger to persons in and about such buildings. All brittle elements either should be

*This discussion is largely from Ref. 27.

permitted to move freely within the structure or should be expected to fail, in which case they should be so designed and detailed as to protect building occupants and people on the streets. It should be noted that walls or partition elements floating free of the frame or which fall out under minor distortions do not contribute beneficial damping values or energy absorption.

Rigid bracing members or shear walls in a glass-walled multistory building stiffen the structure against mild earthquakes and wind but also invite greater seismic shears. Unless designed for more shear than code values generally prescribe, such rigid elements should be considered expendable and other provisions made for severe emergencies. Where a ductile moment-resisting frame cannot be provided to resist all the design lateral forces, the design can be accomplished by (1) providing rigid elements to carry the shear for moderate shocks of quite frequent occurrence, and also to supply the necessary rigidity to prevent excessive drift due to normal wind force; and (2) providing a ductile frame to control flexibility, absorb energy, and prevent building collapse in a severe but possible earthquake (of, say, 50-year frequency) in which the rigid elements may fail.

20. Core Walls Core walls enclosing elevators, stairways, duct shafts, etc., at the interior of a building which is not too tall or slender will be stiffer than the framework. While it may be permissible by code to design the core walls for the lateral forces and the framework for the vertical loads if symmetry indicates no torsion, there is an interaction between the elements even without torsion. The polar moment of inertia of buildings in which core walls offer the only significant lateral-resistant elements may be too small to resist accidentally induced torsion, and any structure of considerable height or slenderness requires lateral strength and rigidity where it will do the most good, which is generally at or near the perimeter. For buildings over 13 stories or 160 ft in height, the 1975 SEAOC Code requires a complete space frame capable of resisting at least 25 percent of the seismic load. These limits should generally be followed as a minimum requirement with the realization that frame interaction throughout seismic response must be carefully evaluated.

In buildings which have a combination of frame and shear wall, the interaction between different types of construction must be considered. The difference in pattern of displacement of a frame and a shear wall is shown in Fig. 15. The partition of shear between them must be such as to produce equal deflections of the two. Because of the difference in shape of the curves, it appears that the shear wall will take more than the total shear near the base but will be restrained relatively, in the opposite direction, by the frame in the upper part of the structure. Provision for the interaction must be made if the structure is to behave properly. Moreover, if the shear wall fails during the deflection of the structure, the change in configuration and energy-absorbing capacity must be considered in assessing the overall behavior of the composite structure.

Possible solutions are many and varied. Sometimes the walls are not considered in determining the resistance. Sometimes the connections between wall and frame are designed to permit relative motion, or a complete interaction may be provided for by modifying the shear stiffness of each story, taking into account both frame and wall. As a general rule, it is suggested that tall slender walls preferably be located between columns that can serve as flanges of a vertical girder.

21. Parts of Buildings Three factors are important in the design of attached elements of buildings:

1. The element itself usually has a short period of vibration which may tune in to the high spectral peak accelerations.

2. The damping is often small and the local structural system may completely lack the desirable quality of being statically indeterminate.

3. Many elements, such as parapet walls, are subjected to a high-level building motion of increased accelerations, rather than to the ground motion.

The design of attached elements involves a good deal of difficulty in analysis, but calculational techniques are available. Some of these are described in Ref. 7, where a design simplification is involved in which the response of the attachment is related to the modal response of the structure. This response is affected by the mass of the attachment relative to that of the structure. Where the relative mass is infinitesimal, the response is affected primarily by the damping of the structure and the equipment, but for a finite, even though small, relative mass an effective relative damping is involved which is related to the square root of the equipment-to-structure effective mass ratio.

The studies reported in Ref. 7 and other research indicate that, in general, the maximum response of a light-equipment mass attached to a structure, even when the equipment mass is tuned to the same frequency as the structure, will not exceed the basic response spectrum to which the primary structure responds multiplied by an amplification factor AF defined by

$$AF = \frac{1}{\beta_e + \beta_s + \sqrt{\gamma}} \tag{27}$$

where β_e = proportion of critical damping for equipment
β_s = proportion of critical damping for structure
γ = ratio of generalized mass of equipment to generalized mass of structure with the mode displacement vector taken to have a unit participation factor

The generalized mass for the nth mode $\overline{M}_n$ is defined for either the equipment or the structure as

$$\overline{M}_n = u_n^T M u_n \tag{28}$$

in which M is the mass matrix and u_n the modal displacement vector (for either the equipment or the structure) normalized to a unit participation factor.

Examination of the results of Eq. (27) will show that a mass ratio for equipment to structure of 0.0001 corresponds to an equivalent added damping factor of 1 percent and a mass ratio of 0.001 to an added factor of about 3.2 percent.

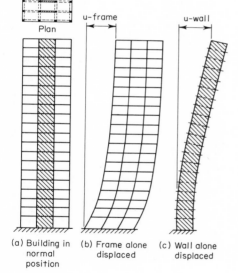

Plan
u-frame
u-wall

Fig. 15 Tall building with moment-resistant frame and shear walls in center interior bay.

(a) Building in normal position
(b) Frame alone displaced
(c) Wall alone displaced

Codes usually specify much larger design coefficients for parts and appendages than for the structure as a whole. These parts must be joined and connected so that there will be strength and inherent ductility whatever the direction of motion. Thoroughness of planning and detailing and the manner in which the parts are joined, or intentionally separated, can do a great deal toward improving seismic resistance.

22. Effects of Soil Conditions The response of a structure to earthquake motions depends on the manner in which it is supported on or in the soil. Interaction between the structure and the soil may allow energy to be absorbed or may result in a motion of the base which differs somewhat from that of the surrounding ground. Thus there is some loss of energy between the ground and the structure, so that the structure generally is not subjected to accelerations as large as those in the earth.

It is shown in Ref. 28 that longer-period motions (say 1 sec or longer) are primarily due to surface waves such as Rayleigh waves or Love waves. It is quite likely, however, that for moderate distances, beyond those corresponding to the depth of focus, surface waves have an important effect even for moderately short-period structures, and more complex motions than those due only to horizontal shears propagated vertically upward must be considered. Moreover, vertical motions cannot be accounted for by a simple horizontal-shear-wave model.

Considerations of variation in intensity of motion with depth beneath the surface are very complex. There are few data that directly relate surface motions to motions beneath the surface; observational data include two or three small earthquakes in Japan. These and other limited data indicate some reduction of surface-motion intensity with depth, but for large motions or high intensities they do not support the contention that variations in intensities of motion with depth can be computed accurately by methods involving only the vertical propagation of a horizontal shear wave.

In regions where unusual types of ground motion can be expected because of oscilla-tions of the soil over deeply buried rock, modifications to the response spectrum must be considered. This is particularly essential in places like Mexico City, where amplification of ground motions in the range of periods from about 2 to 2.5 sec occurs because of the natural frequency of the bowl of soft soil on which most of the city is founded. The Latino Americana Tower in Mexico City was designed for a base shear of the order of 500 metric tons, corresponding to an earthquake of modified Mercalli intensity VIII but taking into account the amplification of motions corresponding to the natural period of vibration of the soil on which the building rests.[18] Shortly after construction was completed, a major earthquake occurred, corresponding to the intensity for which the building was designed. Records were obtained on instruments which had been installed in the building. The values recorded were almost precisely those which had been considered in the design as being consistent with the probable values predicted by a modal analysis.

The forces transmitted to a structure and the feedback to the foundation are complex in nature, and modify the free-field motions. A number of methods for dealing with soil-structure interaction have been proposed. They involve (1) procedures similar to those applicable to a rigid block on an elastic half space; and (2) finite-element or finite-difference procedures corresponding to various forcing functions acting on the combined structure-soil complex.

However one makes the calculation, one normally determines the fundamental fre-quency and higher frequencies of the soil system which interacts with the structure, and the effective damping parameters for the soil system taking into account radiation and material damping. Both these quantities are necessary to obtain rational results, and procedures that emphasize one but not the other cannot give a proper type of interaction.

In general, consideration must be given to the influence of local soil and geologic conditions as they affect the site ground motions, in terms of both intensity and frequency content. Soft soil conditions, for example, may preclude the development of high accelera-tions or velocities within the foundation materials. Consideration must also be given to possible development of unstable conditions such as soil liquefaction, slope instability, or excessive settlements. Further, because of the nature of formation of soil deposits and their lack of uniformity, attention must be given to the determination of the in situ properties and the methods of sampling and testing used to infer these properties. Because of variations in properties and the difficulty of determining them accurately, some variation in the basic parameters used in the calculations must be taken into account.

The method of calculation used should avoid as much as possible the introduction of spurious results arising from the calculational technique. For example, it is often neces-sary to avoid "reflecting" or "hard" boundaries where these do not actually exist.

It is not entirely rational to depend only on calculational methods to modify earthquake motions from some deep layer or bedrock to the surface. It would be desirable to base inferences about site intensity modification on actual observations of surface motions as well as on calculations until measurements of motion become available from actual earthquakes at various depths beneath the surface for a number of different foundation conditions.

23. Detailing and Quality Control Ductility, involving deformations into the inelastic range, is a necessity if structures designed by accepted seismic design procedures are to be capable of resisting earthquakes of the intensity corresponding to those which have been recorded in the United States and elsewhere. Therefore, particular attention must be given to stress concentrations, the choice of types of framing and connection details, and similar matters, in order to ensure that the required ductility can be achieved. The only alternative is to design the structure for greater forces.

Additional ductility or consideration of other energy-absorption capacity may be indicated for special structures, for more severe earthquake risks, or in the upper stories of slender buildings. These ductility ratios can be achieved in most structural materials, but in order to ensure that they will be achieved, attention must be given to appropriate control and inspection procedures during construction, as well as to the factors cited.

The items to which particular care must be given in the design and detailing of reinforced-concrete structures are described in Ref. 14. The following matters need particular attention:

1. In order to avoid compressive failure or crushing of concrete in flexural compression, either a limit must be placed on the amount of tensile reinforcement or compressive reinforcement must be used to give additional strength and ductility on the compressive side. In essence, the difference between the percentages of tensile and compressive steel in a flexural member should not exceed 2 percent of the net section. When compressive steel is used it should be tied into the beam to avoid buckling failures. Such ties are furnished by adequately arranged shear reinforcement.

2. In general, failure of concrete in shear or diagonal tension also involves low ductility, so that appropriate reinforcement must be provided in beams, by either stirrups or inclined bars. However, owing to the fact that flexures in earthquakes generally reverse themselves in direction as the building oscillates, bent-up bars are not usually acceptable unless they are bent up and bent down so as to give reinforcement in the two directions.

3. Further attention is required for concrete columns where the flexure is relatively small compared with the compression. Restraint of the concrete provided by containment, such as that given by spiral reinforcement or closely spaced ties, adds to the compressive strength as well as to the ductility in compression and can generally be used to provide for greater resistance.

4. Appropriate bond and anchorage of reinforcement must be provided in all cases by sufficient lapping of splices, or mechanical connections, welded connections, etc., to avoid failure by loss of anchorage.

5. Tensile forces in reinforced-concrete columns can cause serious difficulties. Such tensions can arise from overall flexure of the building caused by overturning moments. Special reinforcement to resist overturning flexure may be needed in the outer columns of narrow buildings. Recommendations in this regard are given in ACI 318-77, Appendix A.[21]

In steel, problems similar in principle but different in detail must be considered. Connections should be designed to avoid tearing or fracture and to ensure an adequate path for stress to travel across the connection. Because of the possibility of instability by buckling, particularly when steel deforms into the inelastic range, adequate stiffness and restraint of outstanding legs of members must be provided, and the thickness of unrestrained flanges or other elements must be appropriately determined in accordance with applicable plastic-design provisions of building codes such as Part 2 of the AISC Specification and strength-design approaches generally.

In general, columns in the lower stories of either steel or reinforced-concrete buildings should have a reserve of compressive strength for dead load to provide for the additional flexural and overturning compression in the columns under lateral loading in order to ensure appropriate ductility.

Items which do not lend themselves readily to analytical consideration may have an important effect on the response of structures and facilities to earthquake motions and must be considered in the design. Among these are such matters as details and material properties of the elements and components, and inspection and quality control in the construction procedure. The details of connection of the structure to its support or foundations, as well as of the various elements or items within the structure or component, are of major importance. Failures often occur at connections and joints because of inadequacy of these to carry the forces to which they are subjected under dynamic conditions. Inadequacies in properties of material are often encountered and may lead to brittle fracture, even though energy absorption may have been counted on in the design and may be available under static loading conditions. Some of the aspects of these topics are considered in detail for reinforced concrete in Refs. 14 and 29 and for steel in Refs. 30 and 31. Both the designer and the constructor must take into account the requirements for attaining strength and ductility in buildings designed to resist earthquakes.

Surveys of earthquake-damaged buildings throughout the world clearly show that careful attention to the quality of materials and construction significantly enhances the probability that a structure will withstand earthquakes.

24. Cost   The cost of providing earthquake resistance is not a direct function of zone rating or of the seismic coefficient. Earthquake-resistant design, properly done, often provides for wind requirements at little or no additional cost.

REFERENCES

1. Newmark, N. M., and E. Rosenblueth: "Fundamentals of Earthquake Engineering," Prentice-Hall, Inc., Englewood Cliffs, N.J., 1971.
2. Newmark, N. M., and A. S. Veletsos: Design Procedures for Shock Isolation Systems of Underground Protective Structures, vol. III, Response Spectrum of Single-Degree-of-Freedom Elastic and Inelastic Systems, Report for Air Force Weapons Laboratory, by Newmark, Hansen, and Associates under Subcontract to MRD Division, General American Transportation Corporation, RTD TDR 63-3096, 1964.
3. Veletsos, A. S., N. M. Newmark, and C. V. Chelapati: Deformation Spectra for Elastic and Elasto-Plastic Systems Subjected to Ground Shock and Earthquake Motions, *Proc. 3d World Conf. Earthquake Engineering,* New Zealand, vol. 2, 1965.
4. Newmark, N. M., J. A. Blume, and K. K. Kapur: Seismic Design Spectra for Nuclear Power Plants, *J. Power Div. ASCE,* November 1973.
5. Newmark, N. M., and W. J. Hall: Procedures and Criteria for Earthquake Resistant Design, "Building Practices for Disaster Mitigation," National Bureau of Standards, Building Science Series 46, vol. 1, February 1973.
6. Newmark, N. M., and W. J. Hall: Seismic Design Criteria for Nuclear Reactor Facilities, *Proc. 4th World Conf. Earthquake Engineering,* Santiago, Chile, vol. II, 1969.
7. Newmark, N. M.: Earthquake Response Analysis of Reactor Structures, *Nucl. Eng. Des.,* vol. 20, no. 2, July 1972.
8. Newmark, N. M.: Current Trends in the Seismic Analysis and Design of High-Rise Structures, chap. 16 in R. L. Wiegel (ed.), "Earthquake Engineering," Prentice-Hall, Inc., Englewood Cliffs, N.J., 1970.
9. Veletsos, A. S., and N. M. Newmark: Effects of Inelastic Behavior on the Response of Simple Systems to Earthquake Motions, *Proc. 2d World Conf. Earthquake Engineering,* Tokyo, vol. II, 1960.
10. *Proc. 3d World Conf. Earthquake Engineering,* New Zealand National Committee on Earthquake Engineering, New Zealand Institution of Engineers, Wellington, New Zealand, 1965.
11. *Proc. 4th World Conf. Earthquake Engineering,* Chilean Association on Seismology and Earthquake Engineering, Cassillan 2777, Santiago, Chile, 1969.
12. *Proc. 5th World Conf. Earthquake Engineering* (EDIGRAF—Editrice Libraria, 00137, Rome, Italy—Via Giuseppe Chiarini, 6), 1973.
13. Newmark, N. M.: Current Trends in the Seismic Analysis and Design of High-Rise Structures, *Proc. Symp. Earthquake Engineering,* University of British Columbia, Vancouver, September 1965.
14. Blume, J. A., N. M. Newmark, and L. H. Corning: "Design of Multi-Story Reinforced Concrete Buildings for Earthquake Motions," Portland Cement Association, Chicago, Ill., 1961.
15. Jacobsen, L. S., and R. S. Ayre: "Engineering Vibrations," chaps. 8, 9, McGraw-Hill Book Company, New York, 1958.
16. Timoshenko, S., D. H. Young, and W. Weaver, Jr.: "Vibration Problems in Engineering," 4th ed., John Wiley & Sons, Inc., New York, 1974.
17. Blume, J. A.: Period Determinations and Other Earthquake Studies of a Fifteen-Story Building, *Proc. World Conf. Earthquake Engineering,* Berkeley, Calif., 1956.
18. Zeevaert, L., and N. M. Newmark: Aseismic Design of Latino Americana Tower in Mexico City, *Proc. World Conf. Earthquake Engineering,* EERI, Berkeley, Calif., 1956.
19. Derecho, A. T., D. M. Schultz, and M. Fintel: Analysis and Design of Small Reinforced Concrete Buildings for Earthquake Forces, *Eng. Bull. Portland Cement Assoc.,* Skokie, Ill., 1974.
20. Paulay, T.: Some Aspects of Shear Wall Design, *Bull. N. Z. Soc. Earthquake Eng.,* vol. 5, no. 3, September 1972.
21. Building Code Requirements for Reinforced Concrete (ACI 318-77), American Concrete Institute, Detroit, Mich.
22. Uniform Building Code, International Conference of Building Officials, Whittier, Calif., 1976.
23. Recommended Lateral Force Requirements and Commentary, Seismology Committee, Structural Engineers Association of California, 1975.

24. Recommended Comprehensive Seismic Design Provisions for Buildings, Applied Technology Council, San Francisco, Calif., 1977.
25. Newmark, N. M.: Design Criteria for Nuclear Reactors Subjected to Earthquake Hazards, *Proc. IAEA Panel on Aseismic Design and Testing of Nuclear Facilities*, Japan Earthquake Engineering Promotion Society, Tokyo, 1969.
26. Newmark, N. M., and W. J. Hall: Special Topics for Consideration in Design of Nuclear Power Plants Subjected to Seismic Motion, *Proc. IAEA Panel on Aseismic Design and Testing of Nuclear Facilities*, Japan Earthquake Engineering Promotion Society, Tokyo, 1969.
27. Blume, J. A.: Structural Dynamics in Earthquake Resistant Design, *Trans. ASCE*, vol. 125, 1960.
28. Hanks, T. C.: Strong Ground Motion of the San Fernando, Calif., Earthquake: Ground Displacements, *Bull. Seismol. Soc. Am.*, vol. 65, no. 1, 1975.
29. Newmark, N. M., and W. J. Hall: "Dynamic Behavior of Reinforced and Prestressed Concrete Buildings under Horizontal Forces and the Design of Joints (Including Wind, Earthquake, Blast Effects)," Preliminary Publications, 8th Congress, International Association Bridge and Structural Engineering, New York, 1968.
30. Popov, E. P., and V. V. Bertero: Cyclic Loading of Steel Beams and Connections, *J. Struct. Div. ASCE*, June 1973. (See also Errata, *J. Struct. Div. ASCE*, December 1973.)
31. Krawinkler, H., V. V. Bertero, and E. P. Popov: Shear Behavior of Steel Frame Joints, *J. Struct. Div. ASCE*, November 1975.

Section **4**

Fatigue and Brittle Fracture

W. H. MUNSE
Professor of Civil Engineering, University of Illinois, Urbana

FATIGUE OF STRUCTURAL STEEL

1. Significance of Fatigue "The process of progressive localized permanent structural change occurring in a material subjected to conditions which produce fluctuating stresses and strains at some point or points and which may culminate in cracks or complete fracture after a sufficient number of fluctuations."[*]

Fatigue is often considered to be an old-age disease, with failure occurring after a member or structure has been subjected to many applications of loads that may be considerably below the static-load capacity of the member or structure. Thus, the design of a structure for fatigue may be based on stresses that are markedly different from those used for static loadings.

Failures of steel in fatigue are generally thought to be progressive failures starting with dislocations or slip in the crystalline structure of the material, followed by the development of a crack which gradually progresses through the material. The initiation of such failures and the rate at which they propagate vary considerably, depending upon the intensity of stress and a number of other related factors.

The point of initiation of fatigue failures can generally be identified through a careful examination of the fracture surface. They initiate with very little deformation and appear to be of a brittle nature. They generally propagate slowly and intermittently, producing characteristic "oystershell" or "beach" markings (Fig. 1).

2. Fatigue of Structural Steels Fatigue failures were noted by engineers as early as 1829. Since these initial encounters many evaluations have been made of the numerous factors that affect the fatigue behavior of a material, member, or structure. Based on these studies, requirements have been provided in design specifications to protect against such failures.

The principal parameters are the stresses and the number of cycles of these stresses necessary to produce failure. The relationship between them may be shown on an S-N, or Wöhler, curve (Fig. 2). Such a diagram, presented on a log-log basis, portrays the behavior of a specific detail for a given loading condition and can be expressed by

$$F_n = S \left(\frac{N}{n} \right)^K \tag{1}$$

[*]Definition by ASTM E206—62T.

where F_n = fatigue strength computed for failure at n cycles
   S = stress which produced failure in N cycles
   K = slope of the straight-line S-N curve

The curve becomes essentially horizontal at about 2,000,000 cycles, generally referred to as the fatigue limit. Such relationships can be used in design, but they are of greater value in the form of a modified Goodman fatigue diagram (Fig. 3), which provides a relatively

Fig. 1 Typical fatigue fracture in a welded plate.

complete picture of the fatigue behavior of a given connection, member, or structure. The ordinate is the maximum stress, the abscissa the minimum stress (either tension or compression as shown). The radial lines indicate the stress ratio. The curves n_1, n_2, etc., represent failure at various lives. Design stresses can then be presented in the form

$$\sigma_{max} = \frac{\sigma}{1 - K_1 R} \lessgtr \sigma_a \tag{2}$$

where σ_{max} = maximum allowable repeated stress
   σ = stress parameter which is a function of the ultimate strength of the steel, the life, and the type of member
   K_1 = coefficient which is a function of life and the stress ratio
   R = stress ratio (ratio of minimum to maximum stress, taken algebraically)
   σ_a = basic allowable static stress for the member under consideration

To further simplify design, K_1 can be taken equal to 1. This corresponds to a constant range of stress for a given life, and is the basis for many design specifications.

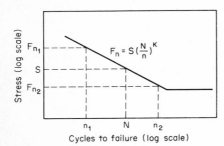

Fig. 2 S-N diagram.

3. Factors Affecting Fatigue Strength   The fatigue behaviors of various types of structures, members, and connections are affected by a variety of factors, many of which produce interrelated effects. Some of the more important parameters are material, rate of cyclic loading, stress variations, residual stresses, size effect, geometry, and prior strain history.

In general, the fatigue resistance of a structural steel is proportional to its ultimate strength. However, this proportionality can vary considerably, or even be nonexistent, depending upon the stress concentrations in the member being considered and the stress cycles to which it is subjected. Small, rotating members of steel (rotating-beam specimens) with polished surfaces generally exhibit a fatigue limit in reversal equal to approximately 50 percent of the ultimate tensile strength of the material. Such a value may

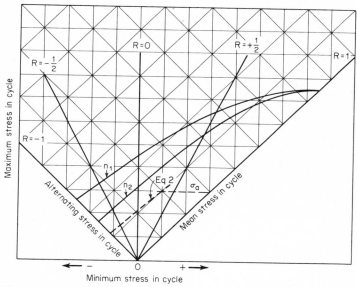

Fig. 3 Modified Goodman diagram for various lives and stress ranges.

be significant in the design of some machine parts. However, when such members contain severe stress concentrations the percentage will be much smaller. Furthermore, corroding environments can be expected to produce further reductions. The effects of geometry and corrosive environment on the relationship between fatigue strength and ultimate strength of steels are shown in Fig. 4.

As-rolled plates and plain rolled members of structural steel also have fatigue strengths proportional to the ultimate strength of the material. This relationship, shown in Fig. 5 for a zero-to-tension loading, may be represented approximately by

$$F_{2,000,000} = 20 + \frac{\text{ultimate tensile strength, ksi}}{4} \tag{3}$$

For other conditions of loading, fatigue resistance may be determined from the following approximate empirical equation:

$$F = \frac{3.0F_r}{1.7 - R} \tag{4}$$

where F_r = fatigue strength for complete reversal of stress
    F = fatigue strength for particular value of R
These equations give an estimate of the basic fatigue strength of the plain material. To relate them to a design or working stress of the form shown in Eq. (2), an appropriate factor of safety must be incorporated.

The rate or frequency of cyclic loading has no significant effect on fatigue strength when the applied stresses are relatively low (long life) and the frequency is less than 3000 cpm. But, if the stresses are high enough to produce plastic deformation with each cycle of loading (low life), an increase in speed of loading will produce an increase in apparent fatigue strength. The magnitude of this effect can be determined only by tests.

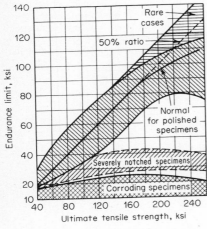

Fig. 4 Relationship between the fatigue limit and ultimate tensile strength of various steels. (*From Ref. 9.*)

Laboratory tests are generally conducted at selected magnitudes of minimum and maximum cyclic stress, but many structures are subjected to maximum and minimum stresses which vary in magnitude from cycle to cycle. Extensive research suggests that many factors affect the manner in which fatigue damage accumulates under this type of loading. Nevertheless, the concept suggested by Palmgren and applied by Miner[3] appears

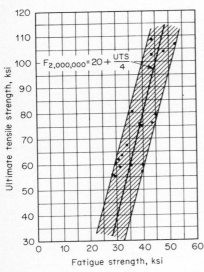

Fig. 5 Relationship between fatigue strength and ultimate tensile strength of plain plates with mill scale on surfaces. Zero-to-tension stressing. (*From Ref. 26.*)

to provide a simple and reasonably reliable prediction of fatigue behavior under random loadings. For failure,

$$\sum_i \frac{n_i}{N_i} = \frac{n_1}{N_1} + \frac{n_2}{N_2} + \frac{n_3}{N_3} + \cdots = 1 \tag{5}$$

where n_i = number of stress cycles at stress level σ_i
N_i = number of stress cycles to produce failure at σ_i

A fourth factor to be considered is the effect of residual stresses. Views differ markedly concerning this factor, and the effect may in fact vary from one instance to another, depending upon the material and geometry of the member, the states and magnitudes of residual and applied stress, and, perhaps, other factors. In general, the introduction of residual compressive stresses at a location that is critical in fatigue can be expected to increase the fatigue resistance at this location for low levels of stress (long life). However, for levels of stress sufficient to cause yielding (short life) the effect of residual stresses will be small or nonexistent. Residual tensile stresses, in general, are found to have little or no effect, unless they can be introduced in such a manner as to reduce the resulting range of stress under service loadings, which can be beneficial. Thus, caution must be exercised in taking advantage of any effect of residual stresses.

Laboratory studies show that the unit fatigue resistance in flexure of relatively small members increases as the diameter decreases; this is a result of the effect of size and the accompanying strain gradient. Under axial loadings the size effect is generally found to be small or nonexistent. In relatively large flexural members (depths greater than 6 in.) the effect of strain gradient is relatively small.

The factor which is generally of greatest importance is the geometry, either internal or external, and the accompanying stress concentration. The magnitude of this effect varies with the sharpness of the change in geometry (the notch radius), the material, and the life or number of cycles to failure. Sharp notches produce very large effective strain concentrations and result in extremely low fatigue resistance. The magnitude of the effective strain concentration for a given detail increases with the ultimate strength of the material (Fig. 4) but decreases as the life decreases, i.e., as the magnitude of the applied stress is increased. Consequently, great care must be used in designing structures that are subjected to repeated loadings, and in the selection of the structural details.

Other variables, such as temperature and rest periods, may also affect the fatigue resistance of a structure. However, these are generally considered to be of secondary importance for most steel structures and are seldom considered in design.

4. Structural Members Fatigue behavior in axially loaded members is markedly different under compressive and tensile loadings. Repeated compressive loadings alone will generally not produce fatigue failures. Fatigue resistance of tensile members is generally controlled by the behavior of their connections. Only if special precautions are taken to protect the connections, or if unusually severe stress concentrations exist in the member itself, will the tension member be critical in fatigue. In this case, fatigue behavior is determined by the factors discussed in Art. 3.

Properly fabricated flexural members, beams, and girders have a high fatigue resistance. However, details such as cover plates, splices, and stiffeners may affect their behavior markedly and must be considered in the design. A summary of the fatigue resistance of rolled beams and built-up beams or girders with various details is given in Table 1 for a zero-to-tension loading cycle. The equations in Arts. 2 and 3 may be used to extend these data to other loading conditions and other lives. Factors of safety must then be applied to obtain design stresses that will protect the structures or members against failure.

5. Riveted Connections Early laboratory studies demonstrated that the fatigue resistance of riveted connections can be relatively low as a result of stress concentrations produced by rivet holes and the bearing of the rivets in the holes. The type of steel generally has little effect, so that there is little or no advantage to high-strength steels under severe fatigue conditions. They may be advantageous only where the number of repeated loads is relatively small and/or the mean stress is relatively high. Variations in joint details and rivet patterns are also found to have little effect.

The clamping force of the fasteners can be quite important. Tight, well-driven rivets are essential to maximum fatigue resistance. They may increase fatigue resistance by as much as 25 percent over that of joints with loose rivets or rivets with minimum clamping force. Nevertheless, design procedures must be based on minimum conditions, and recommendations are based on fatigue strengths of joints with minimum clamping.

6. Bolted Connections The fatigue strength of the ASTM A325 or A490 bolt subjected to direct axial tension is related directly to the clamping force in the fastener. When properly tightened (to the proof load of the fastener) the fatigue limit can be expected to approximate 80 percent of the proof load. However, if the fastener is loose, or loses its clamping force, fatigue resistance may be as low as 20 percent of the proof load.

In many connections in which the fasteners are loaded in tension, there is a prying action which must be taken into account (Sec. 6, Art. 58; Sec. 7, Table 2). The resulting increase in fastener tension may reduce the fatigue limit to as little as 40 percent of the proof load. Thus, eccentricities in the loading of bolted tension connections should be reduced to a minimum, insofar as is practicable.

TABLE 1 Summary of Fatigue Resistance of Beams and Girders*

Member or section	$F_{2,000,000}$, psi	Ratio to plain plate
Structural Carbon Steel (ASTM A7)		
Plain plate (as-rolled condition)....................	31,700	1.0
Rolled I-beam.......................................	31,200	0.98
Rolled I-beam with welded full-length cover plates.....	21,200	0.67
Rolled I-beam with butt-welded splice................	17,200	0.54
Rolled I-beam with riveted partial-length cover plates (based on net section stress)......................	16,900	0.53
Rolled I-beam with welded partial-length cover plates...	10,500	0.33
Welded beam (continuous web-flange welds)...........	23,000–31,500	0.73–1.00
Welded beam with stiffeners.........................	23,000	0.73
Welded beam with butt-welded splice................	20,000	0.63
Welded beam with butt-welded flange transition........	19,000	0.60
Welded beam with welded partial-length cover plate....	12,500	0.39
Low-alloy Steel (ASTM A441)		
Plain plate (as-rolled condition)....................	38,000	1.00
Welded beam (continuous web-flange welds)...........	22,400	0.59
Welded beam with stiffeners.........................	22,000	0.58
Welded beam with butt-welded splice................	22,000	0.58
Welded beam with welded partial-length cover plate....	9,400	0.25
Quenched and Tempered Steel (ASTM A514)		
Plain plate (as-rolled condition)....................	48,100	1.00
Rolled I-beam.......................................	39,000	0.82
Welded beam (continuous web-flange welds)...........	31,800	0.66
Rolled I-beam with butt-welded splice................	21,500	0.45
Rolled I-beam with welded partial-length cover plate...	7,600	0.16

* From Ref. 2.

The clamping force is also important in shear-type connections. Adequate clamping (proof load in A325 or A490 bolts) will provide a fatigue strength in a bolted connection of A36 steel 25 percent greater than that of a similar, properly riveted connection. However, in structures where slip in the connections would be objectionable and must be avoided, frictional resistance rather than fatigue resistance may control the design.

Occasionally it is necessary to use single lap joints rather than butt-type joints. The fatigue resistance of the lap joint is comparable with that of the butt-type joint under tensile loading. However, under compressive loadings the eccentricity in the connection produces a bending which will reduce the fatigue resistance significantly and should be considered in the design.

7. Welded Connections Welded structures can have relatively low resistance to repeated loadings if they are improperly designed and/or fabricated. Extensive data concerning the fatigue resistance of welded structural connections have been accumulated.[2] Of the many factors affecting the fatigue strength of welded structural members and connections, the geometry of the structure or connection and the associated stress concentration is of greatest importance.

The fatigue behavior of welded structural connections varies with the ultimate strength of the materials of which they are fabricated. However, this depends upon the severity of the stress concentrations produced by the welding and the severity of the fatigue loading conditions. Under conditions where a connection will be subjected to high stresses and relatively few applications of loading, the higher-strength materials will prove advantageous. However, under relatively low levels of stress applied a large number of times, and with severe stress concentrations, the benefits of high-strength materials may be relatively small or nonexistent. Thus, the more severe the fatigue conditions the less advantage there is to the use of high-strength material.

An understanding of the flow of stress through the structure or connection is important in designing for resistance to fatigue. Butt-welded joints, because of their relatively smooth flow of stress, provide excellent resistance. However, fillet-welded joints or lap joints, wherein a marked change in direction of stress or load transfer occurs, have relatively low resistance. Thus, care must be exercised in providing proper geometry. Whenever possible, butt-welded connections should be employed and strap plates or doubler plates avoided. Such added material, although it may increase the static strength, will reduce markedly the connection's fatigue resistance. Fatigue resistance of a butt-welded connection may generally be increased by grinding the weld reinforcement flush with the surface of the connected members, provided the weld is sound.

8. Design for Fatigue The principal design specifications[4,5,6] for bridges and buildings include provisions for repeated loadings based on the results of numerous laboratory tests of riveted, bolted, and welded members and connections. For purposes of design, members and connections can be grouped into the 27 details shown in Fig. 6 and described in Table 2.

The fatigue resistances (stress ranges) of the members shown in Fig. 6 are given for a 0.95 level of reliability in Table 3. These values are for constant-cycle loading and can be applied directly only for very limited conditions, since in most instances the load will vary randomly. Four theoretical random-loading conditions are shown in Fig. 7 in terms of a frequency distribution of ratios of range in cyclic stress to the maximum range in cyclic stress. These four conditions are defined in Table 4. The bridge-loading histogram of Fig. 8 is given as an example. Other histograms of highway- and railway-bridge loadings are shown in Refs. 33, 34, and 35.

The allowable fatigue design stress range S_R may be determined from

$$S_R = S_r C_L \qquad (6)$$

where S_r = constant-cycle fatigue stress range for 0.95 reliability and desired life (cycles of loading), Table 3

C_L = loading coefficient to be selected for load type and desired level of reliability, Table 4

Values of C_L are based on a linear log S–log N fatigue relationship, the linear-damage rule of Eq. (5), the loading frequency-distribution functions of Fig. 7, the uncertainty in fatigue data obtained in laboratory tests, and estimated uncertainties for the effects of such other factors as workmanship and fabrication, errors in analysis, and the effects of impact.[36]

The allowable fatigue design stress range is obtained as follows:

1. Identify the type of detail (Fig. 6).

2. Identify the loading frequency distribution which best represents the loading history to which the detail will be subjected during its life (Fig. 7). If no information is available, the type IV distribution can be used and will be conservative.

3. Determine the number of cycles of loading expected during the life of the structure (< 100,000, < 500,000, < 2,000,000, or < 10,000,000).

4. Obtain the basic allowable fatigue design stress range S_r from Table 3.

5. Determine the loading coefficient C_L, based on load type and desired level of reliability, from Table 4.

6. Determine the maximum allowable fatigue design stress range from Eq. (6). This maximum should not exceed the maximum allowable static design stress for the detail in question.

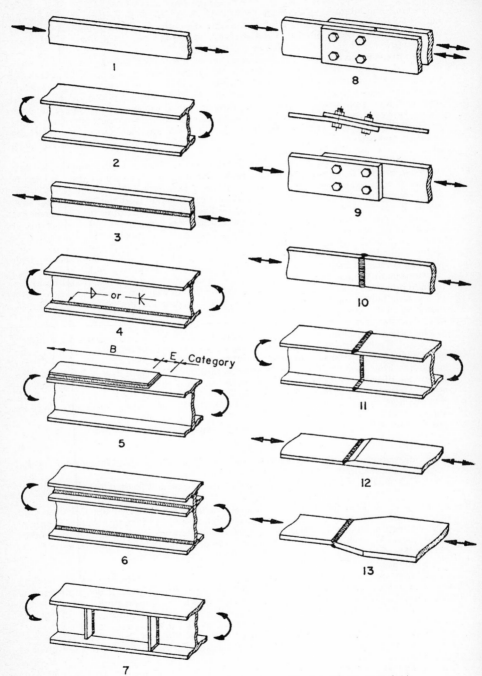

Fig. 6 Structural details for fatigue design requirements of Tables 2 and 3. *(From Ref. 4.)*

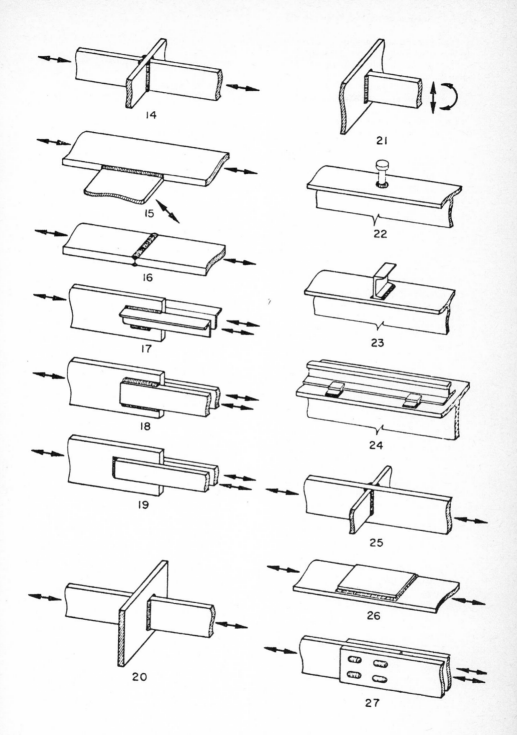

14

15

16

17

18

19

20

21

22

23

24

25

26

27

9. Protecting against Fatigue To protect against fatigue failures in structures subjected to repeated loadings, the designer should:[8]
  1. Avoid details of design that produce severe stress concentrations or poor stress distribution.
  2. Provide gradual changes in the section and avoid reentrant notchlike corners.
  3. Avoid abrupt changes of section or stiffness in members or components.

TABLE 2   Description of Structural Details for Fatigue Design

Type of detail	Example[a] (see Fig. 6)
Plain Material	
• Base metal with rolled or cleaned surface	1, 2
Built-up	
• Base metal and weld metal in members, without attachments, built up of plates or shapes connected by continuous fillet welds or full-penetration groove welds parallel to the direction of applied stress	3, 4, 6
• Base metal at end of partial-length welded cover plates having square or tapered ends, with or without welds across the ends	5
• Calculated flexural stress at toe of welds on girder webs or flanges adjacent to welded transverse stiffeners	7
Mechanically Fastened Connections	
• Base metal at net section of high-strength bolted connections, except bearing-type connections subject to stress reversal, and axially loaded joints which induce out-of-plane bending in connection material	8
• Base metal at net section of other mechanically fastened joints[c]	9[b]
• Shear on fasteners of mechanically fastened joints[c]	9s[d]
Groove Welds	
• Base metal and weld metal, when reinforcement is not removed, in or adjacent to full-penetration groove-welded splices with or without transitions. Transitions must have slopes not greater than 1 to 2½	10, 11, 12, 13
• Base metal or weld metal in or adjacent to full-penetration groove welds in tee or cruciform joints	14
• Base metal at details attached by groove welds subject to transverse and/or longitudinal loading	15
• Weld metal of partial-penetration transverse groove welds, based on effective throat area of the weld or welds	16
Fillet Welded Connections	
• Base metal at ends of intermittent fillet welds	24
• Base metal at junction of axially loaded members with fillet-welded end connections. Welds must be disposed about the axis of the member so as to balance weld stresses	17, 18, 19[b], 20[b]
• Continuous or intermittent longitudinal or transverse fillet welds (except transverse fillet welds in tee joints) and continuous fillet welds subject to shear parallel to the weld axis in combination with shear due to flexure	19s[e], 21
• Transverse fillet welds in tee joints	20s[e]
Miscellaneous Details	
• Base metal adjacent to short (2-in. maximum length in direction of stress) fillet-welded attachments	22, 23, 24, 25
• Base metal adjacent to longer fillet-welded attachments	26
• Base metal at plug or slot welds	27[b]
• Shear on plug or slot welds	27s[e]

[a]These examples are given as guidelines and are not intended to exclude other reasonably similar situations.
[b]This detail provides for stress in the base metal.
[c]Where stress reversal is involved, A307 bolts are not recommended.
[d]This detail provides for shear on the fasteners.
[e]This detail provides for stress on the throat of the weld.

4. Align parts so as to eliminate eccentricities or reduce them to a minimum.
5. Avoid making attachments on parts subjected to severe fatigue loadings.
6. Use continuous welds rather than intermittent welds.
7. Avoid details that introduce high localized constraint.
8. Provide suitable inspection to guarantee proper riveting, adequate clamping in high-strength bolts, and the deposition of sound welds.
9. Provide for suitable inspection during the fabrication and erection of structures.
10. When fatigue cracks are discovered, take immediate steps to prevent their propagation into the structure.

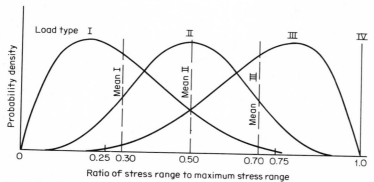

Fig. 7 Loading frequency distributions.

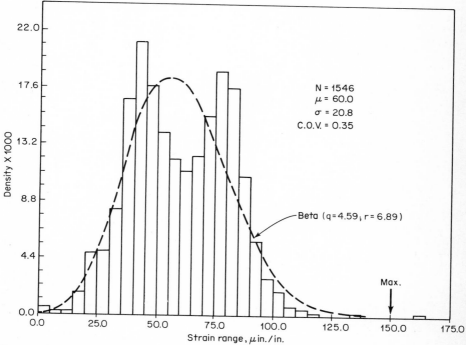

Fig. 8 Histogram for bottom-flange strain range. Beam bridge, heavy truck traffic. *(From Ref. 35.)*

TABLE 3   Basic Allowable Fatigue Design Stress Range S_r, ksi. Constant Cycle, 0.95 Reliability

Detail No. (Fig. 6)	Cycles of loading			
	< 100,000	< 500,000	< 2,000,000	< 10,000,000
1	34.2	29.0	25.2	21.4
2	33.0	25.8	20.8	16.2
3	29.5	22.0	17.1	12.8
4	32.4	18.0	10.9	6.1
5	15.2	9.2	5.9	3.6
6	32.4	18.0	10.9	6.1
7	22.5	14.2	9.6	6.1
8	34.6	28.0	23.4	19.0
9	19.6	15.8	13.1	10.6
9s	27.1	21.8	18.1	14.6
10	23.2	14.4	9.6	6.0
11	25.7	16.9	11.8	7.8
12	20.6	11.8	7.3	4.2
13	27.7	19.4	14.3	10.0
14	21.0	13.2	8.8	5.5
15	15.4	9.7	6.5	4.1
16	18.9	12.2	8.4	5.4
17	15.5	9.7	6.5	4.1
18	9.5	5.0	2.9	1.5
19	16.6	12.7	10.1	7.7
19s	16.4	12.5	9.9	7.6
20	21.7	12.8	8.1	4.8
20s	11.0	7.7	5.7	4.0
21	25.7	20.2	16.4	12.9
22	25.7	14.2	8.5	4.7
23	21.8	13.3	8.7	5.3
24	21.8	13.3	8.7	5.3
25	22.8	12.1	7.0	3.7
26	16.8	10.9	7.5	4.9
27	13.3	9.4	7.0	5.0
27s	14.3	10.0	7.3	5.1

NOTE: Maximum stress must not exceed the allowable static design stress.

TABLE 4   Coefficients C_L for Various Reliability Levels R

Load		R		
Type	Description (see Fig. 7)	0.90	0.95	0.99
I	Primarily light loading cycles: mean range of stress 30% of maximum	2.87	2.50	1.87
II	Medium loading cycles: mean range of stress 50% of maximum	2.07	1.80	1.35
III	Primarily heavy loading cycles: mean range of stress 70% of maximum	1.53	1.33	1.00
IV	Constant loading cycles: stress range constant and equal to 100% of maximum	1.15	1.00	0.75

BRITTLE FRACTURE OF STRUCTURAL STEEL

10. Significance of Brittle Fractures   Brittle failures of structural steels are generally catastrophic, low-ductility fractures that propagate rapidly at relatively low stresses. They can usually be identified by the lack of ductility and the herringbone or chevron appearance of the fracture surface (Fig. 9). The fractures often initiate at nominal stresses equal to or below the yield strength of the material, and propagate at stresses as low as 20 percent of the yield strength.

Fig. 9   Brittle fracture of structural steel.

The most widely publicized brittle fractures have been those of the merchant vessels, liberty ships, and tankers that failed early in World War II.[10] However, many other types of steel structures, primarily of welded construction, have also suffered this type of failure. A brief summary of failures is presented in Table 5. The temperature at the time of failure was generally low, the failures usually initiated at a flaw or some other form of geometrical stress concentration, and the steel was brittle at the service temperature.

Many types of laboratory tests are used to evaluate the susceptibility of structural steels to fracture in a brittle manner. These include the Charpy, drop-weight, notched-tensile, wide-plate, tear, bend, and explosion-bulge tests.[15,16,17] The Charpy V-notch (ASTM E23) and the drop-weight test (ASTM E208) are most widely used for the selection of materials. However, they indicate the behavior of material in a laboratory test, and not necessarily its behavior in a fabricated structure.

The tendency for a material to fracture in a brittle manner is generally defined in terms of *notch toughness* or *transition temperature*. Notch toughness refers to the capacity of a metal to flow plastically under constraint and high local stresses, such as exist at the root of a notch. The transition temperature is the temperature at which a change in fracture behavior takes place; it may be defined in terms of energy absorption, fracture stress, ductility, or fracture appearance (Fig. 10). The ductility transition temperature is generally considered to be a measure of fracture initiation; above this temperature, there will be appreciable plastic flow at the root of a notch before a brittle crack will initiate. The fracture-appearance transition temperature is generally related to crack propagation and is higher than the ductility transition temperature; at this temperature the fracture changes from one which is predominantly cleavage to one which is predominantly a fibrous, shear type.

The temperature at which the transition occurs is also a function of the type of test employed. Figure 11 shows the range of transition temperatures obtained with nine different types of tests on six mild structural steels: a rimmed steel, three semikilled steels, and two fully killed steels.*

---

*Rimmed steels are only slightly deoxidized and are characterized by a rim of low-carbon steel which freezes upon initial contact of the molten metal with the ingot mold, a strong evolution of carbon monoxide gas upon solidification, and a higher-carbon core. In *semikilled steels* the evolution of gas is controlled by limited addition of deoxidizing agents such as silicon and aluminum. *Killed steels* are those in which the carbon-oxygen reaction is completely stopped by the addition of silicon and/or aluminum.

TABLE 5 Examples of Brittle Failures in Structures Other than Ships

Item	Description of structure	Reference
1	Riveted water standpipe at Long Island, New York, 16 to 25 ft diameter by 250 ft high. Oct. 7, 1886. Failed in hydrostatic acceptance test when water reached height of 227 ft	11
2	Riveted molasses tank at Boston, Mass., 90 ft diameter by 50 ft high (overstressed). Jan. 15, 1919	11
3	Welded Vierendeel truss bridge, Albert Canal, Hasselt, Belgium, 245-ft span. Mar. 14, 1938. Failure apparently the result of use of steel with low notch-impact resistance, the existence of high residual stresses, and defective welds with incipient cracks	11
4	Welded plate-girder bridge at Rudersdorf near Berlin, Germany, 17 girder spans totaling 3280 ft. Jan. 2, 1958. Investigators concluded that hardening and residual stresses were responsible	11
5	Welded oil-storage tanks at Fawley, England. Tank 140 ft in diameter by 54 ft high, Feb. 12, 1952. Tank 150 ft in diameter by 48 ft high, Mar. 7, 1952. Failed in hydrostatic tests	11
6	Welded power-shovel dipper sticks, 37 ft long, 20-in.-diameter tubes. 1952. Initiated at stress concentrations caused by abrupt changes in section	11
7	Welded penstock at Anderson Ranch Dam at Boise, Idaho, 15-ft-diameter pipe. Failed during hydrostatic test. Jan. 4, 1950	11
8	Welded gas-transmission lines. Various places in the United States. Individual failures more than 3000 ft long have been reported	11
9	Welded girder bridge, Kings St., South Melbourne, Australia. Multiple-span, multiple-girder bridge. July 10, 1962. Initiated at toe of transverse welds at ends of partial-length cover plates and often extended through entire section	12
10	Caissons on De Long Dock at Thule, Greenland. Material inadequate for service temp. of −40°F. Winter, 1958	13
11	Welded pipe, 150 mm diameter, 45 mm wall thickness. Had been in service 14 years. Failed when temperature dropped from +20 to −45°C	14
12	Eyebar of 1753-ft eyebar-chain suspension bridge. Failed Dec. 15, 1967, after 40 years of service. Temp. 30°F	38
13	Welded plate girder bridge near Pittsburgh, Pa. Crack in 11-ft girder spanning 350 ft. Observed Jan. 28, 1977	39
14	Welded box beam of 66-ft 10-in. span, roof of industrial building in Green Bay, Wis. Failed Jan. 3, 1972, while under construction. Temp. 10°F	

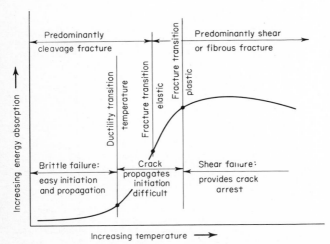

Fig. 10 Transition from ductile to brittle behavior of mild structural steel.

11. Initiation, Propagation, and Arrest   The brittle fracturing of structural steel is often considered in terms of the initiation, propagation, and arrest of the fracture. If initiation can be prevented, there is no problem of propagation or arrest. Initiation occurs when a high stress or strain develops at a point where the material has lost its ductility. The high stress may result from a combination of the applied stress and a local stress concentration,

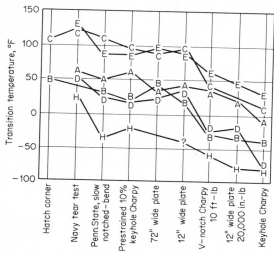

Fig. 11 Transition temperature for six steels. E is a rimmed steel; A, B, and C are semikilled; D and H are fully killed. *(From Ref. 18.)*

residual stresses, and, possibly, dynamic stresses. The loss of ductility or toughness may result from a reduced temperature, a triaxial state of stress, a high strain rate, or an exhaustion of ductility through strain aging, cold working, previous strain history, or hydrogen embrittlement. Various combinations of these factors can be responsible for initiation.

Once initiated, a brittle fracture in mild structural steel may propagate at a tensile stress level as low as 6000 to 8000 psi and at speeds of approximately 4000 to 5000 fps. The elastic energy stored in many structures is of sufficient magnitude to sustain such propagation to complete failure. However, through proper design and the use of appropriate materials, the propagating crack may be arrested.[19] Riveted splices have been used in ships to provide increased deformation capacity and a discontinuity in the crack path. A second procedure is to weld tough strips (strakes) of material in the fracture paths. Either system, if of adequate capacity to absorb the energy released by the propagating crack, will prove effective.

12. Factors Affecting Brittle Fracture   Since the brittle fracturing of structural steels is generally related to a transition temperature, most laboratory studies have been conducted to evaluate the effect of such factors as material, temperature, geometry, state of stress, residual stress, strain rate, strain aging, cold working, strain history, and hydrogen or radiational embrittlement on the transitional behavior.

Material.   The composition of the steel is of major importance and may affect markedly its transition temperature as well as its energy-absorbing capacity. Figure 12 shows the effect of carbon content on the shape and location of the Charpy V-notch transition curve; the lower the carbon content the greater the notch toughness and energy-absorbing capacity. Other elements may also increase or decrease the transition temperature (Fig. 13). For this reason, chemical composition must be considered when selecting a steel for service where brittle fracture is a hazard. Grain size of a material also affects its notch toughness;[22] for a 0.25 percent carbon steel a change from an ASTM fine grain (7 to 8) to

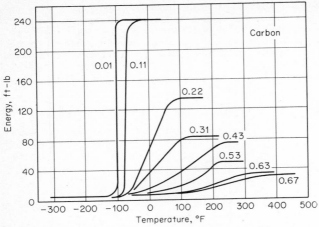

Fig. 12 Effect of carbon on the shape of the Charpy V-notch transition curve. *(From Ref. 20.)*

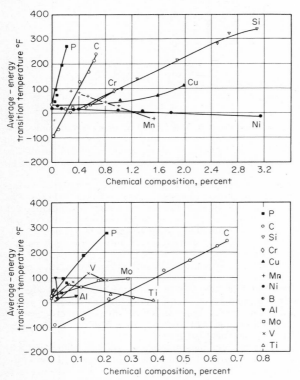

Fig. 13 Effect of chemical composition on average-energy transition temperature. Data adjusted to constant values of carbon, manganese, and silicon except where they are the variables. *(From Ref. 21.)*

coarse grain (2 to 3) increased the transition temperature about 50°F. This accounts for the impairment in notch toughness of the base metal near the heat-affected zone of a weld, and indicates why normalizing at the proper temperature can be used to improve a steel's notch toughness.

Size Effect.   Thick plates require fewer roll passes to produce and the finish rolling temperature is higher than for thin plates. They also cool more slowly. These factors result in a higher transition temperature.

Temperature.   A decrease in temperature increases the strength of a steel and decreases its notch toughness (Fig. 10). Thus, the lowest operating temperature must be known for structures that are to be designed to resist brittle fracture.

Geometry.   The effect of geometry or notches can readily be seen in Fig. 11. The more severe geometries (to the left in the diagram) give the higher transition temperatures; the less severe geometries to the right produce the lowest transition temperatures. Similarly, an increase in width or thickness of a test member generally produces an increase in the transition temperature. Directly associated with the question of geometry is the state of stress. When a high degree of triaxiality is possible (usually thicker material), failure occurs at a lower stress.

Residual Stress.   The failures which have occurred at very low stresses have been in members or structures where there were high residual stresses, generally as a result of welding, in addition to sharp notches or cracks and low temperatures. Although residual stresses per se do not initiate these low-stress brittle fractures, they can have a marked effect on the ease with which the fractures may be initiated. Stress-relief treatments of such members and structures will markedly reduce the residual stresses and improve their low-temperature behavior.[16,23]

Strain Rate.   The rate at which a member is strained will also have a marked effect on its notch toughness. Impact loading of a Charpy V-notched specimen, for example, may provide a transition temperature 100 to 150°F higher than static or slow-bend loading of the same type of member.[24] Consequently, structures subjected to dynamic loadings and low temperatures must be given special consideration in design.

Cold Work.   In general, prior cold working will raise the transition temperature of a steel through the process of strain aging. Mild structural steels strained 5 to 10 percent may have their transition temperatures increased by as much as 90 to 100°F. Thus cold working of these materials is undesirable. Structural steels are listed in Table 6 in descending order of susceptibility to strain aging. Some of these steels are no longer available in the ASTM designations.

TABLE 6   Susceptibility* of Structural Steels to Strain Aging†

Steel	Typical ASTM Grades
Bessemer rimming	A7 (except for bridges)
Open-hearth rimming or semikilled	A7, A30, A36, A113, A131 (A and B), A283, A285
Open-hearth silicon-killed	A201, A212, A7 (over 1½ in.), A36 (over 1½ in.), A373 (over 1 in.), A441
Open-hearth silicon-aluminum-killed	A201, A212, A131C

*In descending order of susceptibility.
†Adapted from Ref. 20.

Strain History.   Since brittle service failures occur either early or late in the life of various structures, their strain history is generally thought to have relatively little effect on their susceptibility to such fractures. However, the failure of the Kings Street Bridge[12] and the failure of laboratory pull heads that had been subjected to a number of loadings indicate that under some conditions repeated loadings may tend to increase the possibility of brittle fracture.

Embrittlement.   Several special treatments have been found to embrittle steel. Although generally of concern only in special applications, hydrogen embrittlement, nitriding, or radiation embrittlements can markedly affect the behavior of steels. Hydrogen will reduce ductility, produce microcracking, and increase the tendency for under-bead cracking;[22,25] nitriding the surface of steel plate will increase the brittleness of the material;[26] and neutron irradiation causes an increase in yield strength and marked decrease in notch toughness.[27] Thus, special precautions may be necessary when the materials in structures are subject to these embrittling processes.

13. Designing against Brittle Fracture   So far as brittle fracture is concerned in steel structures, it is usually sufficient to make sure that the steels meet the requirements of the recognized standard specifications, to proportion the details of the structure with reasonable care, and to insist on good practice in fabrication and erection. However, it is important to note that the lower the expected minimum service temperature, the more urgent these precautions become. Designers who adapt a standard specification to their purposes must remember that whatever assurance the specification gives against brittle fracture is based, in part, on experience with the type of structure and service for which it is written.

Toughness provisions of ASTM A709[1] for bridge steel and the AASHTO specifications ensure suitable materials for the intended service. Materials are obtainable in three classifications, identified as Zones 1, 2, and 3. The corresponding Charpy V-notch requirements are given in Table 7. AASHTO prescribes the following service temperatures:

Zone 1: 0°F and above
Zone 2: −1 to −30°F
Zone 3: −31 to −60°F

These levels of toughness requirements have been established with due consideration of the low strain rates in bridges.

TABLE 7   Supplementary Toughness Requirements: ASTM A709*

	Steel		Zone 1		Zone 2		Zone 3	
Grade	Thickness, in. (mm)		Energy†	Temp.‡	Energy†	Temp.‡	Energy†	Temp.‡
36	To 4 (102) incl.		15(20)	70(21)	15(20)	40(4)	15(20)	10(−12)
50	To 2 (51) incl.		15(20)	70(21)	15(20)	40(4)	15(20)	10(−12)
50	Over 2 to 4 (51 to 102) incl.		20(27)	70(21)	20(27)	40(4)	20(27)	10(−12)
100	To 2½ (64) incl.		25(34)	30(−1)	25(34)	0(−18)	25(34)	−30(−34)
100	Over 2½ to 4 (64 to 102) incl.		35(47)	30(−1)	35(47)	0(−18)	35(47)	−30(−34)

*From Ref. 1.
†Impact-test requirements, ft-lbf (J).
‡Testing temperature, °F (°C).

The AISC specifications do not include specific toughness requirements, since few brittle fractures have occurred in buildings. Some failures which have occurred during construction could have been avoided with tougher steels (or, in some cases, with better design and/or fabrication of details).

Steels for other structures that may be exposed to low temperatures, such as power-transmission poles, oil or gas storage tanks, water storage tanks, and pipelines, should also be selected with appropriate consideration of toughness.

If brittle fracture is considered to be a problem in the case of a structure for which there is no adequate specification or experience, the following recommendations, which are based on Charpy V-notch properties, may be used as a guide.

Temperature.   The minimum and operating temperatures should be determined and used to select a suitable material.

Material.   The material should have adequate strength and notch toughness for the service requirements.

1. To prevent crack initiation (see Fig. 10). Since the design-stress level generally cannot be low enough to prevent initiation at temperatures below the ductility transition, the material should have a ductility transition below the minimum service temperature. The best value of energy absorption to specify for material selection is debatable; however, the 15 ft-lb value is most used. The 15 ft-lb transition can be estimated by the following:[28]

$$T_{15°F} = 168 + 333 \times \%C - 66.6 \times \%Mn - 269 \times \%Si + 210 \times \%Si^2 + 116 \times \%Si$$
$$\times \%Mn - 512 \times \%Al + 2{,}849 \times \%Al^2 + 367 \times \%Al \times \%Si$$
$$- 18.1 \times \text{ferrite grain-size number}$$

2. To provide crack arrest. The fracture-transition plastic (generally about 120°F above the ductility transition) should be below the minimum operating temperature. The usual design-stress levels are generally low enough to permit arrest if this requirement is met. A minimum energy absorption of 45 ft-lb has sometimes been specified at minimum temperature to provide an arresting material.

Design of Details. Since brittle fractures always start at points of stress concentration, the design of details is of great importance. The design may include any or all of the following:

1. Improved geometry. Stress concentrations should be minimized, large radii used at changes in sections, and notches or cracks eliminated.

2. Eliminate flaws or discontinuities. These include laps, seams, cracks, splits, laminations, pits, inclusions, undercut, weld cracks, incomplete fusion, etc.

3. Provide flexibility in the structure to reduce stress concentrations.

4. Reduce residual stresses that increase the possibility of fracture initiation.

5. Use crack arresters. These may be riveted arresters or welded strakes of tough material.

Cracks may initiate at attachments of secondary or nonstructural members and may propagate across a joint and damage the main structural member. For example, some failures in crane booms have originated at attachment of rubbing strips.

Design Stresses. Design stresses based on the elastic properties of structural steels are usually about 60 percent of the yield strength. Such a stress level will generally be adequate to provide arrest if the material toughness requirements noted above for crack arrest are fulfilled.

To protect against crack propagation by stress reduction alone, design stresses should be no greater than 20 percent of the yield strength of the steel (5 to 10 ksi for mild structural steels). For proper design, consideration must be given to all four factors discussed above.

Nil-Ductility Transition Design. An alternative procedure is based on the nil-ductility transition (NDT) as established by the drop-weight test.[29] The NDT temperature is the highest temperature at which a cleavage fracture can be initiated without appreciable deformation at the notch root in a standard drop-weight test (ASTM E208).[1] Table 8 lists the estimated range of nil-ductility temperatures for a variety of steels. The following four design criteria are suggested relative to the fracture-analysis diagram of Fig. 14.[29]

1. NDT temperature criterion. For structures that are not thermally or mechanically stress-relieved, or that may be expected to develop points of local yielding, the lowest service temperature must be above the NDT temperature. This is because, under these conditions, very small flaws may serve as crack initiators.

2. NDT + 30°F criterion. This criterion is based on the crack-arrest temperature (CAT) for stresses on the order of one-half the yield stress. If the lowest service temperature is above the NDT + 30° and the level of stress does not exceed the stated value, flaw size is immaterial, since fractures cannot initiate.

3. NDT + 60°F criterion. This criterion is based on the "fracture-transition elastic" (FTE) temperature, which corresponds to a crack-arrest temperature at yield stress. This criterion applies to high-test pressurization and reactor pressure-vessel service conditions at nozzles due to severe thermal stress.

4. NDT + 120°F criterion. This criterion is based on the "fracture-transition plastic" (FTP) temperature, which corresponds to a crack-arrest temperature at stresses approximating the tensile strength of the steel. It is intended for service requirements involving plastic overload, as in the case of expectation of explosive attack on military structures.

In addition to these notch-toughness criteria, the suggestions given above for the design of details should be followed.

The sharp-crack-propagation concept, originated by Griffith, has been developed to provide a relatively rational design philosophy for the higher-strength, more brittle materials that are considered highly sensitive to flaws.[30,31,37] This approach, generally referred to as fracture mechanics, requires an accurate evaluation of the state of stress, a

TABLE 8   Estimated Ranges in Nil-Ductility Temperatures for Various Steels*

Designation	Description of material	Thickness, in.	Avg or range in NDT, °F†
ASTM A242	High-strength low-alloy struc-tural steel	¾ 1–1½ 2–3	−40 to +30 0 to +70 +20 to +90
ASTM A285 Grade C	Low- and intermediate-tensile-strength carbon-steel plates of flange and firebox qualities	¾–1¼	−20 to +90
ASTM A302 Grade B	Manganese-molybdenum steel plates for boilers and other pressure vessels	2½–4½	+10 to +100
ASTM A441	High-strength low-alloy struc-tural manganese-vanadium steel	¾ 1–1½	−10 to +50 0 to +70
ASTM A514	High-yield-strength, quenched and tempered alloy-steel plates, suitable for welding	2–3 ½–1	+20 to +90 −50 to −100
HY-80 (MIL S-16216)	Quenched and tempered steel (80,000 psi min yield)	1	−120 to −200
HTS (MIL S-16113 Grade HT)	Normalized structural steel (48,000 psi min yield)	¾–1¼	−70 to +40
ABS-C (normalized)	American Bureau of Shipping, hull steel plate	1	−40 to +20

*The values summarized in this table have been reported in various places in the literature. They will vary with chemistry, heat treatment, and thickness of material.[29,32]
†In general the NDT for a normalized material can be expected to be about 50°F lower than that of the as-rolled material.

knowledge of the sizes of all possible critical cracks or flaws, and information regarding the fracture toughness of the material. It has been applied to some special structures,[31,37] but much more work is necessary before it can be used generally.

14. Fabrication The quality of fabrication in steel structures susceptible to brittle fracturing is extremely important. The material must be cut, formed, and assembled with the greatest care.

Cutting:

1. The shearing of plates or punching of holes cold-works the material drastically at the sheared edge or edge of the hole, thus reducing its ductility and notch toughness. The risk is eliminated in punched holes if they are reamed; this requires removing from ¹⁄₁₆ to ¼ in. of the diameter, depending on the thickness of the material and the diameter of the holes.

2. Flame-cut edges of plates, if they are carefully machine-cut, generally provide a satisfactory detail, particularly if the edge is to be welded. However, blowouts can produce harmful irregularities.

3. Notches produced by access openings, changes in section, etc., should be made with as large a radius as is practicable.

Forming:

1. Local forming or cold working can be extremely detrimental. It can cause an exhaustion of ductility and therefore greatly decrease notch toughness.

2. Heating after cold working will often produce strain aging and further enhance the damage from cold work. Steels susceptible to strain aging may be seriously affected if they are heated in the range from about 400 to 850°F, as by galvanizing, adjacent welding, heat straightening, etc., if they have been previously cold-worked by rolling, bending, shearing, punching, straightening, etc.

Assembly:

1. Improper welding can produce harmful flaws and defects.

2. Arc strikes are highly dangerous, and all welders should be made aware of this fact.

3. Small fillet welds on relatively heavy members, like arc strikes, are susceptible to cracking.

4. Preheating can be used effectively to reduce weld cracking and to increase notch toughness of large weldments.

5. Post-heat treatments can be effective in improving notch toughness of large weldments.

6. Peening should generally be avoided. It reduces toughness and impact properties of the weld metal; however, these effects may be eliminated by subsequent weld passes. Peening is not very effective in reducing residual stresses unless the last pass is peened after cooling, but this impairs the impact properties and is not permitted by most specifications.

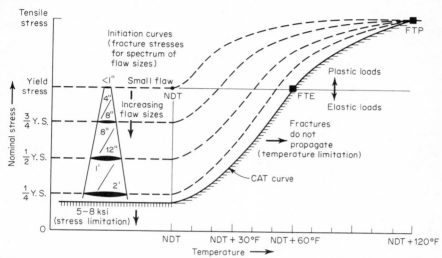

Fig. 14   Generalized fracture-analysis diagram indicating the approximate range of flaw sizes required for fracture initiation at various levels of nominal stress, as referenced by the NDT temperature. *(From Refs. 1 and 29.)*

15. Inspection   Because of the importance of workmanship on the low-temperature performance of structural steels, adequate inspection is essential. Nondestructive techniques that can be used include visual, radiographic, ultrasonic, magnetic-particle, and dye-penetrant procedures (Sec. 6, Art. 32). Destructive methods, such as trepanning, may also be used. However, great care must be exercised in refilling trepan holes.

REFERENCES

1. ASTM Book of Standards: General Testing Methods, Fatigue, Statistical Methods, Appearance of Materials, Sensory Evaluation of Materials and Products, Temperature Measurement; Microscopy, Effect of Temperature on Properties of Metals, part 30, 1976.
2. Munse, W. H.: "Fatigue of Welded Steel Structures," Welding Research Council, New York, 1964.
3. Miner, M. A.: Cumulative Damage in Fatigue, *J. Appl. Mech.*, vol. 12, pp. A159–A164, September 1945.
4. American Institute of Steel Construction: Specification for the Design Fabrication and Erection of Structural Steel for Buildings, 1978.
5. American Railway Engineering Association: Specifications for Steel Railway Bridges, 1977.
6. American Association of State Highway and Transportation Officials: Standard Specification for Highway Bridges, 1977.
7. American Welding Society: Structural Welding Code, D1.1-75.
8. Fatigue of Welded Joints Committee—Welding Research Council: Designing and Making Welded Structural Steel Members for Cyclic Loading, *Welding J. (N.Y.),* vol. 38, *Res. Suppl.,* August 1954.

9. Battelle Memorial Institute: "Prevention of Failure of Metals under Repeated Stress," John Wiley & Sons, Inc., New York, 1946.
10. U.S. Navy Board of Investigation: "The Design and Methods of Construction of Welded Steel Merchant Vessels," Government Printing Office, Washington, D.C., 1947.
11. Shank, M. E.: A Critical Survey of Brittle Failures in Carbon Plate Steel Structures Other Than Ships. *ASTM Spec. Tech. Publ.* 158, 1954.
12. "Report of the Royal Commission into the Failure of Kings Bridge," Melbourne, Australia, 1963.
13. "Fracture of Structural Metals," Watertown Arsenal Laboratories, Watertown, Mass., June 1962.
14. International Institute of Welding: Reports of Brittle Fracture Failures, Commission IX.
15. Parker, Earl R.: "Brittle Behavior of Engineering Structures," John Wiley & Sons, Inc., New York, 1957.
16. Biggs, W. D.: "The Brittle Fracture of Steel," MacDonald and Evans, Ltd., London, 1960.
17. Tipper, C. F.: "The Brittle Fracture Story," Cambridge University Press, New York, 1962.
18. Gensamer, M.: General Survey of the Problems of Fatigue and Fracture of Metals, in W. M. Murray (ed.), "Fatigue and Fracture of Metals," John Wiley & Sons, Inc., New York, 1952.
19. Mosborg, R. J.: An Investigation of Welded Crack Arrèsters, *Welding J. (N.Y.), Res. Suppl.,* January 1960.
20. Shank, M. E. (ed.): "Control of Steel Construction to Avoid Brittle Failure," Welding Research Council, New York, 1957.
21. Rinebolt, J. A., and W. J. Harris, Jr.: Effect of Alloying Elements on Notch Toughness of Pearlitic Steels, *Trans. ASM,* vol. 43, 1951.
22. Stout, Robert D., and W. D'Orville Doty: "Weldability of Steels," Welding Research Council, New York, 1953.
23. Hall, W. J., J. R. Joshi, and W. H. Munse: Studies of Welding Procedures—Part II, *Welding J. (N.Y.), Res. Suppl.,* April 1965.
24. Offenhauer, C. M., and K. H. Koopman: Factors Affecting the Weldability of Carbon and Alloy Steels, *Welding J. (N.Y.), Res. Suppl.,* 1948.
25. Szcepanski, M.: "The Brittleness of Steel," John Wiley & Sons, Inc., New York, 1963.
26. Dvorak, J., and J. Vrtel: Measurement of Fracture Toughness in Low-Alloy Mild Steels, *Welding J. (N.Y.), Res. Suppl.,* June 1966.
27. Faris, F. W.: International Conference on Peaceful Uses of Atomic Energy, Paper 747, Geneva, 1955.
28. Boulger, F. W., and W. R. Hanson: "The Effect of Metallurgical Variables in Ship-plate Steels on the Transition Temperatures in Drop-Weight and Charpy-V-Notch Tests," SSC-145, Dec. 3, 1962.
29. Pellini, W. S., and P. P. Puzak: Fracture Analysis Diagram Procedures for the Fracture-Safe Engineering Design of Steel Structures, *Welding Res. Counc. Bull.* 88, May 1963.
30. Fracture Testing of High Strength Materials, Reports of Special ASTM Committees, Materials Research and Standards ASTM, January 1960; May 1961; November 1961; March 1962; March 1964.
31. Fracture Toughness Testing and Its Applications, *ASTM Spec. Tech. Publ.* 381, April 1965.
32. Pellini, W. S., and P. P. Puzak: Practical Considerations in Applying Laboratory Fracture Test Criteria to the Fracture-Safe Design of Pressure Vessels, *NRL Rept.* 6030, U.S. Naval Research Laboratory, Nov. 5, 1963.
33. Drew, F. P.: Recorded Stress Histories in Railroad Bridges, *J. Struct. Div.* ASCE, December 1968.
34. Cudney, Gene R.: Stress Histories of Highway Bridges, *J. Struct. Div. ASCE,* December 1968.
35. Ruhl, J. A., and W. H. Walker: Stress Histories for Highway Bridges Subjected to Traffic Loading, University of Illinois Urbana, Civil Engineering Studies, Structural Research Series 416, March 1975.
36. Ang, A. H-S., and W. H. Munse: Practical Reliability Basis for Structural Fatigue, ASCE National Structural Engineering Convention, Apr. 14–18, 1975.
37. Rolfe, Stanley T., and John M. Barsom: "Fracture and Fatigue Control in Structures, Applications of Fracture Mechanics," Prentice-Hall, Inc., Englewood Cliffs, N.J., 1977.
38. Collapse of U.S. 35 Highway Bridge, Point Pleasant, W.Va., Dec. 15, 1967, Report NTSB-HAR. 71-1, National Transportation Safety Board, Washington, D.C., Dec. 16, 1970.
39. Cracked Girder Closes I-79 Bridge, *Eng. News-Rec.,* Feb. 10, 1977.

Section **5**

Soil Mechanics and Foundations

Part 1. Soil Mechanics

H. G. LAREW
Professor of Civil Engineering, University of Virginia

NATURE AND CAUSES OF SOIL DEFORMATION

1. Settlement   The principal phenomena which cause footing or foundation movements may be classified as (1) climatic, (2) subsidence, (3) elastic-plastic deformations, (4) shear displacement, and (5) consolidation. Any number of these may be present on a given foundation project, and each is relatively independent of the others; that is to say, each must be considered and dealt with separately. To be safe from one standpoint does not necessarily ensure one's being safe from any of the others.

2. Frost Movements   Climatic effects include movements caused by frost action and by shrinking and swelling of soils. Normally, footing movements caused by frost can be avoided by placing all parts of a foundation below the zone of maximum frost penetration. Except where local conditions such as topography are unusual, Fig. 1 can be used as a *rough guide* in selecting footing depths to avoid foundation movements caused by frost in the United States.

In the case of floor slabs, raft, and footing foundations under cold-storage plants the following precautions or measures are often taken. The cold-storage or locker building is located on natural or imported materials which are not so susceptible to the detrimental effects of frost action. Sites underlain by silts and fine sands are to be avoided if possible. Where this is not feasible, insulating materials are often used between the floor slab, footing system, and the underlying soil. In other instances, a network of perforated or open-jointed pipes is placed in a sand or gravel blanket under the slab and footings, and dry hot air is subsequently and periodically forced through these pipes.

In Arctic regions the depth of freezing may extend several hundred feet into the soil and the placement of foundations below this zone becomes impractical. This has created a multitude of foundation problems in the Arctic, some of which are as yet unsolved. Clean sands and gravels, which are not so susceptible to the detrimental effects of frost action, can be identified from air photographs. Many important buildings in the Arctic are placed on wooden piles which rest in permanently frozen soil. Moreover, these buildings often have an air space between the ground floor and the underlying soil to prevent the thawing of soil around the piles.

3. Shrinkage Another type of soil movement which is normally the result of climatic change occurs in claylike soils which swell in the presence of sufficient moisture and

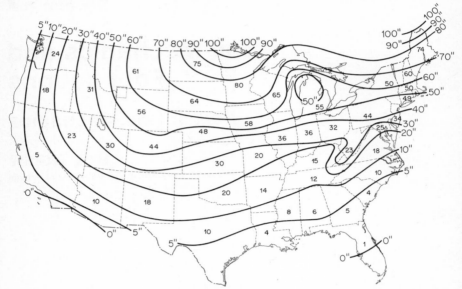

Fig. 1 Extreme frost penetration, inches, based upon state averages. (*National Weather Service, NOAA, U.S. Dept. of Commerce.*)

subsequently shrink upon drying. The problem is usually overcome by placing footings below the zone of seasonal moisture change (3 to 5 ft deep in much of the United States) and by designing for bearing pressures on the soil under the footings which exceed the swell pressure of the clay.

There are some areas in the United States and throughout the world where either the depth of seasonal moisture change exceeds 5 ft and/or the clay's swell pressure exceeds the normal unit loads carried by footings. Certain rather well known areas in the southeastern states, in the Great Plains extending from Texas through Oklahoma and on into Canada, and in most of the western states of the United States are underlain by very highly expansive clays. Frequently these are bentonitic clays which have formed from volcanic ash. These clays are usually characterized by very high plasticity indexes (greater than 30), low shrinkage limits (less than 10), and swell pressures which often exceed 9000 psf. In areas where these highly expansive soils are found, special techniques and foundations have been developed. For example, drilled-in piers, which are founded in soils below the zone of seasonal moisture change and which have their sides isolated from the expansive material which they penetrate, or which employ reinforced bells at the bottom to resist uplift, have been successfully used. Special precautions should be taken to prevent leakage from water lines serving a structure on expansive clay. Moreover, the foundations for furnace rooms on these clays are often structurally isolated from other parts of the building to overcome the detrimental effect of differential settlement caused by the soil's drying and shrinking under these rooms.

4. Subsidence Subsidence caused by the removal of underground oil, water, salt, coal, limestone, and other minerals may cause settlement of structural foundations. Foundation movements caused by subsidence are usually quite difficult and costly to correct. Where there is the slightest evidence to indicate that limestone caverns or mining operations underlie a proposed building site, careful geological reconnaissance, borings, and aerial-photography studies are usually required and justified.

5. Soil Deformation   The construction of fills and structures, lowering of the groundwater table, etc., normally cause an increase in the load carried by an underlying soil deposit. Conversely, excavations and the removal of structures result in a decrease in the intensity of the load. These, or any other changes in loading, will cause such a deposit to deform.

When a soil deposit which is, in part, confined by surrounding soil is subjected to an increasing load from a footing or plate, the soil and footing may settle in a manner similar to one of the curves in Fig. 2. In each of the cases represented, the soil deformation may be characterized by an elastic portion, a plastic portion, and finally a shearing or flow portion. Normally, the elastic portion of soil deformation or settlement is rather small and occurs during construction, so that it can often be neglected. It results from the cumulative elastic deformation of the soil's primary framelike structure and the deformation of individual soil grains comprising this structure. This portion of the total deformation usually can be recovered if the applied load is released. It is of small consequence unless the applied loads are repetitional, as is the case in highway pavements, for example.

As the load is increased beyond the elastic range, a nonrecoverable deformation develops. This portion may be called plastic and can, for many soils, become quite large and significant. Normally, it is larger in the case of soft clays, silts, and loose sands (curve C of Fig. 2). For hard or stiff clays and dense sands it is normally much smaller (curve A).

If the load continues to increase, the deformation and breakup of the soil's natural structure may lead to complete rupture of the soil mass. Present theories of bearing capacity are sufficiently reliable that rupture failures need not occur under building foundations. However, examples of these failures are not uncommon under fills and heavily loaded storage elevators founded on soft clays. Even though two soils (curves A and B of Fig. 2) may not be loaded heavily enough to cause a rupture failure of either, one (curve B) may deform elastically and plastically much more than the other (curve A) even though their ultimate load-carrying capacity is essentially the same.

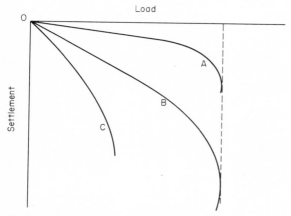

Fig. 2   Load-settlement curves for soils.

6. Time-Dependent Deformation   In the case of soils which are fully saturated (soils located below the groundwater table) there will usually be a time delay between the application of a foundation load and the resulting total elastic-plastic deformation. This delay is caused by the fact that the voids in the soil structure are filled with water, which is relatively incompressible. The water, which undergoes an increase in hydrostatic pressure upon application of the load, cannot quickly escape from a relatively impervious soil. This thwarting of the impending elastic-plastic soil deformation results in a time-dependent deformation which has been called consolidation when the lateral deformation is small in comparison with the vertical movement.

For more permeable materials (clean sands and gravels) the time delay between application of a load and the resulting elastic-plastic deformation is normally insignificant; the deformation occurs during construction. For the more impervious soils (silts and clays) the time required for a consummation of the total elastic-plastic deformation (consolidation) can involve years and even decades. Both the magnitude and time rate of this consolidation can often be estimated by employing principles of soil mechanics.

STRENGTH PROPERTIES OF SOILS

7. Compressive Strength   In many instances engineers need to know the peak strengths, ultimate strengths, and deformation moduli of the soils with which they are working. These strength characteristics should be obtained under the conditions of confinement and loading anticipated during the following stages: (1) preconstruction (excavation or filling), (2) construction, and (3) postconstruction. To reproduce these field conditions exactly in the laboratory is impossible in many instances. This has led to the use of results from simpler tests which are thought to give conservative results. For example, in many practical situations the results of a simple compression test, similar to the test on a concrete cylinder but employing a smaller specimen, may be used to determine the peak strength and deformation modulus of any soil which will cohere enough to retain a given shape (usually cylindrical). The results of such a test may be plotted as shown in Fig. 3 and the peak strength q_u determined. The subscript u indicates that the sides of the sample are not subjected to a confining pressure; i.e., the sample is unconfined. The relative stiffness of the soil may be obtained from the slope of the initial tangent E_i or from the slope of a secant E_s drawn to an arbitrary point on the curve (for example, $q_u/3$).

For claylike soils, Table 1 gives representative values of q_u for six states of consistency. Unconfined values of E_i will usually range from 5 tons/ft² for soft clay to 100 tons/ft² for stiff clays. Values of E_s are commonly 10 to 50 percent less. Normally, the period of loading to failure in these tests ranges from 5 to 20 min. For more rapid or transient rates of loading, both strength and moduli normally show an increase.

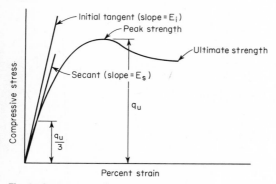

Fig. 3   Stress-strain relationship for soil.

Also shown in Table 1 are the corresponding number of blows N for each of the six states of consistency as obtained from the standard penetration test (Art. 31). On many small- to medium-sized jobs, where the cost of laboratory soil testing may be difficult to justify, engineers frequently use the penetration test as a guide to foundation types and bearing capacities and, at times, the need for additional drilling and laboratory testing.

The unconfined compressive strength of a claylike soil is of practical value in many instances. For example, it can be shown that, for a factor of safety of approximately 3, the allowable bearing capacity q_a of many claylike soils under shallow foundations is essentially equal to q_u, where q_u is an average of several tests run on samples taken within a depth below the footing equal to the width of the footing. This relationship is, for claylike soils, essentially independent of the shape and depth of the footing. However, pressures of this magnitude may cause excessive footing settlements if the clay is normally consolidated, i.e., has never been subjected to a pressure in excess of its present overburden pressure (Art. 8). These normally consolidated clays can usually be identified quite readily, since their natural water contents are closer to their liquid limit than to their plastic limit. The liquid limit is a moisture content for which a remolded soil begins to change from a plastic state into a semiliquid state as its moisture content is increased. The plastic limit is a moisture content for which a remolded soil begins to change from a

TABLE 1   Consistency of Cohesive Soils

Consistency	Unconfined compressive strength q_u, tons/sq ft	Field characteristics	Blows/ft N in standard penetration test
Very soft.........	Less than 0.25	Easily penetrated several inches by fist	Less than 2
Soft.............	0.25–0.50	Easily penetrated several inches by thumb	2–4
Medium.........	0.50–1.0	Can be penetrated several inches by thumb with moderate effort	5–8
Stiff.............	1.0–2.0	Readily indented by thumb but penetrated only with great effort	9–15
Very stiff.........	2.0–4.0	Readily indented by thumbnail	16–30
Hard.............	Greater than 4.0	Indented by thumbnail with difficulty	More than 30

semisolid to a plastic state as the moisture content is increased. These limits are readily obtained by means of the Atterberg limits test. The relationship $q_a = q_u$ is not applicable in the case of claylike soils with unusual secondary structures, i.e., hairline cracks and slickensides (slick internal surfaces). The bearing capacity of these soils is best determined by plate bearing tests (Art. 35).

The unconfined compressive strength is also used to estimate the depth to which temporary vertical cuts can be made in claylike soils. For example, the ultimate or critical depth of a vertical cut H_c in such a soil is given approximately by the formula

$$H_c = \frac{2q_u}{\gamma} \tag{1}$$

where γ is the unit weight of the clay, which will usually range from 115 to 125 pcf. When a factor of safety of 2 is used, the permissible depth of cut is

$$H_A = \frac{q_u}{\gamma} \tag{2}$$

The unconfined compressive strength of claylike soils is often used in analyzing landslides and the stability of slopes with nonvertical sides.

8. Effect of Confining Pressures   Although the confining pressures which are present in an earth mass are often small, or do not affect the strength of the soil appreciably, there are many instances in which they should not be neglected. In other words, the strength determined from an unconfined compressive test may not at all times represent properly the conditions which are present in an earth mass under a structure. For these cases a more realistic laboratory test (triaxial test) has been developed.

Laboratory studies have shown that the shearing and compressive strengths of all soils, except saturated clays in which drainage of water is prevented or does not occur during loading, are affected to a certain degree by the pressures which effectively confine the soil. This is best pictured by plotting the shearing strength of a soil, for various conditions of consolidation pressure, upon a Mohr coordinate field (Fig. 4a). In this figure it is obvious that both the compressive and shearing strengths increase as the confining pressure increases.

The intensity of the effective confining pressure in the earth is not readily determined in many instances. However, it is often related to, and somewhat less than, the effective overburden pressure caused by the weight of the soil directly above the site. Many soils engineers assume that the shearing strength of a soil may be obtained from Fig. 4a by taking AD equal to the full effective overburden pressure. This procedure may not be exact, but it is consistent with the belief that the consolidation pressure σ_3 is often less than the overburden pressure. Where the shearing strength is being evaluated in a homogeneous deposit and along a surface that has a variable depth of overburden, an average depth is often used to determine the overburden pressure and the shearing strength.

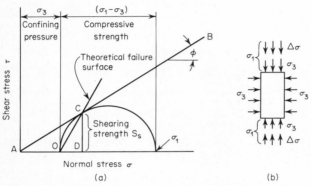

Fig. 4 Strength relationship for sands, silts, and normally consolidated clays.

In the case of granular soils, many silts, and normally consolidated claylike soils which are loaded slowly enough to permit the drainage of water from the soil mass, the relationship between shearing strength and consolidation pressure is essentially linear (AB in Fig. 4a). If a representative sample of soil in a large mass is effectively confined by a pressure σ_3 due to the surrounding soils, an increment of vertical normal stress $\Delta\sigma$ (which might be the result of a foundation load acting some distance above the sample) can be applied before failure is imminent (Fig. 4b). At failure, the magnitude of the resultant stress on the theoretical surface of failure OC is represented by the length of line AC. This resultant stress may be resolved into a normal component σ and a shearing component S_s (AD and CD, respectively). The shearing component can be expressed in terms of the normal component by

$$S_s = \sigma \tan \phi \tag{3}$$

where ϕ = angle of internal friction.

The angle of internal friction is not constant for a given soil or for soils in general but is dependent upon a number of factors, such as void ratio, moisture content, confining pressure, grain shape, grain size, and gradation. In practice, it must be estimated or determined experimentally. For small jobs, the angle of internal friction can be estimated for sands by using Table 2 or the following relationship:

$$\phi = 28.5° + \frac{N}{4}$$

For wet sands, these values should be decreased by 1 to 2°. For gravels and crushed rock with similar states of density, they should be increased from 2 to 6°.

For dry silts and very silty sands values of ϕ are usually 2 to 6° less than those shown in Table 2. This difference is more pronounced as the deposit becomes more dense. For silts and very silty sands below the groundwater table, values of ϕ are, for the great majority of cases, considerably less (one-third to one-half) than the values for dry material. This difference between the dry and saturated silt and silty sands is caused by their low permeability and the buildup of pressure in the pore water during the application of sudden loads to deposits of these soils.

For claylike soils the role played by low permeability and the excess pore pressure caused by superimposed loads is even more pronounced and important. For saturated clays, which have not been preloaded by overburden, or undergone a period of drying,

TABLE 2   Typical Values of ϕ for Dry Sand Composed Primarily of Quartz

Very loose	Loose	Medium	Dense	Very dense
$N < 4$ $\phi < 28.5°$	$5 < N < 10$ $28.5° < \phi < 32°$	$10 < N < 30$ $32° < \phi < 36°$	$30 < N < 50$ $36° < \phi < 41°$	$50 < N$ $41° < \phi < 46°$

the strength relationship is similar to that for sands and silts (Fig. 4a). However, the angle of internal friction for these clays is usually less than that for sand. Where the rate of loading is slow enough to permit full drainage of the pore water and thereby relieve the excess pore-water pressure, ϕ for clays will range from 20 to 30°. Soils under the stacks of an expanding library are believed to be subjected to this type of loading. In most cases, however, the rate of loading of the soil will be more rapid and the excess pore-water pressure will not be fully relieved by internal drainage. In cases of this type the angle of internal friction will range from 10 to 20°.

For clays that have been subjected to loads in excess of their present overburden loads, or for clays which have undergone a period of drying, the strength characteristics may be represented approximately by line AB in Fig. 5. This line is neither exactly straight nor is

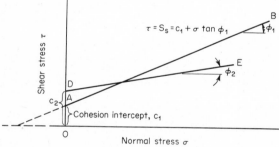

Fig. 5 Strength relationships for preloaded (precompressed) clays.

it fixed in its location even for a given soil. For most practical problems, however, it is sufficiently correct to assume the linear relationship. The position and slope of the line will be influenced by the nature and rate of the loading to be placed upon the soil. Line AB, for example, might represent quite well the strength characteristics of a preloaded or precompressed clay which subsequently is to carry additional increments of building or overburden load which will be applied slowly enough, over a period of years, to permit internal drainage and to prevent the buildup of excess pore-water pressures. The gradual buildup of a fill or slag dump upon a precompressed clay deposit could represent such a loading. If, however, load increments are applied to the same soil more quickly, over a period of weeks or a few months, and drainage is essentially thwarted, line DE would better represent the strength characteristics. The loads applied by most buildings and fills during and immediately after construction are examples of loadings for which DE would best represent the strength characteristics of the underlying precompressed soil.

The two parameters c (cohesion intercept) and ϕ (Fig. 5) are of great importance in determining the strength characteristic of a claylike soil. They are used to estimate the bearing capacity of a soil, to determine the stability of a slope or an earth mass underlying a proposed building site, etc. They are not constant, even for a given soil, and should be evaluated by laboratory tests which duplicate as nearly as possible the most unfavorable field conditions which can be anticipated.

9. Transient and Repeated Loads The structural engineer is often interested in the strength and deformation characteristics of soil under loads which are transient, vibratory, or repeated. Several studies of the effects of each of these loadings have been made in recent years, but no general relationship to cover all possibilities has as yet been developed. However, enough has been learned to indicate the general nature of such a relationship.

Shown in Fig. 6 are typical stress-strain curves for cohesive and cohesionless soils under the action of various types of loading. Gradually applied load curves are generally used in most engineering applications. However, studies have shown that both the stiffness and ultimate strength of soils under the action of repeated loads, where impact is not a factor, are usually less than they are under the action of gradually applied loads. For claylike soils the ultimate strength for repeated loads is usually only 80 to 90 percent of that for gradually applied loads, while for granular materials the range is from 60 to 80 percent. The stiffness of cohesive soils in a repeated-load test, as measured by a slope modulus, ranges from 50 to 95 percent of that obtained in gradually applied load tests. For cohesionless soils the repeated-load modulus is often only about 50 percent of that for gradually applied loads.

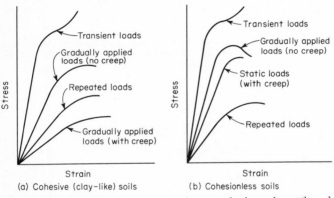

Fig. 6 Probable stress-strain curves for cohesive and cohesionless soils under various loadings.

When the applied loads are sustained over a long period of time, so that creep becomes a factor, there is evidence to indicate that the ultimate strength of saturated claylike soils may be only 30 to 70 percent of that for a gradually applied load which is not sustained. For granular materials these figures are believed to be somewhat greater. The stiffness of saturated clays as obtained from creep tests may range from 20 to 40 percent of that obtained from conventional strength tests in which creep is not a factor.

Where the load duration to failure is measured in a matter of seconds or fractions of seconds, the loads are called transient. Studies thus far have shown that both the stiffness and strength of saturated clays and shales, which are not at the time subject to heavy static loading, will increase as the time of loading to failure decreases. Increases ranging from 0 to 100 percent have been measured for claylike soils. However, the strength and deformation moduli for dry granular materials obtained in transient tests have not been greatly different from those obtained in conventional, gradually applied tests.

FACTORS AFFECTING BEARING PRESSURE

10. Allowable Bearing Pressure The term allowable bearing pressure, as used here, refers to the maximum unit load which can be placed on a soil deposit and not cause any of the following: (1) excessive elastic-plastic deformation of the underlying soil, (2) shear or rupture failure in the soil, and (3) excessive consolidation of the underlying soil, including deeply buried layers of compressible material. These three effects are different and essentially independent phenomena, and each must be considered in determining an allowable bearing pressure. However, in a great many cases involving buildings and similar structures the elastic-plastic soil deformations are found to control the maximum unit load which a soil will carry safely.

11. Permissible Settlement   The amount of differential settlement which can be permitted for various buildings has never been clearly established. Terzaghi and Peck[17] suggest $3/4$ in. as the amount of permissible differential settlement between adjacent columns spaced about 20 ft apart in ordinary buildings. Thornley[46] divides structures into five classes and suggest permissible gross, differential, and net settlements for each (Table 3). Sowers and Sowers[7] have also published a table of limiting settlements based upon structural considerations. Table 4 shows values from that portion of their table which is devoted to differential settlements.

12. Elastic-Plastic Deformation   The source of elastic-plastic deformation normally lies within that depth of soil, below the footing, which is approximately equal to the minimum width of the footing. It is this soil mass which undergoes levels of stress which usually produce the major portion of this type of deformation.

Sand.   On the basis of certain theoretical relationships between settlement, footing width, contact pressures, and the observed settlement of loaded plates and footings placed at the surface of granular deposits, Terzaghi and Peck[6] have prepared a chart which enables one to choose allowable soil pressures which for sands will keep the elastic-plastic settlement to a tolerable magnitude. This chart (Fig. 43) employs the results of the standard penetration-resistance tests (Art. 31). As a rough guide, 10 blows per foot in the standard penetration test produce an allowable pressure of 1 ton/ft^2, 20 blows produce 2 tons/ft^2, etc.

Precompressed Clays.   For most claylike soils which have been precompressed, the elastic-plastic deformations which develop for properly chosen footing loads are likely to be tolerable for most ordinary buildings. The average unconfined compressive strength q_u of samples obtained within a zone equal to the minimum width of the largest footing or foundation is often used as the allowable bearing pressure q_a for these clays. When soil samples are not available for testing, the results of standard penetration tests run in the same zone, though often erratic, are often used to estimate the allowable bearing capacity of precompressed clays. The average of several penetration tests run at intervals of from 2 to 5 ft in depth in the zone of influence below the footing or foundation is used. As a rule, each 8 blows per ft for the penetration-resistance test is equivalent to 1 ton/ft^2 of allowable bearing capacity in precompressed clays. For example, a soil with an average penetration

---

TABLE 3   Structures Classified According to Permissible Foundation Settlements*

Requirements	*Structures*
Class I	
1. Differential settlements under working load must be held within a maximum limit of $1/8$ in.	a. Monumental structures, with interiors or exteriors of marble or other material in which cracking is readily observable
2. Gross settlement under working load, maximum $1/2$ in.	b. Cathedrals and large power plants, steam and hydraulic
As an indication that these requirements will be met—	c. Foundations for heavy machinery, without severe vibration or impact
3. Gross settlement under 200 % working load, maximum $1\frac{1}{2}$ in.	d. Structures designed to remain in serviceable condition for exceptionally long periods, such as monuments
4. Net settlement after application and removal of 200 % working load, maximum limit $1/2$ in.	e. Grain elevators, storage bins, and other structures subject to wide changes in loading lasting over considerable periods
Foundation must develop full capacity by direct bearing without dependence upon transference of load from structure to soil by friction	f. Large concrete tanks
	g. Office buildings, hotels, and stores, all of 10 or more stories in height and all of reinforced-concrete construction or structural steel and concrete construction
	h. Warehouses of multiple-story, heavy-load type of reinforced concrete
	i. Retaining walls

TABLE 3   Structures Classified According to Permissible Foundation Settlements* (Continued)

Requirements	*Structures*

Class II

1. Differential settlements under working load, maximum $\frac{1}{8}$ in.	a. Foundation for machinery causing heavy vibration or impact
2. Gross settlement under working load, maximum $\frac{1}{4}$ in.	b. Buildings housing delicately balanced instruments, such as automatic telephone exchanges, telescopes, testing equipment
As an indication that these requirements will be met—	c. Concrete arches for bridges, hangars, and the like
3. Gross settlement under 200 % working load, maximum $\frac{3}{4}$ in.	
4. Net settlement after 200 % working load released, maximum $\frac{1}{2}$ in.	
Foundation should be capable of developing load capacity by direct bearing without dependence upon friction for transference of load	

Class III

1. Differential settlements under working load, maximum $\frac{1}{4}$ in.	a. Bridges, structural steel or suspension types
2. Gross settlement under working load, maximum $\frac{3}{4}$ in.	b. Steel-frame buildings
As an indication that these requirements will be met—	c. Steel tanks
3. Gross settlement under 200 % working load, maximum $1\frac{1}{2}$ in.	d. Piers and docks
4. Net settlement after 200 % working load has been applied and released, maximum 1 in.	
These structures may be of equal money value with those of Class I and Class II but will be less affected by settlement. They should develop their bearing at least principally by direct loading in or on a noncompressible stratum not underlain by materials of lesser bearing value	

Class IV

1, Differential settlements under working load, maximum $\frac{1}{2}$ in.	a. Factories
2. Gross settlement under working load, maximum $1\frac{1}{2}$ in.	b. Stores
As an indication that these requirements will be met—	c. Apartment buildings
3. Gross settlement under 200 % working load, maximum 2 in.	d. Hotels of less than, say, 15 stories, steel-frame type
4. Net settlement after 200 % working load released, maximum $1\frac{1}{2}$ in.	e. Churches
Loads in these foundation units may be transferred to bearing strata either directly or by friction, or by a combination of both	f. Schools
	g. Warehouses of medium load capacity
	h. Machine shops not housing massive machinery or extra-heavy cranes or gaging devices of high degrees of delicacy
	i. Recreational buildings
	j. Highway structures, grade eliminations

Class V

Permissible settlements vary too widely to tabulate	Temporary structures of all types such as military bridges, falsework for concrete arches, wood-frame buildings, etc., would be included in this class

* From Ref. 46.

resistance of 12 blows per ft would normally be expected to have an allowable bearing capacity of 1.5 tons/ft². Although these rules are approximate, they contain a factor of safety which is normally large enough to limit the maximum elastic-plastic differential settlements to ¾ in. between points approximately 20 ft apart.

Underconsolidated Clays. Settlements of footings placed on normally consolidated or underconsolidated clays will be much larger, and usually excessive, for allowable pres-

TABLE 4   Maximum Allowable Differential Movements for Various Structures*

Type of Structure	*Max Differential Movement*
High continuous brick walls....................	$0.0005-0.001L$
One-story brick mill building, wall cracking.....	$0.001-0.002L$
Plaster cracking (gypsum)......................	$0.001L$
Reinforced-concrete building frame.............	$0.0025-0.004L$
Reinforced-concrete building curtain walls.......	$0.003L$
Steel frame, continuous........................	$0.002L$
Simple steel frame............................	$0.005L$

NOTE: L is the distance between adjacent columns or between any two points that settle differentially.
* From Ref. 9.

sures chosen by the rules for preconsolidated clays. The amount and rate of settlement can frequently be estimated by means of a theory of consolidation described in Arts. 14 and 15. Fortunately, normally consolidated and underconsolidated clays are not often encountered in bridge and building foundation work. They are most likely to be found in coastal areas, former and existing lakes, and alluvial deposits. Where encountered they may be preloaded with fill to reduce future settlement, and vertical sand drains may be employed to speed up the settlement process. In some circumstances an amount of soil which is essentially equal to the weight of the planned structure is removed from the basement area to reduce future settlements to a tolerable amount. In other cases, piles or piers are driven or drilled through the normally consolidated clay layer to reach a more firm and unyielding material below, thereby effectively bypassing the more compressible material.

Silty Soils. For silty soils the allowable bearing capacity, as determined by permissible elastic-plastic deformation, will depend upon the nature of the silt deposit. For medium to dense nonplastic silts or rock flours, the allowable bearing capacity can be determined by assuming that they act much like very fine or silty sands. For medium and stiff plastic silts the allowable bearing capacity can be obtained as for clays. Soft or loose saturated silts are normally unsatisfactory for supporting footings.

13. Shear Failures The second criterion which must be considered when choosing an allowable bearing pressure involves what has been termed a shear or rupture displacement of a mass of soil beneath the footing. The pressure applied to the soil by the footing must not be great enough to cause this displacement. A footing near the surface of a soil deposit and the nature of the soil displacement which occurs when the footing load Q_D becomes excessive are shown in Fig. 7. Both theoretical and experimental studies have

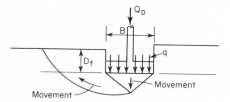

Fig. 7 Shallow continuous footing showing probable mode of shear or rupture displacement.

shown that a small wedge of soil usually forms below a footing and tends to act as a part of the footing. Masses of soil below and on either side of this wedge undergo both lateral and vertical displacement, and this soil displacement is accompanied by a downward movement of the footing when the footing pressures become excessive.

Reasonably reliable solutions have been developed for long, narrow, shallow footings.

For $D_f < 1.5B$, Terzaghi and Peck[17] give the following equations for bearing capacity. For general shear failures, which usually occur in stiff or very stiff clays and dense to very dense sands, the ultimate load Q_D in pounds per foot of footing is

$$Q_D = B(cN_c + \gamma_1 D_f N_q + \tfrac{1}{2}\gamma_2 BN_\gamma) \tag{4}$$

and for local shear failures, which usually occur in very loose sands or soft clays, the ultimate load Q'_D in pounds per foot of footing is

$$Q'_D = B(\tfrac{2}{3}cN'_c + \gamma_1 D_f N'_q + \tfrac{1}{2}\gamma_2 BN'_\gamma) \tag{5}$$

where
B = width of footing, ft
c = cohesive strength of soil, psf
γ_1, γ_2 = unit weights of soil, pcf, above and below the base of footing, respectively
D_f = minimum depth of footing below the adjacent surface, ft
$N_c, N_q,$ and N_γ = dimensionless factors for general shear failures (solid curves in Fig. 8)
$N'_c, N'_q,$ and N'_γ = dimensionless factors for local shear failures (dashed curves in Fig. 8)
For square footings of width B,

$$Q_{ds} = B^2(1.3cN_c + \gamma_1 D_f N_q + 0.4\gamma_2 BN) \tag{6}$$

For circular footings of radius r,

$$Q_{dr} = \pi r^2(1.3cN_c + \gamma_1 D_f N_q + 0.6\gamma_2 rN) \tag{7}$$

In these equations, Q_{ds} and Q_{dr} are total loads for the general shear failure. For a local shear condition these equations must be modified by using $N'_c, N'_\gamma,$ and N'_q values from the dashed curves of Fig. 8 and the factor $\tfrac{2}{3}$ with cohesion values c, as in Eq. (5).

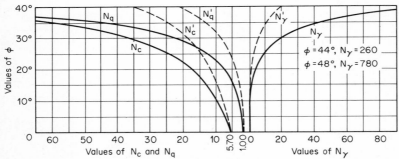

Fig. 8 Relationships between bearing-capacity factors and angle of internal friction. (*After Terzaghi and Peck.*)

A factor of safety of 3 is normally employed when these equations are used to obtain an estimate of the allowable bearing capacity. Although other solutions of this problem have been developed, some of which are slightly more refined, any difference usually becomes academic when one considers the variation in soil strength which can exist in a soil mass, our present methods of evaluating soil strength, and the factor of safety of 3. Moreover, a shear failure is seldom critical when a factor of safety of 3 is used. Only in those cases where a narrow footing ($B < 5$ ft) is at or near the surface of a deposit of loose sand, and for which the groundwater table is at or quite near the surface, is the shear or rupture criterion likely to govern.

14. Consolidation The third criterion which must be satisfied when selecting an allowable bearing pressure involves the time-dependent deformations which may result from the consolidation of soft compressible deposits of soil located below the footing or structure. These consolidation deformations, when combined with any elastic-plastic deformations which may occur, must be kept to a tolerable level.

The amount of settlement or consolidation in a confined layer of normally loaded claylike soil of low to medium sensitivity* may be determined from the following equation:

$$S = \frac{DC_c}{1 + e_0} \log_{10} \frac{p_0 + \Delta p}{p_0} \tag{8}$$

where S = settlement of the compressible layer, ft
  D = thickness of the compressible layer, ft
  C_c = dimensionless compression index of the soil in the compressible layer
  p_0 = present effective overburden pressure, psf, acting on the compressible layer, usually calculated at middepth of the layer
  e_0 = original or field void ratio of the compressible layer where void ratio is the ratio of volume of voids to volume of solids (V_v / V_s)
  Δp = increase in pressure, psf, at the center of the soft compressible layer, caused by footing or building loads, fills, lowering the groundwater table, etc.

Values of D, C_c, and e_0 are obtained by drilling into and through the deposit (noting the thicknesses of various layers) to secure undisturbed samples, and subsequently testing selected and typical samples in the laboratory. The tests usually include the Atterberg limits, natural water content, undisturbed and remolded strength, and consolidation.

The Atterberg limit tests are used to classify the soils and to determine differences in materials at various positions under the structure. The results of these tests, when combined with the natural-moisture-content test results, help to determine whether the soils are normally consolidated or not. For example, normally consolidated clays usually have moisture contents nearer their liquid limit than their plastic limit.

For normally loaded clays which are no more than moderately sensitive, the liquid limit L_w (percent) is related to the compression index C_c by the following equation:

$$C_c = 0.009(L_w - 10) \tag{9}$$

This relationship is used quite often in practice where consolidation tests are not feasible or are thought to be too expensive. It is also used as a check on the value of C_c obtained from the laboratory consolidation test.

Overburden Pressure.   The term p_0 in Eq. (8) is, in the case of normally consolidated clays, the present effective overburden pressure; it is also the maximum pressure p_c to which the deposit has ever been subjected. An estimate of p_c or p_0 can be obtained if one knows the geological history of the area, the unit weight of the soil in each layer, and the location of the groundwater table. A second method of estimating p_c, proposed by Casagrande, involves the following construction on the e–$\log p$ curve of Fig. 9. Point D is located on the laboratory consolidation curve CB where the curvature is maximum. A horizontal line is then drawn from D, and the bisector of the angle between this line and the tangent to the curve at D is found. The straight-line portion of the laboratory curve between D and B is then extended to E, where it intersects the bisector. Directly above E, on the line of original void ratio e_0, the point A can be established, and directly below E the value of p_c can be determined.

Determining Δp.   The term Δp in Eq. (8) represents the increase in pressure at the middepth of the compressible layer caused by lowering of the groundwater table, by proposed building loads, etc. In any normal situation where the groundwater table above the compressible layer is to be lowered, either permanently or for an extended period, Δp will be equal to $h\gamma_w$, where h is the depth in feet that the groundwater table will be lowered and γ_w is the unit weight of water. The commonest type of construction or building load is the uniform load. Where this type of loading can be assumed, it is often sufficiently accurate to estimate Δp by assuming that the surface load is distributed uniformly, at each level below the surface, on an area that becomes larger with depth as determined by the angle θ (Fig. 10). This angle is arbitrarily chosen; the value 30° is often used (Boston Code method) and is believed to be conservative in most cases.

---

*Sensitive soils are those whose undisturbed compressive strength is greater than four times the remolded compressive strength. If this ratio exceeds 8, the soil is called extra-sensitive. Remolding involves the complete breakup and mixing of the soil at unaltered water content.

Taylor[51] gives the following equations for Δp:

For square footings,

$$\frac{\Delta p}{q} = \left(\frac{B/d}{B/d + 2 \tan \theta} \right)^2 \tag{10}$$

For long narrow footings,

$$\frac{\Delta p}{q} = \frac{B/d}{B/d + 2 \tan \theta} \tag{11}$$

Taylor gives curves for these pressures when $\theta = 30°$ (Figs. 11, 12). Either the figure or the equations may be used to estimate the value of Δp upon or within a compressible layer beneath either a single isolated footing or a group of footings, if the pressure beneath each footing is uniformly distributed. The combined effect of a series of isolated footings $\Delta p_1 + \Delta p_2$ (Fig. 10) may be obtained by assuming that the pressure from each footing is dispersed and distributed as shown, and by then combining their separate effects for a selected location in the compressible layer.

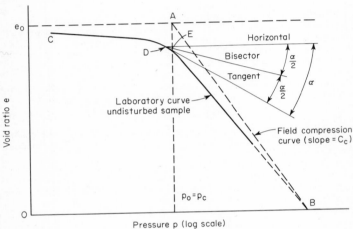

Fig. 9 Typical plot of void ratio vs. log pressure for normally loaded insensitive clay.

More sophisticated methods for estimating Δp, based upon elastic relationships, have been developed by Boussinesq, Westergaard, and others and are widely used in the case of soil deposits. Curves for Δp under the *center* of the loaded area, for both the Boussinesq and Westergaard solutions, are shown in Figs. 11 and 12.

In the preceding discussion of Eq. (8), only the normally consolidated clay, for which the present overburden and preconsolidation pressures are equal, i.e., $p_0 = p_c$, was considered, since it is for this case that the equation is most reliable. For precompressed or overconsolidated soils, where the preconsolidation pressure p_c exceeds the present overburden pressure p_0, settlements predicted by Eq. (8) may be several times greater than the actual settlements depending upon the ratio of $\Delta p/(p_c - p_0)$. Terzaghi and Peck[17] indicate that the actual settlement will be only one-fourth to one-tenth of the predicted value [Eq. (8)] if $\Delta p/(p_c - p_0)$ is less than 50 percent. For the underconsolidated case ($p_c < p_0$), predicted settlements may be considerably smaller than the actual settlements. For these two cases, as well as for the extra-sensitive clays, a quantitative estimate of settlement is difficult to make.

The settlement observed for a given unit pressure under a small-plate load test (Art. 35) will not be the same as that which will develop under a much larger plate or footing on the same soil even though the contact pressures are the same. Therefore, one must extrapolate the test data from the small-plate load test. The following approximate equations may be used:

For sands and gravels,

$$S = S_1 \frac{4B}{B + 2} \tag{12}$$

For more claylike soils,

$$S = S_1 B \tag{13}$$

where S = estimated settlement, in., under footing of width B, ft
S_1 = settlement, in., under a 1-ft-square plate loaded with the design unit load for the footing

The plate load test does not account properly for time-dependent deformation (consolidation) in many cases, namely, where soils are relatively impervious and essentially saturated.

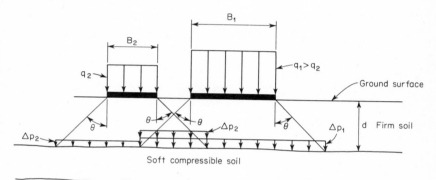

Fig. 10   Pressures on the surface of a buried stratum.

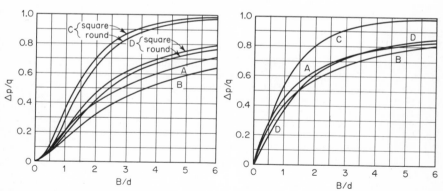

Fig. 11   Pressures on buried strata under round or square footings. (A) Boston Code; (B) Kögler; (C) Boussinesq center stress; (D) Westergaard center stress.

Fig. 12   Pressures on buried strata under long wall footings. (A) Boston Code; (B) Kögler; (C) Boussinesq center stress; (D) Westergaard center stress.

15. Time Rate of Settlement Terzaghi's theory of consolidation may be used to predict time rate of consolidation. This theory assumes (1) one-dimensional consolidation wherein the soft compressible layer is so restrained by the confining layers above and below that it cannot move laterally, i.e., the net deformation within the compressible layer is vertical; moreover, the movement of water from the compressible layer is assumed to be vertical; (2) a completely saturated and homogeneous layer of compressible material; (3) constant values for certain soil parameters which actually vary with pressure; (4) that Darcy's law is valid for the flow of water through soil; (5) an ideal linear relationship between pressure and void ratio; (6) that secondary consolidation which occurs to varying degrees in most soils is negligible; (7) that small samples of soil tested in the laboratory will be representative of large masses of soil in the field; and (8) certain other assumptions which are normally believed to be of lesser importance.

The differential equation governing the time rate of consolidation is

$$\frac{\partial u}{\partial t} = C_v \frac{\partial^2 u}{\partial^2 z} \tag{14}$$

In this equation u is a function which represents the excess pore-water pressure, z is a function which represents the distance to a point in the layer, C_v is the coefficient of consolidation as obtained from a laboratory consolidation test, and t represents time. One solution of primary interest is

$$U = 1 - \sum_{m=0}^{m=\infty} \frac{2}{M^2} e^{-M^2 T} \tag{15}$$

In this expression, U is the average percent consolidation within the layer, T a time factor, and M a dimensionless number which is a function of m, an integer. This has been solved; values of U and T are listed in Table 5. Although the T vs. U relationship is a function of the shape of the initial excess-pore-pressure-distribution curve throughout the depth of the compressible layer, and the conditions of drainage, the values shown in Table 5 are adequate for most typical cases.

TABLE 5   Values of Average Consolidation U and Corresponding Time Factor T

U	0.1	0.2	0.3	0.4	0.5	0.6	0.7	0.8	0.9
T	0.008	0.031	0.071	0.126	0.197	0.287	0.403	0.567	0.848

With the theoretical relationship between the time factor T and the average percent of consolidation U known and evaluated, the time required for a given percent of consolidation to occur in a compressible layer may be calculated by

$$t = \frac{TH^2}{C_v} \tag{16}$$

In this expression, t is the time required to reach a certain average percent consolidation U. The term H is the half thickness of the compressible layer when there are drainage surfaces at both top and bottom of the layer, or the full thickness where a drainage layer is located at only one of the surfaces, top or bottom, of the compressible deposit. The term C_v is the coefficient of consolidation obtained from a laboratory consolidation test. It is not a constant for a given soil but varies with each increment of pressure. T is the dimensionless time factor corresponding to the percent consolidation and may be obtained from Table 5. If time t is expressed in years, H in feet, and C_v in cm²/sec, Eq. (16) becomes

$$t = 29.5 \times 10^{-6} \frac{TH^2}{C_v} \tag{17}$$

Predictions of settlement of typical and critical points under a structure can be made by constructing curves similar to that in Fig. 13. For example, assume that the ultimate settlement predicted by Eq. (8) is 4.3 in.; then, for a settlement of 2.15 in., $U = 0.50$. The corresponding value of T is 0.197 (Table 5). Using this value, and the appropriate value of C_v, in Eq. (17) gives $t = 4.8$ years, which is one point on the time-settlement curve. Additional points may be obtained in a similar manner. Actual measured column footing settlements may be subsequently plotted upon the time-settlement curve, and although the predicted and actual settlement rates may not be the same, the theoretical curve can often be fitted to the actual curve to establish a more reliable projection of the future rate of settlement.

CUTS AND EMBANKMENTS

16. Slope Stability *Cohesionless materials,* such as uncemented clean sands and gravels, will stand at their angle of repose, which is approximately the angle which develops when a pile of the loose material is formed by pouring from a container held a few inches above the pile. Angles of repose of 28 to 35° are commonly used for clean sands and gravels. The angle of repose is essentially independent of time and the depth of excavation or embankment.

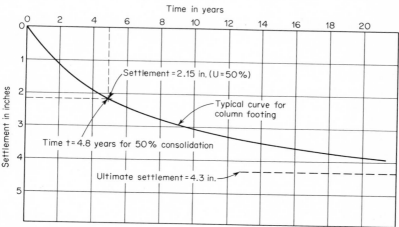

Fig. 13 Time-settlement curve.

Very loose and saturated deposits of granular materials are often quite unstable under the action of transient, blast, and shock loads. Slope angles one-fifth to one-third of those indicated above may be required for stability of these materials, unless the deposit can be made denser by vibration prior to construction. Vibrations caused by pile driving are often used.

For *cohesive materials* such as clays and sandy or silty clays, the concept of angle of repose cannot be employed since the angle or slope to which a cohesive material will stand is a function of the depth of excavation or embankment. The depth to which a cut *above the groundwater table* can be made in a cohesive soil depends upon the shearing strength and unit weight of the soil, the slope angle, and time. The role played by time in the slope-stability problem has not been quantitatively evaluated except for a few isolated deposits. However, it is known that the shearing strength of a soil is affected by time. In some soils (for example, precompressed and certain marine-deposited clays) the strength may decrease with time, while in other cases (for example, certain normally consolidated clays and recent fills) it may increase with time. From the standpoint of stability of the soil it is normally wise to complete construction in a temporary excavation and backfill as quickly as possible. Slopes of permanent excavations, such as highway cuts, are made flatter than those of temporary excavations. Both these practices reflect an appreciation for the influence of time in slope-stability problems.

The stability of slopes of embankments and cuts in homogeneous deposits of cohesive soils bounded by a plane horizontal top surface and a plane inclined slope and underlain by a horizontal surface of a firm stratum (Fig. 14) can be determined by using Fig. 15 or 16. The effects of seepage are not directly accounted for in the solutions on which these figures are based. Given the proposed depth of cut (or height of embankment) H, the overall depth of the deposit to the firm horizontal layer $n_d H$, the unit weight of the soil γ, the slope angle β, and the strength parameters c and ϕ (Art. 8), the stability factor N_s is determined from Fig. 15 or 16.

The cohesive strength required to provide a factor of safety of 1 can be computed from the following equation:

$$c_{\text{reqd}} = \frac{\gamma H}{N_s} \tag{18}$$

This value can then be compared with the available cohesive strength. The factor of safety will depend upon the length of time which the cut must stand open, the threat of losses in life and to adjacent property which might result from failures, the reliability of the soil-strength data, and the probability that the top surface may need to carry live loads. For temporary cuts a factor of safety of from 1.5 to 2.0 is often used in ordinary situations. A larger factor of safety is normally required for permanent cuts since the effects of climate, time, etc., play a greater role.

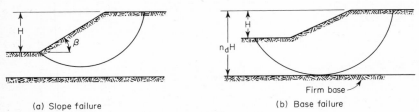

(a) Slope failure   (b) Base failure

Fig. 14 Types and positions of critical slide circles. *(After Terzaghi and Peck.)*

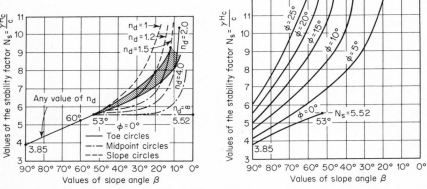

Fig. 15 Relationship between stability factory N_s, slope angle β, and depth factor n_d for a purely cohesive ($\phi = 0$) soil. *(After D. W. Taylor.)*

Fig. 16 Relationship between stability factor N_s, slope angle β and ϕ for a cohesive soil ($\phi = 0$ and $c \neq 0$). *(After D. W. Taylor.)*

Three positions of the failure surface were considered in preparing Figs. 15 and 16: (1) toe surfaces (circles) which pass through the base of the slope or toe, (2) surfaces which break out on the slope above the toe, and (3) midpoint or deep surfaces which break the ground beyond the base or toe of the cut or embankment. When $\phi = 0$ the failure surface will always pass through the toe if $\beta \geqslant 53°$. When $\phi = 0$ and $\beta \leqslant 53°$ any of the three types of failure may occur, depending upon the value of β and the depth factor n_d. Figure 15 shows that the failure surface will be (1) a slope circle for combinations of β and n_d that fall above the shaded area, (2) a toe circle for combinations that fall within the shaded area, and (3) a midpoint or deep circle for combinations that fall below the shaded area. For $\phi > 0$ (Fig. 16) the chance of failures being other than toe failures is quite small; hence the factor n_d loses much of its importance.

The limitations on Figs. 15 and 16 often seriously restrict their use in practice. When unusual site conditions exist which are not covered by these figures, a more involved stability analysis is required.

COMPACTION AND PERMEABILITY

17. Compaction Soils used as backfill around structures and buried utilities, support for pavements and structures, embankments for highways, railroads, parking areas, levees, dams, etc., and linings for canals and reservoirs are compacted so that their strength and stability will be increased, their permeability decreased, their resistance to frost action and erosion enhanced, and their compressibility decreased.

On projects where compaction is anticipated, representative samples of typical soils at the site should be obtained at the time subsurface explorations are being conducted, and subjected to one or more of the following laboratory tests: Atterberg limits, laboratory compaction, natural water content, swell, and permeability.

The laboratory compaction test is performed to determine an optimum moisture content and dry density for each soil. The standard Proctor tests (AASHTO T99, ASTM D698, or British Standard 1377) are generally specified where light to medium rollers are employed. The modified Proctor test (AASHTO T180, ASTM D1557) is usually used where greater compactive effort is required.

Shown in Table 6 are typical values of standard Proctor optimum moisture content and density for various soils. Optimum moisture content for the standard test is usually 2 to 5 percentage points less than a soil's plastic limit, while for the modified Proctor it is usually 5 to 8 percentage points less. Modified Proctor optimum densities are approximately 7 to 10 pcf greater than standard Proctor values.

TABLE 6   Typical Values of Optimum Moisture Content and Density for Various Soils

Unified classification		Range of max dry unit weight, pcf	Range of optimum moisture content, %
Group symbol	Description		
GW	Well-graded clean gravels, gravel-sand mixtures	125–135	11–8
GP	Poorly graded clean gravels, gravel-sand mix	115–125	14–11
GM	Silty gravels, poorly graded gravel-sand-silt	120–135	12–8
GC	Clayey gravels, poorly graded gravel-sand-clay	115–130	14–9
SW	Well-graded clean sands, gravelly sands	110–130	16–9
SP	Poorly graded clean sands, sand-gravel mix	100–120	21–12
SM	Silty sands, poorly graded sand-silt mix	110–125	16–11
SM-SC	Sand-silt clay mix with slightly plastic fines	110–130	15–11
SC	Clayey sands, poorly graded sand-clay mix	105–125	19–11
ML	Inorganic silts and clayey silts	95–120	24–12
ML-CL	Mixture of inorganic silt and clay	100–120	22–12
CL	Inorganic clays of low to medium plasticity	95–120	24–12
OL	Organic silts and silt-clays, low plasticity	80–100	33–21
MH	Inorganic clayey silts, elastic silts	70–95	40–24
CH	Inorganic clays of high plasticity	75–105	36–19
OH	Organic clays and silty clays	65–100	45–21

Shown in Table 7 are typical compaction requirements for various structures based upon the standard Proctor laboratory test. Where highly expansive soils are encountered or when large percentages (exceeding about 20 to 30 percent) of gravel are present in the soil, special compaction requirements will be necessary.

With the start of (and during) construction, field density tests (AASHTO T147) should be made on the following bases:

1. One for every 200 to 500 cu yd of backfill around structures or in trenches, depending upon the total quantity of material involved.
2. One for every 2000 cu yd of material placed for mass earthwork
3. One for every 500 cu yd of highway or airfield pavement subgrade

4. At least one for every 1000 cu yd of material placed in relatively thin sections such as canal or reservoir linings

5. One test whenever there is a definite suspicion of the quality of moisture control and compaction

Close moisture-content control is being practiced if two-thirds of all field values fall within a range ±1 percent of the median moisture content specified. Erratic, yet often satisfactory, control is indicated if only two-thirds of all field values fall in a range ±3 percent.

TABLE 7   Typical Compaction Requirements Based upon the Standard Proctor Laboratory Compaction Test

Purpose of fill	Density required as % of optimum laboratory density		Suggested range in moisture content dry (−) or wet (+) of optimum	Suggested lift thickness after compaction, in. (for hand tamping use ½ of these values)	
	Coarse-grained cohesionless soils	Fine-grained cohesive soils		Coarse-grained cohesionless soils	Fine-grained cohesive soils
Support of structure...	100	98	−1 to +2	10	6
Backfill around structure or in pipe or utility trenches.....	98	95	−2 to +2	8	6
Subgrade of excavation for structure and for earth dams greater than 50 ft high.....	98	98	−1 to +2	12	6
Earth dams less than 50 ft high and for support of highway or airfield pavement.	95	95	−1 to +3	12	6
Canal or small reservoir lining..........	. . .	95	−2 to +2	. . .	6
Drainage blanket or filter..............	98	. . .	+2 to +4	10	
Other fills requiring nominal amount of strength and incompressibility........	93	90	−3 to +3	10	6

Satisfactory compaction procedures are generally being employed if two-thirds of all densities fall within 2 or 3 percent above or below the percent maximum density required. Unsatisfactory or erratic compaction is evidenced where only two-thirds of all field values fall in a range ±5 percent.

Unless test sections have been built or the performance of a given piece of equipment on a particular soil is already known, it is usually better to specify the required percent of density, moisture-content limits, and lift thickness and allow the contractor some leeway in the selection of compaction equipment and methods.

18. Permeability   Permeability refers to the relative ease or difficulty with which a liquid flows through a medium such as soil. The coefficient of permeability k is a velocity of flow under a unit hydraulic gradient and is expressed in terms of distance/time, frequently feet per minute. It is dependent upon the flowing fluid and its temperature. Typical values of k for various soils are shown in Table 8.

The permeability of soil is often of importance in the selection of suitable materials for filter drains, free-draining base courses, and relatively impervious earth cores for dams. It also affects the quantity of groundwater flow into excavations, wells, and underground drains. The size and number of pumps and the layout of a well-point system for drainage will depend upon the permeability of the soils at a given site. Moreover, on vertical sand-drain installations the coefficients of permeability of natural deposits in both the vertical and horizontal directions are required to estimate the relative time rate of consolidation when both radial drainage toward the sand drains and vertical drainage to more pervious horizontal soil layers are involved.

TABLE 8   Typical Values of Coefficient of Permeability k for Soils

Soil Type	Typical Coefficient of Permeability, Ft/Day
Well-graded clean gravels, gravel-sand mixtures...........	75
Poorly graded clean gravels, gravel-sand-silt.............	180
Silty gravels, poorly graded gravel-sand-silt.............	1.5×10^{-3}
Clayey gravels, poorly graded gravel-sand-clay...........	1.5×10^{-4}
Well-graded clean sands, gravelly sands..................	4×10^{-2}
Poorly graded clean sands, sand-gravel mix..............	4×10^{-2}
Silty sands, poorly graded sand-silt mix.................	2×10^{-2}
Sand-silt clay mix with slightly plastic fines.............	3.0×10^{-3}
Clayey sands, poorly graded sand-clay mix..............	7.5×10^{-4}
Inorganic silts and clayey silts.........................	1.5×10^{-3}
Mixture of inorganic silt and clay.....................	3.0×10^{-4}
Inorganic silts of low to medium plasticity..............	1.5×10^{-4}
Organic silts and silt-clays, low plasticity..............	Quite variable
Inorganic clayey silts, elastic silts....................	1.5×10^{-4}
Inorganic clays of high plasticity......................	1.5×10^{-4}
Organic clays and silty clays.........................	Quite variable

1 cm/sec $\approx$ 2,840 ft/day $\approx$ 2 ft/min
1 ft/year $\approx$ 1×10^{-6} cm/sec

   Soil permeability may be determined either in the field by well pumping tests, for example, or in the laboratory on small and, hopefully, representative samples. Field-well pumping tests are relatively expensive and time-consuming, and require specially trained personnel to plan, conduct, and analyze the results. They are seldom justified except where large projects are involved, or where underground water supplies are to be developed by a system of wells. However, the field-well pumping test does give more representative and reliable estimates of permeability, since a much larger mass of the actual soil in an undisturbed condition is involved.

   For most jobs, engineers resort to laboratory tests of small samples to determine k. Two tests, the falling-head test for the more impervious fine sands, silts, and clays, and the constant-head test for more pervious gravels and clean sands, are more commonly employed. These tests hopefully determine k within an order of magnitude, and for most practical situations this is sufficiently accurate.

APPENDIX

   Laboratory Tests. Three volume ratios are used widely in soil mechanics:
   Void ratio is the ratio of volume of voids (air and water) to volume of solid.
   Porosity is the ratio of volume of voids to the total volume of the mass.
   Degree of saturation is the ratio of volume of water to volume of voids.
   The Atterberg limit tests measure the effect of water content on the consistency of fine-grained soils. The *liquid limit* is the moisture content for which a soil begins to change from a plastic to a semiliquid state. It is directly proportional to the compressibility of a soil. The *plastic limit* is the moisture content for which a soil begins to change from a semisolid to a plastic state.
   The *plasticity index* is the numerical difference between the liquid and plastic limits. It is a measure of the range of moisture content through which the soil is plastic, and is inversely proportional to the ease with which water passes through the soil.
   The *shrinkage limit* is the moisture content at which shrinkage ceases to be appreciable as it dries from a saturated condition.
   The *coefficient of permeability* is the velocity of flow under a unit hydraulic gradient. The important laboratory tests are the constant-head permeability test and the falling-head permeability test.
   The *optimum moisture content* is the water content (in percent, dry weight) at which a soil can be compacted to its maximum density for the method of compaction used.

The *unconfined-compression test* is similar to the compression test on a concrete cylinder. It is a rapid-loading test (5 min) which may be used to determine the peak strength q_u and the deformation modulus of any soil which will cohere enough to retain a given shape, usually cylindrical (Art. 7).

The *triaxial-compression test* is used to determine the shearing properties of saturated silts and silty soils. The sample, usually cylindrical, is subjected to a fluid pressure which gives a stress which is essentially uniform over all surfaces. A supplementary axial pressure is then applied and increased until the sample fails. It is the most reliable shear test.

The *direct-shear test* applies normal and shearing forces on a plane surface within the sample to obtain stress-deformation plots and the strength characteristics. The *undrained* test is a quick test which permits practically no change in moisture content so that consolidation or swelling cannot take place. It is commonly used for clays. The *drained* test is a slow test which permits the water content of even a fully saturated soil with low permeability to adapt almost completely to the change in stress. It is commonly used for coarse-grained soils. The *consolidated-undrained* test allows complete consolidation under the normal stress, followed by shear which is applied rapidly enough to allow practically no change in moisture content.

The *consolidation test* measures the time rate of volume decrease in a laterally confined sample subjected to axial load. A laboratory test curve can be converted to a time curve for a layer of any thickness because consolidation time is proportional to the square of the thickness of a stratum. A plot of consolidation vs. void ratio enables a determination of compressibility as well as the pressure to which a soil has been consolidated in its geological history.

The usual laboratory tests for exploratory investigation are given in Table A-1.

TABLE A-1   Laboratory Soil Tests*

Test	Type of soil	Purpose
Specific gravity	All	Determine composition, void ratio
Grain size	Cohesionless (sands, gravels)	Estimate permeability, shear, strength, frost action, compaction
Grain shape	Cohesionless (sands, gravels)	Estimate shear strength
Liquid and plastic limit	Cohesive (silts, clays)	Estimate compressibility, compaction
Water content	Cohesive	Correlate with strength, compressibility, compaction
Void ratio†	Cohesive	Estimate compressibility and strength
Unconfined compression†	Cohesive	Estimate shear strength

*Adapted from Sowers and Sowers, "Introductory Soil Mechanics and Foundations," 3d ed., The Macmillan Company, New York, 1970.

† Samples must be relatively undisturbed.

Part 2. Soil Exploration

THOMAS H. THORNBURN
Consulting Soils Engineer, Las Vegas, Nev.

GENERAL FOUNDATION CONDITIONS

19. Sources of Information Information on the general foundation conditions at a site can be found in a vast variety of sources. For some areas such information is abundant; in others it is practically nonexistent. King's Bibliography of North American Geology enables information on any state or area to be located quickly.[1] The lists of publications of the U.S. Geological Survey[2] and of various state surveys often contain information on reports and maps more recent than those reviewed in the national bibliographies. General geologic information on a particular site can be obtained from various small-scale maps.[3,4,5] The interrelationship between geologic information and various procedures for soil exploration is discussed in Refs. 6, 7, and 8. A particularly good reference list, as well as a review of soils and soil deposits of North America, is contained in Ref. 9.

20. Glacial Materials A major portion of the northern part of North America and Europe was glaciated in the recent geologic past, and the young deposits have been weathered only to a shallow depth. As a consequence, the design of foundations in these areas, especially where the bedrock is covered by a thick deposit of glacial drift, can be correlated to some extent with the type of glacial deposit.

Drift is the general term for all materials transported by glacial ice. Drift deposits may vary in composition from predominantly gravel- and cobble-sized material to accumulations of clay and very fine silt. Because of this variability the types of exploratory procedures which yield the most information will also vary. Deposits should never be assumed to be uniform in character, and the program of soil exploration must be conducted in such a manner as to determine the variability.

Glacial till is an unsorted, nonstratified sediment deposited by glacial ice. Extremely granular tills containing as much as 80 percent coarse sand and gravel are common in the northeastern part of the United States, and it is sometimes difficult to distinguish these materials from the water-sorted, coarse-grained deposits found in the same area. In the central United States, tills containing 80 to 90 percent fine-grained material with clay contents as high as 60 percent are not unusual. If it is not possible to determine in advance the texture of till which exists at a particular site, it is difficult to plan the exploratory program. Tills of medium texture are usually stiff and can be sampled best with a split spoon. For tills which are very granular the standard penetration test or cone penetrometers are more appropriate. Fine-grained tills may be fairly soft and should be sampled with thin-walled tubing.

Glacial outwash is defined as stratified drift deposited by meltwater streams flowing from the glacier. The two major topographic expressions of outwash are the outwash plain and the valley train. If glacial outwash has been laid down on a previously existing plain, it tends to form an apron in front of the ice margin. Closest to the former ice front, the materials will be predominantly gravel and coarse sand. Away from the front, they become increasingly finer in average grain size until they are predominantly fine sand and silt. Where the outwash was confined to a preexisting valley, the bulk of the deposit is usually very granular, except at the surface. Outwash confined to streams flowing within the body of the ice, or in crevasses near the ice front, may be left as a conical-shaped hill (kame), a serpentine ridge (esker) extending more or less perpendicular to the ice front, or a rather blunt-ended, somewhat flat-topped feature called a crevasse filling. Where meltwater streams flowed between an ice lobe and a preexisting valley wall, the outwash was left in the form of terraces. All these deposits are normally coarse-grained, containing mostly boulders, cobbles, gravel, and sand. Because of their topographic characteristics, they can usually be recognized in the field, on topographic maps, or by the study of vertical aerial photographs. In section, bedding of strata composed of various grain sizes is often extremely complex. Along with the variations in size characteristics go variations in porosity, relative density, and, consequently, compressibility. In such deposits, even the most detailed type of soil-exploratory program can be expected to reveal only the general pattern of variability in physical characteristics, and the engineer must adopt a design which is generally more conservative than one which would be used in situations where the deposits are quite uniform.

Within the outwash material itself, the use of the standard penetration test or cone penetrometers is helpful in delineating the variations in relative density. Geophysical methods can help to establish the average thickness of the outwash. In addition, the exploratory program must be completed in such a way as to reveal the characteristics of the materials underlying the outwash where they may be subjected to significant stresses by the structural foundations.

Glacial lacustrine and glacial marine deposits were formed when the ice front stood in or close to a body of impounded water. The coarse-grained fraction was usually deposited in the form of deltas or shore deposits and the fine-grained fraction carried out into the lake or embayment. Deltaic and other shore deposits have the characteristics of typical outwash. The deposits laid down in the bed of the open water form some of the most compressible deposits with which the engineer must deal. In fresh water, the bulk of the coarse silt fraction sedimented out during the period of maximum ice melt and formed a layer of relatively nonplastic silt. The clay and fine-silt fraction was sedimented during the winter when the body of water was relatively undisturbed by currents, thus forming an alternating layer of soft, plastic, silty clay. The two layers represent one annual cycle of deposition, called a varve. Varved clay deposits are extremely common around the shores of the Great Lakes and in many smaller inland lakes in northeastern United States and southern Canada. Glacial lakes also existed in areas of the western United States where few remnants exist today. Here several spectacular foundation failures are attributable to the fact that the characteristics of the substrata were not determined with sufficient precision.

Where the glacial meltwaters flowed into marine embayments the silts and clays tended to be flocculated simultaneously, and thus varving is rare. However, these deposits also tend to be soft, sensitive, and highly compressible.

Foundation exploration in these areas requires exceedingly good undisturbed samples on which physical and chemical tests can be performed. Sampling with at least 3-in.-diameter tubing is recommended. A piston sampler may be necessary to retain samples of varved clay. The vane auger has been used very satisfactorily in the marine deposits.

21. Wind-Laid Materials   *Dunes.*   In desert regions and along beaches sand is usually picked up and reworked by the wind. Continued sorting of the materials takes place, and the resulting dunes are composed of extremely uniform, medium-sized sand. It tends to be loose and to have low stability unless confined. Where the dunes have not been stabilized by vegetation, the windward side may have a fairly dense surface while the lee side will be very unstable. Such deposits can be densified by the driving of compaction piles or stabilized with additives in order to provide satisfactory foundations for many types of structures. The program of exploration should be designed to determine the relative density of the materials in place in order that the proper treatment can be selected. Penetration tests, perhaps combined with field load tests, usually provide the necessary information. Local dune deposits also are found in outwash areas and along the wide valleys of major drainageways which drained glaciated regions. Large-scale soil-survey maps, maps of surficial geology, and aerial photographs indicate the presence of such deposits and may be used to advantage wherever their presence is suspected.

Loess.   Throughout the Missouri and Mississippi drainage basins and in the Pacific Northwest, the surface deposits often consist of a wind-laid accumulation of silt-size particles known as loess. Its textural characteristics vary depending on the distance from the source and the degree of weathering, but it nearly always contains 50 to 60 percent silt-size material. Unless the deposit has been deeply weathered, it has a very high porosity (50 to 60 percent) and a very low natural unit weight. The silt grains are apparently maintained in this loose state by the presence of a clay or mineral bond which fails when its strength is exceeded by overloading or weakened by the presence of excess moisture. Loess tends to have natural planes of vertical cleavage and root holes which extend throughout the depth of the deposit. If slopes are cut quite flat they tend to erode back very rapidly to nearly vertical.

Because of the collapsible structure of loess, it is difficult to obtain samples in which no disturbance has occurred. If wash borings are made by normal procedures, the structure may be destroyed before the sample is obtained. Even with the best undisturbed samples, it is sometimes difficult to predict the performance of these deposits under foundation stresses. For this reason, field load testing must often be relied upon to establish the foundation load which can be applied without destroying the bond strength. Where it is anticipated that loess will be subject to saturation after the structure is built, it is advisable to soak the soil in the test pit while the plate is still under load so that ultimate settlements can be predicted.

22. Water-Laid Materials   In addition to the water-laid deposits associated with glaciation, recent accumulations of water-sorted materials are found in all stream valleys and open bodies of water. Furthermore, accumulations of water-laid materials cover large portions of the North American continent where streams flowing off of highland areas, such as the Rocky and Appalachian Mountains, have deposited a portion of their loads on flat lands. Such deposits are common in the west-central United States from Texas to Montana, in the intermountain basins of Colorado, Nevada, Arizona, and California, and along the shoreline of the Gulf and Atlantic Coast from Texas to Long Island. These deposits are extremely variable, and only a detailed site exploration can hope to reveal the characteristics of a given deposit. However, certain generalizations can be made.

Recent Alluvium and River Terraces.   Characteristics of the alluvium in any particular stream valley depend on two factors, the regimen of the stream and the character of the material from which the stream obtained its load. Youthful streams occur in regions where the gradients are high and the source materials are coarse. The valleys are usually narrow, and the alluvial deposits, which consist predominantly of boulders, cobbles, and coarse sand, usually cover small areas. As the stream decreases its gradient, the channel deposits are normally sands and silts. However, in areas where the stream drains a glaciated region, the surface deposits of medium grain size may be underlain by coarse glacial outwash. Furthermore, preglacial river valleys in such areas may have been filled and subsequently reexcavated so that the present stream flows in a different location from its predecessor. Occasionally, the older deposits may be composed of laminated fine sands, silts, and clay which may compress under high stresses. Explorations in a river valley must be very thorough.

When the stream enters old age, it begins to meander back and forth across its valley, leaving behind an extremely complex deposit of oxbow lakes, filled meanders, natural levees, and backswamp deposits. The floodplain deposits are typically sands and silts, whereas the other deposits vary from coarse to medium sands in the natural levees to highly plastic and compressible organic clays of the backswamp deposits and the filled-in oxbows and meanders. Differential settlements are likely to be excessive, unless the subsurface conditions are thoroughly delineated by a well-planned program of soil exploration and the results carefully utilized in the design of the foundations.

Geophysical methods often yield information which cannot be obtained through normal boring programs except at unusual expense. Vertical aerial photographs have been found to be extremely useful in determining the positions of the various types of deposits within a valley. Continuous tube samples can usually be obtained, though a piston sampler may be required for the more sandy strata. Vane testing can be used to determine the in situ strength of the organic clays.

Interior Plains.   The deposits found in the Great Plains area of west-central United States include those of streams varying in age from youth to early maturity. Sand and gravel layers, with sands and silts between them, are extremely common in old drainageways. Where temporary lakes existed, the deposits may be mostly silt and clay, much of which appears to have been derived from the deposition of volcanic ash. Many of the sand and gravel layers have been partially cemented with calcium carbonate, because the low annual rainfall has prevented leaching. The porosity, as well as the vertical and horizontal permeabilities, may be extremely variable, and careful exploration is required where these properties may influence the structures under consideration.

In the more arid climates where water-laid sediments have accumulated in isolated basins, the variability in texture may be even greater. Near the valley walls, extremely coarse rocky or bouldery fragments may be found. Toward the center of the basin the fineness increases, and where intermittent lakes exist, fine-grained silt and clays are common. Concentrations of free calcium carbonate, as well as alkali salts such as sodium and potassium sulfate, are found. Such salts have a deteriorating effect upon metals and concrete, and the problems of such chemical reactions must be considered in designing and choosing the foundations. These deposits often have high porosity and a particle bond which is destroyed by leaching. Load tests, such as those conducted on loess, may be necessary, especially in areas where irrigation is practiced.

Coastal Plain.   The sediments which cover the Gulf and Atlantic Coastal Plain are usually unconsolidated to a considerable depth, and consist primarily of stratified sands, gravels, clays, and marly deposits. Sometimes the materials are consolidated sufficiently to be classified as sandstone, shale, or limestone. However, such consolidated deposits have a tendency to disintegrate rapidly when exposed to the elements, and some consideration must be given to this effect. Exploration procedures must be flexible so that they can be adapted to the materials encountered at a particular location. Rotary drills capable of sampling either soil or rock are usually most satisfactory, though standard dry-sample boring can also be used.

23. Organic Soils   Local deposits composed predominantly of organic material are common in the Coastal Plain and in the glaciated regions. Such deposits found as fillings in lakes or tidal marshes are usually so highly compressible that they can function as suitable foundation materials only for the lightest and most temporary of structures. The treatment of such deposits is discussed in Ref. 10 and in textbooks on foundation engineering.

Special samplers are usually needed to obtain suitable samples of peat. However, it is usually sufficient to determine the extent of the organic deposit, for most structures must be founded below the peat layer. If the peat is to be preconsolidated, so that only insignificant differential settlements will occur, large-diameter tube samples or carefully hand-carved samples are required for laboratory testing. Field installations of settlement plates and pore-pressure devices are usually required to monitor the progress of consolidation in the field. A prediction of the amount and rate of settlement in these deposits can be made by procedures outlined by Mesri.[11]

24. Residual Soils   The surficial material in many areas is a residual soil developed from the underlying bedrock. Although residual soils often extend to a depth of only a meter or so, in southeastern United States weathered residuum may occur to depths of 50 ft (15 m) or more. It is important to determine not only the characteristics of the surface soil, but also the nature of the rock contact and degree of rock weathering. Since these factors may vary considerably with the nature of the rock, the exploration should reveal the soil-rock profile.

Sedimentary Rock.   Rocks of sedimentary nature exist at or near the surface throughout much of southern United States. They may be broadly classified as sandstone, shale, or limestone. Most rock can support any load which can be carried by the foundation, provided it has not been seriously affected by solution or erosion below the general rock surface. Most rocks have been subjected to some type of crustal movement which has resulted in the development of cracks and joints. The joint spacing may be very wide, or it may be only a few feet. Since rock weathering proceeds most rapidly along these discontinuities, it is essential that the exploratory program provide some information on the joint and crack patterns.

The strength of sandstone depends primarily upon the strength of the binder which cements the grains. If the rock is sound, it is practically impossible for weathering to weaken the foundation seriously during the lifetime of the structure. On the other hand, there is evidence that some weakly cemented sandstone may be roughly the equivalent of loose to medium sand as a result of weathering. Such rock may settle under heavy loads, particularly if the water table rises after construction. Rock coring to a depth of 12 to 25 ft (4 to 8 m) will indicate the degree of weathering and will furnish samples for strength and compressibility testing. If it is necessary to determine the pattern of joint spacing, shallow test pits may be required.

The term shale usually includes all rocks composed of particles varying in size from coarse silt to the finest clay. The strength of shales depends upon not only the character of the cementing agent but also the degree to which the fine-grained sediments have been consolidated. Some shales weather very rapidly when exposed at the surface and may revert to clays within a few years; others are quite resistant to weathering. It is unlikely, however, that weathering beneath the foundation element would cause any significant structural damage within the lifetime of an ordinary structure.

Unless the shale is very soft, rock coring will be required to obtain samples. Where interbedded sandstones and shales are exposed on slopes, consideration must be given to the possibility of sliding along the bedding planes, especially where the rock dips in the same direction as the slope. Seepage may occur where sandstone strata intercept the slope line; this may aggravate the tendency to slide. Thus, the exploratory program should reveal the dip slope, arrangement of strata, and, if possible, the hydraulic conditions.

Limestone is the sedimentary rock most susceptible to weathering. Weathering by solution proceeds rapidly along the vertical joints and also laterally along some of the weaker strata. Where limestone occurs at the surface in humid regions, sinkholes indicating the presence of underground cavities are nearly always present. Rock boring in limestone areas should be carried deep enough to establish whether or not voids are present within the depth which will be stressed by the foundation. If cavernous conditions are found to exist, it is advisable to bore at the location of each foundation element. Geophysical procedures and aerial-photographic interpretation can be used to advantage to delineate rock structure and locate underground cavities. In the interior of the United States, coal and rock salt may be interbedded with the commoner sediments. In mineralized areas, all available mining records should be examined and rock coring carried to sufficient depth to intercept commercially valuable strata.

Crystalline Rock.   Most igneous and metamorphic rocks are hard and relatively resistant to weathering. The most obvious exception to this general statement is marble, which although harder than limestone, is also susceptible to solution. Thus, exploration in marble areas should be conducted as in limestone regions. In crystalline-rock areas, there is a preponderance of steeper slopes, and problems connected with seepage and sliding must be expected. Rocks having a schistose structure and a fairly high mica content tend to weather most rapidly of the crystalline materials. The weathering product is detritus, which has fairly high porosity and compressibility. Water enters this material readily and causes flow slides even on relatively gentle slopes. The exploratory programs should furnish samples of the detritus for mineralogical examination and reveal the thickness of the weathered zone. Aerial photographs will usually show the location of old slide scars and indicate the degree of danger from sliding.

Some crystalline rocks, especially igneous intrusions, tend to weather at a different rate from the adjoining country rock (the older rock into which the igneous material was forced). Thus, abrupt differences in depth of residuum may exist where dikes or similar intrusive features are exposed at or near the surface. Aerial photographs are helpful in delineating such boundaries and should be used wherever possible in planning exploratory programs in crystalline-rock areas. In areas of extrusive igneous rock, layers of basalt are often interbedded with strata composed of coarse volcanic ejecta and volcanic ash. The fragmental material may have weathered to plastic clay, even though the overlying basalt appears to be sound. Exploration of such deposits is often exceedingly difficult. Geophysical methods should help to locate highly permeable or compressible strata which may affect the stability of foundations.

25. Frozen Soils   Frozen soils present special problems not ordinarily encountered in normal engineering practice. Before undertaking soil exploration in such areas, one should be familiar with the types of permanently frozen ground and the problems which may be anticipated with each. Reference 12 is a condensed guide to the descriptions of various types of permanently frozen ground and gives examples of the types of field investigations and records which are necessary for proper identification.

EXPLORATORY PROCEDURES

The classic treatise on soil exploration is Ref. 13. Generalized treatments are given in Refs. 6 and 14. Reference 15 contains a wealth of information on in situ measurement techniques.

Direct Methods

26. Test Pits   Test pits allow direct inspection of the soil or rock in section. Samples carefully hand-carved from the walls or floor may be tested for comparison with samples obtained through other procedures. Ordinarily, it is not economical to conduct a full program of exploration by means of test pits, but in instances where a deposit is extremely variable direct inspection will furnish a truer impression of its nature than can be obtained from many borings. Test pits will also be required where there is a need for load testing of the soil in situ. Excavations for basements and foundations should be inspected to provide a check of the subsurface conditions as predicted from borings and other exploratory procedures. Large-diameter calyx drill holes enable a man to descend to inspect the rock profile.

27. Borings   Borings are the commonest procedure of subsurface exploration. Helical augers 1 to 2½ in. (25 to 60 mm) in diameter or Iwan (post-hole) augers from 3 to 6 in. (75 to 150 mm) in diameter are most commonly used for hand operation. Such augers are usually limited to small jobs where it is not necessary to bore more than about 25 ft (8 m). Although depths as great as 50 ft (15 m) may sometimes be attained, other procedures will probably be more economical. Generally, augers are unsatisfactory where sandy strata are encountered, since the hole will not remain open without support. Casing may be driven to provide support, but hand methods usually become uneconomical if casing is required. In soils possessing some cohesion, augers yield samples which are in a badly disturbed condition. These may be used for classification tests but have no value for determining the structural characteristics of the natural soil.

Mechanical augers 2 to 12 in. (50 to 300 mm) or more in diameter are frequently used. Some of these may be used in continuous flights to make borings to depths of 25 ft (8 m) or more. Usually, the material which comes up on the top of the hole bears no relationship to the material at the bottom. For this reason, this type of equipment cannot be depended upon to yield even representative samples of the various strata encountered. Augers are not ordinarily recommended for large jobs.

Wash borings produce a hole by forcing a wash pipe or hollow drill rod into the ground while jetting and chopping. A chopping bit with water ports is fastened to the end of the pipe, and water is circulated down through the bit and up to the surface, carrying the soil material removed by the chopping process. The water carrying the solids flows from an overflow pipe attached to the top of the casing into a container where the coarser particles settle out. Samples obtained from the fluid are completely disturbed and altered in composition. They should never be used as a basis for design. Changes in color and texture do indicate to the boring foreman changes in substrata encountered, and make it possible to stop the washing process to obtain samples with other tools. Generally, some casing is carried with the wash-boring equipment and a short section is driven at the top of the hole. If the hole will not stand unsupported, casing may be driven the full depth. Figure 17 illustrates the principal components of the wash-boring process.

Dry-sample borings are usually made with the same equipment as wash-sample borings except that samples are obtained intermittently by driving a sampling tool into the soil at the bottom of the hole. The hole is washed a distance of 3 to 5 ft (1 to 2 m), and the chopping bit is replaced with a sampling tool which is driven 12 to 18 in. (about 0.5 m) and then pulled out. The chopping bit is attached again, and washing proceeds until the next sample is taken.

Hollow-stem auger borings can usually be made more rapidly than wash borings combined with dry sampling where heavy truck-mounted equipment can be used. Continuous flights of augers with hollow stems can be drilled through most types of soil to depths of over 200 ft (60 m). A removable plug attached to a center rod prevents entry of soil into the stem until the desired sampling depth is reached. When the plug is removed, a cased hole is provided through which various types of sampling tools may operate. Augers with inside diameters of 2½ to 3⅜ in. (64 to 86 mm) are most commonly used, but a 6-in. (150-mm) size is available. Hollow-stem augers may change the relative density of cohesionless soils by compaction near the mouth or through material being forced prematurely into the stem as the plug is withdrawn below the water table.

Rock Drilling.   The percussion drill rig consists of a string of tools supported by a cable which is alternately raised and dropped against the bottom of the hole by a spudding arrangement on the drilling machine. A small amount of water is poured in the hole, and under the pounding action of the cable tool a slurry is formed with the broken rock particles. These are cleaned out as necessary with a bailing tool. The samples obtained from the bailer are badly disturbed, but the kind of rock material can usually be determined by visual inspection and simple tests. Sampling devices may be attached to the bottom of the cable and cores obtained for inspection when the rock changes. Generally, this method of drilling and sampling is slower than the rotary method, and if the rock is to be sampled to a considerable depth it is usually uneconomical.

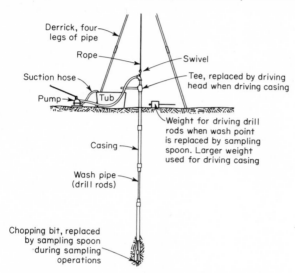

Derrick, four
legs of pipe

Rope

Swivel

Suction hose

Tee, replaced by driving
head when driving casing

Pump

Tub

Weight for driving drill
rods when wash point
is replaced by sampling
spoon. Larger weight
used for driving casing

Casing

Wash pipe
(drill rods)

Chopping bit, replaced
by sampling spoon
during sampling
operations

Fig. 17 Apparatus for making wash boring. *(After H. A. Mohr.)*

Rotary drilling has been adapted for both soil and rock coring. In some aspects, rock-core drilling is similar to the wash-boring processes. In place of a chopping bit, a specially hardened rotary bit or a core bit set with industrial diamonds is used. A suspension of bentonite in water is used for circulating fluid. The characteristics of the fluid are adjusted with additives to make it thixotropic. Whenever drilling or pumping is stopped, the circulating fluid turns to a gel and keeps particles of rock in suspension, thus preventing jamming of the bit. Although drilling can be done with solid bits, with cores taken intermittently, it is much more common to obtain continuous cores if possible. Bits are attached to lengths of flush-joint drill rods. The rods are pressed downward and rotated simultaneously through a special chuck on the drilling machine.

28. Sampling Exploratory borings are made to take samples which may be examined or tested in the laboratory or to permit the access of equipment to test the characteristics of the soil in situ. Samples may usually be carved by hand in test pits, but if the soil material is only slightly cohesive, special techniques must be used to obtain relatively undisturbed samples. For sampling cohesive soils, a simple procedure is to excavate around a pedestal of soil and do the final trimming carefully with a sharp knife or a wire saw. The sample may then be broken off, sealed in paraffin, and shipped to the laboratory. To protect the sample, it is best to invert the open end of a box over the soil pedestal and then, using a spade, extract the sample carefully and turn it upside down. After it has been upended, it may be sealed for shipment by pouring hot paraffin in the box. Another procedure is to press an open-ended cylinder slowly into the soil while trimming away the excess material. Using this procedure, the sample fits snugly inside the cylindrical tube and suffers little or no disturbance. To improve the quality of a sample, a cutting shoe may be attached to the bottom of the tube.

Samples obtained as a by-product of the wash-boring or percussion-drilling processes are of little value except to indicate changing strata. Auger samples, if properly taken, can be used for some identification tests but are too disturbed to provide strength data.

The split spoon is one of the most commonly used tools of soil sampling (Fig. 18). Split-barrel samplers attached to the end of the wash pipe or drill rod are driven into the bottom of the drill hole with a 140-lb (64-kg) weight which is allowed 30 in. (75 mm) free fall on a driving head. Prior to the insertion of the sampling spoon, the bottom of the hole should be cleaned out. When the sampler rests on the bottom, the upper end of the wash pipe or drill rod is marked in 6-in. (150-mm) intervals beginning at the top of the casing. The drill

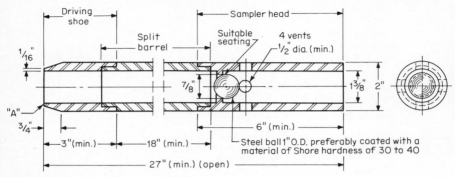

Fig. 18 Standard split-barrel sampler assembly. *(After ASTM.)*
NOTE 1—Split barrel may be 1½ in. inside diameter provided it contains a liner of 16-gage wall thickness.
NOTE 2—Core retainers in the driving shoe to prevent loss of sample are permitted.
NOTE 3—The corners at A may be slightly rounded.

rod is driven to the first 6 in. (150 mm) and the number of blows recorded. The rod is then driven an additional 12 in. (300 mm), or until 100 blows have been applied. The total number of blows required to drive the sampler through the second and third 6 in. of penetration is called the standard penetration resistance N. If it is not possible to obtain 1 ft (300 mm) of penetration, the log should indicate the number of blows and the fraction of the foot penetrated.

After penetration the sampler is brought to the surface and opened for inspection and retrieval of the sample. Such samples are always disturbed, sometimes seriously, by the driving of the spoon and by squeezing of the soil caused by the thick walls of the sampler.

Tube Sampling.   If the subsurface materials are soft or compressible, it is good practice to use samplers with relatively thin walls compared with the split spoon. Samples are usually taken continuously, if possible. The details of the procedure are described in ASTM Method D1587.[16] Table 9 gives the dimensions of commonly used thin-walled tubing.

TABLE 9   Suitable Thin-Walled Steel Sample Tubes

Outside diameter, in. (mm)	2 (50)	3 (75)	5 (125)
Wall thickness:			
BWG	18	16	11
in. (mm)	0.049 (1.24)	0.065 (1.65)	0.120 (3.05)
Tube length, in. (m)	36 (0.9)	36 (0.9)	54 (1.5)
Clearance ratio, %	1	1	1

After ASTM D1587.

Studies of Chicago clay show that the strength of 2-in. (50-mm) tube samples tested in unconfined compression is approximately 75 percent of the strength of carefully hand-carved samples tested in the same manner.[17] Samples of normally loaded clay of ordinary sensitivity will have undrained strengths closely equivalent to that which exists in the ground only if the strains during sampling are not large enough to cause any significant microstructural change.[18] Only carefully hand-carved samples are likely to meet this criterion.

In order to obtain samples several feet long in a sampling device, the inside friction must be reduced. Thus, the diameter of the cutting edge D_e is made slightly smaller than the inside diameter of the tube D_i. The ratio $(D_i - D_e)/D_e$ is defined as the inside clearance ratio. If this ratio is too large, the sample expands excessively as it comes into the tube, changing the microstructure of the clay so that the sample strength may be considerably decreased. For this reason, it is essential to use clearance ratios of about 1 percent.

Tube sampling can be conducted in any stable borehole. Casing or drilling fluid may be required if the soils are sandy or contain sandy strata. Tube sizes other than those listed in Table 9 may be used provided that the ratio of wall thickness to outside diameter and the inside clearance ratios are similar. However, tubes smaller than 2 in. (50 mm) are generally considered unsatisfactory for obtaining undisturbed samples. If the sample is to remain in the tube for any length of time, the inside of the tube should be coated to prevent corrosion and excessive disturbance of the sample when it is ejected. Furthermore, the tube should be reasonably round without bumps, dents, and scratches. Prior to sampling, the hole should be cleaned. The least disturbance of the sample occurs when the tube is pushed with one fairly rapid and continuous motion without impact or twisting. This may be accomplished by placing a heavy weight at the top of the drill rods, by using a special cable and anchor arrangement to pull the rods into the ground, or by exerting hydraulic pressure with a drill rig. Where sand strata or exceptionally hard clay is encountered it may be necessary to drive the sampler. This will cause disturbance of the sample, and driving should not be permitted except on the authority of the supervising engineer.

After the tube is pulled from the hole, the ends are cleaned and sealed with a thin metal disk and a layer of wax. Finally, the ends of the tube should be covered with strong cloth or metal caps taped in place to prevent damage during shipment. It is desirable to make samples as long as possible so that cleanouts and insertions can be reduced to a minimum. On the other hand, the longer the sample the greater the friction between the sample and the inside of the tube. If friction is slight, it is likely that a portion of the sample will be lost as the tube is withdrawn unless special tools are employed.

Piston Sampler.   If soft cohesive soils or cohesionless sands are sampled, it is frequently found that a large percentage of the sample is lost during the process of extracting the tube from the borehole. Recovery can often be increased by the use of a piston sampler. In its simplest form, the piston sampler is a seamless-tube sampler through which has been inserted a piston rod with a plug attached to its lower end. As the sampler is lowered the piston rod is extended through the drill rods and locked to them in such a way as to keep the plug in the mouth of the sampling tube. When sampling depth is reached, the piston rod is locked in position, usually to the casing, and the drill rods are then pushed smoothly and rapidly to advance the sampling tube past the piston. After sampling is complete, the piston rod and drill rods are locked together and withdrawn. If there is a tendency for the sample to slide out of the tube, a slight vacuum is created between the top of the sample and the face of the piston. This tends to retain the sample and to increase the recovery ratio.

If this procedure is used in cohesionless sands below the water table, it is necessary to use drilling mud instead of water. Sampling with a mud containing 3 parts bentonite and 1 part barite is discussed in Ref. 19. Special precautions must be taken in extracting such samples from the tubes, and also in avoiding vibrations during transportation. Laboratory tests on sands obtained with the piston sampler indicate that the average change in density caused by sampling is controlled by the initial relative density value.[20] If the relative density is greater than about 77 percent, the sand tends to decrease in density during sampling, but at smaller relative densities there is a tendency toward density increase. The tests also revealed that the clearance ratio is extremely important in this type of sampling. Clearances of 0.5 or 0.25 percent give better recovery in fine sand. Samples of cohesionless sands below the water table can also be obtained by a technique in which the sampling tube is withdrawn from the soil immediately into a chamber from which the water has been removed by compressed air.[21] The air-water interface at the bottom of the sample tends to hold the sample in the tube by arching.

Good sampling requires a sampler with a smooth and hard inner surface and with moderate or small clearance ratios, a moderate to moderately high punching speed, and a continuous sampling stroke without interruptions. The time lag between pushing the sampler to the bottom of the borehole and the sampling process should be as short as possible. Furthermore, the time between extraction of the sampling tube from the boring and the testing of the samples should be short. Once the samples are obtained they must be protected against frost and shock if good test data are to be obtained.

Core Barrels. Rock samples can be obtained in single- or double-tube core barrels. The essential parts of a single-tube core barrel are the barrel itself, the reamer shell, the core lifter, and the coring bit. The bit is set with diamonds, tungsten carbide, or similar hard materials appropriate to the hardness of the rock or soil being drilled. The double-tube barrel (Fig. 19) differs from the single tube in that it contains an inner tube which remains stationary while the outer tube rotates with the drill rods and the coring bit. It generally produces higher recovery ratios and less fracturing in the samples. Recovery ratio is defined as the percentage ratio between the length of core recovered and the length of core drilled on a given run.

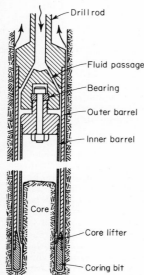

When it is necessary to take large-diameter samples of hardpan, hard clays, highly cemented soils, or extremely stiff deposits, the Denison core barrel may be employed. This is a double-tube core barrel designed particularly for soil sampling. It is manufactured in outside diameters of 3½, 4, 5½, and 7¾ in. (90, 100, 140, and 200 mm) and in lengths to recover either 2- or 5-ft (0.6- or 1.5-m) samples. Drilling mud is commonly used as the circulating fluid. This sampler differs from the normal double-tube barrels in that the inner barrel protrudes below the cutting bit. Thus, the sample is undisturbed by the cutting action of the rotating bit and contact with the drilling fluid. The Denison sampler may also be used with soft soils to provide large-diameter samples which can be trimmed down to give a test sample of about the same quality as a hand-carved sample.

Wireline Drilling. When deep holes are cored into rock, a great deal of time is consumed in removal and replacement of the string of rods in order to extract the core. The use of wireline equipment permits the removal of the inner core barrel through specially designed drill rods. The special assembly which grasps the core barrel can also lower it back down the hole so that drilling can be resumed. The wireline equipment can also be modified to include a longitudinally split third tube nested within the inner core barrel. When the inner barrel is removed, the split tube is extracted by hydraulic pressure and the core is retained, essentially undisturbed, in the split tube. This triple-tube core barrel is useful for sampling brittle or fractured rock.

Fig. 19 Schematic diagram of double-tube core barrel. (*After Peck, Hanson, and Thornburn.*)

Although wireline equipment is more expensive, its timesaving features may more than make up for the extra cost. This is apt to be especially true for deep holes in rather soft rock where the drill bits do not wear rapidly and thus do not require frequent extraction of the rods for replacement.

29. Borehole Cameras If the recovery ratio from rock coring is extremely low, it is impossible to tell whether the poor recovery is due to faulty drilling methods, strata of extremely soft rock, or cavities in the rock. Several types of film and television cameras have been developed which can fit into boreholes as small as NX (3 in. inside diameter) size; descriptions and limitations are given in Ref. 22. Generally, these cameras permit viewing the rock walls in such a way that the dip of the strata as well as the orientation of voids and other discontinuities can be determined.

30. Boring Reports The detailed boring log of each borehole should include the following information:

1. A complete identification of the site on the basis of a legal description or reference to permanent landmarks

2. A plat of the site showing the north arrow, the location of the boring, and all other borings made at the site

3. The date of the boring

4. The elevation of the ground surface with respect to a permanent bench mark

5. The elevation of the water level, including date and time of the observations and procedure used (whether by bailing the hole or filling the hole)

6. The elevation of the upper boundary of each successive stratum of soil or rock

7. A field classification of the strata encountered

8. Values of penetration resistance or other measures of consistency which may have been obtained

9. Elevations or depths at which drilling-water return was lost or artesian pressure observed

10. Type, size, and design of sampling tool, together with size and length of all casing and any observed movements of the casing

11. Length of sample obtained when taking continuous tube or core barrel samples, together with the recovery ratio

12. Any changes in the character of the drilling water or mud

13. Any details regarding the procedure required to advance the hole or keep it open, especially notes with regard to the type of tools used and any changes made, including reasons for change

Borings which cannot be completed often are as important as those which can, and information should be included for the full length of all borings. The final boring report should include a summary of the field information for each boring and data from any laboratory tests and identifications of the samples. It should summarize not only what is known about the deposit, but also what is not known. Only then can the degree of reliability of the conclusions be estimated.

Semidirect Methods

31. Standard Penetration Tests   The commonest procedure for recording the characteristics of the soil in situ is the standard penetration test, which is conducted in connection with sampling by the split-spoon method. Controversy exists with regard to the meaning of the standard penetration test and the relationship of the number of blows N (Art. 28) to the engineering characteristics of the soil. It is generally agreed that the N values bear a much more reliable relationship to the relative density of sand deposits than they do to the consistency of clay deposits. They are extremely suspect when gravel or silt deposits are encountered.

Penetration resistance increases with an increase in either relative density or overburden pressure. Therefore, in cohesionless sands the effect of overburden pressure must be taken into account. The equation

$$C_N = 0.77 \log_{10} \frac{20}{\overline{p}}$$

may be used to compute the correction factor C_N to be applied to the field N value.[6] If the effective overburden pressure $\overline{p}$ at the elevation of the penetration test is in the range of 0.8 to 1.2 tons/ft² (80 to 120 kN/M²), the correction can be ignored without serious error. The equation should not be used for values of $\overline{p}$ less than 0.25 ton/ft² (24 kN/m²), and C_N should be given a maximum value of 2. This correction should be considered inexact and, for the most part, conservative. However, in deposits of very fine sand or silty sand the error may be on the unsafe side.

Cone penetrometers have been popular in Europe for many years and are finding increasing use in the United States. Dynamic penetration tests are similar to the standard penetration test in that the cone is driven into the ground. Variations in the size and shape of the cone as well as the weight and height of fall of the hammer make it necessary to calibrate the blow counts with other tests. A static penetration test is conducted by pushing the cone into the ground and measuring the resistance to penetration. A review and analysis of all types of penetration testing is given in Ref. 23 and a theoretical analysis of the static test in Ref. 15.

The most widely used static penetrometer is the Dutch cone, which utilizes a 60° cone with a base area of 10 cm². The cone is attached to a rod which is protected by casing from the surrounding soil. The cone resistance q_p is defined as the force required to advance the cone, divided by its base area, when the cone is advanced at a rate of 2 cm/sec. The approximate relationship between the Dutch cone resistance in kg/cm^2 and the standard penetration test N value is given in Table 10.[23]

Although penetration tests have obvious drawbacks with respect to calibration and variability related to the field techniques employed, they usually provide the best information that can be economically obtained for granular deposits. They are used in most routine investigations because the cost of obtaining undisturbed samples of sand or of conducting load tests in open pits cannot usually be justified.

TABLE 10   Approximate Ratios of Dutch Cone to Standard Penetration Resistances*

Soil type	q_p/N
Silts, sandy silts, slightly cohesive silt-sand mixtures	2
Clean fine to medium sands and slightly silty sands	3-4
Coarse sands and sands with little gravel	5-6
Sandy gravels and gravels	8-10

*From Ref. 23.

32. In Situ Strength Tests   Several types of tests are available for determining the stress-deformation characteristics of soils in situ. All have advantages and drawbacks, and experience is the key ingredient in obtaining valid and useful results.

Vane Shear Device.   With the exception of the standard penetration test, the use of the vane auger to determine the in situ consistency of clay is probably the commonest semidirect method of soil exploration. Figure 20 shows a typical vane auger. The vane is pushed to the desired depth, and the maximum torque required to turn it is measured by a torque wrench or other suitable device. Because the vanes are relatively thin, insertion disturbs the soil only slightly. Thus, the shearing resistance of a cylinder which surrounds the outside edges of the vanes is measured with a high degree of accuracy. The vane can be turned through several revolutions after the initial reading is taken, and a value of the remolded strength of the soil derived from a second test. In this manner, the sensitivity of the clay may be determined directly in the ground. When a test is completed, the vane may be pushed farther into the soil and another test performed. The vane cannot normally be used where stiff soil layers or sandy or gravelly strata are encountered, since it may be damaged when penetrating the strata.

Values of cohesion obtained from in situ vane tests in soft clays generally are higher than those computed from unconfined or triaxial tests on samples. Ratios as high as 2:1 are not uncommon. A part of this difference is undoubtedly due to disturbance caused by sampling, but there is also a suggestion that the vane may give misleading results if even thin layers or laminations of sand or dense silt are present. Frictional resistance of the vane rods against the casing, the bushing, or the soil is difficult to evaluate under test conditions and may also account for part of the apparently higher shear strength. Table 11 suggests the reliability of field vane tests for clay.

Pressuremeter.   The pressuremeter consists of a volumeter and manometer connected to a cylindrical expansion device, the pressuremeter probe, which is inserted into the borehole. The probe is a steel tube surrounded by two flexible rubber membranes which may be inflated by CO_2 gas applied through leads from the ground surface. A central cell is designed to apply radial pressure on the wall of the borehole and simultaneously measure the increase in diameter of the hole. Two outer guard cells also expand when the pressure is applied, which reduces end effects on the central cell. Volumetric strain is measured by the drop in liquid level in the volumeter at the surface. Interpretation of the data obtained is greatly dependent upon the quality of the test. However, with good pressure-volume curves it is possible to compute the undrained shear strength, an elastic modulus of deformation, and a stress-strain curve. It appears, nevertheless, that proper interpretation of the test results requires considerable experience with the equipment in various kinds of soils.[15]

Borehole Shear Device.   The borehole shear device consists basically of three parts: a shear head, a pulling device, and a measurement console. The shear head is comprised of a gas-operated piston that can be controlled to expand or contract against two curved shear plates which bear against opposite sides of a borehole. After a given normal force is

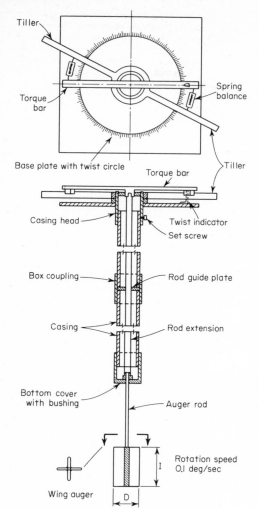

Fig. 20 Schematic diagram of the Swedish Geotechnical Institute vane auger. *(After Hvorslev.)*

TABLE 11 Application and Reliability of Field Vane Tests*

Type of problem	Type of clay	
	Normally consolidated	Overconsolidated
Short-time bearing capacity (no drainage).........	Correct	Correct
Long-time bearing capacity......................	Conservative	Unsafe?
Slopes of temporary cuts (no drainage)...........	Correct	Correct
Slopes of long-time cuts........................	Unsafe (swelling)	Unsafe
Natural slopes.................................	Conservative to correct	Unsafe

* Cadling and Lindskog, ASTM, STP 193, p. 63.

applied to the plates, movement of the whole shear head is actuated by a pulling device operating through steel rods. The pulling device includes a plate resting on the ground surface, a worm gear and a screw arrangement for pulling the rods, and two hydraulic cylinders with a pressure gage for measuring the upward shear force. The console includes a CO_2 bottled-gas supply, a pressure regulator, and a pressure gage. The device as commonly used is inserted into a hole which has been reamed to exactly 3 in. (76 mm) diameter. A series of tests are then performed at gradually increasing normal stresses from about 4 to 15 psi (30 to 100 kN/m²), while allowing 5 to 20 min for consolidation before each shear measurement. Thus the information necessary to plot a Mohr failure envelope is obtained. Present information indicates that the results are usually close to those obtained from consolidated drained triaxial tests on samples but may vary considerably. Thus the borehole shear test, like the pressuremeter test, appears to require considerable experience before reasonable interpretation of the results is assured.

Indirect Methods

33. Correlations with Surficial Geology   The information obtainable from surface geologic maps, agricultural soil reports, the study of aerial photographs, and other earth-science material is of special value to the engineer when it has been correlated with engineering data. This information may often be used, at least in preliminary considerations of a site. Thornburn gives a comprehensive review of the literature on the uses of geology and pedology in highway soil engineering.[24] Many of the items are also applicable to foundation engineering, particularly in the design of shallow foundations.

Aerial Photographs.   Vertical aerial photographs can be a useful aid in the prediction of general terrain conditions at a site. They are especially valuable and often available in areas where even rudimentary knowledge of the geology is lacking. Various soil conditions can usually be delineated on the photographs with some precision by a skilled interpreter, thus greatly simplifying the planning of the exploratory program. Sometimes knowledge of the photographic patterns may even permit an interpreter to recommend construction on one site in preference to another, without any ground exploration. Deposits of highly organic soils can nearly always be recognized and may often be avoided in subsequent project planning.

Other sources of information which are similar to photographs but may be more useful in some situations are available. Remotely sensed information about the earth's surface is regularly transmitted by the earth-resources technological satellite, LANDSAT-1. The imagery is available from the EROS Data Center in Sioux Falls, S. Dak. Radar and infrared scanners can also provide special information often not visible in panchromatic or color photographs. A description of the various techniques and their uses is given in Ref. 25. For most foundation-engineering purposes, however, the photograph will continue to be the most useful tool for terrain analysis. An excellent guide to site selection utilizing aerial photographic interpretation is given by Way.[26]

34. Geophysical Surveys   Geophysical methods of exploration are especially useful in locating the position of a soil-rock contact. Usually, however, some borings are required to interpret the data. For this reason, geophysical methods are most valuable where a fairly large area must be explored rapidly and at low cost. Where only a small area is involved, borings alone will usually suffice unless the position of the soil-rock contact is critical to the design and quite erratic.

Seismic Surveys.   The seismic method of exploration is based on the fact that elastic waves travel at different velocities in different materials. Velocities tend to be low in soft clays, intermediate in dense sand and gravel, and high in sound rock. In its simplest form, the seismic method uses equipment to produce an elastic wave (small explosive charge), a series of detectors (geophones) which are spaced at intervals along a line from the point of origin of the wave, and a mechanism (oscillograph) to record the arrival time of the wave at each detector. Portable equipment has been developed especially for civil-engineering purposes where the maximum depth of exploration does not exceed 50 to 100 ft (15 to 30 m). This equipment can delineate strata contacts from waves induced by a sledgehammer blow, which makes possible the use of seismic surveys in populated areas.

The need to design structures resistant to dynamic loads has increased the importance of in situ investigations by seismic methods. Well-planned, well-executed, and knowledgeably interpreted seismic investigations are essential to the determination of reliable values of elastic moduli. Seismic surveys performed in boreholes where the shot points are established at various elevations in one boring and the geophones arrayed at corresponding elevations in one or more other borings appear to have the greatest all-around utility.[15]

Resistivity Surveys. Subsurface exploration by the electrical-resistivity method is accomplished by measuring changes in resistivity of the earth as the electrode spacing is varied. Usually, four electrodes are arranged in a line at equal intervals. A voltage is impressed across the two outer (current) electrodes, and the amount of current flowing is measured in amperes. The voltage induced across the two inner (potential) electrodes is then measured in volts. In making a survey, the electrode spacing is changed along a given line of traverse after a resistance reading has been obtained. In a relatively confined area several traverses may be made, each oriented in a different direction. The data obtained from each traverse are plotted and compared with standard curves derived from theoretical interpretations. Such interpretations require the services of an expert in geophysics. Resistivity depends upon the presence of chemical salts in solution within the soil and rock mass, so that it is greatly influenced by the water content. However, it is also related to the soil texture. Thus, clay is low in resistivity, sand moderate, gravel fairly high, and sound rock very high. The resistivity method is particularly useful in locating permeable strata in contact with relatively impermeable ones, and zones of weathered rock in contact with sound rock.

Field Testing

A complete program of subsurface investigation must sometimes include field tests and observations in addition to those which have been discussed. Often these are carried on in conjunction with the construction program, but sometimes they will be required before it is possible to complete the final design. Investigations, such as plate load tests, load tests on foundation elements, field permeability tests, pore-pressure measurement, settlement observations on structures after or during construction, determination of slope stability, and tests of dynamic soil characteristics, are representative. Generally, such field observations are not included in the original exploratory program, but the need for special information may be apparent once the initial program has been completed. An authoritative discussion of field measurements is given in Ref. 9.

35. Load Tests The principal difficulty with plate load tests is that the area covered by the plate is usually much smaller than the foundation elements. Thus, at the same intensity of load, the foundation elements stress the soil to a much greater depth than that which is stressed by the bearing plate. If the subsoil consists of stiff clay over soft clay, the load tests may indicate adequate bearing for a structure exerting a high load intensity. When the foundation elements are stressed to their normal working loads, however, settlement may take place in the underlying soft clay layer that was not stressed by the plate. A similar situation may exist where dense granular deposits overlie looser ones. As long as the engineer realizes these limitations, load tests provide a convenient method of predicting the allowable bearing capacity of cohesionless granular materials, of loess, and of soils which contain numerous planes of weakness. Loess has a natural structure which is often destroyed during the sampling process, so that a design based on sample strengths may be considerably in error unless the samples are hand-carved. Loess tends to compress abruptly when some critical load is reached, and one of the best methods of determining the bearing capacity of such strata is the plate loading test. Vane shear tests are not generally accepted in determining the shearing strength of loessial soils.

The plate load test gives information on the soil to a depth equal to about two times the diameter of the bearing plate, but only partially takes into account the effect of time. Figure 21 shows two common assemblies for conducting a load test to determine bearing capacity of soil for static load on spread footings. One of the important considerations is that the width of the test pit be a minimum of four times the diameter of the loaded plate. Terzaghi and Peck suggest that the minimum width be five times the diameter of the loaded plate, in order to prevent the confining effect of surcharge.[17]

The essential elements of the test are the load plate, a jacking post, a reaction beam, and a hydraulic jack. The beam may support a platform loaded with pig iron, concrete blocks, bags of sand, etc., or may react against anchor rods secured in the ground at a sufficient distance from the test pit that they will not affect the test. The load plate is instrumented with dial gages mounted independently of the jacking post to measure the average settlement of the plate. The load is usually applied in increments not exceeding about 10 percent of the estimated allowable load. Each load increment is maintained until the rate of settlement becomes practically imperceptible. Time intervals may be selected, as required, for all settlement to cease, or for the rate of settlement to become uniform, but any time interval so selected must be maintained for each load increment in all tests of any series.

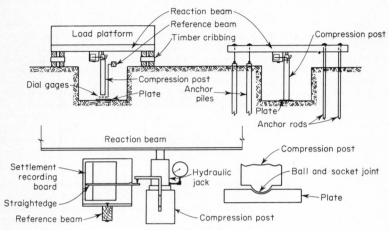

Fig. 21 Typical setup for conducting static-load tests. *(After ASTM.)*

Each test should be continued until a peak load is reached or until the ratio of load increment to settlement increment reaches a minimum steady magnitude. If sufficient load is available, the test should be continued until the total settlement reaches at least 10 percent of the plate diameter, unless a well-defined failure is observed. Measurements of the settlement by dial gages must be made continuously, so that as soon as possible after the test is completed a settlement-load record may be plotted.

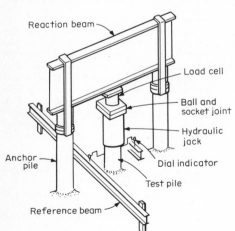

Fig. 22 Arrangement for pile load test using anchor piles. *(After Peck, Hanson, and Thornburn.)*

Pile Testing. Although there are numerous formulas for predicting the bearing capacity of a pile on the basis of its point resistance, it is general practice (most building codes require it) to use test piles to check the design loads where high capacities, especially those in excess of 40 tons per pile, are contemplated. The test pile is usually loaded by jacking against a reaction beam yoked to two or more anchor piles installed at not less than 5 ft (1.5 m) or 5 butt diameters from the test pile (Fig. 22). It may also be loaded directly by means of a box-type platform centered over the pile, or by jacking against a platform resting on blocks three or more diameters from the pile. The load is applied in increments, usually of about 25 percent of the anticipated working load. Under each increment the settlement of the head of the pile is observed until the rate of settlement becomes very small, about 0.3 mm/hr. Additional increments are applied until a total static test load of at least twice the anticipated working load is reached, or until failure occurs. The total load should be maintained for at least 24 hr and rebound readings taken as the pile is unloaded. Further details of standard testing procedures are given in ASTM Designation D1143 and in various building codes.

The driving of a pile not only alters the structure of the surrounding soil but also may cause excess pore pressures which will dissipate with time. Because of these effects it is customary to allow some time to elapse after driving before the test is conducted. For friction piles in cohesive soils or endbearing piles in dense silty sand, a minimum of 3 to 7 days should be allowed. On the other hand, piles in clean granular materials or those driven to point bearing in rock may be tested after 24 hr. Judgment must be used in extrapolating the behavior of a single pile under load to the prospective behavior of a pile group. Special attention must be paid to the possibility of failure of friction-pile groups in cohesive soils at loads below those which would be predicted by considering only individual pile behavior. Valuable information on the performance of friction piles is given by Peck.[27,28]

In some cases, piles or pile groups are required to resist lateral loads or uplift forces. Special testing is required in order to determine the working loads in these instances. Suggested procedures for determining deflections of vertical piles and batter-pile frames under lateral loads are given in Ref. 29. The lateral resistance of piles and methods of analysis are summarized by Davisson.[30] Ireland shows the need for pulling tests to establish uplift resistance of piles and discusses the results of such tests in Ref. 31.

36. Special Observations Although some indication of the location of the groundwater table can be determined from boreholes 24 or more hours after the boring is completed, such observations have little or no meaning unless the soil contains permeable granular strata. If the soil consists of fine-grained silts and clays, the permeability is so low that the true position of the groundwater table cannot be determined in such a short time, and special installations for measuring pore pressures are required.

Permeability of Granular Materials. Because of the difficulties in obtaining undisturbed and representative samples of granular materials, their permeabilities usually cannot be determined with any degree of precision on the basis of laboratory tests. Generally, field permeabilities for any given strata exceed the normal laboratory determinations two to four times.

Where it is necessary to lower the groundwater table or to estimate the amount of seepage into an excavation, field permeability tests must usually be performed. The most reliable procedure for determining the permeability of sand or gravel strata located below the water table is field pumping, in which a central pumping well is surrounded by two or more observation wells. Water is pumped from the pumping well at a constant rate until the water level in the well has become nearly stationary. Once this state has been established, the elevation of the water in the observation wells can be measured and the total average flow across the boundary of any cylindrical cross section computed. According to Terzaghi and Peck, pumping tests require the construction of one test well 10 to 12 in. (250 to 300 mm) in diameter and at least eight observation wells located on two straight lines through the mouth of the test well. One line should be located approximately in the direction of groundwater flow and the other at right angles to it.[17]

Permeability of Fine-Grained Soil. Observations in open boreholes cannot be relied upon for accurate information on the pore-water stress. Even when special piezometer tips are installed at the bottom of the boring it may be extremely difficult to seal off water-bearing or water-taking strata in such a manner as to get reliable results. Therefore, special techniques for measuring the pore-water pressures in such soils must be used. Such measurements may be needed to determine the progress of dewatering for a construction operation, or they may be required for keeping accurate control of excess pore pressures so that the shear strength of an embankment, a slope, or a foundation will not be exceeded during the construction process. The Casagrande piezometer or some slight modification of it is commonly used (Fig. 23). A principal difficulty in obtaining accurate piezometric measurements is the formation of gas bubbles which prohibit complete saturation within the tip. A two-tube piezometer which allows gas-free water to be flushed through the tip prior to the making of measurements permits the use of smaller-diameter tubing and also horizontal runs to convenient gage locations.

The installation of any piezometer produces a temporary change in the pore pressure. Furthermore, the excess pressure produced varies with the soil type, the piezometer type, and the method of installation, so that it cannot be predicted. Thus, there is a variable time lag before the pore pressures at the piezometer tip come to equilibrium with those in the soil. Further discussion of the use of piezometers is given in Ref. 9.

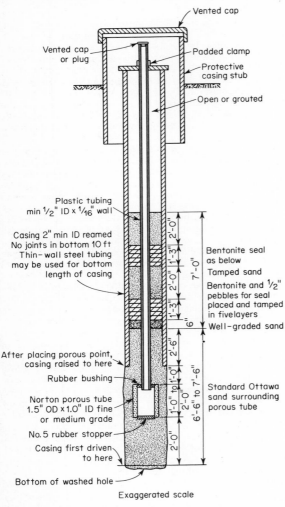

Fig. 23 Casagrande porous-tube piezometer. *(After Leonards.)*

Measurement of Ground Movements.   During and after construction it is often advantageous to measure the vertical and horizontal movements of the earth and the structure. When settlement is expected, vertical movements should be observed to compare their magnitudes with the predicted values. If structures are built on either natural or artificial slopes on which there is a possibility of failure, both vertical and horizontal movements should be observed. From these observations, planes of weakness can be discovered and measures taken to prevent serious failure. Descriptions of various types of field installations designed to monitor ground movements and examples of such movements are given in Refs. 9, 32, and 33.

Dynamic Characteristics. Dynamic loading may produce a wide range of deformations of soils. Such loadings are caused by events varying from severe earthquakes or explosions to relatively minor, but cumulative, repetitions of load from machinery vibrations. The techniques for evaluating the dynamic properties of earth materials are by no means standardized, and test methods are continually being devised or refined.[15,29] Dynamic analysis and design is discussed in Refs. 34, 35, and 36.

Exploratory Program

There is no standard procedure that can be followed in setting up a program of soil exploration. Because of differences in natural soil deposits, types of structures, and peculiarities of construction conditions at any given site, standardized programs usually provide either more information than is necessary or too little pertinent information. Therefore, no more than general guidelines can be given.

37. Preliminary Site Inspection Although a visual examination of the site cannot be considered to be absolutely necessary in carrying out the program of soil investigation, such a visit is nearly always made during the preliminary planning stage for any major structure. At that time certain site characteristics can be observed with little difficulty.

Surface Conditions. If possible, the condition of the site should be observed before it has been disturbed in any way by construction activity. It is particularly important to note any indications of a seasonal high water table. If the site has been previously improved, the location of former improvements and of any structural foundation elements which may be present should be noted. Visual inspection will often reveal whether the area has been filled. If possible, the type of fill should be noted. Fill material, whether rubble, rubbish, or soil, will nearly always be more compressible than the underlying natural soil. Thus, if shallow foundations are to be used, it is absolutely necessary to determine the locations and depths of such materials.

Topography. Sketches of the original site topography can be very helpful in planning the program of exploration, especially where the structure is to cover a large area. It is important that the relationship between the topography of the site and that of adjacent areas be carefully noted. For example, if the site is located on a hillside, or in a valley, a high water table may be expected and should be taken into consideration even though the exploratory program may be made at a time of the year when the water table is low.

Adjacent Structures. The relationship of the proposed project to nearby structures should be carefully noted. Not only may this have a direct influence on the location of the structure within the available site area, but also it will indicate the possible effect of the construction on adjacent structures. If possible, adjacent structures should be inspected for any signs of damage which appear to be due to settlement of their foundations, such as open or repaired cracks in the walls, and distorted lintels, sills, or frames of windows and doors. Such signs nearly always indicate inadequate foundations. However, when no signs of distress are apparent it cannot be assumed that the same type of foundation can be used, since subsurface conditions may change radically in a relatively small distance, both horizontally and vertically.

38. Fact-Finding Survey An absolutely essential initial step in any well-planned program of soil exploration is the fact-finding survey. Where adjacent structures exist, information should be sought regarding types of foundations and soil conditions. If severe stress is evident in existing structures, settlement records may be available.

It is nearly always possible to obtain some information on the local geology. If the engineer is familiar with geologic conditions in the area, a site inspection may permit identification of the local deposits. In any event, reference to appropriate bibliographic sources should be made, and the surficial and areal geology of the site determined.

39. Borings Generally, unless the subsurface conditions are quite well known, the investigation should be started with dry sample borings and the standard penetration test to procure split-spoon samples. Once the first boring has been completed and some detailed information is available with regard to the stratification, the type of sampling can be modified so as to produce the essential information in the most economical manner.

The number of borings to be made at any particular site depends upon the size of the site, the character of the soil conditions, and the critical characteristics of the structure. In the exploration of areas of limited extent, as for individual buildings and bridge piers, the initial borings should preferably be located close to the corners of the area, and at least three borings should be made unless subsurface conditions are known to be very uniform. These preliminary borings must be supplemented by intermediate borings, as required by the extent of the area and the subsurface conditions encountered. Boring spacings of 50 to 100 ft (15 to 30 m) for multistory buildings and 100 to 300 ft (30 to 90 m) for one-story manufacturing plants are often used.[7] If initial borings indicate uniform regular soil conditions, the above spacings may be doubled; for irregular conditions they could be halved.

Depth of Exploration.   Although the depth of exploration depends to some extent on the size and type of the proposed structure, it also depends on the character and sequence of the subsurface strata. Unless the stratigraphy of the area is known, it is impossible to estimate the required depth. It is especially important that the first reconnaissance boring be carried to fully adequate depths. The following general rules are suggested by Hvorslev.[13] All borings should be extended to strata of adequate bearing capacity and should penetrate all deposits which are unsuitable for foundation purposes, e.g., unconsolidated fill, peat, organic silt, and very soft compressible clay. Soft strata should be penetrated even when they are covered with a surface layer of higher bearing capacity, since they may still contribute to excessive settlements. Where settlement of the structure becomes the primary consideration, the boring should penetrate the compressible strata completely or be extended to such a depth that the stress increase for still deeper strata is reduced to values so small that the corresponding consolidation of these strata will not materially influence the settlement of the proposed structure. Except in the case of very heavy loads, or when seepage or other considerations govern, the borings may be stopped when rock is encountered. Boring may also stop after a short penetration into soil strata of exceptional bearing capacity and stiffness, provided it is known from explorations in the vicinity or from the general stratigraphy that these strata have adequate thickness or are underlain by still stronger formations. If the latter conditions are not fulfilled, some of the borings must be extended until it has been established that the stiff strata have adequate thickness, irrespective of the character of the underlying material. If the structure is to be founded on rock, it must be verified that bedrock and not a boulder bed has been encountered. It is advisable to extend one or more borings from 15 to 25 ft (5 to 8 m) into sound rock in order to determine the extent and character of the weathered zone.

Cost.   In general, a well-conducted program of underground exploration will involve expenditures of about 0.1 to 1 percent of the total cost of a project, with exceptional cases as high as 2 percent or as low as 0.03 percent. In situations where the subsurface conditions are unusually complex the exploratory program may cost 5 to 10 percent of the total.

A program of soil exploration must be undertaken except when the designer is thoroughly familiar with the area, or when greater overall economy is achieved if the program is omitted and the foundation elements designed on an ultraconservative basis. Except for lightweight residential structures, these exceptions are extremely rare. Even single-story manufacturing and warehouse structures have suffered serious damage where the exploratory program was neglected.

Even a well-executed program may fail to reveal some important detail of subsurface conditions which may affect the structure. Therefore, the designer should take every opportunity to observe excavations during the construction phase.

Part 3. Retaining Structures and Foundations

HERBERT O. IRELAND
Professor of Civil Engineering, University of Illinois, Urbana

EARTH PRESSURES

40. Stresses in Earth Mass The states of stress in an earth mass may be conveniently represented on a Mohr rupture diagram (Fig. 24) wherein the normal stresses on any plane in a mass of soil are plotted as abscissas and the corresponding shear stresses as ordinates. In this diagram the principal stresses plot on the horizontal axis, which also represents the planes on which the principal stresses act, since, by definition, a principal plane is one on which there are no shear stresses. In a semi-infinite mass of soil with a horizontal surface, the normal stress on any horizontal plane is a principal stress equal to the weight of the overlying material. A second principal plane is oriented at 90° to the former, which corresponds to a vertical plane in this case.

A Mohr circle gives the stresses on every plane through a point in the soil mass. The major and minor principal stresses are denoted by p_v and p_h, respectively. The major principal stress is given by

$$p_v = \gamma z \tag{19a}$$

where γ = unit weight of soil and z = depth. The minor principal stress is given by

$$p_h = k_0 p_v \tag{19b}$$

where k_0 is an empirical constant known as the coefficient of earth pressure at rest. Values of k_0 range from about 0.40 for loose sand to 0.50 for dense sand, and may be increased to about 0.8 by tamping in layers. Overconsolidated clays may have a value k_0 greater than unity, in which case p_v becomes the minor principal stress.

Any change in the value of k for a particular mass of soil must result from either expansion or compression of the entire mass in a horizontal direction. The vertical pressure p_v does not change, because the weight of soil above any horizontal section is unaffected by these deformations. However, the horizontal pressure decreases if the mass of soil expands, and increases if it is compressed. The lower limiting value $p_h = k_a p_v$ (Fig. 24) is controlled by the properties of the material as they influence the location of the rupture line, which is the locus of failure stresses. The rupture line makes the angle ϕ with the horizontal, where ϕ is the angle of internal friction (Art. 8). The corresponding stress circle C represents a failure condition, and any further expansion is associated with sliding along two plane surfaces within the soil mass. These failure surfaces intersect the plane on which the major principal stress acts at an angle of $45° + \phi/2$.

The maximum value $p_h = k_p p_v$ is also controlled by the location of the rupture line. The corresponding stress conditions are represented by circle D. Any further compression results in sliding along two sets of plane surfaces which make an angle of $45° + \phi/2$ with the plane on which the major principal stress acts. In this case, the horizontal pressure is the major principal stress, so that the plane on which it acts is vertical.

The parameters k_a and k_p are called the coefficients of active and passive earth pressure, respectively. Since the stresses corresponding to the active and passive states represent a failure condition, these conditions are said to result in a state of plastic equilibrium. Every intermediate state, including the state of rest, is referred to as a state of elastic equilibrium. It can be demonstrated from the geometry in Fig. 24 that

$$k_a = \frac{1 - \sin \phi}{1 + \sin \phi} = \tan^2 \left(45 - \frac{\phi}{2}\right) \tag{20}$$

$$k_p = \frac{1 + \sin \phi}{1 - \sin \phi} = \tan^2 \left(45 + \frac{\phi}{2}\right) \tag{21}$$

For a soil where the rupture line intercepts a shear strength at zero normal pressure, called cohesion c (Fig. 5), these relationships must be modified as shown below.

$$k_a = \tan^2 \left(45 - \frac{\phi}{2}\right) - \frac{2c}{\gamma z} \tan \left(45 - \frac{\phi}{2}\right) \tag{22}$$

$$k_p = \tan^2 \left(45 + \frac{\phi}{2}\right) + \frac{2c}{\gamma z} \tan \left(45 + \frac{\phi}{2}\right) \tag{23}$$

41. Rankine's Theory Rankine's earth-pressure theory corresponds to the stress and deformation conditions for the states of plastic equilibrium described in Art. 40. The resultant *active* pressure on a vertical plane of height h through a semi-infinite mass of soil whose surface is inclined at an angle β to the horizontal is

$$P_a = \frac{1}{2} \gamma h^2 \left(\cos \beta \, \frac{\cos \beta - \sqrt{\cos^2\beta - \cos^2\phi}}{\cos \beta + \sqrt{\cos^2\beta - \cos^2 \phi}}\right) \tag{24}$$

The resultant force P_a is parallel to the ground surface and acts through the lower third point of the vertical plane. For the special case when the ground surface is horizontal, this equation simplifies to

$$P_a = \frac{1}{2} \gamma h^2 \frac{1 - \sin \phi}{1 + \sin \phi} = \frac{1}{2} \gamma h^2 k_a \tag{25}$$

If a vertical plane rotates about a point at depth z in a soil mass, a wedge-shaped zone develops the state of plastic equilibrium corresponding to Rankine's theory. This wedge of soil is bounded by plane surfaces. Rankine's theory deals strictly with the equilibrium conditions of such a wedge when the shear resistance of the soil is fully mobilized. Theoretically, it is applicable to retaining walls only if the wall does not interfere with the formation of any part of the wedge on either side of the vertical surface which passes through the point of rotation. Nevertheless, Rankine's theory is commonly used to compute the earth pressure against a vertical plane through the heel of a wall.

42. Coulomb's Theory The Coulomb earth-pressure theory for the active state of stress yields the resultant pressure against the back of a retaining wall. It assumes that the soil slides on the back of the wall and mobilizes the shearing resistance between the wall and the soil. The resultant is inclined at an angle δ, the angle of wall friction, with respect to the normal. The other side of the wedge is a failure surface in the soil, which implies that the wall yields sufficiently to develop an active state of stress. It is easily demonstrated that the surface of rupture must be curved in order for the forces involved to be in static equilibrium.

Coulomb's formula for the resultant active earth pressure against the back of a retaining wall is

$$P_a = \frac{1}{2} \gamma h^2 \frac{\cos^2 (\phi - \omega)}{\cos^2 \omega \cos (\delta + \omega) \left[1 + \sqrt{\dfrac{\sin (\delta + \phi) \sin (\phi - \beta)}{\cos (\delta + \omega) \cos (\omega - \beta)}}\right]^2} \tag{26}$$

where ω = angle between vertical plane and back of wall
β = slope of backfill with respect to horizontal
h = vertical projection of pressure surface

43. Equivalent-Fluid Method The equivalent-fluid method is a modification of the Rankine active earth-pressure theory for the case of the backfill with a horizontal ground surface [Eq. (25)]. The equivalent fluid pressure is assigned a value $\gamma_f = \gamma(1 - \sin \phi)/(1 + \sin \phi)$. Typical values for γ_f range from 30 to 45 pcf. A specific value within this range is often used as a standard in many offices. Although the equivalent-fluid method is widely used, it has little to commend it. There is indeed a danger that it may be misused for the case where the ground surface is not horizontal. Furthermore, it leads to misunderstanding and a tendency to obscure the real nature of the earth-pressure problem.

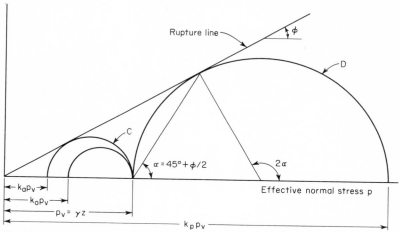

Fig. 24 Shear-stress relationships for earth pressures.

44. Trial-Wedge Method The practical problem often involves boundary conditions which cannot be readily incorporated into formulas based on the Rankine or the Coulomb theories. For these cases, as well as for the usual case, a graphical solution known as the trial-wedge method can be used. The method can be adapted to give earth pressures in accordance with either the Rankine or Coulomb theory, depending upon the assumptions introduced at the outset. One of its principal advantages is that every force involved is represented in both direction and magnitude. Thus, the designer is forced to contemplate each of them independently, and the influence of the shear strength of the backfill material can hardly pass unnoticed. The trial-wedge method is applicable for backfills of soils possessing cohesion, internal friction, or both, for backfills having any configuration of ground surface, and for surcharges located in any position on the backfill.

In the trial-wedge method plane surfaces of rupture may be assumed in all cases. The error introduced by assuming a plane surface when the actual surface of rupture is curved is not significant in active earth-pressure computations.

Rankine's theory is used in the trial-wedge example of Fig. 25. A surface of rupture, such as $BF4$ in a, is assumed. This surface is vertical through the depth of cracking h_0, which is given by

$$h_0 = \frac{2c}{\gamma} \tan \left(45° + \frac{\phi}{2}\right) \tag{27}$$

(Cracking occurs only in the case of a cohesive soil.) The weight of the wedge of soil between the vertical plane AB and the trial rupture surface is computed for a unit length of wall. This weight is held in equilibrium by the forces acting on the plane AB and on the surface of rupture. The active earth pressure P_a must be parallel to the ground surface to satisfy the conditions of static equilibrium. If the ground surface is broken within the active zone, as shown in Fig. 25a, the American Railway Engineering Association (AREA) has adopted the following procedure for approximating the direction of P_a: Point b is located on the ground surface at a distance $2h$ measured horizontally from the back edge of the top of the wall. This point is connected to the back of the wall at a by a straight line, which is assumed to represent the direction of the active earth pressure. The block of soil between the wall and the vertical plane on which the Rankine pressure acts is usually considered in the analysis as a part of the wall.

On the surface of rupture through the backfill the resultant R of the normal reaction and

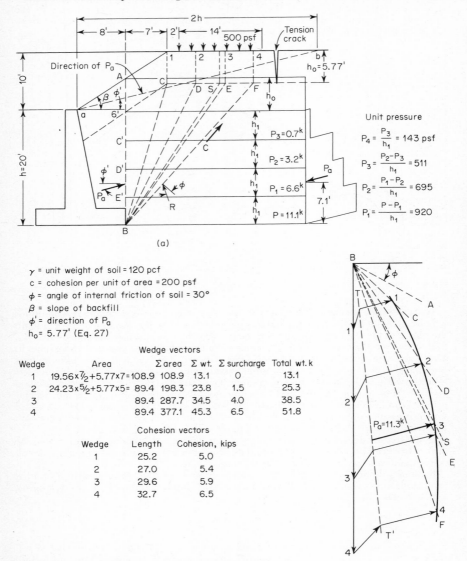

(a)

γ = unit weight of soil = 120 pcf
c = cohesion per unit of area = 200 psf
ϕ = angle of internal friction of soil = 30°
β = slope of backfill
ϕ' = direction of P_a
h_0 = 5.77' (Eq. 27)

Wedge vectors

Wedge	Area	Σ area	Σ wt.	Σ surcharge	Total wt. k
1	19.56×⁷⁄₂+5.77×7=108.9	108.9	13.1	0	13.1
2	24.23×⁵⁄₂+5.77×5= 89.4	198.3	23.8	1.5	25.3
3		89.4	287.7	34.5	4.0	38.5
4		89.4	377.1	45.3	6.5	51.8

Cohesion vectors

Wedge	Length	Cohesion, kips
1	25.2	5.0
2	27.0	5.4
3	29.6	5.9
4	32.7	6.5

Fig. 25 Trial-wedge method for earth-pressure computations. *(From Ref. 37.)*

the friction is inclined at an angle ϕ (Fig. 25a). To this must be added the shearing resistance C due to cohesion along the same surface. Since the directions of all the forces, and the magnitudes of all but the active earth pressure and the resultant force on the surface of rupture are known, it is possible to draw a force polygon from which the active earth pressure can be obtained. Thus, for the wedge $ABF4$ the force polygon is $B4T'4B$, where $B4$ is the weight of the wedge, $4T'$ the cohesion C, $T'4$ the active pressure P_a, and $4B$ the resultant R on the surface of rupture. The procedure is repeated for other trial surfaces, usually by superimposing the force polygons, and the locus of the computed earth pressures is plotted. The maximum value obtained from this locus is the active earth pressure against the vertical surface AB. In this method, surcharge loads can be readily added to the appropriate weight vectors as they are encountered by different trial surfaces of rupture.

Location of Resultant Pressure. Because of the influence of surcharge loads and an irregular ground surface, the point of application of the resultant pressure cannot be arbitrarily taken at the lower third point. Its location can be determined as follows:

1. Divide the pressure surface $6'B$ below the zone of tension cracks $6'CF$ into four equal parts of height h_1 (Fig. 25a).

2. Determine the active pressures P, P_1, P_2, and P_3 as if the base of the wall were, in turn, at B, E', D', and C'. This requires three additional trial-wedge polygons corresponding to Fig. 25b.

3. Compute the increments of pressure P-P_1, P_1-P_2, etc., and divide these differences by the height h_1 of the strips to determine the average unit pressure in each strip.

4. Determine the elevation $\bar{h}$ of the centroid of the resulting approximate-pressure diagram. This is the approximate point of application of the resultant earth pressure.

The foregoing procedure results in the following formula:

$$\bar{h} = \frac{h_1}{3P}(4P_3 + 2P_2 + 4P_1 + P) \tag{28}$$

Thus, for practical applications, Eq. (28) eliminates the calculations in 3 and 4 above. Although the formula is applicable only when $6'B$ is divided into four equal parts, use of a larger number of divisions does not appear to be justified.

RETAINING WALLS

45. Types and Behavior There are five types of retaining wall: gravity, semigravity, cantilever, counterfort, and crib.

The *gravity wall* (Fig. 26a) depends entirely upon the weight of masonry, and any soil resting thereon, for stability. It must be of sufficient thickness to resist the forces acting on it without developing tensile stresses. Concrete gravity walls usually contain a nominal amount of reinforcement near the exposed surfaces to control temperature cracking. Because of the small amount of reinforcement and the relatively simple formwork, gravity walls may prove to be economical for heights up to about 10 ft.

The *semigravity wall* (Fig. 26b) has largely supplanted the gravity wall. It is somewhat more slender and more conservative of materials. However, it requires vertical reinforcement along the inner face and into the footing to resist the rather small tensile forces that develop in these locations. A nominal amount of temperature steel must be provided to control surface cracking. It may be economical for heights up to about 20 ft.

The *cantilever wall* (Fig. 26c) is the commonest type. It consists of a base slab and a stem which are fully reinforced to resist the moments and shears to which they are subjected. The cantilever wall is relatively thin and economical of materials.

The *counterfort wall* (Fig. 26d) consists of a relatively thin concrete face slab which is supported at intervals on the back side by vertical counterforts connected to the base. The loads on the face slab are carried into the counterforts and in turn to the base slab. The space between the counterforts and above the base slab is part of the backfill. All the elements of a counterfort wall are fully reinforced. The primary reinforcement is longitudinal in the face slab, both longitudinal and transverse in the base, and extends in several directions in the counterforts. The counterfort wall requires considerable formwork but has a considerable economic advantage when relatively high walls of some length are required.

Crib Walls. Retaining walls are sometimes formed of stacked rectangular elements to form cells which are filled with soil (Fig. 26e). Their stability depends on the weight of the crib units and their filling, and on the strength of the filling material. Crib walls are relatively inexpensive. They are usually made of reinforced concrete, although both timber and fabricated metal crib walls are not uncommon.

The behavior of a retaining wall or abutment is a problem in soil-structure interaction. The lateral loads depend on the nature of the soil in contact with the structure and the deformations that occur. The backfill is an essential part of a retaining structure, and the properties and characteristics of the backfill are just as important as the properties of the concrete in the structure proper. A retaining structure must support its backfill without

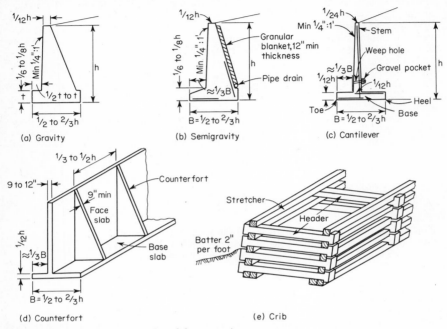

Fig. 26 Retaining-wall types (with trial dimensions).

detrimental lateral movement, and the surface of the backfill must not settle unduly. These deformations depend to a large extent on the properties of the backfill and the subsoil. If the subsoil contains compressible layers, these are likely to consolidate under the weight of the backfill and lead to differential settlements. Since the pattern of settlement tends to be bowl-shaped, it is not uncommon for settlement to result in the retaining wall tilting toward the backfill. More commonly the wall rotates forward or away from the backfill as a result of the lateral forces involved. The settlement at the surface of the backfill includes not only the consolidation that takes place in the subsoil but any reduction in thickness of the backfill as well. Settlement within the backfill may be minimized by careful control during construction.

The *reinforced-earth wall*, which may be considered to be a variety of gravity wall, is patented. It consists of a clean sand backfill containing reinforcing strips, called ties, that are bolted to a face element or skin. The ties are of galvanized steel on the order of 2 mm (about 0.08 in.) thick. Their width depends on their spacing. They must be of adequate length to develop their pullout resistance beyond the boundaries of the conventional active earth-pressure zone. They are often spaced 25 cm (10 in.) vertically and about 1 m (3 ft) horizontally. Since the face elements serve only to prevent loss of the backfill from between ties, they may be of very light-gage metal or they may be precast-concrete architectural elements.[48]

46. Determining Earth Pressure The properties of the retained soil must be known in order to determine the magnitude of the loads that the retaining structure must be designed to resist. It is usually possible to learn in advance the general type of backfill material and to classify it into one of the five categories shown in Table 12. For these classifications, charts for estimating the pressure of the backfill against retaining walls less than about 20 ft in height were prepared by Terzaghi and Peck (Figs. 27 and 28). These charts, based partly on theory and partly empirical, have been adopted by the AREA. For the cases covered, reliable estimates of earth pressures can be made without recourse to earth-pressure theories.

For walls higher than about 20 ft, and for cases where Figs. 27 and 28 are not applicable (for example, surcharged backfill), properties of the backfill should be determined by laboratory tests, and earth pressures computed by methods described previously.

TABLE 12   Type of Backfill for Retaining Walls

Type	Description
1	Coarse-grained soil without admixture of fine soil particles, very free-draining (clean sand, gravel, or broken stone)
2	Coarse-grained soil of low permeability due to admixture of particles of silt size
3	Fine silty sand; granular materials with conspicuous clay content; or residual soil with stones
4	Soft or very soft clay; organic silt; or soft silty clay
5	Medium or stiff clay that may be placed in such a way that a negligible amount of water will enter the spaces between the chunks during floods or heavy rains

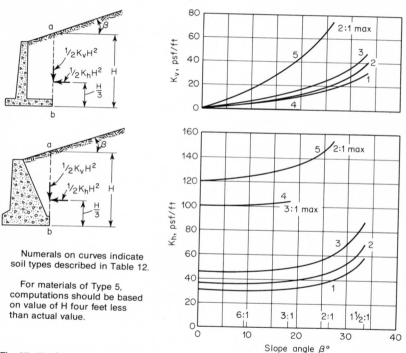

Numerals on curves indicate soil types described in Table 12.

For materials of Type 5, computations should be based on value of H four feet less than actual value.

Fig. 27 Earth-pressure charts for retaining walls less than 20 ft high. See Table 12 for soil types. (*From Ref. 37.*)

Base Pressures.   Figure 29 shows a cross section through a cantilever retaining wall and the forces acting thereon. The earth pressure is denoted by P_a. The forces W_s and W_c are the weight of soil and the weight of concrete, respectively. It is customary to analyze a 1-ft length of retaining wall except for the case of the counterfort wall, where a length equal to the spacing of the counterforts is commonly analyzed. Since a retaining wall is subjected to lateral forces, the foundation is usually eccentrically loaded. The eccentricity e of the resultant force on the base, measured from the center of the base, is determined by taking moments at some convenient point in the plane of the base, such as a, of all the forces acting above the plane of the base.

If the resultant lies within the middle third of the base the pressures on the base are

$$p_{max} = \frac{\Sigma V}{BL}\left(1 + \frac{6e}{B}\right) \qquad p_{min} = \frac{\Sigma V}{BL}\left(1 - \frac{6e}{B}\right) \tag{29}$$

where L = length of wall considered. If the resultant is outside the middle third, the soil-pressure diagram is triangular across a length $3(B/2 - e)$ of the base (Fig. 29b) and the maximum soil pressure is

$$p_{max} = \frac{2\Sigma V}{3L(B/2 - e)} \tag{30}$$

47. Bases on Piles or Piers   In order to calculate the pile or pier reactions, the moment of inertia of the piles and the center of gravity of the pile group must be determined. It is

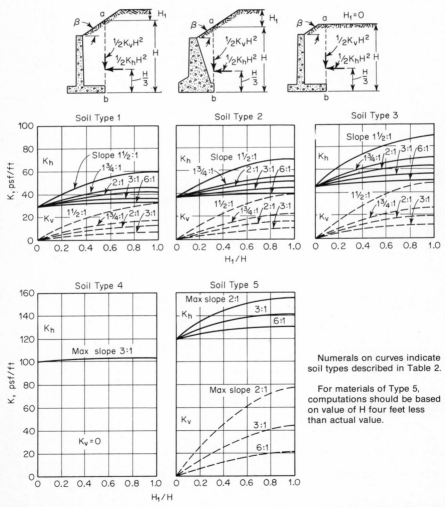

Numerals on curves indicate soil types described in Table 2.

For materials of Type 5, computations should be based on value of H four feet less than actual value.

Fig. 28 Earth-pressure charts for retaining walls less than 20 ft high. See Table 12 for soil types. (*From Ref. 37.*)

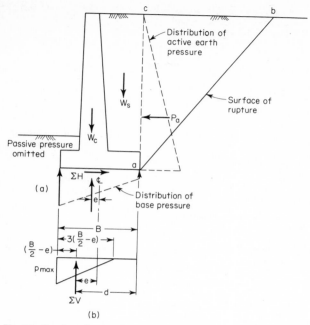

Fig. 29   Resultant force on base of retaining wall.

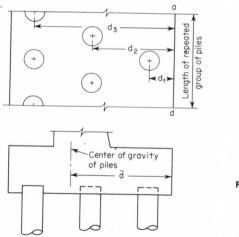

Fig. 30

necessary to consider a length of wall equal to the spacing of a repetitional group of foundation piles. If they are not arranged in a repetitional pattern, the entire length of wall should be analyzed. To determine the center of gravity of the pile group a convenient reference such as aa in Fig. 30 is selected. The piles are located at distances d_1, d_2, d_3 from this base line. Let N_1 equal the number of piles at distance d_1, N_2 the number at d_2, etc. The distance from the selected reference line to the center of gravity of the group is

$$\bar{d} = \frac{N_1 d_1 + N_2 d_2 + N_3 d_3}{\Sigma N} \tag{31}$$

The total reaction on any pile may be found by using the equivalent of the formula $s = P/A \pm Mc/I$. Corresponding to P/A, the axial load per pile is $\Sigma V/N$, where N is the number of piles in the group. The moment M is $e\Sigma V$, while c is the distance from the center of gravity of the group to the pile under consideration. The moment of inertia of the pile group about its center of gravity is $I = \Sigma d^2$.

Stability Requirements

48. Bearing Capacity In many instances involving the construction of embankments, overpasses, or bridge approaches it is necessary to construct a retaining wall backfilled to a considerable elevation above the existing ground surface. In these circumstances, precautions must be taken to ensure that a base failure beneath the weight of the fill does not occur. If the subsoil consists of sands or gravels there is no likelihood of such a failure. However, if the subsoil consists of clays or clayey silts it is necessary to check their supporting capacity. The ultimate bearing capacity of a clay subsoil is $q_d = 2.5\,q_u$, where q_u is the average unconfined compressive strength within the questionable layer (Art. 7). Since the actual bearing pressure $q_a = \gamma H$, where H is the height of fill and γ the unit weight of the fill—normally between 100 and 120 pcf—the factor of safety can be readily computed by

$$\text{F.S.} = \frac{q_d}{q_a} = \frac{2.5 q_u}{\gamma H} \tag{32}$$

Alternatively, the terms in this equation may be rearranged to compute a height of fill that can be built with a particular factor of safety. It is prudent to use a factor of safety of at least 2 for this condition.

Bearing Capacity under Base. In order to protect the completed structure against unsatisfactory foundation behavior, an adequate factor of safety must be provided with respect to the bearing capacity of the subsoil when subjected to the maximum base pressures as determined in Art. 46. This can be assured if the maximum pressure does not exceed the allowable soil pressure (Arts. 12, 13).

49. Stability *Sliding.* It is customary to require a minimum factor of safety of 1.5 against failure by sliding. The forces available to resist sliding are the friction and adhesion that can be developed between the base of the wall and the subsoil. Sometimes it is necessary to include the passive pressure of the soil in front of the wall in order to get a satisfactory factor of safety, which should then be somewhat greater than 1.5. If the passive resistance is used to ensure lateral stability, precautions must be taken to ensure that the subsoil in front of the wall not be removed by scour or other means, and that this material is present before the backfill is completed. The force causing a tendency to slide is the summation of the horizontal components of the earth pressures acting on the retaining wall. Accordingly

$$\text{F.S.} = \frac{\Sigma V \tan \phi + cB}{\Sigma H} \tag{33}$$

where $\tan \phi$ and c represent the coefficient of friction and the adhesion (assumed equal to the cohesion), respectively, and B is the base width of the wall. In the absence of better information, the coefficient of friction may be assigned values of 0.55 for sands and gravels, 0.45 for silty sands and gravels, and 0.35 for silt. It is prudent not to include the passive resistance, but if this be necessary it should be entered into the numerator of this equation.

Keys are sometimes provided as projections beneath the center, the heel, or the toe to provide added resistance to sliding. They are very desirable for walls founded on bedrock or other hard indurated surfaces. For most soil types, however, it can be demonstrated that they are likely to be of little value irrespective of their location.

Overturning. It is usually required that the factor of safety against overturning be at least 1.5. This is commonly determined by taking moments about the toe of all forces acting on the wall above the plane of the base. The factor of safety is the ratio of the moment of the forces resisting overturning to the moment of the forces tending to cause overturning. It is generally not necessary to consider stability against overturning, since this is assured if the resultant soil pressure is located within the middle third, as is usually required for a soil-supported wall, or if the resultant lies within the middle half of the wall, as is sometimes permitted where a retaining structure is founded on rock.

Miscellaneous Requirements

50. Drainage Commonly, drains consist of weep holes about 3 or 4 in. in diameter which extend through the stem of the wall just above the ground surface on the front side. Each is protected on the back by a pocket of gravel, which often does not exceed about 1 cu ft in quantity. Weep holes are commonly spaced at intervals of about 10 ft horizontally and, for relatively high walls, at about 10-ft intervals vertically. Weep holes are most efficient when the backfill is relatively permeable. The pocket of gravel often becomes clogged, and in freezing weather the weep hole is frequently covered with ice. Thus, although they are relatively inexpensive, weep holes may become inoperative. Therefore, a continuous back drain is always preferable when conditions permit its use.

The back drain consists of a perforated metal pipe or a pervious concrete pipe which must be surrounded by an appropriate filter and have adequate facilities for discharge. It should also be accessible for cleaning. In order to facilitate drainage, a relatively permeable backfill material, such as a clean sand and gravel, is most desirable; a clay is least desirable as clay soils can hardly be drained even if drains are provided. It is considered good practice to provide a continuous body of free-draining material between the wall and a clay backfill. This often consists of a layer of pea gravel about 1 ft thick deposited against the back of the retaining wall as the fill is placed (Fig. 26b). Clays are undesirable as backfill for other reasons; they are likely to experience alternate swelling and shrinking with the seasons, and may lose much of their shearing strength if their water content should increase. A reduction in the shear strength of a backfill causes an increase in the lateral earth pressure.

51. Other Considerations The base of a retaining wall or abutment should be located below the frost line or zone of seasonal volume change and, if adjacent to a stream, should be below the anticipated maximum depth of scour. Otherwise, piles or piers should be used for support.

The usual earth pressures that act on a retaining structure can be expected to cause a slight outward tilt at the top. For purposes of appearance it is common to provide a batter on the front of the wall of about ½ to 12 so that it will not give the appearance of overturning when movement occurs. It should be recognized that a relatively high fill placed over a compressible subsoil may cause consolidation which can lead to larger settlement near the heel than at the toe and cause the wall to tilt backward rather than forward. Procedures for estimating settlement are presented in Art. 14.

It is essential that semifluid materials which often accumulate behind a bridge abutment be removed prior to backfilling. If left in place they exert a lateral hydrostatic pressure equal to the vertical pressure that comes on them. This can result in lateral pressures far greater than those assumed in the design (Fig. 31).

Structural Design Requirements

52. Gravity Walls Gravity walls are so proportioned that any horizontal section through the wall is subjected to compressive stresses across the entire section. The front face of the gravity wall is battered for aesthetic purposes, while the back is sloped as required to eliminate tensile stresses at any point. The magnitude, direction, and point of application of the resultant forces on any horizontal section through a gravity wall can be readily determined and the maximum and minimum stresses computed for comparison with permissible values, by the procedures described previously. Temperature steel is required to control cracking.

Crib walls (Art. 45) may be built with open- or closed-faced cribbing. They consist of a front and back row of stretchers connected by a header. The lower stretchers of the crib wall are placed directly on the subsoil after it has been carefully leveled and thoroughly tamped at both the toe and heel. Batter at the front of the crib wall of about 2 in./ft is provided by setting the toe and heel stretchers at slightly different elevations on the prepared subsoil. Crib walls are backfilled with gravel, crushed stone, rock, or other coarse granular material. The crib wall makes a relatively economical retaining structure, useful for many applications where the height of wall required is less than 24 ft, which is the maximum recommended height.

53. Semigravity Walls Although the width of the base for a semigravity wall is comparable with that of a gravity section, the stem at the top of the base is considerably thinner and is often governed by the requirement that its width at the top of the footing be at least one-fourth its height. Because of the thinner stem, the base may project far enough so that some reinforcing steel must be provided in the toe. In a semigravity wall there are tensile forces near the back side of the stem. These must be resisted by providing steel reinforcement. The quantity can be determined in accordance with usual design procedures for cantilever beams. However, a simplified procedure is sometimes used, in which reinforcement is provided in area sufficient to supply a tensile force equal to the computed tension resultant in the concrete, where this tension is computed as for a plain concrete member. Nominal temperature reinforcing should be provided in the front face to control cracking.

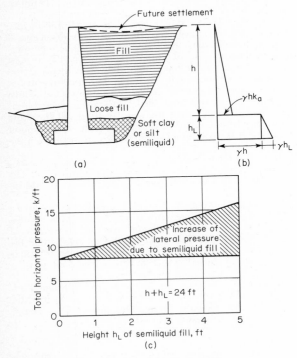

Fig. 31   Influence of a semiliquid under backfill.

54. Cantilever Walls   The cantilever wall consists of three structural elements: the stem, the toe, and the heel, each acting as individual cantilevers. The maximum shear and moment on the toe occurs on a vertical section located at the face of the stem (Fig. 32). In computing the shear and moment, it is customary to omit the soil on top of the toe. The maximum shear and moment on the heel occur on a vertical section at the back of the stem. The downward force due to the weight of backfill on the heel is the major force to be considered, and the reinforcement must therefore be placed at the top of the heel across the critical section. With respect to the stem, the maximum section will be required at the top of the base. The principal force acting on the stem is the earth pressure of the backfill. Thus, reinforcement is required at the back of the stem (Sec. 11, Example 12). Precautions should be taken during construction to ensure that lateral loads in a direction opposite to the earth pressure do not come on the stem. The face of the stem is customarily provided with temperature steel to control cracking. This is sometimes specified as not less than 0.25 in.^{2} of horizontal reinforcement per foot of height.

The width of the base of the cantilever retaining wall is usually 0.4 to 0.65 times the height of the wall. The weaker the backfill and the steeper its slope the wider the base required. The stem is located so that the toe projects about one-third the width of the base beyond the face of the stem. The thickness of the base is normally about one-tenth to one-twelfth of its width. It is generally satisfactory to select a thickness of stem at the base equal to one-tenth to one-twelfth the height of the wall for purposes of a preliminary analysis. The design of a retaining wall is a trial-and-error procedure; several sections may

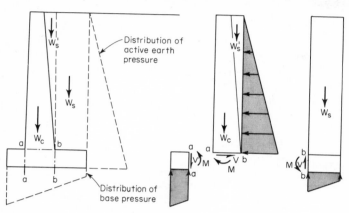

Fig. 32 Critical sections of cantilever wall.

have to be investigated before a suitable one can be selected for a particular site. In many instances, the geometry of the wall must differ considerably from these suggestions because of specific site requirements. For instance, it may be necessary in some cases to construct a cantilever retaining wall with little or no heel.

55. Counterfort Walls The face slab of a counterfort wall is usually designed as a longitudinal beam strip continuous across the counterforts. A strip 1 ft in height loaded uniformly with the normal component of the earth-pressure diagram at the corresponding depth is used. It is assumed that there is no shear between adjacent strips. Thickness of the slab is usually comparable with that of the cantilever wall, i.e., about one-twelfth the height. The slab is commonly designed for equal positive and negative moments of $\frac{1}{12} WL^2$, where L is the clear distance between counterforts.

Spacing of the counterforts is governed by economy and commonly varies from one-third to one-half the height of the wall. The thickness of the counterfort should not be less than 12 in.; it is usually governed by the thickness required to place the reinforcement in the back.

The heel slab is also designed in longitudinal strips 1 ft wide loaded with the weight of the backfill on top and the soil pressure on the bottom. The heel slab functions similarly to the face slab and may be designed for equal positive and negative moments of $\frac{1}{12} WL^2$. The toe of the base acts structurally as a cantilever and is designed as is the toe of the cantilever wall.

The counterfort serves as a tension member between the face and heel slabs. Therefore, it must contain horizontal reinforcement to tie it to the face slab and vertical reinforcement to tie it to the base slab. The tension in the counterfort is carried by the main reinforcing steel located near the back of the counterfort. The tension can be determined at any elevation by taking moments, about a point located at the center of the face slab, of the forces acting above the horizontal section.

The structural interaction in a counterfort retaining wall is relatively complex, and it is recommended that Huntington's presentation be studied before designing a major installation.[38]

56. Joints Joints (Fig. 33) are usually required in retaining walls between successive units of construction and may be required in other locations to minimize the effects of temperature changes, shrinkage, and possible differential settlement. Design codes differ in their requirements for spacing of vertical joints in a long wall. However, joints capable of permitting expansion and contraction are usually not over 100 ft apart and often as close as 50 or 60 ft. Joints are usually provided between the wing walls and the abutments of bridges and should be provided wherever there is an abrupt change in the nature of the subsoil.

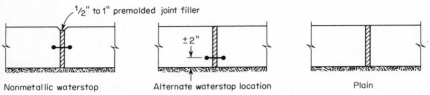

Nonmetallic waterstop   Alternate waterstop location   Plain

Fig. 33 Typical retaining-wall joints.

ABUTMENTS

With the exception of the crib wall, any retaining-wall section can be used for an abutment if it is provided with a bridge seat (Fig. 34a). In order to protect the backfill, bridge abutments usually have wing walls. The U abutment is one in which the wing walls are at 90° with the face of the abutment. Where the additional clearance provided by a vertical wall is not required, spill-through abutments consisting of two or more vertical columns or pilasters carrying a beam, which serves as the bridge seat, are sometimes used. The backfill is placed between and in front of the vertical columns on a natural slope.

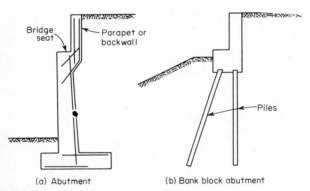

(a) Abutment   (b) Bank block abutment

Fig. 34 Abutments.

The passive resistance of the soil in front of a spill-through abutment can be only partially mobilized because the deformation conditions for a satisfactory structure can never reach a magnitude great enough to develop the passive resistance. Therefore, it is recommended that the passive pressure be limited to a value equal to the active pressure for the height of the openings. The spill-through abutment may be designed to withstand the active earth pressure exerted on the parapet wall above the openings. Because of the beneficial effect of the soil in front of the spill-through abutment, the opening should be made as wide as practicable in order to minimize the earth pressures acting on the wall. The fill should be compacted with equal care on both sides of the wall. However, if there is any possibility that the soil in front of the pilasters may be removed, the pilasters should be proportioned to resist the lateral pressures that would exist if there were no openings.

A bank-block abutment (Fig. 34b) is a spill-through abutment where the vertical columns are either piles or piers installed through previously placed fill to support the bridge seat.

BULKHEADS

Bulkheads are usually waterfront structures which serve the same purpose as the retaining wall. They differ from the retaining wall in their method of support. A bulkhead is usually formed with a vertical wall of sheeting driven into the subsoil and fastened at the top to provide lateral stability (Fig. 35a). A conventional bulkhead is made of either steel or concrete sheeting driven to form a wall and provided with a wale and tie rods to an anchor located so that it can provide the required lateral support. The anchor block may get all its support from the soil, or it may be provided with batter-pile supports to resist the lateral loads.

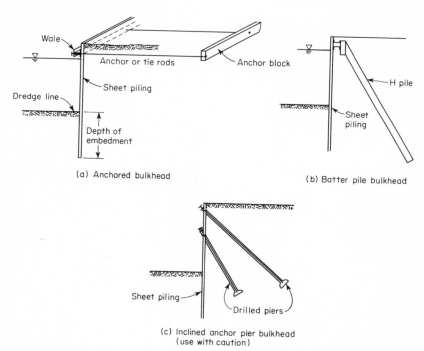

(a) Anchored bulkhead

(b) Batter pile bulkhead

(c) Inclined anchor pier bulkhead
(use with caution)

Fig. 35 Bulkheads.

In the batter-pile bulkhead (Fig. 35b) batter piles are substituted for the tie-rod and anchor system of the conventional bulkhead. Its proponents argue that there is no tie rod to deteriorate or be cut by later construction activity and, therefore, that it is more satisfactory than the conventional anchored bulkhead with tie rods.

A recent innovation is the use of inclined anchor piers to provide the lateral resistance (Fig. 35c). This differs from the batter-pile bulkhead only in the type of structural element used to resist the lateral loads. The inclined anchor piers are usually machine-drilled and may be belled out at the lower end to provide increased anchorage. This will depend primarily on the nature of the subsoil materials involved. The inclined-anchor pier bulkhead is gaining acceptance as a means of protecting the slopes of deep excavations to eliminate the bothersome cross-lot bracing normally required.

57. Forces on Bulkheads   The distribution of earth pressure on the back of any retaining structure is a function of the deformations. If insufficient deformation occurs to develop the active state of stress, the resulting pressures will always be greater than the active pressures. It is likely that any type of anchored bulkhead that can be built will yield sufficiently in its anchorage to develop the active state of stress in the backfill.

Before the forces acting on a bulkhead can be evaluated the soil conditions must be investigated by means of a program of subsurface exploration at the site. The standard penetration resistance of all cohesionless material should be determined in order to ascertain whether these are loose or dense. The unconfined compressive strength or the vane shear strength of all cohesive soils should be determined. The soils should be classified into one of the categories shown in Table 13.

The forces acting upon the inner face of the bulkhead include the active earth pressure produced by the weight of the backfill, the lateral pressure resulting from a uniformly distributed surcharge, the unbalanced water pressure, and the lateral pressures resulting from line loads applied at the surface. The active earth pressure produced by the weight of the backfill can be determined for each stratum in the soil profile by use of the appropriate coefficients from Table 13. Whether the soil is located above or below the water table should be noted, as the submerged weight must always be used in the latter case. The lateral pressure at a particular depth, resulting from a uniformly distributed surcharge load q per unit of area, is equal to q times the appropriate value of k_a. The unbalanced water pressure is the result of tidal action or drawdown following a flood and is included in the forces as a hydrostatic pressure. On the front of the wall, embedment below the dredge line provides passive resistance against an outward movement of the buried part of the sheetpiling. Figure 36 shows the distribution of active and passive pressures on sheet piles driven into sands and into clays.

58. Penetration of Piles In order to determine the depth of sheet-pile penetration, moments are taken about the anchor point at A (Fig. 36). Passive earth pressures should be used with a factor of safety G_s depending upon the nature of the material below the dredge line. For a cohesionless material G_s should be 2 to 3. For silt or clay, the unconfined compressive strength should be divided by a factor of safety of 1.5 to 2. The requirement that the sum of moments about the anchor point be zero usually involves a third-degree equation because both the total pressure and its moment arm must be entered in terms of the required depth of penetration D, which is the unknown. Terzaghi recommends that the computed depth of sheet-pile penetration obtained by taking moments about the anchor point always be increased by 20 percent, irrespective of the subsoil conditions, as insurance against the effects of unintentional excess dredging, unanticipated local scour, and the presence of weak materials in the zone of passive earth pressures which were not revealed by the borings.

The maximum bending moment in the sheet piles may be determined by assuming a hinge at the first hard layer below the dredge line and computing bending moments in the conventional manner. In the event that the computed moments are incorrect because of faulty assumptions, the sheeting may be overstressed locally. However, it is unlikely to fail if it is of a ductile material, provided both the anchor support and the depth of penetration are adequate. Although there might be local yielding, the pressures would be redistributed to other portions of the wall by arch action.

59. Anchorage The anchor pull is computed on the basis that the summation of the horizontal forces must equal zero, the active and passive pressures having the same values as those used in computing the required depth of penetration. The computed depth of penetration without the 20 percent increase is used. The anchor pull A_p is given by the formula

$$A_p = \Sigma P_H L \qquad (34)$$

where ΣP_H = sum of all horizontal forces, per ft length of bulkhead
$\quad\quad L$ = anchor-rod spacing

Particular care should be exercised in designing the anchor and anchor rods because, in reality, the anchor pull may be somewhat greater than computed. Therefore, it is prudent to be conservative in choosing allowable stresses in the anchor rods, and to make sure that the full cross section of the anchor rod is provided in every detail. The anchor block must be located far enough from the bulkhead to be outside the zone of the active earth pressure, and it must develop sufficient passive resistance to withstand the anchor pull. The passive resistance of the anchor block may be computed by using suitable soil constants from Table 13. If the anchor block is located below a line inclined at an angle ϕ with the horizontal and rising from the bottom of the sheeting (Fig. 37), the zone of passive earth pressure will not interfere with the zone of active earth pressure in a

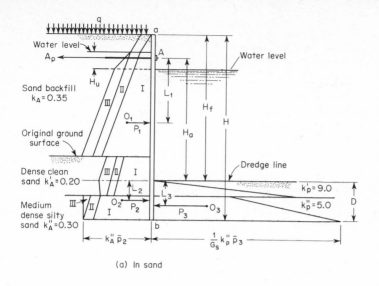

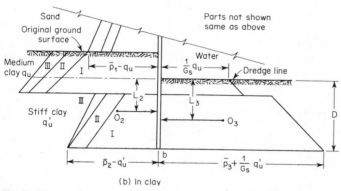

Fig. 36 Forces acting on bulkheads consisting of sheet piles driven into sand and into clay. *(From Ref. 39.)*

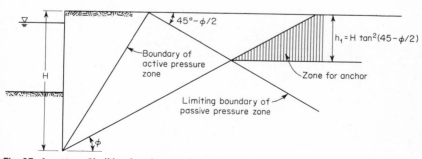

Fig. 37 Location of bulkhead anchor in cohesionless soil.

TABLE 13 Unit Weights of Soils and Coefficients of Earth Pressure*

Type of soil	Weight of moist soil γ, pcf		Weight of submerged soil γ', pcf		Coefficient of active earth pressure k_a		Friction angle		Coefficient of passive earth pressure k_p	Friction angle	
	Min	Max	Min	Max	For backfill	For soils in place	Internal ϕ	On wall δ	For soils in place	Internal ϕ	On wall δ
Clean sand:											
Dense................	110	140	65	78	...	0.20	38	20	9.0	38	25
Medium..............	110	130	60	68	...	0.25	34	17	7.0	34	23
Loose................	90	125	56	63	0.35	0.30	30	15	5.0	30	20
Silty sand:											
Dense................	110	150	70	88	...	0.25	...	...	7.0		
Medium..............	95	130	60	68	...	0.30	...	...	5.0		
Loose................	80	125	50	63	0.50	0.35	...	...	3.0		
Silt and clay†......	$\dfrac{165(1+w)}{1+2.65w}$		$\dfrac{103}{1+2.65w}$		1.00	$1 - \dfrac{q_u}{\bar{p}+\gamma z}$	...	...	$1 + \dfrac{q_u}{\bar{p}+\gamma z}$		

* After K. Terzaghi.

† $\bar{p}$ = effective unit pressure on top surface of stratum, q_u = unconfined compressive stress, w = natural water content in percent of dry weight, z = depth below top surface of stratum. Use γ' where applicable.

cohesionless soil, provided the anchor is located within a depth $h_1 = H \tan^2 (45° - \phi/2)$ from the ground surface, where H is the height of the sheet-pile wall. For a clay, the anchor block should be at a distance of at least $2H$ from the wall.

In many instances it is necessary to provide pile support for the anchorage. If the ground surface behind the anchored bulkhead is to be subjected to heavy surcharge loads, it is advisable to enclose the anchor rods in pipes or conduits to prevent their being subjected to vertical loads due to the effects of differential settlement in the backfill.

The batter-pile bulkhead can be analyzed in a similar fashion. However, instead of providing anchor rods, the anchor pull is resisted by batter piles. The approximate resistance of a batter pile to tensile forces can be estimated by computing the shearing resistance on the embedded portion of the pile resulting from the adhesion between the soil and the pile. This resistance can be realistically determined only by conducting pulling tests on typical piles at the site. The horizontal component of the axial resistance of the batter pile must be adequate to develop the anchor pull with a factor of safety of at least 2.

Although the batter-pile anchorage eliminates some of the problems with conventional anchor rods, the margin of safety may be unusually low. It is customary to assume that the adhesion between a pile and the adjacent soil is equal to the cohesion of the soil. However, if the soil is relatively stiff this assumption may seriously overestimate the adhesion. Furthermore, in the event a batter-pile anchorage approaches its ultimate capacity, there is no reserve strength once movement has started. The connection of the batter-pile anchorages to the sheet-pile wall is relatively complicated and should be carefully planned to ensure that every detail can properly transmit the loads that may come on it.

FOOTING FOUNDATIONS

Spread footings and column footings (isolated footings) usually support a single column, while combined footings usually support more than one. The combined footing is sometimes called a cantilever footing. However, the cantilever footing usually consists of two spread footings connected by a beam, with the column supported at or near the outside edge of one of the footings. A continuous (wall) footing usually extends for a considerable length and generally supports a load-bearing wall. With the exception of some lightly loaded footings, most are constructed of reinforced concrete. Typical examples are shown in Sec. 11, Fig. 39.

Ordinarily, footings are located at the highest level where adequate supporting material may be found, but precautions should be taken to ensure that they are located below the depth of seasonal volume change as a consequence of frost penetration or, in semiarid regions, seasonal changes in moisture that cause appreciable shrinkage and swelling of the soil. Even in the more humid regions soil-shrinkage problems frequently occur where foundations have been established at too shallow a depth, particularly if adjacent vegetation, such as trees, may place a demand on the available soil moisture during times of drought. The extreme depth of frost penetration in the United States is shown in Fig. 1, but local experience is often the best criterion.

The size of a footing foundation for the support of a given load depends on the bearing capacity or settlement characteristics of the underlying subsoil. Since different procedures are used for clays, in contrast to sands, it is customary to classify the subsoil into one of these two categories.

The use of presumptive bearing pressures that appear in most building codes is not recommended except as a general guide or for unimportant structures where the cost of an engineering subsoil investigation would be disproportionate to the cost of the structure.

60. Footings on Clay   The ultimate bearing capacity for a shallow footing on clay is

$$q_d = 2.5q_u \left(1 + \frac{0.2B}{L}\right) \left(1 + \frac{0.2D_f}{B}\right)$$

where q_u = average unconfined compressive strength within a depth equal to the width of the largest footing
B = width of footing
D_f = depth of footing
L = length of footing

A factor of safety of at least 3 should be provided; thus, the allowable bearing pressure is

$$q_a = 0.83q_u \left(1 + \frac{0.2B}{L}\right)\left(1 + \frac{0.2D_f}{B}\right) \tag{35}$$

It follows that q_u is often taken as the allowable bearing pressure for a square footing at shallow depths. Although correlations have been presented that relate q_u to the results of the standard penetration test (Art. 7), such correlations are relatively crude and should not be used to determine the bearing capacity of footing foundations on clay.

If the footing is subjected to a resultant load at an eccentricity e, the width B should be replaced by an effective width $B' = B - 2e$. If there is also eccentricity about the second axis, it is sufficiently accurate to determine similarly an effective length L'. It is assumed that the total vertical load is uniformly distributed over the correspondingly reduced effective area.

At a factor of safety of 3, the stresses in the underlying clay are not likely to cause differential settlement between adjacent footings in excess of the generally accepted tolerable value of ¾ in., provided the footings are sufficiently far apart so that there is no overlap in the underlying stress patterns (Figs. 38 and 39) and provided that the underlying clay has an unconfined compressive strength greater than 2 tons/sq ft.

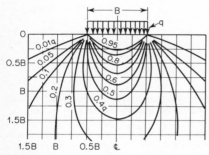

Fig. 38 Stresses beneath continuous footing.

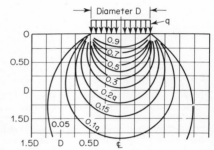

Fig. 39 Stresses beneath circular footing.

To forecast the magnitude of settlement or, more importantly, the differential settlement for foundations underlain by compressible clays generally requires laboratory consolidation tests on undisturbed samples. The settlement is given by

$$S = \frac{\Delta e}{1 + e_0} H \tag{36}$$

where Δe is the change in void ratio at the center of the compressible layer as a consequence of an increase in the original effective overburden pressure p_0 by an amount Δp, e_0 is the initial void ratio of the clay, and H is the thickness of the layer. Δe is best taken from the e-log p curve (Fig. 9) obtained from a laboratory consolidation test. However, if the clay has a sensitivity (ratio of undisturbed to remolded strength) less than 4 and has never been subjected to an overburden pressure greater than at present, i.e., is normally consolidated, a useful relationship is

$$\Delta e = C_c \log \frac{p_0 + \Delta p}{p_0} \tag{37}$$

The compression index C_c may be determined from a statistical relationship $C_c = 0.009$ $(L_w - 10$ percent$)$, where L_w is the liquid limit. For a normally consolidated clay the natural water content is generally closer to the liquid limit than the plastic limit.

The value of Δp is sometimes estimated by considering the surface load to be distributed uniformly on an area limited by a boundary that makes an angle of 30° with the vertical (Fig. 10). However, a better procedure is to use the graphical representation of

Boussinesq's equation, which is based on the theory of elasticity, commonly known as Newmark's influence chart (Fig. 40). This chart is prepared in such a way that a uniform load q covering any one of the influence areas will change the pressure at a depth AB below the center of the chart by $0.005q$. A drawing of the footing plan is prepared on transparent paper to a scale so that the distance AB shown on the chart is equal to the distance from the base of the foundation to the depth where the change in pressure is

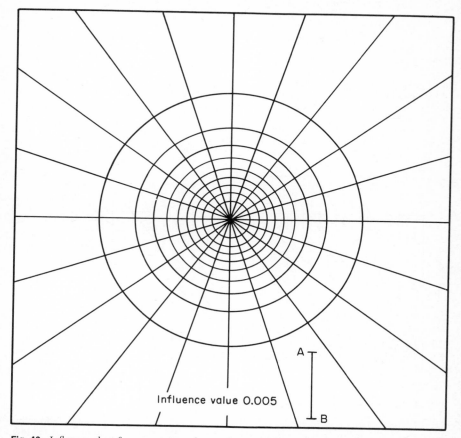

Influence value 0.005

Fig. 40 Influence chart for computation of vertical pressure. *(After N. M. Newmark.)*

desired. The point at the ground surface where settlement is to be determined is then placed over the center of the chart, and the number of influence areas covered by loaded areas is determined separately for each. Then $\Delta p = \Sigma(0.005q$ times the number of influence areas involved). If the pressure is desired at a different depth, a new drawing must be prepared to the appropriate scale $AB =$ depth.

The loads that should be considered to determine the size and probable settlement of a foundation on clay are sustained loads that cause a change in stress beneath the foundation. Exceptional values of live load that may be present for a short time, such as wind loads, snow loads, and earthquake forces, should not be included. However, the factor of safety should not be less than 2 for the most unlikely combinations of loading specified in the building code.

The time required for a predicted settlement to develop is difficult to determine with much accuracy, as it depends to a large extent on the distance between drainage layers in situ. A rate-of-consolidation theory is discussed in Art. 15, but it usually suffices for practical purposes to use

$$t_{\text{field}} = t_{\text{lab}} \frac{H^2_{\text{field}}}{H^2_{\text{lab}}} \tag{38}$$

where t_{field} = time required in the field to obtain the same degree of consolidation as represented by the corresponding time in the laboratory
t_{lab} = time in laboratory from time vs. degree of consolidation curve
H_{lab} = thickness of laboratory sample
H_{field} = thickness between drainage layers in the field

This assumes that the sample and the stratum in the field have drainage at both the top and bottom. Otherwise, H should be taken as the greatest vertical-flow path required for water to reach a drainage surface in both sample and stratum.

61. Footings on Sand (See also Art. 12.) The allowable soil pressure for a shallow footing on sand or other granular material depends on the relative density of the sand, the least width B of the footing, the depth of the foundation D_f, and the position of the groundwater table. The relative density is most readily investigated by means of the standard penetration test (Art. 31). The standard penetration resistance N has been empirically related to the bearing capacity as the basis for Fig. 41. The value of N to be

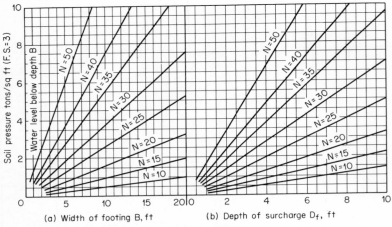

Fig. 41 Bearing-capacity charts for sand. (*From Ref. 6.*)

used in these charts is the average value in the poorest boring within a depth below the largest footing equal to its width B. However, for very fine or silty sands which are below the water table this value must be reduced as follows:

$$N' = 15 + \frac{1}{2}(N - 15) \qquad \text{for } N > 15$$

If the depth of the water table is within a depth B below the base of the footing, the values obtained from Fig. 41a must be multiplied by a correction factor which can be obtained from Fig. 42, where D_w is the distance from the base of the footing to the groundwater table as shown. If the groundwater level is above the base of the footing, the values of Fig. 41a should be multiplied by 0.5 and the values from Fig. 41b should be reduced by a correction factor from Fig. 42. The allowable bearing capacity is the sum of the corrected values from Fig. 41a and b.

The settlement of a footing on sand is similarly predicted from empirical procedures based upon the N value as a measure of the relative density in situ. Figure 43 shows the relationship between the N value and width of foundation for a limiting settlement of 1 in. It is based on the condition that the groundwater is below the base of the footing a distance greater than the width B. Otherwise, the allowable soil pressure obtained from

Fig. 43 must be corrected in the same fashion as previously described for Fig.41*a*. If the allowable soil pressure based on settlement of the largest footing is used for proportioning all other foundations in the same structure, it is anticipated that the differential settlement between adjacent footings will not exceed three-fourths of the settlement of the largest footings, i.e., ¾ in. If a larger differential settlement is tolerable, the values taken from Fig.

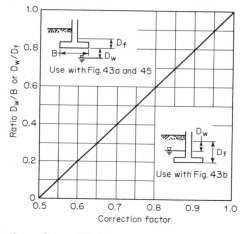

Fig. 42 Water-table correction chart.

43 may be modified by assuming that the settlement varies linearly with the intensity of the bearing pressure. Thus, if a differential settlement of 1½ in. is tolerable, which corresponds to a maximum settlement of the largest footing of about 2 in., the values from Fig. 43 may be doubled.

Since the deformation of sand occurs almost instantaneously with application of load, settlement should be based upon the full dead load plus normal live load and snow or wind load, whichever is greater. This should be the maximum load that may ever reasonably be expected to act upon the footing. The same loads are suitable for the bearing-capacity determination, except that a factor of safety of 2 should be used under the most severe combination of loadings.

In the majority of cases the settlement of a foundation on sand will govern the design, as the corresponding factor of safety against a bearing-capacity failure will be quite large. The usual exceptions involve lightly loaded or wall footings which are often relatively narrow. In these cases, the soil pressure corresponding to a 1-in. maximum settlement for the largest footing should be used to proportion all the smaller footings, with the precaution that the narrow footings be checked against the safe pressure obtained from Fig. 41. Whenever the allowable bearing capacity is smaller than the bearing pressure obtained from Fig. 43, the footings involved should be proportioned on the basis of the smaller value.

In earthquake regions, foundations should not be established on sand with an N value less than 25 blows per ft, because of the likelihood of spontaneous liquefaction due to seismic shocks. However, shallow foundations may prove satisfactory if the N value is at least 15 blows per ft, but these criteria are based on limited information and should be considered tentative.

62. Footings on Silt and Loess   Footings on silt may be proportioned by assigning the silt to the category of either a sand or a clay (Art. 12) and proceeding with the corresponding method for determining the foundation size. For important structures, a more elaborate procedure may be indicated wherein the shear strength of the silt is determined by either field vane tests (Art. 32) or appropriate laboratory triaxial tests.

The behavior of loessial deposits may be quite different from that of either sand or clay, and the final foundation design requires a program of standard load tests (Art. 35). The allowable soil pressure should not exceed one-third the failure load, as represented by the poorest load-settlement curve, or one-half the load at which ½ in. of settlement was obtained in the load test, whichever is smaller.

Inasmuch as loessial deposits usually have a relatively loose structure which is likely to collapse under moderate loads if the natural water content increases, unusual precautions must be taken to ensure adequate surface drainage to prevent, insofar as possible, changes in the water content. A procedure reported by Kezdi as developed in Hungary is based upon a so-called index of collapsibility:

$$i = \frac{\Delta e}{1 + e_1} \tag{39}$$

where Δe is the change in void ratio in a consolidation test as a consequence of adding water when the consolidation pressure is 3 tons/sq ft, and e_1 is the void ratio just prior to

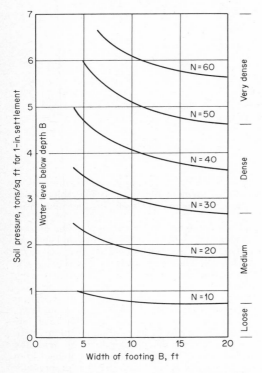

Fig. 43 Settlement chart for foundations on sand. *(From Ref. 6.)*

adding water (Fig. 44). Experience suggests that for values of i less than 0.02 the structure of the loess is not dangerous for the support of footing foundations, but if it exceeds 0.02 precautions are necessary. This may require the use of deep foundations.

RAFT FOUNDATIONS

A raft foundation is essentially a large combined footing which covers the entire area beneath a structure. It is likely to be more economical than individual footings whenever the total area of the latter exceeds about half the plan area of the building. The raft is designed as an inverted-slab system spanning between columns and walls and carrying the building weight as a load assumed to be uniformly distributed over the soil. Since this method does not take into account the stresses associated with differential settlements, and since every soil has variations in compressibility, the actual distribution of soil pressure is likely to be erratic. It is considered good practice to provide more reinforcement than required by the analysis and to provide the same percentage of reinforcement in both top and bottom of the raft. Where a raft foundation is intended to bridge local soft spots in the subsoil, it may be necessary to stiffen it to prevent unusual differential settlements between adjacent columns. Stiffening may be accomplished by providing ribs either above or below the mat and continuous with it so that they function as T-beams. In some instances these may be incorporated into interior partitions.

63. Raft on Clay The allowable bearing pressure for a raft on clay is the same as for a footing [Eq. (35)]. The value of q_a in this expression is the allowable increase in pressure at the base of the raft. Soil that is excavated can be replaced with building load without changing the pressure. Therefore, by increasing the depth of excavation the total building

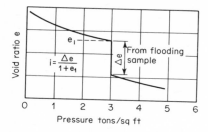

Fig. 44 Index of collapsibility for loess. (*After Kezdi.*)

load that can be supported is correspondingly increased. This can be accomplished by increasing the depth and number of basements. However, the depth of excavation in soft clays must be limited to prevent a base failure. In general, the weight of adjacent surcharge loads plus the soil above the bottom of the excavation should not exceed $2q_u$.

The settlement of a raft on clay is predicted in accordance with the method for footing foundations.

64. Rafts on Sand The width B of a raft foundation is not much greater than the width of the structure and cannot be less. Furthermore, a raft foundation is usually established at a minimum depth of about 8 ft. As a consequence, the allowable pressure based on bearing capacity is always very large and the design soil pressure is governed by settlement. Because of the large size of rafts and their structural integrity, differential settlements are likely to be less than those of a footing foundation at the same soil pressure. Therefore, larger design soil pressures are permissible. Since the soil pressure corresponding to a given settlement is nearly independent of the width of a foundation when the latter is greater than about 20 ft, the allowable design soil pressure for a differential settlement of ¾ in. can be written

$$q_a = \frac{N - 3}{5} \tag{40}$$

where q_a is the allowable bearing pressure in tons per sq ft, and N is the average penetration resistance in blows per ft obtained from the standard penetration test. The value of q_a must be reduced when the water table is within a depth B below the raft. This reduction is obtained from the water-table correction chart (Fig. 42).

The loads that should be used to determine the actual bearing pressure are the dead load of the structure including the raft and the maximum live load that is reasonably expected to act at any time. This load should be reduced by the weight of the soil excavated below the level of the adjacent ground surface in order to obtain the change in stress in the subsoil.

PIER FOUNDATIONS

Pier foundations are often used where it is necessary to carry the foundation to a considerable depth in order to reach a suitable bearing stratum. The principal difference between a footing and a pier foundation is that the depth of the latter is generally greater than $4B$. In order to reach the bearing stratum a variety of construction procedures have been developed. These are important because the behavior of a pier foundation may depend primarily on the details of construction.

65. Open Excavations Perhaps the commonest method of pier construction is in an open excavation. This is often accomplished by making a braced excavation to the bearing stratum and forming a pier therein. Care must be taken in the design of the bracing to provide adequate support for the earth pressures and added hydrostatic pressures that may exist. In the case of large piers which are surrounded by water, excavation is frequently accomplished within the protection of a cellular cofferdam.

A common method of pier construction by hand excavation was developed in Chicago, where it is inappropriately called a caisson. A circular shaft corresponding to the final dimensions of the pier is excavated and lined with vertical boards, or lagging, which are held in place by circular steel rings that are placed as the excavation advances. At the bearing stratum the bottom may be enlarged or belled out to increase the bearing area. On completion the entire excavation is filled with concrete; the rings and lagging are normally left in place. This method is suitable only for clays which do not contain water-bearing seams or pockets. Where such materials are encountered the sheeting is some-times driven ahead of the bottom of the excavation to seal off the water. Although timber sheetpiling is often used for this purpose, the Gow method may be more suitable. Circular steel shells about 6 ft long are driven into the ground and the soil within the shell is excavated. Succeeding sections about 2 in. smaller in diameter are then driven and the process is repeated until the water-bearing layers are passed. The excavations can then be advanced as in the Chicago method, including the formation of a bell. If it is necessary to drive sheeting or a casing to advance through a water-bearing stratum, it is essential that this protection remain in place as the concrete shaft is cast. The pier must extend far enough into the bearing stratum to ensure that the soil in which the bell is to be formed can stand without support until the concrete can be placed.

66. Drilled Piers   Several different types of machines have been developed which can drill holes varying in diameter from about 12 in. to as much as 10 ft. When the bearing stratum has been reached a special belling attachment can be used to enlarge the bottom of the hole and form a bell that may be up to three times the diameter of the pier shaft. For particularly large bells, the final finishing may be done by hand. Under ideal conditions a shaft 50 ft deep can be drilled in about 30 min and a 50- to 75-ft pier is easily completed in one 8-hr shift. When it is necessary to provide lateral support to prevent the intrusion of soft clays or silts, a steel lining may be used in conjunction with the drilling process.

Special techniques have been developed for advancing drilled excavations through running sands located below the water table. Although it is common practice to salvage the lining, this may be somewhat hazardous to the integrity of the pier shaft because of the possibility that the running sand and silt, or soft clay, may squeeze the pier shaft, or actually intrude completely across it should there be momentary arching of the concrete within the casing as the latter is removed. The presence of an intrusion is usually not known until the structure begins to show signs of distress. Therefore, if a lining is required to prevent the intrusion of cohesionless or very soft soils, it should generally be left in place to become a part of the completed pier. Care should be exercised that the bottom of the bell is free of loose or compressible materials which may otherwise cause settlement of the completed pier.

67. Piers on Clay   The allowable bearing pressure q_a for a pier founded on clay is

$$q_a = 1.25 q_u \left(1 + \frac{0.2B}{L}\right) \tag{41}$$

where q_u = average unconfined compressive strength within a depth below pier equal to
      its width
   B = width of base
   L = length of base

This provides a factor of safety of 3 against a bearing-capacity failure. No additional support should be assigned because of the shearing resistance that might develop between the pier shaft and the adjacent soil.

The settlement of a pier foundation in clays may be predicted by the procedure for footing foundations. However, in general, a bearing stratum that is adequate for the support of pier foundations is likely to be overconsolidated and not cause differential settlement between adjacent footings in excess of ¾ in.

68. Piers on Sand   The bearing pressure for a pier on sand is usually governed by considerations of settlement because the effect of depth is to increase the factor of safety against a bearing-capacity failure. Experience has shown that the settlement of a pier is about half that of a footing of the same dimensions at the same unit pressures, provided the relative densities of the subsoil are equal. Therefore, the allowable soil pressure can be determined from Fig. 43 appropriately corrected for the position of the water table. As a consequence of the smaller settlement of a pier compared with a footing, these values may be doubled unless there is a possibility that scour may remove most of the materials above the base of the pier.

69. Caisson Foundations A caisson is a special type of pier consisting of a hollow shell that is sunk into position to form a major part of the completed foundation. There are three principal types.

Box caissons are open at the top and closed at the bottom. They are usually constructed on land, floated to the site, and sunk on a previously prepared bearing surface. Sinking is generally accomplished by filling the box with concrete, stone masonry, or, exceptionally, sand and gravel.

Open caissons are open at both top and bottom and are sunk by dredging out the enclosed material. They are usually the size of the completed foundation and must be provided with a number of wells, extending from top to bottom, large enough to provide easy passage for the excavating buckets. Its construction is started directly over the area where it is to be permanently located. If the ground surface is below water it may be raised by forming a sand island and starting the caisson on the fill. As the caisson is sunk, the upper portion is built up with concrete.

Caissons must be fairly massive in order to overcome the side friction, and occasionally it is necessary to reduce side friction by jetting. When the founding elevation has been reached, the bottom is carefully cleaned, sometimes with the assistance of a diver, and a concrete seal of thickness sufficient to resist uplift of the external water is cast prior to pumping out the caisson. After dewatering, the surface of the seal is inspected before concreting the remainder of the caisson. Open caissons have been sunk to depths in excess of 200 ft.

Pneumatic caissons are closed at the top and open at the bottom and are filled with compressed air. They are generally used where the depth below water is between 40 and 110 ft, which is about the maximum depth men can work advantageously under compressed air. Air locks must be provided for passage of men and materials, and the bottom must have clearance sufficient to provide adequate headroom in the working chamber for the muckers. The pneumatic caisson can be controlled more precisely than the open caisson because selective excavation is possible and the bearing surface can be more suitably prepared and inspected before concrete is placed.

70. Foundation Requirements The allowable bearing pressures for caisson foundations are essentially the same as those for piers. However, because of their size and importance, caisson foundations are generally founded on very hard soils or rock where the strength of the subsoil may be secondary in importance to the allowable stresses permitted in the concrete of the caisson itself.

PILE FOUNDATIONS

Bearing pile foundations are columns that transmit load to some depth in soil. Piles are classified as shown in Table 14. They may be used singly or in groups of several, although it should be recognized that the capacity of a group is not necessarily that of a single pile times the number of piles in the group. Because the final location of any pile may be 3 in. or more from its desired location, it is not good practice to use fewer than three piles for the support of a column unless lateral structural framing capable of withstanding the bending moments due to the possible eccentricity is provided. Piles supporting walls are customarily driven in pairs or staggered. The minimum center-to-center spacing of piles is usually 30 in., although they are customarily driven at a spacing of about three times the butt diameter.

71. Pile-Driving Equipment Table 15 shows a number of hammers and their principal characteristics. Drop hammers are frequently used for relatively small jobs where their slowness is not important. However, single-acting hammers are commoner on larger projects. These are operated by either steam or compressed air and differ from double-acting hammers in that their ram falls by gravity whereas that of the double-acting hammer is forced down by the operating gas under pressure. The double-acting hammer is somewhat faster. However, the energy of its blow depends on the gas pressure and is therefore somewhat less certain than that of a single-acting hammer. However, there is considerable uncertainty as to the energy that gets to the pile in either type.

The energy delivered to the cushion block by a diesel hammer is difficult to evaluate in the field because the stroke of the hammer varies with the driving resistance. Therefore, care is required in its selection and use if it is necessary to evaluate the driving energy for use in a pile-driving formula without supplementary load tests.

TABLE 14   Classification of Bearing Piles
Load-transfer action: friction, point bearing

Attitude	
Battered	Vertical
Between 1 horizontal to 6 vertical and about 5 horizontal to 12 vertical	

Placement		
Driven	Vibrated	Bored
Soil compacted by vibration and displacement Jetted Spudded Preexcavated	Minimal compaction or vibration of adjacent soil	No compaction or vibration

Material of construction					
Timber,* sawn† or uncut		Concrete		Steel	Composite
Treated‡	Untreated§	Precast	Cast-in-place	Pipe	BP sections	
Douglas fir Red oak Southern pine	Cedar Cypress Douglas fir Oak Pine Spruce	Tapered Parallel sides Cylinder (require reinforcing)	With shells Driven Dropped in Without shells Drilled Rammed Pedestal Mixed-in-place	Open end Closed end		Timber with cast-in-place concrete Timber with precast concrete Pipe with cast-in-place concrete

* Subject to attack by marine borers.
† Not common in the United States.
‡ Coal-tar creosote (16 pcf or refusal).
§ Must be permanently below water table.

In general, the hammer should be as large as can be safely used without damaging the pile; it is better to select a hammer on the heavy side. Hammers can be compared by computing ultimate pile capacities (Table 16) for various assumed final driving resistances. The computed capacity may be used to evaluate the driving stresses in the pile, which should not exceed 40 percent of its yield-point strength.

Vibratory equipment is a relatively recent pile-driving development. There are two systems: one vibrates the pile at 10 to 30 cps; the other, which is called a resonant driver, vibrates it at the natural frequency of the pile (average about 100 cps). The vibrators

TABLE 15   Pile-Driving Equipment

Drop Hammers

1,000- to 8,500-lb ram.   Operated from crane with swinging leads.
Height of fall is variable

Description	Weight of ram, lb	Energy, ft-lb	Rate, blows/min
Single-acting Hammers			
Raymond Hammer core.........	......	12,000	
Raymond No. 1................	5,000	15,000	60
Raymond 1S..................	6,500	19,500	58
Raymond 0...................	7,500	24,375	50
Raymond 00	10,000	32,500	50
Raymond 000.................	12,500	40,600	50
Raymond 0000................	15,000	48,750	46
Raymond 00000..............	17,500	56,875	44
McKiernan-Terry:			
No. S-3....................	3,000	9,000	65
S-5.....................	5,000	16,250	60
S-8....................	8,000	26,000	55
S-10....................	10,000	32,500	55
S-14....................	14,000	37,500	60
S-20....................	20,000	60,000	60
Vulcan No. 2.................	3,000	7,260	70
1.................	5,000	15,000	60
0.................	7,500	24,375	80
06.................	6,500	19,500	60
08.................	8,000	26,000	50
010...............	10,000	32,500	50
014..............	14,000	42,000	60
020..............	20,000	60,000	60
Double and Differential-acting Hammers			
Vulcan:			
No. 18C....................	1,800	3,600	150
30C....................	3,000	7,260	133
50C....................	5,000	15,000	120
65C....................	6,500	19,200	117
80C....................	8,000	24,450	111
140C...................	14,000	36,000	103
200C...................	20,000	50,200	98
400C...................	40,000	113,488	100
Raymond:			
No. 50C...................	5,000	15,000	120
65C...................	6,500	19,500	100–110
150C..................	15,000	48,750	
15M..................	5,000	15,060	85
McKiernan-Terry:			
No. 9-B-3.................	1,600	8,750	145
10-B-3.................	3,000	13,100	105
11-B-3.................	5,000	19,150	95
C 5..................	5,000	16,000	100–110
Union Iron Works:			
No. 1½ A..................	1,500	8,680	100–125
1 A...................	1,600	10,020	120
1...................	1,850	13,100	130
0...................	3,000	19,850	110
0A...................	5,000	22,050	90
00...................	6,000	54,900	85

TABLE 15   Pile-Driving Equipment (Continued)

Description	Weight of ram, lb	Energy, ft-lb	Rate, blows/min
		Diesel Hammers	
Link Belt			
No. 105....................	1,460	7,500	90–98
312....................	3,855	18,000	100–105
520....................	5,070	30,000	80–84
McKiernan-Terry:			
No. DE 20.................	2,000	12,000 mean	48–52
		16,000 max	
DE 30................	2,800	16,800 mean	48–52
		22,400 max	
DE 40................	4,000	24,000 mean	48–52
		32,000 max	
Delmag D-5.................	1,100	9,100	50–60
D-12.................	2,750	22,500	50–60
D-22.................	4,850	39,700	50–60

Vibratory Drivers

Description	Weight, lb	Available hp	Frequency, cps f	Max generated force/f, (kips/cycles/sec)
L. B. Foster (Vibrodriver, France):				
No. 2-17......................	6,200	34	18–21	
2-35......................	9,100	70	15–18	62–19
2-50......................	11,200	100	12–17	101–17
Bodine No. B.................	22,000	800	0–140	63/100 to 175/100
Menck (Germany):				
MVB 6.5-30.................	2,000	7.5	....	14/
MVB 22-30.................	4,800	50	....	48/
MVB 44-30.................	8,600	100	....	97/
Muller (Germany):				
MS 26.....................	9,600	72	....	
MS 26D....................	16,100	145	....	
(Russia) BT-5.................	2,900	37	42	48/42
100.................	4,000	37	13	44/13
VPP-2.................	4,900	54	25	49/25
VP.................	11,000	80	6.7	35/6.7
VP-4.................	25,900	208	....	198/
Uragi (Japan):				
VHD-1.....................	8,400	40	16.3–19.7	43/19.7
VHD-2.....................	11,900	80	16.3–19.7	86/19.7
VHD-3.....................	15,400	120	16.3–19.7	129/19.7

TABLE 16 Selected Pile-Driving Formulas

Engineering News (poor, should be discontinued)

$$Q_u = \frac{WH}{s+c}$$

c = 1.0 for gravity hammer

c = 0.1 for other hammers

Recommended factor of safety = 6

Hiley (good)

$$Q_u = \frac{e_f WH}{s + \frac{1}{2}(C_1 + C_2 + C_3)} \times \frac{W + n^2 W_p}{W + W_p}$$

n = coefficient of restitution

e_f = efficiency of hammer

C_1, C_2, C_3 = temporary compression of pile cap and head, pile, and soil, respectively, in inches

C_1 = 0.02 to 0.50, C_2 = 0.05 to 0.5, C_3 = 0 to 0.10
(for guide to selection see Chellis, Ref. 52)

Recommended factor of safety = 2.75

Janbu (very good)

$$Q_u = \frac{WH}{k_u s}$$

$$k_u = C_d \left[1 + \sqrt{1 + \frac{\lambda}{C_d}} \right]$$

$$C_d = 0.75 + 0.15 \frac{W_p}{W}, \quad \lambda = \frac{WHL}{AEs^2}$$

k_u = (approximate) from chart

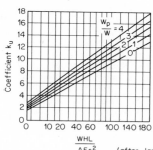

$$\frac{WHL}{AEs^2} \qquad \text{(after Janbu, 1953)}$$

Recommended factor of safety = 3

Pacific Coast (fair)

$$Q_u = \frac{E_n \dfrac{W + k W_p}{W + W_p}}{s + \dfrac{Q_u L}{AE}}$$

k = 0.25 for steel piles, k = 0.1 for others

Recommended factor of safety = 4

Vibrated piles* (tentative)

$$Q_u = \frac{550 \text{ H.P.}}{r_p + f s_t}$$

H.P. = horsepower, ft-lb/sec

r_p = rate of penetration, in./sec

f = frequency, cycles/sec

s_t = temporary compression loss factor
= 0.02 to 0.07 in. (determine from load test to failure)

Q_u = ultimate capacity, lb

W = weight of pile hammer, lb

H = drop of hammer, in.

s = final pile penetration, in.

E_n = energy of driving (WH), in.-lb

L = length of pile, in.

    * From M.T. Davisson

W_p = weight of pile, lb

A = net cross-sectional area
of driven pile, sq in.

E = modulus of elasticity of pile, psi

    timber = 1.2×10^6

    concrete = 2.5×10^6

    steel = 30×10^6

utilize counterrotating shafts to obtain a positive vertical force and to balance the lateral forces. The vertical force generated by the vibratory driver is transmitted to the pile, which will accept more power if the force is increased or if the frequency of the vibrator approaches the natural frequency of the pile in longitudinal vibration. To transmit this force, the vibratory unit must be firmly attached to the pile. Vibratory methods can drive piles at a rate many times faster than is possible with conventional equipment. They are much quieter as the only noise is from the motor and the hum of the vibrator, and ground vibrations generally extend only a few feet away. However, mechanical problems frequently require considerable time for maintenance, which may materially offset the high driving rate when total production is considered.

Conventional pile driving is often assisted by supplemental operations. Jetting is used to loosen the soil around the pile and facilitate penetration into or through medium to dense sands. It requires a discharge pipe of 1½ to 3 in. in diameter, a 1- to 1¼-in. nozzle, and a pumping capacity of about 500 gal/min at a pressure of 100 to 300 psi. Jetting should always be followed by driving to recompact the loosened granular materials.

Spudding is the driving of a heavy mandrel into the soil and withdrawing it to minimize the driving required on the pile placed in the spudded opening. Although it is not often used, spudding may facilitate the penetration of resistant strata because of its more efficient transfer of driving energy.

Preexcavation, which is accomplished with augers or specially developed equipment to remove soil representing the bulk of the volume of the pile to be driven, may facilitate driving, but it is primarily used to control the displacements and heave associated with displacement piles.

72. Pile-Driving Formulas Pile-driving formulas (Table 16) are intended to relate the driving resistance to the supporting capacity under static loading. Every formula (and there are many) includes one or more factors that must be determined experimentally. Contrary to popular opinion, experience has demonstrated that pile-driving formulas are entirely misleading when the soil consists of clay, and they should not be used unless they have been adjusted by experience in a particular locality. However, any pile-driving formula should be considered only as a guide to judgment. For example, tests in Michigan (1965) show that the work done on a load cell located at the top of a steel pile ranged from 0.26 to 0.65 times the rated energy input of the hammers. The average value was 0.45.

The wave equation can be used with reasonable confidence to predict impact stresses in a pile during driving and to estimate its static capacity.[50] The theoretical model requires the evaluation of all factors that may influence the transmission of a stress wave initiated by the pile hammer ram. Thus the properties of the hammer ram, cushion blocks, cap block, pile, and soil must be known. The solution requires an electronic computer and is best presented as curves relating ultimate resistance and maximum pile stress to the driving resistance. The ultimate capacity should be adjusted for freeze or relaxation on the basis of a pile load test to failure.

73. Pile Tests It is essential to distinguish between test piles and pile load tests. Test piles are prototypes driven to determine the lengths of piles required at a particular site. The driving resistance is used as the basis for ordering piles of appropriate length. Test piles should be located near a soil boring representative of the poorest conditions known at the site, and a record of the penetration resistance (blows per foot) should be made for the entire length of the test pile. The driving record should also indicate the characteristics of the driving equipment including the cushion block.

Pile load tests (Art. 35) are used to determine their design capacity. The test load is seldom carried to a sufficient intensity to determine the ultimate capacity. It is usually required that the test load be carried to at least twice the design load and that the resulting load-settlement curve be corrected for the elastic shortening of the pile before an analysis of the data is attempted. The elastic shortening is of particular importance for relatively long point-bearing piles.

Many procedures are used to interpret the results of a pile load test. If the pile has been loaded to failure, the design load should be based on a factor of safety of at least 2.5. A failure load Q_u may be determined as the load at which the load-settlement curve approaches a vertical tangent or as the intersection of tangents to the load-settlement curve near the point where the settlement per unit load increases markedly (Fig. 45). Perhaps the commonest criterion is that the design load shall not exceed one-half the test load which causes a settlement of 0.01 in./ton of test load. Where a failure load has been reached, the design load should be the smaller obtained from these two criteria.

Although pile load tests are often made at the start of construction to verify the design assumption, it is desirable that they be made early enough for the data to serve as a design criterion. It is always desirable to conduct at least two tests in any load-testing program and to base the analysis on the most unsatisfactory result. Averaging should be avoided.

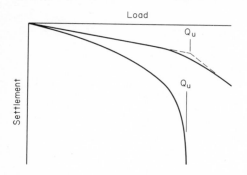

Fig. 45 Failure loads—pile-load tests.

If piles are to be driven through a soft layer and into a firm stratum, the support provided by the soft layer may disappear as the layer consolidates, so that all the load is transmitted to the firm stratum. The capacity of the latter may be determined by conducting a pair of load tests, one on a pile driven into the firm stratum and the second, which must be carried to failure, on a pile stopping 2 to 3 ft above the firm stratum. By subtracting the load carried in the soft layer from the load for the pile penetrating the firm stratum, the support of the latter is obtained. A careful analysis of these conditions is required. If the soft layer is likely to settle as a consequence of the new construction, it may ultimately tend to drag down the piles with a load equivalent to the temporary support it originally provided. This is known as negative skin friction and further reduces the usable capacity of the pile support provided by the firm stratum.

74. Piles in Sand Piles driven into sand generally act in friction, point bearing, or in combination. Their ultimate capacity can be estimated from a formula such as Janbu's (Table 16) or, in the case of friction piles, by

$$Q_{ult} = (\tfrac{1}{2}\gamma l^2 k - \tfrac{1}{2}\gamma_w l_w^2)\pi d \tan \phi_f \tag{42}$$

where Q_{ult} = ultimate load capacity neglecting point resistance
γ = saturated unit weight of soil
l = length of pile
γ_w = unit weight of water
l_w = length of pile below water table
d = pile diameter
k = coefficient of lateral earth pressure
$\tan \phi_f$ = coefficient of friction between sand and pile

Values of k and $\tan \phi_f$ are difficult to evaluate. For driven or vibrated piles, k may be taken as 1.0, but it should not exceed 0.4 if the piles are jetted into the sand. The coefficient of friction may be selected from Table 17.

It should be recognized that these procedures are not precise and, where possible, pile load tests should be conducted as a more reliable indication of pile behavior.

75. Piles in Clay The capacity of a pile driven into clay cannot be reliably predicted from a pile-driving formula. It is best to use the results of a pile load test. However, tests are rather expensive, and the following somewhat less satisfactory procedure may be employed. The ultimate bearing capacity of a single cylindrical friction pile is approximately equal to the adhesion that can be developed between the pile and the surrounding soil times the embedded surface area of the pile. The adhesion is difficult to estimate as it may range from about one-half to one-tenth of the unconfined compressive strength. In general, the stiffer the clay the smaller the relative adhesion factor (Fig. 46). If Fig. 46 is used, a safety factor of 3 should be used for the maximum probable loading. However, a safety factor of at least 2 should be used for the extreme maximum loading of the pertinent building-code load requirement. The relatively small contribution from end bearing in the case of friction piles is generally neglected in the analysis of a single pile. If the subsoil contains well-defined layers of varying strength, the adhesion in each layer should be independently determined. Because of the uncertainty in determining the adhesion in stiff to hard clays, pile load tests should always be used to check the design load whenever friction piles are driven into such deposits.

TABLE 17   Coefficient of Friction

Pile material	$\tan \phi_f$	
	Driven piles	Jetted or drilled piles
Timber.........................	0.65	0.53
Steel: pipe, BP, fluted...........	0.45	0.36
Concrete......................	0.65	0.53
Corrugated steel shells..........	0.70	0.58

For most foundations, friction piles are driven in groups. Therefore, it is necessary to investigate the ultimate load capacity of the group to be certain that it is not less than the product of the load on a single pile and the number of piles, irrespective of how the single-pile load may have been determined. The design capacity is the minimum of these two procedures.

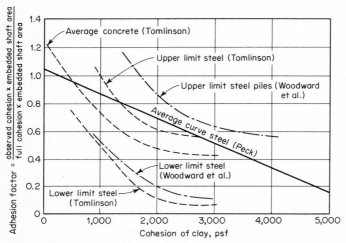

Fig. 46 The adhesion on piles compared with the cohesive strength of clay. *(After Tomlinson.)*

The load capacity of a group of friction piles consists of the shearing resistance on the surface perimeter of the pile group plus the bearing capacity on the plan area of the group at the level of the pile tips. Thus

$$Q_d = (B + L)lq_{u1} + 2.5q_{u2}\left(1 + \frac{0.3B}{L}\right) BL \tag{43}$$

where Q_d = ultimate load on group, tons
$\quad B$ = width of pile group, out to out, ft
$\quad L$ = length of pile group, out to out, ft
$\quad l$ = pile length, ft
$\quad q_{u1}$ = average unconfined compressive strength of clay within length l, tons/sq ft
$\quad q_{u2}$ = average unconfined compressive strength within a distance B below the pile tips, tons/sq ft

The maximum probable loading on the group should not exceed $Q_d/3$. Although the base bearing capacity may contribute considerably to the capacity of the pile group, it should be noted that the greatest benefit of a friction pile foundation is obtained with the longest piles possible within the limits of economy. The longer the pile the smaller the settlement in most instances.

Piles driven into clays that increase materially in strength with depth may be analyzed as for friction piles. However, the point resistance, which may represent a sizable proportion of the pile capacity, can be determined only by means of loading tests. The safe load on a group may be taken as the safe load per pile as determined from load tests times the number of piles in the group. In some localities pile-driving formulas have been adjusted to indicate safe loads corresponding to the pile-driving resistance. Such formulas should never be used outside the geological region in which they were developed.

Piles driven through relatively soft materials to a stiff or hard clay act in point bearing. The load capacity of a group of such piles is equal to the product of the number of piles in the group and the safe load per pile without regard to their spacing. However, these conditions are ideal for the development of negative skin friction (Art. 73), which may be a sizable proportion of the pile load capacity. The magnitude of the negative skin friction can be determined from pile load tests (Art. 35). It may also be estimated as the average shearing resistance of the soft material multiplied by the surface area of the embedded piles.

76. Settlement of Pile Foundations Any pile foundation which has a compressible stratum located below the pile tips is likely to settle, and the magnitude of the settlement should be predicted. It is computed in the same manner as for footings on clay except that the change in pressure Δp is determined somewhat differently depending upon whether the piles act in point bearing or as friction piles.

For a point-bearing pile foundation, the load on the pile group is assumed to be applied to the subsoil at the level of the pile tips on an area equal to the plan area of the pile group. Below the tips, it is considered to be spread uniformly at an angle of 30° from the vertical.

The settlement of a group of friction piles is computed in a similar manner. However, the level of the application of the load to the subsoil is less certain, as load is transferred through much of the length of the piles. A commonly used approximate procedure is based on the assumption that the load is applied at the lower third point of the piles. The load is assumed to spread at an angle of 30° from the vertical, and any compressible material below the lower third point is assumed to contribute to the settlement of the group.

77. Laterally Loaded Piles Where a pile-supported structure is subjected to lateral loads, the vertical piles may provide more lateral resistance than is commonly realized. Prevailing rules of thumb commonly permit an arbitrary lateral load per pile—often 1000 lb—without any consideration as to the type of pile or the soil in which it is driven. Since a pile-supported structure does not transmit load directly to the soil beneath the pile cap, frictional resistance should not be assumed between the base of the structure and the underlying soil. Therefore, the piles must be adequate to resist all lateral loads.

The ultimate lateral bearing pressure per unit length of pile at a given depth in clay is

$$Q_d = 9cB = 4.5q_u B \tag{44}$$

and in sands

$$Q_d = 3B\gamma' z \frac{1 + \sin \phi}{1 - \sin \phi} \tag{45}$$

where Q_d = ultimate load per unit length of pile, lb/ft
$\quad c$ = cohesion, psf
$\quad q_u$ = unconfined compressive strength, psf
$\quad B$ = width of pile, ft

γ' = effective unit weight of soil, pcf
z = depth, ft
ϕ = angle of shearing resistance

The working load should not exceed $Q_d/2$ beyond a depth of $4B$ under any circumstance. For lateral loads smaller than $Q_d/2$, the soil reaction at any depth is given by

$$w = \frac{C_w Q_{hg}}{T} \tag{46}$$

where w = soil reaction, lb/ft
C_w = soil reaction coefficient from Fig. 47
Q_{hg} = shear at ground surface, lb
T = relative stiffness of pile = $\sqrt[5]{EI/n_h}$, in.
EI = flexural stiffness of pile, lb-in.^{2}
n_h = constant of horizontal subgrade reaction, lb/in.^{3}

The modulus of subgrade reaction $k = w/y$ is the ratio of the total reaction per unit length of pile to the corresponding deflection. For granular soils k is directly proportional

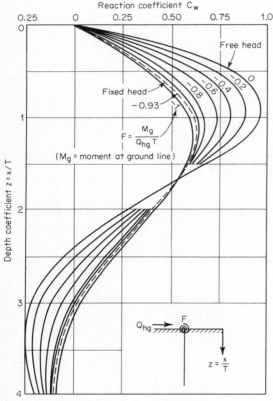

Fig. 47 Coefficients for soil reaction, laterally loaded piles. *(From Prakash, 1961.)*

to the depth x, and it has been shown that k is also proportional to depth for normally consolidated clays and silts. However, for overconsolidated clays k is usually assumed to be a constant and the corresponding relative stiffness is $\sqrt[4]{EI/k}$. Since k is proportional to depth for most soils of interest, the case of constant k is not considered here.

The constant of horizontal subgrade reaction is $n_h = k/x$. Typical values of n_h are presented in Table 18. Actual values can be determined experimentally by driving two instrumented piles relatively close together and jacking them apart. By measuring the loads and deflections at various depths (with a tiltmeter), k can be determined.

The lateral deflection y, in inches, given by

$$y = C_y \frac{Q_{hg} T^3}{EI} \tag{47}$$

where C_y = deflection coefficient from Fig. 48, may also be used to evaluate k from a jacking test as well as to determine anticipated deflections.

The moment in a laterally loaded pile (in.-lb) is

$$M = C_m Q_{hg} T \tag{48}$$

where C_m = moment coefficient from Fig. 49. It is noted in Figs. 47, 48, and 49 that a fixed-head pile is generally more favorable than one with a free head. The fixity depends

TABLE 18 Typical Values of n_h*

Soil type	N	n_h, lb/in.³	
		Dry	Submerged
Sand:			
Loose...........................	<10	9.4	5.3
Medium........................	10–30	28	19
Dense..........................	>30	75	45
Very loose under repeated loading	<5	....	1.5
Silt, very soft, organic............	<3	....	0.4–1.0
Clay:			
Very soft......................	<3		
Static loads...................	.....	....	2
Repeated loads...............	.....	....	1

* M. T. Davisson, 1963.

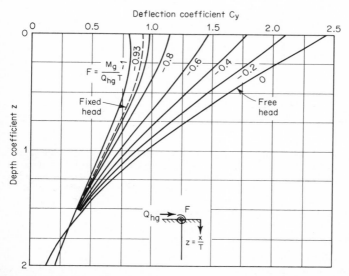

Fig. 48   Deflection coefficients for laterally loaded piles. (*From Matlock and Reese, 1961.*)

on the structural characteristics of the pile cap and the connection between the pile and the pile cap. The latter is attained by embedding the pile at least 24 in. into the pile cap.

The preceding analyses are for single piles, which corresponds to a minimum spacing of eight pile diameters in the direction of the lateral load and three pile diameters normal to the direction of load. Closer spacings cause a reduction in the modulus of subgrade reaction and an increase in the relative stiffness. From the limited information available, it is recommended that a minimum spacing of $3B$ be maintained normal to the load and that T be increased linearly to a limiting value of $1.3T$ as the pile spacing in the direction of the load decreases from $8B$ to $2.5B$.

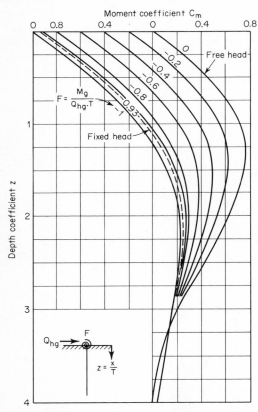

Fig. 49 Moment coefficients for laterally loaded piles. *(From Prakash, 1961.)*

78. Batter Piles Batter piles are often used where the lateral loads exceed the lateral resistance of vertical piles. It is commonly assumed that a batter pile has the same axial load capacity as a vertical pile of the same size driven to the same stratum. In the design of a batter-pile foundation, the sum of the horizontal components of the allowable axial loads must equal or exceed the required lateral resistance.

The vertical component Q of load in a batter pile is given by

$$Q = \frac{\Sigma V}{n} \pm \frac{\Sigma M d}{\Sigma d^2} \tag{49}$$

where ΣV = sum of vertical loads acting on foundation
n = number of piles in group
ΣM = sum of moments about center of gravity of pile group
d = distance from center of gravity of group to pile

A convenient analysis is to assume a batter for the piles and check the corresponding pile reactions by means of a force polygon as shown in Fig. 50. Thus, by trial and error, the number of piles to be battered and the required batter can be determined.

79. Lateral Stability of Poles   The depth of embedment required for the lateral support of poles such as lampposts, sign supports, lighting poles, and pole-type buildings is a rather commonly encountered problem. For a single pole embedded in the ground and subjected to a lateral load at some height above the ground, the proper depth of embedment depends primarily on the location of the point of rotation of the pole. The analysis of the rotational resistance of a pole generally presumes that the pole is stiff enough to be considered rigid. Hence, the rotational resistance is governed entirely by the soil in which the pole is embedded.

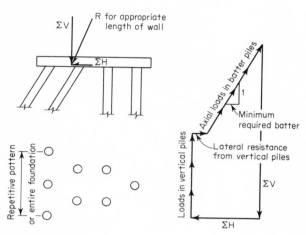

Fig. 50   Batter-pile analysis.

This is a relatively complex problem which has been reviewed in some detail in Ref. 43. A procedure based on empirical data and certain simplifying assumptions concerning the distribution of soil resistance is used extensively (Fig. 51). The required depth of embedment, based on the soil-stress distribution shown in the figure, is

$$D = \frac{1.18P}{bS_1}\left(1 + \sqrt{1 + \frac{1.88bS_1H}{P}}\right) \qquad (50)$$

where b is the diameter, in feet, of a round pole or the width, in feet, normal to S_1, of a rectangular pole. The other terms are defined in Fig. 51, where b is shown in inches rather than feet. A nomographic solution of Eq. (50) is given in the figure.

D should not be less than 4 ft. The author recommends that S_1 be used, instead of the pull-out test values indicated in Fig. 51, and be taken as the unconfined compressive strength for silts and clays and 1000 psf for granular soils. If the free-water table is within the depth of embedment, S_1 for granular soils should be reduced to 500 psf. If the auger test is used, it should be conducted at 1-ft intervals to a depth of at least 6 ft, beginning at a depth of 18 in. below ground surface. The auger hole may be cleaned out between tests to a depth not greater than 6 in. above the next test location.

The maximum bending stress f_b in the post is given by

$$f_b = \frac{P(H + 0.34D)}{S} \qquad (51)$$

where S = section modulus.

80. Guy Anchors   The analysis of guy anchors is a complex problem as the mode of failure is not known except in a very general way. However, an approximate analysis has been developed based upon the shearing resistance that can be mobilized on the base and

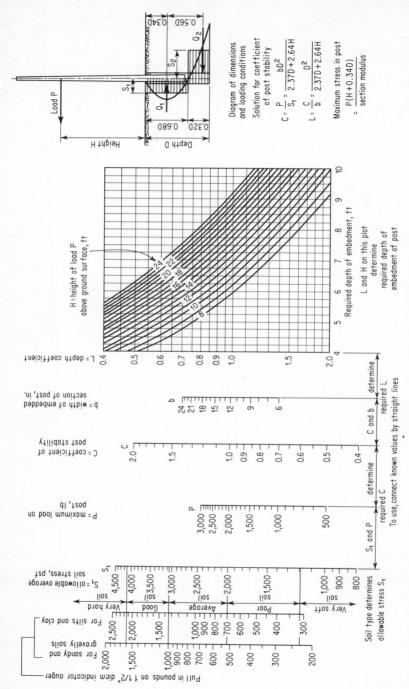

Fig. 51 Embedment of poles, based on ½-in. motion at ground surface. (*Courtesy of Outdoor Advertising Association of America, Inc., New York, N.Y.*)

ends of the anchor block, the weight of the soil above the anchor, the weight of the anchor block itself, and the respective active and passive earth pressures that are likely to act. The forces involved are shown in Fig. 52. The design procedure is one of trial and error, but once dimensions have been selected and a depth established it is not particularly difficult to determine the adequacy of the tentative design.

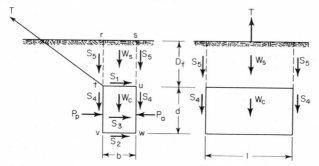

Fig. 52 Analysis of anchor piers.

The design must satisfy the following requirements:

$$T_v \gtreqless W_s + W_c + S_4 + S_5 \gtreqless \gamma b D_f l + \gamma_c b d l$$

$$+ 2\gamma d k \left(D_f + \frac{d}{2}\right)(b + l) \tan \phi + \gamma k D_f^2 (b + l) \tan \phi$$

$$T_h \gtreqless P_p - P_a + S_1 + 2 S_3 \gtreqless \left(D_f + \frac{d}{2}\right)\left(\gamma d N_\phi l - \gamma d \frac{l}{N_\phi} + 2\gamma b d k \tan \phi\right) + \gamma D_f b l \tan \phi$$

where T_v, T_h = vertical, horizontal, components of guy tension (use twice the actual values for safety factor of 2)

W_s = weight of soil above pier $(rstu)$
W_c = weight of pier $(tuvw)$
S_1 = shearing resistance at top of anchor
S_2 = shearing resistance at bottom of anchor (neglected)
S_3 = shearing resistance on ends of anchor
S_4 = shearing resistance on perimeter of anchor $(tuvw)$
S_5 = shearing resistance on soil above anchor $(rstu)$
P_a = active earth pressure on back of anchor (uw)
P_p = passive earth pressure on front of anchor
N_ϕ = $\tan^2 (45° + \phi/2)$
ϕ = angle of shearing resistance
k = coefficient of earth pressure at rest
γ = unit weight of soil (reduce by 60 pcf if located below water table)
γ_c = unit weight of concrete (reduce by 60 pcf if located below water table)

Some care is required in the selection of the appropriate soil constants. Inasmuch as anchor blocks are frequently constructed in oversized excavations, the appropriate soil properties are those of the backfill. Since the relative compaction around such an anchor is always a matter of some uncertainty, it may be assumed that the angle of shearing resistance of the soil is 20° when the backfill material consists of silts or clays and 30° for more granular materials. Corresponding values of the coefficient of earth pressure at rest may be taken as 0.4 and 0.5, respectively.

81. Foundations Subjected to Uplift Many foundations are subjected to loads that act upward. This is particularly true of the foundations for self-supporting towers which are subject to overturning forces as a consequence of wind pressures. For this purpose, pedestal-type piers are frequently used if the subsoil has adequate strength to support the compressive loads involved. Inasmuch as pedestal piers are constructed in open excavations, the soil that participates in resisting the uplift forces is all backfill material. It is generally not possible to obtain pertinent soil parameters from a program of laboratory tests as placement of backfill in relatively restricted locations probably receives less care than any other construction operation. The most satisfactory backfill materials are relatively clean sands and gravels.

The uplift resistance of the foundation in a cohesionless soil may be represented by the equation

$$Q_u = \tfrac{1}{2}\gamma' D_f^2 kP \tan \phi + W_c + W_s \tag{52}$$

where γ' = effective unit weight of soil
D_f = depth of foundation
k = coefficient of lateral earth pressure
P = perimeter of base of foundation
ϕ = angle of shearing resistance on assumed surface of failure
W_c = weight of concrete in pier
W_s = weight of soil over base of pier

Even at sites where cohesive soils prevail, it is desirable to backfill with selected granular materials, if a local source is available, rather than with the locally excavated soil. It is recommended that $\phi = 30°$ and $k = 0.5$ be used for granular backfill and that $\phi = 20°$ be assigned as an equivalent shear-strength parameter for silts and clays, with $k = 0.4$. Higher values of either ϕ or k are not recommended because of the uncertainties inherent in normal backfill construction. They should be increased only if adequate construction control can be assured.

The effect of a high water table on the uplift resistance of foundations must often be considered. Uplift forces due to wind are not likely to occur at their full design value except for short intervals of time, probably not more than a few minutes and certainly not more than a few hours. Whether the uplift of water on the base of a submerged footing can act depends primarily on the rate at which the groundwater can flow through the underlying soil. If an upward force lifts the footing a small amount, a space is created beneath the footing. If this space is not filled with water instantaneously the hydrostatic uplift is destroyed. If the soil beneath the footing is pervious, water will flow rapidly into the space, and buoyancy will be restored and should be considered in the design. On the other hand, studies have shown that days or weeks may be required to provide the necessary volume of water in impervious soils, so that it is not necessary to design the footing for the uplift effect.

Buoyancy can be taken into account by subtracting from Q_u in Eq. (52) the weight W_w of water represented by the block of soil and concrete located below the groundwater table and above the bottom of the pier. The final decision regarding uplift must be based on estimates of the permeability of the intact formation underlying the proposed foundation. This will depend to a large extent on laboratory inspection and tests of undisturbed samples.

If drilled piers must be relied upon to resist uplift forces, particular care is required to evaluate the uplift resistance because the adhesion between the soil and the pier shaft may be considerably less than the cohesion of the materials in which they are built. This is particularly true where piers are constructed in relatively stiff clays. The effective adhesion may be judged from the data in Fig. 46. However, if the drilled pier has an enlarged bottom and is appropriately reinforced, the uplift resistance may be analyzed in the same fashion as for pedestal piers.

The uplift resistance of pile foundations may be evaluated on the basis that the piles act as individual friction piles.

82. Improvement of Subsoil If the subsoil is unusually soft and compressible or consists of organic soils to a relative shallow depth, it is frequently economical to remove these undesirable soils and backfill with select granular materials compacted in layers. If the cost of excavation and replacement is prohibitive, the soft compressible materials can frequently be improved by consolidating them with a surcharge fill. Where permanent fill is required this portion of the fill should be compacted to a height adequate to include the predicted settlements that will occur. Additional fill, equivalent in magnitude to the proposed structural loads, may then be placed to force consolidation of the underlying soft materials. When the desired degree of consolidation has been accomplished, the excess fill may be removed and the structure built with the assurance that the majority of the settlement has already taken place under the surcharge load. However, this takes time.

Vertical sand drains are often installed to accelerate the consolidation process. After removing the vegetation, a pervious gravel or sand blanket of adequate thickness to support construction equipment is placed on the surface of the compressible soil. Since this previous blanket must function as a drain and also serve as a working platform, it should be at least 2 ft thick. The vertical sand drains are then installed by driving a mandrel of 18 to 24 in. diameter through the compressible soil, filling it with relatively clean sand, and withdrawing the casing while the sand is forced out by the application of gas pressure inside the mandrel. The piles are driven in a pattern at a spacing of 8 to 15 ft. As fill is placed over the drainage blanket, the added load forces water from the subsoil into the sand drains, from which it escapes through the drainage blanket.

Sometimes granular soils can be improved by injecting them with grout. However, grouting is often attempted with materials that have little likelihood of penetrating the material to be improved. In general the penetration of grouts into natural soils depends upon the effective size D_{10}* of the material to be grouted. Table 19 shows the relationship between various types of grout and the limiting size of the material to be grouted.

TABLE 19 Criteria for Grout Injection

Grout	Material to be grouted	Finest size that can be grouted, mm
Cement............	Fissures	0.10
	Loose sand	$D_{10} = 0.5$
	Dense sand	$D_{10} = 1.4$
	Soil	$D_{10} = 0.65$
Silicates...........	Soil	$D_{10} = 0.1$
AM-9.............	Soil	$D_{10} = 0.013$
Siroc.............	Soil	$D_{10} = 0.07$

Since loose soils are far more readily grouted than dense ones, the grouting of soils approaching the finest sizes indicated in Table 19 should be approached with some skepticism. Inasmuch as they penetrate the void spaces of the soil, grouts are of principal value in decreasing the soil permeability and are often used only for this purpose. Some of the more desirable chemical grouts are relatively expensive and therefore find use only in special or unusual circumstances.

Loose free draining sand may be improved by making it denser. Under suitable circumstances this may be accomplished with explosives and satisfactory results can be obtained with compaction piles. More commonly, vibroflotation is used. This is a patented process in which a large vibrator unit, known as a vibroflot, is lowered into the sand with the assistance of water jetting. Under the combined action of the jetting and vibration, the sand is compacted to form a crater at the ground surface which is filled with sand. The vibroflot is usually inserted at a spacing of 6 to 8 ft in the area to be treated. Supplemental tests are required after treatment to determine its effectiveness. However, it is difficult to verify compliance with a specification requiring that a certain percentage of relative density must be achieved because indirect methods are usually required. It is probably more realistic to determine, in the field, the results that can be achieved for a particular vibroflot spacing and proceed with foundation design on the basis of the actually achieved results.

83. Construction Problems If it is necessary to excavate below the water table to construct footing foundations in sands, the water table should be lowered in advance of construction by a system of well points, particularly if the water level must be lowered more than 3 or 4 ft. Otherwise, pumping from sumps is likely to lead to quick conditions in the excavated areas with a consequent loosening of the subsoil. The flow that comes toward the excavation from the sides is also likely to cause trouble if this water is permitted to remove fines from the adjacent soils. The removal of fines may gradually increase along some path of least resistance, forming a small underground channel. The corresponding loss of ground associated with this phenomenon, known as piping, may be

*The particle diameter for which 10 percent of the particles in a sample are finer.

particularly important if it should occur under adjacent building foundations or below the foundation elevation for other elements of the proposed structure. A properly planned dewatering system will prevent any loss of ground and will completely eliminate the danger of a quick condition. However, if the dewatered zone is underlain by compressible clay, the lowered water table may cause some settlement as a consequence of consolidation of the clay.

A pier in sand below the water table is often established by sinking a caisson of either an open or pneumatic type. Otherwise, it is generally necessary to dewater, by means of either a multiple-stage well-point installation or deep-well pumps.

The construction of drilled piers and hand-excavated piers protected with rings and lagging is often accompanied by loss of ground. If advanced through relatively soft clays, there is a lateral movement or squeeze into the excavation if the sides remain unsupported even for short periods of time, together with a rise of the bottom of the excavation as it is advanced. In general, these losses of ground in clays are relatively local in their effect on adjacent settlements. However, where such a hand-dug pier encounters a saturated zone of cohesionless materials, even only a few inches thick, these may flow into the caisson and frequently cause settlements of adjacent structures to distances in excess of 100 ft. This is a particularly common problem in the Chicago area, where cohesionless soils are often found immediately overlying bedrock and the consequent piping into the excavation removes soil from beneath the overlying hardpan. Such problems can generally be minimized by careful selection of the construction method and control of the hydraulic conditions to prevent flow into the excavation.

There are many construction problems associated with pile driving. The selection of the most appropriate type of pile should include a consideration of how the pile is expected to function and the possible influence of the construction method on this function. One of the commoner problems is that of heave. This is particularly important where displacement piles are being driven to point bearing, as heave may lift previously driven piles from the bearing stratum. Therefore, it is necessary to take observations so that any piles that have been lifted from the bearing stratum may be redriven. Whether or not observation of the elevation at the butt of the pile will adequately indicate what is happening at the tip will depend to a large extent on the particular type of pile involved. In some instances lightly loaded adjacent foundations may be heaved by pile driving, although the magnitude of the heave can frequently be reduced by preexcavation of most of the volume of the pile. In some materials, displacement may tend to collapse previously driven pile shells. This can sometimes be remedied by using a heavier shell.

Some types of cast-in-place piles are constructed by placing concrete as a casing or soil-filled auger is withdrawn from the subsoil. It is particularly important that the subsoil conditions be carefully analyzed and the necessary precautions taken to prevent the possibility of a discontinuous pile shaft being formed because of an intrusion of soil into the excavated shaft during its construction. One advantage of most cast-in-place and pipe piles is that they can be inspected for attitude and integrity prior to filling with concrete. It is sometimes required that piles be driven within 2 percent of the vertical; thus the tip of a 50-ft pile could be displaced laterally not more than 1 ft relative to the butt.

Under certain conditions it may be necessary to provide special points on piles. For instance, where steel piles are driven through relatively soft soils to bearing on rock which dips rather steeply, it may be necessary to provide a special point that can bite into the rock and develop a good seat. It has also been found necessary on occasion to provide pipe piles with a special point. One type consists of a flat steel plate reinforced with steel plates in the form of a cross mounted within a short cylindrical sleeve welded inside the pile. Special points are often required when pipe piles are driven into granular materials containing some boulders if the piles are to be driven to a relatively high capacity. During casting of concrete piles, diligent inspection is required to ensure that the concrete filling is continuous and free of voids. Placement of a low-slump concrete may be associated with arching within the pile and the formation of extensive voids.

REFERENCES

1. King, Ruth R., et al.: Bibliography of North American Geology, 1959, *U.S. Geol. Surv. Bull.* 1145, Washington, D.C., 1961.
2. Publications of the Geological Survey, 1974, U.S. Geological Survey, Washington, D.C., 1975.

3. Geologic Map of the United States, Scale 1:2,500,000, U.S. Geological Survey, Washington, D.C., 1932.
4. Glacial Map of North America, Scale 1:4,555,000, Geological Society of America, New York, 1945.
5. Glacial Map of the United States East of the Rocky Mountains, Scale 1:1,750,000, Geological Society of America, New York, 1959.
6. Peck, R. B., W. E. Hanson, and T. H. Thornburn: "Foundation Engineering," 2d ed., John Wiley & Sons, Inc., New York, 1974.
7. Sowers, G. B., and G. F. Sowers: "Introductory Soil Mechanics and Foundations," 3d ed., The Macmillan Company, New York, 1970.
8. Tschebotarioff, G. P.: "Foundations, Retaining and Earth Structures," McGraw-Hill Book Company, New York, 1973.
9. Leonards, G. A. (ed.): "Foundation Engineering," McGraw-Hill Book Company, New York, 1962.
10. Survey and Treatment of Marsh Deposits, *Highw. Res. Board Bibliography* 15, 1954.
11. Mesri, G.: Coefficient of Secondary Compression, *J. Soil Mech. Found. Div. ASCE*, January 1973.
12. Pihlainen, J. A., and G. H. Johnston: Guide to a Field Description of Permafrost, *National Research Council of Canada, Associate Committee on Soil and Snow Mechanics, Tech. Mem.* 79, 1963.
13. Hvorslev, M. J.: "Subsurface Exploration and Sampling of Soils for Civil Engineering Purposes," Waterways Experiment Station, Vicksburg, Miss., 1949.
14. Acker, W. L., III: "Basic Procedures for Soil Sampling and Core Drilling," Acker Drill Co., Inc., Scranton, Pa., 1974.
15. In Situ Measurement of Soil Properties, vols. 1 and 2, *Proc. ASCE Geotechnical Specialty Conf.*, New York, 1976.
16. American Society for Testing and Materials: "Annual Book of Standards, Pt. 19, Natural Building Stones; Soil and Rock; Peats, Mosses and Humus," Philadelphia, Pa., 1976.
17. Terzaghi, Karl, and R. B. Peck: "Soil Mechanics in Engineering Practice," 2d ed., John Wiley & Sons, Inc., New York, 1967.
18. Skempton, A. W., and V. A. Sowa: The Behavior of Saturated Clays during Sampling and Testing, *Geotechnique*, vol. 13, no. 4, pp. 269–290, December 1963.
19. Undisturbed Sand Sampling below the Water Table, *Waterways Expt. Sta. Bull.* 35, Vicksburg, Miss., 1950.
20. Density Changes of Sand Caused by Sampling and Testing, *Waterways Expt. Sta. Potomology Investigations, Rept.* 12-1, Vicksburg, Miss., 1952.
21. Bishop, A. W.: A New Sampling Tool for Use in Cohesionless Sands below Ground Water Level, *Geotechnique*, vol. 1, no. 2, pp. 125–131, December 1948.
22. Lundgren, R., F. C. Sturges, and L. S. Cluff: General Guide for Use of Borehole Cameras, *ASTM Spec. Tech. Publ.* 479, 1970.
23. Sanglerat, G.: "The Penetrometer and Soil Exploration," Elsevier Publishing Company, Amsterdam, 1972.
24. Thornburn, T. H.: Geology and Pedology in Highway Soil Engineering, *Rev. Eng. Geol.*, vol. 2, Geological Society of America, 1968.
25. Reeves, R. G., A. Anson, and D. Landen (eds.): "Manual of Remote Sensing," 2 vols., American Society of Photogrammetry, Falls Church, Va., 1975.
26. Way, D. S.: "Terrain Analysis," Dowden, Hutchinson and Ross, Inc., Stroudsburg, Pa., 1973.
27. Peck, R. B.: A Study of the Comparative Behavior of Friction Piles, *Highw. Res. Board Spec. Rept.* 36, 1958.
28. Peck, R. B.: Records of Load Tests on Friction Piles, *Highw. Res. Board Spec. Rept.* 67, 1961.
29. Special Procedures for Testing Soil and Rock for Engineering Purposes, *ASTM Spec. Tech. Publ.* 479, Philadelphia, Pa., 1970.
30. Davisson, M. T.: Lateral Load Capacity of Piles, *Highw. Res. Board Hwy. Res. Rec.* 333, pp. 104–112, 1970.
31. Ireland, H. O.: Pulling Tests on Piles in Sand, *Proc. 4th Int. Conf. Soil Mech. Found. Eng.*, London, vol. 2, pp. 43–45, 1957.
32. *Proc. Specialty Conf. Performance on Earth and Earth-Supported Structures*, ASCE, New York, 1973.
33. Wilson, S. D.: The Use of Slope Measuring Devices to Determine Movements in Earth Masses, *ASTM Spec. Tech. Publ.* 322, 1962.
34. Richart, F. E., Jr., J. R. Hall, Jr., and R. D. Woods: "Vibration of Soils and Foundations," Prentice-Hall, Inc., Englewood Cliffs, N.J., 1970.
35. Richart, F. E., Jr.: Some Effects of Dynamic Soil Properties on Soil-Structure Interaction, *J. Geotech. Eng. Div. ASCE*, December 1975.
36. Newmark, N. M., and E. Rosenblueth: "Fundamentals of Earthquake Engineering," Prentice-Hall, Inc., Englewood Cliffs, N.J., 1971.
37. American Railway Engineering Association: "Manual of Recommended Practice," part 5, Retaining Walls and Abutments.
38. Huntington, W. C.: "Earth Pressures and Retaining Walls," John Wiley & Sons, Inc., New York, 1957.

39. Terzaghi, K.: Anchored Bulkheads, *Trans. ASCE*, vol. 119, 1954.
40. Peck, R. B., and H. O. Ireland: Backfill Guide, *J. Struct. Div. ASCE*, July 1957.
41. "Timber Piles and Construction Timbers," ASCE Manuals of Engineering Practice, no. 17, New York.
42. Davisson, M. T.: Estimating Buckling Loads for Piles, *Proc. 2d Panamerican Conf. Soil Mech. Found. Eng.*, Brazil, vol. 1, 1963.
43. Davisson, M. T., and Prakash Shamsher: A Review of Soil-Pole Behavior, *Highw. Res. Rec.* 39, 1963.
44. Dunham, Clarence W.: "Foundations of Structures," 2d ed., McGraw-Hill Book Company, New York, 1962.
45. Meyerhoff, G. G.: Some Recent Research on the Bearing Capacity of Foundations, *Can. Geotech. J.*, vol. 1, no. 1, September 1963.
46. Thornley, J. H.: "Foundation Design and Practice," Columbia University Press, New York, 1959.
47. Tomlinson, M. J.: "Foundation Design and Construction," John Wiley & Sons, Inc., New York, 1963.
48. Lee, Kenneth L., Bobby Dean Adams, and Jean-Marie J. Vagneron: Reinforced Earth Retaining Walls. *J. Soil Mech. Found. Div. ASCE*, October 1973.
49. Matlock, H., and L. C. Reese; Foundation Analysis of Off-Shore Pile-Supported Structures, *Proc. 5th Int. Conf. Soil Mech. Found. Eng.*, Paris, vol. 2, 1961.
50. Hirsch, T. J., L. L. Lowery, H. M. Coyle, and C. H. Samson, Jr.: Pile-Driving Analysis by One-Dimensional Wave Theory: State of the Art, *Highw. Res. Rec.*, vol. 333, pp. 33–54, 1970.
51. Chellis, R. D.: "Pile Foundations," 2d ed., McGraw-Hill Book Company, New York, 1961.

Design of Steel Structural Members

WILLIAM J. LEMESSURIER
Senior Partner, LeMessurier Associates/SCI, Consulting Engineers, Cambridge, Mass.

HANS WILLIAM HAGEN
Partner, LeMessurier Associates/SCI, Consulting Engineers

LEE C. LIM
Associate, LeMessurier Associates/SCI, Consulting Engineers

1. Design Procedures Two procedures are available for the design of steel structures. In *elastic design* (also called allowable-stress or working-stress design) a member is selected so that under various combinations of the service loads, the computed maximum stress, based on elastic analysis, will not exceed a specified allowable value. In *load-factor design* (also called *plastic design* and *load-and-resistance-factor design*) members are selected so that under various combinations of the service loads, each multiplied by a *load factor*, member forces* do not exceed specified ultimate strengths. Member forces are based on elastic analysis of the structure in some applications and plastic analysis in others. Also in some applications calculated member strengths are multiplied by *resistance factors* to account for variations in the actual strengths.

The time lag between the publication of new research data and their basic incorporation into specifications is often quite long, particularly at the municipal level. Thus, in order for the designer to take advantage of the latest developments, he may have to file a special appeal with the building department of the municipality which has jurisdiction over the particular project. Depending upon the locality, this procedure is frequently unsuccessful.

2. Types of Steel There are basically three groups of structural steels available for use in bridges and buildings:

---

*Member stresses are used in some applications, for example, transmission towers.

CARBON STEELS:
ASTM A36, Structural Steel
HIGH-STRENGTH LOW-ALLOY STEELS:
ASTM A242, High-Strength Low-Alloy Structural Steel
ASTM A440, High-Strength Structural Steel
ASTM A441, High-Strength Low-Alloy Structural Manganese Vanadium Steel
ASTM A572, High-Strength Low-Alloy Columbium-Vanadium Steels of Structural
    Quality
ASTM A588, Corrosion Resistant High-Strength Low-Alloy Structural Steel
A number of proprietary grades of steel meeting the requirements of A572 are available. The specification covers six grades: 42, 45, 50, 55, 60, and 65, for each of which the minimum yield point in ksi is the grade number.
    HEAT-TREATED CONSTRUCTIONAL ALLOY STEELS:
ASTM A514, High Yield Strength Quenched and Tempered Alloy Steel Plates,
    Suitable for Welding
This group consists of many proprietary grades of low-carbon quenched and tempered steels with specified minimum yield points ranging from 90 to 100 ksi and ultimate strengths of 105 to 135 ksi.
    The scope of applications of various major steels is given in Table 1. Properties are given in Table 2.

TABLE 1   Scope of ASTM Carbon and High-Strength Steels

ASTM grade	Bridges			Buildings			General or special purposes		
	R	B	W	R	B	W	R	B	W
A36	X	X	X^a	X	X	X^a	X	X	X^a
A242 ^{b,c}	X	X	X^d	X	X	X^d	X	X	X^d
A440 ^{e,f}	X	X		X	X		X	X	
A441 ^{f,g}	X	X	X	X	X	X			
A514 ^{h}	X	X	X^i	X	X	X^i	X	X	X^i
A572	X	X	$X^{a,j}$	X	X	X^a	X	X	X^a
A588 ^{b,c}	X	X	X^a	X	X	X^a	X	X	X^a

R = riveted; B = bolted; W = welded.
[a]Welding procedure must be suitable for the steel and the intended service.
[b]Intended primarily for use as structural members where savings in weight or added durability are important.
[c]Enhanced atmospheric corrosion resistance equal to two times that of carbon structural steels with copper.
[d]Welding characteristics vary according to the type of steel furnished.
[e]Intended primarily for use where saving in weight is important.
[f]Atmospheric corrosion resistance is approximately twice that of structural carbon steel.
[g]Intended primarily for use in welded bridges and buildings where savings in weight or added durability are important.
[h]Intended primarily for use in welded bridges and other structures.
[i]Welding procedures must be suitable for the materials being welded.
[j]In grades 42, 45, and 50 only.

3. Shapes   Structural-steel shapes are manufactured to certain tolerances with respect to dimensional variations such as camber, cross section, diameter, squareness, flatness, length, straightness, sweep, thickness, weight, and width. The specific limitations are contained in ASTM A6, General Requirements for Delivery of Rolled Steel Plates, Shapes, Sheet Piling, and Bars for Structural Use.
    Quite often alterations or additions must be made to existing structures which contain structural-steel shapes that are no longer produced. The AISC book "Iron and Steel Beams—1873 to 1952" contains member properties, ASTM specification requirements for yield point and tensile strength, and the AISC working-stress recommendations.

TENSION MEMBERS
Tension members with bolted or riveted connections fail by fracture on a cross section through one or more fastener holes. However, they may also become unserviceable

because of excessive, permanent elongation due to yielding of the gross cross section throughout the length.

Net-section strength is influenced by a number of factors, such as:[2,17]

1. Ductility of the steel (strength of a highly ductile material may be 15 to 20 percent more than for a material with relatively low ductility).

2. Method of making holes (punched holes may reduce strength by as much as 15 percent compared with drilled holes).

3. Ratio of gage to fastener diameter.

4. Ratio of net area to area in bearing on fasteners.

5. Length of connection and location of shear plane or planes relative to member cross section (shear lag). This may reduce the efficiency of the net section by as much as 30 percent.

4. Concentrically Loaded Tension Members   The required area is given by

$$A = \frac{P}{F_t} \tag{1}$$

where F_t is the allowable tensile stress. AASHTO permits $F_t = 0.55F_y$ for members without holes and the smaller of $0.55F_y$ or $0.46F_u$ on the net area for members with holes. AREA permits $F_t = 20$ ksi on the net area for A36 steel and $F_t = 0.55F_y$ for high-strength steel.

In the AISC specification, A in Eq. (1) is the *gross* area A_g, using $F_t = 0.6F_y$, or the *effective* area A_e, using $F_t = 0.5F_u$. The effective area is defined as the net area corrected for shear lag.

Net Area.   AISC, AREA, and AASHTO require that the deduction for each hole equal the fastener diameter plus ⅛ in. A fastener hole is normally punched or drilled ¹⁄₁₆ in. oversize; the extra ¹⁄₁₆ in. is to allow for damage caused by punching and for installation adjustments. In the case of a chain of holes extending across a part in a diagonal or zigzag line, the net width of the part is obtained by deducting from the gross width the sum of the diameters of all the holes in the chain and adding, for each gage space in the chain, the quantity $s^2/4g$, where s = longitudinal spacing (pitch) and g = transverse spacing (gage) of any two consecutive holes. The critical net section is obtained from that chain which gives the least net width.

Effective Area.   Shear lag can be accounted for by computing the effective area

$$A_e = A_n \left(1 - \frac{\bar{x}}{L} \right) \tag{2}$$

where L = length of the connection (distance from the first fastener to the last one) and $\bar{x}$ = distance from a fastener plane to the centroid of the portion of the cross section tributary to it (Fig. 1). Thus, for a single-plane connection $\bar{x}$ is the centroidal coordinate of the entire cross section, while for a symmetrical double-plane connection it is the distance from each connector plane to the centroid of the half cross section.

To account for shear lag, AISC prescribes effective areas as percentages of net for

TABLE 2   **ASTM Carbon and High-Strength Steels**

ASTM grade	Thickness, in.	Min yield point, ksi	Tensile strength, ksi	Min elongation in 8 in., %
A36:				
Shapes	All	36	58–80[a]	20[b]
Plates	To ¾ incl.	36	58–80	20[b,c]
	Over ¾ to 1½ incl.	36	58–80	20[c]
	Over 1½ to 2½ incl.	36	58–80	20[c]
	Over 2½ to 4 incl.	36	58–80	20[c]
	Over 4 to 8 incl.	36	58–80	20[c]
	Over 8	32	58–80	20[c]
Bars	To ¾ incl.	36	58–80	20[b]
	Over ¾ to 1½ incl.	36	58–80	20
	Over 1½ to 4 incl.	36	58–80	20
	Over 4	36	58–80	20

TABLE 2   ASTM Carbon and High-Strength Steels   (Continued)

ASTM grade	Thickness, in.	Min yield point, ksi	Tensile strength, ksi	Min elongation in 8 in., %
A242:				
Shapes	Groups 1, 2[h]	50	70	18[b]
	Group 3[h]	46	67	18
	Groups 4, 5[h]	42	63	18
Plates and Bars	To ¾ incl.	50	70	18[b,c]
	Over ¾ to 1½ incl.	46	67	18[c]
	Over 1½ to 4 incl.	42	63	18[c]
A440:				
Shapes	Groups 1, 2[h]	50	70	18[b]
	Group 3[h]	46	67	18
	Groups 4, 5[h]	42	63	18
Plates and bars	To ¾ incl.	50	70	18[b,c]
	Over ¾ to 1½ incl.	46	67	18
	Over 1½ to 4 incl.	42	63	18
A441:				
Shapes[d]	Groups 1, 2[h]	50	70	18[b,c]
	Group 3[h]	46	67	18
	Groups 4, 5[h]	42	63	18
Plates and bars[d]	To ¾ incl.	50	70	18[b,c]
	Over ¾ to 1½ incl.	46	67	18[c]
	Over 1½ to 4 incl.	42	63	18[c]
	Over 4 to 8 incl.	40	60	18[c]
A572:[e]				
Shapes	All	42	60	20
	All	45	60	19
	Groups 1, 2, 3, 4[h]	50	65	18
	Groups 1, 2, 3, 4 to 426 lb/ft[h]	55	70	17
	Groups 1, 2[h]	60	75	16
	Group 1[h]	65	80	15
Plates and bars	To 4 incl.	42	60[c]	20
	To 1½ incl.	45	60[c]	19
	To 1½ incl.	50	65[c]	18
	To 1½ incl.	55	70[c]	17
	To 1 incl.	60	75[f]	16
	To ½ incl.	65	80[f]	15
A588:				
Shapes	All	50	70	18[b]
Plates and bars	Over 5 to 8 incl.	42	63	g
	Over 4 to 5 incl.	46	67	g
	To 4 incl.	50	70	18[b,c]

[a]For W's over 426 lb/ft, tensile strength of 58 ksi only applies.
[b]For material under 5/16 in. in thickness or diameter reduce 1.25% for each 1/32 in. less than 5/16 in.
[c]Reduce two percentage points for plates wider than 24 in.
[d]When the material is normalized, minimum yield point and tensile strength shall be reduced 5 ksi.
[e]In Grades 42, 45, 50, 55, 60, and 65 with minimum yield points of 42, 45, 50, 55, 60, and 65 ksi, respectively.
[f]Reduce three percentage points for plates wider than 24 in.
[g]21 percent in 2-in.
[h]Group 1: W24 × 55, 62; W21 × 44 to 57 incl.; W18 × 35 to 71 incl.; W16 × 26 to 57 incl.; W14 × 22 to 53 incl.; W12 × 14 to 58 incl.; W10 × 12 to 45 incl.; W8 × 10 to 48 incl.; all W6, W5, and W4. M to 37.7 lb/ft incl. S to 35 lb/ft incl. C to 20.7 lb/ft incl. MC to 28.5 lb/ft incl. Angles to ½ in. incl.

Group 2: W36 × 135 to 210 incl.; W33 × 118 to 152 incl.; W30 × 99 to 211 incl.; W27 × 84 to 178 incl.; W24 × 68 to 162 incl.; W21 × 62 to 147 incl.; W18 × 76 to 119 incl.; W16 × 67 to 100 incl.; W14 × 61 to 132 incl.; W12 × 65 to 106 incl.; W10 × 49 to 112 incl.; W8 × 58, 67. S over 35 lb/ft. HP to 102 lb/ft incl. C over 20.7 lb/ft. MC over 28.5 lb/ft. Angles 9/16 to ¾ in. incl.

Group 3: W36 × 230 to 300 incl.; W33 × 201 to 241 incl.; W14 × 145 to 211 incl.; W12 × 120 to 190 incl. HP over 102 lb/ft. Angles over ¾ in.

Group 4: W14 × 233 to 550 incl.; W12 × 210 to 336 incl.

Group 5: W14 × 605 to 730 incl.

various types of members with unconnected elements such as those of Fig. 1, but with a shear-lag analysis as an option.

For a single angle connected by one leg, or two angles back to back on the same side of a gusset plate, AREA and AASHTO specify net area as the area of the connected leg (or legs) plus one-half the area of the unconnected leg(s), but require no reduction, other than holes, if the member consists of two angles connected to opposite sides of a gusset plate.

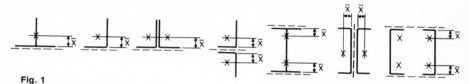

Fig. 1

Eccentricity of fastener gage lines in the direction parallel to a gusset plate, or other element to which the member may be connected, are usually ignored.

Lug angles (Fig. 2) are sometimes used to reduce the length of a single- or double-angle connection. The lug angle should be short and positioned at the beginning of the connection. This will help minimize the effects of unequal fastener shear. A lug angle has dubious efficiency because its only connection to the main angle is through its outstanding leg. AASHTO requires that lug angles be connected with one-third more fasteners than required by the stress to be carried by the lug.

5. Threaded Members The allowable tension on bolts and other threaded parts may be specified for (1) the body (unthreaded) area, (2) the area based on the outside diameter of the thread (which is the same as the body area except for upset rods and similar members), (3) the area at the root of the thread, and (4) the so-called tensile-stress area given by

$$A_s = \frac{\pi}{4}\left(D - \frac{0.9743}{n}\right)^2 \tag{3}$$

where D = nominal size, in., and n = number of threads per inch.

All AISC allowables are for areas according to 1 or 2.

6. Member Type and Selection The type of tension member is governed largely by its method of connection to adjacent portions of the structure. To minimize shear-lag effects, the material should be arranged so that as large a portion as possible is connected directly to the gusset or splice plates. The number and disposition of fasteners will influence the choice. Specified fastener edge distances and

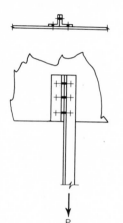

Fig. 2 Lug-angle connection.

pitch must be satisfied. Designers sometimes require 100 percent efficiency in long, heavily loaded bolted or riveted members. This can be achieved by increasing the size of the member at the connection, as in upset rods, or by using a higher-strength steel at the connection (Sec. 18, Art. 30).

To guard against excessive lateral movement (slapping or vibration), AISC recommends that slenderness ratios be kept below the values given in Table 3. This table also gives the mandatory slenderness limits of the AASHTO and AREA specifications.

7. Truss Members The double angle, the double channel, and W shapes are frequently used in trusses for buildings. All connections and members should be symmetrically disposed about the central plane of the truss. Tension members should be selected so that they are sufficiently rigid (Table 3) but not so rigid that large secondary stresses are set up.

AISC, AASHTO, and AREA have placed minimum values on the strength of a connection, requiring it to be designed for the applied load but for not less than 50, 75, and 100 percent, respectively, of the strength of the member. Long connections minimize shear lag, but fasteners in long connections will not be uniformly loaded under service loads. This is caused by unequal deformation of the member relative to the gusset plate between successive fasteners. End fasteners will be more heavily loaded than fasteners in the center of the connection. Yielding of the end fasteners will distribute the loading uniformly between all fasteners in a short connection before failure occurs, but the end fasteners in long connections usually fail prematurely (Fig. 31). Design specifications assume equal load partition between fasteners irrespective of connection length, but AISC specifies a 20 percent reduction in the allowable shear in bearing connections if the distance between the end fasteners in a line parallel to the direction of the axial force exceeds 50 in.

TABLE 3   Maximum Slenderness Ratio—Tension Members

	AISC	AASHTO	AREA
Main	240	200	200
Bracing and secondary	300	240	

Example 1 Determine the net width of the plate shown in Fig. 3. Fasteners are ⅞ in.

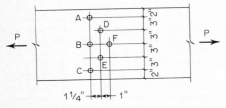

Fig. 3 Example 1.

Chain *ABC*:
$$w_n = 16 - 3 = 13$$

Chain *ABEC* (= chain *ADEC*):
$$w_n = 16 - 4 + 2 \times \frac{1.25^2}{4 \times 3} = 12.26$$

Chain *ADBEC*:
$$w_n = 16 - 5 + 4 \times \frac{1.25^2}{4 \times 3} = 11.52$$

Chain *ADFEC*:
$$w_n = 16 - 5 + 2 \times \frac{1.25^2}{4 \times 3} + 2 \times \frac{1^2}{4 \times 3} = 11.43$$

Example 2 Select a double-angle member for a truss diagonal 12 ft long carrying an axial tension of 85 kips. The width of the truss is limited to 6.5 in. A36 steel, ¾-in. A325 bolts, AISC specification.

$$F_t = 0.5 \times 58 = 29 \text{ ksi} \qquad A_e = \frac{85}{29} = 2.93 \text{ in.}^2$$

$$F_t = 0.6 \times 36 = 22 \text{ ksi} \qquad A_g = \frac{85}{22} = 3.86 \text{ in.}^2$$

Possible sizes are:

t	Equal legs	Unequal legs	A_g	A_n*	A_{eff}†
⁵⁄₁₆		3 ½ × 3 × 6.6 lb	3.87	3.32	2.82
⁵⁄₁₆		4 × 3 × 7.2 lb	4.18	3.63	3.08
⅜	3 × 3 × 7.2 lb		4.22	3.56	3.02

*Deduction for 2 holes = $2 \times 0.875t$.
†$A_{eff} = 0.85 A_n$ (AISC 1.14.2.2).

Either the 4 × 3 × ⁵⁄₁₆ angles or the 3 × 3 × ⅜ angles have the required area.

Try the $3\frac{1}{2} \times 3 \times \frac{5}{16}$ angles, using Eq. (2) instead of $A_e = 0.85A_n$. With $\frac{5}{16}$-in. gusset plates the allowable values for the A325 bolts are:

$$\text{Shear} = 2 \times 0.44 \times 21 = 18.5 \text{ kips}$$
$$\text{Bearing} = 0.75 \times 0.3125 \times 1.5 \times 58 = 20.4 \text{ kips}$$
$$n = \frac{85}{18.5} = 4.6 \qquad \text{Use 5 bolts at } 2\frac{1}{2}\text{-in. spacing}$$

The $3\frac{1}{2}$-in. legs must be back to back to keep the width of the member within the specified 6.5-in. limit. Then $\bar{x} = 0.808$ in. and $L = 4 \times 2.5 = 10$ in.

$$A_{\text{eff}} = A_n \left(1 - \frac{\bar{x}}{L}\right) = 2.82 \left(1 - \frac{0.808}{10}\right) = 3.05 \text{ in.}^2 > 2.93$$

Thus an analysis for shear lag according to Eq. (2) enables the lighter $3\frac{1}{2} \times 3 \times \frac{5}{16}$ angles to be used. For this member, $r_{\min} = 1.10$ and $L/r = 12 \times 12/1.10 = 131 < 240$.

COMPRESSION MEMBERS

8. Column Strength   The principal factor influencing the strength of straight centrally loaded steel columns is the magnitude and distribution of residual stresses within the column cross section.[1,2] Residual stresses result from cooling after hot rolling or welding, or from fabrication operations such as flame cutting, cold bending, or cambering.

Other factors that influence the strength of a concentrically loaded column are shape of cross section, annealing, out-of-straightness, cold straightening, and length. Because of the high heat penetration due to welding of built-up columns, residual stresses tend to be large. In most cases the tensile residual stresses at the weld approach the yield point. Welded columns built up of universal mill plates have somewhat lower strength than comparable rolled shapes. Welded box columns are slightly stronger than welded H columns. On the other hand, welded columns built up from flame-cut plates tend to have higher strength because of a favorable residual stress distribution.

The residual stress distribution in a heavy shape (defined as a shape whose thinnest component is more than $1\frac{1}{2}$ in. thick) differs from that in a small shape in two major respects: the magnitude may be considerably larger and there is a considerable variation through the thickness.[10,11] These account for the reduced strengths of such columns. Out-of-straightness is not an important factor for columns of low slenderness ratio (L/r from 50 to 120); an out-of-straightness of the order of $\frac{1}{4}$ in. in a 20-ft column, as permitted by AISC specifications, will reduce the column strength about 25 percent below that indicated by the SSRC basic column curve.

9. Concentrically Loaded Columns   The following allowable stresses are specified by the AISC:

$$F_a = \left[1 - \frac{(KL/r)^2}{2C_c^2}\right] \frac{F_y}{\text{F.S.}} \qquad \frac{KL}{r} \leqslant C_c \qquad (4)$$

$$F_a = \frac{12\pi^2 E}{23(KL/r)^2} = \frac{149,300}{(KL/r)^2} \text{ ksi} \qquad \frac{KL}{r} \geqslant C_c \qquad (5)$$

where $L = $ length, $r = $ least radius of gyration, $E = $ modulus of elasticity, $K = $ effective-length coefficient, $C_c = \pi\sqrt{2E/F_y}$, and

$$\text{F.S.} = \text{factor of safety} = \frac{5}{3} + \frac{3KL/r}{8C_c} - \frac{(KL/r)^3}{C_c^3}$$

Eq. (4) covers the range of inelastic buckling and Eq. (5) the elastic range. The factor of safety varies from 1.67 at $KL/r = 0$ to 1.92 at $KL/r = C_c$ and remains constant at 1.92 for $KL/r > C_c$. The factor of safety for bracing and secondary members for which L/r exceeds 120 is reduced by dividing the allowable stress given by the appropriate formula [Eq. (4) or (5), with $K = 1$] by $1.6 - (L/r)/200$.

The allowable stresses of the AASHTO are given by

$$F_a = \left[1 - \frac{(KL/r)^2}{2C_c^2}\right] \frac{F_y}{\text{F.S.}} \qquad \frac{KL}{r} \leqslant C_c \qquad (6a)$$

$$F_a = \frac{\pi^2 E}{\text{F.S.} \, (KL/r)^2} \qquad \frac{KL}{r} \geqq C_c \qquad (6b)$$

where F.S. = 2.12. Equation ($6a$) may be written as follows:

$$F_a = C_1 - \frac{C_2}{1000}\left(\frac{KL}{r}\right)^2$$

Values of C_1 and C_2 are given in Table 4.

TABLE 4

F_y, psi	C_1, psi	C_2, psi	Limiting KL/r
36,000	16,980	0.53	126
50,000	23,580	1.03	107
90,000	42,450	3.33	79.8
100,000	47,170	4.12	75.7

The allowable stresses of the AREA are:

$$\frac{KL}{r} \geqq 15 \qquad F_a = 20,000 \text{ psi} \qquad (7a)$$

$$15 \geqq \frac{KL}{r} \geqq 143 \qquad F_a = 21,500 - 100\left(\frac{KL}{r}\right) \qquad (7b)$$

$$\frac{KL}{r} \geqq 143 \qquad F_a = \frac{147,000,000}{(KL/r)^2} \qquad (7c)$$

For high-strength steels up to 60,000 psi yield strength, the corresponding equations are

$$\frac{KL}{r} \leqq \frac{3388}{\sqrt{F_y}} \qquad F_a = 0.55 F_y \qquad (8a)$$

$$\frac{3388}{\sqrt{F_y}} \leqq \frac{KL}{r} \leqq \frac{27,111}{\sqrt{F_y}} \qquad F_a = 0.60F - \left(\frac{F_y}{1662}\right)^{3/2}\frac{KL}{r} \qquad (8b)$$

$$\frac{KL}{r} \geqq \frac{27,111}{\sqrt{F_y}} \qquad F_a = \frac{147,000,000}{(KL/r)^2} \qquad (8c)$$

where $K = 0.75$ for riveted, bolted, and welded connections and 0.875 for pinned connections.

10. Effective Length Values of K for certain idealized end conditions are given in Fig. 4. Values of K for columns in frames depend on the flexural rigidity of adjoining members and the manner in which sidesway is resisted. The values given in Fig. 5 are based on the assumption that all the columns in the frame reach their individual critical loads simulta-

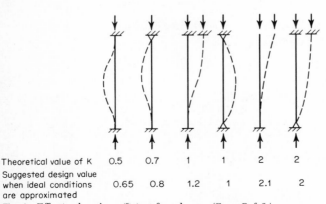

Theoretical value of K	0.5	0.7	1	1	2	2
Suggested design value when ideal conditions are approximated	0.65	0.8	1.2	1	2.1	2

Fig. 4 Effective length coefficient for columns. *(From Ref. 3.)*

neously. Figure 5a is for frames where sidesway is inhibited by diagonal bracing, shear walls, or floor slabs or roof decks secured in the horizontal plane by walls or bracing systems parallel to the plane of the frame. Figure 5b should be used for frames whose lateral stability depends upon the stiffness of the frame itself.

Values of K for compression members in trusses are ordinarily taken as unity. However, rotational restraint from adjoining members may be taken into account, and corresponding reduced effective lengths are given in Chap. 7 of Ref. 4.

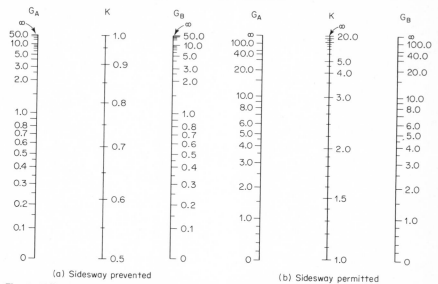

(a) Sidesway prevented   (b) Sidesway permitted

Fig. 5 Effective length coefficient for columns. (*Courtesy of Jackson & Moreland Division of United Engineers & Constructors, Inc.*) Subscripts A and B refer to the joints at the ends of the column. G is defined as

$$G = \frac{\Sigma I_c/L_c}{\Sigma I_g/L_g}$$

where Σ denotes the summation of all members in the plane of buckling which are rigidly connected to the joint, I = moment of inertia, L = length, and subscripts c and g denote column and girder, respectively.

For pinned column bases G is theoretically infinite but should be taken as 10. For fixed column bases G is theoretically zero but should be taken as 1.

Girder stiffness I_g/L_g should be multiplied by the following factor for the stated condition: 1.5 for far end hinged, Fig. 5a; 2 for far end fixed, Fig. 5a; 0.5 for far end hinged, Fig. 5b; 0.67 for far end fixed, Fig. 5b.

11. Amplification Factors and Frame Stability Column design based on the effective-length concept (Art. 10) is satisfactory if all columns in a building are moment-connected so that, together with the girders, they provide lateral stability. However, in most buildings all columns are not rigidly connected, primarily for reasons of economy. The columns that are not part of the lateral stabilizing frames are often designed as pin-ended ($K = 1$). The columns that are parts of the lateral stabilizing frames must resist lateral loadings as well as provide stability to the whole building. Several approximate methods for the design of this type of column are presented in Refs. 5 and 6. The stabilizing frames should be designed to resist the P-Δ forces in addition to all lateral loadings, which can be done by amplifying the building drift and the resultant forces in the frame components. Two sets of amplification factors are given in Table 5. Definitions of the terms in the formulas of this table are:

ΣP = total factored vertical load on columns
$\Sigma P_L = (H/\Delta_{ov})\Sigma V$ = summation of column stiffnesses under lateral load

Δ_{ov} = first-order elastic story drift due to lateral load ΣV

ΣV = total lateral load at story

H = story height

$C_L = P_L/P_e - 1$ = stiffness-reduction factor due to axial load, obtained from second-order analysis

$P_e = \pi^2 EI/(KL)^2$ = Euler load

The value of C_L ranges from 0.0 to 0.216 (Fig. 6). Also plotted in Fig. 6 are the values of β from which P_L can be determined. The term $\Sigma(C_L P)$ is small in practical cases, that is, when G_A and G_B are greater than 2. The maximum error in ignoring this term is less than 4 percent if the amplification factor at factored load is less than 1.5. The omission of $\Sigma(C_L P)$ in LeMessurier's equations results in identical amplification factors for both approaches.

TABLE 5   Amplification Factors

	LeMessurier[5]	P-Δ method[6]
Drift	$\dfrac{1}{1 - \dfrac{\Sigma P + \Sigma(C_L P)}{\Sigma P_L}}$	$\dfrac{1}{1 - \dfrac{\Sigma P}{\Sigma V} \dfrac{\Delta_{ov}}{H}}$
Forces	$\dfrac{1}{1 - \dfrac{\Sigma P}{\Sigma P_L - \Sigma(C_L P)}}$	$\dfrac{1}{1 - \dfrac{\Sigma P}{\Sigma V} \dfrac{\Delta_{ov}}{H}}$

In lieu of Fig. 6, P_L may be computed from

$$\Sigma P_L = \frac{6(G_A + G_B) + 36}{2(G_A + G_B) + G_A G_B + 3} \times \frac{EI}{H^2} \tag{9}$$

The point of inflection of the column measured from the bottom is

$$\frac{G_A + 3}{G_A + G_B + 6} H \tag{10}$$

Values of G_A and G_B do not vary significantly in a multistory building. Thus assuming $G_A = G_B = G_S$, Eq. (9) becomes

$$\Sigma P_L = \frac{12}{1 + G_S} \times \frac{EI}{H^2}$$

and for this case

$$C_L = \frac{0.216}{(1 + G_S)^2}$$

The elastic-buckling load of a story is[5]

$$\Sigma P_e = \Sigma P_L \frac{\Sigma P}{\Sigma P + \Sigma(C_L P)} \tag{11}$$

The buckling load of the ith column in the stabilizing frame is

$$P_{ei} = \frac{P_i}{\Sigma P} \Sigma P_e = \frac{P_i \Sigma P_L}{\Sigma P + \Sigma(C_L P)} \tag{12}$$

where P_i = factored vertical load on ith column, and its effective length K_i is

$$K_i = \sqrt{\frac{\pi^2 EI_i}{P_i H^2} \times \frac{\Sigma P + \Sigma(C_L P)}{\Sigma P_L}} \tag{13}$$

Alternatively[7]

$$K_i = K_{oi} \sqrt{1 + n} \tag{14}$$

where K_{oi} is the value of K determined by the method of Art. 10 and n is the ratio of the total vertical load on the columns that are not part of the lateral stabilizing frames to the total vertical load on the columns that are part of the lateral stabilizing frames. Substituting Eq. (11) into Eq. (13) gives

$$K_i = \sqrt{\frac{\pi^2 E I_i}{P_i H^2} \times \frac{\Sigma P}{\Sigma P_e}} \tag{15}$$

For a simple frame with only one stabilizing column (see Example 8), $\Sigma P_e = P_{ei}$, in which case Eq. (15) becomes

$$K_i = K_{oi} \sqrt{\frac{\Sigma P}{P_i}} \tag{16}$$

By definition of the term n, $\sqrt{\Sigma P/P_i} = \sqrt{1 + n}$, and Eqs. (13) and (14) become identical. Examples are given in Art. 21.

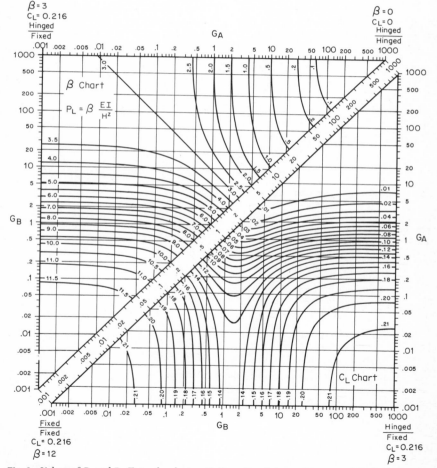

Fig. 6   Values of C_L and P_L. Enter the charts with the column's larger G as G_B and the smaller one as G_A. If the column G's are equal enter each chart with G along the diagonal. (*Ref. 5.*)

12. Proportioning   Radius of gyration and area of cross section are the only factors governing the strength of a concentrically loaded compression member of given length and material. If the variation of allowable stress with slenderness is parabolic, it can be shown that the efficiency of a shape is related to A/r^2, where by efficiency is meant the ratio of the allowable load for a given slenderness ratio to the allowable load for $L/r = 0$. The smaller A/r^2 the more efficient the member. This is particularly significant for slender or lightly loaded members. Approximate values of A/r^2, assuming equal effective lengths in both principal planes, are

Thin-walled square tube—24 × wall thickness ÷ tube depth
Thin-walled pipe—8π × wall thickness ÷ diameter
Solid square—12
Solid circle—4π
W shape—2 to 9 (the lightest W in a given series will be the most efficient)

Material should be positioned as far as possible from the center of gravity of the cross section. End restraints are often different in the two principal planes, so that different moments of inertia in these planes are sometimes desirable to achieve approximately equal slenderness ratios.

Common types of compression members are illustrated in Fig. 7. Single angles are rarely used, except in light roof trusses, because of eccentricities at the connections. Tees are often used in roof trusses. The W shape is used for columns in buildings and truss members. The pipe section is the most efficient but is suitable only for medium loads, and its connection details can be difficult.

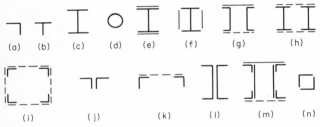

Fig. 7   Types of compression members.

Approximate radii of gyration for various cross sections are given in Table 6.

13. Local Buckling   To prevent local buckling, AISC specifies the following maximum width-thickness ratios b/t:

Unstiffened compression elements
  Single-angle struts; double-angle struts with separators $76/\sqrt{F_y}$
  Double angles in contact and angles or plates
    projecting from compression members; compression
    flanges of beams; stiffeners on plate girders $95/\sqrt{F_y}$
  Stems of tees $127/\sqrt{F_y}$

Stiffened compression elements
  Flanges of square and rectangular box sections of
    uniform thickness $238/\sqrt{F_y}$
  Perforated cover plates $317/\sqrt{F_y}$
  All other uniformly compressed stiffened elements $253/\sqrt{F_y}$

where F_y = yield strength of steel in ksi. Provisions are made for design with elements exceeding these slenderness limits (Art. 16).

AASHTO prescribes the following limiting values of b/t in terms of the calculated compressive stress f_a (psi):

1. Plates supported on one side, outstanding legs of angles, and perforated plates at the perforations: $1625/\sqrt{f_a}$ but not to exceed 12 for main members and 16 for secondary members.

2. Webs of main segments connected by cover plates to form a box section: $4000/\sqrt{f_a}$ but not to exceed 45.

TABLE 6 Approximate Radii of Gyration*

$r_x = 0.29h$ $r_y = 0.29b$	$r_x = 0.42h$ $r_y = 0.42b$	$r_x = 0.31h$ $r_y = 0.48b$
$r_x = 0.40h$ h = mean h	r_y = same as for 2 L	$r_x = 0.37h$ $r_y = 0.28b$
$r_x = 0.25h$	$r_x = 0.42h$ r_y = same as for 2 L	$r_x = 0.31h$
$r = \sqrt{\dfrac{H^2 + h^2}{16}}$ $r = 0.35 H_m$	$r_x = 0.39h$ $r_y = 0.21b$	$r_x = 0.31h$
$r_x = 0.31h$ $r_y = 0.31h$ $r_z = 0.197h$	$r_x = 0.45h$ $r_y = 0.235b$	$r_x = 0.40h$ $r_y = 0.21b$
$r_x = 0.29h$ $r_y = 0.32b$ $r_z = 0.18\dfrac{h+b}{2}$	$r_x = 0.36h$ $r_y = 0.45b$	$r_x = 0.38h$ $r_y = 0.22b$
$r_x = 0.31h$ $r_y = 0.215b$ $= b(0.21 + 0.02s)$	$r_x = 0.36h$ $r_y = 0.60b$	$r_x = 0.39h$
$r_x = 0.32h$ $r_y = 0.21b$ $= b(0.19 + 0.02s)$	$r_x = 0.36h$ $r_y = 0.53b$	$r_x = 0.35h$
$r_x = 0.29h$ $r_y = 0.24b$ $= b(0.23 + 0.02s)$	$r_x = 0.39h$ $r_y = 0.55b$	$r_x = 0.435h$ $r_y = 0.25b$
$r_x = 0.30h$ $r_y = 0.17b$	$r_x = 0.42h$ $r_y = 0.32b$	$r_x = 0.42h$
$r_x = 0.25h$ $r_y = 0.21b$	$r_x = 0.44h$ $r_y = 0.28b$	$r_x = 0.42h$
$r_x = 0.21h$ $r_y = 0.21b$ $r_z = 0.19h$	$r_x = 0.50h$ $r_y = 0.28b$	$r_x = 0.285h$ $r_y = 0.37b$
$r_x = 0.38h$ $r_y = 0.19b$	$r_x = 0.39h$ $r_y = 0.21b$	$r_x = 0.42h$ $r_y = 0.23b$

* J. A. L. Waddell, "Bridge Engineering," John Wiley & Sons, Inc., New York, 1925. Reproduced by permission.

3. Solid cover plates and webs, supported on two edges, connecting main segments of box sections: $5000/\sqrt{f_a}$ but not to exceed 50.

4. Perforated cover plates supported on two edges: $6000/\sqrt{f_a}$ but not to exceed 55.

For cases 2 and 3 above, AREA prescribes limiting values of 32 and 40, respectively, for $F_y = 36,000$ psi, and $6000/\sqrt{F_y}$ and $7500/\sqrt{F_y}$ for high-strength steels.

Plate elements of built-up members may buckle locally between adjacent fasteners. Typical specification requirements to control this are those of the AISC, where for nonstaggered fasteners, the maximum pitch is $127/\sqrt{F_y}$ times the plate thickness but not to exceed 12 in., and for staggered fasteners, $190/\sqrt{F_y}$ times the plate thickness but not to exceed 18 in.

14. Lacing and Perforated Cover Plates   The open sides of built-up compression members must be provided with lacing (Fig. 8a,b,c), battens (tie or stay plates, Fig 8e), or perforated cover plates (Fig. 8f) to prevent local buckling of the components and to ensure

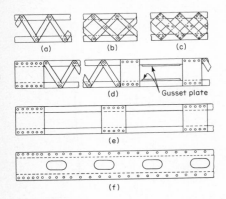

(a)   (b)   (c)

(d)   Gusset plate

(e)

(f)

Fig. 8

that they act as a unit. According to the AISC, lacing must be proportioned to resist a shear, normal to the axis of the member, equal to 2 percent of the axial compression. The AASHTO specifies a shear given by

$$V = \frac{P}{100}\left(\frac{100}{L/r + 10} + \frac{L/r}{3,300,000/F_y}\right) \qquad (17)$$

to which must be added shear due to weight of the member and any external force, other than P, acting on it. In this equation, P is the allowable axial load and r is the radius of gyration about the axis normal to the plane of the lacing or perforated plate. The shear V is divided equally among all parallel planes of shear-resisting elements.

TABLE 7   Geometrical Requirements for Tie Plates

	AISC	AREA and AASHTO
Min length of end plate	B	$1\frac{1}{4}B$
Min length of intermediate plate	$\frac{1}{2}B$	$\frac{3}{4}B$
Min thickness	$\dfrac{B}{50}$	B/50: main members B/60: secondary members

B = clear distance between centerlines of fasteners or welds connecting the plate.

The shear specified by AREA is given by Eq. (17) with the denominator of the second term in parentheses equal to 100 for $F_y = 36,000$ psi and $3,600,000/F_y$ for high-strength steels.

Lacing bars must be spaced so that the segments of the member will not buckle locally. To assure this, AISC requires that the slenderness ratio of the portion of the flange included between adjacent connections of the bars not exceed the slenderness ratio of the member itself. According to AASHTO and AREA, the slenderness ratio of the segment must not exceed 40 or two-thirds the slenderness ratio of the member. The inclination of lacing bars to the axis of the member should be about 45° for double lacing and 60° for single lacing. Lacing bars must be designed to resist both tension and compression.

Tie plates are required at the ends of lacing planes and at any other point where the lacing must be interrupted (Fig. 8d). Tie plates fulfill the dual function of spacing the components of the member and of distributing the load between them. Dimensions of tie plates are given in Table 7.

TABLE 8   Geometrical Requirements for Perforated Cover Plates

	AISC	AASHTO
Ratio of length of hole in direction of load to width	2	2
Clear distance between holes	B	B
Min radius of hole corners	1½ in.	1½ in.
Min distance from edge of end perforation to end of cover plate	. . . . . .	1¼B
Min distance from edge of perforation, measured at its centerline, to nearest line of connecting fasteners	. . . . . .	Art. 13, Case 1

B = clear distance between centerlines of fasteners or welds connecting the plate.

Perforated cover plates have two advantages over lacing: (1) the net width of the plate may be considered to be part of the cross section, and (2) they permit easy access to the interior of the member for maintenance. Geometrical requirements are given in Table 8.

Example 3   Design a square column consisting of four angles (Fig. 9). The effective length KL is 30 ft and the load $P = 740$ kips. A36 steel, AISC specification.

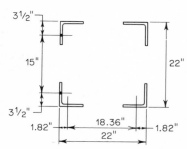

Fig. 9   Example 3.

From Table 6,

$$r = 0.42 \times 22 = 9.24 \text{ in.}$$
$$\frac{L}{r} = \frac{360}{9.24} = 39$$
$$F_a = 19.27 \text{ ksi} \qquad A = \frac{740}{19.27} = 38.4 \text{ in.}^2$$

Try four angles $6 \times 6 \times \frac{7}{8}$, $A = 38.92$ in.²

$$I_x = I_y = 4(31.9 + 9.73 \times 9.18^2) = 3407 \text{ in.}^4$$
$$r_x = r_y = \sqrt{\frac{3407}{38.92}} = 9.35 \approx 9.24 \text{ in.} \qquad \text{Column O.K.}$$

LACING.   For 3½-in. gage on angles, the distance between rows of fasteners is 15 in. For single lacing at 60° with the axis of the member, the distance between lacing points is $15 \times 2 \cot 60° = 17.3$ in. For the 6×6 angle between lacing points,

$$\frac{L}{r} = \frac{17.3}{1.17} = 14.8 < 39$$

The shear on each plane of lacing is

$$V = 0.02 \times 740/2 = 7.4 \text{ kips}$$

The corresponding force in one lacing bar is

$$\frac{7.4}{\cos 30°} = 8.55 \text{ kips}$$

For a maximum permissible L/r of 140

$$r = \frac{17.3}{140} = 0.29t$$
$$t = 0.43 \text{ in. Use } \tfrac{7}{16}$$
$$r = 0.29 \times 0.437 = 0.126 \qquad \frac{L}{r} = \frac{17.3}{0.126} = 137$$
$$F_a = 8.70 \qquad A = \frac{8.55}{8.70} = 0.982 \text{ in.}^2$$
$$b = \frac{0.982}{0.437} = 2.25 \text{ in.}$$

Use $2\tfrac{1}{4} \times \tfrac{7}{16}$ lacing, $\tfrac{3}{4}$-in. rivets.

$$A_n = (2.25 - 0.88)0.437 = 0.6 \text{ in.}^2$$
$$\text{Allowable tension} = 0.6 \times 22 = 13.2 > 8.55$$

TIE PLATES. Minimum length = 15 in., minimum thickness = $\tfrac{15}{50}$ = 0.3. Use 15 × $\tfrac{3}{8}$ tie plates.

15. Tapered Columns The elastic-buckling load for a tapered column can be found by multiplying the Euler buckling load by a factor

$$P = \frac{\pi^2 E I_0 \mu}{(KL)^2} = \frac{\pi^2 E A_0 r_0^2 \mu}{(KL)^2} \qquad (18a)$$

where I_0 = moment of inertia at small end (Fig. 10)
A_0 = area at small end
r_0 = radius of gyration at small end
Values of μ for I and box sections of uniformly tapered depth d and constant flange width and flange and web thickness, and of four-legged tower or boom sections with constant leg area and uniformly tapered depth d are given in Table 9. Buckling is about the strong axis.[8]

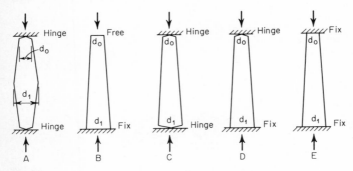

Fig. 10

A formula for inelastic buckling is obtained by replacing E in Eq. (18a) by the tangent modulus E_t:

$$P = \frac{\pi^2 E_t A_0 r_0^2 \mu}{(KL)^2} \qquad (18b)$$

This equation is correct if the area is constant, as in the four-legged tower section, but it is approximate for the I and box sections since the area increases with increase in depth of the section. The increase is relatively small, however, since it is confined to the web (or webs). Furthermore, the formula is on the safe side for the latter sections since the member yields progressively from the smaller end to the larger, rather than uniformly over the whole length.

Equations (18a) and (18b) show that the buckling stress P/A_0 for a tapered member can be found by using an equivalent radius of gyration $r_0 \sqrt{\mu}$ to determine the slenderness ratio. Furthermore, since specification allowable-stress formulas are obtained by dividing buckling-stress formulas by a factor of safety, the equivalent slenderness ratio $KL/(r_0 \sqrt{\mu})$ can be used in allowable-stress formulas.

TABLE 9 Values of μ and K in Eqs. (18)*

Case		d_1/d_0					
(Fig. 10)	K	1	2	3	4	5	6
A	1	1	2.6	4.9	7.7	11.1	14.9
B	2	1	2.6	4.9	7.7	11.1	14.9
C	1	1	2.1	3.4	4.9	6.6	8.3
D	0.7	1	2.0	3.3	4.7	6.2	7.9
E	0.5	1	2.0	3.3	4.6	6.0	7.6

*From Ref. 8.

Based on work reported in Ref. 9, AISC gives a procedure for determining the allowable stress for uniformly tapered I-shaped columns, in which the rotational restraint of connecting beams can be accounted for in determining an effective length KL. Charts for determining the effective-length coefficient (denoted by K_γ in the specification) are given.

Buckling about the weak axis for the tapered members discussed in this article should be computed for the weak-axis radius of gyration at the small end, using $\mu = 1$.

Example 4 Compute the allowable load for the A36-steel column shown in Fig. 11 if it is supported against buckling in the weak direction. Use the AISC allowable-stress formula with the slenderness ratio based on K and μ from Table 9.

$$I_0 = 2 \times 6 \times 3.62^2 + \frac{1}{4} \times \frac{6.5^3}{12} = 157 + 6 = 163 \text{ in.}^4$$

$$A_0 = 2 \times 6 + \frac{1}{4} \times 6.5 = 13.6 \text{ in.}^2$$

$$r_0 = \sqrt{\frac{163}{13.6}} = 3.46$$

$$\frac{d_1}{d_0} = 2 \qquad \text{Case } C, \mu = 2.1$$

$$\frac{KL}{r_0 \sqrt{\mu}} = \frac{1 \times 240}{3.46 \sqrt{2.1}} = 48$$

$$F_a = 18.5 \qquad P = 18.5 \times 13.6 = 252 \text{ kips}$$

The AISC procedure based on K_γ gives the same result.

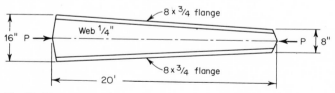

Fig. 11 Example 4.

16. Slender Compression Elements Compression elements whose slenderness exceeds the limits given in Art. 13 are not fully effective. This is because they buckle at stresses below the yield point. Elements of this type are common in cold-formed construction (Sec. 9). To provide for the occasional situation in which they may be used in structural members covered by the AISC specification, formulas and procedures similar to those discussed in Sec. 9 are given in Appendix C of the specification.

BEAMS

The design of rolled sections is normally governed by allowable bending stresses or allowable deflections. Shear will govern only for short heavily loaded spans. Members are usually chosen so that they are symmetrical about the plane of loading, so as to eliminate unsymmetrical bending and torsion.

The I or W is the most efficient rolled section. It has excellent flexural strength, and relatively good lateral strength for its weight. Channels have reasonably good flexural strength, but poor lateral strength, and require horizontal bracing or some other lateral support. Tees and angles are used only for light loads. The flexural strength of a rolled section can be improved by the addition of flange plates (Fig. 12). A plate girder may be used if the loadings are too heavy or the spans too long for a standard rolled section. Plate girders are most commonly fabricated as I-beams but may be fabricated to any shape or depth. A box section (Fig. 12g) may be used if depth is restricted or if lateral stability is a problem.

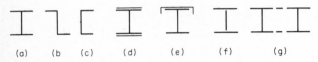

(a)   (b   (c)   (d)   (e)   (f)   (g)

Fig. 12 Typical beam cross sections.

17. Allowable Stresses The required section modulus is given by

$$S = \frac{M}{F_b}$$

where F_b = allowable bending stress.

The AISC divides sections which are symmetrical about the plane of loading into two categories, noncompact and compact. The basic allowable stresses F_b are $0.60F_y$ for noncompact sections and $0.66F_y$ for compact sections. To qualify for these stresses, the requirements of Table 10 must be met. In general these allowable stresses apply for beams bent about the major axis. For the compact members, except those of A514 steel, of doubly symmetric I and H shapes, bent about the minor axis, $F_b = 0.75\ F_y$.

For members, except hybrid girders and members of A514 steel, for which $65/\sqrt{F_y} < b/t_f < 95/\sqrt{F_y}$, F_b may be interpolated between $0.66F_y$ and $0.60F_y$.

For doubly symmetric members of I and H shape, bent about the minor axis, for which $65/\sqrt{F_y} < b/t < 95/\sqrt{F_y}$, F_b may be interpolated between $0.75F_y$ and $0.6F_y$.

Lateral Buckling.   The allowable compression on the extreme fiber of noncompact flexural members, and of members not meeting all the requirements to qualify as compact sections, is given by the following equations:

$$\text{If } \sqrt{\frac{102,000C_b}{F_y}} \leq \frac{L}{r_T} \leq \sqrt{\frac{510,000C_b}{F_y}}$$

$$F_b = \left[\frac{2}{3} - \frac{F_y(L/r_T)^2}{1530 \times 10^3 C_b}\right] F_y \geq 0.6F_y \tag{19a}$$

$$\text{If } \frac{L}{r_T} \leq \sqrt{\frac{510,000C_b}{F_y}}$$

$$F_b = \frac{170,000C_b}{(L/r_T)^2} \geq 0.6F_y \tag{19b}$$

If the compression flange is solid and approximately rectangular in cross section and its area is not less than that of the tension flange, Eq. (20) should be used instead of Eqs. (19) if it gives a larger value of F_b:

$$F_b = \frac{12,000C_b}{Ld/A_f} \geq 0.6F_y \tag{20}$$

In these equations

L = unbraced length of compression flange. For a cantilever free to twist (except at the support) L is the actual length

r_T = radius of gyration of a section comprising the compression flange and one-third of the compression web area about plane of web

A_f = area of compression flange

$C_b = 1.75 + 1.05\ M_1/M_2 + 0.3\ (M_1/M_2)^2 \leqq 2.3$

In the equation for C_b, M_1 is the smaller and M_2 the larger of the bending moments at the ends of the unbraced length. The ratio M_1/M_2 is negative for single-curvature bending and positive for reversed-curvature bending. If the bending moment at any point within the unbraced length exceeds both M_1 and M_2, M_1/M_2 is to be taken as unity, i.e., $C_b = 1$.

TABLE 10   **Requirements for Beam Geometry**

	Allowable compression, extreme fiber	
	Noncompact section $F_b = 0.60F_y$	Compact section $F_b = 0.66F_y$*·†
Unstiffened elements of compression flange, b/t	$\dfrac{95}{\sqrt{F_y}}$	$\dfrac{65}{\sqrt{F_y}}$
Stiffened elements of compression flange, b/t_f	$\dfrac{253\ddagger}{\sqrt{F_y}}$	$\dfrac{190}{\sqrt{F_y}}$
Web, d/t	None	$\leqq \dfrac{640}{\sqrt{F_y}}\left(1 - 3.74\dfrac{f_a}{F_y}\right)$ if $\dfrac{f_a}{F_y} \leqq 0.16$ $= \dfrac{257}{\sqrt{F_y}}$ if $\dfrac{f_a}{F_y} > 0.16$
Circular sections, diameter/thickness	Not specified	$\dfrac{3300}{F_y}$
Distance between lateral supports (compression flange): *(a)* Box ($d \geqq 6b,\ t_f \geqq 2t_w$)	None	$\left(1950 + 1200\dfrac{M_1}{M_2}\right)\dfrac{b\S}{F_y}$ (need not be less than $1200b/F_y$)
(b) Other	Eqs. (19)	$\geqq \dfrac{76b_f}{\sqrt{F_y}}$ and $\geqq \dfrac{20,000}{(d/A_f)F_y}$

*Except hybrid girders and members of A514 steel.
†Flanges must be continuously connected to web or webs and beam must be symmetrical about and loaded in plane of minor axis.
‡238 if section thickness is uniform; 317 for perforated plates.
§Flange thickness not more than two times web thickness.

It should be noted that it is always conservative to use $C_b = 1$.

Provided they satisfy the requirements for a compact section, continuous beams and beams rigidly connected to columns may be proportioned for 90 percent of the gravity-load negative moments at supports. However, the maximum positive moment must then be increased by 10 percent of the average of the negative moments.

AISC permits $F_b = 0.60F_y$ on an unsymmetrical cross section (except channels), provided the compression flange is supported laterally at intervals not exceeding $76b_f/\sqrt{F_y}$. In the case of the channel, the allowable compression F_b is to be determined from Eq. (20) unless the channel is continuously supported against lateral buckling.

It should be noted that beam cross sections which are unsymmetrical about the vertical axis are usually supported laterally so that lateral buckling is not a consideration.

AASHTO The allowable compressive stress for a beam supported laterally its full length by embedment in concrete is $0.55F_y$. Otherwise,

$$F_b = 0.55F_y\left[1 - \frac{(L/r)^2}{2C_c^2}\right] = 0.55F_y\left[1 - \frac{6(L/b)^2}{C_c^2}\right] \qquad (21)$$

where $r = b/\sqrt{12}$ = radius of gyration of compression flange
L = length of unsupported flange

The limiting value of L/b in Eq. (21) is $L/b \gtrsim \pi\sqrt{E/3F_y}$.

The width-thickness ratio b/t of a projecting flange element must not exceed $1625/\sqrt{f_b}$, where f_b = calculated compressive bending stress, but in no case may it exceed 12.

Continuous or cantilever beams may be proportioned for negative moment at interior supports for a stress 20 percent larger than that permitted by Eq. (21), but not more than $0.55F_y$.

AREA The allowable compressive stress for beams and girders is the larger of

$$F_b = 0.55F_y \left[1 - \frac{F_y}{1.8 \times 10^9} \left(\frac{L}{r_y} \right)^2 \right] \quad \text{psi} \qquad \frac{L}{r_y} \gtrsim \frac{29{,}900}{F_y}$$

$$F_b = \frac{10{,}500{,}000}{Ld/A_f} \gtrsim 0.55F_y \qquad \text{psi}$$

The specification defines r_y as the radius of gyration of a section comprising the compression flange and the part of the web in compression. A_f is defined as the area of the smaller flange.

Example 5 Determine the AISC allowable bending stress for an A36 W24 × 131 simply supported beam on a span of 48 ft, laterally supported at midspan and ends.

SOLUTION. $d/A_f = 1.98$, $b_f = 12.855$, $A_w = 13.82$ in.^{2}, $A_f = 12.34$ in.^{2}, and $r_T = 3.40$ in.

$$\frac{b_f}{2t_f} = \frac{12.855}{2 \times 0.960} = 6.70 < \frac{65}{\sqrt{F_y}} = 10.8$$

$$\frac{d}{t} = \frac{24.48}{0.605} = 40 < \frac{640}{\sqrt{F_y}} = 107$$

$$\frac{20{,}000}{(d/A_f)F_y} = \frac{20{,}000}{1.98 \times 36} = 281 \text{ in.} < \frac{48 \times 12}{2} = 288 \text{ in.}$$

The beam satisfies all the requirements for a compact section except for lateral support of the compression flange. Therefore, the allowable stress must be determined by Eqs. (19) and (20).

$$F_b = \frac{12{,}000C_b}{Ld/A_f} = \frac{12{,}000 \times 1.75}{288 \times 1.98} = 36.8 > 24 \text{ ksi}$$

In the formula for C_b [see Eq. (20)], $M_1/M_2 = 0$. Therefore, $C_b = 1.75$. It is not necessary to compute F_b by Eq. (19) because the value by Eq. (20) exceeds $0.66F_y$ and the larger of the allowables by these equations governs.

It should be noted that $C_b = 1$ if this beam does not have lateral support at the center.

Example 6 An A36 beam carrying a uniform load of 3.87 klf is continuous over two 36-ft spans and has continuous lateral support. Determine the lightest W section. AISC specification.

Assume weight of beam = 130 plf

$$w = 3.87 + 0.13 = 4.0 \text{ klf}$$

$$M = -\frac{wL^2}{8} = -\frac{4 \times 36^2}{8} = -648 \text{ ft-kips}$$

$$M = +\frac{9wL^2}{128} = \frac{9 \times 4 \times 36^2}{128} = +365 \text{ ft-kips}$$

$$V = \frac{5wL}{8} = \frac{5 \times 4 \times 36}{8} = 90 \text{ kips}$$

Assume a compact section, $F_b = 24$ ksi

$$M = -648 \times 0.9 = -583 \text{ ft-kips}$$

$$M = +365 + 0.1\frac{648 + 0}{2} = +397 \text{ ft-kips}$$

$$S_x = \frac{583 \times 12}{24} = 292 \text{ in.}^3$$

Try W30 × 108, $\dfrac{b_f}{2t_f} = \dfrac{10.475}{2 \times 0.760} = 6.9 < \dfrac{65}{\sqrt{F_y}}$

$$\frac{d}{t} = \frac{29.83}{0.548} = 54 < \frac{640}{\sqrt{F_y}}$$

Therefore, the section is compact as assumed.

18. Biaxial Bending If the plane of loading passes through the shear center but is not parallel to a principal axis, biaxial bending results. This is a special case of the AISC provision for axial compression and bending [Eq. (29)] with $f_a = 0$, which gives

$$\frac{f_{bx}}{F_{bx}} + \frac{f_{by}}{F_{by}} \lessgtr 1 \tag{22}$$

The following formula may be used to obtain a trial section:

$$S_x = \frac{M_x}{F_b}\left(1 + B\,\frac{M_y}{M_x}\right) \tag{23}$$

where $B = S_x/S_y$. Values of B for commonly used shapes are given in Table 11. The smaller numbers in each group of W shapes in this table are for those with wide flanges.

TABLE 11   Approximate Values of $B*$

Shape	Depth d, in.	$B\dagger$
W	8–16	3–8
W	16–24	5–10
W	24–36	7–12
I	6–8	d
I	10–18	$0.75d$
I	20 and 24	$0.6d$
⊏	7 and under	$1.5d$
⊏	8–10	$1.25d$
⊏	12 and 15	d

*From Ref. 2.
†Values in this table do not apply to shapes classified as miscellaneous or light, or to junior beams and channels.

If the plane of loading does not contain the shear center, bending is accompanied by torsion. Torsion is often ignored if the beam has restraints which limit the twist. For example, a beam which is part of a floor structure may be restrained by the floor slab, so that twisting will be self-limiting if the slab is sufficiently rigid and adequately anchored to the compression flange. However, if such restraints do not exist, torsion should be considered. Simplifying assumptions are usually made for W shapes. If the resultant of the loads on the compression flange is at an angle to the web, e.g., the loads on a crane girder, the horizontal component is usually considered to be resisted by the compression flange in lateral bending and the vertical component by the gross section in simple bending (Fig. 13a). An eccentric load parallel to the web (Fig. 13b) may be resolved into a vertical force in the plane of the web and equal and opposite horizontal forces on the flanges.

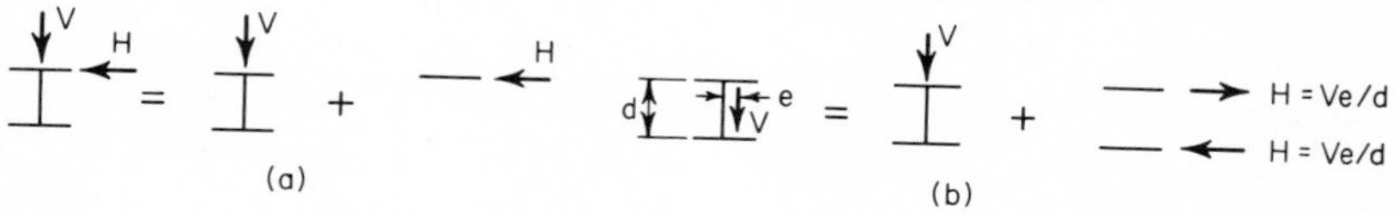

(a)   (b)

Fig. 13

19. Shear The shearing stress f_v in any section subjected to bending stress is given by

$$f_v = \frac{VA\bar{y}}{It} \tag{24}$$

where V = shear force in principal plane, A = area of that part of the cross section above (or below) the plane of shear, $\bar{y}$ = distance between centroid of A and neutral axis, and t = width of section at plane of shear. The shearing stress is very nearly uniform over the web of I-shaped sections loaded in the plane of the web, so that

$$f_v(\text{av}) = \frac{V}{A_w} \tag{25}$$

For biaxial bending it may be necessary to investigate the shearing stress at several points in the cross section.

Allowable shear stresses F_v are as follows: AISC, $0.40F_y$; AASHTO, $0.33F_y$; AREA, $0.35F_y$. However, these values must be reduced for large values of web slenderness (Art. 22).

20. Deflection   AISC.   Live-load deflections of beams and girders supporting plastered ceilings must not exceed 1/360 of the span. The Commentary on the Specifications recommends that the depth-to-span ratio of fully stressed beams in floors be not less than $F_y/800$, and for fully stressed purlins not less than $F_y/1000$.

Roof systems are considered stable under ponding conditions if

$$C_p + 0.9\,C_s \leq 0.25 \qquad \text{and} \qquad I_d \geqslant 25\ S^4/10^6$$

where $C_p = \dfrac{32 L_s L_p^4}{10^7 I_p}$ and $C_s = \dfrac{32 S L_s^4}{10^7 I_s}$

  L_p = column spacing in direction of girder, ft (length of primary members)
  L_s = column spacing perpendicular to direction of girder, ft (length of secondary member)
  S = spacing of secondary members, ft
  I_p = moment of inertia for primary members, in.⁴
  I_s = moment of inertia for secondary member, in.⁴
  I_d = moment of inertia of the steel deck supported on secondary members, in.⁴

I_s for trusses and steel joists must be reduced by 15 percent for use in the above equation. Metal deck is considered to be a secondary member if it is supported by the primary members. Total bending stress due to dead loads, live loads, and ponding should not exceed $0.8F_y$ for primary and secondary members.

AASHTO.   Deflection of simple or continuous spans due to live load plus impact must not exceed 1/800 of the span. The ratio of depth to span of beams and girders should preferably be not less than 1:25.

AREA.   Deflection due to live load plus impact must not exceed 1/640 of the span.

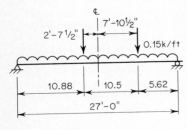

Fig. 14   Example 7.

Example 7   A crane runway beam spans 27 ft. The crane has a capacity of 60 kips, the crane bridge weighs 30 kips and the trolley and hoist 7 kips, and the wheels are 10.5 ft on centers. Design a channel reinforced W shape. Maximum depth 30 in. A36 steel, AISC specification.

MAXIMUM LOAD ON ONE WHEEL:

Vertical: ½(60 + ½ × 30 + 7) = 41
                25 percent impact = 10.3
                            Total = 51.3 kips
Lateral: ½ × 0.20(60 + 7) =   6.7 kips
        Estimated weight of beam =   0.15 klf

MAXIMUM STRONG-AXIS BENDING MOMENT (Fig. 14):

$$M_x = 2 \times 51.3 \times \frac{10.88^2}{27} + \frac{1}{2} \times 0.15 \times 10.88 \times 16.12 = 463 \text{ ft-kips}$$

MAXIMUM WEAK-AXIS BENDING MOMENT:

$$M_y = 2 \times 6.7 \times \frac{10.88^2}{27} = 58.7 \text{ ft-kips}$$

$$F_b = 22 \text{ ksi tension}$$
$$F_b = 20 \text{ ksi compression (assumed)}$$
$$S_x/S_y = 9 \text{ (assumed)}$$

From Eq. (23),

$$S_x = \frac{463 \times 12}{20}\left(1 + 8\frac{58.6}{462.6}\right) = 559 \text{ in.}^3$$

Try W30 × 99 with MC18 × 42.7.
  SECTIONAL PROPERTIES:

W30 × 99: d = 29.65 in., b = 10.450 in., t = 0.67 in.
             w = 0.520 in., A = 29.1 in.^{2}, I_x = 3990 in.^{4}
             I_y = 128 in.^{4}
MC × 42.7: d = 18.00 in., b = 3.95 in., t = 0.625 in.
             w = 0.45 in., A = 12.6 in.^{2}, I_x = 554 in.^{4}
             I_y = 14.4 in.^{4}, $\bar{x}$ = 0.877 in.
Neutral axis: y $= \dfrac{12.6 \times 0.877 + 29.1 \times 15.27}{12.6 + 29.1} = 10.92 \text{ in.}$
             I_x = 3990 + 29.1(15.27 − 10.92)^2 + 14.4 + 12.6(10.92 − 0.877)^2 = 5826 in.^{4}
             S_{xc} = 533.5 in.^{3} S_{xt} = 303.8 in.^{3}

FOR COMPRESSION FLANGE:

$$I_y = \frac{128}{2} + 554 = 618 \text{ in.}^4$$
$$S_{yc} = 68.7 \text{ in.}^3$$
$$A_f = 12.6 + 10.450 \times 0.67 = 19.60 \text{ in.}^2$$

$$r_t = \sqrt{\frac{618}{19.60 + \dfrac{1}{3}(10.92 - 0.45 - 0.67)0.520}} = 5.38$$

$$\frac{L}{r_t} = \frac{27 \times 12}{5.38} = 60.2$$

$$F_b = \left[\frac{2}{3} - \frac{36(60.2)^2}{1530 \times 10^3}\right]36 = 20.93 \text{ ksi}$$

Compression-flange stress:

Vertical load: $463 \times \dfrac{12}{533.5} = 10.39$

Lateral load: $58.7 \times \dfrac{12}{68.7} = 10.25$

Tension flange: $f_b = 20.64 < 20.93 \text{ ksi}$

Vertical load: $463 \times \dfrac{12}{303.8} = 18.28 < 22 \text{ ksi}$

SHEAR:

Vertical load: $51.3\left(1 + \dfrac{16.5}{27}\right) + 0.15 \times \dfrac{27}{2} = 82.6 + 2.0 = 84.6 \text{ kips}$

$$f_v = \frac{V}{A_w} = \frac{84.6}{29.65 \times 0.520} = 5.49 < 0.4 \times 36 = 14.4 \text{ ksi}$$

Lateral load: $6.7\left(1 + \dfrac{16.5}{27}\right) = 10.8 \text{ kips}$

$$A_w = 18 \times 0.45 + 10.450 \times 0.67 = 15.10 \text{ in.}^2$$
$$f_v = \frac{10.8}{15.10} = 0.71 < 14.4 \text{ ksi}$$

21. Combined Bending and Compression   The analysis of a beam column is complicated by the secondary moment, which is the product of the axial load and the deflection.

Considerations of lateral-torsional buckling, local buckling, and residual stresses complicate the analysis further.

The stress in a beam column is given by

$$f = \frac{P}{A} + \frac{Py}{S} + \frac{M}{S}$$

where Py is the secondary moment. With certain simplifying assumptions, this equation can be written in the form

$$\frac{f_a}{F_a} + \frac{f_b}{F_b} \frac{C_m}{1 - f_a/F'_e} \lessgtr 1 \tag{26}$$

where $f_a = P/A$
   $f_b = M/S$
   F_a = allowable stress for axial load alone
   F_b = allowable stress for bending alone
   F'_e = Euler stress divided by factor of safety
   C_m = coefficient which depends on shape of moment diagram

The multiplier of f_b/F_b in Eq. (26) accounts for the stress due to the secondary moment, i.e., the term Py/S, and is sometimes called an amplification factor. If the axial force is small relative to the moment, so that f_a/F_a is relatively small, the amplification factor also tends to be small and may be neglected, leaving only

$$\frac{f_a}{F_a} + \frac{f_b}{F_b} \lessgtr 1 \tag{27}$$

For biaxial bending, with moments M_x and M_y, Eqs. (26) and (27) become

$$\frac{f_a}{F_a} + \frac{f_{bx}}{F_{bx}} \frac{C_{mx}}{1 - f_a/F'_{ex}} + \frac{f_{by}}{F_{by}} \frac{C_{my}}{1 - f_a/F'_{ey}} \lessgtr 1 \tag{28}$$

$$\frac{f_a}{F_a} + \frac{f_{bx}}{F_{bx}} + \frac{f_{by}}{F_{by}} \lessgtr 1 \tag{29}$$

AISC specifies Eq. (28) but allows Eq. (29) to be used instead if $f_a/F_a \lessgtr 0.15$.

Maximum stress may occur at points of support, in which case AISC requires the following additional equation to be satisfied:

$$\frac{f_a}{0.6F_y} + \frac{f_{bx}}{F_{bx}} + \frac{f_{by}}{F_{by}} \lessgtr 1 \tag{30}$$

Values of C_m for the beam column with no relative translation of its ends are as follows:
   1. With end moments M_1 and M_2, where $M_1 < M_2$,

$$C_m = 0.6 - \frac{0.4M_1}{M_2} \gtrless 0.4$$

where M_1/M_2 is negative for single-curvature bending and positive for reversed-curvature bending.
   2. With transverse loads between supports

$$C_m = 1 + \frac{f_a}{F'_e}\left(\frac{\pi^2 \delta_0 EI}{M_0 L^2} - 1\right)$$

where δ_0 = maximum deflection due to transverse load
   M_0 = maximum moment, between supports, due to transverse load

Thus, if a beam column fixed at both ends supports a uniform load w, in addition to P, values of δ_0 and M_0 for determining C_m are

$$\delta_0 = \frac{1}{384} \frac{wL^4}{EI} \qquad M_0 = \frac{wL^2}{24}$$

and the expression in parentheses becomes -0.38. Following are values of C_m for six cases:

Uniform load:
  Simple supports
  Propped cantilever
  Fixed supports
Concentrated load at midspan:
  Simple supports
  Propped cantilever
  Fixed supports

1
$1 - 0.3f_a/F'_e$
$1 - 0.4f_a/F'_e$

$1 - 0.2f_a/F'_e$
$1 - 0.4f_a/F'_e$
$1 - 0.6f_a/F'_e$

Values of C_m for the beam column with relative translations of the ends (as in the column in a frame subjected to sidesway) are more difficult to determine. The AISC recommends for all such cases the value $C_m = 0.85$. This value has been found to be unconservative for certain types of frames.[5]

AASHTO specifies Eqs. (28) and (30) for combined axial and bending stresses. AREA specifies Eq. (28) with $C_m = 1$, Eq. (29), and Eq. (30).

Example 8 A frame rigidly connected at C is loaded as shown in Fig. 15. The drift index is limited to $\Delta_{ov}/H = \frac{1}{250}$. Full bracing is provided on line ACE. A36 steel. Design the frame to AISC specification except for moment amplification of Table 5.

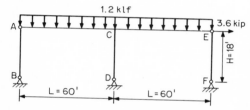

Fig. 15   Example 8.

Design the girder for 0.9 gravity-load moment (AISC 1.5.1.4.1).

$$M = 0.9 \times 1.2 \times \frac{60^2}{8} = 486 \text{ ft-kips}$$

$$S = \frac{486 \times 12}{24} = 243 \text{ in.}^3$$

Try W27 × 94: $S_x = 243$ in.³, $I_x = 3{,}270$ in.⁴
  First-order wind moment = $3.6 \times 18 \times \frac{1}{2} = 32.4$ ft-kips

$$\Sigma P_L = \Sigma V \frac{H}{\Delta_{ov}} = \frac{3.6}{0.004} = 900 \text{ kips}$$

$$\Sigma P = 1.2 \times 120 \times \frac{3}{4} \times \frac{23}{12} = 164 \text{ kips}$$

The (3/4) (23/12) is the factor of safety (load factor) for columns [Eq. (5)], taking into account the ⅓ increase in allowable stress for wind.

The moment-amplification factor, ignoring C_L, is

$$\frac{1}{1 - \Sigma P/\Sigma P_L} = \frac{1}{1 - 164/900} = 1.3$$

Amplified wind moment = $32.4 \times 1.3 = 42.1$ ft-kips

$$0.75\,(486 + 42.1) = 396 \text{ ft-kips} < 486 \qquad \text{Use W27} \times 94.$$

COLUMN:

$$G_C = \frac{I_C/216}{2 \times 0.5 \times 3270/720} = \frac{I_C}{981} \qquad \text{(Fig. 5 footnote)}$$

$$G_D = 10$$

From Eq. (9),

$$\Sigma P_L = \frac{6(I_C/981 + 10) + 36}{2(I_C/981 + 10) + 10I_C/981 + 3} \times \frac{29{,}000 I_C}{216^2}$$

$$900 = \frac{6I_C + 94{,}176}{12I_C + 22{,}563} \times \frac{29{,}000 I_C}{216^2}$$

$$I_C = 412 \text{ in.}^4$$

Try W14 $\times$ 43, $A = 12.6$ in.^{2}, $S_x = 62.7$ in.^{3}, $I_x = 429$ in.^{4} $r_x = 5.82$ in., $r_y = 1.89$ in.

$$\Sigma P_L = \frac{6 \times 429 + 94{,}176}{12 \times 429 + 22{,}563} \times \frac{29{,}000 \times 429}{216^2} = 931 \text{ kips}$$

$$G_C = \frac{429}{981} = 0.44$$

From Fig. 6, $C_L = 0.117$

Check column CD: Total load $= 1.2 \times 120 = 144$ kips

$$\text{Load on column } CD = \frac{5}{8} \times 144 = 90 \text{ kips}$$

From Eq. (13),

$$K = \sqrt{\frac{\pi^2 \times 29{,}000 \times 429}{90 \times 216^2} \times \frac{144 + 0.117 \times 90}{931}} = 2.20$$

$$\frac{K_x L_x}{r_x} = \frac{2.20 \times 216}{5.82} = 81.8$$

$$\frac{K_y L_y}{r_y} = \frac{1 \times 216}{1.89} = 114 \qquad F_a = 11.1 \text{ ksi}$$

Compute amplification factor:

$$\Sigma C_L P = 0.117 \times 90 \times \frac{23}{12} \times \frac{3}{4} = 15.1 \text{ kips}$$

$$\frac{1}{1 - \Sigma P/(\Sigma P_L - \Sigma C_L P)} = \frac{1}{1 - 164/(931 - 15.1)} = 1.22$$

Point of inflection in column, from bottom:

$$\frac{G_A + 3}{G_A + G_B + 6} H = \frac{0.44 + 3}{0.44 + 10 + 6} H = 0.209H$$

Wind moment $= (1 - 0.209) \times 216 \times 3.6 = 615$ in.-kips

$$f_b = \frac{615 \times 1.22}{62.7} = 12.0 \text{ ksi}$$

$$f_a = \frac{90}{12.6} = 7.1 \text{ ksi}$$

$$L = \frac{20{,}000}{(d/A_F)F_y} = \frac{20{,}000}{3.24 \times 36} = 172 \text{ in.} < 216 \text{ in.}$$

$$L = \frac{76 b_f}{\sqrt{F_y}} = \frac{76 \times 8.0}{\sqrt{36}} = 101 \text{ in.} < 216 \text{ in.}$$

Therefore, the column does not qualify as a compact section.

$$\frac{L}{r_T} = \frac{216}{2.14} = 101$$

$$C_b = 1.75 + 1.05 \left(\frac{M_1}{M_2}\right) + 0.3 \left(\frac{M_1}{M_2}\right)^2$$

$$= 1.75 + 1.05 \left(\frac{0.209}{1 - 0.209}\right) + 0.3 \left(\frac{0.209}{1 - 0.209}\right)^2 = 2.048$$

$$\sqrt{\frac{102 \times 10^3 C_b}{F_y}} = \sqrt{\frac{102 \times 10^3 \times 2.048}{36}} = 76.2$$

$$\sqrt{\frac{510 \times 10^3 C_b}{F_y}} = \sqrt{\frac{510 \times 10^3 \times 2.048}{36}} = 170.3$$

$$F_b = \left[\frac{2}{3} - \frac{36(100.9)^2}{1530 \times 10^3 \times 2.048}\right] 36 = 19.79 \text{ ksi [Eq. (19a)]}$$

$$0.75 \left(\frac{7.14}{11.09} + \frac{12.67}{19.79}\right) = 0.963 < 1.0$$

Alternately check column strength using Fig. 5b nomograph and modified K (Eq. 14) with $C_m = 1.0$.

$$K_{oi} = 1.76$$

$$K_i = K_{oi}\sqrt{1+n} = 1.96\sqrt{1 + \frac{54}{90}} = 2.23$$

$$\frac{K_i L_x}{r_x} = \frac{2.23 \times 216}{9.82} = 82.8$$

$$F_{ex} = 21.78 \text{ ksi}$$

$$\frac{C_m}{1 - \dfrac{f_a}{F'_{ex}}} = \frac{1}{1 - \dfrac{7.14}{21.78 \times \frac{4}{3}}} = 1.326$$

$$f_b = 1.326 \times \frac{615.1}{62.7} = 13.01 \text{ ksi}$$

$$0.75\left(\frac{7.14}{11.09} + \frac{13.01}{19.79}\right) = 0.976 < 1.0$$

Note that the term $\dfrac{C_m}{1 - f_a/F'_{ex}} = 1.326$ is close to the amplification factor of 1.292. Use W14 × 43 for column *CD*.

Example 9 The preliminary sizes for the 14th-story column and the 14th- and 15th-floor girders of a 32-story building frame are shown in Fig. 16*a*. The floor plan is shown in Fig. 16*b*. The service-core area is bounded by grid lines 3, *C*, 8, and *D*. The roof plan is similar except that it has a mechanical

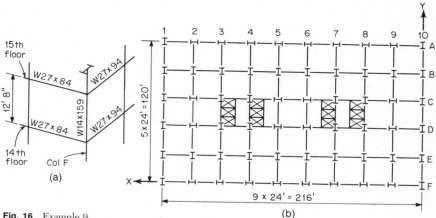

Fig. 16 Example 9.

penthouse in the core area. The frame is not braced against joint translation, but lateral stabilizing frames are provided on grid lines *A*, *C*, *D*, and *F* in the *x* direction and 1, 3, 4, 7, 8, and 10 in the *y* direction. The loads (dead + live) are:

Typical floor*	95 psf
Core area	160 psf
Roof	90 psf
Mechanical penthouse floor	210 psf
Mechanical penthouse roof	140 psf
Facade	380 plf

*With live-load reduction for extent of area loaded.

A computer analysis gave column loads and moments as follows:

Load	Axial load, kips	M_x, ft-kips	M_y, ft-kips
Gravity *G*	433	30	25
Wind W_x normal to *x* axis	120	120	
Wind W_y normal to *y* axis	40		90

Determine whether the A36 W14 × 159 column is satisfactory. AISC specification, using the procedure of Art. 11.

W14 × 159:

$$I_x = 1,900 \text{ in.}^4 \qquad S_x = 254 \text{ in.}^3 \qquad r_x = 6.38 \text{ in.} \qquad A = 46.7 \text{ in.}^2$$
$$I_y = 748 \text{ in.}^4 \qquad S_y = 96.2 \text{ in.}^3 \qquad r_y = 4.00 \text{ in.} \qquad d/A_f = 0.809$$
$$b_f = 15.56 \text{ in.} \qquad t_f = 1.190 \text{ in.} \qquad d = 14.98 \text{ in.} \qquad t_w = 0.745 \text{ in.}$$

W27 × 84: $I_x = 2,850 \text{ in.}^4$ W27 × 94: $I_x = 3270 \text{ in.}^4$

Let ΣP = sum of loads on stabilizing-frame columns

ΣQ = sum of loads on columns not part of stabilizing frames. These columns may be designed for $K=1$ (Art. 11).

LOAD ON TYPICAL FLOOR:

$$\Sigma P + \Sigma Q = 0.095\,(216 \times 120 - 24 \times 120) + 0.16\,(24 \times 120 - 4 \times 12 \times 24)$$
$$+ 0.38 \times 2\,(216 + 120) = 2721 \text{ kips}$$

For stabilizing frames on grid lines A, C, D, and F:

Col. row B   Facade
$$\Sigma Q_x = 2(0.095 \times 216 \times 24 + 0.38 \times 2 \times 24) = 1021 \text{ kips}$$
$$\Sigma P_x = 2721 - 1021 = 1700 \text{ kips}$$

For stabilizing frames on grid lines 1, 3, 4, 7, 8, and 10:

Col. row 2   Col. row 5   Col. row 5
$$\Sigma Q_y = 2(0.095 \times 24 \times 120) + \quad 0.095 \times 24 \times 96 + 0.160 \times 24 \times 24$$
Facade
$$+ 0.38 \times 4 \times 24) \qquad = 1243 \text{ kips}$$
$$\Sigma P_y = 2721 - 1243 = 1478 \text{ kips}$$

LOAD ON ROOF:

$$\Sigma P + \Sigma Q = 0.090(216 \times 120 - 24 \times 120) + 0.210(120 \times 24 - 4 \times 12 \times 24)$$
$$+ 0.38 \times 2(216 + 120) = 2692 \text{ kips}$$

For stabilizing frames on grid lines A, C, D, and F:

Col. row B   Facade
$$\Sigma Q_x = 2(0.090 \times 216 \times 24 + 0.38 \times 2 \times 24) = 970 \text{ kips}$$
$$\Sigma P_x = 2692 - 970 = 1722 \text{ kips}$$

For stabilizing frames on grid lines 1, 3, 4, 7, 8, and 10:

Col. row 2   Col. row 5   Col. row 5
$$\Sigma Q_y = 2(0.090 \times 24 \times 120 + 0.090 \quad \times 24 \times 96 + 0.210 \times 24 \times 24$$
Facade
$$+ 0.38 \times 4 \times 24) = 1248 \text{ kips}$$
$$\Sigma P_y = 2692 - 1248 = 1444 \text{ kips}$$

LOAD ON PENTHOUSE ROOF:

$$\Sigma P + \Sigma Q = 0.140 \times 24 \times 120 = 403 \text{ kips}$$

For stabilizing frames on grid lines A, C, D, and F:

$$\Sigma Q_x = 0$$
$$\Sigma P_x = 403 \text{ kips}$$

For stabilizing frames on grid lines 1, 3, 4, 7, 8, and 10:

Col. row 5
$$\Sigma Q_y = 2 \times 0.140 \times 24 \times 24 = 161 \text{ kips}$$
$$\Sigma P_y = 403 - 161 = 242 \text{ kips}$$

Total load on 14th-story columns is from 18 typical floors, the roof, and the penthouse roof: For stabilizing frames on grid lines A, C, D, and F:

$$\Sigma(\Sigma P_x) = 18 \times 1700 + 1722 + 403 = 32,725 \text{ kips}$$
$$\Sigma(\Sigma Q_x) = 18 \times 1021 + 970 = 19,348 \text{ kips}$$
$$n = \frac{\Sigma(\Sigma Q_x)}{\Sigma(\Sigma P_x)} = \frac{19,348}{32,725} = 0.591$$

For stabilizing frames on grid lines 1, 3, 4, 7, 8, and 10:

$$\Sigma(\Sigma P_y) = 18 \times 1478 + 1444 + 242 = 28,290 \text{ kips}$$
$$\Sigma(\Sigma Q_y) = 18 \times 1243 + 1248 + 161 = 23,783 \text{ kips}$$

$$n = \frac{\Sigma(\Sigma Q_y)}{\Sigma(\Sigma P_y)} = \frac{23{,}783}{28{,}290} = 0.841$$

Effective-length coefficients from Fig. 5b:

$$\frac{I_c}{L_c} = \frac{1900}{152} = 12.50 \qquad \frac{I_g}{L_g} = \frac{3270}{288} = 11.35$$

$$G_A = G_B = 2 \times \frac{12.50}{11.35} = 2.20 \qquad K_{ox} = 1.64$$

$$K_{nx} = K_{ox}\sqrt{1 + n} = 1.64\sqrt{1.841} = 2.23$$

$$\frac{I_c}{L_c} = \frac{745}{152} = 4.90 \qquad \frac{I_g}{L_g} = \frac{2830}{288} = 9.83$$

$$G_A = G_B = 2 \times \frac{4.90}{9.83} = 1.00 \qquad K_{oy} = 1.32$$

$$K_{ny} = K_{oy}\sqrt{1 + n} = 1.32\sqrt{1.591} = 1.66$$

Slenderness ratios:

$$\frac{K_{nx}L}{r_x} = 2.23 \times \frac{152}{6.40} = 53.0 \qquad \frac{K_{ny}L}{r_y} = 1.66 \times \frac{152}{4.00} = 63.1$$

From AISC Manual,

$$F_a = 17.13 \qquad F'_{ex} = 53.16 \qquad F'_{ey} = 37.50$$

Check W14 × 159 for lateral support to qualify as compact section (Art. 17)

$$L = \frac{20{,}000}{F_y d/A_f} = \frac{20{,}000}{36 \times 0.809} = 687 \text{ in.} > 152 \text{ in.}$$

$$L = \frac{76 b_f}{\sqrt{F_y}} = \frac{76 \times 15.56}{\sqrt{36}} = 197 \text{ in.} > 152 \text{ in.}$$

Lateral support O.K. Check web and flange,

$$\frac{d}{t_w} = \frac{14.98}{0.745} = 20.1 \qquad \frac{257}{\sqrt{36}} = 42.8 \text{ allowed}$$

$$\frac{b_f}{2t_f} = \frac{15.56}{2 \times 1.190} = 6.54 \qquad \frac{65}{\sqrt{36}} = 10.8 \text{ allowed}$$

Therefore, $F_{bx} = 0.66 F_y = 24$ ksi, $F_{by} = 0.75 F_y = 27$ ksi

Since allowable stresses may be increased one-third for wind, load combinations are as follows:

Load combination	P, kips	M_x, ft-kips	M_y, ft-kips
1. Gravity G	433	30	25
2. $0.75(W_x + G)$	415	113	19
3. $0.75(W_y + G)$	354	23	86

The resulting stresses are:

Load combination	f_a	f_{bx}	f_{by}
1	9.31	1.42	3.13
2	8.92	5.36	2.38
3	7.61	1.09	10.77

By inspection, load combination 3 governs. From Eq. (28), with $C_{mx} = C_{my} = 1$,

$$\frac{7.61}{17.13} + \frac{1.09}{24(1 - 7.61/53.16)} + \frac{10.77}{27(1 - 7.61/37.50)} = 0.997 < 1$$

From Eq. (29):

$$\frac{7.61}{22} + \frac{1.09}{24} + \frac{10.77}{24} = 0.79 < 1$$

PLATE GIRDERS

Many situations require the use of beams built up from plates or from combinations of plates with shapes. Depth limitations may require built-up sections. The built-up section may be more economical than a rolled section if the weight saving is enough to offset fabrication costs.

Where adequate lateral bracing can be provided, the single-web I-shaped girder is used. Material may be saved by reducing flange thickness as moment drops off or, alternatively, by varying the depth of the section. The tapered girder using unstiffened webs is particularly advantageous for long-span, lightly loaded roof girders. The web is then deep in the region of large moment and small shear and shallow in the region of small moment and large shear.

When lateral bracing is impractical, the box-section girder may be employed. The greatly increased torsional stiffness and the increased moment of inertia about the minor axis make it possible to develop the full strength of the material in compression.

22. Web   For minimum material in a built-up I-shaped section the sum of the areas of the flanges equals the area of the web. The corresponding required depth d is approximately

$$d = \sqrt{\frac{3}{2}\frac{M}{ft}} \tag{31a}$$

or

$$d = \sqrt[3]{\frac{3}{2}\frac{M}{f}\frac{h}{t}} \tag{31b}$$

where h = unsupported (clear) depth of web and t = web thickness. If two webs are used, the fraction $\frac{3}{2}$ becomes $\frac{3}{4}$. These equations may be used in proportioning whenever the depth is not limited by other requirements.

Webs may be classified as (1) unstiffened, (2) stiffened for shear, and (3) stiffened for shear and bending. An unstiffened web must be thick enough to resist the shear with an adequate factor of safety against shear buckling. Shear-stiffened webs may be designed to resist the shear with some margin of safety with respect to shear buckling, or they may be designed as tension-field webs. The web stiffened for both shear and bending uses a longitudinal stiffener to increase bend-buckling resistance of the web in the region of the compressive bending stresses.

AISC   The ratio h/t of depth to thickness of unstiffened webs may not exceed 260. The allowable shear stress on unstiffened webs is given by AISC Eq. (1.10-1) as

$$F_v = \frac{83{,}000}{(h/t)^2} \qquad \frac{h}{t} \leq \frac{547}{\sqrt{F_y}} \tag{32a}$$

$$F_v = \frac{152\sqrt{F_y}}{h/t} \qquad \frac{547}{\sqrt{F_y}} \leq \frac{h}{t} \leq \frac{380}{\sqrt{F_y}} \tag{32b}$$

$$F_v = 0.4F_y \qquad \frac{h}{t} \leq \frac{380}{\sqrt{F_y}} \tag{32c}$$

If the h/t ratio exceeds 260 or the allowable shear exceeds the permissible value, intermediate stiffeners are required. A plate girder with a stiffened web has postbuckling strength which results from the tension field that develops after a web panel reaches the shear-buckling stress. The AISC allowable stress is based on the shear strength of a web panel bounded by vertical stiffeners at the spacing a:

$$F_v = \frac{F_y}{2.89}\left[C_v + \frac{1-C_v}{1.15\sqrt{1+(a/h)^2}}\right] \qquad C_v \lesssim 1 \tag{33a}$$

$$F_v = \frac{F_y}{2.89}C_v \qquad C_v \gtrsim 1 \tag{33b}$$

in which

$$C_v = \frac{F_{v(cr)}}{F_{ys}}$$

$$F_{v(cr)} = \frac{26{,}000k}{(h/t)^2} = \text{critical (buckling) shearing stress}$$

$$k = 4 + \frac{5.34}{(a/h)^2} \qquad \frac{a}{h} \lessgtr 1$$

$$k = 5.34 + \frac{4}{(a/h)^2} \qquad \frac{a}{h} \geq 1$$

$$F_{ys} = \text{yield stress in shear} = \frac{F_y}{\sqrt{3}}$$

The second term in brackets in Eq. (33a) is the contribution of the diagonal tension.

The proportional limit in shear is assumed to be $0.8F_{ys}$. This requires two formulas for C_v, which are given by the AISC as

$$C_v = \frac{45{,}000k}{F_y(h/t)^2} \qquad C_v \lessgtr 0.8$$

$$C_v = \frac{190}{h/t}\sqrt{\frac{k}{F_y}} \qquad C_v \gtrless 0.8$$

The limiting value for a/h is either $\left(\dfrac{260}{h/t}\right)^2$ or 3. The spacing between stiffeners at end panels, and at panels containing large holes and panels adjacent thereto, must be such that f_v does not exceed the value given in Eq. (33b).

The upper limit of slenderness for a shear-stiffened web to prevent buckling of the compression flange in the plane of the web is

$$\frac{h}{t} = \frac{14{,}000}{\sqrt{(F_y + 16.5)F_y}} \tag{34}$$

except that it may be $2000/\sqrt{F_y}$ if transverse stiffeners are provided at a spacing not more than 1½ times the girder depth.

The possibility of buckling of the web as a result of load supported directly by the flange must also be investigated. The compressive stress in the web due to load distributed uniformly over the length of a panel equals the load intensity divided by the web thickness. For concentrated loads between stiffeners (i.e., not supported by bearing stiffeners) the stress is computed by dividing the load by at or dt, whichever is smaller. For distributed loads the stress is computed by dividing the load by the web thickness. The resulting stresses must not exceed

$$\left[5.5 + \frac{4}{(a/h)^2}\right] \frac{10{,}000}{(h/t)^2} \tag{35a}$$

for flanges restrained against rotation, or

$$\left[2 + \frac{4}{(a/h)^2}\right] \frac{10{,}000}{(h/t)^2} \tag{35b}$$

for flanges not restrained against rotation.

The AISC specification does not provide for longitudinally stiffened webs. Instead, the loss in bending strength of the web which results from bend buckling is compensated for by increasing the area of the compression flange. This is accomplished by reducing the allowable stress in the compression flange to

$$F_b' = F_b\left[1 - 0.0005\frac{A_w}{A_f}\left(\frac{h}{t} - \frac{760}{\sqrt{F_b}}\right)\right] \tag{36}$$

when web slenderness is large enough for bend buckling to be significant, i.e.,

$$\frac{h}{t} > \frac{760}{\sqrt{F_b}}$$

AASHTO The ratio h/t of unstiffened webs may not exceed 150. The allowable shear stress on unstiffened webs is

$$f_v = \frac{56.25 \times 10^6}{(h/t)^2} \qquad \frac{h}{t} \leq \frac{13{,}000}{\sqrt{F_y}} \tag{37a}$$

$$f_v = 0.33F_y \qquad\qquad \frac{h}{t} \lessgtr \frac{13,000}{\sqrt{F_y}} \qquad\qquad (37b)$$

The allowable shear for webs with intermediate transverse stiffeners is given by Eq. (33a) with the denominator 2.89 replaced with 3 and with a formula for C_v (denoted by C) valid for all a/h. The upper limit of slenderness is

$$\frac{h}{t} = \frac{23,000}{\sqrt{f_b}} \lessgtr 170$$

where f_b is the calculated bending stress.

The upper limit of web slenderness for a shear-stiffened web with a longitudinal stiffener located at the distance $h/5$ from the compression flange is

$$\frac{h}{t} = \frac{46,000}{\sqrt{f_b}} \lessgtr 340$$

AREA The ratio h/t of unstiffened webs may not exceed 60 for A36 steel or $11,400/\sqrt{F_y}$ for high-strength steel. The allowable shear stress on unstiffened webs is 12,500 psi for A36 steel and $0.35F_y$ for high-strength steel.

The postbuckling shear strength of shear-stiffened webs is not taken into account. The upper limit of slenderness is 170 for A36 steel and $32,500/\sqrt{F_y}$ for high-strength steel. However, if the stress in the compression flange is less than the allowable, the ratio may be multiplied by $\sqrt{p/f}$, where p = allowable fiber stress and f = actual fiber stress.

23. Flanges AISC, AASHTO, and AREA require that plate girders be checked for bending by the formula $M = fI/c$ rather than by the approximate expression $M = fd(A_f + A_w/6)$. However, the latter formula is useful for obtaining a trial section. In the case of riveted and bolted girders, in both AASHTO and AREA the calculated tensile stress is based on the moment of inertia of the entire net section (deducting for holes on each side of the neutral axis), and the compressive stress on the moment of inertia of the entire gross section. The AISC allows the entire gross section to be used in the computation for both tensile and compressive bending stress, except that if the reduction in area of either flange exceeds 15 percent of the flange area the excess must be deducted.

Each flange of a welded plate girder should consist of a single plate, which may consist of a series of plates joined end to end by full-penetration butt welds. Flange connection to the web is usually by fillet welds, which must provide at least VQ/I lb/in. In addition, minimum weld size in relation to thickness of material joined must be provided. Although continuous welds are highly desirable, intermittent welds may be used for economy.

24. Lengths of Flange Plates The theoretical cutoff point of flange or cover plates is found by calculating the moment capacity of the reduced section and locating the corresponding position on the moment diagram. The plate must be extended beyond the theoretical cutoff point so that it can develop its share of force at that point. The required force F to be developed is $F = MQ/I$, where M = moment at point of cutoff. In the riveted or bolted girder sufficient connectors must be provided to develop the force F. Special requirements of the AISC for welded cover plates, based on tests, are summarized in Fig. 17. In each of the cases shown the weld capacity beyond the theoretical cutoff must be at least equal to F, which may require extensions greater than those shown. AASHTO specifies b and c of Fig. 17, regardless of weld size, or extension to a section where the stress in the beam flange is equal to the allowable fatigue stress adjacent to fillet welds, whichever is greater.

25. Lateral Buckling Since most plate girders have relatively deep webs, resistance to lateral buckling is provided primarily by column strength of the compression flange, so that Eq. (19a) will usually control in the case of the AISC specification. Since only relatively deep beams and girders are the rule in bridges, AASHTO and AREA use formulas of the same type as Eq. (19a) (Art. 17).

26. Requirements for Stiffeners Since the postbuckling (tension-field) strength of plate-girder webs is not taken into account in the AREA specifications, stiffeners are spaced close enough to give a margin of safety with respect to shear buckling.* The stiffener spacing a must not exceed

*For consistency in notation, some symbols in this section differ from those used by AASHTO and AREA.

$$a = \frac{10,500t}{\sqrt{f_r}} \gtrless h \tag{38}$$

The size of the shear stiffener itself in the case of the web designed solely as a shear web is determined by the stiffness it needs to confine shear buckles to the adjacent panels. The AASHTO requirement is given by

$$I = \frac{a_0 t^3 J}{10.92} \tag{39}$$

where I = required moment of inertia of stiffener
   a_0 = distance between stiffeners
   $J = 25h^2/a^2 - 20 \gtrless 5$
   a = required distance between stiffeners
The corresponding formula of the AISC specification is

$$I = \left(\frac{h}{50}\right)^4 \tag{40}$$

AREA prescribes width b and thickness t of the stiffener: $b \gtrless 2$ in. plus $\frac{1}{30}$ of the girder depth, and $b/t \lessgtr 16$.

Fillet weld $\gtrless \frac{3}{4}$ cover plate thickness

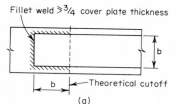

(a)

Fillet weld $< \frac{3}{4}$ cover plate thickness

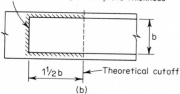

(b)

Fillet welds, sides only

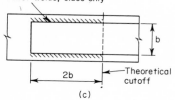

(c)

Fig. 17 Cover-plate extensions, AISC.

Shear stiffeners serve a dual role in the tension-field web; in addition to confining shear buckles they equilibrate the vertical component of the tension field. The required area A_{st} of stiffener which is needed to develop the tension field on which Eqs. (33) are based is given by the AISC formula

$$A_{st} = \frac{1 - C_v}{2}\left[\frac{a}{h} - \frac{(a/h)^2}{\sqrt{1 + (a/h)^2}}\right] YDht \tag{41}$$

where Y = ratio of yield point of web steel to that of stiffener steel
   D = 1 for stiffeners in pairs, 1.8 for single-angle stiffeners, and 2.4 for single-plate stiffeners

If the stiffeners are in pairs, A_{st} is the area of the two. The area may be reduced proportionately if the shear stress is less than the allowable [Eq. (33a)].

The required moment of inertia of a longitudinal stiffener at the distance $h/5$ from the compression flange is given by AASHTO as

$$I = ht^3 \left(\frac{2.4a_0^2}{h^2} - 0.13 \right) \tag{42}$$

27. Combined Bending and Shear The interaction between shear and bending in plate girders, according to the AISC specification, is shown in Fig. 18. In any panel where the tensile bending stress $f_b \lesssim 0.75F_b$ the allowable shearing stress is $0.4F_y$, and in any panel

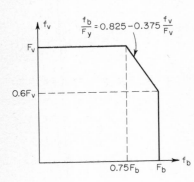

Fig. 18

where $f_v \lesssim 0.6\,F_v$ the allowable bending stress (in the web) is $F_b = 0.6F_y$. Otherwise, the bending stress f_b must satisfy

$$f_b \lesssim \left(0.825 - \frac{0.375f_v}{F_v} \right) F_y \tag{43}$$

Example 10 Design a welded plate girder to span 60 ft and to support a uniform load of 16 klf on its top flange. The flange width is limited to 2 ft, and the depth to the story height of 12 ft. Bracing is provided at 10-ft centers. A36 steel, AISC specification.

Assume girder weight = 0.4 kip/ft; then

$$M = \frac{16.4 \times 60^2}{8} = 7380 \text{ ft-kips}$$

For $F_b = 22$ ksi, the optimum depth from Eq. (31a) is

$$d = \sqrt{\frac{3 \times 7380 \times 12}{2 \times 22t}} = \frac{77.6}{\sqrt{t}}$$

Three possible solutions are:

$t = \tfrac{1}{2}$ in.	$d = 110$ in.	$h/t = 220$
$t = \tfrac{7}{16}$ in.	$d = 118$ in.	$h/t = 270$
$t = \tfrac{3}{8}$ in.	$t = 127$ in.	$h/t = 338$

The maximum permissible web slenderness by Eq. (34) is 322, which rules out the ⅜-in. thickness unless the corresponding depth is reduced to 120 in.

$$V = 16.4 \times 30 = 492 \text{ kips}$$

For the $110 \times \tfrac{1}{2}$ web,

$$f_v = \frac{492}{110 \times \tfrac{1}{2}} = 8.95 < 14.5 \text{ ksi}$$

Check weight of girder with $2A_f = A_w$ (Art. 22)

$$2 \times 3.4 \times 110 \times \tfrac{1}{2} = 374 < 400 \text{ lb/ft}$$

Use $110 \times \tfrac{1}{2}$ web.

BOTTOM FLANGE PLATE:

$$A_f = 0.5A_w = 0.5 \times 110 \times \tfrac{1}{2} = 27.5 \text{ in.}^2$$
$$b = 24 \text{ in.} \qquad t = \frac{27.5}{24} = 1.14$$

Use $24 \times 1\tfrac{1}{8} = 27 \text{ in.}^2$
TOP FLANGE PLATE:

$$\frac{h}{t} = 220 > \frac{760}{\sqrt{22}} = 162$$

From Eq. (36) with $A_w = 55$ and $A_f = 27.5$,

$$F_b' = \left[1 - 0.0005 \times \frac{55}{27.5}(220 - 162) \right] 22 = 20.7 \text{ ksi}$$

Approximate flange area $= 27.5 \times \dfrac{22}{20.7} = 29.2$

$$b = 24 \text{ in.} \qquad t = \frac{29.2}{24} = 1.22$$

Use $24 \times 1\tfrac{1}{4} = 30 \text{ in.}^2$
Allowable stress in top flange, Eqs. (19) and (36)

$$I_y = \frac{1.25 \times 24^3}{12} = 1440 \text{ in.}^4$$

$$r_T = \sqrt{\frac{1440}{30 + 55 \times 0.5/3}} = 6.06 \text{ in.}$$
$$\frac{L}{r_T} = \frac{10 \times 12}{6.06} = 19.8$$
$$F_b = 0.6\,F_y = 22 \text{ ksi}$$
$$F_b = \frac{12{,}000 \times 30}{120 \times 112.4} = 26.7 > 22$$
$$F_b' = [1 - 0.0005 \times 55/30(220 - 162)]\,22 = 20.8 \text{ ksi}$$

Part	Area	y	Ay	Ay^2	I_0
Top flange	30	55.62	1668	92,810	
Bottom flange	27	−55.56	−1500	83,350	
Web	55	0	0	0	55,500
	112		168	176,160	55,500

Neutral axis at $\dfrac{168}{112} = 1.50$ in. above web center

$$I = 176{,}160 + 55{,}500 - 112 \times 1.50^2 = 231{,}400 \text{ in.}^4$$
$$c_{\text{top}} = 55 - 1.50 + 1.25 = 54.75 \text{ in.}$$
$$c_{\text{bot}} = 55 + 1.50 + 1.12 = 57.62 \text{ in.}$$
$$f_{\text{top}} = \frac{7380 \times 12 \times 54.75}{231{,}400} = 20.9 \approx 20.8$$
$$f_{\text{bot}} = \frac{7380 \times 12 \times 57.62}{231{,}400} = 22.0 = 22.0$$

STIFFENER SPACING:
   Max shear $f_v = 8.95$ ksi

$$C_v = \frac{2.89}{F_y} F_v = \frac{2.89 \times 8.95}{36} = 0.718$$

$$C_v = \frac{45{,}000k}{F_y\,(h/t)^2} = \frac{45{,}000k}{36 \times 220^2} = \frac{k}{38.7}$$

$$k = 0.718 \times 38.7 = 27.8$$

$$k = 4.0 + \frac{5.34}{(a/h)^2}$$

$$\frac{a}{h} = \sqrt{\frac{5.34}{27.8 - 4.0}} = 0.477$$

$$a = 52.5 \text{ in.}$$

Other points:

$$\frac{a}{h} \gtrless \left(\frac{260}{h/t}\right)^2 = \left(\frac{260}{220}\right)^2 = 1.4$$

$$\frac{a}{h} \gtrless 3$$

SHEAR STRESS IN SECOND PANEL:

$$V = 492 - \frac{52.5}{12} \times 16.4 = 420 \text{ kips}$$

$$f_v = \frac{420}{55} = 7.64 \text{ ksi}$$

Assume top flange rotationally restrained. From Eq. (35a),

$$\left[5.5 + \frac{4}{(a/h)^2}\right] \frac{10,000}{220^2} = \frac{16}{12 \times \frac{1}{2}}$$

$$\frac{a}{h} = 0.735 \text{ (governs)}$$

$$a = 0.735 \times 110 = 81 \text{ in.}$$

Use stiffener spacing $52 + 8 \times 77 + 52 = 720$ in.

For $a = 77$ in., $\dfrac{a}{h} = \dfrac{77}{110} = 0.7$

$$k = 4.0 + \frac{5.34}{0.7^2} = 14.9$$

$$C_v = \frac{k}{38.7} = 0.385$$

$$F_v = \frac{36}{2.89}\left(0.385 + \frac{0.615}{1.15\sqrt{1 + 0.7^2}}\right) = 10.2 \text{ ksi}$$

COMBINED BENDING AND SHEAR: At quarter point of span $f_b = 0.75f_{b(\max)} = 0.75F_b$ because of uniform load. Also, $f_v = \frac{1}{2}f_{v(\max)} = \frac{1}{2} \times 8.95 = 4.48$ ksi, which is less than $0.6F_v$. Therefore, shearing stress is less than $0.6F_v$ anywhere in the middle half of span, so that bending stress is nowhere limited by interaction with shear stress (Fig. 18).

STIFFENER SIZE:

$$I = \left(\frac{h}{50}\right)^4 = \left(\frac{110}{50}\right)^4 = 23.5 \text{ in.}^4$$

Max projection $= \dfrac{95t}{\sqrt{36}} = 15.8t$

$$\frac{t(15.8t)^3}{3} = \frac{23.5}{2}$$

$$t = 0.307 \text{ in.} \qquad \text{Try 5/16}$$

$$b = 15.8t = 4.94$$

Try pair of stiffeners $5 \times \frac{5}{16}$

$$A_{st} = 2 \times 5 \times \frac{5}{16} = 3.12 \text{ in.}^2$$

Since there is no tension field in end panel, second panel determines required area of stiffener. Substituting C_v and a/h in Eq. (41),

$$A_{st} = \frac{1 - 0.385}{2}\left(0.7 - \frac{0.7^2}{\sqrt{1 + 0.7^2}}\right) \times 1 \times 1 \times 110 \times \frac{1}{2} = 5.05 \text{ in.}^2$$

$$2\,bt = 2 \times 15.8t^2 = 5.05 \qquad t = 0.40$$

Use $6 \times \frac{7}{16}$ stiffeners, $A_{st} = 2 \times 6 \times \frac{7}{16} = 5.25$ in.^{2}

WEB TO FLANGE WELD:

$$V = 492 \text{ kips}, \quad w = 16 \text{ kips/ft} = 1.33 \text{ kips/in.}$$

For top flange:

$$Q = 30 \times 55.62 = 1670 \text{ in.}^3$$
$$\frac{VQ}{I} = \frac{492 \times 1670}{231,400} = 3.55 \text{ kips/in.}$$

Force per inch $= \sqrt{1.33^2 + 3.55^2} = 3.79$ kips/in. Min fillet weld for $1\frac{1}{4}$-in. thickness = $\frac{5}{16}$ in. For E60 electrodes, allowable load on two $\frac{5}{16}$-in. welds is

$$18 \times 0.707 \times 2 \times \tfrac{5}{16} = 7.95 \text{ kips/in.}$$

Max spacing of welds (AISC Specification 1.18.2.3)

$$\frac{127 t_w}{\sqrt{F_y}} = \frac{127 \times \frac{1}{2}}{\sqrt{36}} = 10.6 \text{ in.}$$

Allowable shear on web plate = $0.5 \times 14.4 = 7.2$ kips/in.

Length of weld at 10.5-in. spacing $= 3.79 \times \dfrac{10.6}{7.2} = 5.6$ in. Use $\frac{5}{16} \times 5\frac{1}{2}$ in. at 10 in. on centers each side.

Stiffener to web weld, AISC Specification 1.10.5.4,

$$f_{vs} = h \frac{f_v}{F_v} \sqrt{\left(\frac{F_y}{340}\right)^3} = 110 \frac{7.64}{10.22} \sqrt{\left(\frac{36}{340}\right)^3} = 2.83 \text{ kips/in.}$$

Min fillet weld for $\frac{7}{16}$-in. plate = $\frac{3}{16}$ in. For E60 electrodes, allowable load on four $\frac{3}{16}$-in. welds is

$$18 \times 0.707 \times 4 \times \tfrac{3}{16} = 9.54 \text{ kips/in.}$$

Max clear spacing of intermittent welds $= 16t = 16 \times 0.5 = 8.0$ in.

Therefore, 1 in. of weld can provide $9.54/2.83 = 3.37$ in. of shear transfer. Clear spacing of 1-in. length of weld $= 3.37 - 1.0 = 2.37$ in. For 8 in. max. clear spacing, weld length $= 8/2.37 = 3.38$ in. Use $\frac{3}{16}$ in. $\times$ $3\frac{1}{2}$ in. weld at 11 in. on centers.

END-BEARING STIFFENERS: Use 11-in. plates to extend to flange edges (AISC 1.10.5.1).

$$r = \frac{2 \times 11}{\sqrt{12}} = 6.35 \text{ in. approximate}$$
$$\frac{L}{r} = \frac{3}{4} \times \frac{110}{6.35} = 13.0 \qquad F_a = 21.0 \text{ ksi}$$
$$A = \frac{492}{21} = 23.4 \text{ in.}^2$$
$$A = 12 \times \tfrac{1}{2} \times \tfrac{1}{2} = 3 \text{ in.}^2 \text{ provided by web}$$
$$t = \frac{23.4 - 3.0}{2 \times 11} = 0.93 \qquad \text{Try } \tfrac{15}{16} \text{ in.}$$
$$\frac{b}{t} = \frac{11}{0.94} = 11.7 < \frac{95}{\sqrt{F_y}}$$
$$r = \sqrt{\frac{(15/16) \times 22.5^3/12}{3 + 2 \times 11 \times 15/16}} = 6.14 \text{ in.}$$
$$\frac{L}{r} = 13.4 \qquad F_a = 21.0 \text{ ksi} \qquad \text{O.K.}$$

Use two end stiffeners $11 \times \frac{15}{16}$. Connect with four lines of $\frac{5}{16}$-in. welds $2\frac{1}{4}$ in. long at $8\frac{1}{4}$ in. The girder layout is shown in Fig. 19.

WELDED CONNECTIONS

28. Welding Processes   There are some forty welding processes, which may be categorized into eight major groups: brazing, flow, forge, induction, thermit, gas, resistance, electroslag, electrogas, and arc welding. Shielded metal-arc welding ("stick" welding) and submerged-arc welding are the two processes most commonly used in the building-construction industry. Considerable progress has been made in gas metal-arc welding with carbon dioxide shielding, and the AWS has sanctioned its use in the construction of both buildings and bridges. Gas welding, although still used to a limited extent, is rapidly disappearing from the building-construction scene, and the provisions governing it are no longer included in the AWS code. Resistance welding has been used to a large extent in Europe but has not been used for any significant structural applications in the United States. Electroslag and electrogas welding processes are gradually gaining acceptance in the construction industry.

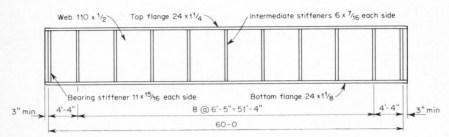

Fig. 19   Example 10.

Shielded-metal-arc welding is a fusion process which is limited primarily to manual application. The filler metal is provided by a consumable electrode. The combination of materials in the electrode coating varies with the metal to be welded and the results desired. Slag-forming fluxes and alloying elements may be included to refine the weld metal. Fluxes with melting points lower than the weld metal provide molten coverings over the weld metal; this allows the weld metal to cool at a slower rate and results in more ductile welds. After cooling, the slag which has floated to the surface of the weld may be removed easily by chipping. The bead shape of the weld can also be controlled by certain materials.

Submerged-arc welding may be performed manually, semiautomatically, or automatically. The filler metal is provided by either a consumable electrode or a supplementary welding rod. The shielding function is performed by a granular, fusible welding composition, commonly called flux, which blankets the welding zone. The bare welding electrode is usually fed automatically from a coiled reel through a mound of flux. The arc is not visible; hence the term "submerged arc."

The flux may incorporate constituents to refine and alloy the weld metal. Sometimes the slag does not have to be removed by mechanical means because the rates of cooling of the slag and the weld metal may be sufficiently different to allow the slag to break away independently.

This process is particularly adapted for use with high-intensity welding currents—on the order of six times as high as those used in manual shielded-metal-arc welding—because of the excellent insulating qualities of the flux. Consequently, the heat is concentrated in a relatively small zone to produce deep-penetration welds, and a saving in the quantity of filler metal may be realized. Estimates indicate that speeds up to four times faster than those with standard manual welding methods may be obtained.

Gas shielded-arc welding is a general term which covers gas tungsten-arc welding and gas metal-arc welding. In the latter process, coalescence is obtained by utilizing the heat of an electric arc between a consumable filler-metal electrode and the base metal to be welded. Shielding of the arc and weld region from atmospheric contamination is provided by a gas, a gas mixture (which may contain an inert gas), or a combination of a gas and a flux. The shielding gas has a beneficial effect on the arc, the weld metal transfer, the weld penetration, the width of fusion, the surface shape patterns, the undercut tendency, and the welding speed. Since only inert gases were used at first, the process was sometimes referred to as Mig welding.

Carbon dioxide is the most widely used gas for arc shielding in gas metal-arc welding of carbon and low-alloy steels. There are three methods of welding with CO_2 depending upon whether the filler metal is supplied by a solid, uncoated electrode, a composite (flux-cored) electrode, or a solid electrode with magnetic flux. All these may be performed semiautomatically, automatically, or by a combination. Regardless of the specific method used, gas metal-arc welding with carbon dioxide shielding consistently gives sound weld deposits, as long as the filler metal is designed for use with CO_2 and contains the proper quantities of deoxidizers.

Gas welding, relative to structural work, usually refers to an oxyacetylene welding operation. It is virtually obsolete in structural work.

The electroslag welding process is a single-pass vertical method for joining plates. The machine consists of two water-cooled copper shoes which press against the sides of the two vertical plates to be joined to form a container in which the welding proceeds vertically upward. Electrodes are fed down into the container and melted by the electric resistance of the welding current. The bare metal of the plates is also melted. Flux-cored electrodes are frequently used. There is no arc in this process, which can handle plates of 1 in. thickness and thicker.

The electrogas welding process is used primarily for welding thin plates. An electric arc is maintained between the flux-cored electrode and the surface of the weld. The arc and weld are shielded by a gas introduced through the sliding copper shoes.

Oxyacetylene flame cutting is extremely useful in the preparation of joints to be welded. An extraordinary variety of shapes and sizes of cuts is possible as long as the proper technique is used. Materials as thick as 30 in. can be cut. Tolerances as low as $\frac{1}{16}$ in. can be maintained in the normal cutting operation of plates up to 6 in. thick; this can be reduced to the order of $\frac{1}{32}$ in. when machine-guided flames are used.

Of utmost importance is the effect which the high flame temperatures have on the metal adjacent to cut edges. These heat-treated zones develop a thin surface layer of brittle material with properties that are determined by the cooling rate. Thick elements usually do not pose any problem, since cooling is adequately slow because of the relatively large quantities of absorbed heat required for cutting.

Resistance welding refers to a group of processes which depend upon the proper combination of electrical energy and mechanical pressure. The work to be welded forms part of the electric circuit. Electrical resistance generates sufficient heat to cause a small region to become plastic. Local coalescence of the pieces is finally effected by the application of pressure. Resistance welding is not presently feasible in the United States for major structural work except in the fabrication of open-web steel joists.

29. Weld Classification Welds may be classified in terms of the position of the weld during welding, type of weld, type of joint, or the magnitude and type of forces to be transmitted. Only a combination of these categories will define a weldment adequately.

Welding position refers to the relative location of the weld during the actual operation. In a *flat position* welding is executed from the top side and the weld face is nearly horizontal. In the *vertical position* the longitudinal axis of the weld is approximately vertical. *Overhead position* requires working from the underside of a joint. Fillet welding in the *horizontal position* is done on the top side of an approximately horizontal surface against an approximately vertical surface. Groove welding in the horizontal position is performed with the longitudinal axis of the weld approximately horizontal and the weld face approximately in the vertical plane.

Four basic weld types used in structural work are fillet, groove, plug, and slot. A *fillet weld* cross section is roughly triangular in shape and serves to join surfaces which are approximately at 90° to each other. A *groove weld* is made in the groove between two members to be joined. If there is inadequate space available for fillet welds, *plug* or *slot welds* may be used to supplement any strength deficiencies and also to prevent buckling of lapped parts. Plug welds are made by depositing weld metal in circular holes cut in one of two lapped members. Similarly, slot welds are made by using elongated holes. In both cases, the holes may be partially or completely filled. Fillet welds in holes or slots are not considered to be plug or slot welds. Before final welds are made, auxiliary *tack welds* are sometimes used for the purpose of holding the parts in proper alignment. Should particular weldment properties or dimensions be required, a *surfacing weld*—one or more string beads or weave beads—is deposited on an unbroken surface.

There are five basic types of welded joints: butt, corner, tee, lap, and edge. The standard *butt joint* is made at the juncture of two parts lying approximately in the same plane. A joint between two members located approximately at right angles to each other is called a *corner joint* if the parts form an L and a *tee joint* if they form a T. An *edge joint* joins the edges of two or more parallel or nearly parallel parts. The joint between two overlapping parts comprises a *lap joint*.

Welds may be classified as *primary* or *secondary* depending upon whether the entire member force, or a fraction thereof, is transferred through a joint.

30. Weldability   This is a general term used to describe the metallurgical compatibility of the base metal and the weld metal, the relative facility and the specific costs involved in fabricating a suitable weldment, and the performance of the weldment in actual service.

The weldability of structural steel is principally controlled by its carbon content. While this element is beneficial to strength, it is detrimental to ductility. A high carbon content combined with the intense heat generated during welding may cause a brittle zone in which weld cracks will develop. A carbon content of approximately 0.20 percent results in a very weldable steel, and good weldability is obtained with an upper limit of about 0.25 percent.

ASTM A36 steel has a carbon content that may be controlled at less than 0.25 to 0.29 percent, depending upon the material thickness, and is considered weldable as long as appropriate techniques are used.

ASTM A242 is primarily a strength specification with specified minimum yield points given on the basis of material thickness. Overall chemical requirements are not strict, although the carbon content is limited to 0.22 percent to enable economical welding. However, special welding techniques may be required because several other elements are often included to furnish strength and to enhance corrosion resistance. Specific instructions must therefore be given when A242 suitable for economical welding is desired.

A440 steel was developed to provide an economical substitute for A242 in riveted and bolted structures. Instead of using expensive alloying elements to acquire the desired strength, the carbon and manganese contents are increased above those of A242. A440 steel is not recommended for welding, and special precautions are therefore required if welding is to be attempted.

ASTM A441 specifies strength requirements identical to those for A242 and A440. It is a high-strength steel which can be welded economically. Limitations on the carbon and manganese levels are the same as those for A242, but about 0.02 percent of vanadium is added to obtain the strength required, in lieu of more expensive alloys.

Quenched and tempered plates produced to the A514 specification are weldable. However, the procedure must be suitable for the type being welded.

A572 steel is weldable in all grades. However, only Grades 42, 45, and 50 are suitable for welded construction of bridges.

A588 steel is weldable. However, the welding process must be suitable for the steel.

In order to verify the weldability of structural steel, the mill-test report should be studied. Weldability should be determined on the basis of the actual rather than the specified chemistry.

31. Electrodes   Each combination of weldable steel, welding condition, and welding position requires the use of a specific electrode. Electrodes are classified on the basis of the mechanical properties of the deposited weld metal, the welding position of the electrode, the type of coating or covering, and the type of current required. Each electrode is identified by a code number EXXXXX, where E stands for electrode and each X represents a number. The first two (or three) numbers indicate the minimum tensile strength, in ksi, of the deposited metal in the as-welded condition. The next number refers to the position in which the electrode is capable of making satisfactory welds: 1 means all positions (flat, vertical, overhead, and horizontal), 2 means flat and horizontal fillet welds, and 3 means flat only. The last X represents a digit which indicates the current to be used and the type of coating or covering on the electrode, e.g., (1) high-cellulose sodium coating for use with direct-current reverse polarity (electrode positive) only; or (2) iron powder, titania coating for use with direct current, either polarity, or alternating current. Hence, E6018 implies a mild-steel arc-welding electrode with a minimum tensile strength of 60 ksi. It is an iron-powder low-hydrogen electrode which may be used in all positions, provided an alternating or direct current, reverse polarity, is supplied.

Electrodes for shielded-metal-arc welding are manufactured in sizes ranging from $\frac{1}{16}$ to $\frac{3}{8}$ in. in diameter and in lengths of 9 to 18 in. Coatings serve as a medium for incorporating alloying elements which affect the tensile strength, hardness, corrosion resistance, and other physical properties of the weld metal.

In submerged-arc welding, appropriate combinations of bare electrodes and granular fusible fluxes are selected to produce the specified properties in the deposited weld metal. Choices are governed by the welding procedure, the type of joint, and the composition of the base metal. The designations are Grade SAW-1 and Grade SAW-2, which imply weld-metal yield points of 45 and 50 ksi, respectively. Clean and shiny bare rods or wires are used as electrodes in order to facilitate relatively high welding currents. The commonly used wires range in size from $\frac{3}{32}$ to $\frac{1}{2}$ in. in diameter. Composition of the welding wires includes a wide variety of steel grades. Similarly, fluxes are made in many different particle sizes and in accordance with several chemical specifications. If particular properties are desired in the weld metal, alloying elements may be included in either the electrodes or the fluxes, or both.

Two levels of weld-metal yield points are specified for the gas metal-arc welding process, Grade GMAW-1 (55 ksi) and Grade GMAW-2 (60 ksi). An appropriate combination of electrode and shielding which satisfies the mechanical-property requirements for Grade GMAW-1 or GMAW-2 may be used, respectively, in lieu of submerged-arc welding Grade SAW-1 or SAW-2. The filler wires used in gas metal-arc welding are small compared with those in other welding processes, the average being $\frac{1}{16}$ in.; the maximum diameter allowed by AWS is $\frac{5}{32}$ in. The surface-to-volume ratio is high, and if the wire surface is not kept clean, an inordinate amount of foreign matter may become part of the weld deposit. In general, the composition of the filler wires matches that of the base metal as nearly as is practicable. Sometimes, however, the filler-metal wires must be designed to negate the detrimental effects of some alloys included in the base metal.

32. Inspection   The production of satisfactory welds is governed by many factors, some of which are the joint (type, preparation, fit-up, included angle, root opening, etc.); the welding position (flat, horizontal, vertical, overhead); the electrode (type, size, inclination, rate of travel); the welding current (type, polarity, amount); the arc (length, speed); the preheating of the base metal; the rate of cooling; and the uniformity of operation.

All surfaces to be welded should be free of paint, slag, loose scale, rust, and other foreign material; mill scale which remains after being subjected to vigorous wire brushing is permitted. In order to counterbalance distortion and minimize shrinkage stresses in the fabrication of welded parts, specific instructions pertaining to the procedure and sequence of welding should be clearly conveyed to the operator. Joints to be welded must be so designed that they are easily accessible to correctly positioned electrodes. Whenever it is feasible, work should be welded in the flat position.

A perfect weld surface does not imply the absence of interior imperfections. There are basically three types of weldment defects in arc welding: dimensional discrepancies (warpage, incorrect weld size, etc.); structural discontinuities (cracks, cavities, etc.); and inadequate properties. Warpage or distortion can usually be avoided by using appropriate jigs, rational welding sequences, or preforming prior to welding. If warpage cannot be prevented, it is sometimes possible to rectify the fault by flame shrinking.

Among the nondestructive methods which may be employed to ascertain the severity of defects are visual, magnetic-particle, penetrant, radiographic, and ultrasonic inspection.

A magnifying glass ($\times 10$ or less) is helpful in the visual examination of welds. Dimensional accuracy and warpage or distortion may be determined by using scales and gages. Although scrutiny of the welds by visual means is obviously limited, a good deal of meaningful information can be obtained by thorough observation and appraisal of the surface conditions. When used in conjunction with other methods, it can be quite effective in deciding the final acceptability of welds.

Magnetic-particle inspection (magnaflux process) determines weld defects such as surface and crater cracks, seams, subsurface porosity, slag inclusions, incomplete fusion, and other discontinuities in magnetic materials. A strong magnetic field is induced in a short section of ferromagnetic material. Discontinuities in the weld create leakages in the magnetic field, and poles are established at these locations. Iron powder sprinkled on the test area will migrate immediately to these poles, forming patterns which approximate the outline of the discontinuities. However, the method has certain drawbacks; its depth penetration is limited and small discontinuities may not be detected.

Penetrant inspection is used to detect small surface defects. There are two types, fluorescent and dye-penetrant inspection. In the former, a highly fluorescent liquid is applied to and absorbed by the surface under inspection. After the excess is removed, a developer is added which draws the penetrant from the discontinuities. Upon exposure to black light, fluorescent indications make it possible to detect surface imperfections. Dye-penetrant inspection is quite similar. The section under inspection is sprayed with a dye penetrant which seeps into surface irregularities. Any excess is removed, and a developer is then applied by spraying. The developer is stained red by the dye, which rises from the surface defects by capillary action.

Radiographic inspection using short-wave radiations, such as x-rays, may be used to establish the existence of interior defects in welds. The amount of absorbed radiation is a function of the weld density and thickness. When a defect is present, the beam of radiation encounters less resistance than it would through a sound weld. Radiation is recorded on a ray-sensitive film producing an image picture (radiograph) which clearly shows the flaws in the weld. The procedure is quite reliable and used extensively.

Ultrasonic inspection is an efficient method capable of determining the presence of surface and subsurface weld defects. Electrically timed high-frequency sound waves or vibrations are propagated through the area under inspection and reflected back by discontinuities and density changes. Reflected signals may be either transmitted as patterns on an oscilloscope or punched on a tape. These results can then be compared with established data.

Proper interpretation and evaluation of test results are of utmost importance. The more sensitive procedures may discover defects which are relatively unimportant in terms of service requirements. In the final analysis, acceptability should be based on a clear understanding of the functional requirements of the weldment.

33. Fillet-Welded Joints   Because of relatively large fit-up tolerances and greater operational flexibility, the fillet weld is probably the most commonly used type to connect lapping and intersecting parts. Ideally, its cross section forms a 45° isosceles triangle with the connected parts. There is some difficulty in obtaining a perfectly flat weld face, so that convexity or concavity normally results. Although the possibility of cracking due to shrinkage is reduced in a convex fillet weld, it results in excess weld metal which does not enhance the strength. On the other hand, a concave fillet weld is more likely to crack as a result of shrinkage, particularly when used to join high-carbon steels.

To obtain a satisfactory fillet weld, the electrode should bisect the angle between the intersecting surfaces to be welded and, for manual welding, should lean about 20° in the direction of travel. The most favorable shop-weld positions are the flat and the horizontal. The flat position is preferable by far because weld metal can be placed with greater speed and ease. The time required to deposit weld metal in the horizontal position may be up to one and a half times as long, the longer time corresponding to the larger weld sizes. Similarly, welding in the vertical or overhead position may well take two and a half times as long as in the flat position.

The size of a fillet weld is expressed as the leg length of the largest isosceles right triangle which can be inscribed within its cross section. Legs of equal size are normal, although unequal legs are sometimes required. On a convex fillet weld the leg length is the critical dimension, as is the throat thickness on a concave shape. The effective stress area is taken to be the throat thickness multiplied by the weld length, regardless of the direction of load.

Stress-carrying fillet welds are usually limited by specifications to a minimum of $\frac{1}{8}$ in. Common sizes increase up to $\frac{1}{2}$ in. by $\frac{1}{16}$ in. and thereafter by $\frac{1}{8}$ in. The most frequently used sizes range from $\frac{3}{16}$ to $\frac{5}{16}$ in., since they can be formed by a single pass of the electrode; appropriate sizes are obtained simply by varying the rate of travel. Small welds also afford a better ratio of effective stress area to volume of weld metal, since the weld metal required increases as the square of the leg. Larger welds are also relatively uneconomical because the slag must be removed, after each pass has cooled, before laying the next pass.

Fillet-weld dimensions are also influenced by the magnitude of the force to be transmitted, the clearance provided for the welding operation, the length of the joint, and the thickness of the parts to be joined. With regard to material thickness, special consideration must be given to selection of the weld size. If relatively thick parts are connected by small fillets, the possibility of obtaining uncracked welds is small. The thick material accelerates the cooling rate of the weld and tends to embrittle it. Also, the thick material does not expand sufficiently to allow the weld to contract completely, so that a residual tensile stress is induced. A combination of these two effects may cause cracking. The AWS code specifies the minimum size of fillet weld to avoid cracked welds (Table 12). However, the weld need not be larger than the thickness of the thinner part joined, unless strength requirements indicate a larger size.

Standard specifications recommend that the maximum size of fillet weld along edges of connected material be the thickness of the material if it is less than $\frac{1}{4}$ in., or $\frac{1}{16}$ in. less than the thickness of the material if it is $\frac{1}{4}$ in. or more, unless specific instructions are given on the drawings to build up the weld to ensure full throat thickness. In any case, the weld must be of such a size so as not to overstress the adjacent base material.

TABLE 12   Minimum Weld Sizes

Thickness of thicker part joined, in.	Minimum leg size of fillet weld, in.	Minimum effective throat of partial-penetration groove weld, in.
To $\frac{1}{4}$ incl.	$\frac{1}{8}$	$\frac{1}{8}$
Over $\frac{1}{4}$ to $\frac{1}{2}$	$\frac{3}{16}$	$\frac{3}{16}$
Over $\frac{1}{2}$ to $\frac{3}{4}$	$\frac{1}{4}$	$\frac{1}{4}$
Over $\frac{3}{4}$	$\frac{5}{16}$	—
Over $\frac{3}{4}$ to $1\frac{1}{2}$		$\frac{5}{16}$
Over $1\frac{1}{2}$ to $2\frac{1}{4}$		$\frac{3}{8}$
Over $2\frac{1}{4}$ to 6		$\frac{1}{2}$
Over 6		$\frac{5}{8}$

Surfaces of parts to be connected by fillet welds should be made to contact as closely as possible. A separation not greater than $\frac{3}{16}$ in. is permitted by specifications. The size of the fillet weld must be increased by the amount of the separation if it is $\frac{1}{16}$ in. or larger.

The ends of a fillet weld are tapered, so that the full length of the weld cannot be considered effective. As a precaution against the occurrence of craters, or "short welds," standard practice requires that detail drawings specify the net length of the weld.

End returns are used to relieve stress concentrations at corners. Returns must extend around the corner for a distance of at least twice the nominal weld size. End returns are usually disregarded in determining the effective length of the weld, even though the AWS code, for example, allows their inclusion. It is strongly recommended that end returns be required on all welded joints subject to eccentricity, impact, or stress reversal.

34. Groove-Welded Joints These are commonly referred to as "butt-welded joints." Strictly speaking, butt joints are limited to those which are made between two members lying approximately in the same plane; this definition does not cover weldments such as tee and corner joints. Groove welds are not used as often as fillet welds because of the closer fit-up tolerances required. In terms of strength, groove welds are superior to fillet welds, since the stresses are transferred directly.

In order to obtain joints with a predictable degree of weld penetration, the AWS standards recommend values for the groove angle, the root opening, and the land. Complete joint penetration means that the filler metal and the base metal are fused throughout the depth of the joint.

Most of the structural-steel welded joints outlined by the AWS and the AISC are considered to be prequalified, provided the welding is done in accordance with the requirements. The term prequalified implies that a joint may be used without performing welding-procedure qualification tests. If a joint differs from those covered in the specifications, its adequacy must be verified by tests such as those prescribed in the AWS Standard Qualification Procedure.

The system of groove-weld classification—not to be confused with the ideographic weld-identification system—adopted by the AWS uses the following notation:

1. Joint type: B = butt, C = corner, T = tee.
2. Material thickness and efficiency:
   L = limited thickness with complete penetration
   U = unlimited thickness with complete penetration
   P = partial penetration
3. Groove type:

1—square	4—single bevel	7—double U
2—single vee	5—double bevel	8—single J
3—double vee	6—single U	9—double J

4. Welding process:
   S = submerged arc
   No symbol indicates manual shielded metal arc.
For example, the designation TC-U4 means (1) the joint may be used for either a tee or a corner, (2) it may be used on an unlimited range of material thicknesses, (3) the groove is a single-bevel, and (4) the welding will be performed by the manual shielded-metal-arc process.

A square-groove weld is most economical since no chamfering is required; unfortunately, it can be used for only a limited range of material thicknesses. To weld thicker materials the edges must be chamfered in order to provide accessibility and to ensure adequate fusion throughout the joint. To minimize the amount of weld metal, joints should be chosen with an appropriate combination of root opening and groove angle to afford satisfactory accessibility for sound welds. J- and U-groove joints are useful in this regard, particularly for the larger material thicknesses, as long as the more expensive chamfering operations result in a net saving.

It should be noted that the strength of groove welds joining members of different cross section is reduced considerably by the presence of stress concentrations at reentrant corners. The transition between such sections should be gradual so as to minimize the stress concentration. Fillet welds are sometimes added to these groove welds to improve the stress distribution. Tests have indicated that the strength of properly transitioned groove-welded joints of variable section is approximately the same as that of constant-section joints.

35. Concentrically Loaded Connections Although some eccentricity usually exists in connections, it may be neglected in certain cases. For instance, in end connections of single-angle, double-angle, and similar members, not subjected to repeated variation in stress, some eccentricity can be tolerated. Similarly, the eccentricity of the applied force in fillet-welded lap joints may be neglected if the thickness of the connected parts is small relative to the weld length.

Axially loaded members may be connected by any of the general types of welds—groove, fillet, plug, and slot—or combinations thereof. Groove welds are generally the most efficient, although not necessarily the most economical.

Example 11   Design a welded connection to sustain the allowable tensile load of a double-angle diagonal truss member consisting of two angles 5 × 3½ × ⁷⁄₁₆ subjected to repeated loading (Fig. 20). A36 steel and E60 electrodes.

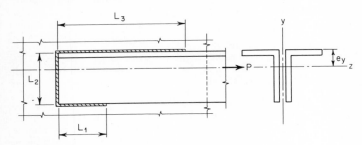

Fig. 20   Example 11.

$$P = 0.60F_yA = 22 \times 7.05 = 155 \text{ kips}$$
$$\frac{P}{2} = \frac{155}{2} = 77.5 \text{ kips}$$
$$a = t - \tfrac{1}{16} = \tfrac{3}{8} \text{ in.}$$
$$q = 12.73a \text{ kips/in.} \qquad 12.73 \times \tfrac{3}{8} = 4.77 \text{ kips/in.}$$

WELD LENGTH:

$$L = L_1 + L_2 + L_3 = \frac{P}{q} = \frac{77.5}{4.77} = 16.2 \text{ in.}$$
$$Q_1L_2 + \frac{Q_2L_2}{2} - Pe_y = 0$$

$$Q_1 = \frac{Pe_y}{L_2} - \frac{Q_2}{2} \qquad L_1 = \frac{1}{2q}\left(\frac{2Pe_y}{L_2} - Q_2\right)$$

For E60 electrodes

$$L_1 = \frac{1}{2 \times 4.77}\left(\frac{2 \times 77.5 \times 1.63}{5} - 4.77 \times 5\right) = 2.8 \text{ in.}$$
$$L_3 = L - L_1 - L_2 = 16.2 - 2.8 - 5 = 8.4 \text{ in.}$$

If no end weld is used, $Q_2 = 0$ and

$$L_1 = \frac{Pe_y}{qL_2} = \frac{77.5 \times 1.63}{4.77 \times 5} = 5.3 \text{ in.}$$
$$L_2 = L - L_1 = 16.2 - 5.3 = 10.9 \text{ in.}$$

36. Beam Seat Connections This type of flexible connection involves a minimum of shop and field welding and is often used (Fig. 21). The AISC recognizes it as Type 2 construction (Art. 39).

Since the seat angle is used only with relatively light loads, welds along the two ends of the vertical leg of the angle are adequate. As a precaution against the formation of craters, these welds should be returned across the top of the seat for a distance of twice the weld size.

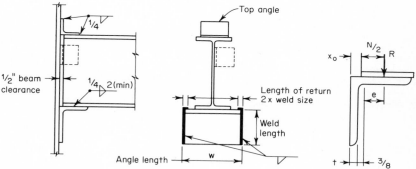

Fig. 21 Beam seat.

The top angle must be small enough to allow simple-beam end rotation. Hence, a 4 × 4 × ¼ angle, field-welded only along its toes, should be used. Generally, a ¼-in. fillet weld, not built out, is called for regardless of the thickness of the material to which the angle is attached.

Example 12 Design a welded flexible beam seat for a W18 × 46 with a reaction of 25 kips. Use A36 steel and E60 electrodes.

$$b = 6.06 \text{ in.} \qquad t_w = 0.360 \text{ in.} \qquad k = 1\tfrac{1}{16} \text{ in.}$$

Bearing length (AISC 1.10.10):

$$N = \frac{R}{t_w \times 0.75F_y} - k = \frac{25}{0.360 \times 0.75 \times 36} - 1.06 = 1.51 \text{ in.}$$

Angle Thickness t:

Bending moment at critical section $M = Re$ (Fig. 21)
$$e = x_0 + \frac{N}{2} - t - \frac{3}{8}$$
$$\frac{wt^2}{6} = \frac{M}{F_b}$$
$$\frac{wt^2 F_b}{6R} + t - \left(x_0 + \frac{N}{2} - \frac{3}{8}\right) = 0$$
$$w = b + \text{(allowance for beam field weld)} = 6.06 + 1.00 = 7.06 \text{ in.}$$

Try $w = 8$ in. $F_b = 0.75 \times 36 = 27$ ksi.

$$x_0 + \frac{N}{2} - \frac{3}{8} = \frac{1}{2} + \frac{1.51}{2} - \frac{3}{8} = 0.88 \text{ in.}$$

$$\frac{8 \times 27}{6 \times 25} t^2 + t - 0.88 = 0 \qquad t = 0.508 \text{ in.}$$

Try $t = \frac{1}{2}$ in.

OUTSTANDING LEG OF ANGLE:

$$\text{OSL} = x_0 + N = \frac{1}{2} + 1.51 = 2.01 \text{ in.} \qquad \text{Use } 3\frac{1}{2} \text{ in.}$$

VERTICAL WELDS:

$$q_y = \frac{R}{2L} \qquad q_x' = \frac{M_z}{S_z} = \frac{Re_x}{2(L^2/6)}$$

$$e_x = x_0 + \frac{N}{2} = \frac{1}{2} + \frac{1.51}{2} = 1.26 \text{ in.}$$

$$q_r = \sqrt{\left(\frac{3Re_x}{L^2}\right)^2 + \left(\frac{R}{2L}\right)^2}$$

$$q_r = \frac{R}{2L^2}\sqrt{36e_x^2 + L^2}$$

$$q = 12.73a \text{ kips/in. allowable}$$

$$a = \frac{R}{25.46L^2}\sqrt{36e_x^2 + L^2}$$

Try $L = 6$ in. $a = \dfrac{25}{25.46 \times 6^2}\sqrt{36 \times 1.39^2 + 6^2} = 0.280$ in.

Use $\frac{5}{16}$ in. fillet weld.

NOTE: With $L = 5$ in. and $a = \frac{3}{8}$ in. approximately 20 percent more weld metal would be required.

37. Stiffened Beam-Seat Connections   The welded stiffened beam seat is generally composed of either a structural tee or two plates welded in a T shape (Fig. 22). The latter offers greater design flexibility. The beam web may be either parallel or perpendicular to the stem. The vertical stem is often trimmed to allow for clearances.

Fig. 22 Stiffened beam seat.

Welded beam seats are sometimes used to support beams with small reactions. This may be occasioned by the lack of adequate clearances for the proper welding of flexible beam seats. For example, a lightly loaded beam which frames into a column web might ordinarily be supported by an unstiffened beam seat. However, it may prove awkward to make the vertical welds adjacent to the column flanges.

Several factors must be considered in arriving at the configuration of the connection. It is recommended that the thickness of the beam-seat plate, or tee flange, be at least equal to the thickness of the stem. Beam-seat width should be approximately equal to the beam flange if bolts are to be used in attaching the beam to the seat, and slightly wider if field welding in the flat position is to be utilized. The minimum length of the beam seat, as well as the outstanding dimension of the vertical stiffener, must be equal to the standard beam setback plus minimum bearing length required to prevent beam-web crippling, unless the beam has end stiffeners. The thickness of the stiffener should be no less than the web thickness of the supported beam, and the width-thickness ratio of the stiffener should be within the limits set by the specifications for outstanding elements in compression. These minimums can be checked by computing the stress on a section mm (Fig. 23a) which intersects the outstanding edge at midheight, i.e.,

$$f = \frac{R_n}{t_x d_s} + \frac{6Re}{t_s d_s^2} \tag{44}$$

The stiffener thickness may be checked by computing the stress on a section perpendicular to the line of action of the load. Then

$$f = \frac{R}{t_s W \cos^2\alpha} \left(1 + \frac{6e}{W}\right) \tag{45}$$

Fillet welds are normally used to connect the stiffener and the beam seat to the supporting member. When the bracket is composed of two separate plates, the stiffener should be fitted to bear against the underside of the seat, and the two plates joined by welds which are equivalent to or stronger than the horizontal welds used in attaching the beam seat to the supporting element. The vertical welds run the full length on both sides of the stiffener, and a minimum length of the horizontal beam-seat weld of four-tenths of the vertical dimension is recommended (Fig. 23b). The horizontal weld should preferably be placed on the underside of the seat plate. In this position it will not interfere with the beam seating in case of an overrun in the beam length, it enhances the torsional stiffness of the connection, and it serves as end returns for the top of the vertical welds. Based on the AISC specification, these welds should not be larger than three-quarters of the stiffener thickness.

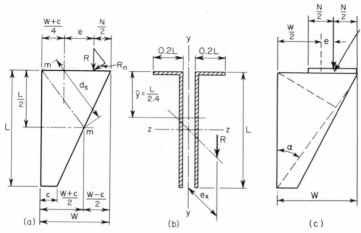

Fig. 23

If the beam web is in the same plane as the stiffener, the reaction is usually assumed to act at the midpoint of the minimum bearing length required to prevent beam-web crippling, measured from the outer edge of the beam seat. When the beam web is perpendicular to the vertical stem, the eccentricity of the reaction is measured from the vertical centerline of the beam web.

A relatively thin (¼-in. minimum) top angle or one-sided clip, fillet-welded along the toes, is sufficient to provide lateral support for the supported beam.

The standard beam setback is ½ in. Calculations based on a setback of ¾ in. allow for mill underrun in the beam length.

Example 13   Design a welded stiffened beam seat for a W18 × 46 with a reaction R = 45 kips. A36 steel and E60 electrodes.

$$b = 6.06 \text{ in.} \qquad t_w = 0.360 \text{ in.} \qquad k = 1 \text{ in.}$$

Bearing length (AISC 1.10.10):

$$N = \frac{R}{0.75 F_y t_w} - k = \frac{45}{0.75 \times 36 \times 0.360} - 1 = 3.63$$
$$W = 3.63 + \tfrac{1}{2} = 4.13 \qquad \text{Use 5 in. (Fig. 22)}$$

Make horizontal seat weld $= 0.4L$ (Fig. 23b). Assume weld of unit width.

$$\bar{y} = \frac{2L \times L/2}{2L + 0.4L} = \frac{L}{2.4}$$

$$I = \frac{2L^3}{12} + 2L \left(\frac{L}{2} - \frac{L}{2.4}\right)^2 + 0.4L \left(\frac{L}{2.4}\right)^2 = \frac{L^3}{4}$$

$$S_{top} = \frac{I}{\bar{y}} = \frac{L^3/4}{L/2.4} = 0.6L^2$$

$$A = 2L + 0.4L = 2.4L$$

$$q_y = \frac{R}{2.4L} \qquad q_x = \frac{Re_x}{0.6L^2}$$

$$q_r = \sqrt{q_x^2 + q_y^2} = \frac{R}{2.4L^2}\sqrt{16e_x^2 + L^2}$$

$$q = 12.73a \text{ (allowable)}$$

$$a = \frac{R}{30.55L^2}\sqrt{16e_x^2 + L^2}$$

$$e_x = W - \frac{N}{2} = 5 - \frac{3.63}{2} = 3.18 \text{ in.}$$

Try $L = 10$ in.

$$a = \frac{45}{30.55 \times 10^2}\sqrt{16 \times 3.18^2 + 10^2} = 0.238 \text{ in.}$$

Use ¼-in. fillet welds 10 in. long.
Stiffener thickness t_s (Fig. 23a)

$$W = 5 \text{ in.} \qquad c = 1 \text{ in.} \qquad e_x = 3.18 \text{ in.}$$

$$e = 3.18 - \frac{5 + 1}{4} = 1.68 \text{ in.}$$

$$d_s = \sqrt{\left(\frac{L}{2}\right)^2 + \left(\frac{W + c}{2}\right)^2} = \sqrt{5^2 + 3^2} = 5.83$$

$$R_n = R \frac{(W + c)/2}{d_s} = 45 \frac{3}{5.83} = 23.2 \text{ kips}$$

From Eq. (44), with $f = 22$ ksi,

$$22 = \frac{23.2}{5.83t_s} + \frac{6 \times 45 \times 1.68}{5.83^2 t_s}$$

$$t_s = 0.181 + 0.607 = 0.788 \text{ in.} \qquad \text{Use } t_s = \text{⅞ in.}$$

If the alternative solution, Eq. (45) is used

$$\cos \alpha = \frac{10}{11.18} = 0.894$$

$$t_s = \frac{45}{22 \times 5 \times 0.894^2}\left(1 + \frac{6 \times 0.68}{5}\right) = 0.930 \text{ in.} \qquad \text{Use } t_s = 1 \text{ in.}$$

Seat width

$$B = b + \text{allowance for field weld} = 7.48 + 1 \qquad \text{Use } 8\text{½ in.}$$

38. Framed Beam Connections This type of connection usually consists of two angles shop-welded to the web of the beam and field-welded to the face of the supporting member. The AISC classifies this connection as Type 2 construction (Art. 39), since it is reasonable to assume that the end of the beam is rotationally unrestrained. The thickness of the framing angles must be limited in order to provide the necessary flexibility. It follows that the connection need be designed only to transfer the beam reaction. It may be desirable to use a combination of bolts and welds, e.g., angles shop-welded to the beam web and outstanding legs bolted to the supporting member.

The connection may be designed by assuming the length of the connection angles and determining the weld size, or vice versa. To minimize the effect of eccentricity the width of the legs connected to the web should be small—3 in. in general and perhaps 4 in. for the larger beams. It is recommended that, when straight electrodes are used for the shop welds, the maximum length of angle be the beam depth less 2½ in. for a 3-in. angle and 3¼ in. for a 4-in. angle. A length equal to the clear depth T of the web may be used if curved electrodes are available. The common range of angle lengths is from one-half to two-thirds the beam depth. The minimum length should be about one-half the clear depth T to provide lateral stability.

Each shop weld is proportioned to resist one-half the beam reaction plus the moment $Re_x/2$ (Fig. 24b). These welds must be sized so as not to overstress the adjacent base material of the beam web. For example, the AISC specification limits the fillet-weld leg size to $\frac{9}{16}$ of the beam-web thickness when A36 steel is used in combination with E60 electrodes.

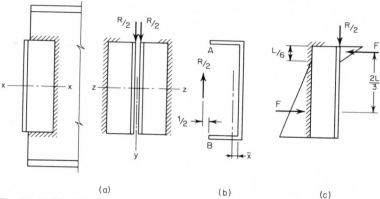

Fig. 24 Welded framed beam connection.

Each field weld should be proportioned to sustain one-half of the beam reaction plus the moment $Re_x/2$ (Fig. 24c). Resistance to the moment is assumed to be provided by the top of the angles bearing against the beam web and by stress in the weld. It is assumed that the neutral axis is located a distance one-sixth the length of the angle from the top of the angle. Sometimes the size of the field welds may be governed by the shear capacity of the supporting member. It may be noted that the effective length of the field weld may be taken as the overall length of the full-sized fillet including returns, although they are often omitted to simplify calculations.

Since connection flexibility is of primary importance, the angle should be relatively thin, but thick enough to accommodate the welds properly. The AISC suggests a minimum thickness of ¼ in. and a maximum thickness of $\frac{1}{16}$ in. greater than the larger leg of either the field weld or the shop weld, unless the weld is to be built out to obtain full throat thickness. Holes for erection bolts, if they are needed, should be placed near the lower end of the outstanding angle legs so as to minimize the rigidity of the connection.

Example 14   Design a welded framed beam connection for a W18 × 46 with a reaction of 25 kips. A36 steel and E60 electrodes.

$$d = 18.06 \text{ in.} \qquad b = 6.06 \text{ in.} \qquad t_w = 0.360 \text{ in.}$$

Try two angles, 3 × 3 × 10.

Shop weld to beam web. Assume weld to be of unit width.

$$\bar{x} = \frac{2 \times 2.5 \times 1.25}{2 \times 2.5 + 10} = 0.42 \text{ in.}$$

$$I_x = 2 \times 2.5 \times 5^2 + \frac{10^3}{12} = 208$$

$$I_y = 2 \times \frac{2.5^3}{12} + 2 \times 2.5 \times 0.83^2 + 10 \times 0.42^2 = 7.81$$

$$J = I_x + I_y = 208 + 7.81 = 216 \text{ in.}^4/\text{in.}$$

Maximum shear stress occurs at points A and B (Fig. 24b).

$$q_y = \frac{25/2}{2 \times 2.5 + 10} = 0.83 \text{ kip/in.}$$

$$M_z = \frac{R}{2} e_x = \frac{25}{2} (3 - 0.42) = 32.2 \text{ in.-kips}$$

$$q'_x = \frac{M_z c_y}{J} = 32.2 \times \frac{5}{216} = 0.75 \text{ kip/in.}$$

$$q'_y = \frac{M_z c_x}{J} = 32.2 \times \frac{2.08}{216} = 0.31 \text{ kip/in.}$$

$$q_r = \sqrt{0.75^2 + (0.83 + 0.31)^2} = 1.36 \text{ kips/in.}$$
$$q = 12.73a \text{ kips/in. allowable}$$

$$a = \frac{q_r}{12.73} = \frac{1.36}{12.73} = 0.107 \text{ in.}$$
$$D \gtrless \text{\%}_{16} t_w \gtrless 0.56 \times 0.360 = 0.188$$

Use $\%_{16}$-in. shop weld to beam web.
Field weld (Fig. 24c):

$$q_y = \frac{25/2}{10} = 1.25 \text{ kips/in.}$$

$$M_x = \frac{R}{2} e_z = 3\frac{25}{2} = 37.5 \text{ in.-kips}$$

$$F = \frac{M_x}{2L/3} = \frac{37.5}{6.67} = 5.63 \text{ kips}$$

$$q'_z = \frac{2F}{5L/6} = \frac{2 \times 5.63}{8.33} = 1.35 \text{ kips/in.}$$

$$q_r = \sqrt{1.25^2 + 1.35^2} = 1.84 \text{ kips/in.}$$
$$a = \frac{q_r}{12.73} = \frac{1.84}{12.73} = 0.145 \text{ in.}$$

Use $\%_{16}$-in. field weld.
Angle thickness t

$$t = \frac{R/2}{F_v L} = \frac{25/2}{14.5 \times 10} = 0.086 \text{ in.}$$
$$t = \%_{16} + \text{\%}_{16} = \text{\%}_4$$

Use two angles $3 \times 3 \times \text{\%}_4 \times 10$.

39. Moment-Resistant Beam Connections Three basic types of construction are permitted by the AISC specification: rigid (continuous) framing (Type 1), simple framing (Type 2), and semirigid framing (Type 3). It is assumed that beam-to-column connections in Type 1 construction are rigid enough to allow virtually no change in the angles between the members. Connections in Type 2 construction are designed as shear connections; i.e., rotational restraint is assumed to be negligible. Semirigid connections permit end rotation intermediate between rigid and simple connections and must develop a predictable moment.

One of the major goals in semirigid construction is to realize overall economy due to recognition of partial continuity. For example, the end moment for a fully fixed, uniformly loaded beam is $wL^2/12$. However, if the end connections can develop only 75 percent of the moment required for a fully fixed beam, the end moments, as well as the midspan moment, will be $wL^2/16$. Consequently, beam material would be used more efficiently with such connections, since the required section modulus is only three-quarters of that required for the fixed-end condition.

Experimental research on semirigid connections for the purpose of ascertaining moment-rotation data has been carried out only to a limited extent. Therefore, if they are to be used with any reliability, it is of utmost importance that sufficient evidence be available to document the design assumptions.

Even if sufficient data were available, strict compliance with the philosophy of semirigid design leads to further complications. Although the basic method of frame analysis remains unaltered, modifications must be made in fixed-end moments, stiffness factors, and carryover factors. All things considered, then, design based on a "pure" approach to semirigidly connected frames would seem to be somewhat impractical.

Semirigid construction has been used to a considerable extent, especially in frames designed for wind. Usually the moment connection is arbitrarily designed for the wind moment only and the beam is proportioned for the larger of either (1) the simple-span positive moment due to gravity loads or (2) three-quarters of the wind moment at the connection—this recognizes the increase in allowable stress for wind forces. An alternate method is arbitrarily to design the moment connection for the wind moment plus a portion of the fixed-end gravity moment. In this case, depending upon the designer's judgment, the beam size may be chosen on the basis of any of the following: (1) the maximum positive or negative moment due to gravity loads, (2) the simple-span positive moment due to gravity loads, or (3) three-quarters of the combined gravity and wind moments.

The design procedure may be rationalized as suggested in Ref. 12, "If the connections are such that they can undergo large inelastic deformations without failure, it will be found that, if the connections are adequate to resist wind stress moments alone at design stresses computed on the assumption that the connections are elastic, they are also adequate for the combination of gravity loads and wind loads."

The ultimate test, of course, is performance under actual conditions. To this extent, experience has demonstrated the validity of these design methods.

The most direct rigid connection of beam to column flange is that shown in Fig. 25. Here the beam flanges are groove-welded and the web fillet-welded to the column flange. In the connection in Fig. 26 the web is field-welded to a plate which is shop-welded to the column flange. This allows a more lenient setback of the web and also eliminates the angle seat in the connection of Fig. 25. In the connection of Fig. 27 high-strength bolts are used instead of field fillet welds in the web-to-plate connection. A similar connection of beam to column web is shown in Fig. 28. Tolerances must be closely controlled in these connections; otherwise considerable difficulty may be encountered in the field.

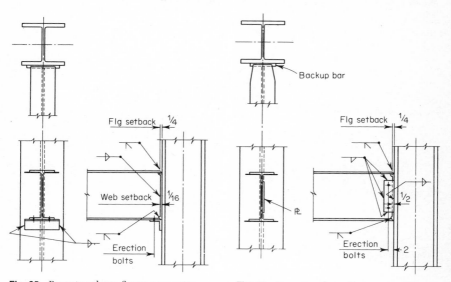

Fig. 25 Beam-to-column flange connection.   **Fig. 26** Beam-to-column flange connection.

More lenient tolerances in beam length can be allowed with a connection of the type shown in Fig. 29. The top plate is shipped loose and positioned after the beam is in place. A vertical-plate web connection as in Fig. 27 may be used instead of the seat bracket, in which case a bottom flange plate can be shop-welded to the column.

Column stiffeners may be required, as in Fig. 30.

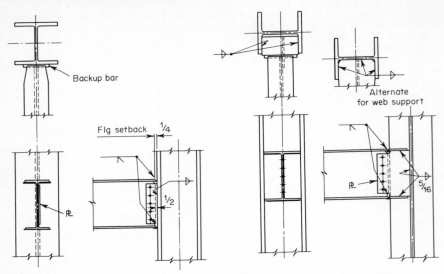

Fig. 27 Beam-to-column flange connection.   **Fig. 28** Beam-to-column web connection.

Example 15 Design a 28-ft beam with semirigid welded top-plate connections, attached to the flanges of W10 × 33 columns, to support a gravity load of 1.50 kips/ft and a wind moment of 67.5 ft-kips (Fig. 29). AISC specification, A36 steel, E60 electrodes.

BEAM: Design for simple-span positive moment due to gravity load.

$$M = \frac{wL^2}{8} = 1.5 \times \frac{28^2}{8} = 147 \text{ ft-kips}$$

$$S = 147 \times \frac{12}{24} = 73.5 \text{ in.}^4$$

Use W18 × 46.
Top plate—reduced section:

$$T = \frac{M_w}{d} = 67.5 \times \frac{12}{18.06} = 44.9 \text{ kips}$$

$$A' = \frac{44.9}{1.33 \times 22} = 1.53 \text{ in.}^2$$

Use $b' = 4\frac{1}{4}$ in., $t = \frac{3}{8}$ in.
Length of $\frac{5}{16}$-in. fillet weld

$$L = \frac{T}{12.73a} = \frac{44.9}{12.73 \times 0.312} = 11.3 \text{ in.}$$

Use $\frac{5}{16}$-in. fillet weld across end and $3\frac{5}{8}$ in. each side
Unwelded length of reduced section = 1.5 × 4.25 = 6.38; use 7 in.
Top plate—full section:

$$b = b' + 2 \times (1\frac{1}{2} \text{ in. radius}) = 4\frac{1}{4} + 3 = 7\frac{1}{4} \text{ in.}$$

W18 × 46 $b = 6.06$ in. $t_f = 0.605$ in. $t_w = 0.360$ in. $k = 1$ in.

FILLET WELD TO BRACKET:
Try $a = t_f - \frac{1}{16} = \frac{7}{16}$ in.

$$L = \frac{T}{12.73a} = \frac{44.9}{12.73 \times 0.438} = 8.05 \text{ in.}$$

Use $\frac{7}{16}$-in. fillet welds 4 in. each side.

$$b + 2a = 6.06 + 2 \times 0.438 = 6.94 \text{ in.}$$

Use WT9 × 32.5 bracket:

$$b = 7.590 \text{ in.} \qquad d = 9.175 \text{ in.} \qquad t_f = 0.750 \text{ in.} \qquad t_w = 0.450 \text{ in.}$$

Bearing length:

$$R = \tfrac{1}{2} \times 28 \times 1.5 = 21 \text{ kips}$$
$$N = \frac{R}{0.75 F_y t_w} - k = \frac{21}{0.75 \times 36 \times 0.360} - 1 = 1.16 \text{ in.}$$

Seat length:

$$W = \text{beam clearance} + \text{weld length} = \tfrac{1}{2} + 4 = 4.5 \qquad \text{Use 5 in.}$$

MINIMUM STEM THICKNESS:

$$W = 5 \text{ in.} \qquad L = 8.00 \text{ in.} \qquad \cos \alpha = 0.872$$
$$e = \frac{1}{2} + \frac{4.5}{2} - 2.5 = 0.25 \text{ in.}$$
$$t = \frac{R}{f_{max} W \cos^2 \alpha}\left(1 + \frac{6e}{W}\right) = \frac{21}{22 \times 5 \times 0.872^2} \times \left(1 + \frac{6 \times 0.25}{5}\right) = 0.326 \text{ in.} < 0.450 \text{ in.}$$
$$M = R e_x + T y = 21 \times 2.94 + 44.9 \times 1.92 = 147.9 \text{ in.-kips}$$
$$f_x = \frac{44.9}{9.55} + \frac{147.9}{10.1} = 19.35 < 1.33\, F_b$$
$$f_y = \frac{44.9}{0.450 \times 9.175} = 10.9 < F_v$$

BRACKET WELD:

$$\text{Flange, } t = (\tfrac{9}{16}) t_f = 0.422 \text{ in.} \qquad \text{Use } \tfrac{7}{16} \text{ in.}$$
$$\text{Stem, } t = (\tfrac{9}{16}) t_w = 0.253 \text{ in.}$$
$$q_x = 19.35 \times \frac{0.450}{2} = 4.35 \text{ kips/in.}$$
$$q_y = 10.9 \times \frac{0.450}{2} = 2.45 \text{ kips/in.}$$
$$q = \sqrt{4.35^2 + 2.45^2} = 4.99 \text{ kips/in.}$$
$$a = \frac{4.99}{1.33 \times 12.73} = 0.295 \text{ in.}$$

Use $\tfrac{5}{16}$-in. fillet.

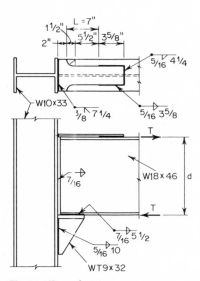

Fig. 29 Example 15.

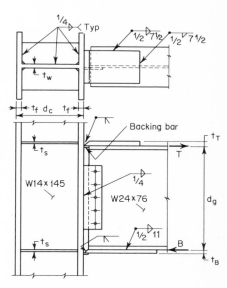

Fig. 30 Example 16.

Example 16 Design a rigid welded connection between a W24 × 76 wind girder and a W14 × 145 column (Fig. 30). AISC specification, A36 steel, E70 electrodes, and A325 bolts for shear connection (friction-type) with oversize holes.

Load	Shear, kips	Moment, ft-kips
DL	22.1	100
LL	8.3	50
DL + LL	30.4	150
WL	24.7	280
DL + LL + WL	55.1	430
¾(DL + LL + WL)	41.3	322

W24 × 76:

$d = 23.91$ in. $b = 8.985$ in.

Horizontal force due to girder moment

$$T = \frac{M}{d} = 322 \times \frac{12}{23.91} = 162 \text{ kips}$$

TOP FLANGE PLATE (complete-penetration groove weld to column flange): Assume $b_T = b - 1.5 = 8.985 - 1.5 = 7.50$ in.

$$A = \frac{T}{0.60F_y} = \frac{162}{22} = 7.35 \text{ in.}^2$$
$$t = \frac{7.35}{7.50} = 0.983 \text{ in.}$$

Use 7½ × 1 top flange plate.
  Length of fillet weld:
Try $a = ½$ in. $L = T/14.85a = 162/(14.85 \times ½) = 21.8$ in. Use ½-in. fillet weld 7½ in. across end and 7½ in. each side.
  BOTTOM FLANGE PLATE (complete-penetration groove weld to column flange): Assume $b_B = b + 1.5 = 8.985 + 1.5 = 10.50$ in.

$$A = 7.35 \text{ in.}^2 \qquad t = \frac{7.35}{10.5} = 0.701 \text{ in.}$$

Use 10½ × ¾ bottom flange plate.
  Length of fillet weld:
Use same as top plate: ½-in. fillet weld 11 in. each side.
  SHEAR PLATE:
Try ⅞-in. bolts, 0.601 × 15 = 9.02 kips.

$$n = \frac{41.3}{9.02} = 4.58$$

Use five ⅞-in. A325 bolts.

$$L = 4 \times 3 + 2 \times 1¼ = 14.5 \text{ in.}$$
$$t = \frac{41.3}{LF_v} = \frac{41.3}{14.5 \times 14.5} = 0.196 \text{ in.}$$

Use 4 × ¼ × 1—2½ shear plate.
  Shear-plate weld:

$$a = 0.25 \times \frac{14.5}{2 \times 14.85} = 0.122 \text{ in. max for balanced design.}$$

But $a = ¼$ in. minimum for column flange of $t = 1.090$ in. (AISC 1.17.2)

$$L = \frac{V}{14.85a} = \frac{41.3}{14.85 \times 0.25} = 11.1 \text{ in.}$$

Use ¼-in. welds 10 in. each side.
  COLUMN WEB STIFFENERS (AISC 1.15.5)
W14 × 145:

$$d = 14.78 \text{ in.} \qquad t_f = 1.090 \text{ in.} \qquad t_w = 0.680 \text{ in.} \qquad k = 1.75 \text{ in.}$$

AISC Eq. (1.15.1):

$$A_{st} = \frac{\frac{4}{3} \times 162 - 36 \times 0.680(1 + 5 \times 1.75)}{36} = \text{negative}$$

Therefore, no stiffeners required unless by AISC Eqs. (1.15-2) and (1.15-3):

$$d_c = \frac{4100 \times 0.680^3\sqrt{36}}{\frac{4}{3} \times 162} = 36 > 14.88 - 2k$$

No stiffeners are needed opposite the compression flange.

$$t_f = 0.4 \sqrt{\frac{\frac{4}{3} \times 162}{36}} = 0.98 < 1.128$$

No stiffeners are needed opposite the tension flange.

RIVETED AND BOLTED CONNECTIONS

40. Rivets   Low-carbon structural rivet steel ASTM A502 Grade 1 is normally used with ASTM A36 structural steel. The A502 Grade 2 carbon-manganese rivet may be used to connect high-strength steels.

The AISC recommends that hot-driven rivets be heated to a uniform temperature not higher than 1950°F and installed at a temperature not lower than 1000°F. If the rivet is worked at too low a temperature, there is a possibility of a brittle formation which is detrimental to a joint. A "cherry-red" hot-driven rivet is essentially in a plastic state and as such fills the hole entirely. Residual tensile stresses in the rivet and corresponding forces between the connected parts may result from cooling of the rivet. The intensity of these forces is erratic, and they are usually neglected in design assumptions. Holes will be filled more completely with short-grip rivets than with long-grip rivets.

Rivet Sizes.   The nominal sizes of rivets used in structural work increase in ⅛-in. increments from ½ to 1½ in. in diameter. Smaller and larger sizes are available for exceptional cases. In the interests of fabrication and erection, a minimum of different fastener diameters should be used.

Rivet Heads.   The commonly used structural rivet is "headed" by two full hemispherical buttons. A high-button or acorn shape characterizes the manufactured head, which obtains a shape similar to the driven head after driving. Flattened and countersunk heads may be fashioned. The finished heads should be fully and uniformly sized, as well as concentric with the holes.

Holes for rivets are punched or drilled ¹⁄₁₆ in. larger than the nominal rivet diameters. Holes may be punched successfully when the thickness of the material is not greater than the nominal diameter of the rivet plus ⅛ in. Thicker material must be either drilled from the solid or subpunched and reamed.

41. High-Strength Bolts   High-strength bolts produce large and predictable tensions when tightened. Initial tensioning of high-strength bolts also results in more rigid joints, more satisfactory stress distributions, and greater assurance against nut loosening.

The A325 bolt is furnished in the following three types:

Type 1. Bolts of medium-carbon steel, in sizes ½ to 1½ in. diameter, inclusive.

Type 2. Bolts of low-carbon martensite steel, in sizes ½ to 1 in. diameter, inclusive (not to be hot-galvanized).

Type 3. Bolts having atmospheric corrosion resistance and weathering characteristics comparable with those of A588 and A242 steels, in sizes ½ to 1½ in. diameter, inclusive. Type 1 is furnished if the type is not specified.

The A325 bolt is identified by the manufacturer's symbol and the legend A325 on the head. In addition, the type 2 bolt head is inscribed with three radial lines, and the legend A325 is underlined on the type 3 bolt. A490 bolts are identified by the manufacturer's symbol and the legend A490.

Dimensional requirements for the A325 and the A490 bolt are identical and must conform to the ANSI Standard B18.2 for heavy hexagon structural bolts, which have shorter thread lengths than other standard bolts.

It is possible to exclude bolt threads from all shear planes since the body length of the bolt is the control dimension. However, bolt lengths must be modified to account for incremental stock lengths, which results in thread runout into the grip as follows: up to ⅜ in. for ½-, ⅝-, ¾-, ⅞-, 1¼-, and 1½-in. bolts; and up to ½ in. for 1-, 1⅛-, and 1⅜-in. bolts. Although it is permissible to have some thread runout in shear planes in bearing-type connections, extreme care should be exercised to provide a sufficient amount of thread for nut tightening lest the nut jam on the thread runout.

Required bolt lengths may be determined in the following manner: add the grip (total thickness of connection material) and the stock length adjustment and the washer thickness (⁵⁄₃₂ in. for a flat hardened washer and ⁵⁄₁₆ in. for a bevel washer), if any. These bolt lengths should be rounded off to the next longer ¼ in.

Nuts.   Different grades of high-strength bolts are combined with various nuts which guarantee failure by bolt yielding rather than by stripping of the nut threads. Heavy semifinished hexagon nuts, with dimensions conforming to the requirements of the ANSI Standard B18.2, are used with both the A325 and the A490 bolts. Nuts satisfying ASTM A325 Specification may be identified either by three circumferential marks on at least one face or by the inscription 2, 2H, D, DH, or 3, and the manufacturer's symbol. Nuts to be used with the A490 bolts are marked 2H or DH and the manufacturer's identification.

Loosening of nuts is usually not a problem in high-strength bolted assemblies because the bolt tension creates high bearing pressures between the nut and the connected part or the washer. However, where bolted joints are subject to shock and vibration, or where the use of power tools is impractical, it may be advantageous to use proprietary nuts which have special self-locking devices.

TABLE 13   Properties of High-Strength Structural Bolts

ASTM Designation	Type name	Bolt diam, in.	Tensile strength on stress area,* ksi	Proof load on stress area,*,† ksi
A325—74	High-strength bolts for structural-steel joints	½–1 1⅛–1½	120 105	85 74
A490—74	High-strength alloy steel bolts for structural-steel joints	½–1½	150–170‡	120‡

*Stress area = $0.785(D - 0.9743/n)^2$: D = nominal bolt size, n = threads per inch.
†Ratio of proof load to tensile strength is roughly 0.70 for A325 bolts and A490 bolts.
‡Same as A354—BD bolts.

42. Installation of High-Strength Bolts   The procedures used in tightening A325 and A490 bolts are essentially the same. Although pneumatic powered impact wrenches are preferred, long-handled manual torque wrenches or electrical wrenches may be used.

The diameter of round bolt holes ("standard" holes) must not be more than ¹⁄₁₆ in. larger than the nominal diameter of the bolt. Holes may be punched in material not thicker than the nominal bolt diameter plus ⅛ in. Holes must be either drilled or subpunched and reamed in thicker material. Oversize, short-slotted, and long-slotted holes may be used if approved by the designer. Table 14 summarizes the sizes and other important features of oversize and slotted holes permitted by AISC.

All contact surfaces in a connection, including those associated with the bolt heads, nuts, and washers, should be free of scale, burrs, dirt, and other foreign matter tending to inhibit uniform seating of the joint components. Tight mill scale is an exception and may be left in place. The light residual oil coating on bolts, nuts, and washers is not injurious to a joint and does not have to be removed even in a friction-type connection.

To attain the required bolt tightening in the recommended time of approximately 10 sec, it is necessary to use impact wrenches. Sometimes, out of necessity, joints are designed too compactly, which leads to inadequate erection clearances. Universal or straight extension adapters which reduce the torquing power by as much as 20 percent must be used here. However, it is possible to build straight extensions into impact wrenches without sacrificing power. Bolts should be tightened beginning with the most rigid portion of the joint and proceeding toward the free edges.

Bolt Tension.   The calibrated-wrench method or the turn-of-nut method is required to control high-strength bolt tension. All fasteners in a joint must be tightened to a tension equal to or greater than the proof load specified in Table 15.

Because of the considerable benefits derived from high initial bolt tension, and in the interest of uniform installation practice, initial tension values identical to those required for friction-type joints are also recommended for the bearing type.

TABLE 14   Oversize and Slotted Holes

Type of hole	Nominal bolt size, in.	Overall dimension of hole, in.	Hardened washer	Remarks
Oversize	$5/8$–$7/8$	Nominal size + $3/16$	One in outer ply	May be used in friction-type connection
	1	$1\frac{1}{4}$		
	$1\frac{1}{8}$	$1\frac{7}{16}$		
	$1\frac{1}{4}$	$1\frac{9}{16}$		
	$1\frac{3}{8}$	$1\frac{11}{16}$		
	$1\frac{1}{2}$	$1\frac{13}{16}$		
Short-slotted	$5/8$	$\frac{11}{16} \times \frac{7}{8}$	One in outer ply	May be used in friction-type or bearing-type connections. Slot must be normal to direction of loading in bearing-type connection
	$3/4$	$\frac{13}{16} \times 1$		
	$7/8$	$\frac{15}{16} \times 1\frac{1}{8}$		
	1	$1\frac{1}{16} \times 1\frac{5}{16}$		
	$1\frac{1}{8}$	$1\frac{3}{16} \times 1\frac{1}{2}$		
	$1\frac{1}{4}$	$1\frac{5}{16} \times 1\frac{5}{8}$		
	$1\frac{3}{8}$	$1\frac{7}{16} \times 1\frac{3}{4}$		
	$1\frac{1}{2}$	$1\frac{9}{16} \times 1\frac{7}{8}$		
Long-slotted	$5/8$–$1\frac{1}{2}$	Width: nominal diameter plus $1/16$. Length: not to exceed $2\frac{1}{2}$ times bolt diameter	$5/16$-in. plate washer with standard hole, and hardened washers if required	May be used in only one of the connected parts of either friction-type or bearing-type connection. Slot must be normal to the direction of loading in bearing-type connection

TABLE 15   Bolt Proof Loads

Bolt size	Min bolt tension,* kips	
	A325 bolts	A490 bolts
$1/2$	12	15
$5/8$	19	24
$3/4$	28	35
$7/8$	39	49
1	51	64
$1\frac{1}{8}$	56	80
$1\frac{1}{4}$	71	102
$1\frac{3}{8}$	85	121
$1\frac{1}{2}$	103	148

*Equal to 70 percent of specified tensile strength.

Hardened washers are required for bolts in oversize and slotted holes (Table 14).

A hardened washer must be used under either the head or the nut, whichever is turned in tightening, with A325 bolts in standard holes, tightened by the calibrated wrench. The washer is not needed in tightening by the turn-of-the-nut method.

Clamping forces produced by A490 bolts are significantly larger than those of A325 bolts. To guard against indenting and galling the connected parts, washer requirements for A490 bolts are dependent upon the yield point of the connected material. When they connect low-strength steels ($F_y < 40$ ksi) a hardened washer must be used under both the head and the nut. Only one washer is required, under the item being turned, if $F_y \geq 40$ ksi. These requirements apply regardless of the tightening method used.

In order to provide a choice of inspection methods, it is a good idea to call for a hardened washer under the element being turned even though it may not be required by the governing code. This suggestion is based on the fact that without a hardened washer galling may take place, which may register an erratic torque reading if the installation is checked with a torque wrench. Because of this, some fabricators and erectors always use a hardened washer under the turned element, regardless of the tightening method specified.

Calibrated-Wrench Method. This method utilizes torque control to assure attainment of appropriate bolt tensions. Adjustable power impact wrenches and manual torque wrenches must be calibrated to induce bolt tensions of 5 percent in excess of the proof-load values indicated in Table 15 for each grade and size of bolt to be used in a given installation. Every wrench is calibrated by having it tighten a minimum of three bolts, of the same diameter, in a hydraulic tension-measuring device. Calibration should be repeated whenever a wrench is required to tighten a different size bolt, or at least once each working day if there is no change in the bolt size. Impact wrenches should be set so as to stall or cut out at the torque effort corresponding to the prescribed fastener tension. When manual torque wrenches are used, the wrench-indicator torque value corresponding to the calibrating tension should be determined and taken as the job standard. Torque measurements must be read while the turned element is in tightening motion.

Subsequent tightening of bolts in any particular assembly may loosen bolts which have already been installed. Therefore, all bolts must be "touched up."

Turn-of-Nut Method. Adequate bolt tensions are achieved with the turn-of-nut method by controlling bolt elongations. Because of this strain control, it is possible to obtain bolt tensions well beyond the specified proof loads. This does not create problems as such, since, in the plastic range, large changes in bolt strains cause small changes in bolt tensions. Therefore, higher clamping forces are provided consistently with their attendant favorable effect on joint tightness.

Either standard power impact wrenches or ordinary spud wrenches may be used to produce the specified bolt tensions. It is recommended that power wrenches be tested frequently in a manner similar to the procedure outlined for the calibrated-wrench method.

Initially, a sufficient number of bolts must be "snugged up" to bring the connection components into full contact. Subsequently, the remaining bolts in the joint should also be brought to snug tightness. This starting position is referred to as the "snug-tight" condition—the point at which the turned element ceases to rotate freely and the impact wrench begins to impact. Or, if common spud wrenches are employed, this starting position is signaled by the full effort of a man. When this phase is completed, all bolts in the joint are then given additional rotation depending on the bolt length and the type of connection (Table 16). If the finger-tight condition is used as a starting point, an extra full turn will correspond approximately to one-half turn from the snug-tight condition. To provide a visual means of determining the additional differential rotation required, it is helpful to matchmark nuts and projecting bolt points with keel or paint. In any case, the final tightening sequence should be performed in an orderly fashion, proceeding from the most rigid portion of the joint toward the free edges.

43. Inspection of High-Strength Bolts Unless inspection procedures are sufficiently outlined in contract documents, some reasonable accord should be reached between the architect/engineer, the contractor, and the inspection agency before lost time begins to accumulate.

As a bare minimum, the inspector should at least verify that (1) appropriate fasteners are being used, (2) joint contact surfaces are prepared so as to permit proper seating of the parts, and (3) correct installation methods are practiced. If the turn-of-nut method is specified, the inspector should note that the turned element is given the additional turn required. When bolts are to be tightened by the calibrated-wrench method, he must observe the on-site calibration tests.

It is recommended that an inspection wrench be used to test bolt installations. Although an impact wrench can certainly be used as the tester, a manual-indicator torque wrench probably gives the inspecting group greater mobility. These inspection wrenches should be calibrated together with the wrenches which will perform the actual tightening. The torque value or the torque effort to be used as the job inspection standard should be that determined by the actual condition of the application. It should be noted that, if washers are not included under the element being turned in tightening, torque readings will be fairly high and quite erratic.

TABLE 16   Nut Rotation from Snug-Tight Condition*

Bolt length (as measured from underside of head to extreme end of point)	Disposition of outer faces of bolted parts		
	Both faces normal to bolt axis	One face normal to bolt axis and other face sloped not more than 1:20 (bevel washer not used)	Both faces sloped not more than 1:20 from normal to bolt axis (bevel washers not used)
Up to and including 4 diameters	⅓ turn	½ turn	⅔ turn
Over 4 diameters but not exceeding 8 diameters	½ turn	⅔ turn	⅚ turn
Over 8 diameters but not exceeding 12 diameters†	⅔ turn	⅚ turn	1 turn

*Nut rotation is relative to bolt, regardless of the element (nut or bolt) being turned. For bolts installed by ½ turn and less, the tolerance should be plus or minus 30°; for bolts installed by ⅔ turn and more, the tolerance should be plus or minus 45°.

†Required rotation must be determined by tests in a tension device simulating the actual conditions.

44. ASTM A307 (Unfinished) Bolts The A307 bolt is known by a variety of names—unfinished, rough, common, ordinary, and machine. Grade A bolts are intended for general application and will be furnished automatically by suppliers unless Grade B bolts, used in piping-system joints, are specified.

A minimum tensile strength of 55 ksi for all sizes, ¼ to 4 in. inclusive, is guaranteed. Dimensions and strengths of nuts are based on the axial proof-load method of testing; satisfactory performance implies that bolt capacities will be reached prior to nut failure. Nuts and bolt heads classified as Grade A are manufactured with a regular square shape, making them easily recognizable. Identification markings on the bolt heads by the manufacturer are optional.

A307 bolts are tightened securely by using long-handled manual wrenches; hardened steel washers are generally not used. They are widely used for connecting relatively light members. The advisability of using them in connections subject to slip or vibration is questionable because of the tendency of nuts to loosen. Nuts with special locking features can be used to limit the effect of this disadvantage.

45. Turned Bolts The term turned bolt applies to all bolts which have regular semifinished heads and strict tolerances on the nominal diameter (0 over and 0.006 in. under). These bolts have been replaced by the high-strength bolt.

46. Ribbed Bolts The ribbed bolt is a proprietary, interference-type fastener made from carbon steel with strengths equal to or greater than that of the A502 Grade 1 rivet. It combines features of a rivet and an unfinished bolt. It has a standard rivet head, button-type or countersunk, a fluted shank with triangular-shaped ribs, and a self-locking nut. The bolt is locked in position when it is driven into any standard punched or drilled hole because the longitudinal ribs deform to a "body-bound" fit. It is ideally suited for connections of very high structures, such as television towers, where it is difficult and also impractical to attempt power installation of rivets or high-strength bolts. Installation requires a hand hammer and an ordinary manual wrench, although power wrenches may be used.

47. Bearing Bolts The bearing bolt incorporates the advantages of the ribbed bolt and the strength of the A325 bolt. It has a flattened button head, a shank with interrupted ribs, effecting a knurled pattern, and a heavy nut with or without a washer.

Bolt ribs are designed with a small taper to facilitate progress through the hole. However, the bolt is somewhat larger than the punched or drilled hole which finally houses it. It is claimed that, while driving, the interrupted ribs will not peel, pack under the head, or break off; they will cut grooves in the sides of the hole in the process of filling it. In this manner full bearing is created along the entire grip length. This body-bound fit affords not only a fully effective cross-sectional bolt area but also a rigid joint with an inherent resistance to slippage. Therefore, the bolt will transmit load primarily by shear and bearing, and the initial tensioning normally required for other high-strength bolts may be considered unnecessary. However, before the decision is made to accept minimal tightening by an ordinary hand wrench, the behavior of the joint should be considered, and if nut loosening is likely, a nut with locking features or proof-load tensioning methods should be used.

48. High-Strength Tension Control and Tension Set Bolts   These bolts, commonly known as TC and TS bolts, are manufactured in Japan. Both are similar in performance. Fasteners equivalent to ASTM A325 and A490 bolts in sizes ranging from ⅝ to 1 in. in diameter can be supplied. Each bolt has a round head with a torque-control notch at its tip. Bolts are installed with an electric wrench which engages both the bolt and the nut in the outer and the inner sleeves of the wrench until the notch of the bolt shears off. The bolt does not rotate during fastening. The notch is designed to shear off when the fastener is loaded to a predetermined tension. The advantage of these bolts is that the completion of bolt fastening can be confirmed by the shear-off of the notch, thus eliminating the use of an inspection wrench.

49. Friction-Type Connections   In high-strength bolt assemblies, joint forces may be transferred either by friction or by bearing. The term "friction-type" derives from the

TABLE 17   Allowable Stresses for Fasteners, ksi[a] (Ref. 13)

Load condition	Hole type	ASTM A325 bolts	ASTM A490 bolts
Applied tension[b]	Standard, oversize, or slotted	44.0	54.0
Shear: friction-type connection	Standard	17.5[c]	22.0[c]
	Oversize	15.0[c]	19.0[c]
	Short-slotted	15.0[c]	19.0[c]
	Long-slotted	12.5[c]	16.0[c]
Shear: bearing-type connection:			
Threads in any shear plane	Standard or slotted	21.0[d]	28.0[d]
No threads in shear plane	Standard or slotted	30.0[d]	40.0[d]
Bearing[e]	Standard, oversize, or slotted	$\dfrac{LF_u}{2d}$ or $1.5F_u$ (whichever is smaller)	

[a]The tabulated stresses, except for bearing stress, apply to the nominal area of bolts used in any grade of steel.

[b]For allowable stresses when bolts are subjected to fatigue loading in tension, see subsection 4(b)2, Ref. 13.

[c]Applicable for contact surfaces with clean mill scale. When the designer has specified special treatment of the contact surfaces in a *friction-type* connection, values in Table 18 may be substituted.

[d]In *bearing-type* connections whose length between extreme fasteners measured parallel to the line of an axial force exceeds 50 in., tabulated values shall be reduced by 20%.

[e]L is the distance in inches measured in the line of force from the centerline of a bolt to the nearest edge of an adjacent bolt or to the end of the connected part toward which the force is directed; d is the diameter of the bolt; and F_u is the lowest specified minimum tensile strength of the connected parts.

assumed behavior of the joint; shear forces are transmitted by static friction between the contact surfaces of the connected parts. This resistance is directly related to the magnitude of the bolt clamping force, the area of contact, and the nature of the faying surfaces. Allowable stresses are given in Tables 17 and 18.

This type of shear connection should be used where slip might be objectionable and where stress reversal, excessive stress variation, impact, or vibration is encountered.

TABLE 18   Allowable Stresses, ksi, for Friction-Type Shear Connections* (Ref. 13)

Class	Surface condition of bolted parts	Standard holes A325	Standard holes A490	Oversize holes and short-slotted holes A325	Oversize holes and short-slotted holes A490	Long-slotted holes A325	Long-slotted holes A490
A	Clean mill scale	17.5	22.0	15.0	19.0	12.5	16.0
B	Blast-cleaned carbon and low-alloy steel	27.5	34.5	23.5	29.5	19.5	24.0
C	Blast-cleaned quenched and tempered steel	19.0	23.5	16.0	20.0	13.5	16.5
D	Hot-dip galvanized and roughened	21.5	27.0	18.5	23.0	15.0	19.0
E	Blast-cleaned, organic zinc-rich paint	21.0	26.0	18.0	22.0	14.5	18.0
F	Blast-cleaned, inorganic zinc-rich paint	29.5	37.0	25.0	31.5	20.5	26.0
G	Blast-cleaned, metallized with zinc	29.5	37.0	25.0	31.5	20.5	26.0
H	Blast-cleaned, metallized with aluminum	30.0	37.5	25.5	32.0	21.0	26.5
I	Vinyl wash	16.5	20.5	14.0	17.5	11.5	14.5

*Values from this table are applicable *only* when they do not exceed the lowest appropriate allowable stresses for *bearing-type* connections, taking into account the position of threads relative to shear planes and, if required, the 20% reduction due to joint length. (See Table 17.)

50. Bearing-Type Connections   It is assumed that this connection transfers shear on the connection by shear and bearing. The bearing-type connection takes greatest advantage of the high-strength bolt. Contact surfaces may be painted, since slip is permitted.

51. Behavior of Riveted and Bolted Connections   In a shear connection it is assumed that a concentrically applied load is distributed equally among all the fasteners. While this assumption is realistic for short joints, it is not realized in long connections. The efficiency of the fasteners decreases with an increase in the number of fasteners in the line of the applied load; Fig. 31 shows the results of tests to failure of high-strength bolted connections. However, it is customary to assume for purposes of design that concentric load is shared equally by all fasteners in the connection.

If fasteners are too close to the edge perpendicular to the direction of the load, shear-out failure may occur in the plates. This type of failure is avoided by specification limits on edge distances.

52. Allowable Stresses   Allowable stresses for A325 and A490 bolts in friction-type connections depend on the hole type and the surface conditions of the bolted parts.

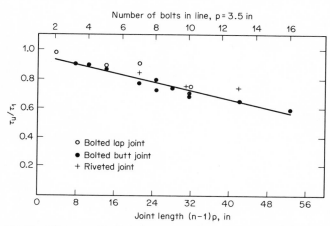

Fig. 31   Efficiency of bolts in long connections. τ_u = average bolt stress when first bolt shears. τ_1 = ultimate shear strength of single bolt of same lot. *(From Ref. 14.)*

Allowable stresses in bearing-type connections depend on the presence or absence of threads in the shear plane. Recommended design values are given in Tables 17 and 18.

53. Eccentrically Loaded Connections—Fasteners in Shear The approximate number of fasteners n in one row of m rows, $n > m$ (Fig. 32a), is

$$n = \sqrt{\frac{6M}{pmR}} \tag{46}$$

where R = allowable force per fastener.

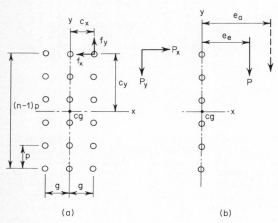

(a)   (b)

Fig. 32

The shearing stresses in the fasteners are given by

$$f_x = \frac{M_z c_y}{JA} \qquad f_y = \frac{M_z c_x}{JA}$$
$$J = I_x + I_y$$

where I_x and I_y are given by[2]

$$I_x = np^2(n^2 - 1)\frac{m}{12}$$
$$I_y = mg^2(m^2 - 1)\frac{n}{12}$$

To account in part for plastic behavior, the AISC suggests, for the case shown in Fig. 32b that the moment M_z be based on the following "effective eccentricities":
For fasteners equally spaced in a single line,

$$e_e = e_a - \frac{1 + 2n}{4}$$

For fasteners equally spaced in two or more gage lines,

$$e_e = e_a - \frac{1 + n}{2}$$

Example 17 Determine the allowable beam reaction for the one-sided shear-plate connection shown in Fig. 33. A325 ¾-in. bolts, friction connection, oversize holes, class A surface

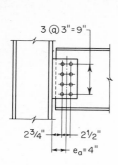

3 @ 3" = 9"

2¾"   2½"

$e_a = 4"$

Fig. 33 Example 17.

$$I_x = \frac{4 \times 3^2(4^2 - 1)2}{12} = 90$$

$$I_y = \frac{2 \times 2.5^2(2^2 - 1)4}{12} = 12.5$$

$$J = I_x + I_y = 102.5$$

$$e_e = 4 - \frac{1 + 4}{2} = 1.50 \text{ in.}$$

Vertical shear:

$$f_y = \frac{P}{\Sigma A} = \frac{P}{8 \times 0.442} = \frac{P}{3.53}$$

Twisting shear (upper right fastener):

$$f'_x = \frac{Pe_e y}{JA} = \frac{P \times 1.5 \times 4.5}{102.5 \times 0.442} = \frac{P}{6.72}$$

$$f'_y = \frac{Pe_e x}{JA} = \frac{P \times 1.5 \times 1.25}{102.5 \times 0.442} = \frac{P}{24.2}$$

Resultant critical shear:

$$15 = \sqrt{\left(\frac{P}{6.72}\right)^2 + \left(\frac{P}{3.53} + \frac{P}{24.2}\right)^2}$$

$$P = 42 \text{ kips}$$

If the eccentricity e_a to the fastener-group centroid had been used, P would be only 27 kips. If it had been neglected, the unconservative P would be 53 kips.

Example 18 Determine the number of ¾-in. A325 bolts for a bearing-type connection with threads in the shear plane to support the eccentrically loaded column-flange bracket of Fig. 34. Assume that the bracket will be adequately sized so as not to affect the choice of fastener.

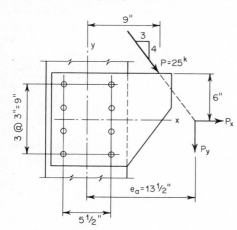

9"

3
4

$P = 25^k$

6"

P_x

P_y

x

y

3 @ 3" = 9"

$e_a = 13\frac{1}{2}"$

5½"

Fig. 34 Example 18.

Estimate the number of bolts needed:

$$M_z = P_y e_a = \tfrac{4}{5} \times 25 \times 13.5 = 270 \text{ in.-kips}$$
$$R = F_v A = 21 \times 0.601 = 12.6 \text{ kips per bolt}$$

$$n = \sqrt{\frac{6 \times 270}{3 \times 2 \times 12.6}} = 4.6$$

Try four bolts per row.

$$e_e = 13.5 - \frac{1+4}{2} = 11.0 \text{ in.}$$

$$M_z = P_y e_e = \tfrac{4}{5} \times 25 \times 11 = 220 \text{ in.-kips}$$

$$I_x = \frac{np^2(n^2-1)m}{12} = \frac{4 \times 3^2(4^2-1) \times 2}{12} = 90$$

$$I_y = \frac{mg^2(m^2-1)n}{12} = \frac{2 \times 5.5^2(2^2-1) \times 4}{12} = 60.5$$

$$J = I_x + I_y = 151 \text{ in.}^4$$

Maximum stress is in fastener at upper right.

$$\text{Horizontal shear } R_x = \frac{P_x}{nm} = \frac{0.6 \times 25}{4 \times 2} = 1.88$$

$$\text{Vertical shear } R_y = \frac{P_y}{nm} = \frac{0.8 \times 25}{4 \times 2} = 2.50$$

$$\text{Twisting shear } R_x' = \frac{M_z c_y}{J} = \frac{220 \times 4.5}{151} = 6.56$$

$$R_y' = \frac{M_z c_x}{J} = \frac{220 \times 2.75}{151} = 4.01$$

Resultant: $f_r = \sqrt{8.44^2 + 6.51^2} = 10.68 < 12.6 \text{ kips}$

54. Eccentrically Loaded Connections—Fasteners in Tension   The approximate number of fasteners n, in one row of m rows, is given by Eq. (46). The tension in the fastener, neutral axis at center of gravity of fasteners, is

$$T_m = \frac{My}{\Sigma y^2} = \frac{6M}{np(n+1)} \approx \frac{6M}{pn^2} \qquad (47)$$

These equations are based on elastic behavior of connections with initial tension in fasteners, assuming that contact pressure between faying surfaces is not lost upon loading the connection.

For connections with no initial tension in fasteners, and for those with initial tension where ultimate-load behavior is considered,

$$n \approx 0.8 \sqrt{\frac{6M}{pmR}}$$

$$f_t = \frac{My}{I_z} \qquad (48)$$

The distance $\bar{y}$ from the compression end of the connection to the neutral axis will be one-sixth to one-seventh of the connection length.

These formulas neglect prying action (Art. 58)

Example 19   Determine the number of $\tfrac{7}{8}$-in. A325 bolts required to support the eccentrically loaded bracket which consists in part of two vertical framing angles with 4-in. legs outstanding (Fig. 35a). Friction-type connection with standard holes.

$$M_z = Pe_x = 40 \times 16 = 640 \text{ in.-kips}$$

$$T = F_t A = 44 \times 0.601 = 26.5 \text{ kips}$$

$$n = \sqrt{\frac{6 \times 640}{2 \times 3 \times 26.5}} = 4.92 \qquad [\text{Eq. (46)}]$$

Try 5 in each row.

$$\Sigma y^2 = 4(3^2 + 6^2) = 180$$

$$T_m = \frac{640 \times 6}{180} = 21.3 \text{ kips} < 26.5 \qquad [\text{Eq. (47)}]$$

Alternative solution:

$$T_b = 39 \text{ kips proof load (Table 15)}$$

$$f_i = \frac{5 \times 39}{4 \times 15} = 3.25 \text{ ksi} \qquad (\text{Fig. } 35b)$$

$$f_m = \frac{M_z}{S} = \frac{6 \times 640}{2 \times 4 \times 15^2} = 2.13 \text{ ksi}$$

The bolt tension T_m is equivalent to the change in contact pressure:

$$T_m = 3.25 \times 3 \times 4 - \frac{1.12 + 1.97}{2} \times 3 \times 4$$
$$= 39 - 18.5 = 20.5 \text{ kips}$$

Bolt shear:

$$f_v = \frac{40}{10 \times 0.601} = 6.65 \text{ ksi} < F_v = 17.5$$

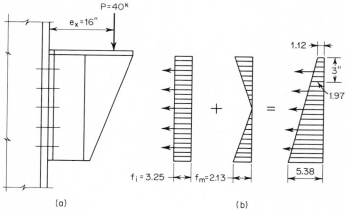

(a)   (b)

Fig. 35   Example 19.

55. Flexible Beam-Seat Connections   This type of connection consists of a seat angle which is usually shop-connected to the supporting member, to facilitate erection, and a top angle or a one-sided clip angle on the beam web to provide lateral and torsional rigidity (Fig. 36). The unstiffened seat is suitable only for relatively small loads in the order of 40 kips. The AISC classifies this connection as Type 2 construction (Art. 39).

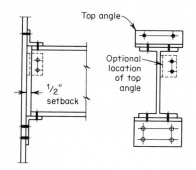

Fig. 36   Seated beam connection.

This connection is somewhat stiffer than the framed beam connection (Art. 57) but still sufficiently flexible to accommodate simple-beam rotations. The rotation capacity is dependent upon the beam depth, the stiffness of the fasteners connecting the top angles to the supporting member, and the stiffness of the element to which the top angle is connected.

The length of the beam bearing N must be sufficient to prevent web crippling. Bearing pressure is assumed to be uniformly distributed, so that the beam reaction acts at the midpoint of the bearing length N.

Flexibility of the connection is assured by using an allowable flexural stress of $F_b = 0.75F_y$ for the seat angle, which may be justified on the basis of restrained cantilever action and the higher shape factor for rectangular cross sections.

A relatively thin ($\frac{1}{4}$ in. minimum) top-angle one-sided clip, with no more than two fasteners in each of its legs, is adequate, since its only function is to provide lateral support for the beam. The standard beam clearance, or setback, is $\frac{1}{2}$ in. Calculations based on a setback of $\frac{3}{4}$ in. allow for mill underrun in the beam length. The spacing of fasteners and gages in seat angles need not be standard as long as the requirements of AISC Secs. 1.16.4 and 1.16.5 are followed.

Example 20   Design a flexible beam-seat connection for a 30-kip reaction from a W18 × 46. A36 steel, $\frac{7}{8}$-in. A325 bolts in a bearing-type connection with threads excluded. (Fig. 36).

$$t_w = 0.360 \text{ in.} \qquad k = 1\frac{1}{16} \text{ in.} \qquad b_f = 6.06 \text{ in.}$$

SEAT ANGLE:

$$N = \frac{R}{t_w \times 0.75F_y} - k = \frac{30}{0.360 \times 27} - 1.06 = 2.03 \text{ in.}$$

Assume k for angle = $1\frac{1}{4}$ in.

$$M = R\left(\frac{3}{4} + \frac{N}{2} - k\right) = 30\left(\frac{3}{4} + \frac{2.03}{2} - 1.25\right) = 15.5 \text{ in.-kips}$$

For 8-in. length of seat and $F_b = 0.75F_y = 27$ ksi.

$$t = \sqrt{\frac{6M}{bF_b}} = \sqrt{\frac{6 \times 15.5}{8 \times 27}} = 0.66 \text{ in.} \qquad \text{Use } \frac{3}{4} \text{ in.}$$

BOLTS IN VERTICAL LEG:

$$\text{Single shear} = 0.601 \times 30 = 18 \text{ kips}$$
$$\text{Bearing} = \frac{(2.5 - 0.50)\,58}{2 \times 0.875} \times 0.75 \times \frac{7}{8} = 43.5 \text{ kips}$$

Use 30/18 = 2 bolts

OUTSTANDING LEG:

$$\text{Minimum length} = \frac{3}{4} + 2.32 = 3.07$$

Use 5 × $3\frac{1}{2}$ × $\frac{3}{4}$ × 8-in. angle with two $\frac{3}{4}$-in. A325 bolts in the vertical leg and two A307 field bolts in the outstanding leg.

Top angle: $3\frac{1}{2}$ × $3\frac{1}{2}$ × $\frac{1}{4}$ × 8-in. with two $\frac{3}{4}$-in. A307 field bolts in each leg.

56. Stiffened Beam-Seat Connections   The makeup of this connection includes all the components of the flexible beam-seat connection, plus single or double vertical stiffeners tightly fitted against the underside of the outstanding leg of the seat angle (Fig. 37). A stiffened beam seat is rarely used to support a beam whose web lies in the same plane as the web of the column to which it connects. Instead, the framed beam connection is preferred, especially since it does not intrude on architectural space.

In terms of analysis, stiffened beam seats are essentially the same as large rigid brackets, their basic difference being related to the magnitude of the eccentricity of the applied load. The beam reaction is usually assumed to act at the midpoint of the outstanding leg of the stiffener. An external moment is developed in the plane of the outstanding leg of the stiffener which requires the upper fasteners to resist tension as well as the usual vertical shear. The flexural compression consists of bearing pressures between the lower end of the stiffener and the face of the supporting element. In addition, if there is a single-angle stiffener or if double stiffeners are spaced too far apart, the fasteners will also be subjected to shears caused by moment in the plane of the connected legs. If the fastener gage of the connected leg is $2\frac{1}{2}$ in. or less, or if the outstanding legs are joined, the twisting is usually neglected.

The length of the outstanding leg of the seat must at least be equal to the standard beam setback plus the minimum bearing length required to prevent web crippling, unless the beam has end stiffeners. Minimum edge distance for fasteners is also a factor. The seat angle must be wide enough to accommodate the fastener arrangement between the stiffener and the supporting member. As a minimum, the seat is generally as wide as the beam flange. The thickness of the outstanding leg of the stiffener is usually determined by the allowable bearing stress. The top of the stiffener angle is often clipped so as to clear the seat-angle fillet, whose radius varies from $\frac{3}{8}$ to $\frac{5}{8}$ in. This results in an effective contact length equal to the length of the outstanding stiffener leg minus the fillet radius, which is normally assumed to be $\frac{1}{2}$ in. Local buckling of the outer edge of the stiffener may be prevented by satisfying the applicable width-thickness ratio for outstanding elements in compression. It is recommended that the stiffener thickness be no less than the web thickness of the supported beam.

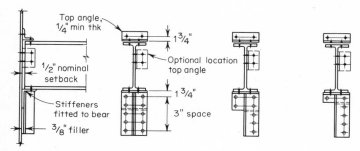

Fig. 37 Stiffened seated beam connection.

A relatively thin ($\frac{1}{4}$ in. minimum) top angle or one-sided clip, with no more than two fasteners in each leg, is sufficient to provide lateral support for the beam.

The standard beam clearance, or setback, is $\frac{1}{2}$ in. Calculations based on a setback of $\frac{3}{4}$ in. allow for mill underrun in the beam length.

Example 21   Design a double-angle stiffened beam-seat connection for a W27 × 84 with a reaction of 80 kips. A36 steel and $\frac{7}{8}$-in. A325 bolts in a bearing connection, threads not excluded.

$$b_f = 9.960 \text{ in.} \qquad t_w = 0.460 \text{ in.} \qquad k = 1\frac{3}{8} \text{ in.}$$

$$N = \frac{R}{t_w \times 0.75 F_y} - k = \frac{80}{0.460 \times 0.75 \times 36} - 1.375 = 5.07 \text{ in.}$$

Use stiffener leg = 5 in. AISC Sec. 1.5.1.5—bearing stiffeners $F_p = 0.90 F_y$.

$$t_s = \frac{80}{2(5 - \frac{1}{2})0.90 \times 36} = 0.274 \text{ in.}$$

Try $t_s = \frac{5}{16}$ in. AISC Sec. 1.9.1.

$$\frac{b}{t_s} \gtrless \frac{95}{\sqrt{F_y}} = 16 \qquad \frac{5}{\frac{5}{16}} = 16$$

Assume column gage = $5\frac{1}{2}$ in. According to AISC standards, the maximum allowable separation between stiffener legs is $2(k_b - t_s) = 2(1\frac{3}{8} - \frac{5}{16}) = 2\frac{1}{8}$ in. Standard gage for 3-in. angle leg is $1\frac{3}{4}$ in. Required stiffener separation = $5\frac{1}{2} - 2 \times 1\frac{3}{4} = 2 < 2\frac{1}{8}$ in. Try two 5 × 3 × $\frac{5}{16}$.
    Seat angle

$$\text{OSL} = \text{beam setback} + N = \frac{3}{4} + 5.07 = 5.82 \text{ in.}$$

Use 6 × 6 × $\frac{3}{8}$ × 10 with $\frac{3}{8}$-in. filler plate as needed.
    BOLTS FOR ONE STIFFENER ANGLE: Assume beam reaction acts at center of stiffener bearing area (Fig. 38a).

$$e_x = 5 + \frac{3}{8} - \frac{5.07}{2} = 2.04 \text{ in.}$$

$$M_z = e_x \frac{R}{2} = 2.84 \times \frac{80}{2} = 114 \text{ in.-kips}$$

$$M_x = e_z \frac{R}{2} = 1.75 \times \frac{80}{2} = 70 \text{ in.-kips}$$

Try 5 bolts in each angle. From Eq. (47),

$$T = \frac{6M_z}{np(n+1)} = \frac{6 \times 114}{5 \times 3 \times 6} = 7.60 \text{ kips}$$

$$\text{Vertical shear} = \frac{R/2}{n} = \frac{40}{5} = 8 \text{ kips}$$

$$J = 2(3^2 + 6^2) = 90 \text{ in.}^4$$

$$\text{Twisting shear} = \frac{M_x c_y}{J} = \frac{70 \times 6}{90} = 4.66 \text{ kips}$$

$$\text{Resultant shear} = \sqrt{8^2 + 4.66^2} = 9.26 \text{ kips}$$

$$f_v = \frac{9.26}{0.6} = 15.4 \text{ ksi} \qquad f_t = \frac{7.60}{0.6} = 12.7 \text{ ksi}$$

From AISC 1.6.3,

$$F_t = 55 - 1.8 f_v = 55 - 1.8 \times 15.4 = 27.3 \text{ ksi} > 12.9$$

A connection with four bolts is adequate.

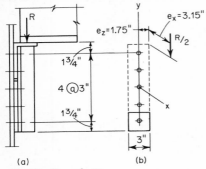

(a)   (b)

Fig. 38   Example 21.

57. Framed Beam Connections   Generally this type of connection consists of two angles shop-connected to the web of the supported member and field-connected to the supporting member (Fig. 39). The AISC classifies this as Type 2 construction (Art. 39). Hence, the connection is designed only for shear due to the beam reaction.

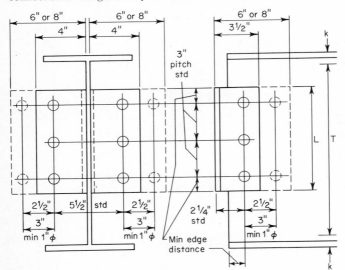

Fig. 39   Bolted framed beam connection.

The double-angle beam-web connection does develop a small amount of end moment which tends to limit the end rotation of the supported member. To assure the least resistance to rotation, the angle thickness should be as small as possible and the gage in the outstanding legs should be large. The eccentricity of the beam reaction may be neglected in all AISC standard framed beam connections.

It is recommended that the length of the connection angles be not less than half the T distance of the beam web in order to provide adequate stiffness and stability to the supported member.

Example 22   Design a framed beam connection for a W16 × 31 with a reaction of 38 kips. A36 steel, ¾-in. A325 bolts in a bearing-type connection with threads in the shear plane.

CONNECTION TO BEAM WEB: From Table 17,
Allowable shear per bolt = 2 × 21 × 0.442 = 18.6 kips

Allowable bearing stress:
$$\frac{LF_u}{2d} = \frac{(3 - 0.438)58}{2 \times 0.75} = 99 \text{ ksi}$$
$$1.5F_u = 1.5 \times 58 = 87 \text{ ksi}$$

Allowable bearing per bolt = 87 × 0.75 × 0.275 = 17.9 kips

$$n = \frac{38}{17.9} > 2 \qquad \text{Use 3 bolts}$$

Length of connection with 3-in. pitch

$$L = 2 \times 3 + 2 \times 1\tfrac{1}{4} = 8\tfrac{1}{2} \text{ in.} > T/2 = 7 \text{ in.}$$

THICKNESS OF CONNECTION ANGLES:
Shear on gross area of vertical section through angles:

$$t = \frac{R/2}{0.4F_y L} = \frac{19}{0.4 \times 36 \times 8.5} = 0.16 \text{ in.}$$

Use ¼ in., which is minimum available

$$\text{Bearing stress} = \frac{38}{2 \times 3 \times 0.75 \times 0.25} = 33.8 \text{ ksi}$$
$$L = \frac{2d}{F_v} F_p = \frac{2 \times 0.75}{58} \times 33.8 = 0.87 \text{ in.} \qquad \text{O.K.}$$

Bolts in outstanding legs:

$$\text{Shear per bolts} = 21 \times 0.442 = 9.28 \text{ kips}$$
$$n = \frac{38}{9.28} = 4.1 \qquad \text{Use 6 bolts (AISC standard connection)}$$

Example 23   Design framed beam connections for a W16 × 36 (span = 22 ft) and a W27 × 94 (span = 25 ft) which frame on opposite sides of a W30 × 99 girder web. (Fig. 40). A36 steel, ¾-in. A325 bearing-type connection with threads in shear planes. AISC recommends that, if contract drawings do not show

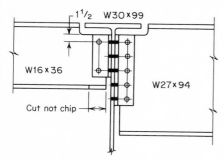

Fig. 40   Example 23.

beam reactions, the connections be designed to support half the total uniform-load capacity of the beam for the given span.

$$R = \frac{W}{2} = \frac{8M}{2L} = \frac{4F_b S}{L}$$

$$W16 \times 36: \quad R = \frac{4 \times 24 \times 56.5}{22 \times 12} = 20.5 \text{ kips}$$

$$W27 \times 94: \quad R = \frac{4 \times 24 \times 243}{25 \times 12} = 77.8 \text{ kips}$$

Bolts in W16 $\times$ 36 web:
Allowable shear per bolt = $2 \times 21 \times 0.442 = 18.6$ kips

$$n = \frac{20.5}{18.6} = 1.1 \qquad \text{Use 2}$$

$$\text{Bearing stress} = \frac{20.5}{2 \times 0.75 \times 0.295} = 46.3 \text{ ksi} < 1.5 F_u$$

$$\text{Min top edge distance} = \frac{46.3 \times 2 \times 0.75}{58} = 1.24 \text{ in.}$$

Use edge distance = $1\frac{1}{2}$ in.

Bolts in W27 $\times$ 94 web:

$$n = \frac{77.8}{18.6} = 4.2 \qquad \text{Use 5}$$

$$\text{Bearing stress} = \frac{77.8}{5 \times 0.75 \times 0.490} = 42.3 \text{ ksi} < 45.7$$

Edge distance of $1\frac{1}{2}$ in. satisfactory

Bolts in W30 $\times$ 99 web:

$$\text{Allowable bearing on web} = 1.5 \times 58 \times 0.75 \times 0.520 = 34.0 \text{ kips}$$

Bolts in W16 $\times$ 36 connection:

$$n = \frac{20.5}{21 \times 0.442} = 2.2 \qquad \text{Use 6 (Fig. 40)}$$

Allowable load per bolt in W27 = allowable bearing on web of W30 − load per bolt in W16 = 34 − 20.5/6 = 30.6 kips

$$\text{Allowable single shear} = 9.28 \text{ kips} < 30.6$$

$$n = \frac{77.8}{9.28} = 8.4 \qquad \text{Use 10}$$

58. Moment-Resistant Beam Connections   The riveted or bolted rigid (Type 1) connection (Art. 39) may take any of several forms, such as built-up brackets, T-stub connections (Fig. 41), flange plates welded directly to the column (Figs. 42 and 43) and end-plate connections (Fig. 44). A vertical connection for beam shear is always a part of the connection.

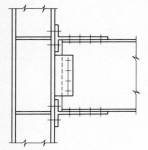

Semirigid connections may consist of seat and top angles, with or without a web connection (Fig. 45), or of structural tees instead of top and bottom angles.

The flexibility of the flange in a T-stub or similar connection may result in a significant "prying" action on the connection. Figure 46 shows the reactive (prying) forces Q between a flange and the part to which it is connected. Thus,

$$T = F + Q$$

Fig. 41   Moment connection.   The magnitude of Q is primarily a function of flange stiffness and would be zero for an infinitely stiff flange. It can be computed from the following formulas:[15]

Connections with A325 bolts:

$$Q = F \frac{100 b d_b^2 - 18 \; w t_f^2}{70 \; a d_b^2 + 21 \; w t_f^2} \tag{49}$$

Connections with A490 bolts:

$$Q = F \frac{100 \; b d_b^2 - 14 \; w t_f^2}{62 \; a d_b^2 + 21 \; w t_f^2} \tag{50}$$

where a = distance from bolt line to edge of flange, not to exceed $2t_f$, in.

b = distance from bolt line to point $\frac{1}{16}$ in. from near face of web, in.

t_f = thickness of T-stub or column flange, whichever is thinner; use average thickness for I-flanges

w = length of T-stub flange tributary to bolt

d_b = nominal bolt diameter, in.

F = externally applied load per bolt

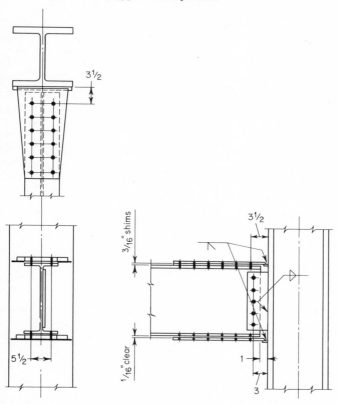

Fig. 42   Moment connection.

Bending moments in the T-stub flange at the critical sections (Fig. 46) are

$$M_1 = (F + Q)b - Q(a + b) \tag{51}$$
$$M_2 = Qa \tag{52}$$

The resulting bending stress, based on the gross width w of the section, should not exceed $0.75F_y$.

Example 24   Design a moment-resistant T-stub connection for a W21 × 62 wind girder which resists the following moments and vertical shears: M_{D+L} = 82 ft-kips, V_{D+L} = 30 kips, M_W = 40 ft-kips, V_W = 10 kips. A36 steel and A325 bolts, friction-type connections, oversize holes, Class A surface (Fig. 41).

$$M = \frac{3}{4}(M_{D+L} + M_W) = \frac{3}{4}(82 + 40) = 91.5 \text{ ft-kips} > 82$$
$$V = \frac{3}{4}(30 + 10) = 30 \text{ kips}$$

W21 × 62:

$$d = 20.99 \text{ in.} \qquad b_f = 8.240 \text{ in.}$$

T-stub: Try ST12 × 52.95 with four $\frac{3}{4}$-in. A325 bolts.

$$b_f = 7.875 \text{ in.} \qquad t_f = 1.102 \text{ in.} \qquad t_w = 0.625 \text{ in.}$$

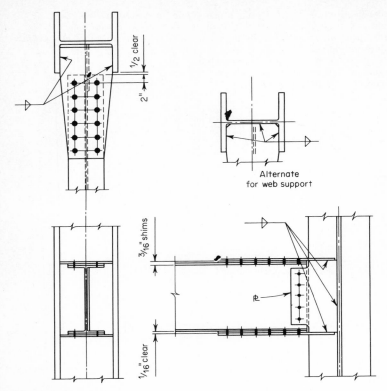

Fig. 43 Moment connection.

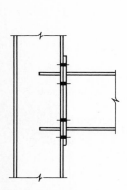

Fig. 44 End-plate moment connection.

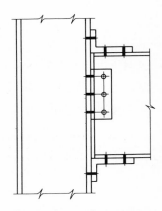

Fig. 45 Semirigid connection.

$$g = 4.5 \text{ in.} \qquad w = (5\tfrac{1}{2} + 2 \times 1\tfrac{1}{4})/2 = 4.25 \text{ in.}$$

$$a = \frac{7.875 - 4.5}{2} = 1.69 \text{ in.} < 2t_f$$

$$b = \frac{4.5 - 0.625}{2} - \frac{1}{16} = 1.88 \text{ in.}$$

$$\frac{Q}{F} = \frac{100 \times 1.88(\tfrac{3}{4})^2 - 18 \times 4.25 \times 1.10^2}{70 \times 1.69(\tfrac{3}{4})^2 + 21 \times 4.25 \times 1.10^2} = 0.076$$

$$F = \frac{91.5 \times 12}{4 \times 20.99} = 13.1 \text{ kips per bolt}$$

$$Q + F = 1.076 \times 13.1 = 14.1 \text{ kips} < 19.4 \, (= 44 \times 0.44)$$

$$Q = 0.076 \times 13.1 = 1.0 \text{ kip}$$

BENDING STRESSES IN T-STUB FLANGE:

At bolt line: $\qquad\qquad M_2 = 1.0 \times 1.69 = 1.69 \text{ in.-kips}$

At fillet: $\qquad\qquad\;\; M_1 = 13.1 \times 1.88 - 1.0 \times 3.57 = 21.1 \text{ in.-kips}$

$$f_b = \frac{6 \times 21.1}{4.25 \times 1.102^2} = 24.5 \text{ ksi}$$

Bolts through T-stub web:
Try ⅞-in.

$$n = \frac{13.2 \times 4}{15 \times 0.601} = 5.8 \qquad \text{Use 6}$$

Bolts through girder web:

$$n = \frac{30}{15 \times 0.601} = 3.3 \qquad \text{Use 4}$$

Example 25 Design an end-plate moment connection (Fig. 47) for a W18 wind girder framed to a W14 × 176 column. The girder moments and shears are: $M_{D+L} = 70$ ft-kips, $V_{D+L} = 25$ kips, $M_W =$

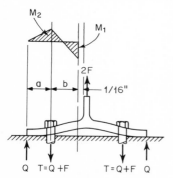

Fig. 46

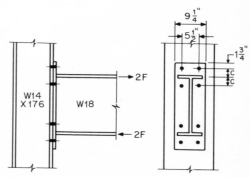

Fig. 47 Example 25.

200 ft-kips, $V_W = 20$ kips. Use $F_y = 50$ ksi steel and A490 bolts, friction-type, standard hole, Class A surface. E70 electrodes.

$$\tfrac{3}{4}(M_{D+L} + M_W) = \tfrac{3}{4}(70 + 200) = 202.5 \text{ ft-kips} > 70 \text{ ft-kips}$$

$$\tfrac{3}{4}(V_{D+L} + V_W) = \tfrac{3}{4}(25 + 20) = 33.75 \text{ kips} > 25 \text{ kips}$$

$$S = \frac{202.5 \times 12}{33} = 73.6 \text{ in.}^3$$

Assume W18 × 50:

$$S = 88.9 \text{ in.}^3 \qquad b_f = 7.50 \text{ in.} \qquad t_f = 0.570 \text{ in.}$$

$$d = 17.99 \text{ in.} \qquad t_w = 0.355 \text{ in.}$$

$$\text{Force in flange } 2F = \frac{202.5 \times 12}{17.99 - 0.57} = 139.5 \text{ kips}$$

Allowable flange force $7.5 \times 0.57 \times 33 = 141.1 > 139.5$ kips.

Use W18 × 50

$$\text{Flange weld} = \frac{139.5}{(2 \times 7.5 + 2 \times 0.57 - 0.355) \times 0.928} = \text{\%-in. fillet}$$

END PLATE: Use 9¼-in. plate (Fig. 47). Try four 1⅛-in.-diameter bolts at the top flange. Minimum bolt clearance = 1⁷⁄₁₆ in.

$$c = \frac{0.57}{2} + 0.625 + 1\tfrac{7}{16} = 2.35 \text{ in.} \qquad \text{Use } 2\% \text{ in.}$$

$$b = 2.375 - \frac{0.57}{2} - \frac{1}{16} = 2.028 \text{ in.}$$

Assuming $M_1 = M_2$ (Fig. 46) in Eqs. (51) and (52),

$$Qa = (F + Q)b - Q(a + b) \qquad Q = \frac{Fb}{2a}$$

$$M_2 = M_1 = Qa = \frac{Fb}{2}$$

$$S = \frac{wt^2}{6} = \frac{Fb}{2f_b}$$

$$t = \sqrt{\frac{3Fb}{wf_b}} = \sqrt{\frac{3 \times 69.7 \times 2.028}{9.25 \times 37.7}} = 1.11 \text{ in.}$$

Try 1⁵⁄₁₆-in. thick plate
From Eq. (50),

$$\frac{Q}{F} = \frac{100 \times 2.028 \times 1.125^2 - 14 \times 4.625 \times 1.313^2}{62 \times 1.75 \times 1.125^2 + 21 \times 4.625 \times 1.313^2} = 0.476$$

$$F = \frac{139.5}{4} = 34.9 \text{ kips} \qquad Q = 0.476 \times 34.9 = 16.6 \text{ kips}$$

$$T = Q + F = 34.9 + 16.6 = 51.5 \text{ kips} < 53.7 \text{ kips} \, (= 54 \times 0.994)$$

Check bending stress in end plate:

$$M_2 = Qa = 16.6 \times 1.75 = 29.1 \text{ in.-kips}$$
$$M_1 = (Q + F)b - Q(a + b)$$
$$= 51.5 \times 2.03 - 16.6 \times 3.78 = 41.8 \text{ in.-kips}$$
$$f_b = \frac{6 \times 41.8}{4.625 \times 1.313^2} = 31.5 \text{ ksi} < 37.5 \text{ ksi}$$

Use end plate 9¼ × 1 × 2—1¾.
Shear on bolts:

$$f_v = \frac{33.75}{8 \times 0.994} = 4.24 \text{ ksi} < 22 \text{ ksi}$$

Since the wind moment is reversible and large in comparison with the gravity moment, provide four 1⅛-in. A490 bolts at the bottom flange.

WELDS: Use ⅝-in. fillet weld all around the top and bottom flanges.
Use ⁵⁄₁₆-in. fillet weld both sides of web

$$\text{Length of weld} = \frac{33.75}{2 \times 21 \times 0.707 \times \tfrac{5}{16}} = 3.63 \text{ in.}$$

$$\text{Web length} = \frac{33.75}{14.5 \times 0.358} = 6.50 \text{ in.}$$

Use ⁵⁄₁₆-in. fillet × 7-in. length both sides
COLUMN WEB STIFFENERS:

W14 × 176 $t_f = 1.310$ in. $t_w = 0.830$ in.
$\phantom{\text{W14 × 176}}$ $d_c = 15.22$ in. $k = 2.0$ in.

Opposite tension flange (AISC 1.15.5.2, 1.15.5.3):

$$P_{bf} = \frac{\tfrac{4}{3}(200 + 70) \times 12}{17.99 - 0.57} = 248 \text{ kips}$$

$$A_{st} = \frac{248 - 50 \times 0.830 (1.25 + 5 \times 2)}{36} = \text{negative}$$

$$0.4\sqrt{248/50} = 0.89 \text{ in.} < t_f$$

No stiffeners required.

Opposite compression flange (AISC 1.15.5.3):

$$\frac{4100 \times 0.830^3 \sqrt{50}}{248} = 66.8 \text{ in.} > d_c = 15.22 - 2 \times 2$$

No stiffeners required.

$$\text{Shear in column web} \approx \frac{139.4}{15.22 \times 0.83} = 11.0 \text{ ksi} < F_v$$

59. Pinned Connections The primary reason for using pinned connections is to allow relatively free end rotation of connected members. Pinned joints in bascule bridges, crane booms, etc., must permit relatively large rotations and must be sufficiently lubricated to preclude rusting and minimize wear. Pinned joints where the associated rotations, normally caused by elastic deformations, are relatively small are found in light bracing systems (clevis pins), hinged arches, links between the cantilever span and the suspended span in cantilever systems, and main supports—sometimes integrated with expansion joints and rocker supports—of heavy trusses and girders.

Steel pins—forged, cast, or cold-rolled and machined to the desired dimensions—generally range in size from 1¼ to 10 in. in diameter; sizes up to 24 in. are also available. Small pins—up to about 4 in. in diameter—may be made with cotter-pin holes at each end, or with a thin flat head at one end and a cotter-pin hole at the other. Pins up to about 10 in. in diameter usually have threaded end projections for recessed retainer nuts. Pins over 10 in. in diameter are usually held in position by a recessed cap at each end and secured by a bolt passing completely through the caps and pin. It should be noted that the designer is not limited to standard sizes, since machining to nonstandard sizes can readily be done.

Pins are cylindrical beams, and bending requirements normally govern. However, the associated failure modes are basically the same as those of mechanically fastened joints. Because of the shape factor of the circular cross section, the allowable bending stress for pins is higher than for ordinary shapes. Allowable shear stresses are about the same as for rivets.

To ensure a uniform distribution of bearing stress over the area of contact, the pin must fit snugly in the pinhole. The AISC specification requires that the diameter of the pinhole be not more than ¹⁄₃₂ in. greater than the diameter of the pin. It may be noted that the required thickness of the parts in contact is inversely proportional to the pin diameter. Thus, a relatively thin bearing may require an excessively large pin. To avoid this situation, members may be reinforced with pin plates at the pinhole.

The allowable bearing stress for pins is somewhat less than that for rivets and bolts; in effect, this allows freer rotation. When rotation is a factor, a high bearing stress would produce undesirable wear. Where large rotations are expected, the AASHTO specification reduces the usual allowable pin bearing stress by 50 percent.

Net sections across and beyond the pinhole must be considered. In addition to net-section failure at the pin, dishing of the plate beyond the pin may occur. Specifications suggest empirical rules which control the geometry of the components near the pinhole. Thus, the AISC specification requires that the allowable tensile stress on the net section transverse to the axis of the member be reduced 25 percent at pinholes in pin-connected plates or built-up members. The net section beyond the pinhole, parallel to the axis of the member, must be not less than two-thirds of the transverse net section at the pinhole. Somewhat different requirements pertain to the design of eyebars. The AASHTO specification requires that the net section across the pinhole be not less than 140 percent and the net section back of the pinhole not less than 100 percent of the net section of the body of the member. The AREA has similar requirements. Furthermore, in order to prevent the use of noncompact sections and large pin diameters, the AISC, AASHTO, and AREA specifications all stipulate, in effect, that the transverse net width at the pinhole be not more than eight times the thickness of the member or element.

Pin plates must be extended beyond the pinhole and connected to the main member by sufficient fasteners or welds. Usually, a pin plate is assumed to transmit a fraction of the main member force proportional to its thickness.

The design of a pinned connection is a trial-and-error procedure, since the pin diameter depends upon the thickness and location of the connected parts, which in turn are governed by the size of the pin. The first step is generally to arrange the members efficiently on the pin (packing) so as to minimize bending and shear. This may have to be done in conjunction with the detailing of other joints in the structure. Clearances must be provided to allow for erection, free rotation, and general maintenance. It is generally assumed that the applied loads act at the centers of the bearing thicknesses. If the parts are relatively thick, this may result in a grossly conservative design, and it may be desirable to consider the loads as distributed.

Example 26 Design a pin connection for the end panel point of a truss (Fig. 48). The diagonal consists of two 15-in. 50-lb channels with the toes out and carries 500 kips. The bottom chord consists of

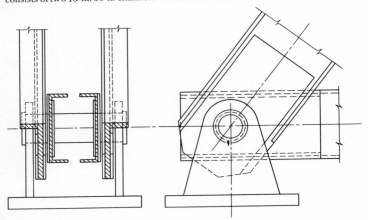

Fig. 48 Example 26.

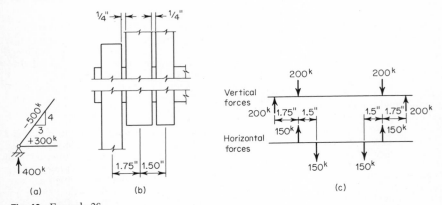

Fig. 49 Example 26.

two 12-in. 25-lb channels with the toes in and carries 300 kips. Use A235 Class E steel ($F_y = 37.5$ ksi) for pin and A36 steel for other components. AISC specification.

C15 × 50:	$t_w = 0.716$ in.	
C12 × 25:	$t_w = 0.387$ in.	$A = 7.35$ in.²

Moment and shear on pin

$$F_V = 400 \text{ kips} \qquad F_H = 300 \text{ kips} \qquad \text{(Fig. 48a)}$$

Assume 1.75 in. between centerlines of shoe and 15-in. channel web and 1.5 in. between centerlines of webs of 15- and 12-in. channels (Fig. 49b and c).

$$M_V = 200 \times 1.75 = 350 \qquad M_H = 150 \times 1.5 = 225 \text{ in.-kips}$$
$$M = \sqrt{350^2 + 225^2} = 416 \text{ in.-kips}$$
$$V = 200 \text{ kips}$$

Pin diameter:

$$F_b = 0.90 F_y = 0.90 \times 36.0 = 32.4 \text{ ksi}$$
$$F_v = 0.40 F_y = 0.40 \times 37.5 = 15.0 \text{ ksi}$$

$$d = \sqrt[3]{\frac{32M}{\pi F_b}} = \sqrt[3]{\frac{32 \times 416}{32.4\pi}} = 5.08 \text{ in.}$$

$$d = \sqrt{\frac{4V}{\pi F_v}} = \sqrt{\frac{4 \times 200}{15\pi}} = 4.11 \text{ in.}$$

Try 5-in. pin.

Shoe and pin plate:

Bearing on A36:
$$F_p = 0.90 \times 36 = 32.4 \text{ ksi}$$

Shoe:
$$t = \frac{200}{32.4 \times 5} = 1.24 \text{ in.} \qquad \text{Use } 1\tfrac{1}{4} \text{ in.}$$

Diagonal:
$$t = \frac{250}{32.4 \times 5} = 1.54 \text{ in.}$$

Pin plates:
$$t = 1.54 - 0.716 = 0.82$$

Try $\tfrac{7}{16}$-in. pin plates, one each side of web.

Bottom chord:
$$t = \frac{150}{32.4 \times 5} = 0.927$$

Pin plates:
$$t = 0.927 - 0.387 = 0.540 \text{ in.}$$

Try $\tfrac{5}{16}$-in. pin plates, one each side of web.

Check 5-in. pin.

Diagonal: At pinhole $t = 0.716 + 0.875 = 1.591$ in.
Bottom chord: At pinhole $t = 0.387 + 0.625 = 1.012$ in.

Assume $\tfrac{1}{4}$-in. clearance.

$$\text{Shoe and diagonal c.c.} = \frac{1.25}{2} + 0.25 + \frac{1.591}{2} = 1.67 \approx 1.75 \text{ in.}$$

$$\text{Diagonal and chord c.c.} = \frac{1.591}{2} + 0.25 + \frac{1.012}{2} = 1.55 \approx 1.50 \text{ in.}$$

Bottom chord at pinhole:

$$F_t = 0.45 F_y = 0.45 \times 36 = 16.2 \text{ ksi}$$
$$A = \frac{150}{16.2} = 9.26 \text{ in.}^2 \text{ net}$$
$$\text{Max width inside plate} = T = 9\tfrac{7}{8} \text{ in.}$$

Make inside plate $9\tfrac{3}{4}$ in., outside plate 12 in.

$$A = 7.32 + 9.75 \times 0.312 + 12 \times 0.312 - 5.03(0.387 + 2 \times 0.313)$$
$$= 9.01 < 9.26 \text{ in.}^2$$

Make outside plate $12 \times \tfrac{3}{8}$, inside plate $9\tfrac{3}{4} \times \tfrac{5}{16}$.

$$A = 9.50 > 9.26 \text{ in.}^2$$

Edge distance:

$$\text{Edge distance} \gtrless 4t \qquad \text{(AISC 1.14.5)}$$
$$12 - 5.03 = 6.97 \text{ in.} < 2 \times 4(0.375 + 0.387 + 0.312) = 8.60 \text{ in.}$$

Net section beyond hole:

$$\tfrac{2}{3} \times 9.26 = 6.17 \text{ in.}^2 \text{ required}$$
$$\frac{6.17}{0.375 + 0.387 + 0.312} = 5.74 \text{ in.}$$
$$5.74 + \frac{5.03}{2} = 8.25 \text{ in.}$$

Use $8\tfrac{1}{2}$-in. end distance.

BEARING PLATES AND SPLICES

60. Beam Bearing Plates These are normally provided under beams which are supported directly on masonry or concrete; in certain cases it is possible to eliminate them. Erection of beams is also facilitated since the bearing plates can be positioned and grouted to specific elevations beforehand. This accounts for their use even where they are not required for stress. The distribution of bearing pressures is complex and is usually assumed to be distributed uniformly over the bearing area.

The plate area must be large enough to keep the bearing pressures within allowable limits, thick enough to maintain the maximum bending stress within the allowable value, and long enough to satisfy web-crippling requirements. The AISC method assumes the critical section for plate bending to be at the distance k from the center of the web, and any pressures exerted on the plate by the beam flanges are neglected.

Beam-anchorage details will depend upon the type of construction used. If masonry extends above the beam support, a government anchor (a bent rod) or short clip angles may be attached to the web. Anchor bolts, bonded to the masonry or concrete, may be used if longitudinal loads are anticipated.

Example 27 Using the AISC recommendations, design a bearing plate for a W24 × 76 beam with a reaction of 47 kips. The available length C for longitudinal bearing is 5 in. Use A36 steel and 3000-psi concrete. Assume the base plate to bear on full area of concrete support (Fig. 50).

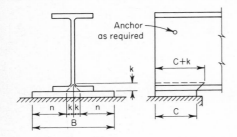

Fig. 50   Beam bearing plate.

Allowable stresses:

$$F_b = 0.75F_y = 0.75 \times 36 = 27.0 \text{ ksi}$$
$$F_p = 0.35f'_c = 0.35 \times 3 = 1.05 \text{ ksi}$$
$$A = \frac{R}{F_p} = \frac{47}{1.05} = 44.8 \text{ in.}^2$$
$$B = \frac{A}{C} = \frac{44.8}{5} = 9 \text{ in.}$$

Actual bearing pressure

$$f_p = \frac{R}{B \times C} = \frac{47}{9 \times 5} = 1.04 \text{ ksi}$$
$$n = \frac{B}{2} - k = \frac{9}{2} - 1\tfrac{1}{4} = 3.25 \text{ in.}$$
$$t = \sqrt{\frac{3f_p n^2}{F_b}} = \sqrt{3 \times 1.04 \times \frac{3.25^2}{27}} = 1.10 \text{ in.}$$

Check web crippling on length $C + k$

$$\frac{R}{(C + k)t_w} \gtrless 0.75F_y = \frac{47}{(5 + 1.25)0.440} = 17.1 < 27 \text{ ksi}$$

Use plate 5 × 1 × 0—9.

61. Column Bases Column bases may be classified in three basic categories: (1) axial load only; (2) axial load plus a relatively small moment; and (3) axial load plus a relatively large moment. In the first case, and sometimes in the second case, the loading produces only compressive stresses between the foundation and the base plate. It may be noted that, with this type of distribution, some degree of fixity is realized, and anchorage requirements are essentially a matter of judgment. When compressive stresses are not present over the entire area of the base plates, the resulting tensile forces may be quite small, so that nominal anchorage, properly oriented, is usually more than adequate to sustain the tensile forces. When the tensile forces are large, the reinforced-concrete beam analogy affords a practical design approach; the bearing pressure at the edge of the compressive zone is taken equal to the allowable bearing pressure of the foundation, and the tension determined accordingly.

Although base plates can be attached to the columns in the shop, they are usually shipped separately and threaded over preset anchor bolts before grouting. Small column bases may be shop-welded to the columns, in which case greater care must be exercised in handling.

Leveling techniques vary depending upon the weight of the base plates. For example, if the plates weigh 400 lb or less, ordinary shim packs or thin leveling plates are generally used. However, if the plates weigh more than 400 lb, nut-and-bolt-type leveling devices are normally required to allow for the necessary adjustments.

Typical column bases are shown in Fig. 51.

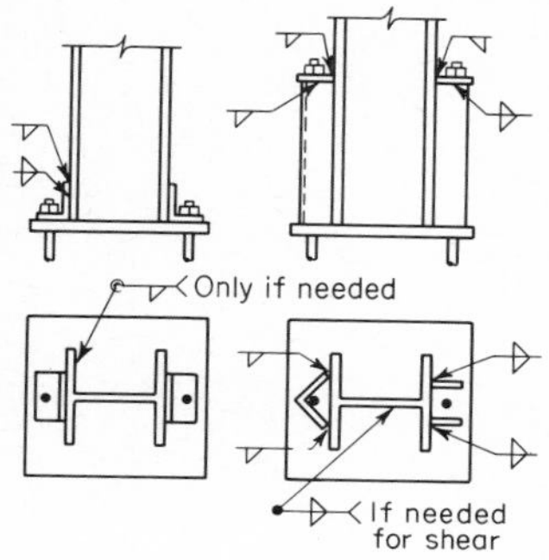

Fig. 51   Column-base-plate details.

Example 28   Design a moment-resistant anchorage and base plate for a W12 × 96 column with an axial load of 15 kips and a moment of 200 ft-kips. The base plate will rest on a 3000-psi concrete foundation. AISC specification, A36 steel, and E60 electrodes.

W12 × 96: $d = 12.71$ in. $b = 12.16$ in.

Assume 4-in. stiffener plates with anchor bolts located 2.5 in. from outside face of column flanges (Fig. 52).

Assume $\dfrac{B}{C} = \dfrac{0.80b_f}{0.95(d + 2 \times 2.5)} = \dfrac{0.80 \times 12.16}{0.95(12.71 + 5)} = 0.578$

Assume $C = d + 2 \times 2.5 + 2 \times 5 = 27.71$ in.

$B = 0.578 \times 27.71 = 16$ in.

Try $B = 16$ in., $C = 28$ in.

STRESS IN ANCHOR BOLT:

Assume $A_2/A_1 > 4$ (AISC 1.5.5).

Allowable pressure $f_1 = 0.7f'_c = 0.7 \times 3 = 2.1$ ksi

$$\Sigma F = 0 = T + 15 - 0.5f_1kdB \qquad T = 16.8kd - 15$$

$$\Sigma M = 0 = 8.88T + 0.5f_1kdB\left(\frac{C}{2} - \frac{kd}{3}\right) - 200 \times 12$$

$$= 8.88T + 235.2kd - 5.6(kd)^2 - 2400 = 0$$

$$kd = 7.38 \text{ in.} \qquad T = 109 \text{ kips}$$

Anchor bolts, $F_t = 20$ ksi

$$A = \frac{109}{20} = 5.45 \text{ in.}^2$$

Use two 2-in. bolts, $A = 6.3$ in.^{2}

BASE PLATE: Assume critical moment at bolt line (Fig. 52).

$$m = 14 - \frac{12.71}{2} - 2.5 = 5.15 \text{ in.}$$

$$f_2 = \frac{5.15f_1}{7.38} = 1.46 \text{ ksi}$$

$$M = \frac{1}{2}f_1 \times 5.15^2 \times \frac{2}{3} + \frac{1}{2}f_2 \times \frac{5.15^2}{3} = 25.0 \text{ in.-kips/in.}$$

$$M = 25.0B \times 16 = 400 \text{ in.-kips}$$

Net width $B' = B - 2 \times 2.75 = 10.5$ in.

$$t^2 = \frac{6 \times 400}{10.5 \times 0.75F_y} = 8.47 \qquad t = 2.91 \text{ in.}$$

Use $16 \times 3 \times 2$—4 plate

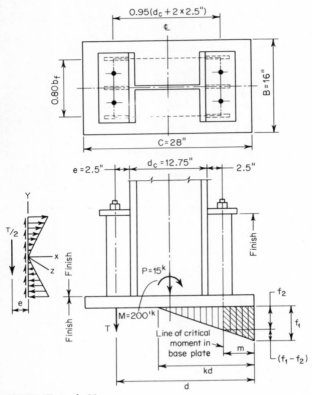

Fig. 52 Example 28.

STIFFENER WELDS: Load on center stiffener (Fig. 52) $\approx T/2 = 54.5$ kips. Assume unit-width fillet welds, length L

$$M = 54.5 \times 2.5 = 136 \text{ in.-kips}$$

$$\text{Shear due to flexure } q_s = \frac{136}{2L^2/6} = \frac{408}{L^2} \text{ kips/in.}$$

$$\text{Vertical shear } q_v = \frac{54.5}{2L} = \frac{27.3}{L} \text{ kips/in.}$$

$$\text{Resultant shear} = \sqrt{0.2q_x^2 + q_v^2} \qquad \text{(Ref. 16)}$$
$$q = 12.73 \text{ kips/in. allowable}$$

$$12.73D = \sqrt{0.2\left(\frac{408}{L^2}\right)^2 + \left(\frac{27.3}{L}\right)^2}$$

Column flange thickness $= 0.900$ in., minimum fillet weld $= \frac{5}{16}$ in.

$$q = \frac{5}{16} \times 12.73 = 3.98 \text{ kips/in.}$$

$$3.98 = \sqrt{0.2\left(\frac{408}{L^2}\right)^2 + \left(\frac{27.3}{L}\right)^2}$$
$$L = 8.66 \text{ in.}$$

Use ⁵⁄₁₆-in. fillet weld each side of stiffener plate of length 9 in.

SIZE OF STIFFENER: Equating shear on stiffener to shear on welds:

$$F_v t = 2 \times 12.73 D$$

$$t = 2 \times 12.73 \times \frac{0.3125}{14.5} = 0.549 \text{ in.}$$

Try ⁹⁄₁₆ in., $b/t = 4/0.5625 = 7.1 < 16$

Use $4 \times \text{⁹⁄₁₆} \times 0{-}9$ stiffeners, three each side.

62. Compression-Member Splices If the ends of axially loaded compression members which are to be spliced are finished to obtain uniform bearing between the two, the connection serves primarily to hold the two in line. It is customary to specify that there be enough rivets, bolts, or welds to hold the parts securely in place, or to require that the splice be proportioned for a specified portion of the load.

Tier buildings are normally erected in two-story lifts with splices about 2 ft above the finished floor. This allows for clearance of beam-to-column connections and facilitates erection. When such columns are finished to bear, it would seem reasonable to take advantage of the potential friction at the contact surfaces in considering the horizontal shear.

Typical tier-building splices are shown in Fig. 53.

Example 29 Design a tier-building column splice between a W12 × 65 and a W12 × 96 whose contact surfaces are to be finished to bear. The design forces and moments at girder-column intersections are:

	DL	LL	WL
P, kips	133	70	10
M_x, ft-kips	40	20	100

The story height is 15 ft, with the splice 3 ft above the girder centerline (Fig. 54). A36 steel, A325 bolts, friction-type connection, standard holes, Class A surface condition. AISC specification.

W12 × 65:
W12 × 96: $\quad d = 12.12 \quad\quad b = 12.0 \quad\quad t_f = 0.605$
$ \quad t_w = 0.390 \quad\quad A = 19.11 \text{ in.}^2 \quad S_x = 87.9 \text{ in.}^3$
$ \quad d = 12.71$

MOMENT SPLICE:
Gravity load:

$$\text{Moments at splice} = \frac{54}{90} \times \text{moments at floor}$$

$$M_D + M_L = 24 + 12 = 36 \text{ ft-kips}$$
$$P_D + P_L = 133 + 70 = 207$$
$$f_G = -\frac{207}{19.1} \pm \frac{36 \times 12}{87.9} = -10.84 \pm 4.91$$
$$= -15.75, \ -5.93 \text{ ksi (no tension)}$$

Gravity plus wind (AISC 1.15.8):

$$M_D + M_L + M_W = 24 + 12 + 60 = 96 \text{ ft-kips}$$
$$0.75 P_D - P_W = 0.75 \times 133 - 10 = 90 \text{ kips}$$
$$f_{GW} = -\frac{90}{19.1} \pm \frac{96 \times 12}{87.9} = -4.71 \pm 13.11$$
$$= -17.82, \ +8.40 \text{ ksi}$$

Locate neutral axis (Fig. 54).

$$c_T = d \frac{f_2}{f_1 + f_2} = \frac{12.12 \times 8.38}{17.80 + 8.38} = 3.88 \text{ in.}$$
$$f_i = \frac{8.38(3.88 - 0.605)}{3.88} = 7.07 \text{ ksi}$$

Tension in flange = ½ (7.07 + 8.38) × 12 × 0.605 = 56.18
Tension in web = ½ × 7.07(3.88 − 0.605) × 0.390 = 4.51

$$T = 60.69 \text{ kips}$$

To account for ⅓ increase in allowable stress, $T = \text{¾} \times 60.69 = 45.5$ kips.

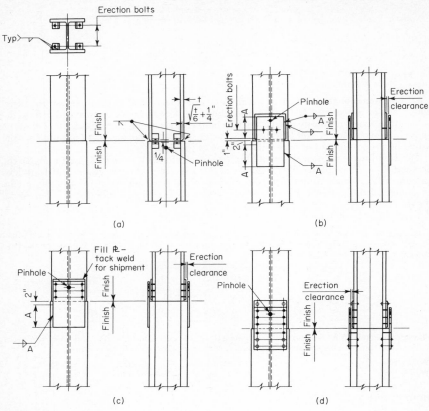

Fig. 53 Column splices.

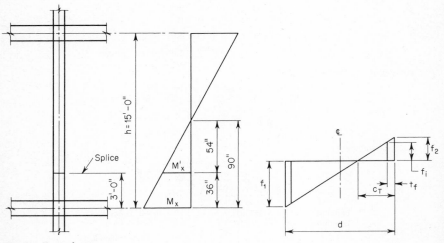

Fig. 54 Example 29.

Try ¾-in. A325 bolts (friction-type connection).

$$F_v A_b = 17.5 \times 0.442 = 7.73 \text{ kips}$$
$$n = \frac{45.5}{7.73} = 5.9 \qquad \text{Use } 6$$

FLANGE SPLICE PLATES:

$$A = \frac{45.5}{22} = 2.07 \text{ in.}^2$$

Min width $b = 8$ in. for $g = 5\frac{1}{2}$ in.
Net width $8 - 2 \times \frac{7}{8} = 6.25$ in.

$$t = \frac{2.07}{6.25} = 0.331 \text{ in.}$$

Use two $8 \times \frac{3}{8} \times 1$—6 splice plates (Fig. 55).

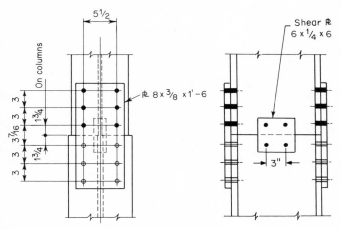

Fig. 55 Example 29.

FILLER PLATES:

$$t = \frac{1}{2}(12.75 - 12.12) - \frac{1}{16} = 0.253 \text{ in.}$$

Use two $8 \times \frac{1}{4} \times 0$—9 filler plates; tack weld to upper shaft.

SHEAR SPLICE
    Gravity load:

$$\text{Frictional resistance } F_G = \frac{\text{slip coefficient} \times \text{axial load}}{\text{factor of safety}}$$
$$= \frac{0.25(133 + 70)}{1.65} = 30.8 \text{ kips}$$
$$\text{Shear} = \frac{M_D + M_L}{h/2} = \frac{40 + 20}{7.5} = 8.0 < 30.8$$

    Gravity plus wind:

$$F_{GW} = \frac{0.25(0.75 \times 133 - 10)}{1.65} = 13.6 \text{ kips}$$
$$\text{Shear} = \frac{3}{4} \times \frac{M_D + M_W}{h/2} = \frac{3}{4} \times \frac{40 + 100}{7.5} = 14.0 < 13.6$$

Use two $6 \times \frac{1}{4} \times 6$ shear plates connected to the webs with four ¾-in. A325 bolts.

Example 30   Design a tier-building column splice between two W14 × 145's to transfer 50 percent of the computed stress. Neglect friction. The design forces and moments at girder-column intersections are:

	DL	LL	WL_x	WL_y
P, kips	300	100	100	15
M_x, ft-kips	50	30	180	
M_y, ft-kips	25	15		55

Story height is 12 ft 8 in., and the splice is 32 in. above the girder centerline. A36 steel, E60 electrodes, AISC specification. Calculations are shown for the critical case, viz., dead load plus live load plus wind with respect to the x axis.

W14 × 145: $\qquad A = 42.7$ in.$^2 \quad d = 14.78 \quad b = 15.50$
$\qquad\qquad\qquad S_x = 232$ in.$^3 \quad t_w = 0.680 \quad t_f = 1.090$
$\qquad\qquad\qquad S_y = 87.3$ in.^{3}

Forces at splice for x-axis loading:

$$\text{Moments at splice} = \frac{44}{76} \times \text{moments at floor}$$

$$M_D + M_L + M_{Wx} = (50 + 30 + 180)\frac{44}{76} = 150.4 \text{ ft-kips}$$

$$P_D + P_L + P_{Wx} = 300 + 100 + 100 = 500 \text{ kips}$$

To account for the ⅓ increase in allowable stress, multiply loads by ¾.

$$f = -\frac{3}{4} \times \frac{500}{42.7} \pm \frac{3}{4} \times \frac{150.4 \times 12}{232} = -8.78 \pm 5.84$$
$$= -14.62, -2.94 \text{ ksi}$$

Flange force F_1 (Fig. 56):

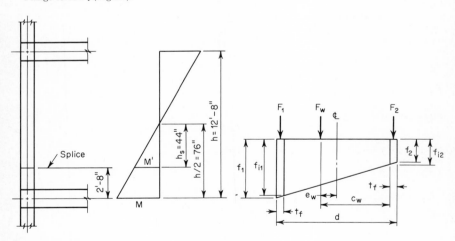

Fig. 56   Example 30.

$$f_{i1} = 14.62 - 11.68 \times \frac{1.090}{14.78} = 13.76 \text{ ksi}$$

$$F_1 = \frac{14.62 + 13.76}{2} \times 15.50 \times 1.090 = 242 \text{ kips}$$

Flange force F_2:

$$f_{i2} = 2.94 + 11.68 \times \frac{1.090}{14.78} = 3.80 \text{ ksi}$$

$$F_2 = \frac{2.94 + 3.80}{2} \times 15.50 \times 1.090 = 56.9 \text{ kips}$$

Axial force F_w in web:

$$F_w = \frac{13.76 + 3.80}{2} \times 12.60 \times 0.680 = 75 \text{ kips}$$

Check: $242 + 57 + 75 = 374$ kips $0.75 \times 500 = 375$ kips

Eccentricity e_w of F_w (Fig. 56)

$$c_w = \frac{12.60}{3} \times \frac{2 \times 13.76 + 3.80}{13.76 + 3.80} = 7.49$$

$$e_w = 7.49 - 6.30 = 1.19 \text{ in.}$$

Shear force S_w in web

$$S_w = \frac{3}{4} \times \frac{50 + 30 + 180}{12.67/2} = 30.8 \text{ kips}$$

Forces at splice for y-axis loading

$$\text{Moments at splice} = \frac{44}{76} \times \text{moments at floor}$$

$$M_D + M_L = (25 + 15)\frac{44}{76} = 23.2 \text{ ft-kips}$$

Shear force S_f in each flange

$$S_f = \frac{1}{2} \times \frac{3}{4} \times \frac{25 + 15}{12.67/2} = 2.37 \text{ kips}$$

FLANGE SPLICE:
    Design for 50 percent:

$$V_f = 0.5F_1 = 0.5 \times 239 = 120 \text{ kips}$$
$$H_f = S_f = 2.37 \text{ kips}$$
$$M_f = 0.5 \times 0.75 \times 23.2 = 8.7 \text{ ft-kips}$$

Assume 12-in. splice plate 23 in. long (Fig. 57).

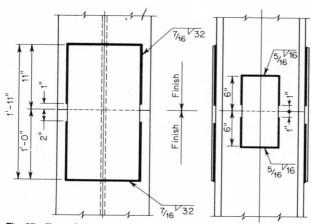

Fig. 57 Example 30.

Weld group on upper shaft (Fig. 58).
Assume unit-width welds.

$$\bar{y} = \frac{2 \times 10 \times 5}{2 \times 10 + 12} = 3.13 \text{ in.}$$

$$I_x = 2 \times \frac{10^3}{12} + 2 \times 10 \times 1.87^2 + 12 \times 3.13^2 = \quad 354$$

$$I_y = \frac{12^3}{12} + 2 \times 10 \times 6^2 \qquad\qquad\qquad = \quad 864$$

$$J = \overline{1218} \text{ in.}^4/\text{in.}$$

Maximum stress occurs at A.

$$q_x = -\frac{H_f}{32} = -\frac{2.37}{32} = -0.074 \text{ kip/in.}$$

$$q_y = \frac{V_f}{32} = \frac{120}{32} = 3.75 \text{ kips/in.}$$

$$M_z = M_f - 7.87 H_f = 8.7 \times 12 - 2.37 \times 7.87 = 85.7 \text{ in.-kips}$$

$$q'_x = \frac{M_z c_y}{J} = 85.7 \times \frac{6.87}{1218} = 0.48 \text{ kip/in.}$$

$$q'_y = \frac{M_z c_x}{J} = 85.7 \times \frac{6}{1218} = 0.42 \text{ kip/in.}$$

$$q = \sqrt{(-0.074 + 0.48)^2 + (3.75 + 0.42)^2} = 4.2 \text{ kips/in.}$$

$$q = 12.73a \text{ allowable}$$

$$a = \frac{4.2}{12.73} = 0.33 \text{ in.} \qquad \text{Use } \tfrac{3}{8}\text{-in. welds}$$

Try two flange splice plates $12 \times \frac{5}{8} \times 1$—11.

$$f_a = \frac{120}{12 \times 0.625} = 16.0 \text{ ksi}$$

$$f_b = \frac{8.7 \times 12 \times 6}{0.625 \times 12^2} = 6.96 \text{ ksi}$$

$$\frac{f_a}{0.6F_a} + \frac{f_b}{F_b} = \frac{16.0}{22} + \frac{6.96}{27} = 0.98 < 1.0$$

$$f_v = \frac{2.37}{12 \times 0.625} = 0.32 \text{ ksi} < F_v$$

Use two flange plates $12 \times \frac{5}{8} \times 1$—11.

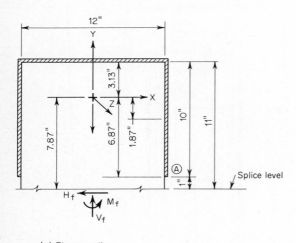

(a) Flange splice   (b) Web splice

Fig. 58   Example 30.

Web splice (Figs. 57 and 58)

$$V_w = 0.5F_w = 0.5 \times 77 = 38.5 \text{ kips}$$

$$H_w = S_w = 30.8 \text{ kips}$$

$$M_w = 0.5F_w e_w = 0.5 \times 77 \times 1.20 = 46.2 \text{ in.-kips}$$

Assume 6-in. plates 12 in. long. Weld group one side of web.

$$\bar{y} = \frac{2 \times 5 \times 2.5}{2 \times 5 + 6} = 1.56 \text{ in.}$$

$$I_x = 2 \times \frac{5^3}{12} + 2 \times 5 \times 0.94^2 + 6 \times 1.56^2 = \quad 44$$

$$I_y = \frac{6^3}{12} + 2 \times 5 \times 3^2 \qquad\qquad\qquad = 108$$

$$J = \overline{152} \ \text{in.}^4/\text{in.}$$

Maximum stress occurs at B.

$$q_x = \frac{H_w}{2A} = \frac{30.8}{32} = 0.96 \ \text{kip/in.}$$

$$q_y = \frac{V_w}{2A} = \frac{38.5}{32} = 1.20 \ \text{kip/in.}$$

$$M_z = 0.5 \times 30.8 \times 4.44 - 0.5 \times 46.2 = 45.3 \ \text{in.-kips}$$

$$q'_x = \frac{M_z c_y}{J} = 45.3 \times \frac{3.44}{152} = 1.03 \ \text{kip/in.}$$

$$q'_y = \frac{M_z c_x}{J} = 45.3 \times \frac{3}{152} = 0.89 \ \text{kip/in.}$$

$$q = \sqrt{(0.96 + 1.03)^2 + (1.20 + 0.89)^2} = 2.89 \ \text{kip/in.}$$

$$a = \frac{2.89}{12.73} = 0.227 \ \text{in.} \qquad \text{Use } \tfrac{1}{4} \text{ in.}$$

Use two web splice plates $6 \times \frac{5}{16} \times 1$—0.

$$f_a = \frac{38.5}{2 \times 6 \times 0.312} = 10.27 \ \text{ksi}$$

$$f_{bx} = \frac{46.2 \times 6}{2 \times 0.312 \times 6^2} = 12.32 \ \text{ksi}$$

$$f_v = \frac{30.8}{2 \times 0.312 \times 6} = 8.21 < F_v$$

$$\frac{10.27}{27} + \frac{12.32}{27} = 0.92 < 1$$

Use two web splice plates $6 \times \frac{5}{16} \times 1$—0.

REFERENCES

1. Tall, L., et al: "Structural Steel Design," 2d ed., The Ronald Press Company, New York, 1974.
2. Gaylord, E. H., and C. N. Gaylord: "Design of Steel Structures," 2d ed., McGraw-Hill Book Company, New York, 1972.
3. Structural Stability Research Council: "Guide to Stability Design Criteria for Metal Structures," 3d ed., John Wiley & Sons, Inc., New York, 1976.
4. Bleich, F.: "Buckling Strength of Metal Structures," McGraw-Hill Book Company, New York, 1952.
5. LeMessurier, W. J.: A Practical Method of Second Order Analysis, *AISC Eng. J.*, vol. 13, no. 4, 1976, vol. 14, no. 2, 1977.
6. Winter, G., et al.: "Stability," vol. SB, chap. 16, Structural Design of Tall Steel Buildings, ASCE-IABSE International Conference on the Planning and Design of Tall Buildings, 1972.
7. Lim, L. C., and R. McNamara: Stability of Novel Building Systems, *Rept.* 5, chap. 16 of Ref. 6.
8. Gere, J. M., and W. O. Carter: Critical Buckling Loads for Tapered Columns, *J. Struct. Div. ASCE*, February 1962.
9. Lee, G. C., M. T. Morrell, and R. L. Ketter: Design of Tapered Members, *WRC Bull.* 173, June 1972.
10. Tebedge, N., and L. Tall: Residual Stresses in Structural Steel Shapes—A Summary of Measured Values, *Fritz Eng. Lab. Rept.* 337.34, February 1973.
11. Alpsten, G. A., and L. Tall: Residual Stresses in Heavy Welded Shapes, *Welding J.*, vol. 49, April 1970.
12. Sourochnikoff, B.: Wind Stresses in Semirigid Connections of Steel Framework, *Trans. ASCE*, vol. 115, p. 382, 1950.
13. "Structural Joints Using ASTM A325 or A490 Bolts," Research Council on Riveted and Bolted Structural Joints of the Engineering Foundation, Feb. 4, 1976.
14. Bendigo, R. A., R. M. Hansen, and J. L. Rumpf: Long Bolted Joints, *J. Struct. Div. ASCE*, December 1963.
15. Nair, R. S., P. Birkemoe, and W. H. Munse: "Behavior of Bolts in Tee-Connections Subjected to Prying Action," Structural Research Series 353, University of Illinois, September 1969.

16. Archer, F. E., H. K. Fischer, and E. M. Kitchem: Fillet Welds Subjected to Bending and Shear, *Civ. Eng. Public Works Rev.*, vol. 54, p. 634, London, April 1959.
17. Munse, W. H., and E. Chesson, Jr.: Riveted and Bolted Joints: Net Section Design, *J. Struct. Div. ASCE*, February 1963.
18. American Institute of Steel Construction: Specification for the Design Fabrication and Erection of Structural Steel for Buildings, 1978.
19. American Railway Engineering Association: Specifications for Steel Railway Bridges, 1977.
20. American Association of State Highway and Transportation Officials: Standard Specification for Highway Bridges, 1977.
21. American Welding Society: Structural Welding Code D1.1-75.

Section **7**

Plastic Design of Steel Frames

LYNN S. BEEDLE
Director, Fritz Engineering Laboratory, Lehigh University

T. V. GALAMBOS
H. D. Jolley Professor of Civil Engineering, Washington University

1. Inelastic Bending Figure 1a shows a typical tensile stress-strain curve for steel. Figure 1b shows that portion which is utilized in plastic design. An idealized curve consists of an elastic portion having a slope equal to E, the modulus of elasticity, and a portion having zero slope.

The moment-curvature relationship can be determined by assuming that each fiber will strain according to the idealized stress-strain curve. Such a curve is shown in Fig. 2a for a W8 × 31 member bent about its strong axis. The initial elastic response (stress distribution in Fig. 2b) results in the straight-line segment OA for which the curvature is proportional to the applied moment ($\phi = M/EI$). Initiation of yielding terminates this region of proportionality, after which increasingly larger curvatures result from each equal increment of moment. The corresponding stress distribution is shown in Fig. 2c. Finally, when one-half the cross section has yielded in compression and the other half in tension (Fig. 2d), a maximum moment is reached. This moment, termed the *plastic moment* M_p, is the ultimate moment which this section can maintain under the assumption of an ideal elastic-plastic stress-strain law. Upon attainment of the plastic moment the section continues to rotate at a constant moment, and the member is said to have formed a *plastic hinge*. It is customary to use an idealized moment-curvature curve which consists of an elastic portion OB and a flat, inelastic portion BC.

The value of M_p is given by

$$M_p = \sigma_y Z \tag{1}$$

where Z is called the *plastic section modulus*. Values of Z for rolled shapes are listed in Ref. 1. For shapes built up of steels having the same yield point and having at least one axis of symmetry, Z is given by

$$Z = \frac{Aa}{2} \tag{2}$$

where a = distance between centroids of two equal areas lying above and below the neutral axis and A = cross-sectional area.

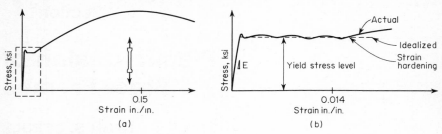

Fig. 1 Typical stress-strain curve.

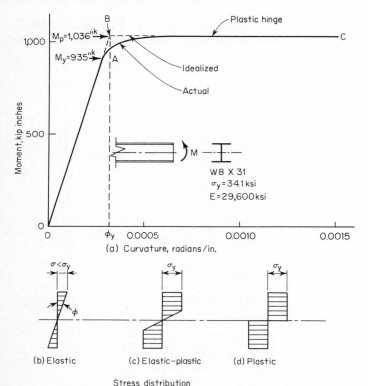

Stress distribution

Fig. 2 Moment-curvature plot for W8 $\times$ 31

The ratio of the plastic moment, M_p to the moment M_y at the initiation of yielding is

$$f = \frac{M_p}{M_y} = \frac{Z\sigma_y}{S\sigma_y} = \frac{Z}{S} \tag{3}$$

where f is called the *shape factor*. For wide-flange sections f is approximately 1.12 for strong-axis bending. Shape factors and formulas for plastic moduli are listed for several sections in Table 1.

2. Indeterminate Structures   Although the more efficient utilization of material which results from plastic design stems in part from the use of M_p rather than M_y as a measure of bending strength, the principal advantage lies in the redistribution of moment which develops in statically indeterminate structures. This is illustrated in Fig. 3.

The moment diagram (Fig. 3a) shows that the maximum moment occurs at the left support as long as the beam remains elastic. If the moment-curvature relationship is

TABLE 1   Plastic Moduli and Shape Factors

Section	Plastic modulus Z	Shape factor f
	$\dfrac{bd^2}{4}$	1.5
	x − x axis $bt(d-t) + \dfrac{w}{4}(d-2t)^2$	1.12 (approx.)
	y − y axis $\dfrac{b^2t}{2} + \dfrac{1}{4}(d-2t)w^2$	1.55 (approx.)
	$\dfrac{d^3}{6}$	1.70
	$\dfrac{d^3}{6}\left[1-(1-\dfrac{2t}{d})^3\right]$ td^2 for $t \ll d$	$\dfrac{16}{3\pi}\left[\dfrac{1-(1-\frac{2t}{d})^3}{1-(1-\frac{2t}{d})^4}\right]$ 1.27 for $t \ll d$
	$\dfrac{bd^2}{4}\left[1-(1-\dfrac{2w}{b})(1-\dfrac{2t}{d})^2\right]$	1.12 (approx.) for thin walls
	$\dfrac{bd^2}{12}$	2

idealized, a plastic hinge will form at the left support as soon as the moment reaches M_p (Fig. 3b). No additional moment can be carried by the cross section at this hinge point, but additional load can be carried by the structure, which now behaves as a propped cantilever. The next plastic hinge is reached under the load point (Fig. 3c). However, the structure has not yet "failed," because still more load can be carried, with the right two-thirds of the beam acting as a cantilever. The ultimate load is attained when the moment at the right support reaches M_p, so that a mechanism is formed (Fig. 3d, e), with the plastic hinges at A and B continuing to rotate without any change in M_p until the moments become fully redistributed. For this beam a 33 percent increase of load is realized by accounting for the reserve strength of the beam after formation of the first hinge. The load-deflection curve in Fig. 4 illustrates the complete history of the beam.

The behavior of more complex structures is similar. Results of tests on many full-scale structural frames have shown excellent agreement with the predicted ultimate load based on the formation of a mechanism.[2]

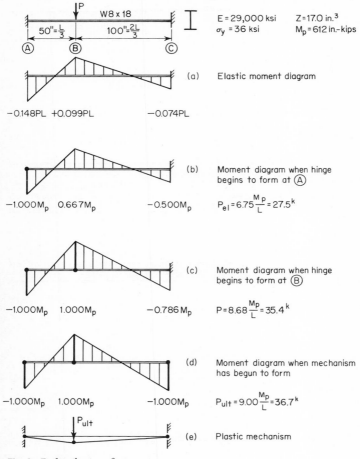

$E = 29{,}000$ ksi $Z = 17.0$ in.[3]
$\sigma_y = 36$ ksi $M_p = 612$ in.-kips

$-0.148PL$ $+0.099PL$ $-0.074PL$

(a) Elastic moment diagram

$-1.000M_p$ $0.667M_p$ $-0.500M_p$

(b) Moment diagram when hinge begins to form at (A)

$P_{el} = 6.75\dfrac{M_p}{L} = 27.5^k$

$-1.000M_p$ $1.000M_p$ $-0.786M_p$

(c) Moment diagram when hinge begins to form at (B)

$P = 8.68\dfrac{M_p}{L} = 35.4^k$

$-1.000M_p$ $1.000M_p$ $-1.000M_p$

(d) Moment diagram when mechanism has begun to form

$P_{ult} = 9.00\dfrac{M_p}{L} = 36.7^k$

(e) Plastic mechanism

Fig. 3 Redistribution of moment.

ANALYSIS

3. Theorems Although the ultimate strength of steel frames can be determined by the step-by-step procedure illustrated in Fig. 3, several much simpler methods are available. The ultimate load can be shown to lie between the loads computed by the upper-bound and lower-bound theorems.

Upper-Bound Theorem. A load computed on the basis of an assumed mechanism will always be greater than, or at best equal to, the true ultimate load.

Lower-Bound Theorem. A load computed on the basis of an assumed moment distribution which is in equilibrium with the applied loading and where no moment exceeds M_p is less than, or at best equal to, the true ultimate load.

If the loads computed by the two methods coincide, the true and unique ultimate load has been found. At ultimate load the following conditions must be met:
1. The applied loads must be in equilibrium with the internal forces.
2. There must be a sufficient number of plastic hinges for the formation of a mechanism.
3. The plastic moment must not be exceeded at any point in the structure.
4. Statical Method The statical method of solution, which is based on the lower-bound theorem, is particularly suited for the analysis of simple frames and continuous beams with variable or uniform stiffness. The procedure is illustrated in Examples 1, 2, 3, and 4.
1. Select the redundant moments.
2. Draw the statically determinate moment diagram.
3. Superimpose the redundant moments on the determinate moment diagram and determine peak moments.
4. Set peak moments equal to M_p and check that the number of plastic hinges is sufficient to form a mechanism.
5. Compute the corresponding ultimate load by statics.
5. Mechanism Method The principle of virtual work can be used in the mechanism method, which is based on the upper-bound theorem. The virtual-work principle can be stated as follows:
If a system of forces in equilibrium is subjected to a virtual displacement, the work done by the external forces is equal to the work done by the internal forces.
In plastic analysis, virtual work is generated by subjecting the frame to virtual displacements consistent with the motion of the mechanism. The external work is equal to the sum of the products of all the forces times the displacements through which they move, while the internal work is equal to the sum of the products of the plastic moments times their hinge rotations. No internal work is performed by the moments between the hinges, since there is no change in the curvature in the elastic parts of the structure.
It is convenient to classify mechanisms as *independent* mechanisms and *combined* mechanisms. In the rectangular frame of Example 6 the independent mechanisms are the beam mechanism and the sidesway mechanism. If more than two members meet at a joint, as in Example 7, an additional independent mechanism is the *joint* mechanism. The number of independent mechanisms is equal to the number of possible plastic hinges minus the number of redundancies of the structure.

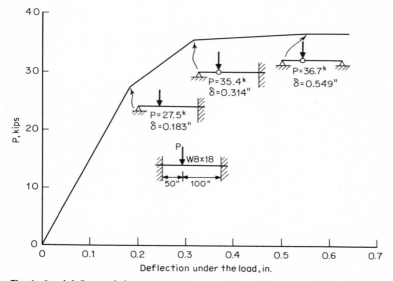

Fig. 4 Load-deflection behavior.

Various combinations of independent mechanisms can be made. Several common combined mechanisms are given in Examples 6 through 9. Combinations should be chosen so as to make the external work as large as possible and the internal work small, since this is the condition which will determine the ultimate load. Therefore, one tries combinations which eliminate as many plastic hinges as possible and which involve displacements of as many loads as possible.

Example 1. Two-span continuous beam. Determine P_u, A36 steel.
P may occupy either span or both.

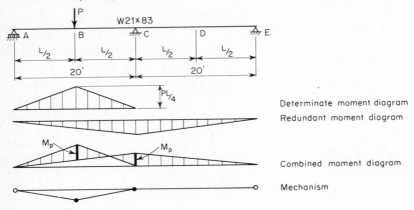

Equilibrium:

At B : $\dfrac{PL}{4} - \dfrac{M_p}{2} = M_p$; $P_u = \dfrac{6 M_p}{L}$

Equilibrium:

At B and D : $\dfrac{PL}{4} - \dfrac{M_p}{2} = M_p$; $P_u = \dfrac{6 M_p}{L}$

For W21×83, $Z = 196$ in.^{3}

$M = Z\sigma_y = 196 \times 36 = 7,060^{"k}$

$P_u = \dfrac{6 \times 7,060}{240} = 176^k$

Examples 5 through 7 illustrate the determination of the ultimate load by the mechanism method. The procedure is as follows:

1. Determine the location of all possible plastic hinges. Hinges are likely to form at points of support, at joints, and under concentrated loads; that is, at all locations where the moment diagram could have peak values.

2. Determine the degree of indeterminacy and sketch the independent mechanisms.

3. Select those independent mechanisms and any suitable combinations thereof which might give the lowest value of the ultimate load, and compute the corresponding ultimate loads.

4. Since the mechanism method is based on the upper-bound theorem, the load corresponding to any mechanism will be either larger than or at least equal to the correct ultimate load. Because the critical mechanism may be overlooked in step 3, it is necessary to subject the mechanism leading to the lowest load to a moment check. If this check indicates that M_p is nowhere exceeded, then the correct mechanism, and thus the correct ultimate load, is found.

Although all possible combinations were investigated in Examples 5 and 6, so that the moment check is not needed, it is shown for illustrative purposes.

6. Moment Check   The usual design problem of determining member sizes for a frame to support given loads is illustrated in Example 7. Three of the four possible independent

Example 2. Two-span continuous beam. Determine member sizes, A36 steel.

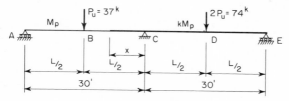

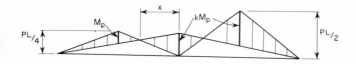

Equilibrium:

At D : $kM_p + \dfrac{kM_p}{2} = \dfrac{PL}{2}$; $kM_p = \dfrac{PL}{3} = \dfrac{37 \times 360}{3} = 4{,}440''^k$

At B : $M_p + \dfrac{kM_p}{2} = \dfrac{PL}{4}$; $M_p = \dfrac{PL}{4} - \dfrac{kM_p}{2} = \dfrac{37 \times 360}{4} = -2{,}220 = 1{,}110''^k$

Splice location (at M=0):

$M_x = \dfrac{Px}{2} + \dfrac{kM_p x}{L} - kM_p = x\left(\dfrac{P}{2} + \dfrac{P}{3}\right) - \dfrac{PL}{3} = 0$; $x = \dfrac{2L}{5} = \dfrac{2 \times 30}{5} = 12'$

Member sizes:

$kM_p = Z\sigma_y$; $Z = \dfrac{4{,}440}{36} = 123.2$ in.^{3} ;   Use W24×55 (Z=134 in.^{3})

$M_p = Z\sigma_y$; $Z = \dfrac{1{,}110}{36} = 30.8$ in.^{3} ;   Use W14×22 (Z=33.2 in.^{3})

A W18×60 and a W18×35 can be used if a uniform depth is required.

mechanisms (the joint mechanism leads only to the equation of equilibrium of moments at the joint) and one combined mechanism are investigated. Therefore, it is important that a moment check be performed, since it is possible that a combination giving a larger value of M_p may have been overlooked. The structure is found to be statically indeterminate after the critical mechanism (beam mechanism in the right span) has formed. However, it

Example 3. Rigid frame. Determine member sizes, A36 steel.

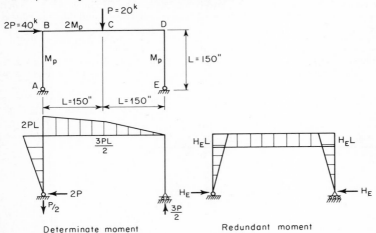

Determinate moment   Redundant moment

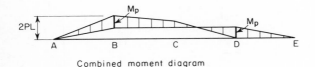

Combined moment diagram

Equilibrium:

At B : $2PL - M_p = M_p$; $M_p = PL = 20 \times 150 = 3,000^{"k}$

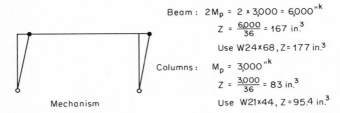

Mechanism

Beam : $2M_p = 2 \times 3,000 = 6,000^{"k}$

$Z = \dfrac{6,000}{36} = 167$ in.^{3}

Use W24×68, Z= 177 in.^{3}

Columns : $M_p = 3,000^{"k}$

$Z = \dfrac{3,000}{36} = 83$ in.^{3}

Use W21×44, Z =95.4 in.^{3}

NOTE: Members must be checked for secondary effects.

is not necessary to determine the exact moment distribution; if any distribution can be found which satisfies equilibrium and for which the moment nowhere exceeds M_p, the mechanism is the correct one.

The remaining redundancy after a mechanism has formed can be computed by

$$I = X - (M - 1) \tag{4}$$

where I = number of redundancies remaining
$\quad\quad X$ = number of redundancies in the original structure
$\quad\quad M$ = number of hinges necessary to develop the mechanism under investigation

In Example 7 six moments are still unknown at the formation of the mechanism. Four are assumed equal to their plastic values and the remaining two are computed by statics.

Example 4. Gabled frame. Determine member sizes, A36 steel.

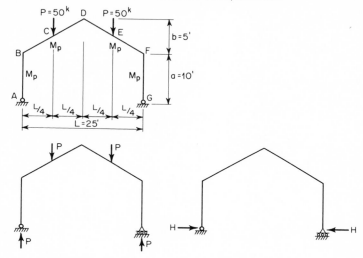

Determinate structure   Redundant structure

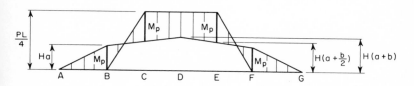

Equilibrium:

At B : $M_p = Ha$; $H = \dfrac{M_p}{a}$

At C : $H\left(a + \dfrac{b}{2}\right) + M_p = \dfrac{PL}{4}$; $M_p = \dfrac{PL}{2(4 + b/a)} = \dfrac{50 \times 300}{2(4 + 5/10)} = 1{,}667 \text{"}^k$

$\quad\quad\quad Z = \dfrac{1{,}667}{36} = 46.3 \text{ in.}^3$

Use W14 x 30 ; $Z = 47.3 \text{ in.}^3$ (Check for secondary effects.)

7. Instantaneous Center   If a structure contains sloping members, as in a gabled frame, the instantaneous center (I.C.) may be used to determine mechanism displacements. This is illustrated in the single-bay gabled frame of Example 8. In addition to the beam and panel mechanisms a, b, and c a fourth type of independent mechanism, sometimes called a gable mechanism, is shown in d. One combined mechanism is shown in e. The determination of the hinge rotations for mechanisms d and e is simplified if use is made of the fact that each rotates about three points whose locations are known (A, C, and I.C.). The instantaneous center is determined by the intersection of two straight lines, each originating at a known nontranslating point (A and C) and passing through the adjacent hinge. The structure is subjected to a virtual displacement by rotating it through an angle θ about any one of the pivotal points. Rotations and translations of other points are then determined from the geometry of the structure. It will be noted that the vertical and horizontal components of the displacement of a load on the rafter are equal to the products of the rotation at one of the pivotal points and the horizontal and vertical distances, respectively, between the point and the load.

Example 5. Fixed-end beam. Determine P_u, A36 steel.

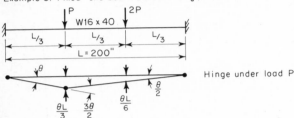

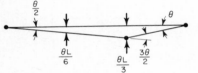

Hinge under load P

External work: $W_e = P\dfrac{\theta L}{3} + 2P\dfrac{\theta L}{6} = \dfrac{2P\theta L}{3}$

Internal work: $W_i = M_p\theta + \dfrac{3M_p\theta}{2} + \dfrac{M_p\theta}{2} = 3M_p\theta$

$W_e = W_i:$ $P_u = \dfrac{9M_p}{2L}$

Hinge under load 2P

$W_e = P\dfrac{\theta L}{6} + 2P\dfrac{\theta L}{3} = \dfrac{5PL\theta}{6}$

$W_i = 3M_p\theta$

$W_e = W_i:$ $P_u = \dfrac{18M_p}{5L} < \dfrac{9M_p}{2L}$ $\therefore$ this is correct mechanism

Moment check

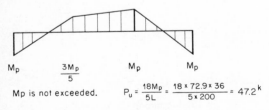

M_p $\dfrac{3M_p}{5}$ M_p M_p

M_p is not exceeded. $P_u = \dfrac{18M_p}{5L} = \dfrac{18 \times 72.9 \times 36}{5 \times 200} = 47.2^k$

Example 9 illustrates the application of the instantaneous-center method for a two-bay gabled frame. Although there are two instantaneous centers in this case, the mechanism motion is determined by prescribing the rotation at only one of the five pivotal points.

8. Distributed Loads   If the loads are distributed, the exact location of the hinge within the span is not generally known. In some cases, such as the propped cantilever in Example 10, it is possible to obtain an exact solution by maximizing the moment expression. This is not a practical solution in case there are three or more unknown hinge locations in the structure, and it is better to replace the distributed loads by judiciously placed equivalent concentrated loads, as is illustrated in Example 10. This is always conservative because the actual maximum moment in the beam is less than, or at most equal to, the moment due to the concentrated loads. In this example the exact and the approximate moment diagrams agree very well, even though only three concentrated loads are used. The required beam size is the same in either case.

A number of diagrams and other aids are available in Refs. 3 to 6. Various analysis procedures are reviewed and classified in Ref. 2. Many frame problems as well as the plastic analysis of arches, plates, and shells are discussed in Refs. 3 to 10. It is possible to obtain optimum designs (minimum weight) of rigid frames by the use of linear programming.[5,7,8,11,12]

Example 6. Rigid frame. Determine P_u, A36 steel.

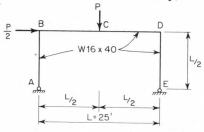

Possible hinges	3
Redundancies	1
Independent mechanisms	2

Independent mechanisms

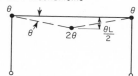

Beam mechanism

$$W_e = \frac{P\theta L}{2}$$

$$W_i = 4M_p\theta$$

$$P_u = \frac{8M_p}{L}$$

Sidesway (panel) mechanism

$$W_e = \frac{P\theta L}{4}$$

$$W_i = 2M_p\theta$$

$$P_u = \frac{8M_p}{L}$$

Combined mechanism

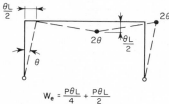

$$W_e = \frac{P\theta L}{4} + \frac{P\theta L}{2}$$

$$W_i = 4M_p\theta$$

$$P_u = \frac{16M_p}{3L}$$

Moment check for $P_u = \frac{16M_p}{3L}$

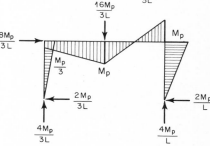

Moment check OK: $P_u = \frac{16M_p}{3L} = \frac{16 \times 72.9 \times 36}{3 \times 300} = 46.7^k$

9. Moment Balancing　This procedure can be used to make the moment check discussed in Art. 6 and as a direct technique for design. The procedure consists in adjusting a set of initial moments to obtain equilibrium. The initial moments may be those obtained from a mechanism analysis, in which the problem is to determine whether equilibrium can be established without violating the yield condition, i.e., without exceeding the plastic moment at any point. On the other hand, the initial moments may be obtained from the basic independent mechanisms, in which case the problem is to bring them into equilibrium at the joints so that a mechanism results without violating the yield condition.

Figure 5 shows five incremental moment diagrams which are useful in the moment-balancing procedure. These diagrams show the effect of a unit moment at a given point in a span upon the moments at other points in the span. The sign convention is shown in f. Diagrams d and e are useful in developing combined beam and panel mechanisms, d with wind from the left and e with wind from the right.

For purposes of design, any set of moments in equilibrium with the vertical loads can be used as initial moments for the beams in a frame. Usually these will be the moments corresponding to beam mechanisms. Any set of column end moments in equilibrium with

Example 7. Rigid frame. Determine member sizes, A36 steel.

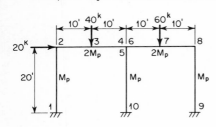

Possible hinges	10
Redundancies	6
Independent mechanisms	4

Independent mechanisms:

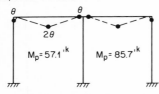

$M_p = 57.1^{'k}$　　$M_p = 85.7^{'k}$

Two beam mechanisms

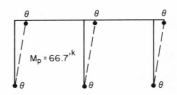

$M_p = 66.7^{'k}$

Sidesway mechanism

Joint mechanism

Beam mechanism 6-7-8

$W_e = 60^k \times 10' \times \theta$

$W_i = 2M_p\theta + 2M_p \times 2\theta + M_p\theta$

$M_p = 600/_7 = 85.7^{1k}$

$W_e = 20^k \times 20' \times \theta$

$W_i = 6M_p\theta$

$M_p = 400/_6 = 66.7^{1k}$

Combined mechanisms:

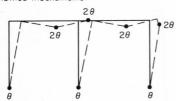

Possible combined mechanism

$W_e = 20^k \times 20' \times \theta + 40^k \times 10' \times \theta + 60^k \times 10' \times \theta$

$W_i = 3M_p\theta + 3 \times 2M_p \times 2\theta + M_p \times 2\theta$

$M_p = 1400/_{17} = 82.4^{'k}$

Largest M_p was found for beam mechanism 6-7-8: $M_p = 85.7^{lk}$

Moment check:

Original redundancies $\bar{X} = 6$

$M - I$ $= 2$

Remaining redundancies $= 4$ (Eq.4)

Known moments: $M_6 = -171.4^{lk}$, $M_7 = +171.4^{lk}$, $M_8 = -85.7^{lk}$

Assume four of the six unknown moments to be equal to or less than M_p:

$M_1 = -85.7^{lk}$, $M_4 = -171.4^{lk}$, $M_9 = 85.7^{lk}$, $M_{10} = -85.7^{lk}$

Compute remaining unknown moments by statics

$M_2 = 57.2^{lk}$, $M_3 = 142.9^{lk}$, $M_5 = 0$

M_p is nowhere exceeded and beam mechanism 6-7-8 is critical

$$Z = \frac{85.7 \times 12}{36} = 28.6 \text{ in.}^3 \text{ columns}$$

$Z = 2 \times 28.6 = 57.2$ in.^{3} beams

Columns: W12x22 ($Z = 29.3$ in.^{3})

Beams: W18x35 ($Z = 66.5$ in.^{3})

Moment diagram (not necessarily correct, but statically admissible):

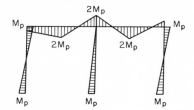

the horizontal shear in a story may be used as initial moments for the columns; usually these will be those corresponding to a panel mechanism.

A convenient tabular form for recording the moments is shown in Fig. 6. Entries in the joint table are moments on the ends of the members meeting at the joint. Entries in the sway table are unbalances in the sway moments; that is, each entry is the algebraic sum of the column end moments.

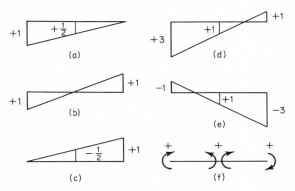

Fig. 5 Incremental diagrams for moment balancing.

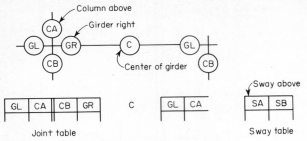

Fig. 6 Suggested tabular form for moment balancing.

Example 8. Gabled frame. Determine member sizes, A36 steel.

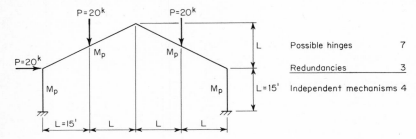

Possible hinges	7
Redundancies	3
Independent mechanisms	4

Independent mechanisms:

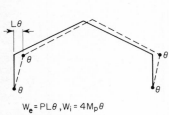

$W_e = PL\theta , W_i = 4M_p\theta$
$P_u = 4M_p/L$

(a)

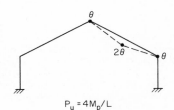

$P_u = 4M_p/L$

(b)

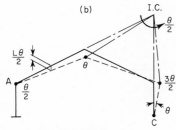

$W_e = PL\theta , W_i = 4M_p\theta$
$P_u = 4M_p/L$

(c)

$W_e = PL\theta/2 + PL\theta/2$
$W_i = M_p(\theta/2 + \theta + 3\theta/2 + \theta)$
$P_u = 4M_p/L$

(d)

Combined mechanism:

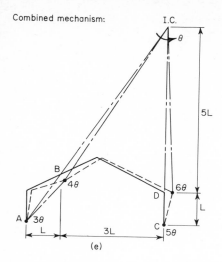

Joint rotations:

IC: θ

A: $\dfrac{3L\theta}{L} = 3\theta$

B: $3\theta + \theta = 4\theta$

C: $\dfrac{5L\theta}{L} = 5\theta$

D: $5\theta + \theta = 6\theta$

$W_e = P \times 3\theta \times L + P \times 3\theta \times L + P\theta L = 7P\theta L$

$W_i = M_p\theta(3+4+5+6) = 18M_p\theta$

$P_u = 18M_p/7L$

(e)

A moment check (not shown) reveals that M_p is nowhere exceeded for this mechanism:
Therefore:

$$M_p = \frac{7PL}{18} = \frac{7 \times 20 \times 15}{18} = 117^{lk}$$

$$Z = 117 \times 12 / 36 = 39.0 \text{ in.}^3$$

Use W14 x 26 ($Z = 40.2$ in.^{3}). Check for secondary effects.

Example 11, which is adapted from Chap. 2 of Ref. 10, illustrates the procedure. The initial moments (beam mechanisms) for Case 1 loading are determined and entered in the table as multiples of M_p. The corresponding sum of the moments at the column tops is zero. The joints are then balanced by adding appropriate moments at the column tops, and the zero sum is entered in the sway column. The final moments are in equilibrium at the joints and with respect to shear in the story, resulting in simultaneous beam mechanisms 4-5-6 and 8-9-10.

The panel mechanism determines the moments at the column tops for Case 2 loading. These are entered, together with the beam-mechanism moments, in the first row of the table as multiples of $M_p/3$. The total unbalance at the joints is the sway moment, -8. Balance can be achieved in a number of ways. However, this loading tends to produce a combined beam and panel mechanism, so that distribution d of Fig. 5 should be used. This distribution can be assigned to beam 8-9-10, as in Trial 1. Since it is the end moments that affect the column tops, the required distribution can be determined, i.e., +6, +2, +2, for which the sum of the end moments is +8. Following this the joints are balanced and the totals obtained. In Trial 2 the correction distribution is divded equally between beams 4-5-6 and 8-9-10, after which the joints are balanced and totals obtained. This results in a combined mechanism such as that shown in Example 7.

The comparison of weights of Trials 1 and 2 is based on the assumption that the weight per foot of beam is directly proportional to the plastic moment it furnishes. It will be noted that Case 1 loading governs the design of this frame. Further discussion of the moment-balancing method may be found in Refs. 2, 5, 13, and 17, and its application to the design of planar multistory frames in Refs. 10 to 17.

DEFLECTION ANALYSIS

The primary requirement in the design of a structure is that it must carry the assumed load; deflection limitations are secondary. Estimates of deflections fall into two categories:

Example 9. Two – bay gabled frame, all members same size.
Analysis of one combined mechanism.

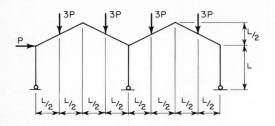

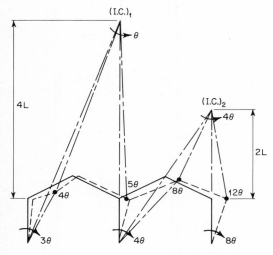

$W_e = P \times 3\theta \times L + 3P \times 3\theta \times \dfrac{L}{2} + 3P \times \theta \times \dfrac{L}{2} + 3P \times 4\theta \times \dfrac{L}{2} + 3P \times 4\theta \times \dfrac{L}{2} = 21P\theta L$

$W_i = M_p \theta (4 + 5 + 8 + 12) = 29\, M_p \theta$

$P_u = \dfrac{29 M_p}{21L} = 1.38\dfrac{M_p}{L}$ (Moment check must be made)

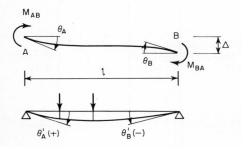

Fig. 7 Nomenclature and sign convention for Eq. (5).

determination of the deflection at ultimate load, and estimates of the deflection at working load.

10. Deflection at Ultimate Load   Deflection calculations are based on the assumptions of the simple plastic theory, the most important of which is the idealized moment-curvature relationship (Fig. 2). This concept assumes unlimited rotations at plastic hinges at the moment $M = M_p$, with each segment between hinges retaining its flexural rigidity EI.

Examination of mechanisms as they form in a structure reveals an additional important concept that simplifies what otherwise would be a formidable task, and permits deflection calculations to be made in a straightforward manner. Discontinuities (kinks) exist at the hinges which develp during the formation of the mechanism, except for the hinge which forms last. This is because the structure attains its ultimate load as soon as the moment reaches M_p at the last hinge to form, so that inelastic rotation is not required at this point.

Consider the load-deflection plot (Fig. 4) of the beam of Fig. 3. The deflection corresponding to point a is determined by an elastic analysis of the fixed-end beam AC. The increment of deflection ab is calculated for the corresponding increment of load acting on the propped cantilever AC, and the increment bc for the corresponding load increment acting on the cantilever BC. This suggests that deflection at ultimate load must be computed step by step. However, the following procedure is simpler:

1. Obtain the ultimate load and the corresponding moment diagram.
2. Compute the rotations θ (Fig. 7) of the various frame segments. Continuity at the last hinge to form requires that $\theta_{ij} = \theta_{jk}$.

Example 10. Propped cantilever. Determine size, A36 steel.

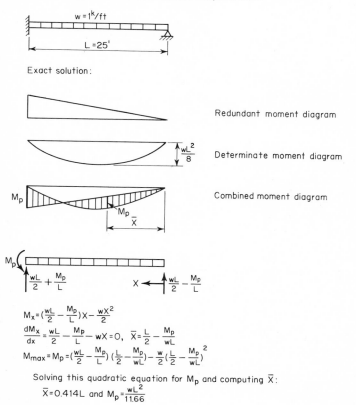

Exact solution:

Redundant moment diagram

Determinate moment diagram

Combined moment diagram

$$M_x = \left(\frac{wL}{2} - \frac{M_p}{L}\right)X - \frac{wX^2}{2}$$
$$\frac{dM_x}{dx} = \frac{wL}{2} - \frac{M_p}{L} - wX = 0, \quad \bar{X} = \frac{L}{2} - \frac{M_p}{wL}$$
$$M_{max} = M_p = \left(\frac{wL}{2} - \frac{M_p}{L}\right)\left(\frac{L}{2} - \frac{M_p}{wL}\right) - \frac{w}{2}\left(\frac{L}{2} - \frac{M_p}{wL}\right)^2$$

Solving this quadratic equation for M_p and computing $\bar{X}$:
$$\bar{X} = 0.414L \text{ and } M_p = \frac{wL^2}{11.66}$$

Approximate solution

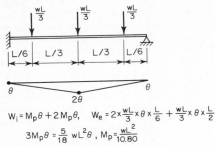

$$W_i = M_p \theta + 2M_p \theta, \quad W_e = 2 \times \frac{wL}{3} \times \theta \times \frac{L}{6} + \frac{wL}{3} \times \theta \times \frac{L}{2}$$

$$3M_p \theta = \frac{5}{18} wL^2 \theta, \quad M_p = \frac{wL^2}{10.80}$$

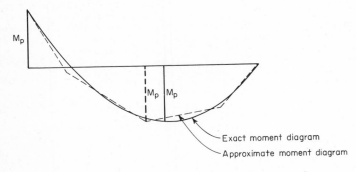

Exact moment diagram

Approximate moment diagram

$$M_p = \frac{1 \times 25^2}{11.66} = 53.6^{\,ik} \qquad Z = \frac{53.6 \times 12}{36} = 17.9 \,in.^3 \text{ exact}$$

$$M_p = \frac{1 \times 25^2}{10.80} = 57.9^{\,ik} \qquad Z = \frac{57.9 \times 12}{36} = 19.3 \,in.^3 \text{ approx.}$$

Use W12×16.0 (Z = 20.6 in.³)

3. Solve the resulting equations for the deflection(s) Δ.

In the beam of Fig. 3 it is obvious that the last hinge forms at C. If the location of the last hinge is not known, the calculations must be repeated, assuming each hinge in turn to be the last to form. The correct deflection is the largest value so obtained. Alternatively, an incorrect assumption as to the last hinge to form results in a "kink" in the wrong direction in the mechanism; if the kink is removed through a mechanism displacement and the resulting deflections are added to those determined in step 3, the correct deflections are obtained.[4]

A convenient form of the slope-deflection equation is given in Eq. (5), the nomenclature being shown in Fig. 7. Clockwise moments and rotations are taken as positive. Thus

$$\theta_{AB} = \theta'_{AB} + \frac{\Delta}{l} + \frac{l}{3EI}\left(M_{AB} - \frac{M_{BA}}{2}\right) \tag{5}$$

where θ'_{AB} is the slope at end A due to the loads between A and B acting on a simply supported beam AB. Example 12 illustrates the procedure.

11. Deflection at Working Load Where deflections are important, primary interest will probably be in an estimate of deflection at working load, which will usually be in the elastic range. Thus, one obvious procedure is to make an elastic-deflection analysis.

A simpler approach gives an upper limit to the true deflection at working load. The deflection at ultimate load is calculated and divided by the load factor F. Thus

$$\delta'_w = \frac{\delta_u}{F} \tag{6}$$

As is evident from Fig. 8, δ'_w may be more than double the working-load deflection δ_w. However, comparison of δ'_w with an allowable limit will indicate whether more refined calculations are needed.

DESIGN REQUIREMENTS

12. Specifications Plastic design of members in steel buildings is regulated by the provisions of Part 2 of the AISC Specification for the Design, Fabrication and Erection of Structural Steel for Buildings.[15] The 1978 edition permits plastic design for simply supported and continuous beams, and braced and unbraced planar frames.

Example 11. Moment balancing for two-bay frame.

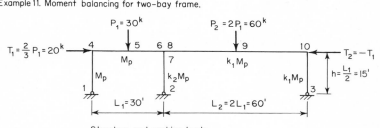

Structure and working loads

Loading conditions:

Case 1 (DL+LL)	F=1.7	$P_1 = 51^k, P_2 = 102^k$
Case 2 (DL+LL+T_1)	F=1.3	$P_1 = 39^k, P_2 = 78^k, T_1 = 26^k$
Case 3 (DL+LL+T_2)	F=1.3	$P_1 = 39^k, P_2 = 78^k, T_2 = 26^k$

Plastic moment ratios:

$$M_p = \frac{P_1 L_1}{8} \; ; \; k_1 M_p = \frac{P_2 L_2}{8} \; ; \; k_1 = \frac{P_2 L_2}{P_1 L_1} = 2 \times 2 = 4$$

Interior column: $M_7 = -M_6 - M_8$

$$k_2 M_p = M_p - k_1 M_p = -3M_p$$
$$k_2 = 3$$

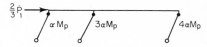

Panel mechanism:

$$\alpha M_p (1+3+4) = T_1 h = \frac{2}{3} P_1 \frac{L_1}{2}$$
$$8\alpha \frac{P_1 L_1}{8} = \frac{P_1 L_1}{3} \; ; \; \alpha = \frac{1}{3}$$

Case 1 – Trial design

CB	GR	C	GL	CA	CB	GR	C	GL	CA	CB		SA	SB
4-1	4-5	5	5-6		7-2	8-9	9	9-10		10-3			
0	-1	+1	+1		0	-4	+4	+4		0			0
+1					+3					-4			0
+1	-1	+1	+1		+3	-4	+4	+4		-4			0

Starting moment x $\frac{1}{M_p}$
Balance
Total

Required plastic moments:

$$M_{p4-6} = M_p = \frac{P_1 L_1}{8} = \frac{51 \times 30}{8} = 191^{ik}$$

$$M_{p8-10} = k_1 M_p = 4 \times 191 = 764^{ik}$$

$$M_{p2-7} = k_2 M_p = 3 \times 191 = 573^{ik}$$

Case 2 – Trial design No. 1

	CB	GR	C	GL	CA	CB	GR	C	GL	CA	CB		SA	SB
	4-1	4-5	5	5-6	–	7-2	8-9	9	9-10	–	10-3			
Starting moment $\times \frac{3}{M_p}$	-1	-3	+3	+3		-3	-12	+12	+12		-4			-8
Inc. moment							+6	+2	+2					
Balance	+4					+6					-10			0
Total	+3	-3	+3	+3		+3	-6	+14	+14		-14			-8

Case 2 – Trial design No. 2

	CB	GR	C	GL	CA	CB	GR	C	GL	CA	CB		SA	SB
	4-1	4-5	5	5-6	–	7-2	8-9	9	9-10	–	10-3			
Starting moment $\times \frac{3}{M_p}$	-1	-3	+3	+3		-3	-12	+12	+12		-4			-8
Inc. moment		+3	+1	+1			+3	+1	+1					
Balance	+1					+8					-9			0.
Total	0	0	+4	+4		+5	-9	+13	+13		-13			-8

Compare weight of designs 1 and 2

Design 1: $(3+3+14)\frac{L}{2} + 3L + 14 \times 2L = 41L$

Design 2: $(0+5+13)\frac{L}{2} + 4L + 13 \times 2L = 39L$

Required plastic moments, design 2 :

$$M_{p4-6} = 4\frac{M_p}{3} = \frac{4}{3}\frac{P_1 L_1}{8} = \frac{39 \times 30}{6} = 195^{ik}$$

$$M_{p8-10} = \frac{13}{4} M_{p4-6} = 633^{ik}$$

$$M_{p2-7} = \frac{5}{4} M_{p4-6} = 244^{ik}$$

Plastic design is also used for structures which are intended to absorb energy released by bomb blasts or earthquakes. It is not recommended for members subjected primarily to fluctuating loads, such as crane girders. However, rigid-frame bents supporting crane runways may be considered as coming within its scope.

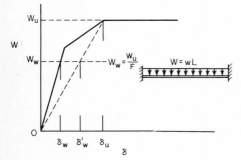

Fig. 8 Approximate working-load deflection.

Example 12. Deflections for frame of Example 6.

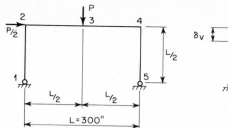
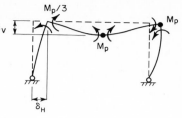

Slope deflection equations (Eq. 5):

$$\theta_{AB} = \theta'_{AB} + \frac{\Delta}{l} + \frac{1}{3EI}\left[M_{AB} - \frac{M_{BA}}{2}\right]$$

$$\theta_{21} = 0 + \frac{\delta_H}{L_{/2}} + \frac{L_{/2}}{3EI}\left[0 - \frac{M_p}{3}\right] = \frac{2\delta_H}{L} - \frac{M_p L}{18EI}$$

$$\theta_{23} = 0 + \frac{\delta_V}{L_{/2}} + \frac{L_{/2}}{3EI}\left[\frac{M_p}{3} + \frac{M_p}{2}\right] = \frac{2\delta_V}{L} + \frac{5M_p L}{36EI}$$

$$\theta_{32} = 0 + \frac{\delta_V}{L_{/2}} + \frac{L_{/2}}{3EI}\left[-M_p - \frac{M_p}{6}\right] = \frac{2\delta_V}{L} - \frac{7M_p L}{36EI}$$

$$\theta_{34} = 0 - \frac{\delta_V}{L_{/2}} + \frac{L_{/2}}{3EI}\left[M_p - \frac{M_p}{2}\right] = -\frac{2\delta_V}{L} + \frac{M_p L}{12EI}$$

$$\theta_{43} = 0 - \frac{\delta_V}{L_{/2}} + \frac{L_{/2}}{3EI}\left[M_p - \frac{M_p}{2}\right] = -\frac{2\delta_V}{L} + \frac{M_p L}{12EI}$$

$$\theta_{45} = 0 + \frac{\delta_H}{L_{/2}} + \frac{L_{/2}}{3EI}\left[-M_p + 0\right] = \frac{2\delta_H}{L} - \frac{M_p L}{6EI}$$

Since joint 2 is elastic, $\theta_{21} = \theta_{23}$

$$\frac{2\delta_H}{L} - \frac{M_p L}{18EI} = \frac{2\delta_V}{L} + \frac{5M_p L}{36EI} \qquad \delta_H = \delta_V + \frac{7M_p L^2}{72EI}$$

Assume last hinge forms at 3: $\theta_{32} = \theta_{34}$

$$\frac{2\delta_V}{L} - \frac{7M_p L}{36EI} = -\frac{2\delta_V}{L} + \frac{M_p L}{12EI} \qquad \delta_{V3} = \frac{5M_p L^2}{72EI}$$

Assume last hinge forms at 4: $\theta_{43} = \theta_{45}$

$$-\frac{2\delta_V}{L} + \frac{M_p L}{12EI} = \frac{2\delta_H}{L} - \frac{M_p L}{6EI} \qquad \delta_{V4} = \frac{M_p L^2}{72EI}$$

Since $\frac{1}{72} < \frac{5}{72}$, hinge forms last at 3

$$\delta_V = \frac{5M_p L^2}{72EI} = \frac{5 \times 72.8 \times 36 \times 300^2}{72 \times 29,000 \times 518} = 1.09''$$

$$\delta_H = \frac{M_p L^2}{6EI} = 2.62''$$

Estimate of working load deflection for $F = 1.70$

$$\delta_{V(W)} = 1.09 / 1.7 = 0.6''$$

$$\delta_{H(W)} = 2.62 / 1.7 = 1.5''$$

Steels suitable for plastic design are those which exhibit the necessary ductility and strain-hardening properties. The AISC specification recommends the following steels: ASTM A36, A242, A441, A529, A572, and A588.

13. Loads and Forces Prescribed design or working loads are multiplied by an appropriate load factor F to obtain the ultimate loading. Members are selected so that the frame will just support the most critical combinations of these ultimate loads. The following combinations must be investigated: dead load plus live load, and dead load plus live load plus wind or earthquake load.

It is assumed that these loads are static and that they remain proportional during the entire loading history. Although this assumption is not fulfilled in most cases, studies have shown that the usual load fluctuations existing in buildings will not cause a significant reduction of the ultimate load computed by assuming proportional loading. Plastic design should not be used if load fluctuations are severe enough and repeated often enough to cause fatigue failure.

The following load factors are prescribed in the AISC specification:

$F = 1.70$ for dead load plus live load
$F = 1.30$ for dead load plus live load plus earthquake or wind forces

14. General Design Procedure Plastic design of a building frame is performed in the following steps:

1. Determine the dimensions of the structure, working loads, and ultimate loads.
2. Assume the relative sizes of the members, type of steel, types of connections, bracing systems, etc.
3. Analyze the tentative structure and modify the relative sizes of the members so that the structure will fail under the most critical combination of the ultimate loads.
4. Select sections based on the moment diagram obtained in step 3.
5. Design the beam columns.
6. Check the selected sizes for adequacy against local buckling, lateral buckling, excessive deflection, and shear.
7. Design the connections.
8. Consider problems of fabrication and erection.

15. Preliminary Design The purpose of the preliminary design is to make initial decisions as to detail, and to establish tentative ratios of the plastic moments of the members making up the framework. Although the latter step requires a considerable amount of judgment, the following guides may be helpful:

1. If the critical mechanism is a local one (that is, a portion of the structure remains redundant at failure) the material in the redundant portions is not used to its full capacity. A more efficient choice of moment ratios can be made in the next trial cycle.
2. Minimum sections for continuous beams are obtained if all spans fail simultaneously as independent mechanisms.
3. The absolute minimum section for beams under gravity loads is obtained if each beam fails as an independent beam mechanism with all the necessary hinges forming in the beam itself. Similarly, minimum column sections are obtained under the action of horizontal forces alone when full plastic moments are developed at their ends.
4. Maximum economy is not necessarily associated with the lightest possible sections; it is also necessary to consider costs of fabrication.

16. Analysis Following the assumption of the relative values of the various M_p's the structure is analyzed by any suitable method, and the design is checked to ensure that a mechanism has formed and that the assumed value of M_p is nowhere exceeded. If the assumed structure does not fulfill these conditions, adjustments are made and a new analysis is performed. The choice of the best analysis procedure is a matter of experience; however, the following may prove helpful:

1. For continuous beams of uniform section and for single-story single-bay frames, both the statical method and the mechanism method work with equal advantage.
2. For continuous beams of nonuniform section the statical method is best.
3. For regular framed structures of few stories and bays the mechanism method is best.
4. For multibay single-story gabled and lean-to frames the use of charts is most efficient.

5. For relatively complex frames the moment-balancing procedure has merit.

6. With digital computers either an elastic-plastic step-by-step analysis[2] or a linear-programming approach[11,12] is recommended.

Example 13. Continuous beam with cover plates. A36 steel.

Use same size member throughout with cover plates in middle span.

$F = 1.7$

$w_{1u} = 1.70^k/ft$, $w_{2u} = 2.55^k/ft$

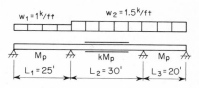

Span 1 (see Example 10):

$$M_p = \frac{w_1 L_1^2}{11.66} = \frac{1.7 \times 25^2}{11.66} = 91.1^{'k}$$

Working loads shown. Floor slab provides lateral support.

Span 3:

$$M_p = \frac{w_2 L_3^2}{11.66} = \frac{2.55 \times 20^2}{11.66} = 87.5^{'k}$$

$$Z = \frac{91.1 \times 12}{36} = 30.4 \text{ in.}^3$$

Try W14×22 ($Z = 33.2$ in.^{3}, $M_p = 99.6^{'k}$)

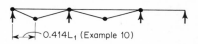

$0.414L_1$ (Example 10)

Shear:

$V = 0.55\sigma_y wd = 0.55 \times 36 \times 0.23 \times 13.72$
$= 62.5^k > 38.2^k$ O.K.

Local buckling

$$\frac{d}{w} = \frac{13.72}{0.23} = 60 < 70 \text{ O.K.}$$

$$\frac{b}{t} = \frac{5/2}{0.335} = 7.5 < 8.5 \text{ O.K.}$$

$0.414L_3$

Span 2:

$$\theta M_p (1 + 2k + 1) = w_2 L_2 \times \frac{1}{2} \times \frac{\theta L_2}{2}$$

$$kM_p = \frac{w_2 L_2^2}{8} - M_p = \frac{2.55 \times 30^2}{8} - 91.1$$

$$= 175.4^{'k}$$

Cov. pls. $\Delta M_p = kM_p - M_p = 175.4 - 99.6$
$= 75.8^{'k}$

$$A = \frac{\Delta M_p}{\sigma_y d} = \frac{75.8 \times 12}{36(13.72 + 0.5)} = 1.8 \text{ in.}^2$$

Use 4 × ½ pls.

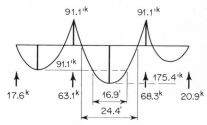

$91.1^{'k}$ $91.1^{'k}$

$91.1^{'k}$

$175.4^{'k}$

17.6^k 63.1^k $16.9'$ 68.3^k 20.9^k

$24.4'$

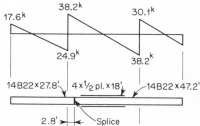

17.6^k 38.2^k 30.1^k

24.9^k 38.2^k

14B22×27.8'   4 × ½ pl. × 18'   14B22 × 47.2'

$2.8'$ Splice

SECONDARY DESIGN CONSIDERATIONS

With the information from the plastic analysis of the frame for the critical loading combination, the selection of member sizes and details can begin. However, in addition to furnishing the required plastic modulus, each member must meet certain additional requirements which assure its capacity to maintain the plastic moment through large enough rotations to permit all the hinges in the failure mechanism to form. These secondary checks are

1. Adjustment of member size for axial force
2. Design of the lateral bracing system

3. Local buckling
4. Shear force
5. Frame stability
6. Rotation capacity and deflection

17. Axial Force   If, in addition to bending moment, an appreciable axial force is present in a member, the plastic moment is reduced and the available inelastic-rotation capacity may also be reduced because of inelastic instability. The end moment which can be developed in the presence of an axial force P is always less than M_p. The maximum value depends on M_p, P, $P_y = \sigma_y A$, the slenderness ratio L/r_x, and the ratio of the end moments. The following two equations closely approximate the maximum beam-column capacity when failure is by inelastic instability in the plane of the applied loads:[2,14]

$$\frac{P}{P_{cr}} + \frac{M}{M_p}\frac{C_m}{1 - P/P_e} \gtrsim 1.0 \tag{7}$$

$$\frac{M}{M_p} = 1.18\left(1 - \frac{P}{P_y}\right) \le 1.0 \tag{8}$$

where P = axial load
M = numerically larger end moment

$$P_{cr} = P_y(1 - 0.25\lambda^2) \qquad \text{for } \lambda \le \sqrt{2} \tag{9a}$$

$$P_{cr} = \frac{P_y}{\lambda^2} \qquad \text{for } \lambda \ge \sqrt{2} \tag{9b}$$

$$\lambda = \frac{1}{\pi}\frac{L}{r_x}\sqrt{\frac{\sigma_y}{E}} \tag{10}$$

$$P_e = \frac{P_y}{\lambda^2} \tag{11}$$

$$P_y = A\,\sigma_y \tag{12}$$

$$C_m = 0.6 - 0.4\frac{M_1}{M_2} \ge 0.4 \tag{13}$$

In Eq. (13), M_1 is the numerically smaller of the end moments M_1 and M_2. The ratio M_1/M_2 is positive if the member is bent in reverse curvature and negative if it is bent in single curvature. Both Eq. (7) and Eq. (8) must be checked.

Equations (7) and (8) apply only if (1) lateral bracing is provided between the ends of the beam column to prevent lateral buckling (Art. 18) and (2) the frame is braced against sidesway buckling (Art. 21).

If the lateral bracing requirements of Art. 18 are not satisfied, Eq. (7) must be modified as follows:

1. P_{cr} is determined for the larger of L_x/r_x and L_y/r_y, where L_x and L_y are unbraced lengths.

2. M_p in the denominator of the second term is replaced by the maximum moment M_m the unbraced member can support in the absence of axial force. The AISC specification[15] gives the following approximate formula for column-type wide-flange sections:

$$M_m = \left[1.07 - \frac{(L/r_y)\sqrt{\sigma_y}}{3160}\right] M_p \le M_p \tag{14}$$

The authors suggest the following for beam-type wide-flange sections:

$$M_m = (5/3)S_x F_b \tag{15}$$

where S_x is the elastic section modulus and F_b is the allowable bending stress determined according to Sec. 1.5.1.4.6a of the AISC specification with $C_b = 1.0$.

In case of biaxial bending the following interaction equations apply:[14]

$$\left(\frac{M_x}{M_{pcx}}\right)^\alpha + \left(\frac{M_y}{M_{pcy}}\right)^\alpha = 1 \tag{16a}$$

$$\left(\frac{C_{mx}M_x}{M_{ucx}}\right)^{\beta} + \left(\frac{C_{my}M_y}{M_{ucy}}\right)^{\beta} = 1 \tag{16b}$$

where M_x and M_y are applied end moments, C_{mx} and C_{my} are the corresponding equivalent moment factors from Eq. (13), and

$$M_{pcx} = 1.18 \, M_{px} \left(1 - \frac{P}{P_y}\right) \leq M_{px} \tag{17a}$$

$$M_{pcy} = 1.19 \, M_{py} \left[1 - \left(\frac{P}{P_y}\right)^2\right] \leq M_{py} \tag{17b}$$

$$\alpha = 1.6 - \frac{P/P_y}{2 \log_e(P/P_y)} \tag{18}$$

$$M_{ucx} = M_{ux}\left(1 - \frac{P}{P_{cr}}\right)\left(1 - \frac{P}{P_{ex}}\right) \tag{19a}$$

$$M_{ucy} = M_{uy}\left(1 - \frac{P}{P_{cr}}\right)\left(1 - \frac{P}{P_{ey}}\right) \tag{19b}$$

$$\beta = 1 \qquad \text{if} \, \frac{b_f}{d} \gtrsim 0.3 \tag{20a}$$

$$\beta = 0.4 + \frac{P}{P_y} + \frac{b_f}{d} \gtrsim 1 \qquad \text{if} \, \frac{b_f}{d} \gtrsim 0.3 \tag{20b}$$

In these equations b_f is the flange width and d the member depth, and M_{ux} and M_{uy} are the moment capacities in the absence of axial force about the x and y axes, respectively, M_{uy} is the fully plastic moment about the weak axis and M_{ux} is the value of M_m from Eq. (14) or (15). P_{cr} is the critical axial load in the absence of flexure and P_{ex} and P_{ey} are the elastic buckling loads about the major and the minor axes, respectively. Both Eqs. (16a) and (16b) must be checked.

Additional references and design aids for beam columns are given in Refs. 2, 10, and 14.

18. Lateral Bracing In order to develop a mechanism, the members of a frame must undergo relatively large inelastic rotations at the hinges which form early. Lateral buckling must not interfere with the hinge development; this is accomplished by providing lateral bracing. Rolled wide-flange members are able to deliver a plastic hinge of sufficient rotation capacity if the lateral bracing is designed to fulfill the following minimum requirements:

1. Lateral braces are provided at the hinge location and at distances L on either side of the hinge given by

$$\frac{L}{r_y} \gtrsim \frac{1375}{\sigma_y} + 25 \qquad \text{for} + 1 \gtrsim \frac{M}{M_p} \gtrsim -0.5 \tag{21a}$$

$$\frac{L}{r_y} \gtrsim \frac{1375}{\sigma_y} \qquad \text{for} -0.5 > \frac{M}{M_p} \gtrsim -1.0 \tag{21b}$$

In Eqs. (21), σ_y is the specified yield stress in ksi, r_y is the radius of gyration of the member to be braced, and M is the lesser of the moments at the ends of the braced segment. The ratio M/M_p is positive when the segment is bent in double curvature and negative when it is bent in single curvature.

2. Lateral bracing members must be attached to the compression flange of the braced member or just below it. Where this is not possible, as is sometimes the case in the negative-moment region, knee braces must be provided.

3. Vertical web stiffeners must be provided at plastic hinge locations.

4. The minimum required area of the bracing member is (conservatively) 0.04 times the area of the member to be braced.

The above requirements need not be met in the region of the last hinge to form (since inelastic rotation is not needed at the last hinge) or in regions not adjacent to hinges. However, the possibility of elastic lateral buckling in these regions must be checked (Sec.

1.5.1.4.6a of AISC specification). The location of the last hinge to form is not determined in the usual analysis, so that all hinges must be braced. Some economy in bracing may be achieved by determining the sequence of hinge formation. Further discussion of lateral bracing is given in Art. 6.3 of Ref. 2.

19. Local Buckling Local buckling may also inhibit hinge formation; this can be precluded by limiting the width-thickness ratios of the plate elements which comprise the member. If these ratios are less than the limits listed below, the average strains in the compressed plate elements near plastic hinges can reach strain hardening before local buckling sets in. Rotations corresponding to these strains are sufficient to permit the development of the mechanism. The maximum width-thickness ratios are

1. Projecting elements (flanges of rolled shapes, or similar outstanding elements, stiffeners, etc.), $b/t \lessgtr 52/\sqrt{\sigma_y}$, where b is the width of the element, t its thickness, and σ_y the yield stress in ksi

2. Flange plates in box sections, and similar elements, $b/t \lessgtr 190/\sqrt{\sigma_y}$
3. Webs of rolled or built-up shapes

$$\frac{d}{w} \lessgtr \frac{412}{\sqrt{\sigma_y}} \left(1 - 1.4\,\frac{P}{P_y}\right) \qquad \text{for } \frac{P}{P_y} \lessgtr 0.27 \qquad (22a)$$

$$\frac{d}{w} = \frac{257}{\sqrt{\sigma_y}} \qquad \text{for } \frac{P}{P_y} > 0.27 \qquad (22b)$$

(but need not be less than $257/\sqrt{\sigma_y}$, where d is the depth and w the thickness of the web).

20. Shear Although theoretical considerations suggest that the plastic moment may be reduced in the presence of large shearing forces, tests show that this is not true for wide-flange and I shapes. Since large shear means a steep moment gradient, yielding under these conditions is localized so that the beneficial effects of strain hardening enable the beam to reach its full plastic moment. It is sufficient to satisfy the requirement that the maximum shear not exceed

$$V = 0.55\sigma_y wd \qquad (23)$$

where d is the depth of the beam and w the web thickness. If this condition is not satisfied, a section with a larger web thickness may be used or w increased by a doubler plate.

21. Frame Instability A frame may become unstable as a whole, especially if relatively high axial forces are present and no sway bracing is provided. Frame instability is essentially a magnification of the lateral deflections to the point where uncontrolled sway occurs with no increase in load.

Stability can be controlled by the stiffness of the rigid framing system itself or by providing a vertical bracing system (e.g., diagonal or K bracing or shear walls). Frames with a vertical bracing system are called *braced frames* and those whose stability depends on their own stiffness *unbraced frames.*

The analysis and design of braced and unbraced multistory frames requires special considerations involving the determination of second-order story shear forces arising from the product of the gravity load and the story sway deflection (P-Δ analysis). These stability requirements are defined in Sec. 2.3 of the AISC specification.[15] Methods for considering frame stability in braced multistory frames are presented in Refs. 2, 10, and 16. The latter is devoted entirely to this topic. Stability design requirements of unbraced plastically designed multistory steel frames are discussed in Refs. 2, 10, and 14.

The second-order effect need not be considered for one- or two-story frames for which the methods of plastic analysis and design given in this chapter apply. If such frames are braced by a vertical system, consideration of frame stability is not required. For unbraced rigid frames stability considerations are met by modifying certain terms in Eqs. (7) and (8) as follows:

1. P_{cr} is determined for the larger of KL_x/r_x or L_y/r_y, where L_x and L_y are the unbraced lengths.
2. P_e is determined for KL/r_x.
3. M_p is replaced by M_m from Eq. (14) or Eq. (15).
4. $C_m = 0.85$.

The in-plane effective length factor K is determined by elastic-stability analysis as discussed in Sec. 6. Example 15 illustrates the treatment of stability in a two-story frame.

CONNECTIONS

Plastic hinges often form at the intersection of two or more members. Since the connecting devices are located at these points, they are subjected to severe loading conditions. The basic requirement for most connections is the same: they must be designed so that they will develop the plastic moments of the members joined and must have an adequate reserve of ductility in order to assure that all required plastic hinges will form throughout the structure.

22. Corner Connections Ignoring the shear and axial forces on the knee of the simple beam-to-column connection shown in Fig. 9, the force in the flange of the beam is given by $T = M_p/d_b$. Considering equilibrium of the top flange, the maximum possible value of T corresponds to the condition in which the web is yielded all along the flange. Thus, $T_w = \tau_y w d_c$, and since $\tau_y = \sigma_y/\sqrt{3}$,

$$\frac{M_p}{d_b} = \frac{\sigma_y}{\sqrt{3}} \, w d_c$$

from which

$$w = \frac{\sqrt{3} M_p}{\sigma_y d_b d_c} \tag{24}$$

This analysis may be used in the design of many types of connections. Frequently, as in the case of a connection that fails to meet the design requirement without stiffening, the required size of the stiffener may be determined by considering the combined resistance at full yield of the several parts which act together. Thus, in a diagonally stiffened corner connection, the required area of stiffener can be determined by considering the combined resistance at full yield of the web and the diagonal stiffener.

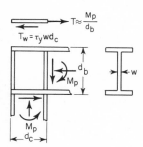

Fig. 9 Corner connection.

Unless properly braced, haunched knees may exhibit poor rotation capacity. Therefore, bracing must be supplied at the haunch extremities and the intersections of the compression flange, and the compression flange must be checked between these braced points to assure that it does not buckle. In the event of inadequacies, additional bracing must be supplied or the proportions of the connection changed.

Curved haunches must be checked for the same factors as tapered haunches, and in addition the inner compression flange must be checked for "cross bending." This is necessary because the flange tends to bend across the web. The flange thickness must be greater than $b^2/2R$ for satisfactory performance, where b is the flange width and R the haunch radius. For curved knees of normal proportions an increase of one-third in flange thickness over that of the flanges in the adjoining member is normally adequate to assure that the plastic moment will be sustained throughout the haunch.

Table 2 lists formulas for proportioning corner connections. Applications are given in Examples 14 and 15, which are adapted from Ref. 4.

23. Interior Beam-to-Column Connections If beams attached to a column transmit unequal moments, the column web must resist the unbalance and must be proportioned to keep shear deformation within suitable limits. In addition, the connection must also resist the thrusts of the flanges of the adjoining beams. Formulas are given in Table 2 and an illustration in Example 15.

In proportioning the elements of connections it is important to keep in mind the local buckling requirements (Art. 19).

TABLE 2 Formulas for Design of Connections

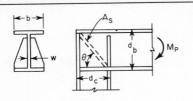

Corner connection

$$w \geq \frac{\sqrt{3}\, M_p}{\sigma_y d_b d_c} \quad \text{without diagonal stiffener}$$

$$A_s = \frac{1}{\cos\theta}\left(\frac{M_p}{\sigma_y d_b} - \frac{w d_c}{\sqrt{3}}\right) \quad \text{if needed}$$

or use doubler plates to increase w.

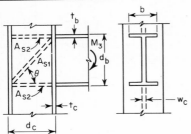

Beam-to-column connections

$$w_c \geq \frac{\sqrt{3}\, M}{\sigma_{yc} d_b d_c} \quad \text{without diagonal stiffener}$$

$$A_{s1} = \frac{1}{\cos\theta}\left(\frac{M}{\sigma_{yc} d_b} - \frac{w_c d_c}{\sqrt{3}}\right) \quad \text{if needed}$$

where $M = M_2 - M_1$ for interior connection

$\qquad M = M_3$ for exterior connection

$\qquad d_b$ = depth of larger beam

$$w_c \geq \frac{b t_b}{t_b + 5 k_c}\, \frac{\sigma_{yb}}{\sigma_{yc}}$$

$$w_c \geq \frac{d_c}{30} \quad \begin{array}{l}\text{(interior}\\ \text{conn. only)}\end{array}$$

$$t_c \geq 0.4\,\sqrt{b t_b \frac{\sigma_{yb}}{\sigma_{yc}}}$$

$\left.\phantom{\begin{array}{c}1\\2\\3\\4\\5\end{array}}\right\}$ without flange stiffener

$$A_{s2} = b t_b - w_c\,(t_b + 5 k_c) \quad \text{if needed}$$

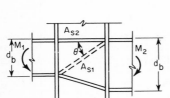

TABLE 2. Formulas for Design of Connections (*Continued*)

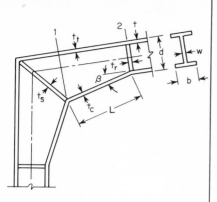

Tapered haunch

$Z = \dfrac{M}{\sigma_y}$ at sections 1 and 2

$t_t = t$ if $L < 4.8b$

$t_t = t\left[1 + 0.1\left(\dfrac{L}{b} - 4\right)\right]$ if $L > 4.8b$

$t_c = \dfrac{t_t}{\cos \beta}$

$\left.\begin{array}{l} t_s = \sqrt{2}\,t - 0.82\,\dfrac{wd}{b} \\[2mm] t_s = \sqrt{2}\,t\,(1 - \tan \beta) \end{array}\right\}$ use larger value

$t_r = t_c \sin \beta$

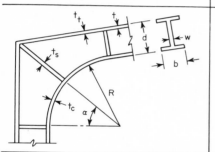

Curved haunch

$R \leqslant \dfrac{4b}{\alpha}$

$t_t = \dfrac{4}{3}t$

$t_c = \dfrac{4}{3}t$ but $> \dfrac{b^2}{2R}$

$t_s = \sqrt{2}\,t - 0.82\,\dfrac{wd}{b}$

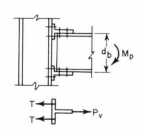

Bolted tee connection

Shear P_v, beam flange to tee:

$$P_v = \dfrac{M_p}{d_b}$$

Tension T: tee to column flange:

$$T = \dfrac{P_v}{n}$$

TABLE 2.   Formulas for Design of Connections (*Continued*)

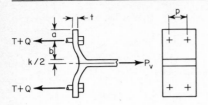

Tension due to prying action:

$$\frac{Q}{T} = \frac{\frac{1}{2} - \frac{pt^4}{30\,ab^2 A_b}}{\frac{a}{b}\left(\frac{a}{3b}+1\right)+\frac{pt^4}{6\,ab^2 A_b}}$$

where A_b = nominal area of bolt.

If $a > 1.25b$, use $a = 1.25b$.

If column flange is thinner than tee stub flange provide stiffeners.

Check bolt forces for $T+Q \lessgtr$ proof load.

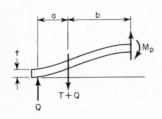

Tee-stub flange:

Compute bending moment at bolt line

$$M = Qa$$

and at edge of fillet radius

$$M = Tb - Qa$$

Moments must not exceed

$$M = \frac{\sigma_y\,pt^2}{4}$$

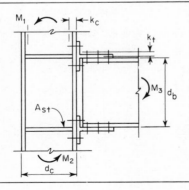

Column web w_c:

$$w_c \gtrless \frac{P_v}{\sigma_{yc}(k_t + 5k_c)} \quad \text{without stiffener}$$

$$A_{st} = \frac{P_v}{\sigma_{ys}} - w_c(k_t + 5k_c)\frac{\sigma_{yc}}{\sigma_{ys}} \quad \text{if needed}$$

Web thickness for shear:

$$w_c = \frac{\sqrt{3}\,M_3}{\sigma_y\,d_b d_c}$$

TABLE 2.   **Formulas for Design of Connections (*Continued*)**

	Shear, ksi					α	
	Fric-tion type	Bear-ing type	Bearing	Tension	Combined tension and shear	Shear through shank	Shear through threads
			High-strength bolts at ultimate load				
A325	20	50	Less than tensile strength	$1.15 \times$ P.L.	$V^2/\alpha^2 + T^2 = (1.1 \times \text{P.L.})^2$	0.8	0.65
A490	30	68		P.L.	$V^2/\alpha^2 + T^2 = (\text{P.L.})^2$	0.75	0.55

Welds at ultimate load

	Fillet welds, shear on throat		Tension on butt welds
	ksi	lb per in. per $\frac{1}{16}$ in.	
E60	30	1300	σ_y of base metal
E70	35	1550	

Example 14 Design of gabled frame. A36 steel.

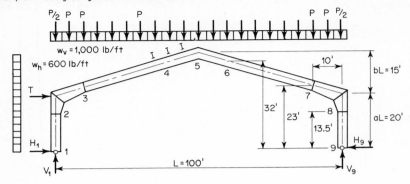

Structure and working loads
Purlins 5' o.c.

Case 1. DL + LL: $w_u = 1.7 \times 1.0 = 1.7^{k}/ft$, $P_u = 5 \times 1.7 = 8.5^{k}$

Case 2. DL + LL + wind: $w_u = 1.3 \times 1.0 = 1.3^{k}/ft$, $P_u = 5 \times 1.3 = 6.5^{k}$

$$w_h = 1.3 \times 0.6 = 0.78^{k}/ft$$

$$T = \frac{w_h(a+b)^2 L^2}{2a} = \frac{0.78 \times 35^2}{2 \times 20} = 23.9^{k} = 3.67P$$

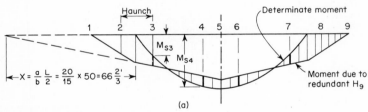

(a)

Assume mechanism shown in (b)

$$M_{S3} = 10 \times 10P - 5P - 10P/2 = 90P$$

$$M_{S4} = 40 \times 10P - 20 \times 7P - 40P/2 = 240P$$

$$M_3 = 23H_1 - M_{S3} = M_4 = M_{S4} - 32H_1 = M_p$$

$$H_1 = \frac{M_{S3} + M_{S4}}{23 + 32} = \frac{330P}{55} = 6P$$

$$M_p = 240P - 32 \times 6P = 48P$$

$$M_{p(col)} = 13.5H_1 = 13.5 \times 6P = 81P$$

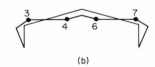

(b)

Moment diagram (a) shows all $M \lessgtr M_p$

Case 2 analysis (not given) shows Case 1 is critical.

$\therefore$ $M_p = 48P = 48 \times 8.5 = 408^{\,\text{ik}}$ girder

$M_p = 81P = 688.5^{\,\text{ik}}$ column

$H_1 = H_9 = 6P = 51^{\,k}$

$V_1 = V_9 = 10P = 85^{\,k}$

Girder: $Z = \dfrac{M_p}{\sigma_y} = \dfrac{408 \times 12}{36} = 136$ in.^{3} Use W21×68 $\begin{cases} A = 20.0 \quad b = 8.27 \\ d = 21.13 \quad t = 0.685 \\ w = 0.430 \ r_y = 1.80 \end{cases}$
$Z = 160$

Column: $Z = \dfrac{688.5 \times 12}{36} = 230$ in.^{3} Use W24×94 $\begin{cases} A = 27.7 \quad b = 9.06 \\ d = 24.31 \quad t = 0.875 \\ w = 0.515 \ r_y = 1.98 \end{cases}$
$Z = 254$

Axial force:

Column: $\dfrac{P}{P_y} = \dfrac{V_1}{\sigma_y A} = \dfrac{85}{36 \times 27.7} = 0.085 < 0.15$

Girder: $\tan \theta = 15/50$, $\theta = 16°40'$

$P = H_9 \cos \theta + (V_9 - 2.5P) \sin \theta$ (at 3,7)

$= 51 \times 0.96 + 63.8 \times 0.288 = 67.2^{\,k}$

$\dfrac{P}{P_y} = \dfrac{67.2}{36 \times 20.0} = 0.093 < 0.15$

No increase in size required because of axial force

Shear:

Column: $V_{all} = 0.55 \, \sigma_y wd = 0.55 \times 36 \times 0.515 \times 24.31 = 248^{\,k}$

$V_{max} = H_1 = 51^{\,k}$ O.K.

Girder not critical

Local buckling:

Girder: $b/t = 4.14/0.685 = 6 < 8.5$; $d/w = 21.13/0.43 = 50 < (412/\sqrt{36})(1 - 1.4 \times 0.093) = 59.7$

Column: $b/t = 4.53/0.875 = 5.2 < 8.5$; $d/w = 24.31/0.515 = 47 < (412/\sqrt{36})(1 - 1.4 \times 0.085) = 60.5$

Lateral bracing:

Girder: purlins 5' o.c. $\dfrac{L}{r_y} = \dfrac{5 \times 12}{1.80} = 33.3 < \dfrac{1,375}{36} = 38$

Column: provide brace midway between sections 1 and 2

$\dfrac{L}{r_y} = \dfrac{6.75 \times 12}{1.98} = 41 < \dfrac{1,375}{36} + 25 = 63$

Bracing details :

Haunches : provide bracing to inner (compression) flange at each end (sections 7 and 8) and at center.

Peak : provide bracing to inner flange.

Purlins : must be adequately braced in order to provide lateral support to rafters.

Connection details :

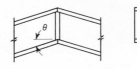

Peak : proportion stiffener to transmit flange thrust.

$\sigma_y A_s = 2\sigma_y A_f \sin\theta$

$bt_s = 2bt \sin\theta$

$t_s = 2 \times 0.685 \times 0.288 = 0.40"$ Use $\frac{1}{2}$ pl.

Haunch   (Table 3):

Use $\beta \approx 13°$

$d_1 = 44"$

$Z_1 = bt(d_1 - t) + \frac{W}{4}(d_1 - 2t)^2$

$= 8.27 \times 0.685 (44 - 0.685)$

$+ \frac{0.5}{4}(44 - 1.37)^2 = 472$ in.^{3}

$\frac{M_h}{\sigma_y} = \frac{20 H_9}{\sigma_y} = \frac{20 \times 51 \times 12}{36} = 340$ in.^{3} O.K.

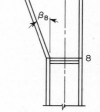

Taper width from $8\frac{1}{4}$ to 9

Flange :

Width the same as W21 × 68 , use $8\frac{1}{4}"$

Length of compression flange = 108"

$L_{cr} = 4.8b = 4.8 \times 8.25 = 40.8"$

$t_t = t\left[1 + 0.1\left(\frac{L}{b} - 4\right)\right] = 0.685\left[1 + 0.1\left(\frac{108}{8.25} - 4\right)\right] = 1.31$

$t_c = t_t / \cos\beta = 1.31 / 0.98 = 1.34$. Use $1\frac{3}{8}"$ pl.

Transverse stiffeners :

$t = t_c \sin\beta = 1.375 \times 0.284 = 0.39"$

$t = b/8.5 = 4/8.5 = 0.47$. Use $\frac{1}{2}"$ pl.

Diagonal :

$t_s \gtrless \sqrt{2}\, t - 0.82\, wd_1/b = \sqrt{2} \times 1.375 - 0.82 \times 0.5 \times 44 / 8.25 =$ negative

$t_s \gtrless \sqrt{2}\, t(1 - \tan\beta) = \sqrt{2} \times 0.685 \times 0.769 = 0.74"$. Use $\frac{3}{4}"$ pl.

Example 15. Design of two-bay two-story unbraced frame. A36 steel

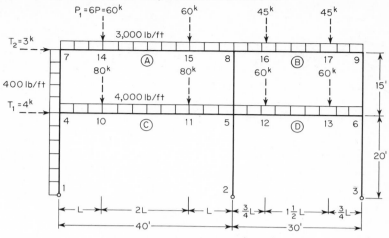

Structure and working loads: $P = 10^k$, $L = 10'$

Replace roof and floor loads by concentrated loads at quarter points
Replace wind load by concentrated loads at roof and floor

$$T_1 = \frac{400 \times 20 \times 10}{20} = 4^k \qquad T_2 = \frac{400 \times 15 \times 7.5}{15} = 3^k$$

Plastic moment ratios:
    For simultaneous mechanisms in spans A and B use plastic
moments in ratio of squares of spans.

$k_B \qquad = 1 \qquad k_D = (60/45)\, k_B = 1.33$
$k_A = (4/3)^2\, k_A = 1.78 \qquad k_C = (4/3)^2\, k_D = 2.37$

Use $k = k_A$ for column $1-4-7$, $k = k_B$ for columns $2-5-8$ and $3-6-9$

23 possible hinges $\left\{\begin{array}{ll} \text{hinge at } 7,14,15,16,17,9,10,11,12,13 & =10 \\ 3 \text{ hinges at } 4,6,8 & = 9 \\ 4 \text{ hinges at } 5 & = 4 \end{array}\right.$
 9 redundants
14 independent mechanisms

8 beam mechanisms: 7–14–8, 7–15–8, 8–16–9, etc.
4 joint mechanisms: 4, 5, 6, 8
2 panel mechanisms

| Typical beam mechanism | Typical joint mechanism | Panel mechanism | Panel mechanism |

Case 1 loading: DL+LL			
Mechanism	Internal work $\dfrac{W_i}{M_p\theta}$	External work $\dfrac{W_e}{P\theta L}$	$\dfrac{M_p}{P_u L}$
1 4/3 1/3 (beam)	$k_A(1+\frac{4}{3}+\frac{1}{3})=\frac{8}{3}\times1.78=4.74$	$6\times1\times1+6\times\frac{1}{3}\times1=8$	1.69
1/3 4/3 1 (beam)	Same as above		1.69
1 4/3 1/3 (beam)	$k_B(1+\frac{4}{3}+\frac{1}{3})=\frac{8}{3}\times1=2.67$	$4.5\times1\times\frac{3}{4}+4.5\times\frac{1}{3}\times\frac{3}{4}=4.5$	1.69
Remaining beam mechanisms similar Panel mechanisms not critical			1.69 —
1 4/3 4/3 4/3 4/3 (combined) k_A, k_B, k_C, k_D	$k_A(\frac{4}{3}+\frac{4}{3})+k_B(\frac{4}{3}+\frac{4}{3})$ $+k_C(\frac{4}{3}+\frac{4}{3})+k_D(\frac{4}{3}+\frac{4}{3})$ $=\frac{8}{3}(k_A+k_B+k_C+k_D)$ $=\frac{8}{3}(1+1.78+2.37+1.33)$ $=17.28$	$6(1+\frac{1}{3})+4.5(1+\frac{1}{3})\times\frac{3}{4}$ $+8(1+\frac{1}{3})+6(1+\frac{1}{3})\times\frac{3}{4}$ $=8+4.5+10.67+6$ $=29.17$	1.69
Case 2 loading: DL+LL+ wind from left			
Same as combined mechanism above for Case 1	Same as above $=17.28$	Same as above plus work due to wind forces $=29.17+0.3\times1\times3.5+0.4\times1\times2$ $=31.02$	1.80
Case 3 loading: DL+LL+ wind from right. Same as Case 2			1.80

Case 1 loading   Case 2 loading

$P_u = 1.70\,P = 1.70 \times 10 = 17.0^k$ $P_u = 1.30\,P = 1.30 \times 10 = 13^k$

$M_P = 1.69\,P_u L = 1.69 \times 17.0 \times 10 = 287^{lk}$ $M_P = 1.80\,P_u L = 1.80 \times 13 \times 10 = 234^{lk}$

Case 1 controls: $M_{PA} = 1.78\,M_P = 1.78 \times 287 = 511^{lk}$ $M_{PC} = 2.37\,M_P = 681^{lk}$

                 $M_{PB} = 1 \times M_P = 287^{lk}$ $M_{PD} = 1.33\,M_P = 382^{lk}$

Number of remaining redundancies (Eq.4): $I = X - (M-1) = 9 - (8-1) = 2$

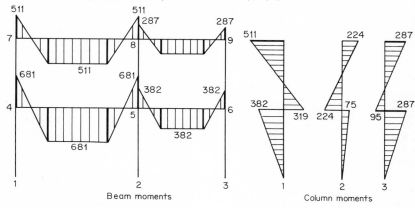

Beam moments   Column moments

Case 1 loading

Moment check:

Joint 8: $M_{85} = -M_{87} - M_{89} = 511 - 287 = +224^{lk}$

Joint 6: assume $M_{63} = +M_P (= +M_{PB})$

     $M_{69} = -M_{65} - M_{63} = 382 - 287 = +95^{lk}$

Sway equilibrium of top story: assume $M_{58} = M_{85} = 224^{lk}$

     $M_{47} = -M_{74} - M_{58} - M_{85} - M_{69} - M_{96}$

          $= -511 + 224 + 224 + 95 + 287 = 319^{lk}$

Joint 4: $M_{41} = -M_{45} - M_{47} = -681 + 319 = -362^{lk}$

Sway equilibrium of bottom story:

     $M_{41} + M_{52} + M_{63} = +362 - 75 - 287 = 0$

Moment check shows all $M \lessgtr M_P$

Reactions for Case 1:

$V_1 = 6P + 8P = 14P_u = 14 \times 17.0 = 238^k$

$V_2 = 6P + 8P + 4.5P + 6P \quad = 417^k$

$V_3 = 4.5P + 6P \quad = 179^k$

From column moment diagrams:

$H_1 = 287/20 = 14.4^k$ (maximum possible value $= 511/20 = 25.6^k$)

$H_2 = 75/20 = 3.8^k$ (maximum possible value $= 287/20 = 14.4^k$)

$H_3 = 287/20 = 14.4^k$

Member		M, ft-kips	P, kips	Section *	M_p, ft-kips	P_y, kips	G_T, **	G_B, **	K **	KL/r_x	M, ft-kips Eq.(8)	Σ Eq.(7)
	1-4	511	238	W18×106	690	1120	2.7	10	2.2	67	641	0.96
	4-7	511	102	W18×86	558	911	2.1	2.7	1.7	39	585	0.91
	2-5	287	417	W12×120	558	1271	0.9	10	1.7	74	442	0.96
	5-8	287	179	W12×79	357	835	0.6	0.9	1.2	40	330	0.94
	3-6	287	179	W12×96	441	1015	1.9	10	2.0	88	429	0.90
	6-9	287	77	W12×79	357	835	1.4	1.9	1.5	51	357	0.80
	7-8	511	—	W24×68	531							
	8-9	287	—	W21×50	330							
	4-5	681	—	W27×84	732							
	5-6	382	—	W24×55	402							

*Local buckling requirements were checked and found to be OK; continuous lateral bracing was assumed to be provided for all members.

**Effective length calculations performed according to AISC Specification, Sec. 1-8

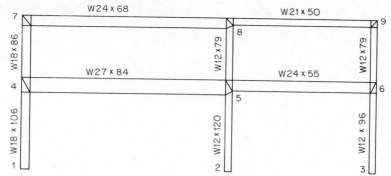

Corner connection joint 9 (Table 3):

$$w = \frac{\sqrt{3}\ M_p}{\sigma_y\, d_b d_c} = \frac{\sqrt{3} \times 287 \times 12}{36 \times 20.83 \times 12.38} = 0.643 > 0.470$$

$$A_s = \frac{1}{\cos\theta}\left(\frac{M_p}{\sigma_y\, d} - \frac{w d_c}{\sqrt{3}}\right) = \frac{24.2}{12.38}\left(\frac{287 \times 12}{36 \times 20.83} - \frac{0.47 \times 12.38}{\sqrt{3}}\right)$$

$$= 2.42\ \text{in.}^2$$

For 3" plates: $t = 2.42/6 = 0.403$, Use $\tfrac{1}{2}$"

$b/t = 3/0.5 = 6 < 8.5$

Use 3 × 1/2 diagonal stiffeners

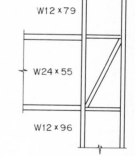

Side connection joint 6 (Table 3):

$$w = \frac{\sqrt{3}\ M_p}{\sigma_y\, d_b d_c} = \frac{\sqrt{3} \times 382 \times 12}{36 \times 23.57 \times 12.38} = 0.756 > 0.470$$

$$A_{s1} = \frac{1}{\cos\theta}\left(\frac{M_p}{\sigma_y\, d_b} - \frac{w d_c}{\sqrt{3}}\right) = \frac{26.60}{12.38}\left(\frac{382 \times 12}{36 \times 23.57} - \frac{0.47 \times 12.38}{\sqrt{3}}\right)$$

$$= 4.40\ \text{in.}^2$$

For $3\tfrac{3}{4}$" plates: $t = 4.40/7.5 = 0.59$, use $\tfrac{5}{8}$"

Use $3\tfrac{3}{4}$ × $\tfrac{5}{8}$ diagonal stiffeners

Flange stiffener:

$$w = \frac{b t_b}{t_b + 5 k_c} = \frac{7.0 \times 0.505}{0.505 + 5 \times 1.44} = 0.46 < 0.470$$

No stiffeners needed, but use nominal 3 × $\tfrac{3}{8}$ stiffener

REFERENCES

1. "Manual of Steel Construction," American Institute of Steel Construction, New York, 8th ed., 1979.
2. "Manual of Engineering Practice," no. 41, American Society of Civil Engineers, New York, 1971.
3. "Plastic Design in Steel," American Institute of Steel Construction, New York, 1959.
4. Beedle, L. S.: "Plastic Design of Steel Frames," John Wiley & Sons, Inc., New York, 1958.
5. Neal, B. G.: "The Plastic Methods of Structural Analysis," John Wiley & Sons, Inc., New York, 1956.
6. Baker, J. F., M. R. Horne, and J. Heyman: "The Steel Skeleton," vol. II, "Plastic Behavior and Design," Cambridge University Press, New York, 1956.
7. Hodge, P. G.: "Plastic Analysis of Structures," McGraw-Hill Book Company, New York, 1959.
8. Massonnet, C., and M. Save: "Plastic Analysis and Design," vol. I, Blaisdell Publishing Company, New York, 1965.
9. Disque, R. O.: "Applied Plastic Design in Steel," Van Nostrand Reinhold, New York, 1971.

10. "Plastic Design of Multi-Story Frames," Summer Conference Lecture Notes, Lehigh University, 1965.
11. Toakley, A. R.: Optimum Design Using Available Sections, *J. Struct. Div. ASCE*, May 1968.
12. Bigelow, R. H., and E. H. Gaylord: Design of Steel Frames for Minimum Weight, *J. Struct. Div. ASCE*, December 1967.
13. Rubinstein, M. F., and J. Karagozian: Building Design Using Linear Programming, *J. Struct. Div. ASCE*, December 1966.
14. Structural Stability Research Council: "Guide to Design Criteria for Metal Compression Members," 3d ed., John Wiley & Sons, Inc., New York, 1976.
15. Specification for the Design, Fabrication and Erection of Structural Steel for Buildings, American Institute of Steel Construction, New York, 1978.
16. American Iron and Steel Institute: "Plastic Design of Braced Multi-Story Steel Frames," Washington, D.C., 1968.
17. Gaylord, E. H.: Plastic Design by Moment Balancing, *AISC Eng. J.*, October 1967.

Fabrication and Erection of Structural Steel

Part 1. Fabrication*

C. F. HARRIS

Chief Engineer—Drawing Rooms, United States Steel Corporation, Pittsburgh, Pa.

The fabrication of structural steel consists of manufacturing the parts required to construct the steel frames of bridges, buildings, and related structures. It involves interpreting design drawings and specifications, making the shop and erection drawings, procuring the necessary materials, cutting, forming, assembling, and fastening the material into units, and shipping it to the point of construction.

1. Drawing-Room Operations In addition to design drawings and specifications the drawing room needs the following information: source of approval of drawings, source of missing information, basis of payment (unit price or lump sum), material items to be included and excluded, erection to be included or excluded, required sequence of delivery, and schedule of delivery dates.

For large structures it is often desirable to divide the structure into sections or divisions to facilitate the scheduling and control of material orders and deliveries, preparation of drawings, and fabrication.

The normal sequence of operations is as follows:

Material Orders. The first step is the preparation of a preliminary bill of material from which the material can be procured while drawings are being made. Fabricators usually stock only a limited amount of commonly used grades and sections of steel, suitable for small connections or emergency needs; therefore, main material must be ordered to exact lengths. Short pieces are ordered in multiples of the exact lengths to supply all needs with a minimum of waste. For simple structures it may be possible to order all material directly from the design drawings, but for more complicated structures, or when designs are not

---

*Dana M. Taylor, Manager of Works (retired), Bethlehem Steel Corporation, Pottstown, Pa., authored Part 1 of this section for the first edition. Much of the material has been retained.

fully developed, it is necessary to make layout or preliminary drawings showing proposed connections and dimensions. These drawings are submitted to the owner's engineers for approval prior to ordering material.

Special items that may require extra time for procurement, such as forgings, castings, or bronze plates, are ordered as early as possible.

Erection diagrams comprise a complete set of plans with elevations and cross sections which locate all pieces and provide essential dimensions and all other necessary erection information. They also establish dimensions required for the detail drawings. For complicated structures, particularly those having curved or irregular alignments, extensive calculations and numerous layout drawings are often required to develop the location of working points and dimensions before work on the diagrams or detail drawings can proceed.

For large structures that require shipment in the sequence of erection, the drawing room arranges with the erector or owner a division into areas known as shipping installments. These are numbered and shown on the erection diagram or on a separate drawing.

Shop drawings are sometimes called detail drawings because they show in detail all the information required to make the component parts and assemble them into shipping pieces. Each component part is assigned an assembly mark for identification, and each shipping piece is assigned an erection mark. Shipping pieces are made as large as can be conveniently fabricated, shipped, and erected. Before the drawings are made for unusually large pieces, their dimensions and weights must be referred to railroads or truckers to determine if they can be cleared to the destination.

The shop drawings are submitted to the owners or designers for approval of details and compliance with designs and specifications, including strength of the connections.

A *bill of material* (shop bill) is written for each shop drawing, listing the assembly mark, quantity, section, dimension, and specification of each part shown on the drawing. The item numbers assigned on the preliminary orders are applied to the bills of material for identification of the material from which the components are to be cut. Bills of miscellaneous material for items such as field rivets, bolts, weld, washers, shims, or other components needed in erection must also be prepared as soon as the required information is available.

A *bill of finished parts* (shipping bill) is prepared for each shop drawing, listing the shipping pieces detailed thereon. The weight of each piece, calculated in accordance with specification requirements, and the shipping installment to which it belongs are also shown. These bills are used for recording production and shipment, and as a source of data for shipping documents.

2. Shop Operations   The size and arrangement of a structural-steel fabrication shop depend upon the type, range, and volume of work to be scheduled. Inasmuch as the handling of material is a substantial part of the cost of fabrication, it is very important that machinery, cranes, and other handling facilities be arranged to provide for an efficient flow of material to and between all operations. It is also important that machinery and handling equipment be adaptable to a wide range of work. This has limited the application of automatic equipment.

The *receiving yard* is most frequently an open crane runway. A contract may require the procurement of dozens of different sections of steel, originating at several different mills with rollings spaced over a period of several weeks. As shipments of steel arrive at the receiving yard, they are checked for specification, size, and quantity, and marked for identification and stored to await shop needs. Stock material is also stored in the receiving yard.

Template Shop. The location of holes, cuts, bends, etc., for steel sections is determined by the use of some form of wood or cardboard template, unless individual layouts made directly on the steel, or some other process, is deemed more economical. After working drawings are issued to the shop, all templates that will be required in the course of fabrication are made. For small pieces, such as connection angles or gusset plates, full-sized cardboard templates are made, locating all holes, cuts, and bends and carrying notations for any other required operations. For irregularly shaped or bent pieces, full-sized wood or cardboard templates are extensively used. Spacing templates are made on long wooden or steel strips to show the longitudinal spacing of holes for plates or sections that can be punched on machines equipped with spacing tables. Full-sized layouts are often made on the template floor for the development of dimensions in roof trusses and curved or irregularly shaped bridge members.

Cutting. Small pieces, often called detail material, are sheared from long plates or angles, using paper or wooden templates to determine the required dimensions. Beams, channels, or other shapes are sawed or flame-cut to length.

Many forms of flame-cutting machines carrying one or more torches and following a track or template by mechanical, magnetic, electronic, or numerically controlled (N/C) guidance are available. Additional heating of the edges may be specified as a means of avoiding embrittlement of flame-cut edges of certain grades of high-strength steel. Machine flame cutting may also be used to remove rough edges from sheared plates in lieu of planing or milling.

Laying out consists of locating working lines, cuts, and holes on the piece manually by the use of tape line, rule, square, or template. It is used wherever no mechanical device is well suited for the purpose. Lines for cuts or bends are marked with talc crayon, and the location of holes is stamped with a steel center punch for accurate centering of the punching or drilling tool.

Punching and Drilling. Rivet and bolt holes are ordinarily formed by punch-press machines adapted to the various types of sections normally used. Detail material is punched through full-sized cardboard templates, one hole at a time. Material on which holes have been located by laying out are punched on a single- or odd-hole punching machine. Long angles, plates, and beams may be passed through multiple-punching machines, equipped with roller tables and a carriage, which space the required holes on one or more parallel gage lines at intervals determined by wood or steel spacing templates. Various degrees of automatic operation may be used to control the movement of the carriage and the operation of the punching machine.

Where the thickness or hardness of the material is too great for satisfactory punching, holes are drilled. Single-spindle drills are generally used, but multiple-spindle machines may be used when the pattern of holes is repeated. The location of holes is determined either by laying out or by template.

Both punching machines and drilling machines may be equipped with spacing equipment having numerical controls. Rectangular coordinates locating all holes are processed by a computer and transferred to a perforated paper tape. The tape is fed into an electronic control unit which automatically regulates the movement of material and/or the operation of the machine to locate accurately and punch or drill the required holes.

Straightening and Bending. Steel-mill tolerances for straightness of structural shapes and plates permit greater variations than is acceptable for many uses. Any material which is not straight enough for proper fabrication must be straightened before fitting.

Long plates are flattened and straightened by passing them through a leveling roll before punching. Side sweep in narrow plates is corrected by applying greater pressure near the inside of the curve. Specially designed rolls curve angles and plates used in the fabrication of cylindrical members such as pipe and tanks. Hydraulic or mechanical presses of various types are used for straightening or bending angles, beams, and assembled sections. Curved or bent sections occasionally require hot bending or forming by special dies, or hand forging. Gutters, curbs, and similar bent shapes are formed from flat plates on a press brake machine. Such work is generally performed without heating, but the radius of the bend must be great enough to avoid cracking the plate along the bend line. Low-yield-stress grades of steel are preferred for such use.

Limited straightening or bending of plates or shapes may also be effected by local applications of heat with a gas torch. The result of locally heating a small area that is surrounded by unheated steel is that upon cooling the heated area becomes smaller and thicker. The successful application of this principle requires considerable experience in order to plan the extent and location of heats needed to achieve the desired results.

Fitting is the most important and varied operation in the structural fabricating shop. All the components required to form an assembly are brought together, fitted, and secured by temporary or permanent bolts, or tack welds. There may be successive fitting operations for large pieces, with small assemblies combined into larger ones until the final size of the shipping piece is attained.

Each shipping piece is marked with paint to show the contract number, drawing number, and erection mark, and also a direction mark if required by the erector.

Reaming.   When assemblies are fitted, there is usually some misalignment of matching holes. This is corrected by reaming all unfair holes with fluted reamers. Specifications for bridges usually require that rivet or bolt holes of all important connections be punched or drilled with holes ¼ in. smaller in diameter than the final size, and reamed to size after the components are assembled. This is known as "subpunched reaming." Holes for shop connections are reamed with manually guided, single-spindle drilling arms attached to a traveling gantry. Holes for field connections are reamed to final size with matching members assembled, or through steel templates.

Riveting.   Structural rivets are heated in gas, oil, or electric furnaces to a cherry-red color. Machine-driven rivets are less costly than those driven by a pneumatic hammer, and are generally used for all groups of rivets accessible to a machine. Cold-driven rivets are used for some applications but are not generally used by the industry. Many fabricators no longer have riveting equipment; so rivets are seldom used.

Bolting.   Bolts are of two types, regular machine and high-strength. The machine bolt is generally used only for minor connections. The high-strength bolt is available in different grades and sizes and is used extensively for both shop and field fasteners. Special installation requirements must be observed and monitored to assure that high-strength fasteners are installed with proper surface preparation and tension.

Welding.   Practically all welding in structural-steel fabrication is by some form of the electric-arc process, either manual or automatic. In manual arc welding the arc length, rate of metal deposit, and quality of the weld depend to a large extent upon the skill and care of the operator. This process has the great advantage of adaptability, in that it can be used for applications that are too varied or inaccessible for the use of automatic equipment.

Automatic welding is performed by a mechanically guided electric arc. Regulation of electrical characteristics, feed of the electrode, and other factors are mechanically controlled. Two or more arcs having the same or varied characteristics may be operated in tandem along a seam to increase the rate of production. This process has the advantage of better quality control and lower production cost, but cannot be adapted to many fabrication requirements. Semiautomatic welding is a modified form of automatic welding wherein the movement of the arc along the seam is controlled by hand. This process is more adaptable than the automatic, but less adaptable than manual welding.

Finishing and Machining.   Machine finishing is required for the ends of compression members to provide a smooth surface which will bear uniformly against the abutting surface. Machine finishing is also required for pins and pinholes, and bearing surfaces on bridge shoes and bearing plates.

The edges of web plates or gusset plates may require finishing to remove the irregularities resulting from the shearing process. This is accomplished by a planer or milling machine or by machine flame cutting.

Chipping with a pneumatic hammer is used to remove excess material from countersunk rivet heads and to cut out welds or to smooth irregular surfaces. This is a slow and expensive process. Electric or pneumatic portable grinders are used to smooth rough edges or ends of material.

Bridge Assembling.   The specifications for large bridges or other complicated structures often require that adjoining main members be assembled to ensure proper fit and for reaming the subpunched splice holes. This operation is usually performed in an open runway served by heavy-capacity cranes and equipped with portable machinery for drilling or reaming from the top, bottom, and sides of the assembly. Very satisfactory results with bridge trusses are obtained by assembling the top and bottom chords separately for reaming the splices, and by reaming the connections for the verticals, diagonals, and bracing to accurately located steel templates. Matching steel templates are used for reaming the connections on the verticals, diagonals, and bracing members.

Members whose connections have been drilled with numerically controlled equipment usually have the holes drilled full size. The extreme accuracy of this equipment eliminates the necessity of assembly and reaming; however, it is usual to make representative test assemblies to prove the accuracy.

Painting.   The application of one coat of primer paint is ordinarily required. The type of paint, method of surface preparation, and method of application is specified by the owner. Shop painting is generally performed in the shipping yards following all other shop operation and final inspection.

The owner should give consideration to the total paint system in specifying the primer and surface preparation since the performance of this system is a design consideration and thus is outside the scope of responsibility of the fabricator or fabricator-erector. Reference is made to the Steel Structures Painting Council (SSPC) publications for recommendations of several painting systems.

Shipping. The shipping yard handles material for painting, sorts it into installments, and stores it until time for loading. Local shipments are usually transported by trucks, whereas more distant shipments are transported on railroad cars or river barges. Larger pieces, exceeding the length of a single car, may be loaded on two or more cars. The dimensions of unusually high, wide, or long loads must be kept within previously determined clearance limits. Continuity of erection requires that material be segregated into the proper installments and delivered according to schedule.

3. Suggestions for Design Engineers Many engineers cannot be fully informed regarding the problems of steel fabrication and, as a result, may include features in the design that add unnecessarily to the cost. The following suggestions are offered as a guide:

1. Follow the AISC Code of Standard Practice.

2. Visits to fabrication shops and consultations with fabricators in advance of design will supply information not available elsewhere, and should prove valuable, particularly for the design of unusual structures.

3. A separate set of design drawings showing all structural steel fully dimensioned for location and elevation will permit the drawing room to work more effectively. When this information is incorporated on the architectural drawings it is often necessary for the drawing room to make plans and obtain approval of dimensions before the material can be ordered. This delays fabrication and frequently leads to errors resulting from misinterpretation of the design.

4. Avoid revisions after work on a structure has been started. They add to the cost and delay fabrication to a greater extent than one might expect. Changes in the section or length of material are likely to involve ordering new material and stocking or scrapping the old. Since most material is ordered to cut length, its value is greatly depreciated for other applications. Partial releases of design drawings are also to be avoided. This delays fabrication and results in extra costs for which the fabricator is entitled to recompense. If complete information cannot be supplied for normal handling of the fabrication, it is frequently more economical to make the necessary changes during erection rather than to interrupt or delay fabrication.

5. Uniformity of span lengths for bridges and of bay lengths for buildings will result in less costly fabrication. The design of a structure should be kept as uniform as its function permits. For example, when an expansion joint is required in a building, the bay length on each side of the joint should be that of the typical bay, thereby avoiding special-length material which entails additional drawings and fabrication cost.

6. Specify machine finishing only where necessary, and use liberal machining tolerances. Machine finishing is always expensive, and the cost is increased as the tolerances are decreased. For example, designs for bridge bearings often require machining on surfaces where it is unnecessary, or require closer finish tolerances than are needed in service.

7. Critical dimensions and/or tolerances should be identified, preferably on the drawings.

8. Minor changes in size or section of material to obtain the lightest possible weight are not always economical. Consideration should be given to the design of connections and the possible repetitive nature of the pieces, since the overall cost is the primary objective.

Part 2. Erection*

D. B. REES

Director—Construction Procedures, American Bridge Division,
United States Steel Corporation

The following discussion of erection problems is intended primarily for the engineer, the architect, the detailer, and the many design agencies of the various government offices who have to do with creating useful structures in a competitive economy.

4. Drawings General design drawings indicating governing dimensions, loading, and special information, and stress sheets showing sections of all members are required. For tier buildings, column and beam schedules for each tier and floor are sufficient. On plans for bridges and unusual structures, stresses and required sections should be noted on the individual members. Explanatory notes should be clear and concise.

If there is a general and detailed location plan, as in the case of bridges, it should show all surrounding physical conditions, contours, roads, public facilities, power lines, and other structures. Available boring charts and other information important to the erector should be shown. Special requirements or restrictions concerning the site or use of other facilities should be noted.

Detail Drawings. Standard details and connections, as in beam-and-column work, should be indicated by reference to the applicable code or manual. Typical details of the American Institute of Steel Construction Manual should always be permitted when possible. Large joints and connections should be detailed sufficiently to give the fabricator and erector a true picture of what is required. Detailed and dimensioned shop drawings are made by fabricators to conform to their methods, shop practices, and erection requirements. Special or unusually heavy types of details greatly increase the cost of detailing, fabrication, and erection.

Connections and other details affecting field assembly, such as erection clearances and arrangement of parts to permit easy assembly, are usually standardized as far as possible. Multiple, overlapping, and inaccessible connections are to be avoided. Weight and length of members and location of field splices must be considered. The fabricator and erector should be permitted to locate and arrange field splices and connections, consistent with structural requirements and architectural appearance.

5. Specifications Specifications should be complete enough to give the erector the information required for estimating costs of field work, including any unusual items. Reference to "standard" or "usual," or such phrases as "satisfactory to the engineer" or "in a workmanlike manner" should not be used. Any unusual requirements for tightening and testing high-strength bolts, inspecting or testing field welding, etc., should be clearly specified.

Specifications should clearly state the required standards of workmanship and excellence, but not the methods used to obtain them. While consulting engineers have the responsibility of protecting the client's interest in obtaining an economical and safe

---

*E. L. Durkee, Engineer of Erection (deceased), Bethlehem Steel Corporation, authored Part 2 of this section for the first edition. Much of the material has been retained.

structure, they should permit erectors to use their ingenuity to the greatest extent possible consistent with reasonable safety. Temporary structures, such as falsework, should not be subject to the same requirements as the permanent structure, unless so stated in the specifications.

6. Consultation with Erector Prior to advertising for bids, or at any time during the preparation of the plans, the designer may choose to consult informally with a fabricator and erector on items such as maximum sizes and weights for shipping, handling in the field, and erecting with standard equipment. Site conditions frequently govern and limit the scheme of erection, and this in turn may control the location, type, and number of splices and the arrangement of details in general.

It may not be possible to incorporate all the ideas and suggestions of various contractors as to schemes of erection. In such cases, only the necessary and essential structural data should be given on the design drawings. Typical details may be shown, but as much latitude as is possible without affecting strength or appearance should be permitted in rearranging them.

Frequently, the prescribed sections and makeup of members may be subject to discussion and change to suit fabrication and erection. Improvements in welding and other fabricating processes may suggest designs which are economical in weight and have architectural and aesthetic advantages but which may prove much more expensive to erect because of additional temporary supports and falsework that may be required.

7. Budget Cost Estimates The designer must usually make an estimate of the cost of the erected steelwork for budget purposes. For certain standard types of construction the estimate may be based on the cost of similar structures of recent construction. For more unusual or complicated structures, or where erection conditions are entirely different and more difficult, the engineer may release a more or less complete set of preliminary plans to a few selected firms for an approximate cost estimate. Such figures should be kept confidential, of course, and the engineer should not hold the contractor to the preliminary figure when submitting the formal bid later. A high and a low figure might be asked for, without necessitating a detailed survey and cost estimate for which the contractor cannot be reimbursed.

8. Bidding When soliciting firm bids, the designer should give as much time as possible for the bidders to prepare their estimates. Frequently they are forced to ask for an extension of time because of the extent and complication of the structure and pressure of other work. Generally, the more time that can be given, the more accurate and realistic the bids will be.

9. Schedules It is customary to require some kind of schedule for the various phases of the construction to be submitted with the bid, even though the completion date of the entire contract is usually specified. Preparation of detailed and elaborate schedules is time-consuming and costly, and should be required of the successful bidder only. This applies particularly to computerized scheduling aids, such as the critical-path-method chart covering all individual operations.

The engineer should have a realistic overall schedule of the various construction phases from excavation, driving of piles, construction of foundation and piers, and on through to the finished structure. A contractor should not be required to begin a phase of the work before another has completed operations at least sufficiently for the work to continue without interruption and undue interference. Foundations and piers should be scheduled for completion in proper order and in time for steel erection to proceed as planned, especially for bridges. If this is not economical or desirable, erectors should be permitted to postpone the start of their work until they can proceed according to the original plan.

DESIGN CONSIDERATIONS

10. Economy in Design Economy in steel construction starts with the designer's concept of the final structure. Problems of fabrication, transportation, and field assembly must be considered during the design stages to achieve the most satisfactory and economical result. The ability to conceive, analyze, and design a structure does not necessarily imply a knowledge of the problems that may be encountered in fabricating and erecting it, and it is possible to design a structure that would be impractical, if not impossible, to build. Of course, this does not mean that new ideas should not be tried out. However, the designer should, consistent with any unusual requirements, keep the structure as simple as possible and still meet utilitarian requirements and functions. Unusual or "trick" designs should be avoided unless they serve a valid purpose, as in the case of exhibition and show buildings, or where it is desired to demonstrate feasibility of new types of construction.

Composite construction and orthotropic design frequently result in lighter and, theoretically, more economical structures, but they have at the same time increased costs of fabrication and have raised serious problems for the erector.

Greatest economy in any structure will be realized when there is as much duplication of the various units and parts as is possible. Structures in which every bay and every span are even slightly different will result in increased costs of shop drawings, of fabrication, and probably of erection.

The designer should not specify unusual and unnecessary tolerances. When close tolerances are required, as for elevator guides and connection of prefabricated siding, window frames, etc., adjustable guides and supporting members should be furnished. It is impractical and expensive to require machine-shop accuracy and workmanship in ordinary structural members.

The greatest overall usefulness and economy in steel structures can be obtained only if the designer, fabricator, and erector work together as a team—hence the desirability of consultation as early as possible, especially for unusual and monumental structures. Out of such conferences have come many advances in design, materials of construction, and the development of new equipment and methods for fabrication and erection. Such cooperation promotes development in the structural field and eliminates expensive and impractical ideas before they have advanced to the stage where they are difficult to revise or change. However, both fabricator and erector should be prepared to undertake new kinds of construction which may require new methods and equipment if this is what is wanted.

The designer should always remember that the least weight does not necessarily mean least cost. Not only should the initial cost of material be considered, but also the difficulty of handling long, limber pieces during fabrication, shipping, and erection. This handling cost is often more significant than the total cost of the material. In using composite construction on bridges and the floor systems of multistory buildings, the designer should not overlook the difficulties encountered by the fabricator/erector with the long, flexible members. This is not to say that weight should not be saved whenever possible, either by design or by changing the grade of material and utilizing a higher working stress, but rather that a full understanding of all effects is necessary. A good question to ask is "Will this member, because of its limberness, require temporary stiffening elements, expensive falsework or additional erecting cranes to provide its lateral stability during erection?" If the answer is "yes," consideration should be given to increasing the size of the member or its elements.

Where weight is a consideration, as in long-span roofs and bridges, a saving in weight in some members may result in equivalent or even greater weight savings in other parts of the structure. In this case, every expedient should be used to save weight. This consideration has led to the development of the higher-strength steels, which justify their increased cost in many cases by the savings they permit in the lighter members elsewhere.

Shear connectors, usually in the form of welded plates, shapes, studs, or spiral coils, present another problem in composite construction. Some specifications require them to be shop-welded, which is unquestionably more economical than field welding. However, they are a hazard in shipping and handling, susceptible to damage, and constitute a serious hazard and expense during erection. They are a danger to workmen who must walk the top flange. In bridge construction where there is an erection traveler operating and running on the top flange, a special track or support must be devised to clear the connectors. If the girders are erected by a crane operating on the ground, this is not a problem. However, shear connectors are best welded in the field and the contractor should be at least permitted to make a choice. Local and state laws and applicable union rules should be reviewed to be certain shop application of these devices is permitted.

11. Lateral Stability   Lateral stability is a major problem for the erector in long-span plate girder bridge construction, particularly if the girder is one section of a long continuous span. Continuous-span girders are somewhat lighter than those for simple spans but lose some of their economy if they have to be haunched over the piers or supports, and if they have to be spliced. These items are an additional expense both to fabricate and to erect. Such girders often require heavier-capacity erection equipment and, usually, longer booms. If erected from the ground this is not so serious, but when erection derricks must operate over the deck of a span the floor members may have to be strengthened, or special devices employed, to spread the heavier reactions over a greater number of beams.

Long, slender girders must frequently be stiffened until other members have been placed. Sometimes a stiffening beam can be bolted flatwise to the top flange until the girder can be permanently braced, or a lateral stiffening truss, especially fabricated for the purpose, may have to be used to obtain the necessary stiffness. Unstable girders may also require the use of an additional costly crane to hold on while an adjacent girder and the cross framing are erected.

Whenever a new design or innovation results in lighter and less stable members, the designer should call this to the erector's attention and provide proper field supervision. Most reliable erectors have large and experienced engineering staffs and are aware of such problems.

12. Girder Bridges When long-span continuous girders have to be spliced, a temporary support or falsework must be used and some of the savings over simple-span construction are lost. Designs of shorter, simple-span, parallel-flange girders with additional piers should be compared with that of the larger, haunched girders. Deck-truss or through-truss spans, while not considered so desirable aesthetically, may also be cheaper.

For bridges with wide roadways, deck-girder construction, consisting of two main girders with transverse floor beams and longitudinal stringers, may be excessively heavy and expensive to fabricate and ship. A bridge with longitudinal girders 6 to 10 or 12 ft apart requires lighter erection equipment and may be faster and cheaper to erect.

13. Splices Girders and other members 120 ft or longer are readily fabricated and shipped in one piece, and have been made in single lengths up to 180 ft and weighing over 100 tons. When this can be done and the girder is handled in the field, it will save a temporary falsework support and may be economical. A field splice may be economical for lengths over 120 to 150 ft, since shop splices in the individual sections will be required anyway. It is almost always cheaper to use a bolted (HS) field splice than a welded field splice. The bolted field splice can more readily accommodate fabrication tolerances of the two sections being joined, and the tightness of the bolts can be inspected with less skilled inspectors. The field-welded splice requires closer fabrication tolerances at the abutting ends of the joining members, welded erection aids for aligning members prior to start of welding, control of the electrodes, and finally, a qualified ultrasonic or radiograph inspector. Furthermore, repair of an unacceptable flaw in the weld is very difficult and expensive.

14. Geometric and Cambered Shapes The following terms are generally used to denote the shape of a truss or structure under various loading conditions. The *geometric shape* or outline is that indicated on the drawings. Unless otherwise stated, it is the shape which the structure is intended to assume under dead load. For some long-span truss bridges, it is specified as the shape under dead load plus one-half the design live load.

The *cambered shape*, also called the "no-load" outline, is that which the structure will have when the members are under no stress. This condition could exist in a truss assembled on its side in the shop or while being assembled in the field and supported on falsework at every panel point.

Cambering for a specified loading means that the truss or girder is to be fabricated so as to have a specified shape under that loading condition. To do this, the expected deflections for the specified loading are determined. Camber is built into a girder in the shop. For a truss, the tension members are shortened and compression members lengthened by the amounts of the calculated deformations, giving what are called the cambered lengths of the members. This introduces a problem in both fabrication and erection, and a decision must be made as to how the truss is to be built in the shop and assembled in the field. On a skewed structure with the cross frames or floor beams framing normal to the main carrying girder or truss, it should be recognized that difficulties will arise at times in the erection of these transverse members because of the differential in elevation due to the cambered position of the two framing or connecting points.

Reaming Assembled. Many specifications require trusses to be reamed while assembled in the shop. If the members have been cambered, this will give the desired overall shape, but the angles of the web members with the chords will be slightly different from the angles computed for the geometric shape. When the truss is assembled in the field, the connections will fit readily and the erector has no difficulty in fitting up and bolting them. Numerically controlled drilling which eliminates shop assembly will also provide the desired camber. When the span is completed and swung, and subjected to the specified loading, it will deflect to its geometric shape, but in so doing the changed lengths of the members under stress will cause angular changes in the connections at the truss joints. Since they are rigidly pinned or bolted, this will result in a slight bending in the members, thus inducing secondary stresses.

If these secondary stresses are within the prescribed limits of the specification, there is no objection to this method of reaming assembled. However, if they are to be minimized, a different fabrication procedure is required. The gusset plates must be laid out and the connection holes for the web members drilled or reamed to the geometric angles. This can best be accomplished by drilling the holes full size through a metal template laid out to the geometric angles, or by the use of numerically controlled equipment. If reaming while assembled is still required, one end of each web member must be disconnected and set to the correct geometric angle with the other chord and the connection to that chord drilled or reamed. This procedure is then repeated for the other end. This method of cambering will result in a small angular misalignment as the truss is assembled in the field, whether on falsework or cantilevered, since the members are under virtually no stress and still have their cambered length. In fitting up these connections, heavy pinning is required to bring the holes to match, which results in slight S-shaped bends in the various members. The chords usually do not bend noticeably, but the web members, being lighter, are likely to bend enough to be noticeable, especially if they are not too rigid in the plane of the truss. When the truss is then subjected to the specified loading, the changed lengths of the members will result in the theoretical geometric shape, and the reverse secondary stresses induced during field assembly will be eliminated.

Admittedly, this condition cannot be fully attained, but it is the only practical approach to the problem. If the web members cannot be fully sprung by heavy pinning, there will still be a small misalignment of the connections, which may have to be reamed out. However, this is usually not necessary, and the connections can be bolted or riveted with proper pinning.

Designers should understand this problem and make clear in their notes and specifications exactly what end results are desired. If secondary stresses are of no concern, small trusses and other assemblies may be put together and all connections reamed while so assembled. For larger and more important structures, if elimination of secondary stresses to the greatest possible extent is desirable, all members and connections should be fabricated to the geometric shape. Modern shop equipment and methods result in highly accurate fabrication, and shop assembly is usually unnecessary. Perfect matching of field connections can be obtained by drilling and reaming to metal templates, or by the use of numerically controlled drilling. It may be desirable to assemble bridge chords in long lengths, set them to the proper geometric shape, and ream them with all milled ends in full bearing. They are easily sprung to the cambered shape during field assembly.

15. Stress Participation The assumptions made in analysis sometimes result in no calculated stress in certain members of a structure, or ignore their secondary sharing of load with other members. For example, changes in length, under load, of the chord members of the main trusses of a bridge force the floor systems, including the lateral bracing, to participate; this is not ordinarily taken into account. In many cases, this may not be serious, but the designer must recognize when these conditions may exist and make provision for them if they may be objectionable.

The situation can be partly relieved in many instances by delaying the final fitting up and bolting of critical field connections until as much as possible of the final structure and dead load is in place. While this is good in theory, it can be very objectionable in practice, since it delays the erector and may entail considerable expense. In addition, the structural integrity of the structure may dictate completion of bracing connections because of extraneous loading during erection, such as wind. In building construction it could mean interference with and delay to other trades as well. In bridge construction, floor-system stringers and bracing should be finally secured before the deck is placed to avoid expensive scaffolding for later access to these connections. In general, design and details of structures should permit erectors to complete all bolting, welding, or other fastening of a structure before they leave the work.

16. Expansion and Construction Joints Construction joints and expansion joints are common in long-span bridge decks to avoid participation of the floor system in chord stresses, and may be required every four or five panels. If they are not used, and the chords are fabricated to a cambered length and the stringers to the geometric length (which is the usual case), there will be serious mismatching of the connections as the stringers are erected. This will result in lateral bowing of the floor beams, if the connections can be made at all. A seat and slotted-hole type of connection will avoid this difficulty. The difference in panel length between a cambered chord and a geometric-length stringer may be as much as $\frac{3}{16}$ to $\frac{1}{4}$ in., and a series of floor expansion joints is the best answer.

17. Subpunching and Reaming At one time many designers and specifications required certain connections to be subpunched or subdrilled in the shop and reamed to full size in the field after erection, in the belief that a more accurately aligned joint would be obtained and possible secondary stresses avoided. The fallacy of this requirement is that the erector must match the connections as they are, simply in order to erect and hold the members in place. The only criterion is to match the holes as they exist. If they are subpunched or subdrilled, the erector pins and bolts them as they are and reams them to full size, a job which the shop can do much cheaper.

Another fallacy is that of full-sized holes in one member and subpunched holes in the matching member. The erector has no criterion for proper fitting up, and if the holes are not centered the reamer will center itself in the small hole and enlarge part of the larger hole, making it egg-shaped, after which it may have to be reamed oversize to get a round hole.

18. Bolting, Welding, and Other Fasteners Most steel structures are fabricated by either bolting or welding in the shop and bolting or welding in the field. Field bolting with high-strength bolts is by far the commonest. The engineer should specify the methods of bolt tightening that will be acceptable, and the method for checking the tension after the bolts have been tightened. The engineer's inspectors should check each connection as soon as the bolts are tightened or the other fasteners are placed and before the crew has left the point, particularly where the location is not readily accessible and requires scaffolding.

In the case of field-welded structures, some suitable means for temporarily supporting, aligning, and holding each member as it is erected must be devised. Such connections are usually bolted sufficiently for this purpose and to take any additional erection stress that may occur before they can be conveniently welded. In plate work, such as large storage tanks, special erection and adjusting devices have been developed to hold and align each ring or segment, after which welding proceeds to a sufficient stage so that erection can continue.

All things considered, field bolting is the primary kind of field connection and the most desirable from the erector's point of view. However, for certain structures welding can be less costly overall owing to reduced shop fabrication. When the design permits, the option of bolting or welding should be left to the contractor. High-strength bolts do introduce some erection problems. The faying surfaces of the joint should be left unpainted to give a high friction coefficient, but a thin coat of lacquer or the equivalent should be permitted if rusting may be a problem. At splices and other connections the erector requires a small clearance, $\frac{1}{32}$ to $\frac{1}{16}$ in. for each faying surface, for entering the spliced material. This may not be permissible if there are many surfaces involved, and it may prevent tight clamping of the assembly. In such cases, the splice may be redesigned, as by making it of the continuous overlapping type, or fillers may be provided or, possibly, bolts added to compensate. As a last resort, it may be necessary to ship most of the splice material loose and assemble it in place, which increases the time of erection, the number of field bolts, and the cost of erection. The designer should understand the problem and keep field connections and splices as simple as possible.

19. Weighing Reactions Some specifications for continuous-span construction have required weighing of the reactions and adjustment of the elevations at the piers to meet the required reactions more closely. Actually, relatively large differences in elevations at the piers produce very small changes in stress in a continuous structure having reasonably long spans, which is usually the case. The weight of any structure is difficult to estimate accurately, and weighing equipment of the necessary capacity is difficult and expensive to calibrate and is probably not accurate within less than 5 percent. Weighing usually results in no adjustment being made anyway, because excessive movements would be required to produce significant changes in the reactions. Moreover, such changes in elevation would usually result in objectionable changes in vertical alignment.

Reactions or loads occasionally are weighed in building construction where a portion of a column is being removed and its load must be carried by a girder to two adjacent columns. Here it is desirable to weigh the reactions by jacking and hold the jacking load while the new connection is made.

In continuous structures, one of the members may be kept inoperative while erection proceeds over falsework or other temporary supports, until a certain stage at which continuity is to be developed. In this case, the calculated stress for that condition is jacked into the member, which is then drilled or reamed and connected. In this way, several simple spans may be joined to act as a continuous structure for subsequent loading. For continuity under live load only, such members are connected under no stress and no jacking is required.

20. Strain-Gage Measurements An occasional requirement that might well be eliminated from erection specifications is that requiring strain-gage measurements on various members of the erected structure for various loading conditions. It is practically impossible to set up a pattern of strain gages or gage holes on a built-up fabricated member that will permit accurate determination of the average stress on the cross section. Whittemore and similar strain gages are limited in their accuracy and require many sets of readings. The SR4 strain gage requires an elaborate electrical installation and console for taking the readings. The gages must be applied in profusion, they are easily damaged, and they are not weatherproof, even when varnished. Strain-gage measurements in the field are of doubtful value. Certainly they should never be required as a criterion for any phase of the erection procedure. If designers insist on such measurements, they should arrange to do it with their own engineering staffs in such a way as to interfere with or delay the field work as little as possible.

21. Erection Stresses Erection stresses were of little consequence in most building construction until the advent of composite floor systems and other innovations, for which the transfer of derrick loads and wind now require analysis. The erector may have to increase sizes and connections of beams and girders to support derricks, other equipment, and floor loads. This is solely the erector's responsibility, subject to the engineer's approval.

In long-span bridge construction, particularly where erection is by cantilevering over piers or falsework supports, stresses from the weight of the erection equipment and the structure itself are a major consideration and will often control sizes of the truss members. The designer should obtain a reasonable estimate of the weight of erection equipment, temporary tracks, or construction of any kind, and provide for it in the design. If the erection method or procedure is not obvious, this may not be possible. However, in many cases, as with bridges over railroad yards, busy streets, congested areas, and deep navigable streams, it is practically certain that cantilever erection will be used.

If the erector must increase the cross section of certain members to accommodate the scheme of erection, permission to do so should be given subject to the engineer's approval. Frequently a member can be increased in strength sufficiently by using a higher-strength steel with no change in cross section. Such changes are always at the erector's expense.

22. Maintenance Exposed steelwork should be readily accessible for maintenance and painting. The steelwork in most bridges is completely exposed and must be reasonably accessible in every detail for cleaning, painting, general inspection, and maintenance, particularly for long-span and multiple-span girder or truss bridges.

Expansion shoes and rollers often become clogged with dirt and debris and rust quickly, requiring repair or even replacement. The designer should provide means for jacking up the span at each shoe, usually in the form of a heavy end floor beam over each pier, with suitable bearing areas on the pier. For structures that have no end floor beam, other provisions, such as the addition of temporary jacking wing plates or brackets, can be made should repairs to the expansion shoes become necessary.

Walkways and ladders should be provided for ready access to all parts of the structure, particularly those below the deck. Most large bridges have a continuous inspection walkway with handrails, with access by ladders over the side of the deck. Lateral walks give access to the shoes at each pier and to intermediate expansion joints and other critical points, such as navigation lights.

For box girders and towers of cellular construction, access to the interior must be provided through manholes whose covers can be readily removed from the inside as well as the outside. Ladders or other means of access should be available for the full height of enclosed areas.

23. Storage and Shipment   The fabricator starts work in the shop long before erection begins. For economy, all members of the same kind will usually be fabricated in one operation. Plain beam sections and other simple members will usually be fabricated first and must be shipped and stored. A considerable quantity of steel may have to be fabricated and stored before erection begins, to ensure a continuous field operation.

The erector may have to rent storage and working space near the site. Whenever possible, the engineer should arrange for such areas for the various contractors, particularly for large bridge projects in congested locations.

The method of shipment may determine the largest, heaviest, and longest pieces that can be fabricated. All carriers have limitations on weight and length, and extra charges for members in excess of certain minimums. State highway departments limit the width and weight of loads which may be hauled. The designer should keep these considerations in mind and limit the necessity for oversize members as far as possible.

EQUIPMENT AND METHODS

24. Equipment   Although it is not necessary for designers to have a detailed knowledge of erection equipment, they should know something about the various kinds of cranes and derricks, their uses and limitations.

Truck cranes and crawler cranes have a very short mast and a long boom mounted on a revolving platform having a control cab and a heavy counterweight to balance the load. The truck crane can travel over the highways under its own power. Crawler cranes must be hauled on special low-body carriers. They are heavier and generally have greater capacity than truck cranes. The upper assembly of these cranes can be rotated a full 360°.

The truck-crane body is equipped with outrigger beams which can be extended and blocked up from the ground to give greater capacity by increasing stability against overturning. Crawler cranes depend entirely on the long and widely spaced caterpillar tracks for stability. Special crane attachments such as increased mast lengths and additional counterweights are sometimes used to enable heavier loads to be picked, within the limits of the boom capacity and the hoisting equipment.

The maximum rated capacity of a truck or crawler crane is based on the shortest length of boom at minimum reach. A 50-ton-capacity crawler crane can pick 50 tons only on a 40- or 50-ft boom at the minimum radius of about 12 ft, and usually not over the side. It is always a good idea to consult a crane load rating chart when heavy, high, or long-radii lifts are to be made, in order to assure a safe operation.

Guy derricks are characterized by a very long or high mast, generally about 10 ft longer than the boom itself, and supported by wire-rope guys from the structure on which it stands. Its principal advantage is the ability to handle capacity or near-capacity loads at long reaches. The guy derrick can rotate a full 360° with the boom against the mast and swinging under the guys. The usual guy derrick has about a 100-ft mast and a capacity of 20 to 40 tons, but they have been built to 70 tons or more for building construction. They are used principally for the erection of high tier buildings.

The *stiffleg derrick* has a moderate length of mast, supported by two stifflegs at the top which are connected to the outer end of two horizontal sills on the ground. Mast heights of 25 to 50 ft are common, and the boom may be two or three times longer. Capacities vary, but 50 to 60 tons is not unusual. Many derricks of capacities from 85 to 120 tons have been built, and a few of as much as 500 tons. The derrick capacity is limited not only by the strength of its members but also by the counterweight or the strength of tie-downs at the ends of the sills to resist overturning. The stifflegs are generally set at either 60 or 90° with each other, which limits the swing of the boom to 300 or 270°. They are generally used in a fixed position or location and are not mobile as such.

For building or bridge construction, a stiffleg derrick can be mounted on a suitable tower of any reasonable height, even up to 150 ft. The tower may be fixed, or it may be equipped with trucks and wheels for lateral mobility. Stiffleg derricks can also be mounted on a platform and by use of overhead cathead beams and hoisting rigging be raised vertically for tier building erection.

Hammerhead tower cranes consisting of variable-height towers and horizontal rotating beams are especially adaptable to tier building construction. These cranes can be raised vertically, or tower-extension sections can be added as the height of the structure increases.

Luffing boom tower cranes are also available. The tower and climbing capabilities are similar to those of the hammerhead crane.

Derrick boats are useful for structures built in or over waterways. The larger and more elaborate boats have a whirley-type (rotating) crane permanently mounted on a special steel barge. Any crawler crane may be mounted on a suitable barge, or a stiffleg derrick may be supported on one barge, or on two barges rigidly tied together, to make a temporary derrick boat. Frequently a stiffleg derrick is mounted on a steel tower supported on two barges for erecting bridges high over the water.

Locomotive cranes have a revolving crane mounted on a specially designed railroad body, and have their own propelling machinery for moving at low speeds for short distances. The revolving body is heavily counterweighted to give boom capacities of as much as 60 to 75 tons, and the car body is equipped with outrigger beams. They are used mostly for work in railroad yards and for erection of small railroad girder bridges.

Derrick cars generally consist of a short, heavy, A-frame mast supported by short stifflegs and heavy wire-rope tiebacks, and are mounted on a railroad-car body. The hoist helps act as counterweight, and outrigger beams are provided for lateral stability. Boom lengths are generally 60 to 80 ft. They are used for erection of railroad girder bridges in locations inaccessible to other equipment. They are generally equipped so that they can travel short distances at slow speed under their own power.

Traveler is a term used to designate any kind of derrick mounted on a supporting underframe which has either skids or wheels to permit it to be moved over a structure. The derrick is usually a standard piece of equipment. The underframe is specially designed to fit the structure to be erected, and must be arranged to distribute the derrick reactions properly to the structure without overstressing any of the permanent members. Sometimes these members must be reinforced and heavier connections provided.

Whenever possible, bridge travelers are designed to be supported on the permanent floor framing, running over the floor stringers and being blocked up over the floor beams in working positions. Occasionally the traveler is designed to run over the top chord of a bridge, or over arch ribs, in which case provision must be made to level the traveler frame and plumb the mast in each working position. The designer should check the traveler reactions on the structure but should remember that the traveler itself is the sole responsibility of the erector.

25. Responsibility of Engineer and Contractor   Designers seldom have responsibility for the erection of a structure beyond making sure that it is constructed to accurate alignment according to plans and specifications, and seeing that no members are overstressed during erection. A designer should make sure that the erector has enough qualified engineers and field personnel, ample equipment of necessary capacity, and a safe and suitable scheme of erection. A designer who believes that some equipment is inadequate or unsafe, or a scheme unduly dangerous and hazardous, should warn the erector, but only in extreme cases should the erector be stopped for such reasons. Erectors may require additional equipment and personnel to be employed if they deem it necessary to ensure proper progress to complete the work on schedule.

Specifications customarily require the contractor to submit for approval the scheme of erection, including the design of any special devices or equipment. This does not relieve the erector of responsibility. Most contractors welcome such checks.

26. Erection of Buildings   Erection methods for most types of building construction are fairly well standardized. Tier buildings and tall structures of any kind are usually erected by derricks or tower cranes. One guy derrick with a 90-ft boom can economically cover an area of about 100 by 100 ft. For larger areas, more derricks, or a longer boom, may be required.

Reasonable tolerances for plumbing tall building frames should be specified by the designer, usually in accordance with a standard specification such as that of the AISC.

Closer plumbing tolerance may be required for elevator wells and guides, and for the supports for window framing and special wall panels and enclosures.

Column-core buildings have a composite type of construction which, while not common, is still occasionally used. Its advantages (principally some saving in weight of the steel columns) would seem to be few compared with the difficulties and dangers of erection. The steel columns are relatively lightweight sections which are later encased in concrete. They are spaced and aligned by light steel struts, there being no steel floor framing. Lugs on the columns support the flat-slab concrete floors. Only a tier or two at a time can be erected, aligned, and temporarily braced until the floor slabs are placed. The erection is thus interrupted for every tier, besides which it is extremely slow, dangerous, and expensive. The designer would do well to avoid this type of construction.

Manufacturing and mill buildings, assembly plants, warehouses, and a wide variety of commercial buildings are most generally erected by means of truck or crawler cranes, operating directly on the ground or on the building floor slabs. Erection procedures present no problems to the designer. Some mill buildings, such as those for steel plants and other heavy industries, may be of considerable height and support heavy overhead traveling cranes. These will require large-capacity crawler or locomotive cranes with long booms for their erection, and sometimes involve complicated erection problems. Stiffleg derricks mounted on towers and wheels rolling on tracks are used to erect extremely high structures where clearance for swinging the boom is limited and relatively heavy loads are handled.

27. Erection of Bridges   There are so many types and sizes of bridges, which must be constructed in different locations and under such entirely different conditions, that it is impossible to generalize as to methods of erection. The following covers the more usual types and some common methods and kinds of equipment used.

Beam and Girder Spans.   Short beam and girder spans that are accessible from the ground can be erected by crawler or truck cranes operating on the ground. This is fast and economical but is limited to heights of about 240 ft. Above that height, some kind of derrick or traveler supported on and moving over the structure or a cableway will most likely be needed. Over water, a crane or derrick mounted on a barge can be used for low crossings. For bridges of greater height, a tower derrick boat will give a better reach and avoid the use of a very long boom.

Long beam and girder spans present special problems and hazards to the erector. To be economical they must be fabricated in long lengths, resulting in heavy and usually laterally unstable loads to be handled at great heights or long reaches. After it is landed, a girder must be temporarily braced until the next one can be erected and braced to it. A temporary beam is sometimes bolted to the top flange for this purpose, or a light, framed stiffening truss may be used. Lateral bracing in the form of wire ropes is not suitable for this purpose, and its use should not be permitted.

In truss-type bridges the truss members are generally shorter and lighter than girders, and easier and safer to handle. Long truss spans are usually erected in place, member by member, by what is known as semicantilevering. The first of a series of spans is erected on falsework supports, following which subsequent spans are cantilevered by tying back to the preceding one and cantilevering over only one or two temporary supports to the next pier.

Chords are usually fabricated in two-panel lengths of from 50 to 80 ft. Chord splices must be located just beyond the falsework support in the direction of erection. This may result in unsymmetrical location of splices, which the designer should permit.

For spans high over water it may be possible to use a tower derrick boat or long-boom crane with or without special attachments mounted on a suitable barge. If this is not practicable, a traveler running on the deck would be used. In some cases it may be possible to run the traveler over the top chords, resulting in a shorter boom and simplifying the erection since all members are then below the traveler, as with a deck span. Two-panel-length chords are the rule, except for unusually heavy spans. Diagonal truss members are usually in one length, even up to 120 ft, unless the erection procedure makes a splice at a subpanel point desirable. The designer should permit changes in location and number of splices to suit the erector.

Arches.   There are many types of arches. Those having pins or hinges at the crown, or at the abutments, or both, offer no unusual problems. Fixed arches should be erected only on accurately located and constructed abutments, finished to the correct angle. The ribs must be fabricated to a high degree of accuracy as to length and curvature, and care should be taken in erection to avoid locking in dead-load and erection stresses for which the ribs were not designed. The designer may require the thrust at the crown to be measured by weighing with hydraulic jacks placed between the two half-rib sections. While such measurements may confirm the stress calculations to a very close degree, and allay any fears the designer may have, they should not be used as a criterion for making small adjustments in either the length or angularity of the arch rib at the crown. Field adjustments to correct inaccurate surveys or fabrication should not be required.

Small arches can be erected by cranes from the ground, or from the water, where there is access. Where this is not practical, a traveler may be used. For long, heavy arch spans, particularly where access for delivery of materials from directly below to the erection rig is impossible, cableways are adaptable. Arches are usually supported on falsework until they can be joined and swung free. Where falsework cannot be used, some kind of overhead tieback system with suitable anchorages must be used. This will probably be more expensive.

If possible, the designer should try to determine, by consultation with the erector, what method of erection is likely to be used, as this may govern some of the design features and details. Otherwise, changes may have to be permitted later.

The engineer should be responsible for the original surveys but may require the erector to check the location and elevation of all piers and abutments.

28. Falsework   The term falsework is used to denote any kind of temporary support required or used to hold a structure in place until it can be completed and made self-supporting. Falsework may be of any kind of material such as timber, steel, wire ropes, or concrete. In many cases, members of the structure itself which are not required to be erected until later may be used temporarily as falsework. The designer should permit such use, where the members will not be overstressed or damaged, with any necessary changes in design or details at the erector's expense.

Falsework will not affect the design of most structures, particularly buildings and short-span bridges. But for long-span bridges of either the girder or truss type the reaction on the span at a falsework support may be large, and special details may be required to develop this load and distribute it properly into the bridge structure. The designer should permit this, but should check the revised details.

Complete falsework to support a structure at every panel point is seldom used. Most erectors will use as few bents or supports as possible, cantilevering the structure over them for the greatest possible length before using another support. This cantilevered length is generally determined by the stress in critical members. It is often desirable or necessary for the erector to reinforce or redesign some members to permit a greater cantilevered length to suit a better falsework location. This requires close cooperation between designer and erector, since changes in design of one member may affect stresses in other members.

29. Erector's Responsibility   For some monumental, long-span structures, involving high falsework carrying very heavy loads, the designer may require the erector to submit falsework design and details, including the footings, for review and approval. This should be stated in the specifications, but such review or approval does not relieve the erector of responsibility for the design and safety of the falsework.

Since falsework is of a temporary nature, carrying its maximum design load for a very short period of time, the erector should be permitted to use more lenient specifications and higher unit stresses than for the permanent structure. Loads on piling or other footings may be much greater than would be desirable for a permanent structure where settlement could not be tolerated.

When falsework is located in a waterway, many hazards exist, as from shipping, floods, floating ice, and debris. The designer should help to obtain any permits necessary for putting falsework obstructions in the channel. Permits are usually arranged for by the erector. However, where a channel has to be obstructed during erection, the designer can often ascertain the minimum width of channel and other navigation requirements when obtaining permits for the permanent structure.

Section **9**

Design of Cold-Formed Steel Structural Members

GEORGE WINTER
Professor of Engineering, emeritus, Cornell University

Cold-formed steel structural members are produced from steel sheet, strip, plate, or bar stock. Forming in press brakes is economical for moderate production runs of limited quantities of a given shape. This is so because, in the semimanual use of press brakes, only a minimum change of tooling is needed to accommodate the fabrication of a great variety of shapes. The production of large quantities of identical shapes is best accomplished by cold roll forming. Once the special set of rolls needed for each shape is at hand, this fully mechanized, high-speed process is superior for true mass production. While it is possible by cold forming to produce shapes up to ½ and even 1 in. thick, cold-formed steel construction is generally restricted to the thinner plates and to the sheet thicknesses given in Table 1.

Sheet and strip steel is ordered to decimal thicknesses, rather than by U.S. Standard gage numbers. However, the use of gage thicknesses is so ingrained that its informal use continues in trade and profession.

1. Materials A considerable variety of sheet and strip steels is available for use in cold-formed construction. To be suitable, the material must possess adequate and reliable strength and ductility. The specified minimum yield point is the primary criterion for strength under static loading, and allowable unit stresses are generally expressed in terms of it. The tensile strength is of secondary importance provided there is a reasonable spread between it and the yield point. This is so because fatigue strength and resistance to brittle fracture, which relate chiefly to tensile strength rather than yield point, are rarely of consequence in cold-formed steel construction, the former because the loading of such structures is seldom of the repetitive type, and the latter because the very thinness of the material almost always precludes the development of those triaxial stress conditions which appear to be essential for the initiation of brittle fracture. Ductility (measured as permanent elongation in a 2-in. gage length) is required to enable the material to be cold-formed to relatively sharp radii without cracking and, in the formed member, to provide plastic stress relief in localized regions of stress concentration, particularly in connections.

Elongation in a 2-in. gage length in combination with the ratio of tensile to yield strength, F_u/F_y, can be used as a measure of ductility. Generally, sufficient ductility is available if F_u/F_y is not smaller than 1.08 and elongation in 2 in. not less than 10 percent. Steels which do not satisfy these requirements can be used, but only for shapes which merely require mild cold forming and which are used without highly stressed connections.

TABLE 1   Thicknesses and Weights of Uncoated Steel Sheets

U.S. Standard gage No.	Weight, psf	Approx thickness, in.
4	9.375	0.2242
6	8.125	0.1943
8	6.875	0.1644
10	5.625	0.1345
12	4.375	0.1046
14	3.125	0.0747
16	2.500	0.0598
18	2.000	0.0478
20	1.500	0.0359
22	1.250	0.0299
24	1.000	0.0239
26	0.750	0.0179
28	0.625	0.0149
30	0.500	0.0120

The following ASTM specifications define the qualities of sheet and strip steels suitable for cold-formed construction:

ASTM A446: Steel Sheet, Zinc Coated (Galvanized) by the Hot-Dip Process

ASTM A570: Hot-Rolled Carbon Steel Sheets and Strip, Structural Quality

ASTM A606: Steel Sheet and Strip, Hot-Rolled and Cold-Rolled, High-Strength, Low-Alloy, with Improved Corrosion Resistance

ASTM A607: Steel Sheet and Strip, Hot-Rolled and Cold-Rolled, High-Strength, Low-Alloy Columbium and/or Vanadium

ASTM A611: Steel Cold-Rolled Sheet, Carbon, Structural

Table 2 summarizes the specified minimum mechanical properties of these various steels.

Apart from sheet and strip steel, steel plate and bar of the applicable ASTM grades (Sec. 6, Table 2) are used for cold-formed construction.

In addition to material ordered and produced to these ASTM specifications, other sheet and strip steels are being used. To ensure the safety of such use, the minimum physical properties of such steels should be specified in a manner similar to that of ASTM steels, and their structural suitability established by stipulating an adequate spread between yield and tensile strengths, sufficient ductility, and, where needed, weldability. Adequate controls by tests, analyses, etc., should be instituted by producer or purchaser to verify the specified properties.

Steels with yield points of 30 ksi or less are rarely used structurally. The bulk of usage lies in the range of 33 to 42 ksi, with the high-strength steels of 50 ksi yield and over generally restricted to special situations.

The static strength of steel structural members depends not only on the yield point f_y but also on the shape of the initial portion of the stress-strain diagram. Sheet and strip steels and structural members made of them exhibit one of the two types of diagram of Fig. 1. Curve a shows the behavior of a sharp-yielding steel,

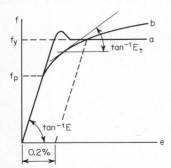

Fig. 1 Initial portions of stress-strain curves of (*a*) sharp-yielding steel and (*b*) gradual-yielding steel.

with a practically straight line up to yielding, a definite horizontal yield plateau, and, in most cases, a distinct upper and lower yield point. In such steels it is the lower yield point which determines the performance of the member. Curve b shows the behavior of a gradually yielding steel. For such steels the yield strength is defined either by a 0.2 percent permanent elongation, as shown, or by a stipulated amount of total elongation.

The strength of members which fail by buckling depends not only on Young's modulus E but also on the tangent modulus E_t (Fig. 1), determined at the magnitude of stress at

TABLE 2 Properties of Sheet and Strip Steels

ASTM standard	Grade	Yield point, ksi	Tensile strength, ksi	Elongation,* % in 2 in.
A446-71	A	33	45	20
	B	37	52	18
	C	40	55	16
	D	50	65	12
	E†	80	82	——
	F	50	70	12
A570-70	A	25	45	23–27
	B	30	49	21–25
	C	33	52	18–23
	D	40	55	15–21
	E	42	58	13–19
A606-71:				
Cold-rolled		45	65	22
Hot-rolled	Cut	50	70	22
	Coils	45	65	22
	Annealed	45	65	22
A607-70	45	45	60	22–25
	50	50	65	20–22
	55	55	70	18–20
	60	60	75	17–18
	65	65	80	15–16
	70	70	85	14
A611-70	A	25	42	26
	B	30	45	24
	C	33	48	22
	D	40	52	20
	E†	80	82	——

*Where a range of elongations is given, values depend on thickness or production method. For details, consult the ASTM standard.
†Grade E is a fully hard product of low ductility, for roofing and other applications which require only mild cold forming.

which buckling occurs. Current design procedures are formulated for steels whose proportional limit f_p is not lower than about 70 percent of the specified minimum yield point. This precludes the use of the AISI specification[1] for such materials as stainless steels in which the proportional limit is often considerably lower. However, AISI has promulgated a special specification and manual for the design of stainless-steel cold-formed structural members.[3]

2. Shapes and Uses In view of the relative ease of producing a great variety of cold-formed shapes, structural sections of this type have not been standardized. A number of fabricators have developed their individual lines of members, and their catalogs should be consulted for specific information. At the same time it is possible for designers to devise their own shapes for particular jobs. Shapes can be divided into two categories: framing members and surface members.

A number of *framing members* are shown in Fig. 2. Some of these have the general outline of well-established hot-rolled sections. The depth of such sections ranges from about 3 to 12 in. and the thickness from about 8 to 18 gage. These dimensions frequently result in plane elements of much larger ratios of flat width to thickness than are found in hot-rolled shapes. Such thin elements are frequently stiffened with lips or other edge stiffeners to forestall premature local buckling. Special shapes are common for particular uses. Thus, *g*, *h*, and *i* are used as flexural members and compression members, and for chords of trusses. Shapes such as *j* and *k* can also be employed for chord members. Girts and eaves struts can be made in shapes *l* and *m*, respectively. Closed sections such as *r*

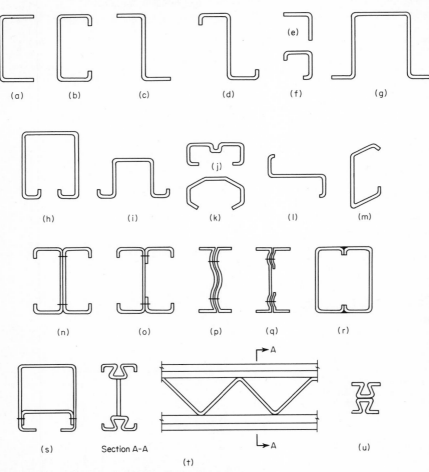

Fig. 2 Cold-formed framing members. *(Adapted from Ref. 6.)*

and *s* are particularly favorable for columns. Nailability can be provided by the use of projection welding as in *p* and *q*, or by appropriate shape, such as in the cold-formed chords of the open-web joist *t* and the stud *u*. Using special, hardened nails, it is also possible to nail directly into the flanges of cold-formed members of moderate thickness.

Properties of several of the simpler framing members are tabulated for a wide range of dimensions in the AISI Manual.[1] These are not standardized stock sections, although many of them are in current production with minor dimensional variations. The Manual gives them primarily as a guide to designers to indicate economical ranges of dimensions, properties, and capacities which can be obtained, but not to restrict them either to the tabulated shapes or to their specific dimensions.

A number of *surface members* are shown in Fig. 3. These can be defined as load-resisting shapes which also constitute useful surfaces. Generally they fall into one of the four illustrated categories, for each of which only a few of many current or possible shapes are shown. Ribbed-steel roof deck is generally made of 18 to 22 gage sheet, 1½ to 2 in. deep, with ribs spaced at 6 in., and is used for spans of about 5 to 10 ft. Some of these sections, in the inverted ribs-up position, are also used as reinforcing forms for concrete

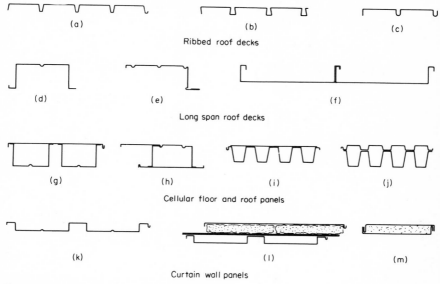

Ribbed roof decks

Long span roof decks

Cellular floor and roof panels

Curtain wall panels

Fig. 3 Light-gage surface members. *(Adapted from Ref. 6.)*

floors, e.g., in high-rise buildings. In this case the deck constitutes the slab form during construction and the positive slab reinforcement in the finished structure. The long-span roof decks utilize 0.05- to 0.075-in. sheet, are 3 to 8 in. deep, and span up to 30 ft. Cellular panels in the same range of dimensions as long-span decks are used for both floors and roofs. For floors they are usually topped by a concrete fill at least 2-in. thick which carries the flooring proper. The cells are utilized for a variety of purposes: electrical conduits, telephone and other wiring, recessed lighting, and air ducts for heating and air conditioning. Wall panels, insulated or not, are generally used as curtain walls which need only resist wind pressures, but some forms are also employed as bearing walls, mostly for one-story structures.

While framing members are optimized to produce maximum strength and stiffness per unit weight of material, surface members must satisfy a variety of functional requirements of which optimum strength is only one. Another is coverage, i.e., the amount of surface area provided by one unit, and specific requirements in regard to such functions as floor electrification and conduction of hot and cold air. The design aims at shapes which best serve these multifunction requirements.

A variety of uses of cold-formed light-gage steel structures and structural members are discussed and illustrated in Ref. 4.

3. Shear Diaphragms Surface members are generally designed to resist loads acting perpendicular to the surface, such as gravity loads on floors and wind loads on walls. However, surfaces consisting of such members also can be made to resist sizable loads acting in the plane of the surface. For that purpose it is only necessary adequately to interconnect the individual panels, by seam welding or otherwise, and to connect the diaphragm so obtained to the main structural framing. Horizontal floor and roof diaphragms of this type have long been used successfully to resist wind and seismic loads in single and multistory buildings in the same manner as concrete slabs are used for receiving such loads and carrying them off to shear walls or braced bays.

The action of a light-gage steel shear diaphragm is illustrated in Fig. 4, which shows the conventional steel framing of a one-story structure with roof panels in place. The framing is designed for vertical load only. Horizontal wind, earthquake, or other loads in the transverse direction are represented by the load w in the plane of the roof deck. If the panels are merely fastened to the frame, but not interconnected, they offer little resistance

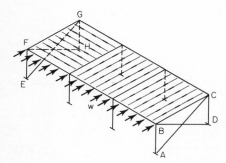

Fig. 4 Shear diaphragm bracing against wind and seismic forces (schematic).

to such loads, and special bracing or other measures must be provided. However, if they are interconnected by intermittent seam welds or otherwise, as schematically shown, then the diaphragm so obtained, together with the longitudinal framing members BF and CG, acts similarly to a plate girder supported at BC and FG by the shear walls $ABCD$ and $EFGH$ (schematically indicated by the cross bracing). A difference compared with normal plate-girder action consists in the fact that such diaphragms, though possessing sizable shear strength, do not resist any significant flexural compression or tension in their own plane. Therefore, the entire moment is resisted by the perimeter members acting as flanges, the diaphragm supplying only the shear resistance. Shear stresses being equal in sections at right angles to each other, it is irrelevant whether individual panels are placed longitudinally or transversely (Fig. 4).

The shear strength of such diaphragms depends mainly on the configuration of the panels, their thickness, and the type and arrangement of connections. In view of the great variability of these factors, one relies chiefly on the results of large-scale tests, of which great numbers have been carried out for many shapes. Test results and recommended design values can be obtained from the manufacturers of the particular panels or decks. The results of tests made at Cornell University on diaphragms composed of shapes very similar to b, c, g, and h of Fig. 3 and of corrugated sheet are very briefly summarized in Table 3 to indicate the range of strengths which can be obtained. The table gives the ultimate shear strength along lines of support such as BC and FG of Fig. 4; among the variables, only shape and gage thickness are indicated. Details of connections, panel depths, deflection, and other pertinent information are given in Refs. 7 and 11. The ultimate shear strengths are used in design by multiplying the design loads by load factors which range from 2.0 to 2.3 for dead load, 2.7 to 3.0 for live load, and 2.1 to 2.5 for wind or earthquake load, depending on types of connections. Current practice in diaphragm testing and design is given in Ref. 12.

Apart from furnishing resistance to horizontal loads, the in-plane strength and rigidity of such diaphragms has other structural uses. One of them, the use in shell-type roofs, is described below. Another is the utilization of in-plane rigidity for the bracing of structural members, such as columns and beams, against buckling in the direction of the wall, roof, or floor diaphragms connected to these members. This often permits a substantial reduction in member size. Details of design methods are given in Ref. 13.

4. Folded-Plate and Shell Roofs   Shear-rigid diaphragms are also used in shell structures for lightweight roofs. The types of shells that can be so constructed are limited by two requirements. (1) Surface members of the type of Fig. 3 are fabricated straight; they can be slightly warped in place but are not produced with longitudinal curvature except for special, proprietary systems. Hence, they can be used only for shells which have straight generators. (2) Since such diaphragms, in their own planes, resist shear but not normal stress, they can be used only for shells where the surface proper is essentially subject to shear and where special members can be provided to resist normal forces of tension or compression. Two types of shells satisfy both these conditions; folded plates and hyperbolic paraboloids.

A simple folded-plate roof is shown in Fig. 5a. Folded-plate action, in general, recognizes the fact that any constituent plate is very rigid in its own plane and quite flexible perpendicular to its plane. Therefore, any load such as P applied at a fold line is resisted

TABLE 3   Representative Shear Strengths of Light-Gage Steel Diaphragms

Shape	Gage	Range of ultimate shear strength, plf
b	16	2,600–3,600
	18	2,300
	20	1,000–1,330
c	18	1,700–2,600
g	16	3,500
	18	2,300
h	16	1,800–3,200
	18	1,500–2,400
Corrugated sheet	22	520
$2\frac{2}{3} \times \frac{1}{2}$ in.	24	440
	26	370
	28	330

by the two abutting plates in the manner shown in Fig. 5b. Under such loading each plate deflects in its own plane, and the actual deflection ab of the fold line can be found from the construction of Fig. 5c. In customary reinforced-concrete folded-plate structures, the plates themselves resist the longitudinal flexural tension and compression stresses as well as the shear stresses. In light-gage steel structures, the plates resist shear only, and it is

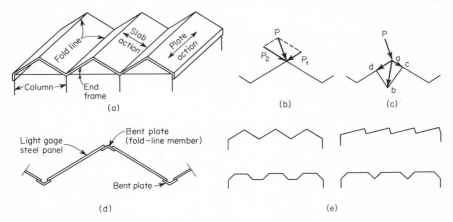

Fig. 5   Folded-plate roof structures. *(From Ref. 8.)*

necessary to provide special fold-line members to resist the longitudinal forces (Fig. 5d). They consist of simple bent plates, usually ¼ in. thick and of minimum width (about 3½ in.) for proper bearing and connection of the transverse panels. Various other folded-plate roof shapes are shown in Fig. 5e. Round structures with radially arranged folds are also possible.

The structure acts in the following manner: Each inclined or horizontal plane is made up of panels which span transversely between fold-line members and are designed as simple spans. At any fold-line member the two reactions are combined into a single force, such as P in Fig. 5b, and this force is decomposed in the direction of the two abutting plates as shown. Thus, in general, any one plate is loaded by one such load in its own plane at each of its two longitudinal edges and is designed for this loading as a beam spanning between the end frames. The longitudinal forces in the fold-line members are obtained by dividing the bending moment by the distance between these members. Thus, each fold-line member, acting as a flange, receives a longitudinal force from each abutting plate and is designed for the sum of these two forces. The shear caused by the two in-plane edge loads is carried by the diaphragm, which is attached to the end frames by shear-resistant connections, generally puddle welds or fillet welds.

Because the plate action in such structures is not continuous across fold lines, as it is in reinforced-concrete folded plates, small differences in fold-line deflections do not affect moments and shears, which makes superfluous the type of deflection analysis which is frequently necessary in reinforced concrete. Also, because of the longitudinal flexibility of such diaphragms (bellows effect of panel deformations) strain continuity of the plates meeting at one fold line need not be considered. These two features make the design of light-gage steel folded-plate structures much simpler than that of reinforced concrete. Details and full-scale test confirmation are given in Ref. 8. This type of construction is used for spans up to about 100 ft.

Hyperbolic paraboloids are doubly curved shells made up by two families of straight generators intersecting at right angles (Fig. 6). When properly connected to edge mem-

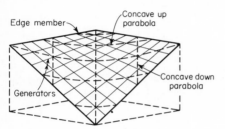

Fig. 6   Hyperbolic-paraboloid shell.

bers, such a shell, according to elementary theory, is subject to tensile membrane stresses along the concave-down parabolas, and to equal compression membrane stresses along concave-up parabolas, i.e., along lines at 45° to the edges (Sec. 20, Art. 26). This state of stress is equivalent to pure shear in the directions of the generators. Consequently, if panels are laid in the direction of one set of generators, and interconnected to form a diaphragm, under uniform load this diaphragm is subject to shear only, a state of stress to which it is well suited. The shear is transferred to the edge members by suitable connections; these members are then stressed in tension or compression, depending on the manner in which the shell is supported at the corners. Details and test confirmation are given in Ref. 9.

DESIGN

Many of the design procedures which apply to structures made of hot-rolled shapes or plates are equally applicable to cold-formed light-gage structures. However, differences in behavior under load between the two types of construction are sufficiently pronounced to necessitate corresponding differences in design methods. The main reasons for this are the following: (1) The design procedures for hot-rolled construction have been developed chiefly around the relatively few structural shapes and forms germane to that type of structure. In contrast, in cold-formed construction the variety of shapes which can be fabricated is almost unlimited. For this reason, design methods must be so general that they apply to almost any possible shape, extant or yet to be developed. (2) The so-called *flat-width ratio*, i.e., the ratio of width to thickness, of any of the various plane elements of which shapes are made up is frequently much larger in light-gage than in hot-rolled construction. For such thin-walled members it is necessary to employ special and more elaborate design procedures to safeguard against local buckling of elements subject to compression, shear, or in-plane bending. Also, the relatively smaller wall thickness results in torsional stiffness which, in relation to flexural stiffness, is much smaller than in similar thick-walled members. This is important in regard to torsional-flexural buckling of beams and columns. (3) The production and fabrication processes peculiar to the two types of steel construction affect in different ways the effective mechanical properties of the material. Thus, hot-rolling causes residual cooling stresses in structural shapes, which may strongly affect buckling strength of members. On the other hand, the strain hardening resulting from cold forming of light-gage members changes the mechanical properties of the sheet or strip, particularly in and near corners.

For those reasons, among others, the AISI specification[1,2] and related material applicable to this type of construction have been developed. Design procedures are given in the Specification for the Design of Light Gage Cold-Formed Steel Structural Members issued by the American Iron and Steel Institute as Part I of its "Cold-formed Steel Design Manual."[1] The Manual contains a commentary by the author which explains the background of the various design provisions and contains an extensive bibliography of research documentation.[2]

5. Basic Design Stresses   The AISI specification provides a safety factor of 1.67 in cases where failure would be initiated by simple yielding. Correspondingly, the basic design stress F in tension, compression, or bending is $F = 0.6F_y$, where F_y is the minimum specified yield strength before forming to shape. If yielding is caused by shear rather than by normal stress, the allowable value is $F_v = 2F/3 = 0.4F_y$. For the range of customary yield points, Table 4 gives the pertinent basic allowable stresses.

TABLE 4   Basic Design Stresses, F, ksi

Yield point F_y	Tension, compression, bending, F
25	15
30	18
33	20
37	22
40	24
45	27
50	30
Other	$0.6F_y$

The increase in yield strength brought about by strain hardening due to cold forming is concentrated in those regions of the section which are strongly cold-worked (corners, etc.) and is generally disregarded. However, for relatively stocky sections (i.e., shapes with relatively small flat-width ratios) and under special conditions of control by test, design may be based on the higher yield point of the member as formed. Details of taking advantage of this increased strength are discussed in Art. 15.

The basic design stresses in Table 4 apply only if the strength of the section or element is not governed by factors other than simple yielding. Where necessary, they must be modified for local or general buckling, web crushing, and other possible influences.

6. Section Properties   The calculation of areas, moments of inertia, and other properties of intricately shaped thin-wall sections is lengthy and tedious unless appropriate simplifications are made. Two of these are particularly helpful.

In many shapes the inside radius of right-angle corners is made approximately equal to the sheet thickness. For such sharp corners, particularly if the total corner material is a small fraction of the entire cross section, the actual rounded transitions can be replaced by square corners for purposes of calculation, with negligible error. More generally, section properties of any thin-wall shape are best computed by the so-called linear method. One first assumes the area of the section to be concentrated along the centerline of the sheet, so that the various area elements which compose the section are replaced by straight- or curved-line elements. Calculating the total length, moment of inertia, etc., of this middle line, one then obtains the appropriate property of the actual section by multiplying by the sheet thickness. Figure 7 gives the properties of variously shaped and disposed line elements for use in calculations by the linear method.

It should be noted that the actual area of thin compression elements, such as compression flanges, must frequently be replaced by a reduced, effective area for calculating effective cross-sectional properties (Art. 7).

7. Thin Compression Elements The main structural characteristic of the flat element is its flat-width ratio w/t, where w = flat width measured from end of corner transition radius and t = sheet thickness. Elements with large w/t ratios, when in compression, are subject to local buckling. The compression element of the flexural member of Fig. 8a,

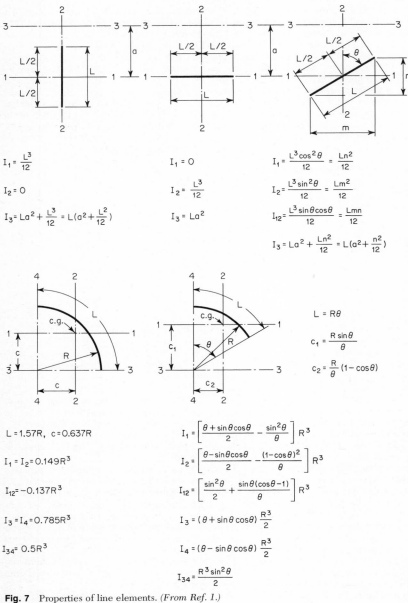

$$I_1 = \frac{L^3}{12}$$

$$I_2 = 0$$

$$I_3 = La^2 + \frac{L^3}{12} = L\left(a^2 + \frac{L^2}{12}\right)$$

$$I_1 = 0$$

$$I_2 = \frac{L^3}{12}$$

$$I_3 = La^2$$

$$I_1 = \frac{L^3\cos^2\theta}{12} = \frac{Ln^2}{12}$$

$$I_2 = \frac{L^3\sin^2\theta}{12} = \frac{Lm^2}{12}$$

$$I_{12} = \frac{L^3\sin\theta\cos\theta}{12} = \frac{Lmn}{12}$$

$$I_3 = La^2 + \frac{Ln^2}{12} = L\left(a^2 + \frac{n^2}{12}\right)$$

$$L = 1.57R, \quad c = 0.637R$$

$$I_1 = I_2 = 0.149R^3$$

$$I_{12} = -0.137R^3$$

$$I_3 = I_4 = 0.785R^3$$

$$I_{34} = 0.5R^3$$

$$L = R\theta$$

$$c_1 = \frac{R\sin\theta}{\theta}$$

$$c_2 = \frac{R}{\theta}(1-\cos\theta)$$

$$I_1 = \left[\frac{\theta + \sin\theta\cos\theta}{2} - \frac{\sin^2\theta}{\theta}\right]R^3$$

$$I_2 = \left[\frac{\theta - \sin\theta\cos\theta}{2} - \frac{(1-\cos\theta)^2}{\theta}\right]R^3$$

$$I_{12} = \left[\frac{\sin^2\theta}{2} + \frac{\sin\theta(\cos\theta-1)}{\theta}\right]R^3$$

$$I_3 = (\theta + \sin\theta\cos\theta)\frac{R^3}{2}$$

$$I_4 = (\theta - \sin\theta\cos\theta)\frac{R^3}{2}$$

$$I_{34} = \frac{R^3\sin^2\theta}{2}$$

Fig. 7 Properties of line elements. *(From Ref. 1.)*

being held straight only along the longitudinal edge where it is joined to the web, buckles locally at a lower stress and, once buckled, shows more pronounced distortions than an element of the same w/t ratio, but of the type of Fig. 8*b*. The latter is stiffened by webs along both its longitudinal edges, which increases its buckling strength. Elements of the type of Fig. 8*a* are known as *unstiffened compression elements*, while those of Fig. 8*b* are called *stiffened compression elements.* Because of their different behavior in local buckling, different design methods apply to the two types.

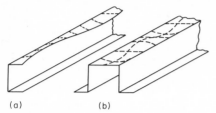

(a)   (b)

Fig. 8 Local buckling of *(a)* unstiffened compression flange, and *(b)* stiffened compression flange. *(From Ref. 2.)*

In order to avoid excessively deformable members and safeguard the integrity of cross sections, the AISI specification limits the w/t ratio for stiffened elements to 500 when both longitudinal edges are connected to other stiffened elements, 60 when one edge is stiffened by a simple lip, 90 when one edge is stiffened by any other kind of stiffener. For unstiffened elements, the maximum w/t ratio is 60.

Stiffeners.   Compression elements can be stiffened either by webs along both longitudinal edges (e.g., top flange of Fig. 2*g*) or by an edge stiffener such as the lip of Fig. 2*b*. For very wide elements, intermediate stiffeners are frequently provided, as in Fig. 3*d*. To be fully effective the *minimum moment of inertia of an edge stiffener* is

$$I_{min} = 1.83t^4 \sqrt{\left(\frac{w}{t}\right)^2 - \frac{4000}{F_y}} \geqslant 9.2t^4 \tag{1}$$

where I_{min} = minimum permissible moment of inertia of stiffener about its centroidal axis
            parallel to stiffened compression element
     w/t = flat-width ratio of stiffened element
If an edge stiffener consists of a simple lip bent at right angles to the stiffened element, the overall depth d_{min} of the lip may be determined by

$$d_{min} = 2.8t \sqrt[6]{\left(\frac{w}{t}\right)^2 - \frac{4000}{F_y}} \geqslant 4.8t \tag{2}$$

A simple lip is not permitted as an edge stiffener for any element whose w/t exceeds 60.

The moment of inertia of an intermediate stiffener must be not less than twice the value of I_{min} by Eq. (1).

Recent research tends to show that these AISI stiffener requirements may be somewhat unconservative. Until improved requirements are promulgated, it may be advisable to dimension stiffeners so that their actual moment of inertia is 1.5 to 2.0 times that given by Eq. (1), and 1.2 to 1.6 times that given by Eq. (2), respectively, the lower multiples applying to smaller w/t and F_y values and the higher multiples to larger w/t and F_y values.

Stiffened compression elements develop slight and imperceptible buckling waves at stresses which decrease with increasing w/t ratio. Such slight buckling does not impair their carrying capacity; it merely means that the previously uniformly distributed compression stress now becomes concentrated in the less buckled regions of the plate, i.e., in the two strips adjacent to the stiffened edge. The more highly distorted central portion is not capable of resisting increasing stress as the load on the member is increased. Therefore, once buckling starts, the longitudinal compression stresses in a stiffened compression element, such as the top flange of Fig. 8b, are distributed across the width nonuniformly as shown in Fig. 9. With increasing load on the member, the maximum compression stress at the edges increases and the nonuniformity of distribution becomes more pronounced.

In order to account for this nonuniform stress distribution in design, the actual width w of the stiffened compression element is replaced by a reduced effective width b which is

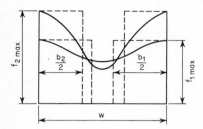

Fig. 9 Stress distribution and effective width of stiffened compression element. *(From Ref. 2.)*

determined in such a manner that the total area under the actual stress-distribution curve is equal to the combined area of the two rectangular stress distributions whose combined width is b (Fig. 9). This effective width depends on the flat-width ratio w/t and on the maximum edge stress f (ksi). The element fails when the edge stress becomes equal to the yield point. At design load, the edge stress is smaller, and the effective width is therefore larger than at failure. For this reason it is necessary to distinguish between effective widths to be used for determining allowable loads, i.e., on the one hand, failure loads divided by the safety factor, and on the other those to be used for computing deflections at design loads. Thus, according to the AISI specification (1968):

1. For safe-load determination elements are fully effective $(b = w)$ up to $(w/t)_{\lim} = 171/\sqrt{f}$. For larger values of w/t,

$$b = \frac{253}{\sqrt{f}} \left[1 - \frac{55.3}{(w/t)\sqrt{f}} \right] t \qquad (3a)$$

2. For deflection determination elements are fully effective to $(w/t)_{\lim} = 221/\sqrt{f}$. For larger values of w/t,

$$b = \frac{326}{\sqrt{f}} \left[1 - \frac{71.3}{(w/t)\sqrt{f}} \right] t \qquad (3b)$$

The AISI specification gives slightly larger effective widths for closed square and rectangular tube, because of their greater stability of cross-sectional shape.

The charts of Figs. 10 and 11 give effective widths according to Eqs. (3). More detailed and elaborate charts are presented in the AISI Manual.

Once the effective widths of all stiffened compression elements of a section are determined, its cross-sectional properties are determined in the usual manner, but for the effective rather than the actual area. Thus, the effective areas of the hat section used as a flexural member and as a column are shown in Fig. 12a and b, respectively.

The efficiency of stiffened compression elements of large w/t ratio can be greatly improved by providing additional intermediate stiffeners between edges, such as the longitudinal stiffening ribs in Fig. 3d, e, h, k. Such elements are known as *multiple stiffened elements* and, for purposes of computation, are divided into subelements, i.e., flat portions between intermediate stiffeners or between an intermediate stiffener and the adjacent edge stiffener. If the w/t ratio of such a subelement is smaller than 60, its effective width is computed as for a stiffened element. If it is larger than 60, its effective width is

$$b' = b - 0.10 \left[\frac{w}{t} - 60 \right] t \qquad (4)$$

where b is the effective width determined as above for the particular w/t ratio. In this case, for calculating section properties, the area of the intermediate stiffener(s) is also reduced, such that

$$A_{\text{eff}} = \alpha A_{\text{full}} \tag{5}$$

where for w/t between 60 and 90:

$$\alpha = \left(3 - 2\frac{b'}{w}\right) - \frac{1}{30}\left(1 - \frac{b'}{w}\right)\frac{w}{t} \tag{6a}$$

and for w/t greater than 90:

$$\alpha = \frac{b'}{w} \tag{6b}$$

Unstiffened compression elements (Fig. 8a) develop buckling waves at considerably lower stresses than stiffened elements of the same w/t ratio, and these distortions may be somewhat more pronounced because of lack of restraint along one of the longitudinal

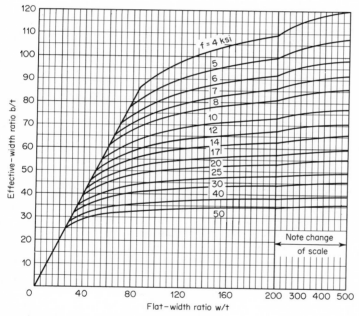

Fig. 10 Effective width of stiffened elements for safe-load determination [Eq. (3a)].

edges. Because of this, the allowable stress on unstiffened elements is reduced with increasing w/t to prevent failure by local buckling and reduce possible distortion at service loads. According to the AISI specification (1968), the allowable stress on unstiffened compression elements is

For $w/t \lesssim 63.3/\sqrt{F_y}$:

$$F_c = 0.6F_y \tag{7a}$$

For $63.3/\sqrt{F_y} \lesssim w/t \lesssim 144/\sqrt{F_y}$:

$$F_c = F_y\left(0.767 - 0.00264\frac{w}{t}\sqrt{F_y}\right) \tag{7b}$$

For $144/\sqrt{F_y} \lesssim w/t \lesssim 25$:

$$F_c = \frac{8000}{(w/t)^2} \tag{7c}$$

For $25 \leqslant w/t \leqslant 60$: $\qquad\qquad F_c = 19.8 - 0.28\,\dfrac{w}{t}$ $\hfill (7d)$

except that Eq. (7c) shall be used for angle struts.

Equations (7) are plotted in Fig. 13. It is seen that allowable stresses for w/t in the range of about 30 to 60 are so low as to make this range uneconomical for most design situations.

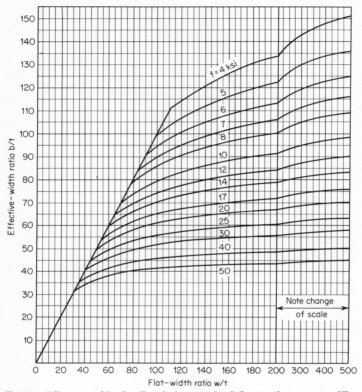

Fig. 11 Effective width of stiffened elements for deflection determination [Eq. (3b)].

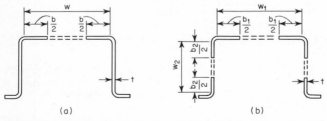

Fig. 12 Actual width w and effective width b in (a) flexural member and (b) compression member.

The 1968 AISI specification method of dealing with stiffened compression elements by means of a reduced effective width, and with unstiffened compression elements by means of a reduced allowable stress, leads to certain inconsistencies and overconservatisms, particularly for the higher-strength steels. Recent research[13] has shown that the effective-width method is also applicable to unstiffened elements and results in greater consistency and economy.

The ratio of effective to actual width w/b can be written in general terms as

$$b/w = \sqrt{\sigma_{cr}/f}\,(1 - 0.22\sqrt{\sigma_{cr}/f}) \qquad (a)$$

where σ_{cr} is the critical buckling stress of compressed plates (see, for example, chap. 4 of Ref. 15). Introducing the safety factor and other transformations, the following two equations are obtained from Eq. (a):

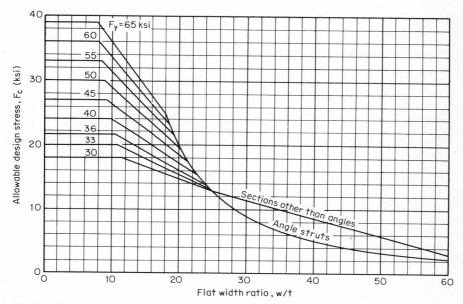

Fig. 13 Allowable stresses on unstiffened compression elements [Eqs. (7)].

For safe-load determinations, when

$$w/t > (w/t)_{\text{lim}} = 85.6\sqrt{k}/\sqrt{f}$$

$$b = \frac{126.4\sqrt{k}}{\sqrt{f}}\left[1 - \frac{27.6\sqrt{k}}{(w/t)\sqrt{f}}\right]t \qquad (b)$$

and for deflection determinations, when

$$w/t > (w/t)_{\text{lim}} = 110.8\sqrt{k}/\sqrt{f}$$

$$b = \frac{163.3\sqrt{k}}{\sqrt{f}}\left[1 - \frac{35.6\sqrt{k}}{(w/t)\sqrt{f}}\right]t \qquad (c)$$

Here k is the numerical coefficient in the plate-buckling equation, which depends primarily on edge conditions.[13]

These equations are perfectly general. For stiffened compression elements, assuming conservatively $k = 4$, Eqs. (b) and (c) become Eqs. ($3a$) and ($3b$), previously given for stiffened elements.

For unstiffened elements, again assuming somewhat conservatively $k = 0.5$, one obtains the following effective-width equations:

1. For safe-load determination elements are fully effective ($b = w$) up to

$$\left(\frac{w}{t}\right)_{\text{lim}} = \frac{60.5}{\sqrt{f}}$$

For elements with w/t larger than $(w/t)_{\text{lim}}$ the effective width

$$b = \frac{89.4}{\sqrt{f}}\left[1 - \frac{19.5}{(w/t)\sqrt{f}}\right]t \qquad (8a)$$

2. For deflection determination elements are fully effective ($b = w$) up to

$$\left(\frac{w}{t}\right)_{\text{lim}} = \frac{78.3}{\sqrt{f}}$$

For elements with w/t larger than $(w/t)_{\text{lim}}$ the effective width

$$b = \frac{115.4}{\sqrt{f}}\left[1 - \frac{25.2}{(w/t)\sqrt{f}}\right]t \qquad (8b)$$

The somewhat conservative values $k = 4$ and $k = 0.5$, usually assumed in design, correspond to simple support along the stiffened edges. For the actual rotational restraints usually provided by the stiffening components (e.g., webs) correspondingly larger values of k, if determined by rational analysis, can be used in Eqs. (b) and (c), resulting in moderately larger effective widths.

The stresses in unstiffened compression elements in the postbuckling range are distributed as shown in Fig. 14. The stress is largest at the supported edge and smallest at the outer, unsupported edge, which is understandable from the manner of postbuckling deformation (Fig. 8a). Correspondingly, the effective width in unstiffened elements is adjacent to and extends outwards from the stiffened edge as shown in Fig. 14.

Design charts for unstiffened elements, based on Eqs. (8a) and (8b), are given in Figs. 15 and 16.

8. Compression Members Thin-walled compression members can fail in one of the following ways: by simple column buckling, by local buckling of the component elements, by a combination of local and column buckling, and by torsional-flexural buckling, i.e., by simultaneous twisting and bending. The last three of these modes are more consequential for thin-walled than for more stocky members because the resistance to both local and torsional buckling depends approximately on the square of the thickness.

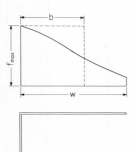

Fig. 14 Stress distribution and effective width of unstiffened compression element. *(From Ref. 14.)*

Closed sections, such as in Fig. 2r and s, will not buckle torsionally. This and their favorable ratio of I/A in both principal directions make them very efficient as compression members. Doubly symmetrical sections (Fig. 2n) and point-symmetrical sections (Fig. 2c and d), whose shear center coincides with the centroid, also do not buckle torsional-flexurally. The chief singly symmetrical sections which need checking for this buckling mode are angles, channels, C-sections, and hat sections (Fig. 2e, f, a, b, g). For these sections the three charts of Fig. 17 permit one to ascertain whether an axially compressed member of given length and cross section will fail flexurally or by combined torsion and flexure. A given member corresponds to a point in the pertinent plot. If that point falls to the left of the applicable curve, member buckling will be flexural and torsion need not be checked. If it falls to the right, the member will buckle in the torsional-flexural mode at a load which may be considerably below that for flexural buckling. In this case it is advisable to redesign the member to eliminate the torsional mode. In regard to channel, C-, and hat sections it is seen that torsional-flexural buckling cannot occur unless the flange width b is at least 0.6 to 0.8 the web dimension a.

In general, if shape and length of a member are such that torsional-flexural buckling can occur, the member will always buckle at a lower stress than a member of the same slenderness, KL/r, but not subject to torsion. Thus, where possible, it is economical to

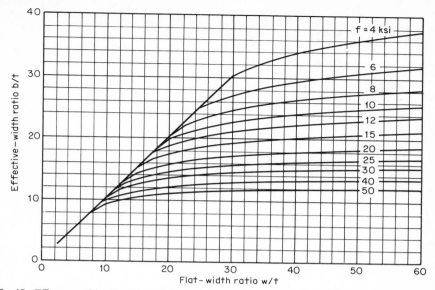

Fig. 15 Effective width of unstiffened elements for safe-load determination [Eq. (8a)].

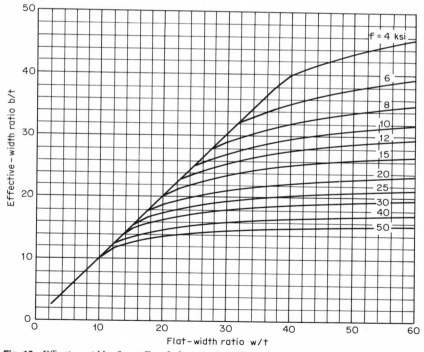

Fig. 16 Effective width of unstiffened elements for deflection determination [Eq. (8b)].

9-17

design in a manner which will prevent the torsional-flexural mode. Twisting is often easily prevented by available bracing, such as in wall studs where adequate attachment to exterior and interior wall sheathing will prevent torsion. If members are not braced against twist, it is often possible to select cross-sectional shapes and dimensions so that the buckling mode is flexural rather than torsional-flexural.

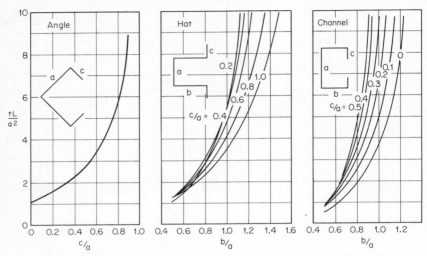

Fig. 17 Type of buckling for angle, hat, or channel compression members. Members which plot to left of pertinent curve buckle in simple flexure; members which plot to right in combined torsion and flexure. *(From Ref. 10.)*

If the w/t ratios of the cross-sectional elements of compression members exceed certain limits, local buckling can combine with overall buckling, which results in a lower carrying capacity. Local buckling of compression members can be dealt with by using a shape factor Q, which is the ratio of the load at which a short length of member fails by local buckling to the load at which it would fail in simple yielding if local buckling were prevented. It is calculated as follows:

If all elements of the section are stiffened (Fig. 2b, r):

$$Q = \frac{A_{\text{eff}}}{A_{\text{full}}} \qquad (9)$$

If all elements of the section are unstiffened (Fig. 2e):

$$Q = \frac{F_c}{F} \qquad (10)$$

If the section is composed of both stiffened and unstiffened elements (Fig. 2a, c):

$$Q = \frac{F_c}{F} \frac{A_{\text{eff}}}{A_{\text{full}}} \qquad (11)$$

where F_c = allowable compression stress for the unstiffened element with largest w/t ratio [Eqs. (7)]

A_{eff} = reduced effective area of all stiffened elements plus unreduced area of unstiffened elements, if any

The effective width for calculating A_{eff} is determined from Eqs. (3), using $f = F$ in Eq. (10) and $f = F_c$ in Eq. (11).

Strength calculations for compression members which are subject to torsional-flexural buckling are fairly lengthy and complex. Part I of the AISI Manual gives the necessary design formulas and Part V contains charts which greatly simplify the calculations.

The discussion of compression-member design which follows deals only with simple flexural buckling. The allowable axial compression stress F_a, taking account of both local and overall buckling, is determined from the following equations:

For $KL/r < 765/\sqrt{QF_y}$:
$$F_a = 0.522 \, QF_y - \left(\frac{QF_y KL/r}{1494}\right)^2 \qquad (12a)$$

For $KL/r \geq 765/\sqrt{QF_y}$:
$$F_a = \frac{151,900}{(KL/r)^2} \qquad (12b)$$

where $F_a = P/A$ = allowable compression stress on the full, unreduced area, ksi
  K = effective length factor
  r = radius of gyration of the full, unreduced section

The chart of Fig. 18 gives allowable stresses F_a for $F_y = 33$ ksi. It can be used for steels with other yield points in the manner indicated in the caption. The AISI specification

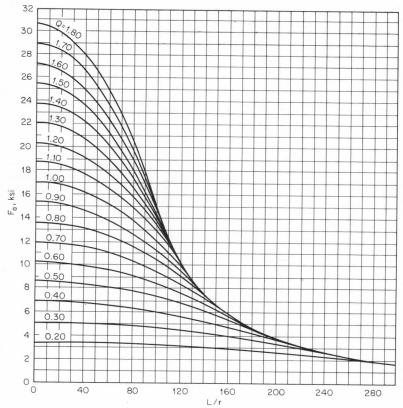

Fig. 18 Allowable stresses $F_a = P/A$ for compression members. Chart is drawn for 33-ksi yield point; it can be used for other yield points F_y by multiplying Q by $F_y/33$. Curves for Q greater than 1.00 do not apply to steels with 33-ksi yield point. *(From Ref. 1.)*

(1968) makes special provisions for relatively thicker stocky sections, as follows: When $Q = 1$, the steel is 0.09 in. or more in thickness, and $KL/r < 765/\sqrt{QF_y}$, the somewhat more liberal provisions of the AISC specification can be used instead of Eq. (12a).

Example The light box-shaped compression member of Fig. 19 is produced by spot-welding a flat plate to a cold-rolled hat section. (Shapes such as this have the advantage of large rigidities in both principal directions, and of high torsional rigidity which eliminates the possibility of torsional-flexural buckling. A typical use is in industrial storage racks.) Determine the allowable axial load P for $L = 6$ ft and for $L = 10$ ft, assuming $F_y = 50$ ksi and $k = 1$.

The basic design stress (Art. 5) is $F = 0.6 \times 50 = 30$ ksi. The section properties for the full, unreduced section are

$$A = 0.585 \text{ in.}^2 \qquad \bar{y} = 0.815 \text{ in.}$$
$$I_x = 0.390 \text{ in.}^4 \qquad I_y = 0.555 \text{ in.}^4$$

$$r_x = \sqrt{\frac{I_x}{A}} = \sqrt{\frac{0.390}{0.585}} = 0.82 \text{ in.}$$

It is assumed that spot welds are placed at appropriately close intervals so that the lines of welds act as lines of stiffening for the flat plate. Requirements for such weld spacing are given in Sec. 4.4 of Ref. 1; the main requirement is that, in order to prevent buckling of the compressed sheet between welds,

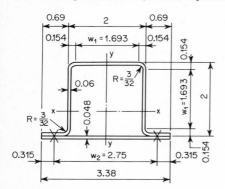

Fig. 19

the weld spacing is not to exceed $s = 200t/\sqrt{f}$, where f = design stress in the flat plate. From Fig. 19 the flat widths of all three sides of the hat section are the same $w_1 = 1.693$ in., and the flat width of the plate between lines of welds $w_2 = 2.75$ in. The flat-width ratios of these stiffened elements are then $w_1/t = 1.693/0.06 = 28.2$; $w_2/t = 2.75/0.048 = 57.3$.

The effective design widths for these two types of stiffened elements are calculated for $F = 30$ ksi, from Eq. (3a) or Fig. 10. They are, respectively,

$$b_1 = w_1 = 1.693 \text{ in.}$$
$$b_2 = 38.0t_2 = 38.0 \times 0.048 = 1.83 \text{ in.}$$

It is simplest to calculate the effective area by subtracting from the full, unreduced area the ineffective portions $w - b$ of the stiffened elements. Thus,

$$A_{\text{eff}} = 0.585 - (2.75 - 1.83)0.048 = 0.541 \text{ in.}^2$$

For the unstiffened elements, i.e., the flat portions projecting beyond the weld lines, $w/t < 8.95$ so that no stress reduction is required [Eq. (7a)]. Therefore, the shape factor is given by Eq. (9):

$$Q = \frac{A_{\text{eff}}}{A_{\text{full}}} = \frac{0.541}{0.585} = 0.925$$

To determine which of Eqs. (12) applies, one determines

$$\frac{L}{r} = \frac{765}{\sqrt{QF_y}} = \frac{765}{\sqrt{0.925 \times 50}} = 112$$

For an unbraced length of 10 ft:

$$\frac{L}{r} = \frac{120}{0.82} = 146$$
$$F_a = \frac{151,900}{146^2} = 7.13 \text{ ksi}$$
$$P = 7.13 \times 0.585 = 4.17 \text{ kips}$$

For an unbraced length of 6 ft:

$$\frac{L}{r} = \frac{72}{0.82} = 87.8 \text{ in.}$$

$$F_a = 0.522 \times 0.925 \times 50 - \left(\frac{0.925 \times 50 \times 87.8}{1494}\right)^2 = 16.75 \text{ ksi}$$

$$P = 16.75 \times 0.585 = 9.80 \text{ kips}$$

The use of the shape factor Q to account for the interaction of local and overall buckling is approximate and may lead to significant errors, on both the conservative and the unconservative side. A more accurate and consistent method has been developed,[14] using Eqs. (12) with $Q = 1$ and sectional properties calculated for the effective cross section rather than the full cross section. The effective area A and the corresponding radius of gyration r are calculated at the stress F_a, using the effective-width equations for load determination, Eqs. (3a) and (8a), where F_a is the allowable compression stress on the effective area A. At the outset F_a is unknown; so successive approximation must be used. A good first approximation is obtained by using properties of the unreduced cross section in computing KL/r to determine an initial value of F_a from Eq. (12) or from Fig. 18, using $Q = 1$.

9. Flexural Members *Unstiffened Compression Flanges.* If the compression flange of a flexural member consists of unstiffened elements, then according to the current (1968) AISI specification[1] section properties are computed for the full, unreduced section. The allowable compression stresses in such flanges are then those given by Eqs. (7). However, if reduced effective widths instead of reduced allowable stresses are used for unstiffened elements, the allowable stress is $F = 0.6F_y$, and if $w/t > (w/t)_{\lim}$, effective section properties are calculated using the effective width as determined from Eqs. (8).

Stiffened Compression Flanges. If the compression flange consists of stiffened or multiple stiffened elements, the allowable stress is F in both compression and tension. However, if $w/t > (w/t)_{\lim}$, the reduced effective design width [Eqs. (3)] must be employed to calculate effective section properties.

If the neutral axis is closer to the tension flange than to the compression flange, compression governs and the effective width is computed from Eq. (3a) using $f = F$. If the neutral axis is closer to the compression flange, tension governs and the location of the axis depends on the magnitude of the compression stress f, calculated on the basis of the reduced, effective properties. It is possible to determine the position of the neutral axis and other properties by writing equations which incorporate the formulas for effective width. While this is often advantageous for computer calculations, successive approximations are usually preferable for ordinary design calculations when only one or a few shapes are being investigated. One estimates $\overline{y}$, the distance from the neutral axis to the compression flange, keeping in mind that the compression flange is not fully effective, and computes the effective width of the compression flange for the stress $f = F\,\overline{y}/h - \overline{y})$, where h is the depth of the section. The distance $\overline{y}$ is then calculated on the basis of the effective area for the stiffened compression elements and the full area for all other elements, if present (1968) AISI procedures are used. However, if Eqs. (8) are used for unstiffened flanges, $\overline{y}$ is calculated using effective areas for all compression elements and full areas for all others. The calculation is repeated until satisfactory agreement is obtained.

Section properties based on effective widths for load determination can also be used to compute deflections. This gives a slight overestimate, and if a more precise calculation is desired, the section properties are recomputed, using Eq. (3b) and, if applicable, (8b).

Example The panel shown in Fig. 20 is used as a flexural member. Determine its moment capacity and the moment of inertia to be used for deflection calculations for steel with $F_y = 33$ ksi, for which $F = 20$ ksi.

MOMENT CAPACITY. Since the effective width of the top flange depends on the stress in that flange, successive approximations are necessary.

FIRST APPROXIMATION. The flat-width ratio of the top flange is $w/t = 8.76/0.06 = 146$. Assume $f_c = 11$ ksi. From Eq. (3a) or Fig. 10, the effective width, exclusive of corners, is $b = 67.6t = 4.056$ in.

Cross sectional properties, using linear method:

Element	Element length L, in.	Distance y from top fiber, in.	Ly	Ly^2	I_0
Top flange........	4.056	0.030	0.1217		
Bottom flanges....	$2 \times 0.630 = 1.260$	1.470	1.8522	2.7227	
Webs............	$2 \times 1.260 = 2.520$	0.750	1.8900	1.4175	0.3334
Top corners.......	$2 \times 0.141 = 0.282$	0.063	0.0178	0.0011	0.0002
Bottom corners....	$2 \times 0.141 = 0.282$	1.437	0.4052	0.5823	0.0002
Summation Σ......	8.400		4.2869		0.3338

$$y_{cg} = \frac{4.2869}{8.400} = 0.510 \text{ in.}$$

$$f_c = 20\,\frac{0.510}{1.500 - 0.510} = 10.3 \text{ ksi}$$

SECOND APPROXIMATION. On the basis of the first approximation, assume

$$y_{cg} = 0.500 \text{ in.} \qquad f_c = 10 \text{ ksi}$$

for which Eq (3a) or Fig. 10 gives the effective top-flange width $b = 70.4t = 4.224$ in. The corresponding values of Ly and Ly^2 are 0.1267 and 0.0038. All other entries in the table of cross-

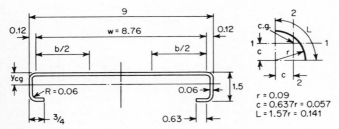

Fig. 20

sectional properties for the first approximation remain the same. Inserting the corrected top-flange width and the corresponding values of Ly and Ly^2 in the table, the following summations are obtained:

$$\Sigma L = 8.568 \qquad \Sigma Ly = 4.2919 \qquad \Sigma Ly^2 = 4.7274$$

Therefore,

$$y_{cg} = \frac{4.2919}{8.568} = 0.501 \text{ in.}$$

which agrees with the assumed value 0.500 in. Then

$$I'_x = \Sigma Ly^2 + \Sigma I_0 - y_{cg}^2 \Sigma L = 4.7274 + 0.3338 - 2.1506 = 2.9106$$
$$I_x = I'_x t = 2.9106 \times 0.060 = 0.1746 \text{ in.}^4$$
$$S_x = \frac{I_x}{c} = \frac{0.1746}{1.500 - 0.501} = 0.1748 \text{ in.}^3$$
$$M = 0.1748 \times 20 = 3.50 \text{ in.-kips}$$

DEFLECTION. For reasons discussed in Art. 7, the effective width for deflection calculations is different from that for allowable moment. Assume that the deflection at allowable load is to be determined.

FIRST APPROXIMATION. Assume $f = 10$ ksi. From Eq. (3b) or Fig. 11, the effective width, exclusive of corners, is $b = 87.1t = 5.226$ in.

Cross-sectional properties

Element	Element length L, in.	Distance y from top fiber, in.	Ly	Ly^2	I_0
Top flange........	5.226	0.030	0.1568	0.0047	
Bottom flanges....	$2 \times 0.630 = 1.260$	1.470	1.8522	2.7227	
Webs.............	$2 \times 1.260 = 2.520$	0.750	1.8900	1.4175	0.3334
Top corners.......	$2 \times 0.141 = 0.282$	0.063	0.0178	0.0011	0.0002
Bottom corners....	$2 \times 0.141 = 0.282$	1.437	0.4052	0.5823	0.0002
Summation Σ......	9.570		4.3220	4.7283	0.3338

$$y_{cg} = \frac{4.322}{9.570} = 0.452$$
$$I'_x = \Sigma Ly^2 + \Sigma I_0 - y_{cg}^2 \Sigma L = 4.7283 + 0.3338 - 1.953 = 3.1091$$
$$I_x = I'_x t = 3.1091 \times 0.060 = 0.1865 \text{ in.}^4$$
$$f_c = \frac{3.50 \times 0.452}{0.1865} = 8.47 \text{ ksi}$$

This stress is significantly smaller than the assumed value 10 ksi.

SECOND APPROXIMATION. Assume $f_c = 8$ ksi. From Eq. (3b) or Fig. 11, the effective top-flange width is $b = 95.2t = 5.712$ in. The corresponding values of Ly and Ly^2 are 0.1714 and 0.0051. All other entries in the table of cross-sectional properties for the first approximation remain the same. The corrected summations are

$$\Sigma L = 10.056 \qquad \Sigma Ly = 4.3366 \qquad \Sigma Ly^2 = 4.7287$$
$$y_{cg} = \frac{4.3366}{10.056} = 0.431 \text{ in.}$$
$$I'_x = \Sigma Ly^2 + \Sigma I_0 - y_{cg}^2 \Sigma L = 4.7287 + 0.3338 - 1.8680 = 3.1945$$
$$I_x = I'_x t = 3.1945 \times 0.060 = 0.1917 \text{ in.}^4$$
$$f_c = \frac{3.50 \times 0.431}{0.1917} = 7.86 \text{ ksi}$$

This is close enough to the assumed value 8 ksi. Therefore, $I_x = 0.1917$ in.^{4} should be used in the deflection calculations. This is about 10 percent larger than the value 0.1746 in.^{4} which governed load-carrying capacity. Hence, the latter would be conservative, but uneconomical, for deflection determination.

10. Beam Webs Webs of beams must be checked for local buckling caused by shear stresses and by bending compressive stresses. According to the AISI specification (1968), the allowable maximum average shear stress $v = V/ht$ on the gross area of flat webs is given by

For $h/t \lesssim 547/\sqrt{F_y}$: $$F_v = \frac{152\sqrt{F_y}}{h/t} \leq 0.40\, F_y \text{ ksi}$$ (13a)

For $h/t > 547/\sqrt{F_y}$: $$F_v = \frac{83,200}{(h/t)^2} \text{ ksi}$$ (13b)

where t = web thickness
h = clear distance between flanges

When the web consists of two sheets, each sheet is considered separately, carrying its share of the total shear.

The allowable maximum compressive stress in the web due to bending is

$$F_{bw} = \frac{520,000}{(h/t)^2}$$ (14)

In cases where large bending moments and large shears occur in the same cross section, such as at interior supports of continuous beams, webs must be proportioned to satisfy the interaction equation

$$(f_{bw}/F_{bw})^2 + (f_v/F_v)^2 \leq 1 \tag{15}$$

where f_{bw} and f_v are, respectively, the maximum compression stress at the flange junction due to bending and the actual average shear stress. To avoid excessively deformable webs, the AISI specification limits the b/t ratio to 150 if no stiffening devices are employed at points of concentrated loads or reactions, and to 200 if such devices are provided.

Crippling.   To guard against crippling of flat webs which have no stiffeners at supports or other points of concentrated load, safe loads are established by the following equations, which hold for single webs with inside corner radii equal to or less than the sheet thickness and for $F_y = 33$ ksi: Permissible end reaction $P_{\max}$ per web, kips:

$$P_{\max} = t^2[98 + 4.2(N/t) - 0.022(N/t)(h/t) - 0.011(h/t)] \tag{16a}$$

Permissible interior reaction or concentrated load $P_{\max}$ per web, kips:

$$P_{\max} = t^2[305 + 2.3(N/t) - 0.009(N/t)(h/t) - 0.5\,(h/t)] \tag{16b}$$

where N = length of bearing, in., but not greater than h.

For larger corner radii and for other grades of steel the value of $P_{\max}$ from Eq. (16a) is multiplied by

$$k(1.15 - 0.15n)(1.33 - 0.33k)$$

and the value from Eq. (16b) by

$$k(1.06 - 0.06n)(1.22 - 0.22k)$$

where $k = F_y/33$

n = ratio of inside bend radius to web thickness

For sections which furnish a high degree of restraint against rotation of the web, as in Fig. 2n and o, the value of $P_{\max}$ per web, kips, is given by

$$P_{\max} = t^2 F_y(4.44 + 0.558\sqrt{N/t}) \tag{17a}$$

for end reactions, and

$$P_{\max} = t^2 F_y(6.66 + 1.446\sqrt{N/t}) \tag{17b}$$

for interior reactions or concentrated loads.

When concentrated loads or reactions, which may cause web crippling, are located in regions with high bending moments, the following interaction equation must be satisfied:

$$(M/M_{\max}) + (P/P_{\max}) \leq 1.3 \tag{18}$$

where M and P are the actual moment and concentrated load or reaction at the section, respectively, and $M_{\max}$ and $P_{\max}$ are the allowable values for these quantities in the absence of the respective other quantity.

It is generally not economical in cold-formed construction to reinforce webs against local crippling by providing nonintegral stiffeners, as is done in hot-rolled construction. However, methods have been developed to form end-reaction integral stiffeners in webs of cold-rolled shapes.

11. Lateral Buckling of Beams   This is not likely to occur in most finished structures; for example, a floor joist cannot usually bend out of its vertical plane because the floor it supports constitutes a rigid diaphragm which braces it. However, buckling can occur if a beam is unbraced over significant portions of its length or, more often, is temporarily unbraced during construction. In such cases allowable stresses must be adjusted to forestall buckling. Additionally, this information is needed when dealing with combined compression and bending, such as in eccentrically loaded columns (Art. 12).

Because of the variety of cold-formed beam shapes, calculation of allowable stresses for unbraced beams is slightly more complex than for hot-rolled I-shaped sections. The slenderness ratio of such cold-formed sections, which determines the lateral buckling strength, can be expressed as $L^2 S_{xc}/dI_{yc}$. Apart from this slenderness, the allowable stress depends on the bending-moment distribution along the member; this is expressed by the constant C_b. Correspondingly, the allowable bending stress F_b (ksi) depends on the parameter

$$\eta = \frac{L^2 S_{xc}}{d I_{yc}} \frac{1}{C_b} \tag{19}$$

where L = unbraced length, in.

I_{yc} = moment of inertia of compression portion of section about section gravity axis parallel to web, in.[4]

S_{xc} = compression section modulus of entire section

d = depth of section

C_b can conservatively be taken as unity or calculated from

$$C_b = 1.75 + 1.05 \frac{M_1}{M_2} + 0.3 \left(\frac{M_1}{M_2}\right)^2 \le 2.3$$

where M_1 is the smaller of the two bending moments M_1 and M_2 at the ends of the unbraced length. M_1/M_2 is positive for reverse-curvature bending and negative for single-curvature bending. When the moment at any point along L is larger than both M_1 and M_2, $C_b = 1.0$. Also, for members subject to combined axial load and bending, $C_b = 1.0$.

For bending about the centroidal axis perpendicular to the web, the AISI provisions can be written as follows: For

1. I-beams symmetrical about the web and symmetrical channels:

When $\dfrac{10.5 \times 10^4}{F_y} < \eta \le \dfrac{52.4 \times 10^4}{F_y}$

$$F_b = \frac{2}{3} F_y - \frac{F_y^2}{157 \times 10^4} \eta \le 0.6 F_y \tag{20a}$$

When $\eta \ge \dfrac{52.4 \times 10^4}{F_y}$

$$F_b = \frac{17.5 \times 10^4}{\eta} \tag{20b}$$

2. Point-symmetrical Z-beams, bent about centroidal axis perpendicular **to web:**

When $\dfrac{5.24 \times 10^4}{F_y} < \eta \le \dfrac{26.2 \times 10^4}{F_y}$

$$F_b = \frac{2}{3} F_y - \frac{F_y^2}{78.6 \times 10^4} \eta \le 0.6 F_y \tag{21a}$$

When $\eta \ge \dfrac{26.2 \times 10^4}{F_y}$

$$F_b = \frac{8.73 \times 10^4}{\eta} \tag{21b}$$

Box Beams.   In contrast to the single-web open section, some commonly used thin-walled sections are highly resistant to lateral buckling. Thus, the torsional constant J for box beams exceeds that of the comparable open section by several orders of magnitude, and this often holds for I_y relative to I_x. With any reasonable dimensions, such beams are unlikely to buckle laterally. Therefore, the AISI specification provides that lateral buckling need not be considered for closed box-type beams if the length-width ratio L/b does not exceed 75.

Hat Section.   The torsional constant J is of the same order as that of the comparable I, but I_y is much larger relative to I_x. Thus, the hat section is relatively more stable than the I, but less stable than the box. For hat sections bent about the weak axis, i.e., $I_x < I_y$ where x is the axis perpendicular to the webs, lateral buckling will not occur. For the opposite case, $I_x > I_y$, a conservative estimate can be made by regarding the compression portion of the hat section as a column tending to buckle about the y axis, in which case the allowable stress can be obtained from Eq. $(11b)$.

12. Combined Bending and Axial Compression   For shapes which, under axial loading, are not subject to torsional-flexural buckling, the interaction of bending and axial compression is basically the same for thin-walled, cold-formed members as for the generally thicker hot-rolled or welded shapes. This applies chiefly to doubly symmetrical I-shaped members and closed box shapes. For these, the equations governing combined loading, including definitions, are the same as in the AISI specification as in the AISC specification, which are given in Sec. 6, Art. 21.

The AISI specification contains special provisions for cases met with primarily in cold-formed construction. These include singly symmetric shapes which in axial compression may be subject to torsional-flexural buckling, singly symmetric shapes loaded other than in the plane of symmetry, and singly symmetric shapes which also may be subject to local buckling.[1,4]

13. Bracing of Channels and Z's *Beams.*   Channels and Z's are the simplest two-flange shapes which can be cold-formed from a single piece of sheet steel. Because neither shape is doubly symmetrical, when used as a beam loaded in the plane of the web it not only deflects in the loading plane but also twists. Therefore, it is essential that these shapes be braced not only to prevent lateral buckling (Art. 11) but also so that twisting can be limited to insignificant amounts.

When channels are used as beams they can be connected in pairs to form I-sections (Fig. 21a) or they can be used individually. In the latter case, special braces must be used to counteract twisting (Fig. 21b). Let Q be the load per channel on a portion of length s

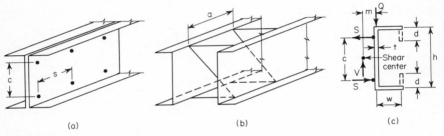

(a)   (b)   (c)

Fig. 21

(Fig. 21a) or a (Fig. 21b), and let this portion be centered on a cross section braced by welds (Fig. 21a) or braces (Fig. 21b). Equilibrium of that portion requires that the moment of forces be zero, so that (Fig. 21c)

$$S = \frac{Qm}{c} \qquad (22)$$

where S = required strength of the connection and m = distance to shear center. For simple channels

$$m = \frac{w^2}{2w + h/3}$$

while for C-shaped channels with stiffening lips (Fig. 21c)

$$m = \frac{wht}{4I_x} \left[wh + 2d \left(h - \frac{4}{3} \frac{d^2}{h} \right) \right]$$

It is seen from Eq. (22) that the strength S depends entirely on the local load Q over the portion of the beam extending the distance $a/2$ or $s/2$ on either side of the bracing. For beams designed for uniform load q per unit length of one channel, the actual load intensity may vary considerably over the span. To account for this, the AISI specification stipulates $Q = 3qs$ for two channels welded back to back at the relatively close spacing s, and $Q = 1.5qa$ for single channels braced at the larger spacing a. The further limitations $s_{max} = L/6$ and $a_{max} = L/4$ are prescribed for safety as well as to prevent significant deformations between braced sections.

Z-Sections. The centroid and the shear center coincide for this shape, so that a load in the plane of the web does not cause a primary twisting moment about the shear center.

However, unless bracing is provided, twisting will develop because the principal axes are inclined with respect to the plane of the loads. For this case

$$S = Q\frac{I_{xy}}{I_x} \tag{23}$$

where x and y are centroidal axes perpendicular and parallel to the web, respectively, and I_{xy} is the product of inertia.

Columns. If the two-channel member of Fig. 21a is used as a compression member it is essential that the column buckling strength of the single channel between spot welds be larger than the buckling strength of the entire member of length L. This is assured by making the slenderness ratio s/r_2 of the individual channel one-half the ratio L/r_1 of the member. This gives the maximum permissible spacing

$$s_{max} = \frac{Lr_2}{2r_1} \tag{24}$$

where r_1 is the governing radius of gyration of the I-section and r_2 that of the individual channel about its axis parallel to the web.

14. Connections Connections used in cold-formed steel construction include fusion welds, resistance welds, regular bolts, high-strength bolts with or without prestressing by controlled torquing, various types of metal screws, cold or blind rivets, and other special devices.

Fusion welds are used for on-site welding to connect cold-formed to cold-formed or cold-formed to structural members. Welding of thin steel, which is often galvanized, requires weld shapes and welding techniques which are often different from those in ordinary weldments. The puddle weld, which is the standard way of connecting floor or roof deck to structural framing, is made by burning through the deck and then filling the hole with weld metal. This gives the same type of connection as the plug weld in structural-steel members, where the hole is predrilled or prepunched. The AISI specification provides permissible shear stresses on the throat of fillet or plug welds which depend on the particular combinations of steel yield strength and class of electrode. The throat is taken to be half the thickness of the pertinent sheet. AWS D1.3-77 contains detailed provisions on allowable weld stresses and on welding procedures, control, inspection, etc., and should be consulted on all matters concerning fusion welding of cold-formed steel construction.

Resistance welding, in the form of spot welds, is mostly used for shop fabrication of light-gage members. Allowable shear forces per weld, based on a safety factor of about 2.5, are given in Table 5. Values for thicknesses not listed can be obtained by interpolation.

TABLE 5   Allowable Shear for Spot Welding

Sheet thickness, in.	Allowable shear per spot, lb
0.010	50
0.020	125
0.040	350
0.060	725
0.080	1,075
0.125	2,000
0.250	6,000

Bolted connections in cold-formed steel behave somewhat differently from those in heavier construction. The chief reason for this lies in the relative dimensions of sheet and bolt, the sheet thickness generally being a small fraction of the bolt (or head or nut) diameter. Extensive testing has shown that bolted, light-gage steel connections can fail in any of four modes:

1. When the edge distance in the direction of stress is small, by shearing of the sheet along two lines parallel to the direction of stress; this occurs at a shear stress in the sheet, on the two tearing planes, of about $0.55F_u$, where F_u = tensile strength of sheet.

2. When the edge distance in the direction of stress is large, by piling of the sheet material in front of the bolt in the direction of stress; this occurs at a bearing stress of about 3.0 to $3.6F_u$.

3. By tearing of the sheet in the reduced net section perpendicular to the direction of applied tension; such tearing was found to occur for relatively close bolt spacing when the stress in the net section is equal to the tensile strength F_u in the sheet. However, for large bolt spacing perpendicular to the direction of stress, because of the corresponding stress concentration, tearing was found to occur at a smaller stress in the net section, viz.,

$$f_{net} = (1.0 + 0.9r + 3rd/s)F_u \leq F_u$$

where d = bolt diameter, s = bolt spacing, and r = force transmitted by bolt or bolts at section, divided by tension force in member at that section.

4. By shearing of the bolt, at a shear stress on the root area of the thread of about 0.6 times the tensile strength of the bolt material.

The 1968 edition of the AISI specification gives allowable stresses and edge distances based on the above information and on a safety factor of about 2.2. However, this information was obtained from connection tests with washers under both head and nut of each bolt. Since then, connections with only one washer per bolt, or none at all, have come into use. Tests have shown that this can reduce the tension strength on the net section and the bearing strength by 10 to 20 percent.

In regard to the strength of screws, cold or blind rivets, and other special but frequently used connectors, information on strength characteristics should be obtained from the particular manufacturer because the great variety of such devices makes it impossible to develop information of more general validity.

15. Effects of Cold Forming on Steel Properties   Cold working has marked effects on the mechanical properties of ductile metals, caused by a combination of strain hardening, strain aging, and the Bauschinger effect. In general terms, it can be said that the yield strength can be raised considerably by cold working, the tensile strength is affected to a much smaller degree, and the ductility as measured by permanent elongation is reduced. When shapes are made from flat sheet or plate by cold forming, different amounts of cold work are performed in different parts of the section. The largest effects are produced in the corners and are the more pronounced the sharper the corner, i.e., the smaller the r/t ratio, r being the inside radius. When cold forming is done in press brakes, the metal in the flat portions is little affected, but when roll forming is employed, the yield strength of the flat portions can also be raised significantly. It follows that the stress-strain curve of an as-formed member differs from that of the flat material before forming. This stress-strain curve represents the weighted average of the individual curves for corner materials and for the flats. It can be obtained directly by performing tension or compression tests on short pieces of the entire member (so-called full-section tests), rather than on isolated coupons. The yield strengths determined in such tests are higher than those of the virgin material, in both tension and compression. The magnitude of the increase depends, among other things, on the fraction of the total material which is located in corners. In investigations at Cornell University, increases in yield strength ranging from 25 to 45 percent have been observed for roll-formed sections with relatively large percentages of corner material, and from 3 to 12 percent for press-braked sections with smaller percentages of corner material. It is certain that even larger increases will be found in more highly deformed sections than those which have been tested.

Allowable stresses for panels and decks are mostly based on the steel properties before forming, without taking advantage of cold-forming effects. This is so because the design of such members is mostly governed by deflection rather than strength requirements. On the other hand, for structural shapes (posts, studs, joists, purlins, etc.) increasing advantage is being taken of the raised strength properties obtained in the cold-forming process. The AISI specification permits the determination of the effective yield point of the as-formed member by two methods. One is to make full-section tension or compression tests on short specimens cut from the member. The stress-strain curves so obtained for the entire section are then used to establish the yield point by standard ASTM procedures. The other method takes advantage of the fact that, based on extensive research, it is possible to calculate without tests the yield point of corners F_{yc} from the geometry of the corner and the strength properties of the virgin material before forming. The weighted average yield point F_{ya} of the entire section is

$$F_{ya} = CF_{yc} + (1 - C)F_{yf} \qquad (25)$$

where C = ratio of total corner area to total area of full section of compression member, or of full flange area for flexural members

F_{yf} = average yield point of flat portions of section, established by coupon tests, or virgin yield point if no such tests are made

F_{yc} = yield point of corners = $B_c F_y/(R/t)^m$

$B_c = 3.69(F_u/F_y) - 0.819(F_u/F_y)^2 - 1.79$

$m = 0.192(F_u/F_y) - 0.068$

R = inside bend radius

t = sheet thickness

It is seen that the corner yield point depends on the sharpness of bend R/t and on the ratio of tensile to yield strength F_u/F_y of the virgin material. The formula does not hold for steels with relatively low ductility, i.e., values of F_u/F_y less than 1.2, or for R/t larger than 7.

The AISI specification contains detailed provisions on the test procedures which may be necessary when taking advantage of the higher steel strength obtained in the process of cold forming. The cost of such testing is generally small compared with the economy achieved through the utilization of the higher material strength.

16. Test Determination of Structural Performance   The cold-forming processes employed to shape light-gage steel members make it possible to produce an infinity of different shapes. It is not feasible to develop design procedures to the point where they would be applicable to any conceivable shape, extant or not yet thought of. Because of this, on occasion it may be necessary to employ tests rather than accepted methods of calculations to define structural performance. To avoid misinterpretation, it is not suggested either here or in the AISI Manual that tests should ever be substituted for calculations which can be made by accepted methods. The established practice is to make such tests only when methods of calculations suitable to the particular structure or component do not exist.

By appropriate measurements of deflections and strains it is possible to calculate flexural rigidity, moment of inertia, section modulus, and other effective properties from experimental data. Likewise, load-carrying capacity may be determined from tests. The AISI specification provides that (1) where practicable, tests shall be carried out in triplicate; (2) for determining allowable loads the member or assembly shall (*a*) be capable of sustaining without failure twice the live load plus one-and-one-half the dead load and (*b*) not show harmful distortions when under one-and-one-half the live load plus one dead load.

In evaluating such test determinations it is essential that the actual strength of the tested material be considered. For example, assume that a given new shape or assembly is intended for fabrication from an ASTM steel with a minimum specified yield strength of 33 ksi. Since, as a rule, actual strengths of sheet steels exceed specified minima sizably, the members or assemblies on which test determinations are being made may easily have been formed from steel with, say 42 ksi yield strength. Thus, their carrying capacity will be larger than if they had been made of steel with exactly the specified minimum value of 33 ksi. However, the excess strength cannot be counted upon since it resulted from an accidentally higher yield strength. To deduce, then, the carrying capacity which can be counted upon, it is necessary (1) to determine by test the properties (yield strength, etc.) of the particular steel from which the test members or assemblies have been made, and (2) to adjust the strength values obtained by testing the structure for the differences between the actual and the specified yield strength.

It should be noted that the strength of a member is not always directly proportional to the strength of the material from which it is made. For instance, the carrying capacity of columns of considerable slenderness is entirely independent of the strength of the material and depends only on E and L/r [see Eq. (12b)]. Therefore, the evaluation of tests to determine structural behavior requires considerable skill and theoretical knowledge.

REFERENCES

1. "Cold-Formed Steel Design Manual," American Iron and Steel Institute, Washington, D.C., 1977.
2. Winter, George: Commentary on Specification for the Design of Cold-Formed Steel Structural Members, 1968, part II of Reference 1.

3. "Stainless Steel Cold-Formed Structural Design Manual," American Iron and Steel Institute, Washington, D.C., 1974.
4. Yu, Wei Wen: "Cold-Formed Steel Structures," McGraw-Hill Book Company, New York, 1973.
5. Winter, George: Cold-Formed, Light Gage Steel Construction, *J. Struct. Div. ASCE*, November 1959.
6. Winter, George: Development of Cold-Formed Light Gage Steel Structures, American Iron and Steel Institute Regional Technical Meeting Paper, 1959.
7. Nilson, A. H.: Shear Diaphragms of Light Gage Steel, *J. Struct. Div. ASCE*, November 1960.
8. Nilson, A. H.: Folded Plate Structures of Light Gage Steel, *J. Struct. Div. ASCE,* October 1961.
9. Gergely, Peter, and George Winter: Experimental Investigation of Thin-Steel Hyperbolic Paraboloid Structures, *J. Struct. Div. ASCE*, October 1972; Banavalkar, P. V., and Peter Gergely: Analysis of Thin-Steel Hyperbolic Paraboloid Shells, *J. Struct. Div. ASCE*, November 1972.
10. Chajes, A., and George Winter: Torsional-Flexural Buckling of Thin-Walled Members, *J. Struct. Div. ASCE*, August 1965; Chajes, A., P. Fang, and George Winter: Elastic and Inelastic Torsional-Flexural Buckling, *Cornell Eng. Res. Bull.* 66-1, 1966.
11. Luttrell, L. D.: Strength and Behavior of Light Gage Steel Shear Diaphragms, *Cornell Eng. Res. Bull.* 67-1, 1967.
12. "Design of Light Gage Steel Diaphragms," American Iron and Steel Institute, Washington, D.C., 1967.
13. Errera, S. J., and T. V. S. R. Apparao: Design of I-Shaped Beams with Diaphragm Bracing, *J. Struct. Div. ASCE*, April 1976; Design of I-Shaped Columns with Diaphragm Bracing, *J. Struct. Div. ASCE*, September 1976.
14. Kalayanaraman, V., T. Pekoz, and G. Winter: Unstiffened Compression Elements, *ASCE Preprint* 2807, Cold-Formed Steel Structures, 1976.
15. Structural Stability Research Council, B. G. Johnston (ed.): "Guide to Stability Design Criteria for Metal Structures," 3d ed., John Wiley & Sons, Inc., New York, 1976.

Section **10**

Design of Aluminum Structural Members

JOHN W. CLARK

Manager, Engineering Properties and Design Division,
Alcoa Laboratories, Alcoa Center, Pa.

MATERIALS AND SPECIFICATIONS

1. Shapes Aluminum-alloy products are available in the form of shapes, plate, sheet, bar, tube, pipe, rod, wire, forgings, rivets, bolts, screws, nails, and various kinds of castings. Structural shapes are produced by either extruding or rolling. Sizes up to 10 × 10 × ¼-in. angles, 12-in. I-beams, and 15-in. channels are available as standard items, and the commoner sizes are produced in lengths up to 85 ft. Sheet and plate are available in thicknesses up to 6 in. and in widths up to 200 in. Lengths up to 45 ft are produced in some thicknesses of plate.

The extrusion process permits the fabrication of many useful shapes that are impractical to produce by rolling. Structural shapes with stiffening lips or bulbs to support outstanding flanges, stiffened sheet panels, hollow tubular shapes, sections with integral backup strips and bevels for welding, interlocking shapes, and many other types of sections are readily produced by extrusion. Maximum economy of metal and fabricating cost can often be achieved by designing members especially adapted to a given purpose.

2. Codes and Specifications American Society for Testing and Materials specifications that cover aluminum structural products are listed in Table 1. Most of these products are also covered by military specifications. Specifications[1] and engineering data[2] for aluminum structures have been published by the Aluminum Asociation. These specifications are based to a large extent on earlier specifications published by the American Society of Civil Engineers.

3. Characteristics of Aluminum Alloys Some aluminum alloys are heat-treatable; others are not. A widely used heat-treated wrought aluminum alloy is 6061-T6. The number 6061 identifies the alloy composition; the use of four digits in this number distinguishes it as a wrought rather than a cast alloy; the -T shows that the metal has been heat-treated; and the final 6 indicates the type of heat treatment.

An example of a non-heat-treatable alloy is 3003-H14. The designation 3003 identifies alloy and wrought condition as in the heat-treated alloys; the letter -H signifies that temper is produced by strain hardening rather than heat treatment; the final number 14 indicates the specific temper, which in turn relates to a definite set of mechanical properties for the alloy in question. The letter -O following an alloy number designates the annealed temper.

Aluminum alloys in general have excellent resistance to corrosion, and most of the alloys that are considered as general-purpose structural alloys are used in structures without paint or other protection. Some of the high-strength heat-treated alloys, however,

TABLE 1 ASTM Specification Numbers for Some Aluminum Alloys

Alloy	Sheet and plate	Rolled rod and bar	Structural shapes, rolled or extruded	Extruded rod, bar, shapes, tubes	Drawn tube General	Drawn tube Heat exchangers	Extruded tube and pipe
2014	B209	B211	....	B221			
Alclad 2014	B209						
3003	B209	B211	....	B221	B210	B234	B241
3004	B209	....	....	B221			
5083	B209	....	....	B221	B210	....	B241
5086	B209	....	....	B221	B210	....	B241
5454	B209	....	....	B221	....	B234	B241
5456	B209	....	....	B221	B210	....	B241
6061	B209	B211	B308	B221	B210	B234	B241
6063	....	....	....	B221	B210	....	B241

need to be protected against corrosion by painting, anodizing, or, in the case of sheet and plate, coating with a thin, integral layer of a more corrosion-resistant alloy. The latter product is designated as "alclad." Among the alloys listed in the tables, only 2014 is given paint protection in normal environments.

Both the weight and the modulus of elasticity of aluminum alloys are roughly one-third of the corresponding values for steel. Typical physical properties of a number of aluminum alloys are listed in Table 2. Minimum mechanical properties for the most widely used thicknesses of some typical alloys and products appear in Table 3. These tables include several of the alloys that are currently used most widely in structural applications. However, many other alloys are available with characteristics especially suited for different applications.

All the usual fabrication operations can be performed readily on aluminum. Ordinary

TABLE 2 Typical Physical Properties of Some Aluminum Alloys

Alloy and temper	Weight, lb/cu in.	Modulus of elasticity Avg tensile and compression[a]	Modulus of elasticity Shear[b]	Coefficient of thermal expansion,[c] per °F, 68 to 212°F	Electrical conductivity[d]	Thermal conductivity at 25°C, cgs units
2014-T6	0.101	10,700	4,000	0.0000128	40	0.37
3003[e]	0.099	10,000	3,750	0.0000129	46	0.42
3004[e]	0.098	10,000	3,750	0.0000133	42	0.39
5083[e]	0.096	10,300	3,850	0.0000132	29	0.28
5086[e]	0.096	10,300	3,850	0.0000132	32	0.30
5454[e]	0.097	10,300	3,850	0.0000131	34	0.32
5456[e]	0.096	10,300	3,850	0.0000133	29	0.28
6061-T6	0.098	10,000	3,750	0.0000131	43	0.40
6063-T5	0.098	10,000	3,750	0.0000130	55	0.50
6063-T6	0.098	10,000	3,750	0.0000130	53	0.48

[a]The compressive modulus is 100 ksi higher than the average, and the tensile modulus is 100 ksi lower.

[b]Poisson's ratio for aluminum alloys is one-third.

[c]Values of coefficient of thermal expansion for cold-worked tempers may vary slightly from lot to lot.

[d]Percent of the International Annealed Copper Standard.

[e]Values apply to all tempers.

TABLE 3 Minimum Mechanical Properties for Some Aluminum Alloys

Alloy and temper	Product	Thickness range, in.	Tension			Compression	Shear		Bearing	
			TS	YS	El	YS	US	YS	US	YS
2014-T6	Shapes	≤1.999†	60	53	7	52	35	31	114	85
Alclad 2014-T6	S and P	0.020–0.499†	63	55	*	56	38	32	120	88
3003-H14	S and P	0.009–1.000‡	20	17	*	14	12	10	40	25
3004-H34 or 24	S and P	0.009–1.000‡	32	25	*	22	19	14	64	40
5083-H111	Shapes	≤5.000	40	24	12	21	23	14	78	38
5083-H321	S and P	0.188–1.500§	44	31	12	26	26	18	84	53
5086-H111	Shapes	≤5.000	36	21	12	18	21	12	70	34
5086-H32, H116, H117	S and P	0.020–2.000	40	28	*	26	24	16	78	48
5454-H111	Shapes	≤5.000	33	19	12	16	19	11	64	30
5454-H34	S and P	0.020–1.000‡	39	29	*	27	23	17	74	49
5456-H111	Shapes	≤5.000	42	26	12	22	24	15	82	42
5456-H321, H116, H117	S and P	0.188–1.251§	46	33	12	27	27	19	87	56
6061-T6	Shapes	≤1.000§	38	35	10	35	27	20	80	56
6061-T6	S and P	0.006–4.000	42	35	*	35	27	20	88	56
6063-T5	Shapes	≤0.500§	22	16	8	16	13	9	46	26
6063-T6	Shapes	≤1.000	30	25	8	25	19	14	63	40

*Depends on thickness.
†Strengths are higher for some thicknesses within this range.
‡Thicker plate is available in softer tempers.
§Thicker plate is available in this temper with slightly reduced strength.
   TS = tensile strength, ksi
   YS = yield strength, ksi
   El = elongation, percent in 2 in.
   US = ultimate strength in shear or bearing, ksi
S and P = sheet and plate
Minimum tensile properties are specified values. All other quantities in the table are corresponding minimum values but are not specified.

presses, brakes, and rolls are suitable for forming, but the surfaces of the tools that come in contact with the aluminum should be smooth and free from toolmarks, dents, or rough edges which would tend to tear or score the metal. Similarly, shearing, sawing, and machining operations are performed using the same methods and equipment that are used for other metals. Tools should have keen cutting edges and, preferably, more side and top rake than is usual for steel. High blade speeds are desirable in sawing.

DESIGN

The basic principles of design for aluminum are the same as for other ductile metals. Specific design procedures may differ for different materials, however, and this section is concerned principally with design information that applies particularly to aluminum. Where it is customary in designing with other materials to allow increased allowable stresses for specific loading conditions, such as wind loads, similar increases can be allowed for aluminum.

Tension Members

4. Yielding and Fracture For a member loaded in axial tension, two strength values are important to the designer: the load at which the member begins to yield or take appreciable permanent set, and the load at which the ultimate tensile strength is reached. Yielding usually begins to become significant when the average stress on the minimum net section reaches the tensile yield strength. Ultimate failure may be expected when the average stress on the minimum net section is equal to the tensile strength. The net section of an aluminum tension member is computed in the same way as for a steel member.

The stress-strain curve for aluminum alloys does not have a flat spot or sharp break at the yield strength. Yield strength is defined as the stress at which the material undergoes a permanent set of 0.2 percent.

Factor of Safety. It is customary to permit a smaller factor of safety against yielding than against fracture or collapse of a structure. Factors of safety that are recommended for two major classifications of aluminum-alloy structures are as follows:[1]

For bridge structures and other structures that would be designed in steel in accordance with the AASHTO, AREA, or similar specifications, design stresses should be selected to provide factors of safety of at least 1.85 on yielding and 2.2 on ultimate strength.

For building structures and other structures that woud be designed in steel according to the AISC or similar specifications, design stresses should be selected to provide factors of safety of at least 1.65 on yielding and 1.95 on ultimate strength.

Based on the foregoing factors of safety, the specifications published by the Aluminum Association[1] prescribe allowable tensile stresses for alloy 6061-T6 of 17 ksi for bridge structures and 19 ksi for building structures.

5. Welded Tension Members The heat of welding softens aluminum alloys in the strain-hardened or heat-treated tempers, and the resulting reduction in strength must be considered in the design of welded members (Art. 25). If only part of the cross section is affected, as is generally the case for members welded longitudinally, the material adjacent to the weld is reinforced by the higher-strength, unaffected material at a greater distance from the weld. The tensile strength (per unit of area) of such a part can be calculated as the weighted average of the tensile strength of the material in a zone adjacent to the weld and the tensile strength of the material outside this zone.[5] For this purpose, the material should be considered to have the strength shown in Table 13 within a zone extending a distance of 1 in. on either side of the weld, and the material outside this region should be considered to have parent-metal properties. This heat-affected zone is measured from the centerline of the weld for butt welds and from the root of the fillet for fillet welds.

The foregoing procedure can also be used to calculate the yield strength of welded parts. The yield strength in the heat-affected zone is considered to be the value measured on a 10-in. gage length across a butt weld. Outside this zone, the material is considered to have the yield strength of unaffected parent metal. The yield strength for the entire cross section is the weighted average of the yield strength for the material in the heat-affected zone and the parent-metal yield strength outside this zone. (In the case of welded parts with longitudinal welds, the actual yield strength in the heat-affected zone may be

0.210	552	0.0158	0.0105	736	0.0210	0.0140	920	0.0263	0.0175	0.292	0.248	0.876
0.220	574	0.0165	0.0110	766	0.0220	0.0147	957	0.0275	0.0183	0.305	0.260	0.870
0.230	596	0.0173	0.0115	795	0.0230	0.0153	994	0.0288	0.0192	0.319	0.271	0.864
0.240	618	0.0180	0.0120	824	0.0240	0.0160	1030	0.0300	0.0200	0.333	0.283	0.858
0.250	639	0.0188	0.0125	853	0.0250	0.0167	1066	0.0313	0.0208	0.347	0.295	0.853
0.260	660	0.0195	0.0130	880	0.0260	0.0173	1101	0.0325	0.0217	0.361	0.307	0.847
0.270	681	0.0203	0.0135	908	0.0270	0.0180	1135	0.0338	0.0225	0.375	0.319	0.841
0.280	701	0.0210	0.0140	935	0.0280	0.0187	1169	0.0350	0.0233	0.389	0.330	0.835
0.290	721	0.0218	0.0145	962	0.0290	0.0193	1202	0.0363	0.0242	0.403	0.342	0.829
0.300	741	0.0225	0.0150	988	0.0300	0.0200	1234	0.0375	0.0250	0.416	0.354	0.823
0.310	760	0.0233	0.0155	1013	0.0310	0.0207	1267	0.0388		0.430	0.366	0.817
0.320	779	0.0240	0.0160	1038	0.0320	0.0213	1298	0.0400		0.444	0.378	0.811
0.330	797	0.0248	$>0.75\rho_b$	1063	0.0330	$>0.75\rho_b$	1329	0.0413	$>0.75\rho_b$	0.458	0.389	0.805
0.340	815	0.0255		1087	0.0340		1359	0.0425		0.472	0.401	0.799
0.350	833	0.0263		1111	0.0350		1389	$>0.75\rho_b$		0.486	0.413	0.794
0.360	851	0.0270		1134	0.0360		1418			0.500	0.425	0.788
0.370	868	0.0278		1157	0.0370		1446			0.514	0.437	0.782

*Adapted from Ref. 10.

†For $f'_c = 5000$ psi, the given c/d must be multiplied by 0.85/0.80.

‡Above these lines $\rho < 200/f_y$

TABLE 6   Areas of Groups of Bars, in.^{2}

Bar No.	Number of bars											
	1	2	3	4	5	6	7	8	9	10	11	12
4	0.20	0.40	0.60	0.80	1.00	1.20	1.40	1.60	1.80	2.00	2.20	2.40
5	0.31	0.62	0.93	1.24	1.55	1.86	2.17	2.48	2.79	3.10	3.41	3.72
6	0.44	0.88	1.32	1.76	2.20	2.64	3.08	3.52	3.96	4.40	4.84	5.28
7	0.60	1.20	1.80	2.40	3.00	3.60	4.20	4.80	5.40	6.00	6.60	7.20
8	0.79	1.58	2.37	3.16	3.95	4.74	5.53	6.32	7.11	7.90	8.69	9.48
9	1.00	2.00	3.00	4.00	5.00	6.00	7.00	8.00	9.00	10.0	11.0	12.0
10	1.27	2.54	3.81	5.08	6.35	7.62	8.89	10.2	11.4	12.7	14.0	15.2
11	1.56	3.12	4.68	6.24	7.80	9.36	10.9	12.5	14.0	15.6	17.2	18.7
14	2.25	4.50	6.75	9.00	11.2	13.5	15.7	18.0	20.2	22.5	24.7	27.0
18	4.00	8.00	12.0	16.0	20.0	24.0	28.0	32.0	36.0	40.0	44.0	48.0

TABLE 7   Areas of Bars in Slabs, in.^{2}/ft

Bar No.	Spacing, in.											
	3	3½	4	4½	5	5½	7	8	9	10	11	12
4	0.80	0.69	0.60	0.53	0.48	0.44	0.40	0.34	0.30	0.27	0.24	0.22	0.20
5	1.24	1.06	0.93	0.83	0.74	0.68	0.62	0.53	0.46	0.41	0.37	0.34	0.31
6	1.76	1.51	1.32	1.17	1.06	0.96	0.88	0.75	0.66	0.59	0.53	0.48	0.44
7	2.40	2.06	1.80	1.60	1.44	1.31	1.20	1.03	0.90	0.80	0.72	0.65	0.60
8	3.16	2.71	2.37	2.11	1.90	1.72	1.58	1.35	1.18	1.05	0.95	0.86	0.79
9	4.00	3.43	3.00	2.67	2.40	2.18	2.00	1.71	1.50	1.33	1.20	1.09	1.00
10	5.08	4.35	3.81	3.39	3.05	2.77	2.54	2.18	1.90	1.69	1.52	1.38	1.27
11	6.24	5.35	4.68	4.16	3.74	3.40	3.12	2.67	2.34	2.08	1.87	1.70	1.56

TABLE 8   Minimum Beam Widths, ACI Code

Deduct ¾ in. if stirrups not required

Maximum aggregate size ¾ of clear space between bars

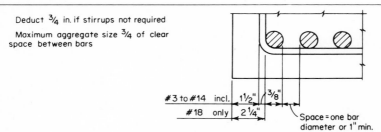

#3 to #14 incl.   1½"   ⅜"

#18 only   2¼"

Space = one bar diameter or 1" min.

Size of bars	No. of bars in single layer of reinforcement							Add for each added bar
	2	3	4	5	6	7	8	
No. 4	5¾	7¼	8¾	10¼	11¾	13¼	14¾	1½
No. 5	6	7¾	9¼	11	12½	14¼	15¾	1⅝
No. 6	6¼	8	9¾	11½	13¼	15	16¾	1¾
No. 7	6½	8½	10¼	12¼	14	16	17¾	1⅞
No. 8	6¾	8¾	10¾	12¾	14¾	16¾	18¾	2
No. 9	7¼	9½	11¾	14	16¼	18½	20¾	2¼
No. 10	7¾	10¼	12¾	15¼	17¾	20¼	23	2⅝
No. 11	8	11	13¾	16½	19½	22¼	25	2⅞
No. 14	9	12¼	15¾	19	22½	25¾	29¼	3⅜
No. 18	10½	15	19½	24	28½	33		

support to a much deeper one at midspan, a check on the compressive capacity may also be necessary.

Example 2 Design for flexure a rectangular, continuous, reinforced-concrete beam of three equal 20-ft spans, with ends simply supported, carrying 2 klf live and 1 klf dead load (Fig. 4a), f'_c = 4000 psi, f_y = 60,000 psi. The load factors are 1.4 for dead load and 1.7 for live load (Art. 5).

SOLUTION. The three-moment equation gives $M_1 l_1 + 2M_2 (l_1 + l_2) + M_3 l_2 = -w_1 l_1^3/4 \; -w_2 l_2^3/4$.

CASE I. Maximum positive moment at center of interior span. Since $M_1 = 0$,

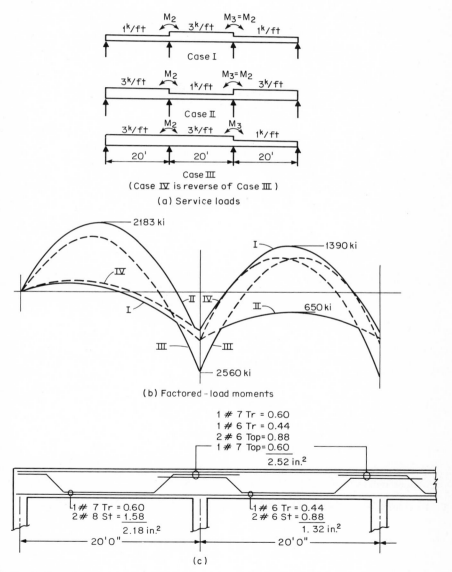

Fig. 4 Example 2.

$$2M_2(20 + 20) + 20M_2 = -\frac{1.4 \times 1 \times 20^3}{4} - \frac{(1.4 + 1.7 \times 2)20^3}{4}$$

$$M_2 = -\frac{(1.4 + 4.8) \times 20^3 \times 12}{4(2 \times 40 + 20)} = -1490 \text{ kip-in.}$$

$$M = 4.8 \times 20^2 \times 12/8 - 1490 = +1390 \text{ kip-in. at midspan}$$

CASE II. Minimum moment at center of interior span.

$$M_2 = -\frac{(4.8 + 1.4) \times 20^3 \times 12}{4(2 \times 40 + 20)} = -1490 \text{ kip-in.}$$

$$M = 1.4 \times 20^2 \times 12/8 - 1490 = -650 \text{ kip-in. at midspan}$$

CASE III. Maximum negative moment at interior support.

$$2M_2 \times 40 + M_3 \times 20 = -(4.8 + 4.8) \times 20^3 \times 12/4$$
$$M_2 \times 20 + 2M_3 \times 40 = -(4.8 + 1.4) \times 20^3 \times 12/4$$

from which $M_2 = -2560$ kip-in., $M_3 = -1216$ kip-in.

The moments of resistance must encompass the envelopes of all the moment curves (Fig. 4b). The maximum negative moment is 2,560,000 in.-lb, and from Table 5, with a reinforcement ratio of 0.010, $M_u/\phi bd^2$ is 547. Therefore, $bd^2 = 2,560,000/(0.9 \times 547) = 5200$, which can be met by the following values for b and d:

b	8	10	12	14	16
d	25.5	22.8	20.8	19.3	18.0

Using the 12×20.8 beam size and adding ⅜ in. for the radius of a longitudinal bar, ⅜ in. for a stirrup leg, and 1½ in. of cover gives 23.05 in. total depth. Then, using a 23-in. depth with $d = 20.7$ in.,

$$k_u = M_u/\phi bd^2 = 2,560,000/(0.9 \times 12 \times 20.7^2) = 553$$

for which $\rho = 0.0101$ (Table 5). Then $A_s = 0.0101 \times 12 \times 20.7 = 2.51$ in.²

1 No. 6 bent up in this beam	= 0.44
1 No. 7 bent up from end span	= 0.60
2 No. 6 top	= 0.88
1 No. 7 top	= 0.60
	2.52 in.²

At the center of the interior span, $M_u/\phi bd^2 = 650,000/(0.9 \times 12 \times 20.7^2) = 140$, for which Table 5 shows ρ to be less than the minimum $200/f_y = 200/60,000 = 0.00333$. Therefore, $A_s = 0.00333 \times 12 \times 20.7 = 0.83$ in.² Use two No. 6 = 0.88 in.², which gives $\rho = 0.00354$. The corresponding value of k_u is 207, for which $M_u = 958,000$ kip-in. Therefore, the two No. 6 bars are adequate for the negative moment over the central portion of the interior span of length l_0 given by $(1490 - 650)(l_0/20)^2 = 958 - 650$, which gives $l_0 = 12.1$ ft. The No. 7 top bar at each interior support must extend $(20 - 12.1)/2 = 4$ ft into the interior span, plus the necessary development length.

For maximum positive moment in the interior span, $M_u/\phi bd^2 = 1,390,000/(0.9 \times 12 \times 20.7^2) = 300$; so $\rho = 0.0052$ and $A_s = 0.0052 \times 12 \times 20.7 = 1.29$ in.² Use one No. 6 truss bar and two No. 6 straight bars = 1.32 in.²

For maximum positive moment in the end span, $M_u/\phi bd^2 = 2,183,000/(0.9 \times 12 \times 20.7^2) = 472$; so $\rho = 0.0085$ and $A_s = 0.0085 \times 12 \times 20.7 = 2.11$ in.² Use one No. 7 truss bar and two No. 6 straight bars = 2.18 in.²

9. Doubly Reinforced Beams Since strength design assumes a high intensity of compression distributed over a relatively small area, compression reinforcement is seldom necessary for resisting stress. Compression steel is one of the most effective ways to reduce deflection and is much more likely to be supplied for that purpose. It is then the designer's option to include such compression steel in computations for load-carrying capacity.

The ultimate resisting moment is given by

$$M_u = \phi \left[(A_s - A_s')f_y \left(d - \frac{a}{2} \right) + A_s' f_y (d - d') \right] \tag{5}$$

where $a = (A_s - A_s') f_y/0.85 f_c' b$ and the use of f_y with A_s' ignores the concrete displaced by A_s'.

The equation is based on the assumption that both tension steel and compression steel are at yield stress when the beam fails. To guarantee this condition requires that

$$\rho - \rho' \gtrless \frac{0.85\beta_1 f'_c}{f_y} \frac{d'}{d} \frac{87,000}{87,000 - f_y} \tag{6}$$

In order to assure failure by yielding of the tension reinforcement, rather than by crushing of the concrete, the steel ratio must also satisfy the requirement $\rho - \rho' \lesssim 0.75\rho_b$, where ρ_b is given by Eq. (4).

Example 3 A beam required to develop an ultimate moment of 7000 in.-kips is limited to 12×24 in., $f'_c = 4000$, $f_y = 60,000$ psi. Determine the reinforcement. Assume two rows of tensile reinforcement with $d = 20$ in.

SOLUTION. Using $\rho = \rho_{max}$, the capacity of a singly reinforced section is determined from Eq. (4) with $\beta_1 = 0.85$:

$$\rho_b = 0.85 \times 0.85 \frac{4}{60} \times \frac{87}{87 + 60} = 0.0285$$
$$\rho_{max} = 0.75 \times 0.0285 = 0.02$$
$$A_s = 0.02 \times 12 \times 20 = 4.80 \text{ in.}^2$$
$$a = \frac{4.80 \times 60}{0.85 \times 4 \times 12} = 7.05 \text{ in.} \qquad d - \frac{a}{2} = 16.48$$
$$M_u = 0.90 \times 4.80 \times 60 \times 16.48 = 4270 < 7000 \text{ kip-in.}$$

Assuming one layer of compressive reinforcement, the moment arm of the steel couple will be about $24 - 4.0 - 2.5 = 17.5$ in. Therefore, to gain the 2730 kip-in. required, add

$$A_s = A'_s = \frac{2730}{0.9 \times 60 \times 17.5} = 2.89 \text{ in.}^2$$

Try 3 No. 9 = 3.00 in.² for A'_s. Then $A_s = 4.80 + 2.89 = 7.69$ in.²
Try 3 No. 11 + 3 No. 9 = 7.68 in.² For the arrangement shown in Fig. 5, $d = 20.53$ in. and $d' = 2.5$ in.

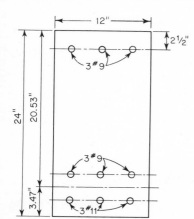

Fig. 5 Example 3.

Then

$$\rho - \rho' = \frac{7.68 - 3.00}{12 \times 20.53} = 0.0190$$

which satisfies the requirements

$$\rho - \rho' \lesssim 0.75\rho_b \lesssim 0.75 \times 0.0285 = 0.021$$
$$\rho - \rho' \gtrless 0.85 \times 0.85 \times \frac{4}{60} \times \frac{2.5}{20.53} \times \frac{87}{87 - 60} = 0.0189$$

10. Tee Beam Figure 6a shows an isolated tee beam such as might be used for a crane girder or similar freestanding beam. At b, the crosshatched area shows the theoretical outline of a tee beam made up of a portion of a monolithic beam-and-slab floor system, while c is taken from a concrete-joist floor system. Since the compression is not uniformly distributed across a tee of extreme width, the maximum symmetrical flange for assumed uniformly distributed stress is the smallest of (1) one-quarter of the beam span, (2) a projection of eight slab thicknesses on each side of the stem, and (3) one-half the distance to the next beam on either side.

In one-sided beams, the limits are a projection beyond the stem of $L/12$ or of $6t$, or one-half the distance to the next beam.

Tee beams are of three varieties. If the neutral axis is in the flange or at its bottom, the

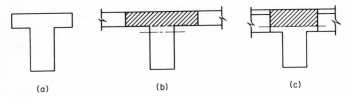

(a)   (b)   (c)

Fig. 6   Tee beam.

beam can be designed as a rectangular cross section of width equal to the width of the flange. If the neutral axis is in the stem and the flange is not too thin (say, $t \geqslant 0.15d$), the resistance of the small piece of stem between bottom of flange and neutral axis may be neglected. If the neutral axis is in the stem and the flange is thin, compression in the stem should be considered.

Rectangular beams can be reinforced with 2 percent or more of tension steel and still be below 75 percent of balanced reinforcement. As a result, tee beams are not often necessary with ultimate-strength design. However, formulas are available.

If the flange thickness exceeds $1.18qd/\beta_1$, the tee beam is, in effect, rectangular. If it is less,

$$M_u = \phi \left[(A_s - A_{sf})f_y \left(d - \frac{a}{2} \right) + A_{sf}f_y(d - 0.5t) \right] \tag{7}$$

where $A_{sf} = 0.85(b - b') \, tf'_c/f_y$ = steel area needed to develop compressive strength of overhanging flanges

$a = (A_s - A_{sf})f_y/0.85f'_c b$

The quantity $\rho_w - \rho_f$ must not exceed $0.75\rho_b$, where $\rho_w = A_s/b'd$ and $\rho_f = A_{sf}/b'd$.

Example 4   Redesign the beam of Example 3 as a tee with a flange 4 in. thick.

SOLUTION.   The flange must replace strength of $A'_s = 2.89$ in.^{2} as found in Example 3. A flange 4 in. thick would have equal compressive strength if

$$b - b' = \frac{2.89 \times 60}{0.85 \times 4 \times 4} = 12.8 \text{ in.}$$
$$a = \frac{60(7.75 - 2.89)}{0.85 \times 4 \times 12} = 7.15 \text{ in.}$$

This is slightly on the safe side, since the centroid of the flange lies 0.5 in. higher than A'_s. Check M_u provided by the web and flange (Fig. 7):

$$M_u = 0.90 \times 0.85 \times 4[12 \times 7.15(20.53 - 3.58) + 2 \times 7 \times 4 \times 18.53] = 7625 \text{ kip-in.} > 7000$$

Example 5   With a 3-in. slab and beams spanning 20 ft at 6 ft on centers, design an interior tee beam for a live load of 3 klf and a dead load of 1 klf. $f'_c = 4000$, $f_y = 60,000$ psi, maximum depth of beam 20 in. (Fig. 8a).

Assume a maximum positive bending moment of $wL^2/16$ and a maximum negative moment of $wL^2/11$ as suggested in ACI 318-77. Sketch the moment envelopes (Fig. 8b).

POSITIVE MOMENT

$$M_{LL} = 3.0 \times 1.7 \times 20^2 \times 12/16 = 1530$$
$$M_{DL} = 1.0 \times 1.4 \times 20^2 \times 12/16 = \underline{420}$$
$$M_{LL+DL} = 1950 \text{ kip-in.}$$

NEGATIVE MOMENT

$$M_{LL} = 3.0 \times 1.7 \times 20^2 \times 12/11 = -2225$$
$$M_{DL} = 1.0 \times 1.4 \times 20^2 \times 12/11 = \underline{-610}$$
$$= -2835 \text{ kip-in.}$$

The flange width is governed by ¼ the beam span = 60 in. (Fig. 8). With a 20-in. maximum depth, d = 17.6 in.,

$$k_u = M_u/\phi bd^2 = \frac{1,950,000}{0.90 \times 60 \times 17.6^2} = 117$$

For negative moment, with a stem width of 14 in. and a depth of 17.6 in.,

$$k_u = M_u/\phi bd^2 = \frac{2,835,000}{0.90 \times 14 \times 17.6^2} = 727$$

To obtain reinforcement, refer to Table 5. For positive moment, $k_u = 117$, for which ρ is less than the minimum value 200/60,000 = 0.0033. Therefore, $A_s = 0.0033 \times 17.6 \times 60 = 3.52$ sq in.

$$\begin{aligned} 2 \text{ No. 7 truss bars} &= 1.20 \text{ sq in.} \\ 2 \text{ No. 10 straight bars} &= \underline{2.54} \text{ sq in.} \\ +A_s &= 3.74 \text{ sq in. (Fig. 9)} \end{aligned}$$

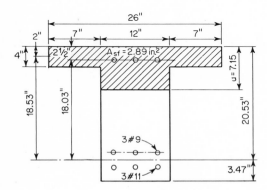

Fig. 7   Example 4.

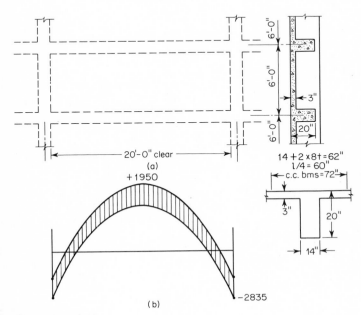

Fig. 8   Example 5.

For negative moment, $k_u = 727$. From Table 5, $\rho = 0.0138$ and $A_s = 0.0138 \times 17.6 \times 14 = 3.40$ sq in.

$$
\begin{array}{ll}
\text{2 No. 7 this span} & = 1.20 \text{ sq in.} \\
\text{2 No. 7 continuing span} & = 1.20 \text{ sq in.} \\
\text{1 No. 10 top} & = \underline{1.27} \text{ sq in.} \\
& A_s = 3.67 \text{ sq in. (Fig. 9)}
\end{array}
$$

11. Special Beam Shapes Strength design is adaptable to beams of any shape. In this case, as for rectangular beams, (1) the neutral axis must result in a total tension equal to total compression, (2) the tension steel (at least the bars farthest from the neutral axis) must

Fig. 9 Reinforcing for beam of Example 5.

be at yield stress (usually above minimum yield strain), (3) the compression strain farthest from the neutral axis must be taken as 0.003, and (4) the compression area can be divided into rectangles or triangles.

12. Shear and Diagonal Tension According to ACI, the mean intensity of shear as a measure of diagonal tension is $v_u = V_u / \phi b_w d$, computed at a distance d from the support, where b_w is the width of the web of a T-beam or rectangular beam. In the absence of torsion, the capacity v_c of an unreinforced web to resist diagonal tension is $2\sqrt{f_c'}$ or, if a more detailed analysis is made, $1.9\sqrt{f_c'} + 2500\, \rho_w V d / M$ but not to exceed $3.5\sqrt{f_c'}$, where $\rho_w = A_s / b_w d$.

With web reinforcement (and torsion not important) v_u must not exceed $10\sqrt{f_c'}$. Stirrup spacing must not exceed $d/2$ for $v_u \lesssim 6\sqrt{f_c'}$ or $d/4$ for $6\sqrt{f_c'} < v_u \lesssim 10\sqrt{f_c'}$. Wherever stirrups are required, minimum $A_v = 50 b_w s / f_y$, where A_v is the total area of web reinforcement within the distance s.

The required area A_v of vertical stirrups is

$$
A_v = \frac{V_s s}{\phi f_v d} = \frac{v_s b s}{f_y} \tag{8}
$$

where $v_s = v_u - v_c$.

For a single bent bar or a single group of parallel bars at the angle α with the longitudinal axis of the member, all bent at the same distance from the support,

$$
A_v = \frac{V_s}{\phi f_v \sin \alpha} \tag{9}
$$

For a series of parallel bars or groups of bars bent at different distances from the support,

$$
A_v = \frac{V_s s}{\phi f_v d\,(\sin \alpha + \cos \alpha)} \tag{10}
$$

In these formulas, $V_s =$ shear carried by web reinforcement.

Spacing of stirrups in a triangular shear diagram of base a (Fig. 10) is

$$
s = a\,\frac{\sqrt{n} - \sqrt{n - 0.5}}{\sqrt{n}} \qquad a\,\frac{\sqrt{n} - \sqrt{n - 1.5}}{\sqrt{n}} \quad \cdots \quad a\,\frac{\sqrt{n} - \sqrt{0.67}}{\sqrt{n}} \tag{11}
$$

where $n =$ number of stirrups in distance a. This locates the stirrups at midlength of the trapezoidal segments of the excess shear prism, except that the last one is at the centroid of the triangular segment. Maximum spacing often controls near the small end of the triangle.

Web reinforcement must be provided to account for three portions of the shear diagram:

1. To continue the size and spacing of such reinforcement as is required at distance d from the face of the support back to the face of the support (the crosshatched area in Fig. 10)

2. To locate a stirrup at or near the centroid of each equal volume into which the excess shear prism (the stippled area in Fig. 10) is divided

3. A length with minimum stirrups beyond area 3 in Fig. 10, within which $v_u \gtrless v_c/2$.

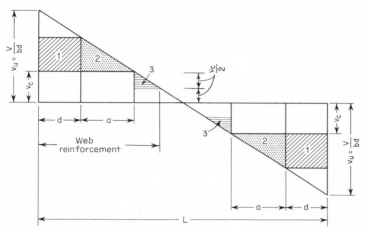

Fig. 10   Location of web reinforcement.

Example 6   For the continuous beam of Example 2, b = 12 in., d = 20.7 in., design interior-span stirrups for f_y = 40,000 psi, neglecting any truss bars.

SOLUTION.   Plot the shear envelope curves of Fig. 11a from the data of Example 2 and Fig. 4b:

CASE I $V_u = wL/2 = (1.4 \times 1 + 1.7 \times 2)10 = 48$ kips

CASE II $V_u = 1 \times 1.4 \times 10 = 14$ kips

CASE III V_{uL} = Case I + $(M_L - M_R)/L = 48 + (-1216 + 2560)/240 = 53.6$ kips

$V_{uR} = 48 - 5.6 = 42.4$ kips

CASE IV V_{uL} and V_{uR} are reversed from Case III.

With a maximum V_L of 53.6 kips, b = 12, and d = 20.7 in.,

$$v_u = V_u/\phi bd = 53,600/(0.85 \times 12 \times 20.7) = 254 \text{ psi}$$

At distance d = 20.7 in. = 1.72 ft from the face of the support,

$$v_u = (53.6 - 1.72 \times 4.8)1000/(0.85 \times 12 \times 20.7) = 215 \text{ psi} > 2\sqrt{f_c} = 126 \text{ psi}$$

Thus web reinforcement is required, and must be extended at least to the point where $v_u = 126/2 = 63$ psi.

The unit shear diagram, shown in Fig. 11b, has a slope of 1.89 psi/in. The shaded areas represent (1) two lengths (AB and BC) where stirrups are required because $v_u > v_c$ and (2) the length CD where minimum stirrups are required because $v_u > v_c/2$.

Assume No. 3 U stirrups (smallest bar size) for length AB and calculate spacing:

$$s = \frac{A_v f_y}{(v_u - v_c)b} = \frac{0.22 \times 40,000}{89 \times 12} = 8.23 \text{ in., say 8 in.}$$

as dimensioned in Fig. 11b, with the first stirrup at 8/2 = 4 in. from support.*

For length BC, the spacing can be increased as the shear decreases; compute the spacing at the fourth stirrup:

$$v_s = 89 - (28 - 20.7)1.89 = 75 \text{ psi} \qquad s = \frac{0.22 \times 40,000}{75 \times 12} = 9.8 \text{ in.}$$

Alternate calculation: $s = (89/75)8.23 = 9.8$ in.

*See footnote to Example 7, p. 11–19.

Maximum spacing = $d/2$ = 20.7/2 = 10.3 in., say 10 in. The maximum 10-in. spacing governs both here and to within $s/2$ = 5 in. of point D. Next locate D: L_{AD} = 191/1.89 = 101 in.

Use 1 No. 3 U at 4 in., 4 at 8 in., 6 at 10 in., each end.

Example 7   Given the service loads in Fig. 12a, b = 12 in., d = 22.6 in., design stirrups for f'_c = 3750 psi, f_y = 40,000 psi.

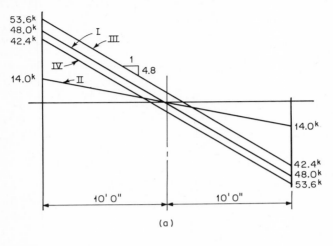

(a)

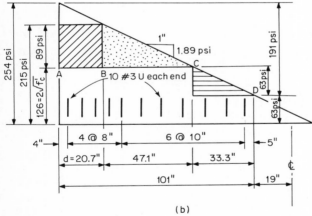

(b)

Fig. 11   Example 6.

SOLUTION.   The factored-load shear diagram is shown in Fig. 12b. Since the contribution of truss bars is to be disregarded, it will be sufficiently accurate to use the shear diagram of Fig. 12b for either a continuous or simple-span beam. The controlling ultimate unit shears are:

$$v_{\text{end}} = \frac{62,000}{0.85 \times 12 \times 22.6} = 270 \text{ psi} \qquad v_c = 2\sqrt{f'_c} = 2\sqrt{3750} = 122 \text{ psi}$$

$$v_d = v_{22.6} = \frac{62,000 - 1.88 \times 2800}{0.85 \times 12 \times 22.6} = 246 \text{ psi}$$

$$v_s = v_u - v_c = 246 - 122 = 124 \text{ psi}$$

$$v_c = \frac{34,000}{0.85 \times 12 \times 22.6} = 147 \text{ psi}$$

$$v_s = 147 - 122 = 25 \text{ psi}$$

For distance $d = 22.6$ in. from the support (shown crosshatched), try No. 3 U for which $A_v = 0.22$ sq in.

$$s = A_v f_y / v_s b = 0.22 \times 40{,}000/(124 \times 12) = 5.9 \text{ in.}$$

say 5.5 in. for 4 spaces which then reaches to 2.4 in. beyond point B.*
At $2.4 + s =$ say 8.4 in. beyond B, $v_s = 124 - 8.4 \times 1.025 = 115$ psi (where 1.025 is the slope of the shear diagram).

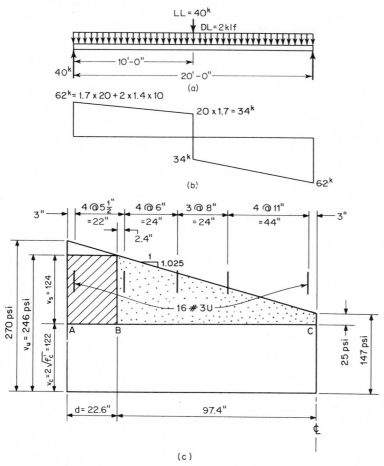

Fig. 12 Example 7.

$$s = 0.22 \times 40{,}000/(115 \times 12) = 6.38 \text{ in.}$$

say 6 in. for 4 spaces to 26.4 in. past B. At $26.4 + 8$ (estimated) $= 34.4$ in. beyond B, $v_s = 124 - 34.4 \times 1.025 = 89$ psi.
Since s varies as $1/v_s$, $s = 6.38 \times 115/89 = 8.2$ in., say 8 in. for 3 spaces, to $26.4 + 24 = 50.4$ in. past B. At $50.4 + 11 = 61.4$ in. beyond B, $v_s = 124 - 61.4 \times 1.025 = 61$ psi, and $s = 6.38 \times 115/61 = 12.0$ in.,

*Each stirrup theoretically is placed at the centroid of the v_s area it serves. Hence the stirrups just used provide for $s/2$ beyond the last stirrup (to $2.4 + 5.5/2 \approx 5+$ in. beyond B) and the estimated s used next should theoretically be 5.5/2 plus half the next s to be calculated. This is also why on Fig. 12c the first stirrup was placed $s/2 \approx 3$ in. from the support.

which is greater than the maximum $d/2 = 11.3$ in.; say 4 spaces at 11 in. (Note that the 3 in. left to midspan might be made anything from zero to 11/2 in. in similar cases.)

Since stirrups are required over the entire span, requirement No. 3 of Art. 12 does not apply; hence use 16 No. 3 stirrups each end, as shown on Fig. 12c.

Example 8 Determine maximum live shears with a movable live load w_L on the beam of Fig. 13, which shows the load position, the corresponding shear, and the fact that live-load shear at midspan is one-quarter the live shear at the end. It also shows that the assumption of a linear variation from end to center of span is on the safe side.

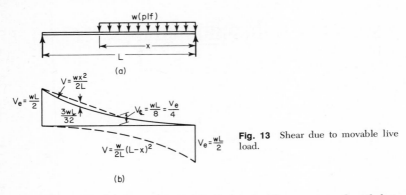

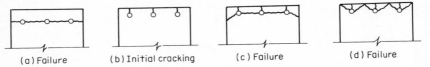

Fig. 13 Shear due to movable live load.

13. Development and Anchorage of Reinforcement

The necessary length between the point of maximum stress in a bar and the end of the bar is called its development or anchorage length. An inadequate length can result in a splitting failure of the concrete, the split running along the bar as indicated in the cross sections sketched in Fig. 14. Since

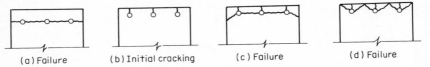

Fig. 14 Development length splitting.

such a failure is normally brittle, conservatism is necessary. The Code development length attempts to assure that such a failure is less probable than some other strength limitation.

Closely spaced bars can split off the cover, as in Fig. 14a. If the cover is thin, say less than a half (clear) bar space, the weakness may initially develop as in Fig. 14b and bring failure either as shown in Fig. 14c for closely spaced bars or in Fig. 14d for widely spaced bars.

Neither closely spaced bars (closer than possibly 3 in. clear) nor widely spaced bars in mass concrete are mentioned in the Code; the designer should be aware of the deficiencies of the first and the advantages of the second.

Code development lengths are given in Table 9. These are lengths of straight bars but are also used for trussed bars bent at 45° or flatter. The development length of standard hooks (180, 135, or 90°) is determined by substituting f_h for f_y in Table 9, where $f_h = \xi\sqrt{f_c'}$ and both f_h and $\sqrt{f_c'}$ are in units of psi. Values of ξ are given in Table 10.

The development length may be a combination of the equivalent embedment length of a hook or mechanical anchorage plus additional embedment of the reinforcement.

Research shows that very little tension develops beyond a hook and only the length in front is effective as part of the needed development. The specified extension beyond a hook is primarily to assure that the hook is restrained against straightening under tension.

Hooks in compression bars are ineffective.

TABLE 9 Development Length l_d, in., of Deformed Bars and Deformed Wire*

Deformed Bars and Deformed Wire in Tension

(a) l_d = (basic development length) × (modification factor) $\geqslant$ 12 in.

(b) Basic development length†
    No. 11 bars or smaller $\geqslant 0.04 A_b f_y/\sqrt{f_c'}$ and $0.0004 d_b f_y$

    No. 14 bars $\geqslant 0.085 f_y/\sqrt{f_c'}$

    No. 18 bars $\geqslant 0.11 f_y/\sqrt{f_c'}$

    Deformed wire $\geqslant 0.03 d_b f_y/\sqrt{f_c'}$

(c) Modification factor‡
1. Top reinforcement§   1.4
2. Bars with $f_y > 60,000$ psi $2 - 60,000/f_y$
3. Reinforcement located at least 3 in. from   0.8
   side face of member and spaced laterally
   at least 6 in. on center
4. Reinforcement in excess of that required $A_{s(req)}/A_{s(act)}$
5. Bars enclosed within a spiral which is   0.75
   not less than ¼ in. in diameter and not
   more than 4 in. pitch

Deformed Bars in Compression

(d) Development length $l_d = 0.02 f_y d_b/\sqrt{f_c'} \geqslant$
0.0003 $f_y d_b \geqslant$ 8 in. This length may be
modified according to provisions of (c) 4
and (c) 5

Bundled Bars

(e) l_d = (factor below) × (development length
of individual bar)
    Three-bar bundle   1.2
    Four-bar bundle   1.33

*Adapted from Ref. 6.

†This basic development length must be increased 33 percent for all-lightweight concrete and 18 percent for sand–lightweight concrete. Use linear interpolation for partial sand replacement. Alternatively, the basic development length may be multiplied by $6.7\sqrt{f_c'}/f_{ct} \geqslant 1$ when f_{ct} is specified and concrete is proportioned in accordance with Code 4.2. The factors in (c) also apply.

‡All modification factors can be applicable in a given case, e.g., for top tensile reinforcing in excess of that required, located at least 3 in. from side face and spaced at least 6 in. on center and with $f_y > 60,000$ psi, the basic development length from (b) is multiplied by $1.4 × 0.8 × (2 - 60,000/f_y)A_{s(req)}/A_{s(act)}$.

§Horizontal reinforcement with more than 12 in. of concrete in the member below the bars.

TABLE 10 Values of ξ in $f_h = \xi\sqrt{f_c'}$*†

Bar No.	$f_y = 60$ ksi		$f_y = 40$ ksi All bars
	Top bars	Other bars	
3–5	540	540	360
6	450	540	360
7–9	360	540	360
10	360	480	360
11	360	420	360
14	330	330	330
18	220	220	220

*From Ref. 6.

†Values of ξ may be increased 30 percent where enclosure is provided perpendicular to plane of hook. Enclosure may consist of external concrete or internal closed ties, spirals, or stirrups.

Example 9 A 10×18-in. cantilevered beam has 2 No. 7 bars in the top. Determine the embedment required to anchor and fully develop these bars. $f'_c = 4000$ psi, $f_y = 60,000$ psi.

SOLUTION. Since these are top bars, the appropriate formula from Table 9 is modified in accordance with (c). A_b for the No. 7 bar is 0.6 in.² Hence

$$l_d = 1.4 \left(\frac{0.04 A_b f_y}{\sqrt{f'_c}} \right) = 1.4 \frac{0.04 \times 0.6 \times 60,000}{\sqrt{4000}} = 31.9 \text{ in.}$$

Part of this embedment will be made up by a hook. This hook for No. 7 to No. 11 bars, inclusive (Table 10) has a value of $360\sqrt{f'_c} = 22,800$ psi. It needs a lead-in length to care for $60,000 - 22,800 = 37,200$ psi, which requires $31.9 \times 37,200/60,000 = 19.8$ in. This requires an anchorage-member thickness of $19.8 + 3.5$ (for hook) + 2 cover = 25.3 in., more than would usually be available unless the beam is one that continues beyond the support. Theoretically, a larger-radius hook should give more resistance, but no Code guidance is available for this. The Code does not forbid a 90° hook plus extra anchorage beyond the bend, but this combination is not recommended, for reasons given in Art. 13. Therefore, the design needs revision.

One solution would be the use of 3 No. 7 bars, equivalent to lowering the bar stress to 40,000 psi. With the same 22,800 psi on the hook the necessary lead-in becomes $(17,200/60,000)31.9 = 9.1$ in. and the required anchorage-member thickness only $9.1 + 3.5 + 2 = 14.6$ in.

Bars could also be made smaller, such as 4 No. 5, furnishing 1.24 in.² at a stress of $(1.20/1.24)60,000 = 57,100$ psi. The hook cares for the same 22,800 psi, and lead-in length for $57,100 - 22,800 = 34,300$ psi is $1.4 \times 0.04 \times 0.31 \times 34,300/\sqrt{4000} = 9.4$ in. Total thickness needed would be $9.4 + 2.50 + 2 = 13.9$ in.

Example 10 Design the longitudinal reinforcement for a 10×18-in. beam carrying a uniform load of 2500 plf live load and 2000 plf dead load on a simple span of 10 ft and determine embedment lengths. $f'_c = 4000$, $f_y = 60,000$ psi.

SOLUTION. $d = 18 - 1\frac{1}{2} - \frac{3}{8} - \frac{1}{2} = 15.6$ in. To determine the reinforcing steel:

$$\text{DL} = 2000 \times 1.4 = 2800 \text{ plf}$$
$$\text{LL} = 2500 \times 1.7 = \underline{4250 \text{ plf}}$$
$$7050 \text{ plf}$$
$$M = wL^2/8 = 7050 \times (10^2 \times 12)/8 = 1,058,000 \text{ lb-in.}$$
$$M/\phi bd^2 = \frac{1,058,000}{0.90 \times 10 \times 15.6 \times 15.6} = 483$$

From Table 5, $\rho = 0.0088$, $A_s = 0.0088 \times 10 \times 15.6 = 1.37$ sq in. Use 1 No. 7 bar and 1 No. 8 bar = 1.39 sq in.

Since the bond resistance is increased by a compressive reaction (Code 12.2.3), the l_d provided must not exceed

$$1.3 M_t/V_u + l_a = 1.3(1.39 \times 60,000 \times 15.6)/35,250 + l_a$$
$$= 48 \text{ in.} + l_a$$

where l_a is the additional length beyond the theoretical reaction point. For the No. 8 bar (the more critical) from Table 9,

$$l_d = 0.04 A_b f_y/\sqrt{f'_c} = 0.04 \times 0.79 \times 60,000/\sqrt{4000} = 30 \text{ in.}$$

which is considerably less than the 48-in. limit for M_t/V_u just calculated, even without considering the probable l_a extension of the No. 8 bar past the center of the support.

14. Splices Because reinforcing bars are delivered in manageable lengths, longer runs must be spliced. Splices may be made by lapping (if not larger than No. 11), butt welding, mechanical coupling, or other device. Lapped bars may be in contact or spaced far enough apart to allow more wet concrete between them. The following are Code requirements.

Lap Splices in Compression. The minimum lap length is l_d from (d) of Table 9, but not less than $0.0005 f_y d_b$ for $f_y \le 60,000$ psi or $(0.0009 f_y - 24)d_b$ for $f_y > 60,000$ psi, or 12 in., where d_b = diameter of bar. This lap must be increased by one-third for $f'_c < 3000$ psi.

The lap length for spiral compression members, and for tied compression members with ties having an effective area of at least $0.0015hs$ throughout the lap, where h = member thickness and s = tie spacing, may be taken at 0.75 and 0.83, respectively, of the lap length defined above, but not less than 12 in.

Splices in Tension. Lap splices are not permitted in tension tie members; instead, the bars must be fully connected by welding or by mechanical connections, and the splices must be staggered at least $1.7 l_d$. Lap splices may be used in other members.

If the area A_s of the steel at the splice is less than twice that required by analysis, lap splices must have a lap of $1.3l_d$ if no more than half the bars are spliced within the lap length and $1.7l_d$ if more than half are spliced, and in either case not less than 12 in.

If the area A_s of the steel at the splice is more than twice that required by analysis, lap splices must have a lap of l_d if no more than three-quarters of the bars are spliced within the lap length and $1.3l_d$ if more than three-quarters are spliced, and in either case not less than 12 in.

The embedment lengths l_d are those for the full f_y (Table 9).

The lap splice of a circumferential tension bar in a round tank need not be considered as being in a tension tie, although it is especially desirable that such splices be well staggered.

Butt-welded or coupled splices must provide positive connections capable of transmitting at least 125 percent of the specified yield strength of the bar. Splices that will be called upon only to transmit compression may have ends of bars sawed square within 1.5° and held in true alignment with a welded sleeve or splice or a mechanical coupler.

Example 11   The tension tie between bottom hinges of a three-hinged arch is 150 ft long and is to consist of deformed bars buried in concrete below the floor, to carry a tension of 220 kips, of which 120 kips is dead load and 100 kips live load. Design the tie, the intermediate splices, and the end anchorages. $f'_c = 4000$ psi, $f_y = 60,000$ psi.

SOLUTION.

$$
\begin{aligned}
DL &= 120,000 \times 1.4 = 168,000 \\
LL &= 100,000 \times 1.7 = \underline{170,000} \\
&\qquad\qquad\qquad\quad 338,000
\end{aligned}
$$

This being a tension tie, ϕ equals 0.90 and $\phi f_y = 0.90 \times 60,000 = 54,000$ psi. Hence the area of bars should be

$$A_s = 338,000/54,000 = 6.28 \text{ in.}$$
$$\text{Use 4 No. 11 bars} = 6.24 \text{ sq in.}$$

Use mechanical splices, such as Cadweld, or butt-welded splices; lap splices are not allowed by Code.

The four bars must be spliced at points staggered $1.7l_d$ along the tie. Since $l_d = 0.04A_b f_y/\sqrt{f'_c} = 0.04 \times 1.56 \times 60,000/\sqrt{4000} = 59.2$ in., this requires 101 in.

ANCHORAGE.   Length of anchorage is l_d for bottom bars and $1.4l_d$ for top bars, or 59.2 and 83 in., respectively. These lengths give real problems, since little tension can be carried beyond the usual 90° bend. The bars could be extended through the vertical member into an anchorage block, as in Fig. 15a,

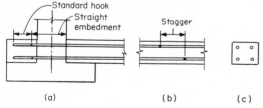

(a)   (b)   (c)

Fig. 15   Example 11.

more bars of smaller diameter could be used to reduce the required l_d, or the bars might be carried through the vertical member and welded to steel anchor plates.

As a further precaution the tie should be buried deep enough or placed on a sufficiently soft layer that loads on the floor above will not damage it from beam action.

Example 12   The wall thickness of the vertical stem of a cantilevered retaining wall varies from 8 to 14 in. in a net height of 12 ft. For $f'_c = 3000$ psi, $f_y = 60,000$ psi, design the dowels from the base to the stem and the vertical bars in the stem (Fig. 16). There is a surcharge of 300 psf on the backfill. The service-load moment and shear in the stem are $M = 60(h_x + 3)^3$ and $V = 15(h_x + 3)^2$ per foot length of wall, where h is in feet.

SOLUTION.   Computed shears, moments, effective depths, and areas of steel are given in Table 11, using a load factor of 1.7 for horizontal loads. The shears and moments at all levels below 8 ft are given in the schedule. At lesser depths, their values are not critical.

Although the bars shown in the table are all that are required for stress, it would be well to provide at least a mat of No. 4 bars 12 in. c/c horizontal and vertical for temperature reinforcement in the exterior face of the wall.

Development length of No. 5 = $0.04 A_b f_y / \sqrt{f'_c}$ = 0.04 × 0.31 × 60,000/$\sqrt{3000}$ = 13.6 in. Therefore, the dowel must extend 14 in. into the footing. However, many designers would prefer to have it terminate in a semicircular hook.

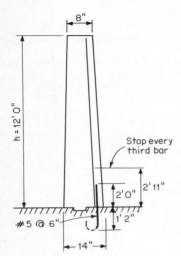

Fig. 16   Example 12.

The projection into the stem acts as a splice, which requires $1.7 l_d$ = 1.7 × 13.6 = 23.1 in.

The vertical bars in the wall can theoretically be reduced to No. 5 at 9 in. at a depth of 10 ft from the top, but the Code requires an extension of d = 10.5 in. or $12 d_b$ (7.5 in.). Every third bar can then be cut off at 2 ft + 10.5 in. from the top of the base, say 2 ft 11 in. Alternatively, every third bar could be omitted if every third dowel projected 2 ft 11 in. and thus acted as the third bar for the wall.

TABLE 11   Example 12

h_x, ft	V, lb	M, lb-in.	$M/\phi bd^2$		d, in.	ρ Table 5	A_s, in.²	Bars
8	3080	135,800	$\dfrac{135,800}{0.90 \times 12 \times 9.5 \times 9.5}$	= 139	9.5			
9	3670	176,300	$\dfrac{176,300}{0.90 \times 12 \times 10 \times 10}$	= 163	10			
10	4310	223,900	$\dfrac{223,900}{0.90 \times 12 \times 10.5 \times 10.5}$	= 188	10.5	0.0033	0.42	No. 5 at 9
10.5	4650	251,000	$\dfrac{251,000}{0.90 \times 12 \times 10.75 \times 10.75}$	= 201	10.75	0.0035	0.45	No. 5 at 8
11	5000	280,000	$\dfrac{280,000}{0.90 \times 12 \times 11 \times 11}$	= 214	11	0.0037	0.49	No. 5 at 7.5
12	5740	344,200	$\dfrac{344,200}{0.90 \times 12 \times 11.5 \times 11.5}$	= 241	11.5	0.0042	0.58	No. 5 at 6

15. Bar Cutoffs and Bend Points

Example 13 A 10 × 18 in. beam is carrying a factored load of 2550 plf on a simply supported span of 20 ft. Design the tension bars and determine the cutoff for each. f'_c = 4000, f_y = 60,000 psi. (Fig. 17.)

SOLUTION. If the steel is in two layers, d = 18 − 1½ − ⅜ − ½ − ½ = 15.1 in.

$$M_u = wL^2/8 = 2550 \times 20^2 \times 12/8 = 1,530,000 \text{ in.-lb}$$

$$\frac{M}{\phi b d^2} = \frac{1,530,000}{0.90 \times 10 \times 15.1 \times 15.1} = 745$$

From Table 5 ρ = 0.0142, A_s = 0.0142 × 10 × 15.1 = 2.14 sq in. Use 5 No. 6 bars = 2.20 sq in.

Code 12.2.1 requires at least one-third of these bars to extend 6 in. into the support, or 2 No. 6 × 21 ft 0 in. Of the remaining bars, one can theoretically be cut off or bent up at $\sqrt{1/5}$ of the distance from midspan to support, i.e., $\sqrt{1/5}$ × 10 ft = 4.47 ft. However, bars should extend the depth d = 15.1 in. = 1.26 ft (or 12 bar diameters if greater) past the theoretical cutoff point, which is 4.47 + 1.26 = 5.73 ft from midspan.

The next bottom bar can be spared, cut off or bent up at $\sqrt{2/5}$ of the distance from midspan to support, i.e., 6.32 ft plus the same 1.26 ft for anchorage = 7.58 ft. The third bar can theoretically be spared at $\sqrt{3/5}$ of the distance from midspan, i.e., 7.74 ft + 1.26 ft = 9.00 ft, but might be detailed the same length as the full-length bottom bars.

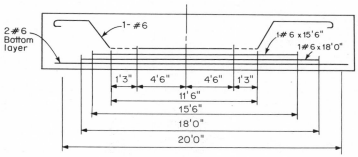

Fig. 17   Example 13.

16. Deflection   Safety consists largely in keeping service-load stresses within the maximum allowable or, in strength design, factored loads within stated margins. Serviceability means creating a structure that will not misbehave visibly, i.e., will not settle, crack, warp, or sag to an undesirable degree.

Short-time elastic deflection can be approximated quite adequately by the usual methods. However, due consideration must be given to end restraints (e.g., a simply supported, uniformly loaded beam deflects five times as much as the same beam with ends fully fixed).

ACI limits short-term live-load elastic deflection to $L/180$ for flat roofs not supporting or attached to nonstructural elements likely to be damaged by large deflections, and to $L/360$ for floors under the same restrictions. In roof or floor construction supporting or attached to nonstructural elements not likely to be damaged by large deflections, the deflection which occurs after attachment is limited to $L/240$. This includes long-term deflections due to all sustained load plus the immediate deflection from any additional live load; camber may be deducted. The corresponding limit for members supporting or attached to elements which are likely to be damaged, such as partitions subject to cracking, is $L/480$ except where adequate countermeasures are taken to avoid damage.

Because concrete shrinks as it ages, and creeps or yields plastically under continuous pressure, the compression side of a reinforced-concrete beam shortens with age while the tension side undergoes very little change. Thus, deflections increase with age somewhat in proportion to the amount of compression constantly maintained in the concrete. The amount of such increase depends upon a variety of factors, one of the most important of which is the amount of compressive reinforcement. Table 12 gives fairly workable approximations of the *increase* in the short-term elastic deflection caused by loads that are more or less constantly in place, such as dead load and portions of the live load.

The part of the live load which is likely to be in place most of the time is, to a considerable extent, a question of the designer's judgment. Of the specified live load, only a small part (perhaps a third or so) might be considered permanent for churches or meeting rooms, while practically all would be considered permanent for storage warehouses.

Below certain span-depth ratios, deflections need not be computed. Table 13 lists such ratios, with footnotes indicating modifications required for lightweight concrete or for steel yield strengths other than 60,000 psi.

The ACI procedures are adequate for the control of deflections and may predict actual deflections in monolithic structures reasonably well. However, when tight fits are

TABLE 12   Increase in Short-Time Deflection

Duration of load	$A_s' = 0$	$A_s' = A_s/2$	$A_s' = A_s$
1 month...............	0.6	0.4	0.3
6 months...............	1.2	1.0	0.7
1 year.................	1.4	1.1	0.8
5 or more years.........	2.0	1.2	0.8

involved, e.g., glass panels under a spandrel beam, it is prudent to provide clearance or adjustment beyond the estimated deflection.

17. Column Design—Combined Compression and Bending   Columns rarely carry axial load alone. Because of loads applied off center, moments introduced at joints, or lateral· loading in unbraced frames, column design must be for combined direct load and flexure.

TABLE 13   Minimum Thickness of Beams by Code 9.5*

	Min thickness h, in			
	Simply supported	One end continuous	Both ends continuous	Cantilever
Member	Members not supporting or attached to partitions or other construction likely to be damaged by large deflections			
Solid one-way slabs†	$l/20$	$l/24$	$l/28$	$l/10$
Beams or ribbed one-way slabs	$l/16$	$l/18.5$	$l/21$	$l/8$

*From Ref. 6.
†The span length l is in inches.

The values given in this table shall be used directly for nonprestressed reinforced-concrete members made with normal-weight concrete ($w = 145$ pcf) and Grade 60 reinforcement. For other conditions, the values shall be modified as follows:

(a) For structural lightweight concrete having unit weights in the range 90 to 120 lb/cu ft, the values in the table shall be multiplied by $1.65 - 0.005w$ but not less than 1.09, where w is the unit weight in lb/cu ft.

(b) For nonprestressed reinforcement having yield strengths other than 60,000 psi, the values in the table shall be multiplied by $0.4 + f_y/100,000$.

Effective Column Length.   Figure 18a shows the deflected shape of a column in a braced frame and Fig. 18b one in an unbraced frame. When columns are long, the secondary moment $P\Delta$ can be substantial, especially in unbraced frames. Joint rotations are controlled by the relative stiffnesses $\Sigma EI/l_u$ of the columns and $\Sigma EI/l$ of the beams at the joints. The effective length of the column is kl_u, where l_u is the unsupported length of the column. Values of k can be determined from Fig. 5 of Sec. 6.

By Code definition a column in an unbraced frame is short if $kl_u/r < 22$ and in a braced frame if $kl_u/r < 34 - 12M_1/M_2$, where M_1 and M_2 are the column end moments, with $M_1 < M_2$. The ratio M_1/M_2 is positive for members in single curvature and negative for members in double curvature.

Since $k = 1$ and $M_1/M_2 = 1$ are on the safe side for braced frames, $kl_u/r = 34 - 12 \times 1 =$

22, the same as for unbraced frames. Also, since $r = 0.3h$ is a good approximation for rectangular cross sections, $l/h = 0.3 \times 22 = 6.6$ is near the changeover point from short column to long column for braced and unbraced frames.

Design aids. Because of the many variables involved, either interaction diagrams or tables are a practical requirement for the design of compression members. A typical

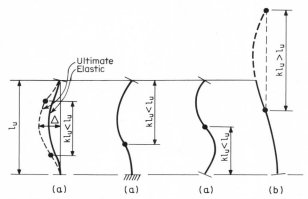

Fig. 18 Effective column length. (*a*) Braced frame; (*b*) unbraced frame.

interaction diagram is shown in Fig. 19. Point P_0 represents concentric compression ($M = 0$) and point M_0 pure bending ($P = 0$). At point a failure is by simultaneous crushing of the concrete and yielding of the steel, which gives a balanced design ($M = M_b$ and $P = P_b$). The segment P_0a represents failure which begins with crushing of the concrete and the portion M_0a failure which begins by yielding of the steel.

The Code specifies $\phi = 0.70$ for tied columns and 0.75 for spiral-reinforced columns.

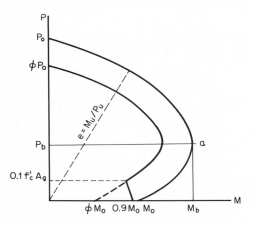

Fig. 19 Column interaction diagram.

Since $\phi = 0.9$ for beams, a linear transition from the value for beams to the value for columns is specified from $P = 0$ to $P = 0.10 f'_cA_g$ (Fig. 19).

The Code also specifies the following maximum values of the load P_u for concentrically loaded, nonprestressed members:

Members with spiral reinforcement:

$$P_u = 0.85\phi[0.85f'_c(A_g - A_{st}) + f_yA_{st}] \tag{12}$$

Members with tied reinforcement:

$$P_u = 0.80\phi[0.85f'_c(A_g - A_{st}) + f_y A_{st}] \tag{13}$$

A_{st} is the total area of the longitudinal reinforcement. These equations replace the minimum accidental-eccentricity requirements for concentrically loaded members in earlier editions of ACI 318. The reduction in load is about the same as was achieved by the specified minimum eccentricities.

Interaction diagrams for $f_y = 60,000$ psi and concrete of any strength up to 4000 psi are given in Figs. 20 to 31. They can also be used for steel with other yield strengths without substantial error, at least for preliminary design. In the designation pq for each curve, $q = f_y/0.85 f'_c$.

Example 14 Design a square, tied, short column with bars on two opposite faces for a factored load $P_u = 730$ kips, $M_u = 200$ kip-ft, $f'_c = 4$ ksi, $f_y = 60$ ksi.

SOLUTION. $e = M_u/P_u = 200 \times 12/730 = 3.29$ in. Assume $\rho = 0.03$, $h = 17$ in., $\gamma = 0.7$. Then $e/h = 3.29/17 = 0.19$. $q = f_y/0.85 f'_c = 60/(0.85 \times 4) = 17.6$, $q\rho = 17.6 \times 0.03 = 0.53$. On the chart of Fig. 21 these e/h and $q\rho$ values intersect at $K = 0.84$. Therefore,

$$bh = \frac{P_u}{\phi f'_c K} = \frac{730}{0.7 \times 4 \times 0.84} = 310 \text{ in.}^2$$

Try 18×18 in., for which

$$\frac{e}{h} = \frac{3.29}{18} = 0.18 \qquad K = \frac{730}{0.7 \times 4 \times 18^2} = 0.80$$

With these values $q\rho = 0.45$. Therefore, $\rho = 0.45/17.6 = 0.026$, which is close enough to the assumed value 0.03.

Instead of entering Fig. 21 with e/h and $q\rho$, one may assume values of b and h and proceed as follows:

Assume $b = h = 18$ in. and $\gamma = 0.7$.

$$K = \frac{P_u}{\phi f'_c bh} = \frac{730}{0.7 \times 4 \times 17 \times 17} = 0.80$$

$$K\frac{e}{h} = \frac{M_u}{\phi f'_c bh^2} = \frac{200 \times 12}{0.7 \times 4 \times 17 \times 17^2} = 0.15$$

Entering Fig. 21 with these values gives $q\rho = 0.45$, from which $\rho = 0.45/17.6 = 0.026$ as before.

Then $A_s = 0.026 \times 18^2 = 8.42$ in.^{2} Use 4 No. 11 and 2 No. 10, for which $A_s = 8.78$ in.^{2} For simplicity, many designers would use 6 No. 11 = 9.36 in^2

$$\gamma = \frac{h - 2d'}{h} = \frac{18 - 2(1.50 + 0.5 + 0.70)}{18} = 0.70, \text{ assumed } 0.70.$$

Table 14 shows that there is no bar-spacing problem.

Use 18×18-in. column, 4 No. 11 + 2 No. 10 bars (No. 11 in corners).

Code 7.12.3 requires at least No. 4 ties be used with No. 11 bars, with maximum tie spacing the smallest of $16d_b = 16 \times 1.41 = 22.5$ in., 48 tie diameters = 24 in., or the least dimension of the column = 18 in., the last governing. Also, no bar should be more than 6 in. clear on either side from a braced bar. On a center-to-center basis, these bars are spaced 6.3 in., but not less than 6 in. clear. Crossties are not required, but to be conservative, will be used.

Use No. 4 closed ties at 18 in. around No. 11 corner bars plus No. 4 crossties around No. 10 bars, also at 18-in. spacing (Fig. 32).

Example 15 Design a short round spiral column for a factored load $P_u = 1000$ kips, $M_u = 400$ kip-ft, $f'_c = 4$ ksi, $F_y = 60$ ksi, reinforcement ratio ρ about 0.04.

SOLUTION. $e = 400 \times 12/1000 = 4.8$. Assume $h = 22$ in. $e/h = 4.8/22 = 0.22$
$q = 60/(0.85 \times 4) = 17.6$ $pq = 0.04 \times 17.6 = 0.70$. Assume γ
                                                       $= 0.8$.

From Fig. 30, $K = 0.60 = \dfrac{1000}{0.75 \times 4h^2}$ $h = 23.6$ in.

Try $h = 24$, $e/h = 0.20$ $K = \dfrac{1000}{0.75 \times 4 \times 24^2} = 0.58$

From Fig. 30, $pq = 0.55$. $\rho = 0.55/17.6 = 0.031$ $A_s = 0.031 \times 0.785 \times 24^2 = 14.13$ in.^{2}
Use 10 No. 11 = 15.62 in.^{2} or 12 No. 10 = 15.19 in.^{2}
Either can be used, even if the bars are lap-spliced side by side (Table 15).

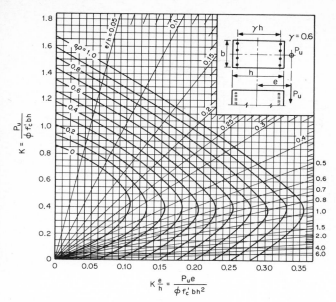

Fig. 20 Column interaction diagram—rectangular section, $\gamma = 0.9$, $f'_c \leq 4$ ksi, $f_y = 60$ ksi. (*Adapted from Ref. 3.*)

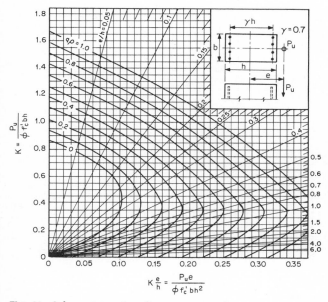

Fig. 21 Column interaction diagram—rectangular section, $\gamma = 0.8$, $f'_c \leq 4$ ksi, $f_y = 60$ ksi. (*Adapted from Ref. 3.*)

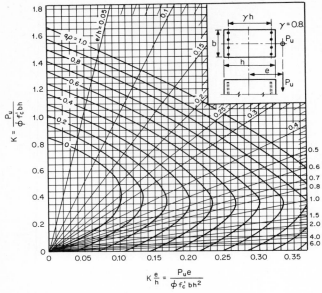

Fig. 22 Column interaction diagram—rectangular section, $\gamma = 0.7$, $f'_c \leq 4$ ksi, $f_y = 60$ ksi. (*Adapted from Ref. 3.*)

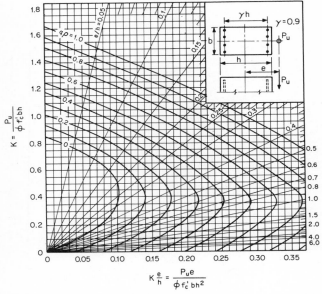

Fig. 23 Column interaction diagram—rectangular section, $\gamma = 0.6$, $f'_c \leq 4$ ksi, $f_y = 60$ ksi. (*Adapted from Ref. 3.*)

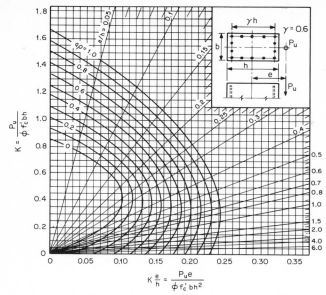

Fig. 24 Column interaction diagram—rectangular section, $\gamma = 0.9$, $f'_c \leq 4$ ksi, $f_y = 60$ ksi. (*Adapted from Ref. 3.*)

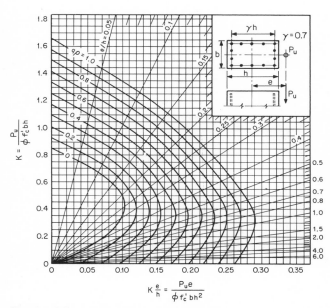

Fig. 25 Column interaction diagram—rectangular section, $\gamma = 0.8$, $f'_c \leq 4$ ksi, $f_y = 60$ ksi. (*Adapted from Ref. 3.*)

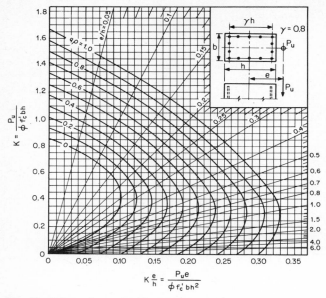

Fig. 26 Column interaction diagram—rectangular section, $\gamma = 0.7$, $f'_c \leq 4$ ksi, $f_y = 60$ ksi. (*Adapted from Ref. 3.*)

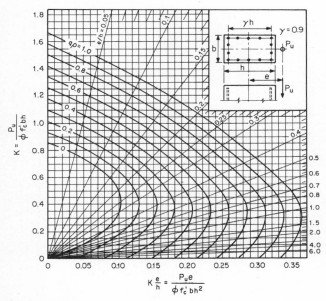

Fig. 27 Column interaction diagram—rectangular section, $\gamma = 0.6$, $f'_c \leq 4$ ksi, $f_y = 60$ ksi. (*Adapted from Ref. 3.*)

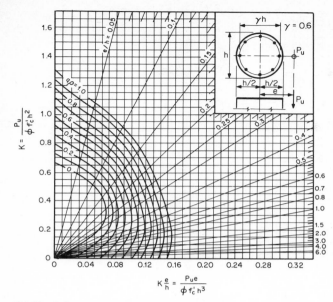

Fig. 28 Column interaction diagram—circular section with spiral, $\gamma =$ 0.9, $f'_c \le 4$ ksi, $f_y = 60$ ksi. (*Adapted from Ref. 3.*)

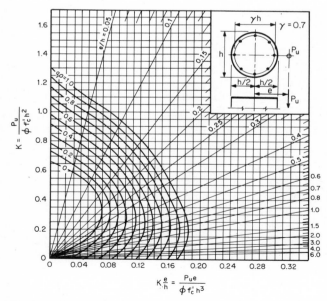

Fig. 29 Column interaction diagram—circular section with spiral, $\gamma =$ 0.8, $f'_c \le 4$ ksi, $f_y = 60$ ksi. (*Adapted from Ref. 3.*)

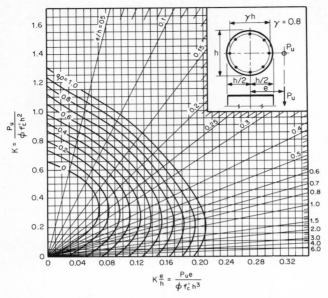

Fig. 30 Column interaction diagram—circular section with spiral, $\gamma =$ 0.7, $f'_c \leq 4$ ksi, $f_y = 60$ ksi. (*Adapted from Ref. 3.*)

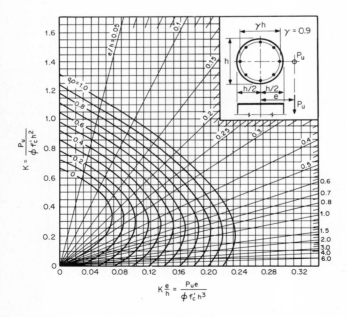

Fig. 31 Column interaction diagram—circular section with spiral, $\gamma =$ 0.6, $f'_c \leq 4$ ksi, $f_y = 60$ ksi. (*Adapted from Ref. 3.*)

TABLE 14   Maximum Number of Bars in One Face of Square Tied Columns*

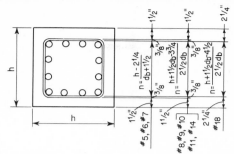

#11,14, and 18
require 1/2-in. ties

Column size C	No. of bars of a size n								
	No. 5	No. 6	No. 7	No. 8	No. 9	No. 10	No. 11	No. 14	No. 18
10	3	3	3	3	2	2	2		
11	4	3	3	3	3	2	2	2	
12	4	4	4	3	3	3	2	2	
13	5	4	4	4	3	3	3	2	
14	5	5	4	4	4	3	3	3†	
15	6	5	5	5	4	4	3	3	2
16	6	6	5	5	4	4	4	3	2
17	6	6	6	5	5	4	4	3	2
18	7	7	6	6	5	5	4	3	2
19	7	7	7	6	6	5	4	4	3†
20	8	7	7	7	6	5	5	4	3
21	8	8	7	7	6	6	5	4	3
22	9	8	8	7	7	6	5	4	3
23	9	9	8	8	7	6	6	5	3
24	10	9	9	8	7	6	6	5	4†
25	10	10	9	9	8	7	6	5	4
26	11	10	10	9	8	7	6	5	4
27	11	11	10	9	8	7	7	6	4
28	12	11	10	10	9	8	7	6	4
29	12	11	11	10	9	8	7	6	4
30	13	12	11	11	9	8	8	6	5
31	13	12	12	11	10	9	8	7	5
32	14	13	12	11	10	9	8	7	5
33	14	13	12	12	10	9	8	7	5
34	14	14	13	12	11	10	9	7	5
35	15	14	13	13	11	10	9	7	6
36	15	15	14	13	12	10	9	8	6
37	16	15	14	13	12	11	9	8	6
38	16	15	15	14	12	11	10	8	6
39	17	16	15	14	13	11	10	8	6
40	17	16	16	15	13	12	10	9	6

*Modified from Ref. 1.
†If this number of bars is used in each of four faces, ρ will exceed 8 percent.

TABLE 15 Maximum Number of Spliced Column Verticals in Various Patterns That Can Be Accommodated in Single Ring within Column Spirals*

Butt-welded

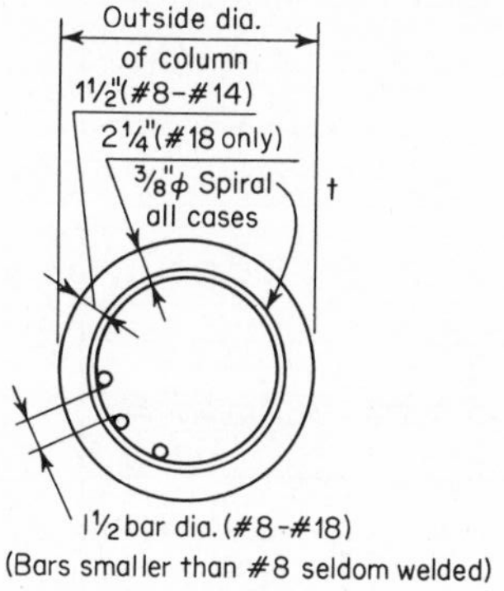

- Outside dia. of column
- 1½"(#8–#14)
- 2¼"(#18 only)
- ⅜"φ Spiral all cases
- 1½ bar dia. (#8–#18)
- (Bars smaller than #8 seldom welded)

Radially lapped

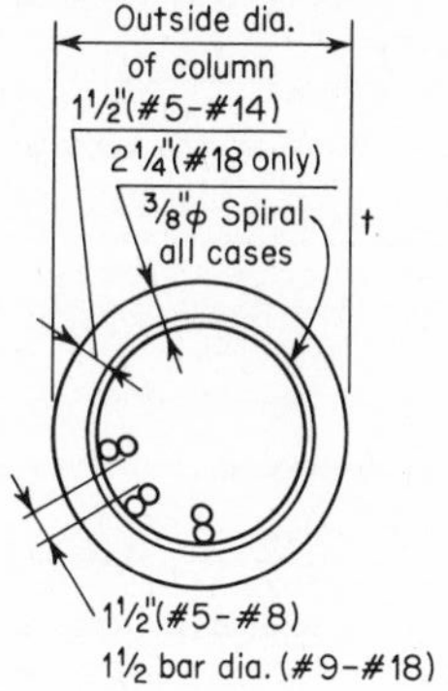

- Outside dia. of column
- 1½"(#5–#14)
- 2¼"(#18 only)
- ⅜"φ Spiral all cases
- 1½"(#5–#8)
- 1½ bar dia. (#9–#18)

Circumferentially lapped

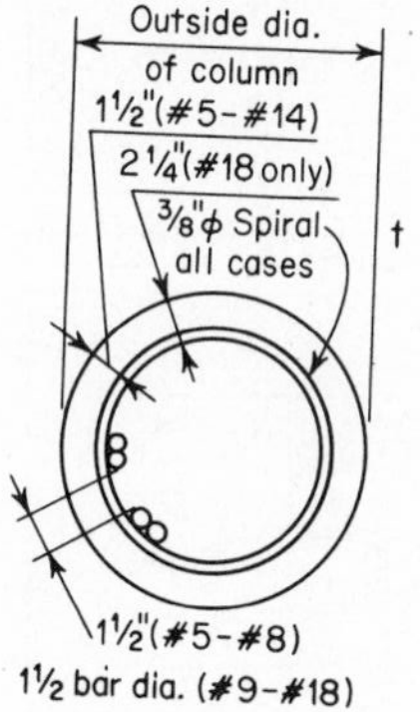

- Outside dia. of column
- 1½"(#5–#14)
- 2¼"(#18 only)
- ⅜"φ Spiral all cases
- 1½"(#5–#8)
- 1½ bar dia. (#9–#18)

Diam of column	Butt-welded						Radially lapped									Circumferentially lapped								
	No. 8	No. 9	No. 10	No. 11	No.‡ 14	No.‡ 18	No. 5	No. 6	No. 7	No. 8	No. 9	No. 10	No. 11	No.‡ 14	No.‡ 18	No. 5	No. 6	No. 7	No. 8	No. 9	No. 10	No. 11	No.‡ 14	No.‡ 18
10†	6						6									6								
11†	7	6					8	7	6							7	6							
12†	9	7	6				9	8	7	6						8	7	6						
13†	10	9	7	6			10	9	8	7	6					9	7	7	6					
14	11	10	8	7			12	11	10	9	8	6				11	9	8	7	6				
15	12	11	9	8	6		13	12	11	10	8	7	6			12	11	10	9	8	6			
16	14	12	10	9	7		15	13	12	11	9	8	6			13	12	11	10	8	7	6		
17	15	13	11	10	8		16	15	14	12	11	9	7	6		14	13	11	11	9	7	6	6	
18	16	14	12	11	9		18	16	15	14	12	10	8	6		15	14	12	11	10	8	7	6	
19	17	15	13	12	10		19	18	16	15	13	11	9	7		16	15	13	12	11	9	8	7	

	C1	C2	C3	C4	C5	C6	C7	C8	C9	C10	C11	C12	C13	C14	C15	C16	C17	C18	C19	C20	C21	C22	C23	C24
20		7	9	10	12	13	14	16	17		8	10	12	14	16	18	19	21	6§	10	13	14	16	19
21		8	10	11	12	14	15	17	18		8	11	13	15	17	19	20	22	7§	11	14	15	18	20
22	6	8	10	12	13	15	16	18	20		9	11	14	16	18	20	22	23	7§	12	15	16	19	21
23	6	9	11	12	14	16	17	19	21	6	10	12	15	17	20	21	23	25	8§	13	15	17	20	22
24	6	9	11	13	15	17	18	20	22	7	11	13	16	18	21	22	24	26	9§	13	16	18	21	24
25	7	10	12	14	16	18	19	21	23	7	11	14	17	19	22	24	26	28	9§	14	17	19	22	25
26	7	10	13	14	16	19	20	22	24	8	12	15	18	20	23	25	27	29	10	15	18	20	23	26
27	8	11	13	15	17	20	21	23	25	8	13	16	19	21	25	26	28	31	11	16	19	21	24	28
28	8	11	14	16	18	20	22	24	26	9	13	17	20	22	26	28	30	32	11	16	20	22	25	29
29	8	12	15	16	19	21	23	25		9	14	17	21	23	27	29	31	33	12	17	21	23	26	30
30	9	12	15	17	20	22	24	26		10	15	18	22	25	28	30	32	35	12	18	22	24	28	31
31	9	13	16	18	20	23	25	27		11	16	19	23	26	29	31	34		13	18	23	25	29	33
32	10	14	16	19	21	24	26			12	16	20	24	27	31	33			14	19	23	26	30	34
33	10	14	17	19	22	25	27			12	17	21	25	28	32	35			14	20	24	27	31	35
34	10	15	18	20	23	26				12	18	22	26	29	33				15	21	25	28	32	36
35	11	15	18	21	23	27				13	19	23	26	30	34				15	21	26	29	33	38
36	11	16	19	22	24					13	19	24	27	31					16	22	27	30	34	
37	12	16	20	23	25					14	20	25	29	33					16	23	28	31	35	
38	12	17	21	24	26					14	20	26	29	33					17	24	29	32	37	
39	12	17	22	24	27					15	22	27	31	35					17	24	30	33	38	
40	13	18	22	25						15	22	28	31						18	25	31	34		
41	13	18	23	26						16	23	29	33						19	26	33	35		
42	14	19	23	26						16	24	30	33						19	27	34	36		
43	14	19	24	27						17	25	31	35						20	27	35	37		
44	14	20	25							18	25	31							20	28	36	38		
45	15	20	25							18	26	33							21	29	35			
46	15	21	26							19	27	33							21	30	36			
47	16	22	27							19	28	34							22	30	37			
48	16	22	27							20	28	35							22	31	38			

* From Ref. 1.

† ⅜-in. spiral too large to meet all code requirements in 10- to 13-in. columns, but ¼-in. spiral not standard.

‡ Lapped splices not recommended for No. 14 and No. 18 bars.

§ Limited to number of bars that provide a maximum of 8 percent vertical reinforcement.

$$\gamma = \frac{h - 2d'}{h} = \frac{24 - 2 \times 1.5 - 2 \times 0.375 - 1.27}{24} = 0.79, \text{ assumed } 0.8 \qquad \text{O.K.}$$

The spiral must replace the strength of the concrete cover. Code 10.9.2 requires the ratio ρ_s of the volume of spiral steel to the volume of the core out-to-out of spiral to be at least

$$\rho_s = \frac{0.45 f'_c (A_g/A_c - 1)}{f_y} = \frac{0.45 \times 4 (24^2/21^2 - 1)}{60} = 0.0092$$

Table 16 shows that a 2¼-in. pitch gives $\rho_s = 0.0093$.
Figure 33 shows the column cross section.

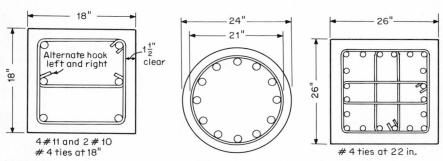

Fig. 32 Example 14.   **Fig. 33** Example 15.   **Fig. 34** Example 16.

Example 16   Design a square tied column, reinforced on all four faces, for a factored load $P_u = 1600$ kips, $M_u = 800$ kip-ft, $f'_c = 4$ ksi, $f_y = 60$ ksi.

SOLUTION. $e = 6$ in. Assume $h = 28$ in.; then $e/h = 0.21$. Try $\rho = 0.04$ $q\rho = 17.6 \times 0.04 = 0.70$. Assume $\gamma = 0.8$. From Fig. 26, $K = 0.85$.

$$\frac{P_u}{\phi f'_c bt} = \frac{1600}{0.7 \times 4h^2} = 0.85 \qquad h = 25.9 \text{ in.} \qquad \text{Try } 26 \text{ in.}$$

$$\frac{e}{h} = \frac{6}{26} = 0.23 \qquad K = \frac{1600}{0.7 \times 4 \times 26^2} = 0.85$$

From Fig. 26, $\rho q = 0.80$ $\rho = \dfrac{0.80}{17.6} = 0.045.$

$$A_s = 0.045 \times 26^2 = 30.4 \text{ in.}^2 \qquad \text{Use 20 No. 11 or 14 No. 14}$$

Table 15 shows that either six No. 11 or five No. 14 can be placed on one face. Therefore, either arrangement can be used. Use 20 No. 11.
Check γ:

$$d' = 1.5 + 0.5 + 0.71 = 2.71 \text{ in.}$$
$$\gamma = \frac{26 - 2 \times 2.71}{26} = 0.79, \text{ assumed } 0.8 \qquad \text{O.K.}$$

Ties must brace each bar that is as much as 6 in. clear from neighboring braced bars. Distance center to center of corner bars is $26 - 2 \times 1.5 - 1.41 = 20.6$ in. Spacing of six bars is $20.6/5 = 4.12$ in. Therefore, must tie around alternate bars.
Use No. 4 ties at 22 in. ($14d_b = 22.6$ in.) vertically in sets of 3 pieces: one around all bars, one closed around two center bars on each face (Fig. 34).

18. Column Splices   If bars must be offset between columns, the slope of the inclined portion with the axis of the column must not exceed 1 in 6, and the portion of the bar above and below the offset must be parallel to the axis of the column. Horizontal support at the offset bends can be provided by the floor construction or or by metal ties or spirals placed not more than 6 in. from the point of bend. The devices must be designed to resist a horizontal thrust equal to one and one-half times the horizontal component of the nominal force in the inclined portion of the bar.
When a column face is offset as much as 3 in., the bars must be spliced with separate

TABLE 16 Spirals as Percentages of Core Volume (Out to Out of Spirals)

Core diam, in.	⅝ in. diam Pitch, in.							½ in. diam Pitch, in.							⅜ in. diam Pitch, in.						
	2	2¼	2½	2¾	3	3¼	3½	2	2¼	2½	2¾	3	3¼	3½	1¾	2	2¼	2½	2¾	3	3¼
11															2.29						
12	5.17							3.33							2.09	1.83					
13	4.77							3.08							1.93	1.69					
14	4.44	3.94						2.86	2.54						1.80	1.57	1.40				
15	4.13	3.68	3.31					2.67	2.37	2.13					1.68	1.47	1.30	1.17			
16	3.88	3.45	3.10					2.50	2.22	2.00					1.57	1.38	1.22	1.10			
17	3.65	3.24	2.92	2.65				2.35	2.09	1.88	1.71				1.48	1.29	1.15	1.03	0.94		
18	3.45	3.06	2.76	2.51	2.30			2.22	1.97	1.78	1.62	1.48			1.40	1.22	1.09	0.98	0.89	0.81	
19	3.26	2.90	2.61	2.37	2.18			2.11	1.87	1.68	1.53	1.40			1.32	1.16	1.03	0.93	0.84	0.77	
20	3.10	2.76	2.48	2.25	2.07	1.91		2.00	1.78	1.60	1.45	1.33	1.23		1.26	1.10	0.98	0.88	0.80	0.73	0.68
21	2.95	2.63	2.36	2.15	1.97	1.82	1.69	1.90	1.69	1.52	1.38	1.27	1.17	1.09	1.20	1.05	0.93	0.84	0.76	0.70	0.64
22	2.82	2.51	2.26	2.05	1.88	1.74	1.61	1.82	1.62	1.45	1.32	1.21	1.12	1.04	1.14	1.00	0.89	0.80	0.73	0.67	0.62
23	2.70	2.40	2.16	1.96	1.80	1.66	1.54	1.74	1.55	1.39	1.26	1.16	1.07	0.99	1.09	0.96	0.85	0.76	0.70	0.64	0.59
24	2.59	2.30	2.07	1.88	1.72	1.59	1.48	1.67	1.48	1.33	1.21	1.11	1.03	0.95	1.05	0.92	0.81	0.73	0.67	0.61	0.56
25	2.48	2.21	1.98	1.80	1.65	1.53	1.42	1.60	1.42	1.28	1.16	1.07	0.98	0.91	1.01	0.88	0.78	0.70	0.64	0.59	0.54
26	2.39	2.12	1.91	1.73	1.59	1.47	1.36	1.54	1.37	1.23	1.12	1.03	0.95	0.88	0.97	0.85	0.75	0.68	0.62	0.56	0.52
27	2.30	2.04	1.84	1.67	1.53	1.41	1.31	1.48	1.32	1.19	1.08	0.99	0.91	0.85	0.93	0.82	0.72	0.65	0.59	0.54	0.50
28	2.21	1.97	1.77	1.61	1.48	1.36	1.27	1.43	1.27	1.14	1.04	0.95	0.88	0.82	0.90	0.79	0.70	0.63	0.57	0.52	0.48
29	2.14	1.90	1.71	1.55	1.43	1.32	1.22	1.38	1.23	1.10	1.00	0.92	0.85	0.79	0.87	0.76	0.67	0.61	0.55	0.51	0.47
30	2.07	1.84	1.65	1.50	1.38	1.27	1.18	1.33	1.19	1.07	0.97	0.89	0.82	0.76	0.84	0.73	0.65	0.58	0.53	0.49	0.45

* Ref. 1.

dowels. Lap splices of No. 14 and No. 18 bars are not acceptable under the Code, in either tension or compression, except for dowels into footings.

For bars carrying compression only, end bearing of bars, held together by an acceptable device, is permitted by Code 7.7.2. However, Code 7.10.5 requires a minimum *tensile* strength at each face of a column equal to 25 percent of the vertical reinforcement area times f_y. Also, Code 7.10.3 requires that bars with stresses ranging between f_y in compression and not more than $f_y/2$ in tension must have splices (or continuing bars) providing a tension resistance of at least twice the computed tension. If the tension calculated exceeds $f_y/2$, lap splices designed for f_y in tension, full-welded splices, or full positive connections are required by Code 7.10.4.

19. Columns with Biaxial Bending   For round columns, biaxial bending does not require special design methods because resistance about any axis is the same. One simply designs for $\sqrt{M_x^2 + M_y^2}$, where M_x and M_y are moments about the x and y axes.

For square columns, it is more than safe (somewhat wasteful) to design for the sum of the eccentricities in the two directions, although it is not quite safe to design separately for moment about each axis.

Bresler proposed the following formula for the design of rectangular cross sections under biaxially eccentric load:[7]

$$\frac{1}{P_u} = \frac{1}{P_{x0}} + \frac{1}{P_{y0}} - \frac{1}{P_0} \tag{14}$$

where P_u = ultimate load with eccentricities e_x and e_y
  P_{x0} = ultimate load for e_x only
  P_{y0} = ultimate load for e_y only
  P_0 = ultimate concentric load

Comparisons of the results of this formula with tests, and with ultimate loads for various cross sections computed according to the standard assumptions for ultimate-strength analysis, have shown it to be satisfactory.[8]

Charts and tables to facilitate the design of columns under biaxially eccentric load have been prepared.[2,9]

Figures 20 to 31 can also be used to evaluate Eq. (14).

Example 17   Compute the design load for the column of Fig. 35 for load eccentricities $e_x = 12$ in. and $e_y = 4$ in., with $f_c' = 4$ ksi, $f_y = 60$ ksi.

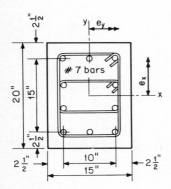

Fig. 35   Example 17.

SOLUTION. $A_s = 10 \times 0.6 = 6$ in.² $\rho = 10/300 = 0.0333$
          $q = 60/0.85 \times 4 = 17.6$ $q\rho = 0.59$
  For the x axis:

$$\gamma = 15/20 = 0.75 \qquad e/h = 12/20 = 0.6$$

From Figs. 25 and 26, $K = 0.36$ and 0.38 for $\gamma = 0.7$ and 0.8, respectively. Therefore,

$$P_{x0} = 0.37 \times 4 \times 15 \times 20 = 444 \text{ kips}$$

For the y axis:

$$\gamma = 10/15 = 0.67 \qquad e/b = 4/15 = 0.27$$

From Figs. 24 and 25, $K = 0.62$ and 0.66 for $\gamma = 0.6$ and 0.7, respectively. Therefore,

$$P_{y0} = 0.65 \times 4 \times 15 \times 20 = 780 \text{ kips*}$$

The ultimate concentric load is

$$P_0 = 0.85 \times 4(15 \times 20 - 10) + 60 \times 10 = 1586 \text{ kips}$$

Substituting into Eq. (14) gives

$$\frac{1}{P_u} = \frac{1}{444} + \frac{1}{780} - \frac{1}{1586} = 0.00290 \qquad P_u = 345 \text{ kips}$$
$$P = 0.7 \times 345 = 241 \text{ kips}$$

20. Stairs Concrete stairs are usually constructed after the floors are placed, using dowels to tie them in place. They may be framed in a variety of ways (Fig. 36). They are

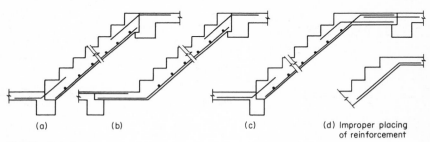

(a)   (b)   (c)   (d) Improper placing
of reinforcement

Fig. 36 Stair framing.

usually designed as simple spans, since they are a small part of the structure and the care and expense involved in locating negative-moment bars are not justified. The design is simple. The span is taken as the horizontal distance L between supports and the live load in psf of horizontal projection (Fig. 37). The dead load is calculated as for a slab of length L and thickness z plus one-half the height of riser, plus weight of any finish. Slab dimensions t and d are usually measured at the heel of the step.

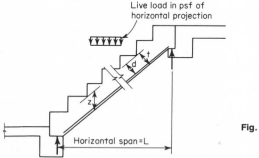

Live load in psf of horizontal projection

Horizontal span $=L$

Fig. 37

Typical reinforcement is shown in Fig. 36. Reinforcement at the upper landing should always be spliced and lapped as shown; reinforcing bars in tension should never be bent around a reentrant corner as in d.

21. Wall Footings The width of wall footings is determined by the area required to distribute the load. The allowable soil pressure is usually stated for unfactored service

---

*Comparisons of Eq. (14) with analyses and results of the tests in Ref. 7 were based on P_0 as computed here. According to Eq. (13), P_0 would be 80 percent of this value, which would give a less conservative result ($P = 255$ kips).

loads. Depth is determined by structural requirements for factored loads. Cross reinforcement is required if the flexural stress exceeds the allowable tension for plain concrete. Longitudinal reinforcement is desirable for spacing cross reinforcement.

The bending moment is taken at the face of a concrete wall, column, or pedestal and halfway between the middle and the face for masonry walls.

Example 18 Design a footing for a 12-in. concrete wall carrying a live load of 10,000 plf and a dead load of 5000 plf. Allowable soil pressure = 5000 psf, f'_c = 3000 psi, f_y = 60,000 psi. Neglect weight of footing.

SOLUTION. (Fig. 38). Width of footing = 15,000/5000 = 3 ft. Try a 12-in. depth with d = 8½ in. to allow 3 in. of concrete cover.

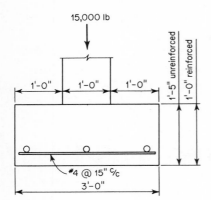

Fig. 38 Example 18.

Factored load w = 1.4 × 5000 + 1.7 × 10,000 = 24,000 plf = 8000 psf.

$$M_u = (8000 \times 1)0.5 = 4000 \text{ lb-ft} = 48,000 \text{ lb-in.}$$

Assume internal lever arm = $0.9d$ = 7.65 in.

$$A_s = M_u/(7.65 \, \phi f_y) = 48,000/(0.9 \times 60,000 \times 7.65) = 0.116 \text{ in.}^2/\text{ft}$$

Even though slabs of uniform thickness are excluded in Code 10.5.1, check the minimum reinforcement ratio:

$$\rho_{min} = 200/f_y = 200/60,000 = 0.0033 \text{ min} \qquad A_s = \rho bd = 0.0033 \times 12 \times 8.5 = 0.34 \text{ in.}^2$$

This minimum is not required where A_s is over one-third more than required for the given moment. Therefore, use 1.33 × 0.116 = 0.154 in.²/ft.

Use No. 4 at 15 in., A_s = 0.20 × 12/15 = 0.16 in.²/ft. Required development l_d = $0.04A_b \, f_y /\sqrt{f'_c}$ = 0.04 × 0.20 × 60,000/$\sqrt{3000}$ = 8.8 in.

From face of wall the bar projects 12 in. minus end cover > 8.8 in., O.K. Shear at d from face of concrete wall (Code 11.2.2) = 8000(12 − 8.5)/12 = 2330 lb. $v = V_u/(\phi bd) = 2330/(0.85 \times 12 \times 8.5) = 27$ psi < $2\sqrt{3000}$ = 109 psi. Longitudinal spacer bars may be 2 No. 3 or 2 No. 4. The latter are easier to place and are of the same diameter as moment bars, which may be convenient.

22. Column Footings Individual footings for columns may be square, rectangular, polygonal, round, or irregular to suit the space available (Fig. 39). Square footings are the simplest. Theoretically, footings could taper in depth to a working minimum at the periphery, but as they are frequently constructed by placing a prepared mat of reinforcing bars in a neatly excavated pit and filling to the required level with concrete, such tapering involves more labor than the value of the concrete saved.

Combined footings (Fig. 39d, e) may be rectangular or trapezoidal in plan. The trapezoidal footing is used when it is impossible to extend the footing sufficiently beyond the more heavily loaded column. The centroid of the bearing pressure should coincide as nearly as is practicable with that of the loads. In the absence of specific recommendations to the contrary by a soils engineer, the soil pressure may be assumed uniform over the contact surface. Cantilevered footings (Fig. 39g) may be used in place of the combined footing.

The main requirement for footings is that they be safe, simple, and economical. They should be proportioned to produce substantially equal settlements of the supported columns.

Example 19 Design a square footing to support a column load of 200 kips dead + 260 kips live load on 5000-psf soil. $f_c' = 3000$ psi, $f_y = 60,000$ psi, column 18 in. square.

SOLUTION. Column load + estimated weight of footing = 460 + 40 = 500 kips. Required size of footing = 500/5 = 100 sq ft = 10 ft square.

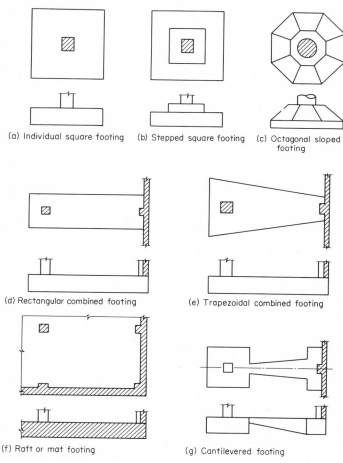

(a) Individual square footing   (b) Stepped square footing   (c) Octagonal sloped footing

(d) Rectangular combined footing   (e) Trapezoidal combined footing

(f) Raft or mat footing   (g) Cantilevered footing

Fig. 39 Typical column footings.

Try a footing 1 ft 10 in. deep with 3 in. clear cover for placement on earth. With bars in two directions, d for the upper layer will be about 17.5 in.

Effective soil pressure $w = (200 \times 1.4 + 260 \times 1.7)/100 = 7.22$ ksf. Except on long narrow footings, punching shear around the column usually determines necessary depth. This is calculated around a perimeter located $d/2$ from the column face (Code 11.10b).

$$\text{Perimeter} = b_0 = 4(18 + 2 \times 17.5/2) = 4 \times 35.5 = 142 \text{ in. or } 2.96 \text{ ft per side}$$

$$V_u = 7.22(10^2 - 2.96^2) = 659 \text{ kips}$$
$$v_u = V_u/(\phi b_0 d) = 659{,}000/(0.85 \times 142 \times 17.5) = 312 \text{ psi}$$

Allowable $v_u = 4\sqrt{3000} = 219$ psi = 0.7 of that needed.
Noting that b_0 also increases as d is increased, try $d = 22.5$ in. = 1.88 ft, thickness = 27 in.

$b_0 = 4(18 + 22.5) = 162$ in. or 3.37 ft per side
$V_u = 7.22(10^2 - 3.37^2) = 640$ kips $v_u = 640,000/(0.85 \times 162 \times 22.5) = 207$ psi < 219 O.K.

Beam-type shear per foot of width of footing on plane $d = 22.5$ in. from column:

$$V_u = 7.22 \times 1(5 - 0.75 - 1.87) = 17.2 \text{ kips}$$

$$v_u = 17,200/(0.85 \times 12 \times 22.5) = 75 \text{ psi} < 2\sqrt{f_c'} = 109 \text{ psi}$$

Check M_u at column face $= 7.22(5 - 0.75)^2/2 = 65.2$ kip ft/ft

$$M_u/(\phi bd^2) = 65,200 \times 12/(0.9 \times 12 \times 22.5^2) = 143$$

From Table 5, $\rho = 0.0025$, min $\rho = 200/f_y = 200/60,000 = 0.0033 = (4/3)0.0025.$*
$$A_s = 0.0033 \times 12 \times 22.5 = 0.90 \text{ in.}^2$$
Use 15 No. 7 at 8 in. each way ($A_s = 0.90$ in.^{2}).

$$l_d = 0.04A_b f_y/\sqrt{f_c'} = 0.04 \times 0.60 \times 60,000/\sqrt{3000} = 26.3 \text{ in.}$$

Since footing extends over 4 ft from column face, development is no problem.

The assumed weight of 40,000 lb can now be checked as $10 \times 10 \times 2.25 \times 150 = 33,800$ lb. If a higher degree of precision seems desirable, a recomputation can be made. No substantial savings over the present design appear likely.

Example 20   Design a combined footing for a 12-in. exterior wall whose outside face is on a property line, 20 ft from the center of a 20×20-in. interior column (Fig. 40). The load on the wall is 200 kips and that on the column 325 kips. Allowable soil pressure is 5000 psf, $f_c' = 4000$ psi, and $f_y = 60,000$ psi. Assume average load factor as 1.65 on both wall and column.

SOLUTION. At service loads, the centroid of the loads is $0.5 + 325 \times 19.5/525 = 12.57$ ft from the building line, making the length of a concentric, rectangular footing $2 \times 12.57 = 25.14$ ft, or, say, 25 ft.

Load on exterior wall   = 200,000
Load on interior column   = 325,000
Footing (at 8 to 10 percent of load) = 50,000
                                     575,000 lb

$b = 575,000/(5000 \times 25) = 4.6$ ft $= 4$ ft 8 in. Net soil pressure $= 525,000/(4.67 \times 25) = 4500$ psf at service loads or $4500 \times 1.65 = 7420$ psf under factored loads.

The distance to the point of zero shear from exterior face is $200,000 \times 1.65/(4.67 \times 7420) = 9.52$ ft. Max negative $M_u = -330(9.52 - 0.5) + 34.7 \times 9.52^2/2 = -1410$ kip ft

Clearances or other constraints sometimes demand minimum member size, but economy of the member itself more often goes with ρ smaller than the maximum. Try $\rho = 0.015$, for which $M_u/\phi bd^2 = 781$. Then

$$781 = 1410 \times 1000 \times 12/(0.90 \times 4.67 \times 12d^2)$$
$$d = 20.7 \text{ in}$$

Punching shear seems improbable with this d plus the 20-in. column using up 41 in. of the total 56-in. width. Hence, check flexural shear first. Shear usually controls footing depth, since heavy stirrups in wide members are awkward to place. Flexural shear, by inspection, is maximum near the inside face of the wall or the inside face of the column. Net soil pressure on 56-in. width is $4.67 \times 7.42 = 34.7$ kips/ft

$$V_{wall} = 330 - 34.7 \times 1 = 295 \text{ kips}$$

Summing forces from left and including the column,

$$V_{col} = 34.7(5.0 + 10/12) - 1.65 \times 325 = 202 - 536 = -334 \text{ kips}$$

Critical flexural shear is at d from column or wall.

$$\text{Crit. } V_{col} = -334 + 34.7 \times 19/12 = -334 + 55 = -279 \text{ kips}$$

$$v_u = V_u/(0.85 \ bd) = 279,000/(0.85 \times 56 \times 21) = 278 \text{ psi} > 2\sqrt{4000} = 126 \text{ psi}$$

The d of 21 in. could be used with this shear, but the authors prefer not over half the shear to be carried by stirrups in footings, that is, $v_u \lesssim 4\sqrt{f_c'} = 252$ psi. Try $d = 24$ in.

$$\text{Crit } V_{col} = -334 + (34.7 \times 24/12) = -265 \text{ kips}$$
$$v_u = 265,000/(0.85 \times 56 \times 24) = 232 \text{ psi. Stirrups required. (Fig. 40}d\text{.)}$$

Crit. V_{col} on extension beyond column $= 34.7(5 - 0.83 - 2) = 75.3$ kips $< \frac{1}{3} \times 265$ kips (231 psi) on other side of column. No stirrups are needed here.

---

*Whether footings are excluded (as slabs of uniform thickness) from this minimum ρ might be debated legally, but high shear and low ρ are always a bad combination.

Crit. $V_{\text{wall}} = 295 - 34.7 \times 2 = 226$ kips

$v_u = 226{,}000/(0.85 \times 56 \times 24) = 198$ psi $> 2\sqrt{f_c'}$. Stirrups required.

Punching shear, on the boundary of a square $d/2$ beyond column, i.e., a square of $20 + 24 = 44$ in. (3.67 ft) each side gives $b_0 = 4 \times 44 = 176$ in.

Crit. $V_u = 325 \times 1.65 - 7.42 \times 3.67^2 = 536 - 100 = 436$ kips

$v_u = 436{,}000/(0.85 \times 176 \times 24) = 121$ psi $< 4\sqrt{f_c'} = 4\sqrt{4000} = 252$ psi   O.K.

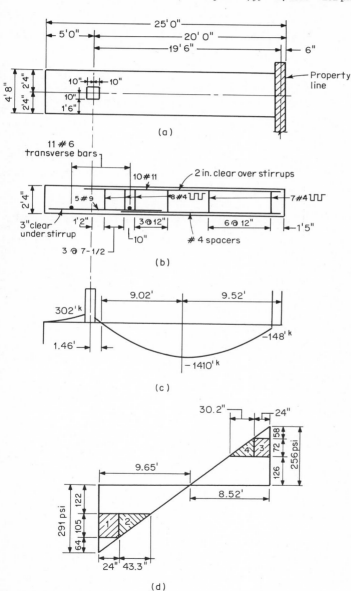

Fig. 40 Example 20.

Stirrups will be designed after flexural reinforcement is established.
Use thickness = 28 in., d = 24 in.
Footing weight = $25 \times 4.67 \times 2.33 \times 150$ = 40,800 lb vs. 50,000 lb assumed. O.K.
TOP BARS, LONGITUDINAL.

$$\frac{M_u}{\phi b d^2} = \frac{1,410,000 \times 12}{0.9 \times 56 \times 24^2} = 583$$

Table 5 gives ρ = 0.0108. A_s = $0.0108 \times 56 \times 24$ = 14.52 in.^{2} Use 10 No. 11 = 15.6 in.^{2}

Required l_d with 1.4 factor for top bars = $1.4 \times 0.04 \times 156 \times 60,000/\sqrt{4000}$ = 82.9 in. = 6.9 ft.
Continue all bars to end under wall.
BOTTOM BARS, LONGITUDINAL. At exterior face of column, M_u = $34.7(5 - 0.83)^2/2$ = 302 kip-ft

$$\frac{M_u}{\phi b d^2} = \frac{302,000 \times 12}{0.9 \times 56 \times 24^2} = 125$$

Table 5 gives ρ = 0.0027. Minimum ρ = 200/60,000 = 0.0033

$$A_s = 0.0033 \times 56 \times 24 = 4.44 \text{ in.}^2 \qquad \text{Use 5 No. 9} = 5.00 \text{ in.}^2$$

Required l_d = $0.04 \times 1.00 \times 60,000/\sqrt{4000}$ = 37.9 in. = 3.1 ft. Continue all No. 9 to left end of footing.
    Draw moment diagram to detail bars to right of column (Fig. 40c). At right of column, M_u = 34.7(5 + 0.83)^2/2 − 1.65 × 325 × 0.83 = +145 kip-ft. M_u at left face of wall = −0.5(330 − 34.7) = 148 kip-ft.
    For the top bars, the development requirements at the point of inflection (Fig. 40c) are the same as for positive-moment bars except for the 1.4 factor for top bars. Therefore, l_d = $1.3M_t/V_u + l_d$, where 1.3 represents the 30 percent increase for reinforcement confined by a compressive reaction (Code 12.2.3).
    $M_t \approx M_u$ = 1410 kip-ft
    V_u = 34.7 × 9.02 = 313 kips (since V = 0 at point of maximum moment)
    Allowable l_d = 1.3 × 1410 × 12/313 + l_a = 70.3 + l_a

Required l_d for No. 11 bars = $1.4(0.04 \times 1.56 \times 60,000/\sqrt{4000})$ = 83 in. Development is O.K. if bars are extended l_a = 13 in. beyond the point of inflection.
    Code 12.3.3 requires at least one-third of the negative-moment reinforcement to extend at least d beyond the point of inflection, which is 24 in. (since $12d_b$ and $\frac{1}{16}$ of clear span are each less). However, the compression from the column modifies the conditions on which this requirement is based, and a designer who understands these various complications can be less conservative, running five bars to the center of the column and the other five to the farther face of the column.
    BOTTOM BARS, TRANSVERSE. Use bottom transverse bars centered under the column over a length equal to width of footing = 4.67 ft. The footing projects beyond the face of the columns 0.5(4.67 − 1.67) = 1.50 ft

$$M_u = 4.67 \times 7420 \times 1.50^2/2 = 39,000 \text{ lb-ft total}$$

A_s = $M_u/(\phi f_y \times 0.9d)$ = 39,000 × 12/(0.9 × 60,000 × 0.9 × 23) = 0.42 in.^{2}
    Min. ρ = 200/60,000 = 0.0033 A_s = 0.0033 × 4.67 × 12 × 23) = 4.29 in.^{2} Use 10 No. 6 = 4.42 in.^{2}

l_d = $0.04 \times 0.44 \times 60,000/\sqrt{4000}$ = 16.7 in. vs. 16 in. available.
Try 11 bars, l_d = (4.29/4.86) × 16.7 = 14.7 in.   O.K.   Use 11 No. 6 bottom transverse bars.
    STIRRUPS. The unit-shear diagram is shown in Fig. 40d. The shaded areas require stirrups. For the d = 24-in. length adjacent to the face of the column,

$$A_v = \frac{105 \times 56 \times 24}{60,000} = 2.35 \text{ in.}^2$$

Use 3 No. 4 double-U stirrups at 7.5 in., which gives 4 × 0.2 × 24/7.5 = 2.55 in.
    At x = 36 in. from column face,
    v_s = 105 − 12 × 2.52 = 75 psi
    s = 0.8 × 60,000/(75 × 56) = 11.4 in.   Use 11 in.
    At x = 48 in., v_s = 75 − 11 × 2.52 = 47 psi
    Maximum s = $d/2$ = 12 in. controls
    For area 3, with No. 4 double-U stirrups, s = 0.8 × 60,000/(72 × 56) = 11.9 in. Therefore, use maximum allowable (12 in.) for areas 3 and 4.
    STIRRUP DEVELOPMENT. Code 12.13.1. A straight leg must have a basic development length l_d above or below middepth of member, with a minimum of 24 stirrup diameters (here 12 in.). Therefore, a hook is required at the top, its 2 in. inside diameter (Code 7.1.3.1) using a total of 1.5 in. plus cover of 2 in. from start of hook, which leaves 24/2 − 2.5 = 8.5 in. for the 0.5 l_d required in addition to the hook.

0.5 l_d = $0.5(0.04 \times 0.20 \times 60,000/\sqrt{4000})$ = 3.79 in. <8.5 in. available

Use No. 4 longitudinal spacers in each corner of stirrups where main longitudinal bars are not available.

Example 21 Design a cantilever footing for a 12-in. exterior wall on the property line, which is 20 ft from the center of a 20-in.-square column (Fig. 41). Wall load = 160 kips, column load = 300 kips, average load factor on each = 1.65, allowable soil pressure = 6000 psf, f_c = 3750 psi, f_y = 60,000 psi.

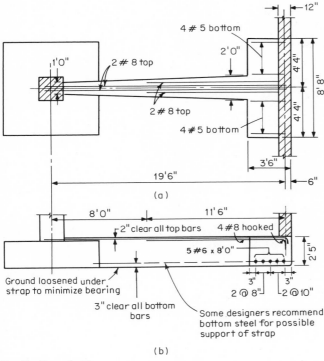

Fig. 41 Example 21.

SOLUTION. Assume the footing under the exterior wall to be 3.5 ft wide. Distance from center of this footing to center of wall is then 1.25 ft and to the center of the interior column is 20 − 1.75 = 18.25 ft. Neglecting weight of the strap beam, the uplift on the interior column (a downward load on the wall footing) is 160,000 × 1.25/18.25 = 10,960 lb.

Wall load	= 160,000 lb
Downward reaction	= 10,960 lb
Footing (assumed)	= 11,000 lb
Total	= 182,000 lb at service loads

Required area of wall footing = 182,000/6000 = 30.3 ft².

Use footing 3 ft 6 in. by 8 ft 8 in. = 30.3 ft²

Net soil pressure = (182,000 − 11,000)/30.3 = 5640 psf at service loads or 5640 × 1.65 = 9310 psf with factored loads.

Assume a strap 2 ft wide where it joins the wall footing, leaving the footing projecting 3.33 ft on each side. If the footing supported a column instead of a wall, it would act as balanced cantilevers beyond each side face of the column and strap beam. The wall, however, delivers its load along the length of the footing but is not interconnected to provide a moment reaction for footing strips perpendicular to it. On the outside strip parallel to the wall it is on the safe side to assume all the net soil pressure carried to the strap beam, creating the worst design shear situation there. This ignores the fact that some reaction is carried directly to the wall because of diagonal strips between wall and strap. It also ignores the torsion developed because footing deflection along the wall is near zero and largest in the corner. If the footing projected much farther from the wall, a more exact analysis of combined shear and torsion would be indicated. Here it is assumed the conservative shear calculation offsets the neglected torsion.

On a 1 ft-strip at the edge* of strap

$$V_u = 9310(4.33 - 1.0) = 31{,}000 \text{ lb}$$

Required $d_v = V_u/\phi v_c b = 31{,}000/(0.85 \times 2\sqrt{3750} \times 12) = 24.8$ in. Use $h = 29$ in., $d = 25.6$ in.

The footing may be sloped or stepped down from the strap toward the outer edges; however, a uniform depth will probably be more economical.

$M_u = 9310 \times 3.33^2/2 = 51{,}600$ lb-ft/ft

$A_s = M_u/\phi f_y 0.9 d = 51{,}600 \times 12/(0.90 \times 60{,}000 \times 0.9 \times 25.6) = 0.50$ in.^{2}/ft

Minimum $\rho = 200/f_y = 200/60{,}000 = 0.0033$†

Minimum $A_s = 0.0033 \times 25.6 \times 12 = 1.01$ in.$^2 > (4/3)0.50 = 0.67$ in.^{2}/ft

Use 5 No. 6 × 8 ft 0 in. spaced from inner edge as shown on Fig. 41a.

$l_d = 0.04 \times 0.44 \times 60{,}000/3750 = 17.2$ in. < available 37 in.

Nominal reinforcing is also desirable perpendicular to wall.

Use on each side of strap beam 4 No. 5 at 10 in., total of 8 bars.

Weight of footing = 3.5 × 8.67 × 2.42 × 150 = 11,000 lb, as assumed.

DESIGN OF STRAP BEAM. Neglecting weight of strap and the limited soil reaction between footings, the point of zero shear is 160,000 × 1.65/(8.67 × 9300) = 3.28 ft from the building line.

$M_u = -(160{,}000 \times 1.65)(3.28 - 0.50) + 8.67 \times 9300 \times 3.28^2/2 = -300{,}200$ lb-ft

$A_s = 300{,}100 \times 12/(0.90 \times 60{,}000 \times 0.9 \times 26.5) = 2.80$ in.^{2}, say 4 No. 8 = 3.16 in.^{2}

Development of bars is calculated as explained in Example 20. With 4 No. 8 bars, $M_t = 3.16 \times 60{,}000 \times 0.9 \times 26.5/12 = 377{,}000$ lb-ft. At face of wall $V_u = 160{,}000 \times 1.65 - 8.67 \times 9310 \times 1 = 183{,}000$ lb. The bars extend 4 in. beyond the centerline of the wall; so $l_a = 0.33$ ft. Max usable $l_d = 1.3 \times 377{,}000/183{,}000 + 0.33 = 3.01$ ft = 36 in. Required l_d for No. 8 top bars = 1.4(0.04 × 0.79 × 60,000/$\sqrt{3750}$) = 43.3 in. > 36 in. The deficiency in development length available can be made up by adding hooks at the end. An alternate would be to use 5 No. 7 = 3.00 in.^{2} without hooks.

Critical V_u = load transferred from column to wall footing = 10,960 × 1.65 = 18,100 lb.

$$v_u = V_u/\phi bd = 18{,}100/(0.85 \times 24 \times 26.5) = 33.5 \text{ psi} < 2\sqrt{f_c'} = 122 \text{ psi}$$

Strap can be narrowed to 12 in., say at face of column.

Bars may be terminated (Code 12.1.4) when moments permit, provided one of the following conditions is satisfied: (1) shear at the point is not more than half the allowable, (2) extra stirrups are provided, or (3) the continuing bars provide double the area required. The first condition is satisfied over most of the strap beam. Half of the bars can theoretically be cut off where $M_u = 0.5\ M_t$ already calculated = 0.5 × 377,000 = 188,000 lb-ft = $V_u x = 18{,}100x$, where x is the distance from center of column. $x = 10.22$ ft. Code 12.14 also requires the bars to be carried an extra length, the larger of d or $12d_b = 26.5$ in. = 2.20 ft.

Stop 2 No. 8 at 8 ft from center of column.

COLUMN FOOTING. The design of the column footing can be the same as for an isolated footing, usually using the column load reduced by the shear transferred to it by the strap beam. Some modifications could be made in the zone under the beam, but this is usually not done.

If settlement is likely to be such as to compress the soil under the strap, the designer should consider a combined footing.

23. Walls Concrete walls may be plain or reinforced, bearing or nonbearing, cast in forms or in neatly excavated trenches without forms. Design is semiempirical, experience being an even larger factor than tests. The most critical points in a wall are openings or abrupt changes in cross section or outline. Wherever openings occur, there is danger of diagonal tension cracks at reentrant corners (Fig. 42a). Reinforcement, equivalent to two or three No. 5 bars in walls of usual thickness, should be supplied as in Fig. 42b and c.

The capacity of bearing walls is increased considerably by the stiffening effects of intersecting walls, offsets, pilasters, floors, and the like. Unsupported walls, such as the high back wall of a theater stage, require special study. Where loads on reinforced bearing walls fall within the middle third of the thickness, the walls may be proportioned for the allowable compression given by Code 14.2:

$$P_u = 0.55\phi f_c' A_g[1 - (l_c/40h)^2] \tag{15}$$

where $\phi = 0.7$ and l_c is the vertical distance between supports and h is the overall thickness, which must be at least $\frac{1}{25}$ of the unsupported length or width, whichever is

---

*Not at d from the face because the lack of an external compressive reaction at the top of the strap makes the shear more critical.

†See footnote on page 11-44.

shorter. Walls in which the load falls outside the middle third must be designed for combined direct compression and flexure.

If the reinforcement is designed, placed, and anchored as for tied columns, the allowable loads may be computed as for tied columns.

The design of walls under tension or tension and flexure is discussed in Sec. 22, Arts. 13 and 14.

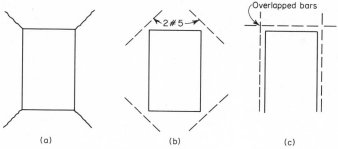

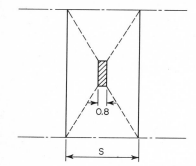

Fig. 42   Reinforcement of wall openings.

24. Slabs   One-way solid slabs supported on masonry walls, or on steel or concrete beams, are designed as rectangular beams, isolating a 12-in. width of strip for study.

One-way solid slabs may be required to support concentrated loads rather than uniformly distributed ones. The simplest example is a wheel load. Analysis and tests indicate that if a single wheel is placed on a slab of single span, the width of strip which will support it is about as shown in Fig. 43. The entire slab assumes a saucer shape. Those slab

Fig. 43   Concentrated load on slab.

elements directly under the load are most highly stressed, but for a considerable width on either side, elements will participate. A satisfactory scheme is to assume that the radiating lines flare at an included angle of $\tan^{-1} 0.6\pm$. For more precise results, careful analyses are required.

Example 22   A simply supported concrete floor slab spans 10 ft over a factory tunnel. If the slab is designed for a live load of 200 psf, can it carry a truck wheel concentration of 6000 lb?

SOLUTION.   The live-load moment on a strip 1 ft wide = $1.7 \times 200 \times 10 \times 11 \times 12/8 = 56{,}100$ lb-in. The corresponding moment of a wheel placed in midspan and assumed distributed over a strip $0.8 + 2 \times 0.6 \times 5 = 6.8$ ft wide is $M = 1.7 \times 6000 \times 11 \times 12/4 \div 6.8 = 49{,}500$ lb-in.; so unless impact and vibration exist, a wheel concentration of $6000 \times 56{,}100/49{,}500 = 6800$ lb would be about equivalent.

Ordinarily, such slabs are cast upon temporary formwork or centering of lumber, plywood, or panels. Sometimes formed metal sheets are used for such supports, the necessary reinforcing bars and concrete placed, and the sheets left permanently in place. In other instances, the formed sheets are bonded to the concrete with welded cross wires or other devices so as to replace the bars and become the slab reinforcement.

Another variation is to key or anchor the supporting beams, whether of steel or precast concrete, to the slab with shear developers (composite construction). Precast-concrete slabs or planks, with or without voids, are available.

Holes in floors are common because of stairs, elevators, ducts, pipes, and the like. Reentrant corners are points of weakness and require supplementary reinforcement in the way of corner bars or overlapped side and end bars. One rough rule is to make the tensile resistance of the added bars at least equal to that of the concrete removed by the opening. Obviously, the stress bars must not be cut off, or the capacity is reduced; rather the full amount of required steel must be grouped, some at either side of the opening. In addition, bars at right angles are needed to divert the stresses around the openings.

Two-Way Solid Slabs. As distinguished from two-way flat slabs or flat plates, two-way solid slabs span in two directions (usually at right angles to each other) onto supporting beams or girders, possibly of steel or precast concrete, but frequently of concrete cast monolithically with the slab. The nearer the panels are to being square, the more effective they become. The moment and shear in Appendix A of the 1963 ACI Code are recommended as giving adequate information for the design of two-way solid slabs. The comments in the previous paragraphs regarding holes and openings apply equally well here.

Two-Way Flat Slabs. The two-way flat slab (Fig. 44) is supported on columns, usually

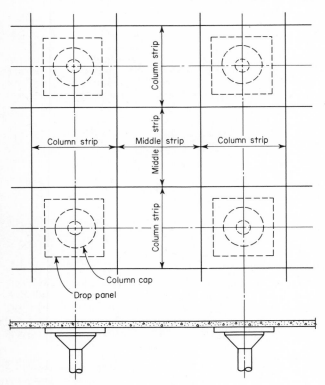

Fig. 44 Two-way flat slab.

arranged in square or rectangular patterns. Flaring heads on the columns and plinthlike dropped panels are an integral part of the system. Methods for design are set out in detail in ACI 318. Numerous variations, such as columns in triangular pattern, four-way reinforcement of approximately square panels, and structural steel or heavy bar shear heads in lieu of column capitals, have been successfully built.

Two-Way Flat Plates.   Two-way flat plates are quite similar to flat slabs except that the drop panels and column capitals are omitted. Columns may be increased in size to keep the shear within allowable limits. Flat plates are much used in high-rise apartment or hotel construction, especially when the underside of the plate can also serve as the ceiling of the room below. Columns may be irregularly spaced to suit the location of corners of closets or bathrooms. Columns are usually uniform in size from bottom to top, with the amount of reinforcement varied from maximum to minimum. The design is an adaptation of ACI 318.

One-Way Concrete Ribbed Slabs.   Since the concrete in a solid slab below and near the neutral axis is greatly in excess of what is needed to resist shear, much of it can be saved, and dead weight eliminated, by forming voids with permanent steel forms or removable forms of metal, wood, plastic, or other material (Fig. 45). The joists are

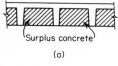

Surplus concrete

(a)

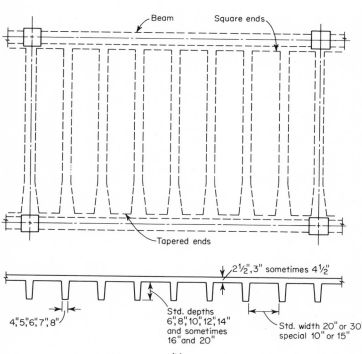

Beam   Square ends

Tapered ends

2½",3" sometimes 4½"

4",5",6",7",8"

Std. depths
6",8",10",12",14"
and sometimes
16"and 20"

Std. width 20" or 30"
special 10" or 15"

(b)

Fig. 45   One-way ribbed slab.

designed as regular tee beams. It is important that form depths, joist widths, tapered ends, etc., all conform to the standard forms that it is proposed to use (see U.S. Department of Commerce Simplified Practice Recommendation R87). Maintaining the same sizes throughout a project makes for maximum economy because of the many reuses of a minimum amount of form material.

Two-Way Joist Construction.   Much of the unnecessary concrete in flat-slab construction can be eliminated by using waffle slabs or two-way joist construction (Fig. 46). ACI 318 recommends designing for the shears and moments used in flat-slab design.

25. Structural Framing Systems One of the first decisions to be made by the designer is the selection of a suitable framing system for any structure. The determining factors are many. Compliance with all laws, codes, and ordinances (or a special permit to depart for

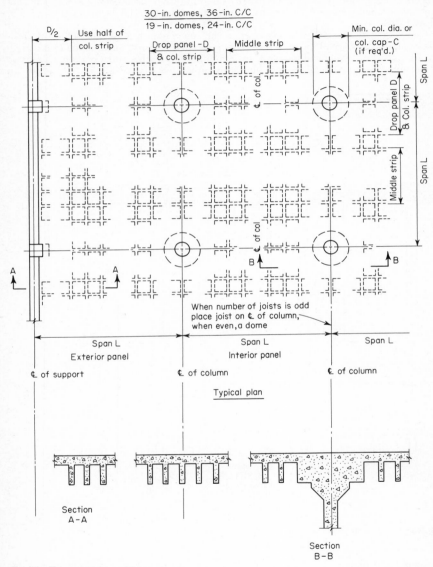

Fig. 46 Two-way joist system.

good reason) is the first essential. Economy, not merely in the structural frame itself but, more importantly, overall economy in the finished structure, including mechanical equipment, is the second most important factor. Following well-understood local practices and customs results in better bids. After these factors are duly considered, studies are usually made of various possibilities, but the following observations are pertinent. Two-way flat slabs are appropriate for heavy loads on moderate spans. Loads as low as 100 or 150 psf can be accommodated, but as the loads rise to 300 or 400 psf or higher, the more suitable the flat slab. While column spacings can be as low as 15 or 18 ft, those of 20, 25, and even 30 ft are common and economical. Waffle slabs are appropriate in lieu of flat slabs when the loads are modest, say 100 to 200, or even 300 psf, or the spans considerable, say 25 to 40 or more feet, because dead weight becomes an increasingly important consideration. Both flat slabs and waffle slabs are particularly suitable for industrial plants, garages, and shopping centers.

Flat plates are most natural for relatively light loads, say 20 or 30 up to 80 or 100 psf on spans of 15 or 16 up to 20 or 25 ft. By elimination of drops and caps, they provide much more acceptable framing in apartments and hotels.

Solid slabs have worked particularly well for relatively low, several-storied dormitories spanning 15 to 18 ft from exterior walls to corridor bearing wall and across the corridor. Required wall thicknesses have limited such structures to five or six stories. The underside of the slab can be painted to become the ceiling of the room below, but all ducts, pipes, etc., must be run in vertical shafts or in suspended ceilings over the corridors.

Whatever system is chosen, rough framing sketches with approximate sizes should be checked repeatedly with all interested parties, for clearances, story heights, interferences, economy, simplicity, and likelihood of obtaining favorable bids before a final choice is made, because resulting computational work makes later changes costly.

REFERENCES

1. "Design Handbook," vol. II, Concrete Reinforcing Steel Institute, Chicago, 1965.
2. "CRSI Handbook," Concrete Reinforcing Steel Institute, Chicago, 1975.
3. Everard, Noel J., and Edwin Cohen: "Ultimate Strength Design of Reinforced Concrete Columns," American Concrete Institute Publication SP-7, Detroit, 1964.
4. "Manual of Standard Practice for Detailing Reinforced Concrete Structures," ACI 315-74, American Concrete Institute, Detroit, 1974.
5. Ferguson, Phil M.: "Reinforced Concrete Fundamentals," 3d ed., John Wiley & Sons, Inc., New York, 1973.
6. "Building Code Requirements for Reinforced Concrete," ACI 318-77, American Concrete Institute, Detroit, 1975.
7. Bresler, B.: Design Criteria for Reinforced Concrete Columns under Axial Load and Biaxial Bending, *J. ACI,* November 1960.
8. Ramamurthy, L. N.: "Investigation of Ultimate Strength of Square and Rectangular Columns under Biaxial Eccentric Loads, Symposium on Reinforced Concrete Columns," ACI Publication SP-13, 1966.
9. Parme, A. L., J. M. Nieves, and A. Gouwens: Capacity of Rectangular Columns Subject to Biaxial Bending, *J. ACI,* September 1966
10. "Design Handbook," ACI Publication SP-17, 1973.

Section **12**

Design of Prestressed-Concrete Structural Members*

T. Y. LIN

Professor Emeritus of Civil Engineering, University of California, Berkeley

PAUL ZIA

Professor of Civil Engineering, North Carolina State University

NOTATION

a	= lever arm between centers of C and T
A	= sectional area
A_c	= net cross-sectional area of concrete
A_{ps}	= area of steel
A_t	= transformed area
C	= total compressive force in concrete
C'	= ultimate compressive force in concrete
c_b	= distance from c.g.c. to bottom fiber
C_c	= coefficient of creep = δ_t/δ_i
c.g.c.	= centroid of concrete
c.g.s.	= centroid of steel
c_t	= distance from c.g.c. to top fiber
e	= eccentricity of c.g.s.
E_c	= modulus of elasticity of concrete
E_s	= modulus of elasticity of steel
F	= effective prestress (after losses)
F_1, F_2	= total prestress at points 1 and 2, respectively
F_i	= total initial prestress before transfer
F_0	= total prestress just after transfer
f	= unit stress in general
f_r	= modulus of rupture of concrete
f_c	= unit stress in concrete
f'_c	= compressive strength of concrete, 28 days

*Much of the material for this section was taken from T. Y. Lin, "Design of Prestressed Concrete," 2d ed., John Wiley & Sons, Inc., New York, 1963.

f'_{ci} = compressive strength of concrete at transfer
$f'_{c\infty}$ = strength of concrete at time infinity
f_{se} = effective prestress after deducting all losses
f_i = initial prestress before transfer
f_0 = prestress just after transfer
f_s = steel stress in general
f_{pu} = ultimate unit stress in steel
f_{ps} = stress in steel at ultimate load on section
f_{py} = yield strength of prestressing steel
f'_t, f_b = tensile stress at top (bottom) fiber
f_y = yield strength of nonprestressed reinforcement
h = sag of cable, overall depth
I_t = moment of inertia of transformed section
K = wobble coefficient, per ft
k = coefficient for depth of compression
k' = k at ultimate load
k_b, k_t = kern distances from c.g.c. to bottom (top)
M_u = ultimate resisting moment
m = load factor or factor of safety
n = E_s/E_c
ρ_p = A_{ps}/bd
r = radius of gyration of cross section
S_t = principal tensile stress
T = total tension in prestressing steel
T' = ultimate tension in prestressing steel
V_c = total shear carried by concrete
V_s = total shear carried by steel
δ = unit elastic shortening due to transfer of prestress
δ_i = initial unit strain in concrete due to elastic shortening
δ_s = unit strain in steel due to shrinkage of concrete
δ_t = final unit strain in concrete including effect of creep but not of shrinkage
Δ_a = deformation of anchorage
Δf_s = loss of prestress
μ = coefficient of friction between tendon and surrounding material

MATERIALS

1. Concrete   Higher-strength concrete is usually required for prestressed than for reinforced work. In the United States, 28-day cylinder strength of 4000 to 5000 psi is generally specified and is often the most economical mix for prestressed concrete, while a strength of 6000 psi is commonly used in the plant. A low-slump 5000-psi concrete can be obtained with a water-cement ratio of about 0.45 by weight and a cement content under 7 sacks per cu yd of concrete, provided good internal or external vibration is used.

High early strength is desirable for fast turnover in a pretensioning plant, or fast removal of formwork at the jobsite. With steam or hot-air curing, a transfer strength of 3500 psi is often attained in less than 24 hr. A transfer strength of over 3000 psi can be attained in 3 or 4 days by using high-early-strength cement without special curing.

Shrinkage and creep characteristics affect the behavior and efficiency of prestressed concrete. The total amount of creep strain at the end of 20 years ranges from one to five times the instantaneous elastic deformation under load (averaging about three times), the low values occurring for moist storage and for limestone aggregates. Of the total amount of creep strain, about one-fourth takes place within the first 2 weeks after application of prestress, another one-fourth within 2 to 3 months, another one-fourth within a year, and the last one-fourth in the course of many years. Upon removal of the sustained stress, roughly 80 to 90 percent of the creep will be recovered during the same length of time that it has developed.

Shrinkage is primarily dependent on time and on moisture conditions, but not on stresses. The magnitude of shrinkage strain varies with many factors; it may range from 0.0000 to 0.0010 and beyond. At least a portion of the shrinkage strain resulting from drying is recoverable upon the restoration of lost water. The amount of shrinkage is somewhat proportional to the amount of water employed in the mix. Larger size of aggregates needing a smaller amount of cement paste and harder and denser aggregates produce smaller shrinkage. The chemical composition of the cement also affects the amount of shrinkage. For the purpose of design, shrinkage strain is assumed to be about 0.0002 to 0.0004 for the usual concrete mixtures employed in prestressed construction.

The rate of shrinkage depends chiefly on weather conditions. If the concrete is left dry, most of the shrinkage will take place during the first 2 or 3 months. When stored in air at 50 percent humidity and 70°F, the rate of shrinkage is comparable with that of creep.

Lightweight concrete has been successfully used in prestressed-concrete construction. It has a lower modulus of rupture than normal-weight concrete and slightly less favorable shrinkage and creep characteristics. However, with the better aggregates these properties are comparable with those of normal-weight concrete.

2. Steel   High-tensile steel for prestressing usually takes one of three forms: wires, strands, or bars. Wires for prestressing generally conform to ASTM Specification A421 for Uncoated Stress-Relieved Wire for Prestressed Concrete. They are made from rods produced by the open-hearth or electric-furnace process. After being cold-drawn to size, wires are mechanically straightened and stress-relieved by a continuous low-temperature (about 700°F) heat treatment to produce the prescribed mechanical properties.

The tensile strength and the minimum yield strength (measured by the 1.0 percent total-elongation method) are prescribed in Table 1 for the various sizes of wires. Currently, the ¼-in. wire is the most commonly used.

TABLE 1   Tensile and Yield Strength for Prestressing Wires*

Nominal diam, in.	Remarks	Area, sq in.	Min tensile strength, psi		Min yield strength, psi	
			Type WA	Type BA	Type WA	Type BA
0.192	Gage 6	0.02895	250,000		200,000	
0.196	5 mm	0.03017	250,000	240,000	200,000	192,000
0.250	¼ in.	0.04909	240,000	240,000	192,000	192,000
0.276	7 mm	0.05983	235,000		188,000	

*Type WA for wedge-type anchorage; type BA for button-type anchorage.

A typical stress-strain curve for a stress-relieved ¼-in. wire conforming to ASTM A421 is shown in Fig. 1, with a typical modulus of elasticity between 28,000,000 and 30,000,000 psi. The specified minimum elongation in 10 in. is 4 percent. Typical elongation at

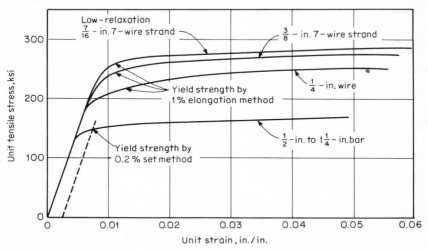

Fig. 1   Typical stress-strain curve for prestressing steels.

rupture is likely to be from 5 to 6 percent.

Strands for prestressing generally conform to ASTM Specification A416 for Uncoated Seven-Wire Stress-Relieved Strand for Prestressed Concrete. While these specifications were intended for pretensioned, bonded construction, they are applicable to posttensioned construction, whether of the bonded or the unbonded type. These strands have a guaranteed minimum ultimate strength of 250,000 psi, with the properties listed in Table 2. A higher-strength steel known as 270K grade, which is more commonly used, has a guaranteed minimum ultimate strength of 270,000 psi (Table 2).

A typical stress-strain curve for a stress-relieved ⅜-in. seven-wire strand (ASTM A416 grade) is shown in Fig. 1, which is also typical for strands of all sizes. For approximate

TABLE 2   Seven-Wire Uncoated Stress-Relieved Strands

Nominal diam, in.	Weight per 1000 ft, lb	Approx area, in.²	Ultimate strength, lb	Yield strength, lb
		250K grade		
¼	122	0.036	9,000	7,650
⁵⁄₁₆	197	0.058	14,500	12,300
⅜	272	0.080	20,000	17,000
⁷⁄₁₆	367	0.108	27,000	23,000
½	490	0.144	36,000	30,600
0.6	737	0.216	54,000	45,900
		270K grade		
⅜	290	0.085	23,000	19,550
⁷⁄₁₆	390	0.115	31,000	26,350
½	520	0.153	41,300	35,100
0.6	740	0.217	58,600	49,800
		270K grade (low relaxation)		
⅜	292	0.085	23,000	20,700
⁷⁄₁₆	400	0.115	31,000	27,900
½	532	0.153	41,300	37,170
0.6	737	0.215	54,000	48,600
		Dyform strand		
⁵⁄₁₆	230	0.069	20,000	17,000
⅜	330	0.099	28,000	23,800
⁷⁄₁₆	450	0.134	38,000	32,300
½	600	0.174	47,000	39,950
0.6	860	0.253	65,000	55,250

calculations, a modulus of elasticity of 27,000,000 psi is often used for ASTM A416 grade and 28,000,000 psi for 270K grade. The specified minimum elongation of the strand is 3.5 percent in a gage length of 24 in. at initial rupture, although typical values are usually in the range of 6 percent. When these strands are galvanized, they are about 15 percent weaker.

Another type of seven-wire strand of 270K grade is the "stabilized" strand, which is produced by a combined process of low-temperature heat treatment and high tension. Because of this special process, the yield strength of the strand is raised and its relaxation is substantially reduced. This type is also called low-relaxation strand.

A "Dyform" seven-wire stress-relieved strand, originated in Great Britain, differs from the regular seven-wire strand in that the outer wires of the Dyform strand are deformed, being run through a die after the stranding operation. It has the advantage of having a greater steel area than that of a regular strand with the same nominal diameter, thus resulting in a larger ultimate tensile strength.

Three-wire stress-relieved strands up to ⅜ in. in diameter and four-wire stress-relieved strand of ⁷⁄₁₆ in. in diameter are also available. These strands have the same diameter, steel area, and ultimate-strength requirements as the corresponding seven-wire strands.

High-strength alloy bars for prestressing generally conform to ASTM Specification A722. These bars are usually proof-stressed (cold-stretched) to at least 80 percent of the guaranteed ultimate strength, which has a minimum value of 150,000 psi. A typical stress-strain curve for these bars is shown in Fig. 1, which shows that a constant modulus of elasticity exists only for a limited range (up to about 80,000 psi) with a value between 25,000,000 and 28,000,000 psi.

The yield strength of high-tensile bars is often defined by the 0.7 percent extension method or the 0.2 percent offset method, as indicated in Fig. 1. Most specifications call for a minimum yield strength of 85 percent of the guaranteed ultimate strength, though actual values are often higher. Minimum elongation at rupture in 20 diameters length is specified at 4 percent, with minimum reduction of area at 20 percent. Common sizes and properties of high-tensile bars for prestressing are listed in Table 6.

3. Grouting For bonding the tendons to the concrete after tensioning (in the case of posttensioning), cement grout is injected, which also serves to protect the steel against corrosion. Entry for the grout into the cableway is provided by means of holes in the anchorage heads and cones, or pipes buried in the concrete members. The grout can be injected at one end of the member until it is forced out of the other end. For longer members, it can be applied at both ends until forced out of a center vent. Either ordinary portland cement or high-early-strength cement may be used for the grout. Coarse sand is preferred for bond and strength, but sufficient fineness is necessary considering the limited space through which the grout has to pass. To ensure good bond for small conduits, grouting under pressure is desirable; however, care should be taken to ensure that the pressure on the walls of the cable enclosure can be safely resisted. Machines for mixing and injecting the grouts are commercially available.

Where larger space between the wires is obtained, such as in a Magnel cable, a 1:1 cement-sand mix is often used with a water-cement ratio of about 0.5 by volume, and a pressure of a few psi may be sufficient for short cables. Where the space is limited, as in a Freyssinet or Prescon cable, neat cement paste with about the same water-cement ratio should be employed. Admixtures are generally used to increase workability, reduce bleeding and shrinkage, or provide expansion. When it is desired to save cement, fine sand of $\frac{1}{64}$-in. grain size can be added. The water:cement:sand proportion should be about 1.0:1.3:0.7 by volume with water-cement ratio of no more than 0.45 by weight. Grouting pressure generally ranges from 80 to 100 psi. After the grout has discharged from the far end, that end is plugged and the pressure is again applied at the injecting end to compact the grout. It is also good practice to wash the cables with water before grouting is started, the excess water being removed with compressed air. When tendons are unbonded, they must be properly greased and wrapped to prevent corrosion.

A minimum grout temperature of 60°F is generally recommended. The temperature of members at time of grouting should be maintained above 35°F until job-cured 2-in. cubes of grout reach a minimum compressive strength of 800 psi. Test cubes should be cured under temperature and moisture conditions as close as possible to those of the grout in the member. During mixing and pumping, the grout temperature should not exceed 90°F. Otherwise, difficulties may be encountered in pumping.

When tendons are unbonded, they must be properly greased and wrapped to prevent corrosion. Plastic shielded wires and strands up to 0.6 in. in diameter, prepacked with corrosion inhibitor, are commercially available for posttensioned work. Such tendons are claimed to have a very low coefficient of friction during tensioning.

METHODS AND SYSTEMS OF PRESTRESSING

4. Tensioning Methods Methods of tensioning tendons can be classified into four groups: (1) mechanical prestressing by means of jacks, (2) electrical prestressing by application of heat, (3) chemical prestressing by means of expansive cements, (4) others.

Mechanical stressing of the tendons is by far the commonest method for both posttensioning and pretensioning. In posttensioning, hydraulic jacks are used to pull the steel against the hardened concrete; in pretensioning, to pull it against bulkheads or molds. The capacity of these jacks varies from about 3 tons up to 200 tons or more. The Clifford-Gilbert system in England employs a small screwjack weighing about 20 lb pulling one wire at a time. The B.B.R.V. and Prescon systems employ jacks of various capacities to fit cables of different sizes. The Leonhardt system in Germany employs reinforced-concrete jacks tensioning hundreds of wires at one time. In all cases, both the jack gage pressure and the tendon elongation are measured to determine the amount of prestress.

Tendons can be lengthened by heating with electricity. Originated in the United States for the posttensioning process, it has not proved to be commercially applicable. A combination of electrical and mechanical stressing, known as the electrothermal method, has been developed in the U.S.S.R. for pretensioning.

Chemical prestressing utilizes expansive cements that expand chemically after setting and during hardening. When these cements are used to make concrete with embedded steel, the steel is elongated by the expansion of the concrete. Thus, the steel is prestressed in tension, which in turn produces compressive prestress in the concrete, resulting in chemically prestressed or self-stressed concrete. Modern development of expansive cement started in France in 1940 (Lossier cement). Its use for self-stressing has been investigated intensively in the U.S.S.R. since 1953. At the University of California, Berkeley, the use of calcium sulfoaluminate admixtures for expansive cements was developed by A. Klien in 1956. Since 1963, a number of structures have been built using his shrinkage-compensating cement, which has an expansion of about 0.05 to 0.10 percent, intended to compensate the expected amount of shrinkage strain.

The Preflex method in Belgium consists of prebending a high-tensile steel beam and encasing its tensile flange in concrete: releasing the bending places the concrete under compression, thus enabling it to take tension.

5. Pretensioning   Pretensioning in the United States is usually accomplished in the plant by the long-line process. Tendons are stretched between two bulkheads held against the ends of a stressing bed several hundred feet long. Concrete is then placed along the bed between steel, timber, or concrete forms. When the concrete has set sufficiently to carry the prestress, the tendons are freed from the bulkheads and the prestress is transferred to the members, generally through bond between steel and concrete. Since strands anchor themselves much better than wires, they are widely used.

Devices for gripping the tendons to the bulkheads are usually made on the wedge and friction principle. Quick-release grips for holding strands are employed.

In order to improve the behavior of prestressed beams, their tendons are often bent to given profiles. In the long-line process, this is achieved by deflecting the tendons up and down along the length of the bed, known as harping or draping. When individual molds are used for pretensioning, complicated patterns of tendon arrangement can be accomplished, such as are carried out in the U.S.S.R. by their continuous prestressing process whereby the tendons are mechanically fed under a controlled tension force and woven around pegs fixed to the mold.

6. Posttensioning Systems   There are hundreds of patents and systems for posttensioning. A partial list is given in Table 3. Other systems are described in publications of the Post-tensioning Institute, Chicago. Patent royalties are indirectly included in the bid price for supplying tendons and anchorages. The bid sometimes includes the furnishing of equipment and technical supervision of jacking operations. Tables 4 through 7 give data for a few prestressing systems commonly used in the United States. Typical end anchorages are shown in Fig. 2.

Loss of Prestress

Loss of prestress results from the following:
1. Immediate elastic shortening of concrete under compression
2. Creep of concrete under sustained compression
3. Shrinkage of concrete due to drying
4. Relaxation in steel under tension
5. Slippage and slackening of tendons during anchoring
6. Frictional force between tendon and concrete during tensioning

7. Elastic Shortening of Concrete   As prestress is transferred to the concrete, shortening of the concrete results in loss of prestress. Considering the axial shortening of concrete produced by pretensioning, we have

$$\text{Unit elastic shortening } \delta = \frac{f_c}{E_c} = \frac{F_0}{A_c E_c}$$

where F_0 is the total prestress just after transfer, that is, after the shortening has taken place. Loss of prestress in steel is

TABLE 3 Linear Prestressing Systems

Type	Classification	Description		Name of system	Country of origin
Pretensioning	Methods of stressing	Against buttresses or stressing beds		Hoyer	Germany
		Against central steel tube		Shorer, Chalos	U.S., France
		Continuous stressing against molds		Continuous wire winding	U.S.S.R.
		Electric current to heat steel		Electrothermal	U.S.S.R.
	Methods of anchoring	During pre-stressing	Wires	Various wedges	
			Strands	Strandvise, Supreme	U.S.
		For transfer of prestress	Bond, for strands and small wires		Europe, U.S.
			Corrugated clips, for big wires	Dorland	U.S.
Posttensioning	Methods of stressing	Steel against concrete		Most systems	
		Concrete against concrete		Leonhardt	Germany
				Billner	U.S.
		Expanding cement		Lossier	France
		Electrical prestressing		Billner	U.S.
		Bending steel beams		Preflex	Belgium
	Methods of anchoring	Wires, by frictional grips		Freyssinet	France
				Magnel	Belgium
				Morandi	Italy
				Holzmann	Germany
				Preload	U.S.
				Kelly	U.S.
				WCS	U.S.
		Wires, by bearing		B.B.R.V.	Switzerland
				INRYCO	U.S.
				WCS	U.S.
				Prescon	U.S.
				Texas P.I.	U.S.
		Wires, by loops and combination of methods		Billner	U.S.
				Monierbau	Germany
				Huttenwerk Rheinhausen	Germany
				Leoba	Germany
				Leonhardt	Germany
		Bars, by bearing and by grips		Lee-McCall	England
				Stressteel	U.S.
				Stress rods	U.S.
				Finsterwalder	Germany
				Dywidag	Germany
				Karig	Germany
				Polensky and Zollner	Germany
				Wets	Belgium
				Bakker	Holland
		Strands, by bearing		Roebling	U.S.
				Wayss and Freytag	Germany
		Strands, by friction grips		CCL	England
				Freyssinet	U. S., France
				Anderson	U.S.
				Atlas	U.S.
				VSL	Switzerland, U.S.
				Prescon	U.S.
				CCS	U.S.
				CONESCO	U.S.
				Continental Structures	U.S.
				CONA	Switzerland, U.S.
				PTS/Howlett	U.S.
				PTI	U.S.
				Kelly	U.S.
				WCS	U.S.

$$\Delta f_s = E_s \delta = \frac{E_s F_0}{A_c E_c} = \frac{n F_0}{A_c} \tag{1}$$

where $n = E_s/E_c$.

The value of F_0 may not be known exactly. However, the value of the initial prestress F_i before transfer is usually known; hence another solution can be obtained. Using the transformed-section method, with $A_t = A_c + nA_{ps}$,

$$\delta_i = \frac{F_i}{A_c E_c + A_{ps} E_s}$$

$$\Delta f_s = E_s \delta_i = \frac{E_s F_i}{A_c E_c + A_{ps} E_s} \tag{2}$$

$$= \frac{n F_i}{A_c + n A_{ps}}$$

$$\Delta f_s = \frac{n F_i}{A_t}$$

TABLE 4   Typical Tendons for B.B.R.V. System*

No. of ¼-in. wires	1	14	28	40
Section area of wires, in.²	0.04909	0.687	1.3744	1.963
Max force after anchoring (70% of ultimate), lb	8250	115,500	231,000	330,000
Max jacking force (80% of ultimate), lb	9420	131,880	263,760	376,800
Ultimate strength, lb	11,780	164,920	329,840	471,200
Baseplate, B.B.R.V., in.		6¾ × 6¾	9¼ × 9¼	11 × 11
Baseplate, Prescon, in.		6 × 8½	7 × 12	

*Almost any number of ¼-in. wires up to about 192 for either system.

TABLE 5   Freyssinet System. Cable Characteristics

Wires (0.196 and 0.276 in. diameters)

Cable size	12/0.196	18/0.196	12/0.276
Nominal steel area, in.²	0.362	0.543	0.718
Ultimate strength, lb	90,000	135,000	168,500
Max jacking force, lb (80% ultimate)	72,000	108,000	135,000
Max force after anchoring, lb (70% ultimate)	63,000	94,500	117,950
Cable weight—sheath not included, lb/ft	1.23	1.85	2.45
Recommended hole diam, in.	1⅛	1½	1½
Anchorage diam, in.	3⅞	4⅞	4⅞
Anchorage length, in.	4	4⅞	4⅞

12-strand (½-in. 7-wire Strands)

Nominal steel area, in.²	1.73
Ultimate cable strength (1.73 × 250,000 psi), lb	432,000
Max jacking force (80% of ultimate), lb	345,600
Max force after anchoring (70% of ultimate), lb	302,400
Cable weight (sheath not included) lb/ft	5.93
Recommended hole diam (ID), in.	2⅝
Anchorage diam, in.	8¼

TABLE 6 Prestressing Bars

			Ultimate strength guaranteed min		Initial tensioning load, $0.7f_{pu}$*		
			Regular†	Special‡	Regular	Special	
			All values in units of 1000 lb				Anchorage plate, in.
Stressteel bars (smooth)							
¾	1.50	0.442	66	71	46	50	4 × 4
⅞	2.04	0.601	90	96	63	67	4½ × 5
1	2.67	0.785	118	126	82	88	5 × 5½
1⅛	3.38	0.994	149	159	104	111	6 × 6
1¼	4.17	1.227	184	196	129	137	6 × 7
1⅜	5.05	1.485	223	238	156	166	7 × 7½
Dywidag bars (threaded)							
⅝	0.98	0.28	43.5		30.5		3 × 3
1	3.01	0.85	127.8	136.3	89.5	95.4	5 × 5½
1¼	4.39	1.25	187.5	200.0	131.0	140.0	6 × 7
1⅜	5.56	1.58	234.0	163.8	249.6	174.7	7 × 7½

*Losses due to creep, shrinkage of concrete, and steel relaxation should be deducted from this value. Overtension to $0.8f_{pu}$ is permitted to account for friction loss and/or wedge seating loss.

†Regular is for minimum ultimate strength of 150,000 psi.

‡Special is for minimum ultimate strength of 160,000 psi.

TABLE 7 VSL Posttensioning System Using ½-in. 7-Wire Strands of 270K Grade

Unit	No. of strands	Steel area, in.²	Weight, lb/ft	Max temp. force, kips	Initial force, kips	Sheath diam, in.		Bearing plate, in.
						Flexible tubing	Rigid tubing	
E5-3	2	0.31	1.05	66	58	1¼	1½	5¼ × 5¼
	3	0.46	1.58	99	87	1½	1¾	
E5-4	4	0.61	2.10	132	116	1⅝	1¾	6⅛ × 6⅛
E5-7	5	0.77	2.63	165	145	1¾	2¹⁄₁₆	8 × 8
	6	0.92	3.15	198	174	1⅞	2¹⁄₁₆	
	7	1.07	3.68	231	202	2	2¼	
E5-12	8	1.22	4.20	264	231	2	2¼	10½ × 10½
	9	1.38	4.73	297	260	2⅛	2⁷⁄₁₆	
	10	1.53	5.25	330	289	2¼	2⁷⁄₁₆	
	11	1.68	5.78	363	318	2⅜	2⁷⁄₁₆	
	12	1.84	6.30	397	347	2½	2¹³⁄₁₆	
E5-19	13	1.99	6.83	430	376	2⅝	3	13¼ × 13¼
	14	2.14	7.35	463	405	2⅝	3	
	15	2.30	7.88	496	434	2¾	3³⁄₁₆	
	16	2.45	8.40	529	463	2⅞	3³⁄₁₆	
	17	2.60	8.93	562	492	3	3⁹⁄₁₆	
	18	2.75	9.45	595	520	3	3⁹⁄₁₆	
E5-22	19	2.91	9.98	628	549	3⅛	3⁹⁄₁₆	14⅜ × 14⅜
	20	3.06	10.50	661	578	3¼	3¾	
	21	3.21	11.03	694	607	3¼	3¾	
	22	3.37	11.55	727	636	3⅜	3¾	
E5-31	23	3.52	12.08	760	665	3½	3¹⁵⁄₁₆	17 × 17
	24	3.67	12.60	793	694	3½	3¹⁵⁄₁₆	
	25	3.83	13.13	826	723	3⅝	3¹⁵⁄₁₆	
	26	3.98	13.65	859	752	3⅝	3¹⁵⁄₁₆	
	27	4.13	14.18	892	781	3¾	4⁵⁄₁₆	
	28	4.28	14.70	925	810	3⅞	4⁵⁄₁₆	
	29	4.44	15.23	958	838	3⅞	4⁵⁄₁₆	
	30	4.59	15.75	991	867	4	4½	
	31	4.74	16.28	1024	896	4	4½	
E5-55	55	8.42	28.88	1818	1590	5½	6	23 × 23

For posttensioning, the problem is different. If we have only a single tendon in the member, the concrete shortens as that tendon is jacked against the concrete. Since the force in the cable is measured after the elastic shortening of the concrete has taken place, no loss due to that shortening need be accounted for.

If we have more than one tendon and the tendons are stressed in succession, then the prestress is gradually applied to the concrete, the shortening of concrete increases as each

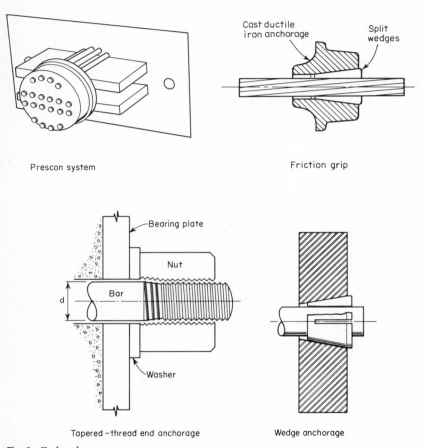

Prescon system

Friction grip

Tapered−thread end anchorage

Wedge anchorage

Fig. 2   End anchorages.

cable is tightened against it, and the loss of prestress due to elastic shortening differs in the tendons. The tendon that is first tensioned would suffer the maximum amount of loss due to the shortening of concrete by the subsequent application of prestress from all the other tendons. The tendon that is tensioned last will not suffer any loss due to elastic concrete shortening. For practical purposes, it is accurate enough to determine the loss for the first cable and use half that value for the average loss of all the cables. If each tendon is tensioned to a value above the specified initial prestress by the magnitude of the expected loss, no loss from elastic shortening need be considered.

8. Creep   The shortening due to creep ranges from one to five times the instantaneous elastic shortening, averaging about three times for normal-weight concrete. A simple method to estimate creep is therefore to use the elastic shortening and multiply it by a suitable creep coefficient C_c, realizing that this coefficient depends on many factors. Thus, the loss of prestress Δf_s due to creep is given by

$$\Delta f_s = C_c n f_c \qquad (3a)$$

where f_c = compressive stress in the concrete.

To estimate the creep coefficient the approach of the European Concrete Committee may be used. For the usual range of mixes at stress-strength ratios not exceeding 0.35,

$$C_c = \eta_b \eta_c \eta_d \eta_e \eta_t \qquad (3b)$$

Values of η are given in Table 8.

TABLE 8a   Values of η_b in Eqs. (3b) and (4b)

W/C ratio	Cement content in mix, lb/yd³			
	340	505	675	840
0.3				0.65
0.4		0.6	0.8	1.0
0.5	0.6	0.88	1.2	1.5
0.6	0.75	1.13	1.6	
0.7	1.0	1.5		

TABLE 8b   Values of η_c^* in Eq. (3b)

Relative humidity %	40	50	60	70	80	90	100
η_c	3.1	2.85	2.6	2.3	1.9	1.5	1.0

TABLE 8c   Values of η_d^* in Eq. (3b)

Concrete age at loading, days	1	3	7	28	90	360
Normal cement	1.8	1.6	1.4	1.0	0.7	0.5
High-early-strength cement	1.7	1.4	1.1	0.7	0.5	0.3

*For concrete hardening at a constant temperature of 68°F. If the concrete hardens at a temperature other than 68°F, the age at loading should be computed as $D = \Sigma[\Delta t(T - 14)/54]$ in which Δt is the number of days during which hardening has taken place at $T°F$.

TABLE 8d   Values of η_e in Eq. (3b)

Volume/surface ratio, in.	1	2	4	6	9	10
η_e	1.2	1.0	0.85	0.75	0.72	0.7

9. Shrinkage Loss of prestress resulting from shrinkage of the concrete is given by

$$\Delta f_s = \delta_s E_s \qquad (4a)$$

The shrinkage strain δ_s is commonly taken to be 0.0003. If a closer estimate is required, one may use the approach of the European Concrete Committee,

$$\delta_s = \eta_b \eta_t \eta_e' \eta_\rho \epsilon_h \qquad (4b)$$

In the above expression, $\eta_p = 1/(1 + n\rho_p)$, in which ρ_p is the percentage of longitudinal bonded reinforcement A_{ps} with respect to the cross-sectional area A_c of member. The values of η_b and η_t are the same as for predicting creep, and the values of η_e' and ϵ_h are given in Table 9.

10. Relaxation in Steel Stress relaxation in steel is the loss of stress when it is maintained at a constant strain for a period of time. Relaxation varies with steels of different compositions and treatments, but its approximate characteristics are known for most of the

TABLE 8e   Values of η_t in Eqs. (3b) and (4b)

Loading duration	Volume/surface ratio, in.				
	1	2	4	8	16
3 days	0.2	0.1			
7	0.3	0.18	0.1		
14	0.4	0.25	0.15		
28	0.5	0.36	0.2	0.1	
90	0.7	0.6	0.4	0.2	0.1
180	0.85	0.75	0.55	0.33	0.15
1 year	0.95	0.85	0.7	0.5	0.25
2	0.98	0.92	0.85	0.68	0.35
5	0.99	0.95	0.95	0.85	0.65
10	1.0	0.99	0.99	0.95	0.85
20		1.0	1.0	0.99	0.95

prestressing steels. In general, the percentage increases with increasing stress, and when a steel is under low stress, relaxation is negligible. Typical curves giving the relation between relaxation and initial stress level in three types of steel wires are shown in Fig. 3.

Relaxation in stress-relieved seven-wire strands has characteristics similar to those of stress-relieved wires, and can be expressed as

$$\Delta f_s = f_{si}\left[\frac{\log_{10}t}{10}\left(\frac{f_{si}}{f_{py}} - 0.55\right)\right]$$

where f_{si} is the initial stress in the prestressing strand, f_{py} is the yield strength of the strand at 1 percent elongation, and t is the time in hours. An expression for stabilized strands is

$$\Delta f_s = f_{si}\left[\frac{\log_{10}t}{45}\left(\frac{f_{si}}{f_{py}} - 0.55\right)\right]$$

Both formulas are applicable only for $f_{si} \geq 0.60f_{py}$.

For high-tensile bars stressed to about $0.60f_{pu}$, relaxation is about 3 percent, which is also the average loss assumed for relaxation of most tendons stressed to the usual allowable values.

TABLE 9a   Values of η_e' in Eq. (4b)

Volume/surface ratio, in.	1	2	4	6	8	10
η_e'	1.20	1.00	0.80	0.65	0.55	0.50

TABLE 9b   Values of ϵ_h in Eq. (4b)

Relative humidity %	40	50	60	70	80	90	100
ϵ_h, 10^{-6} in./in.	420	380	330	275	210	115	0

11. Slippage of Tendons during Anchoring For most systems of posttensioning, when the jack is released and the prestress transferred to the anchorage, the tendon tends to slip slightly. The amount of slippage depends on the type of wedge and the stress in the wires, an average value being around 0.1 in. For direct bearing anchorages, the heads and nuts are subject to a slight deformation at the release of the jack, an average value for such deformation being about 0.03 in. If long shims are required to hold elongated wires in place, there will be a deformation in the shims. A shim 1 ft long may deform 0.01 in.

A general formula for computing the loss of prestress due to deformation Δ_a at anchoring is

$$\Delta f_s = \frac{\Delta_a E_s}{L} \tag{5}$$

where L is the length of tendon in inches. Since this loss of prestress is caused by a fixed total amount of shortening, the percentage of loss is higher for short wires than for long

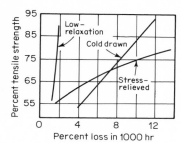

Fig. 3   Relaxation of prestressing wires.

ones. Hence it is quite difficult to tension short wires accurately, especially for systems of prestressing whose anchorage losses are relatively large. On the other hand, in the long-line process of pretensioning, this type of loss is insignificant and is not taken into consideration in design.

12. Friction   The stress in a tendon tends to decrease with distance from the tensioning end, because there is friction between the tendon and its surrounding concrete or sheathing. This frictional loss can be conveniently considered in two parts: the length effect and the curvature effect. The length effect is the amount of friction that would be encountered if the tendon were straight. Since in practice the duct for a straight tendon will not be perfectly straight, some friction will exist between the tendon and its surrounding material. This is sometimes described as the wobbling effect of the duct and is dependent on the length of and stress in the tendon, the coefficient of friction, and the workmanship and method used in aligning and forming the duct.

The loss of prestress due to curvature effect results from the intended curvature of the tendons. This loss is also dependent on the coefficient of friction and the pressure exerted by the tendon on the concrete. The coefficient of friction depends on the nature of the surfaces in contact, the amount and nature of lubricants, and sometimes the length of contact. The pressure between the tendon and concrete depends on the stress in the tendon and the change in angle.

The force F_2 at any point on the tendon is given by

$$F_2 = F_1 e^{-\mu\theta - Kx} \tag{6}$$

where F_1 = force at the jacking end
      θ = angle between F_1 and F_2, radians
      x = length between points 1 and 2, ft
      μ = coefficient of friction between tendon and surrounding material
      K = wobble coefficient, per ft

When $\mu\theta + Kx$ is less than 0.3, the following approximate formula can be used:

$$F_2 = F_1(1 - \mu\theta - Kx) \tag{7}$$

Table 10 may be used to estimate values of μ and K. Actual values may differ greatly from those given and can only be obtained from experience. For example, values depend a great deal on the care exercised in construction. Tendons well greased and carefully wrapped in plastic tubes will offer little friction, but if mortar leaks through openings in the tube, the tendons may become tightly stuck.

There are several methods for overcoming the frictional loss in tendons. One method is to overtension them. Jacking from both ends is another. Lubricants can be used to advantage for unbonded tendons. For bonded tendons, water-soluble oils can be used to reduce friction while tensioning; the lubricant is flushed off with water afterward.

13. Effective Prestress   The effective prestress is obtained by deducting the losses from the initial prestress. The loss of prestress varies with many factors. For a close estimate it is necessary to consider the amount of various losses in successive time intervals such as before transfer of prestress, during transfer of prestress, first year after transfer of prestress, and from first year to the end of service life of the structure. However, for the average case, the values in Table 11 are representative.

TABLE 10   Coefficients for Frictional Loss

	Wobble coefficient, K	Curvature coefficient, μ
Grouted tendons in metal sheathing:		
Wire tendons	0.0010–0.0015	0.15–0.25
High-strength bars	0.0001–0.0006	0.08–0.30
7-wire strand	0.0005–0.002	0.15–0.25
Unbonded tendons:		
Mastic-coated		
Wire tendons	0.001–0.002	0.05–0.15
7-wire strand	0.001–0.002	0.05–0.15
Pregreased		
Wire tendons	0.0003–0.002	0.05–0.15
7-wire strand	0.0003–0.002	0.05–0.15

TABLE 11   Prestress Losses

	Pretensioning, %	Posttensioning, %
Elastic shortening of concrete	4	2
Creep of concrete	6	5
Shrinkage of concrete	7	6
Relaxation in steel	6	6
Total loss not including frictional loss	23	19

14. Elongation of Tendons   If a tendon has uniform stress F along its entire length L, the total elongation is given by

$$\Delta_s = \frac{FL}{E_s A_{ps}} \tag{8}$$

For a tendon with uniform curvature, considering frictional loss throughout its length L, the total elongation is given by

$$\Delta_s = \frac{F_2 L}{E_s A_{ps}} \frac{e^{\mu\theta + KL} - 1}{\mu\theta + KL} \tag{9}$$

For an approximate solution,

$$\Delta_s = \frac{F_1 + F_2}{2} \frac{L}{E_s A_{ps}} \tag{10}$$

ANALYSIS FOR FLEXURE

15. Basic Concepts   Three different concepts may be used to explain and analyze the behavior of prestressed concrete.

1. Consider a simple rectangular beam, eccentrically prestressed by a tendon (Fig. 4). The prestress F produces a stress f at any cross section:

$$f = \frac{F}{A} \pm \frac{Fey}{I} \tag{11}$$

where A = area of concrete section
e = eccentricity of tendon
y = distance from centroidal axis (c.g.c.)
I = moment of inertia of the section

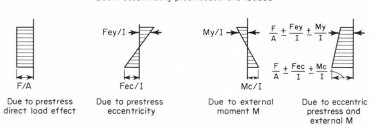

Fig. 4   Stress distribution across an eccentrically prestressed concrete section.

If M is the moment at a section due to external load, the stress at any point on that section is

$$f = \frac{F}{A} \pm \frac{Fey}{I} \pm \frac{My}{I} \tag{12}$$

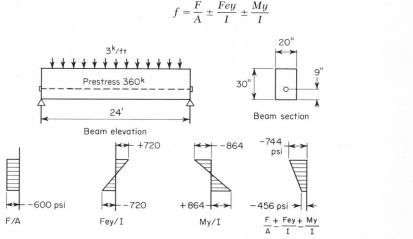

Fig. 5   Example 1.

If prestress eccentricities and external moments exist on both principal axes

$$f = \frac{F}{A} \pm \frac{Fe_y y}{I_x} \pm \frac{Fe_x x}{I_y} \pm \frac{M_x y}{I_x} \pm \frac{M_y x}{I_y} \tag{13}$$

Example 1   A prestressed-concrete rectangular beam 20 × 30 in. has a simple span of 24 ft and is loaded by a uniform load of 3 kips/ft which includes its own weight (Fig. 5). The prestressing tendon is located as shown and produces an effective prestress of 360 kips. Compute fiber stresses in the concrete at the midspan section.

SOLUTION. $F = 360$ kips, $A = 20 \times 30 = 600$ in.^{2} (neglecting any hole due to the tendon), $e = 6$ in., $I = bd^3/12 = 20 \times 30^3/12 = 45,000$ in.^{4}; $y = 15$ in. for extreme fibers.

$$M = 3 \times \frac{24^2}{8} = 216 \text{ kip-ft}$$

$$f = \frac{-360,000}{600} \pm \frac{360,000 \times 6 \times 15}{45,000} \pm \frac{216 \times 12,000 \times 15}{45,000}$$

$$= -600 + 720 - 864 = -744 \text{ psi for top fiber}$$

$$= -600 - 720 + 864 = -456 \text{ psi for bottom fiber}$$

The resulting stress distribution is shown in Fig. 5.

2. The internal resisting couple C-T is shown in Fig. 6. When the prestress F and the

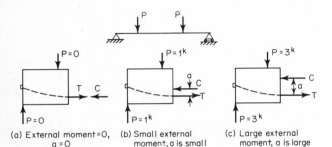

(a) External moment=0,   (b) Small external   (c) Large external
       a =0   moment, a is small   moment, a is large

Fig. 6 Variation of a.

external moment M are known and since $C = T = F$, the lever arm a, which locates the center of the compressive force C, is given by

$$a = \frac{M}{F} \tag{14}$$

Once C is located, the stress at any point on the section is given by

$$f = \frac{C}{A} \pm \frac{Ce'y}{I} \tag{15}$$

where e' = eccentricity of C with respect to c.g.c.

Various stress distributions in the concrete can be obtained, depending upon the location of C (Fig. 7). Thus, if C is at the bottom kern point, the triangular distribution b results. Similarly, if C is at the top kern point, distribution e is obtained. The kern distances k_b and k_t are given by

$$k_b = \frac{r^2}{c_t} \qquad k_t = \frac{r^2}{c_b}$$

where r = radius of gyration of the cross section
c_t = distance from c.g.c. to top fiber
c_b = distance from c.g.c. to bottom fiber

Example 2 Same data as Example 1.

SOLUTION. $M = 3 \times 24^2/8 = 216$ kip-ft. The internal couple furnished by the forces $C = T = 360$ kips (Fig. 8) has the lever arm

$$a = \frac{216}{360} \times 12 = 7.2 \text{ in.}$$

Since T acts 9 in. from the bottom, C lies 16.2 in. from the bottom, and $e' = 16.2 - 15 = 1.2$ in. Then

$$f = \frac{-360,000}{600} \pm \frac{360,000 \times 1.2 \times 15}{45,000}$$
$$= -600 \pm 144$$
$$= -744 \text{ psi for top fiber}$$
$$= -456 \text{ psi for bottom fiber}$$

3. By this concept, one visualizes prestressing as an attempt to balance a portion of the external loads on a beam. In its simplest form, one assumes a parabolic tendon in a simple beam prestressed so as to exert a uniform upward force on the beam. If this beam supports an external downward load of equal intensity, then the net transverse load is zero, so that there is a uniform compressive stress $f = F/A$ at any cross section. If the external load is not balanced by the upward force, the moment M of the *unbalanced* load produces an additional stress $f = My/I$.

The prestress F required to balance a uniform load w lb/ft is given by

$$F = \frac{wL^2}{8h}$$

where L = span of beam, ft
h = sag of cable, ft

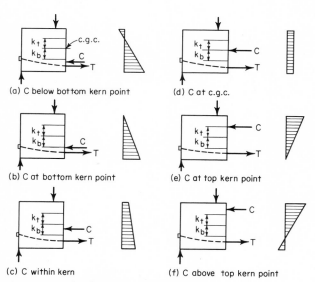

(a) C below bottom kern point

(d) C at c.g.c.

(b) C at bottom kern point

(e) C at top kern point

(c) C within kern

(f) C above top kern point

Fig. 7   Stress distribution (elastic theory).

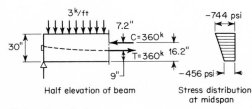

Half elevation of beam

Stress distribution at midspan

Fig. 8   Example 2.

Example 3 A 20×30-in. concrete beam is prestressed with a parabolic cable located as shown in Fig. 9. Compute the prestress required to balance an external load of 2.5 kips/ft.

SOLUTION. From Fig. 9, $h = 6$ in.

$$F = \frac{wL^2}{8h} = \frac{2.5 \times 24^2}{8 \times 0.5} = 360 \text{ kips}$$

Under this prestress, the beam will be uniformly stressed to $360,000/600 = 600$ psi. (The horizontal component of the prestress force should be used if greater accuracy is desired.)

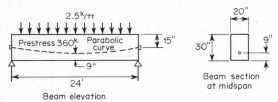

Beam elevation

Fig. 9 Example 3.

The beam will be under no bending when subjected to an external load of 2.5 kips/ft. If the external load is 3 kips/ft (Examples 1 and 2), the unbalanced load is 0.5 kip/ft, and the midspan moment

$$M = \frac{wL^2}{8} = \frac{0.5 \times 24^2}{8} = 36 \text{ kip-ft}$$
$$f = \pm \frac{My}{I} = \frac{36 \times 12,000 \times 15}{45,000} = \pm 144 \text{ psi}$$

The stresses resulting from prestress and the external load of 3 kips/ft are

$$f = -600 \pm 144$$
$$= -744 \text{ psi for top fiber}$$
$$= -456 \text{ psi for bottom fiber}$$

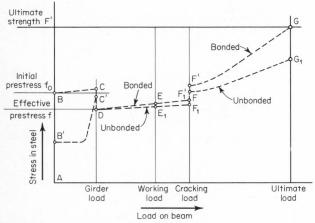

Fig. 10 Variation of steel stress with load.

16. Stress in Steel Variation in stress in the steel of a prestressed-concrete beam is shown in Fig. 10. For a posttensioned beam, as the tendons are tensioned, the steel stress increases from A to B. Simultaneously, the prestress is transferred to the beam. If the beam is heavy, its full dead load will not come into play until after its falsework is removed, causing a slight increase in steel stress from B to C. If the beam is relatively light, it usually begins to camber before the steel stress reaches B, and its dead load comes into play immediately. Thus the steel stress may vary from an intermediate point, say B' to C'. Because of the camber the tendons shorten slightly so that their stress is slightly lower, as represented by C'. In a pretensioned beam, the steel stress generally varies from B to C' upon transfer of prestress.

Assuming the losses of prestress take place before the application of superimposed dead and live loads, the steel stress is reduced from C or C' to D. Only minor changes in the range DE are induced by superimposed dead and live loads.

At cracking, stress in the steel at the crack jumps from F to F', after which it continues to increase until the ultimate load G is reached. Unbonded tendons would be forced to slip except for frictional resistance. This slippage would allow any strain in the unbonded tendon to distribute throughout its entire length. Consequently, as load increases, the tendon stress will increase more slowly than that in a bonded tendon. The line $DE_1F_1F_1'G_1$ shows the stress variation in unbonded tendons, assuming the same effective prestress before addition of external load.

17. Cracking Moment   The external moment producing first hair cracks in a prestressed-concrete beam is known as the cracking moment. It is a measure of the serviceability of the beam. It occurs when the tensile stress in the extreme fiber of concrete reaches its modulus of rupture f_r (Fig. 11).

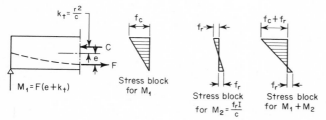

Fig. 11   Cracking moment.

$$-\frac{F}{A} - \frac{Fec}{I} + \frac{Mc}{I} = f_r$$

Transposing, we have the value of cracking moment,

$$M = Fe + \frac{FI}{Ac} + \frac{f_r I}{c} \tag{16a}$$

This can also be written

$$M = F\left(e + \frac{r^2}{c}\right) + \frac{f_r I}{c}$$
$$= F(e + k_t) + \frac{f_r I}{c} \tag{16b}$$

which is shown as $M = M_1 + M_2$ in Fig. 11. Unless the value of f_r is known from tests, it is commonly specified as $7.5\sqrt{f_c'}$ psi.

18. Ultimate Moment   *Underreinforced Beams.*   The steel in an underreinforced bonded beam is usually stressed to its ultimate strength f_{pu} under the action of the ultimate moment. Thus, the ultimate tension force is $T' = A_{ps}f_{pu}$. The compression force $C' = T'$ can be located approximately using a rectangular stress block, with an average stress of $0.85f_c'$ and a depth of $k'd$ (Fig. 12) such that the area of concrete A_c' within the depth $k'd$ satisfies the relation $A_c' = C'/0.85f_c'$. Equating T' and C' results in the equation

$$k' = \frac{A_{ps}f_{pu}}{0.85f_c'db} \tag{17}$$

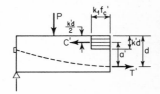

Fig. 12   Ultimate moment.

These formulas apply even though the cross section is not rectangular, provided that the compression flange has a uniform width b for the depth $k'd$.

The ultimate resisting moment $T'a'$ (Fig. 12) is

$$M_u = A_{ps}f_{pu}d\left(1 - \frac{k'}{2}\right) \tag{18}$$

Most building codes and bridge specifications give a slightly different and more conservative formula for rectangular sections

$$M_u = A_{ps}f_{ps}d\left(1 - 0.6\frac{\rho_p f_{ps}}{f'_c}\right) \tag{19}$$

where f_{ps} = stress in steel at failure and $\rho_p = A_{ps}/bd$.

For flanged sections in which the neutral axis falls outside the flange (usually where the flange thickness t is less than $1.4d\rho_p f_{ps}/f'_c$),

$$M_u = A_{sr}f_{ps}d\left(1 - \frac{k'}{2}\right) + 0.85f'_c(b - b')\,t\left(d - \frac{t}{2}\right) \tag{20}$$

where b is the effective width of the flange, b' is the width of the web, and

$$A_{sr} = A_{ps} - A_{sf}$$
$$A_{sf} = 0.85f'_c(b - b')t/f_{ps}$$

Unless the value of f_{su} is determined from detailed analysis, the following is often specified:

Bonded members: $$f_{ps} = f_{pu}\left(1 - 0.5\frac{\rho_p f_{pu}}{f'_c}\right) \tag{21a}$$

Unbonded members: $$f_{ps} = f_{se} + 10{,}000 + \frac{f'_c}{100\rho_p}$$
$$\leq f_{se} + 60{,}000 \leq f_{pu} \tag{21b}$$

where f_{se} is the effective stress in the prestressing steel after losses, in psi, and f_{pu} is the specified yield strength of the prestressing steel.

In unbonded members, it is desirable to use a moderate amount of bonded nonprestressed reinforcement which will help distribute cracks and improve the postcracking stiffness and the ultimate strength of the beam. The minimum amount of such reinforcement required by the ACI Code is $A_s = 0.004A$, where A is the area of that part of the cross section between the flexural tension face and the center of gravity of the cross section.

Overreinforced Beams.   It is safe to assume that a rectangular section is underreinforced if $\rho_p f_{ps}/f'_c < 0.30$. If $\rho_p f_{ps}/f'_c > 0.30$, the beam is usually considered to be overreinforced, and it is often specified that the ultimate flexural strength be taken not greater than $M_u = 0.25f'_c bd^2$. More accurate predictions can be made with the following procedure, considering both equilibrium and strain compatibility.

The maximum compressive strain in the concrete at failure is taken to be 0.003. Assuming plane sections remain plane and using a trial value of $k'd$, the strain in the steel at rupture of the beam is given by $e_s = e_{s1} + e_{s2}$, where e_{s1} is the strain at the time when the concrete compressive strain on the top fiber is zero, and $e_{s2} = 0.003\,(1 - k')/k'$ (Fig. 13). The value of f_{ps} can then be obtained from the stress-strain diagram. If f_{ps} so determined is near the ultimate value f_{pu}, the section is not overreinforced. However, if f_{ps} is appreciably lower than f_{pu}, the section is overreinforced.

Knowing the value of f_{ps}, the equilibrium condition $C' = T'$ should be checked. If the condition is not satisfied, a new trial value of $k'd$ should be assumed and the procedure repeated until $C' = T'$. The moment capacity is then computed by Eq. (19) or Eq. (20).

Example 4 RECTANGULAR SECTION. Given a beam of rectangular cross section 12 in. wide by 24 in. deep. The c.g.s. of the prestressing wires is 4 in. above the bottom of the beam. Area of wires is 1.5 in.^{2}, $f_{pu} = 240{,}000$ psi, $f_0 = 150{,}000$ psi, $f'_c = 5000$ psi. Compute the ultimate resisting moment.

SOLUTION. Assuming that the beam is underreinforced, the stress in the steel at the ultimate moment is $f_{ps} = f_{pu} = 240{,}000$ psi. The depth $d = 24 - 4 = 20$ in., and from Eqs. (17) and (18).

$$k' = \frac{1.5 \times 240{,}000}{0.85 \times 5000 \times 20 \times 12} = 0.353$$

$$M_u = 1.5 \times 240,000 \times 20 \left(1 - \frac{0.353}{2}\right) = 5,930,000 \text{ in.-lb}$$

Using the more conservative formula, Eq. (19), the reinforcement ratio $\rho_p = 1.5/(12 \times 20) = 0.00625$, and from Eq. (21a)

$$f_{ps} = f_{pu} \left(1 - 0.5 \frac{\rho_p f_{pu}}{f_c'}\right) = 240,000 \left(1 - 0.5 \frac{0.00625 \times 240}{5}\right)$$
$$= 204,000 \text{ psi}$$

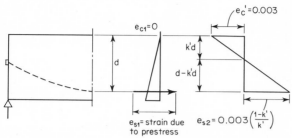

Beam elevation   Strains due to prestress   Strains due to loading and girder weight

Fig. 13 Strains at rupture.

From Eq. (19),

$$M_u = 1.5 \times 204,000 \times 20 \left(1 - 0.6 \times \frac{0.00625 \times 204}{5}\right) = 5,184,000 \text{ in.-lb}$$

This result is about 13 percent smaller than the previous value. Now

$$\rho_p \frac{f_{ps}}{f_c'} = 0.00625 \times \frac{204,000}{5000} = 0.255$$

which, since it is less than 0.3, suggests that the beam is underreinforced as assumed. If a more accurate value of M_u is desired, assume $k' = 0.31$ and the maximum strain in the concrete to be 0.003 (Fig. 13). Then

$$e_{s2} = 0.003 \times \frac{1 - 0.31}{0.31} = 0.0067$$

With $f_0 = 150,000$ psi, assume the effective prestress $f_e = 125,000$ psi. The corresponding strain $f_e/E_s = 0.0042$ and the total strain at failure is $0.0067 + 0.0042 = 0.0109$. From Fig. 1, this corresponds to a stress of 210,000 psi. Therefore,

$$k' = \frac{1.5 \times 210,000}{0.85 \times 5000 \times 20 \times 12} = 0.309$$

which is very close to the originally assumed value of 0.31. Therefore, with $f_{ps} = 210,000$ psi,

$$M_u = 1.5 \times 210,000 \times 20 \left(1 - 0.6 \times \frac{0.00625 \times 210}{5}\right)$$
$$= 5,308,000 \text{ in.-lb}$$

which is about 10 percent less than the first computed value.

T-SECTION. Assume a T-section 24 in. deep whose flange is 20 × 3 in. and web is 5 in. thick. The c.g.s. of the prestressing wires is 4 in. above the bottom of the beam. Area of wires is 1.5 in.^{2}, $f_{pu} = 240,000$ psi, $f_0 = 150,000$ psi, $f_c' = 5000$ psi. Compute the ultimate resisting moment.

SOLUTION. The required area in compression is (assuming $f_{ps} = f_{pu}$)

$$\frac{1.5 \times 240,000}{0.85 \times 5000} = 85 \text{ in.}^2$$

The flange furnishes 60 in.^{2}, leaving 25 in.^{2} to be supplied by the web. Therefore, the neutral axis is $25/5 = 5$ in. below the bottom of the flange.

For an approximate solution compression in the web can be neglected and the center of compression assumed at middepth of the flange. The lever arm is $24 - 4 - 1.5 = 18.5$ in. and $M_u = 360,000 \times 18.5 = 6,660,000$ in.-lb.

For a more exact solution, the centroid of the compressive area, including the web, can be determined. Thus, the distance from middepth of flange to the centroid is $4 \times 25/(25 + 60) = 1.2$ in., and the lever arm $18.5 - 1.2 = 17.3$ in. The resulting moment is $M_u = 360,000 \times 17.3 = 6,220,000$ in.-lb., which is about 6 percent less than the approximate value.

19. Composite Sections Figure 14 shows a composite section at the midspan of a simply

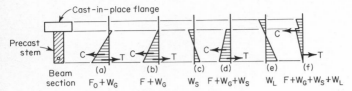

Fig. 14 Stress distribution for a composite section.

supported beam, whose stem is precast and lifted into position with the top slab cast in place directly on the stem. If no temporary intermediate support is furnished, the weight of both the slab and the stem will be carried by the stem acting alone. After the slab concrete has hardened, the composite section will carry live or dead load that may be added to it. The following stress distributions are shown for various stages of working-load conditions.

a. Owing to the initial prestress F_0 and the weight of the stem W_G there will be heavy compression in the lower fibers and possibly some small tension in the top fibers. The tensile force T in the steel and the compressive force C in the concrete form a resisting couple with a small lever arm between them.

b. After losses have taken place, the effective prestress F together with the weight of the stem will result in a slightly lower compression in the bottom fibers and some small tension or compression in the top fibers. The C-T couple will act with a slightly greater lever arm.

c. Addition of the slab of weight W_s produces additional moment and stresses as shown. Stresses resulting from differential creep and shrinkage between the slab and the stem are neglected.

d. Adding *b* to *c*, smaller compression is found to exist at the bottom fibers and some compression at the top fibers. The lever arm for the C-T couple increases further.

e. Stresses resulting from live load W_L are shown, the moment being resisted by the composite section.

f. Adding *d* to *e*, we have the stress block *f*. The C-T couple now acts with an appreciable lever arm.

The cracking moment and ultimate moment can be determined using methods similar to those previously described for noncomposite sections. An illustration of composite design is given in Example 12.

DESIGN FOR FLEXURE

20. Preliminary Design Preliminary design for flexure cn be based on the C-T couple. Under working load, the lever arm a varies between $0.30h$ and $0.85h$, where h = total depth of section, and averages about $0.60h$. Hence, the required effective prestress F can be estimated from the equation

$$F = T = \frac{M_T}{0.60h} \tag{22}$$

where M_T = total external moment produced by the working load.

The depth h for a prestressed section varies between 50 and 80 percent of that of an equivalent reinforced-concrete section, and may be taken at 70 percent for a first trial. Having estimated the force F, the area of steel is computed by

$$A_{ps} = \frac{F}{f_s} \qquad (23)$$

where f_s depends on the steel but usually equals about 150 ksi.

The area of concrete required is estimated by

$$A_c = \frac{F}{f_{av}} \qquad (24)$$

where f_{av}, the average precompression in the concrete, varies from 700 to 1300 psi for I- and T-beams and from 250 to 500 psi for solid slabs.

The load-balancing method, Concept 3 of Art. 15, can also be used for preliminary design (Art. 23).

21. Elastic Design Concept 1 of Art. 15 can be used for design by the elastic theory. However, it is often more convenient to use Concept 2. Thus:

Case 1. Girder moment $M_G = 0$. Allowing no tension in the concrete, either at transfer or at maximum load, the permissible moment is (Fig. 15a)

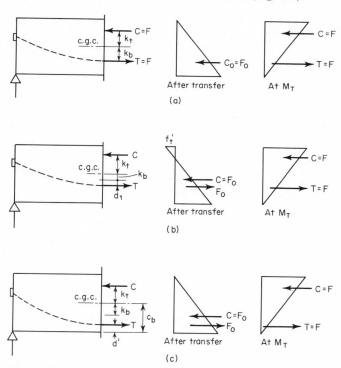

Fig. 15 Elastic design for different ratios of M_G/M_T: (a) $M_G = 0$; (b) small M_G/M_T; (c) large M_G/M_T.

$$M_T = F(k_t + k_b) \qquad (25)$$

where k_t and k_b are the kern distances defined in Art. 15. If tensile stress f_b' is allowed for the bottom fibers, an additional moment $f_b'I/c_b$ can be carried:

$$M_T = F(k_t + k_b) + \frac{f_b'I}{c_b} \qquad (26)$$

If tensile stress f_t' is allowed for the top fibers at transfer, another additional moment $f_t'I/c_t$ can be carried:

$$M_T = F(k_t + k_b) + \frac{f_b'I}{c_b} + \frac{f_t'I}{c_t} \tag{27}$$

Case 2. When M_G is small, so that the c.g.s. cannot be located at its lowest possible position, as determined by the required concrete protection d', the distance d_1 from the bottom kern point to the c.g.s. is given by

$$d_1 = \frac{M_G + f_t'I/c_t}{F_0} \tag{28}$$

where f_t' is the allowable tension in the top fiber at transfer. The permissible moment is (Fig. 15b)

$$M_T = F(k_t + k_b + d_1) \tag{29}$$

If tensile stress f_b' is allowed for the bottom fibers

$$M_T = F(k_t + k_b + d_1) + \frac{f_b'I}{c_b} \tag{30}$$

Case 3. When M_G is large so that c.g.s. is located at the lowest possible position as determined by the required concrete protection d' (Fig. 15c),

$$M_T = F(k_t + c_b - d') \tag{31}$$

If tensile stress f_b' is allowed for the bottom fibers

$$M_T = F(k_t + c_b - d') + \frac{f_b'I}{c_b} \tag{32}$$

After the prestress F has been determmined the area of steel is computed by

$$A_s = \frac{F}{f_s}$$

and the extreme fiber stresses in the concrete are computed under M_G and under M_T, using the formulas in Art. 15. If the stresses are not satisfactory, the section is revised. Direct design formulas and computer programs are available.

Allowable Stresses. The ACI318-77 allowable stresses are as follows:

18.4. Permissible stresses in concrete—flexural members
  18.4.1. Flexural stresses immediately after transfer, before losses, shall not exceed the following:
    (a) Compression $0.60f_{ci}'$
    (b) Tension except as permitted in (c) $3\sqrt{f_c'}$
    (c) Tension at ends of simply supported members $6\sqrt{f_c'}$
  Where the calculated tension stress exceeds this value, reinforcement shall be provided to resist the total tension force in the concrete computed on the assumption of an uncracked section.
  18.4.2. Stresses at service loads, after allowance for all prestress losses, shall not exceed the following:
    (a) Compression $0.45f_c'$
    (b) Tension in precompressed tensile zone $6\sqrt{f_c'}$
    (c) Tension in precompressed tensile zone in members, other than in two-way slab systems, where computations based on the transformed cracked section and on bilinear moment-deflection relationships show that immediate and long-term deflections comply with requirements of ACI Section 9.5 $12\sqrt{f_c'}$
  18.4.3. The permissible stresses in Sections 18.4.1 and 18.4.2 may be exceeded when it is shown experimentally or analytically that performance will not be impaired.
18.5. Permissible stresses in steel
  18.5.1. Due to jacking force $0.80f_{pu}$ or $0.94f_{py}$ whichever is smaller, but not greater than the maximum value recommended by the manufacturer of the steel or of the anchorages
  18.5.2. Pretensioning tendons immediately after transfer, or posttensioning tendons immediately after anchoring $0.70f_{pu}$

---

22. Ultimate Design   The ultimate moment capacity of the section must be not less than the working load moment multipled by a load factor m, usually 1.8 for buildings and 2.0 for bridges. For underreinforced sections, the ultimate lever arm will be around $0.9d$, where d = effective depth. The area of steel required is

$$A_{ps} = \frac{mM_T}{0.9df_{pu}} \tag{33}$$

Assuming that the concrete on the compressive side is stressed to $0.85f_c'$, the required area under compression is

$$A_c = \frac{mM_T}{0.9d \times 0.85f_c'} \tag{34}$$

When designed by the above formulas, the section should be checked by the equations in Art. 18. In addition, compressive stresses at transfer must be investigated for the tension flange, usually by the elastic theory, and checks for excessive camber, deflection, and cracking may be required.

23. Balanced-Load Design   Concept 3 of Art. 15 can be conveniently used for design. Balancing of a uniformly distributed load by a parabolic cable was described in Art. 15. Figure 16 illustrates the balancing of a concentrated load by bending the tendon at

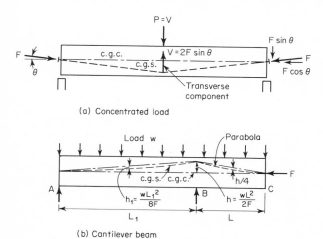

(a) Concentrated load

(b) Cantilever beam

Fig. 16   Load balancing for beams.

midspan, creating an upward component $V = 2F \sin \theta$. If this V exactly balances a concentrated load P, the fiber stress in the beam at any section (except for local stress concentrations) is given by $f = (F \cos \theta)/A_c = F/A_c$ for small values of θ. Any loading in addition to P will cause bending in an elastic homogeneous beam (up to point of cracking), and the additional stresses can be computed by $f = Mc/I$, where M is the moment produced by the additional load.

Now consider a cantilever beam (Fig. 16b). The conditions for load balancing become slightly more complicated, because any vertical component at the cantilever end C will upset the balance unless there is an externally applied load at that end. To balance a uniformly distributed load w, the tangent to the c.g.s. at C must be horizontal. The parabola for the cantilever portion is located by computing $h = wL^2/2F$, and the parabola for the anchor arm by $h_1 = wL_1^2/8F$.

Example 5   A double cantilever beam is to be designed so that its prestress will exactly balance a total uniform load of 1.6 kips/ft on the beam (Fig. 17a). Design the beam using the least amount of prestress, assuming that the c.g.s. must have a concrete protection of at least 3 in. If a concentrated load $P = 14$ kips is added at midspan, compute the maximum fiber stresses.

SOLUTION. In order to balance the load in the cantilever, the c.g.s. at the tip must be located at the c.g.c. with a horizontal tangent. To use the least amount of prestress, the eccentricity over the support should be a maximum, that is, $h = 12$ in. or 1 ft. The prestress required is

$$F = \frac{wL^2}{2h} = \frac{1.6 \times 20^2}{2 \times 1} = 320 \text{ kips}$$

In order to balance the load on the center span, using the same prestress, the sag for the parabola must be

$$h_1 = \frac{wL_1^2}{8F} = \frac{1.6 \times 48^2}{8 \times 320} = 1.44 \text{ ft} = 17.3 \text{ in.}$$

Hence the c.g.s. is located as shown in Fig. 17b.

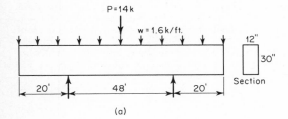

(a)

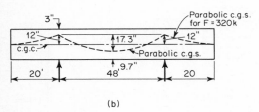

(b)

Fig. 17 Example 5.

Under the combined action of the uniform load and the prestress, the beam has no deflection anywhere and is under uniform compressive stress of

$$f = \frac{F}{A_c} = \frac{320,000}{360} = -889 \text{ psi}$$

Owing to $P = 14$ kips, the moment M at midspan is

$$M = \frac{PL}{4} = \frac{14 \times 48}{4} = 168 \text{ kip-ft}$$

and the extreme fiber stresses are

$$f = \frac{Mc}{I} = \frac{6M}{bd^2} = \frac{6 \times 168 \times 12,000}{12 \times 30^2} = \pm 1120 \text{ psi}$$

The resulting stresses at midspan are

$$f_{\text{top}} = -889 - 1120 = -2009 \text{ psi compression}$$
$$f_{\text{bot}} = -889 + 1120 = +231 \text{ psi tension}$$

Note that the actual cable placement may not possess the sharp bend shown over the supports, and the effect of any deviation from the theoretical position must be investigated accordingly. Also note that $F = 320$ kips is the effective prestress, so that under the initial prestress there will be a slight camber at midspan and either a camber or a deflection at the tips which can be computed.

For better stress conditions under the load P, it would be desirable to relocate the c.g.s. so that it would have more sag at midspan. Then a balanced condition would not exist under the uniform load w.

24. Deflections While controlled deflections resulting from prestress can be advantageously used to produce desired cambers and to offset load deflection, excessive camber can cause serious trouble. Deflection owing to prestress can be computed as in Example 6.

Example 6 A concrete beam of 32-ft simple span (Fig. 18) is posttensioned with 1.2 sq in. of high-tensile steel to an initial prestress of 140 ksi immediately after prestressing. Compute the initial deflection at midspan due to prestress and the beam's own weight, assuming $E_c = 4,000,000$ psi.

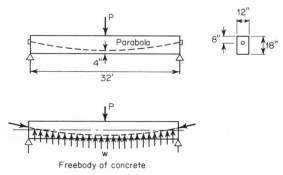

Freebody of concrete

Fig. 18 Example 6.

Estimate the deflection after 3 months, assuming a creep coefficient $C_c = 0.8$ and an effective prestress of 120 ksi at that time.

SOLUTION. The parabolic tendon with 6-in. midordinate is replaced by a uniform load acting along the beam with intensity

$$w = \frac{8Fh}{L^2} = \frac{8 \times 140,000 \times 1.2 \times 6}{32^2 \times 12} = 655 \text{ plf}$$

The prestress force is eccentric 1 in. at each end of the beam, producing a moment of $140,000 \times 1.2 \times 1/12 = 14,000$ ft-lb.

Since the weight of the beam is 225 plf, the net uniform load on the concrete is $655 - 225 = 430$ plf, which produces an upward deflection at midspan given by

$$\Delta = \frac{5wL^4}{384EI} = \frac{5 \times 430 \times 32^4 \times 12^3}{384 \times 4,000,000 \times (12 \times 18^3)/12} = 0.434 \text{ in.}$$

The end moments produce a downward deflection given by

$$\Delta = \frac{ML^2}{8EI} = \frac{140,000 \times 1.2 \times 1 \times 32^2 \times 12^2}{8 \times 4,000,000 \times (12 \times 18^3)/12} = 0.133 \text{ in.}$$

Thus the net deflection due to prestress and beam weight is

$$0.434 - 0.133 = 0.301 \text{ in. upward}$$

To calculate the estimated deflection at 3 months, the prestress deflection is reduced proportionately for loss of prestress, and the resulting net deflection is multiplied by 1.8 to allow for creep. The camber due to initial prestress is $0.434 \times 655/430 - 0.133 = 0.528$ in. The initial deflection due to beam weight is $0.434 \times 225/430 = 0.227$ in. Thus the required deflection is

$$\Delta = 1.8 \left(0.528 \times \frac{120}{140} - 0.227\right) = 0.407 \text{ in. upward}$$

Deflections resulting from external load are calculated in the usual manner for homogeneous beams, provided the concrete is not cracked. If the beam is bonded, the moment of inertia should be computed on the basis of the transformed section, but it can be approximated by using the gross concrete section. If the beam is unbonded, it is close enough for practical purposes to use the moment of inertia of the gross section of concrete. For computing instantaneous deflection due to live load, the following approximate values of E_c may be used:

Age of concrete	E_c, ksi	
	Hard rock	Lightweight
1 day...........	4,000	2,500
7 days...........	4,500	3,000
30 days..........	5,000	3,400
1 year..........	5,500	3,800

If live load is of long duration, creep must be considered. Also, if the load produces cracking, the elastic theory can be used only as an approximation. Accurate data concerning deflection after cracking are not available. However, the ultimate deflection can be computed accurately if moment-curvature relationships are known.

It is difficult to predict camber, because it varies not only with E_c and creep of concrete, but also with age, support conditions, temperature and shrinkage differential between top and bottom fibers, and variations in properties of the concrete. It is usually necessary to have experience with the product of a particular plant before accurate prediction can be made. Lacking such experience, camber computations for 1-day strength of 4000 psi may be based on $E_c = 4,000,000$ psi for hard rock and 2,500,000 psi for lightweight concrete. These values may then be modified for loss of prestress and creep, which can be approximated roughly by the following table:

Age of concrete	Ratio of effective to initial steel stress, %	Coefficient of flexural creep	
		Hard rock	Lightweight
1 day..............	94	1.0	1.0
7 days.............	89	1.6	1.3
30 days............	86	2.0	1.5
1 year.............	83	2.5	1.8

It is good practice to balance the dead-load deflection by camber whenever possible. If this is done, flexural creep and variation in E_c will have little effect on camber or deflection.

Formulas for the calculation of prestress camber are given in Fig. 19. In these formulas, moments M_1 and M_2 are determined by multiplying the prestress F (more accurately, its horizontal component) by the corresponding ordinate y.

SHEAR, BOND, AND BEARING

25. Principal Tension Under service-load conditions a prestressed-concrete beam is generally uncracked and behaves elastically. The principal tensile stress in the web is computed as follows:

1. From the total external shear V at the section, deduct the shear V_s carried by the tendon to obtain the shear V_c carried by the concrete

$$V_c = V - V_s \tag{35}$$

Occasionally, though rarely, $V_c = V + V_s$; this happens when the cable inclination is such that it adds to the shear on the concrete.

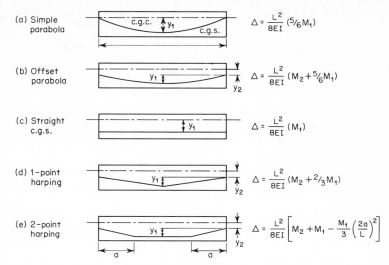

(a) Simple parabola $\quad\Delta = \dfrac{L^2}{8EI}(5\!\!/\!\!_6 M_1)$

(b) Offset parabola $\quad\Delta = \dfrac{L^2}{8EI}(M_2 + 5\!\!/\!\!_6 M_1)$

(c) Straight c.g.s. $\quad\Delta = \dfrac{L^2}{8EI}(M_1)$

(d) 1-point harping $\quad\Delta = \dfrac{L^2}{8EI}(M_2 + 2\!\!/\!\!_3 M_1)$

(e) 2-point harping $\quad\Delta = \dfrac{L^2}{8EI}\left[M_2 + M_1 - \dfrac{M_1}{3}\left(\dfrac{2a}{L}\right)^2\right]$

Fig. 19   Formulas for computing midspan camber due to prestress (simple beams).

2. Compute the distribution of V_c by

$$v = \frac{V_c Q}{Ib} \tag{36}$$

where v = shearing unit stress at any given level

Q = statical moment of the cross-sectional area above (or below) that level about the centroidal axis

b = width of section at that level

3. Compute the fiber stress distribution for that section due to external moment M and the prestress F

$$f_c = \frac{F}{A} \pm \frac{Fec}{I} \pm \frac{Mc}{I} \tag{37}$$

4. The maximum principal tensile stress S_t is given by

$$S_t = \sqrt{v^2 + \left(\frac{f_c}{2}\right)^2} - \frac{f_c}{2} \tag{38}$$

The greatest principal tensile stress does not necessarily occur at the centroidal axis, where the maximum vertical shearing stress exists. At some point, where f_c is diminished, a higher principal tension may exist even though v is not a maximum.

For sections without web reinforcement, S_t is often limited to $2\sqrt{f_c'}$ or less.

In composite construction the shearing stress v between precast and in-place portions is computed from Eq. (36) with V = the total shear applied after the in-place portion has been cast. The allowable value of v varies from about 40 psi for a smooth surface without ties to 160 psi for a roughened surface with adequate ties. Ties are not usually required for composite slabs or panels with large contact areas. For beams with a narrow top flange composite with in-place slabs, ties are almost always needed.

Push-off tests at the Portland Cement Association indicate an ultimate stress for composite action of about 500 psi for a rough bonded surface and 300 psi for a smooth bonded surface. About 175 psi may be added for each 1 percent stirrup reinforcement crossing the joint.

26. Web Reinforcement   Tension cracks which may develop under combined flexure and shear reduce moment capacity unless web reinforcement is provided. ACI318-77 gives an empirical method, based on ultimate-strength test results at the University of Illinois, for design of web reinforcement. The yield strength of the stirrups extending over the length d is assumed to be effective in transmitting the shear (Fig. 20). Thus

$$\frac{A_v f_y d}{s} = V_u - V_{ci} \tag{39}$$

$$\frac{A_v f_y d}{s} = V_u - V_{cw} \tag{40}$$

where A_v = area of one stirrup
f_y = yield stress of stirrup, not more than 60,000 psi
d = effective depth of section, in.
s = spacing of stirrups, in.
V_u = shear at ultimate load
V_{ci} = shear at section when the vertical flexure crack starts to develop into an inclined one
V_{cw} = shear at section when web cracking starts without prior flexural cracking
The critical section is taken at $h/2$ from the theoretical point of maximum shear, where h is the overall depth of the beam.

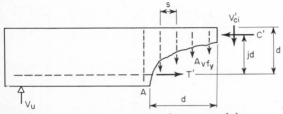

Fig. 20   Ultimate design for combined moment and shear.

V_{ci} in Eq. (39) is given by

$$V_{ci} = \frac{M_{cr}}{M/V} + V_d + 0.6 b_w d \sqrt{f_c} \tag{41}$$

where M_{cr} = cracking moment [Eq. (43)]
M = maximum moment at section due to superimposed loads
V = shear coincident with M
V_d = dead-load shear at section
b_w = width of web
d = distance from extreme fiber in compression to centroid of prestressing tendons or $0.8h$, whichever is larger
This equation is identical in form with ACI318-77, Eq. (11-11). The last term accounts for the additional shear required to cause the vertical flexural crack to develop into an inclined one. Test results show that V_{ci} need not be taken less than $1.7 b_w d \sqrt{f_c'}$.
V_{cw} in Eq. (40) is given by

$$V_{cw} = (3.5\sqrt{f_c'} + 0.3 f_{pc}) b_w d + V_p \tag{42}$$

where f_{pc} = compressive stress in concrete at centroid of section (or at junction of web and flange if centroid is in flange) due to effective prestress
V_p = vertical component of effective prestress at section
This is Eq. (11-13) of ACI318-77. Instead of the value given by this equation, V_{cw} may be taken as the shear at the section for the multiple of dead load plus live load that produces a principal tensile stress of $4\sqrt{f_c'}$ at the centroid, or at the junction of the web and flange if the centroid is in the flange.

The cracking moment is given by

$$M_{cr} = \frac{I}{c_t} (6\sqrt{f_c'} + f_{pe} - f_d) \qquad (43)$$

where I = moment of inertia of beam section
  c_t = distance from c.g.c. to extreme fiber in tension
  f_{pe} = effective prestress at extreme fiber which is in tension due to superimposed load
  f_d = dead-load stress at extreme fiber which is in tension due to superimposed load
  Procedures for design of web reinforcement are the same as for nonprestressed members (Sec. 11, Art. 12) except that, if the effective prestress force is equal to at least 40 percent of the tensile strength of the flexural reinforcement, the minimum area of web reinforcement may be determined by

$$A_v = \frac{A_{ps}}{80} \frac{f_{pu}}{f_y} \frac{s}{d} \sqrt{\frac{d}{b_w}} \qquad (44)$$

instead of the formula in Sec. 11.

Example 7   Assume for the cantilever beam shown in Fig. 21: $f_c' = 5000$ psi, $f_y = 40,000$ psi (stirrups), effective prestress = 390 kips. Design the web reinforcement by ACI318-77.

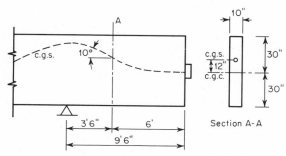

Fig. 21   Example 7.

SOLUTION. The weight of the beam is $10 \times 60 \times 150/144 = 625$ lb/ft. The critical section is at $h/2 = 30$ in. from the support, and the design shears and moments at the section are

$$V_d = 1.4 \times 0.625 \times 6 = 5.25 \text{ kips}$$
$$M_d = 3V_d = 15.75 \text{ ft-kips}$$
$$V = 1.7 \times 160 = 272 \text{ kips}$$
$$M = 1.7 \times 160 \times 6 = 1632 \text{ ft-kips}$$

The section modulus $I/c_t = 10 \times 60^2/6 = 6000$ in.^{3} and the cross-sectional area $= 60 \times 10 = 600$ in.^{2} Then

$$f_d = \frac{15.75 \times 12}{6000} = 0.0315 \text{ ksi}$$
$$f_{pc} = \frac{390}{600} = 0.65 \text{ ksi}$$

The distance from c.g.c. to c.g.s. is 12 in. (Fig. 21). Therefore,

$$f_{pe} = f_{pc} + \frac{390 \times 12}{6000} = 1.43 \text{ ksi}$$

With $\sqrt{f_c'} = \sqrt{5000} = 70.7$ psi $= 0.0707$ ksi, Eq. (43) gives $M_{cr} = 6000 (6 \times 0.0707 + 1.43 - 0.0315) = 10,940$ in.-kips $= 911$ ft-kips
  The vertical component V_p of the tendon tension is $390 \sin 10° = 67$ kips. Since $d = 42$ in. and $0.8h = 48$ in., use $d = 48$ in. From Eqs. (41) and (42),

$$V_{ci} = \frac{911}{1632/272} + 5.25 + 0.6 \times 10 \times 48 \times 0.0707 = 177 \text{ kips}$$
$$V_{cw} = (3.5 \times 0.0707 + 0.3 \times 0.65) \, 10 \times 48 + 67 = 279 \text{ kips}$$

With the required capacity-reduction factor $\phi = 0.85$, Eq. (7) of Sec. 11 gives

$$\frac{\phi A_v f_y d}{s} = V_u - \phi V_{ci} = 5.25 + 272 - 0.85 \times 177 = 127 \text{ kips}$$

Using ⅝-in. U stirrups, $A_v = 0.62$ in.^{2} and

$$s = \frac{0.85 \times 0.62 \times 40 \times 48}{127} = 8.0 \text{ in.}$$

27. Prestress Transfer Bond   Pretensioned tendons usually transfer their stress to the surrounding concrete through bond. The length of transfer varies with many factors. For plain wires, it averages about 100 diameters; for seven-wire strands it is often taken as 50 to 75 diameters.

If a flexural crack should occur near the end of a pretensioned member, there is a tendency for the tendons to be pulled out as a result of bond slippage. For seven-wire strands without mechanical anchorage, the following formula gives the minimum length of embedment L in inches, required to prevent such slippage:

$$L = (f_{ps} - \tfrac{2}{3} f_{se})D$$

where f_{ps} = stress in steel to be developed at ultimate moment, ksi
f_{se} = effective prestress, ksi
D = diameter of strand, in.

28. Anchorage   For tendons with end anchorages, where the prestress is transferred to the concrete by direct bearing, the stress may be transmitted by steel plates, steel blocks, or reinforced-concrete blocks. Stress analysis for anchorages is complicated, and as a result they are often designed by experience, tests, and usage. Since they are usually supplied by the prestressing companies, the engineer does not ordinarily have to design them.

The allowable bearing stress depends on several factors, such as the amount of reinforcement at the anchorage, the ratio of bearing area to the total area, and the assumptions made in computing the stress. The value $0.60f'_c$ is commonly used, assuming uniform bearing over the contact area. Many codes and bridge design specifications prescribe the allowable bearing stress as $0.6f'_{ci}\sqrt[3]{A'/A_b} < f'_{ci}$, where A_b is the bearing area of the anchor plate and A' is the area of the maximum portion of the anchorage surface that is geometrically similar to and concentric with the area A_b.

Because of strict economy in the design of end anchorages, it has not been unusual for poor concrete to fail under application of prestress. Therefore, concrete must be of high quality and must be carefully placed at the anchorages.

End Block.   The portion of a prestressed member surrounding the anchorages of the tendons is often called the end block. The theoretical length of the block, sometimes called the lead length, is the distance required to transfer the prestress and distribute it throughout the entire beam cross section. End blocks are required for beams with posttensioning tendons but are not needed where all tendons are pretensioned wires or seven-wire strand.

In posttensioned members a closely spaced grid of vertical and horizontal bars must be placed near the end face of the end block to resist bursting, and closely spaced vertical and horizontal reinforcement is required throughout the length of the block.

In pretensioned beams, vertical stirrups acting at a unit stress of 20,000 psi to resist at least 4 percent of the total prestressing force should be placed within the distance $d/4$ of the end of the beam, with the end stirrup as close to the end of the beam as is practicable. The Portland Cement Association Laboratories developed an empirical equation for the design of stirrups to control horizontal cracking in the ends of pretensioned I-girders,

$$A_t = 0.021 \frac{T}{f_s} \frac{h}{l_t}$$

where A_t = required total cross-sectional area of stirrups at the end of girder, to be uniformly distributed over a length equal to one-fifth of the girder depth

T = total effective prestress force, lb
f_s = allowable stress for the stirrups, psi
h = depth of girder, in.
l_t = length of transfer, assumed to be 50 times the strand diameter, in.

TYPICAL SECTIONS

29. Beam Sections   Cross sections commonly used for prestressed-concrete beams are the rectangle, the symmetrical I, the unsymmetrical I, the T, the inverted T, and the box. The suitability of these shapes will depend on the simplicity, availability, and reusability of formwork, ease in placing concrete, functional and aesthetic requirements, and theoretical considerations. The rectangular shape is easiest to form but uneconomical in material. The T is suitable for high ratios of M_G/M_T, where there is little danger of overstressing at transfer and where the concrete is effectively concentrated at the compressive flange. The inverted T is good for low ratios of M_G/M_T (to avoid overstressing at transfer) but does not have a high ultimate moment. The I and box have more concrete near the extreme fibers and are efficient both at transfer and under ultimate loads, but have weaker webs and require more complicated forming.

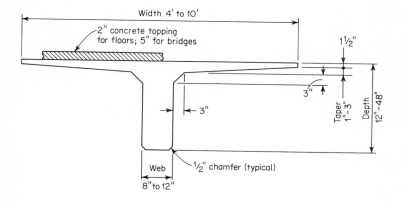

Table of properties
Width = 6'; web = 8"; taper = 2"; no topping

Depth, in.	Area, in.²	I, in.⁴	c_b, in.
12	265	2,360	9.2
16	297	6,170	11.8
20	329	11,730	14.5
24	361	19,700	17.0
28	393	30,400	19.5
32	425	44,060	21.9
36	457	61,000	24.2

Fig. 22   Single T-section.

Typical T-sections are shown in Figs. 22 and 23. Properties of T-, I-, and box sections are given in Table 12. Typical bridge sections are shown in Sec. 18, Figs. 84, 85, and 86.

30. Span-Depth Ratios   For reasons of economy and aesthetics, higher span-depth ratios are almost always used for prestressed concrete than for reinforced concrete. Higher ratios are possible because deflection can be much better controlled. On the other hand, when these ratios get too high, camber and deflection become quite sensitive to variations in

TABLE 12   Properties of Sections

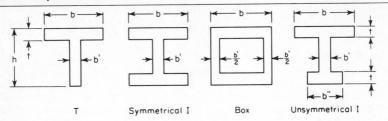

T	Symmetrical I	Box	Unsymmetrical I

T-section

b'/b	t/h	A	c_b	c_t	I	r^2	k_t	k_b
0.1	0.1	$0.19bh$	$0.714h$	$0.286h$	$0.0179bh^3$	$0.0945h^2$	$0.132h$	$0.333h$
0.1	0.2	0.28	0.756	0.244	0.0192	0.0688	0.0910	0.282
0.1	0.3	0.37	0.755	0.245	0.0193	0.0520	0.0689	0.212
0.1	0.4	0.46	0.735	0.265	0.0202	0.0439	0.0597	0.165
0.2	0.1	0.28	0.629	0.371	0.0283	0.1010	0.161	0.272
0.2	0.2	0.36	0.678	0.322	0.0315	0.0875	0.129	0.272
0.2	0.3	0.44	0.691	0.309	0.0319	0.0725	0.105	0.234
0.2	0.4	0.52	0.684	0.316	0.0320	0.0616	0.090	0.195
0.3	0.1	0.37	0.585	0.415	0.0365	0.0985	0.169	0.237
0.3	0.2	0.44	0.626	0.374	0.0408	0.0928	0.148	0.248
0.3	0.3	0.51	0.645	0.355	0.0417	0.0819	0.127	0.231
0.3	0.4	0.58	0.645	0.355	0.0417	0.0720	0.112	0.203
0.4	0.1	0.46	0.559	0.441	0.0440	0.0954	0.171	0.216
0.4	0.2	0.52	0.592	0.408	0.0486	0.0935	0.158	0.229
0.4	0.3	0.58	0.609	0.391	0.0499	0.0860	0.141	0.220
0.4	0.4	0.64	0.612	0.388	0.0502	0.0785	0.128	0.205
1.0	1.0	1.00	0.500	0.500	0.0833	0.0833	0.167	0.167

Symmetrical I- and Box Sections

b'/b	t/h	A	c_b	c_t	I	r^2	k_t	k_b
0.1	0.1	$0.28bh$	$0.500h$	$0.500h$	$0.0449bh^3$	$0.160h^2$	$0.320h$	$0.320h$
0.1	0.2	0.46	0.500	0.500	0.0671	0.146	0.292	0.292
0.1	0.3	0.64	0.500	0.500	0.0785	0.123	0.246	0.246
0.2	0.1	0.36	0.500	0.500	0.0492	0.137	0.274	0.274
0.2	0.2	0.52	0.500	0.500	0.0689	0.132	0.264	0.264
0.2	0.3	0.68	0.500	0.500	0.0791	0.117	0.234	0.234
0.3	0.1	0.44	0.500	0.500	0.0535	0.121	0.243	0.243
0.3	0.2	0.58	0.500	0.500	0.0707	0.122	0.244	0.244
0.3	0.3	0.72	0.500	0.500	0.0796	0.111	0.222	0.222
0.4	0.1	0.52	0.500	0.500	0.0577	0.111	0.222	0.222
0.4	0.2	0.64	0.500	0.500	0.0725	0.113	0.226	0.226
0.4	0.3	0.76	0.500	0.500	0.0801	0.105	0.211	0.211

TABLE 12 Properties of Sections *(Continued)*

Unsymmetrical I-sections

b''/b	b'/b	t/h	A	c_b	c_t	I	r^2	k_t	k_b
0.3	0.1	0.1	$0.21bh$	$0.650h$	$0.350h$	$0.0260bh^3$	$0.1236h^2$	$0.190h$	$0.354h$
0.3	0.1	0.2	0.32	0.675	0.325	0.0345	0.1080	0.160	0.332
0.3	0.1	0.3	0.43	0.672	0.328	0.0387	0.0900	0.134	0.274
0.3	0.2	0.1	0.29	0.610	0.390	0.0316	0.1090	0.179	0.280
0.3	0.2	0.2	0.38	0.647	0.353	0.0378	0.0994	0.153	0.282
0.3	0.2	0.3	0.47	0.655	0.345	0.0402	0.0856	0.131	0.248
0.5	0.1	0.1	0.23	0.597	0.403	0.0326	0.1420	0.238	0.352
0.5	0.1	0.2	0.36	0.611	0.389	0.0464	0.1288	0.210	0.331
0.5	0.1	0.3	0.49	0.606	0.394	0.0535	0.1090	0.180	0.274
0.5	0.2	0.1	0.31	0.572	0.428	0.0373	0.1204	0.210	0.282
0.5	0.2	0.2	0.42	0.595	0.405	0.0488	0.1160	0.195	0.286
0.5	0.2	0.3	0.53	0.599	0.401	0.0540	0.1020	0.170	0.254
0.5	0.3	0.1	0.39	0.557	0.443	0.0430	0.1103	0.198	0.250
0.5	0.3	0.2	0.48	0.582	0.418	0.0510	0.1065	0.183	0.255
0.5	0.3	0.3	0.57	0.592	0.408	0.0553	0.0970	0.164	0.238
0.7	0.1	0.1	0.25	0.554	0.446	0.0381	0.1525	0.276	0.342
0.7	0.1	0.2	0.40	0.560	0.440	0.0560	0.1391	0.248	0.316
0.7	0.1	0.3	0.55	0.557	0.443	0.0651	0.1182	0.212	0.267
0.7	0.2	0.1	0.33	0.540	0.460	0.0425	0.1290	0.239	0.280
0.7	0.2	0.2	0.46	0.552	0.448	0.0578	0.1258	0.228	0.281
0.7	0.2	0.3	0.59	0.553	0.447	0.0657	0.1113	0.202	0.249
0.7	0.3	0.1	0.41	0.534	0.466	0.0467	0.1140	0.214	0.244
0.7	0.3	0.2	0.52	0.546	0.454	0.0598	0.1150	0.210	0.254
0.7	0.3	0.3	0.63	0.550	0.450	0.0663	0.1051	0.191	0.234

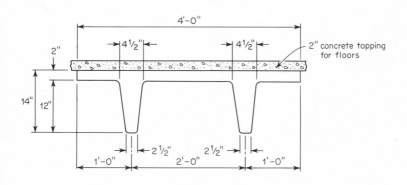

Table of properties

	Area, in.^{2}	I, in.^{4}	c_b, in
With 2" topping	276	4,456	11.74
Without topping	180	2,862	10.00

Fig. 23 Double T-section.

loadings, in properties of materials, in magnitude and location of prestress, and in temperature. Furthermore, the effects of vibration become more pronounced. Care should be taken with cantilever beams, since they are particularly sensitive to deflection and vibration.

Span-depth ratio limitations should vary with the nature and magnitude of the live load, the damping characteristics, the boundary conditions, the shape and variations of the section, the modulus of elasticity, and the span. If the structure is carefully investigated for camber, deflection, and vibration, there is no reason to adhere to any given ratio.

The limiting values in Table 13 may be used as a preliminary guide for building design. In general, with span-depth ratios some 10 percent below the tabulated values, problems of camber, deflection, and vibration are not likely to develop unless the loadings are extremely heavy and vibratory in nature. On the other hand, these ratios can be exceeded by 10 percent or more if careful study ensures acceptable behavior. The ratios are intended for both hard-rock concrete and lightweight concrete but should be reduced by about 5 percent for lightweight concrete having E_c less than 3,000,000 psi. For long spans (say, in excess of about 70 ft) and for heavy loads (say, live loads over 100 psf) the values should be reduced by 5 to 10 percent. For in-place concrete in composite action with precast elements, the total depth may be considered in computing span-depth ratios.

Experience with prestressed-concrete railway bridges is not sufficient to establish span-depth ratios. Usual ratios have been in the range of 10 to 14 for box sections up to 100 ft or more. For simple-span highway bridges of the I-beam type, up to about 200 ft, a span-depth ratio of 20 is considered conservative, 22 to 24 is normal, while 26 to 28 would be the critical limit. Box sections can have ratios about 5 to 10 percent higher than I-beams, while T-sections spaced far apart should have ratios about 5 to 10 percent lower than I-beams. Again, there is no reason to believe that a fixed span-depth ratio will apply to all cases.

31. Cable Layouts Typical cable layouts for pretensioned and posttensioned simple spans are shown in Figs. 24 and 25, respectively. Layouts for single and double cantilevers are shown in Figs. 26 and 27.

32. Tendon Protection and Spacing Minimum concrete protection for tendons is governed by requirements for fire resistance and for corrosion protection. ACI318-77 specifies the following minimum thickness (inches) of concrete cover for prestressing steel, ducts, and nonprestressed steel:

	Min cover, in.
Cast against and permanently exposed to earth	3
Exposed to earth or weather:	
Wall panels, slabs, and joists	1
Other members	1½
Not exposed to weather or in contact with the ground:	
Slabs, walls, joists	¾
Beams, girders, columns:	
Principal reinforcement	1½
Ties, stirrups, or spirals	1
Shells and folded-plate members:	
Reinforcement ⅝ in. and smaller	⅜
Other reinforcement	Nominal diameter but not less than ¾

Minimum spacing of tendons is governed by several factors. First, the clear spacing between tendons, or between tendons and side forms, must be sufficient to permit easy placing of concrete. A minimum of 1⅓ times the size of the maximum aggregate is recommended. Second, to develop the bond between steel and concrete properly, the clear distance between bars should be at least the diameter of the bars for special anchorage and 1½ times the diameter for ordinary anchorage, with a minimum of 1 in. These limitations may not be necessary for small wires and strands, which are often bundled together.

ACI318 calls for a minimum clear spacing at each end of the member of four times the diameter of individual wires or three times the diameter of strands, in order to develop the transfer bond properly.

33. Partial Prestress Partial prestress in prestressed concrete may mean one or more of the following:
   1. Tensile stresses are permitted in concrete under working loads.
   2. Nonprestressed reinforcements, whether of mild steel or high-tensile steel, are employed in addition to tendons.
   3. Tendons are stressed to a lower level than usual.

TABLE 13   Approximate Limits for Span-Depth Ratios

	Continuous spans		Simple spans	
	Roof	Floor	Roof	Floor
One-way solid slabs.........................	52	48	48	44
Two-way solid slabs (supported on columns only)..	48	44	44	40
Two-way waffle slabs (3-ft waffles)..............	40	36	36	32
Two-way waffle slabs (12-ft waffles).............	36	32	32	28
One-way slabs with small cores.................	50	46	46	42
One-way slabs with large cores.................	48	44	44	40
Double tees and single tees (side by side)........	44	40	40	36
Single tees (spaced 20-ft centers)................	36	32	32	28
Cantilever solid slab: roof 20, floor 18				

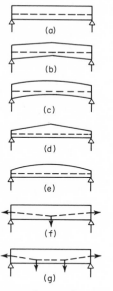

Fig. 24 Layouts for pretensioned beams.

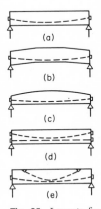

Fig. 25 Layouts for posttensioned beams.

Figure 28 shows load-deflection curves for a bonded beam with its concrete subjected to varying degrees of prestress. The case of full prestress permitting no tension under working load is represented by curve *b*. Partial prestressing permitting tension up to the modulus of rupture is shown by curve *c*. A nonprestressed concrete beam is shown in curve *d*. An overprestressed region is indicated between curves *a* and *b*.

The advantages of partial prestress compared with full prestress are (1) better control of camber, (2) saving in prestressing steel, (3) saving in end anchorages, (4) possible greater resilience in the structure, and (5) economical utilization of mild steel.

The disadvantages of partial prestress are (1) earlier appearance of cracks, (2) greater deflection under overload, (3) higher principal tensile stress under working load, and (4) slight decrease in ultimate flexural strength for the same amount of steel.

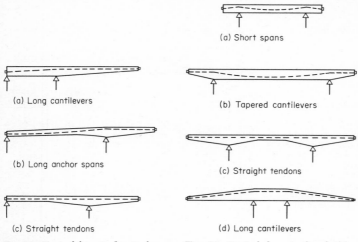

(a) Short spans

(a) Long cantilevers

(b) Tapered cantilevers

(b) Long anchor spans

(c) Straight tendons

(c) Straight tendons

(d) Long cantilevers

Fig. 26 Typical layouts for single cantilevers.

Fig. 27 Typical layouts for double cantilevers.

Nonprestressed reinforcements can be placed at various positions in a prestressed beam to improve its behavior and strength at different stages. Frequently, one set of reinforcement can serve to strengthen the beam in several ways:

1. To provide strength immediately after transfer of prestress:
   a. Along the compression flange, which may be under tension at transfer
   b. Along the tension flange, which may be under high compressive stress at transfer
2. To reinforce certain portions of the beam for special or unexpected loads during handling, transportation, and erection
3. To distribute cracks under working loads
4. To increase ultimate capacity of the beam
5. To help carry high compression in the concrete
6. To reinforce the concrete along directions which are not prestressed: web, end block, and flange slab reinforcements

When nonprestressed reinforcements are used to carry compression, the compressive stress in the steel is generally quite high because of shrinkage and creep in the concrete. When they are used to carry tension, the reinforcements cannot function effectively until the concrete has cracked. However, the design of tension reinforcements is usually made on the assumption that they will be stressed to the usual allowable values (e.g., 20,000 psi for intermediate-grade steel) and that their total tension will replace the tension in the portion of concrete which might be lost as a result of cracking.

Figure 29 shows the stresses and strains produced in various reinforcements under different stages of loading. It is to be noted that nonprestressed reinforcements will be stressed very little under working loads but will be effective at ultimate load, especially for underreinforced beams.

34. Combination of Prestressed and Reinforced Concrete While a combination of prestressed and reinforced concrete is represented in the use of nonprestressed reinforcement, the flexural strength is essentially supplied by the tendons, with the nonprestressed steel playing a minor role. For certain types of construction, a full combination of prestressed and reinforced concrete may be the best design, making use of the advantages

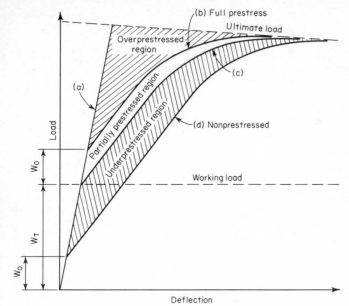

Fig. 28 Load-deflection curves for varying degrees of prestress.

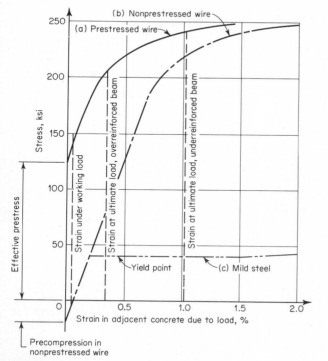

Fig. 29 Stress-strain diagrams.

of both. Reinforced concrete has the advantage of simplicity in construction, monolithic behavior, no camber, less creep, and reasonably high ultimate strength. Prestressed concrete utilizes high-strength steel economically, produces a favorable distribution of stress under certain conditions of loading, and controls deflection and cracking.

Certain structural elements and systems favor reinforced concrete, others prestressed concrete, still others partially prestressed concrete. Some will be best designed with a combination of reinforced and prestressed concrete having the nonprestressed steel carrying perhaps 50 percent or more of the total ultimate load.

One occasion for the use of this combination is the case of high live-load to dead-load ratio, when prestressing alone may produce excessive camber. Another is the case of high added dead load requiring prestressing in stages, which may be cumbersome. A third case is the requirement of high ultimate strength or resilience to resist dynamic loadings. There is also reason to believe that a heavy amount of nonprestressed steel used in conjunction with unbonded tendons will result in economy and in developing a high ultimate stress in the tendons.

For precast columns, prestressing will help control cracking during transportation and erection and will contribute to the bending strength. Nonprestressed steel will increase both the axial load and the flexural capacity. Hence a combination may be the best solution for certain cases. The use of nonprestressed reinforcement for joineries and continuity is, of course, often a simple and economical solution.

Nonprestressed steel does not act until the concrete cracks, and does not contribute toward the precracking strength. Hence if cracking could result in a primary or a secondary failure, nonprestressed steel may be of no help. The possibility of corrosion of the prestressing steel if the member cracks too early or too often should also be investigated.

CONTINUOUS BEAMS

Continuous beams may be fully cast in place with tendons continuous from one end to the other (Fig. 30). They may also be precast in smaller elements which are made continuous by special posttensioning arrangements (Fig. 31).

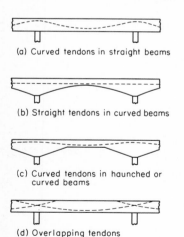

(a) Curved tendons in straight beams

(b) Straight tendons in curved beams

(c) Curved tendons in haunched or curved beams

Fig. 30 Layouts for fully continuous beams.

(d) Overlapping tendons

No support reactions are induced by prestressing a statically determinate system. In a continuous beam, or any statically indeterminate system, support reactions are generally induced by the application of prestress, because the bending of the beam due to prestress may tend to deflect the beam at its supports. These reactions produce secondary moment in the beam.

35. Continuous-Beam C Lines Under the action of prestress alone, neglecting the weight of the beam and all other external loads, the C line (the line of pressure in the concrete) in a simple beam coincides with the c.g.s. line. In a continuous beam, the C line departs from the c.g.s. line in amounts required to resist the secondary moments. When

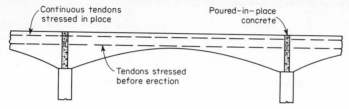

(a) Continuous tendons stressed after erection

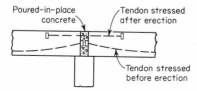

(b) Short tendons stressed over supports

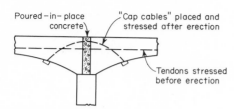

(c) Cap cables over supports

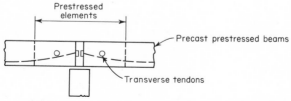

(d) Continuous elements over supports transversely prestressed

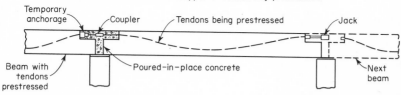

(e) Couplers over supports

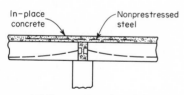

(f) Nonprestressed steel over supports

Fig. 31 Layouts for partially continuous beams.

12-41

the c.g.s. line is coincident with the C line, it is termed a "concordant cable." All simple beams and some continuous beams have concordant cables, while most continuous beams have nonconcordant cables.

While a nonconcordant cable usually gives a more economical solution, the concordant cable is sometimes preferred because it induces no external reactions. The concordant cable works better in a precast continuous beam. Several methods have been proposed for obtaining concordant cables. A simple rule is the following: Every real moment diagram for a continuous beam on nonsettling supports, produced by any combination of external loadings, whether transverse loads or moments, plotted to any scale, is one location for a concordant cable in that beam. This theorem is easily proved. Since the moments in a continuous beam are computed on the basis of no deflection over the supports, and since any c.g.s. line following the corresponding moment diagram will produce a similar moment diagram, that c.g.s. line will also produce no deflection over the supports; hence it will induce no reactions and is a concordant cable.

The C line can be located from the c.g.s. line by a linear transformation. Linear transformation is defined as the location of the C line from the c.g.s. line by displacements at the interior supports without changing its intrinsic shape within each span. Since the C line deviates from the c.g.s. line on account of the moments produced by the induced reactions, and since such moments vary linearly within the span, it follows that the C line can be linearly transformed.

A simple method of determining the C line is shown in Examples 8 and 9.

Example 8   A continuous prestressed-concrete beam with bonded tendons is shown in Fig. 32a. The c.g.s. is eccentric at A, bent sharply at D and B, and has a parabolic curve for the span BC. Locate the line of pressure (the C line) in the concrete due to prestress alone. Consider a prestress of 250 kips.

SOLUTION. The primary moment diagram due to prestress is shown in b. The corresponding shear diagram is shown in c, from which the loading diagram is drawn in d. The fixed-end moments are computed for this loading. At A there is the additional moment $0.2 \times 250 = 50$ kip-ft resulting from the prestress eccentricity. Moment distribution is performed in e.

The eccentricity of the line of pressure at B is $246/250 = 0.98$ ft. The line of pressure for the beam can be computed by plotting its moment diagram and dividing the ordinates by the value of the prestress. But this is not necessary; since the line of pressure deviates linearly from the c.g.s. line, it is only necessary to move the c.g.s. line so that it passes through the points located over the supports, as shown in f. The line of pressure at D is translated upward by the amount $(0.98 - 0.4) 30/50 = 0.35$ ft and is now located $0.80 - 0.35 = 0.45$ ft below the c.g.c. line. At midspan of BC, the line of pressure is translated upward by the amount $(0.98 - 0.4) 25/50 = 0.29$ ft and is now located 0.61 ft below the c.g.c. line.

Example 9   A uniform load of 1.2 kips/ft is applied to the beam of Example 8. Compute the stresses in the concrete at section B, where $I = 39,700$ in.^{4} and $A_c = 288$ in.^{2} (Fig. 33).

SOLUTION. The moment diagram for distributed load is plotted in c. Dividing these moments by the prestress of 250 kips gives the C line in d. Adding d to f of Example 8, the pressure line for both prestress and external load is given in e.

The resulting moment at section B is $250 \times 0.52 = 130$ kip-ft, from which

$$f = \frac{-250}{288} \pm \frac{130 \times 12 \times 18}{39,700} = -0.867 \pm 0.707 \text{ ksi}$$
$$= -160 \text{ psi top fiber}$$
$$= -1574 \text{ psi bottom fiber}$$

36. Load-Balancing Method   This method is convenient for both the analysis and the design of prestressed continuous beams. When the external load is exactly balanced by the transverse component of the prestress, the beam is under a uniform stress $f = F/A_c$ across any section. For any change in load from that balanced condition, only the effects of change need be computed. For example, if the additional moment is M, the additional stresses are given by $f = My/I$. Thus, after load balancing, the analysis of prestressed continuous beams is reduced to the analysis of nonprestressed continuous beams. Since such analysis will be applied to only the unbalanced portion of the load, approximate methods may often prove sufficient.

Design by the load-balancing method gives a different visualization of the problem. It becomes a simple matter to lay out the cables in an economical manner and to compute the required prestress and the corresponding fiber stresses in concrete. This is illustrated in Example 10.

In the load-balancing method it is often assumed that the dead load of the structure is balanced by the effective prestress. Therefore, a slight amount of camber may exist under the initial prestress. It is not always necessary to balance all the dead load, since such balancing may require too much prestress, and a limited amount of deflection may not be objectionable. On the other hand, when the live load is large compared with the dead load, it may be necessary to balance some of the live load in addition to the dead load.

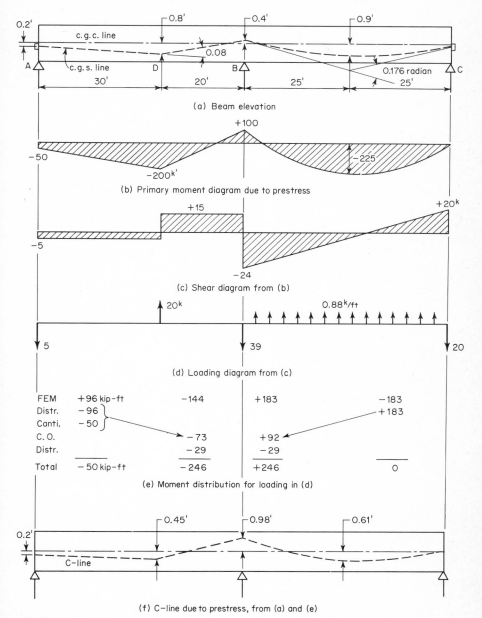

Fig. 32 Example 8.

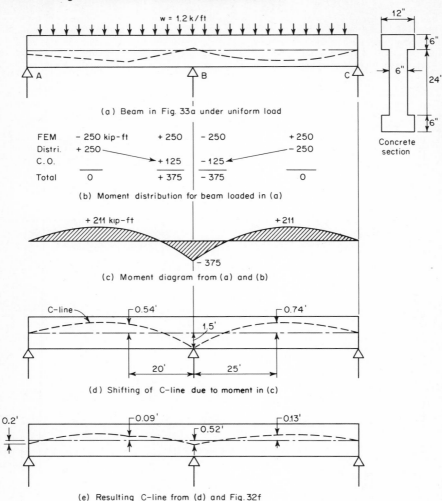

(a) Beam in Fig. 33a under uniform load

(b) Moment distribution for beam loaded in (a)

(c) Moment diagram from (a) and (b)

(d) Shifting of C-line due to moment in (c)

(e) Resulting C-line from (d) and Fig. 32f

Fig. 33   Example 9.

Example 10   For the continuous beam in Fig. 34, determine the prestress F required to balance a uniform load of 1.03 kips/ft, using the most economical location of cable. Assume a concrete protection of at least 3 in. for the c.g.s. Compute the midspan section stresses and the reactions for the effect of prestress and an external load of 1.6 kips/ft.

SOLUTION.   The most economical cable location is one with the maximum sag so that the least amount of prestress will be required to balance the load. A 3-in. protection is given to the c.g.s. over the center support and at midspan (a theoretical parabola based on these clearances will have slightly less than 3 in. at a point about 20 ft from the exterior support). The c.g.s at the beam ends should coincide with the c.g.c. and cannot be raised, not only because such raising will destroy the load balancing but because it will not help to increase the efficiency of the cable, since unfavorable end moments will be introduced.

The cable now has a sag of 18 in., and the prestress F required to balance the load of 1.03 kips/ft is

$$F = \frac{wL^2}{8h} = \frac{1.03 \times 50^2}{8 \times 1.5} = 214 \text{ kips}$$

The fiber stress under this balanced-load condition is

$$f = \frac{F}{A_c} = \frac{214,000}{360} = -593 \text{ psi}$$

Owing to the additional load of $1.6 - 1.03 = 0.57$ kip/ft the negative moment at the center support is

$$M = \frac{wL^2}{8} = 0.57 \times \frac{50^2}{8} = -178 \text{ kip-ft}$$

and

$$f = \frac{Mc}{I} = \frac{6 \times 178 \times 12,000}{12 \times 30^2} = \pm989 \text{ psi}$$

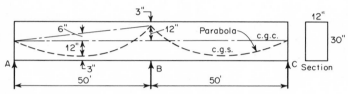

Fig. 34   Example 10.

The resulting stresses at the center support are

$$f = -593 + 989 = +396 \text{ psi tension top}$$
$$f = -593 - 989 = -1582 \text{ psi compression bottom}$$

The reactions due to 1.03 kips/ft can be computed from the vertical components of the cable and are, very closely,

Exterior support: $R_A = 1.03 \times 25 - \dfrac{214}{50} = 25.8 - 4.3 = 21.5 \text{ kips}$

Interior support: $R_B = 51.6 + 2 \times 4.3 = 60.2 \text{ kips}$

Under the action of 0.57 kip/ft load, the reactions are, by the elastic theory,

Exterior support: $R_A = 10.6 \text{ kips}$
Interior support: $R_B = 35.6 \text{ kips}$

Hence the total reactions due to 1.6 kips/ft load and the effect of $F = 214$ kips are

Exterior support: $R_A = 21.5 + 10.6 = 32.1 \text{ kips}$
Interior support: $R_B = 60.2 + 35.6 = 95.8 \text{ kips}$

37. Ultimate Strength of Continuous Beams   The ultimate strength of prestressed continuous beams can be estimated by limit analysis. Plastic hinges form at points of maximum moment in underreinforced beams. Complete plastic hinges may not develop at cross sections where shear is large and in overreinforced beams, in which case the action will be only partly plastic.

Cracking in prestressed continuous beams can be computed by the elastic theory; cracking begins when the tensile fiber stress reaches the modulus of rupture of the concrete.

DESIGN EXAMPLES

Example 11   Design a precast, prestressed roof T panel with the cross section shown in Fig. 35. Given: simple span 100 ft, roofing and piping 6 psf, live load 20 psf, and with $f'_{ci} = 3500$ psi, $f'_c = 5000$ psi, $E_c = 4,000,000$ psi.

Allowable tension: $f_b = 6\sqrt{f'_c} = 424$ psi at full dead load plus live load
Allowable compression: $= 0.45f'_c = 2250$ psi at full dead load plus live load
    $= 0.60f'_{ci} = 2100$ psi at transfer

Pretensioning tendon ½-in. 7-wire strand (Table 5). Posttensioning ¼-in. buttonhead wires (Table 4).

$f_{pu} = 250,000$ psi, $f_0 = 175,000$ psi, $f_{se} = 145,000$ psi, $f_v' = 40,000$ psi. Properties of T (Fig. 22): $A_c = 457$ in.^{2}, $I_c = 61,000$ in.^{4}, $c_b = 24.2$ in.

SOLUTION.

$$w_G = 457 \times \frac{150}{144} = 476 \text{ plf} \qquad w_s = 6 \times 6 = 36 \text{ plf}$$

$$w_L = 6 \times 20 = 120 \text{ plf} \qquad w_T = 632 \text{ plf}$$

$$M_G = 476 \times \frac{100^2}{8} = 595$$

$$M_S = 36 \times \frac{100^2}{8} = 45$$

$$M_L = 120 \times \frac{100^2}{8} = \underline{150}$$

$$M_T = 790 \text{ kip-ft}$$

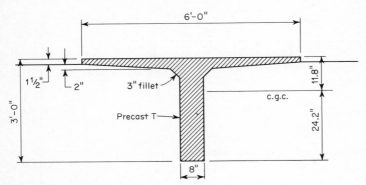

Fig. 35 Example 11.

AMOUNT OF PRESTRESS. Since M_G is large, we can probably locate c.g.s. at the lowest position assuming concrete protection $d' = 4$ in. to the c.g.s. at midspan (Fig. 36). From Eq. (32).

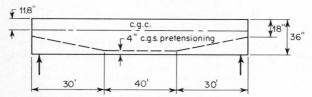

Fig. 36 Location of c.g.s. for girder pretensioning.

$$M_T = F(k_t + c_b - d') + \frac{f_b' I}{c_b}$$

$$k_t = \frac{61,000}{457 \times 24.2} = 5.52 \text{ in.}$$

$$790 \times 12,000 = F(5.5 + 24.2 - 4) + 424 \times \frac{61,000}{24.2}$$

$$F = 327,000 \text{ lb}$$

The corresponding top fiber stress is [Eq. (12)]

$$f_t = \frac{F}{A} - \frac{Fec}{I} + \frac{Mc}{I}$$

$$= \frac{-327,000}{457} + \frac{327,000 \times 20.2 \times 11.8}{61,000} - \frac{790 \times 12,000 \times 11.8}{61,000}$$

$$= -716 + 1276 - 1832 = -1272 < 2250 \text{ psi}$$

The stress in the steel at transfer is somewhat lower than f_0 and may be taken as 165,000 psi. Thus

$$F_0 = \frac{165}{145} \times 327,000 = 372,000 \text{ lb}$$

Then, with $M_G = 595$ kip-ft, we have

$$f_t = \frac{-372,000}{457} + \frac{372,000 \times 20.2 \times 11.8}{61,000} - \frac{595 \times 12,000 \times 11.8}{61,000}$$

$$= -814 + 1455 - 1380 = -739 \text{ psi}$$

$$f_b = -814 - \frac{24.2}{11.8}(1455 - 1380)$$

$$= -814 - 154 = -968 \text{ psi}$$

These fiber stresses at transfer indicate a fairly uniform stress distribution at the midspan section, which is desirable.

PRETENSIONING TENDONS. To supply 327 kips, using ½-in. 7-wire strands with $A_{ps} = 0.1438$ in.² and $F = 145 \times 0.1438 = 20.8$ kips requires $327/20.8 = 15.7$ strands. Use 16 strands arranged as in Fig. 37.

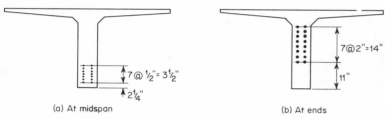

(a) At midspan   (b) At ends

Fig. 37 Arrangement of pretensioning tendons.

CAMBER AT TRANSFER. The camber at transfer can be computed from Eq. (e) of Fig. 19, to which is added the deflection due to the weight of the beam

$$\Delta = \frac{L^2}{8EI}\left[M_2 + M_1 - \frac{M_1}{3}\left(\frac{2a}{L}\right)^2\right] - \frac{5}{48}\frac{M_G L^2}{EI}$$

$$= \frac{100^2 \times 144}{8 \times 4000 \times 61,000}\left[372 \times 6.2 + 372 \times 14 - \frac{372 \times 14}{3}\left(\frac{2 \times 30}{100}\right)^2 - \frac{5}{6} \times 595 \times 12\right]$$

$$= 0.69 \text{ in. upward}$$

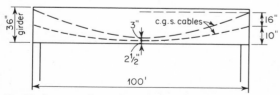

Fig. 38 Location of c.g.s. for girder posttensioning.

POSTTENSIONING. If the girder is posttensioned (rather than pretensioned) with ¼-in. wires with $A_{ps} = 0.049$ in.², the effective prestress is $0.049 \times 145 = 7.1$ kips per wire. The number of wires required is $327/7.1 = 46$. Two parabolic tendons of 23 or 24 wires each can be used (Fig. 38).

CRACKING. Cracking strength can be determined from Eq. (16b), assuming the modulus of rupture $f_r = 7.5\sqrt{f_c} = 530$ psi.

$$M = F(e + k_t) + \frac{f_r I}{c}$$

$$= \frac{327(20.2 + 5.52)}{12} + \frac{530 \times 61,000}{24.2 \times 1000 \times 12}$$
$$= 812 \text{ kip-ft}$$

This corresponds to a uniform load $w = 8 \times 812/100^2 = 0.650$ klf, or a live load of $650 - 476 - 36 = 138$ plf $= 138/6 = 23$ psf.

ULTIMATE STRENGTH (ART. 18). Assuming that the ultimate strength of the steel is fully developed, and with $A_{ps} = 16 \times 0.1438 = 2.3$ in.^{2},

$$T' = 2.3 \times 250 = 575 \text{ kips}$$
$$A' = \frac{575}{0.85 \times 5} = 135 \text{ in.}^2$$

The area A' is supplied by the 6-ft flange with $k'd$ about 2 in. The moment arm is $36 - 4 - 1 = 31$ in. and the ultimate moment

$$M_u = 2.3 \times 250 \times 31/12 = 1486 \text{ kip-ft}$$

This corresponds to a uniform load $w' = 8 \times 1486/100^2 = 1.19$ klf. The ACI required ultimate load capacity is $U = 1.4D + 1.7L = 1.4 (476 + 36) + 1.7 \times 120 = 0.921$ klf < 1.19 klf.

STIRRUPS. The maximum shear at design load is $632 \times 50 = 31.6$ kips. Thus

$$v = \frac{VQ}{Ib} = \frac{31,600 \times 8 \times 24.2^2/2}{61,000 \times 8} = 152 \text{ psi}$$

At c.g.c., $f = 327,000/457 = 716$ psi. The principal tension is [Eq. (38)]

$$S_t = \sqrt{152^2 + 358^2} - 358 = 32 \text{ psi}$$

which is very low.

Combined moment and shear failure will not occur in a simple beam such as this, where the tendons are relatively low in the beam. Hence, only nominal stirrups are needed. The minimum requirement is given by Eq. (44).

$$A_v = \frac{A_{ps}}{80} \frac{f_{pu}}{f_y} \frac{s}{d} \sqrt{\frac{d}{b_w}} = \frac{2.3}{80} \times \frac{250}{40} \times \frac{s}{32} \sqrt{\frac{32}{8}} = 0.0112s \text{ in.}^2$$

For single No. 4 bars, $A_v = 0.20$ in.^{2} and $s = 0.20/0.0112 = 17.9$ in. According to the ACI Code, s cannot exceed 24 in. or $\frac{3}{4}d$, whichever is smaller. Therefore, use No. 4 single stirrups at 18 in.

Example 12 The top flange of a composite section 4 in. thick by 60 in. wide is to be cast in place. Design a precast section with a total depth of 36 in. (including the thickness of the slab) for the following moments:

$$M_t = 320 \text{ kip-ft} = \text{total moment on section}$$
$$M_c = 220 \text{ kip-ft} = \text{moment in composite section}$$
$$M_p = 100 \text{ kip-ft} = \text{moment on precast portion}$$
$$M_g = 40 \text{ kip-ft} = \text{girder load moment}$$

The allowable stresses are

$$f_b = 1.80 \text{ ksi} = \text{compression in precast section at transfer}$$
$$f_t' = 0.30 \text{ ksi} = \text{tension in precast section at transfer}$$
$$f_t = 1.60 \text{ ksi} = \text{compression in composite section at working load}$$
$$f_b = 0.16 \text{ ksi} = \text{tension in composite section at working load}$$

Initial prestress $= 150$ ksi; effective prestress $= 125$ ksi.

SOLUTION. Assume the lever arm $= 0.65h$ for M_T.

$$F = \frac{M_T}{0.65h} = \frac{320 \times 12}{0.65 \times 36} = 164 \text{ kips}$$
$$F_0 = 164 \times 150/125 = 197 \text{ kips}$$

The concrete area required for an inverted T can be approximated by

$$A_c = \frac{1.5F_0}{f_b} = \frac{1.5 \times 197}{1.8} = 164 \text{ in.}^2$$

The resulting trial section is shown in Fig. 39. Section properties for the precast portion are

$$\begin{array}{rll}
4 \times 14 = & 56 \times & 2 = & 112 \\
28 \times 4 = & 112 \times & 18 = & 2016 \\
\hline
A_c = & 168 & & 2128 \div 168 = 12.7 \text{ in.} = c_b
\end{array}$$

$$56 \left(\frac{4^2}{12} + 10.7^2 \right) = 6{,}500$$

$$112 \left(\frac{28^2}{12} + 5.3^2 \right) = 10{,}450$$

$$I_c = \overline{16{,}950} \text{ in.}^4$$

Section properties for the composite section are

$$4 \times 60 = 240 \times 2 \quad = \quad 480$$

$$\frac{168 \times 23.3}{408} = \frac{3920}{4400} \div 408 = 10.8 \text{ in.} = c_t$$

$$240 \left(\frac{4^2}{12} + 8.8^2 \right) \quad = 18{,}800$$

$$168 \times 12.5^2 \quad = 26{,}250$$

$$I \text{ of precast portion} = 16{,}950$$

$$I' = \overline{62{,}000} \text{ in.}^4$$

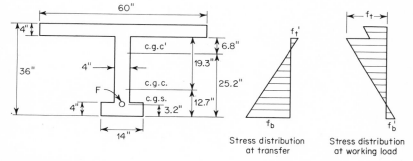

Fig. 39 Example 12.

1. The c.g.s. of the composite section must be located for optimum capacity to resist moment, but not so low as to overstress the precast portion at transfer. Equating the tensile stress in the top fiber to the allowable value gives

$$f_t' = -\frac{F_0}{A_c} + \frac{F_0 e c}{I_c} - \frac{M_G c}{I_c}$$

$$0.30 = -\frac{197}{168} + \frac{197 \times e \times 19.3}{16{,}950} - \frac{40 \times 12 \times 19.3}{16{,}950}$$

$$= -1.17 + 0.224e - 0.545 \qquad (a)$$

$$e = 9.0 \text{ in.}$$

Thus, the c.g.s. can be located $12.7 - 9 = 3.7$ in. above the bottom fiber.

2. The required value of F is determined by equating the tensile stress in the bottom fiber of the composite section at working load to the allowable value

$$f_b' = -\frac{F}{A_c} - \frac{Fec}{I_c} + M_p \frac{c}{I_c} + M_c \frac{c'}{I'}$$

$$0.16 = -\frac{F}{168} - \frac{F \times 9 \times 12.7}{16{,}950} + \frac{100 \times 12 \times 12.7}{16{,}950} + \frac{220 \times 12 \times 25.2}{62{,}000}$$

$$= -F(0.00595 + 0.00675) + 0.90 + 1.07 \qquad (b)$$

$$F = 143 \text{ kips}$$

$$F_0 = 143 \times 150/125 = 172 \text{ kips}$$

Since the eccentricity was computed for $F_0 = 197$ kips in Eq. (a), it must be revised for $F_0 = 172$ kips. Thus,

$$0.30 = -1.02 + 0.196e - 0.545$$

$$e = 9.5 \text{ in.}$$

which locates the c.g.s. $12.7 - 9.5 = 3.2$ in. above the bottom fiber. Correcting Eq. (b) for $e = 9.5$ in. gives

$$0.16 = -F(0.00595 + 0.00712) + 0.90 + 1.07$$
$$F = 139 \text{ kips}$$
$$F_0 = 139 \times 150/125 = 167 \text{ kips}$$

3. With this value of F_0 and the assumed trial cross section, the compressive stress (bottom fiber) in the precast section at transfer is

$$f_b = -\frac{167}{168} - \frac{167 \times 9.5 \times 12.7}{16,950} + \frac{40 \times 12 \times 12.7}{16,950}$$
$$= -1.0 - 1.19 + 0.36$$
$$= -1.83 \text{ ksi}$$

which is close enough to the allowable value 1.80 ksi.

4. The compressive stress in the top fiber of the precast portion at working load

$$f_t = -\frac{139}{168} + \frac{139 \times 9.5 \times 19.3}{16,950} - \frac{100 \times 12 \times 19.3}{16,950} - \frac{220 \times 12 \times 6.8}{62,000}$$
$$= -0.83 + 1.51 - 1.37 - 0.29$$
$$= -0.98 \text{ ksi}$$

which is less than the allowable value 1.60 ksi.

The stress in the top fiber of the cast-in-place flange can be computed by using the appropriate values in $f = Mc/I$. However, this stress will not be critical (Fig. 39).

Section **13**

Concrete Construction Methods

FRANCIS A. VITOLO

President, Corbetta Construction Company, Inc., White Plains, N.Y.

The principal goal of a designer is to obtain the most economical structure while maintaining its basic utilitarian functions and its architectural integrity, all without sacrificing quality. The owner's choice of the numerous forms of contract will vary both the designer's and the contractor's roles in attempting to achieve this goal. When the designer and the contractor are part of the same organization or when they are part of a team, the open line of communications permits them to share the effort.

The actual construction contract can assume one of three basic forms.

With a *guaranteed maximum price* contract, contractors agree to perform the work at a guaranteed maximum price for a fixed fee. As an inducement in keeping costs down, they share in any savings. Under this arrangement, the contractor will naturally strongly influence the designer. On a *cost plus fixed fee* contract, the contractor agrees to perform the work at a fixed fee, without guaranteeing the maximum price. The major incentive to the contractor is to maintain low costs, in order to establish, or perpetuate, a reputation that will promote future contracts. The owner and the designer must share the desire for economy. These forms of contract permit work to start before final drawings are prepared, offering completion months sooner.

The *lump sum* form of contract is the most common because it offers a high degree of competition. This gives contractors the incentive to devote their skills, ingenuity, and inventiveness to reducing costs in order to maintain a fair margin of profit. By law, except in rare instances, all public works contracts must be lump sum.

Contractors are constantly planning ahead, weighing which forthcoming projects will best fit their bidding schedules. Therefore, prospective bidders should be made aware of imminent projects as early as possible. When a firm bid date has been established, the availability of bid documents should be advertised at least a month in advance. When feasible, bids should be solicited from recognized qualified contractors. Since many contractors are preparing bids for other contracts, conflicting bid dates should be avoided. (Public works agencies receive bids on known specific days of the week or month.) During the last hours, indeed minutes, before bid time the general contractor is swamped with late proposals from prospective subcontractors and material suppliers, with comparing qualified bids, with bargaining, etc. Therefore, bid openings should not be scheduled on Mondays or on the day following a holiday; afternoon is preferable to morning.

The time allotted for bid preparations should include a reasonable allowance for obtaining and distributing bid documents, for quantity take-offs and analyses of the results, for the planning, devising, and evaluation of construction methods and of proper sequencing, and finally for estimating the costs of the various elements and assembling them into a bid package. Some additional time should be allowed for the issuance of addenda or of clarifications so as not to face the necessity for granting a postponement.

The designer should be responsive to the questions raised by prospective bidders. All too often designers create an unhealthy climate by treating contractors as adversaries. The bid documents should contain all pertinent information in clear, concise form, without resort to exculpatory phrases. A properly conducted prebid conference, where contractors' questions are given straightforward answers, will often eliminate misunderstandings, promote good relationships, and benefit all parties.

Bid documents should be made as simple as possible. Alternate and unit prices should be solicited only if they will be of significant value and are explicit for all trades involved.

The specification writer should refrain from naming proprietary articles, which inhibit competitive pricing and later may cause the contractor to relinquish control of deliveries, etc.

A reasonable time should be allowed for completion of the project, taking into consideration such factors as urgency, local climatic conditions, accessibility, and availability of materials, equipment, and manpower (especially if they are unusual).

Finally, the designer should encourage rather than inhibit contractor resourcefulness. Specific procedures, sequences, etc., should be given only if they are a function of the design.

1. General Considerations   The period of original layout and preliminary general design establishes the basic pattern for the structure from which the design details are developed. Even at this early stage, the designer should be mindful of the construction point of view and guide the design accordingly.

It is not suggested that the engineer deviate from improved or advanced ideas of design merely for the purpose of facilitating construction. It is suggested, however, regardless of the design or the use for which the structure is to be constructed, that various principles and practices can be incorporated which will reduce construction costs.

Reductions in costs or time are inherent in mass production, which requires repetition. A detail may be complicated, or of an unusual configuration, and still be economically mass-produced, as long as sufficient repetition is involved to warrant the planning and equipment required. In the project layout, this will involve a repeated, geometric pattern, possibly of column, beam, or girder spacings, which in turn will produce slabs of similar sizes and dimensions. In order to meet architectural, design, or other standards of the owner, the repetition may consist of sections of the structure rather than individual units. While many variations might occur in individual elements within the section, the repeated sections would be similar in every respect.

Architectural or mechanical features which cause variations throughout an otherwise repetitive series of elements or sections may, in many instances, be congregated in one area to take advantage of the repetition in the remainder of the series.

2. Formwork   Formwork, a major item of cost in concrete construction, accounts for 35 to 60 percent of the cost of the concrete work.

The cost of form fabrication, per unit of concrete surface formed, is inversely proportional to the number of reuses of the form, i.e., the number of times the form can be used without refabrication or changes between uses. With repetitive use, the form is fabricated only once. As the number of reuses is increased, the quality of the form can be economically increased. Sufficient reuse will justify plastic, steel, or concrete forms for areas which normally are economically restricted to wood forms.

Forms are preferably fabricated in a central location with stationary, powered equipment, with all the advantages of assembly-line mass production. However, mass production requires a quantity of identical units.

Columns.   Increasing the size of several smaller odd-size columns to the dimensions of a greater number of larger columns will increase the quantity and, therefore, the cost of the concrete. However, the savings in form costs will, in many instances, exceed the cost of additional concrete and result in a decrease in the overall cost of the structure. This cost reduction will also apply to spread footings, pile caps, buttresses, beams, girders, and, in some circumstances, slabs.

In multistory structures consideration should be given to repetition of column sizes from floor to floor. A minor reduction in column dimensions from floor to floor, due to smaller loadings, not only causes refabrication of the column form panels but also requires changes to the top of the form, since the slabs, girders, and beams will normally retain the same dimensions from floor to floor. Minor reduction in column dimensions from floor to floor should be accumulated, and one comparatively large reduction made after several floors. When making the accumulated reduction, consideration should be given to a reduction in only one dimension of the column, thereby requiring the refabrication of only two sides of the column forms.

From the above, it will be noted that when it is undesirable to use the same column cross section throughout one floor or sectional area, it will be advantageous to use rectangular columns with one dimension equal. This will vary only two of the four column side panels.

Beams and Girders. The same general principles apply with respect to beam and girder design. These require form sides, bottoms, and shores. Variation in beam widths affects the slab forms and the beam bottom form. A variation in the height affects two beam side forms and the shores (as well as reshores). Obviously, where a change from a standard is required, changing only the width involves dimensional variations in a much smaller percentage of the formwork and shores.

When selecting the original dimensions of a series of beams or girders a width equal to that of the column face into which they frame will reduce the form details required at the head of the column. Similarly, selection of a beam depth equal to that of intersecting beams will reduce form details at beam intersections and will produce greater quantities of shoring of equal lengths.

Slabs. The length and width of a slab are not particularly significant, except for their effect on the lengths of beams and girders into which the slab frames. Variation of the slab depth in small areas of lighter loadings, however, may seriously affect the forms of the adjacent framing. This requires beams with special sides, special column top details (unless the beams are equal to or wider than the column width), and special-length shores and reshores for this slab area.

Walls. For wall forms the side-panel dimensions become important only with reference to the length of adjacent framing members. Variation in the thickness of the wall, however, does affect considerably the fabrication and installation of box-outs for openings and sleeves and the side forms of the adjacent framing members.

Foundation walls which require forms sloped at the bottom are undesirable. They are more economically placed with stepped forms.

Tie Beams. For tie beams placed monolithically with slabs on ground, side forms are not used for depths below slab up to 2 or 3 ft, depending on soil conditions. The cost of sloping trench sides and placing additional concrete is less than the cost of the forms. It will be noted that when forms are required, they cannot be recovered. This means not only a form cost based upon no reuse, but lost material that cannot be refabricated for use elsewhere. Such lost forms are expensive and should be avoided whenever possible.

Accessories. Form costs include the cost of form accessories, such as ties, and in the case of fireproofing of structural steel, form hangers. These accessories are wholly, or in part, used only once as they become embedded in the concrete. The standardization of girder, beam, and column dimensions, and wall and slab thicknesses, all helps to reduce the number of sizes and lengths of form accessories to be stored, handled, and placed.

3. Reinforcing Steel The contractor uses assembly-line mass-production methods for the cutting, bending, and handling of reinforcing steel. This is done in a central yard, equipped with heavy-duty powered equipment. When possible, jigs and templates are established, near the cutting yard, for the assembly-line fabrication of column and beam cages, footing and pile-cap mats, and other assemblies. As in the form yard, unit costs are lowered with a reduction in the numbers of bar sizes, lengths of bars of the same size, types of bend to be processed, and the number of cage or mat variations to be assembled.

Columns. In many cases, simplifying the formwork (for example, increasing smaller odd-size columns to the size of larger columns) automatically changes the column reinforcing to a standard size. A small increase in the bar size of isolated columns will often eliminate a quantity of various cage sizes and result in only a negligible increase in the weight of the reinforcing steel.

The use of overall rectangular column ties and beam stirrups combined with interior C-type ties, in lieu of double interlocking rectangular ties, will facilitate the assembly of column and beam cages.

Long Bars.   Except for special designs, bar lengths which exceed standard mill lengths should be avoided. Bars exceeding approximately 20 ft in length which require bends on both ends are awkward to handle in the bending yard. After one end is bent, they must be turned around to bend the other end. Furthermore, yards are not normally set up with sufficient vacant area to turn the bars in the vicinity of the bending machine. In this situation, the contractor will prefer to bend one end and splice a short, bent bar to the other. Long bars requiring a sharp bend toward the center are awkward to bend, handle, store, and place. Again, it is preferable to splice two bars, one straight and the other with a bend on one end.

Obviously, the use of spliced bars to eliminate bending does increase the weight and cost of the reinforcing material. However, a review of bending costs, and the higher handling and placing costs of some unusual bent bars, will usually show a saving in labor costs which exceeds the cost of the additional steel used for splicing. This cost differential can be considerable in areas of high labor rates.

Concentrations of Steel.   The problem of concentrations of reinforcing steel which interfere with the concrete placement, or with the placement of the reinforcement itself, should be a matter of routine checking during the design. When corrective measures must be taken during the preparation of shop drawings, delays result during the period required by the designer to investigate them. Further delays may be encountered in changing the drawings to conform to the designer's corrective measures. When the situation is encountered during placement of the steel, the resultant delays and/or additional labor can be very costly.

Concentrations occur mainly in heavily reinforced members or at the framing connections of these members. However, they may also occur in framing systems which are considered as lightly reinforced. They are not readily apparent in the latter systems and, not being suspected, are often overlooked during the design and preparation of shop drawings. These concentrations may be the result of the reinforcing from several beams or girders intersecting over a column which, when added to the column reinforcement, may restrict or prohibit the concrete from passing into the column forms. The intersection of beam and girder reinforcing, combined with considerable slab reinforcing which is continuous over the beams, can also create the same problem. The latter situation becomes more serious when the slab reinforcing is fanned, with the vertex passing over the beam or intersection.

It should be remembered that dowels which are to be placed prior to concreting may have been noted but not pictured on the drawings, and are easily overlooked. These dowels may introduce concentrations.

Lap splices double the steel area within the splice length. Although not shown or indicated on the contract drawings, additional splices may be required in continuous bars which exceed stock lengths, or in the case of long bent bars which cannot be placed in one piece.

Occasionally there are concentrations known to the engineer which, for design reasons, are not readily avoidable by changing bar sizes or arrangements or by increasing the size of the member. In these cases, difficulties in concrete placement may be overcome by using a smaller aggregate. Care should be exercised in using this solution, however, because of possible difficulties in vibrating the concrete, possible segregation of the concrete, and the added cost of using an additional concrete mix.

Thin concrete members, or members of special design, may require closer placement tolerances to ensure compliance with design computations. Any such requirement should be specifically noted, and detailed if necessary, in the specifications or on the drawings.

Inclusion on the drawings of specific information covering dowel lengths and locations, end-anchorage requirements, and required and permissible splice locations and details will avoid misunderstandings and delays in the preparation of shop drawings and in the field work.

4. Concrete   The designer specifies the strength and other qualities which the concrete should have to serve its purpose. The experienced concrete contractor can produce the specified concrete and will accept this responsibility when given control of the entire operation, from the design of mix through the curing operations. In other words, the specification should give, in as much detail as is necessary, the strength and other qualities required. However, the means and methods used to achieve these results, including materials, design mix, mixing, transporting, placing and curing, should be left to the contractor. Of course, the engineer must retain overall control. This is best accomplished simply by requiring compliance with recognized standards such as the municipal building code and the various standards of the industry. The reliable contractor will use these standards as minimum requirements, supplementing them when necessary to produce the desired results. Any additional requirements, limitations, and restrictions serve only to increase the cost of the work and to transfer responsibility from the contractor to the engineer or owner.

The selection of the raw materials, and the combining of them into a concrete mix, are done by the contractor on the basis of knowledge of the materials themselves, and of available facilities. A rigidly specified mix may be difficult to place in restricted form openings caused by thin sections, concentrations of reinforcing, or the presence of embedded items. Under these conditions segregation and excessive honeycombing are likely to occur. With control of the mix design, the contractor can, without difficulty or loss of time, readily adjust the concrete supply to overcome these problems. The answer may be merely a matter of using smaller aggregate, or increasing the slump (with the other adjustments necessary to ensure strength, wearability, and durability) above that normally required.

The use of high-early-strength cement should, in most cases, be permitted at the contractor's option. When it is used by those who understand it, its performance and results cannot be questioned. Although more expensive than regular cement, it is often the secret of success in cutting costs when it is used to accelerate the release of formwork, to bring a project back to schedule, and to anticipate delays from inclement (or freezing) weather.

Some projects are adapted to the use of paving machines, some to transit-mixed concrete, and others centrally mixed concrete, or a combination of mixing and transportation methods. The contractor can best judge the method which will prove most economical and, at the same time, produce the required results.

Testing. The testing of concrete is basically an inspection control. Therefore, the testing laboratory should be an independent agency retained by the owner. The contractor's reputation can only be adversely affected by problems resulting from the placement of substandard concrete, and he must have the fair and honest tests which can be assured by a reliable, experienced laboratory.

Removing Forms. It is generally specified that forms be left in place for various lengths of time depending on their location, i.e., columns, walls, beam bottoms, beam sides, slabs, etc. These requirements are for the purpose of ensuring that the concrete has attained sufficient strength to warrant removal of the forms, and are based upon the theoretical strengths of concrete that should have developed in the specified period. These strengths are not confirmed until cylinders have been tested, usually not sooner than 28 days after the concrete is placed. Ordinarily, this is a considerable time after the forms have been removed.

Stripping on strength requires that the concrete strength be supported by test cylinders prior to form removal in lieu of the theoretical strengths as noted above. Specifications which permit stripping of forms on the basis of attained strength, rather than time, may result in savings from quicker turnover of formwork and a reduction in completion time of the structure. Although this requires additional test cylinders, since the 28-day tests must be made to meet building-code requirements, the additional cost of testing is negligible compared with the saving that can result. However, when stripping on the basis of proved strength is permitted by the specifications, stripping on the basis of time should be optional in the event of damaged test cylinders or questionable test results.

Curing. Curing affects not only the strength and quality of the concrete but also, depending on the method used, the properties of the surface. Some commercial curing agents will adversely affect the bonding of paint and other surface finishes and the color stability of paints. Curing agents may leave surfaces not scheduled to receive further treatment with an undesirable color. Here, again, it is suggested that the desired results be explicitly noted, and the means and methods of accomplishment be left to the contractor.

5. Embedded Items Items to be embedded in the concrete include inserts, hangers, nailers, reglets, anchors, sleeves, frames, and structural steel. Too often, the importance of studying the details of embedded items is overlooked, resulting in a missed opportunity to reduce costs or avoidable difficulties in the field.

Hangers, anchors, and similar embedded items which protrude from the face of the concrete must necessarily pierce the forms. Unless these items occur in the same form location in succeeding repetitive sections, forms must be either patched or refabricated. Furthermore, they impede stripping of the form and may, depending upon their size and number, interfere with placement of the form supports and shoring. In the case of hooked anchors, the form must be fabricated with a separate waste panel for each use in order to minimize form damage upon stripping. The waste panel represents an additional operation and increases both forming and stripping costs. Embedded items which are attached to the form face, without penetrating the form itself, are preferred.

Architectural metal and delicate steel items with attached anchors are placed in one operation prior to placement of the concrete. When using embedded inserts in lieu of attached anchors, this placement requires two operations. However, the cost of supporting these items with attached anchors in the forms, and their protection during concreting and subsequent construction, may more than offset the cost of the additional operation when embedded inserts are used. When contractors are given the option of placement by either method, their experience will indicate the most economical method, and the possibility of damage in the event the items are installed in the formwork. It also helps to avoid concreting delays, in the event of late delivery of the item.

The question of whether to embed heavy castings, structural-steel members, and other assemblies in the formwork prior to concreting, or to "box out" for these items and place them later should also be left to the contractor's option.

Contract documents should show in exact detail the number, location, position, size, and type of each embedded item, with details of attached or detached anchorages where either is permissible.

Each embedded item should be studied with respect to its effect on the reinforcing steel, concrete cover required, and any changes resulting from its proximity to openings, columns, beams or girders, and mechanical-trade items.

6. Special Designs During the design stage, the engineer is minutely familiar with all phases of the project. During this state of the work, problems of a special or unique nature may arise which will be of vital concern to the contractor in bidding, planning, and executing the work. Their details should be properly documented and noted in the contract documents.

The following paragraphs indicate the type of information that should be noted but which is normally made available to the contractor only after a question is raised and the information requested. The latter practice may result in delays in planning and may on occasion require expensive changes in planning or in the field.

The contract drawing notation of the load used in the design of vehicle passageways such as bridges, elevated bypasses, ramps, and entrances informs contractors as to the extent they may use these facilities during construction.

Specific information should be noted concerning framing designed for composite action, deflection to be controlled through camber, and those special structures in which a definite sequence of form release is essential to avoid a reversal of stress for which the framing is not designed.

Definite and specific information on the sequence of concreting, shoring, and reshoring is essential where columns or beams are required to assume gradually increasing increments of load and deflection as the concreting progresses upward in the structure.

Exact elevations to which formwork is to be installed should be given for elevated highways, bridges, and other structures where deflection must be controlled so that finished elevations will conform to proper slope for drainage and to required grade for high-speed roadways. The formwork elevations are also required for concrete framing members which are cambered to provide for deflection that will result from hung framing or mechanical equipment.

Formwork for long-span arches is normally required to be released in predetermined sectional sequences to assure that all the arch ring remains in compression during the operation. Since this often establishes limitations in the design and operation of the falsework, form design and job planning will be facilitated if this information is made available with the contract documents. The contractor's attention should be drawn to those structures, or parts of structures, which depend upon other parts of the construction for their stability. These notes should indicate the degree of completion required to assure stability.

In the case of cantilevered or other similar unbalanced construction, the contractor should, if possible, be given the option to locate construction joints within permissible areas noted on the drawings.

The above indicates the type of information desired by the contractor, and it should be anticipated and furnished in the design of folded plates, domes, arches, thin shell, very long spans, or other structures in which construction details involve more than routine construction procedures.

7. Tolerances It has always been understood that concrete construction, being an on-site operation, cannot meet the close tolerances required of manufactured items of other materials. Prior to the advent of precast concrete, tolerances, as such, were not specified, but were matters of judgment between the engineer and contractor. Depending upon the member involved and its intended use, the degree of accuracy was a rule-of-thumb consideration of workmanship. Precast concrete has come to be recognized as a manufactured item, and tolerances are more frequently specified and adhered to.

Dimensional and strength tolerances in concrete work have been defined and standardized by the industry, particularly by the American Concrete Institute and the American Society for Testing and Materials. The adoption of these standards by the industry has minimized questions of acceptability of concrete work on the basis of dimensional and strength accuracy. However, this acceptability extends only to the finished concrete work and not the completed structure. Confusion still results from the engineer's acceptance of these tolerances, on the one hand, while on the other he details metal frames for openings, installation of masonry units, mechanical installations, etc., to tolerances which cannot be guaranteed with the accepted standards. Failure of the engineer to recognize and to provide for this situation creates unnecessary problems for the concrete contractor and others concerned.

Tolerances for warp, camber, and concrete finishes have not reached the standardization and recognition of other tolerances and should be more closely checked and provided for in the specifications.

8. Shop Drawings Upon the execution of the contract between the owner and the contractor, the owner's engineer assumes two obligations: (1) the approval of shop, or working, drawings and material samples submitted by the contractor, and (2) the inspection and supervision of the actual field work.

Shop drawings are prepared by the contractor for the purpose of expediting construction. They are, with few exceptions, generally an amplification of the information contained in the contract drawings and specifications. They clarify the work to be accomplished by means of (1) larger-scale details, (2) detailed dimensions computed from overall dimensions shown on the contract drawings, (3) combining details which may appear on several contract drawings, (4) combining on the shop drawing information from both the contract drawings and specifications, (5) transforming tables, schedules, and written information into a visual form of sketches and diagrams, and (6) giving the field personnel, on one or several drawings, all the information, but only that, required by them to accomplish the particular portion of the work with which they are immediately concerned. Shop drawings also serve to "proofread" the contract drawings, for it is during preparation of the shop drawings that most errors, inconsistencies, and missing information are discovered.

Shop drawings are submitted to the designer for approval. This provides a review of the work and an opportunity to correct errors and to make alterations prior to any expensive preparatory work in the field. They also show designers the contractor's interpretation of the work to be performed, and give them an early opportunity for clarification in the event that the contractor's interpretation differs from that intended.

Needless to say, designers should review the shop drawings carefully. If it is not their intention to do so, the requirement that they be submitted to them should be deleted from the contract for reasons of economy. This, in turn, will alert the contractor to the fact that the drawings are not being checked.

9. Material Samples Material samples are submitted to the designer for approval, for the purpose of ensuring a mutual understanding and interpretation of the material-requirement sections of the specifications. These sample-submission requirements of the specifications should be limited to those materials which the engineer plans to test, or about which a question might arise as to the expected color shade, finish, or quality of fabrication. The submission of samples of trade-name materials which are specifically identified in the specifications is a waste of time and money for both the engineer and the contractor.

INSPECTION

10. The Resident Engineer The engineer's contact with the actual construction is through the resident engineer. The resident engineer was once viewed as a policeman, hired by the engineer, whose main duties were to see that the contractor complied with the terms of the contract. Through a broader understanding by both engineer and contractor this viewpoint has changed, and the resident engineer's position has assumed its proper perspective.

The resident engineer is the field supervisor and coordinator who expedites the completion of the construction work. With the understanding that teamwork between engineer and contractor is essential to both parties, resident engineers are impartial, even though employed by, and reporting to, the engineer. They are impartial in that they protect the owner by making sure that the contractor fulfills the contractual obligations and at the same time protects the contractor from improper requests or orders which are beyond the scope of the contract. Their experience and authority permit them to make on-site decisions, and corrections and changes which may be needed to expedite the work. They should be aware of the limits of their authority and instructed as to the nature or extent of such corrections or changes as are beyond that authority and which should be referred to the engineer. Only by familiarity with the contract documents, and a knowledge of the intent of the designer are they able to interpret the documents and make those decisions required to produce a more satisfactory project. They should be sufficiently experienced to realize that there are often several ways of accomplishing a result, and that the method should be the option of the contractor provided the proposed method will produce the desired result.

CONTRACT DOCUMENTS

The contract documents, other than the proposal form, consist of the contract drawings and specifications. Both are interdependent and inseparable, and together they inform the contractor exactly what the owner and engineer require of the contractor. It is upon these documents that the contractor has based the proposal to construct the project.

11. Preparation Contractors cannot anticipate the intentions of the engineer or owner, and must assume that the contract documents represent what is required. Under competitive bidding, they cannot do otherwise and expect to receive the contract award. At the same time, where ambiguities are found during the short time prior to bidding and sufficient time is not available for clarification, contractors must increase their proposals to cover any possible additional expenses which may be involved.

A review of the contract documents gives the contractor a fairly good indication of the problems and progress to be expected during the planning and preparatory stages and during the period of actual construction. Well-prepared documents which appear to be accurate, complete, and concise, while fully covering essential details, indicate that the design was carefully executed and that a minimum of delay and problems can be expected to develop during the contract. Incomplete or carelessly prepared documents suggest a long construction period with many delays in the preparation of shop drawings, and in the actual construction work, pending detailed information, corrections, and clarification. They also suggest the probability of discussions and possible disagreements as to the extent of the work included in the contract price. This applies to the contract specifications as well as to the drawings, for since they complement each other, the sum is only as good as the least effective part. A well-prepared set of drawings can come to naught when accompanied by hastily or carelessly prepared specifications.

12. Specifications The specifications inform the contractor of those requirements and obligations which are more effectively given in words, or which cannot be satisfactorily transmitted by diagrams or notes on the drawings. As instructions or information, they, like the drawings, should be exact, complete, and yet concise. They should be prepared by competent personnel, and completely checked to avoid conflicts between specifications and drawings.

The specifications should be prepared specifically for the individual project. The inclusion of extraneous material, copied from specifications for other projects, may cause confusion in an otherwise clear presentation. In an attempt to cover the engineer and owner from every conceivable angle, specification writers sometimes create more problems than they had hoped to avoid.

13. Intent   The construction contract implies to the contractor that the owner knows what is wanted, is willing to pay for what is ordered, and has retained a competent designer who has fulfilled his obligations and will accept full responsibility for the design. The contractor is concerned when items appear in the specifications or on the contract drawings which attempt to transfer the designer's responsibility to the contractor. The following are examples of this transfer of responsibility which, in every case, requires the requested work to be done at no cost to the engineer or owner:

1. The contractor agrees that all foundation work required for mechanical equipment, located but not detailed, shall be included in the contract price.

2. The intent of the contract is to construct facilities complete for their intended use. The contractor agrees to furnish and install any and all items required to complete these facilities, whether or not shown or indicated by the contract documents.

3. The contractor agrees that the contract documents have been reviewed and found to be complete and correct in every respect, including but not limited to the structural adequacy.

The contractor, upon viewing items such as these in the contract, can only conclude that those responsible for the preparation of the documents are, at best, uncertain as to their completeness or accuracy, and must decide whether to increase his proposal to cover eventualities, or to forgo submitting a proposal.

14. Scope of Work   The "scope of work" section of the contract specifications establishes the extent of the work covered by the contract. It is that part of the contract documents which enumerates the items desired by the owner and upon which the contractor's proposal is based.

As this section is the nucleus of the entire contractual agreement, there should be no question or doubt, on the part of anyone concerned, as to the meaning or intent of the section or any of its parts. The drawings and remainder of the specifications furnish details and further information on the items included in the scope of work section but are not construed as changing the limitations of the contract work as set forth in that section.

In contracts wherein one contractor is to construct the entire project, this section may be very short and general in nature. In contracts involving several contractors working concurrently under separate contracts, and in contracts involving additions or alterations to existing structures, the construction of only a portion of a project, a portion of the work covered by the contract drawings, etc., the section on scope of work must be more explicit and may of necessity be voluminous by comparison.

15. Drawings   The contract drawings are the picture half of the contract documents and usually show the work required by the written specifications. They should show all the information, in sufficient detail, which, when combined with the specifications, will enable the contractor to construct the project as it is intended. They should be legible, well correlated as to sections and details, and accurately drawn to scale.

On larger projects, the architectural, structural, and mechanical drawings may be independent sets prepared by separate design firms. The better the correlation of the various sets of drawings prior to their being issued for proposals, the less the likelihood of confusion, delay, and expense during construction. When conflicts are found during construction, work stops and labor is idled until such time as the inconsistency is clarified or corrected. Checking drawings in the field is very expensive for the contractor, designer, and owner.

Contract drawings inevitably contain errors. Revised drawings are required to correct these errors and to incorporate changes required for various other reasons. Once a change is made, the revised drawing should be issued promptly. While the revision may not affect work currently being done in the field, it may seriously affect planning, scheduling, and other items preparatory to future work. In many instances, it is advisable to notify the contractor verbally of a pending revision to avoid unnecessary preparatory work.

Of immediate concern to the contractor upon receipt of a revised drawing is the date and details of the revision. A description of the revision in the drawing title box is seldom sufficient. A more satisfactory procedure is to outline the revised dimension, detail, or section with an irregular, heavy pencil line on the reverse side of the tracing. On the front side, in heavy print, the irregular line is identified with an arrow and a revision number. The revision is then readily apparent, and will not require a detailed comparison of the entire drawing with the previous one to determine the extent of the revision. This procedure not only facilitates finding the revision, but also helps to guard against the overlooking of parts of a revision when several details on various sections of the drawing are revised under one date and revision number.

Whenever dimensions are changed, the drawing should be corrected to scale to avoid a distorted picture. Where this is not practical because of the size or complexity of the drawing, a note under the revision number should state "not to scale." In either case, revisions should be thoroughly checked to ensure correlation with details not changed, and with the architectural and mechanical drawings, to ensure that the correction of one error has not introduced another.

References to drawings in correspondence should always include the drawing title, number, and date, to avoid any possible confusion as to the exact issue of the drawing involved.

Section **14**

Design of Composite Beams and Girders

W. H. FLEISCHER, D. C. FREDERICKSON, I. M. VIEST
Bethlehem Steel Corporation, Bethlehem, Pa.

W. C. HANSELL
Wiss, Janney, Elstner and Associates, Northbrook, Ill.

1. Definitions Composite beams and girders comprised of a steel beam and a rein-forced-concrete slab so interconnected that the component elements act together as a unit are treated in this section. Beams with both full and partial shear connection are included.

The steel beam may be fully encased in concrete. If the concrete encasement is cast continuously with the slab, meets certain thickness requirements, and is properly rein-forced, the natural bond between the steel and the concrete can be relied on to provide composite action. Otherwise the composite action is assured by small pieces of steel bars or shapes welded to the top flange of the steel beam and embedded in the concrete of the slab.

The *steel beam* may be a rolled beam, a rolled beam with cover plates, or a built-up section. An unsymmetrical section, such as a rolled beam with a cover plate on the bottom flange, is often economical.

The *reinforced-concrete slab* acts as a very effective cover plate if it is on the compres-sion side of the steel beam. In negative-moment regions, where the concrete slab is on the tension side of the steel beam, adequately anchored slab reinforcing steel parallel to the steel beam can be made to act like a tension cover plate. Slab dimensions are usually dictated by the beam spacing and the required capacity for load transfer to the beams. The design of the slab is independent of the composite action and is carried out in the same manner as for noncomposite floors.

The *shear connectors* or the *concrete encasement* provide the necessary connection between the slab and the beam. Their function is to transfer horizontal shear from the slab to the beams and to force the concrete and steel parts to act as a unit.

NOTATION

a_c = depth of compression-stress block in slab at ultimate load
A_b = area of bottom flange
A_B = area of rolled beam

A_c = effective area of concrete at working load = bt/kn
A_p = area of cover plate
A_{pe} = effective cover-plate area = $f_{yp}b_p t_p/f_y$
A_r = area of longitudinal slab reinforcement within effective width at section of maximum negative moment
A_{re} = effective area of slab reinforcement = $A_r f_{yr}/f_y$
A_s = area of steel section
A_t = area of top flange
A_w = area of web
b = effective slab width
b' = projecting steel flange width
b_b = width of bottom flange
b_p = width of cover plate
b_t = width of top flange
d = depth of rolled steel beam or plate girder
d_s = stud diameter, in.
d_w = depth of web
e, e', e'' = moment arms, Eqs. (9)
e''' = moment arm = $e_r + 0.5d$
e_c = distance from top of steel section to center of gravity of effective slab
e_r = distance from top of steel section to center of gravity of longitudinal slab reinforcement
E = modulus of elasticity
E_c = modulus of elasticity of concrete
E_s = modulus of elasticity of steel
f_c' = compressive strength of concrete, ksi
f_r = range of fluctuating stress in slab reinforcement over the support, ksi
f_y = yield point of steel section, ksi
f_{yp} = yield point of tension cover plate, ksi
f_{yr} = yield point of reinforcing bars, ksi
g = distance of top layer of bars from centroidal axis of longitudinal slab reinforcement
h_r = height of rib or haunch
H = horizontal shear force, Eqs. (19)
H' = horizontal shear force corresponding to N'
H_r = range of horizontal shear force per unit of length
H_s = stud length including head
H_u = horizontal shear force at ultimate load, Eqs. (7)
I = moment of inertia
I_B = moment of inertia of rolled beam
I_c = moment of inertia of composite section
I_{eff} = effective moment of inertia for composite beam with partial shear connection
I_r = moment of inertia of longitudinal slab reinforcement about own centroidal axis
I_s = moment of inertia of steel section
I_w = moment of inertia of web
k = numerical factor depending on type of loading; $k = 1$ for transient loads, $k = 2$ or 3 for sustained loads
K_c = $A_c/(A_c + A_s)$
K_r = $A_r/(A_s + A_r)$
K_s = A_p/A_s
L = span length
m = statical moment of steel section about neutral axis of composite beam $A_s \bar{y}_c$
M = moment
M_{Dc} = moment caused by dead loads resisted by composite beam
M_{DL} = dead-load moment
M_{Ds} = moment caused by dead loads resisted by steel beam
M_{LL} = live-load moment
M_{L+I} = live-load moment including impact
$M_{\max}$ = maximum positive moment
M_u = ultimate moment
n = modular ratio E_s/E_c
n_i = number of ribs with i studs
n_r = number of studs in one rib
n_t = total number of ribs
N = number of connectors for full shear connection
N' = number of connectors provided for partial shear connection
N_c = number of additional connectors at point of contraflexure
q = allowable static load for one shear connector, kips
q_i = allowable static load per rib with i studs
q_r = allowable fluctuating load on one connector

q_u = ultimate static shear strength for one connector
R = reduction coefficient, Eq. (20)
s = spacing of connectors along the beam
S = stringer spacing, ft
S_{bc} = section modulus, bottom steel fiber, composite section
S_{bs} = section modulus, bottom fiber, steel section
S_{cc} = section modulus, top concrete fiber, composite section
S_{eff} = effective section modulus for composite beam with partial shear connection
S_{rc} = section modulus, top layer of longitudinal slab reinforcement, composite section
S_s = section modulus, bottom and top fiber, symmetrical steel section
S_{tc} = section modulus, top steel fiber, composite section
S_{ts} = section modulus, top fiber, steel section
t = slab thickness above haunch or rib
t_b = thickness of bottom flange
t_f = average flange thickness of channel, in.
t_p = thickness of cover plate
t_t = thickness of top flange
t_w = web thickness, in.
V_{L+I} = live-load vertical shear including impact
V_{max} = maximum vertical shear caused by live load (including impact)
V_{min} = minimum vertical shear caused by live load (including impact)
w = length of channel connector, in.
w_c = air-dry unit weight of concrete, pcf
w_r = average width of rib or haunch
y_{bc} = distance from neutral axis to bottom steel fiber, composite section
y_{bs} = distance from neutral axis to bottom fiber, steel section
y_{cc} = distance from neutral axis to top concrete fiber, composite section
y_{ts} = distance from neutral axis to top fiber, steel section

$\overline{y}_c$ = shift in neutral axis from addition of concrete slab or slab reinforcement

$\overline{y}_s$ = shift in neutral axis from addition of steel cover plate or unsymmetrical flanges
Z_B = plastic section modulus, rolled beam
Δ = deflection at midspan of a simple beam

2. Elastic Properties of Cross Section Flexural stresses and deflections up to initial yielding are based on elastic properties of the cross section. Formulas for the elastic properties are given below for unsymmetrical steel sections and for composite sections.

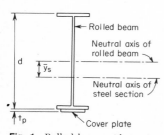

Fig. 1 Rolled beam with tension cover plate.

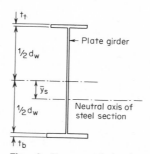

Fig. 2 Unsymmetrical plate girder.

Those for composite sections are derived for complete interaction. They may be used for all beams with full shear connection. Elastic properties for beams with partial shear connection are described in Art. 9.

    1. Rolled beam with tension cover plate (Fig. 1)

$$\overline{y}_s = 0.5(d + t_p)K_s \tag{1a}$$

$$I_s = 0.5(d + t_p)\overline{y}_s A_B + I_B \tag{1b}$$

$$S_{bs} = \frac{I_s}{0.5d + t_p - \overline{y}_s} \tag{1c}$$

$$S_{ts} = \frac{I_s}{0.5d + \overline{y}_s} \tag{1d}$$

2. Unsymmetrical welded plate girder (Fig. 2)

$$\overline{y}_s = \frac{0.5(d_w + t_b)A_b - 0.5(d_w + t_t)A_t}{A_s} \tag{2a}$$

$$I_s = 0.25(d_w + t_t)^2 A_t + 0.25(d_w + t_b)^2 A_b + I_w - A_s(\overline{y}_s)^2 \tag{2b}$$

$$S_{bs} = \frac{I_s}{0.5d_w + t_b - \overline{y}_s} \tag{2c}$$

$$S_{ts} = \frac{I_s}{0.5d_w + t_t + \overline{y}_s} \tag{2d}$$

3. Composite beam—neutral axis below the slab (Fig. 3). The properties of the composite cross section are a function of the modular ratio n and a numerical factor k. The

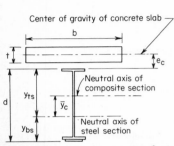

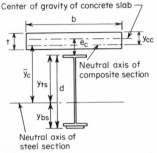

Fig. 3 Composite beam, neutral axis below slab.

Fig. 4 Composite beam, neutral axis in slab.

factor k is used to account for the effects of creep. For loads of short duration, such as live loads for bridges, no creep effects are present ($k = 1$). For sustained loads, such as dead loads, the effects of creep may be accounted for by taking k larger than 1 (Art. 7).

$$\overline{y}_c = (y_{ts} + e_c)K_c \tag{3a}$$

$$I_c = (y_{ts} + e_c)\overline{y}_c A_s + I_s + \frac{A_c t^2}{12} \tag{3b}$$

$$S_{bc} = \frac{I_c}{y_{bs} + \overline{y}_c} \tag{3c}$$

$$S_{tc} = \frac{I_c}{y_{ts} - \overline{y}_c} \tag{3d}$$

$$S_{cc} = \frac{I_c}{y_{ts} - \overline{y}_c + e_c + 0.5t} \tag{3e}$$

4. Composite beam—neutral axis in the slab (Fig. 4). If the neutral axis is located in the slab, only the portion of the slab located above the neutral axis is considered effective in resisting stresses. However, Eqs. (3) give sufficiently accurate results even if the neutral axis is located in the slab as long as the following condition is satisfied.

$$\frac{d}{t} \le \frac{1}{3}\frac{A_c}{A_s}$$

Otherwise, Eqs. (4) must be used.

$$y_{cc} = \frac{A_s t}{A_c}\left[\sqrt{1 + \frac{2A_c}{A_s t}(y_{ts} + e_c + 0.5t)} - 1\right] \tag{4a}$$

$$\overline{y}_c = y_{ts} + e_c + 0.5t - y_{cc} \tag{4b}$$

$$I_c = A_c \frac{y_{cc}^3}{3t} + A_s \overline{y}_c^2 + I_s \tag{4c}$$

$$S_{bc} = \frac{I_c}{\overline{y}_c + y_{bs}} \tag{4d}$$

$$S_{tc} = \frac{I_c}{y_c - y_{ts}} \tag{4e}$$

$$S_{cc} = \frac{I_c}{y_{cc}} \tag{4f}$$

5. Composite beam with slab in tension (Fig. 5). Equations (3) and (4) apply to cross sections subjected to positive moments. At negative-moment sections the slab is stressed

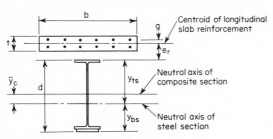

Fig. 5 Composite beam, slab in tension.

in tension; assuming that the concrete resists no tensile stresses, the following formulas apply:

$$\bar{y}_c = (y_{ts} + e_r)K_r \tag{5a}$$

$$I_c = (y_{ts} + e_r)\bar{y}_c A_s + I_s + I_r \tag{5b}$$

Formulas for section moduli S_{bc} and S_{tc} are the same as Eqs. (3c) and (3d), and the section modulus of the top layer of reinforcing bars is

$$S_{rc} = \frac{I_c}{y_{ts} - \bar{y}_c + e_r + g} \tag{5c}$$

3. Plastic Strength of Cross Section The ultimate strength of a composite beam is computed from the fully plastic distribution of stresses.[1,2] Formulas for the plastic bending strength of composite sections are given below.

If the steel section is a hybrid beam, it is convenient to convert the component steel areas to one yield point. For example, if the yield point of the rolled beam is f_y and the yield point of the tension cover plate is f_{yp}, the equivalent area of the cover plate is

$$A_{pe} = \frac{f_{yp}}{f_y} b_p t_p \tag{6}$$

At ultimate load, the horizontal shear H_u at the interface of the slab and the beam is equal to the smallest of (1) the compressive strength of the slab, (2) the tensile strength of the steel section, and (3) the shear strength of the connectors provided between the sections of maximum and zero moments. Accordingly, the horizontal shear is equal to the smallest value given by the following equations:

$$H_u = 0.85 f'_c bt \tag{7a}$$
$$H_u = f_y(A_B + A_{pe}) \tag{7b}$$
$$H_u = N'q_u \tag{7c}$$

The depth of the compression-stress block may be computed as

$$a_c = \frac{f_y(A_B + A_{pe})}{0.85 f'_c b} \tag{8}$$

Equation (8) always results in a conservative value of the moment arms given by Eqs. (9).

1. Composite beam with full steel section in tension (Fig. 6). This condition governs when the horizontal shear is given by Eq. (7b). Designating

$$e' = e_c + 0.5t - 0.5a_c \tag{9a}$$
$$e = e' + 0.5d \tag{9b}$$
$$e'' = e' + d + 0.5t_p \tag{9c}$$

the ultimate moment is

$$M_u = f_y A_B e + f_y A_{pe} e'' \tag{10}$$

2. Composite section with plastic centroid in top steel flange (Fig. 6). This condition governs when $f_y(A_B + A_{pe}) \geqslant H_u \geqslant f_y(A_w + A_{pe})$. The ultimate moment may be computed from the following equation:

$$M_u = f_y(0.5A_B + A_{pe})(d - t_t) + 0.5f_y A_{pe}(t_t + t_p) + H_u(e' + 0.5t_t) \tag{11}$$

Eq. (11) is approximate: the computed M_u is always slightly smaller than the exact value.*

3. Composite beam with plastic centroid in the web (Fig. 6). When the horizontal shear is no more than $f_y(A_w + A_{pe})$, the ultimate moment is given by

$$M_u = f_y Z_B + 0.5f_y A_{pe}(d + t_p) + H_u e - \frac{(H_u - f_y A_{pe})^2}{4t_w f_y} \tag{12}$$

4. Composite beam with slab in tension (Fig. 7). Equations (10) through (12) apply when the slab is in compression. When the slab is in tension, only the slab reinforcement

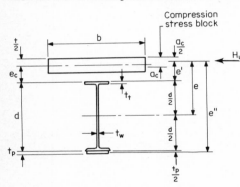

Fig. 6 Composite beam, slab in compression.

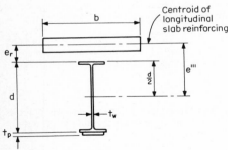

Fig. 7 Composite beam, slab in tension.

contributes to the plastic strength of the cross section. It is convenient to convert the area of the reinforcement with yield point f_{yr} to an equivalent area with yield point f_y

$$A_{re} = A_r \frac{f_{yr}}{f_y} \tag{13}$$

Normally, A_{re} is less than $(A_w + A_{pe})$ so that the plastic centroid is in the web of the steel section. The ultimate moment capacity is then

$$M_u = f_y Z_B + 0.5f_y A_{pe}(d + t_p) + f_y A_{re} e''' - \frac{f_y(A_{re} - A_{pe})^2}{4t_w} \tag{14}$$

*If the depth of the compression zone in the top flange is designated a_t, the neglected term in Eq. (11) is $f_y b_t a_t(t_t - a_t)$. Since a_t varies from 0 to t_t and is always small in comparison to both d and e', it is evident that the neglected term is much smaller than M_u given by Eq. (11).

4. Shear Connectors Stud and channel shear connectors shown in Fig. 8 are in common use in the United States.

The *stud shear connector* is a short length of round steel bar welded to the steel beam at one end and having an anchorage at the other end. The commonest diameters are ½, ⅝, ¾, and ⅞ in. The most commonly used lengths are 3 and 4 in., although longer studs are needed when the slab has a deep haunch over the steel beam. The anchorage is provided in the form of a round head (Fig. 8); the head thickness is usually ⅜ or ½ in. and the diameter ½ in. larger than the stud diameter. Stud connectors are end-welded. The joint

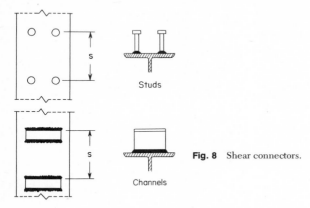

Studs

Channels

Fig. 8 Shear connectors.

encompasses the whole cross-sectional area of the stud and, when properly executed, is stronger than the steel of the stud. Stud connectors are used in both building and bridge structures.

The *channel shear connector* is a short length of rolled channel with one flange welded to the beam and the other providing an anchorage for the slab (Fig. 8). Channels C3 × 4.1 and C4 × 5.4 are most commonly used. They are usually welded with continuous fillet welds front and back of the channel. Channels are used in bridge construction.

The capacities of stud and channel connectors meeting the following conditions are given in Arts. 11 and 16.

1. The studs are made with ASTM A108, AISI Grades C1010, C1015, C1017, or C1020 cold-drawn steel having a minimum tensile strength of 60 ksi, and a minimum elongation of 20 percent in 2-in. gage length as specified in the AWS Structural Welding Code D1.1-75.[3]

2. The ratio of the overall length to the diameter of the stud is no less than 4.

3. The channels are made of structural-grade steel ASTM A36 or high-strength steel.

4. Concrete aggregates conform to ASTM C33 or C330 for normal weight or light-weight concretes, respectively.

Capacity of connectors other than studs and channels should be determined by tests. When used primarily for static loads, connectors may be designed for full or partial shear connection. Full shear connection is capable of developing the full plastic flexural capacity of the composite section. Then the horizontal shear at ultimate load is limited by Eqs. (7a) and (7b). For partial shear connection, flexural capacity of the composite section varies with the number of connectors provided. For moving loads, consideration must be given also to the fatigue strength of the connectors (Art. 16).

When the slab is haunched over the beam, the connectors should extend into the main body of the slab. Unless the stud is located directly over the web of the steel beam, the stud diameter should not exceed 2.5 times the beam flange thickness.

5. Unsymmetrical Steel Sections As the concrete slab performs the function of a heavy top steel plate, substantial savings of steel may be accomplished through the use of unsymmetrical steel sections. In built-up beams the bottom flange is made of a larger plate than the top flange; as this involves no additional fabrication costs, the use of unsymmetrical, rather than symmetrical, built-up beams is always economical. With rolled beams, the unsymmetrical steel section is obtained by welding a cover plate on the bottom flange. In this case the economy depends on the relative costs of welding the cover plate and of the steel saved.

6. Negative-Moment Sections   Negative-moment sections may be designed assuming that (1) the slab reinforcement acts compositely with the steel beam or (2) the steel beam alone resists the negative moment. The first assumption is usually preferable, since longitudinal slab reinforcement is always needed to control the tensile cracking of the slab.

When the slab reinforcement is assumed to act compositely with the steel beam, shear connectors must be provided throughout the negative-moment region. If the beam is subjected to repeated loading, the allowable tensile stress for the top flange should be selected with consideration of fatigue (Sec. 4).

When the external moment is assigned to the steel section alone, no connectors are required in the negative-moment regions. However, if the beam is subjected to fatigue loading and the slab reinforcement is continuous, additional connectors must be provided as discussed in Art. 16.

7. Deflections and Vibrations   In computing deflections, it is necessary to account for the method of construction and for the effects of creep of concrete. When no temporary supports are used during casting and curing of the concrete slab, the dead loads are resisted by the steel beam alone; but if effective temporary supports are used, the dead loads are resisted by the composite section.

If loads of long duration, such as dead loads, are resisted by the composite section, they cause creep of the concrete. Creep may be accounted for in computing deflections by increasing the conventional value of the modular ratio n. The multiplication factor $k = 2$ may be used for building designs, while $k = 3$ is recommended for bridge designs.

Live loads are always resisted by the composite section. If the loads are of short duration, the live-load deflections are computed with the conventional modular ratio ($k = 1$).

Dead-load deflections may cause excessive thickening of the slab in beams built without shores and excessive dishing of the slab in beams built with shores. Provisions should be made in the construction to compensate for large dead-load deflections.

Formulas for deflections are given in Sec. 1, Table 4.

Vibrations are either steady-state or transient. Steady-state vibrations are usually of no direct concern in structural design because they can and should be eliminated by insulating their source. On the other hand, transient vibrations cannot be eliminated; so the structure must be designed to accommodate them without impairment of its own function. Commonest transient vibrations are those caused by walking on a floor and by vehicles moving across a bridge.

The classical method of guarding against annoying vibrations is by limiting the slenderness of the load-carrying members. The following maximum depth-to-span ratios have been recommended for composite beams:

1:24 for ordinary building applications
1:20 for building applications where vibrations and shock are present
1:25 for bridges

These ratios originated in noncomposite design; their use in composite design was based on the overall depth including the steel beam and the concrete slab.

In general, composite designs have proved satisfactory. As with other types of construction, several cases of annoying vibrations that have been reported tended to be associated with large, open floor areas.

Studies have indicated that the damping characteristics of the structure play a very important role in the human reaction to transient vibrations.[4,5] Most types of buildings provide large amounts of damping through such construction features as walls, partitions, and flooring. On the other hand, in a few types of buildings, such as large clear-span schoolrooms, department stores, and churches, the inherent damping is very low, so that particular attention must be given to providing very stiff floors.

Studies have also indicated that under transient vibrations even noncomposite beams respond as composite. This would suggest the desirability of using the above ratios in conjunction with the depth of the steel beam alone. Such a conservative approach may be warranted in the design of floors which have large open areas, since they are particularly susceptible to annoying vibrations.

BUILDING DESIGN

8. Assumptions   Two techniques are available for the design of composite floor systems for buildings: working-stress design (WSD) and load and resistance-factor design (LRFD). These are discussed in Sec. 6, Art. 1. Design by WSD is explained in this section.

For design purposes, a composite floor is assumed to consist of a series of T-beams. Two types of composite beams are used in buildings: beams fully encased in concrete and beams without concrete encasement. In a fully encased beam, natural bond between the two materials may be assumed to provide composite action. When the concrete slab rests on the steel beam and there is only a partial encasement, or none, composite action may be assumed only if the slab is connected to the beam with shear connectors.

Encased Beams.   In order to qualify for recognition as a composite beam, an encased beam must be surrounded on its sides and soffit by 2 in. of concrete or more. The top of the beam must be at least 1½ in. below the top of the slab and at least 2 in. above the bottom of the slab. The encasement must be adequately reinforced throughout the whole depth and across the soffit of the beam, and the encasement must be cast integrally with the supported slab.

In proportioning encased beams, distinction must be made between the loads resisted by the steel section alone and those resisted by the composite section. Unless temporary supports are used during the casting and curing of concrete, the weight of the steel and concrete is supported by the steel beam alone and must be so considered in the design. Dead load applied after the concrete has hardened and all live loads are resisted by the composite section. The effective slab width, the allowable stresses, and the modular ratio (the effects of creep need not be considered) are the same as for unencased composite beams for buildings. Alternatively, the AISC specification[6] permits proportioning of the steel beam alone to resist all loads, live and dead, using an allowable bending stress of $0.76f_y$; in this case temporary shoring is not required.

Beams with Shear Connectors.   In proportioning composite beams with shear connectors, all loads may be assumed to be resisted by the composite section regardless of whether the steel section is supported on temporary shores during construction. The method of proportioning for this case is described in Arts. 9 and 10, and illustrated in Art. 12. Since flat soffit concrete slabs have been almost universally supplanted by slabs cast on formed steel decks, the design tables of Art. 10 and the examples of Art. 12 are based on this condition.

In addition to the stresses caused by dead and live loads, composite beams are subjected also to stresses caused by creep, shrinkage, expansion of concrete, and differential temperature changes. Except in unusual cases, none of these need be considered in the proportioning of composite beams for buildings. However, the effects of creep should be accounted for in deflection computations (Art. 7).

9. Design of Composite Beams   Composite beams are designed by the theory of the transformed section.

The width of the slab assumed effective as the flange of an interior composite T-beam is taken as the smallest of the following:

1. One-fourth of the span of the beam
2. The distance center to center of beams
3. The width of the top flange of the steel section plus sixteen times the least thickness of the slab.

For an edge beam, the effective width of the slab projection on either side of the steel section must not exceed:

1. One-twelfth of the span of the beam
2. Six times the least thickness of the slab

Furthermore, the effective width on the outside of the steel section must not exceed the actual width of the overhang, and the effective width on the inside must not extend beyond the centerline between the edge beam and the adjacent interior beam.

In negative-moment areas, the reinforcement of the slab parallel with the beam within the effective width of the slab may be included in computing the properties of the composite section, provided shear connectors are furnished in accordance with the requirements of Art. 11.

The effective width b of the concrete slab is transformed to the width of an equivalent steel plate by dividing by the modular ratio n. For concretes with unit weight w_c from 90 to 145 pcf the modular ratio may be expressed as

$$n = \frac{27,800}{w_c^{1.5} f_c'^{0.5}} \tag{15}$$

Integer accuracy for n is sufficient. The transformed width of the slab is then b/n.

The required section modulus is based on the maximum moment caused by all dead and live loads. The AISC allowable bending stress for the steel (usually the governing factor) is $0.66f_y$ for compact sections and $0.6f_y$ for noncompact sections. At positive-moment sections the compact values apply to both tension and compression flanges of rolled wide-flange beams because the concrete slab provides effective support against lateral buckling of the steel beam. The allowable stress for concrete in flexural compression is $0.45f_c'$. It should be noted that the concrete stress is caused only by loads applied subsequent to the time when the concrete has reached 75 percent of its required strength in unshored construction, and by the total dead and live load in shored construction.

In determining section properties for stress computations for composite beams with lightweight concrete (ASTM C330 aggregates) slabs, the effective width of the slab is assumed to be the same as for normal-weight concrete (ASTM C33 aggregates) of the same strength. For deflection calculations, the section properties are based on the appropriate ratio n computed from Eq. (15) for the specified strength and weight concrete.

After the steel section has been selected (Art. 10), the properties of the cross section may be computed from the pertinent equations given in Art. 2. If the beams are to be built without effective temporary shores, the section modulus of the composite section used in stress calculations must not exceed the following:

$$S_{bc} \leqslant \left(1.35 + 0.35 \frac{M_{LL}}{M_{DL}} \right) S_{bs} \tag{16}$$

For beams in positive bending, the section modulus S_{bs} refers to the bottom flange; for beams in negative bending, both flanges have to be investigated. The one-third increase in allowable stress for wind and seismic loads is not applicable for unshored beams. The stresses in the steel beam caused by loads supported by the steel beam alone must not exceed the allowable values for a bare steel beam.

For beams with partial shear connection, the effective section modulus may be taken as

$$S_{\text{eff}} = S_s + (H'/H)^{0.5}(S_{bc} - S_s) \tag{17}$$

and the effective moment of inertia for deflection computations may be taken as

$$I_{\text{eff}} = I_s + (H'/H)^{0.5}(I_c - I_s) \tag{18}$$

In some cases, it is convenient in design computations to replace the shear ratio H'/H in Eqs. (17) and (18) with the stud ratio N'/N.

Certain additional requirements have to be observed when the concrete slab is cast on a formed steel deck. Two cases must be considered: beams running perpendicular to the deck ribs and beams which run parallel to the deck ribs. In both cases the effective width of the concrete flange is based on the total slab thickness including the rib height. For beams running perpendicular to the deck ribs concrete below the top of the steel deck is neglected in computing the section properties, but for beams parallel to the deck ribs it may be included. It is satisfactory to assume the total amount of concrete in the ribs of the steel deck to be a flat soffit addition of the same amount to the slab above the top of the steel deck.

10. Selection of Steel Section   The required steel section for a composite beam made of a wide-flange beam and a concrete slab may be selected directly from Tables 1 through 8 on the basis of the required section modulus. The tables are for slabs with 2½ and 3¼ in. of concrete topping on steel decks 2 and 3 in. deep. For each slab-beam combination, the tables include the section modulus for the bottom fiber S_{bc} and the distance of the neutral axis from the bottom fiber y_{bc} for several values of b/n.

11. Design of Shear Connectors   The design of a full shear connection for a composite beam subjected essentially to static loading is based on the requirement that the composite beam be capable of developing its fully plastic flexural strength. Where it is not possible or economical to provide full shear connection (N connectors) between the slab and its supporting beam, the number of shear connectors may be reduced to N' (partial shear connection). The number of connectors N' should not be less than 25 percent of N.

Should the beams be subjected to fatigue loading, it is recommended to design the shear connectors according to the procedure described in Art. 16.

Number of Shear Connectors.   The total horizontal shear to be resisted at the interface of the concrete slab and the steel beam between the section of maximum positive moment and the adjacent section of zero moment may be taken as the smallest of the values given by the following formulas:

$$H = 0.425f'_c bt \qquad (19a)$$
$$H = 0.5A_s f_y \qquad (19b)$$
$$H' = N'q \qquad (19c)$$

For slabs cast on a formed steel deck, the slab thickness t in Eq. (19a) should be treated as explained in Art. 9 for computation of section properties.

When it is assumed that the slab reinforcement in the negative-moment region acts compositely with the steel beam, the total horizontal shear to be resisted by shear connectors between the sections of maximum positive and maximum negative moment may be taken as the sum of the smallest of the values given by Eqs. (19a), (19b), and (19c) and

$$H = 0.5A_r f_{yr} \qquad (19d)$$

The sum of the allowable loads for all shear connectors provided between the slab and the beam must be at least equal to H. Therefore, the required number of connectors is $N = H/q$. The allowable load q for one connector may be computed from Tables 9, 10, and 11.

Allowable Loads q.   The allowable static shear loads for stud connectors embedded in flat soffit slabs of concrete made with ASTM C33 aggregates are given in Table 9. These loads are applicable to slabs with a concrete haunch over the steel beam as long as the connectors have at least 1 in. lateral cover.

For slabs made with rotary-kiln-produced aggregates, the allowable shear loads are obtained by multiplying the values from Table 9 by the appropriate coefficient from Table 10. For slabs cast on a formed steel deck with the ribs perpendicular to the beams, the allowable loads are obtained by multiplying the values for flat-soffit slabs by the reduction factor

$$R = \frac{0.85}{\sqrt{n_r}} \frac{w_r}{h_r} \left(\frac{H_s}{h_r} - 1.0 \right) \le 1.0 \qquad (20)$$

For reduction-factor calculations, H_s should not be taken more than $h_r + 3$ in. and n_r should not be taken more than 3. When the ribs are parallel to the beams, Eq. (20) is modified by replacing $0.85/\sqrt{n_r}$ with 0.6 and is used only for decks with w_r/h_r less than 1.5. Values of R are listed in Table 11.

Equation (20) is applicable only to formed steel decks with nominal depth no more than 3 in. and having studs ¾ in. or less in diameter extending at least 1.5 in. above the top of the ribs. Where the ribs of the steel decks are parallel to the supporting beams, the deck may be split over the supporting member to form a haunch. When the ribs are 1.5 in. or more in height, the average width of a haunch or rib over the supporting member should not be less than 2 in. for the first stud in a transverse row plus four stud diameters for each additional stud. When the deck ribs are perpendicular to the steel beam, the slab must be anchored to the steel beam to resist uplift by welded studs or a combination of welded studs and puddle welds or other types of vertical anchorages. The anchor spacing should not exceed 16 in. and the stud spacing should not exceed 32 in.

Connector Placement.   Connectors may be spaced uniformly between sections of maximum and zero moments in an area of positive bending unless there are sizable abrupt changes in the shear diagram between the two sections. In the latter case the percentage of shear connectors between any concentrated load in that region and the nearest section of zero moment should be determined as

$$\frac{S_{bc}M/M_{\max} - S_s}{S_{bc} - S_s} \times 100 \text{ percent} \qquad (21)$$

TABLE 1 Section Properties of Composite Beams with Formed Steel Deck 2½-in. Topping on 2-in. Deck Perpendicular to Beam

Steel section	b/n = 0 S_{bs} in.³	y_{bs} in.	b/n = 2 S_{bcs} in.³	y_{bcs} in.	b/n = 4 S_{bc} in.³	y_{bc} in.	b/n = 6 S_{bcs} in.³	y_{bcs} in.	b/n = 8 S_{bcs} in.³	y_{bcs} in.	b/n = 10 S_{bcs} in.³	y_{bcs} in.	b/n = 12 S_{bcs} in.³	y_{bcs} in.
W36 × 160	542	18.01	588	20.05	620	21.73	643	23.15	661	24.35	675	25.39	686	26.29
W36 × 150	504	17.93	551	20.08	582	21.83	604	23.29	621	24.52	634	25.57	644	26.49
W36 × 135	439	17.78	485	20.13	515	22.01	536	23.54	552	24.82	563	25.90	573	26.82
W33 × 141	448	16.65	491	18.79	519	20.51	539	21.92	554	23.11	566	24.12	575	24.99
W33 × 130	406	16.55	448	18.83	476	20.64	495	22.12	509	23.34	519	24.36	528	25.24
W33 × 118	359	16.43	402	18.91	428	20.83	445	22.37	458	23.63	468	24.67	475	25.56
W30 × 116	329	15.01	368	17.33	393	19.14	409	20.57	421	21.74	430	22.71	436	23.54
W30 × 108	299	14.92	339	17.39	362	19.27	378	20.75	389	21.94	397	22.92	403	23.75
W30 × 99	269	14.83	308	17.48	330	19.45	345	20.97	355	22.19	362	23.18	368	24.00
W27 × 94	243	13.46	278	16.02	298	17.89	310	19.33	319	20.47	325	21.39	330	22.15
W27 × 84	213	13.36	248	16.14	266	18.13	277	19.61	285	20.77	290	21.69	294	22.45
W24 × 94	222	12.16	256	14.51	274	16.24	286	17.57	294	18.61	300	19.46	304	20.16
W24 × 84	196	12.05	229	14.63	246	16.46	256	17.83	263	18.90	268	19.75	272	20.44
W24 × 76	176	11.96	207	14.74	222	16.65	232	18.06	238	19.13	243	19.98	246	20.67
W24 × 68	154	11.87	185	14.86	199	16.89	207	18.32	213	19.40	217	20.24	220	20.92
W24 × 62	131	11.87	162	15.13	176	17.23	184	18.70	189	19.79	193	20.62	195	21.28
W24 × 55	114	11.79	145	15.33	157	17.52	164	19.01	168	20.09	171	20.91	174	21.55
W21 × 68	140	10.57	169	13.33	182	15.17	189	16.49	195	17.47	198	18.24	201	18.85
W21 × 62	127	10.50	154	13.44	167	15.35	174	16.69	178	17.67	181	18.43	184	19.03
W21 × 57	111	10.53	139	13.71	151	15.69	157	17.05	161	18.04	164	18.79	166	19.38
W21 × 50	94.5	10.42	121	13.88	132	15.95	137	17.32	141	18.29	143	19.02	145	19.59
W21 × 44	81.6	10.33	107	14.10	116	16.23	121	17.61	124	18.56	126	19.26	128	19.80
W18 × 55	98.3	9.06	123	11.96	133	13.75	139	14.97	142	15.85	145	16.52	147	17.05
W18 × 50	88.9	9.00	113	12.10	122	13.95	127	15.18	130	16.05	132	16.71	133	17.21
W18 × 46	78.8	9.03	102	12.35	111	14.26	116	15.49	118	16.36	120	17.00	122	17.50
W18 × 40	68.4	8.95	90.4	12.58	97.8	14.55	102	15.78	104	16.62	105	17.24	107	17.71
W18 × 35	57.6	8.85	78.5	12.80	84.9	14.81	88.1	16.02	90.0	16.84	91.3	17.42	92.3	17.86

Shape														
W16 × 50	81.0	8.13	104	11.02	113	12.74	117	13.88	120	14.69	122	15.30	124	15.77
W16 × 45	72.7	8.07	94.5	11.16	102	12.92	106	14.06	109	14.86	111	15.45	112	15.90
W16 × 40	64.7	8.01	85.0	11.35	91.8	13.17	95.2	14.30	97.4	15.08	98.8	15.65	99.9	16.08
W16 × 36	56.5	7.93	76.0	11.51	82.1	13.36	85.1	14.48	86.9	15.24	88.2	15.78	89.1	16.19
W16 × 31	47.2	7.94	65.7	11.90	70.9	13.79	73.4	14.90	74.9	15.63	75.9	16.14	76.7	16.52
W16 × 26	38.4	7.85	55.4	12.22	59.6	14.12	61.5	15.18	62.7	15.86	63.5	16.33	64.1	16.68
W14 × 38	54.6	7.05	73.7	10.23	79.8	11.91	82.9	12.95	84.8	13.65	86.0	14.16	87.0	14.55
W14 × 34	48.6	6.99	66.5	10.40	71.8	12.11	74.4	13.13	76.0	13.82	77.0	14.30	77.9	14.67
W14 × 30	42.0	6.92	58.9	10.59	63.5	12.32	65.7	13.32	67.0	13.97	67.9	14.43	68.6	14.77
W14 × 26	35.3	6.96	51.3	10.98	55.2	12.72	57.1	13.70	58.2	14.33	59.0	14.76	59.6	15.08
W14 × 22	29.0	6.87	43.5	11.27	46.7	13.01	48.1	13.93	49.0	14.51	49.6	14.90	50.1	15.19
W12 × 30	38.6	6.17	54.6	9.59	58.9	11.18	60.9	12.11	62.2	12.71	63.0	13.14	63.7	13.46
W12 × 26	33.4	6.11	48.1	9.81	51.6	11.41	53.3	12.31	54.3	12.88	55.0	13.28	55.5	13.57
W12 × 22	25.4	6.16	39.8	10.25	42.9	11.86	44.3	12.72	45.2	13.26	45.8	13.62	46.3	13.89
W12 × 19	21.3	6.08	34.5	10.49	37.0	12.07	38.1	12.88	38.8	13.38	39.4	13.71	39.8	13.95
W12 × 16	17.1	6.00	29.1	10.76	31.1	12.28	32.0	13.03	32.6	13.49	33.0	13.77	33.4	13.99
W12 × 14	14.9	5.96	25.8	10.98	27.5	12.46	28.3	13.16	28.8	13.58	29.2	13.85	29.5	14.04
W10 × 26	27.9	5.17	42.4	8.50	45.8	9.94	47.4	10.75	48.4	11.26	49.1	11.62	49.6	11.88
W10 × 22	23.2	5.09	36.4	8.71	39.1	10.14	40.4	10.90	41.2	11.38	41.8	11.70	42.2	11.94
W10 × 19	18.8	5.12	31.4	9.06	33.7	10.48	34.8	11.21	35.5	11.65	36.0	11.95	36.4	12.17
W10 × 17	16.2	5.06	27.9	9.21	29.9	10.60	30.8	11.29	31.4	11.70	31.9	11.98	32.2	12.18
W10 × 15	13.8	5.00	24.6	9.38	26.3	10.72	27.1	11.37	27.7	11.75	28.1	12.00	28.4	12.19
W10 × 12	10.9	4.94	20.1	9.73	21.3	10.98	21.9	11.56	22.3	11.89	22.7	12.12	22.9	12.29
W8 × 21	18.2	4.14	30.7	7.45	33.1	8.71	34.3	9.38	35.0	9.79	35.6	10.07	36.0	10.27
W8 × 18	15.2	4.07	26.4	7.64	28.4	8.87	29.3	9.49	29.9	9.87	30.4	10.12	30.8	10.30
W8 × 15	11.8	4.06	22.2	7.92	23.8	9.11	24.6	9.69	25.1	10.03	25.6	10.26	25.9	10.44
W8 × 13	9.9	4.00	19.3	8.09	20.6	9.23	21.3	9.76	21.7	10.07	22.1	10.29	22.3	10.45
W8 × 10	7.8	3.95	15.3	8.46	16.2	9.50	16.7	9.95	17.1	10.23	17.3	10.43	17.5	10.58

TABLE 2 Section Properties of Composite Beams with Formed Steel Deck 2½-in. Topping on 2-in. Deck Parallel to Beam

Steel section	b/n = 0		b/n = 2		b/n = 4		b/n = 6		b/n = 8		b/n = 10		b/n = 12	
	S_{bs} in.³	y_{bs} in.	S_{bc} in.³	y_{bc} in.	S_{bc} in.³	y_{bc} in.	S_{bc} in.³	y_{bc} in.	S_{bc} in.³	y_{bc} in.	S_{bc} in.³	y_{bc} in.	S_{bc} in.³	y_{bc} in.
W36 × 160	542	18.01	598	20.70	633	22.77	656	24.41	673	25.75	686	26.86	696	27.80
W36 × 150	504	17.93	560	20.75	594	22.90	616	24.58	632	25.94	644	27.06	653	28.00
W36 × 135	439	17.78	495	20.85	526	23.13	547	24.88	561	26.26	572	27.39	580	28.33
W33 × 141	448	16.65	499	19.44	530	21.53	549	23.16	563	24.45	574	25.51	582	26.40
W33 × 130	406	16.55	457	19.53	486	21.71	504	23.38	517	24.69	526	25.76	534	26.64
W33 × 118	359	16.43	410	19.65	437	21.94	454	23.66	465	25.00	474	26.06	480	26.93
W30 × 116	329	15.01	376	18.02	401	20.16	416	21.76	427	23.00	434	23.99	440	24.79
W30 × 108	299	14.92	346	18.11	370	20.33	384	21.95	394	23.20	401	24.18	406	24.98
W30 × 99	269	14.83	315	18.23	337	20.53	351	22.19	359	23.44	366	24.42	371	25.21
W27 × 94	243	13.46	284	16.73	303	18.90	314	20.45	322	21.61	327	22.51	331	23.23
W27 × 84	213	13.36	253	16.90	271	19.17	281	20.74	287	21.90	292	22.78	295	23.48
W24 × 94	222	12.16	260	15.16	279	17.16	289	18.58	296	19.65	301	20.48	305	21.14
W24 × 84	196	12.05	233	15.32	249	17.40	259	18.85	265	19.91	269	20.73	272	21.37
W24 × 76	176	11.96	211	15.46	226	17.62	234	19.08	239	20.13	243	20.93	246	21.55
W24 × 68	154	11.87	188	15.64	202	17.87	209	19.33	214	20.37	217	21.15	220	21.75
W24 × 62	131	11.87	166	15.93	179	18.23	186	19.70	190	20.73	193	21.49	195	22.07
W24 × 55	114	11.79	148	16.17	159	18.52	165	19.99	169	20.99	172	21.72	173	22.27
W21 × 68	140	10.57	172	14.02	184	16.05	191	17.38	195	18.33	198	19.04	200	19.58
W21 × 62	127	10.50	157	14.16	169	16.24	175	17.57	178	18.50	181	19.19	183	19.72
W21 × 57	111	10.53	142	14.45	152	16.59	160	17.93	161	18.85	164	19.52	166	20.03
W21 × 50	94.5	10.42	124	14.66	133	16.84	138	18.16	141	19.05	143	19.69	144	20.17
W21 × 44	81.6	10.33	109	14.91	117	17.11	122	18.41	124	19.26	126	19.87	127	20.32
W18 × 55	98.3	9.06	125	12.62	134	14.53	139	15.72	142	16.53	144	17.12	146	17.57
W18 × 50	88.9	9.00	114	12.78	122	14.72	127	15.90	129	16.70	131	17.27	133	17.70
W18 × 46	78.8	9.03	104	13.05	112	15.03	116	16.20	118	16.98	120	17.53	121	17.94
W18 × 40	68.4	8.95	91.7	13.31	98.2	15.30	101	16.44	103	17.18	105	17.70	106	18.08
W18 × 35	57.6	8.85	79.6	13.54	85.2	15.53	87.9	16.63	89.5	17.33	90.8	17.81	91.7	18.17

Shape														
W16 × 50	81.0	8.13	105	11.64	113	13.44	117	14.53	120	15.26	122	15.79	123	16.19
W16 × 45	72.7	8.07	95.8	11.79	103	13.61	106	14.69	108	15.40	110	15.90	111	16.28
W16 × 40	64.7	8.01	86.0	12.01	92.0	13.84	94.9	14.89	96.8	15.57	98.1	16.05	99.2	16.40
W16 × 36	56.5	7.93	76.9	12.18	82.1	14.01	84.7	15.03	86.3	15.68	87.5	16.13	88.5	16.46
W16 × 31	47.2	7.94	66.3	12.58	70.8	14.41	73.0	15.39	74.3	16.00	75.4	16.42	76.2	16.72
W16 × 26	38.4	7.85	55.8	12.90	59.4	14.69	61.1	15.60	62.3	16.16	63.1	16.53	63.8	16.80
W14 × 38	54.6	7.05	74.5	10.82	79.8	12.49	82.4	13.44	84.1	14.05	85.4	14.47	86.4	14.79
W14 × 34	48.6	6.99	67.1	11.00	71.6	12.67	73.9	13.59	75.4	14.17	76.5	14.57	77.3	14.86
W14 × 30	42.0	6.92	59.3	11.19	63.2	12.84	65.2	13.72	66.5	14.27	67.4	14.64	68.2	14.91
W14 × 26	35.3	6.96	51.6	11.58	55.0	13.22	56.7	14.06	57.8	14.57	58.6	14.91	59.3	15.17
W14 × 22	29.0	6.87	43.7	11.86	46.4	13.44	47.8	14.22	48.7	14.68	49.4	14.99	50.0	15.23
W12 × 30	38.6	6.17	54.8	10.12	58.5	11.65	60.4	12.46	61.6	12.96	62.6	13.30	63.4	13.55
W12 × 26	33.4	6.11	48.2	10.34	51.2	11.84	52.8	12.60	53.9	13.07	54.7	13.38	55.4	13.62
W12 × 22	25.4	6.16	39.9	10.78	42.6	12.24	44.0	12.96	45.0	13.39	45.7	13.69	46.2	13.92
W12 × 19	21.3	6.08	34.5	11.00	36.7	12.40	37.9	13.06	38.7	13.46	39.3	13.74	39.7	13.96
W12 × 16	17.1	6.00	29.0	11.22	30.8	12.54	31.9	13.14	32.5	13.51	33.0	13.78	33.4	13.99
W12 × 14	14.9	5.96	25.7	11.42	27.3	12.67	28.2	13.23	28.8	13.59	29.2	13.85	29.5	14.04
W10 × 26	27.9	5.17	42.4	8.96	45.5	10.29	47.0	10.97	48.1	11.39	48.9	11.68	49.5	11.91
W10 × 22	23.2	5.09	36.3	9.15	38.8	10.44	40.1	11.07	41.0	11.46	41.7	11.73	42.2	11.95
W10 × 19	18.8	5.12	31.3	9.49	33.4	10.74	34.6	11.33	35.4	11.70	36.0	11.97	36.4	12.17
W10 × 17	16.2	5.06	27.7	9.61	29.6	10.81	30.7	11.37	31.4	11.73	31.9	11.98	32.2	12.18
W10 × 15	13.8	5.00	24.5	9.75	26.1	10.88	27.0	11.42	27.6	11.76	28.1	12.00	28.4	12.19
W10 × 12	10.9	4.94	19.9	10.04	21.2	11.07	21.9	11.57	22.3	11.89	22.7	12.12	22.9	12.29
W8 × 21	18.2	4.14	30.5	7.80	32.8	8.92	34.1	9.47	34.9	9.82	35.6	10.08	36.0	10.27
W8 × 18	15.2	4.07	26.2	7.96	28.1	9.03	29.2	9.54	29.9	9.88	30.4	10.12	30.8	10.30
W8 × 15	11.8	4.06	22.0	8.22	23.7	9.22	24.6	9.72	25.1	10.03	25.6	10.26	25.9	10.44
W8 × 13	9.9	4.00	19.1	8.35	20.5	9.30	21.3	9.77	21.7	10.07	22.1	10.29	22.3	10.46
W8 × 10	7.8	3.95	15.2	8.65	16.2	9.52	16.7	9.95	17.1	10.23	17.3	10.43	17.5	10.58

TABLE 3 Section Properties of Composite Beams with Formed Steel Deck 3¼-in. Topping on 2-in. Deck Perpendicular to Beam

Steel section	$b/n = 0$		$b/n = 2$		$b/n = 4$		$b/n = 6$		$b/n = 8$		$b/n = 10$		$b/n = 12$	
	S_{tbs} in.^{3}	y_{tbs} in.	S_{bcs} in.^{3}	y_{bcs} in.	S_{bcs} in.^{3}	y_{bcs} in.	S_{bcs} in.^{3}	y_{bcs} in.	S_{bcs} in.^{3}	y_{bcs} in.	S_{bcs} in.^{3}	y_{bcs} in.	S_{bcs} in.^{3}	y_{bcs} in.
W36 × 160	542	18.01	602	20.63	640	22.69	666	24.35	685	25.71	699	26.85	710	27.81
W36 × 150	504	17.93	564	20.69	601	22.82	626	24.52	643	25.91	657	27.06	667	28.03
W36 × 135	439	17.78	499	20.79	533	23.05	556	24.82	572	26.24	584	27.41	594	28.38
W33 × 141	448	16.65	504	19.39	537	21.48	559	23.12	575	24.45	586	25.54	596	26.46
W33 × 130	406	16.55	461	19.47	493	21.66	513	23.35	528	24.70	538	25.80	547	26.72
W33 × 118	359	16.43	414	19.59	444	21.90	463	23.65	476	25.02	485	26.13	493	27.04
W30 × 116	329	15.01	380	17.98	408	20.14	425	21.77	437	23.05	446	24.08	453	24.93
W30 × 108	299	14.92	350	18.07	377	20.31	393	21.98	404	23.27	412	24.30	418	25.14
W30 × 99	269	14.83	319	18.19	344	20.52	359	22.23	369	23.53	376	24.56	382	25.39
W27 × 94	243	13.46	288	16.71	310	18.92	323	20.52	332	21.73	338	22.68	343	23.45
W27 × 84	213	13.36	257	16.88	277	19.19	289	20.83	296	22.05	302	22.99	306	23.73
W24 × 94	222	12.16	265	15.15	286	17.20	298	18.67	306	19.80	312	20.67	317	21.38
W24 × 84	196	12.05	238	15.32	256	17.46	267	18.97	274	20.09	279	20.96	283	21.65
W24 × 76	176	11.96	215	15.47	232	17.68	242	19.21	248	20.33	252	21.19	256	21.86
W24 × 68	154	11.87	193	15.65	208	17.95	217	19.49	222	20.60	226	21.44	229	22.09
W24 × 62	131	11.87	170	15.95	185	18.33	192	19.88	197	20.98	201	21.80	203	22.43
W24 × 55	114	11.79	152	16.20	165	18.65	172	20.20	176	21.28	179	22.07	181	22.67
W21 × 68	140	10.57	176	14.05	191	16.16	198	17.57	203	18.59	207	19.35	209	19.94
W21 × 62	127	10.50	161	14.20	175	16.36	182	17.78	186	18.78	189	19.53	191	20.11
W21 × 57	111	10.53	146	14.50	158	16.73	165	18.15	169	19.15	171	19.88	174	20.44
W21 × 50	94.5	10.42	128	14.72	139	17.00	144	18.42	147	19.38	150	20.08	152	20.61
W21 × 44	81.6	10.33	113	14.98	123	17.31	127	18.70	130	19.63	132	20.30	133	20.80
W18 × 55	98.3	9.06	129	12.69	140	14.70	146	15.98	149	16.87	152	17.52	153	18.01
W18 × 50	88.9	9.00	118	12.86	128	14.92	133	16.19	136	17.06	138	17.68	140	18.16
W18 × 46	78.8	9.03	108	13.14	117	15.24	122	16.51	124	17.36	126	17.97	128	18.43
W18 × 40	68.4	8.95	95.4	13.42	103	15.54	107	16.78	109	17.60	111	18.18	112	18.60
W18 × 35	57.6	8.85	83.1	13.68	89.5	15.81	92.6	17.01	94.5	17.79	95.8	18.32	96.8	18.72

Shape														
W16 × 50	16.67	130	16.22	129	15.64	127	14.83	124	13.65	119	11.73	110	8.13	81.0
W16 × 45	16.78	118	16.36	116	15.80	115	15.01	112	13.84	108	11.90	99.8	8.07	72.7
W16 × 40	16.93	105	16.54	104	16.00	102	15.25	100	14.10	96.9	12.14	89.9	8.01	64.7
W16 × 36	17.02	93.7	16.64	92.8	16.14	91.5	15.42	89.7	14.30	86.7	12.32	80.5	7.93	56.5
W16 × 31	17.31	80.7	16.97	79.9	16.50	78.8	15.82	77.3	14.74	74.9	12.75	69.7	7.94	47.2
W16 × 26	17.43	67.7	17.12	67.0	16.70	66.1	16.07	64.9	15.06	62.9	13.10	58.8	7.85	38.4
W14 × 38	15.34	92.0	14.99	90.9	14.51	89.6	13.83	87.7	12.78	84.6	10.97	78.3	7.05	54.6
W14 × 34	15.44	82.3	15.11	81.4	14.66	80.3	14.01	78.7	12.99	76.1	11.17	70.7	6.99	48.6
W14 × 30	15.51	72.7	15.21	71.9	14.79	70.9	14.17	69.5	13.19	67.3	11.39	62.7	6.92	42.0
W14 × 26	15.79	63.2	15.51	62.5	15.12	61.7	14.54	60.5	13.60	58.6	11.80	54.7	6.96	35.3
W14 × 22	15.87	53.3	15.62	52.7	15.27	52.0	14.74	51.0	13.87	49.5	12.12	46.4	6.87	29.0
W12 × 30	14.16	67.8	13.88	67.0	13.49	66.1	12.92	64.7	12.01	62.6	10.33	58.3	6.17	38.6
W12 × 26	14.25	59.3	13.99	58.6	13.63	57.7	13.10	56.6	12.24	54.9	10.58	51.3	6.11	33.4
W12 × 22	14.55	49.6	14.31	49.0	13.98	48.3	13.50	47.3	12.68	45.8	11.05	42.7	6.16	25.4
W12 × 19	14.60	42.7	14.37	42.2	14.07	41.6	13.63	40.7	12.87	39.5	11.31	37.0	6.08	21.3
W12 × 16	14.64	35.9	14.42	35.5	14.14	35.0	13.74	34.3	13.06	33.3	11.57	31.2	6.00	17.1
W12 × 14	14.70	31.7	14.49	31.4	14.23	31.0	13.85	30.4	13.21	29.5	11.80	27.7	5.96	14.9
W10 × 26	12.52	53.5	12.29	52.8	11.96	52.0	11.49	50.8	10.71	49.1	9.21	45.6	5.17	27.9
W10 × 22	12.57	45.7	12.35	45.1	12.06	44.4	11.62	43.4	10.89	42.0	9.44	39.2	5.09	23.2
W10 × 19	12.81	39.5	12.59	39.0	12.31	38.4	11.91	37.5	11.23	36.3	9.81	33.9	5.12	18.8
W10 × 17	12.82	35.0	12.61	34.6	12.34	34.0	11.97	33.3	11.33	32.2	9.97	30.1	5.06	16.2
W10 × 15	12.84	30.8	12.64	30.5	12.39	30.0	12.03	29.4	11.43	28.4	10.13	26.6	5.00	13.8
W10 × 12	12.95	24.9	12.77	24.6	12.53	24.3	12.19	23.8	11.66	23.1	10.48	21.7	4.94	10.9
W8 × 21	10.90	39.4	10.69	38.9	10.42	38.2	10.04	37.3	9.41	36.0	8.13	33.4	4.14	18.2
W8 × 18	10.94	33.7	10.74	33.3	10.49	32.8	10.13	32.0	9.55	30.9	8.32	28.8	4.07	15.2
W8 × 15	11.08	28.4	10.90	28.1	10.65	27.6	10.32	27.0	9.78	26.0	8.62	24.3	4.06	11.8
W8 × 13	11.11	24.5	10.93	24.3	10.70	23.9	10.38	23.4	9.88	22.6	8.79	21.1	4.00	9.9
W8 × 10	11.24	19.2	11.08	19.0	10.88	18.8	10.58	18.4	10.12	17.8	9.15	16.7	3.95	7.8

TABLE 4 Section Properties of Composite Beams with Formed Steel Deck 3¼ in. Topping on 2-in. Deck Parallel to Beam

Steel section	$b/n = 0$		$b/n = 2$		$b/n = 4$		$b/n = 6$		$b/n = 8$		$b/n = 10$		$b/n = 12$	
	S_{bos} in.³	y_{bos} in.	S_{bcs} in.³	y_{bcs} in.	S_{bcs} in.³	y_{bcs} in.	S_{bcs} in.³	y_{bcs} in.	S_{bcs} in.³	y_{bcs} in.	S_{bcs} in.³	y_{bcs} in.	S_{bos} in.³	y_{bos} in.
W36 × 160	542	18.01	611	21.24	650	23.62	675	25.44	692	26.87	705	28.04	715	29.00
W36 × 150	504	17.93	573	21.32	610	23.77	634	25.63	650	27.08	662	28.24	671	29.20
W36 × 135	439	17.78	507	21.46	542	24.04	563	25.95	578	27.42	589	28.58	597	29.53
W33 × 141	448	16.65	511	20.01	545	22.39	566	24.17	580	25.54	591	26.64	599	27.54
W33 × 130	406	16.55	468	20.12	500	22.59	519	24.41	532	25.80	542	26.89	549	27.78
W33 × 118	359	16.43	420	20.28	450	22.86	468	24.71	479	26.11	488	27.20	494	28.07
W30 × 116	329	15.01	386	18.61	413	21.02	429	22.75	440	24.04	448	25.05	454	25.86
W30 × 108	299	14.92	356	18.73	382	21.21	397	22.96	407	24.25	414	25.25	419	26.04
W30 × 99	269	14.83	324	18.88	348	21.44	362	23.21	371	24.50	378	25.48	382	26.25
W27 × 94	243	13.46	292	17.35	313	19.76	325	21.41	333	22.60	338	23.50	342	24.21
W27 × 84	213	13.36	261	17.56	280	20.06	290	21.71	297	22.88	302	23.76	305	24.44
W24 × 94	222	12.16	269	15.74	289	17.97	300	19.48	307	20.58	312	21.41	316	22.06
W24 × 84	196	12.05	241	15.94	259	18.24	268	19.76	274	20.84	279	21.65	282	22.27
W24 × 76	176	11.96	218	16.11	234	18.47	243	19.99	248	21.05	252	21.84	255	22.44
W24 × 68	154	11.87	195	16.32	210	18.73	217	20.25	222	21.29	225	22.04	228	22.62
W24 × 62	131	11.87	172	16.64	186	19.11	193	20.62	197	21.64	200	22.37	203	22.92
W24 × 55	114	11.79	154	16.92	166	19.42	172	20.90	175	21.88	178	22.58	180	23.10
W21 × 68	140	10.57	178	14.65	192	16.86	198	18.24	203	19.18	206	19.87	208	20.40
W21 × 62	127	10.50	163	14.81	175	17.05	182	18.42	185	19.35	188	20.02	190	20.52
W21 × 57	111	10.53	148	15.14	159	17.42	165	18.78	168	19.69	171	20.33	173	20.82
W21 × 50	94.5	10.42	129	15.38	139	17.68	144	19.00	147	19.87	149	20.48	151	20.93
W21 × 44	81.6	10.33	114	15.65	123	17.95	127	19.24	129	20.06	131	20.63	133	21.05
W18 × 55	98.3	9.06	131	13.25	140	15.29	145	16.50	148	17.30	151	17.87	153	18.30
W18 × 50	88.9	9.00	119	13.44	128	15.49	132	16.68	135	17.46	137	18.00	139	18.40
W18 × 46	78.8	9.03	109	13.73	117	15.80	121	16.98	124	17.73	125	18.25	127	18.64
W18 × 40	68.4	8.95	96.0	14.01	103	16.08	106	17.21	108	17.91	110	18.40	111	18.76
W18 × 35	57.6	8.85	83.5	14.26	89.2	16.31	92.1	17.38	93.9	18.04	95.3	18.49	96.5	18.81

W16 × 50	81.0	8.13	111	12.25	119	14.17	123	15.27	126	15.99	128	16.49	129	16.87
W16 × 45	72.7	8.07	100	12.43	108	14.34	111	15.42	114	16.11	116	16.59	117	16.94
W16 × 40	64.7	8.01	90.2	12.67	96.5	14.57	99.6	15.61	102	16.27	103	16.72	105	17.04
W16 × 36	56.5	7.93	80.7	12.85	86.3	14.74	89.0	15.74	90.9	16.36	92.3	16.78	93.5	17.09
W16 × 31	47.2	7.94	69.8	13.28	74.5	15.14	76.8	16.09	78.4	16.66	79.7	17.05	80.6	17.35
W16 × 26	38.4	7.85	58.7	13.61	62.5	15.40	64.5	16.28	65.9	16.79	66.9	17.16	67.6	17.44
W14 × 38	54.6	7.05	78.5	11.44	84.2	13.18	87.0	14.12	89.0	14.70	90.5	15.10	91.7	15.40
W14 × 34	48.6	6.99	70.7	11.64	75.6	13.36	78.1	14.26	79.8	14.81	81.2	15.18	82.2	15.47
W14 × 30	42.0	6.92	62.6	11.84	66.8	13.53	69.0	14.38	70.6	14.89	71.7	15.25	72.6	15.53
W14 × 26	35.3	6.96	54.5	12.25	58.2	13.90	60.1	14.70	61.5	15.18	62.5	15.53	63.2	15.79
W14 × 22	29.0	6.87	46.2	12.54	49.2	14.10	50.8	14.84	51.9	15.29	52.7	15.62	53.3	15.87
W12 × 30	38.6	6.17	58.1	10.74	62.1	12.30	64.3	13.08	65.8	13.56	66.9	13.90	67.8	14.17
W12 × 26	33.4	6.11	51.1	10.97	54.5	12.48	56.3	13.21	57.6	13.67	58.6	14.00	59.3	14.25
W12 × 22	25.4	6.16	42.5	11.42	45.5	12.87	47.1	13.56	48.2	14.00	49.0	14.31	49.6	14.55
W12 × 19	21.3	6.08	36.8	11.64	39.3	13.01	40.7	13.66	41.6	14.07	42.2	14.37	42.7	14.60
W12 × 16	17.1	6.00	31.0	11.86	33.1	13.14	34.3	13.75	35.0	14.14	35.5	14.42	35.9	14.64
W12 × 14	14.9	5.96	27.5	12.05	29.4	13.26	30.4	13.85	31.0	14.23	31.4	14.49	31.7	14.70
W10 × 26	27.9	5.17	45.4	9.54	48.7	10.89	50.6	11.55	51.9	11.98	52.8	12.29	53.5	12.52
W10 × 22	23.2	5.09	38.9	9.74	41.7	11.03	43.3	11.65	44.4	12.06	45.1	12.35	45.7	12.57
W10 × 19	18.8	5.12	33.6	10.08	36.1	11.32	37.5	11.92	38.4	12.31	39.0	12.59	39.5	12.81
W10 × 17	16.2	5.06	29.9	10.21	32.1	11.39	33.3	11.97	34.0	12.34	34.6	12.61	35.0	12.82
W10 × 15	13.8	5.00	26.4	10.34	28.4	11.46	29.4	12.03	30.0	12.39	30.5	12.64	30.8	12.84
W10 × 12	10.9	4.94	21.5	10.63	23.1	11.67	23.8	12.19	24.3	12.53	24.6	12.77	24.9	12.95
W8 × 21	18.2	4.14	33.1	8.35	35.8	9.48	37.3	10.05	38.2	10.42	38.9	10.69	39.4	10.90
W8 × 18	15.2	4.07	28.5	8.51	30.8	9.58	32.0	10.13	32.8	10.49	33.3	10.74	33.7	10.94
W8 × 15	11.8	4.06	24.1	8.77	26.0	9.79	27.0	10.32	27.6	10.65	28.1	10.90	28.4	11.08
W8 × 13	9.9	4.00	21.0	8.90	22.6	9.88	23.4	10.38	23.9	10.70	24.3	10.93	24.5	11.11
W8 × 10	7.8	3.95	16.7	9.20	17.8	10.12	18.4	10.58	18.8	10.88	19.0	11.08	19.2	11.24

TABLE 5 Section Properties of Composite Beams with Formed Steel Deck 2½-in. Topping on 3-in. Deck Perpendicular to Beam

Steel section	$b/n = 0$ S_{bs}, in.³	y_{bs}, in.	$b/n = 2$ S_{bc}, in.³	y_{bc}, in.	$b/n = 4$ S_{bc}, in.³	y_{bc}, in.	$b/n = 6$ S_{bc}, in.³	y_{bc}, in.	$b/n = 8$ S_{bc}, in.³	y_{bc}, in.	$b/n = 10$ S_{bc}, in.³	y_{bc}, in.	$b/n = 12$ S_{bc}, in.³	y_{bc}, in.
W36 × 160	542	18.01	595	20.14	632	21.91	658	23.39	678	24.65	694	25.73	706	26.68
W36 × 150	504	17.93	558	20.18	593	22.02	618	23.54	637	24.83	652	25.94	664	26.89
W36 × 135	439	17.78	492	20.24	526	22.21	550	23.81	567	25.15	580	26.29	591	27.25
W33 × 141	448	16.65	498	18.89	530	20.70	553	22.19	570	23.44	583	24.50	594	25.41
W33 × 130	406	16.55	455	18.95	487	20.85	508	22.40	524	23.68	536	24.76	545	25.68
W33 × 118	359	16.43	408	19.03	438	21.06	458	22.67	473	23.99	483	25.09	492	26.02
W30 × 116	329	15.01	375	17.46	403	19.36	422	20.88	435	22.11	445	23.14	453	24.00
W30 × 108	299	14.92	346	17.53	372	19.51	390	21.07	402	22.33	412	23.37	419	24.23
W30 × 99	269	14.83	315	17.62	340	19.70	357	21.31	368	22.59	376	23.64	383	24.51
W27 × 94	243	13.46	285	16.17	307	18.16	322	19.68	331	20.89	339	21.86	344	22.67
W27 × 84	213	13.36	254	16.31	275	18.41	288	19.99	297	21.21	303	22.19	308	22.99
W24 × 94	222	12.16	262	14.66	284	16.51	297	17.92	307	19.03	313	19.94	319	20.68
W24 × 84	196	12.05	235	14.79	255	16.75	267	18.21	275	19.34	281	20.25	285	20.99
W24 × 76	176	11.96	213	14.92	231	16.96	242	18.46	249	19.61	254	20.51	258	21.24
W24 × 68	154	11.87	191	15.08	207	17.22	217	18.75	223	19.90	228	20.80	231	21.51
W24 × 62	131	11.87	168	15.34	184	17.59	193	19.15	199	20.31	203	21.20	206	21.90
W24 × 55	114	11.79	150	15.57	164	17.91	172	19.49	177	20.64	181	21.51	183	22.20
W21 × 68	140	10.57	174	13.53	190	15.50	199	16.91	205	17.97	209	18.80	212	19.45
W21 × 62	127	10.50	160	13.66	175	15.71	183	17.14	188	18.19	192	19.01	194	19.65
W21 × 57	111	10.53	144	13.94	158	16.07	166	17.52	170	18.58	174	19.39	176	20.02
W21 × 50	94.5	10.42	127	14.14	139	16.35	145	17.82	149	18.87	152	19.65	154	20.26
W21 × 44	81.6	10.33	112	14.38	123	16.67	129	18.14	132	19.17	134	19.92	136	20.50
W18 × 55	98.3	9.06	129	12.19	141	14.13	147	15.45	151	16.41	154	17.13	156	17.69
W18 × 50	88.9	9.00	118	12.36	129	14.36	135	15.68	138	16.63	141	17.34	142	17.88
W18 × 46	78.8	9.03	108	12.62	118	14.68	123	16.02	126	16.96	129	17.65	130	18.19
W18 × 40	68.4	8.95	95.2	12.88	104	15.01	108	16.34	111	17.25	113	17.92	114	18.42
W18 × 35	57.6	8.85	83.0	13.13	90.6	15.30	94.2	16.62	96.4	17.50	97.9	18.13	99.0	18.60

W16 × 50	81.0	8.13	109	11.27	120	13.14	125	14.38	129	15.27	131	15.93	133	16.44
W16 × 45	72.7	8.07	99.7	11.43	109	13.35	114	14.59	117	15.46	119	16.10	120	16.60
W16 × 40	64.7	8.01	89.9	11.65	98.1	13.63	102	14.86	105	15.71	106	16.33	107	16.80
W16 × 36	56.5	7.93	80.7	11.83	87.9	13.84	91.4	15.07	93.5	15.89	95.0	16.48	96.0	16.93
W16 × 31	47.2	7.94	70.0	12.26	76.1	14.32	79.0	15.52	80.7	16.31	81.9	16.87	82.7	17.29
W16 × 26	38.4	7.85	59.2	12.61	64.1	14.69	66.4	15.84	67.7	16.58	68.6	17.10	69.3	17.47
W14 × 38	54.6	7.05	78.7	10.54	86.0	12.38	89.6	13.52	91.8	14.29	93.3	14.85	94.4	15.28
W14 × 34	48.6	6.99	71.1	10.74	77.5	12.61	80.5	13.73	82.3	14.48	83.6	15.02	84.5	15.42
W14 × 30	42.0	6.92	63.2	10.95	68.7	12.85	71.2	13.95	72.8	14.66	73.8	15.17	74.6	15.55
W14 × 26	35.3	6.96	55.2	11.37	59.9	13.29	62.0	14.36	63.3	15.05	64.2	15.52	64.8	15.87
W14 × 22	29.0	6.87	47.0	11.71	50.7	13.61	52.4	14.63	53.4	15.27	54.1	15.70	54.6	16.01
W12 × 30	38.6	6.17	59.0	9.95	64.1	11.72	66.5	12.74	68.0	13.41	69.0	13.88	69.7	14.23
W12 × 26	33.4	6.11	57.0	10.20	56.3	11.98	58.3	12.97	59.4	13.60	60.2	14.04	60.8	14.36
W12 × 22	25.4	6.16	43.4	10.69	47.1	12.47	48.7	13.42	49.7	14.01	50.4	14.42	50.9	14.71
W12 × 19	21.3	6.08	37.7	10.97	40.7	12.71	42.0	13.61	42.7	14.16	43.3	14.53	43.7	14.79
W12 × 16	17.1	6.00	32.0	11.27	34.3	12.96	35.3	13.79	35.9	14.29	36.4	14.62	36.8	14.85
W12 × 14	14.9	5.96	28.4	11.53	30.4	13.16	31.2	13.94	31.8	14.40	32.2	14.70	32.5	14.92
W10 × 26	27.9	5.17	46.5	8.90	50.6	10.51	52.5	11.41	53.7	11.98	54.4	12.38	55.0	12.67
W10 × 22	23.2	5.09	40.1	9.15	43.4	10.75	44.9	11.60	45.8	12.13	46.4	12.50	46.9	12.76
W10 × 19	18.8	5.12	34.7	9.53	37.5	11.12	38.8	11.94	39.6	12.43	40.1	12.77	40.5	13.01
W10 × 17	16.2	5.06	31.0	9.71	33.3	11.26	34.4	12.04	35.0	12.50	35.5	12.81	35.9	13.03
W10 × 15	13.8	5.00	27.4	9.91	29.4	11.41	30.3	12.14	30.9	12.57	31.3	12.85	31.6	13.06
W10 × 12	10.9	4.94	22.4	10.31	23.9	11.72	24.5	12.37	25.0	12.74	25.3	12.98	25.6	13.17
W8 × 21	18.2	4.14	34.5	7.90	37.4	9.33	38.7	10.09	39.5	10.55	40.1	10.87	40.6	11.10
W8 × 18	15.2	4.07	29.8	8.12	32.1	9.52	33.2	10.23	33.8	10.66	34.3	10.94	34.7	11.15
W8 × 15	11.8	4.06	25.2	8.45	27.1	9.81	27.9	10.46	28.5	10.85	28.9	11.11	29.2	11.30
W8 × 13	9.9	4.00	21.9	8.66	23.5	9.95	24.2	10.56	24.7	10.91	25.0	11.15	25.3	11.32
W8 × 10	7.8	3.95	17.4	9.09	18.4	10.27	19.0	10.79	19.3	11.09	19.6	11.30	19.8	11.46

TABLE 6 Section Properties of Composite Beams with Formed Steel Deck 2½-in. Topping on 3-in. Deck Parallel to Beam

Steel section	b/n = 0		b/n = 2		b/n = 4		b/n = 6		b/n = 8		b/n = 10		b/n = 12	
	S_{bts} in.³	y_{bts} in.	S_{bcs} in.³	y_{bcs} in.	S_{bcs} in.³	y_{bcs} in.	S_{bcs} in.³	y_{bcs} in.	S_{bcs} in.³	y_{bcs} in.	S_{bcs} in.³	y_{bcs} in.	S_{bcs} in.³	y_{bcs} in.
W36 × 160	542	18.01	611	21.13	652	23.47	678	25.27	696	26.72	710	27.89	720	28.87
W36 × 150	504	17.93	573	21.21	612	23.62	637	25.46	654	26.92	667	28.10	676	29.08
W36 × 135	439	17.78	507	21.34	544	23.89	566	25.79	582	27.27	593	28.45	602	29.42
W33 × 141	448	16.65	512	19.90	547	22.25	569	24.02	584	25.41	595	26.53	603	27.44
W33 × 130	406	16.55	469	20.01	502	22.45	522	24.27	536	25.67	546	26.79	554	27.69
W33 × 118	359	16.43	421	20.16	452	22.72	471	24.58	483	25.99	492	27.10	499	28.00
W30 × 116	329	15.01	387	18.51	415	20.90	433	22.64	444	23.95	452	24.98	458	25.81
W30 × 108	299	14.92	357	18.63	384	21.09	400	22.85	410	24.17	418	25.19	423	26.01
W30 × 99	269	14.83	325	18.78	351	21.33	365	23.11	375	24.42	381	25.43	386	26.23
W27 × 94	243	13.46	294	17.26	316	19.67	328	21.33	336	22.55	342	23.48	346	24.21
W27 × 84	213	13.36	262	17.47	282	19.96	293	21.64	300	22.85	305	23.76	309	24.47
W24 × 94	222	12.16	270	15.66	291	17.89	303	19.42	311	20.55	316	21.40	320	22.08
W24 × 84	196	12.05	242	15.85	261	18.16	271	19.71	278	20.83	282	21.66	286	22.32
W24 × 76	176	11.96	220	16.03	237	18.40	246	19.96	251	21.05	255	21.87	258	22.50
W24 × 68	154	11.87	197	16.24	212	18.67	220	20.23	225	21.30	228	22.09	231	22.69
W24 × 62	131	11.87	173	16.56	188	19.06	195	20.61	200	21.67	203	22.43	206	23.01
W24 × 55	114	11.79	155	16.84	168	19.38	174	20.91	178	21.93	181	22.66	183	23.21
W21 × 68	140	10.57	180	14.58	194	16.82	201	18.24	206	19.22	209	19.94	212	20.49
W21 × 62	127	10.50	165	14.75	178	17.02	184	18.44	188	19.40	191	20.10	193	20.63
W21 × 57	111	10.53	149	15.07	161	17.39	167	18.80	171	19.75	173	20.43	175	20.94
W21 × 50	94.5	10.42	130	15.32	141	17.67	146	19.04	149	19.95	151	20.59	153	21.07
W21 × 44	81.6	10.33	115	15.60	125	17.96	129	19.30	132	20.16	134	20.77	135	21.21
W18 × 55	98.3	9.06	132	13.21	143	15.29	148	16.55	151	17.39	153	17.99	155	18.44
W18 × 50	88.9	9.00	121	13.40	130	15.51	135	16.74	138	17.56	140	18.13	141	18.56
W18 × 46	78.8	9.03	110	13.69	119	15.83	123	17.05	126	17.84	128	18.40	129	18.81
W18 × 40	68.4	8.95	97.4	13.98	105	16.12	108	17.30	110	18.05	112	18.56	113	18.94
W18 × 35	57.6	8.85	84.7	14.25	90.9	16.36	94.0	17.49	95.7	18.19	97.1	18.67	98.2	19.02

W16 × 50	81.0	8.13	112	12.23	121	14.19	125	15.34	128	16.10	130	16.63	132	17.03
W16 × 45	72.7	8.07	102	12.41	110	14.38	114	15.51	116	16.23	118	16.74	119	17.12
W16 × 40	64.7	8.01	91.7	12.65	98.3	14.63	102	15.72	104	16.41	105	16.89	106	17.24
W16 × 36	56.5	7.93	82.1	12.85	88.0	14.81	90.8	15.86	92.7	16.52	94.1	16.97	95.2	17.29
W16 × 31	47.2	7.94	71.0	13.29	76.0	15.23	78.4	16.23	79.9	16.84	81.1	17.26	82.1	17.56
W16 × 26	38.4	7.85	59.9	13.63	63.8	15.51	65.8	16.44	67.1	16.99	68.2	17.37	68.9	17.65
W14 × 38	54.6	7.05	79.9	11.45	86.0	13.26	89.0	14.24	90.9	14.86	92.3	15.29	93.5	15.60
W14 × 34	48.6	6.99	72.1	11.65	77.2	13.45	79.8	14.39	81.5	14.98	82.8	15.38	83.9	15.68
W14 × 30	42.0	6.92	63.9	11.87	68.3	13.63	70.5	14.53	72.0	15.08	73.2	15.45	74.1	15.73
W14 × 26	35.3	6.96	55.7	12.29	59.5	14.02	61.4	14.87	62.8	15.38	63.8	15.74	64.5	16.00
W14 × 22	29.0	6.87	47.2	12.60	50.3	14.25	51.9	15.03	53.0	15.50	53.8	15.83	54.4	16.08
W12 × 30	38.6	6.17	59.4	10.78	63.6	12.41	65.8	13.25	67.3	13.76	68.4	14.11	69.3	14.37
W12 × 26	33.4	6.11	52.2	11.02	55.8	12.61	57.6	13.40	58.9	13.87	59.9	14.20	60.6	14.46
W12 × 22	25.4	6.16	43.6	11.49	46.6	13.03	48.2	13.76	49.4	14.20	50.2	14.52	50.8	14.76
W12 × 19	21.3	6.08	37.7	11.73	40.2	13.19	41.6	13.86	42.5	14.28	43.2	14.58	43.7	14.81
W12 × 16	17.1	6.00	31.8	11.97	33.9	13.33	35.1	13.95	35.8	14.35	36.4	14.64	36.8	14.85
W12 × 14	14.9	5.96	28.3	12.18	30.1	13.46	31.1	14.06	31.7	14.44	32.2	14.71	32.5	14.92
W10 × 26	27.9	5.17	46.6	9.61	50.1	11.04	51.9	11.74	53.2	12.18	54.2	12.49	54.9	12.73
W10 × 22	23.2	5.09	40.0	9.82	42.9	11.19	44.4	11.85	45.5	12.26	46.3	12.56	46.8	12.78
W10 × 19	18.8	5.12	34.6	10.18	37.1	11.50	38.5	12.12	39.4	12.52	40.0	12.80	40.5	13.02
W10 × 17	16.2	5.06	30.8	10.32	33.0	11.58	34.2	12.17	34.9	12.55	35.5	12.83	35.9	13.03
W10 × 15	13.8	5.00	27.2	10.47	29.1	11.66	30.2	12.23	30.8	12.60	31.3	12.86	31.6	13.06
W10 × 12	10.9	4.94	22.2	10.78	23.7	11.87	24.5	12.40	25.0	12.74	25.3	12.98	25.6	13.17
W8 × 21	18.2	4.14	34.2	8.46	36.9	9.66	38.4	10.24	39.4	10.62	40.1	10.90	40.6	11.11
W8 × 18	15.2	4.07	29.5	8.64	31.7	9.77	33.0	10.33	33.8	10.69	34.3	10.95	34.7	11.15
W8 × 15	11.8	4.06	24.9	8.91	26.8	9.98	27.8	10.52	28.5	10.86	28.9	11.11	29.2	11.30
W8 × 13	9.9	4.00	21.6	9.06	23.3	10.07	24.1	10.59	24.7	10.91	25.0	11.15	25.3	11.32
W8 × 10	7.8	3.95	17.2	9.38	18.4	10.32	19.0	10.79	19.3	11.09	19.6	11.30	19.8	11.46

TABLE 7 Section Properties of Composite Beams with Formed Steel Deck 3¼-in. Topping on 3-in. Deck Perpendicular to Beam

Steel section	$b/n = 0$		$b/n = 2$		$b/n = 4$		$b/n = 6$		$b/n = 8$		$b/n = 10$		$b/n = 12$	
	S_{bs} in.³	y_{bs} in.	S_{bs} in.³	y_{bs} in.	S_{bs} in.³	y_{bs} in.	S_{bs} in.³	y_{bs} in.	S_{bs} in.³	y_{bs} in.	S_{bs} in.³	y_{bs} in.	S_{bs} in.³	y_{bs} in.
W36 × 160	542	18.01	611	20.75	654	22.91	683	24.64	704	26.07	720	27.26	732	28.27
W36 × 150	504	17.93	573	20.82	614	23.05	642	24.83	662	26.28	677	27.48	688	28.50
W36 × 135	439	17.78	507	20.93	546	23.30	572	25.15	590	26.64	603	27.86	613	28.88
W33 × 141	448	16.65	512	19.53	550	21.72	575	23.44	593	24.83	606	25.98	616	26.94
W33 × 130	406	16.55	469	19.62	505	21.91	528	23.69	545	25.11	557	26.26	566	27.23
W33 × 118	359	16.43	422	19.75	456	22.17	477	24.01	492	25.45	502	26.61	510	27.57
W30 × 116	329	15.01	388	18.14	420	20.41	440	22.13	453	23.48	463	24.57	471	25.46
W30 × 108	299	14.92	358	18.24	388	20.60	407	22.36	419	23.72	428	24.81	435	25.69
W30 × 99	269	14.83	327	18.38	355	20.83	372	22.63	384	24.00	392	25.09	398	25.96
W27 × 94	243	13.46	296	16.90	321	19.24	336	20.93	346	22.22	353	23.22	358	24.03
W27 × 84	213	13.36	265	17.09	288	19.54	301	21.27	309	22.56	315	23.55	320	24.35
W24 × 94	222	12.16	273	15.34	297	17.51	311	19.09	320	20.28	327	21.21	332	21.97
W24 × 84	196	12.05	245	15.52	267	17.80	279	19.41	287	20.60	293	21.52	297	22.26
W24 × 76	176	11.96	223	15.69	242	18.05	253	19.68	260	20.87	265	21.78	269	22.49
W24 × 68	154	11.87	200	15.89	217	18.34	227	19.99	233	21.17	237	22.05	241	22.75
W24 × 62	131	11.87	176	16.21	193	18.74	202	20.40	208	21.57	212	22.44	215	23.12
W24 × 55	114	11.79	158	16.48	173	19.09	181	20.75	185	21.90	189	22.74	191	23.38
W21 × 68	140	10.57	183	14.29	200	16.55	209	18.06	215	19.15	219	19.97	222	20.61
W21 × 62	127	10.50	168	14.46	184	16.77	192	18.30	197	19.37	200	20.17	203	20.79
W21 × 57	111	10.53	152	14.78	167	17.16	174	18.69	179	19.76	182	20.54	184	21.14
W21 × 50	94.5	10.42	134	15.03	146	17.47	153	18.99	156	20.02	159	20.77	161	21.34
W21 × 44	81.6	10.33	119	15.31	130	17.81	135	19.30	138	20.30	140	21.01	142	21.55
W18 × 55	98.3	9.06	136	12.97	149	15.15	155	16.53]	159	17.48	162	18.18	164	18.72
W18 × 50	88.9	9.00	125	13.17	136	15.39	142	16.76	145	17.70	147	18.37	149	18.89
W18 × 46	78.8	9.03	114	13.47	125	15.73	130	17.10	133	18.02	135	18.68	136	19.17
W18 × 40	68.4	8.95	101	13.77	110	16.07	114	17.41	117	18.29	118	18.91	120	19.37
W18 × 35	57.6	8.85	88.1	14.06	95.6	16.37	99.1	17.67	101	18.50	103	19.08	104	19.51

Shape														
W16 × 50	17.39	140	16.91	138	16.28	136	15.40	132	14.12	127	12.04	116	8.13	81.0
W16 × 45	17.53	127	17.07	125	16.46	123	15.61	120	14.34	116	12.23	106	8.07	72.7
W16 × 40	17.70	113	17.27	112	16.69	110	15.87	108	14.63	104	12.49	95.5	8.01	64.7
W16 × 36	17.80	101	17.40	99.8	16.85	98.4	16.06	96.3	14.85	92.9	12.70	85.7	7.93	56.5
W16 × 31	18.12	87.0	17.75	86.1	17.24	84.9	16.50	83.2	15.32	80.4	13.17	74.4	7.94	47.2
W16 × 26	18.26	73.0	17.93	72.3	17.47	71.3	16.79	70.0	15.68	67.8	13.56	63.0	7.85	38.4
W14 × 38	16.12	99.6	15.73	98.5	15.21	97.0	14.47	94.8	13.32	91.3	11.34	83.9	7.05	54.6
W14 × 34	16.23	89.2	15.87	88.2	15.38	87.0	14.67	85.1	13.56	82.2	11.57	75.8	6.99	48.6
W14 × 30	16.33	78.8	15.99	77.9	15.53	76.9	14.86	75.3	13.79	72.8	11.81	67.4	6.92	42.0
W14 × 26	16.63	68.6	16.32	67.9	15.89	66.9	15.26	65.6	14.23	63.5	12.26	59.0	6.96	35.3
W14 × 22	16.73	57.9	16.45	57.3	16.07	56.5	15.49	55.5	14.54	53.8	12.62	50.2	6.87	29.0
W12 × 30	14.98	74.0	14.67	73.1	14.24	72.1	13.61	70.6	12.61	68.2	10.76	63.1	6.17	38.6
W12 × 26	15.08	64.7	14.80	63.9	14.40	63.0	13.82	61.8	12.87	59.9	11.04	55.7	6.11	33.4
W12 × 22	15.40	54.3	15.14	53.7	14.78	52.9	14.25	51.8	13.35	50.2	11.55	46.6	6.16	25.4
W12 × 19	15.46	46.8	15.22	46.2	14.90	45.6	14.41	44.7	13.57	43.3	11.84	40.5	6.08	21.3
W12 × 16	15.51	39.4	15.28	39.0	14.99	38.4	14.55	37.7	13.79	36.6	12.15	34.3	6.00	17.1
W12 × 14	15.58	34.8	15.36	34.5	15.08	34.0	14.67	33.4	13.97	32.4	12.41	30.5	5.96	14.9
W10 × 26	13.36	59.0	13.10	58.3	12.74	57.4	12.21	56.2	11.34	54.3	9.67	50.2	5.17	27.9
W10 × 22	13.42	50.4	13.18	49.8	12.86	49.1	12.37	48.0	11.56	46.5	9.94	43.2	5.09	23.2
W10 × 19	13.66	43.7	13.43	43.2	13.13	42.5	12.68	41.6	11.92	40.3	10.35	37.5	5.12	18.8
W10 × 17	13.68	38.8	13.46	38.3	13.18	37.7	12.76	36.9	12.05	35.8	10.53	33.4	5.06	16.2
W10 × 15	13.71	34.2	13.50	33.8	13.23	33.3	12.84	32.6	12.18	31.6	10.73	29.6	5.00	13.8
W10 × 12	13.83	27.6	13.64	27.3	13.39	27.0	13.03	26.5	12.45	25.7	11.12	24.2	4.94	10.9
W8 × 21	11.74	44.1	11.52	43.6	11.23	42.8	10.80	41.9	10.09	40.4	8.64	37.5	4.14	18.2
W8 × 18	11.79	37.8	11.58	37.4	11.31	36.8	10.92	35.9	10.26	34.7	8.88	32.4	4.07	15.2
W8 × 15	11.94	31.9	11.75	31.5	11.49	31.0	11.13	30.4	10.53	29.3	9.21	27.4	4.06	11.8
W8 × 13	11.98	27.6	11.79	27.3	11.55	26.9	11.20	26.3	10.65	25.5	9.41	23.9	4.00	9.9
W8 × 10	12.13	21.6	11.96	21.4	11.74	21.1	11.42	20.7	10.93	20.1	9.83	18.9	3.95	7.8

TABLE 8 Section Properties of Composite Beams with Formed Steel Deck 3¼-in. Topping on 3-in. Deck Parallel to Beam

Steel section	$b/h = 0$ S_{bts} in.³	y_{bts} in.	$b/h = 2$ S_{bcs} in.³	y_{bcs} in.	$b/h = 4$ S_{bcs} in.³	y_{bcs} in.	$b/h = 6$ S_{bcs} in.³	y_{bcs} in.	$b/h = 8$ S_{bcs} in.³	y_{bcs} in.	$b/h = 10$ S_{bcs} in.³	y_{bcs} in.	$b/h = 12$ S_{bcs} in.³	y_{bcs} in.
W36 × 160	542	18.01	625	21.68	669	24.30	697	26.26	715	27.79	729	29.00	740	30.00
W36 × 150	504	17.93	586	21.78	629	24.48	655	26.47	672	28.00	685	29.22	695	30.20
W36 × 135	439	17.78	520	21.96	559	24.78	583	26.82	598	28.36	610	29.57	618	30.54
W33 × 141	448	16.65	524	20.47	562	23.09	585	24.99	601	26.45	612	27.59	620	28.52
W33 × 130	406	16.55	481	20.60	516	23.32	538	25.26	552	26.71	562	27.85	569	28.76
W33 × 118	359	16.43	432	20.79	466	23.61	485	25.59	497	27.04	506	28.16	513	29.05
W30 × 116	329	15.01	398	19.11	429	21.75	446	23.59	458	24.94	466	25.98	472	26.81
W30 × 108	299	14.92	367	19.25	396	21.96	413	23.81	423	25.16	430	26.18	436	26.99
W30 × 99	269	14.83	335	19.43	362	22.21	377	24.08	387	25.42	393	26.42	398	27.20
W27 × 94	243	13.46	303	17.89	326	20.51	339	22.25	347	23.49	353	24.41	357	25.13
W27 × 84	213	13.36	271	18.13	292	20.83	303	22.57	310	23.78	315	24.67	319	25.36
W24 × 94	222	12.16	279	16.25	302	18.68	314	20.28	321	21.43	327	22.28	331	22.94
W24 × 84	196	12.05	251	16.47	270	18.97	281	20.58	287	21.70	292	22.53	296	23.16
W24 × 76	176	11.96	227	16.68	245	19.23	254	20.83	260	21.92	264	22.72	267	23.33
W24 × 68	154	11.87	204	16.92	220	19.51	228	21.10	233	22.16	236	22.92	239	23.50
W24 × 62	131	11.87	180	17.27	195	19.91	203	21.48	208	22.52	211	23.25	213	23.80
W24 × 55	114	11.79	161	17.57	174	20.24	181	21.77	185	22.76	188	23.46	190	23.98
W21 × 68	140	10.57	187	15.22	202	17.60	209	19.05	214	20.03	217	20.73	220	21.25
W21 × 62	127	10.50	171	15.41	185	17.81	191	19.25	196	20.19	199	20.87	201	21.37
W21 × 57	111	10.53	155	15.75	168	18.20	174	19.61	178	20.54	180	21.19	183	21.67
W21 × 50	94.5	10.42	136	16.02	147	18.47	152	19.84	155	20.72	158	21.33	160	21.78
W21 × 44	81.6	10.33	121	16.33	130	18.76	134	20.09	137	20.91	139	21.48	141	21.90
W18 × 55	98.3	9.06	138	13.83	149	16.03	154	17.30	158	18.12	160	18.70	162	19.12
W18 × 50	88.9	9.00	126	14.05	136	16.25	141	17.49	144	18.28	146	18.82	148	19.23
W18 × 46	78.8	9.03	115	14.36	124	16.57	129	17.79	132	18.55	134	19.08	135	19.46
W18 × 40	68.4	8.95	102	14.67	109	16.86	(113	18.02	116	18.74	117	19.22	119	19.58
W18 × 35	57.6	8.85	88.8	14.96	95.2	17.10	98.3	18.20	100	18.86	102	19.31	103	19.65

Shape														
W16 × 50	81.0	8.13	117	12.84	127	14.90	131	16.05	135	16.79	137	17.30	139	17.67
W16 × 45	72.7	8.07	107	13.04	115	15.09	119	16.21	122	16.91	124	17.39	126	17.75
W16 × 40	64.7	8.01	96.0	13.30	103	15.33	107	16.40	109	17.07	111	17.52	112	17.86
W16 × 36	56.5	7.93	86.2	13.51	92.3	15.51	95.5	16.53	97.6	17.16	99.2	17.59	101	17.91
W16 × 31	47.2	7.94	74.6	13.97	79.8	15.92	82.5	16.89	84.3	17.47	85.7	17.87	86.8	18.18
W16 × 26	38.4	7.85	63.0	14.33	67.2	16.19	69.4	17.08	71.0	17.61	72.1	17.99	72.9	18.28
W14 × 38	54.6	7.05	84.2	12.06	90.6	13.92	93.9	14.89	96.1	15.49	97.9	15.90	99.2	16.21
W14 × 34	48.6	6.99	76.0	12.28	81.5	14.11	84.3	15.03	86.3	15.59	87.8	15.99	89.0	16.29
W14 × 30	42.0	6.92	67.4	12.51	72.1	14.28	74.7	15.16	76.4	15.69	77.7	16.07	78.7	16.36
W14 × 26	35.3	6.96	58.8	12.94	62.9	14.66	65.2	15.48	66.7	15.99	67.7	16.36	68.6	16.64
W14 × 22	29.0	6.87	50.0	13.25	53.3	14.88	55.2	15.64	56.4	16.12	57.3	16.46	57.9	16.73
W12 × 30	38.6	6.17	62.9	11.39	67.5	13.04	70.0	13.85	71.7	14.35	73.0	14.71	73.9	14.99
W12 × 26	33.4	6.11	55.4	11.64	59.2	13.22	61.4	13.99	67.8	14.47	63.9	14.82	64.7	15.09
W12 × 22	25.4	6.16	46.3	12.12	49.7	13.63	51.6	14.35	52.8	14.81	53.6	15.15	54.3	15.40
W12 × 19	21.3	6.08	40.1	12.36	43.0	13.78	44.6	14.46	45.5	14.90	46.2	15.22	46.8	15.46
W12 × 16	17.1	6.00	34.0	12.59	36.4	13.92	37.6	14.57	38.4	14.99	39.0	15.28	39.4	15.51
W12 × 14	14.9	5.96	30.2	12.79	32.3	14.06	33.4	14.68	34.0	15.08	34.5	15.36	34.8	15.58
W10 × 26	27.9	5.17	49.8	10.18	53.7	11.62	55.9	12.32	57.3	12.78	58.3	13.10	59.0	13.36
W10 × 22	23.2	5.09	42.8	10.41	46.1	11.76	47.9	12.43	49.0	12.87	49.8	13.18	50.4	13.41
W10 × 19	18.8	5.12	37.1	10.77	40.0	12.07	41.5	12.72	42.5	13.13	43.2	13.43	43.7	13.66
W10 × 17	16.2	5.06	33.1	10.91	35.6	12.15	36.9	12.78	37.7	13.18	38.3	13.46	38.7	13.68
W10 × 15	13.8	5.00	29.3	11.05	31.5	12.24	32.6	12.84	33.3	13.23	33.8	13.50	34.2	13.71
W10 × 12	10.9	4.94	23.9	11.35	25.5	12.47	26.5	13.03	27.0	13.39	27.3	13.64	27.6	13.83
W8 × 21	18.2	4.14	37.0	9.00	40.1	10.21	41.8	10.83	42.8	11.23	43.6	11.52	44.1	11.74
W8 × 18	15.2	4.07	32.0	9.18	34.6	10.33	35.9	10.92	36.8	11.31	37.4	11.58	37.8	11.79
W8 × 15	11.8	4.06	27.1	9.46	29.3	10.56	30.3	11.13	31.0	11.49	31.5	11.75	31.9	11.94
W8 × 13	9.9	4.00	23.6	9.60	25.5	10.66	26.3	11.20	26.9	11.55	27.3	11.79	27.6	11.98
W8 × 10	7.8	3.95	18.8	9.93	20.1	10.93	20.7	11.42	21.1	11.74	21.4	11.96	21.6	12.13

where M represents the moment at a concentrated load point. Within each interval between individual concentrated loads, the shear connectors may be spaced uniformly.

Connectors required in the region of negative bending of a continuous beam may be uniformly distributed between the sections of maximum and zero moments.

In placing the shear connectors, the minimum center-to-center spacing of studs should

TABLE 9   Allowable Loads for Connectors in Flat-Soffit Slabs of Stone Concrete

Diameter of stud, in.	Allowable load q, kips*		
	$f'_c \times 3.0$ ksi	3.5 ksi	$\geqslant 4.0$ ksi
½	5.1	5.5	5.9
⅝	8.0	8.6	9.2
¾	11.5	12.5	13.3
⅞	15.6	16.8	18.0

*Applicable when $H_s \geqslant 4d_s$.

TABLE 10   Reduction Coefficients for Connectors in Lightweight Concrete for Buildings

	Air-dry unit weight, pcf						
	90	95	100	105	110	115	120
Coefficient, $f'_c \leqslant 4.0$ ksi	0.73	0.76	0.78	0.81	0.83	0.86	0.88
Coefficient $f'_c \geqslant 5.0$ ksi	0.82	0.85	0.87	0.91	0.93	0.96	0.99

TABLE 11   Reduction Coefficients for Stud Connectors on Steel Deck

n_r	w_r/h_r	H_s/h_r					
		1.5	2.0	2.5	3.0	3.5	4.0
1	0.67	0.29	0.57	0.85	1.00	1.00	1.00
	1.0	0.43	0.85	1.00	1.00	1.00	1.00
	1.5	0.64	1.00	1.00	1.00	1.00	1.00
	2.0	0.85	1.00	1.00	1.00	1.00	1.00
	2.5	1.00	1.00	1.00	1.00	1.00	1.00
2	0.67	0.20	0.40	0.60	0.81	1.00	1.00
	1.0	0.30	0.60	0.90	1.00	1.00	1.00
	1.5	0.45	0.90	1.00	1.00	1.00	1.00
	2.0	0.60	1.00	1.00	1.00	1.00	1.00
	2.5	0.75	1.00	1.00	1.00	1.00	1.00
	3.0	0.90	1.00	1.00	1.00	1.00	1.00
	3.5	1.00	1.00	1.00	1.00	1.00	1.00
3	0.67	0.16	0.33	0.49	0.66	0.82	0.99
	1.0	0.25	0.49	0.74	0.98	1.00	1.00
	1.5	0.37	0.74	1.00	1.00	1.00	1.00
	2.0	0.49	0.98	1.00	1.00	1.00	1.00
	2.5	0.61	1.00	1.00	1.00	1.00	1.00
	3.0	0.74	1.00	1.00	1.00	1.00	1.00
	3.5	0.86	1.00	1.00	1.00	1.00	1.00
	4.0	0.98	1.00	1.00	1.00	1.00	1.00

be six diameters along the longitudinal axis and four diameters transverse to the longitudinal axis of the supporting composite beam. The maximum center-to-center spacing should not exceed eight times the total thickness of the slab.

12. Example 1   Design beams B1 and B2 of the floor shown in Fig. 9 as composite beams with stud shear connectors, using A36 steel and 110-pcf lightweight 3-ksi concrete. The floor loading is as follows: live load, 250 psf; 2½ in. lightweight concrete over 3-in. steel deck, 41 psf; fire-rated ceiling, 10 psf; miscellaneous and mechanical, 4 psf; which gives a total superimposed load of 305 psf. The beams are to be designed simply supported. The average sustained live load is equal to 50 percent of the total live load. Thus, in computing deflections, for one-half of the live load $k = 1$ and for the other half $k = 2$. The steel deck shown in Fig. 10 spans perpendicular to beam B1 and parallel to beam B2. Wide-flange beams without cover plates are to be used.

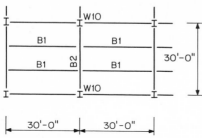

Fig. 9 Floor layout for Example 1.

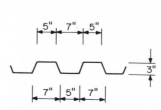

Fig. 10 Deck geometry for Example 1.

TOTAL SLAB THICKNESS:

$$3 + 2.5 = 5.5 \text{ in.}$$

ALLOWABLE STRESSES:
  Steel: $0.66f_y = 24$ ksi
  Concrete: $0.45f_c' = 1.35$ ksi
Modular ratio for stress calculations [Eq. (15)]:

$$n = \frac{27,800}{145^{1.5} \times 3^{0.5}} = 9$$

Modular ratio for deflection calculations:

$$n = \frac{27,800}{110^{1.5} \times 3^{0.5}} = 14$$

BEAM B1: Effective slab width:

$$b = 0.25 \times 30 \times 12 = 90 \text{ in.}$$
$$b = 10 \times 12 = 120 \text{ in.}$$
$$b = 16 \times 5.5 + 8 = 96 \text{ in.}$$
Governing: $b = 90$ in.

Transformed slab width for stress calculations:

$$b/kn = 90/(1 \times 9) = 10.0 \text{ in.}$$

Transformed slab width for transient-load deflection calculations:

$$b/kn = 90/(1 \times 14) = 6.43 \text{ in.}$$

Transformed slab width for sustained-load deflection calculations:

$$b/kn = 90/(2 \times 14) = 3.21 \text{ in.}$$

Estimated beam weight:

$$70 \text{ plf} = 7 \text{ psf}$$

Total uniform load:

$$305 + 7 = 312 \text{ psf}$$

Total moment:

$$M = 0.125 \times 0.312 \times 10 \times 30^2 = 351 \text{ ft-kips}$$

Required section modulus:

$$S_{bc} = 351 \times 12/24 = 176 \text{ in.}^3$$

From Table 5, the required steel section for $b/kn = 10$ is a W21 × 62: $S_{bs} = 127$ in.^{3}, $I_s = 1330$ in.^{4}. The section modulus of the composite section is $S_{bc} = 192$ in.^{3} and $y_{bc} = 19.01$ in. Check Eq. (16):

$$\left(1.35 + 0.35\frac{250}{62}\right) \times 127 = 351 \text{ in.}^3 > 176 \text{ in.}^3$$

Therefore, the actual transformed section modulus may be used in stress computations.
STRESSES:
Bottom fiber (total load):

$$M/S_{bc} = 351 \times 12/192 = 21.9 \text{ ksi} < 24.0 \text{ ksi}$$

Top of concrete (Live load only):

$$\frac{250}{312} \times \frac{7.48}{19.01} \times \frac{21.9}{9} = 0.77 \text{ ksi} < 1.35 \text{ ksi}$$

SHEAR CONNECTORS: use ¾-in.-diameter stud connectors, 4½ in. high.
From Eqs. (19):

$$H \leq 0.425 \times 3.0 \times 90.0 \times 2.5 = 287 \text{ kips}$$
$$H \leq 0.5 \times 18.3 \times 36.0 = 329 \text{ kips}$$

Assuming two studs per rib, from Tables 9, 10, and 11:

$$q = 11.5 \times 0.83 \times 0.60 = 5.73 \text{ kips per stud}$$
and
$$q_2 = 2 \times 5.73 = 11.46 \text{ kips per rib}$$
$$N = H/q = 287/5.73 = 50 \text{ studs for half beam length}$$

However, this number of connectors is not needed since beam B1 is not stressed up to the allowable stress when fully composite. Thus a partial shear connection will be satisfactory.
From Eq. (17):

$$H' = 287\left(\frac{176 - 127}{192 - 127}\right)^2 = 163 \text{ kips}$$
$$N' = \frac{163}{5.73} = 29$$

The available number of ribs is $n_t = 14$. Therefore, two studs must be used in some ribs and three in others.
The capacity q_3 of three studs per rib is

$$q_3 = 3 \times 11.5 \times 0.83 \times 0.49 = 14.03 \text{ kips}$$

Then n_3 may be computed from

$$n_3 q_3 + n_2 q_2 = H'$$
$$n_3 + n_2 = n_t$$

Therefore,

$$n_3 = \frac{163 - 14 \times 11.46}{14.03 - 11.46} = 1$$

One rib at each end of the beam will have three studs and the remaining 26 ribs will have two studs each for a total of 58 studs per beam. Thus partial shear connection saves $(2 \times 50 - 58) = 42$ studs without any increase in beam size.
Check:

$$1 \times 14.03 + 13 \times 11.46 = 163 \text{ kips} \gtreqless 163 \text{ kips}$$

DEFLECTION:

$$\Delta = \frac{5ML^2}{48EI} \times 1728 = \frac{180ML^2}{EI} = \frac{3ML^2}{500I}$$

with M in ft-kips, L in ft, and I in in.^{4}
Dead load, $b/kn = 0$ (Table 5): $I_s = 1330$ in.^{4}

$$M = \frac{62}{312} \times 351 = 69.7 \text{ ft-kips}$$
$$\Delta = \frac{3 \times 69.7 \times 30^2}{500 \times 1330} = 0.28 \text{ in.}$$

Transient live load, $b/kn = 6.43$ (Table 5):

$$S_{bc} = 184 \text{ in.}^3 \qquad y_{bc} = 17.39 \text{ in.}$$
$$I_{eff} = 1330 + \left(\frac{163}{287}\right)^{0.5}(184 \times 17.39 - 1330) = 2740 \text{ in.}^4$$

$$M = \frac{125}{312} \times 351 = 141 \text{ ft-kips}$$

$$\Delta = \frac{3 \times 141 \times 30^2}{500 \times 2740} = 0.28 \text{ in.}$$

Sustained live load, $b/kn = 3.21$ (Table 5):

$$S_{bc} = 170 \text{ in.}^3 \qquad y_{bc} = 14.99 \text{ in.}$$

$$I_{\text{eff}} = 1330 + \left(\frac{163}{287}\right)^{0.5}(170 \times 14.99 - 1330) = 2250 \text{ in.}^4$$

$$M = \frac{125}{312} \times 351 = 141 \text{ ft-kips}$$

$$\Delta = \frac{3 \times 141 \times 30^2}{500 \times 2250} = 0.34 \text{ in.}$$

Total deflection:

$$0.28 + 0.28 + 0.34 = 0.90 \text{ in.} = \text{span}/400$$

BEAM B2; Effective slab width:

$$b = 0.25 \times 30 \times 12 = 90 \text{ in.}$$
$$b = 30 \times 12 = 360 \text{ in.}$$
$$b = 16 \times 5.5 + 11 = 99 \text{ in.}$$
$$\text{Governing: } b = 90 \text{ in.}$$

As the effective slab width for beam B2 is the same as for beam B1, the transformed slab widths b/kn are also the same. Estimated beam weight:

$$120 \text{ plf} = 4 \text{ psf}$$

Total uniform load:

$$305 + 7 + 4 = 316 \text{ psf}$$

Total moment:

$$M = 0.333 \times 0.316 \times 10 \times 30^2 = 948 \text{ ft-kips}$$

Required section modulus:

$$S_{bc} = 948 \times 12/24 = 474 \text{ in.}^3$$

From Table 6, the required steel section for $b/kn = 10$ is a W33 × 118: $S_{bs} = 359 \text{ in.}^3$, $I_s = 5900 \text{ in.}^4$. The composite section modulus is $S_{bc} = 492 \text{ in.}^3$ and $y_{bc} = 27.10 \text{ in.}$ Check Eq. (16):

$$\left(1.35 + 0.35 \frac{250}{66}\right) \times 359 = 961 \text{ in.}^3 > 474 \text{ in.}^3$$

Therefore, the actual transformed section modulus may be used in stress computations.
STRESSES:
Bottom fiber (total load):

$$M/S_{bc} = 948 \times 12/492 = 23.1 \text{ ksi} < 24.0 \text{ ksi}$$

Top of concrete (live load only):

$$\frac{250}{316} \times \frac{11.26}{27.10} \times \frac{23.1}{9} = 0.84 \text{ ksi} < 1.35 \text{ ksi}$$

SHEAR CONNECTORS: use ¾-in.-diameter stud connectors, 4½ in. high. No reduction of stud capacity because of deck geometry is necessary, since $w_r/h_r = 2 > 1.5$.
From Eqs. (19):

$$H \leq 0.425 \times 3.0 \times 90.0 \times 4.0 = 459 \text{ kips}$$
$$H \leq 0.5 \times 34.7 \times 36.0 = 625 \text{ kips}$$

From Tables 9 and 10:

$$q = 11.5 \times 0.83 = 9.55 \text{ kips}$$
$$N = H/q = 459/9.55 = 48 \text{ studs for half beam length}$$

Because of the presence of concentrated loads, beam B2 must satisfy Eq. (21). Accordingly, all studs must be placed between the concentrated load and the nearest end of the beam.
Minimum longitudinal spacing of studs: $6 \times 0.75 = 4.5$ in. Maximum number of studs:

$$3 \times \left(\frac{10 \times 12 - 7.5}{4.5} + 1\right) = 78 > 48$$

Therefore, beam B2 can be built with full shear connection. However, partial shear connection may be used instead, since the maximum bending stress of 23.1 ksi is less than the allowable 24 ksi. The required number of studs is computed from Eq. (17).

$$N' = 48 \left(\frac{474 - 359}{492 - 359}\right)^2 = 36 \text{ studs per outer third of span}$$

Thus partial shear connection saves $(48 - 36) \times 2 = 24$ studs per beam without any increase of girder size. Additional studs must be placed between the concentrated loads at the maximum spacing of 32 in.; a single line of studs will be sufficient. The total number of studs in beam B2 is $2 \times 36 + 5 = 77$.

DEFLECTION:

$$\Delta = \frac{23ML^2}{216EI} \times 1728 = \frac{23ML^2}{3750I}$$

Dead load, $b/kn = 0$ (Table 6): $I_s = 5900$ in.^{4}

$$M = \frac{66}{316} \times 948 = 198 \text{ ft-kips}$$

$$\Delta = \frac{23 \times 198 \times 30^2}{3750 \times 5900} = 0.19 \text{ in.}$$

Transient live load, $b/kn = 6.43$ (Table 6):

$$S_{bc} = 474 \text{ in.}^3 \qquad y_{bc} = 24.91 \text{ in.}$$

$$I_{\text{eff}} = 5900 + \left(\frac{36}{48}\right)^{0.5} (474 \times 24.91 - 5900) = 11{,}000 \text{ in.}^4$$

$$M = \frac{125}{316} \times 948 = 375 \text{ ft-kips}$$

$$\Delta = \frac{23 \times 375 \times 30^2}{3750 \times 11{,}000} = 0.19 \text{ in.}$$

Sustained live load, $b/kn = 3.21$ (Table 6):

$$S_{bc} = 442 \text{ in.}^3 \qquad y_{bc} = 21.81 \text{ in.}$$

$$I_{\text{eff}} = 5900 + \left(\frac{36}{48}\right)^{0.5} (442 \times 21.81 - 5900) = 9140 \text{ in.}^4$$

$$M = \frac{125}{316} \times 948 = 375 \text{ ft-kips}$$

$$\Delta = \frac{23 \times 375 \times 30^2}{3750 \times 9140} = 0.23 \text{ in.}$$

Total deflection:

$$0.19 + 0.19 + 0.23 = 0.61 \text{ in.} = \text{span}/590$$

BRIDGE DESIGN

13. Assumptions Specifications for the design of composite-beam highway bridges[7] include two methods: working-stress design (WSD) and load-factor design (LFD). The LFD method, which assures greater consistency in the live-load capacity of bridges, is discussed here (see Ref. 8 for WSD).

In composite-beam bridges, the concrete slab rests on the top of the steel beams and there is no encasement of the steel. For design purposes, the structure is assumed to consist of a series of T-beams. The slab must be connected to the steel beams with mechanical shear connectors.

Three types of loading conditions are considered in computing the maximum strength of a noncompact member and in checking the stresses and deflections at service loads: (1) live loads, (2) dead loads resisted by the steel section, and (3) dead loads resisted by the composite section (effects of creep of concrete included). Shrinkage, expansion of concrete, and differential temperature changes need to be considered only in exceptional cases. On the other hand, no distinction is made between the two types of dead load in computing the maximum strength of compact members.

The live loads are always carried by the composite section. Since they are of short duration, live-load stresses and deflections are computed with $k = 1.0$.

The steel beams, concrete slab, diaphragms, wearing surface, and other parts of the bridge superstructure constitute dead load. All dead loads placed after the concrete slab has attained at least 75 percent of its 28-day strength may be assumed to be carried by the composite section. In unshored construction, all other dead loads should be considered to be carried by the steel beams alone. If at least three temporary supports (at the quarter points) are provided between the end points of a span, it is sufficiently accurate for design purposes to assume that all dead loads are carried by the composite section.

Dead loads acting on the composite section set up stresses in the concrete slab that cause creep. The effect of creep of concrete on stresses in the steel beam may be accounted for approximately by multiplying the modular ratio n by $k = 3$ in computing the stresses due to dead loads carried by the composite section.

14. Design of Composite Beams The design of composite beams for bridges is based on the strength of the cross section and on the maximum stress under an overload. The strength of a compact beam is computed from the fully plastic stress distribution (Art. 3). The strength of a noncompact beam and the maximum stress under an overload are computed using elastic properties of the transformed section (Art. 2).

The width of the slab assumed effective as the flange of an interior T-beam must not exceed any of the following:

1. One-fourth of the span of the beam
2. The distance center to center of beams
3. Twelve times the least thickness of the slab

For an edge beam, the effective width on either side of the steel section must not exceed:

1. One-twelfth of the span of the beam
2. Six times the least thickness of the slab

Furthermore, the effective width on the outside of the steel section must not exceed the actual width of the overhang, and the effective width on the inside must not extend beyond the centerline between the edge beam and the adjacent interior beam.

In negative-moment areas, the reinforcement of the slab parallel with the beam and within the effective width of the slab may be included in computing the properties of the composite section, provided shear connectors are furnished in accordance with the requirements of Art. 16.

For computation of elastic properties the effective width b of the concrete slab is transformed to the width of an equivalent steel plate as discussed in Art. 9.

The cross section of a *compact beam* is governed either by the ultimate moment, which must not be less than the sum of the factored external moments

$$M_u \geq 1.3 \left(M_{DL} + \frac{5}{3} M_{L+I} \right) \tag{22}$$

or by the maximum stress under an overload equal to M_{L+I} multiplied by $\frac{5}{3}$. The maximum stress caused by all dead loads and the overload must not exceed $0.95 f_y$:

$$\frac{M_{Ds}}{S_{bs}} + \frac{M_{Dc}}{S_{bc}} + \frac{5}{3} \frac{M_{L+I}}{S_{bc}} \leq 0.95 f_y \tag{23}$$

The overload condition usually governs the design. The cross section of a *noncompact beam* is governed by the maximum stress under the factored external loads. The maximum stress should not exceed the yield stress:

$$1.3 \left(\frac{M_{Ds}}{S_{bs}} + \frac{M_{Dc}}{S_{bc}} + \frac{5}{3} \frac{M_{L+I}}{S_{bc}} \right) \leq f_y \tag{24}$$

In Eqs. (23) and (24), the section modulus of the composite beam S_{bc} should include the effects of creep when computing stresses caused by composite dead loads ($k = 3$).

15. Steel-Member Selection The first step in the design of a composite beam is the selection of the steel section. This is usually accomplished on the basis of comparable earlier designs. In the absence of such information, one may start with a section a few sizes smaller than that required for a noncomposite design and proceed with successive trials on the basis of the applicable equations from Art. 14.

In general, rolled structural shapes are used for spans up to about 70 ft. Cover plating the rolled members extends their use to spans of about 90 ft. Over 90 ft, plate girders are generally used. The span lengths refer to simple beams. For continuous designs, the distance between the points of contraflexure should be substituted for the span length.

16. Design of Shear Connectors In bridge beams shear connectors must be able to develop the fully plastic strength of the composite section and to withstand the repeated passages of vehicles without failure.

Fatigue Strength. Tests of stud and channel shear connectors have demonstrated that fatigue failure occurs in the connecting welds[9] and that the number of cycles to failure can be predicted with a satisfactory accuracy from the range of the horizontal shear stress alone. The allowable repeated loads for four sizes of stud connectors and for channel connectors attached with $3/16$-in. fillet welds are listed in Table 12. Although only $3/4$- and $7/8$-

TABLE 12   Allowable Range of Horizontal Shear for Connectors under Repeated Loads

Type and size of connector	Allowable repeated load q_r, kips*		
	100,000 cycles	500,000 cycles	2,000,000 cycles
$1/2$-in. stud	3.3	2.7	2.0
$5/8$-in. stud	5.1	4.1	3.1
$3/4$-in. stud	7.3	6.0	4.4
$7/8$-in. stud	10.0	8.1	6.0
Channels	$4.0w$	$3.0w$	$2.4w$

*Applicable to studs with $H_s \geq 4d_s$ and to channels with at least $3/16$-in. continuous fillet welds at the heel and toe.

in.-diameter studs are used in bridge construction, the two smaller sizes are included since they are used occasionally in buildings.

The fluctuating load acting on connectors is the range of horizontal shear caused by live loads (including impact) at the top surface of the steel beam,

$$H_r = \frac{(V_{max} - V_{min})m}{I_c} \tag{25}$$

where H_r = range of horizontal shear
V_{max}, V_{min} = maximum and minimum vertical shear caused by live load (including impact) at the section considered; if of opposite sign, V_{min} should be taken as a negative quantity
$m = A_s \bar{y}_c$

The spacing s of connectors is given by

$$s = \frac{\Sigma q_r}{H_r} \tag{26}$$

where Σq_r = sum of allowable fluctuating loads on all connectors placed at the same beam cross section.

As the governing load in the design of shear connectors against fatigue is the range of horizontal shear, the variation of spacing in any beam is small. This, and the experimental basis of the design, do not warrant extreme refinements in spacing the connectors. Accordingly, a uniform spacing can be used through long lengths of beams.

The maximum spacing of connectors should not exceed 24 in. except over the supports of continuous beams where wider spacing often avoids placing connectors at locations of high stresses in the tension flange. A clear depth of concrete cover over the tops of connectors of at least 2 in. is recommended. The connectors should extend at least 2 in. above the bottom of the slab.

In continuous bridges, the longitudinal slab reinforcement over the supports is always subjected to large fluctuating stresses since it ordinarily does not resist the weight of the beams and the slab. When the external moment is assigned to the steel section alone, additional shear connectors are needed at the points of contraflexure to avoid premature fatigue failure of the connectors in the regions of positive moments. This additional number of connectors, N_c, may be computed from the following formula:

$$N_c = \frac{A_r f_r}{q_r} \tag{27}$$

where f_r is the range of stress due to live load, including impact, in the slab reinforcement over the support. In lieu of more accurate computations, f_r may be taken as 10 ksi. It is recommended that these additional connectors be placed on both sides of the dead-load point of contraflexure within a distance of one-third of the effective slab width.

Ultimate Strength. The total number of connectors must not be less than that required for the fully plastic stress distribution. This total number is computed by the same procedure as described in Art. 11 except that the horizontal shear H_u at ultimate load replaces the horizontal shear H and the reduced strength $0.85q_u$ of a connector replaces the allowable load q. Accordingly, the required number of connectors is $N = H_u/0.85q_u$.

The horizontal shear H_u is computed from Eqs. (7a) and (7b) for the regions of positive moment and from the formula

$$H_u = A_r f_{yr} \tag{28}$$

for the regions of negative moments. The strength of stud connectors is given by the formula

$$q_u = 0.41d_s^2(w_c f_c')^{0.75} \tag{29}$$

and that of channel connectors may be computed from the formula

$$q_u = 17.4(t_f + 0.5t_w)w f_c'^{0.5} \tag{30}$$

The static strengths of commonly used stud and channel connectors embedded in flat-soffit slabs of concrete made with ASTM C33 aggregates are listed in Table 13. The loads are applicable to slabs with a concrete haunch over the steel beam as long as the connectors have at least 1 in. lateral cover. For slabs made with rotary-kiln-produced aggregates, the allowable shear loads are obtained by multiplying the values from Table 13 by the appropriate coefficient from Table 14.

TABLE 13   Static Strength of Connectors in Flat-Soffit Slabs of Stone Concrete

Type and size	Static strength q_u of one connector, kips*		
	$f_c' = 3.0$ ksi	3.5 ksi	4.0 ksi
½-in. stud	9.7	10.9	12.1
⅝-in. stud	15.2	17.1	18.9
¾-in. stud	21.9	24.6	27.2
⅞-in. stud	29.8	33.5	37.0
C3 × 4.1	10.8w	11.7w	12.5w
C4 × 5.4	11.7w	12.6w	13.5w
C5 × 6.7	12.5w	13.5w	14.4w

*Applicable to studs with $H_s \geq 4d_s$.

TABLE 14   Reduction Coefficients for Connectors in Lightweight Concrete in Bridges

Air-dry unit weight, pcf	90	95	100	105	110	115	120
Coefficient	0.70	0.73	0.76	0.78	0.81	0.84	0.87

17. Example 2 The procedures described in Arts. 13 through 16 are illustrated with the design of an interior stringer for a two-span continuous rolled-beam bridge assuming composite action only in the regions of positive moments. The stringer is fabricated from two rolled sections, one in the positive-moment regions and the other over the center support. Both spans are 70 ft long and are designed for the standard HS20-44 truck and 500,000 cycles. The designs are made for 4-ksi concrete and A588 steel; accordingly, $n = 8$ and $f_y = 50$ ksi. The girders are spaced 8.33 ft and the bridge is to be built without temporary supports. The slab is 7.5-in.-thick reinforced concrete, but the thickness of the structural slab is assumed as 7 in.

Dead load on steel beam:

Slab $7.5/12 \times 8.33 \times 0.145 = 0.75$
Beam and framing details $= \underline{0.15}$
$\qquad\qquad\qquad\qquad\qquad\quad 0.90$ kip/ft

Dead load on composite beam:

Curb and railings $\qquad = 0.165$ kip/ft

Moments and shears are calculated by an elastic analysis. Generally, the first analysis is made assuming a constant moment of inertia for the entire length of the beam even though the composite section in the span is stiffer than the noncomposite section over the center support. With the AASHTO[7] wheel-load distribution of $S/5.5$, such an analysis results in the moments and shears listed in Table 15. After the preliminary design, the moments should be recomputed accounting for the differences in the moments of inertia in the regions of positive and negative moments. The effect of the nonuniform section on shears is usually small and need not be considered.

When the section of the beam is compact, AASHTO specifications[7] permit a 10 percent reduction of the maximum negative moment provided that the positive moments are increased proportionally. This is accomplished by superimposing on the original moment diagram positive moments varying linearly from zero at the exterior support to 10 percent of the maximum negative moment at the interior support. The maximum positive moment is located at the 0.4 point or 28 ft from the end support. Thus it must be increased 0.04 times the maximum negative moment. Accordingly, the redistributed moments are as follows:

Maximum negative moments:

$$M_{Ds} = 0.90 \times 551.3 = 496.2 \text{ ft-kips}$$
$$M_{Dc} = 0.90 \times 101.1 = 91.0 \text{ ft-kips}$$
$$M_{L+I} = 0.90 \times 601.1 = 541.0 \text{ ft-kips}$$

Maximum positive moments:

$$M_{Ds} = 308.7 + 0.04 \times 551.3 = 330.8 \text{ ft-kips}$$
$$M_{Dc} = 56.6 + 0.04 \times 101.1 = 60.6 \text{ ft-kips}$$
$$M_{L+I} = 750.5 + 0.04 \times 601.1 = 774.5 \text{ ft-kips}$$

These redistributed moments are used in calculations at overload and at maximum strength.

STEEL-MEMBER SELECTION. For rolled-beam design, it is a reasonable assumption that the overload condition governs. In accordance with Art. 15, for positive moment choose a section a few sizes smaller than required for noncomposite design. Assuming that the resulting section is compact, AASHTO 1.7.136(A) gives

$$S_s = \frac{330.8 + 60.6 + \tfrac{5}{8} \times 774.5}{0.8 \times 50} \times 12 = 505 \text{ in.}^3$$

Try W30 $\times$ 108 with $S = 299$ in.^{3}, $d = 29.83$ in.,

$$b_t = 10.475 \text{ in.}, t_t = 0.760 \text{ in.}, \text{ and } t_w = 0.545 \text{ in.}$$
$$b' = \tfrac{1}{2}(10.475 - 0.545) = 4.97 \text{ in.}$$
$$b'/t_t = 4.97/0.760 = 6.54 < 7.2$$
$$d/t_w = 29.83/0.545 = 54.7 < 59$$

Thus the section in the positive-moment regions is compact.

For the negative-moment region, the required section modulus is

$$S_s = \frac{496.2 + 91.0 + \tfrac{5}{8} \times 541.0}{0.8 \times 50} \times 12 = 447 \text{ in.}^3$$

Try W30 $\times$ 173 with $S = 539$ in.^{3}, $d = 30.44$ in.,

$$b_b = 14.985 \text{ in.}, t_b = 1.065 \text{ in.}, \text{ and } t_w = 0.655 \text{ in.}$$
$$b' = \tfrac{1}{2}(14.985 - 0.655) = 7.17$$
$$b'/t_b = 7.17/1.065 = 6.73 < 7.2$$
$$d/t_w = 30.44/0.655 = 46.5 < 59$$

Thus the section over the support is compact if it is adequately braced.

MAXIMUM POSITIVE MOMENT—28 ft from end support

Effective slab width: $b = 12 \times 7 = 84$ in.

Properties of trial section are computed from Eqs. (3). For W30 $\times$ 108, $d = 29.83$ in., $A_s = 31.7$ in.^{2}, $I_s = 4470$ in.^{4} and $S_s = 299$ in.^{3} Assume $e_c = 5.0$ in.

Composite section, $k = 3$

$$A_c = \frac{84 \times 7}{3 \times 8} = 24.5 \text{ in.}^2 \qquad K_c = \frac{24.5}{24.5 + 31.7} = 0.436$$

$$\bar{y}_c = (14.91 + 5.0)0.436 = 8.68 \text{ in.}$$
$$I_c = 19.91 \times 8.68 \times 31.7 + 4470 + 24.5 \times 49/12 = 10{,}050 \text{ in.}^4$$
$$S_{bc} = \frac{10{,}050}{14.91 + 8.68} = 426 \text{ in.}^3$$

Composite section, $k = 1$

$$A_c = \frac{84 \times 7}{8} = 73.5 \text{ in.}^2 \qquad K_c = \frac{73.5}{73.5 + 31.7} = 0.699$$
$$\bar{y}_c = (14.91 + 5.0)0.699 = 13.91 \text{ in.}$$
$$I_c = 19.91 \times 13.91 \times 31.7 + 4470 + 73.5 \times 49/12 = 13{,}550 \text{ in.}^4$$
$$S_{bc} = \frac{13{,}550}{14.91 + 13.91} = 470 \text{ in.}^3$$

Stresses at overload [Eq. (23)]
    Allowable stress: $0.95 \times 50 = 47.5$ ksi

$$\frac{330.8 \times 12}{299} + \frac{60.6 \times 12}{426} + \frac{5}{2} \times \frac{774.5 \times 12}{470} = 47.9 \text{ ksi} \approx 47.5 \text{ ksi}$$

Maximum strength, Eqs. (7), (8), (9), (10), and (22)

$$H_u \leq 0.85 \times 4 \times 84 \times 7 = 1999 \text{ kips}$$
$$H_u \leq 50 \times 31.7 = 1585 \text{ kips}$$
$$a_c = \frac{1585}{0.85 \times 4 \times 84} = 5.55 \text{ in.} < 7.0 \text{ in.}$$
$$e' = 5.0 + 3.5 - 2.77 = 5.73 \text{ in.}$$
$$e = 5.73 + 14.91 = 20.64 \text{ in.}$$
$$M_u = 50 \times 31.7 \times 20.64 = 32{,}710 \text{ in.-kips} = 2726 \text{ ft-kips}$$

Applied moment:

$$1.3(330.8 + 60.6 + \tfrac{5}{8} \times 774.5) = 2187 \text{ ft-kips} < 2726 \text{ ft-kips}$$

Thus the trial section W30 × 108 is adequate both at the overload and at the maximum strength.

MAXIMUM NEGATIVE MOMENT. The negative moments are resisted by the steel section alone. The properties of the trial section W30 × 173 are: $d = 30.44$ in., $A_s = 50.8$ in.^{2}, $I_s = 8200$ in.^{4}, $S_s = 539$ in.^{3}, and $Z_B = 605$ in.^{3}.
Stresses at overload

Allowable stress: $0.8 \times 50 = 40$ ksi [AASHTO 1.7.136(A)]
$$\frac{496.2 + 91.0 + \tfrac{5}{8} \times 541.0}{539} \times 12 = 33.2 \text{ ksi} < 40 \text{ ksi}$$

Maximum strength

$$M_u = 605 \times 50 = 30{,}250 \text{ in.-kips} = 2520 \text{ ft-kips}$$

Applied moment:

$$1.3(496.2 + 91.0 + \tfrac{5}{8} \times 541.0) = 1936 \text{ ft-kips} < 2520 \text{ ft-kips}$$

Thus the trial section W30 × 173 is adequate both at overload and at maximum strength.

SPLICE LOCATION. Assuming that overload governs, the splice between the two rolled sections is located where the total external moment is equal to the moment resistance of the smaller steel section. Moment resistance of W30 × 108:

$$\frac{40 \times 299}{12} = 997 \text{ ft-kips}$$

External moment (including moment redistribution) at 63 ft from end support

$$297.7 + 54.6 + \tfrac{5}{8} \times 378.6 - 0.09(551.3 + 101.1 + \tfrac{5}{8} \times 601.1) = 834 \text{ ft-kips}$$

External moment (including moment redistribution) at the interior support

$$496.2 + 91.0 + \tfrac{5}{8} \times 541.0 = 1489 \text{ ft-kips}$$

Thus the splice is located within 7 ft of the interior support. Using straight-line interpolation, the distance from the 0.9 point is

$$7 \times \frac{997 - 834}{1489 - 834} = 1.74 \text{ ft}$$

Locate the splices at 5 ft 3 in. both sides of the interior support.

With the trial sections for both the positive and negative moments and the splice location known, one should recompute the moments to account for the variable moment of inertia. That step is omitted here, but connector spacing, based on the uncorrected moments, is determined to illustrate the procedure.

SHEAR CONNECTORS. Use groups of three studs $\frac{7}{8}$ in. in diameter. For 500,000 cycles, the strength of the group of three studs is (Table 12)

$$\Sigma q_r = 3 \times 8.1 = 24.3 \text{ kips}$$

The spacing of the groups of connectors is computed at several points along the beam from Eqs. (25) and (26) with

$$I_c = 13,550 \text{ in.}^4$$
$$m = 31.7 \times 13.91 = 441 \text{ in.}^3$$
$$s = \frac{24.3 \times 13,550}{(V_{max} - V_{min})441} = \frac{747}{V_{max} - V_{min}}$$

where V_{max} and V_{min} are values of V_{L+I} from Table 15. The spacing is computed in Table 16. Note in Table 15 that the dead-load point of contraflexure is approximately 52.5 ft from the end support.

TABLE 15   **Moments and Shears for HS20-44 Vehicle, Example 2**

Distance from end support ft	Tenth point	M_{Ds}, ft-kips	M_{Dc}, ft-kips	M_{L+I}, ft-kips Pos.	Neg.	V_{L+I}, kips Pos.	Neg.
0	0	0.0	0.0	0.0	0.0	57.0	6.0
7	1	143.3	26.3	341.8	42.1	49.2	6.2
14	2	242.5	44.5	571.2	84.1	41.4	10.1
21	3	297.7	54.6	695.3	126.2	33.9	15.5
28	4	308.7	56.6	750.5	168.3	26.7	23.0
35	5	275.6	50.5	733.8	210.4	19.8	31.2
42	6	198.4	36.4	661.2	252.4	13.6	39.0
49	7	77.2	14.1	515.2	294.5	8.1	45.9
56	8	−88.2	−16.2	316.3	336.6	3.8	52.1
63	9	−297.7	−54.6	80.5	378.6	1.6	57.5
70	10	−551.3	−101.1	0.0	601.1	0.0	62.1

TABLE 16   **Spacing of Shear Connectors, Example 2**

Distance from end support, ft	Tenth point	$V_{max} - V_{min}$, kips	s, in.	Actual spacing
0	0	63.0	11.8	10 × 12 in. = 10 ft 0 in.
7	1	55.4	13.5	
14	2	51.5	14.5	
21	3	49.4	15.1	
28	4	49.7	15.0	36 × 14 in. = 42 ft 0 in.
35	5	51.0	14.6	
42	6	52.6	14.2	
49	7	54.0	13.8	

ULTIMATE STRENGTH. The required number of connectors between the points of zero and maximum positive moment is computed with the aid of Table 13 as

$$N = \frac{1585}{0.85 \times 37} = 50$$

The number of connectors provided between the end support and the section of maximum moment is

$$(10 + 15 + 1) \times 3 = 78 > 50$$

and between the section of maximum moment and the point of dead-load contraflexure is

$$(36 - 15) \times 3 = 63 > 50$$

Thus the requirement of ultimate strength is satisfied.

ADDITIONAL CONNECTORS NEAR POINT OF CONTRAFLEXURE [Eq. (27)]. Assuming the minimum required amount of longitudinal slab reinforcement (1 percent) and taking $f_r = 10$ ksi,

$$N_c = \frac{0.01 \times 84 \times 7 \times 10}{8.1} = 7$$

Add three rows of three connectors spaced at 6 in. Thus the total number of connectors is

$$(1 + 10 + 36 + 3) \times 3 = 150$$

in each span.

REFERENCES

1. Slutter, R. G., and G. C. Driscoll, Jr.: Flexural Strength of Steel-Concrete Composite Beams, *J. Struct. Div. ASCE*, April 1965.
2. Larson, M. A.: Composite Steel-Concrete Construction, *J. Struct. Div. ASCE*, December 1974.
3. Structural Welding Code AWS D1.1-75, American Welding Society, 1975. Also 1976 revisions.
4. Lenzen, K. H.: Vibration of Floor Systems of Tall Buildings, *ASCE-IABSE Int. Conf. Proc.*, vol. II-17, 1973.
5. Wiss, J. F., and R. A. Parmelee: Human Perception of Transient Vibrations, *J. Struct. Div. ASCE*, April 1974.
6. Specification for the Design, Fabrication and Erection of Structural Steel for Buildings, American Institute of Steel Construction, 1978.
7. Standard Specifications for Highway Bridges, American Association of State Highway and Transportation Officials, 1977.
8. Viest, I. M., R. S. Fountain, and R. C. Singleton: "Composite Construction in Steel and Concrete for Bridges and Buildings," McGraw-Hill Book Company, New York, 1958.
9. Slutter, R. G., and J. W. Fisher: "Fatigue Strength of Shear Connectors," Highway Research Board, 1965.

Masonry Construction

WALTER L. DICKEY
Consulting Civil and Structural Engineer, Los Angeles, Calif.

NOTATION

A_g = gross area of masonry cross section
A_s = area of tension reinforcement
A_v = area of stirrup
b = width of beam
d = depth of beam to reinforcement
D = diameter of bar
E_m = modulus of elasticity of masonry
E_s = modulus of elasticity of steel
f_m = compressive stress in extreme fiber
f'_m = approved ultimate compressive stress of masonry
f_s = tensile stress in reinforcement
f_v = tensile stress in stirrup
h = vertical or horizontal distance between supports, also width of bond beams in plane of wall
h' = effective height of wall or column
M_m = resisting moment as limited by masonry
M_s = resisting moment as limited by steel
n = E_s/E_m
Σ_0 = sum of perimeters of reinforcing bars
p = A_s/bd
t = overall dimension of column, also wall thickness
v_m = shear stress in masonry
v' = shear carried by stirrups

MATERIALS

1. Burned-Clay Units Burned-clay units include common and face brick, hollow clay tile, terra-cotta, and ceramic tile. The latter two are not considered to be structural materials.

Solid units are those whose net cross-sectional area in any plane parallel to the bearing surface is not less than 75 percent of the gross area. Units whose net area is less than 75 percent of the gross area are called *hollow units*.

2. Brick Building brick is available in three grades. Grade SW is for use where the brick may freeze when permeated with water, and where a uniform, high degree of resistance to weathering is desired. Grade MW may be used for exposure to temperatures below freezing, where the brick is not likely to be permeated. It is suitable for the face of a wall above grade. Grade NW is intended for backup or interior masonry. Physical requirements are given in Table 1 (ASTM C62).

TABLE 1   Physical Requirements for Building Brick and Facing Brick

Grade	Compressive strength, flat, min, psi		Water absorption, 5-hr boil, max, %		Saturation* coefficient, max, %	
	Avg of 5	Indi-vidual	Avg of 5	Indi-vidual	Avg of 5	Indi-vidual
SW, severe weathering........	3,000	2,500	17.0	20.0	0.78	0.80
MW, moderate weathering.....	2,500	2,200	22.0	25.0	0.88	0.90
NW, no exposure............	1,500	1,250	No limit	No limit	No limit	No limit

* Ratio of 24-hr cold absorption to 5-hr boil absorption.

Facing brick is available in grades SW for high resistance to frost action and MW for moderate resistance to frost action (ASTM C216). Type FBX is intended for use where a high degree of mechanical perfection, narrow color range, and minimum variation in size are required. Type FBS is used where wider variation in color and size is acceptable. Type FBA is manufactured and selected to produce characteristic architectural effects from nonuniformity in size, color, and texture.

3. Structural Clay Tile   There are two grades of load-bearing wall tile. Grade LBX is suitable for general use in masonry construction and for use in masonry exposed to weathering. This grade is suitable for direct application of stucco. Grade LB is intended for masonry not exposed to frost action and for exposed masonry protected with a facing of 3 in. or more of other masonry.

Non-load-bearing tile (Grade NB) includes partition tile, furring tile, and fireproofing tile.

Facing tile is intended for general use in interior and exterior walls and partitions. Type FTX is a smooth-faced tile low in absorption, easily cleaned, and resistant to staining. Type FTS is a smooth or rough tile of moderate absorption, moderate variation in face dimensions, and medium range in color. These are available in two classes: standard and special-duty, the latter having superior resistance to impact and transmission of moisture and greater lateral and compressive load resistance.

Floor tile is available in grades FT1 and FT2, both of which are suitable for use in flat or segmented arches or in combination tile and concrete ribbed-slab construction.

Other types of hollow units, such as brick-block, may be used satisfactorily and should not be ruled out simply because they are not included in ASTM specifications for tile.

Ceramic glazed facing tile, facing brick, and solid masonry units are available in ASTM C126 Grade S for comparatively narrow mortar joints and Grade G where variation in face dimension must be very small.

Physical requirements for structural clay tile are given in Table 2.

4. Concrete Units   Concrete building brick (ASTM C55), solid load-bearing units (ASTM C145), and hollow load-bearing units (ASTM C90) are available in Types I and II. The moisture content of Type I units is controlled while that of Type II is not. The moisture limits are related to the shrinkage characteristics of the units and to the relative humidity of the jobsite (Table 3). This is to control shrinkage and the consequent hazard of cracking due to structural restraint. These limits may be waived if special precaution is taken to prevent such stress, as by control joints, etc. These units are of two grades. Grade N is for general use, as in exterior walls that may or may not be exposed to water penetration, and for interior or back-up use. Grade S is limited to use where not exposed to weather.

Hollow non-load-bearing units (ASTM C129) are available in Types I and II.
Physical requirements for concrete masonry units are given in Table 4.

5. Mortar Bond is more important to the proper functioning of masonry than is the strength of the mortar itself. Mortars for unit masonry may be specified as to proportions, or on the basis of property specifications. Proportions for five different types of mortar are given in Table 5. No strength requirements are stipulated. However, for reinforced masonry, or other masonry where assurance of strength is important, field sampling and testing should be done. The alternate method of proportioning requires mixing of the ingredients to a required flow and meeting the 2-in. cube strengths given in Table 6.

TABLE 2   Physical Requirements for Structural Clay Tile

Type and grade	Absorption, % (1 hr boiling)		Compressive strength, psi (based on gross area)			
			End-construction tile		Side-construction tile	
	Avg of 5 tests	Indi-vidual	Min avg of 5 tests	Indi-vidual	Min avg of 5 tests	Indi-vidual
Load-bearing (ASTM C34):						
LBX..........................	16	19	1,400	1,000	700	500
LB...........................	25	28	1,000	700	700	500
Non-load-bearing (ASTM C56),						
NB..........................	...	28				
Floor tile (ASTM C57):						
FT1..........................	...	25	3,200	2,250	1,600	1,100
FT2..........................	...	25	2,000	1,400	1,200	850
Facing tile (ASTM C212):						
Standard.....................	...	...	1,400	1,000	700	500
Special-duty.................	...	...	2,500	2,000	1,200	1,000
Glazed units (ASTM C126)......	...	...	3,000	2,500	2,000	1,500

TABLE 3   Maximum Moisture Content for Type I Units
(Percent Absorption, Average of Three Units)

Linear shrinkage, %	Conditions at job site		
	Humid*	Intermediate†	Arid‡
0.03 or less	45	40	35
0.03–0.045	40	35	30
0.045–0.065 max	35	30	25

*Average annual humidity above 75 percent.
†Average annual humidity 50 to 75 percent.
‡Average annual humidity less than 50 percent.

REINFORCED MASONRY

6. Materials Mortar and grout for reinforced masonry are covered by ASTM C476, which is a proportion specification. The mortar may consist of 1 part portland cement, ¼ to ½ part lime, and fine aggregate 2¼ to 3 times the sum of the volumes of cement and lime. The ½ part lime (i.e., 1:½:4½) generally gives better bond and water retention as well as better workability. The mortar may also consist of 1 part portland cement and 1 part Type II masonry cement with fine aggregate 2¼ to 3 times the combined volumes of cement. Masonry cement is favored in some areas, but many agencies prefer the lime, in which all

TABLE 4   Requirements for Concrete Masonry Units

| | Compressive strength, min, psi on average gross area | | Water absorption, max, pcf, average of 5 units | | | |
| | | | Oven-dry weight of concrete, pcf | | | |
Product	Average of 5 units	Individual unit	Over 125	105–125	105 or less	85 or less
Concrete building brick (ASTM C55):						
Grades NI, NII	3500*	3000*	13	15	18	
Grades SI, SII	2500*	2000*				20
Solid load-bearing concrete masonry units (ASTM C145):						
Grades NI, NII	1800	1500	13	15	18	
Grades SI, SII	1800	1500				20
Hollow load-bearing concrete masonry units (ASTM C90):						
Grades NI, NII	1000	800	13	15	18	
Grades SI, SII	1000	800				20
Hollow non-load-bearing concrete masonry units (ASTM C129), Types I, II	800	500				

*Brick flatwise.

TABLE 5   Unit-Masonry Mortar. Proportioning by Volume

| | Parts by volume | | | Aggregate measured in damp, loose condition |
Mortar type	Portland cement	Masonry cement	Hydrated lime or lime putty	
M	1	1 (Type II)		Not less than
S	½	I (Type II)		2¼ and not more
N		I (Type II)		than 3 times the
O		1 (Type I or II)		sum of volumes
M	1		¼	of cement and lime
S	1		Over ¼ to ½	
N	1		Over ½ to 1¼	
O	1		Over 1¼ to 2½	
K	1		Over 2½ to 4	

TABLE 6   Unit-Masonry Mortar. Proportioning by Strength

Mortar type	Average 28-day compressive strength, psi
M	2500
S	1800
N	750
O	350
K	75

the ingredients can be known and tested to give predictably better results. Two types of grout are specified (Table 7). A high water-cement ratio is used to obtain fluidity, but it is reduced rapidly by absorption of the masonry units. Recommended size of aggregate is ⅛ in. maximum for grout spaces to 2 in., ⅜ in. maximum for spaces to 3 in., and ¾ in. for wider spaces.

Cold-drawn wire complying with ASTM A82 and reinforcing bars recognized by ACI for reinforced concrete are used. In general, bar or wire under ¼ in. is not deformed. For the few situations in which placement is critical, positive positioners must be used.

7. Design Tests indicate that the structural performance of reinforced masonry is analogous to that of reinforced concrete within the extremely low limits of stress that are permitted. The bond between the units is such that masonry can be assumed to act as a homogeneous material within the range of working stresses. The assumption that masonry carries no tension is ultraconservative in most cases and may lead to erroneous conclusions.

TABLE 7 Grout Proportions by Volume

Parts by volume of portland cement or portland blast-furnace slag cement	Parts by volume of hydrated lime or lime putty	Aggregate, measured in a damp, loose condition	
		Fine	Coarse
Fine grout....	1	0–⅒	2¼–3 times the sum of the volumes of the cementitious materials	
Coarse grout..	1	0–⅒	2¼–3 times the sum of the volumes of the cementitious materials	1–2 times the sum of the volumes of the cementitious materials

The net section, particularly in hollow units, is an important consideration. Mortar joints are sometimes raked for appearance, which results in a greatly reduced effective section. This is especially critical for forces perpendicular to the wall. Design for shear is similar to that for reinforced concrete.

8. Allowable Stresses Two levels of stress are permitted by the Uniform Building Code, depending upon inspection (Table 8). A special inspector employed by the owner or the owner's agent must be present at all times during construction of the masonry if the allowable stresses requiring special inspection are used.

If the value of f'_m is determined by tests (Art. 13), the allowable stresses given in Table 8 may be used.

9. Beams The following procedure is suggested for the design of reinforced-masonry beams. For the allowable stresses f_m and f_s and the corresponding values of E_m and E_s, determine

$$k = \frac{1}{1 + f_s/nf_m} \qquad j = 1 - \frac{k}{3} \qquad K = \frac{1}{2} f_m jk$$

For preliminary design, k and j may be assumed to be 0.30 and 0.90, respectively. Determine b, d, and A_s from

$$M = Kbd^2 \qquad M = A_s f_s jd$$

Values of K as a function of np and f_m are given in Fig. 1.

Bending stress can be checked by

$$f_m = \frac{M}{bd^2} \frac{2}{jk}$$

Shearing stress is checked by $v_m = V/bjd$. If the allowable stress for no web reinforcement is exceeded, web reinforcement must be provided. Stirrup spacing is given by $s = f_v A_v/bv'$.

TABLE 8 Maximum Working Stresses, psi, for Reinforced Solid and Hollow Unit Masonry

Type of stress[a]	Hollow clay units Grade LB or hollow concrete units[b] Grade A		Grouted solid hollow units, concrete Grade A, clay Grade B, or solid units 2500 psi on gross area		Solid units 3000 psi on gross area		Special testing[f] f'_m established by prism tests				
Ultimate compressive strength f'_m	675	1350	750	1500	900	1800	2000	2700	3000	3500	4000
Special inspection required	No	Yes	No	Yes	No	Yes	Yes	Yes	Yes	Yes	Yes
Compression—axial, walls,[h] 0.2 f'_m	135	270	150	300	180	360	400	540	600	700	800
Compression—axial, columns,[i] 0.18f'_m	122	244	135	270	162	324	360	486	540	630	720
Compression—flexural, 0.33f'_m	225	450	250	500	300	600	667	900			
Shear: No shear reinforcement, Flexural,[c] $1.1\sqrt{f'_m}$ Shear walls[d]	25	40	25	42	25	47	49	50			
$M/Vd \geq 1$,[g] $0.9\sqrt{f'_m}$	17	33	17	34	17	34	34				
$M/Vd = 0$, $2\sqrt{f'_m}$	25	50	25	50	25	50	50				
Reinforcing taking all shear: Flexural, $3\sqrt{f'_m}$ Shear walls[d]	75	110	75	115	75	127	134	150			
$M/Vd \geq 1$,[g] $1.5\sqrt{f'_m}$	35	55	35	58	35	64	67	75			
$M/Vd = 0$, $2\sqrt{f'_m}$	60	73	60	77	60	85	89	104	110	118	120
Modular ratio n, 30,000f'_m	44	22	40	20	33	17	15	11	10		

Bearing:										
Full area, $0.25f'_m$	170	340	187	375	225	450	500	675	750	875	1000
$\frac{1}{3}$ or less of area,[e] $0.3f'_m$	200	400	225	450	270	540	600	810	900	1050	1200
Bond—plain bars	30	60	30	60	30	60	60				
Bond—deformed bars	100	140	100	140	100	140	140				

[a] Allowable values according to Uniform Building Code, 1976.

[b] Stresses for hollow unit masonry are based on net section.

[c] Web reinforcement shall be provided to carry the entire shear in excess of 20 psi whenever there is required negative reinforcement for a distance of one-sixteenth the clear span beyond the point of inflection.

[d] Where determinations involve rigidity considerations in combination with other materials or where deflections are involved, the moduli of elasticity and rigidity under columns entitled "Yes" for special inspection shall be used.

[e] This increase shall be permitted only when the least distance between the edges of the loaded and unloaded areas is a minimum of one-fourth of the parallel side dimensions of the loaded area. The allowable bearing stress on a reasonably concentric area greater than one-third, but less than the full area, shall be interpolated between the values given.

[f] Special testing shall include preliminary tests conducted to establish f'_m and at least one field test during construction of walls per each 5000 sq ft of wall but not less than three such field tests for any building.

[g] Use straight-line interpolation for M/Vd values between 0 and 1.

[h] See Eq. (2).

[i] See Eq. (3).

Bond stress is given by $u = V/\Sigma_0 jd$ and development length L of reinforcing bars by $L = f_s D/4u$.

Values of k, j, and $2/kj$ are given in Table 9.

Lintels are designed as beams supporting the triangular portion of the wall bounded by lines at 45° from each support. Concentrated loads from beams framing into the wall above the opening may be assumed to be distributed over a length equal to the base of the trapezoid formed by drawing, from the edges of the beam or bearing plate, lines at 60° with the horizontal. Alternatively, the wall spanning an opening may be designed as a deep beam.

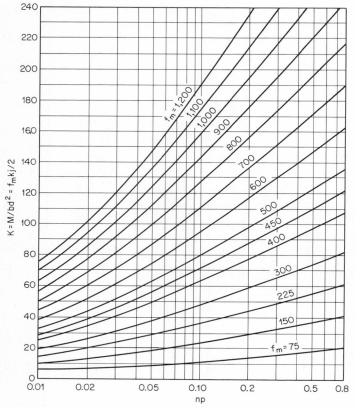

Fig. 1   Values of K for reinforced masonry. *(From Concrete Masonry Association of California.)*

Deep Beams. Walls are often designed to span from caisson to caisson to eliminate grade beams. Methods of design have been developed by the Portland Cement Association and others. Tests have shown that these give large factors of safety; so simplified methods may be used up to ratios of $h/t = 48$. In one approximate method a strip at the bottom of the wall is designed to carry the total load. Another, which is simple and quite conservative, is to proportion the tension steel as in a beam or tied arch; the reactions are assumed to be carried by a vertical strip at each end, which is the width of the end bearing plus twice the wall thickness, with the stress limited by

$$f_m = 0.20 f'_m \left[1 - \left(\frac{h/t}{48} \right)^3 \right]$$

(1)

10. Walls The ratio of height or length of a reinforced-masonry bearing wall to its thickness is limited by the Uniform Building Code to 25. The corresponding ratios for nonbearing walls are 30 and 48 for exterior and interior walls, respectively. The allowable axial stress f_m is given by

$$f_m = 0.20f'_m \left[1 - \left(\frac{h}{40t} \right)^3 \right] \qquad (2)$$

where f'_m = approved ultimate compressive stress, not to exceed 6000 psi. This formula does not take end conditions into consideration. The following effective heights h' are suggested: $h' = 2h$ for cantilever walls, $h' = 1.8h$ for cantilever guided at top, $h' = h$ for pin-ended wall, $h' = 0.75h$ for wall pinned at one end and fixed at the other, $h' = 0.5h$ for

TABLE 9

np	k	j	$2/kj$
0.010	0.131	0.956	15.93
0.020	0.181	0.939	11.76
0.030	0.216	0.927	9.95
0.040	0.245	0.918	8.89
0.050	0.270	0.909	8.14
0.055	0.281	0.906	7.87
0.060	0.291	0.902	7.60
0.065	0.301	0.899	7.38
0.070	0.310	0.896	7.19
0.075	0.319	0.893	7.01
0.080	0.327	0.890	6.85
0.085	0.336	0.888	6.71
0.090	0.343	0.885	6.58
0.095	0.351	0.883	6.45
0.100	0.358	0.880	6.34
0.105	0.365	0.878	6.24
0.110	0.371	0.876	6.14
0.115	0.378	0.873	6.04
0.120	0.384	0.871	5.97
0.125	0.390	0.869	5.89
0.130	0.396	0.867	5.82
0.135	0.401	0.866	5.75
0.140	0.407	0.864	5.68
0.145	0.412	0.862	5.62
0.150	0.417	0.860	5.56
0.155	0.422	0.859	5.51
0.160	0.427	0.857	5.46
0.165	0.432	0.855	5.41
0.170	0.437	0.854	5.36
0.175	0.441	0.852	5.31
0.180	0.446	0.851	5.26
0.185	0.450	0.849	5.22
0.190	0.455	0.848	5.18
0.195	0.459	0.846	5.14
0.200	0.463	0.845	5.11
0.250	0.500	0.833	4.80
0.300	0.530	0.823	4.58
0.350	0.556	0.814	4.41
0.400	0.579	0.806	4.27
0.450	0.600	0.800	4.17
0.500	0.618	0.794	4.07

wall fixed at both ends. However, judgment must be used in evaluation of the degree of fixity or lateral support.

Minimum reinforcement is 0.2 percent of the gross cross-sectional area of the wall, with at least one-third in either direction.

Lateral forces will generally govern the design of a wall. If there are pilasters or intersecting walls, moments can be based on end fixity.

Shear walls must be checked for tie-down and for the condition of no live load as well as full live load.

Typical wall elevations are shown in Fig. 2. The spacing H should not exceed 12 ft for low-shrinkage units and 8 ft for other units. The spacing S should not exceed 8 ft, except in seismic areas for which $S_{max} = 4$ ft. Where reinforcement is continuous $c \geq 2b$.

Arrangement of horizontal reinforcement is usually determined by the position of openings and by code requirements. According to the UBC, horizontal reinforcement is required in the top of footings, at the top of wall openings, at roof and floor levels, and at the top of parapet walls. Only horizontal reinforcement which is continuous in the wall is considered in determining the minimum area of reinforcement.

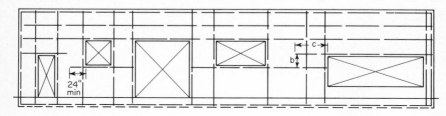

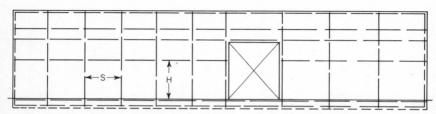

Fig. 2 Wall reinforcing. *(From Concrete Masonry Association of California.)*

Vertical reinforcement is generally determined by design for wind, seismic, and vertical loads. A recommended reinforcement for 8-in. masonry walls is No. 5 at 4 ft.

11. Columns   The allowable load (Uniform Building Code) on reinforced-masonry columns is

$$P = A_g(0.18f'_m + 0.65p_a f_s)\left[1 - \left(\frac{h}{40t}\right)^3\right] \tag{3}$$

The least dimension must be 12 in., except that it may be 8 in. if the column is designed for one-half the allowable stress. The unsupported length must not exceed 20 times the least dimension.

The ratio p_g may not be less than 0.5 percent or more than 4 percent. The minimum number of reinforcing bars is four and the smallest diameter ⅜ in. Lateral ties at least ¼ in. in diameter must be spaced not over 16 bar diameters, 48 tie diameters, or the least dimension of the column. Ties may be in the bed joints or tight around the vertical bars. Additional ties to enclose anchor bolts are advisable at tops of columns. This is especially important if the element supported may be subjected to lateral loads, as from earthquake or wind.

Combined vertical load and bending is evaluated by

$$\frac{f_a}{F_b} + \frac{f_b}{F_b} \leq 1$$

12. Diaphragms   Horizontal diaphragms are often used to distribute lateral forces to shear walls or other bracing systems. In the design of diaphragms in reinforced-masonry buildings, the walls are usually considered as flanges of deep plate girders whose webs are the diaphragms. Longitudinal shear between the web and the flange must be considered in determining the connection to the wall. Horizontal truss systems usually consist of struts and tension-rod diagonals. This type of bracing is not generally recommended for masonry buildings because deflection is usually excessive compared with other systems.

TESTS AND INSPECTION

13. Compressive Strength of Masonry Some codes permit tests to determine the approved ultimate compressive stress f'_m. Two methods are used.

The strength may be established by preliminary tests of prisms built of the same materials under the same conditions and, insofar as is possible, with the same bonding and workmanship as will be used. Prisms for walls are 16 × 16 in. and of the thickness and type of construction of the prototype. Shorter prisms may be needed to fit within the test depth of available machines. However, if the length is less than the width, the results may be inaccurate because the direction of splitting may be changed. Prisms for columns are 8 × 8 in. in plan and 16 in. high. All specimens should have a height-to-thickness ratio not less than 2. If other ratios are used, the value of f'_m is taken as the compression strength of the specimens multiplied by the following correction factor:

h/d	1.5	2.0	2.5	3.0
Factor	0.86	1.00	1.11	1.20

Prisms must be tested under conditions specified by the governing code.

The approved value of f'_m may be established by tests of individual units (Table 10).

Table 10

Compressive Strength of Units, psi	Assumed f'_m, psi
1,000	900
1,500	1,150
2,500	1,550
4,000	2,000
6,000	2,400
8,000	2,700
10,000	2,900
12,000	3,000

DETAILING AND CONSTRUCTION

14. Detailing The cost of masonry can be greatly influenced by the detailing. It is important to detail the units and the reinforcement so that they can be placed with the rhythmic procedure of good laying. Care must be given to modular layouts to minimize field cutting of units. Wide spacing of steel will usually result in economy. Steel in grout spaces must have at least ¼ in. cover for embedment. If spacer bars are used on joint reinforcement, they can serve as chairs to keep the reinforcement clear so that mortar may flow beneath. In general, the thickness of joints should be at least twice the diameter of the embedded bar. However, adequate bond might also be achieved by the use of welded spacers to serve as anchors in transferring stress.

Steel details are similar to those for reinforced concrete, but with the additional limitations of modular spacing and emphasis on wide spacing of reinforcement.

15. Concrete Foundations Horizontal concrete surfaces that are to receive masonry should be clean and preferably slightly damp, with the aggregate exposed on the surface roughened to assure good bond. Grout spaces should be kept clear of mortar and the bottom course grouted solid before additional courses are placed.

Foundation dowels should not be bent to a slope of more than 1 in 6. Incorrectly positioned dowels should be grouted into a cell adjacent to the cell containing a vertical wall-reinforcing bar if necessary. Wall steel need not be tied to the dowels, although this is a good way to keep it in place.

16. Workmanship Since masonry construction consists of assemblages of small units, workmanship is of utmost importance. For grouted masonry, bed joints should be beveled from the inside, rather than furrowed. This may leave an open space on the inside face of the bed joint, but this will be filled with grout and will serve as a mechanical key. Units must be pressed down firmly while the mortar is moist and plastic. Head joints should be shoved tight and full, although in grouted work the back of the joint can be open. Joints in hollow masonry should be full for the thickness of the face shell. One of the advantages of this is that it breaks capillary action. Units must not be tapped to relocate them after the mortar has lost its initial plasticity. Such tapping will break the bond and cause weakness and leakage. If adjustments are necessary, the old mortar should be removed. Mortar fins protruding more than the thickness of the joint into the grout space, from either head or bed joints, should be removed but not allowed to fall into the grout space.

Racking (stepping back successive courses) is acceptable, but toothing is not. Toothed joints cannot be pressed together to make a tight bond. If toothing is permitted, caulking material may be used to seal the joints, or the joints made by pressing mortar onto the surface and tucking mortar tightly into the joint.

Joints should be tooled to compress the mortar against the edges of the masonry to make the joint watertight as well as neat.

Wetting.   To assure proper bonding, clay units should be wetted so that the surface is slightly damp. The absorption rate should not exceed 0.025 oz/sq in./min when placed in water to a depth of ⅛ in. Concrete units should not be wetted before laying except in very hot, dry weather, when the bearing surfaces to receive mortar may be slightly moistened immediately before laying. Wetting should be minimized since it causes slight expansion and subsequent shrinkage which may cause cracking.

Admixtures must be used sparingly and carefully. Generally they are used to reduce the water content. They affect the flow after suction of the mortar and should be checked for their effect on bond.

Section **16**

Timber Structures

KENNETH P. MILBRADT

Associate Professor of Civil Engineering, Illinois Institute of Technology

STRUCTURAL PROPERTIES OF WOOD

1. Anisotropic Nature of Wood Relationships between load, duration of load, deformation, and material strength are usually based on the assumption that the structural material is homogeneous and isotropic, in both elastic constants and strength properties. Wood is neither, because of its cellular structure and growth characteristics.

The growth of trees is by the addition of cells under the bark, at the ends of branches, and at the roots. The thickness and structural characteristics of the new cell layer depend upon many factors such as temperature, moisture, and species. Most cells are oriented along the longitudinal axis of the tree while fewer cells develop radially and usually none tangentially. The growth pattern, in certain climates, leads to the development of annual dark and light rings. These rings exhibit different strength properties, the summer wood being darker in color and stronger than the lighter-colored spring wood. In estimating the strength properties of some woods the number of annual rings per inch is used as a partial guide.

The addition of cells occurs, overall, in a uniform manner and creates the "grain" appearance in wood. Wood is often assumed to be an orthotropic material with the three principal elasticity directions coinciding with the longitudinal, radial, and tangential directions in the tree. This requires nine constants to specify the elastic behavior: three Young's moduli, three shear moduli, and three Poisson's ratios.

2. Elastic Constants The most important elastic constant in design is the modulus of elasticity E along the grain, since the grain is usually parallel to the maximum normal stress. Moisture content considerably influences the modulus of elasticity, which is usually given for a 15 or 19 percent moisture content. The modulus of elasticity of green timber is approximately 25 percent lower than that for a 15 percent moisture content and, within a limited range, may vary about 2 percent for each 1 percent change in moisture content.[1] For short-term loading, E does not change with time. Long durations of load cause creep, which may be approximated by using a secant modulus of one-half (or less) of the modulus E.

The moduli of elasticity E_R and E_T in the radial and tangential directions (across grain) usually are not important in timber construction. However, they are needed to determine the properties of plywood. In general, they are affected more by moisture content than is E, the range of variation being about 2:1 for moisture contents ranging from 7 to 21 percent. Average values are, approximately, $E_T = 0.05E$ and $E_R = 0.10E$. The shear moduli of elasticity are, approximately, $G_{LT} = 0.06E$, $G_{TR} = 0.075E$, and $G_{RT} = 0.018E$, where L, R, and T are axes along the grain, radially, and tangentially, respectively.

3. Directional Strength Properties   As with most materials, there is inherent variability in the strength of small, clear samples of wood under short-time loading. Added to this variability are the effects of duration of load and strength-reducing factors such as knots. In addition, wood exhibits directional properties when subjected to various stress states. The strength properties to consider are associated with normal and shear stresses parallel to the grain, perpendicular to the grain radially, and perpendicular to the grain tangentially. The difference in strength properties in the radial and tangential directions is seldom of significance in design. Thus, it is necessary only to differentiate between directions normal and parallel to the grain.

Tension Parallel to Grain.   It is usual to accept the modulus of rupture in bending as the measure of tensile strength of wood, the tensile strength being two-thirds of the modulus of rupture. Wood exhibits its maximum tensile strength parallel to the grain.

Compression Parallel to Grain.   Wood usually fails under uniaxial compressive stresses by buckling of the fibers. This takes place on a 20 to 30° plane and sometimes is referred to as a shear failure. Upon seasoning from the green state to 15 percent moisture content, compressive strength is increased by 50 to 75 percent in small clear specimens and less in larger cross sections. The increase for larger specimens is limited because of defects introduced by drying.

Compression Perpendicular to Grain.   The cells of the wood, being essentially hollow, exhibit relatively low stiffness and strength when stressed in compression perpendicular to the grain. Seasoning green lumber to 15 percent moisture content increases this strength by 50 percent.

Tension Perpendicular to Grain.   Wood exhibits relatively low tensile strength perpendicular to the grain, and in the presence of defects, such as checks, it is substantially reduced. This strength usually does not enter into design, but abrupt changes in cross section may cause stress concentrations with tension perpendicular to the grain which should be considered in determining load capacity.

Shear Strength Parallel to Grain.   Significant shear stresses may be developed parallel to the grain in beams. Checks, splits, and shakes, being fractures parallel to the grain, significantly reduce the shear strength in this direction. These flaws are associated with larger specimens. Small clear specimens without flaws may exhibit a shear strength two to five times that of the larger specimens.

4. Factors Affecting Strength   Working stresses for timber are approximately 20 to 75 percent of the ultimate strength. The large variation of the factor of safety is due to the variability of the strength properties, which depend on knots, moisture content, grain, density, shakes, splits, checks, and other factors.

Growth Characteristics and Strength.   The effect of knots is to decrease strength because their grain and the surrounding grain is at a large angle to the maximum tensile or compressive stress. The effects in tension are more detrimental than in shear or compression. Thus, for beams it is advantageous to locate the higher-grade sections of the lumber in the maximum-tensile-stress zones. The mass modulus of elasticity for stress-graded lumber is little affected by the presence of knots.

As with many other materials, wood increases its strength properties with an increase in specific gravity, and relationships have been suggested for strength and specific gravity.[1] Density is usually recognized in the stress grading rules for lumber.

Other factors such as sloping grain, decay, insect attack, pitch pockets, shakes, wane, splits, checks, compression failure, tension, and mold affect the strength properties of wood. These are considered in the detailed instructions for stress grading of lumber published by various organizations.

Environmental Conditions.   As the moisture content of wood drops below 30 percent (the approximate fiber-saturation point) its strength properties increase, with the exception of toughness. Table 1 lists the approximate variation of strength properties, the applicable range being from 2 to 25 percent moisture content, approximately. Reduction in water content from the fiber-saturation point to zero is accompanied by radial shrinkage ranging from 4 to 6 percent, tangential shrinkage from 6 to 8 percent, and longitudinal shrinkage from 0.1 to 0.3 percent. The relation between shrinkage and moisture content is linear.

If wood is kept either continuously dry or continuously wet, decay does not occur. Moisture and temperature are the prime factors affecting decay rate. Wood should not be in direct contact with masonry or concrete where excessive moisture will be transferred to the wood. Ventilated air spaces around untreated members or pressure treatment with preservatives retard or prevent decay.[1]

TABLE 1   Effect of Moisture Content on Strength Properties*

	% Change per 1% Change in Moisture Content
Static bending:	
Stress at proportional limit	5
Modulus of rupture	4
Modulus of elasticity	2
Work to proportional limit	8
Work to maximum load	0.5
Impact bending, height of drop causing fracture	0.5
Compression parallel to grain:	
Fiber stress at proportional limit	5
Maximum crushing strength	6
Compression perpendicular to grain:	
Fiber stress at proportional limit	5.5
Shear parallel to grain, maximum strength	3
Tension perpendicular to grain, maximum strength	1.5
Hardness:	
End	4
Side	2.5

* From Ref. 1.

The strength properties of wood at 12 percent moisture content, within a range of 70 to 150°F, may be expected to decrease by ⅓ to ½ percent per 1°F.[1] Temperatures above 150°F leave permanent detrimental effects. Fire resistance of heavy timber construction is recognized as adequate by most building codes. For light timber construction it may be necessary to use fire retardants and reduce the allowable stresses.

Time-Load Effects. The strength properties of wood are affected by the duration of

TABLE 2   Duration of Maximum Load

Duration	% of 10-year strength
Impact	200
Wind or earthquake	133
7 days	125
2 months (as for snow)	115
Permanent	90

loading (Fig. 1). The usual durations of load considered in design, and the corresponding percent of 10-year strength, are given in Table 2.

Long-term deflections may be estimated by using effective moduli of elasticity one-third and one-half the value in Table 3 for unseasoned timber and for seasoned timber kept dry, respectively.

The ability of wood to absorb impact is considerable (Table 2). Meager data are available on strength properties under cyclic loading. For bending, the endurance limit (10⁷ cycles of full reversal) is about 25 percent of the modulus of rupture in bending.[3] Available data indicate that the allowable horizontal shearing stress should be decreased about 20 percent for cyclic loading.

5. Working Stresses for Sawn Lumber   Tables 3 and 4 present allowable stresses for stress-graded lumber of 2 of the 36 species listed by the National Design Specification.[2]

TABLE 3 Allowable Unit Stresses for Structural Lumber, Normal Loading Conditions, Visual Grading[a]

Species and commercial grade	Size classification	Allowable unit stresses, psi							Grading rules agency
		Extreme fiber in bending F_b		Tension parallel to grain F_t	Horizontal shear F_v	Compression perpendicular to grain $F_{c\perp}$	Compression parallel to grain F_c	Modulus of elasticity E	
		Single-member uses[a]	Repetitive-member uses[a]						
		Southern pine (surfaced dry, used at 19% max m.c.)							
Select structural	2 to 4 in. thick, 2 to 4 in. wide	2100	2400	1250	90	405	1600	1,800,000	Southern Pine Inspection Bureau (see footnotes c, i, j, and l)
Dense select structural		2450	2800	1450	90	475	1850	1,900,000	
No. 1		1750	2000	1000	90	405	1250	1,800,000	
No. 1 dense		2050	2350	1200	90	475	1450	1,900,000	
No. 2		1250	1450	725	75	345	850	1,400,000	
No. 2 medium grain		1450	1650	850	90	405	1000	1,600,000	
No. 2 dense		1700	1950	1000	90	475	1150	1,700,000	
No. 3		825	950	475	75	345	600	1,400,000	
No. 3 dense		950	1100	550	90	475	700	1,500,000	
Stud		825	950	475	75	345	600	1,400,000	
Construction	2 to 4 in. thick, 4 in. wide	1050	1200	620	75	345	1100	1,400,000	
Standard		590	700	340	75	345	925	1,400,000	
Utility		275	325	165	75	345	600	1,400,000	
Select structural	2 to 4 in. thick, 6 in. and wider	1800	2050	1200	90	405	1400	1,800,000	
Dense select structural		2100	2400	1400	90	475	1650	1,900,000	
No. 1		1500	1750	1000	90	405	1250	1,800,000	
No. 1 dense		1800	2050	1200	90	475	1450	1,900,000	
No. 2		1050	1200	700	75	345	900	1,400,000	
No. 2 medium grain		1250	1450	825	90	405	1050	1,600,000	
No. 2 dense		1450	1650	975	90	475	1250	1,700,000	
No. 3		725	825	475	75	345	650	1,400,000	
No. 3 dense		850	975	575	90	475	750	1,500,000	

Dense standard factory	2 to 4 in. thick, 2 to 4 in. wide	2000	2300	1200	90	475	1450	1,900,000	Southern Pine Inspection Bureau (see footnotes j and l)
No. 1 factory		1400	1600	825	90	405	1000	1,600,000	
No. 1 dense factory		1650	1900	975	90	475	1150	1,700,000	
No. 2 factory		1400	1600	825	90	405	1000	1,600,000	
No. 2 dense factory		1700	1950	975	90	475	1150	1,700,000	
Dense standard factory	2 to 4 in. thick, 6 in. and wider	1750	2000	1200	90	475	1450	1,900,000	
No. 1 factory		1250	1450	825	90	405	1050	1,600,000	
No. 1 dense factory		1450	1650	975	90	475	1250	1,700,000	
No. 2 factory		1250	1450	825	90	405	1050	1,600,000	
No. 2 dense factory		1450	1650	975	90	475	1250	1,700,000	
Dense structural 86	2 to 4 in. thick	2750	3150	1850	150	475	2050	1,900,000	
Dense structural 72		2300	2650	1550	125	475	1700	1,900,000	

Southern pine (surfaced at 15% m.c. kiln dry, used at 15% max m.c.)

Select structural	2 to 4 in. thick, 2 to 4 in. wide	2250	2600	1350	95	405	1850	1,900,000	Southern Pine Inspection Bureau (see footnotes c, i, j, and l)
Dense select structural		2650	3050	1550	95	475	2150	2,000,000	
No. 1		1900	2200	1100	95	405	1450	1,900,000	
No. 1 dense		2250	2600	1300	95	475	1700	2,000,000	
No. 2		1350	1550	775	80	345	975	1,500,000	
No. 2 medium grain		1550	1800	925	95	405	1150	1,700,000	
No. 2 dense		1850	2150	1050	95	475	1350	1,800,000	
No. 3		875	1000	525	80	345	700	1,500,000	
No. 3 dense		1050	1200	600	95	475	825	1,600,000	
Stud		875	1000	525	80	345	700	1,500,000	
Construction	2 to 4 in. thick, 4 in. wide	1150	1300	670	75	345	1300	1,500,000	
Standard		640	750	375	75	345	1050	1,500,000	
Utility		300	350	175	75	345	700	1,500,000	

TABLE 3 Allowable Unit Stresses for Structural Lumber, Normal Loading Conditions, Visual Grading[a] (Continued)

Species and commercial grade	Size classification	Allowable unit stresses, psi							Grading rules agency
		Extreme fiber in bending F_b		Tension parallel to grain F_t	Horizontal shear F_v	Compression perpendicular to grain $F_{c\perp}$	Compression parallel to grain F_c	Modulus of elasticity E	
		Single-member uses[a]	Repetitive-member uses[a]						
		Southern Pine (surfaced at 15% m.c. kiln dry, used at 15% max m.c.)							
Select structural	2 to 4 in. thick, 6 in. and wider	1950	2250	1300	95	405	1650	1,900,000	Southern Pine Inspection Bureau (see footnotes j and l)
Dense select structural		2250	2600	1500	95	475	1900	2,000,000	
No. 1		1650	1900	1100	95	405	1450	1,900,000	
No. 1 dense		1900	2200	1300	95	475	1700	2,000,000	
No. 2		1150	1300	750	80	345	1050	1,500,000	
No. 2 medium grain		1350	1550	900	95	405	1250	1,700,000	
No. 2 dense		1550	1800	1050	95	475	1450	1,800,000	
No. 3		800	900	525	80	345	750	1,500,000	
No. 3 dense		925	1050	625	95	475	875	1,600,000	
Dense standard factory	2 to 4 in. thick, 2 to 4 in. wide, decking	2200	2550	1300	95	475	1700	2,000,000	
No. 1 factory		1500	1750	900	95	405	1150	1,700,000	
No. 1 dense factory		1800	2050	1050	95	475	1350	1,800,000	
No. 2 factory		1500	1750	900	95	405	1150	1,700,000	
No. 2 dense factory		1800	2050	1050	95	475	1350	1,800,000	
Dense standard factory	2 to 4 in. thick, 6 in. and wider, decking	1900	2200	1300	95	475	1650	2,000,000	
No. 1 factory		1350	1550	900	95	405	1250	1,700,000	
No. 1 dense factory		1550	1800	1050	95	475	1450	1,800,000	
No. 2 factory		1350	1550	900	95	405	1250	1,700,000	
No. 2 dense factory		1550	1800	1050	95	475	1450	1,800,000	
Dense structural 86	2 to 4 in. thick	3000	3450	2000	160	475	2350	2,000,000	
Dense structural 72		2500	2900	1650	135	475	2000	2,000,000	

Douglas fir-larch (surfaced dry or surfaced green, used at 19% max. m.c.)

Dense select structural	2 to 4 in. thick, 2 to 4 in. wide	2450	2800	1400	95	455	1850	1,900,000
Select structural		2100	2400	1200	95	385	1600	1,800,000
Dense No. 1		2050	2400	1200	95	455	1450	1,900,000
No. 1		1750	2050	1050	95	385	1250	1,800,000
Dense No. 2		1700	1950	1000	95	455	1150	1,700,000
No. 2		1450	1650	850	95	385	1000	1,700,000
No. 3		800	925	475	95	385	600	1,500,000
Appearance		1750	2050	1050	95	385	1500	1,800,000
Stud		800	925	475	95	385	600	1,500,000
Construction	2 to 4 in. thick, 4 in. wide	1050	1200	625	95	385	1150	1,500,000
Standard		600	675	350	95	385	925	1,500,000
Utility		275	325	175	95	385	600	1,500,000
Dense select structural	2 to 4 in. thick, 6 in. and wider	2100	2400	1400	95	455	1650	1,900,000
Select structural		1800	2050	1200	95	385	1400	1,800,000
Dense No. 1		1800	2050	1200	95	455	1450	1,900,000
No. 1		1500	1750	1000	95	385	1250	1,800,000
Dense No. 2		1450	1700	950	95	455	1250	1,700,000
No. 2		1250	1450	825	95	385	1050	1,700,000
No. 3		725	850	475	95	385	675	1,500,000
Appearance		1500	1750	1000	95	385	1500	1,800,000
Dense select structural	Beams and stringers	1900		1100	85	455	1300	1,700,000
Select structural		1600		950	85	385	1100	1,600,000
Dense No. 1		1550		775	85	455	1100	1,700,000
No. 1		1300		675	85	385	925	1,600,000
Dense select structural	Posts and timbers	1750		1150	85	455	1400	1,700,000
Select structural		1500		1000	85	385	1200	1,600,000
Dense No. 1		1400		950	85	455	1200	1,700,000
No. 1		1200		825	85	385	1000	1,600,000
Select Dex	Decking	1750	2000			385		1,800,000
Commercial Dex		1450	1650			385		1,700,000

West Coast Lumber Inspection Bureau and Western Wood Products Association (see footnotes b through j)

West Coast Lumber Inspection Bureau (see footnotes b through j)

TABLE 3 Allowable Unit Stresses for Structural Lumber, Normal Loading Conditions, Visual Grading[a] (Continued)

Species and commercial grade	Size classification		Extreme fiber in bending F_b		Tension parallel to grain F_t	Horizontal shear F_v	Compression perpendicular to grain $F_{c\perp}$	Compression parallel to grain F_c	Modulus of elasticity E	Grading rules agency
			Single-member uses[a]	Repetitive-member uses[a]						
			Southern pine (surfaced dry, used at 19% max m.c.)							
Dense select structural	Beams		1900		1250	85	455	1300	1,700,000	Western
Select structural	and		1600		1050	85	385	1100	1,600,000	Wood
Dense No. 1	stringers		1550		1050	85	455	1100	1,700,000	Products
No. 1			1350		900	85	385	925	1,600,000	Association
										(see foot-
Dense select structural	Posts and		1750		1150	85	455	1350	1,700,000	notes b
Select structural	timbers		1500		1000	85	385	1150	1,600,000	through k)
Dense No. 1			1400		950	85	455	1200	1,700,000	
No. 1			1200		825	85	385	1000	1,600,000	
Selected decking	Decking		2000	1650					1,800,000	
Commercial decking									1,700,000	
Selected decking	Decking		2150	1800	Surfaced at 15% max m.c. and used at 15% max m.c.				1,900,000	
Commercial decking									1,700,000	

SOURCE: National Forest Products Association.

[a] Allowable unit stresses for single-member uses are intended for structures where an individual member such as a beam, girder, or post carries its full design load. Stresses for repetitive-member uses are intended for members in bending, such as joists, trusses, rafters, studs, planks, or decking that are spaced not more than 24 in., are not less than three in number, and are joined by floor, roof, or other load-distributing elements adequate to support the design load.

Recommended allowable unit stresses for visually graded lumber are determined in accordance with the provisions of ASTM D245-74, Methods for Establishing Structural Grades and Related Allowable Properties for Visually Graded Lumber.

[b] Recommended design values are applicable to lumber that will be used under dry conditions such as in most covered structures. For 2- to 4-in.-thick lumber the dry surfaced size should be used. In calculating design values, the natural gain in strength and stiffness that occurs as lumber dries has been

taken into consideration as well as the reduction in size that occurs when unseasoned lumber shrinks. The gain in load-carrying capacity due to increased strength and stiffness resulting from drying more than offsets the design effect of size reductions due to shrinkage. For 5-in. and thicker lumber, the surfaced sizes also may be used because design values have been adjusted to compensate for any loss in size by shrinkage which may occur.

[c]Values for F_b, F_b and F_c for the grades of construction, standard, and utility apply only to 4-in. widths. Design values for 2- and 3-in. widths of these grades are available from Northeastern Lumber Manufacturers Association, Redwood Inspection Service, Southern Pine Inspection Bureau, West Coast Lumber Inspection Bureau, Western Wood Products Association, and National Lumber Grades Authority.

[d]Values for 2- to 4-in. thicknesses are based on edgewise use. When such lumber is used flatwise, the recommended values for fiber stress in bending may be multiplied by the following factors:

Width, in.	Thickness, in.		
	2	3	4
2 to 4	1.10	1.04	1.00
6 and wider	1.22	1.16	1.11

[e]Recommended design values of F_b for decking may be increased by 10 percent for 2-in.-thick decking and by 4 percent for 3-in.-thick decking.

[f]When 2- to 4-in.-thick lumber is manufactured at a maximum moisture content of 15 percent and used in a condition where the moisture content does not exceed 15 percent, the values for surfaced dry or surfaced green lumber may be multiplied by the factors in Table 3a.

[g]When 2- to 4-in.-thick lumber is designed for use where the moisture content will exceed 19 percent for an extended period of time, values should be multiplied by the factors in Table 3a.

[h]When lumber 5 in. and thicker is designed for use where the moisture content will exceed 19 percent for an extended period of time, values should be multiplied by the factors in Table 3a.

[i]Horizontal shear values are based on the conservative assumption of the most severe checks, shakes, or splits possible, as if a piece were split full length. When lumber 4 in. and thinner is manufactured unseasoned, the tabulated values should be multiplied by a factor of 0.92. Horizontal shear values for any grade and species of lumber may be established by Table 3b when the length of split or check is known.

[j]Stress-rated boards of nominal 1-, 1¼-, and 1½-in. thickness, 2 in. and wider, are permitted the recommended design values shown for select structural, No. 1, No. 2, No. 3, construction, standard, utility, appearance, clear heart structural, and clear structural grades as shown in the 2- to 4-in.-thick categories here, when graded in accordance with the stress-rated board provisions in the applicable grading rules. Information on stress-rated board grades applicable to the various species is available from the respective grading rules agencies.

TABLE 3 Allowable Unit Stresses for Structural Lumber, Normal Loading Conditions, Visual Grading^a **(Continued)**

^{k}When decking is surfaced at 15 percent moisture content and used where the moisture content will exceed 15 percent for an extended period of time, the tabulated design values should be multiplied by the following factors: F_b, 0.79; E, 0.92.

^{l}When 2- to 4-in.-thick lumber is designed for use where the moisture content will exceed 19 percent for an extended period of time, the values for surfaced green lumber in Table 3 should be used for the corresponding grades of lumber surfaced dry or kiln dried.

TABLE 3a

Footnote	F_b	F_t	F_v	$F_{c\perp}$	F_c	E
f	1.08	1.08	1.05	1.00	1.17 1.15*	1.05 1.04*
g	0.86	0.84	0.97	0.67	0.70	0.97
h	1.00	1.00	1.00	0.67	0.91	1.00

*Redwood only.

TABLE 3b

Length of split on wide face	Multiply value of F_b by	
	2 in. nom.	3 in. and thicker
None	2.00	2.00
½ × wide face	1.67	1.67
¾ × wide face	1.50	
1 × wide face	1.33	1.33
1½ × wide face or more	1.00	1.00

Table 3 is for visually graded lumber while Table 4 is for machine-stress-rated lumber. These are 10-year values for lumber used under continuously dry conditions. For other conditions, see footnotes. Values in Tables 3 and 4 should be adjusted for duration of load according to Table 2.

Compression at Angle to Grain. The allowable stress at an angle to the grain is given by

$$F_\theta = \frac{F_c F_{c\perp}}{F_c \sin^2 \theta + F_{c\perp} \cos^2 \theta} \tag{1}$$

where F_c = allowable compressive stress parallel to grain
$F_{c\perp}$ = allowable compressive stress perpendicular to grain
θ = inclination of F_θ to grain

6. Glued-Laminated Lumber Glued-laminated lumber is fabricated from boards ¾ in. thick and greater. Depths of glued-laminated members are multiples of ¾ or 1½ in. The lamina, being thin, can be seasoned with fewer checks and other detrimental effects associated with seasoning large timber. The ¾-in. lamina are more desirable than 1½-in.

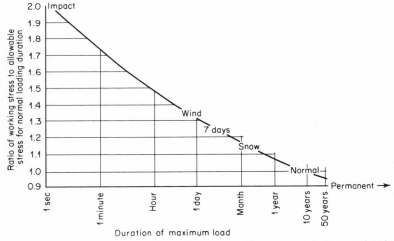

Fig. 1 Allowance for duration of load. *(Derived from Forest Prod. Lab. Rept. R 19 16.)*

lamina because of better distributions of flaws and less detrimental effect of warpage. Table 5 lists allowable stresses for laminated Douglas fir, larch, and southern pine lumber. The usual widths of board are given in the following table:

Nominal width, in.	3	4	6	8	10	12	14	16
Finished width, in.	2¼	3⅛	5⅛	6¾	8¾	10¾	12¼	14¼

Glues and gluing are important since the glue transmits shearing and normal stresses. Casein glues are usually used for dry conditions of use, while resorcinol-formaldehyde or phenol-formaldehyde glues are used for wet or weathering conditions. The moisture content of the lumber at the time of gluing may range from 7 to 16 percent but should approximate the content at which the member is expected to be put into service. Variation in moisture content of the lamina should not exceed 5 percent at the time of gluing.

Scarf joints are used to splice lamina. Scarf-joint efficiency factors (ratio of scarfed-joint stress to basic clear stress) are as follows:

Scarf slope	1:12 or flatter	1:10	1:8	1:5
Efficiency factor	0.85	0.80	0.75	0.60

TABLE 4 Allowable Unit Stresses for Structural Lumber, Normal Loading Conditions, Machine-Stress-Rated[a]

Grading rules agency Grade designation	Size classification	Extreme fiber in bending[c] F_b		Tension parallel to grain F_t	Compression parallel to grain F_c	Modulus of elasticity E
		Single-member uses	Repetitive-member uses			
Western Wood Products Association						
1200f, 1.2E		1200	1400	600	950	1,200,000
1500f, 1.4E		1500	1750	900	1200	1,400,000
1650f, 1.5E	Machine-	1650	1900	1020	1320	1,500,000
1800f, 1.6E	rated lumber	1800	2050	1175	1450	1,600,000
2100f, 1.8E	2 in. thick or	2100	2400	1575	1700	1,800,000
2400f, 2.0E	less, all	2400	2750	1925	1925	2,000,000
2700f, 2.2E	widths	2700	3100	2150	2150	2,200,000
3000f, 2.4E		3000	3450	2400	2400	2,400,000
3300f, 2.6E		3300	3800	2650	2650	2,600,000
900f, 1.0E	Machine-	900	1050	350	725	1,000,000
900f, 1.2E	rated joists 2	900	1050	350	725	1,200,000
1200f, 1.5E	in. thick or	1200	1400	600	950	1,500,000
1350f, 1.8E	less, all	1350	1550	750	1075	1,800,000
1800f, 2.1E	widths	1800	2050	1175	1450	2,100,000
West Coast Lumber Inspection Bureau						
900f, 1.0E		900	1050	350	725	1,000,000
1200f, 1.2E		1200	1400	600	950	1,200,000
1450f, 1.3E	Machine-	1450	1650	800	1150	1,300,000
1500f 1.4E	rated lumber	1500	1750	900	1200	1,400,000
1650f, 1.5E	2 in. thick or	1650	1900	1020	1320	1,500,000
1800f, 1.6E	less, all	1800	2050	1175	1450	1,600,000
2100f, 1.8E	widths	2100	2400	1575	1700	1,800,000
2400f, 2.0E		2400	2750	1925	1925	2,000,000
2700f, 2.2E		2700	3100	2150	2150	2,200,000
900f, 1.0E	Machine-	900	1050	350	725	1,000,000
900f, 1.2E	rated joists 2	900	1050	350	725	1,200,000
1200f, 1.5E	in. thick or	1200	1400	600	950	1,500,000
1500f, 1.8E	less, 6 in.	1500	1750	900	1200	1,800,000
1800f, 2.1E	and wider	1800	2050	1175	1450	2,100,000
Southern Pine Inspection Bureau						
1200f, 1.2E		1200	1400	600	950	1,200,000
1500f, 1.4E		1500	1750	900	1200	1,400,000
1650f, 1.5E	Machine-	1650	1900	1020	1320	1,500,000
1800f, 1.6E	rated	1800	2050	1175	1450	1,600,000
2100f, 1.8E	lumber 2 in.	2100	2400	1575	1700	1,800,000
2400f, 2.0E	thick or less,	2400	2750	1925	1925	2,000,000
2700f, 2.2E	all widths	2700	3100	2150	2150	2,200,000
3000f, 2.4E		3000	3450	2400	2400	2,400,000
3300f, 2.6E		3300	3800	2650	2650	2,600,000
900f, 1.0E	Machine-	900	1050	350	725	1,000,000
900f, 1.2E	rated lumber	900	1050	350	725	1,200,000
1200f, 1.5E	2 in. thick or	1200	1400	600	950	1,500,000
1350f, 1.8E	less, all	1350	1550	750	1075	1,800,000
1800f, 2.1E	widths	1800	2050	1175	1450	2,100,000

Allowable unit stresses, psi[b]

SOURCE: National Forest Products Association.
(Notes on next page.)

Cyclic Loading.   Tests on glued joints indicate an endurance limit (3×10^7 cycles) of 40 percent of the modulus of rupture of the joint.[3] It is recommended that the allowable horizontal shear stress be reduced 15 percent for cyclic loading.

Curved Laminated Members.   Bending of lamina to produce curved members induces stresses along the grain which reduce the strength of the laminated member. The reduction is given by

$$F' = F \left[1 - 2000 \left(\frac{t}{R} \right)^2 \right]$$

(2)

where F = allowable stress for straight member
$\quad\quad\; F'$ = allowable stress for curved member
$\quad\quad\;\; t$ = thickness of lamination
$\quad\quad\; R$ = radius of curvature of member

Recommended limits for t/R in laminated members are $\frac{1}{100}$ for hardwoods and southern pine and $\frac{1}{125}$ for other softwoods.[2]

Radial stresses are induced by bending of a beam which is curved in the plane of the loads. Because of the relatively low strength of wood perpendicular to the grain, these must be checked. The maximum radial stress f_r, is psi, for prismatic rectangular sections is

$$f_r = \frac{3M}{2Rbh}$$

(3)

where M = bending moment, in.-lb
$\quad\quad\; R$ = radius of curvature, in.
$\quad\quad\;\, b$ = width of cross section, in.
$\quad\quad\;\, h$ = depth of cross section, in.

For rectangular beams of varying depth and with the greatest depth at midspan the maximum radial stress is[15]

$$f_r = K_r \frac{6M}{bh^2}$$

(4a)

NOTE: Footnotes *b*, *g*, and *i* to Table 3 apply also to machine-stress-rated lumber.

[a]Recommended allowable unit stresses for machine-stress-rated lumber are determined by nondestructive pretesting of individual pieces to establish the allowable unit working stress for each piece.

[b]Allowable unit stresses for F_v (dry) and F_c (dry) are:

	Douglas fir-larch (WWPA/ WCLIB)	Douglas fir-S (WWPA)	Hem-Fir (WWPA/ WCLIB)	Western hemlock (WWPA/ WCLIB)	Pine[d] (WWPA)	Englemann spruce (WWPA)	Cedars[e] (WWPA/ WCLIB)	Southern pine (SPIB)
F_v	95	90	75	90	70	70	75	90 95[f]
$F_{c\perp}$	385	335	245	280	190	195	265	405

[c]Tabulated values F_b are applicable to lumber loaded on edge. When loaded flatwise, they may be multiplied by the following factors:

Nominal width, in.	3	4	6	8	10	12	14
Factor	1.06	1.10	1.15	1.19	1.22	1.25	1.28

[d]Pine includes Idaho white, lodgepole, ponderosa, or sugar pine.
[e]Cedar includes incense or western red cedar.
[f]Southern pine kiln-dried.

TABLE 5 Allowable Unit Stresses for Structural Glued-Laminated Softwood Timber for Normal Loading Duration (Dry Conditions of Use)

Combination symbol	Number of laminations	Extreme fiber in bending $F_b{}^a$		Tension parallel to grain F_t	Compression parallel to grain F_c	Compression perpendicular to grain $F_{c\perp}$		Horizontal shear F_v	Modulus of elasticity
		Load∥to wide face of laminations[b]	Load⊥to wide face of laminations			Tension face	Compression face		
Douglas fir and larch									
16F	4 or more	See note d	1600	1600	1500	385	385	165	1,600,000
18F	4 or more		1800	1600	1500	385	385	165	1,700,000
20F	4 or more		2000	1600	1500	385^f	385	165	1,700,000
22F^c	4 or more		2200	1600^e	1500^e	450^g	385^g	165	1,800,000
24F^c	4 or more		2400	1600^e	1500^e	450	385	165	1,800,000
26F^c	4 or more		2600	1600^e	1500^e	450	410	165	1,800,000
1^h	4 or more	900	1200	1200	1500	385	385	145^i	1,600,000
2^h	4 or more	1500	1800	1800	1800	385	385	145^i	1,800,000
3^h	4 or more	1900	2200	2200	2100	450	450	145^i	1,900,000
4^h	4 or more	2100	2400	2400	2000	410	410	145^i	2,000,000
5^h	4 or more	2300	2600	2600	2200	450	450	145^i	2,100,000
Southern pine									
18F-1^c	4 or more	See note d	1800	1600^e	1500^e	385	385	200	1,600,000
18F-2^c	12 or more		1800	1600^e	1500^e	385	385	200	1,600,000
20F-1^c	10 or more	See note d	2000	1600^e	1500^e	385	385	200	1,700,000
20F-2^c	8 or more		2000	1600^e	1500^e	450	450	200	1,700,000
20F-3^c	7 or more		2000	1600^e	1500^e	450	450	200	1,700,000
22F-1^c	4 or more	See note d	2200	1600^e	1500^e	450	450	200	1,700,000
22F-2^c	12 or more		2200	1600^e	1500^e	385	385	200	1,700,000
22F-3^c	14 or more		2200	1600^e	1500^e	385	385	200	1,700,000

Combination	Laminations								
24F-1[c]	10 or more	See note d	2400	1600[e]	1500[e]	385	385	200	1,800,000
24F-2[c]	4 or more		2400	1600[e]	1500[e]	450	450	200	1,800,000
24F-3[c]	14 or more		2400	1600[e]	1500[e]	450	450	200	1,800,000
26F-1[c]	11 or more	See note d	2600	1600[e]	1500[e]	385	385	200	1,800,000
26F-2[c]	16 or more		2600	1600[e]	1500[e]	450	450	200	1,800,000
26F-3[c]	11 or more		2600	1600[e]	1500[e]	450	450	200	1,800,000
1[h]	4 or more		900	1600	1400	385	385	165[i]	1,500,000
2[h]	4 or more		1550	2200	1900	385	385	165[i]	1,700,000
3[h]	4 or more		1800	2600	2200	450	450	165[i]	1,800,000
4[h]	4 or more		1900	2400	2100	385	385	165[i]	1,900,000
5[h]	4 or more		2200	2600	2200	450	450	165[i]	2,000,000

SOURCE: National Forest Products Association.

NOTE: The 26F combination may not be readily available, and the designer should check on availability. The 22F and 24F combinations are generally available from all laminators.

[a] The tabulated bending stresses are applicable to members 12 in. or less in depth. For members greater than 12 in. in depth, the requirements of Art. 11 apply.

[b] The allowable stress F_b is applicable to three or more laminations when load is parallel to grain.

[c] For members stressed principally in bending.

[d] For more detail see Standard Specifications for Structural Glued Laminated Timber of Douglas Fir, Western Larch, Southern Pine and California Redwood, AITC 117-71 by American Institute of Timber Construction.

[e] Allowable unit stress may be increased when slope of grain is more restrictive than basic requirement. (See reference in Note d.)

[f] When dense lumber is used in outer laminations in members 4 to 12 laminations deep, $F_{c\perp}$ in outer tension and compression faces is 450 psi for dry conditions of use and 305 psi for wet conditions of use. When close-grain lumber is used in members 4 to 8 laminations deep, $F_{c\perp}$ in tension and compression faces is 410 psi for dry conditions of use and 275 psi for wet conditions of use.

[g] When close-grain lumber is used in outer laminations in members 4 to 10 laminations deep, $F_{c\perp}$ in both tension and compression faces is 410 psi for dry conditions of use and 275 psi for wet conditions of use.

[h] For members stressed principally in axial tension or axial compression or loaded in bending with the direction of the load parallel to the wide face of the laminations.

[i] Allowable horizontal shear value is based on three or more laminations with the load applied parallel to the wide faces of the laminations. When loaded perpendicular to the wide face, the horizontal shear allowed for the bending combinations of the species may be used.

where $K_r = A + B\dfrac{h}{r} + C\left(\dfrac{h}{r}\right)^2$ $\qquad\qquad\qquad\qquad$ (4b)

$\qquad R$ = mean radius at midspan

Values of the coefficients in Eq. (4b) in terms of the slope β of the top of the beam with the horizontal are given in Table 6. Allowable values of f_r in Eq. (4a) depend on the direction of bending, as follows:

For bending which decreases R:

$$f_r = F_{c\perp}$$

For bending which increases R:

$\qquad$ Douglas fir and larch:

$$f_r = \tfrac{1}{3}\,F_v \text{ for wind or earthquake}$$
$$f_r = 15 \text{ psi for other loads}$$

$\qquad$ Other species:

$$f_r = \tfrac{1}{3}\,F_v$$

7. Plywood Plywood is usually fabricated with the grains of adjacent layers at right angles, although layers are sometimes composed of two or more parallel-laminated veneers. This results in a more nearly isotropic material for normal stresses. Plywood demonstrates advantages over timber for biaxial-stress states, such as those associated with pure shear (shear walls and beam webs).

TABLE 6   Coefficients in Eq. (4b)

β	A	B	C
0.0	0.0	0.2500	0.0
2.5	0.0079	0.1747	0.1284
5.0	0.0174	0.1251	0.1939
7.5	0.0279	0.0937	0.2162
10.0	0.0391	0.0754	0.2119
15.0	0.0629	0.0619	0.1722
20.0	0.0893	0.0608	0.1393
25.0	0.1214	0.0605	0.1238
30.0	0.1649	0.0603	0.1115

Construction and industrial plywoods are manufactured from some 70 different species of wood. The American Plywood Association divides these into four strength groups. Two types are manufactured—*Exterior* (with glue that is 100 percent waterproof, and inner-ply grade restrictions) and *Interior* (with glue that is highly moisture-resistant). Exterior type should be specified for all exposed applications. Appearance grades are designated N (intended for natural finish), A, B, C, C(plugged), and D. The double symbol A-A, A-B, etc., describes the face and back, in that order.

Structural I and Structural II are grades of C-D sheathing made with exterior glue. Structural I is limited to Group 1 species, while Structural II may be Group 1, 2, or 3. These grades are intended for applications such as box beams, gusset plates, and stressed-skin panels.

C-D Interior grade is intended for subflooring, wall sheathing, roof decking, etc. It may be made in any species. Structural I, Structural II, and Exterior C-C are identified as to recommended spacing of supports for use as decking. Thus, the number 32/16 means that spacing of supports should not exceed 32 in. for roof decking and 16 in. for subflooring. If the second number is 0, the panel should not be used for subflooring.

Plyform grade, which is intended for concrete forms, is furnished in Class I or Class II, with Class I the stronger and stiffer due to species groups required for each class.

Table 7 gives allowable stress for the construction and industrial plywoods. Section properties are given in Tables 8 and 9. Tables 10 and 11 are guides to the use of these tables.

TABLE 7   Allowable Stresses for Plywood*

| Type of stress | Species group of face ply | Grade-stress level (Table 10) | | | | |
| | | S-1† | | S-2‡ | | S-3§ |
		Wet	Dry	Wet	Dry	Dry only
Extreme fiber in bending F_b	1	1430	2000	1190	1650	1650
	2,3	980	1400	820	1200	1200
Tension in plane of plies F_t (Face grain parallel or perpendicular to span, at 45° to face grain use $^1/_6$)	4	940	1330	780	1110	1110
Compression in plane of plies F_c	1	970	1640	900	1540	1540
	2	730	1200	680	1100	1100
(Parallel or	3	610	1060	580	990	990
perpendicular to face grain, at 45° to face grain use $^1/_3$)	4	610	1000	580	950	950
Shear in plane perpendicular to plies F_v	1	205	250	205	250	210
	2,3	160	185	160	185	160
(Parallel or perpendicular to face grain, at 45° to face grain use $2F_v$)	4	145	175	145	175	155
Shear, rolling, in the plane of plies F_s	Marine and Structural I	63	75	63	75	
(Parallel or perpendicular to face grain, at 45° to face grain use $1^1/_3\,F_s$)	Structural II and 2-4-1	49	56	49	56	55
	All other	44	53	44	53	48
Modulus of rigidity (shear modulus) G in plane of plies	1	70,000	90,000	70,000	90,000	82,000
	2	60,000	75,000	60,000	75,000	68,000
	3	50,000	60,000	50,000	60,000	55,000
	4	45,000	50,000	45,000	50,000	45,000
Bearing on face $F_{c\perp}$	1	210	340	210	340	340
(Perpendicular to plane	2,3	135	210	135	210	210
of plies)	4	105	160	105	160	160
Modulus of elasticity E in bending in plane of plies	1	1,500,000	1,800,000	1,500,000	1,800,000	1,800,000
	2	1,300,000	1,500,000	1,300,000	1,500,000	1,500,000
	3	1,100,000	1,200,000	1,100,000	1,200,000	1,200,000
(Face grain parallel or perpendicular to span)	4	900,000	1,000,000	900,000	1,000,000	1,000,000

SOURCE: American Plywood Association.

*Conforming to U.S. Product Standard PS-1-74 for Construction and Industrial plywood. Normal-load basis in psi.

†To qualify for stress level S-1, glue lines must be exterior and only veneer grades N, A, and C are allowed in either face or back.

‡For stress level S-2, glue lines must be exterior and veneer grade B, C-plugged, and D are allowed on the face or back.

§Stress level S-3 includes all panels with interior or intermediate glue lines.

TABLE 8 Effective Section Properties for Plywood*

Nominal thickness, in.	Approx weight, psf	Effective thickness for shear, in.	Stress applied parallel to face grain				Stress applied perpendicular to face grain			
			A, area, in.²/ft	I, moment of inertia, in.⁴/ft	KS eff. section modulus, in.³/ft	Ib/Q rolling shear constant, in.²/ft	A, area, in.²/ft	I, moment of inertia, in.⁴/ft	KS eff. section modulus, in.³/ft	Ib/Q rolling shear constant, in.²/ft
Unsanded panels										
⁵⁄₁₆-U	1.0	0.283	1.914	0.025	0.124	2.568	0.660	0.001	0.023	
³⁄₈-U	1.1	0.293	1.866	0.041	0.162	3.108	0.799	0.002	0.033	
¹⁄₂-U	1.5	0.316	2.500	0.086	0.247	4.189	1.076	0.005	0.057	2.585
⁵⁄₈-U	1.8	0.336	2.951	0.154	0.379	5.270	1.354	0.011	0.095	3.252
³⁄₄-U	2.2	0.467	3.403	0.243	0.501	6.823	1.632	0.036	0.232	3.717
⁷⁄₈-U	2.6	0.757	4.109	0.344	0.681	7.174	2.925	0.162	0.542	5.097
1-U	3.0	0.859	3.916	0.493	0.859	9.244	3.611	0.210	0.660	6.997
1⅛-U	3.3	0.877	4.621	0.676	1.047	10.008	3.464	0.307	0.821	8.483
Sanded panels										
¼-S	0.8	0.304	1.680	0.013	0.092	2.175	0.681	0.001	0.020	
⅜-S	1.1	0.313	1.680	0.038	0.176	3.389	1.181	0.004	0.056	
½-S	1.5	0.450	1.947	0.077	0.266	4.834	1.281	0.018	0.150	3.099
⅝-S	1.8	0.472	2.280	0.129	0.356	6.293	1.627	0.045	0.234	3.922
¾-S	2.2	0.589	2.884	0.197	0.452	7.881	2.104	0.093	0.387	4.842
⅞-S	2.6	0.608	2.942	0.278	0.547	8.225	3.199	0.157	0.542	5.698
1-S	3.0	0.846	3.776	0.423	0.730	8.882	3.537	0.253	0.744	7.644
1⅛-S	3.3	0.865	3.854	0.548	0.840	9.883	3.673	0.360	0.918	9.032
Touch-sanded panels										
½-T	1.5	0.346	2.698	0.083	0.271	4.252	1.159	0.006	0.061	2.746
¹⁹⁄₃₂-T	1.7	0.491	2.618	0.123	0.337	5.403	1.610	0.019	0.150	3.220
⅝-T	1.8	0.497	2.728	0.141	0.364	5.719	1.715	0.023	0.170	3.419
²³⁄₃₂-T	2.1	0.503	3.181	0.196	0.447	6.600	2.014	0.035	0.226	3.659
¾-T	2.2	0.509	3.297	0.220	0.477	6.917	2.125	0.041	0.251	3.847
(2-4-1)1⅛-T	3.3	0.855	4.592	0.653	0.995	9.933	4.120	0.283	0.763	7.452

SOURCE: American Plywood Association.

*Face plies of different species group from inner plies (includes all product standard grades except those noted in Table 9).

TABLE 9 Effective Section Properties for Plywood*

Nominal thickness, in.	Approx weight, psf	Effective thickness for shear, in.	Stress applied parallel to face grain				Stress applied perpendicular to face grain			
			A, area, in.²/ft	I, moment of inertia, in.⁴/ft	KS eff. section modulus, in.³/ft	Ib/Q rolling shear constant, in.²/ft	A, area, in.²/ft	I, moment of inertia, in.⁴/ft	KS eff. section modulus, in.³/ft	Ib/Q, rolling shear constant, in.²/ft
						Unsanded Panels				
5/16-U	1.0	0.356	2.375	0.025	0.144	2.567	1.188	0.002	0.029	
3/8-U	1.1	0.371	2.226	0.041	0.195	3.107	1.438	0.003	0.043	
1/2-U	1.5	0.403	2.906	0.091	0.318	4.188	1.938	0.007	0.077	2.574
5/8-U	1.8	0.434	3.464	0.155	0.433	5.268	2.438	0.015	0.122	3.238
3/4-U	2.2	0.606	3.672	0.247	0.573	6.817	2.938	0.059	0.334	3.697
7/8-U	2.6	0.776	4.388	0.346	0.690	6.948	3.510	0.192	0.584	5.086
1-U	3.0	1.088	5.200	0.529	0.922	8.512	6.500	0.366	0.970	6.986
1 1/8-U	3.3	1.119	6.654	0.751	1.164	9.061	5.542	0.503	1.131	8.675
						Sanded panels				
1/4-S	0.8	0.342	1.680	0.013	0.092	2.172	1.226	0.001	0.027	
3/8-S	1.1	0.373	1.680	0.038	0.177	3.382	2.126	0.007	0.078	
1/2-S	1.5	0.545	1.947	0.078	0.271	4.816	2.305	0.030	0.217	3.076
5/8-S	1.8	0.576	2.280	0.131	0.361	6.261	2.929	0.077	0.343	3.887
3/4-S	2.2	0.748	3.848	0.202	0.464	7.926	3.787	0.162	0.570	4.812
7/8-S	2.6	0.778	3.952	0.288	0.569	7.539	5.759	0.275	0.798	5.671
1-S	3.0	1.091	5.215	0.479	0.827	7.978	6.367	0.445	1.098	7.639
1 1/8-S	3.3	1.121	5.593	0.623	0.955	8.840	6.611	0.634	1.356	9.031
						Touch-sanded panels				
1/2-T	1.5	0.403	2.698	0.084	0.282	4.246	2.086	0.008	0.082	2.720
19/32-T	1.7	0.567	3.127	0.124	0.349	5.390	2.899	0.030	0.212	3.183
5/8-T	1.8	0.575	3.267	0.144	0.378	5.704	3.086	0.037	0.242	3.383
23/32-T	2.1	0.598	3.337	0.201	0.469	6.582	3.625	0.057	0.322	3.596
3/4-T	2.2	0.606	3.435	0.226	0.503	6.900	3.825	0.067	0.359	3.786

SOURCE: American Plywood Association.
*All plies from same species group (includes Structural I and Marine).

TABLE 10 Guide to Use of Tables 7, 8, and 9

Plywood type and grade	Description and use	Veneer grade			Common thicknesses, in.	Grade stress level (Table 7)	Species group	Section property table
		Face	Back	Inner				
Interior-type plywood								
C-D INT-APA	Unsanded sheathing grade for wall, roof, subflooring, and industrial applications such as pallets and for engineering design, with proper stresses. Also available with intermediate and exterior glue.* For permanent exposure to weather or moisture only exterior-type plywood is suitable	C	D	D	5/16, 3/8, 1/2, 5/8, 3/4	S-3*	See Table 11	Table 8 (unsanded)
Structural I, C-D INT-APA or Structural II C-D INT-APA†	Plywood grades to use where strength properties are of maximum importance, such as plywood-lumber components. Made with exterior glue only. Structural I is made from all Group 1 woods. Structural II allows Group 3 woods	C	D	D	5/16, 3/8, 1/2, 5/8, 3/4	S-2	Structural I, use Group 1; Structural II, use Group 3	Table 9 (unsanded)
Underlayment INT-APA	For underlayment or combination subfloor-underlayment under resilient floor coverings. Available with exterior glue. Touch-sanded. Available with tongue and groove	C plugged	D	C and D	1/2, 19/32, 5/8, 23/32, 3/4	S-3*	As specified	Table 8 (touch-sanded)
C-D plugged INT-APA	For built-ins, wall and ceiling tile backing, *not* for underlayment. Available with exterior glue. Touch-sanded	C plugged	D	D	1/2, 19/32, 5/8, 23/32, 3/4	S-3*	As specified	Table 8 (touch-sanded)
Structural I or II† underlayment or C-D plugged	For higher strength requirements for underlayment or built-ins. Structural I constructed from all Group 1 woods. Made with exterior glue only	C plugged	D	C and D	1/2, 19/32, 5/8, 23/32, 3/4	S-2	Structural I, use Group 1; Structural II, use Group 3	Table 9 (touch-sanded)
2.4.1 INT-APA	Combination subfloor-underlayment. Quality floor base. Available with exterior glue, most often touch-sanded. Available with tongue and groove	C plugged	D	C and D	1 1/8	S-3*	Group 1	Table 8
Appearance grades	Generally applied where a high-quality surface is required. Includes N-N, N-A, N-B, N-D, A-A, A-B, A-D, B-B, and B-D INT-APA grades	B or better	D or better	D	1/4, 3/8, 1/2, 5/8, 3/4	S-3*	As specified	Table 8 (sanded)

Grade	Description				Thickness	Stress level	Species group	Table
C-C EXT-APA	Unsanded sheathing grade with waterproof glue bond for wall, roof, subfloor, and industrial applications such as pallet bins	C	C	C	5/16, 3/8, 1/2, 5/8, 3/4	S-1	See Table 11	Table 8 (unsanded)
Structural I C-C EXT-APA or Structural II C-C EXT-APA†	"Structural" is a modifier for this unsanded sheathing grade. For engineering applications in construction and industry where full exterior-type panels are required. Structural I is made from Group 1 woods only	C	C	C	5/16, 3/8, 1/2, 5/8, 3/4	S-1	Structural I, use Group 1 Structural II, use Group 3	Table 9 (unsanded)
Underlayment EXT-APA and C-C plugged EXT-APA	Underlayment for combination subfloor-underlayment or two-layer floor under resilient floor coverings where severe moisture conditions may exist. Also for controlled atmosphere rooms and many industrial applications. Touch-sanded. Available with tongue and groove	C plugged	C	C	1/2, 19/32, 23/32, 5/8, 3/4	S-2	As specified	Table 8 (touch-sanded)
Structural I or II† underlayment EXT-APA or C-C plugged EXT-APA	For higher-strength underlayment where severe moisture conditions may exist. All Group 1 construction in Structural I Structural II allows Group 3 woods	C plugged	C	C	1/2, 19/32, 5/8, 23/32, 3/4	S-2	Structural I, use Group 1 Structural II, use Group 3	Table 9 (touch-sanded)
B-B Plyform Class I or II†	Concrete-form grade with high reuse factor. Sanded both sides, mill-oiled unless otherwise specified. Available in HDO. For refined design information on this special-use panel see APA publication "Plywood for Concrete Forming" (form V345). Design using values from this specification will result in a conservative design	B	B	C	5/8, 3/4	S-2	Class I, use Group 1 Class II, use Group 3	Table 8 (sanded)
Marine EXT-APA	Superior Exterior type plywood made only with Douglas fir or western larch. Special solid-core construction. Available with MDO or HDO face. Ideal for boat-hull construction	A or B	A or B	B	1/4, 3/8, 1/2, 5/8, 3/4	A face and back, use S-1, B face or back, use S-2	Group 1	Table 9 (sanded)
Appearance grades	Generally applied where a high-quality surface is required. Includes AA, A-B, A-C, B-B, B-C, HDO, and MDO EXT-APA. Appearance grades may be modified to Structural I. For such designation use Group 1 stresses and Table 9 (sanded) section properties	B or better	C or better	C	1/4, 3/8, 1/2, 5/8, 3/4	A or C face and back use S-1, B face or back use S-2	As specified	Table 8 (sanded)

*When exterior glue is specified, i.e., "interior with exterior glue," stress level 2 (S-2) should be used.
†Check local suppliers for availability of Structural II and Plyform Class II grades.
SOURCE: American Plywood Association.

Example Determine the effective section properties and allowable stresses for a plywood to be used for subflooring on joists spaced 16 in.
SOLUTION. Table 10 shows that C-D INT-APA should be used, at stress level S-3 and with effective section properties from Table 8.
Table 11 shows that 32/16 index is available in ½-in. thickness in species group 1 and ⅝-in. thickness in species group 3.
Table 7 gives, for stress level S-3 and species group 1, F_b = 1650 psi, F_s = 48 psi, and E = 1,800,000 psi.
Table 8 gives, for ½-in. thickness, I = 0.086 in.^{4}, KS = 0.247 in.^{3}/ft, and Ib/Q = 4.189 in.^{2}/ft.

TABLE 11   Species Groups

Thickness, in.	Identification index*						
	12/0	16/0	20/0	24/0	32/16	42/20	48/24
⁵⁄₁₆	4	3	1				
⅜		4	3	1			
½				4	1		
⅝					3	1	
¾						3	1
⅞						4	3

SOURCE: American Plywood Association.
*30/12-⅝ and 36/16-¾ panels also sometimes available; use Group 4 stresses.

Plate, Skin, and Diaphragm Construction. When properly designed as a stiffened skin, and with particular attention to connections, plywood sheets may be used for shear diaphragms, straight or curved panels with various sandwich cores or stiffeners, glued beam webs, folded plates, and geodesic domes. Design procedures for these various structural forms are available.[12]

FASTENERS

Member sizes in wood structures, such as trusses, are often determined by the connections rather than by the structural properties of the members. The relatively low ratios of allowable shear and bearing stresses to allowable stress parallel to the grain lead to fastener problems. The usual fasteners for heavy timber are bolts, split rings, or shear plates.
  8. Bolts The shear-load capacity of a bolt depends upon the ratio of the length l of bolt in the main member and its diameter d, and on the wood. Allowable loads are given in Table 12 for three-member joints.
Load Parallel to the Grain. The tabulated values are for side members whose thickness is at least $l/2$. For thinner side members, l is taken to be twice the thickness of the thinner one.
  For two-member joints (bolt in single shear) the allowable load is half the allowable load for a value of l equal to twice the thickness of the thinner member. For joints with more than three members where the members are of equal thickness, the allowable load per shear plane is one-half the tabulated load for a value of l equal to the thickness of one member.
  Values parallel to the grain may be increased 25 percent if metal side plates are used.
Load Perpendicular to the Grain. For load applied perpendicular to the grain of the main member through either wood or steel side plates, the value of l is the thickness of the main member.
Load Inclined to Grain. The allowable load acting at the angle θ with the grain of the main member is given by Eq. (1), using values of P and Q from Table 12.
  In a two-member joint, loads may act at other than 90° to the bolt axis, subjecting the bolt to combined shear and axial force. The allowable shear component is one-half the tabulated load for a value of l twice the thickness of the thinner member.
Moisture Content. For intermittently wet and continuously wet conditions the allowable loads are 75 and 66 percent, respectively, of the tabulated values.
  If the lumber is at or above the fiber-saturation point and is expected to season in place, the allowable loads of Table 12 may be used for a joint with wood side members having a

single bolt loaded either parallel or perpendicular to the grain, a single row of bolts loaded parallel to the grain, or multiple rows loaded parallel to the grain with separate side members for each row. Allowable loads for other configurations are 40 percent of those in Table 12.

Duration of Load.   Allowable loads in Table 12 should be adjusted for the expected duration of load according to Table 2.

Spacing.   The minimum bolt spacing in a row parallel to the load axis should be four bolt diameters for both parallel-to-grain and perpendicular-to-grain loading. If the bolt

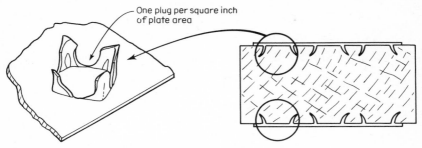

One plug per square inch of plate area

Fig. 2   Truss-plate connector. *(Woodclaw Inc.)*

bearing is less than the allowable for the side members, spacing for the perpendicular-to-grain load may be proportionately reduced.

Spacing between rows of bolts for parallel-to-grain loading should be not less than $1\frac{1}{2}$ bolt diameters.

For perpendicular-to-grain loading with $l/d \lesssim 2$ the spacing should be $2\frac{1}{2}d$ and for $l/d \gtrsim 6$ at least $5d$. For $2 \lesssim l/d \lesssim 6$ the spacing may be linearly interpolated.

End and Edge Distances.   A bolt should be at least $7d$ from the end of a softwood tension member and $5d$ for hardwood. For compression members a distance of $4d$ is required.

The edge distance for bolts with $l/d \lesssim 6$ should be $1\frac{1}{2}d$ for parallel-to-grain loading. For $l/d \gtrsim 6$ the distance should be one-half the spacing between rows. For perpendicular-to-grain loading the loaded edge distance should be at least $4d$.

9. Split Rings and Shear Plates   Split rings and shear plates vary in diameter from $2\frac{1}{2}$ to 4 in. and are rated at higher loads than bolts. Loads and guides to design are given in Ref. 16.

10. Truss Plates   One type of gusset plate for wood trusses is made of sheet steel of 20, 18, or 16 gage. Plates may be nailed, or more often, they have teeth punched out of and perpendicular to the sheet. The teeth are pressed into the wood either by roll pressing or

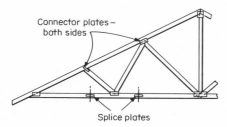

Connector plates — both sides

Splice plates

Fig. 3   Typical truss with truss-plate connectors. *(Woodclaw Inc.)*

by a hydraulic press. The truss plates shown in Fig. 2 have four teeth per plug spaced at 90° in the plug circle. Thus the angle of the grain is not relevant to the load capacity. The rated capacity of this plate for Douglas fir and southern pine is 190 psi for 20-gage and 140 for 16-gage thicknesses.

A typical truss using these connectors is shown in Fig. 3. Trusses fabricated with 2 × 4 to 2 × 8 members and spaced 2 to 4 ft can carry loads of 25 to 55 psf on spans from 20 to 80 ft.

TABLE 12   Allowable Loads in Pounds on One Bolt in Double Shear (Three-Member Joint) (normal loading conditions)

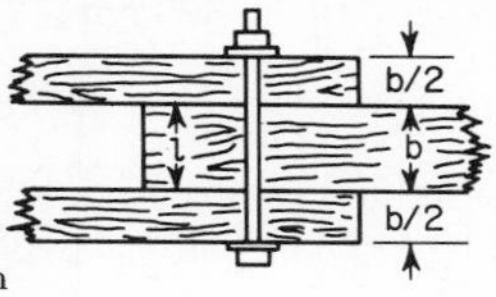

Length of bolt in main member l	Diameter of bolt d	l/d	Projected area of bolt $A = l \times d$	Douglas fir–larch (dense) Southern pine (dense)		Douglas fir–larch Southern pine (med. grain)		Southern pine (open grain)	
				Parallel to grain P	Perpendicular to grain Q	Parallel to grain P	Perpendicular to grain Q	Parallel to grain P	Perpendicular to grain Q
1½	½	3.00	0.750	1120	500	960	430	820	370
	⅝	2.40	0.938	1420	570	1210	490	1030	420
	¾	2.00	1.125	1700	630	1460	540	1240	470
	⅞	1.71	1.313	1990	700	1700	600	1440	520
	1	1.50	1.500	2270	760	1940	650	1650	570
2	½	4.00	1.000	1400	670	1200	570	1020	500
	⅝	3.20	1.250	1860	760	1590	650	1350	560
	¾	2.67	1.500	2260	840	1930	720	1640	630
	⅞	2.29	1.750	2640	930	2260	790	1920	690
	1	2.00	2.000	3030	1010	2590	870	2200	750
2½	½	5.00	1.250	1510	840	1290	720	1100	620
	⅝	4.00	1.563	2190	950	1870	810	1590	710
	¾	3.33	1.875	2780	1060	2370	900	2020	790
	⅞	2.86	2.188	3290	1160	2810	990	2390	860
	1	2.50	2.500	3770	1270	3230	1080	2740	940
3	½	6.00	1.500	1530	1010	1310	860	1110	750
	⅝	4.80	1.875	2350	1140	2010	970	1710	850
	¾	4.00	2.250	3150	1270	2690	1080	2290	940
	⅞	3.43	2.625	3860	1390	3300	1190	2810	1040
	1	3.00	3.000	4500	1520	3840	1300	3270	1130
3½	½	7.00	1.750	1530	1140	1310	980	1110	870
	⅝	5.60	2.188	2380	1330	2030	1130	1730	990
	¾	4.67	2.625	3360	1480	2870	1260	2440	1100
	⅞	4.00	3.063	4290	1630	3670	1390	3120	1210
	1	3.50	3.500	5120	1770	4380	1520	3720	1320
4	½	8.00	2.000	1530	1180	1310	1010	1110	960
	⅝	6.40	2.500	2380	1510	2040	1290	1730	1130
	¾	5.33	3.000	3420	1690	2920	1440	2490	1260
	⅞	4.57	3.500	4560	1860	3900	1590	3310	1380
	1	4.00	4.000	5600	2030	4790	1730	4070	1510
4½	⅝	7.20	2.813	2380	1640	2040	1400	1730	1270
	¾	6.00	3.375	3430	1900	2940	1620	2500	1410
	⅞	5.14	3.938	4640	2090	3970	1790	3370	1560
	1	4.50	4.500	5910	2280	5060	1950	4300	1700
	1¼	3.60	5.625	8170	2670	6990	2280	5940	1990
5½	⅝	8.80	3.438	2380	1650	2040	1410	1730	1380
	¾	7.33	4.125	3430	2200	2930	1880	2490	1720
	⅞	6.29	4.813	4680	2550	4000	2180	3400	1900
	1	5.50	5.500	6080	2790	5200	2380	4420	2080
	1¼	4.40	6.875	9160	3260	7830	2790	6660	2430
7½	⅝	12.00	4.688	2380	1480	2040	1260	1730	1290
	¾	10.00	5.625	3430	2130	2930	1820	2490	1800
	⅞	8.57	6.563	4670	2840	3990	2430	3390	2360
	1	7.50	7.500	6100	3550	5210	3030	4430	2800
	1¼	6.00	9.375	9540	4450	8160	3800	6930	3310

TABLE 12   Allowable Loads in Pounds on One Bolt in Double Shear (Three-Member Joint) (normal loading conditions) (*Continued*)

Length of bolt in main member l	Diameter of bolt d	l/d	Projected area of bolt $A = l \times d$	Douglas fir–larch (dense) Southern pine (dense)		Douglas fir–larch Southern pine (med. grain)		Southern pine (open grain)	
				Parallel to grain P	Perpendicular to grain Q	Parallel to grain P	Perpendicular to grain Q	Parallel to grain P	Perpendicular to grain Q
9½	¾	12.67	7.125	3430	1920	2930	1640	2490	1700
	⅞	10.86	8.313	4680	2660	4000	2270	3400	2260
	1	9.50	9.500	6100	3460	5210	2960	4430	2900
	1¼	7.60	11.875	9530	5210	8150	4450	6920	4140
	1½	6.33	14.250	13740	6480	11750	5530	9990	4830
11½	⅞	13.14	10.062	4680	1980	4000	2060	3400	2170
	1	11.50	11.500	6090	3240	5210	2770	4430	2780
	1¼	9.20	14.375	9530	5110	8150	4360	6930	4270
	1½	7.67	17.250	13730	7200	11750	6150	9980	5740
13½	1	13.50	13.500	6090	2410	5210	2530	4430	2680
	1¼	10.80	16.875	9530	4860	8150	4160	6920	4130
	1½	9.00	20.250	13730	7070	11740	6040	9980	5920

SOURCE: National Forest Products Association.

BEAMS

Glued-laminated and timber beams are used in schools, churches, and one- to three-story industrial buildings. Versatility in shape and size and greater load capacity allows the glued-laminated beam to be used for main structural members with longer spans, while the timber beam is restricted to use as purlins or in small-bay buildings. Their excellent fire resistance is recognized by building codes and insurance underwriters, and the use of glued-laminated beam construction in buildings is widely accepted. Heavy timber beams are impractical because of lack of supply and cost.

For simple spans in roof systems the glued-laminated beam may span to 90 ft and the tapered, pitched, and curved beam to 100 ft. If cantilever or continuous spans are used for roof framing, the column spacing may extend to 100 ft for glued-laminated beams. Beam systems for floors are usually limited to 40 ft whether simple or continuous.

11. Flexure Flexural stresses are calculated from the formula

$$f_b = \frac{M}{SC_d}$$

where S = section modulus
  C_d = depth factor
The depth factor C_d accounts for the increased probability of lower mean strengths with increased volume. For glued-laminated or timber beams 12 in. or more in depth

$$C_d = 0.81 \frac{h^2 + 143}{h^2 + 88} \tag{5a}$$

where h = depth of beam, in.
For I- and box beams

$$C_d = 0.81 \left[1 + \left(\frac{h^2 + 143}{h^2 + 88} - 1 \right) C \right] \tag{5b}$$

where $C = p^2(6 - 8p + 3p^2)(1 - q) + q$
  p = depth of compression flange divided by h
  q = sum of web thicknesses divided by b

An additional form factor accounts for the difference in stress gradients and volume of material at maximum stress in shapes other than rectangular. Thus, beams of circular section are allowed an 18 percent increase in working stress. Square beams with the bending moment in the plane of a diagonal are allowed a 41 percent increase. Thus, beams of circular section have the same strength as square beams of the same area, and square beams have the same strength with the diagonal in a vertical plane as in the normal position.

Table 13 gives properties of sawn timber sections.

TABLE 13   Sectional Properties of Standard Dressed (S4S) Lumber*

Nominal size	Dressed size	A, in.^{2}	I, in.^{4}	S, in.^{3}
1 × 3	¾ × 2½	1.88	0.98	0.78
1 × 4	¾ × 3½	2.63	2.68	1.53
1 × 6	¾ × 5½	4.13	10.40	3.78
1 × 8	¾ × 7¼	5.44	23.82	6.57
1 × 10	¾ × 9¼	6.94	49.47	10.70
1 × 12	¾ × 11¼	8.44	88.99	15.82
2 × 3	1½ × 2½	3.75	1.95	1.56
2 × 4	1½ × 3½	5.25	5.36	3.06
2 × 6	1½ × 5½	8.25	20.80	7.56
2 × 8	1½ × 7¼	10.88	47.64	13.14
2 × 10	1½ × 9¼	13.88	98.93	21.39
2 × 12	1½ × 11¼	16.88	178.0	31.64
2 × 14	1½ × 13¼	19.88	290.8	43.89
3 × 1	2½ × ¾	1.88	0.088	0.23
3 × 2	2½ × 1½	3.75	0.70	0.94
3 × 4	2½ × 3½	8.75	8.93	5.10
3 × 6	2½ × 5½	13.75	34.66	12.60
3 × 8	2½ × 7¼	18.13	79.39	21.90
3 × 10	2½ × 9¼	23.13	164.9	35.65
3 × 12	2½ × 11¼	28.13	296.6	52.73
3 × 14	2½ × 13¼	33.13	484.6	73.15
3 × 16	2½ × 15¼	38.13	738.9	96.90
4 × 1	3½ × ¾	2.63	0.12	0.33
4 × 2	3½ × 1½	5.25	0.98	1.31
4 × 3	3½ × 2½	8.75	4.56	3.65
4 × 4	3½ × 3½	12.25	12.51	7.15
4 × 6	3½ × 5½	19.25	48.53	17.65
4 × 8	3½ × 7¼	25.38	111.1	30.66
4 × 10	3½ × 9¼	32.38	230.8	49.91
4 × 12	3½ × 11¼	39.38	415.3	73.83
4 × 14	3½ × 13¼	46.38	678.5	102.4
4 × 16	3½ × 15¼	53.38	1034.4	135.7

*From Ref. 2.

12. Shear   Shear stress in timber or glued-laminated beams is given by

$$f_v = \frac{VQ}{Ib}$$

For rectangular sections,

$$f_v = \frac{3V}{2A} \tag{6}$$

Because of checks in sawn timber beams, shear stresses are larger than those given by the above equation. However, the allowable shear stress F_v is based on the presence of checks.

If critical shear stresses result from loads adjacent to a support, some increase of the shear strength along the grain may be expected by the development of local radial stress fields; i.e., the load is transferred by radial compressive stresses in addition to shear stresses. The following approximate method for calculating the vertical shear V for rectangular beams accounts for this effect.

1. Neglect all loads adjacent to supports within a distance equal to the beam depth.

2. For moving concentrated loads place the largest at three times the beam depth from the support or at the quarter point, whichever is closer. If the resulting shearing stress is greater than the allowable in Table 14 use the following more accurate equation to determine the shear V:

$$V = \sum \frac{10P_i(L - x_i)(x_i/h)^2}{9L[2 + (x_i/h)^2]}$$

where x_i = distance of P_i from support

TABLE 14   Allowable Shearing Stress, psi, for Moving Concentrated Loads*

	19% m.c.	15% m.c.	Unseasoned
Douglas fir	185	195	170
Southern pine, medium grain	175	185	165
Southern pine, open grain	150	160	140

*From Ref. 2.

Notched Beams.   The allowable shear for a rectangular beam notched on the lower face at the support is given by

$$V = \frac{2F_v bd}{3} \frac{d}{h}$$

where d = depth of beam above notch

13. Bearing   For a bearing of any length at the end of a beam, or for bearings 6 in. or more in length at any location other than the end, the allowable bearing stress is $F_{c\perp}$. For a bearing length L_B less than 6 in., and at least 3 in. from the end of the beam, the allowable bearing stress is

$$\frac{F_{c\perp}(L_B + \frac{3}{8})}{L_B}$$

14. Deflections   The usual equations for deflections are valid using values of E given in Table 3. However, deflection under long-term loading may be approximately double that computed by the elastic theory (Art. 2).

Deflection of a flat roof may cause ponding of water during rainstorms, which can lead to collapse. As water accumulates deflection increases, thereby increasing the capacity of the basin. It can be shown that a flat roof panel, considered as a simple beam of span L, will deflect enough to contain the water that falls on it if[5]

$$\frac{\pi^4}{L^4} EI = w$$

where w = unit weight of water

If the left side of this equation exceeds w, the deck sheds water over the supports. For a factor of safety of 2, the result can be written

$$EI \geq 15L^4 \qquad (7)$$

where L = span, ft
  I = moment of inertia, in.^{4}/in.
  E = modulus of elasticity, psi

Equation (7) gives the required stiffness of the deck. It may be used for a beam (or truss) by using for I the moment of inertia of the beam divided by the beam spacing. Alternatively, Eq. (7) is equivalent to limiting the deflection in a simple span to 1 in. under a load of 10 psf.

To account for ponding, stresses and deflections should be multiplied by

$$\frac{1}{1 - 0.64(L/EI)^4}$$

to check if allowables are exceeded.

Camber of flat roofs should be equal to or greater than 2.5 times the dead-load deflection. Roof slopes of approximately 1:48 or greater assure runoff of water. Drain capacities must be adequate to prevent backup of water on the roof for the highest expected rate and duration of rainfall.

15. Lateral Stability Wood beams are usually rectangular in cross section and, if laterally unsupported, may buckle laterally if the depth is greater than the width. Such buckling will be prevented if the allowable stress F_b' is limited to the following:[6]

$$F_b' = F_b \qquad\qquad\qquad 0 < R < 10$$

$$F_b' = F_b \left[1 - \frac{1}{3}\left(\frac{R}{K}\right)^4 \right] \qquad 10 < R < K$$

$$F_b' = 0.4\,\frac{E}{R^2} \qquad\qquad K < R < 50$$

where F_b' = allowable bending stress laterally unsupported
$\quad\ F_b$ = allowable bending stress laterally supported
$\quad\ K = \sqrt{0.6E/F_b}$

$\quad\ R = \sqrt{L_e h/b^2}$

$\quad\ L_e$ = effective length for loads at gravity axis (Table 15)
In the derivation of these equations it is assumed that the ends of the beam are fixed

TABLE 15   Effective Length L_e

Type of beam	Load	L_e
Simple................	Concentrated at center	$1.37L$*
Simple................	Uniform	$1.63L$*
Simple................	Equal end moments	$1.84L$
Cantilever	Concentrated at end	$1.44L$*
Cantilever.............	Uniform	$0.90L$*

* For loads applied at top increase L_e by $3h$; for loads applied at bottom decrease L_e by h.

against torsional rotation. The value $R = 50$ is recommended as an upper limit on slenderness of laterally unsupported beams.

Beams usually have lateral support sufficient to allow the full bending stress F_b. In lieu of the above criteria the following are approximate rules for adequate lateral support for rectangular beams of ordinary proportions:[2]

1. If $h/b = 2$ lateral support is not required.
2. If $h/b = 3$ the ends should be held in position.
3. If $h/b = 4$ the member should be held in line, as in a well-bolted truss chord member.
4. If $h/b = 5$ one edge should be held in line.
5. If $h/b = 6$ use diagonal bridging at intervals not exceeding 8 ft.
6. If $h/b = 7$ both edges should be held in line.

16. Continuous Spans Continuous beams may be constant in cross section or variable, the latter resulting in economy of material beyond that of the constant cross section. Continuous-beam structures often are made determinate by using hinged splices in every other span to give a cantilevered system. This is based on the principle that, for fixed loads, any determinate system of beams may be made as efficient as a continuous system by proper placement of hinges. Figure 4 illustrates possible hinge locations for uniform loading. For beams of constant cross section the hinges are placed so that the maximum negative moment equals the maximum positive moment. Thus, $L_c = 0.172L$ for two equal spans L uniformly loaded; the maximum moment is $0.0858wL^2$ (Fig. 4a). For three equal spans uniformly loaded (Fig. 4b), $L_c = 0.146L$ and $M_{max} = wL^2/16$.

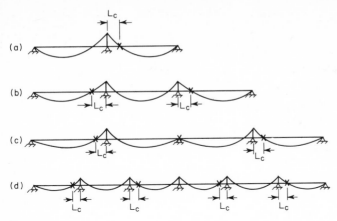

Fig. 4

Increasing the depth of continuous beams at the supports results in greater efficiency of material. Assuming a constant cross section for negative moment and one for positive moment, the ratio of the moment capacities for the most economical solution can be determined.

For a uniformly loaded beam of two equal spans, with resisting moments M_m and kM_m (Fig. 5), $k = 2.2$. For a rectangular beam of uniform width, the corresponding ratio of depth at support to that of the positive-moment section is 1.48. The half length of the midsection is $a = 0.134L$ and the distance to the point of zero moment $L_c = 0.281L$.

For three equal spans, the ratio of depth of the negative-moment sections to that of the positive-moment sections is 1.55. For the middle span, $a = 0.104L$ and $L_c = 0.220L$, while for the end spans $a = 0.134L$ and $L_c = 0.281L$. Hinges may be located in the center span or in the end spans.

A continuous-beam system of uniformly loaded equal spans will be economical of material if the moment capacity at the supports is 2.4 times that of the positive-moment

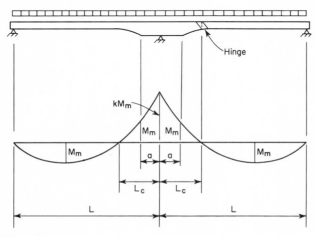

Fig. 5

sections. For this case, $a = 0.116L$ and $L_c = 0.228L$. The ratio of depths for rectangular sections of uniform width is 1.55.

If the cantilevers at the support are tapered, as in Fig. 5, they must be designed according to the procedure in Art. 17.

Deflections must be checked and, for roofs, ponding avoided (Art. 14). In cantilever systems the shedding of water by alternate spans onto adjacent spans should be prevented by designing the spans for nearly equal deflections for uniform loads.

17. Pitched and Tapered Beams—Simple Spans   Glued-laminated beams may be pitched to form roofs and vaulted ceilings. The pitched beam can be straight or tapered, the tapered beam being more efficient in terms of lumber (Fig. 6).

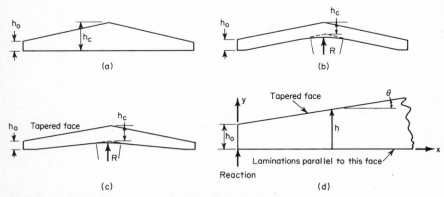

Fig. 6   (a) Tapered beam; (b) pitched beam; (c) pitched and tapered beam.

The design procedure for uniform loading is as follows:[7]
1. Determine the depth at the support by $h_0 = 3V/2bF_v$.
2. Determine the depth h_c of a beam of uniform depth for the same width b.
3. The center depth h_c is given by $h_c = (h_c^2 + h_0^2)/2h_0$.
4. The midspan deflection may be calculated with sufficient accuracy by

$$\Delta = \frac{216wL^4}{Ebh_c^3}(0.85n + 0.42) \qquad n = \frac{h_c}{h_0} \lessgtr 3$$

5. Check the radial stresses by Eq. (3) with $h = h_c$. The radius R should be 32 ft or greater for 1½-in. lamina.
6. Check the stresses by the interaction equation

$$\left(\frac{f_x}{F_b}\right)^2 + \left(\frac{f_y}{F_y}\right)^2 + \left(\frac{f_{xy}}{F_v}\right)^2 \lessgtr 1$$

where f_x = bending stress parallel to x axis
f_y = stress parallel to y axis
f_{xy} = shear stress
F_b = allowable bending stress
$F_y = F_v/3$ for f_y tension
$F_y = F_c$ for f_y compression
The bending stresses f_x are assumed to follow the elementary theory. In general, for any continuously tapered beam[14]

$$f_x = \frac{12M(y - h/2)}{bh^3}$$

from which, at the faces $y = 0$ and $y = h$,

$$f_x = \pm\frac{6M}{bh^2}$$

The shear stress is assumed to vary parabolically at the support. However, at other cross sections it may be maximum at the tapered face. The shear stress is given by

$$f_{xy} = \frac{6M}{bh^2}\frac{dh}{dx}\left[3\left(\frac{y}{h}\right)^2 - 2\frac{y}{h}\right] + \frac{6}{bh}\frac{dM}{dx}\left[\frac{y}{h} - \left(\frac{y}{h}\right)^2\right]$$

from which, at the tapered face $y = h$,

$$f_{xy} = \frac{6M}{bh^2}\frac{dh}{dx} = f_x\frac{dh}{dx}$$

The stress f_y is

$$f_y = \frac{6M}{bh^2}\left(\frac{dh}{dx}\right)^2\left(\frac{y}{h}\right)^2\left(4\frac{y}{h} - 3\right) + \frac{6M}{bh^2}\frac{d^2h}{dx^2}\frac{y}{h}\left(1 - \frac{y}{h}\right)$$
$$+ \frac{12}{bh}\frac{dh}{dx}\frac{dM}{dx}\left(\frac{y}{h}\right)^2\left(1 - \frac{y}{h}\right) + \frac{1}{b}\frac{d^2M}{dx^2}\left[1 - 3\left(\frac{y}{h}\right)^2 + 2\left(\frac{y}{h}\right)^3\right]$$

from which, at the tapered face $y = h$,

$$f_y = \frac{6M}{bh^2}\left(\frac{dh}{dx}\right)^2 = f_x\left(\frac{dh}{dx}\right)^2$$

For uniformly loaded beams with straight tapers the critical combination of stresses is at the section whose depth

$$h = 2h_0\frac{h_0 + L\tan\theta}{2h_0 + L\tan\theta}$$

For beams with straight tapers supporting a concentrated load at midspan, the critical combination of stresses is at the section whose depth $h = 2h_0$ if $h_c \geqslant 2h_0$ and at the section whose depth $h = h_c$ if $\frac{4}{3}h_0 < h_c < 2h_0$. If $h_c < \frac{4}{3}h_0$, various sections should be checked.

COLUMNS

18. Solid Columns Columns may be classified as short, intermediate, and long according to the ratio L/d, where L is the unsupported length and d the least dimension. The allowable compressive stress P/A for pin-ended rectangular columns is given by the following equations:

Short ($L/d \leqslant 10$): $\qquad\qquad\qquad\dfrac{P}{A} = F_c$ (Table 3)

Intermediate ($10 < L/d \leqslant K$): $\qquad\dfrac{P}{A} = F_c\left[1 - \dfrac{1}{3}\left(\dfrac{L/d}{K}\right)^4\right]$

Long ($K < L/d < 50$): $\qquad\qquad\dfrac{P}{A} = \dfrac{0.274E}{(L/d)^2}$

where $K = 0.64\sqrt{E/F_c}$

The National Design Specification does not classify columns as to length but recommends a single formula

$$\frac{P}{A} = \frac{\pi^2 E}{2.727(L/r)^2}$$

where r = least radius of gyration. The value of P/A must not exceed F_c. For rectangular cross sections, this formula becomes

$$\frac{P}{A} = \frac{0.30E}{(L/d)^2}$$

Allowable values are subject to adjustment for duration of load (Table 2).

19. Box Columns The equations of Art. 18 apply to box columns of square cross section with the substitution of $d = \sqrt{d_1^2 + d_2^2}$, where d_1 and d_2 are the outside and inside dimensions of the box.

20. Spaced Columns Spaced columns are formed of two or more timbers separated by spacer blocks and connected to end blocks by timber connectors, e.g., split rings (Fig. 7). This type of member is often used in trusses. L/d for the individual members is limited to 80 and L_2/d to 40. If a single spacer block is used, it should be located within the middle tenth of the column length and the individual members connected with a bolt through the block. If two or more interior spacer blocks are used, connectors are required, and the distance between any two spacers may not exceed one-half the distance between centers of connectors in the end blocks.

Connectors for the end blocks are designed to resist a shear force equal to the cross-sectional area of an individual member of the column times the appropriate constant in Table 16. Spacers and end blocks should have a thickness not less than d and a width and length sufficient to provide required end and edge distances for the connectors.

Allowable unit stresses depend on the ratio L/d and the location of the end blocks; the

TABLE 16 End-Spacer-Block Constants for Connector-Joined Spaced Columns

	End-spacer-block constant	
L/d ratio of individual member in the spaced column	Douglas fir, larch (dense), southern pine (dense)	Douglas fir, larch, southern pine (med. grain)
0–11	0	0
15	38	33
20	86	73
25	134	114
30	181	155
35	229	195
40	277	235
45	325	277
50	372	318
55	420	358
60–80	468	399

latter determines the rotational restraint. Equations are given for two end-block distances a, i.e., $a \lessgtr L/20$, and $L/10 > a > L/20$. According to the Canadian Institute of Timber Construction,[7]

Short columns ($L/d \lessgtr 10$):

$$\frac{P}{A} = F_c$$

Intermediate columns ($10 < L/d < K$):

$$\frac{P}{A} = F_c \left[1 - \frac{1}{3}\left(\frac{L/d}{K}\right)^4 \right]$$

where $K = 1.01 \sqrt{E/F_c}$ if $a \lessgtr L/20$

$K = 1.11 \sqrt{E/F_c}$ if $L/20 < a < L/10$

Long columns ($L/d > K$):

$$\frac{P}{A} = \frac{0.685E}{(L/d)^2} \qquad a \lessgtr \frac{L}{20}$$

$$\frac{P}{A} = \frac{0.822E}{(L/d)^2} \qquad \frac{L}{20} < a < \frac{L}{10}$$

The National Design Specification recommends the following:

$$\frac{P}{A} = \frac{0.75E}{(L/d)^2} \qquad a \leq \frac{L}{20}$$

$$\frac{P}{A} = \frac{0.90E}{(L/d)^2} \qquad \frac{L}{20} < a < \frac{L}{10}$$

provided $L/d > \sqrt{0.3E/F_c}$. For lesser values of L/d, the individual members are designed as simple solid columns.

Spaced columns must be checked for buckling in the direction of the width of the individual members. Values of P/A are subject to adjustment for duration of load (Table 2).

21. Beam Columns   The following equation is suggested in the National Design Specifications for beam columns of rectangular cross section:

$$\frac{M/S}{F_b} + \frac{P/A}{F_c} \leqq 1 \qquad 0 \leqq \frac{L}{d} \leqq \sqrt{\frac{0.3E}{F_c}} \qquad (8)$$

where F_c = allowable compressive stress parallel to grain for axially loaded column with same L/d
        F_b = allowable bending stress if only bending existed

TRUSSES

The basic types of timber roof trusses may be loosely classified into four categories: bowstring, lenticular, pitched, and flat. The bowstring truss is usually the most economical. The circular top chord of the bowstring closely approximates a parabola, so that it is subjected to nearly uniform axial thrust under uniform loading. The lower chord acts as a tie and the web system resists relatively small shear forces. In a flat truss the axial forces in the chords vary directly with the moment and the web members carry the total shear. Figure 8 illustrates some common truss forms.

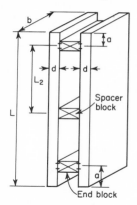

Fig. 7 Spaced column.

The Pratt web system is generally used for parallel-chord trusses. Pratt, Fink, or Belgian web systems are used for pitched trusses. The Warren web system, with or without verticals, is used for bowstring trusses. Bowstring trusses have been largely replaced by either wood joists or truss-plate trusses.

Glued-laminated members, although usually more expensive than solid lumber, permit higher stresses and may be curved to any shape, varied in cross section, and fabricated to any practical shipping length. Sawn timber presents problems because of knots, checks, and splits in large sizes. Although it is often used for web members and lower chords, it is recommended that both chords be made of glued-laminated timber.

22. Proportions   In order to use as few connectors as possible, it is usually economical to minimize the number of pieces meeting at a joint. Joints are costly, so that the number of panels should be kept to a minimum consistent with reasonable sizes of members. The top chord will usually be the determining factor. Panel lengths ordinarily range from 6 to 10 ft for material 2 to 4 in. thick. Bowstring panels vary from about 7 to 12 ft.

Pitched trusses may span up to 80 ft economically, parallel-chord trusses from 40 to 140 ft, and bowstrings from 30 to over 250 ft.

Depth-span ratio should be about 1:5 to 1:6 for pitched trusses, 1:7 to 1:9 for parallel-chord trusses, and about 1:6 to 1:8 for bowstrings. The top-chord radius of the bowstring is often made equal to the span; this results in a depth-span ratio of 0.134:1.

Spacing of heavy trusses with double chord members and split-ring connectors ranges from 10 to 30 ft for spans from 30 to 100 ft or more. The larger spacing is used with glued-laminated purlins. Solid timber is sometimes used for shorter purlins. Closer spacing is required if roof decking spans the trusses. Trusses fabricated with truss-plate connectors are frequently spaced 2 ft on centers with plywood roof decking spanning the trusses.

23. Design of Members   The combinations of load outlined in Table 17 are suggested for investigation in the analysis of the forces in trusses. The analysis is based on the usual assumption of pinned joints, and any continuity of the chord is considered independently. Moments resulting from between-joint loads or from eccentricity of the axial force P in a curved chord are determined for use in Eq. (8). Chord deflection may be significant in trusses with long panels; the additional moment due to P should be added to the primary moment.

Compression members should be designed for an effective length equal to the distance between joints for buckling in the plane of the truss. Distances between purlins, braced

TABLE 17   Recommended Load Combinations*

Description	Orientation		Load combinations		Note	Stress modifying factor
	Windward side	Leeward side	Windward side	Leeward side		
Arches or steep pitched trusses, roof slope 10 in 12 or steeper	Wind → 12 / 10		$DL + LL$ $DL + WL$ $DL + \frac{1}{2}LL + WL$ $DL + LL + \frac{1}{2}WL$	$DL + LL$ $DL + WL$ $DL + \frac{1}{2}LL + WL$ $DL + LL + \frac{1}{2}WL$	1 2 3 4	1.15 1.33 1.33 1.33
Arches, pitched trusses, bowstring trusses, or tied arches	Wind → Wind →		$DL + LL$ $DL + \frac{1}{2}LL + WL$ $DL + LL$ $DL + LL + \frac{1}{2}WL$ $DL + \frac{1}{2}LL + WL$	$DL + LL$ $DL + \frac{1}{2}LL + WL$ $DL + \frac{1}{2}LL$ $DL + \frac{1}{2}LL + \frac{1}{2}WL$ $DL + \frac{1}{4}LL + WL$	5 6 7 8 8	1.15 1.33 1.15 1.33 1.33
Beams, parallel-chord trusses, flat or nearly flat roof slope	Wind →		$DL + LL$ $DL + LL$ $DL + \frac{1}{2}LL + WL$	$DL + LL$ $DL + \frac{1}{2}LL$ $DL + \frac{1}{2}LL + WL$	9 10 11	1.15 1.15 1.33

Notes:
1. Make as a routine check.
2. Occasionally critical along rafter portion.
3. Generally critical for steeper slopes.
4. May be critical for flatter slopes.
5. Usually critical for pitched arches.   Always critical for truss chords and the tie in tied arches; occasionally critical for columns.
6. May be critical for knee braces and columns and occasionally for webs in pitched trusses.
7. Critical for the arch in tied arches, also for bowstring-truss webs if on masonry and generally if on knee-braced columns.
8. Sometimes critical for truss webs if on knee-braced columns.
9. Critical for beams, truss chords, and generally webs.   Occasionally for columns.
10. Occasionally critical for webs.
11. Critical for knee braces and columns and sometimes for webs.
* From Ref. 7.

joints, or other lateral support is the effective length for buckling perpendicular to the plane of the truss.

Tension members are usually determined by the net section requirements at the joints.

If eccentric connections are unavoidable, the moment due to the eccentricity must be considered in design of the members.

Multiple-element members may be used for chords and web system. Whenever possible, members should be symmetrical about the plane of the truss. Heavy trusses are often fabricated with double-member chords and a single-member web system. If the truss is flat or pitched and has single chord members, the Howe web system with steel-rod verticals and wood diagonals may be used. Multiple-connector and bolted joints for web-to-chord or web-to-web connections should be avoided because of shrinkage stresses and associated fractures which result in a loss of load capacity. These effects are due mainly to the angles at which these members meet.

Splices in chord members should be made between panel points and be symmetrical about the center plane of the truss. Wood splice plates and split rings, truss plates, or steel

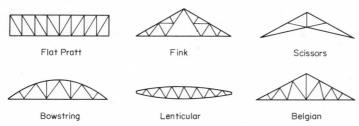

Flat Pratt   Fink   Scissors

Bowstring   Lenticular   Belgian

Fig. 8 Common timber trusses.

splice plates with bolts and rings are used. Usually only two chord splices are required for spans up to 100 ft; longer spans require three or more.

24. Deflections Truss deflections may be determined by virtual work or the Williot-Mohr diagram. Deflections due to dead load, live load, and wind load are calculated and combined to obtain critical values according to deflection criteria, e.g., for nonbearing partitions, roof drainage, glass, etc. Calculated dead-load deflections should be multiplied by 2 for seasoned lumber (15 percent or less moisture content) and by 3 for unseasoned lumber.

In calculating truss deflections, joint slippage must be added to the deformation PL/AE of the member. The slip for split rings and shear plates may be approximated by

$$\Delta_j = 0.07 \text{ in. times the load ratio for unseasoned lumber}$$
$$\Delta_j = 0.05 \text{ in. times the load ratio for seasoned lumber}$$

where load ratio is defined as the actual load on a connector divided by its allowable load. The slip in bolted joints is assumed to be the difference between the drilled hole size and the bolt diameter.

25. Camber The bottom chord of bowstring trusses should be cambered ½ in. for every 10 ft of truss length. For pitched trusses the top chord should be cambered ⅜ in. and the bottom chord ½ in. per 10 ft of length. Alternatively, trusses may be cambered for the live-load deflection plus 3 times the dead-load deflection for seasoned lumber and 4 times the dead-load deflection for green lumber.

Reference 2 suggests the following:

$$\Delta = K_1 \frac{L^3}{H} + K_2 \frac{L^2}{H}$$

where Δ = camber, in., at center of truss
L = span, ft
H = height, ft
K_1 = 0.000032 for any type of truss
K_2 = 0.0028 for flat and pitched trusses or 0.00063 for bowstring trusses, i.e., trusses without top-chord splices

26. Bracing Bracing of trusses for lateral stability during and after erection is required. Each truss must be connected to bracing systems in vertical planes perpendicular to the trusses. The horizontal struts should be snugly fitted and designed to resist tension as well as compression. They may be solid struts or built-up I or T.

Sidewall wind loads may be resisted by end walls and other structural components that develop horizontal shear strength (plywood diaphragm). If end walls are used for this purpose, the reactions from the sidewalls and trusses must be transmitted to the end walls through horizontal X-bracing or a diaphragm roof (plywood). If this is not possible, knee braces or other means of developing continuity between trusses and columns is necessary.

27. Trussed Joists Trussed joists with wood chords are an alternative to steel-bar joists, and have largely replaced wood bowstring trusses for spans to 70 ft and, with special erection procedures, spans to 100 ft. A variety of profiles is available.[17] They may have chords of solid lumber, machine-stress-graded, or of Micro-Lam lumber, the latter manufactured with $\frac{1}{10}$-in.-thick laminations of parallel grain. Finger joints are used for splicing the solid or Micro-Lam chords. Web members are steel tubes of diameters 1 to 2 in. flattened at each end to fit into the chord members. Steel pins ranging from $\frac{5}{8}$ to $1\frac{1}{4}$ in. in diameter are press-fitted to connect the web members to the chords. Wood and/or steel bridging is used for lateral bracing during erection and, if required, left in place. Lateral stability of the compression chord is assured by proper attachment to floor or roof deck.

ARCHES

Timber arches are used extensively in churches, schools, and other public buildings. The Tudor, Gothic, radial, and parabolic are common types. Tudor and Gothic arches are usually three-hinged. The radial and parabolic arches may be three-hinged but are often two-hinged. The usual spans range from 30 to 250 ft and the usual spacing from 16 to 30 ft. The arch rib may be segmental, with mechanical fasteners joining the segments, or laminated. The latter is more common by far.

Trussed arches are used also. They may be three-hinged and are often two-hinged. Table 17 gives suggested load combinations to be investigated.

28. Three-Hinged Tudor Arch The design of the three-hinged Tudor arch (Fig. 9) is relatively simple. Usually, only four to six sections need to be investigated after initial

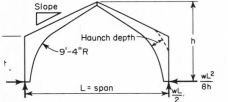

Fig. 9 Tudor arch.

assumptions and the analysis are made. Gravity loads usually govern for roof slopes under 40°. Combinations of gravity and wind load are likely to be critical for greater slopes.

Base Section. Assuming the width b, the depth d is determined from Eq. (6):

$$d = \frac{3V}{2bF_v}$$

The width b should be a standard for glued-laminated lumber (Art. 6). The maximum value of V often results from the combination of dead load, live load, and half wind load acting on the leeward side.

The peak section is usually subjected to relatively low shear forces. Its depth may be assumed to be two-thirds the base depth but not less than b.

Knee Section. The depth of the knee (haunch) at the tangent points is assumed to be four-thirds to three-halves of the base depth. A minimum radius of 32 ft for the haunch is recommended for members of $1\frac{1}{2}$-in. lamina. If desirable or necessary, a radius as small as 9 ft 4 in. may be used for members of $\frac{3}{4}$-in. lamina (Fig. 9). Lamina less than $\frac{3}{4}$ in. are special and more costly. The stress-reduction factors from Eq. (2) are 0.964 and 0.910 for the 32 ft and 9 ft 4 in. radii, respectively.

Stress Check. After tentative sections are chosen the arch is drawn to scale and the tangent points located. Stresses are checked by Eq. (8). Sections suggested for investigation are the roof arm at the tangent point, the midpoint, and the quarter point near the peak. For the vertical leg the horizontal shear at the base and the combined stress at midpoint and tangent point should suffice.

The allowable compressive stress F_c may need to be reduced for (1) buckling normal to the plane of bending (if the roof deck and walls do not provide continuous support) and (2) buckling in the plane of bending.

For (1) the effective length of the roof member is the distance between purlins and for the vertical leg the distance from base to knee.

For (2) the effective length of the leg is the distance from base to point of contraflexure, and the radius of gyration that at the tangent point. If there is no point of contraflexure, and the roof slope is less than 40°, the knee height or length of the roof arm is usually used as the effective length. The corresponding radius of gyration is that at a section one-third up the leg or one-third down the arm from the peak. If there is no point of contraflexure, and the roof slope exceeds 40°, the effective length is usually taken to be the chord distance from base to peak and the radius of gyration that at the tangent point.

The value of F_b must also be adjusted for the depth factor [Eq. (4)] and the curvature factor [Eq. (2)]. Radial stresses at the knee are checked by Eq. (3).

Deflections. Horizontal deflection of the knee and/or vertical deflection of the peak may be controlling factors. If they are critical, base and peak depths of five-eighths the tangent-point depth will reduce them to acceptable limits.

Bracing. If the walls are not of masonry, X-bracing between the legs should be provided in every third bay with every arch connected to the bracing system. The bracing may be designed as a tension system.

Table 18 gives typical dimensions for three-hinged Tudor arches for roof slopes up to 40°. The sections shown are adequate for gravity loading. Wind load may govern for steeper slopes.

29. Two-Hinged Arches Two-hinged arches are of three types: foundation, tied, and buttressed (Fig. 10). They have been used for spans of more than 200 ft. Foundation arches may have tie rods in or under the floor to resist the thrust.

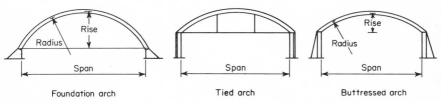

Foundation arch   Tied arch   Buttressed arch

Fig. 10 Two-hinged arches.

Section sizes for two-hinged arches of constant radius and constant section are given in Tables 19 and 20.

Coefficients for calculating the value of moments, thrusts, and reactions caused by dead, live, and wind loads on circular arches are given in Table 2, Sec. 17.

SHELL STRUCTURES

Because of the anisotropic properties of woods, and because of the methods of laminating timber or plywood for shells, elastic-shell analysis should be applied to timber shell structures with caution and, if possible, confirmed by experimental results. Limited success in adapting elastic-shell theory for timber has been achieved for certain types of shells, and some have been constructed in Great Britain, mostly of the hyperbolic-paraboloid and conoid shapes.[8,9]

Some simplified design methods have been developed for timber shells, but there are insufficient experimental data to evaluate predicted load capacities. Therefore, most types of timber shells are, in reality, framed space structures, with the sheathing playing a limited structural role.

30. Domes Domes may have spheroidal, ellipsoidal, or other shape, and may span from 50 to 350 ft, or even greater with low rise-span ratios (radius equal to or greater than the span). Generally, only surfaces of revolution about a vertical axis are used, with glued-laminated members covered by wood sheathing. The supporting timber framework may be built in a geodesic pattern or a triangulated pattern, or it may consist of a network of meridional glued-laminated ribs and circumferential members (Fig. 11). The circumferential members in radial domes and the intermeshing grids of the geodesic and triangulated domes limit bending deflections and associated stresses, so that, except for local

TABLE 18 Typical Haunch Sections, Three-Hinged Tudor Arch*

Slope	H \ w	400	600	800	1000	1200
Span:				30 ft		
3:12	8	3⅛ × 9¾	5⅛ × 9¾	5⅛ × 10½	5⅛ × 12	5⅛ × 15
	10	3⅛ × 11¼	5⅛ × 11¼	5⅛ × 12¾	5⅛ × 14¼	5⅛ × 15¾
	12	5⅛ × 9¾	5⅛ × 12	5⅛ × 14¼	5⅛ × 15¾	5⅛ × 17¼
	14	5⅛ × 10½	5⅛ × 12¾	5⅛ × 15	5⅛ × 16½	5⅛ × 17¼
	16	5⅛ × 11¼	5⅛ × 13½	5⅛ × 15¾	5⅛ × 17¼	5⅛ × 18¾
4:12	8	3⅛ × 9¾	3⅛ × 12	5⅛ × 11¼	5⅛ × 12¾	5⅛ × 14¼
	10	3⅛ × 11¼	5⅛ × 11¼	5⅛ × 12¾	5⅛ × 14¼	5⅛ × 15¾
	12	3⅛ × 12	5⅛ × 12	5⅛ × 14¼	5⅛ × 15¾	5⅛ × 15¾
	14	5⅛ × 10½	5⅛ × 12¾	5⅛ × 15	5⅛ × 16½	5⅛ × 17¼
	16	5⅛ × 11¼	5⅛ × 13½	5⅛ × 15¾	5⅛ × 17¼	5⅛ × 18
6:12	8	3⅛ × 9	3⅛ × 11¼	3⅛ × 12¾	5⅛ × 12	5⅛ × 12
	10	3⅛ × 11¼	5⅛ × 11¼	5⅛ × 12¾	5⅛ × 14¼	5⅛ × 14¼
	12	3⅛ × 12	5⅛ × 12	5⅛ × 14¼	5⅛ × 15¾	5⅛ × 16½
	14	3⅛ × 12¾	5⅛ × 12¾	5⅛ × 15	5⅛ × 16½	5¼ × 18
	16	5⅛ × 11¼	5⅛ × 13½	5⅛ × 15¾	5⅛ × 17¼	5⅛ × 18¾
8:12	8	3⅛ × 10½	3⅛ × 12	5⅛ × 11¼	5⅛ × 12¾	5⅛ × 13½
	10	3⅛ × 12	5⅛ × 11¼	5⅛ × 12¾	5⅛ × 14¼	5⅛ × 15¾
	12	3⅛ × 12¾	5⅛ × 12¾	5⅛ × 14¼	5⅛ × 15¾	5⅛ × 17¼
	14	3⅛ × 12¾	5⅛ × 12½	5⅛ × 15	5⅛ × 17¼	5⅛ × 18¾
	16	5⅛ × 12	5⅛ × 14¼	5⅛ × 15¾	5⅛ × 18	5⅛ × 19½
10:12	8	3⅛ × 10½	3⅛ × 12	5⅛ × 11¼	5⅛ × 12¾	5⅛ × 13½
	10	3⅛ × 11¼	5⅛ × 11¼	5⅛ × 12¾	5⅛ × 14¼	5⅛ × 15
	12	3⅛ × 12¾	5⅛ × 12	5⅛ × 13½	5⅛ × 15¾	5⅛ × 16½
	14	3⅛ × 12¾	5⅛ × 12¾	5⅛ × 15	5⅛ × 16½	5⅛ × 18
	16	5⅛ × 11¼	5⅛ × 13½	5⅛ × 15¾	5⅛ × 17¼	5⅛ × 18¾
				40 ft		
3:12	10	5⅛ × 12	5⅛ × 15	5⅛ × 17¼	5⅛ × 20¼	6¾ × 17¼
	12	5⅛ × 13½	5⅛ × 16½	5⅛ × 19½	5⅛ × 21¾	6¾ × 18¾
	14	5⅛ × 14½	5⅛ × 17¼	5⅛ × 21	6¾ × 19½	6¾ × 19½
	16	5⅛ × 15	5⅛ × 18	6¾ × 18	6¾ × 20¼	6¾ × 21¾
	18	5⅛ × 15¾	5⅛ × 18¾	6¾ × 18¾	6¾ × 21	6¾ × 22½
4:12	10	5⅛ × 12¾	5⅛ × 15	5⅛ × 17¼	5⅛ × 19½	5⅛ × 20¼
	12	5⅛ × 13½	5⅛ × 16½	5⅛ × 18¾	5⅛ × 21	5⅛ × 21
	14	5⅛ × 14¼	5⅛ × 17¼	5⅛ × 20¼	5⅛ × 21	6¾ × 20¼
	16	5⅛ × 15	5⅛ × 18	5⅛ × 21	6¾ × 19½	6¾ × 21¾
	18	5⅛ × 15	5⅛ × 18¾	6¾ × 18	6¾ × 21	6¾ × 22½
6:12	10	5⅛ × 12¾	5⅛ × 14¼	5⅛ × 16½	5⅛ × 18¾	5⅛ × 19½
	12	5⅛ × 12¾	5⅛ × 15¾	5⅛ × 18	5⅛ × 20¼	5⅛ × 20¼
	14	5⅛ × 13½	5⅛ × 16½	5⅛ × 19½	5⅛ × 21	6¾ × 19½
	16	5⅛ × 14¼	5⅛ × 17¼	5⅛ × 20¼	6¾ × 19½	6¾ × 21
	18	5⅛ × 15	5⅛ × 18¾	5⅛ × 21	6¾ × 21	6¾ × 21¾
8:12	10	5⅛ × 11¼	5⅛ × 14¼	5⅛ × 16½	5⅛ × 18	5⅛ × 19½
	12	5⅛ × 12¾	5⅛ × 15¾	5⅛ × 18	5⅛ × 20¼	5⅛ × 21
	14	5⅛ × 14¼	5⅛ × 16½	5⅛ × 19½	5⅛ × 21	6¾ × 20¼
	16	5⅛ × 15	5⅛ × 18	5⅛ × 20¼	5⅛ × 21	6¾ × 21¾
	18	5⅛ × 15¾	5⅛ × 18¾	5⅛ × 21	6¾ × 21	6¾ × 22½
10:12	10	5⅛ × 11¼	5⅛ × 13½	5⅛ × 15¾	5⅛ × 17¼	5⅛ × 18¾
	12	5⅛ × 12	5⅛ × 15	5⅛ × 17¼	5⅛ × 18¾	5⅛ × 21
	14	5⅛ × 12¾	5⅛ × 15¾	5⅛ × 18	5⅛ × 20¼	5⅛ × 21
	16	5⅛ × 13½	5⅛ × 16½	5⅛ × 19½	5⅛ × 21	6¾ × 20¼
	18	5⅛ × 14¼	5⅛ × 17¼	5⅛ × 20¼	6¾ × 19½	6¾ × 21¾

TABLE 18 Typical Haunch Sections, Three-Hinged Tudor Arch* (*Continued*)

Slope	H \ w	400	600	800	1000	1200
Span:				50 ft		
3:12	10	5⅛ × 15	5⅛ × 18	5⅛ × 18¾	5⅛ × 19½	5⅛ × 21¾
	12	5⅛ × 16½	5⅛ × 20¼	5⅛ × 21	6¾ × 21¾	6¾ × 23¼
	14	5⅛ × 18	5⅛ × 19½	6¾ × 21	6¾ × 24	6¾ × 25½
	16	5⅛ × 18¾	5⅛ × 21	6¾ × 22½	6¾ × 24¾	6¾ × 26¼
	18	5⅛ × 19½	6¾ × 20¼	6¾ × 23¼	6¾ × 25½	6¾ × 27¾
4:12	10	5⅛ × 15	5⅛ × 18	5⅛ × 21	6¾ × 19½	6¾ × 21¾
	12	5⅛ × 16½	5⅛ × 20¼	6¾ × 19½	6¾ × 21¾	6¾ × 23¼
	14	5⅛ × 17¼	5⅛ × 21	6¾ × 21	6¾ × 23¼	6¾ × 24
	16	5⅛ × 18	6¾ × 18¾	6¾ × 21¾	6¾ × 24	6¾ × 25½
	18	5⅛ × 18¾	6¾ × 19½	6¾ × 22½	6¾ × 24¾	6¾ × 27
6:12	10	5⅛ × 14¼	5⅛ × 17¼	5⅛ × 19½	6¾ × 18¾	6¾ × 20¼
	12	5⅛ × 15¾	5⅛ × 18¾	5⅛ × 21	6¾ × 20¼	6¾ × 21¾
	14	5⅛ × 16½	5⅛ × 20¼	6¾ × 19½	6¾ × 21¾	6¾ × 24
	16	5⅛ × 17¼	5⅛ × 21	6¾ × 21	6¾ × 23¼	6¾ × 25½
	18	5⅛ × 18	6¾ × 18¾	6¾ × 21¾	6¾ × 24	6¾ × 26¼
8:12	10	5⅛ × 13½	5⅛ × 16½	5⅛ × 19½	5⅛ × 21¾	6¾ × 18¾
	12	5⅛ × 15	5⅛ × 18¾	5⅛ × 21	6¾ × 20¼	6¾ × 21¾
	14	5⅛ × 15¾	5⅛ × 19½	6¾ × 18¾	6¾ × 21	6¾ × 23¼
	16	5⅛ × 17¼	5⅛ × 20¼	6¾ × 20¼	6¾ × 22½	6¾ × 24¾
	18	5⅛ × 17¼	5⅛ × 21¾	6¾ × 21	6¾ × 24	6¾ × 26¼
10:12	10	5⅛ × 13½	5⅛ × 16½	5⅛ × 18¾	5⅛ × 18¾	5⅛ × 21
	12	5⅛ × 14¼	5⅛ × 17¼	5⅛ × 20¼	6¾ × 18¾	6¾ × 21
	14	5⅛ × 15	5⅛ × 18¾	5⅛ × 21	6¾ × 21	6¾ × 22½
	16	5⅛ × 16½	5⅛ × 19½	6¾ × 19½	6¾ × 21¾	6¾ × 24
	18	5⅛ × 17¼	5⅛ × 21	6¾ × 21	6¾ × 23¼	6¾ × 25½
				60 ft		
3:12	12	5⅛ × 19½	6¾ × 20¼	6¾ × 23¼	6¾ × 25½	8¾ × 24
	14	5⅛ × 21	6¾ × 21¾	6¾ × 24¾	6¾ × 27¾	8¾ × 24¾
	16	5⅛ × 20¼	6¾ × 22½	6¾ × 26¼	6¾ × 27	8¾ × 26¼
	18	6¾ × 19½	6¾ × 24	6¾ × 29¼	8¾ × 27	8¾ × 27¾
	20	6¾ × 20¼	6¾ × 24¾	6¾ × 28½	8¾ × 27¾	8¾ × 30
4:12	12	5⅛ × 18¾	6¾ × 19½	6¾ × 22½	6¾ × 25½	8¾ × 23¼
	14	5⅛ × 20¼	6¾ × 21	6¾ × 24¾	6¾ × 27	8¾ × 24
	16	6¾ × 18	6¾ × 22½	6¾ × 25½	6¾ × 29¼	8¾ × 25½
	18	5⅛ × 21	6¾ × 23¼	6¾ × 27	8¾ × 27	8¾ × 27¾
	20	6¾ × 19½	6¾ × 24	6¾ × 27¾	8¾ × 27¾	8¾ × 29¼
6:12	12	5⅛ × 17¼	5⅛ × 21	6¾ × 21	6¾ × 25½	6¾ × 25½
	14	5⅛ × 18¾	6¾ × 19½	6¾ × 22½	6¾ × 25½	6¾ × 27¾
	16	5⅛ × 19½	6¾ × 21	6¾ × 24	6¾ × 27	8¾ × 26¼
	18	5⅛ × 21	6¾ × 21¾	6¾ × 25½	6¾ × 27¾	8¾ × 27¾
	20	6¾ × 18¾	6¾ × 23¼	6¾ × 26¼	8¾ × 26¼	8¾ × 28½
8:12	12	5⅛ × 18	5⅛ × 20¼	6¾ × 21¾	6¾ × 24	6¾ × 24
	14	5⅛ × 18¾	6¾ × 20¼	6¾ × 23¼	6¾ × 25½	6¾ × 27
	16	5⅛ × 20¼	6¾ × 21	6¾ × 24¾	6¾ × 27	8¾ × 25½
	18	5⅛ × 21	6¾ × 21¾	6¾ × 25½	6¾ × 28½	8¾ × 27
	20	5⅛ × 21¾	6¾ × 22½	6¾ × 26¼	6¾ × 29¼	8¾ × 28½
10:12	12	5⅛ × 16½	5⅛ × 19½	6¾ × 19½	6¾ × 21¾	6¾ × 23¼
	14	5⅛ × 17¼	5⅛ × 21	6¾ × 21	6¾ × 23¼	6¾ × 25½
	16	5⅛ × 18	6¾ × 19½	6¾ × 22½	6¾ × 24¾	6¾ × 27
	18	5⅛ × 19½	6¾ × 20¼	6¾ × 23¼	6¾ × 25½	8¾ × 25½
	20	5⅛ × 20¼	6¾ × 21	6¾ × 24¾	6¾ × 27¾	8¾ × 26¼

TABLE 18 Typical Haunch Sections, Three-Hinged Tudor Arch* (Continued)

Slope	w / H	400	600	800	1000	1200
Span:				70 ft		
3:12	12	6¾ × 19½	6¾ × 24	6¾ × 27	8¾ × 27	8¾ × 28½
	14	6¾ × 21	6¾ × 25½	8¾ × 27	8¾ × 29¼	8¾ × 30
	16	6¾ × 21¾	6¾ × 27	8¾ × 27	8¾ × 30	8¾ × 31½
	18	6¾ × 22½	6¾ × 27¾	8¾ × 27¾	8¾ × 31½	8¾ × 33¾
	20	6¾ × 23¼	6¾ × 28½	8¾ × 29¼	8¾ × 32¼	8¾ × 34½
4:12	12	6¾ × 18	6¾ × 22½	6¾ × 25½	8¾ × 25½	8¾ × 27
	14	6¾ × 19½	6¾ × 24	6¾ × 27¾	8¾ × 27	8¾ × 27¾
	16	6¾ × 21	6¾ × 25½	8¾ × 26¼	8¾ × 29¼	8¾ × 30¾
	18	6¾ × 21¾	6¾ × 27	8¾ × 27¾	8¾ × 30¾	8¾ × 32¼
	20	6¾ × 22½	6¾ × 27¾	8¾ × 27¾	8¾ × 31½	8¾ × 33
6:12	12	5⅛ × 21	6¾ × 21¾	6¾ × 24¾	6¾ × 27¾	8¾ × 25½
	14	6¾ × 18¾	6¾ × 23¼	6¾ × 26¼	6¾ × 29¼	8¾ × 27
	16	6¾ × 19½	6¾ × 24	6¾ × 27¾	8¾ × 27¾	8¾ × 29¼
	18	6¾ × 20¼	6¾ × 24¾	8¾ × 25½	8¾ × 28½	8¾ × 30¾
	20	6¾ × 21¾	6¾ × 26¼	8¾ × 27	8¾ × 29¼	8¾ × 32¼
8:12	12	5⅛ × 19½	6¾ × 20¼	6¾ × 23¼	6¾ × 26¼	8¾ × 24
	14	5⅛ × 21	6¾ × 21¾	6¾ × 25½	6¾ × 27¾	8¾ × 27
	16	6¾ × 19½	6¾ × 23¼	6¾ × 27	6¾ × 29¼	8¾ × 29¼
	18	6¾ × 20¼	6¾ × 24¾	6¾ × 27¾	8¾ × 27¾	8¾ × 30¾
	20	6¾ × 21	6¾ × 25½	6¾ × 28½	8¾ × 29¼	8¾ × 31½
10:12	12	5⅛ × 18¾	6¾ × 19½	6¾ × 22½	6¾ × 25½	8¾ × 23¼
	14	5⅛ × 20¼	6¾ × 21	6¾ × 24	6¾ × 27	8¾ × 25½
	16	5⅛ × 21	6¾ × 22½	6¾ × 25½	8¾ × 25½	8¾ × 27¾
	18	6¾ × 19½	6¾ × 23¼	6¾ × 27	8¾ × 27	8¾ × 29¼
	20	6¾ × 20¼	6¾ × 24¾	8¾ × 24¾	8¾ × 27¾	8¾ × 30¾
				80 ft		
3:12	14	6¾ × 23¼	8¾ × 25½	8¾ × 29¼	8¾ × 33	8¾ × 34½
	16	6¾ × 24¾	8¾ × 26¼	8¾ × 30¾	8¾ × 33¾	8¾ × 35¼
	18	6¾ × 25½	8¾ × 27¾	8¾ × 32¼	8¾ × 35¼	10¾ × 32¼
	20	6¾ × 26¼	8¾ × 28½	8¾ × 33	8¾ × 36¾	10¾ × 33¾
	22	6¾ × 26¼	8¾ × 27¾	8¾ × 31½	8¾ × 35¼	10¾ × 35¼
4:12	14	6¾ × 22½	6¾ × 27¾	8¾ × 27¾	8¾ × 31½	8¾ × 32¼
	16	6¾ × 24	8¾ × 29¼	8¾ × 29¼	8¾ × 33	8¾ × 33
	18	6¾ × 24¾	8¾ × 27	8¾ × 30¾	8¾ × 34½	8¾ × 34½
	20	6¾ × 25½	8¾ × 27¾	8¾ × 31½	8¾ × 35¼	10¾ × 33
	22	6¾ × 26¼	8¾ × 27	8¾ × 31½	8¾ × 35¼	10¾ × 34½
6:12	14	6¾ × 20¼	6¾ × 24¾	6¾ × 28½	8¾ × 27¾	8¾ × 29¼
	16	6¾ × 21¾	6¾ × 26¼	8¾ × 27	8¾ × 30	8¾ × 31½
	18	6¾ × 22½	6¾ × 27¾	8¾ × 28½	8¾ × 31½	8¾ × 33
	20	6¾ × 24	6¾ × 29¼	8¾ × 29¼	8¾ × 33	8¾ × 35¼
	22	6¾ × 24¾	8¾ × 26¼	8¾ × 30	8¾ × 33	10¾ × 33
8:12	14	6¾ × 19½	6¾ × 24	6¾ × 27	8¾ × 27	8¾ × 28½
	16	6¾ × 20¼	6¾ × 25½	6¾ × 29¼	8¾ × 28½	8¾ × 30¾
	18	6¾ × 21¾	6¾ × 26¼	8¾ × 27	8¾ × 30	8¾ × 32¼
	20	6¾ × 22½	6¾ × 27¾	8¾ × 28½	8¾ × 31½	8¾ × 33¾
	22	6¾ × 23¼	8¾ × 25½	8¾ × 29¼	8¾ × 32¼	8¾ × 35¼
10:12	14	5⅛ × 21	6¾ × 22½	6¾ × 26¼	8¾ × 25½	8¾ × 27¾
	16	6¾ × 20¼	6¾ × 24	6¾ × 27¾	8¾ × 27	8¾ × 30
	18	6¾ × 21	6¾ × 25½	8¾ × 25½	8¾ × 29¼	8¾ × 31½
	20	6¾ × 21¾	6¾ × 27	8¾ × 27	8¾ × 30	8¾ × 33
	22	6¾ × 23¼	8¾ × 24¾	8¾ × 27¾	8¾ × 31½	8¾ × 34½

SOURCE: Tables by Timber Structures, Inc.

Sections shown are dictated by vertical loading. Wind load may govern on slope steeper than 10:12. H = wall height, ft (Fig. 9); w = vertical load, plf.

TABLE 19   Foundation Arches (Rise Equals One-Third of Span)

Span, ft	Rise	Radius	Section sizes required, in. Design load on horizontal projection, plf of arch span					Max horizontal thrust per 100 lb of design load, lb
			400	600	800	1000	1200	
			Depth increments based on ¾-in. laminations					
50	16 ft 8 in.	27 ft 1 in.	3⅛ × 13½	5⅛ × 13½	5⅛ × 15	5⅛ × 16½	5⅛ × 18	1700
60	20 ft 8 in.	32 ft 6 in.	5⅛ × 15	5⅛ × 16½	5⅛ × 19½	5⅛ × 21	5⅛ × 22	2100
70	23 ft 4 in.	37 ft 11 in.	5⅛ × 16½	5⅛ × 19½	5⅛ × 21	5⅛ × 25½	6¾ × 22	2450
			Depth increments based on 1½- or ¾-in. laminations					
80	26 ft 8 in.	43 ft 3 in.	5⅛ × 18	5⅛ × 24	6¾ × 24	6¾ × 25½	6¾ × 25½	2800
90	30 ft 0 in.	48 ft 9 in.	5⅛ × 21	6¾ × 24	6¾ × 25½	6¾ × 27	6¾ × 28½	3150
100	33 ft 4 in.	54 ft 2 in.	5⅛ × 24	6¾ × 25½	6¾ × 27	6¾ × 30	6¾ × 31½	3500
110	36 ft 8 in.	59 ft 7 in.	6¾ × 24	6¾ × 27	6¾ × 30	6¾ × 33	8¾ × 31½	3850
120	40 ft 0 in.	65 ft 0 in.	6¾ × 25½	6¾ × 28½	6¾ × 31½	8¾ × 31½	8¾ × 33	4200
130	43 ft 4 in.	70 ft 5 in.	6¾ × 27	6¾ × 30	8¾ × 31½	8¾ × 34½	8¾ × 36	4550
140	46 ft 8 in.	75 ft 10 in.	6¾ × 28½	6¾ × 33	8¾ × 34½	8¾ × 36	8¾ × 39	4900
150	50 ft 0 in.	81 ft 3 in.	6¾ × 30	8¾ × 31½	8¾ × 36	10¾ × 36	10¾ × 39	5250
160	53 ft 4 in.	86 ft 8 in.	6¾ × 28½	8¾ × 33	8¾ × 37½	10¾ × 39	10¾ × 40½	5600
170	56 ft 8 in.	92 ft 1 in.	8¾ × 30	8¾ × 36	8¾ × 39	10¾ × 40½	10¾ × 45	5950
180	60 ft 0 in.	97 ft 6 in.	8¾ × 31½	8¾ × 37½	10¾ × 39	10¾ × 42¼	10¾ × 46½	6300

SOURCE: Tables by Timber Structures, Inc.

TABLE 20   Tied and Buttressed Arches (Radius Equals Span)
Depth Increments Based on ¾- or 1½-in. Laminations

Span, ft	Rise	Section sizes required, in. Design load on horizontal projection, plf on arch span					Max horizontal thrust per 100 lb of design load, lb
		400	600	800	1000	1200	
50	6 ft 8⅜ in.	3⅛ × 12	3⅛ × 13½	5⅛ × 13½	5⅛ × 15	5⅛ × 15	4,700
60	8 ft 0⁷⁄₁₆ in.	3⅛ × 13½	5⅛ × 13½	5⅛ × 15	5⅛ × 16½	5⅛ × 18	5,640
70	9 ft 4⁹⁄₁₆ in.	5⅛ × 13½	5⅛ × 16½	5⅛ × 18	5⅛ × 19½	5⅛ × 21	6,580
80	10 ft 8⅝ in.	5⅛ × 15	5⅛ × 18	5⅛ × 19½	5⅛ × 21	5⅛ × 24	7,520
90	12 ft 1⁷⁄₁₆ in.	5⅛ × 16½	5⅛ × 19½	5⅛ × 24	6¾ × 24	6¾ × 24	8,860
100	13 ft 4¾ in.	5⅛ × 19½	5⅛ × 24	6¾ × 24	6¾ × 25½	6¾ × 25½	9,400
110	14 ft 8⅞ in.	5⅛ × 21	6¾ × 21	6¾ × 25½	6¾ × 27	6¾ × 28½	10,340
120	16 ft 0¹⁵⁄₁₆ in.	5⅛ × 24	6¾ × 25½	6¾ × 27	6¾ × 28½	6¾ × 30	11,280

SOURCE: Tables by Timber Structures, Inc.

bending moments, the members are designed to resist only axial forces. For the most part, low-rise domes are used, so that uniform loads produce only compressive stresses in the rib and circumferential members.

For radial-rib domes the axial forces in circumferential members are compression if the angle between the dome surface and horizontal is less than 52°. The compressive forces are always greatest near the crown. The base tension ring may be of steel or prestressed concrete, and should be supported directly on the foundation. Tension-ring stresses increase rapidly as the dome rise-span ratio decreases below 0.13. Radial-rib domes require compression rings at the crown (usually a steel assembly); geodesic or triangulated domes do not. In addition, radial-rib domes require cross bracing while the geodesic and triangulated domes do not. Bracing is usually used to prevent lateral deflection of the ribs and to limit stresses due to unbalanced and concentrated loads. It is frequently provided by crossed steel tension rods.

Design Procedure for Radial-Rib Domes.[10]   The ribs are at maximum stress with full uniform live load plus dead load over the entire dome. The force F_i at any joint is associated with the area defined by the intersections of diagonals in adjacent bays (Fig. 11). Thus, $F_i = (w_L + w_D) \Delta A_i$ where w_L and w_D are the load intensities on the horizontal

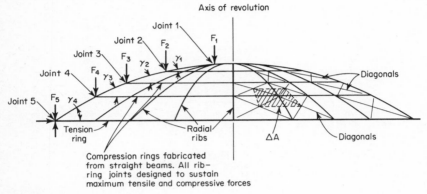

Fig. 11   Radial-rib dome.

projection. The analysis begins at the junction of a rib with the top compression ring (Fig. 12). The load F_1 is resolved into a component $F_1/\sin \gamma_1$ along chord 1-2 and a horizontal component $F_1 \cot \gamma_1$ in the plane of the ring. The angles γ are measured from the horizontal to the chords. With the component on chord 1-2 known, the forces at joint 2 can be determined. Thus, the component on chord 2-3 is $(F_1 + F_2)/\sin \gamma_2$ and the component normal to the second ring is $(F_1 + F_2) \cot \gamma_2 - F_1 \cot \gamma_1$. The compressive force in the chord adjacent to the tension ring is

$$\frac{\sum_{1}^{n-1} F_i}{\sin \gamma_{n-1}}$$

and a trial size for the rib is found by dividing this force by the allowable compressive stress parallel to the grain.

Between the centerline of the rib and the chord connecting adjacent joints is an eccentricity due to the curvature of the rib. If the rib is articulated at the rib-ring intersections, the resulting moment at any point on the rib is Fe, where e is the eccentricity of the rib relative to the chord. If the rib is continuous at the joints, the fixed-end moments are approximately $2Fe'/\pi$, where e' is the eccentricity at $L/2$. The resulting fixed-end moments may be distributed in the usual manner for straight members, and the final moments obtained by linear superposition of the distributed moments and the moments Fe. The rib section is checked by Eq. (8).

Compression Rings.   The components of F_1 in the plane of the ring are resisted by the straight chords of which the rings are built. Thus, from Fig. 12b for the ring at level 1,

$$P_{R1} = \frac{F_1 \cot \gamma_1}{2 \sin (\pi/N)}$$

where P_{R1} = compressive force in ring chord at level 1
       N = number of ribs
At ring 2,

$$P_{R2} = \frac{(F_1 + F_2) \cot \gamma_2 - F_1 \cot \gamma_1}{2 \sin (\pi/N)}$$

where P_{R2} = compressive force in ring chord at level 2, etc.

Ring chords at any level are subjected to maximum tension when the dome above the level is fully loaded, and to maximum compression when the portion of the dome below the level is loaded. In addition to the axial force, each chord supports a uniform load from the sheathing. Assuming it to be simply supported, the resulting moment is $wL^2/8$. The chord is checked for the combined bending and axial force by Eq. (8).

Tension Ring.   The tension ring should be supported on the foundation. If the ring is continuous, the analysis is made for a ring subjected to the radial forces $R_h = \left(\sum_{1}^{n-1} F_i\right) \cot \gamma_{n-1}$ (Sec. 23, Art. 5). However, the ring may be segmented, so that the chords are subjected only to tensile forces $R_h/2 \sin (\pi/N)$.

Diagonal Rods.   For polar symmetric loads, the diagonals are not stressed. Statics alone demonstrates that they may be highly stressed by asymmetrical loading. However, experimental investigations and an examination of the statical analysis show that this is not the case.[13] Most domes are designed for polar symmetrical loads. If asymmetrical loading is to be considered, it is recommended that such additional load as is necessary to produce polar symmetry be added to simplify the analysis. Some designers evaluate wind stresses by assuming a vertical load of 15 to 20 psf in addition to dead load and live load.

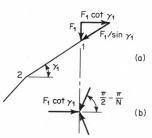

(a)

(b)

Fig. 12

Diagonals are not needed except for purposes of erection; they are often used in alternate bays for this purpose.

A number of computer programs can be used for the analysis of domes. For example, SAP4, NASTRAN, STRESS, and other programs that contain beam elements are suitable for modeling dome framing.

31. Barrel Vaults   Laminated barrel-shell roofs have been built using four or more lamina, two longitudinally and two at 60° to the shell axis.[8,11] Short-span barrels have been built with a single skin or a double skin and an internal rib system. Generally, the single membrane is used for shells with radii less than 16 ft where buckling is not likely to be critical, and the ribbed shell for larger radii.

If a sufficient number of crossed layers are used, the timber barrel vault is more amenable to the isotropic material aproximation and the usual design methods may be used (Sec. 20). It should be noted, however, that the low value of Young's modulus for timber or plywood means that buckling could be a problem. Some solutions are available for buckling of plane plywood sheets subjected to various edge tractions, but the solution for the buckling of a timber or plywood barrel vault has not been obtained. Because of this, vaults of small-radius or ribbed vaults should be used.

Where edge beams are used, care must be taken to ensure a proper connection between edge beams and membrane.

32. Hyperbolic Paraboloids   This type of shell uses materials efficiently. The basic unit is a shell supported by four straight edge members which are warped about their long axes to follow the curvature (Sec. 9, Fig. 6). The usual elementary analysis is sufficiently accurate for timber shells. The ratio of rise to length of side should be 0.2 or more to limit secondary stresses and inhibit buckling.

Two layers of sheathing, usually of 1-in. nominal thickness, are adequate in most cases. The sheathing may be laid with the boards of one layer parallel to one family of parabolas and those of the other parallel to the second family. Alternatively, one layer may be laid in the direction of one set of generators and the other in the direction of the second set. The first method requires bending the lumber, and the second warping it, to fit the curvature of the shell.

The first method has the advantage that the grain is in the direction of the principal stresses in the shell, so that one layer is in tension and the other in compression. The two layers should be nailed together to prevent buckling of the layer in compression. In the second method, the layers are shear panels so that shear connections must be developed.

Edge Members.   The shell should join the edge members at their neutral axes. Only shear stress, tangential to the edge member, is transferred from the shell, and the edge member is designed for axial force. The edge members which are in compression are subject to buckling normal to the surface of the shell, and the compressive stress must be limited accordingly.

REFERENCES

1. Agriculture Handbook 72, U.S. Department of Agriculture, Government Printing Office.
2. National Design Specifications, National Forest Products Association, Washington, D.C., 1973.
3. Lewis, W. C.: Fatigue of Wood and Glued Joints Used in Laminated Construction, *For. Prod. Res. Soc. Proc.*, vol. 5, pp. 221–229, 1951.
4. Gurfinkel, G.: "Wood Engineering," Southern Forest Products Association, New Orleans, La., 1973.
5. Haussler, R. W.: Roof Deflections Caused by Rainwater Pools, *Civil Eng.*, October 1962, pp. 58–59.
6. Hooley, R. F., and B. Madsen: Lateral Stability of Glued Laminated Beams, *J. Struct. Div. ASCE*, June 1964.
7. Canadian Institute of Timber Construction: "Timber Construction Manual," Ottawa, 1963.
8. Tottenham, H.: The Design of Timber Shell Roofs, *Proc. 1st Int. Conf. Timber Eng.*, pp. 123–131, Southampton, 1961.
9. Tottenham, H.: The Analysis of Orthotropic Cylindrical Shells, *Civil Eng. Public Works Rev., Timber Eng. Suppl.*, vol. 54, no. 635, pp. 597–599, May 1959.
10. American Institute of Timber Construction: "Timber Construction Manual," Washington, D.C.
11. Barrel Vault—A Timber Shell Roof for a British Railway Parcel Depot, *Wood*, vol. 26, no. 1, pp. 16–17, January 1961.
12. American Plywood Association: "Plywood Diaphragm Construction," Tacoma, Wash., 1970.
13. Anderson, Paul: Steel Framed Domes—Design and Research, *1959 Proc., AISC Natl. Eng. Conf.*
14. Maki, A. C., and E. W. Kuenzi: Deflection and Stresses of Tapered Wood Beams, U.S. Forest Service Research Paper FPL34, 1965.
15. Foschi, R. O., and S. P. Fox: Radial Stresses in Curved Timber Beams, *J. Struct. Div. ASCE*, October 1970.
16. Timber Engineering Company: "Design Manual for Teco Timber Construction," Washington, D.C.
17. Trus Joist Corporation, Glenview, Ill.

Arches and Rigid Frames

THOMAS C. KAVANAGH
Vice-President (deceased), Iffland Kavanagh Waterbury, New York, N.Y.

ROBERT C. Y. YOUNG
Vice-President, URS/Madigan-Praeger, Inc., New York, N.Y.

Arches and rigid frames are generally characterized by the development under vertical loads of inclined rather than vertical reactions. Mixed construction is designated by the predominant characteristic; thus, Fig. 1 is spoken of as a rigid-frame bridge. Rings are completely enclosed arches or rigid frames.

Cross sections are designed for thrust, moment, and shear, with magnitudes depending on the location of the pressure line (the funicular polygon of applied loads) as shown in Fig. 2. If the pressure line coincides with the axis of the structure (as in a uniformly loaded parabolic arch), all cross sections are subjected to compression, with no moment or shear. If the pressure line falls within the kern of the section, there will exist thrust, bending moment, and shear, but no tension on the cross section. If, finally, as in Fig. 2c and d, the shape of the structure differs (for utilitarian or other reasons) from the pressure line, moment may become dominant. A survey of approximate methods based on sketching deflected shapes is given in Ref. 28.

1. Nomenclature and Classification *Arches* may consist of several parallel *ribs*, or of a single curved sheet or *barrel* (Fig. 3). The *crown* is the highest point of the rib or barrel; the *soffit* and *back* are, respectively, the undersurface and top surface of the arch, and the *springing line* is the intersection of the pier or abutment and the soffit. The *haunch* is the midsection of arch rib or barrel between crown and springing line. *Intrados* and *extrados* are lines of intersection of soffit and back with a vertical plane through the crown and springing lines. The *skewback* is the inclined surface on which the arch rests. The *clear rise* is the height of intrados at crown above the springing line. The *centerline rise* is the height of the axis of the arch at the crown above the axis at the skewback.

In bridges, the space between back of arch and roadway is called the *spandrel*. Depending on whether this space is left relatively open (with columns) or filled with earth, one has either an *open-spandrel arch* (Fig. 3) or a *filled-spandrel arch*.

Tied arches are those in which a structural tie is built between reaction points to take the horizontal thrust.

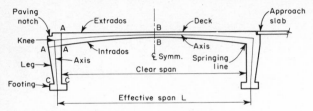

Fig. 1 Rigid-frame bridge.

Types of arches are shown in Table 1.
 Rings.   The nomenclature and classification of rings follows that of arches and rigid frames. Cross sections may be constant or variable, members may be straight or curved, and curvatures may be constant or variable. The ring may rest on point supports or, as in the case of culverts, on a continuous earth support having rigid, elastic, semielastic, or plastic properties.
 Rigid frames comprised of columns supporting transverse members are spoken of as *bents.* Various types are shown in Fig. 4. Frames may have members of constant or variable cross section (Fig. 1). Rigidity of joints (i.e., continuity of members of the frame) is

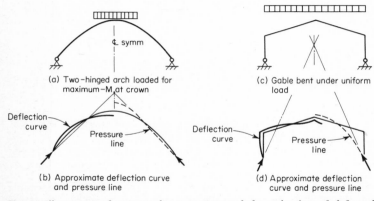

Fig. 2 Illustrations of pressure lines approximated from sketches of deformed structure.

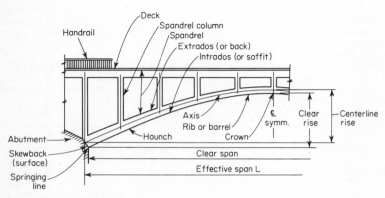

Fig. 3 Open-spandrel arch.

TABLE 1   Arch Classifications

Fixed.....................	Three degrees indeterminate. Requires good foundations. Relatively rigid and theoretically slightly more economical. High-temperature stresses with flat arches. Commonest usage in reinforced concrete
One-hinged................	Two degrees indeterminate. Rarely used
Two-hinged................	One degree indeterminate. Common usage in steel, timber, aluminum. Less rigid than fixed arch, but thrust line is definitely located at abutments. Relatively insensitive to rotations and to moderate differential deflections of supports
Three-hinged...............	Statically determinate. Common usage in steel and timber. Free from temperature stresses and insensitive to foundation settlements
Two- or three-hinged, tied...	Does not require massive abutments. Not affected by settlements of supports
Through and half-through...	Overhead lateral bracing may be objectionable aesthetically. Overcome by using slender ribs with relatively heavier tie girder deck (Langer arch-stiffened girder)
Deck......................	Rib usually carries all load, though occasionally deck girder made to participate (Langer arch-stiffened girder)
Solid rib..................	Loads brought to individual ribs by beams, shells, or other transverse elements
Barrel....................	In bridges, the spandrel fill gives excessive dead weight. Path of loads to barrel uncertain
Braced rib (lattice)	
Open spandrel..............	Clearer path of load to ribs
Spandrel-braced............	Analysis as a trussed framework
Semicircular	
Segmental (arc of circle)	
Multicentered..............	Use of segments of circles facilitates field layout
Parabolic..................	Funicular line for uniformly distributed load
Elliptical	
Catenary..................	Shape of inverted free-hanging string
Gothic	
Flat......................	Large horizontal thrust. Large effect of rib shortening and temperature
High-rise..................	Reduced horizontal thrust. Reactions may approach vertical

a requisite for structural stability. If elasticity or inelasticity of joints must be taken into consideration, special methods of frame analysis for *semirigid* joints are used.

Rigid frames are so shaped as to utilize the interior space better than is possible with arches, as would be the case with a bridge crossing a major thoroughfare with limited vertical clearances. The nearly straight deck, rigidly connected to piers and abutments, emphasizes functional characteristics for aesthetic treatment.

Although moments are increased at the knees of rigid frames, the moments at the center

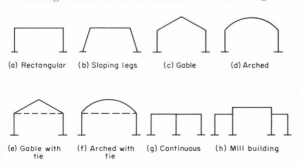

(a) Rectangular   (b) Sloping legs   (c) Gable   (d) Arched

(e) Gable with tie   (f) Arched with tie   (g) Continuous   (h) Mill building

Fig. 4 Types of bents.

of span are small relative to a simple span of the same length. The shallower depth at midspan produces better clearance and appearance, and lowers the approach lengths of the overpassing road, or raises the underpassing road. Embankment excavation or fill is reduced, land acquisition lowered, and the total cost decreased. Maintenance and replacement costs are low, and widening is readily accomplished in the future.

ANALYSIS

2. Assumptions Analysis of a final design is necessary to verify that the shape and cross sections are such as to produce stresses or deformations within permissible values under the given system of loads. While analysis is thus a terminal step, it also provides the basis for many approximations useful in preliminary design.

The analysis of stresses on curved members is based on the assumption that the cross-sectional dimensions of the member are small compared with the radius of curvature, so that the stresses may be calculated by formulas applicable for straight members. For very sharp curvatures, curved-beam formulas for angle changes and stresses must be used. It is also assumed, except where otherwise noted, that the arch may be analyzed by first-order or so-called "elastic" theory; that is, the effect of deformations on the bending moments may be neglected and the principle of superposition will apply. For long spans, and for buckling investigations, deformation theory (second-order theory) must be applied.

3. Kern Relationships In designing or checking cross sections for combined axial loads and bending moments, it is often advantageous to use the kern relationships. The limiting points of the kern are the "kern points," which have eccentricities e_k from the center of gravity given by

$$e_k = \frac{r^2}{c} \tag{1}$$

where r = radius of gyration
c = distance to extreme fiber

The extreme fiber stresses f for an unsymmetrical section (Fig. 5) are given by the following equations:

Upper flange:
$$f_u = \frac{N}{A} + \frac{Mc_u}{I} = N\left(e + \frac{r^2}{c_u}\right)\frac{c_u}{I} = M_{k_u}\frac{c_u}{I} \tag{2}$$

Lower flange:
$$f_l = \frac{N}{A} - \frac{Mc_l}{I} = N\left(e - \frac{r^2}{c_l}\right)\frac{c_l}{I} = M_{k_l}\frac{c_l}{I} \tag{3}$$

Each stress f may be maximized by constructing the influence line for the appropriate M_k.

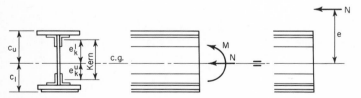

Fig. 5 Kern points.

4. Arches and Closed Rings Energy methods lead to a consistent, mnemonic, and general form of the static equations for solving indeterminate arches, rings, etc. Thus, with a typical fixed arch (Fig. 6), indeterminate to the third degree,

$$\begin{aligned}
1 \cdot \delta_a &= \delta_{a0} + X_a\delta_{aa} + X_b\delta_{ab} + X_c\delta_{ac} + \delta_{at} \\
1 \cdot \delta_b &= \delta_{b0} + X_a\delta_{ba} + X_b\delta_{bb} + X_c\delta_{bc} + \delta_{bt} \\
1 \cdot \delta_c &= \delta_{c0} + X_a\delta_{ca} + X_b\delta_{cb} + X_c\delta_{cc} + \delta_{ct}
\end{aligned} \tag{4}$$

where X_a, X_b, X_c, = the three redundants
δ_a, δ_b, δ_c, = displacements in the direction of X_a, X_b, and X_c, respectively

$\delta_{a0}, \delta_{b0}, \delta_{c0}$ = displacements in the direction of X_a, X_b, and X_c, respectively, due to the applied external loads acting on the determinate base system

δ_{mn} = displacement in direction of X_m due to $X_n = 1$

$\delta_{at}, \delta_{bt}, \delta_{ct}$ = displacements of base system in direction of X_a, X_b, and X_c, respectively, due to uniform temperature change $t°$ and/or a differential temperature $\Delta t°$ between top and bottom fibers

The displacements δ are given by

$$\delta_{mn} = \delta_{nm} = \int \frac{M_m M_n ds}{EI} + \int \frac{N_m N_n ds}{EA} + \int \frac{T_m T_n ds}{GJ} + \int \frac{\kappa V_m V_n ds}{AG} \tag{5}$$

X_a X_c
X_b

Fig. 6   Fixed-ended-arch redundants.

For temperature effects $t°$ and $\Delta t°$,

$$\delta_{mt} = \int \frac{M_m \epsilon_t \Delta t° ds}{h} + \int N_m \epsilon_t t° ds \tag{6}$$

In the above expressions M_m and M_n are the bending moments in the base system due to $X_m = 1$ and $X_n = 1$, respectively. Similarly, N, T, and V are the normal forces, torques, and shears in the base system subjected to the unit redundants designated by the respective subscripts. The temperature coefficient of expansion is ϵ_t. The shape coefficient κ appears in the shear-energy expression. For common arches and loads, the shear-energy effects are neglected and the torsional term does not apply, and the equations reduce to a consideration of flexural effects, normal forces (rib shortening), and temperature.

For electronic-computer solution, Eq. (4) may be written in matrix form:

$$\begin{bmatrix} \delta_{aa} & \delta_{ab} & \delta_{ac} \\ \delta_{ba} & \delta_{bb} & \delta_{bc} \\ \delta_{ca} & \delta_{cb} & \delta_{cc} \end{bmatrix} \begin{bmatrix} X_a \\ X_b \\ X_c \end{bmatrix} = \begin{bmatrix} (\delta_a - \delta_{a0} - \delta_{at}) \\ (\delta_b - \delta_{b0} - \delta_{bt}) \\ (\delta_c - \delta_{c0} - \delta_{ct}) \end{bmatrix}$$

or

$$AX = K$$

which may be solved by inversion,

$$X = A^{-1}K$$

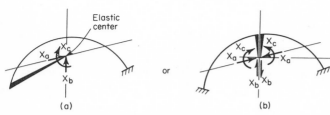

(a)   (b)

Fig. 7   Fixed-ended-arch redundants at elastic center.

In practical arch analysis the solution of the three simultaneous equations is frequently avoided by use of the neutral-point method or its equivalent, column analogy (Art. 17). The equations are reduced to three independent equations in three unknowns by equating to zero all δ terms with dissimilar paired subscripts (e.g., δ_{ab}, δ_{bc}, δ_{ac}). This is accomplished (Fig. 7) by placing the origin of coordinates at the center of gravity of elastic weights (ds/EI) and by making the axes conjugate (axes for which the product of inertia of elastic weights is zero). For symmetrical arches the axes are orthogonal and are principal axes. The elastic-center method is particularly convenient in computing influence lines, as well as for estimating effects of rib shortening, temperature change, and support displacements.

For unsymmetrical arches, the conjugate axes will not generally be perpendicular to each other, and it is simpler to employ orthogonal axes through the neutral point (elastic center) and dispense with the requirement $\delta_{ab} = 0$. For immovable abutments, and considering loads only (i.e., $\delta_t = 0$), the solution of Eq. (4) in both neutral-point and column-analogy symbolism becomes:

$$
X_a = -\frac{\delta_{a0} - \delta_{b0}\delta_{ab}/\delta_{bb}}{\delta_{aa} - \delta_{ab}^2/\delta_{bb}} = -\frac{M_{xx} - M_{yy}I_{xy}/I_{yy}}{I_{xx} - I_{xy}^2/I_{yy}}
$$

$$
X_b = -\frac{\delta_{b0} - \delta_{a0}\delta_{ab}/\delta_{aa}}{\delta_{bb} - \delta_{ab}^2/\delta_{aa}} = -\frac{M_{yy} - M_{xx}I_{xy}/I_{xx}}{I_{yy} - I_{xy}^2/I_{xx}} \qquad (7)
$$

$$
X_c = -\frac{\delta_{c0}}{\delta_{cc}} = -\frac{W}{A}
$$

The second terms in the numerator and denominator of these equations are called "corrections for dissymmetry." The terminology relating to column analogy is as follows:

W = total Mds/EI load on analogous column
M_{xx} = moment of Mds/EI load about XX axis
M_{yy} = moment of Mds/EI load about YY axis
A = total ds/EI area of analogous column
I_{xx} = moment of inertia of ds/EI area about XX axis
I_{yy} = moment of inertia of ds/EI area about YY axis
I_{xy} = product of inertia of ds/EI area about XX and YY axes

Application of Eqs. (7) is illustrated in Example 1, which shows the calculations for influence-line data for a unit load at A.

5. Rigid Frames The energy methods of analysis described above are equally applicable to rigid frames, and in fact form the basis for several useful and extensive compilations of rigid-frame formulas.[1,2] Computer programs such as STRESS and STRUDL are widely used in analysis, and also in section selection.

DESIGN OF ARCHES

6. General Procedure Design of an arch, ring, or rigid frame involves a succession of stages each representing a refinement of the previous one. This is particularly true of indeterminate structures, in which initial, but guided, approximations of design moments, shears, and thrusts are necessary, whereas in determinate systems the statical values may be established precisely at the outset.

The first stage involves planning with respect to character of the spanned opening, clearance requirements, rise, span, loading, architecture, and aesthetics.

The second stage proceeds with determination of the general shape and approximate design moments, shears, and thrusts from which tentative cross sections are determined. Such values are usually derived from approximate influence-line data. Alternately, the cross sections may be assumed from data on similar designs which have proved satisfactory, or from empirical relationships based on experience. Frequently, as in the case of concrete arches and steel rigid frames, an intermediate stage of design refinement is possible at this point, using available design charts or tables, some of which are comprehensive enough to yield a final design requiring very little later adjustment.

The last stage of design (for indeterminate structures) is the formal analysis of the structure, to verify stresses and deformations, and to make minor adjustments if necessary.

7. Preliminary Selection of Shape To reduce bending moments, the arch axis should conform as nearly as practicable to the pressure line or equilibrium polygon of the loads. Such a curve is parabolic for uniformly distributed loads. However, dead load usually increases toward the abutments, resulting in a rising of the pressure line between crown and springing, and an increase in the inclination at the springing line. The pressure line also varies as live load is superimposed on the dead load; this variation will occur with

Example 1   Analysis of fixed unsymmetrical arch for a unit vertical load.*

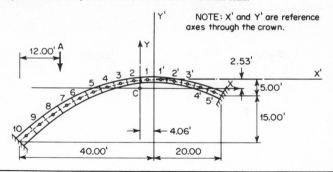

NOTE: X' and Y' are reference axes through the crown.

1	2	3	4	5	6	7	8	9	10	11	12	13	14	15	16	17
						Properties of section								**Unit load at** *A*		
		Given						**Derived**								
	ds	d	x'	y'	$a = \dfrac{ds}{d^2}$	ay'	y	ax'	x	ay^2	ax^2	axy	M_0	M_0a	M_0ay	M_0ax
10	5.57	3.34	−38.0	−18.05	0.149	−2.69	−15.52	−5.66	−33.94	35.89	171.6	+78.5	−10	−1.49	+23.1	+50.6
9	5.23	3.04	−34.0	−14.45	0.186	−2.69	−11.92	−6.32	−29.94	26.43	166.7	+66.4	−6	−1.12	+13.4	+33.5
8	5.00	2.79	−30.0	−11.25	0.230	−2.59	−8.72	−6.90	−25.94	17.49	154.8	+52.0	−2	−0.46	+4.0	+11.9
7	4.78	2.54	−26.0	−8.45	0.292	−2.47	−5.92	−7.59	−21.94	10.23	140.6	+37.9				
6	4.55	2.33	−22.0	−6.05	0.360	−2.18	−3.52	−7.92	−17.94	4.46	115.9	+22.7				
5	4.45	2.16	−18.0	−4.05	0.442	−1.79	−1.52	−7.76	−13.94	1.02	85.9	+9.4				
4	4.26	2.00	−14.0	−2.45	0.533	−1.31	+0.08	−7.46	−9.94	0	52.7	−0.4				
3	4.17	1.83	−10.0	−1.25	0.680	−0.85	+1.28	−6.80	−5.94	1.11	24.0	−5.2				
2	4.05	1.67	−6.0	−0.45	0.870	−0.39	+2.08	−5.22	−1.94	3.76	3.3	−3.5				
1	4.01	1.54	−2.0	−0.05	1.098	−0.05	+2.48	−2.20	+2.06	6.75	4.7	+5.6				
1'	4.01	1.54	+2.0	−0.05	1.098	−0.05	+2.48	+2.20	+6.06	6.75	40.3	+16.5				
2'	4.05	1.67	+6.0	−0.45	0.870	−0.39	+2.08	+5.22	+10.06	3.76	88.0	+18.2				
3'	4.17	1.83	+10.0	−1.25	0.680	−0.85	+1.28	+6.80	+14.06	1.11	134.4	+12.2				
4'	4.26	2.00	+14.0	−2.45	0.533	−1.31	+0.08	+7.46	+18.06	0	173.8	+0.8				
5'	4.45	2.16	+18.0	−4.05	0.442	−1.79	−1.52	+7.76	+22.06	1.02	215.1	−14.8				
Summation					8.463	−21.40		−34.39		119.78	1,571.8	+296.3		−3.07	+40.5	+96.0
Corrections for dissymmetry	$-\Sigma axy \times \Sigma axy/\Sigma ax^2$									−55.86						
	$-\Sigma axy \times \Sigma axy/\Sigma ay^2$										−733.0					
	$-\Sigma M_0ax \times \Sigma axy/\Sigma ax^2$														−18.1	
	$-\Sigma M_0ay \times \Sigma axy/\Sigma ay^2$															−100.2
Summations corrected for dissymmetry										63.9	839			−3.07	+22.4	−4.2

Correction from y' to y: $\dfrac{\Sigma ay'}{\Sigma a} = \dfrac{-21.40}{8.463} = -2.53$

Correction from x' to x: $\dfrac{\Sigma ax'}{\Sigma a} = \dfrac{-34.39}{8.463} = -4.06$

$$X_c = -\frac{-3.07}{8.463} = +0.363$$

$$X_a = -\frac{+22.4}{63.9} = -0.351$$

$$X_b = -\frac{-4.2}{839} = +0.005$$

* Courtesy of Portland Cement Association.

live load over a part of the span, or even with full or partial live load over the entire span. Many designers overlook the fact that uniform live load over the full span has the effect of producing a more nearly parabolic pressure line than does dead load alone.

The AASHTO specifications require the arch axis to be shaped to the equilibrium polygon for full dead load.

Arch ribs are most commonly of variable thickness, increasing in depth from the crown to the springing. A rough approximation is to relate this variation to the secant of the angle of inclination of the arch axis to the horizontal. The assumption that moments of inertia vary directly as the secant of the angle is particularly convenient in developing simple approximate formulas, and it has been found that results are relatively insensitive to moderate deviations from this assumption.

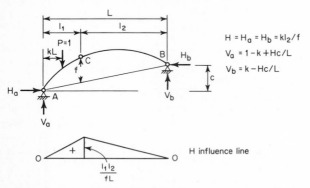

$$H = H_a = H_b = kl_2/f$$
$$V_a = 1 - k + Hc/L$$
$$V_b = k - Hc/L$$

H influence line

$$\frac{l_1 l_2}{fL}$$

Fig. 8 Unsymmetrical three-hinged arch.

8. Approximations for Special Shapes The following approximations are applicable to all constructional (elastic) materials.

Three-Hinged Arch.   The influence line for the horizontal reaction $H_a = H_b$ is shown in Fig. 8.

Symmetrical Two-Hinged Arch.   For a parabolic or flat circular shape ($f/L \leq \frac{1}{8}$) with the variation in moment of inertia given by $I = I_c \sec \phi$, where I_c is the moment of inertia

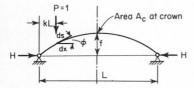

Fig. 9 Symmetrical two-hinged arch.

at the crown and ϕ the inclination of the tangent with the horizontal (Fig. 9), the influence line for the horizontal reaction is given by

$$H = \frac{5L}{8f}(k - 2k^3 + k^4)\nu \tag{8}$$

where

$$\nu = \frac{1}{1 + \frac{15}{8}I_c/A_cf^2}$$

The parameter ν expresses the effect of rib shortening due to axial forces and may be neglected except for very flat arches (for arches with $f/L = \frac{1}{7}$ to $\frac{1}{8}$, $\nu = 0.98$). At midspan, $k = 0.5$, so that $H_{max} = 0.195\nu L/f$.

The above formula may be further approximated by a parabolic influence line

$$H = \frac{3Lk(1 - k)\nu}{4f} \tag{9a}$$

with $H_{\max} = 3L\nu/16f$. For full-span uniform load w,

$$H = \frac{wL^2\nu}{8f} \tag{9b}$$

The formula may also be approximated by a semicubic parabola with central ordinate $H_{\max} = 0.20\nu L/f$, the equation of which is

$$
\begin{aligned}
H &= \frac{L}{5f}[1 - (1 - 2k)^{3/2}]\nu &\quad 0 \leqslant k \leqslant 0.5 \\
H &= \frac{L}{5f}[1 - (2k - 1)^{3/2}]\nu &\quad 0.5 \leqslant k \leqslant 1.0
\end{aligned}
\tag{10}
$$

The semicubic parabola is also a reasonable approximation for H for the two-hinged spandrel-braced arch.

For a uniform temperature change $t°$,

$$H = \frac{\epsilon_t t° E I_c}{\tfrac{8}{15}f^2 + I_c/A_c} \tag{11}$$

or if the normal forces, represented by the term I_c/A_c, are neglected,

$$H = \frac{15}{8}\,\epsilon_t t° E \frac{I_c}{f^2} \tag{12}$$

Table 2 gives values of moment, thrust, and reactions for circular two-hinged arches, of constant moment of inertia, for rise-span ratios f/L of 1:3 and 1:4. The dead load w_D and live load w_L per foot of horizontal projection are assumed constant. The uniform wind load w_W per foot of vertical projection is a uniform-load equivalent to the radial wind-pressure distribution recommended by the ASCE Task Committee on Wind Forces (*Trans. ASCE*, vol. 126, pt. II, 1962). The maximum live-load moment, and concurrent axial-force and reaction components at each point, are calculated for vertical live load over that portion of the span which will produce maximum moment at the point.

Symmetrical Tied Arch. For a parabolic or flat circular shape ($f/L \gtrsim 1{:}8$) with $I = I_c$ sec ϕ, the influence line for H is given by Eq. (8), except that the parameter ν is replaced by

$$\nu' = \frac{1}{1 + (15/8f^2)(I_c/A_T + I_c/A_c)} \approx 1 \tag{13}$$

where A_T = area of tie.

For a uniform temperature rise $t°$ of both arch and tie, $H = 0$. If, however, the tie is warmed to $t°$ and the arch to $t° + \Delta t°$, then

$$H = \frac{\epsilon_t \Delta t° E I_c}{\tfrac{8}{15}f^2 + I_c(1/A_c + 1/A_T)} \tag{14}$$

Symmetrical Fixed Arch. For a parabolic or flat circular shape ($f/L \gtrsim 1{:}8$), with $I = I_c$ sec ϕ, the influence lines for the reactions (Fig. 10) are given by

$$
\begin{aligned}
V_A &= (1 + 2k)(1 - k)^2 \\
H_A &= \frac{15}{4}\frac{L}{f}(k - k^2)^2\nu \\
M_A &= \frac{L}{2}(1 - k)^2(5k^2\nu - 2k)
\end{aligned}
\tag{15}
$$

where $\nu = \dfrac{1}{1 + {}^{45}\!/_4 I_c/A_c f^2}$

Since $\nu \approx 1$, it may ordinarily be neglected.

TABLE 2 Coefficients for Circular Two-Hinged Arch with Constant I*

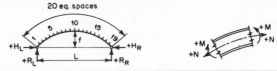

Point	Dead load§		Wind load¶		Max $+ M_L$ and concurrent N, H, and R due to critical live load					Max $- M_L$ and concurrent N, H, and R due to critical live load				
	$\dfrac{M_D}{w_D L^2}$	$\dfrac{N_D\dagger}{w_D L}$	$\dfrac{M_W}{w_W L^2}$	$\dfrac{N_W}{w_W L}$	$\dfrac{M_L}{w_L L^2}$	$\dfrac{N_L\dagger}{w_L L}$	$\dfrac{H\ddagger}{w_L L}$	$\dfrac{R_L}{w_L L}$	$\dfrac{R_R}{w_L L}$	$\dfrac{M_L\dagger}{w_L L^2}$	$\dfrac{N_L\dagger}{w_L L}$	$\dfrac{H\ddagger}{w_L L}$	$\dfrac{R_L}{w_L L}$	$\dfrac{R_R}{w_L L}$
					$f/L = 1:3$									
1	−.00650	.583	.01452	.425	.00265	.196	.055	.222	.032	.00915	.387	.295	.278	.468
2	−.00961	.560	.02396	.442	.00579	.162	.055	.222	.032	.01540	.398	.295	.278	.468
3	−.00994	.530	.02820	.450	.00902	.173	.080	.264	.049	.01897	.357	.270	.236	.451
4	−.00819	.495	.02717	.448	.01185	.184	.109	.304	.070	.02004	.311	.241	.196	.430
5	−.00508	.458	.02090	.437	.01425	.149	.109	.304	.070	.01933	.309	.241	.196	.430
6	−.00134	.424	.01250	.421	.01562	.165	.141	.341	.095	.01696	.258	.209	.159	.405
7	.00235	.393	.00514	.408	.01581	.186	.175	.375	.125	.01346	.208	.175	.125	.375
8	.00542	.370	−.00107	.396	.01489	.211	.209	.405	.159	.00947	.160	.141	.095	.341
9	.00744	.355	−.00606	.387	.01302	.239	.241	.430	.196	.00561	.117	.109	.070	.304
10	.00813	.350	−.00976	.380	.01278	.190	.190	.187	.187	.00463	.160	.160	.313	.313
11	.00744	.355	−.01211	.376	.01302	.239	.241	.430	.196	.00561	.117	.109	.070	.304
12	.00542	.370	−.01308	.374	.01489	.211	.209	.405	.159	.00947	.160	.141	.095	.341
13	.00235	.393	−.01265	.375	.01581	.186	.175	.375	.125	.01346	.208	.175	.125	.375
14	−.00134	.424	−.01084	.378	.01562	.165	.141	.341	.095	.01696	.258	.209	.159	.405
15	−.00508	.458	−.00767	.384	.01425	.149	.109	.304	.070	.01933	.309	.241	.196	.430
16	−.00819	.495	−.00426	.390	.01185	.184	.109	.304	.070	.02004	.311	.241	.196	.430
17	−.00994	.530	−.00175	.395	.00902	.173	.080	.264	.049	.01897	.357	.270	.236	.451
18	−.00961	.560	−.00016	.398	.00579	.162	.055	.222	.032	.01540	.399	.295	.278	.468
19	−.00650	.583	.00048	.399	.00265	.196	.055	.222	.032	.00915	.387	.295	.278	.468
					$f/L = 1:4$									
1	−.00377	.667	.01012	.449	.00420	.205	.086	.236	.037	.00798	.462	.396	.264	.463
2	−.00548	.641	.01665	.460	.00809	.221	.119	.275	.054	.01357	.420	.363	.226	.446
3	−.00557	.612	.01954	.464	.01143	.187	.119	.275	.054	.01701	.425	.363	.226	.446
4	−.00451	.583	.01875	.463	.01390	.208	.157	.311	.074	.01840	.376	.325	.189	.426
5	−.00272	.556	.01431	.456	.01529	.181	.157	.311	.074	.01802	.375	.325	.189	.426
6	−.00064	.531	.00842	.447	.01566	.210	.198	.344	.098	.01630	.321	.284	.156	.402
7	.00139	.510	.00332	.438	.01478	.244	.241	.375	.125	.01339	.266	.241	.125	.375
8	.00306	.495	−.00094	.432	.01290	.282	.284	.402	.156	.00984	.213	.198	.344	.098
9	.00415	.485	−.00433	.426	.01031	.322	.325	.426	.189	.00619	.163	.157	.311	.074
10	.00452	.482	−.00682	.422	.01026	.244	.243	.172	.172	.00574	.238	.238	.338	.338
11	.00415	.485	−.00838	.420	.01031	.322	.325	.426	.189	.00619	.163	.157	.311	.074
12	.00306	.495	−.00901	.419	.01290	.282	.284	.402	.156	.00984	.213	.198	.344	.098
13	.00139	.510	−.00870	.419	.01478	.244	.241	.375	.125	.01339	.266	.241	.125	.375
14	−.00064	.531	−.00744	.421	.01566	.210	.198	.344	.098	.01630	.321	.284	.156	.402
15	−.00272	.556	−.00526	.425	.01529	.181	.157	.311	.074	.01802	.375	.325	.189	.426
16	−.00451	.583	−.00292	.428	.01390	.208	.157	.311	.074	.01840	.376	.325	.189	.426
17	−.00557	.612	−.00121	.431	.01143	.187	.119	.275	.054	.01701	.425	.363	.226	.446
18	−.00548	.641	−.00013	.433	.00809	.221	.119	.275	.054	.01357	.420	.363	.226	.446
19	−.00377	.667	.00030	.434	.00420	.205	.086	.236	.037	.00798	.462	.396	.264	.463

* From Ref. 27.
† All values in this column negative.
‡ $H_L = H_R = w_L L$.
§ $H_L = 0.350 w_D L$
¶ $H_L = -0.394 w_W L$
¶ $H_R = -0.165 w_W L$ $\left.\right\} f/L = 1:3$
¶ $R_L = -0.265 w_W L$
¶ $R_R = -0.363 w_W L$

$H_L = 0.482 w_D L$
$H_L = -0.424 w_W L$
$H_R = -0.268 w_W L$ $\left.\right\} f/L = 1:4$
$R_L = -0.223 w_W L$
$R_R = -0.341 w_W L$

For full-span uniform vertical load w per unit length of span,

$$H = \frac{wL^2\nu}{8f}$$
$$V_A = \frac{wL}{2}$$
$$M_A = M_B = \tfrac{1}{12}wL^2(1 - \nu)$$
$$M_C = \tfrac{1}{24}wL^2(1 - \nu)$$

(16)

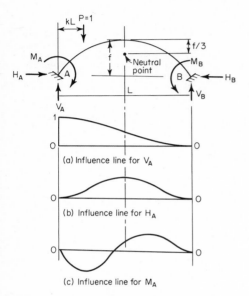

(a) Influence line for V_A

(b) Influence line for H_A

(c) Influence line for M_A

Fig. 10 Symmetrical fixed circular arch.

For uniform full-rise lateral load w per unit vertical rise, acting from left to right,

$$H_A = \frac{11wf}{14} \qquad H_B = \frac{3wf}{14}$$
$$V_A = -\frac{wf^2}{4L} \qquad V_B = \frac{wf^2}{4L}$$
$$M_A = -\frac{51wf^2}{280} \qquad M_B = \frac{19wf^2}{280}$$

(17)

For a temperature change $t°$,

$$H = \frac{45}{4} \frac{\epsilon_t t° EI_c \nu}{f^2}$$
$$V = 0$$
$$M_A = \tfrac{2}{3}fH$$

(18)

Fixed Circular Arch of Constant Section. Exact values of reactions and moments for a fixed circular arch of constant cross section are given in Table 3.

Fixed Concrete Highway Arch Bridges. Certain empirical expressions in common use, also helpful in choosing first trial sections for concrete highway arches, are noted below.

1. Whitney variation of rib thickness (Fig. 11)

$$d_k = d_c c\sqrt[6]{1 + \tan^2 \phi}$$

(19a)

TABLE 3   Coefficients for Circular Fixed Arch Constant I*

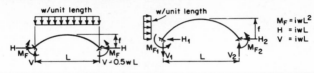

f/L	M_F	H	M_{F_1}	M_{F_2}	H_1	H_2	V_1, V_2
0.10	0.001230	1.2584	0.001831	0.000689	0.07857	0.02143	0.002480
0.12	0.001680	1.0507	0.002647	0.000994	0.09433	0.02567	0.003559
0.14	0.002224	0.9027	0.003598	0.001377	0.10990	0.03010	0.004825
0.16	0.002898	0.7924	0.004709	0.001817	0.12556	0.03444	0.006274
0.18	0.003659	0.7069	0.005981	0.002320	0.14125	0.03875	0.007899
0.20	0.004507	0.6388	0.007404	0.002900	0.15689	0.04311	0.009696
0.22	0.005439	0.5833	0.008989	0.003551	0.17254	0.04746	0.011660
0.24	0.006452	0.5372	0.010734	0.004282	0.18817	0.05183	0.013784
0.26	0.007549	0.4983	0.012643	0.005094	0.20378	0.05622	0.016062
0.28	0.008725	0.4652	0.014719	0.005994	0.21938	0.06062	0.018488
0.30	0.009981	0.4366	0.016963	0.006983	0.23496	0.06504	0.021054
0.32	0.011315	0.4117	0.019380	0.008065	0.25053	0.06947	0.023755
0.34	0.012726	0.3898	0.021972	0.009245	0.26609	0.07391	0.026583
0.36	0.014213	0.3705	0.024740	0.010527	0.28163	0.07837	0.029533

* From Ref. 11.

where d_c = depth at crown

$\quad\quad d_k$ = depth at kL from crown   (19b)

$\quad\quad c = [1 - 2k(1 - I_c/I_s \cos \phi_x)]^{-1/3}$

2. F. F. Wald formula (barrel arches)

$$d_c = \sqrt{L} + \frac{L}{10} + \frac{w_L}{200} + \frac{w_c}{400} \tag{20}$$

where d_c = depth at crown, in.

$\quad\quad L$ = clear span, ft

$\quad\quad w_L$ = live load, psf

$\quad\quad w_c$ = dead load, psf, above arch at crown

Wald suggests that the depth of springing be taken at from 1½ to 3 times d_c, increasing with the flatness of the arch.

3. W. J. Douglas formula (barrel arches)

$$
\begin{array}{lll}
L = 20 \text{ ft} & d_c = 0.03(6 + L) & \text{ft} \\
L = 20 \text{ to } 50 \text{ ft} & d_c = 0.015(30 + L) & \text{ft} \\
L = 50 \text{ to } 150 \text{ ft} & d_c = 0.0001(11,000 + L^2) & \text{ft} \\
L = 150 \text{ ft} & d_c = 0.016(75 + L) & \text{ft}
\end{array} \tag{21}
$$

4. V. A. Cochrane formulas for bridges (Fig. 9)

Open-spandrel arch:

$$y = \frac{8f}{6 + 5f/L}\left(3k^2 + \frac{10k^4 f}{L}\right) \tag{22}$$

$$\tan \phi_s = \frac{8f/L}{6 + 5f/L}\left(3 + \frac{5f}{L}\right) \tag{23}$$

Filled-spandrel arch:

$$y = \frac{4f}{1 + 3f/L}\left(k^2 + \frac{24k^5 f}{L}\right) \tag{24}$$

$$\tan \phi_s = \frac{4f/L}{1 + 3f/L}\left(\frac{1 + 7.5f}{L}\right) \tag{25}$$

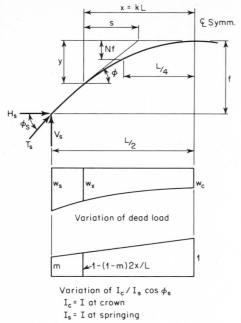

Variation of $I_c / I_s \cos \phi_s$
$I_c = I$ at crown
$I_s = I$ at springing

Fig. 11 Whitney-arch notation.

Table 4 gives dimensions of typical highway arch bridges of reinforced concrete which may be of value in preliminary proportioning of arch bridges.

Circular and Parabolic Arches of Constant Section.   In roofs, bridges, etc., where arch ribs of prismatic section may be desired for appearance and ease of fabrication, exact influence coefficients are given in Tables 5 and 6.

9. Intermediate Design   Tables for circular and ellptical arches of variable thickness are available.[29] For symmetrical concrete arches, Whitney's tables[3] enable a rapid intermediate refinement before proceeding with formal final elastic analysis. Results are given in

TABLE 4   Typical Concrete Highway Arch Bridges

Span, ft	Rise, ft	Road-way, ft	Sidewalks	Crown depth and width	Springing depth and width	Type
50	9	23	2 at 5 ft 0 in.	10 in. × 36 ft 6 in.	2 ft 6 in. × 36 ft 6 in.	Filled spandrel
60	12	20	None	12 in. × 23 ft 0 in.	2 ft 6 in. × 23 ft 0 in.	Filled spandrel
70	14	19	None	1 ft 3 in. × 22 ft 0 in.	3 ft 7½ in. × 22 ft 0 in.	Filled spandrel
80	14	24	....	2 ft 0 in. × 4 ft 0 in.	3 ft 6 in. × 4 ft 0 in.	Open spandrel, 2 ribs
90	21	27	None	1 ft 6 in. × 5 ft 0 in.	3 ft 5 in. × 5 ft 0 in.	Open spandrel, 3 ribs
100	22	20	2 at 5 ft 0 in.	3 ft 0 in. × 3 ft 0 in.	6 ft 0 in. × 3 ft 0 in.	Open spandrel, 2 ribs
120	16	36	2 at 8 ft 0 in.	2 ft 1 in. × 6 ft 0 in.	4 ft 11 in. × 6 ft 0 in.	Open spandrel, 4 ribs
145	21	18	None	3 ft 0 in. × 4 ft 3 in.	5 ft 3 in. × 4 ft 3 in.	Open spandrel, 2 ribs
160	46	18	None	3 ft 2 in. × 4 ft 0 in.	6 ft 4 in. × 4 ft 0 in.	Open spandrel, 2 ribs
190	36	20	1 at 5 ft 0 in.	2 ft 3 in. × 10 ft 0 in.	5 ft 0 in. × 10 ft 0 in.	Open spandrel, 2 ribs
205	31	21	None	4 ft 0 in. × 4 ft 0 in.	8 ft 0 in. × 4 ft 0 in.	Open spandrel, 2 ribs

TABLE 5 Fixed Symmetrical Arches*
Influence values—vertical loading

Influence values for	h/L	Position of load								
		0.1L	0.2L	0.3L	0.4L	0.5L	0.6L	0.7L	0.8L	0.9L
Parabolic arch: Moment at right springing	0.1	0.0112L	0.0317L	0.0464L	0.0471L	0.0307L	0.0002L	-0.0360L	-0.0629L	-0.0602L
	0.2	0.0112L	0.0310L	0.0449L	0.0448L	0.0290L	0.0001L	-0.0340L	-0.0606L	-0.0589L
	0.3	0.0113L	0.0305L	0.0432L	0.0427L	0.0276L	0.0004L	-0.0320L	-0.0580L	-0.0576L
	0.4	0.0113L	0.0300L	0.0423L	0.0408L	0.0261L	0.0005L	-0.0303L	-0.0561L	-0.0566L
	0.5	0.0113L	0.0298L	0.0408L	0.0393L	0.0249L	0.0005L	-0.0293L	-0.0546L	-0.0559L
Vertical reaction at right springing	0.1	0.0285	0.1053	0.2175	0.3530	0.5000	0.6470	0.7825	0.8947	0.9715
	0.2	0.0297	0.1083	0.2210	0.3553	0.5000	0.6447	0.7790	0.8917	0.9703
	0.3	0.0310	0.1113	0.2246	0.3576	0.5000	0.6424	0.7754	0.8887	0.9690
	0.4	0.0319	0.1137	0.2273	0.3596	0.5000	0.6404	0.7727	0.8863	0.9681
	0.5	0.0326	0.1155	0.2297	0.3612	0.5000	0.6388	0.7703	0.8845	0.9674
Horizontal reaction	0.1	0.3075	0.9665	1.6546	2.1554	2.3354	2.1554	1.6546	0.9665	0.3075
	0.2	0.1591	0.4888	0.8282	1.0682	1.1546	1.0682	0.8282	0.4888	0.1591
	0.3	0.1098	0.3323	0.5531	0.7067	0.7610	0.7067	0.5531	0.3323	0.1098
	0.4	0.0849	0.2528	0.4173	0.5255	0.5638	0.5255	0.4173	0.2528	0.0849
	0.5	0.0695	0.2049	0.3327	0.4171	0.4460	0.4171	0.3327	0.2049	0.0695
Circular arch: Moment at right springing	0.1	0.0121L	0.0341L	0.0491L	0.0503L	0.0341L	0.0032L	-0.0331L	-0.0604L	-0.0592L
	0.2	0.0131L	0.0351L	0.0500L	0.0507L	0.0354L	0.0069L	-0.0272L	-0.0548L	-0.0561L
	0.3	0.0155L	0.0386L	0.0532L	0.0554L	0.0404L	0.0165L	-0.0162L	-0.0435L	-0.0496L
	0.4	0.0197L	0.0431L	0.0571L	0.0585L	0.0471L	0.0254L	-0.0021L	-0.0281L	-0.0394L
	0.5	0.0241L	0.0479L	0.0617L	0.0639L	0.0553L	0.0375L	+0.0139L	-0.0097L	-0.0237L
Vertical reaction at right springing	0.1	0.0286	0.1053	0.2176	0.3529	0.5000	0.6471	0.7824	0.8947	0.9714
	0.2	0.0307	0.1100	0.2226	0.3562	0.5000	0.6438	0.7774	0.8900	0.9693
	0.3	0.0348	0.1178	0.2305	0.3610	0.5000	0.6390	0.7695	0.8822	0.9652
	0.4	0.0416	0.1287	0.2406	0.3669	0.5000	0.6331	0.7594	0.8713	0.9584
	0.5	0.0510	0.1423	0.2522	0.3735	0.5000	0.6265	0.7478	0.8577	0.9490
Horizontal reaction	0.1	0.3193	0.9944	1.6817	2.1850	2.3678	2.1850	1.6817	0.9944	0.3193
	0.2	0.1706	0.5099	0.8466	1.0826	1.1668	1.0826	0.8466	0.5099	0.1706
	0.3	0.1305	0.3628	0.5778	0.7304	0.7742	0.7304	0.5778	0.3628	0.1305
	0.4	0.1178	0.2935	0.4443	0.5432	0.5774	0.5432	0.4443	0.2935	0.1178
	0.5	0.1150	0.2523	0.3638	0.4348	0.4591	0.4348	0.3638	0.2523	0.1150

* From J. Michalos, "Theory of Structural Analysis and Design," The Ronald Press Company, New York, 1958.

TABLE 5 Fixed Symmetrical Arches*

Influence values at right springing—horizontal loading

Influence values at right springing for	h/L	Position of load								
		$0.1L$	$0.2L$	$0.3L$	$0.4L$	$0.5L$	$0.6L$	$0.7L$	$0.8L$	$0.9L$
Parabolic arch:										
Moment	0.1	0.0039L	0.0096L	0.0127L	0.0128L	0.0123L	0.0136L	0.0178L	0.0232L	0.0221L
	0.2	0.0078L	0.0189L	0.0246L	0.0247L	0.0238L	0.0261L	0.0345L	0.0449L	0.0434L
	0.3	0.0119L	0.0280L	0.0358L	0.0348L	0.0344L	0.0391L	0.0496L	0.0650L	0.0636L
	0.4	0.0158L	0.0369L	0.0469L	0.0463L	0.0446L	0.0493L	0.0637L	0.0840L	0.0837L
	0.5	0.0199L	0.0460L	0.0575L	0.0569L	0.0545L	0.0597L	0.0777L	0.1026L	0.1033L
Vertical reaction	0.1	0.0099	0.0311	0.0533	0.0695	0.0753	0.0695	0.0533	0.0311	0.0099
	0.2	0.0206	0.0641	0.1088	0.1410	0.1523	0.1410	0.1088	0.0641	0.0206
	0.3	0.0323	0.0989	0.1664	0.2140	0.2310	0.2140	0.1664	0.0989	0.0323
	0.4	0.0444	0.1349	0.2252	0.2883	0.3107	0.2883	0.2252	0.1349	0.0444
	0.5	0.0566	0.1713	0.2847	0.3632	0.3908	0.3632	0.2847	0.1713	0.0566
Horizontal reaction	0.1	0.1071	0.2900	0.4296	0.4902	0.5000	0.5098	0.5704	0.7100	0.8929
	0.2	0.1108	0.2951	0.4310	0.4905	0.5000	0.5095	0.5690	0.7049	0.8892
	0.3	0.1157	0.3007	0.4341	0.4848	0.5000	0.5152	0.5659	0.6993	0.8843
	0.4	0.1185	0.3061	0.4382	0.4905	0.5000	0.5095	0.5618	0.6939	0.8815
	0.5	0.1216	0.3109	0.4397	0.4921	0.5000	0.5079	0.5603	0.6891	0.8784
Circular arch:										
Moment	0.1	0.0045L	0.0084L	0.0124L	0.0133L	0.0127L	0.0140L	0.0188L	0.0250L.	0.0222L
	0.2	0.0099L	0.0223L	0.0277L	0.0281L	0.0272L	0.0293L	0.0366L	0.0463L	0.0466L
	0.3	0.0191L	0.0377L	0.0452L	0.0457L	0.0447L	0.0471L	0.0558L	0.0690L	0.0720L
	0.4	0.0350L	0.0578L	0.0663L	0.0667L	0.0658L	0.0681L	0.0768L	0.0911L	0.1003L
	0.5	0.0581L	0.0822L	0.0905L	0.0915L	0.0908L	0.0927L	0.1002L	0.1140L	0.1272L
Vertical reaction	0.1	0.0101	0.0313	0.0531	0.0687	0.0744	0.0687	0.0531	0.0313	0.0101
	0.2	0.0230	0.0662	0.1074	0.1355	0.1454	0.1355	0.1074	0.0662	0.0230
	0.3	0.0435	0.1072	0.1624	0.1982	0.2105	0.1982	0.1624	0.1072	0.0435
	0.4	0.0724	0.1539	0.2161	0.2552	0.2682	0.2552	0.2161	0.1539	0.0724
	0.5	0.1145	0.2037	0.2673	0.3055	0.3183	0.3055	0.2673	0.2037	0.1145
Horizontal reaction	0.1	0.1167	0.2720	0.4218	0.4911	0.5000	0.5089	0.5782	0.7280	0.8833
	0.2	0.1291	0.3174	0.4403	0.4923	0.5000	0.5077	0.5597	0.6826	0.8709
	0.3	0.1630	0.3450	0.4524	0.4939	0.5000	0.5061	0.5476	0.6550	0.8370
	0.4	0.2110	0.3777	0.4639	0.4955	0.5000	0.5045	0.5361	0.6223	0.7890
	0.5	0.2696	0.4075	0.4733	0.4967	0.5000	0.5033	0.5267	0.5925	0.7304

* From J. Michalos, "Theory of Structural Analysis and Design," The Ronald Press Company, New York, 1958.

tables and diagrams for reactions due to dead and live loads, and to temperature and rib-shortening effects, together with complete influence lines for moments at crown, springing line, and quarter points, and diagrams giving maximum and minimum moments with corresponding thrusts for crown, springing line, and quarter points. The data are presented in terms of two parameters: the *shape coefficient N*, the ratio of the drop of arch axis at quarter point to the rise *f*, and the *form coefficient m*, which characterizes the variation of the cross section of the rib (Fig. 11). These parameters, together with the span, rise, and thickness or moment of inertia of rib at the crown, entirely establish the rib and its reactions under live load and temperature change.

In design, the value of $g = w_s/w_c$ is computed first. In investigating a design already completed, the value of N is computed directly from the ordinates of the rib axis, after which g is computed from

$$g = \frac{1}{2} \cdot \left(\frac{1}{N} - 2\right)^2 - 1 \qquad (26)$$

If this value of g does not correspond with the value computed from the actual weights, Whitney recommends for Table 9 the use of the value $w_c = w_s/g$ based on the actual w_s and the value of g from Eq. (26).

Whitney's tables are set up for values at 10 points along the half arch rib, equally spaced in the horizontal direction (Fig. 12). Table 7 is used to lay out the axis of the rib after N or

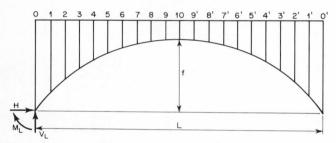

Fig. 12 Whitney-arch load points.

g has been computed. Table 8 enables the tangent to the axis at each point to be determined for use in plotting, or in calculating thrusts on the cross section. Table 9 yields horizontal and vertical reactions H_d and V_d for dead load. The dead-load thrust at any point is then

$$T_d = H_d(1 - \tan^2 \phi)^{1/2} \qquad (27)$$

If the rib is laid out to the dead load, there are no dead-load moments except for rib shortening. Tables 10 and 11 allow the solution of Eq. (19), which completely determines the rib size at all points.

Figures 13 through 16 are influence lines for the horizontal reaction H_s and for moments at crown, springing, and quarter points. Figures 17 through 19 show values of maximum uniform live-load moments and corresponding values of thrust at crown, quarter point, and springing. Figure 20 gives values of thrust due to temperature change, assuming a concrete modulus of 3,000,000 psi.

Shrinkage effects may normally be calculated as equivalent to a 15° temperature drop.

The horizontal thrust due to rib shortening under dead load is usually calculated separately in terms of the horizontal thrust due to dead load (Table 9):

$$H_{RS} = - H_d \frac{I_c}{A_c C'_m C f^2} \qquad (28)$$

where H_{RS} = horizontal thrust due to rib shortening under dead load
H_d = horizontal thrust under dead load
A_c = area of cross section at crown
C = coefficient from Table 12
C'_m = coefficient from Fig. 21

TABLE 7 Arch-Axis Coordinates—Values of y/f*

$g = \dfrac{w_s}{w_c}$	N	Point 0 (springing line)	Point 1	Point 2	Point 3	Point 4	Point 5 (quarter point)	Point 6	Point 7	Point 8	Point 9	Point 10 (crown)
1.000	0.250	1.000	0.8100	0.6400	0.4900	0.3600	0.2500	0.1600	0.0900	0.0400	0.0100	0.0000
1.167	0.245	1.000	0.8060	0.6339	0.4834	0.3539	0.2450	0.1563	0.0878	0.0390	0.0098	0.0000
1.347	0.240	1.000	0.8019	0.6277	0.4769	0.3478	0.2400	0.1527	0.0856	0.0380	0.0095	0.0000
1.543	0.235	1.000	0.7977	0.6214	0.4701	0.3416	0.2350	0.1493	0.0835	0.0370	0.0092	0.0000
1.756	0.230	1.000	0.7934	0.6151	0.4632	0.3353	0.2300	0.1458	0.0814	0.0360	0.0090	0.0000
1.987	0.225	1.000	0.7890	0.6087	0.4563	0.3293	0.2250	0.1424	0.0792	0.0350	0.0087	0.0000
2.240	0.220	1.000	0.7847	0.6022	0.4494	0.3229	0.2200	0.1386	0.0771	0.0340	0.0085	0.0000
2.514	0.215	1.000	0.7801	0.5957	0.4425	0.3167	0.2150	0.1351	0.0749	0.0330	0.0082	0.0000
2.814	0.210	1.000	0.7755	0.5891	0.4355	0.3104	0.2100	0.1315	0.0728	0.0320	0.0080	0.0000
3.141	0.205	1.000	0.7709	0.5824	0.4285	0.3041	0.2050	0.1281	0.0707	0.0310	0.0077	0.0000
3.500	0.200	1.000	0.7662	0.5757	0.4215	0.2978	0.2000	0.1245	0.0686	0.0300	0.0074	0.0000
3.893	0.195	1.000	0.7615	0.5689	0.4145	0.2914	0.1950	0.1209	0.0665	0.0290	0.0072	0.0000
4.324	0.190	1.000	0.7567	0.5621	0.4073	0.2851	0.1900	0.1176	0.0644	0.0281	0.0070	0.0000
4.801	0.185	1.000	0.7518	0.5551	0.4000	0.2787	0.1850	0.1140	0.0623	0.0271	0.0067	0.0000
5.321	0.180	1.000	0.7469	0.5481	0.3927	0.2723	0.1800	0.1106	0.0602	0.0262	0.0065	0.0000
5.898	0.175	1.000	0.7420	0.5410	0.3854	0.2659	0.1750	0.1072	0.0582	0.0252	0.0062	0.0000
6.536	0.170	1.000	0.7367	0.5337	0.3781	0.2595	0.1700	0.1037	0.0562	0.0243	0.0059	0.0000
7.244	0.165	1.000	0.7313	0.5264	0.3707	0.2531	0.1650	0.1002	0.0541	0.0233	0.0057	0.0000
8.031	0.160	1.000	0.7259	0.5190	0.3632	0.2466	0.1600	0.0968	0.0521	0.0224	0.0055	0.0000
8.906	0.155	1.000	0.7205	0.5116	0.3557	0.2399	0.1550	0.0934	0.0501	0.0215	0.0053	0.0000
9.889	0.150	1.000	0.7151	0.5040	0.3480	0.2332	0.1500	0.0901	0.0483	0.0206	0.0050	0.0000

* From Ref. 3.

TABLE 8 Intercepts to Determine Positions of Tangents—Values of $\frac{s}{L/2}$ *

$g = \frac{w_s}{w_c}$	N	Point 0 (springing line)	Point 1	Point 2	Point 3	Point 4	Point 5 (quarter point)	Point 6	Point 7	Point 8	Point 9	Point 10 (crown)
1.000	0.25	0.5000	0.4500	0.4000	0.3500	0.3000	0.2500	0.2000	0.1500	0.1000	0.0500	∞
1.347	0.24	0.4743	0.4311	0.3865	0.3410	0.2943	0.2466	0.1982	0.1492	0.0998	0.0500	∞
1.756	0.23	0.4503	0.4129	0.3734	0.3319	0.2884	0.2432	0.1965	0.1485	0.0996	0.0500	∞
2.240	0.22	0.4279	0.3957	0.3607	0.3229	0.2825	0.2397	0.1946	0.1478	0.0993	0.0499	∞
2.814	0.21	0.4070	0.3792	0.3482	0.3140	0.2765	0.2360	0.1926	0.1469	0.0991	0.0499	∞
3.500	0.20	0.3872	0.3634	0.3360	0.3051	0.2704	0.2323	0.1905	0.1459	0.0988	0.0498	∞
4.324	0.19	0.3686	0.3482	0.3241	0.2962	0.2643	0.2285	0.1884	0.1449	0.0985	0.0498	∞
5.321	0.18	0.3510	0.3335	0.3125	0.2875	0.2583	0.2246	0.1862	0.1439	0.0983	0.0497	∞
6.536	0.17	0.3342	0.3194	0.3011	0.2787	0.2520	0.2206	0.1840	0.1429	0.0980	0.0497	∞
8.031	0.16	0.3182	0.3058	0.2899	0.2700	0.2457	0.2164	0.1817	0.1419	0.0976	0.0496	∞
9.889	0.15	0.3030	0.2926	0.2788	0.2613	0.2393	0.2121	0.1792	0.1407	0.0972	0.0496	∞

* From Ref. 3.

17-18

Moments due to the above rib shortening are calculated from

$$M_{RS} = - H_{RS}y'$$ (29)

where y' is the vertical distance between the elastic center and the point about which moments are taken. Table 13 locates the elastic center at the distance y_c below the crown. Table 14 gives influence-line data for the left vertical reaction.

10. Approximations of Whitney Data Hardy Cross[1] derived average values of maximum moments at crown and springing from Whitney's tables, which can be applied as an alternate, shorter, intermediate refinement of the preliminary design:

$$\int \frac{ds}{EI} = A = \frac{L}{EI_c}\left(\frac{1+m}{2}\right)$$

$$\int x^2 \frac{ds}{EI} = \frac{1}{12} AL^2 \left(\frac{1+3m}{2+2m}\right)$$ (30)

$$\int y^2 \frac{ds}{EI} = \frac{4}{45} Af^2 \left(\frac{1+3m}{2+2m}\right)$$

If ds/I = constant, the Whitney form coefficient m is unity, and the parenthetical multipliers in the above formulas drop out, yielding exact values for a parabolic arch with this particular variation of cross section. For uniform live load, the kern moments are

$$\text{Max } M_s^k = \pm \tfrac{1}{45} w_L L^2$$
$$\text{Max } M_c^k = \pm \tfrac{1}{225} w_L L^2$$ (31)

For max M_s^k,

$$H = 0.10 w_L \frac{L^2}{f}$$

TABLE 9 Values of N, g, H_d, V_d, $\tan \phi_s$[*]

N	$g = \dfrac{w_s}{w_c}$	$C_d = \dfrac{fH_d}{w_c L^2}$	$\dfrac{V_d}{w_c L}$	$\dfrac{L}{f} \tan \phi_s$
0.250	1.000	0.1250	0.5000	4.000
0.245	1.166	0.1285	0.5277	4.120
0.240	1.347	0.1320	0.5566	4.217
0.235	1.543	0.1358	0.5875	4.328
0.230	1.756	0.1397	0.6206	4.442
0.225	1.988	0.1438	0.6554	4.556
0.220	2.240	0.1483	0.6933	4.675
0.215	2.515	0.1530	0.7331	4.793
0.210	2.814	0.1579	0.7761	4.915
0.205	3.141	0.1632	0.8219	5.030
0.200	3.500	0.1687	0.8713	5.165
0.195	3.893	0.1746	0.9242	5.294
0.190	4.324	0.1808	0.9812	5.427
0.185	4.799	0.1875	1.0430	5.561
0.180	5.321	0.1946	1.1092	5.700
0.175	5.897	0.2022	1.1874	5.841
0.170	6.536	0.2104	1.2593	5.985
0.165	7.244	0.2193	1.3443	6.132
0.160	8.031	0.2287	1.4370	6.283
0.155	8.910	0.2389	1.5382	6.440
0.150	9.889	0.2499	1.6496	6.601

[*] From Ref. 3.

TABLE 10 Rib Thickness—Values of $\dfrac{L^2}{f^2}\tan^2\phi$*

$g = \dfrac{w_s}{w_c}$	N	Point 0 (springing line)	Point 1	Point 2	Point 3	Point 4	Point 5 (quarter point)	Point 6	Point 7	Point 8	Point 9	Point 10 (crown)
1.000	0.25	16.000	12.960	10.240	7.840	5.760	4.000	2.560	1.440	0.640	0.160	0
1.347	0.24	17.780	13.846	10.552	7.821	5.583	3.788	2.377	1.316	0.580	0.145	0
1.756	0.23	19.722	14.769	10.852	7.791	5.403	3.578	2.200	1.199	0.523	0.130	0
2.240	0.22	21.841	15.730	11.149	7.748	5.220	3.371	2.029	1.088	0.468	0.116	0
2.814	0.21	24.150	16.731	11.443	7.694	5.034	3.166	1.864	0.983	0.417	0.102	0
3.500	0.20	26.676	17.780	11.734	7.629	4.844	2.964	1.706	0.883	0.369	0.089	0
4.324	0.19	29.441	18.885	12.020	7.552	4.650	2.765	1.554	0.789	0.325	0.076	0
5.321	0.18	32.476	20.047	12.297	7.463	4.450	2.569	1.408	0.701	0.284	0.065	0
6.536	0.17	35.812	21.269	12.564	7.360	4.243	2.376	1.268	0.619	0.246	0.056	0
8.031	0.16	39.494	22.553	12.820	7.238	4.029	2.187	1.135	0.541	0.211	0.048	0
9.889	0.15	43.572	23.899	13.063	7.095	3.809	2.001	1.008	0.467	0.180	0.041	0

* From Ref. 3.

TABLE 11 Rib Thickness—Values of c in Eq. (19b)*

k \ m	Point 0 (springing line)	Point 1	Point 2	Point 3	Point 4	Point 5 (quarter point)	Point 6	Point 7	Point 8	Point 9	Point 10 (crown)
	0.50	0.45	0.40	0.35	0.30	0.25	0.20	0.15	0.10	0.05	0
1.0	1.000	1.000	1.000	1.000	1.000	1.000	1.000	1.000	1.000	1.000	1.000
0.8	1.077	1.068	1.060	1.052	1.044	1.036	1.028	1.021	1.014	1.007	1.000
0.6	1.186	1.160	1.137	1.116	1.096	1.077	1.060	1.044	1.028	1.014	1.000
0.5	1.260	1.221	1.186	1.154	1.126	1.101	1.077	1.056	1.036	1.017	1.000
0.4	1.357	1.295	1.244	1.199	1.160	1.126	1.096	1.068	1.044	1.021	1.000
0.3	1.494	1.393	1.315	1.252	1.199	1.154	1.116	1.082	1.052	1.024	1.000
0.25	1.587	1.454	1.357	1.282	1.221	1.170	1.126	1.089	1.056	1.026	1.000
0.20	1.710	1.529	1.405	1.315	1.244	1.186	1.137	1.096	1.060	1.028	1.000
0.15	1.882	1.621	1.462	1.352	1.268	1.203	1.149	1.103	1.064	1.030	1.000

* From Ref. 3.

For max M_c^k,

$$H = 0.06w_L \frac{L^2}{f}$$

For dead load, approximately,

$$H_c = \frac{1}{8} \frac{WL}{f} \tag{32}$$

$$T_s = H_c \sec \phi_s \tag{33}$$

where W = total dead load.

For temperature,

$$H_t = \frac{\epsilon_t Et°L}{I_y} = \frac{45}{4} \epsilon_t Et° \frac{I_c}{f^2} \approx 200t° \frac{I_c}{f^2} \frac{4}{1+3m} \tag{34}$$

$$M_c = H_t y_c = \frac{15}{4} \epsilon_t Et° \frac{I_c}{f} \approx 67.5t° \frac{I_c}{f} \frac{2+2m}{1+3m} \tag{35}$$

$$f_t = \frac{M_c y}{I} = \frac{15}{8} \epsilon_t Et° \frac{d}{f} \approx 33t° \frac{d_c}{f} \frac{2+2m}{1+3m} \tag{36}$$

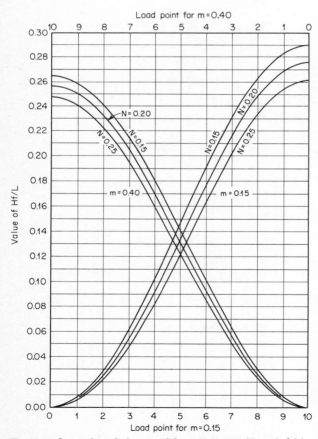

Fig. 13 Influence lines for horizontal thrust H, Fig. 12. *(From Ref. 3.)*

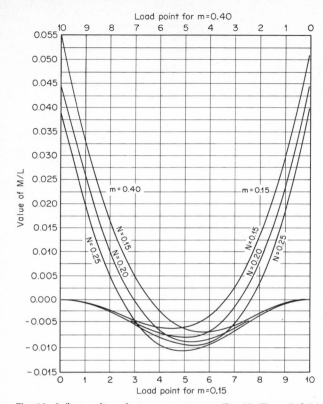

Fig. 14 Influence lines for moment at crown, Fig. 12. *(From Ref. 3.)*

TABLE 12   Parameter C for Rib Shortening*

N	g	$m = 0.15$	$m = 0.20$	$m = 0.25$	$m = 0.30$	$m = 0.40$	$m = 0.50$
0.25	1.000	0.0329	0.0370	0.0410	0.0448	0.0520	0.0588
0.24	1.347	0.0320	0.0361	0.0400	0.0437	0.0509	0.0577
0.23	1.756	0.0312	0.0352	0.0390	0.0428	0.0498	0.0566
0.22	2.240	0.0302	0.0342	0.0380	0.0417	0.0486	0.0553
0.21	2.814	0.0294	0.0332	0.0370	0.0407	0.0475	0.0541
0.20	3.500	0.0285	0.0323	0.0361	0.0396	0.0466	0.0530
0.19	4.324	0.0276	0.0314	0.0351	0.0386	0.0455	0.0519
0.18	5.321	0.0267	0.0305	0.0341	0.0376	0.0443	0.0507
0.17	6.536	0.0258	0.0295	0.0331	0.0366	0.0432	0.0496
0.16	8.031	0.0250	0.0286	0.0322	0.0356	0.0422	0.0484
0.15	9.889	0.0241	0.0277	0.0312	0.0346	0.0411	0.0472

* From Ref. 3.

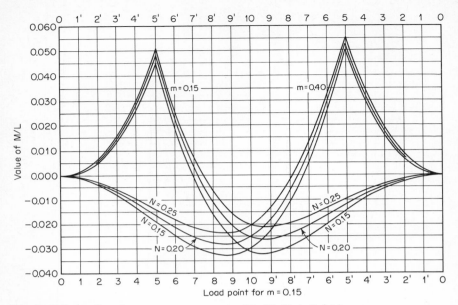

Fig. 15 Influence lines for moment at quarter point, Fig. 12. *(From Ref. 3.)*

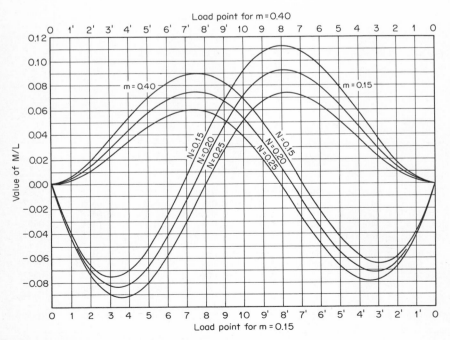

Fig. 16 Influence lines for moment at springing line, Fig. 12. *(From Ref. 3.)*

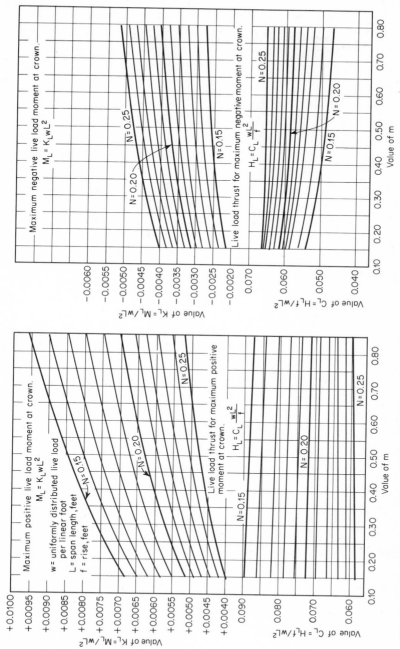

Fig. 17 Maximum crown moments and corresponding thrusts for uniform live loading. *(From Ref. 3.)*

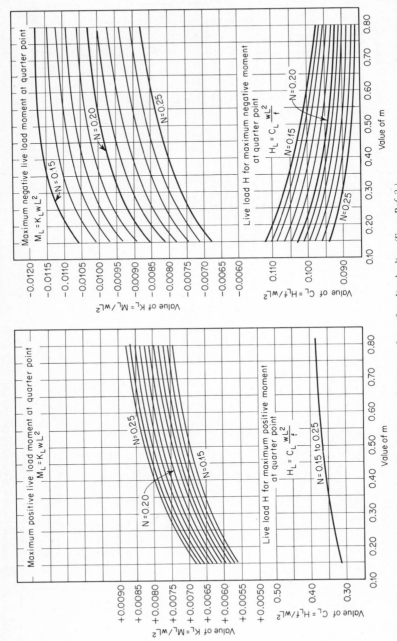

Fig. 18 Maximum quarter-point moments and corresponding thrusts for uniform live loading. (*From Ref. 3.*)

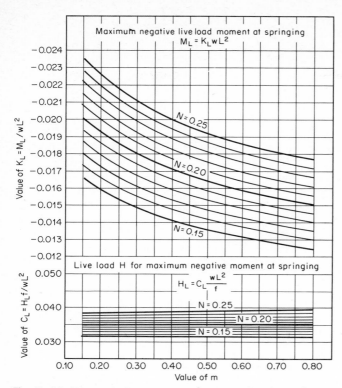

Fig. 19 Maximum springing moments and corresponding thrusts for uniform live loading. *(From Ref. 3.)*

TABLE 13 Value of y_c/f^*

N	g	$m = 0.15$	$m = 0.20$	$m = 0.25$	$m = 0.30$	$m = 0.40$	$m = 0.50$
0.25	1.000	0.2101	0.2222	0.2333	0.2436	0.2619	0.2778
0.24	1.347	0.2044	0.2163	0.2273	0.2374	0.2556	0.2713
0.23	1.756	0.1985	0.2103	0.2212	0.2312	0.2491	0.2647
0.22	2.240	0.1926	0.2043	0.2150	0.2250	0.2427	0.2580
0.21	2.814	0.1867	0.1983	0.2089	0.2187	0.2362	0.2513
0.20	3.500	0.1808	0.1922	0.2027	0.2124	0.2296	0.2446
0.19	4.324	0.1748	0.1861	0.1964	0.2060	0.2230	0.2378
0.18	5.321	0.1688	0.1799	0.1901	0.1996	0.2164	0.2309
0.17	6.536	0.1628	0.1738	0.1838	0.1931	0.2097	0.2240
0.16	8.031	0.1567	0.1675	0.1774	0.1865	0.2029	0.2170
0.15	9.889	0.1506	0.1612	0.1709	0.1799	0.1960	0.2099

*From Ref. 3.

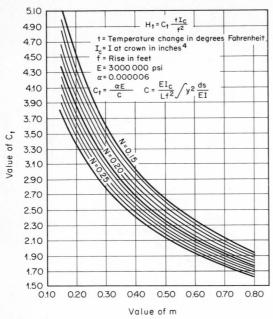

Fig. 20 Temperature thrust H_t. *(From Ref. 3.)*

TABLE 14 Influence-Line Data for Left Vertical Reaction*

Point loaded	$m = 0.15$	$m = 0.20$	$m = 0.25$	$m = 0.30$	$m = 0.35$	$m = 0.40$
0	1.0000	1.0000	1.0000	1.0000	1.0000	1.0000
1	0.9966	0.9960	0.9955	0.9951	0.9948	0.9945
2	0.9844	0.9825	0.9810	0.9797	0.9786	0.9777
3	0.9614	0.9582	0.9555	0.9532	0.9512	0.9496
4	0.9268	0.9222	0.9185	0.9153	0.9126	0.9103
5	0.8804	0.8750	0.8704	0.8668	0.8636	0.8608
6	0.8220	0.8164	0.8119	0.8079	0.8047	0.8018
7	0.7525	0.7476	0.7434	0.7399	0.7369	0.7344
8	0.6741	0.6704	0.6672	0.6645	0.6622	0.6603
9	0.5890	0.5869	0.5852	0.5837	0.5825	0.5814
10	0.5000	0.5000	0.5000	0.5000	0.5000	0.5000
9'	0.4110	0.4131	0.4148	0.4163	0.4175	0.4186
8'	0.3259	0.3296	0.3328	0.3355	0.3378	0.3397
7'	0.2475	0.2524	0.2566	0.2601	0.2631	0.2656
6'	0.1780	0.1836	0.1881	0.1921	0.1953	0.1982
5'	0.1196	0.1250	0.1296	0.1332	0.1364	0.1392
4'	0.0732	0.0778	0.0815	0.0847	0.0874	0.0897
3'	0.0386	0.0418	0.0445	0.0468	0.0488	0.0504
2'	0.0156	0.0175	0.0190	0.0203	0.0214	0.0223
1'	0.0034	0.0040	0.0045	0.0049	0.0052	0.0055
0'	0.0000	0.0000	0.0000	0.0000	0.0000	0.0000

* From Ref. 3.

Where live loads are uniform, the approximate positioning for maximum moments is indicated in Fig. 22.

As a guide in selection of sections for stress, Cross suggested proportioning allowable concrete stress as follows:

Stress due to dead load	50% (range 40–60%)
Stress due to live-load compression	5%
Stress due to live-load flexure	30% (range 20–50%)
Stress due to flexure from rib shortening, shrinkage, and temperature change	15% (range 5–20%)
Total stress	100%

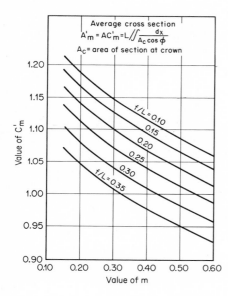

Fig. 21 Parameter C'_m for rib shortening. (*From Ref. 3.*)

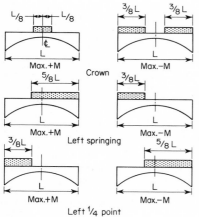

Fig. 22 Approximate uniform-live-load positioning for maximum moments.

11. Final Design The final design of an arch usually comprises a formal analysis to verify the preliminary or intermediate design under various combinations of load, temperature change, shrinkage, rib shortening, etc., with such final corrections to cross section, reinforcement, and shape as are dictated thereby. It should be noted that allowable stresses for combined compression and bending are variable and a function of the relative value of moment to thrust, and reference must be made to the particular interaction or combined-stress formulas of the specifications involved. It is also to be noted that for loading combinations involving temperature change, shrinkage, rib shortening, etc., an increase in permissible stresses is allowed by the various codes.

Preliminary and intermediate design of a highway-bridge concrete arch are illustrated in Example 2.

12. Unsymmetrical Arches The design of unsymmetrical arches is hampered by the lack of simple approximations, which are somewhat difficult to formulate. Whitney[5] has presented an extension to unsymmetrical arches of the methods described in Art. 9. Many designers use a trial-and-error approach, starting from an initial shaping to fit the pressure line for dead load or dead load plus one-half live load.

13. Ultimate Design of Concrete Arches Past reports of ACI Committee 312 have sought to develop a general method of design of concrete arches which will take into account all important factors, such as dead and live loading, effects of volume changes (shrinkage, temperature effects, and plastic flow), abutment movements, and arch deflection.[6] Because of the nonlinear relation between loads and deflection, the Committee has favored a basically ultimate-strength method of design. It recommended two sets of ultimate-load criteria:

$$U = 1.2 \left(B + 2L + \frac{W}{2} \right) + D$$

or

$$U = 1.2 \left(B + \frac{L}{2} + 2W \right) + D$$

(37)

and

$$U = K \left(B + L + \frac{W}{2} + D' \right)$$

or

$$U = K \left(B + \frac{L}{2} + W + D' \right)$$

(38)

where U = ultimate thrust-moment strength of rib
B = thrust-moment strength required by basic load, i.e., by dead load, temperature change, and shrinkage
L = thrust-moment strength required by live load
W = thrust-moment strength required by wind load
D = thrust-moment strength required by rib deformation under ultimate-load conditions
D' = thrust-moment strength required by rib deformation under working stress
K = 2 for arches and columns

Equations (37) provide for overload of the rib. They control section proportions and reinforcing steel of most arch ribs because of the usual occurrence of large dead-load thrusts but small dead-load moments. Equations (38) control the proportioning of those exceptional members in which axial load is so large that load eccentricities are less than six-tenths the depth of the member. They also govern section proportions and steel reinforcement of ribs with large dead-load moments.

The first step of the suggested design procedure consists in determining moments and thrusts in the rib for each of the specified loading conditions, assuming the rib to behave as an elastic member and neglecting influence of deflections on moments. This follows standard methods of elastic analysis.

The second step involves computation of the deformation of the arch rib by standard elastic methods, giving the angular rotations and deflections at all points of the rib. Secondary moments, i.e., horizontal thrust times deflection, are determined, and deflections and rotations calculated again. The procedure is repeated to convergence. The effect of horizontal deflections may also be considered but may be neglected for arches whose rise-span ratio is less than 1:5.

Values of modulus of elasticity of 4,000,000 psi are suggested for volumetric changes and 2,000,000 psi for determination of deflection moments and stability. It is also recommended that standard ultimate-strength design procedures for combined bending and axial thrust be used in the proportioning and design of concrete-arch ribs.

Example 2. Two-ribbed, open-spandrel fixed arch.
Specs. AASHTO, 12th ed., 1977
Concrete 4,000 psi reinforcement grade 60.
Temperature ± 40°, shrinkage–15° equivalent.

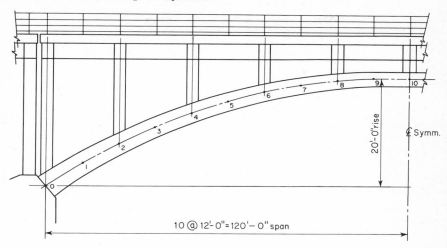

10 @ 12'-0" = 120'-0" span

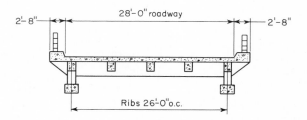

Ribs 26'-0" o.c.

PRELIMINARY DESIGN:

Design of the roadway yields concrete slab, stringer, floor beam, and spandrel column sizes from which the following dead loads are calculated. The arch rib is estimated to be 2'-6" × 1'-8" at crown and 2'-6" × 2'-8" at springing.

Item	DL (one rib)
Slab	200 kips
Curb	41
Steel railing	7
Stringers	64
Floor beams	90
Lateral strut	2
Spandrel columns	53
Arch rib	100
Total	557 = 4.64 k/ft = 55.7 k/panel point

Using a three-hinged arch approximation, dead-load thrust is

$$H_C = \frac{1}{8} \times \frac{4.64 \times 120^2}{20} = 418^k$$

Assume allowable *DL* stress = 20% × f'_c = 800 psi

$$A_c = 418/0.800 = 522 \text{ in.}^2, \text{ use } 30'' \times 20'' = 600 \text{ in.}^2$$

17-31

Depth of springing = 1½ to 2½ × depth at crown. Use 30″ × 32″ springing. Preliminary layout of axis. Spandrel arch dead load is close to uniform, for which axis would be parabolic with $N = 0.25$. Assume $N = 0.23$. From Table 9, tan $\phi = 4.442 \times 20/120 = 0.740$, $\Phi_s = 36°30'$, cos $\Phi_s = 0.804$

$$M = I_c/I_S \cos \phi_S = 24^3/(32^3 \times 0.804) = 0.303$$

Tables 10 and 11, with Eq. (19), yield preliminary depths d_k shown in figure. From scaled lengths, rib weights are calculated for 12-ft panel lengths and entered in the table.

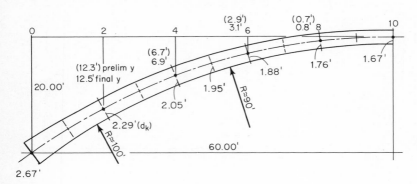

INTERMEDIATE DESIGN:
Refine arch shape to conform to equilibrium polygon for full dead load.

Panel-point Load Summary (12-ft Panels):

Load	0	2	4	6	8	10	Total 1/2 rib	Avg.
Superstructure	20.2	40.4	40.4	40.4	40.4	20.2	202.0^k	3.36 k/ft
Spandrel column	6.6	8.6	5.2	2.9	1.6	0.6	25.5	0.43
Rib	7.0	11.8	9.9	8.7	8.0	3.8	49.2	0.82
Total load at p.p.	33.8	60.8	55.5	52.0	50.0	24.6	276.7	4.61

Shape rib so that y is proportional to DL moment diagram.

Point	0	2	4	6	8	10
P	-33.8	-60.8	-55.5	-52.0	-50.0	-24.6
V	+242.9	+182.1	+126.6	+74.6	+24.6	
M_{DL}	0	+242.9	+425.0	+551.6	+626.2	+650.8 × 12′
Corrected y	0	7.5	13.1	16.9	19.2	20.0 = 20M/6
$20-y$	20	12.5	6.9	3.1	0.8	0

$20 - y = 4.78$ at point 5.

Corrected N = 4.78/20 = 0.239

$$H_D = 650.8 \times 12/20 = 390^k \qquad V_D = 243^k$$
$$\tan \phi_S = 243/390 = 0.623, \cos \phi_S = 0.849$$
$$\text{Corrected } m = 20^3/(32^3 \times 0.849) = 0.29$$

Live-load moments and thrusts

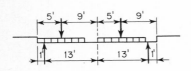

$$\frac{(8 + 22)/26 = 1.16}{}$$

$$I = \frac{50}{125 + 20} = 20.4\%$$

$0.640 \times 1.16 \times 1.204 = 0.893$ k/ft per rib

$$18 \times \frac{2}{120} \times 1.16 \times 1.204 = 0.419$$

$$LL = 1.312$$
$$\times 12 = 15.7^k/\text{p.p.}$$

For N = 0.239 and m = 0.29, Figs. 13 and 14 give:

$$H_C = (0.023 + 0.084 + 0.164 + 0.231 + \frac{1}{2} \times 0.259) \times 2 \times 15.7^k \times 120/20 = 120^k$$

$$M_C = (-0.0028 - 0.0081 - 0.0086 + 0.0051 + \frac{1}{2} \times 0.0407) \times 2 \times 15.7^k \times 120 = 22 \text{ ft-kips}$$

Moment at the springing is found similarly from ordinates in Fig. 16: $M_S = +88$ ft-kips.

Figures 13, 14, and 16 are used to calculate the live load thrusts and moments for truck loading and lane loading. Truck loading governs.

Wheels $(14/120)20 = 2.33$ units apart with span = 20 units
Truck loading $32^k \times 1.16 \times 1.204 = 44.7k$
$8^k \times 1.16 \times 1.204 = 11.2k$
For $N = 0.239$, $m = 0.29$

$+M_{CL} = [44.7(+0.0407) + 11.2(+0.0013)]120 = +220$ ft-kips (load points 10.0 and 12.33)
$H_{CL} = [44.7(+0.259) + 11.2(+0.222)]120/20 = +84k$ (load points 10.0 and 12.33)
$-M_{SL} = [44.7(-0.0855) + 11.2(-0.0543)]120 = -531$ ft-kips load points 3.5 and 5.83)
$H_{SL} = [44.7(+0.066) + 11.2(+0.156)]120/20 = +47k$ (load points 3.5 and 5.83)

With last two, right springing moment $M = +244$ ft-kips
Table 14 gives $V_S = 44.7 \times 0.939 + 11.2 \times 0.824 + (531 + 244)/120 = 58k$

Check sections

	Crown		Springing		
	M	H	M	H	V
D	0	+390	0	+390	+243
1.67 (L + I)	+367	+140	−887	+78	+97
Total	+367	+530	−887	+468	+340
× 1.3 =	+477	+689	−1153	+608	+442

$$T_S = \sqrt{608^2 + 442^2} = 752^k \approx P_S$$

Crown	Springing
$\dfrac{e}{h} = \dfrac{477 \times 12}{689 \times 20} = 0.415$	$\dfrac{e}{h} = \dfrac{1153 \times 12}{752 \times 32} = 0.575$
$\dfrac{P}{\phi f'_c bh} = \dfrac{689}{0.7 \times 4 \times 30 \times 20} = 0.410$	$\dfrac{P}{\phi f'_c bh} = \dfrac{752}{0.7 \times 4 \times 30 \times 32} = 0.280$
$\times \dfrac{e}{h} = 0.170$	$\times \dfrac{e}{h} = 0.161$
$\gamma = \dfrac{20\text{-}6}{20} = 0.70$	$\gamma = \dfrac{32\text{-}6}{32} = 0.813$
From Sec. 11, Fig. 21	From Sec. 11, Figs. 22 and 23
$\mu\rho = 0.23$	$\mu\rho = 0.20, 0.17$

$$\mu = \frac{f_y}{0.85 f'_c} = \frac{60}{0.85 \times 4} = 17.6$$

$p = 0.013$, $A_s = 7.8$ in.²	$p = 0.011$, $A_s = 10.56$ in.²
10 No. 8 = 7.85 in.²	6 No. 9 + 4 No. 10 = 11.06 in.²

CHECK TEMPERATURE, SHRINKAGE, AND RIB SHORTENING:

$$A_c = 30 \times 20 + 7.85 \times 7 = 655 \text{ in.}^2 = 4.55 \text{ ft}^2$$
$$I_c = 30 \times 20^3/12 + 7.85 \times 7 \times 7^2 = 22,690 \text{ in.}^4 = 1.09 \text{ ft}^4$$

Temperature $\pm 40°$ From Fig. 20 for $N = 0.24$, $m = 0.29$

$$H = \frac{C_t t I_c}{f^2} = \frac{2.90 \times 40 \times 22,690}{20^2} = \pm 6.6^k$$

Table 13 gives $y_c = 0.235 \times 20 = 4.70$ ft

$$M_{ct} = 6.6 \times 4.70 = \pm 31 \text{ ft-kips}, \quad M_{st} = 6.6 (20 - 4.70) = \pm 101 \text{ ft-kips}$$

Shrinkage at $-15°$ equivalent $= 15°$ equivalent $= 15°/40° = 0.375$

$$H_{sh} = 2k \quad M_{csh} = +12 \text{ ft-kips} \quad M_{ssh} = -38 \text{ ft-kips}$$

DL rib shortening: From Fig. 21, $C'_m = 1.12$. From Table 12, $C = 0.0430$

$$H_{RS} = -H_D \frac{I_c}{A_cC'_mCf^2} = \frac{390 \times 1.09}{4.55 \times 1.12 \times 0.0430 \times 20^2} = -5k$$

$$M_{CRS} = 5 \times 4.70 = 24 \text{ ft-kips}, \ M_{SRS} = -5(20 - 4.70) = -77 \text{ ft-kips}$$

	Crown		Springing		
	M	H	M	H	V
D	0	+390	0	+390	+243
$L + I$	+220	+84	−531	+47	+58
T	+31	−7	−101	−7	0
R	+12	−2	−38	−2	0
RS	+24	−5	−77	−5	0
Total	+287	+460	−747	+423	+301
× 1.3	+373	+598	−971	+550	+391

$$T = \sqrt{550^2 + 391^2} = 675 \text{ k} \approx P$$

Moments and thrusts all less than for 1.3 $[D + 1.67(L + I)]$. Therefore, final design analysis may proceed. Before doing this it is desirable to fit the intrados to a three-centered circular arc for ease of construction, and to correct the ordinates as necessary. The required radii are shown in the figure.

The plasticity of concrete under high stresses or under stresses of long duration also results in redistribution of moment between sections. Plastic flow of concrete is rapid in the early stages but decreases with time. It causes a decrease in the stresses in the concrete and an increase in the stress in the reinforcement. It is desirable to decenter arches as soon as possible if sufficient reinforcing is provided to prevent overstress in the steel.

The 1932 and 1940 reports of ACI Committee 312 contain considerable data from which probable effects of temperature change, shrinkage, and plastic flow may be estimated. While temperature, shrinkage, and abutment-movement stresses are relieved by plastic flow and cracking of the ribs, deformation stresses are increased. Records show that cracking and plastic flow have caused failure months after decentering. To prevent this, ribs should be proportioned to control initial volumetric and displacement stresses, and sufficient steel should be provided to prevent excessive flow under long time periods. A minimum of 1 percent reinforcement is recommended for important arches.

DESIGN OF FRAMES

14. Steel Frames For buildings constructed of single-span rigid frames, the following will usually provide good economy for average roof loads:[7]

Span, ft	Frame spacing, ft
30–40	16
40–60	18
60–100	20
Over 100	$\frac{1}{5}$–$\frac{1}{6}$ of span

Few structures have bases completely fixed against rotation. Therefore, rigid frames are frequently designed as if hinged at their column bases. Expensive pin-connected base details are rarely required; in most cases, an ordinary flat base plate with a single line of anchor bolts set perpendicular to the span at the column centerline will suffice.

Frames should be properly braced for lateral stability. Korn[7] suggests the use of laced struts, portals, or sway frames as follows:

In spans up to 40 ft, no lateral bracing is necessary where purlins or deep metal deck is used. Rod bracing would be desirable for one or more bays.

In spans from 41 to 60 ft, bracings are required on diagonal centerline of haunch or knee and at ridge. (Intermediate bracing members would be desirable for the 60-ft span.)

In spans over 60 ft, bracings are required at haunches and along ridges, with additional intermediate sway frames spaced at a distance equal to 90 times the least radius of gyration and not to exceed 100 times that dimension.

15. Concrete Rigid-Frame Bridges The following proportions are recommended[8] for frames of the type shown in Fig. 1:

1. Thickness at center of span (B-B) equal to about L/35. This value may be reduced to L/40 when the frame is founded on a practically unyielding foundation; it should be increased where the footings rest on highly compressible soils.

2. Thickness at A-A equal to about L/15.

3. Thickness at base (C-C) equal to about 1 ft 6 in. for 30-ft spans, 2 ft 6 in. for 60-ft spans, and 3 ft 4 in. for 90-ft spans.

16. Design By sketching the deflected shape of the structure under specific combinations of load, one may approximately locate the inflection points, which makes the structure statically determinate. The section near the center of span would be governed by dead plus live plus impact, or by the same combination plus temperature change plus shrinkage, or by the total of all these plus spread of the footings (often specified, or otherwise arbitrarily assumed as ½ in.) The knee section would be governed by dead plus live plus impact plus active lateral earth pressure on the legs, or by this combination plus rise in temperature.

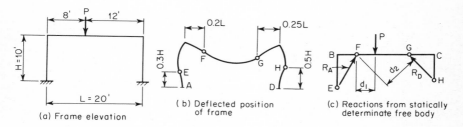

(a) Frame elevation

(b) Deflected position of frame

(c) Reactions from statically determinate free body

Fig. 23 Frame analysis by locating points of inflection.

Example A frame under a concentrated load is shown in Fig. 23. The deflected position is sketched and the inflection points E, F, G, and H located. By considering the partial structure EBFGCH in (c) the reaction $R_D = Pd_1/d_2 = P \times 4/7.8 = 0.51P$. Similarly, $R_A = 0.74P$. Therefore,

$$V_D = \frac{5}{\sqrt{5^2 + 5^2}} \times 0.51P = 0.361P$$

$$H_D = \frac{5}{\sqrt{5^2 + 5^2}} \times 0.51P = 0.361P$$

$$M_D = 0.5 \times 10 \times H_D = 1.80P$$

$$V_A = \frac{7}{\sqrt{7^2 + 4^2}} \times 0.74P = 0.642P$$

$$H_A = \frac{4}{\sqrt{7^2 + 4^2}} \times 0.74P = 0.368P$$

$$M_A = 0.3 \times 10 \times H_A = 1.1P$$
$$M_C = 0.25 \times 20 \times V_D = 1.80P$$

Discrepancies in these reactions are due to approximations in locating the points of inflection.

Approximate locations of the inflection points for rectangular bents are given in Fig. 24. These locations are based on members with constant moment of inertia and equal stiffnesses of columns and beams. If these two conditions are not met, the point of inflection should be moved toward the stiffer joint (by about 5 to 15 percent of the indicated distance to that joint) in order to obtain a better first approximation.

Charts for horizontal reactions in two-legged rigid frames with hinged bases are given in Figs. 25 to 28. Moment and reaction tables in Ref. 27 also list the corresponding lightest W section in A36 steel. Similar aids for first approximation of moments and forces are available in tabular form.[9] Formulas for a large variety of framed structures are also available.[1]

Member sizes and dimensions of certain frames, such as pier bents for grade-separation structures, are proportioned for appearance or for standardization of design. In such cases, a final analysis can be made without recourse to a preliminary design.

A formal analysis should be made to check the preliminary design.

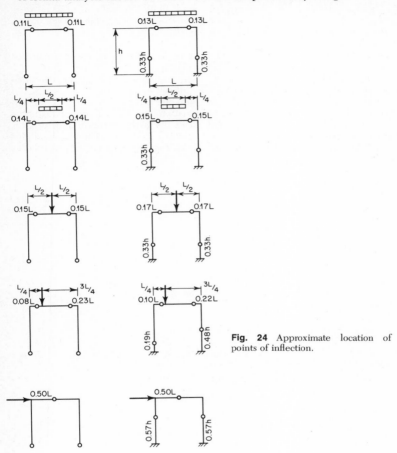

Fig. 24 Approximate location of points of inflection.

Example 3 Design of steel-building rigid frame. A single-story building requires a clear span of approximately 100 ft and a minimum height of 18 ft. Welded A36 steel gabled frames, 20 ft on centers, will satisfy the requirements. The AISC specification will be followed. The dimensions of the frame are shown in Fig. 29. The bases are considered to be hinged.

Loads and load combinations:
  Dead load D:

$$
\begin{array}{ll}
\text{Roofing} & = 4 \text{ psf} \\
\text{Insulation} & = 3 \\
\text{Metal deck} & = 2 \\
\text{Miscellaneous} & = 4 \\
\text{Purlins} & = \underline{4} \\
& \overline{17 \text{ psf of sloping surface}}
\end{array}
$$

$$0.017 \times \frac{52.7}{50} \times 20 = 0.36 \text{ kip/ft of frame}$$

$$
\begin{array}{rl}
\text{Frame} = & \underline{0.10} \\
D = & 0.46 \text{ kip/ft}
\end{array}
$$

Snow load S: At 30 psf, $S = 0.03 \times 20 = 0.6$ kip/ft
Drift snow load Dr: 0.6 kip/ft on half span only
Wind load W: At 20 psf, $W = 0.02 \times 20 = 0.4$ kip/ft

Various combinations of these loads are investigated (Fig. 30). Normally, combination I controls the design of the frame. The remaining combinations may control the design of details. A 33 percent increase in allowable stresses is permitted for combinations III to V.

PRELIMINARY ANALYSIS. Figures 25 and 26 are used to determine the horizontal reactions. It is assumed that $I_2/I_1 = 0.6$. This ratio will be checked later.

$$K = \frac{I_2}{I_1}\frac{h}{m} = 0.6\,\frac{20}{52.7} = 0.228$$
$$Q = \frac{f}{h} = \frac{16.67}{20} = 0.833$$

For uniform vertical load w, Fig. 25 gives

$$C_1 = 0.059 \qquad H_A = H_E = \frac{C_1 w L^2}{h} = 28w$$

For horizontal loads, Fig. 26 gives

$$C_6 = 0.56 \qquad H_E = C_6 h w = 11.2w$$

$H_E = C_1 w l^2/h$ for uniform vertical load on full span
 ($H_E = C_1 w l^2/2h$ when half-span only is loaded)
$H_E = C_2 C_1 Pl/h$ for concentrated load P on span, distant al from either end
 ($H_E = 2C_2 C_1 Pl/h$ for equal concentrated loads P, distant al from both ends)
$H_E = C_3 Pe/h$ for concentrated load P on either bracket of length e
 ($H_E = 2C_3 Pe/h$ for equal concentrated loads P on both brackets

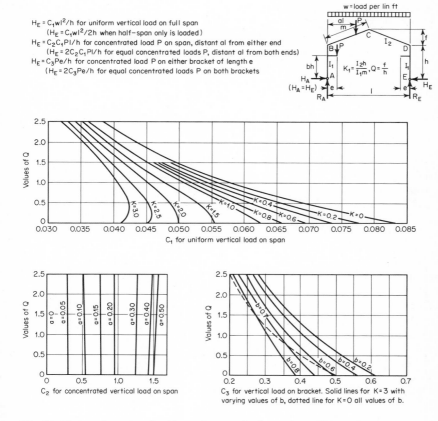

Fig. 25   Horizontal reactions for ridge and rectangular frames, hinged bases. (*Courtesy of American Institute of Steel Construction.*)

Loading	Horizontal reactions
D	$H_A = H_E = 28 \times 0.46 = 12.9\,\text{kips}$
S	$H_A = H_E = 16.8\,\text{kips}$
Dr	$H_A = H_E = 8.4\,\text{kips}$
W	$H_E = 4.5\,\text{kips}$
	$H_A = 36.67 \times 0.4 - 4.5 = 10.2\,\text{kips}$

The moments and forces at various points of the frame may now be computed, and are summarized in Table 15.

DESIGN OF MEMBERS

RAFTER: $M = +\,338\,\text{ft-kips}$
$N = +\,31.1\,\text{kips}$ (not corrected for moment redistribution)
$V = 1.4\,\text{kips}$ (not corrected for moment redistribution)

Unbraced lengths:

$L_x = 44.27\,\text{ft}$ (assume $K_x = 1.5$) $L_y = 6.32\,\text{ft}$

Try W24 × 84: $r_x = 9.79\,\text{in.}$ $\left(\dfrac{KL}{r}\right)_x = 81.5$ $F_a = 15.2\,\text{ksi}$

$H_E = C_4\,wh$, for uniform load against roof
$H_E = C_5\,wh$, for uniform load against vertical side
$H_E = C_6\,wh$, for uniform load against total height
$H_E = Pb(C_7 - b^2 C_8)$, for concentrated load P on one vertical side ($b \lesssim 1.0$)
($H_E = P$, for concentrated load P on both vertical sides at same elevation

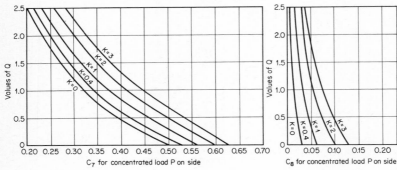

Fig. 26 Horizontal reactions for ridge and rectangular frames, hinged bases. (*Courtesy of American Institute of Steel Construction.*)

$$\frac{L_y d}{A_f} = 263 < 500 \qquad \frac{76b_f}{\sqrt{F_y}} = 114 > L_y \qquad F_b = 24 \text{ ksi}$$

$$f_a = \frac{31.1}{24.7} = 1.26 \qquad f_b = \frac{338 \times 12}{196} = 20.6 \text{ ksi}$$

$$\frac{f_a}{F_a} = \frac{1.26}{15.2} = 0.083 < 0.15$$

$$\frac{f_b}{F_b} = \frac{20.6}{24.0} = \underline{0.859}$$

$$0.942 < 1.0$$

COLUMN

Use $M = -417$ ft-kips (no moment redistribution)
$N = +53.0$ kips
$V = 29.7$ kips

Unbraced lengths: $L_x = 14$ ft $L_y = 14$ ft $K_y = 1.0$

For effective length coefficient K_x,

At top of column: $G_T = \dfrac{I_1/h}{I_2/2m} = \dfrac{1}{0.6}\dfrac{2 \times 52.7}{20} = 8.8$

At bottom of column: $G_B = 10$

$$K_x = 2.9 \qquad \text{(Sec. 6, Fig. 5)}$$

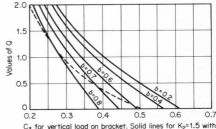

$H_E = C_1 w l^2/h$ for uniform vertical load on full span
($H_E = C_1 w l^2/2h$ when half-span only is loaded)
$H_E = C_2 C_1 P l/h$ for concentrated load P on span, distant al from either end
($H_E = 2C_2 C_1 P l/h$ for equal concentrated loads P, distant al from both ends
$H_E = C_3 Pe/h$ for concentrated load P on either bracket of length e
($H_E = 2C_3 Pe/h$ for equal concentrated load P on both brackets

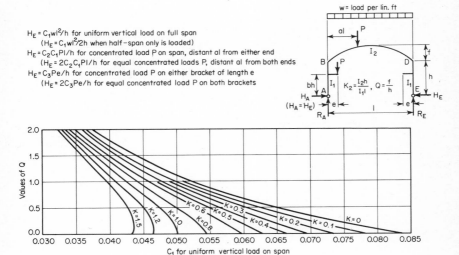

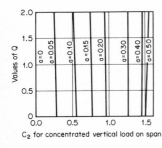

Fig. 27 Horizontal reactions for parabolic and rectangular frames, hinged bases. (*Courtesy of American Institute of Steel Construction.*)

$H_E = C_4$ wh for uniform load against roof
$H_E = C_5$ wh for uniform load against vertical side
$H_E = C_6$ wh for uniform load against total height
$H_E = Pb$ $(C_7 - b^2 C_8)$ for concentrated load P on one vertical side (b≲1.0)
    ($H_E = P_1$ for concentrated load P on both vertical sides at same elevation)

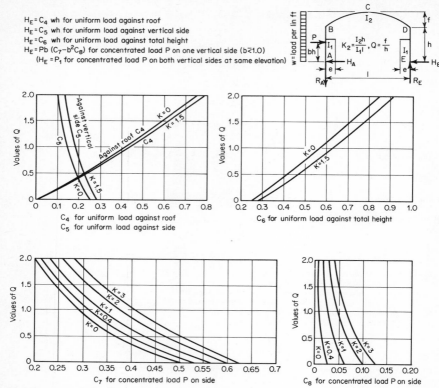

Fig. 28 Horizontal reactions for parabolic and rectangular frames, hinged bases. *(Courtesy of American Institute of Steel Construction.)*

Try W30 × 99:

$$r_x = 11.7 \qquad \left(\frac{KL}{r}\right)_x = 42$$

$$r_y = 2.1 \qquad \left(\frac{KL}{r}\right)_y = 84 \qquad F_a = 14.9 \text{ ksi}$$

$$\frac{76}{\sqrt{F_y}} b = 132 \text{ in.} < 14 \text{ ft} \qquad F_b = 22 \text{ ksi}$$

$$f_a = \frac{53}{29.1} = 1.82 \text{ ksi} \qquad f_b = \frac{417 \times 12}{269} = 18.6 \text{ ksi}$$

$$\frac{f_a}{F_a} = \frac{1.82}{14.9} = 0.12 < 0.15$$

$$\frac{f_b}{F_b} = \frac{18.6}{22.0} = \underline{0.84}$$

$$0.96 < 1.0$$

FINAL ANALYSIS. Check assumed sections:

$$\frac{I_2}{I_1} = \frac{2364.3}{3988.6} = 0.594 \approx 0.6$$

HAUNCH DESIGN (Fig. 31)
Try haunch makeup:

	A	I
2 flange plates 10½ × ¹¹⁄₁₆	14.45	5550
1 web plate 38½ × ½	19.25	2380
	33.7	7930

TABLE 15 Moments, Axial Forces, and Shears for Frame of Fig. 29

Loading	A		1			B			2			3			C		
Points	R	H	M*	N†	V	M	N	V	M	N	V	M	N	V	M	N	V
D	23.0	12.9	−181	23.0	12.9	−258	23.0	12.9	−123	18.4	14.2	+121	13.5	0.6	+102	12.3	4.1
S	30.0	16.8	−236	30.0	16.8	−336	30.0	16.8	−160	24.0	18.5	+158	17.6	0.8	+133	16.0	5.3
Dr (on right half)	7.5	8.4	−118	7.5	8.4	−168	7.5	8.4	−130	10.4	4.4	+ 29	10.4	4.4	+ 67	10.4	4.4
Dr (on left half)	22.5	8.4	−118	22.5	8.4	−168	22.5	8.4	− 29	13.1	12.6	+130	7.1	−5.3	+ 67	5.6	−9.8
W (on right side)	2.7	4.5	− 63	2.7	4.5	− 90	2.7	4.5	− 80	5.2	1.2	− 40	5.2	1.2	− 30	5.2	1.2
W (on left side)	−2.7	−10.2	+104	−2.7	−4.6	+124	−2.7	−2.2	+106	−1.9	−2.3	+ 3	2.3	−3.7	− 30	3.4	4.0
D + S	......	......	−417	53.0	29.7	−594	53.0	29.7	−283	42.4	32.7	+279	31.1	1.4	+235	28.3	9.4
With moment redistribution‡	......	......	......	......	......	−535	......	......	−224	......	......	+338	......	......	+294		

* Moments in ft-kips.
† N in kips; negative sign denotes tension.
‡ AISC Specification, Art. 1.5.1.4.

Use loads at point B (Table 15).
At section eb:

$$M = -542 \text{ kips (Fig. 32)}$$
$$N = +53 \text{ kips}$$
$$V = 29.7 \text{ kips}$$

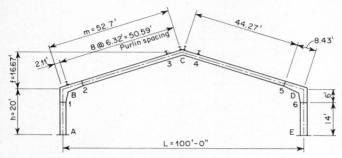

Fig. 29   Frame elevation, Example 3.

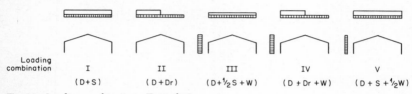

Fig. 30   Loading combinations, Example 3.

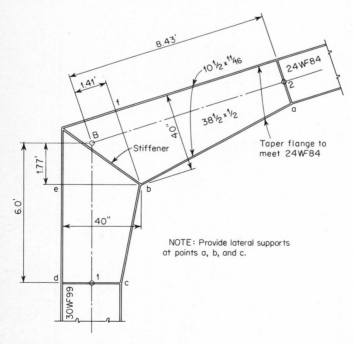

Fig. 31   Haunch detail, Example 3.

Unbraced length, $L = 6.0 - 1.77 = 4.23$ ft $= 51$ in.

AISC Formula (1.5–6a): $r_y^2 = \dfrac{b^2}{12} \dfrac{A_f}{A_f + A_w/6} = 7.52$

$$r_y = 2.74 \qquad \frac{L}{r_y} = 18.6 < \sqrt{\frac{102{,}000\,C_b}{F_y}}$$

AISC Formula (1.5–7): $\dfrac{Ld}{A_f} = 282 < 545 \qquad F_b = 22$ ksi

$$f_b = \frac{N}{A} + \frac{Mc}{I} = \frac{53}{33.7} + \frac{542 \times 12 \times 20}{7930} = 18.0 \text{ ksi}$$

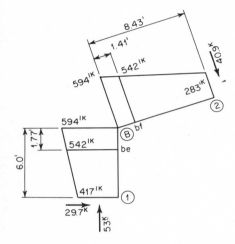

Fig. 32 Moments and forces at haunch.

At section bf, the haunch is checked similarly and found to be satisfactory ($f_b = 17.7$ ksi).

For the diagonal stiffener required at b, the resultant of the flange loads in rafter and column is 117 kips.

$A_s = {}^{117}\!/_{22} = 5.3$ in.^{2}; use 2 plates $5 \times {}^{9}\!/_{16}$.

17. Arched Bents; Continuous Arches on Elastic Piers Arches resting on and integral with relatively flexible columns are used in both bridge and building work. They are common in buildings with thin-barrel concrete-shell roofs, where the arches act as single or continuous end or intermediate frames to take the loads of the roof shell. Building arches are commonly made with constant-section members of regular geometric shape (circular, parabolic, etc.), while bridge arches may be of varying section and shape. Tabular solutions for circular arched bents and gable bents are available.[30,31]

Arched frames may be analyzed by the general method of virtual work. Single-span frames are most easily handled by the column analogy. With continuous arched frames, however, variations of modern numerical procedures, such as those involving moment distribution or slope deflection, are more popular. The moment-distribution procedure is described primarily because it is basically similar to that used for multistory rectangular bents, wherein fixed-end moments are distributed while translation of joints is prevented and corrections are made for the actual translation of the joints. The corrections are made by applying unit horizontal displacements successively at each joint and writing simultaneous equations equal in number to the total of such joints having horizontal freedom of movement, in order to balance the horizontal thrusts. The simultaneous-equation technique is particularly adaptable to cases with several different loading conditions, which advantage is not shared by alternate variations of this procedure (such as by successive balancing of moments and thrusts), and which may even offer difficulties in convergence.

The primary difference in operating with curved members is that fixed-end moments, flexural stiffness, and moment carryover factors are modified. In addition, however, there exist fixed-end thrusts due to loads, a thrust induced by applied moment at one end when the far end is free to rotate, and a moment induced by horizontal displacement when there is no end rotation. The following briefly outlines the use of the column analogy for obtaining these basic constants.

From Fig. 33 and the formula $f = P/A + M_x y_c/I_x + M_y x_c/I_y$,

$$K = \text{moment stiffness} = f_a = \frac{1}{A} + \frac{y_c^2}{I_x} + \frac{x_c^2}{I_y} \tag{39}$$

$$CK = \text{carryover moment} = f_b = \frac{1}{A} + \frac{y_c^2}{I_x} - \frac{x_c^2}{I_y} \tag{40}$$

$$C = \text{carryover factor} = \frac{f_b}{f_a} \tag{41}$$

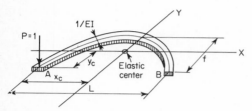

Fig. 33 Analogous column loaded to calculate moment stiffness and carryover.

M_v = moment due to unit vertical displacement (no end rotation)

= f_a for $P = 0$, $M_x = 0$, $M_y = 1$

$$= \frac{x_c}{I_y} \tag{42}$$

M_u = moment due to unit horizontal displacement

= f_a for $P = 0$, $M_x = 1$, $M_y = 0$

$$= \frac{y_c}{I_x} \tag{43}$$

H_m = thrust due to unit moment (far end fixed)

$$= \frac{y_c/I_x}{K} \tag{44}$$

H'_m = thrust due to unit moment (far end pinned)

$$= \frac{H_m}{1 + C} \tag{45}$$

For arches of constant cross section and regular shape, the above values as well as fixed-end moments and thrusts may be taken from Tables 3, 5, and 16. For arches of varying section, Whitney's tables may be used, or the constants calculated by the above formulas.

Example 4 illustrates the application of moment distribution to arched frames. A single-span circular arch frame under uniform vertical load is used only for illustration; it is actually simpler to solve this case by column analogy or other methods. There are two degrees of freedom (points A and B) with respect to horizontal movement. However, because of symmetry, only one equation is needed. Where dissymmetry of loading or of geometry exists, two sidesway corrections would have to be made in a single-bay frame, requiring two simultaneous equations.

A slope-deflection solution of this problem is given in Ref. 11.

SPECIAL TOPICS

18. Second-Order Theory   The methods of analysis discussed in Arts. 4 and 9 are first-order analyses; that is, they neglect the secondary moments resulting from deflection of the arch or frame. For example, if one were to calculate the deflections Δ_1 corresponding to a first-order analysis of the two-hinged arch, increments of moment $dM_1 = H\Delta_1$ could be determined. These, in turn, induce additional deflections and a second increment of moment dM_2, etc. Calling $dM_1/M_1 = \alpha_1$, $dM_2/dM_1 = \alpha_2$, and assuming that succeeding ratios dM_3/dM_2, etc., are equal to α_2, the final moment may be approximated by

$$M = M_1(1 + \alpha_1 + \alpha_1\alpha_2 + \alpha_1\alpha_2^2 + \cdots)$$
$$= M_1\left(1 + \frac{\alpha_1}{1 - \alpha_2}\right)$$

Example 4. Arch bent analysis.

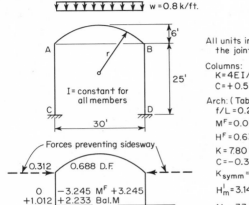

All units in feet and kips. Positive moments rotate the joint clockwise. Thrusts to right are positive.

Columns:
$K = 4EI/L = 0.160\,EI$
$C = +0.5$

Arch: (Tables 3 and 16)
$f/L = 0.20$
$M^F = 0.004507 \times 0.8 \times 30^2 = 3.245$
$H^F = 0.6388 \times 0.8 \times 30 = 15.33$
$K = 7.80\,EI/30 = 0.260\,EI$
$C = -0.357$
$K_{symm} = 0.260\,EI\,(1 + 0.357) = 0.353\,EI$
$H_m^I = 3.14/30 = 0.105$
$M_u = 33.17\,EI/30^2 = 0.0369\,EI$

Moment distribution — no sidesway

Thrusts at A
+15.33
− 0.47

at B
−15.33 $= H^F$
+ 0.47 $=$ in arches due to rotation $= 3.245 - 1.012 = 2.233$ reduction
$\underset{4.466}{\times\quad 2 \text{ ends}}$
$\times 0.105 = H_m^I$
0.47

+ 0.06
+14.92
→

− 0.06 = in columns due to rotation $(1.012 + 0.506)/25 = 0.06$
−14.92 = total thrust preventing sidesway
←

Apply horiz. $\Delta = 1$ at A   Can be done
    ←   simultaneously
         $\Delta = 1$ at B   because of symmetry
         →

Translational stiffnesses
$\text{Cols:} \dfrac{6EI\Delta}{L^2} = \dfrac{6EI \cdot 1}{25^2} = 0.0096EI$
Arches: M_u $= 0.0369EI$

Apply arbitrary M of 36.9 in arch, 9.6 in cols.

Correction moment distribution

Thrusts at A   at B $y_C = 3.96$
  −18.64   +18.64 = thrust due to rotation, $73.800/3.96 = 18.64$
  + 9.28   − 9.28 = in arches due to rotation $29.630 − 73.800 = −44.170$ reducton
                                                                                   2 ends
                                                                                88.340
                                                                                $\dfrac{0.105}{9.28} = H_m^{!}$

  − 1.97   + 1.97 = in columns due to rotation $(29.630 + 19.615)/25 = 1.97$
  −11.33   +11.33 = total thrust at A and B
   ←   →

To eliminate unbalanced thrusts at A and B
  $+14.92 − 11.33X = 0$, $X = 1.317$

Final $M_A = -1.012 + 1.317(+29.630) = +38.0$ ft-kips $= -M_B$
      $M_C = 0.506 + 1.317(-19.615) = -25.3$ ft-kips $= -M_D$
      $H_C = -H_D = -0.06 + 1.317(+1.97) = +2.53$ kips

If M_1 is thought of as an incremental moment, the formula becomes

$$M = M_1 \left(\frac{1}{1-\alpha} \right)$$

Charts for approximating the amplification of stress resulting from secondary moments in flexible steel arches are useful for preliminary design of long-span arches, as well as for avoiding proportions which might prove dangerous because of instability.[12] Presentations of second-order theory are given in Refs. 13 and 14.

TABLE 16   Properties of Arches of Constant Cross Section*

Arch shape	f/L	y_c	Moment stiffness K	Moment carry-over factor C	Thrust due to unit moment (other end free to rotate) H_m'	Moment due to unit horizontal displacement (no end rotation) M_u	Moment due to unit vertical displacement (no end rotation) M_v
Parabolic	0.1	0.006L	8.55EI/L	0.330	6.28L	71.42EI/L²	5.73EI/L²
	0.2	0.129	7.54	0.322	3.17	31.64	5.11
	0.3	0.187	6.46	0.316	2.15	18.27	4.42
	0.4	0.242	5.53	0.312	1.64	11.89	3.80
	0.5	0.295	4.79	0.311	1.33	8.34	3.30
Circular	0.1	0.067	8.71	0.343	6.27	73.23	5.72
	0.2	0.132	7.80	0.357	3.14	33.17	5.02
	0.3	0.196	6.69	0.383	2.10	19.42	4.13
	0.4	0.258	5.59	0.415	1.58	12.52	3.27
	0.5	0.318	4.63	0.450	1.27	8.55	2.55

* From J. Michalos, "Theory of Structural Analysis and Design," The Ronald Press Company, New York, 1958.

19. Interaction of Arch and Deck In tied arches, where the tie is a continuous deck girder, and in spandrel arches, where the deck is continuous (no expansion joints), the moment at any section will be divided between the arch rib and the deck girder. The division may be approximated as though the two elements were a flitched beam, with common deflections and angular changes. This problem is discussed in Refs. 15, 16, and 17.

20. Buckling of Arches Discussion of arch buckling may be found in the paper by Wastlund.[18] Common cases of critical loads are given in the following summary:

1. Circular ring, of radius r and constant section, under radial pressure p per unit length (antisymmetric buckling):

$$\text{Critical axial force } S_{cr} = \frac{3EI}{r^2}$$

2. Two-hinged circular arch with rise f, radius r, central angle θ, and of constant section, under radial pressure p per unit length (antisymmetric buckling):

$$\text{Critical axial force } S_{cr} = \left(\frac{\pi^2}{\theta^2} - 1\right)\frac{EI}{r^2}$$

3. Parabolic arch of rise f and span L:

$$H_{cr} = \frac{\beta EI}{L^2}$$

where β is given in Table 17.

TABLE 17   Values of β for Buckling of Parabolic Arches*

f/L		0	0.1	0.2	0.3	0.4	0.5	Source
3-hinged	$I = $ const.	29.8	28.5	24.9	20.2 (19.8)	15.4 (13.6)	...	Stuessi
	$I = I_c \sec \phi$	29.7	29.4	27.7	25.3 (25.1)	22.6 (19.4)	19.8 (15.0)	Dischinger
2-hinged	$I = $ const.	39.4	35.6	28.4	19.4	13.7	9.6	Lockschin
	$I = I_c \sec \phi$	39.4	37.2	31.6	25.1	19.4	15.0	Dischinger
Fixed	$I = $ const.	80.8	75.8	63.1	47.9	34.8	...	Stuessi
	$I = I_c \sec \phi$	80.8	78.4	70.8	61.1	51.1	41.8	Dischinger

* Values are for symmetrical buckling.   Figures in parentheses are for unsymmetrical buckling.

21. Laterally Loaded Arches and Frames   Laterally loaded arches and frames, such as those subjected to wind, earthquake, etc., may be analyzed by the general method discussed in Art. 4 [Eq. (4)], taking into account the effect of torsion on the sections. Free-standing ribs without bracing are relatively infrequent, since it is customary in bridge work to tie the ribs by lateral struts. Such struts may vary in flexural and torsional stiffness from very flexible elements (which substantially divide the load equally between ribs) to very stiff struts which transform the rib system into a Vierendeel truss fixed at the abutments.

Analysis of such laterally loaded systems is extremely complex. A numerical procedure by Baron and Michalos[19] provides an orderly and systematic calculation. Michalos[20] also gives influence-line data for laterally loaded parabolic arches as a function of the ratios of rise to span and of bending to torsional stiffness. Such data are satisfactory for designing ordinary arches approximating a parabolic shape, and for preliminary design of long-span arches.

22. Skewed Barrel Arches and Rigid-Frame Slabs   A fixed, right barrel arch or rigid-frame slab, analyzed as a plane structure of unit width, is three degrees statically indeterminate. By contrast, the corresponding skewed structure constitutes a space structure of six degrees redundancy, having three unknown internal thrusts and three unknown internal moments at any cut section in the arch (Fig. 34). A skewed two-hinged barrel arch has four redundants. For extreme skews, long spans, or multiple spans, shearing stresses due to torsion become important. Although solution by the method of virtual work is entirely feasible, considerable difficulty is experienced because of the torsional moments.

Modern approaches to skew-barrel analysis evaluate the effects of skew separately as a kind of secondary stress, in which the ordinary stresses (T_x, T_y, M_z) for the rectangular structure of the same right span and the same elastic and geometric properties are used as primary stresses. The skew stresses (T_z, M_x, M_y) are then developed by simple approxi-

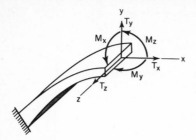

Fig. 34 Skewed barrel arch.

mations based on many theoretical and experimental investigations. Methods of analysis have been developed.[21,22]

CONSTRUCTION AND DETAILS

23. Concrete Arches and Frames *Reinforcement.*   Arch rib and rigid-frame reinforcement usually is provided in the form of parallel intradosal and extradosal bars, with some transverse distribution and temperature reinforcement (say, 0.25 percent of the transverse area) in the case of barrel arches, and stirrups as spacing hoops and for shear reinforcement as needed. Fabricated structural-steel sections, such as single or paired angles, plates, or channels, may also be used as top and bottom longitudinal reinforcement, tied together with latticing or battens to form a truss. Such trusses may be erected as three-hinged arches between abutments, with transverse bracing for stability. Occasionally, the

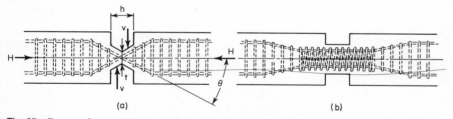

Fig. 35 Concrete hinges. (a) Mesnager; (b) Considére.

trusses are made sufficiently strong and stiff to support the arch forms and the wet concrete without the need for centering.

Falsework or Centering.   Pipe scaffolding and timber or steel trussed centering are common methods of supporting the arch rib or barrel during placing of concrete. Reuse of centering for different ribs or sections, or even for different bridges or building roofs, is facilitated by providing means of dismantling or movement on rails or skids, or by crane, etc. Centering must be carefully designed to allow for camber and for shrinkage and settlement of the centering. Release or striking of the centering should take place only after the concrete has attained its design strength. Release is accomplished by means of jacks, wedges, or other devices permitting a controlled sequence of operations; usually the crown is released first, then the haunches simultaneously on each side.

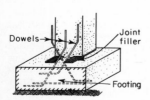

Fig. 36 Typical detail of hinged base for concrete rigid frame. (*Courtesy of Portland Cement Association.*)

Sequence of Placing Concrete.   To minimize shrinkage stresses, concrete arches are poured in alternate lateral sections (voussoirs), preferably with small keyways between the sections, the keyways being placed last. The sequence should take account of deflections of the arch centering, with the abutment sections preferably placed last. Other sections should be placed symmetrically with respect to the center of the bridge span. Where wide barrel arches require a longitudinal joint, the sections on each side of such joint should be poured on independent centering to avoid relative settlements.

Concreting of rigid frames is conventionally done in the sequence: footings, legs, deck. Proper shear keys are placed between successive pours of these elements. In low frames

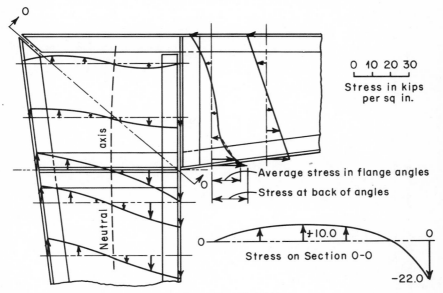

Fig. 37   Normal stresses in square-knee frame. *(From Ref. 24.)*

with heavy legs, or in continuous frames where temperature and shrinkage stresses are important, the deck may be poured with a shrinkage keyway in the spans. This keyway is filled with concrete after the major shrinkage in the deck has occurred.

Drainage and Waterproofing.   Drainage of filled-spandrel arches by adequate tile or pipes to an outlet beyond the abutments is important to eliminate frost damage and the surcharge of accumulated water. Spandrel-filled barrel arch bridges should be treated with an asphaltic membrane or surface-coat waterproofing on extrados and backs of spandrel walls.

Expansion Joints.   The spandrel walls, handrails, deck, and other elements which may participate in the longitudinal action of the arch must be provided with expansion joints. Joints at crown and springing are satisfactory for spans to about 70 ft, but for longer spans five or more such joints should be used.

Deck Participation.   An interaction of the rib with the superstructure occurs, particularly in the case of spandrel walls and the deck near the crown, which stiffens the rib of a spandrel-filled arch. In the case of an open-spandrel system having stiff columns near the crown, the interaction tends to convert the arch into a Vierendeel truss. There is a tendency in practice to treat participation stresses as secondary effects and not to count upon them to contribute to the carrying capacity of the rib. Therefore, measures such as jointing, mentioned previously, must be considered to prevent interaction; otherwise objectionable cracking may occur. It is also possible to form the superstructure after striking the arch centering, but this will remove the participation only with respect to dead load. In longer spans, with relatively less stiff columns, the interaction may be reduced. Frequently, control joints are placed at ends of spandrel columns to produce articulation without impairing the appearance, but some means must be provided to carry traction forces of the deck.

Temporary and Permanent Hinges.   Various devices may be employed to compensate for the stresses produced by the displacement of the pressure line as a result of rib shortening, shrinkage, drop in temperature, and foundation settlements. Most popular on long-span concrete arches, particularly with low rise-span ratios, is the use of temporary hinges at crown and abutments to form a three-hinged arch under dead load, so that on

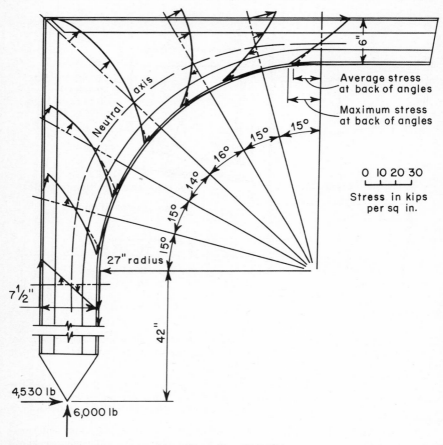

Fig. 38   Normal stresses on curved-knee frame. *(From Ref. 24.)*

removal of centering there is no displacement of the pressure line due to rib shortening. At the same time, by placing the hinges with calculated eccentricities, the effects of shrinkage and temperature drop may be counteracted by stresses of opposite sign induced in the rib. All these produce substantial economies.

All hinges are subsequently concreted solid, so that under live load the structure acts as a hingeless arch. Hinges may be of concrete, using the Mesnager or Considére designs (Fig. 35); or they may be of cast steel similar to standard hinges for steel bridges, but at a considerable increase in cost over concrete hinges. The Mesnager hinge is usually designed with hinge opening h equal to or slightly greater than the vertical thickness t. Hinge bars are limited to an L/r of about 20 to 40 based on the inclined length between faces of concrete. The two inclined hinge bars are designed to take, as a truss, the normal and shear forces from the rib acting at their intersection, with areas selected so as not to exceed 30 percent of the yield stress of the steel. Lateral stirrups and ties must also be provided to resist the tendency of the concrete to burst, and must be fully developed in a distance of 8 bar diameters from the face. Mesnager hinges should not be used for high shear-thrust ratios, and limits of V/H of 0.024, 0.195, and 1.0 have been suggested for angles $\theta = 15°$, 30°, and 45°, respectively.

The Considére hinge is essentially a short section of spiral column, its dimension being such that the resistance to rotation is small compared with that of the main adjoining members. Considére hinges are designed by obtaining the percentage of longitudinal and spiral steel from spiral-column formulas and limiting the deformation in the extreme fibers at maximum rotation to a permissible proportion of the ultimate value. Details and charts for design of both types of hinges are given in Ref. 23.

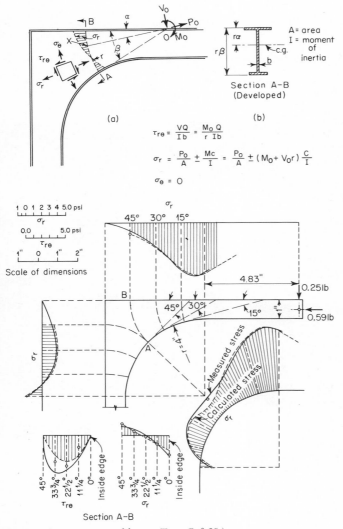

$$\tau_{re} = \frac{VQ}{Ib} = \frac{M_o\,Q}{r\,Ib}$$

$$\sigma_r = \frac{P_o}{A} \pm \frac{Mc}{I} = \frac{P_o}{A} \pm (M_o + V_o r)\frac{C}{I}$$

$$\sigma_\theta = 0$$

Fig. 39 Stresses in curved knees. *(From Ref. 26.)*

Stress compensation may also be effected by a system, developed by Freyssinet,[17] of inserting hydraulic jacks at the crown in a temporary gap to adjust the line of pressure, as desired, before concreting the gap. Other means of compensation used by Freyssinet induce residual stresses by pouring the arch rib in layers.

Details of a hinged base for a concrete rigid frame are shown in Fig. 36.

Foundations.   The importance of stable, rigid foundations for arches cannot be over-emphasized. Factors to be considered include vertical and lateral stability of bearing foundations and pile foundations, protection against erosion and scour, proper anchorage of reinforcement, and proper design of hinges.

24. Steel Arches and Rigid Frames   *Knees.* The knee of a rigid frame cannot be analyzed by elementary beam-flexure theory. Considerable experimentation has been done, both on full-scale knees of steel and by photoelastic analyses of models. Figures 37 and 38 show stresses in tests of an actual structure, and indicate the shift of the neutral axis toward the inside corner.[24]

Analysis of rigid-frame knees of steel may be made by approximation procedures given by Bleich[25] and by Olander.[26] Olander's formulas, based on the *developed* cross sections shown by the circular arcs, approximate the stresses quite closely (Fig. 39).

Wind Bracing.   In bridges, wind bracing must be placed between the ribs for lateral stability. This bracing usually consists of struts rigidly connected to the ribs, to form with longer spans a curved Vierendeel truss capable of resisting the lateral forces. An alternate is the use of a truss of diagonals and struts between the ribs, or of crossed diagonals alone.

Fig. 40   Basket-handle arch bridge, Rio Blanco, Mexico. *(Kavanagh and Piccone.)*

With arch bridges of the Langer-girder type (employing a heavy tie girder and relatively slender rib rising over the roadway), heavy overhead lateral bracing of the truss or Vierendeel type is unsightly. Successful use of the arches to brace themselves, as employed in the 250-ft-span Rio Blanco Bridge (Figs. 40 and 41) and in the 815-ft span of the Fehrmarnsund Bridge on the International Highway Route between Germany and

Denmark illustrates the lightness which such lateral bracing may take. The Rio Blanco bridge consists essentially of two box-type steel arch ribs, open on the underside, with 30 wire-rope hangers and two tie girders. The floor is a reinforced-concrete slab on a gridwork of diagonal floor beams. The framework was prefabricated in sections and field-welded on timber falsework. The bridge has three traffic lanes and two walks.

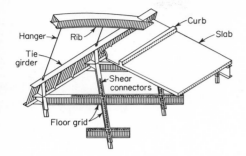

Fig. 41   Details of Rio Blanco Bridge.

25. Economics   Costs of arches and rigid frames are often dictated by practical and functional factors other than those of structural efficiency. For example, costs of long-span arch bridges may be governed by the feasibility and economy of the large falsework requirements.

REFERENCES

 1. Kleinlogel, A.: "Rahmenformeln," Wilhelm Ernst & Sohn, KG, Berlin, 1943.
 2. Leontovich, V.: "Frames and Arches," McGraw-Hill Book Company, New York, 1959.
 3. Whitney, C. S.: Design of Symmetrical Concrete Arches, *Trans. ASCE,* vol. 88, p. 931, 1925.
 4. Cross, H., and N. D. Morgan: "Continuous Frames of Reinforced Concrete," John Wiley & Sons, Inc., New York, 1932.
 5. Whitney, C. S.: Analysis of Unsymmetrical Concrete Arches, *Trans. ASCE,* vol. 99, p. 1268, 1934.
 6. Whitney, C. S.: Plain and Reinforced Concrete Arches, Report of Committee 312, *J. ACI,* vol. 47, p. 681, May 1951; vol. 37, p. 1, September 1940; vol. 28, p. 479, March 1932.
 7. Korn, M. P.: "Steel Rigid Frames Manual—Design and Construction," J. W. Edwards, Publishers, Incorporated, Ann Arbor, Mich., 1953.
 8. "Analysis of Rigid Frame Concrete Bridges," Portland Cement Assoc., Chicago.
 9. Griffiths, J. D.: "Single Span Rigid Frames in Steel," American Institute of Steel Construction, New York, 1948.
10. Griffiths, J. D.: Multiple-Span Gable Frames, *Trans. ASCE,* vol. 121, p. 1288, 1956.
11. Parcel, J., and R. B. B. Moorman: "Analysis of Statically Indeterminate Structures," John Wiley & Sons, Inc., New York, 1955.
12. Rowe, R. S.: Amplification of Stress in Flexible Steel Arches, *Trans. ASCE,* vol. 119, p. 910, 1954.
13. Asplund, S. O.: Deflection Theory of Arches, *Trans. ASCE,* vol. 128, p. 307, 1963.
14. Freudenthal, A.: Deflection Theory for Arches, *Int. Assoc. Bridge Struct. Eng. Mem.,* vol. 3, p. 100, 1935.
15. Garrelts, J. M.: Design of St. Georges Tied Arch Span, *Trans. ASCE,* vol. 108, p. 543, 1943.
16. Hardesty, S., and J. M. Garrelts: Rainbow Arch Bridge over Niagara Gorge: Design, *Trans. ASCE,* vol. 110, p. 6, 1945.
17. Freyssinet, E.: Three Monumental Bridges Built in Venezuela, *Civil Eng.,* March 1953, p. 157.
18. Wastlund, G.: Stability Problems of Compressed Steel Members and Arch Bridges, *Proc. ASCE,* vol. 86, ST6, p. 47, June 1960.
19. Baron, F., and J. P. Michalos: Laterally Loaded Plane Structures and Structures Curved in Space, *Trans. ASCE,* vol. 117, p. 279, 1952.
20. Michalos, J.: Effects of Lateral Loads on Arches, *J. ACI,* vol. 47, p. 377, 1951.
21. Barron, M.: Reinforced Concrete Skewed Rigid-Frame and Arch Bridges, *Trans. ASCE,* vol. 116, p. 999, 1951.
22. Michalos, J.: Analysis of Skewed Rigid Frames and Arches, *J. ACI,* vol. 48, p. 437, 1952.
23. Ernst, G. C.: Design of Hinges and Articulations in Reinforced Concrete Structures, *Trans. ASCE,* vol. 106, p. 862, 1941.
24. Lyse, I., and W. E. Black: An Investigation of Steel Rigid Frames, *Trans. ASCE,* vol. 107, p. 127, 1942.

25. Bleich, F.: "Design of Rigid Frame Knees," American Institute of Steel Construction, 1959.
26. Olander, H. C.: Stresses in the Corners of Rigid Frames, *Trans. ASCE*, vol. 119, p. 797, 1954.
27. "Steel Gables and Arches," American Institute of Steel Construction, New York, 1963.
28. Benjamin, J. R.: "Statically Indeterminate Structures," McGraw-Hill Book Company, New York, 1959.
29. Design Constants for Circular and Elliptical Arches of Variable Thickness, *Adv. Eng. Bull.* 8, Portland Cement Assoc., Chicago, 1963.
30. Design Constants for Circular Arch Bents, *Adv. Eng. Bull.* 7, Portland Cement Assoc., Chicago, 1963.
31. Design Constants for Continuous Gable Frames, *Adv. Eng. Bull.* 15, Portland Cement Assoc., Chicago, 1965.

Bridges

Part 1. Steel and Concrete Bridges

ARTHUR L. ELLIOTT

Bridge Engineer, Sacramento, Calif.

LOADS

1. Loads Bridge structures are designed for such of the following loads as are applicable: dead load, live load, impact or dynamic effect of live load, wind load, longitudinal force, centrifugal force, thermal forces, earth pressure, buoyancy, shrinkage stresses, erection stresses, ice and current pressures, and earthquake forces.

Dead Load. The dead load includes the weight of all parts of the structure, including the supporting members, deck, sidewalks, railing, and an allowance for future wearing surface if applicable. Snow and ice loads are treated as dead load.

Highway Live Loads. The current system of live loading for the design of highway structures is that of the American Association of State Highway and Transportation Officials. Their Standard Specifications for Highway Bridges is the accepted guide for all highway bridge work done by public agencies in the United States. Adherence to these specifications is recommended for all highway bridge work.

Two systems of loading are specified, the H and the HS (Fig. 1). The loading currently used by the majority of the states is the HS20-44, a 20-ton truck with 16-ton semitrailer (Fig. 1b). The variable axle spacing of the HS20-44 loading may approximate more closely vehicles now in use. The variable spacing is also more satisfactory for continuous spans, in that the position of the heavy loads on adjoining spans may be adjusted to produce maximum negative moment. In applying the truck loadings, only one truck per lane is used. Truck trains are represented by a lane loading which approximates a 20-ton truck preceded and followed by 15-ton trucks at 30-ft intervals (Fig. 2).

The truck and lane loadings are assumed to occupy a width of 10 ft. They are placed in "design traffic lanes" having a width $W = W_c/N$, where W = width of design traffic lane, W_c = roadway width between curbs, exclusive of median strip, and N = number of design traffic lanes as follows:

W_c, ft	N	W_c, ft	N
20–30 incl.	2	Over 78– 90 incl.	7
Over 30–42 incl.	3	Over 90–102 incl.	8
Over 42–54 incl.	4	Over 102–114 incl.	9
Over 54–66 incl.	5	Over 114–126 incl.	10
Over 66–78 incl.	6		

The lane loadings or standard trucks are assumed to occupy any position within their individual design traffic lanes which will produce the maximum stress.

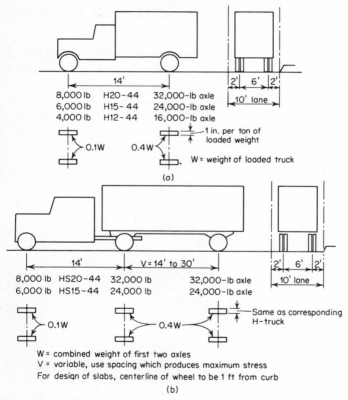

Fig. 1   AASHTO truck loads.

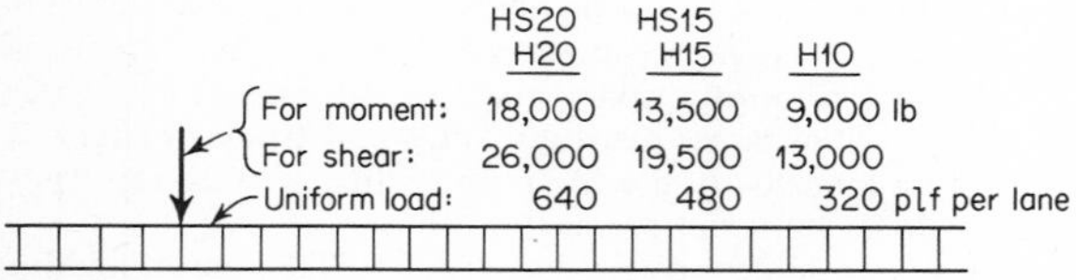

	HS20 H20	HS15 H15	H10
For moment:	18,000	13,500	9,000 lb
For shear:	26,000	19,500	13,000
Uniform load:	640	480	320 plf per lane

Fig. 2   AASHTO lane load. For continuous spans one additional concentrated load of equal weight to be placed in one other span in such a position as to produce maximum negative moment.

Railroad Live Loads. Live loading for railroad structures is usually based on the wheel loads of two locomotives followed by a uniform load representing the weight of the heaviest cars (Table 3). Most designs are made for Cooper's E-80 loading.

2. Maximum Moments and Shears in Simple Spans The maximum moment caused by a group of wheels in a simple span will occur under one of the wheels when the center of gravity of the loads within the given span and the point of application of that wheel are equidistant from the center of the span. This is illustrated for a three-axle truck in Fig. 3.

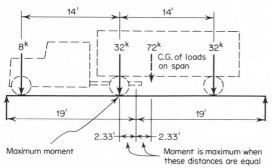

Fig. 3 Simple-span maximum moment under wheel group.

Envelopes of maximum moment in a simple span may be approximated by a second-degree parabola whose ordinate at midspan is the combined live- and dead-load moments. These envelopes are used for determining cutoff points for reinforcing steel in concrete beams and to determine the lengths of cover plates in steel beams. Ordinates to the true moment envelope and the parabolic envelope are given in the following table for a 50-ft simple span for which the dead load is 1 klf and the live load 1.5 wheel lines of the standard HS20-44 truck.

Distance from support, ft	Dead-load moment, ft-kips	Live-load moment, ft-kips	Impact 28.6 %	Total moment, ft-kips	Parabola, ft-kips	% variation
5	113	194	56	363	329	−9
10	200	334	96	630	585	−7
15	263	405	116	784	768	−2
20	300	463	132	895	877	−2
22.7	310	471	135	916	906	−1
25	313	467	134	914	914	0

Highway Loadings. The following formulas give maximum moments and shears resulting from one lane of HS20-44 loading on a simple span of length L, impact not included:

Span, Ft	*Max Moment, Ft-kips*
0–23.9	$8L$
23.9–33.8	$16L + 784/L − 224$
33.8–145.6	$18L + 392/L − 280$
Over 145.6 (equivalent load)	$0.08L^2 + 4.5L$

Span, Ft	*Max Shear, Kips*
0–14	32
14–28	$64 − 448/L$
28–127.5	$72 − 672/L$
Over 127.5 (equivalent load)	$0.32L + 26$

TABLE 1 HS20-44 Live Load, One Lane

Impact Not Included
T = truck loading, L = lane loading

Positive moment, ft-kips

		Beams fixed both ends						Beams fixed one end							
Span in ft	Impact	SBM @ .5L	FEM @ B	FEM @ C	SBM @ .2L	FEM @ B'	FEM @ C	SBM @ .2L	FEM @ B	SBM @ .4L	FEM @ B	SBM @ .6L	FEM @ B	SBM @ .8L	FEM @ B
1	0.300	8.0T	4.0	4.0	5.1T	4.1	1.0	5.1T	3.1	7.7T	5.4	7.7T	6.1	5.1T	4.6
2	0.300	16.0	8.0	8.0	10.2	8.2	2.0	10.2	6.1	15.4	10.8	15.4	12.3	10.2	9.2
3	0.300	24.0	12.0	12.0	15.4	12.3	3.1	15.4	9.2	23.0	16.1	23.0	18.4	15.4	13.8
4	0.300	32.0	16.0	16.0	20.5	16.4	4.1	20.5	12.3	30.7	21.5	30.7	24.6	20.5	18.4
5	0.300	40.0	20.0	20.0	25.6	20.5	5.1	25.6	15.4	38.4	26.9	38.4	30.7	25.6	23.0
6	0.300	48.0	24.0	24.0	30.7	24.6	6.1	30.7	18.4	46.1	32.3	46.1	36.9	30.7	27.6
7	0.300	56.0	28.0	28.0	35.8	28.7	7.2	35.8	21.5	53.8	37.6	53.8	43.0	35.8	32.3
8	0.300	64.0	32.0	32.0	41.0	32.8	8.2	41.0	24.6	61.4	43.0	61.4	49.2	41.0	36.9
9	0.300	72.0	36.0	36.0	46.1	36.9	9.2	46.1	27.7	69.1	48.4	69.1	55.3	46.1	41.5
10	0.300	80.0	40.0	40.0	51.2	41.0	10.2	51.2	30.7	76.8	53.8	76.8	61.4	51.2	46.1
11	0.300	88.0	44.0	44.0	56.3	45.1	11.3	56.3	33.8	84.5	59.1	84.5	67.6	56.3	50.7
12	0.300	96.0	48.0	48.0	61.4	49.2	12.3	61.4	36.9	92.2	64.5	92.2	73.7	61.4	55.3
13	0.300	104.0	52.0	52.0	66.6	53.2	13.3	66.6	39.9	99.8	69.9	99.8	79.9	66.6	59.9
14	0.300	112.0	56.0	56.0	71.7	57.3	14.3	71.7	43.0	107.5	75.3	107.5	86.0	71.7	64.5
15	0.300	120.0	60.0	60.0	76.8	61.4	15.4	76.8	46.1	115.2	80.6	115.2	92.2	76.8	69.1
16	0.300	128.0	64.0	64.0	81.9	65.5	16.4	81.9	49.2	122.9	86.0	122.9	98.3	81.9	73.7
17	0.300	136.0	68.0	68.0	87.0	69.6	17.4	87.0	52.2	130.6	91.4	130.6	104.5	87.0	78.3
18	0.300	144.0	72.0	72.0	94.7	74.0	30.7	94.7	67.7	138.2	96.8	138.2	110.6	94.7	89.3
19	0.300	152.0	76.0	76.0	105.0	80.1	53.2	105.0	93.2	145.9	102.1	145.9	116.7	105.0	106.7
20	0.300	160.0	80.0	80.0	115.2	87.7	72.3	115.2	116.2	153.6	107.5	153.6	122.9	115.2	123.8
21	0.300	168.0	84.0	84.0	125.4	96.4	88.8	125.4	137.0	161.3	112.9	161.3	129.0	125.4	140.8
22	0.300	176.0	88.0	88.0	135.7	105.9	103.1	135.7	156.1	169.0	118.3	169.0	135.2	135.7	157.4
23	0.300	184.0	92.0	92.0	145.9	116.0	115.6	145.9	173.6	176.6	123.7	176.6	141.3	145.9	173.8
24	0.300	192.7	116.3	104.4	156.2	126.6	126.7	156.2	190.0	189.4	141.5	189.4	153.9	156.2	189.9
25	0.300	207.4	122.7	120.5	166.4	137.4	136.5	166.4	205.2	204.8	164.5	204.8	169.6	166.4	205.7
26	0.300	222.2	129.4	135.9	176.6	148.5	145.3	176.6	219.6	220.2	186.4	220.2	185.3	176.6	221.2
27	0.300	237.0	136.3	150.5	186.9	159.8	153.2	186.9	233.1	235.5	207.2	235.5	200.9	186.9	236.4
28	0.300	252.0	143.5	164.5	197.1	171.1	160.4	197.1	246.0	250.9	227.1	250.9	216.4	197.1	251.3
29	0.300	267.0	150.8	178.0	207.4	182.6	166.9	207.4	258.2	266.2	246.3	266.2	231.8	207.4	266.0
30	0.300	282.1	158.3	191.0	217.6	194.0	172.9	217.6	269.9	281.6	264.8	281.6	247.2	217.6	280.5
32	0.300	312.5	173.6	215.9	238.1	216.9	183.6	238.1	292.1	312.3	300.1	312.3	277.6	238.1	308.7
34	0.300	343.5	179.5	236.6	258.6	239.6	192.9	258.6	312.7	343.0	333.4	343.0	307.7	258.6	336.0
36	0.300	378.9	201.4	261.5	280.3	262.3	207.2	280.3	338.3	375.7	366.7	375.7	340.5	280.3	365.9
38	0.300	414.3	223.1	285.6	303.4	285.5	225.3	303.4	368.1	410.2	400.4	410.2	375.9	303.4	398.2
40	0.300	449.8	244.4	309.0	326.4	309.3	241.1	326.4	395.8	444.8	433.0	444.8	410.6	326.4	429.8
42	0.299	485.3	265.5	331.9	349.4	333.3	255.1	349.4	421.7	479.4	464.7	479.4	444.7	349.4	460.8
44	0.296	520.9	286.3	354.2	372.5	357.5	267.5	372.5	446.2	513.9	495.7	513.9	478.3	372.5	491.2
46	0.292	556.5	306.9	376.2	395.5	381.7	278.7	395.5	469.6	548.5	526.0	548.5	511.5	395.5	521.1
48	0.289	592.2	327.3	397.8	418.6	405.9	289.0	418.6	491.9	583.0	555.7	583.0	544.3	418.6	550.4
50	0.286	627.8	347.6	419.1	441.6	430.0	298.3	441.6	513.3	617.6	584.9	617.6	576.7	441.6	579.1
52	0.282	663.5	367.7	440.2	464.6	454.0	307.0	464.6	534.0	652.2	613.7	652.2	608.8	464.6	607.5
54	0.279	699.3	387.6	461.0	487.7	477.8	315.2	487.7	554.1	686.7	642.1	686.7	640.6	487.7	635.4
56	0.276	735.0	407.5	481.5	510.7	501.5	322.8	510.7	573.6	721.3	670.3	721.3	672.1	510.7	662.9
58	0.273	770.8	427.2	501.9	533.8	525.1	330.0	533.8	592.5	755.8	698.0	755.8	703.4	533.8	690.1
60	0.270	806.5	446.8	522.1	656.8	548.5	336.9	556.8	611.1	790.4	725.6	790.4	734.4	556.8	716.9
62	0.267	842.3	466.3	542.2	579.8	571.7	343.4	579.8	629.3	825.0	752.9	825.0	765.3	579.8	743.4
64	0.265	878.1	485.7	562.1	602.9	594.8	349.7	602.9	647.1	859.5	780.0	859.5	796.0	602.9	769.6
66	0.262	913.9	505.1	581.9	625.9	617.7	355.8	625.9	664.6	894.1	806.9	894.1	826.5	625.9	795.6
68	0.259	949.8	524.4	601.6	649.0	640.4	361.7	649.0	681.9	928.6	833.6	928.6	856.8	649.0	821.3
70	0.256	985.6T	543.6	621.2	672.0T	663.0	367.4	672.0T	698.9	963.2T	860.2	963.2T	887.0	672.0T	846.7

TABLE 1 HS20-44 Live Load, One Lane (*Continued*)

Impact Not Included
T = truck loading, L = lane loading

Shear, kips			Negative moment, ft-kips												
Simple span			Beams fixed one end					Beams fixed both ends							
Shear at end	Shear at midspan	Center reaction 2 equal simple spans	SBM @ .42L	FEM @ B	aL ft	SBM @ aL	FEM @ B	SBM @ .33L	FEM @ B	FEM @ C	aL ft	SBM @ aL	FEM @ B	FEM @ C	
32.0T	16.0T	32.0T	4.5L	3.5	0.4	7.8T	6.2	4.1L	2.7	1.4	0.3	7.1T	4.7	2.4	1
32.0	16.0	32.0	9.1	7.2	0.8	15.6	12.3	8.3	5.5	2.9	0.7	14.2	9.5	4.7	2
32.0	16.0	32.0	13.9	11.1	1.3	23.4	18.5	12.6	8.5	4.5	1.0	21.3	14.2	7.1	3
32.0	16.0	32.0	18.8	15.1	1.7	31.2	24.6	17.1	11.5	6.2	1.3	28.4	19.0	9.5	4
32.0	16.0	32.0	23.9	19.3	2.1	39.1	30.8	21.8	14.7	8.0	1.7	35.6	23.7	11.9	5
32.0	16.0	32.0	29.2	23.7	2.5	46.9	37.0	26.6	17.9	9.9	2.0	42.7	28.4	14.2	6
32.0	16.0	32.0	34.6	28.2	3.0	54.7	43.1	31.5	21.3	11.9	2.3	49.8	33.2	16.6	7
32.0	16.0	32.0	40.1	32.8	3.4	62.5	49.3	36.6	24.7	14.1	2.7	56.9	37.9	19.0	8
32.0	16.0	32.0	45.9	37.7	3.8	70.3	55.4	41.8	28.3	16.3	3.0	64.0	42.7	21.3	9
32.0	16.0	32.0	51.7	42.6	4.2	78.1	61.6	47.1	32.0	18.7	3.3	71.1	47.4	23.7	10
32.0	16.0	32.0	57.8	47.8	4.7	85.9	67.7	52.6	35.8	21.1	3.7	78.2	52.2	26.1	11
32.0	16.0	32.0	64.0	53.1	5.1	93.7	73.9	58.2	39.7	23.7	4.0	85.3	56.9	28.4	12
32.0	16.0	32.0	70.3	58.6	5.5	101.5	80.1	64.0	43.7	26.3	4.3	92.4	61.6	30.8	13
32.0	16.0	32.0	76.8	64.2	5.9	109.3	86.2	69.9	47.8	29.1	4.7	99.6	66.4	33.2	14
34.1	16.0	34.8	83.5	70.0	6.3	117.2	92.4	76.0	52.0	32.0	5.0	106.7	71.1	35.6	15
36.0	16.0	37.0	90.3	75.7	6.8	125.0	98.5	82.2	56.3	35.0	5.3	113.8	75.9	37.9	16
37.7	16.0	39.1	97.3	82.0	7.2	132.8	104.7	88.6	60.7	38.1	5.7	120.9	80.6	40.3	17
39.1	16.0	40.9	104.4	88.3	7.6	140.6	110.9	95.0	65.3	41.3	6.0	128.0	85.3	42.7	18
40.4	16.0	42.5	111.7	94.7	8.0	148.4	117.0	101.7	69.9	44.6	6.3	135.1	90.1	45.0	19
41.6	16.0	44.0	119.1	101.3	3.8	112.2	123.9	108.4	74.7	48.0	6.7	142.2	94.8	47.4	20
42.7	16.0	45.3	126.7	108.0	4.1	123.7	140.8	115.4	79.5	51.5	7.0	149.3	99.6	49.8	21
43.6	16.0	46.5	134.5	114.9	4.4	135.7	157.4	122.4	84.5	55.1	5.5	151.8	107.1	96.0	22
44.5	16.0	47.7	142.4	122.0	4.7	148.1	173.8	129.6	89.5	58.9	5.2	156.4	116.4	113.8	23
45.3	16.0	48.7	150.4	129.2	5.0	160.8	190.0	137.0	94.7	62.7	5.1	162.3	126.7	126.6	24
46.1	16.0	49.6	158.6	136.6	5.4	173.8	206.0	144.4	100.0	66.7	5.2	169.3	137.5	136.8	25
46.8	16.0	50.5	167.0	144.1	5.7	187.0	221.7	152.1	105.4	70.7	5.2	177.1	148.5	145.4	26
47.4	16.0	51.3	175.6	151.9	6.1	200.5	237.2	159.8	110.9	74.9	5.3	185.6	159.8	152.9	27
48.0	16.0	52.0	184.2	159.7	6.4	214.1	252.6	167.8	116.5	79.1	5.5	194.7	171.2	159.7	28
48.8	16.6	52.7	193.1	167.7	6.8	227.9	267.8	175.8	122.2	83.5	5.7	203.3	182.6	165.9	29
49.6	17.1	53.3	202.1	175.9	7.2	241.9	282.8	184.0	128.0	88.0	5.9	214.2	194.0	171.7	30
51.0	18.0	54.5	220.6	192.8	7.9	270.2	312.4	200.8	139.9	97.3	6.3	235.2	216.9	182.4	32
52.2	18.8	55.5	239.7	210.3	8.7	299.0	341.5	218.2	152.3	107.0	6.8	257.4	239.6	192.3	34
53.3	19.6	56.4	259.4	228.4	9.4	328.1	370.1	236.2	165.1	117.1	7.1	278.4	262.3	205.8	36
54.3	20.2	57.3	279.7	247.2	8.8	331.5	400.4	254.7	178.3	127.7	7.4	299.1	285.6	224.0	38
55.2	20.8	58.0	300.7	266.6	9.5	362.8	433.3	273.8	192.0	138.7	7.8	321.4	309.3	239.2	40
56.0	21.3	58.7	322.3	286.6	10.3	394.6	465.9	293.4	206.1	150.1	8.2	344.8	333.3	253.1	42
56.7	21.8	59.3	344.5	307.3	11.0	426.8	498.1	313.7	220.6	161.9	8.7	369.3	357.5	266.0	44
57.4	22.3	59.8	367.4	328.4	11.8	459.3	530.0	334.5	235.5	174.2	9.2	394.6	381.7	278.3	46
58.0	22.7	60.3	390.8	350.6	12.6	492.1	561.6	355.8	250.9	186.9	9.7	420.6	405.9	290.0	48
58.6	23.0	60.8	414.9	373.2	13.3	525.1	592.9	377.8	266.7	200.0	10.2	447.3	430.0	301.4	50
59.1	23.4	61.2	439.6	396.5	14.1	558.4	624.1	400.3	282.9	213.5	10.7	474.4	454.1	312.4	52
59.6	23.7	61.6	465.0	420.3	14.9	591.8	655.0	423.4	299.5	227.5	11.3	502.0	478.1	323.3	54
60.0	24.0	62.0T	491.0	444.9	15.7	625.3	685.7	447.0	316.6	241.9	11.8	530.0	502.0	334.0	56
60.4	24.4	63.1L	517.5	470.0	16.5	659.0	716.2	471.2	334.1	256.7	12.4	558.3	525.8	344.6	58
60.8	24.8	64.4	544.8	495.8	17.3	692.8	746.6	496.0	352.0	272.0	13.0	586.9	549.5	355.0	60
61.2	25.2	65.7	572.6	522.3	18.1	726.7	776.8	521.4	370.3	287.7	13.6	615.8	573.1	365.4	62
61.5	25.5	67.0	601.1	549.4	18.9	760.8	806.9	547.3	389.1	303.8	14.2	645.0	596.7	375.7	64
61.8	25.8	68.2	630.2	577.1	19.7	794.9	836.8	573.8	408.3	320.3	14.7	674.3	620.1	386.0	66
62.1	26.1	69.5	659.9	605.5	20.5	829.0	866.7	600.8	427.9	337.3	15.3	703.8	643.5	396.3	68
62.4T	26.4T	70.8L	690.2L	634.5	21.3	863.3T	896.4	628.4L	448.0	354.7	15.9	733.5T	666.7	406.5	70

TABLE   HS20-44 Live Load, One Lane *(Continued)*

Impact Not Included

T = truck loading, L = lane loading

		Positive moment, ft-kips													
		Beams fixed both ends						Beams fixed one end							
Span in ft.	Impact	SBM @ .5L	FEM @ B	FEM @ C	SBM @ .2L	FEM @ B	FEM @ C	SBM @ .2L	FEM @ B	SBM @ .4L	FEM @ B	SBM @ .6L	FEM @ B	SBM @ .8L	FEM @ B
72	0.254	1021L	563	641	695T	685	373	695T	716	998T	887	998T	917	695T	872
74	0.251	1057	582	660	718	708	378	718	732	1032	913	1032	947	718	897
76	0.249	1093	601	679	741	730	384	741	749	1067	939	1067	977	741	922
78	0.246	1129	620	699	764	752	389	764	765	1101	965	1101	1007	764	946
80	0.244	1165	639	718	787	774	394	787	781	1136	991	1136	1036	787	971
82	0.241	1201	658	737	810	796	399	810	797	1171	1017	1171	1066	810	995
84	0.239	1237	677	756	833	817	404	833	813	1205	1043	1205	1095	833	1019
86	0.237	1273	695	775	856	839	409	856	829	1240	1068	1240	1125	856	1044
88	0.235	1308	714	794	879	861	414	879	844	1274	1094	1274	1154	879	1067
90	0.233	1344	733	813	902	882	419	902	860	1309	1120	1309	1183	902	1091
92	0.230	1380	752	832	925	903	424	925	875	1343	1145	1343	1212	925	1115
94	0.228	1416	770	851	948	924	428	948	890	1378	1171	1378	1242	948	1138
96	0.226	1452	789	870	972	945	433	972	906	1412	1196	1412	1271	972	1162
98	0.224	1488	808	888	995	966	438	995	921	1447	1221	1447	1300	995	1185
100	0.222	1524	826	907	1018	987	442	1018	936	1482	1247	1482	1329	1018	1209
105	0.217	1614	873	954	1075	1039	454	1075	974	1568	1310	1568	1401	1075	1266
110	0.213	1704	919	1000	1133	1091	465	1133	1011	1654	1372	1654	1473	1133	1324
115	0.208	1793	965	1047	1190	1142	477	1190	1048	1741	1435	1741	1544	1190	1381
120	0.204	1883	1011	1093	1248	1193	488	1248	1085	1827	1497	1827	1616	1248	1437
125	0.200	1973	1057	1139	1306	1243	499	1306	1121	1914	1560	1914	1687	1306	1493
130	0.196	2063	1103	1185	1363	1294	511	1363	1157	2000	1622	2000	1758	1363	1549
135	0.192	2153	1149	1231	1421	1344	522	1421	1194	2086	1684	2086	1829	1421	1605
140	0.189	2243	1195	1277	1478	1393	533	1478	1230	2173	1746	2173	1900	1478	1660
145	0.185	2333	1241	1323	1536	1443	544	1536	1266	2259	1807	2259	1971	1536	1715
150	0.182	2423	1286	1369	1594	1492	556	1594	1302	2346	1869	2346	2041	1594	1770
155	0.179	2513	1332	1415	1651	1541	567	1651	1337	2432T	1931	2432T	2112	1651	1825
160	0.175	2602T	1378	1461	1709	1590	578	1709T	1373	2657L	2532	2657L	2601	1709	1879
165	0.172	2921L	1823	1823	1766	1639	589	1869L	2463	2804	2677	2804	2748	1766	1933
170	0.169	3077	1924	1924	1824	1688	600	1969	2606	2954	2826	2954	2900	1824	1988
175	0.167	3238	2027	2027	1882	1736	611	2072	2752	3108	2979	3108	3055	1882	2042
180	0.164	3402	2133	2133	1939	1784	623	2177	2903	3266	3136	3266	3214	1939	2096
185	0.161	3571	2242	2242	1997	1833	634	2285	3058	3428	3297	3428	3377	1997	2150
190	0.159	3743	2353	2353	2054	1881	645	2396	3216	3593	3463	3593	3545	2054	2203
195	0.156	3920	2467	2467	2112	1929	656	2508	3379	3763	3632	3763	3716	2112	2257
200	0.154	4100	2583	2583	2170	1977	667	2624	3546	3936	3805	3936	3891	2170	2311
205	0.152	4285	2703	2703	2227	2025	679	2742	3716	4113	3982	4113	4070	2227	2364
210	0.149	4473	2824	2824	2285	2073	690	2863	3891	4294	4163	4294	4254	2285	2418
215	0.147	4666	2949	2949	2342	2120	701	2986	4070	4479	4348	4479	4441	2342	2471
220	0.145	4862	3076	3076	2400	2168	712	3112	4252	4668	4537	4668	4632	2400T	2524
225	0.143	5063	3206	3206	2458	2216	724	3240	4439	4860	4730	4860	4828	3240L	4633
230	0.141	5267	3339	3339	2515	2263	735	3371	4629	5056	4928	5056	5027	3371	4828
235	0.139	5476	3474	3474	2573	2311	746	3504	4824	5256	5129	5256	5230	3504	5027
240	0.137	5688	3612	3612	2630	2358	757	3640	5023	5460	5334	5460	5437	3640	5230
245	0.135	5905	3753	3753	2688	2406	769	3779	5225	5668	5543	5668	5649	3779	5437
250	0.133	6125	3896	3896	2746T	2453	780	3920	5432	5880	5756	5880	5864	3920	5648
255	0.132	6350	4042	4042	4064L	4056	3615	4064	5643	6096	5973	6096	6083	4064	5863
260	0.130	6578	4190	4190	4210	4204	3755	4210	5857	6315	6194	6315	6307	4210	6082
265	0.128	6811	4342	4342	4359	4356	3898	4359	6076	6538	6419	6538	6534	4359	6305
270	0.127	7047L	4495	4495	4510L	4510	4044	4510L	6299	6765L	6648	6765L	6765	4510L	6532

TABLE 1 HS20-44 Live Load, One Lane (*Continued*)

Impact Not Included
T = truck loading. L = lane loading

Shear, kips			Negative moments, ft-kips												
Simple span			Beams fixed one end					Beams fixed both ends							
Shear at end	Shear at midspan	Center reaction 2 equal simple span	SBM @ .42L	FEM @ B	aL ft	SBM @ aL	FEM @ B	SBM @ .33L	FEM @ B	FEM @ C	aL ft	SBM @ aL	FEM @ B	FEM @ C	
62.7T	26.7T	72.1L	721L	664	22.2	898T	926	657L	468	372	16.6	763T	690	417	72
62.9	26.9	73.4	753	694	23.0	932	956	685	489	394	17.2	793	713	427	74
63.2	27.2	74.6	785	725	23.8	966	985	715	511	409	17.8	824	736	437	76
63.4	27.4	75.9	818	757	24.6	1001	1014	745	532	428	18.4	854	759	447	78
63.6	27.6	77.2	851	789	25.4	1035	1044	775	555	448	19.0	884	782	458	80
63.8	27.8	78.5	885	822	26.3	1070	1073	806	577	468	19.6	915	805	468	82
64.0	28.0	79.8	920	855	27.1	1104	1102	838	600	488	20.3	945	828	478	84
64.2	28.2	81.0	955	890	27.9	1139	1131	870	624	509	20.9	976	850	488	86
64.4	28.4	82.3	991	924	28.7	1174	1160	903	648	530	21.5	1006	873	498	88
64.5	28.5	83.6	1028	960	29.6	1208	1189	936	672	552	22.2	1037	896	509	90
64.7	28.7	84.9	1065	996	30.4	1243	1218	970	697	574	22.8	1068	918	519	92
64.9	28.9	86.2	1103	1033	31.2	1278	1247	1004	722	597	23.4	1099	941	529	94
65.0	29.0	87.4	1142	1070	32.0	1313	1276	1039	748	620	24.1	1130	963	539	96
65.1	29.1	88.7	1181	1108	32.9	1347	1304	1075	774	643	24.7	1161	986	550	98
65.3	29.3	90.0	1220	1146	33.7	1382	1333	1111	800	667	25.3	1192	1008	560	100
65.6	29.6	93.2	1322	1246	35.8	1469	1405	1204	868	728	26.9	1269	1064	586	105
65.9	29.9	96.4	1428	1349	37.8	1556	1476	1300	939	792	28.5	1347	1120	611	110
66.2	30.2	99.6	1538	1456	39.9	1643	1548	1400	1012	859	30.2	1426	1175	637	115
66.4	30.4	102.8	1652	1568	42.0	1731	1619	1504	1088	928	31.8	1504	1230	663	120
66.6T	30.6	106.0	1770	1683	44.1	1818T	1690	1611	1167	1000	33.4	1582	1286	689	125
67.6L	30.8	109.2	1891	1802				1722	1248	1075	35.0	1661	1341	715	130
69.2	31.0	112.4	2017	1926				1836	1332	1152	36.6	1740	1396	741	135
70.8	31.2	115.6	2146	2053				1954	1419	1232	38.3	1819T	1450	767	140
72.4	31.4	118.8	2279	2184				2075	1508	1315					145
74.0	31.5	122.0	2416	2320				2200	1600	1400					150
75.6	31.7	125.2	2557	2459				2328	1695	1488					155
77.2	31.8	128.4	2702	2602				2460	1792	1579					160
78.8	31.9	131.6	2851	2750				2596	1892	1672					165
80.4	32.0	134.8	3004	2901				2735	1995	1768					170
82.0	32.2	138.0	3161	3056				2878	2100	1867					175
83.6	32.3	141.2	3321	3216				3024	2208	1968					180
85.2	32.4	144.4	3486	3379				3174	2319	2072					185
86.8	32.5	147.6	3654	3546				3327	2432	2179					190
88.4	32.6	150.8	3827	3717				3484	2548	2288					195
90.0	32.6	154.0	4003	3893				3644	2667	2400					200
91.6	32.7	157.2	4183	4072				3808	2788	2515					205
93.2	32.8	160.4	4367	4255				3976	2912	2632					210
94.8	32.9	163.6	4555	4443				4147	3039	2752					215
96.4	32.9	166.8	4747	4634				4322	3168	2875					220
98.0	33.0	170.0	4942	4829				4500	3300	3000					225
99.6	33.1	173.2	5142	5029				4682	3435	3128					230
101.2	33.1	176.4	5346	5232				4867	3572	3259					235
102.8	33.2	179.6	5553	5439				5056	3712	3392					240
104.4	33.3	182.8	5764	5651				5248	3855	3528					245
106.0	33.3T	186.0	5980	5866				5444	4000	3667					250
107.6	33.4L	189.2	6199	6085				5644	4148	3808					255
109.2	33.8	192.4	6422	6309				5847	4299	3952					260
110.8	34.2	195.6	6649	6536				6054	4452	4099					265
112.4L	34.6L	198.8L	6880L	6767				6264L	4608	4248					270

18-7

Table 1 gives moments, shears, and reactions for the HS20-44 loading for spans ranging from 1 to 270 ft.

Railroad Loadings.   Table 2 gives the position of wheels of the Cooper load which produce maximum moments. Table 3 gives summations of moments about any wheel of any number of wheels on either side of the moment center. Table 4 gives maximum moments, shears, and reactions for spans of 7 to 250 ft.

TABLE 2   Position of Wheels to Produce Maximum Moment with Cooper's Loading

Maximum moment will occur under wheel noted in table for given span and distance from left end of span. Engine normally faces left. If wheel number is overlined, face engine to right.

Example   In an eight-panel truss of 200-ft span, maximum moment at panel point L_1, 25 ft from the left end, occurs with wheel 4 at that point. Maximum moment at L_2 occurs with wheel 7 at that point, etc.

Span, ft	10	15	20	25	30	35	40	45	50	55	60	65	70	80	90	100	110	120	130	140
									Distance from left end of span											
300–260	2	3	3	4	4	5	5	6	7	7	8	9	10	11	12	13	14	15	17	18
250–200	2	3	3	4	4	5	5	6	7	8	8	9	10	11	12	13	14	15	17	18
190–150	2	3	3	4	4	5	5	6	7	8	9	9	11	12	12	13	14	15	17	18
140	3	3	3	4	4	5	5	6	7	8	9	10	11	12	12	13	14	15	17	18
130	3	3	3	4	4	5	5	6	7	8	9	10	11	12	12	13	14	15	17	
120	3	3	3	4	4	5	5	6	7	8	9	10	11	12	13	13	14	15		
110	3	3	3	4	4	5	6	7	7	8	9	10	11	12	13	13	14			
100	3	3	3	4	5	5	6	$\overline{14}$	$\overline{14}$	$\overline{14}$	$\overline{13}$	$\overline{13}$	11	12	13	13				
90	3	3	4	4	5	$\overline{13}$	$\overline{13}$	$\overline{13}$	$\overline{13}$	$\overline{13}$	$\overline{13}$	$\overline{13}$	$\overline{13}$	12	13					
80	3	3	4	4	$\overline{13}$	$\overline{13}$	$\overline{13}$	$\overline{12}$	$\overline{12}$	$\overline{12}$	$\overline{12}$	$\overline{12}$	$\overline{12}$	12						
70	3	3	4	4	$\overline{13}$	$\overline{13}$	$\overline{12}$	$\overline{12}$	$\overline{12}$	$\overline{12}$	$\overline{11}$	$\overline{11}$	11							
65	3	3	4	4	$\overline{12}$	$\overline{12}$	$\overline{12}$	$\overline{12}$	$\overline{12}$	$\overline{11}$	$\overline{11}$	11								
60	3	3	4	4	5	$\overline{13}$	$\overline{12}$	$\overline{11}$	$\overline{11}$	$\overline{11}$	11									
55	12	12	12	4	12	$\overline{13}$	$\overline{12}$	$\overline{12}$	$\overline{13}$	11										
50	12	12	12	12	12	$\overline{13}$	$\overline{13}$	$\overline{13}$	12											
45	12	12	12	12	12	$\overline{13}$	$\overline{13}$	13												
40	3	3	3	12	12	$\overline{13}$	13													
35	3	3	4	4	13	13														
30	3	3	4	4	13															
25	3	3	4	4																
20	$\overline{4}$	3	4																	
15	3	3																		
10	3																			

Example   Calculate the moment of wheels 1 to 15, inclusive, about wheel 15. In Table 3 follow the vertical line through wheel 15 to the stepped line, then to the left, and find 5408.0 ft-kips to the right of the vertical line through wheel 1. Multiply this value by 7.2 to convert to the E-72 loading. Answer: 38,900 ft-kips per track.

Example   Calculate the moments of wheels 17, 16, 15, and 14 about wheel 13. In Table 3 follow the vertical line through wheel 13 to the stepped line, then to the right, and find 427.0 ft-kips to the left of the vertical line through wheel 17. Multiply this value by 7.2. Answer: 3070 ft-kips per track.

Example   Given a 200-ft eight-panel railroad truss, calculate the maximum moment at L_1 and the maximum shear in panel L_0L_1 for one track of E-72 loading. From Table 2, moment at L_1 will be maximum when wheel 4 is at L_1. From Table 3, the distances from wheel 4 to wheel 1 and to the uniform load are 18 and 91 ft, respectively. Therefore, all wheels are on the bridge, and there will be $175 - 91 = 84$ ft of uniform load on the span. The left reaction R_L is found from

$$200R_L = 8182 + 142 \times 84 + 1 \times \frac{84^2}{2}$$

TABLE 3 Moments in Foot-Kips for Class E-10 Engine Loading
One Track of Two Rails

Wheel No.	1	2	3	4	5	6	7	8	9	10	11	12	13	14	15	16	17	18	1 klf
Axle loads	5.0	10.0	10.0	10.0	10.0	6.5	6.5	6.5	6.5	5.0	10.0	10.0	10.0	10.0	6.5	6.5	6.5	6.5	(1 klf)
Spacing, ft	8	5	5	5	9	5	6	5	8	8	5	5	5	9	5	5	6	5	5
Totals from end of train: Kips	142.0	137.0	127.0	117.0	107.0	97.0	90.5	84.0	77.5	71.0	66.0	56.0	46.0	36.0	26.0	19.5	13.0	6.5	0
Feet	109	101	96	91	86	77	72	66	61	53	45	40	35	30	21	16	10	5	0

Moments of wheel loads about:

(about)	1	2	3	4	5	6	7	8	9	10	11	12	13	14	15	16	17	18	1 klf
End of train	8,182.0	7,637.0	6,627.0	5,667.0	4,757.0	3,897.0	3,396.5	2,928.5	2,499.5	2,103.5	1,838.0	1,388.0	988.0	638.0	338.0	201.5	97.5	32.5	
Wheel No.18	7,472.0	6,952.0	5,992.0	5,082.1	4,222.0	3,412.0	2,944.0	2,508.5	2,112.0	1,748.0	1,508.0	1,108.0	758.0	458.0	208.0	104.0	32.5		
17	6,794.5	6,299.5	5,389.5	4,529.5	3,719.5	2,959.5	2,524.0	2,121.0	1,757.0	1,425.5	1,210.5	860.5	560.5	310.5	110.5	39.0		32.5	
16	6,020.5	5,555.5	4,705.5	3,905.5	3,155.5	2,455.5	2,059.0	1,695.0	1,370.0	1,077.5	892.5	602.5	362.5	172.5	32.5		39.0	110.5	
15	5,408.0	4,968.0	4,168.0	3,418.0	2,718.0	2,068.0	1,704.0	1,372.5	1,080.0	820.0	660.0	420.0	230.0	90.0		32.5	104.0	208.0	
14	4,364.0	3,969.0	3,259.0	2,599.0	1,989.0	1,429.0	1,123.5	850.5	616.5	415.0	300.0	150.0	50.0		58.5	149.5	279.5	442.0	
13	3,834.0	3,464.0	2,804.0	2,194.0	1,634.0	1,124.0	851.0	610.5	409.0	240.0	150.0	50.0		50.0	141.0	264.5	427.0	622.0	
12	3,354.0	3,009.0	2,399.0	1,839.0	1,329.0	869.0	628.5	420.5	251.5	115.0	50.0		50.0	150.0	273.5	429.5	624.5	852.0	
11	2,924.0	2,604.0	2,044.0	1,534.0	1,074.0	664.0	456.0	280.5	144.0	40.0		50.0	150.0	300.0	456.0	644.5	872.0	1,132.0	
10	2,316.0	2,036.0	1,556.0	1,126.0	746.0	416.0	260.0	136.5	52.0		80.0	210.0	390.0	620.0	828.0	1,068.5	1,348.0	1,660.0	
9	1,748.0	1,508.0	1,108.0	758.0	458.0	208.0	104.0	32.5		40.0	200.0	410.0	670.0	980.0	1,240.0	1,532.5	1,864.0	2,228.0	
8	1,425.5	1,210.5	860.5	560.5	310.5	110.5	39.0		32.5	97.5	307.5	567.5	877.5	1,237.5	1,530.0	1,855.0	2,219.0	2,615.5	
7	1,077.2	892.5	602.5	362.5	172.5	32.5		39.0	110.5	205.5	475.5	795.5	1,165.5	1,585.5	1,917.0	2,281.0	2,684.0	3,119.5	
6	820.0	660.0	420.0	230.0	90.0		32.5	104.0	208.0	328.0	648.0	1,018.0	1,438.0	1,908.0	2,272.0	2,668.5	3,104.0	3,572.0	
5	415.0	300.0	150.0	50.0		58.5	149.5	279.5	442.0	607.0	1,017.0	1,477.0	1,987.0	2,547.0	2,969.5	3,424.5	3,918.5	4,445.0	
4	240.0	150.0	50.0		50.0	141.0	264.5	427.0	622.0	812.0	1,272.0	1,782.0	2,342.0	2,952.0	3,407.0	3,894.5	4,421.0	4,980.0	
3	115.0	50.0		50.0	150.0	273.5	429.5	624.5	852.0	1,067.0	1,577.0	2,137.0	2,747.0	3,407.0	3,894.5	4,414.5	4,973.5	5,565.0	
2	40.0		50.0	150.0	300.0	456.0	644.5	872.0	1,132.0	1,372.0	1,932.0	2,542.0	3,202.0	3,912.0	4,432.0	4,984.5	5,576.0	6,200.0	
1		80.0	210.0	390.0	620.0	828.0	1,068.5	1,348.0	1,660.0	1,940.0	2,580.0	3,270.0	4,010.0	4,800.0	5,372.0	5,976.5	6,620.0	7,296.0	
Totals from wheel 1: Kips	5.0	15.0	25.0	35.0	45.0	51.5	58.0	64.5	71.0	76.0	86.0	96.0	106.0	116.0	122.5	129.0	135.5	142.0	142.0
Feet	0	8	13	18	23	32	37	43	48	56	64	69	74	79	88	93	99	104	109

TABLE 4 Maximum Moments, Shears, and Reactions for Class E-10 Engine Loading
One Track of Two Rails

Axle loads: (5.0) (10.0) (10.0) (10.0) (10.0) (6.5) (6.5) (6.5) (6.5) (10.0) (10.0) (10.0) (10.0) (5.0)   1 klf ▨ or (12.5) (12.5)

Spacing, ft....: 8 5 5 5 9 5 6 5 5 5 9 5 5 5 8   7

Span, ft	Max moment, ft-kips	Max shear, kips	Max floor-beam reaction, kips	Equivalent uniform load			Span, ft	Max moment, ft-kips	Max shear, kips	Max floor-beam reaction, kips	Equivalent uniform load		
				Moment	Shear	Reaction					Moment	Shear	Reaction
7	21.9	12.5	15.1	3.57	3.57	2.15	26	162.4	29.1	38.8	1.92	2.24	1.49
7½	23.5	13.4	16.0	3.33	3.56	2.12	27	172.3	29.6	40.0	1.89	2.20	1.48
8	25.0	14.0	16.8	3.12	3.51	2.11	28	182.7	30.2	41.2	1.86	2.16	1.47
9	28.1	15.3	18.2	2.78	3.39	2.02	29	194.0	30.8	42.2	1.84	2.12	1.46
10	31.2	16.2	19.2	2.50	3.25	1.92	30	205.2	31.5	43.1	1.82	2.10	1.44
11	34.4	17.0	21.0	2.28	3.09	1.90	31	216.5	32.2	44.3	1.80	2.08	1.43
12	40.0	17.7	23.3	2.22	2.95	1.94	32	227.7	32.9	45.5	1.78	2.05	1.42
13	47.5	18.3	24.6	2.25	2.81	1.90	33	239.0	33.5	46.7	1.75	2.03	1.41
14	55.0	18.8	26.1	2.25	2.68	1.86	34	250.3	34.1	47.8	1.73	2.00	1.40
15	62.5	20.0	27.3	2.22	2.67	1.82	35	261.5	34.6	48.8	1.71	1.98	1.39
16	70.0	21.3	28.5	2.19	2.66	1.78	36	274.3	35.3	49.8	1.69	1.96	1.38
17	77.5	22.4	29.4	2.15	2.63	1.73	37	287.2	35.9	50.7	1.68	1.94	1.37
18	85.0	23.3	30.3	2.10	2.59	1.69	38	300.0	36.5	51.8	1.66	1.92	1.36
19	93.3	24.2	31.5	2.07	2.55	1.66	39	313.3	37.2	52.9	1.65	1.90	1.36
20	103.1	25.0	32.8	2.06	2.50	1.64	40	327.8	37.7	54.0	1.64	1.88	1.35
21	112.9	25.7	34.0	2.05	2.45	1.62	42	356.7	39.2	56.0	1.62	1.87	1.34
22	122.8	26.3	35.1	2.03	2.40	1.60	44	385.8	40.3	58.2	1.60	1.83	1.32
23	132.7	27.0	36.1	2.01	2.34	1.57	46	414.9	41.4	60.3	1.57	1.80	1.31
24	142.6	27.7	37.0	1.98	2.31	1.54	48	443.8	42.4	62.4	1.54	1.77	1.30
25	152.5	28.4	37.8	1.95	2.27	1.51	50	475.5	43.5	64.3	1.52	1.74	1.29

Axle loads (5.0) (10.0) (10.0) (10.0) (10.0) (10.0) (5.0) (8.0) (6.5) (6.5) (6.5) (6.5) (6.5) (6.5) (10.0) (10.0) (10.0) (10.0) (10.0) (10.0)   1 klf /// or (12.5) (12.5) (12.5)

Spacing, ft.... : 8   5   5   9   5   5   8   8   6   5   6   5   5   5   9   5   5   8 ... 7

Span, ft	Max moment, ft-kips	Max shear, kips	Max floor-beam reaction, kips	Equivalent uniform load		
				Moment	Shear	Reaction
52	507.6	44.6	66.7	1.50	1.72	1.28
54	540.5	45.6	69.0	1.48	1.69	1.28
56	576.1	46.5	71.4	1.47	1.66	1.27
58	611.6	47.7	74.0	1.46	1.65	1.28
60	649.5	48.8	76.6	1.44	1.63	1.28
62	688.2	50.0	79.1	1.43	1.61	1.27
64	727.7	51.3	81.5	1.42	1.60	1.27
66	769.7	52.5	83.9	1.41	1.59	1.27
68	811.7	53.9	86.2	1.40	1.59	1.27
70	853.7	55.3	88.5	1.39	1.58	1.26
72	896.7	56.7	90.7	1.38	1.58	1.26
74	939.0	58.1	93.0	1.37	1.57	1.26
76	986.0	59.5	95.2	1.36	1.57	1.25
78	1,032.7	60.9	97.3	1.36	1.56	1.25
80	1,080.0	62.1	99.4	1.35	1.55	1.24

Span, ft	Max moment, ft-kips	Max shear, kips	Max floor-beam reaction, kips	Equivalent uniform load		
				Moment	Shear	Reaction
82	1,128.3	63.5	101.5	1.34	1.55	1.24
84	1,177.7	64.8	103.5	1.34	1.54	1.23
86	1,229.7	66.1	105.4	1.33	1.54	1.23
88	1,282.0	67.4	107.3	1.32	1.53	1.22
90	1,334.7	68.6	109.3	1.32	1.53	1.22
92	1,388.3	69.9	111.2	1.31	1.52	1.21
94	1,442.7	71.2	113.1	1.31	1.52	1.20
96	1,497.3	72.4	115.0	1.30	1.51	1.20
98	1,552.7	73.7	116.8	1.29	1.51	1.19
100	1,609.7	75.0	118.6	1.29	1.50	1.19
125	2,497.7	89.7	140.5	1.28	1.44	1.12
150	3,531.0	103.7	162.7	1.25	1.38	1.08
175	4,676.3	117.3	185.8	1.22	1.34	1.06
200	5,939.0	130.5	209.5	1.19	1.31	1.05
250	8,796.3	156.6	257.6	1.13	1.25	1.03

in which the moment, 8182 ft-kips, and the sum of the weights of the 18 wheels, 142 kips, are found in Table 3. Therefore, $R_L = 118.2$ kips, and the moment at L_1 is

$$M_1 = 118.2 \times 25 - 240 = 2715 \text{ ft-kips}$$

in which the moment, 240 ft-kips, is found in Table 3. For the E-72 loading,

$$M_1 = 7.2 \times 2{,}715 = 19{,}550 \text{ ft-kips per track}$$

The shear in panel L_0L_1 is $V = R_L - 240/25 = 118.2 - 9.6 = 108.6$ kips. For the E-72 loading, $V = 7.2 \times 108.6 = 782$ kips per track.

3. Positive Moment in Continuous Spans Because of the many possible combinations of span, loading, ratio of dead to live load, and degree of fixity, no general formulas or curves can be given for all cases.

Positive Moments in End Span. The envelope of maximum positive moments in the end span of a series of continuous spans on simple supports can be approximated satisfactorily by parabolas. If the length of the end span is 75 to 100 percent of that of the adjacent span, the maximum positive moment occurs at or near the 0.4 point nearest the discontinuous end. The envelope can be approximated by two second-degree parabolas with common vertexes at the point whose ordinate is the maximum moment at the 0.4 point. One parabola passes through the (zero) moment ordinate at the discontinuous end, the other through the moment ordinate at the 0.8 point.

Example Determine the envelope of maximum positive moments for the end span in a continuous beam of three 60-ft spans (Fig. 4a). The dead load is 2 klf, the live load 1.5 wheel lines of HS20-44 load with 27 percent impact.

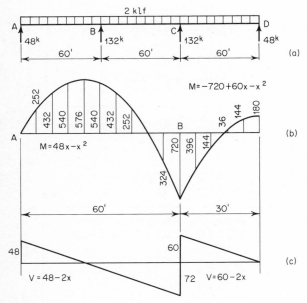

Fig. 4

1. The dead-load moments at B and C are 720 ft-kips, from which the reaction at A is found to be 48 kips. The corresponding dead-load moment diagram is shown in Fig. 4b. Only the moments at the 0.4 and 0.8 points are needed in this example; the remaining ordinates are used in the determination of other maximum-moment envelopes discussed later.

2. From Table 1, with the truck positioned for maximum moment at the 0.4 point, the simple-beam moment and fixed-end moment are, respectively, 790.4 and 725.6 ft-kips. With the truck positioned for maximum moment at the 0.8 point the simple-beam moment and fixed-end moment are,

respectively, 556.8 and 716.9 ft-kips. Converting these values to 1.5 wheel lines per girder, and adding 27 percent impact, the following values are obtained:

	0.4	0.8
SBM	753	531
FEM	692	683

3. With the fixed-end moment, 692 ft-kips, known, the live-load negative moment at B can be determined. When moving loads are being investigated, it is convenient to do this by distributing a unit moment acting at B (Fig. 5). The resulting distribution factors can be used to distribute the fixed-end moment at B in AB for any combination of loads in any of the three spans. In this case, we get

$$\text{FEM}_{BA} = 0.533 \times 692 = 369 \text{ for maximum moment at } 0.4$$
$$\text{FEM}_{BA} = 0.533 \times 683 = 364 \text{ for maximum moment at } 0.8$$

From these values, corrections to the live-load simple-beam moments can be found. Finally, then, the $DL + LL + I$ moments are

$$M_{0.4} = 576 + 753 - 0.4 \times 369 = 1182 \text{ ft-kips}$$
$$M_{0.8} = 0 + 531 - 0.8 \times 364 = 240 \text{ ft-kips}$$

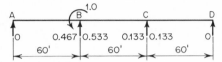

Fig. 5   Moment-distribution factors.

4. The approximate maximum-moment envelope is shown in Fig. 6. The ordinates at intermediate points are easily found from properties of the parabola. Thus

$$M_{0.5} = 1182 - (1182 - 240)(\tfrac{1}{4})^2 = 1123 \text{ ft-kips}$$

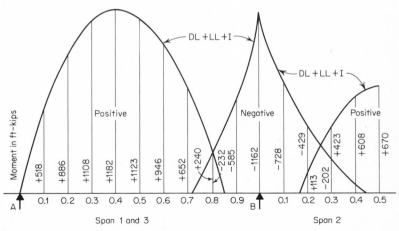

Fig. 6   Maximum moment envelopes, continuous beam, three 60-ft spans.

Positive Moment in Interior Span.   An approximate envelope for maximum positive moment in the interior span is constructed by passing a second-degree parabola through the maximum-moment ordinates at the 0.2 point and the midpoint. For maximum moment at the 0.2 point, the truck heads into the span with the rear wheel at the 0.2 point. Position of the load for maximum midspan moment is shown in Fig. 3. Although the maximum moment is at the 32-kip axle, it is assumed to be at midspan. Fixed-end moments for loads at these positions may be found in Table 1.

Example Construct the approximate envelope for maximum positive moment in the interior span of the beam of Fig. 4a.

The following moments are found in Table 1:

Simple-beam moment at 0.2 point	556.8
Fixed-end moment at B	548.5
Fixed-end moment at C	336.9

Using the distribution factors determined previously (Fig. 5) the negative moments at B and C are

$$M_B = 0.467 \times 548.5 + 0.133 \times 336.9 = 301$$
$$M_C = 0.133 \times 548.5 + 0.467 \times 336.9 = 230$$

From these values and the simple-beam moment 556.8, the moment at the 0.2 point is found to be 270 ft-kips. For 1.5 wheel lines and 27 percent impact, this reduces to 257 ft-kips, which, combined with the dead-load moment, -144 ft-kips, from Fig. 4b, gives the maximum $DL + LL + I$ moment of $+113$ ft-kips.

The maximum moment at midspan, 670 ft-kips, is found similarly, and the maximum moment envelope is determined by constructing the parabola shown in Fig. 6.

4. Negative Moments in Continuous Spans Envelopes of maximum negative moment can be obtained by determining the maximum negative moments at a support and at several points in the adjoining spans. Maximum negative moment at a support occurs when the two adjoining spans are loaded. In the case of equal spans over 41 ft long the equivalent lane loading controls. If the adjoining spans are unequal, this separation point must be determined by trial.

The position of load to obtain maximum negative moment at an interior point of a span can be determined from an influence line, or from a series of trials. However, a satisfactory approximation is obtained by loading the span adjacent to the one under consideration with the heavy wheels of the truck equidistant from the 0.4 point adjacent to the support common to the two spans. Thus, for maximum negative moments in span 1, the loads are positioned as in Fig. 7a, while for maximum negative moments in span 2 they are located as in Fig. 7b.

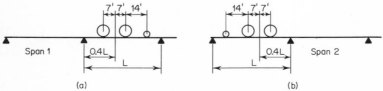

(a)   (b)

Fig. 7 Approximate maximum negative moments in continuous beams. (a) Position for negative moment in span 1; (b) Position for negative moment in span 2.

For equal spans, Table 1 gives the simple-beam moment and the fixed-end moments, for both truck and lane loading, with the loads positioned for maximum negative moments in the adjoining, unloaded span.

Example Construct the approximate envelope of maximum negative moments for the continuous beam of Fig. 4a.

1. Since the spans are equal and over 41 ft long, the maximum live-load negative moment at an interior support is produced by the lane loading. Therefore, the required fixed-end moments from Table 1 are the following: $FEM_{BA} = 495.8$, $FEM_{BC} = 352.0$, and $FEM_{CB} = 272.0$. Using the distribution factors in Fig. 5,

$$M_B = 0.533 \times 495.8 + 0.467 \times 352.0 + 0.133 \times 272.0 = 464$$

For 1.5 wheel lines and 27 percent impact, this gives $M_B = 442$ ft-kips. Adding this to the dead-load moment from Fig. 4b, the $DL + LL + I$ moment is found to be -1162 ft-kips.

2. Maximum negative moments in the end span are produced when the live load is in the second

span. From Table 1, it is found that the truck loading is critical, so that $\text{FEM}_{BC} = 549.5$ and $\text{FEM}_{CB} = 355.0$. Following the procedure in step 1, we get $M_B = 290$ ft-kips. Since the corresponding moment diagram for span AB is a straight line from 0 at A to -290 ft-kips at B, the negative moments at the 0.7, 0.8, and 0.9 points are easily determined (Fig. 8). These live-load moments are given in the following table. The dead-load moments are taken from Fig. 4.

Location	DL	LL + I	DL + LL + I
0.7	+252	−203	+ 49
0.8	0	−232	− 232
0.9	−324	−261	− 585
B	−720	−290	−1,010

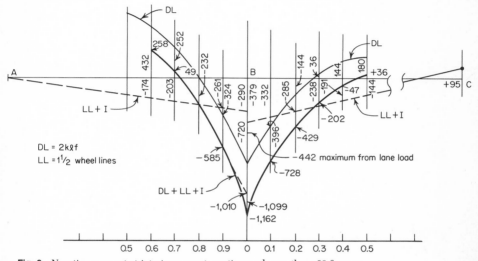

Fig. 8 Negative moment at interior support, continuous beam, three 60-ft spans.

3. Maximum negative moments in the center span are produced when the live load is in the end span. From Table 1, $\text{FEM}_{BA} = 746.6$. Therefore,

$$M_B = 0.533 \times 746.6 \times \tfrac{3}{4} \times 1.27 = -379$$
$$M_C = 0.133 \times 746.6 \times \tfrac{3}{4} \times 1.27 = +95$$

Since the moment diagram in span BC is a straight line, moments at intermediate points are easily found (Fig. 8). The results are given in the following table.

Location	DL	LL + I	DL + LL + I
B	−720	−379	−1,099
0.9	−396	−332	− 728
0.8	−144	−285	− 429
0.7	+ 36	−238	− 202
0.6	+144	−191	− 47
0.5	+180	−144	+ 36

The maximum-moment envelope is shown in Fig. 6. Note that the ordinate at B is the value found in step 1, which exceeds the values found in steps 2 and 3.

5. Shears in Continuous Spans The following approximate limits determine the spans up to which the HS20-44 truck produces maximum shear, and above which the equivalent loading should be used:

Shear and reaction at the end support......... To 140 ft
Shear at left of first interior support.......... To 110 ft
Shear at right of first interior support......... To 110 ft
Reaction at first interior support............. To 58 ft

Shear in End Span. A satisfactory envelope for maximum shear can be found by determining the following: (1) shear at end support, (2) positive and negative shears at the 0.4 point nearest the end support, and (3) negative shear just to the left of the first interior support.

Example Determine the approximate envelope of maximum shears for the beam of Fig. 4a.
1. Maximum shear at A occurs with the truck in the position shown in Fig. 9a. The moment at B is

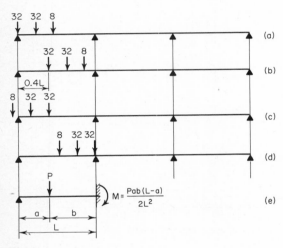

Fig. 9

found by determining the fixed-end moment and using the distribution of Fig. 5. The fixed-end moment, which is easily computed from the known formula (Fig. 9e), is 302 ft-kips. Therefore, $M_B = 0.533 \times 302 = 160$ ft-kips, and $R_A = 58.1$ kips. For 1.5 wheel lines and impact of 27 percent, $R_A = 55.5$ kips.
2. Maximum positive shear at the 0.4 point occurs with the truck in the position shown in Fig. 9b. Proceeding as in step 1, the maximum shear $V = R_A$ is found to be 24.7 kips.
3. Maximum negative shear at the 0.4 point occurs with the truck in the position shown in Fig. 9c. For this case, the maximum shear $V = R_B$ is 21.8 kips.
4. Maximum negative shear to the left of B occurs with the truck in the position shown in Fig. 9d. The maximum shear $V = R_B$ is 61.2 kips.
5. The maximum $LL + I$ values are plotted in Fig. 10, and the envelope of maximum shears found by combining them with dead-load shears shown in the figure.

Shear in Interior Span. The shear envelope for an interior span is found in a manner similar to that of the preceding example. In a three-span symmetrical bridge, the center-span shears need be calculated only at the interior supports and at the midpoint.

6. Impact *Highway Loadings.* Impact is expressed as an increase in live-load stress according to the formula

$$I = \frac{50}{L + 125}$$

where I = impact factor (maximum 30 percent)
  L = length, ft, of the portion of the span which is loaded to produce the maximum stress in the member
The length L is to be taken as follows:
  1. For roadway floors, L = design length of span.
  2. For transverse members, such as floor beams, L = span center to center of supports.
  3. For truckload moments, L = span of member, except that for cantilever arms L = distance from moment center to farthermost axle.
  4. For truckload shears, L = length of loaded portion of span from point under consideration to far reaction, except that for cantilever arms I = 30 percent.
  5. For continuous spans, L = length of span under consideration for positive moment; L = average of two adjacent loaded spans for negative moment.

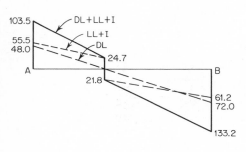

Fig. 10 End-span maximum shear envelope, continuous beam, three 60-ft spans.

Railroad Loadings. According to the AREA specifications, impact on railroad structures is the following percentages of the live load, applied equally and vertically at the top of each rail:
  a. For stringers, girders, floor beams, posts of deck-truss spans carrying floor-beam loads only, and floor-beam hangers:
  L less than 100 ft, $60 - L^2/500 + 100/S$
  L 100 ft or more, $1800/(L - 40) + 10 + 100/S$
  b. For truss spans, $4000/(L + 25) + 15 + 100/S$
  Where the hammer-blow effect is absent, as in the case of diesel locomotives, the percentage of impact is given by
  L less than 80 ft, $40 - 3L^2/1600 + 100/S$
  L 80 ft or more, $600/(L - 30) + 16 + 100/S$
  The impact allowance is also governed by the number of tracks on the structure:
  *a.*Where the load is from two tracks:
  L less than 175 ft, full impact on two tracks
  L from 175 to 225 ft, full impact on one track; on the other $(450 - 2L)$ percent of full impact.
  L over 225 ft, full impact on one track, none on the other
  b. Where there are more than two tracks, full impact is applied to any two tracks for all values of L.
  The length L in these formulas is the following:
  a. For stringers, transverse floor beams without stringers, longitudinal girders, and main members of trusses, L = length, ft, center to center of supports.
  b. For floor beams, floor-beam hangers, subdiagonals of trusses, transverse girders, supports for longitudinal and transverse girders, and viaduct columns, L = length, ft, of the longer adjacent supported stringer, longitudinal beam, girder, or truss.
  The length S is the distance, ft, between centers of longitudinal beams, girders, or trusses, or between supports of floor beams or transverse girders.

Impact for ballasted-deck bridges is 90% of the values for open-deck bridges.

7. Wind   Wind load is assumed to act horizontally from any direction. Wind forces may result in vertical or twisting motions of flexible structures. In the design of suspension bridges, it is prudent to verify the stable aerodynamic characteristics of the stiffening trusses or girders by wind-tunnel tests on models.

Care should be taken in determining the path through which lateral pressures will be transmitted to the ground. Continuous concrete structures may well be stiff enough to carry the lateral loads to the abutments, rather than to intermediate piers. Structures having concrete decks will usually transmit wind loads to the diaphragms or cross frames at the ends of the span, and thence to the bearings. For diaphragm and cross-frame design, the wind load is applied to the top flange of a plate girder.

Wind loads for which various bridge structures are to be designed are prescribed in the AASHTO, AREA, or other applicable specifications.

8. Other Loads   *Earthquake Loads.*   A horizontal load of 2 to 20 percent of the dead load of the structure, applied at the center of gravity, is a simple approximation of earthquake forces on a bridge. Among the factors that must be considered in a more sophisticated approach are the nature of the earthquake, the natural period of vibration of the structure, the natural period of vibration of the material that supports the foundation, and most importantly the details of the structure. In severe earthquakes, structures usually fail because of poor details Joints pull apart, bearings fall off their supports, concrete disintegrates and destroys reinforcing bond, and footings break or pull away from pile supports. Therefore, in earthquake-resistant design the details of a structure should be carefully worked out to withstand severe movement and racking.

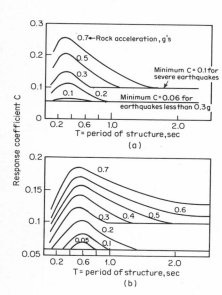

Fig. 11 Response coefficient for earthquake forces.

The design of very large or unusual bridges and bridges adjacent to faults should be based on up-to-date dynamic techniques. Dynamic analyses of a number of smaller and more conventional structures were made by a task force of the Structures Section of the California Department of Transportation to determine equivalent static forces. The following simplified approach, developed by the author, is based on the work of the task force and was reviewed by it.

The equivalent static force EQ, which is to be distributed in proportion to the stiffnesses of the superstructure and supporting members, is given by

$$EQ = CFW$$

where C = response coefficient from Fig. 11
$\quad F$ = framing factor = 1 for bridges on single-column bents and 0.8 for bridges on bents with two or more columns
$\quad W$ = dead-load weight of superstructure

The response coefficient C depends on the depth of the alluvium cover to rocklike material, the peak rock acceleration, and the natural period T of the structure. Values of T in seconds are given by

$$T = 0.32 \sqrt{\frac{W}{P}}$$

where P is the total uniformly distributed lateral force required to produce a 1-in. deflection of the structure.

The above procedure, with some modifications of Fig.11, has been adopted in the AASHTO specifications.[3]

Example Compute the natural period T for a two-span continuous bridge supported at the center on a two-column bent.

The lateral deflection Δ of the bridge at midspan is

$$\Delta = \frac{5PL^3}{384EI_s} - 2 \times \frac{P_cL^3}{48EI_s}$$

where L = total length of bridge
$\quad I_s$ = moment of inertia of superstructure
$\quad P_c$ = lateral force on one column

Assuming the column to be fixed against rotation at top and bottom, the deflection at the top is

$$\Delta = \frac{P_ch^3}{12EI_c}$$

where h = height of column. Eliminating P_c between these two equations and using $\Delta = 1$ in. gives

$$P = \frac{1 + (I_c/2I_s)\,(L/h)^3}{5L^3/384EI_s}$$

Substituting this value of P into the formula for T enables the response coefficient C to be determined from Fig. 11, after which the equivalent static force EQ can be calculated.

Buoyancy. Submerged structures should be designed for the buoyant effect of the water. Although this can be of considerable importance in a large structure, it is of little effect in smaller ones. Tremie seals for spread footings should have a thickness of 0.43 times the hydrostatic head. Minimum thickness of a tremie seal should be 1.5 ft without piles and 2 ft with piles. Uplift on piles for intermittent loads is generally figured at 40 percent of the allowable design load, provided, of course, there is sufficient weight in the land mass engaged by the pile to develop this calculated resistance.

Longitudinal Force. AASHTO requires that provision be made for a longitudinal force equal to 5 percent of the lane load, excluding impact, acting in all lanes carrying traffic headed in the same direction. The center of gravity of the loading is assumed to be 6 ft above the floor. This loading rarely affects a simple-span bridge, but piers, bents, and footings should be checked. Very stiff or continuous structures are sometimes affected by this loading.

Centrifugal force is seldom of consequence in a highway structure, but it must be included in the design of railroad structures. In accordance with the AREA and the AASHTO specifications, the centrifugal force is the percentage $0.00117S^2D = 6.68S^2/R$ of each axle load, excluding impact, where S = design speed, mph, D = curvature, degrees, and R = radius of curve, ft. This force is assumed to act horizontally 6 ft above the rail or roadway surface.

Thermal forces can be very important. The extent to which they are provided for will depend upon the locality and the possible temperature variation. As a general guide, the following variations from a normal of 70°F should be provided for:

	Rise, °F	Fall, °F
Moderate climate:		
Steel structures...........	40	50
Concrete structures........	30	40
Cold climate:		
Steel structures...........	40	80
Concrete structures........	35	45

When design for thermal forces governs, the basic allowable stresses may be increased by 25 percent.

Earth Pressure.   AASHTO requires that structures retaining a fill be designed for an equivalent fluid pressure of at least 30 pcf. This is a good figure for general use. To avoid movement in high walls or abutments, the more conservative figure 36 pcf is recommended in computing toe pressures for high walls or abutments, while for heel pressures 27 pcf may be used.

Shrinkage of concrete creates stresses which must be accommodated. Just as in a complicated weldment where the welding sequence is carefully programmed to minimize shrinkage and distortion, so in a complicated concrete structure must the pouring sequence be carefully worked out and specified to minimize shrinkage effects. Concrete takes most of its shrinkage soon after it sets. Consequently, a delay of 24 hr between pours for thin sections, and of from 3 to 5 days for heavier sections, is usually sufficient to allow for most of the shrinkage. Large-mass concrete structures raise more serious problems, but the average highway or railroad bridge does not have sections massive enough to make shrinkage a problem after 3 to 5 days. Arch rings are poured in intermittent sections and then filled in between, with a final pour at the crown, to minimize shrinkage effects. Complicated rigid frames or interconnected concrete members should be similarly poured in sections and allowed to shrink before the final gap is closed.

The shrinkage coefficient for unreinforced concrete is 0.0002. However, reinforcing bars restrict the concrete so that this shrinkage is seldom realized. Shrinkage is manifested most often by the numerous minute cracks which are inevitable with presently constituted cements. In ordinary concrete beam-and-slab construction, shrinkage stresses may be ignored in the design, provided proper field control of pouring sequences is exercised. In arches and in prestressed construction, the effect of shrinkage is appreciable and must be compensated for in the design of the member.

Erection Stresses.   In certain types of structures and methods of erection, stresses which exist only during erection can be considerable. In some cases, such as cantilever trusses, members needed temporarily during erection may be removed later, or altered to prevent their having an undesirable effect on stresses in other members. It is usual practice for the erector to determine and provide for any necessary strengthening of members during erection.

Pressure of Moving Water.   Portions of structures which are subject to the force of flowing water, floating ice, or debris must be designed to resist these forces. The force of flowing water on piers is given by AASHTO as

$$P = KV^2$$

where P = water pressure, psf
V = velocity of the water, fps
K = shape constant = $1\frac{3}{8}$ for square ends, $\frac{1}{2}$ for angle ends of 30° or less, $\frac{2}{3}$ for circular piers

Ice Pressure.   AASHTO requires that the pressure of ice on a pier be calculated at 400 psi. The thickness of ice to be provided for, and its elevation, must be determined from local records.

9. Design Methods   Either allowable-stress design or load-factor design may be used for bridges. The two methods are discussed in Secs. 6 and 11. AASHTO allows design by either method. Allowable-stress design is used in the examples in this section.

STEEL BRIDGES

10. Floor Systems Floors for steel highway bridges are usually reinforced concrete. The use of other types, such as steel grid, armor type, steel plate, and timber, depends upon dead load, traffic, and the location of the structure. Floors should have a transverse slope and, if possible, a longitudinal slope, to provide adequate drainage. Transverse slope should be at least 1 percent, and preferably more. Adequate drains should be placed at suitable locations to prevent water from collecting on the deck.

Types. Exclusive of the main carrying members, floor systems can be classified as one-, two-, three-, and four-element systems, as follows:

   1. A transversely reinforced concrete slab on main stringers or girders is a one-element system. Transverse-laminated timber floors are in the same category.

   2. A longitudinally reinforced concrete slab, on transverse floor beams supported by girders or trusses is a two-element system.

   3. A transverse slab on longitudinal stringers supported by floor beams framing into the main girders or trusses is a three-element system. The design of the slab in the three-element system is similar to that of the one-element system.

   4. A steel-grid floor on transverse beams resting on stringers supported by floor beams framing into the main girders or trusses is a four-element system.

In addition to the above, the orthotropic plate floor, which consists of a deck plate acting compositely both longitudinally and transversely and, in addition, with the main supporting members, is used.

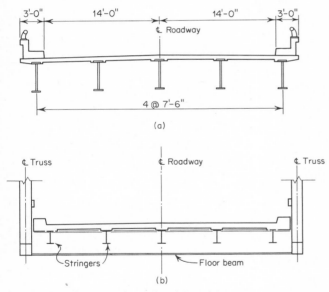

Fig. 12 Typical transversely reinforced floor slab. (*a*) Steel-stringer span; (*b*) steel truss.

Slabs. The dead load consists of the weight of the slab plus any anticipated wearing surface. Bituminous surfacing is occasionally placed over concrete slabs to match the bridge approaches, or to level up or seal cracks in an old concrete deck. Where an added blanket is a definite possibility, it should be provided for in the design. Where a blanket is added after years of service, the higher stresses normally allowed in computing the capacity of existing bridges will usually compensate for the extra weight of the blanket.

11. Concrete Floors Transversely reinforced concrete slabs are commonly used on steel-stringer spans, plate girders, and truss bridges, as well as on several types of reinforced-concrete structures. Typical examples are shown in Fig. 12. A steel-stringer span, in which the slab is a one-element floor system, is shown in *a*. A truss span, in which the slab is part of a three-element floor system, is shown in *b*.

Example   Design a transversely reinforced concrete slab continuous over three or more longitudinal steel stringers, AASHTO specifications. Live load is HS20-44. Allowable stresses f_c = 1200 psi, f_s = 20,000 psi. No allowance will be made for future wearing surface. A typical section is shown in Fig. 13. The assumed slab thickness is 7 in., weight 88 psf.

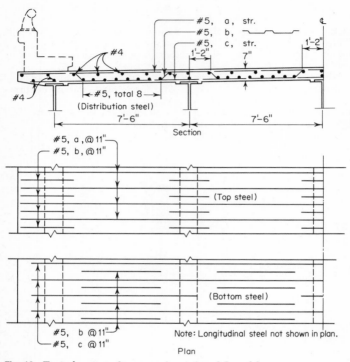

Fig. 13   Typical section of transversely reinforced floor slab.

The design span *S* for a slab continuous over two or more supports is the distance center to center of stringers minus half the flange width, in this case, 7 ft (Fig. 14). The live-load moment per foot width of slab, both positive and negative, is given by $0.8P(S + 2)/32$, where *P* is the wheel load (AASHTO 1.3.2). Impact for a 7-ft span is 30 percent.

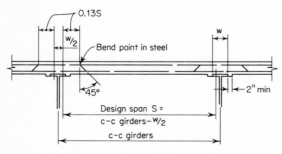

Fig. 14   Transversely reinforced slab on steel stringers.

The design moments per foot width of slab are:

$$DL = 0.088 \times 7^2/10 = 0.43$$
$$LL + I = 1.30 \times 0.8 \times 16(7 + 2)/32 = 4.68$$
$$5.11 \text{ ft-kips}$$

With $K = 198$, $d^2 = 5110/198 = 25.8$, $d = 5.1$ in. Using $d = 5\frac{1}{8}$ in. with $1\frac{7}{8}$ in. from center of steel to top of slab gives a 7-in. slab. Then

$$A_s = \frac{5.11 \times 12}{20 \times 0.88 \times 5.13} = 0.68 \text{ in.}^2$$

Use No. 5 at $5\frac{1}{2}$ in.

The specifications require distribution steel in the bottom of the slab, to be determined as the percentage $220/\sqrt{S}$ of the main steel, but not to exceed 67 percent.

$$\frac{220}{\sqrt{7}} = 83 > 67\%$$
$$0.67 \times 0.68 = 0.46 \text{ in.}^2$$

Use No. 5 at 8 in.

Other than distribution steel, the longitudinal steel in the top and bottom of the slab can be No. 4 with 18 in. minimum spacing. Bars should be placed at the bend points in the transverse steel and at other points where the reinforcing should be adequately tied to hold it in place during pouring operations.

Longitudinally Reinforced Slabs. When used on truss spans the longitudinally reinforced slab is part of a two-element floor system. More commonly it is used as a primary structure supported on piers, bents, or piles. Its use with trusses is limited to comparatively short spans as the dead load is considerably greater than that in the floor system using a transverse slab.

For a slab supported by the steel floor beams of a truss, the effective span (AASHTO) is the span center to center of floor beams less one-half the flange width. Practical procedure is to use the span and, where refinement is necessary or desired, make corrections to the computed moments in accordance with the ratio of effective span to span.

The AASHTO requirements in determining placement, sizes, and length of reinforcing bars in continuous beams are:

1. At the freely supported end of a continuous slab, one-third of the positive-moment reinforcement shall extend beyond the face of the support a distance sufficient to develop one-half the allowable stress in the bars.

2. At the restrained ends of continuous slabs one-fourth of the positive-moment reinforcement shall extend beyond the face of the supports.

3. Between the supports, reinforcement bars shall extend at least 15 diameters, but not less than one-twentieth of the span length, beyond the point where they are no longer needed to resist stress.

Example Design a longitudinally reinforced slab continuous over four supports. AASHTO specifications, HS20-44 live load. Allowable stresses $f_c = 1200$ psi, $f_s = 20,000$ psi. The layout is shown in Fig. 15 which is a longitudinal section of a truss span. Calculations will be based on the 16-ft span, rather than the effective span.

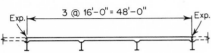

Fig. 15 Longitudinally reinforced floor slab.

The effective width of slab for distribution of one wheel line is $E = 4 + 0.06S = 4 + 0.06 \times 16 = 4.96$ ft. The slab is analyzed as a continuous strip 1 ft wide. The wheel-line concentrations (impact not included) are shown in Fig. 16a and their positions for maximum positive and negative moments in Fig. 16b. Other positions of load, not shown in the figure, are required to determine the envelope of maximum moment shown in Fig. 17. The slab depth is governed by the positive moment, 17.1 ft-kips, in the end span, and $d^2 = 17,100/198$, $d = 9.4$ in. With $d = 9\frac{1}{2}$ in. and $1\frac{1}{2}$-in. cover, total depth of slab is 11 in. A shallower depth could be used by designing the end spans as doubly reinforced slabs.

Positive moment in end span:

$$A_s = \frac{12 \times 17.1}{20 \times 0.88 \times 9.5} = 1.23 \text{ in.}^2$$

	A_s	M
No. 6 at 16.........	0.33	4.6
No. 7 at 16.........	0.45	6.3
No. 7 at 16.........	0.45	6.3
	1.23 in.²	17.2 ft-kips

Negative moment at support:

$$A_s = \frac{12 \times 15.8}{20 \times 0.88 \times 9.0} = 1.20 \text{ in.}^2$$

	A_s	M
No. 6 at 16.........	0.33	4.4
No. 7 at 16.........	0.45	5.9
No. 7 at 16.........	0.45	5.9
	1.23 in.²	16.2 ft-kips

Positive moment in interior span:

$$A_s = \frac{12 \times 12.6}{20 \times 0.88 \times 9.5} = 0.91 \text{ in.}^2$$

	A_s	M
No. 6 at 16.........	0.33	4.6
No. 6 at 16.........	0.33	4.6
No. 6 at 16.........	0.33	4.6
	0.99 in.²	13.8 ft-kips

In the bottom of the slab, transverse distribution steel in the center half of the span is determined as the percentage of the main steel $100/\sqrt{S}$, but not to exceed 50 percent. Distribution steel can be reduced one-half in the outer quarters. Transverse temperature steel in the top of the slab is usually No. 4 bars at 18 in. Details of the slab are shown in Fig. 18.

Lightweight-Concrete Bridge Decks. Lightweight concrete composed of aggregates made from expanded shale, slate, clay, and slag is occasionally used in bridge decks. Concrete in the structural range of 3000 to 4500 psi can be obtained with weights of about 100 to 120 pcf, not including the reinforcing steel. Lightweight-concrete decks are frequently constructed in two courses, the bottom course consisting of the lightweight concrete and, to provide durability, the top course consisting of natural-sand mortar composed of 1 part cement and 3 parts by dry weight of natural sand. Thickness of the top course should be not less than ½ in. or more than ¾ in.

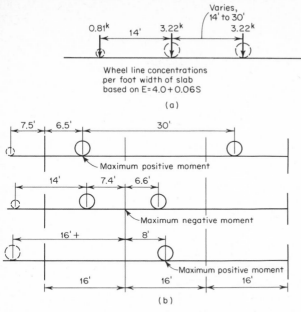

Wheel line concentrations
per foot width of slab
based on E=4.0+0.06S

(a)

Maximum positive moment

Maximum negative moment

Maximum positive moment

(b)

Fig. 16 Wheel positions for maximum moment in floor slab of Fig. 15

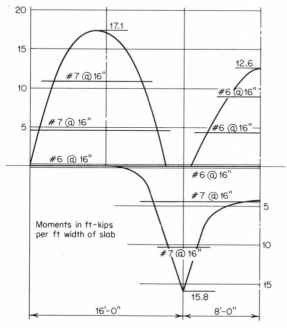

17.1

#7 @ 16"

12.6

#6 @ 16"

#7 @ 16"

#6 @ 16"

#6 @ 16"

#6 @ 16"

#7 @ 16"

Moments in ft–kips
per ft width of slab

#7 @ 16"

15.8

16'-0"   8'-0"

Fig. 17 Maximum moment envelopes for floor slab of Fig. 15.

Design coefficients will vary somewhat. Lightweight concrete has a larger value of n because of its lower modulus of elasticity. There is a decrease in j and an increase in k and p. For balanced reinforcement a higher percentage of steel is required, but there is a reduction in depth of the member because of decreased dead load. Beams will have more deflection owing to the smaller modulus of elasticity. Tests show that, for comparable strengths, working stresses for bond and diagonal tension applicable to conventional sand and gravel concrete may be used for lightweight concrete.

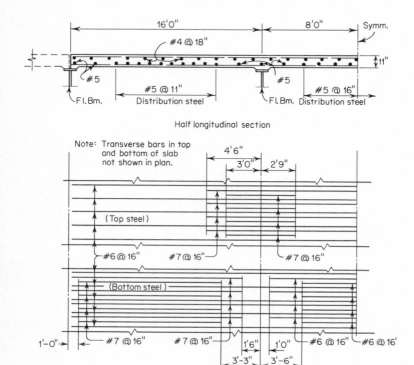

Fig. 18   Reinforcement for floor slab of Fig. 15.

Some of the important items which should be covered in a specification for lightweight concrete are (1) minimum amount of portland cement per cubic yard; (2) desired weight per cubic foot; (3) design strength in flexure; (4) type of lightweight aggregate; (5) grading limits of the aggregate; (6) percentage of unburned or underburned lumps in expanded shale (ASTM Designation C142); (7) hardness of aggregate to conform to a specified test; (8) maximum weight of aggregate in loose, surface-dry condition; (9) limitations on the moisture content prior to mixing; indicate the quantity of aggregate sample to be submitted for test; (10) provision for substitution of natural sand, provided it does not increase the weight beyond the desired maximum; (11) control of use of internal vibrators, so as to guard against flotation of the coarse aggregate; (12) use of air-entraining agents to improve workability and minimize separation of aggregates.

12. Steel Floors   Battledeck floors can be used for spans from ±15 to ±30 ft. They can be used as primary elements, e.g., on pile bents, and in the floor system for plate-girder and truss bridges. Light weight is their distinctive feature. Each section or unit is usually 6 to 9 ft wide, depending upon the number of units required to make up the desired width. Originally, these units consisted of a deck plate welded to either I- or wide-flange beams.

More recently, the units are built up completely from plates (Fig. 19). Transverse stiffeners welded to the floor plate between beams allow a thinner floor plate. Tests* on battledeck floors were performed at Lehigh University and subsequent rules for design developed.†

A durable and satisfactory surface for smooth floor plate is composed of a coal-tar-modified thermosetting epoxy resin, with a cover of aluminum-oxide aggregate added to

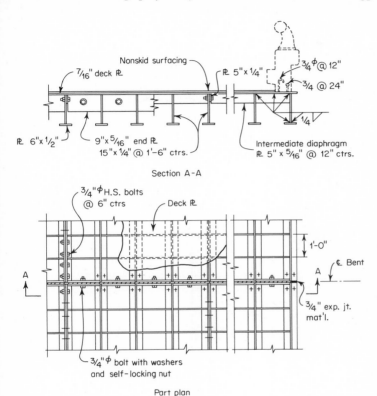

Section A-A

Part plan

Fig. 19 Typical battledeck floor, 26-ft spans.

provide the nonskid surface. Before application it is important that the surface of the steel deck be thoroughly sandblasted. The time for hardening of this surfacing may range from 1 to 2 hr at 90°F to 12 or more hours at 60°F.

A surface of asphaltic concrete material suitably sealed and bonded to the steel plate will work satisfactorily in some locations. It is recommended that its use be limited to decks that have a longitudinal gradient not greater than 2 percent and where maximum temperatures do not exceed 100°F. Asphalt plank, fixed in place with an asphaltic cement, has been used on some structures. In some installations hard stone chips have been pressed into the plank to increase traction.

To improve traction where traffic runs directly on the floor plate, "checkered" or "traffic" plates can be used, preferably those in which the patterns are raised above the

*Lyse, Inge, and I. E. Madsen: Structural Behavior of Battledeck Floor Systems, *Trans. ASCE*, vol. 104, p. 244, 1939.
†Committee on Technical Research, AISC, "The Battledeck Floor for Highway Bridges," 1938.

nominal thickness of the plates. When choosing the type of surface finish, consideration should be given to weather conditions. Battledeck floors are sensitive to weather changes. Because of the free access of air to both sides of the floor plate, freezing or thawing occurs almost simultaneously with air-temperature variations.

Steel-grid floors are used where light deck systems are desired. They have been used on many movable structures, such as vertical-lift spans, bascule spans, and swing bridges. In general, they consist of two types, the I-Beam Lok and the Irving type. Either type can be used as an open grid or filled with concrete. An open deck weighs approximately 15 to 19 psf, depending on the type. Recommended spans for the various types are given in manufacturers' literature.

Both transverse and longitudinal distribution of load is considered in the design of grid flooring. According to AASHTO, a wheel load is distributed, normal to the main bars, over a width equal to 1¼ in. per ton of axle load plus twice the distance center to center of main bars. The portion of the load assigned to each main bar is applied uniformly over a length equal to the rear tire width of the design truck (20 in. for H-20, 15 in. for H-15). This distribution is shown in Fig. 20 for an H-20 wheel.

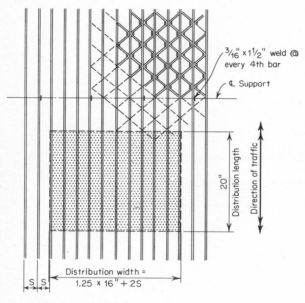

Fig. 20 Steel-grid flooring, showing AASHTO distribution for 16-kip wheel.

In the design of steel-grid floors filled with concrete AASHTO calls for load distribution and bending moments determined in the same manner as for concrete slabs. This also applies to edge beams (longitudinal), to unsupported edges (transverse), and to the design span length. Properties of the composite steel and concrete slab are determined by the transformed area.

Units of deck grating are welded to the supporting elements in accordance with the manufacturer's instructions. A successful method of attaching the Irving decking is to weld the deck to the supporting elements using ³⁄₁₆- by 1½-in. welds on one side of every fourth bar (Fig. 20). Welds are staggered so that each main bar is welded at every fourth support.

Structural Plate Flooring.   For redecking existing bridges the structural plate floor is economical and comparatively easy to erect under traffic. It is rolled in various gage thicknesses in widths of 24 in. Usually it is installed transverse to traffic and welded to the stringers. Depending on the gage of the plate, the stringer spacing can vary from about 2 to 3 ft. The effective span of this flooring on steel stringers is defined for design purposes as the clear distance between flanges plus one-half the stringer flange width. Lateral and longitudinal distributions of a wheel load are not defined in the AASHTO specifications, but the same distribution as for a laminated timber floor can be used. A continuity factor of 0.8 can be applied to the simple-beam moment for both positive and negative moments.

A bituminous material, consisting of two coats, is placed as a wearing surface. The first course is placed to a level slightly above the top of the corrugations. The second course is placed so that after rolling the resulting surface is at least 2 in. above the corrugations at centerline of roadway and 1 in. thick at the edges to provide a crown. A typical installation is shown in Fig. 21.

Orthotropic plate floors are discussed in Part 2 of this section.

13. Floor Beams Steel floor beams may be standard rolled shapes or may be built up from plates. The distribution of wheel loads to floor beams in decks without longitudinal stringers depends on their spacing and the kind of floor. Distribution factors are given in AASHTO 1.3.1. No lateral distribution of wheel loads is assumed.

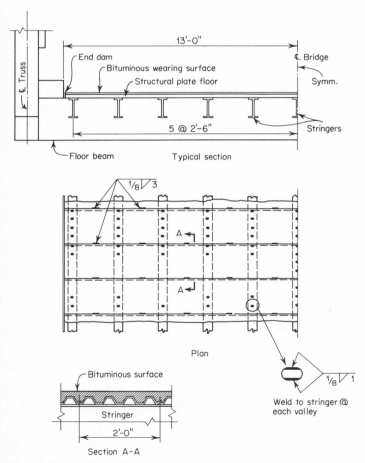

Fig. 21 Structural plate flooring.

In floors with longitudinal stringers, dead load from the stringers can be considered as a uniform load on the floor beam for design purposes without appreciable error. Figure 22a shows a floor beam for a truss bridge having 20-ft panels. If the dead load from the stringers is applied to the floor beam as uniform load, the maximum moment is 240 ft-kips; applying the stringer reactions as concentrated loads gives a maximum moment of 230 ft-kips.

Live loads on the floor beam can be considered on the basis of wheel lines. The live-load reaction per line (Fig. 22b) is $4 \times \frac{6}{20} + 16 + 16 \times \frac{6}{20} = 22.0$ kips.

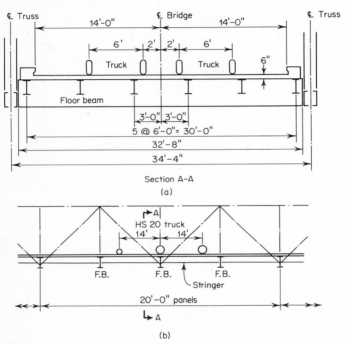

Section A-A
(a)

(b)

Fig. 22

It is assumed that trucks stay within the design lanes. For maximum live-load moment they are placed as shown in Fig. 22a. The resulting moment is

$$2 \times 22 \times 14.33 - 2 \times 22 \times 3 = 44 \times 11.33 = 500 \text{ ft-kips}$$

The total moment is

DL floor beam	=	25
DL stringers	=	240
LL	=	500
Impact 30 percent	=	150
DL + LL + I	=	915 ft-kips

For an allowable stress of 20 ksi, the section modulus is

$$S = 915 \times \frac{12}{20} = 549 \text{ in.}^3$$

Use W36 × 170.

For maximum end reaction of the floor beam one of the trucks should be placed within 2 ft of the curb. The other truck would be placed in the same position as shown for moment in order to stay within the design lane.

BEARING AND EXPANSION DETAILS

14. End Bearings Bearing details are designed to transmit the loads to the foundation and to provide for expansion of the superstructure. They may vary from thin elastomeric bearing pads to huge pin and rocker assemblies. Certain basic requirements must be satisfied. First, the bearing must be adequate for the loading, which may occur simultaneously from several different directions. Second, the bearing must be able to accommodate movements of the structure. These movements, which may be in any direction, result from load, deflection, temperature change, earthquake movement, impact, centrifugal force, etc. The design must restrict movements to reasonable limits. This may be done with keeper plates, lugs, and anchor bolts in slotted holes. Third, bearings must be easy to maintain or else designed so as to require a minimum of maintenance. Many failures have resulted from freezing of expansion details.

Expansion details must allow for a thermal movement of 1¼ in. for each 100 ft of structure. Provision must also be made to accommodate changes in length due to live load. Spans over 300 ft long should also contain provision for independent expansion and contraction of the floor.

Spans having less than 2 in. of anticipated movement are best supported on elastomeric bearing pads. Spans with more than 2 in. of movement and spans over 50 ft not using elastomeric pads may be provided with rollers, rockers, or lubricated sliding plates and must have a type of bearing employing a hinge, curved bearing plates, or a pin to allow for rotation of the upper bearing plate.

Sliding Plates and Small Pedestals. Steel plates of equal area which are intended to slide on each other usually corrode, freeze, and cause trouble. A better detail can be made by using a block or small pedestal of steel for one of the plates, thus keeping the contact area small.

Elastomeric bearing pads may be used for all types of bridges. Figure 23 shows a

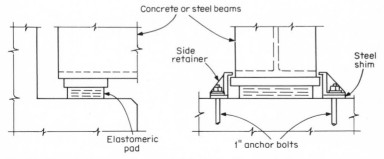

Fig. 23 Typical elastomeric bearing pad detail for short-span bridges.

typical elastomeric bearing detail suitable for short steel spans, concrete T-beams, box girders, and slabs. A durometer hardness of 55 ± 5 is the most satisfactory hardness for bridge service. The thickness of the pad should not exceed one-fifth the width nor should it be less than twice the horizontal movement. The minimum thickness should be 1 in. made up of two ½-in. layers. Maximum pressure on the pad should not exceed 800 psi. The initial vertical deflection from dead load should not exceed 15 percent of the uncompressed thickness of the pad.

The shear force on the pad, which is equal to shearing modulus × area × movement ÷ pad thickness, should not exceed one-fifth the dead load. The modulus of elasticity in shear should be about 135 psi at 70°F. Because the pads are seldom set at the mean temperature, 1½ times the temperature range suitable for the location should be used for the design. For thicknesses greater than 1 in., increments of ½ in. are used. For convenience in fabrication, pad length and width should be specified in increments of 2 in. To reduce bulging of the pad, nonelastic separators are used between the layers. If metal plates are used, the pads must be molded to final size when manufactured, with at least ⅛ in. cover of elastomeric material over the edges and faces of the pad. If fabric separators are used, the pads may be sawed from larger pads, using care to make clean cuts. Quality of material is of crucial importance in elastomeric bearing pads, and a good specification should be used and rigidly followed. Table 5 gives allowable loads for various combinations of length and width at 800 psi, and required thicknesses for various spans.

TABLE 5   Elastomeric Bearing Pads

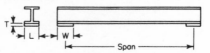

Maximum Load, $DL + LL + I$, Kips, for Allowable Stress = 800 psi

W, in.	L, in.											
	10	12	14	16	18	20	22	24	26	28	30	32
6	48	57	67	76	86	96	105	115	125			
8	64	76	89	102	115	128	141	153	166	179		
10	. . .	96	112	128	144	160	176	192	208	224	240	256
12	. . .	. . .	134	153	172	191	210	230	249	268	287	306

Span, Ft, of Steel and Concrete Beams for Various Thicknesses

T, in.	Min W, in.	Range in temperature					
		Extreme, 120°F		Moderate, 100°F		Mild, 80°F	
		Steel	Concrete	Steel	Concrete	Steel	Concrete
1	6	0–70	0–115	0–80	0–130	G–105	0–155
1½	8	71–105	116–170	81–125	131–195	106–155	156–230
2	10	106–140	171–230	126–165	196–260	156–210	231–310

Figure 24 illustrates an anchor-bolt detail for an elastomeric bearing pad.

Self-lubricating bearings consist of two steel plates with a lubricated bronze plate between them. The bronze plate may have trepanned, drilled, or bored inserts filled with a nonplastic lubricant over at least 25 percent of its area, or it may be an oil-impregnated, sintered, powdered-metal bronze plate with a compressive strength of not less than 15,000 psi. The coefficient of friction between the steel and bronze plates should not exceed 0.1. Unlubricated bronze bearings used on bridges will usually be unsatisfactory because they tend to gall and freeze.

Roller and rocker bearings may be used to advantage for spans of moderate length (50 to 150 ft). Rollers and rockers should be of as large a diameter as is practicable, with a minimum of 6 in. Small rollers and roller nests tend to collect dirt and become rusted, clogged, and inoperative. In all expansion details provision should be made for drainage and thorough cleaning.

Segmental rockers, in which the faces of the rocker are cut to a radius greater than half the rocker height, require less space than do cylindrical rollers of sufficient radius to have the same allowable stress (Fig. 25). They are satisfactory where the movement is not too great. Large movements introduce a component of force which lifts the bridge. Large, single rollers offer a simple, effective solution. Test results seem to indicate that allowable stresses higher than those permitted by the AASHTO specification formula may be justified for such rollers. In all cases it is important that plates and rollers or rockers be geared together with lugs and restrained from moving laterally.

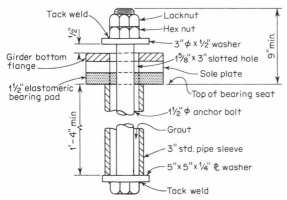

Fig. 24   Anchor-bolt detail for elastomeric bearing pad.

Pedestals and shoes may be cast or made up of welded structural elements. The AASHTO specifications require that the difference in width between the top and bottom bearing surfaces not exceed twice the distance between them. This is to assure uniform bearing and minimize bending in the pedestal or shoe. The web plates and angles which connect built-up units to the base plate should be not less than ⅝ in. thick. The webs should be rigidly connected transversely, if possible. The minimum thickness of metal in cast-steel pedestals should be 1 in.

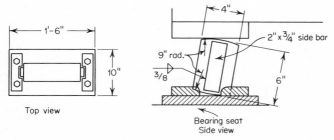

Fig. 25   Typical rocker detail.

The design of bearings is largely a matter of preference and experience. Ease of fabrication should always be kept in mind. Pins should be made as large as is practicable; small pins have a tendency to freeze up. Drain holes should be provided both in and around the bearing detail so that water will not be trapped. If pockets cannot be avoided they should be filled with a mastic. Figures 26 and 27 show typical pedestal and shoe details.

15. Expansion Hangers   In continuous girder spans with intermediate suspended spans, support details must permit movement. For short spans, a small pedestal may be set on a step fabricated on the end of the supporting girder. For longer spans, where there is appreciable movement, pins and hanger links such as are shown in Fig. 28, are quite satisfactory. For long spans and heavy loads it is usually necessary to use pin plates on each side of the web to stiffen it and provide adequate bearing area for the pin. Figure 29 illustrates a pin connection where provision for horizontal movement is not needed.

16. Deck Expansion Joints It is probably safe to say that a completely satisfactory deck expansion joint has not yet been designed. Such joints are subject to warpage during

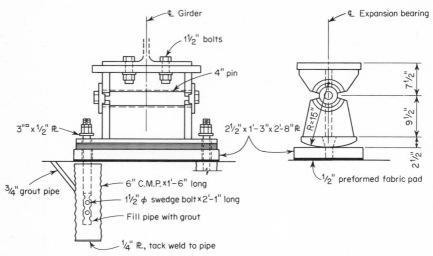

Fig. 26 Expansion bearing.

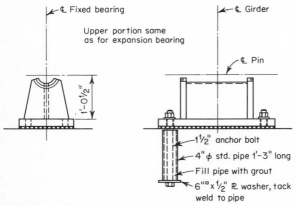

Fig. 27 Fixed bearing.

fabrication or galvanizing and are difficult to place properly. They pound loose, become noisy, and are difficult to repair. They collect water and dirt and do not ride smoothly. A design that would overcome these faults without extensive maintenance would be well worth its cost. If conditions permit, a completely open joint, with steel angles on the exposed corners and the lower portion of the deck undercut to prevent debris jamming in the crack, is to be recommended. An open joint cannot be used where falling water and debris would be objectionable.

Figures 30 and 31 show some examples of deck expansion joints which, if properly fabricated and installed, should give good service.

BEAM AND PLATE-GIRDER BRIDGES

17. Beam Bridges The principal advantage of the rolled-beam bridge is its simplicity. Fabrication is usually simple, and such bridges have been designed so that fabrication consisted only of cutting the beams to length. No falsework is required and erection is simple. The appearance is clean. Rolled beams are economical for spans up to about 60 ft. In competition with other materials and with welded-steel construction, the rolled-beam

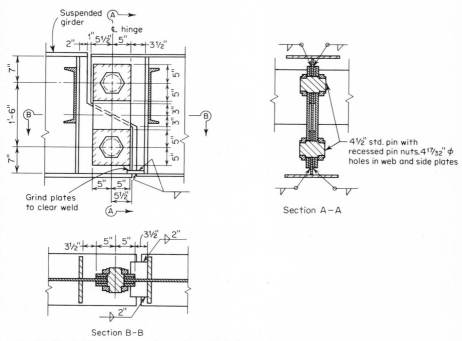

Fig. 28 Typical link expansion hinge for welded plate girder.

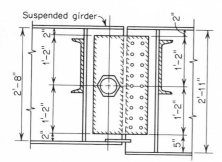

Fig. 29 Typical shear hinge.

bridge often comes out second best. This is because the rolled beam is of uniform cross section, so that there is an uneconomical distribution of steel in relation to the stress requirements.

18. Plate-Girder Bridges Plate girders have the advantage that they can be tailored to fit shear and moment requirements more closely than the rolled beam. They become advantageous at about 60 ft and are commonly used for spans to 300 ft or more. Welded girders made up of three plates compete with rolled beams in all but the shortest spans.

The deck bridge is used most often. It is simple in its details and easy to fabricate. The floor slab may be supported directly on the girders, so that floor beams and stringers are eliminated. Its disadvantage is in the clearance required for the girder. Through-girder

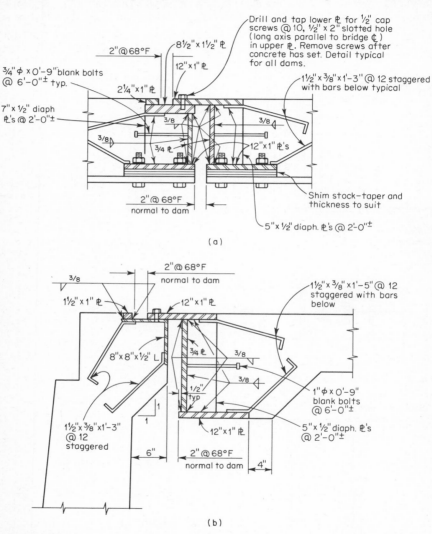

Fig. 30 Deck-expansion details: (*a*) midspan; (*b*) abutment.

bridges, used where clearance is limited, require floor beams or a system of floor beams and stringers to support the deck. Knee braces are used in place of cross frames and outriggers are sometimes used.

The box girder is an efficient shape with superior torsional stiffness. It is suitable for long-span bridges, particularly when it is used in conjunction with an orthotropic steel plate deck.

19. Composite Beam Bridges Composite action incorporates the concrete deck with the top flange of the supporting steel beams, so that the two act as a unit. The union is effected by mechanical bond between the flange and the deck provided by welded lugs, studs, or other devices. Composite action may be of two kinds, depending upon the method of construction. More commonly, the concrete deck is cast using the stringers to support the formwork. In this case, composite action is effective only for live loads and any dead load placed subsequent to construction of the deck. But if the stringers are supported by shores which are not removed until the concrete has set, the deck acts compositely with the beams to support dead load as well as live load.

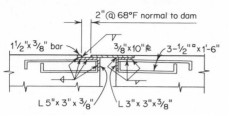

Fig. 31 Sidewalk expansion detail.

Composite construction is most economical for spans 60 ft and longer. The saving in steel for shorter spans will not usually pay for the shear connections. Economy is increased if a cover plate is used on the bottom flange of a rolled beam or, in the case of a welded I, if the bottom flange is larger than the top flange. Composite bridges are also stiffer.

The transformed section is used for analysis, with the cross-sectional area of the concrete flange tansformed to steel. The AASHTO specifications limit the effective width of deck to the smallest of (1) one-fourth the span of the beam, (2) the distance center to center of beams, and (3) twelve times the least thickness of slab.

Shear connectors are designed to develop the horizontal shear between deck and beams. Because imperfections and natural roughnesses of slab and beams contribute to shear resistance, it is likely that current formulas for shear connector resistance are conservative. Connectors extend into the area of deck reinforcing, and if continuous connectors are used, deck reinforcing bars may have to be threaded through—a tedious and costly operation. (See also Sec. 14.)

20. Continuous Spans Although continuous spans will often show economy in weight, requirements of field splicing, erection falsework, and equipment often outweigh savings in cost. A series of simple spans is usually better for bridges whose overall length is less than about 100 ft. However, many variables are involved in the decision, so that each location should be studied in the light of conditions at the site.

21. Spacing Ordinarily, the number of beams or girders in a deck highway bridge should be such as to optimize the cost of the bridge. This involves a comparison of the cost of the beams or girders with that of the deck. If insufficient clearance is available it may be necessary to use more than the number of beams or girders corresponding to the optimum combination.

For spans up to about 100 ft, either three or four girders will usually be economical for a two-lane highway bridge. As the span increases, the depth of girder and its weight rise rapidly, so that for longer spans two girders with a floor system will be economical.

22. Lateral Systems A concrete or steel deck, well secured to the stringers, acts as a horizontal diaphragm and as a tie between stringers. For spans under 125 ft, the AASHTO specifications do not require lateral truss systems for beam bridges and girder bridges which have cross frames or diaphragms and decks adequate to transmit lateral forces. Even in these cases, however, a lateral system may be needed to facilitate erection. If a timber or precast concrete deck is used with minimal connection to the stringers, other provision should be made for stiffening the structure in the horizontal plane, usually by a lateral truss.

Beam-bridge diaphragms should be spaced not more than 25 ft. These diaphragms are simple, usually a channel connected through an angle to the web of the stringer. Similar diaphragms are used at each end. Deck plate-girder bridges must have cross frames at each end, with intermediate cross frames at intervals not to exceed 25 ft. If a lateral truss is required, it is usually placed in the plane of the bottom flanges. It is desirable to space cross frames so that lateral diagonals will be approximately 45° to the girder.

23. Deflection For simple or continuous spans, deflection due to live load plus impact should not exceed $\frac{1}{800}$ of the span. In urban areas where pedestrian traffic is involved, some states prefer deflection limits of $\frac{1}{1000}$ or even $\frac{1}{1200}$ in order to minimize deflection and vibration. Deflection of cantilevers should not exceed $\frac{1}{300}$ of their length. Deflection limitations often defeat the economy of higher-strength steels for short spans.

Live-load deflections for truck loading may be computed by adding deflections for each axle. However, a close approximation may be made by using the simple-beam moment for the truck loading from Table 1. This moment is increased by a factor which varies linearly from 15 percent for a 50-ft span to 9 percent for a 140-ft span (Fig. 32). The corrected moment is converted to an equivalent concentrated load at midspan and the corresponding midspan deflection is computed (Art. 25).

Depth Ratios. The depth-span ratio of plate girders and rolled beams used as girders should be not less than 1:25. For continuous spans, the span is taken as the distance between dead-load points of contraflexure. If this limitation cannot be met, the section should be increased so that the deflection is no greater than would result for the ratio 1:25.

24. Welded Plate Girders Typical details of the welded plate girder are shown in Fig. 33. The American Welding Society Specifications for Welded Highway and Railway Bridges recommend single-plate flanges, butt-welded where a change of section is desired. Flange splices should be staggered at least 1 ft. The effect of flange thickness on the size of fillet required to weld the flange to the web should be kept in mind, since the minimum permissible size of weld increases with thickness of the flange. A $\frac{5}{16}$-in. fillet is usually the largest that can be placed in one pass without special equipment. Larger welds require multiple passes, cleaning, and chipping, and the cost of placing rises much faster than strength as weld size increases.

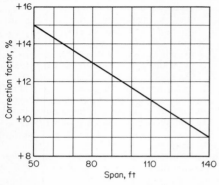

Fig. 32   Correction factor for computing deflection.

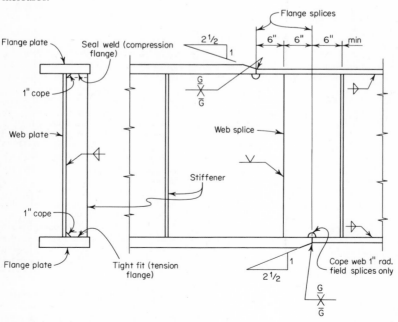

Fig. 33   Typical welded-girder details.

Intermediate transverse stiffeners may be milled or ground to fit, or cut short 1 in. at each end to allow access for painting. Stiffeners not cut short should be coped at the inside corners to clear the flange welds. Bearing stiffeners are milled or ground to bear on both flanges. Fillet welds to the web may be intermittent or continuous and as small as plate thicknesses permit. Because of the difficulty of starting and stopping automatic welding machines, fabricators often prefer the continuous weld. Stiffeners may be welded to the compression flange, but to the tension flange only where the flange stress is less than 75 percent of the allowable. AASHTO permits intermediate stiffeners on only one side of the web, with a consequent saving in weight and cost of fabrication, and improvement in appearance. Stiffeners on one side only are required to be welded to the compression flange.

Longitudinal stiffeners permit the use of thinner webs and may prove economical for deep girders. They may be cut at their intersections with transverse stiffeners.

In field splices, the web-to-flange fillet welds should be omitted for at least 1 ft on each side of the splice and welded after the flange splice is made. This results in better distribution of shrinkage stresses and permits lateral displacement of the web if flange shrinkage causes the ends of the webs to jam.

Fatigue. AASHTO and AREA provisions for fatigue are based on the type of member or detail, the kind of stress to which it is subjected, the range of service-load stress, and the expected number of cycles of stress.

In welded work, probably more than in any other structural fabrication process, workmanship must be of the highest quality. This requires adequate and competent inspection. Although the designer is not usually responsible for inspection, the success of the design depends so heavily upon the quality of welding that the designer should make sure that proper steps are taken to obtain the weld strength anticipated in the design.

25. Example: Design of Welded Plate Girder A two-lane, continuous, two-girder bridge of spans 110,154,110 ft (Fig. 34a) will be designed for the HS20-44 loading, AASHTO specifications. The roadway width is 28 ft. Steel is A36.

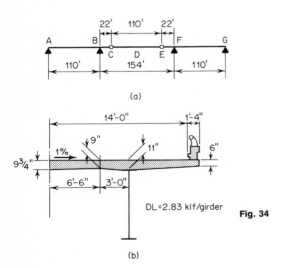

(a)

(b)

DL=2.83 klf/girder

Fig. 34

The dead load is 2.83 klf per girder (Fig. 34b). Dead-load moment and shear diagrams are shown in Fig. 35a and b.

The lane loads are positioned as shown in Fig. 36. The lane load per girder is $(4.5 + 18.5)/19 = 1.21$. Impact for a 110-ft span is 21.3 percent (Table 1). Thus, for one girder, $LL + I = 1.21 \times 1.213 = 1.47$ lanes.

LIVE-LOAD MOMENTS. The maximum moment at the center of the 110-ft span is 1704 ft-kips per lane (Table 1). Therefore, $M = 1.47 \times 1704 = 2510$ ft-kips. The influence line for the moment at B is shown in Fig. 37a. To find the maximum moment we try both the lane loading and the truck loading,

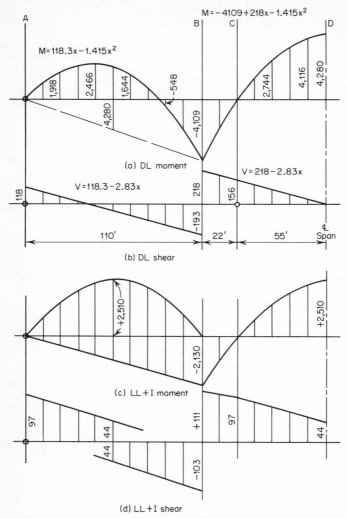

(a) DL moment

$M = 118.3x - 1.415x^2$

$M = -4109 + 218x - 1.415x^2$

$V = 118.3 - 2.83x$

$V = 218 - 2.83x$

(b) DL shear

(c) LL + I moment

(d) LL + I shear

Fig. 35   Shear and moment diagrams for girder of Fig. 34.

the latter with the truck in the suspended span with the trailer axle at hinge C. The truck-load moment is the larger, and $M_{BC} = 1.47 \times 1450 = 2130$ ft-kips.

LIVE-LOAD SHEARS. The maximum shear at A is $1.47 \times 65.9 = 97$ kips (Table 1). For maximum shear at midspan of AB, the truck is positioned with the trailer axle at midspan. This gives $V = 1.47 \times 29.9 = 44$ kips.

To find the shear to the left of B, the lane loading is used in conjunction with the influence line of Fig. 37b. With the 26-kip load at B

$$V_{BA} = 1 \times 26 + 0.64 \times \tfrac{1}{2}(110 \times 1 + 132 \times 0.2) = 69.6 \text{ kips}$$
$$= 1.47 \times 69.6 = 103 \text{ kips per lane}$$

From the influence line for shear to the right of B (Fig. 37c),

$$V_{BC} = 1.47(1 \times 26 + 0.64 \times 77) = 111 \text{ kips per lane}$$

From the influence line for the reaction at B (Fig. 37d),

$$R_B = 1.47 \times {}^{132}\!/_{110}(26 + 121 \times 0.64) = 182 \text{ kips per lane}$$

Moments and shears are summarized in the following table and plotted in Fig. 35c and d.

	Moments		Shears right of B	
	B	D	Load for max shear	Load for max moment
DL...............	4,109	4,280	218	218
LL + I...........	2,130	2,510	111	97
Total..........	6,239	6,790	329	315

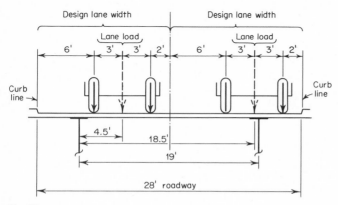

Fig. 36

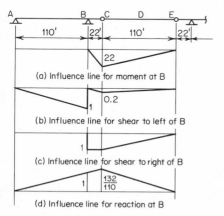

(a) Influence line for moment at B

(b) Influence line for shear to left of B

(c) Influence line for shear to right of B

(d) Influence line for reaction at B

Fig. 37   Influence lines, support B of girder of Fig. 34.

Assuming the envelope of maximum moments for simple spans to be a parabola (Art. 3), the envelopes of maximum moment shown in Fig. 38 are determined from the values in Fig. 35. These should be drawn to scale. A scale of 1 in. = 20 ft and 1 in. = 2000 ft-kips is satisfactory.

In determining the number of variations in cross section, the cost of splicing must be balanced against the cost of the steel it will save. Such considerations usually keep the number of splices to a minimum. A constant depth and thickness of web facilitate fabrication. Also it is desirable to hold the flange width constant and vary the thickness as required.

GIRDER SECTION I. The cross section is proportioned for the maximum positive moment (at D). If no longitudinal stiffener is used, the web must be not thinner than $D/165$, where D is the unsupported depth. Try a $100 \times \frac{5}{8}$ web, for which $100/0.625 = 160$, with $18 \times 1\frac{3}{4}$ flanges:

$$\text{Flange } 18 \times 1\frac{3}{4} \times 5176 = 163{,}040$$
$$\text{Web} \qquad\qquad\qquad\qquad\quad 52{,}080$$
$$\overline{\qquad\qquad\qquad I = 215{,}120 \text{ in.}^4}$$

$$f_b = \frac{Mc}{I} = \frac{6790 \times 12 \times 51.75}{215{,}120} = 19.5 \text{ ksi}$$

$$M = \frac{20 \times 215{,}120}{51.75 \times 12} = 6930 \text{ ft-kips}$$

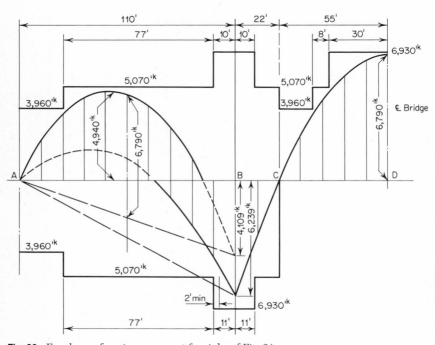

Fig. 38   Envelopes of maximum moment for girder of Fig. 34.

GIRDER SECTION II. The compressive stress in the bottom flange must be investigated for the maximum negative moment $M = 6239$ ft-kips at support B. The allowable stress is

$$F_b = 20{,}000 - 7.5 \left(\frac{L}{b}\right)^2 \qquad \frac{L}{b} \lessgtr 36$$

where L = length of unsupported flange between lateral supports
         b = width of flange

For continuous beams and girders, L may be taken as the distance from interior support to the dead-load point of contraflexure if this distance is less than that defined above. Furthermore, continuous beams and cantilever beams may be proportioned for negative moment at interior supports at an allowable stress 20 percent higher than that given by the formula, but not to exceed 20,000 psi.

To determine L, it is necessary to decide on the pattern and spacing of the diagonal bracing and cross frames. The latter should be located so that bracing is at about 45° to the girders, and spaced evenly throughout the span. For the bracing layout shown in Fig. 39, $L = 11$ ft and $L/b = 11 \times \frac{12}{18} = 7.3$. Therefore, with the 20 percent increase, the allowable stress is 20 ksi and the allowable moment at support B is the same as that at D, 6930 ft-kips.

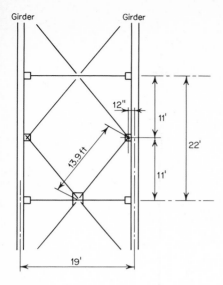

Fig. 39 Lateral bracing for girder of Fig. 34.

GIRDER SECTION III. This section is near midspan of AB. The maximum moment, which may be scaled from Fig. 38 or computed, is 4940 ft-kips. The required section modulus

$$S = 4940 \times \tfrac{12}{20} = 2964 \text{ in.}^3$$
$$I = Sc = 2964 \times 51.25 = \quad 151,900$$
$$I \text{ of } 100 \times \tfrac{5}{8} \text{ web} \qquad \underline{\quad 52,080}$$
$$99,820$$

For $18 \times 1\tfrac{1}{8}$ plates, $I = 103,500$ in.[4] The allowable moment

$$M = \frac{20 \times 155,580}{12 \times 51.13} = 5070 \text{ ft-kips}$$

GIRDER SECTION IV. The minimum sections are at the ends of the suspended span and at the ends of the bridge. The minimum allowable flange thickness is governed by the limitation $b/t = 23$. Use $18 \times \tfrac{3}{4}$ plate:

$$\text{Flange } 18 \times \tfrac{3}{4} \times 5075 \quad = \quad 68,510$$
$$\text{Web} \qquad\qquad\qquad\qquad \underline{\quad 52,080}$$
$$I = \overline{120,590 \text{ in.}^4}$$
$$M = \frac{20 \times 120,590}{12 \times 50.75} = 3960 \text{ ft-kips}$$

The allowable moments for the various sections are plotted on the moment envelope of Fig. 38 to determine where the section changes can be made. The larger of the two flange plates to be joined must be extended at least 1 ft beyond the theoretical point of splice. It is good practice to stagger splices in the upper and lower flanges by at least 1 ft (Fig. 33).

Splices will be located at the following points (Fig. 38):

	Bottom, ft	Top, ft
Right of A.............	22	23
Left of B..............	11	10
Right of B.............	11	10
Right of C.............	16	17
Right of C.............	24	25

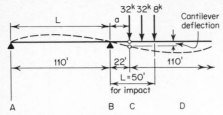

Fig. 40   Load position for cantilever deflection.

DEFLECTION. The load position for maximum deflection of the cantilever is shown in Fig. 40. The loaded length for determining impact is 50 ft. Therefore, $I = 28.6$ percent. With $R = 65.9$ kips (Table 1).

$$R_C = 65.9 \times 1.21 \times 1.286 = 102.5 \text{ kips}$$
$$\Delta = \frac{Pa^2(L + a)}{3EI} = \frac{102.5 \times 22^2 \times (110 + 22) \times 12^3}{3 \times 29,000 \times 215,120} = 0.61 \text{ in.}$$
$$\Delta = \frac{L}{300} = \frac{22 \times 12}{300} = 0.88 \text{ in. allowable}$$

Deflection of the suspended span will be approximated by the method discussed in Art. 23. The loading is shown in Fig. 41.

$$M = 1704 \text{ ft-kips}, I = 21.3 \text{ percent (Table 1)}$$
$$M = 1704 \times 1.21 \times 1.213 = 2500 \text{ ft-kips}$$
$$\text{Equivalent concentrated load} = \frac{4M}{L} = \frac{4 \times 2500}{110} = 91 \text{ kips}$$
$$\Delta = \frac{PL^3}{48EI} = \frac{91 \times (110 \times 12)^3}{48 \times 29,000 \times 215,120} = 0.69 \text{ in.}$$

The correction factor is 1.11 (Fig. 32).

$$\Delta = 0.69 \times 1.11 = 0.77 \text{ in.}$$
$$\Delta = \frac{L}{800} = \frac{110 \times 12}{800} = 1.65 \text{ in. allowable}$$

SPAN BF. Taking the deflection for span BF as the sum of the deflection of the cantilever for its severest loading condition (Fig. 40) and of the suspended span for its severest loading condition (Fig. 41).

$$\Delta = 0.61 + 0.77 = 1.38 \text{ in.}$$
$$\Delta = \frac{154 \times 12}{800} = 2.31 \text{ in. allowable}$$

The calculated deflection is probably about 10 percent low because of the variable moment of inertia, which was neglected in the computations, but it is also high because deflections for two different load

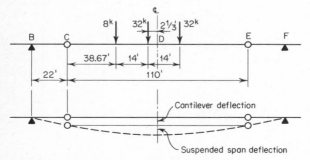

Fig. 41   Load position for suspended-span deflection.

positions were assumed to be additive. If it is critical, deflection of the cantilever should be recomputed, using the load position of Fig. 41.

FLANGE-TO-WEB WELDS. The influence lines for shear at pier B (Fig. 37) show that there is no reversal of live-load shear. The maximum shear is 329 kips (see table, p. 18-41). Since the maximum shear results from lane loading, the number of stress cycles to be assumed for determining the allowable fatigue stress is 500,000 for an average daily truck traffic of 2500 or more (AASHTO 1.7.2). The flange-to-web weld is in stress category B. Therefore, the allowable stress range F_{sr} is 27.5 ksi. This is greater than the basic allowable stress, 12.4 ksi, and since there is no stress reversal, the basic value governs. The minimum fillet weld for a 1¾-in. flange plate is ⅜ in. Therefore,

$$q = 12.4 \times 0.707 \times 0.375 = 3.29 \text{ kli allowable}$$

The stress in each weld is

$$q = \frac{VQ}{I} = \frac{329 \times 31.5 \times 50.88}{2 \times 215,120} = 1.22 \text{ kli}$$

At hinge C, $V_{max} = 156 + 97 = 253$ kips (Fig. 35) and there is no reversal. The minimum weld for a ¾-in. plate is ¼ in. Then

$$q = 12.4 \times 0.707 \times 0.25 = 2.19 \text{ kli allowable}$$

The stress in each weld is

$$q = \frac{253 \times 13.5 \times 50.375}{2 \times 120,590} = 0.71 \text{ kli}$$

BEARING STIFFENERS. The reaction at B is $DL + (LL + I) = 411 + 182 = 593$ kips. Using 8-in. plates,

$$t = \frac{583}{8 \times 29} = 2.55 \text{ in.}$$

and, with four stiffeners, $t = 2.55/4 = 0.64$ in. Therefore, use four 8 × ¾-in. Stiffeners at other points of support are designed similarly.

INTERMEDIATE STIFFENERS (AASHTO 1.7.43). The maximum shear is $V = 329$ kips, at support B.

$$f_v = \frac{329,000}{100 \times 0.625} = 5260 \text{ psi}$$

$D/t = 100/0.625 = 160$, which exceeds the slenderness limit 150 for which no intermediate stiffeners are required. From AASHTO 1.7.43, Fig. 1.7.43DI, the allowable shear for a web with $D/t = 160$ and with stiffeners at the maximum allowable spacing ($d = 1.5D$) is 7900 psi. Therefore, $d = 1.5 \times 100 = 150$ in. Use a spacing of 132 in. to provide a mid-panel stiffener for fastening lateral bracing. (Fig. 39). The required moment of inertia of the stiffener is

$$I = \frac{d_0 t^3 J}{10.92}$$

where $J = 25 (D/d)^2 - 20$ but not less than 5
$\quad d =$ required distance between stiffeners
$\quad d_0 =$ actual distance between stiffeners
The minimum width of stiffener is $2 + 100/30 = 5.3$ in. and the minimum thickness $5.3/16 = 0.331 = ⅜$ in.

$$J = 25 \left(\frac{100}{150}\right)^2 - 20 = \text{negative, use 5}$$

$$I = \frac{132 \times 0.625^3 \times 5}{10.92} = 14.8 \text{ in.}^4$$

Use 5 × ⅜ stiffeners on one side of the web, for which $I = bd^3/3 = 0.375 \times 5^3/3 = 15.6$ in.[4]

LONGITUDINAL STIFFENER. A longitudinal stiffener allows a thinner web to be used (AASHTO 1.7.43). The effect of such a stiffener on the design of the girder will be investigated.

$$t = \frac{100}{330} = 0.303 \text{ in.}$$

Try 18 × 2 flanges with 100 × 5⁄16 web:

$$
\begin{array}{lll}
\text{Flange } 18 \times 2 \times 5202 & = 187,270 \\
\text{Web} & \underline{\quad 26,040} \\
& I = 213,310 \text{ in.}^4
\end{array}
$$

$$f_b = \frac{6790 \times 12 \times 52}{213,310} = 19.9 < 20 \text{ ksi}$$

Intermediate stiffeners:

$$f_v = \frac{329,000}{100 \times 0.312} = 10,600 \text{ psi}$$
$$F_v = 10,700 \text{ psi} \qquad \text{for } d/D = 0.40$$

This is a very close stiffener spacing for a 100-in. girder. With a $100 \times \%$ web with 18×2 flange plates, the bending stress s is reduced to 19.4 ksi while the value of d is increased to 85 in. Use 66 in. to provide a mid-panel stiffener for fastening lateral bracing (Fig. 39).

$$J = 25 \left(\frac{100}{85}\right)^2 - 20 = 14.6$$
$$I = \frac{66 \times 0.375^3 \times 14.6}{10.92} = 4.7 \text{ in.}^4$$

Use $4 \times \%$ stiffeners, for which $I = 8.0$ in.^{4}

The required moment of inertia of the longitudinal stiffener is

$$I = Dt^3 \left(2.4 \frac{d_0^2}{D^2} - 0.13\right)$$

This should be determined near the center of the girder, where shear is small and d large. To provide a stiffener at mid-panel for fastening lateral bracing (Fig. 39), use $d_0 = 66$ in.

$$I = 100 \times 0.375^3 \left(2.4 \frac{66^2}{100^2} - 0.13\right) = 4.83 \text{ in.}^4$$

For a $4 \times \%$ stiffener $I = \% \times 4^3/3 = 8$ in.^{4} A smaller stiffener would not be desirable.

The design using the longitudinal stiffener results in an overall saving of about 19,000 lb, which is about 6 percent of the weight of the structure without the stiffener.

CROSS BRACING. The exposed area of the bridge in elevation (Fig. 42) is 12.1 ft^2/ft. Wind force on the structure at 50 psf is $50 \times 12.1 = 0.61$ klf. Since the wind load is uniform, the reactions at the supports may be determined by proportion from the deadload reactions.

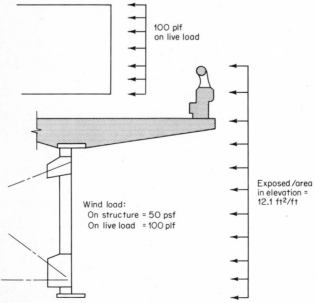

100 plf
on live load

Exposed /area
in elevation =
12.1 ft²/ft

Wind load:
On structure = 50 psf
On live load   = 100 plf

Fig. 42

Wind-load reaction at pier $B = 411 \times 0.61/2.83 = 88$ kips. This load is carried to the end-cross frames by the deck, so that 44 kips acts at each of the upper corners of the frame (Fig. 43).

$$\text{Stress in } CE = 44 \text{ kips } C$$
$$AC = 44 \text{ kips } T$$
$$CF = 56.1 \text{ kips } T = 44 \times \frac{145}{114}$$

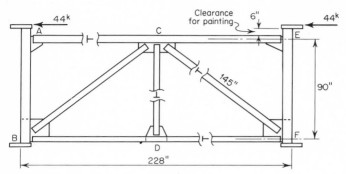

Fig. 43   Cross frame for bridge of Fig. 34.

Member AE: Try WT6 $\times$ 15, $A = 4.40$ in.,^{2} $r_x = 1.75$ in.

$$\frac{L}{r_x} = \frac{228}{1.75} = 130 < 140$$
$$F_a = 16,000 - 0.3 \times 130^2 = 10,930 \text{ psi}$$
$$f_a = \frac{P}{A} = \frac{44,000}{4.40} = 10,000 \text{ psi}$$

Member BC, CF: Try WT6 $\times$ 15, $A = 4.40$ in.,^{2} $r_y = 1.52$ in.

$$\frac{L}{r_y} = \frac{145}{1.52} = 95$$
$$F_a = 16,000 - 0.3 \times 95^2 = 13,290 \text{ psi}$$
$$f_a = \frac{P}{A} = \frac{56,100}{4.40} = 12,750 \text{ psi}$$

Member BF: $P = 0$, $r_{min} = {}^{228}/_{140} = 1.63$ in.
  Use WT6 $\times$ 15. Member CD: $P = 0$, $r_{min} = {}^{90}/_{140} = 0.64$ in.
  Use S3 $\times$ 6.25, $r_y = 0.71$ in., $A = 1.81$ in.^{2}
  Members of the cross frames will be welded to their gusset plates. The connection must be designed for the average of the calculated stress and the strength of the member, but for not less than 75 percent of the strength of the member.
  INTERMEDIATE CROSS FRAMES. Since it is assumed that the wind force is carried to the end cross frames by the deck, members of the intermediate cross frames are chosen for maximum L/r requirements. The slenderness ratios of all but members BC and CF of the end cross frames are only slightly less than the permissible limit, 140. Therefore, all cross frames will be made alike.
  LATERAL BRACING. These members are also free of calculated stress and are proportioned for maximum $L/r = 140$.

$$r = 13.9 \times {}^{12}/_{140} = 1.19$$

Use WT5 $\times$ 11, $r_y = 1.33$ in.
  26. Web Splice   The locations of web splices usually depend upon available lengths of plates, and the actual location is often left to the fabricator. Specifications do not agree on design criteria. The AREA specifications require that they be designed for both (1) the shear strength of the gross area of the web and (2) the combination of moment strength of the net section of the web and the maximum shear at the splice. The AASHTO specifications require that the splice be designed for the average of the calculated stress at the point of splice and its strength, but in no case less than 75 percent of the strength.

27. Field Splices When girders are continuous and when single spans are too long to ship in one piece, field splices are required. Such splices are usually bolted. Field splices may be bolted even in welded work because of the difficulty of welding large girders in the field. In some cases, the flanges may be butt-welded and the web splice bolted.

TRUSS BRIDGES

Trusses are used in highway bridges only for very long spans. However, trusses may be economically used for shorter spans where aesthetics and (in the case of through bridges) safety with high-speed traffic are not critical. Deck-truss bridges are preferable.

28. Proportions The Warren truss is more economical than the Pratt for parallel-chord trusses. For spans greater than about 320 ft, the K-truss is advisable to reduce floor weight and the inclination of the diagonals. Depth-span ratios vary considerably, usually between 1:5 and 1:10. For best results, the depth should be such that the angle of the diagonal falls in the range of 50 to 55° with the horizontal. Long panels require a heavier deck while short panels increase the cost of the truss, so that a compromise must be made, usually in the range of 16 to 32 ft for the panels of a highway-bridge truss.

Gravity axes of truss members should coincide with their working lines. However, a small eccentricity may be used to compensate for the dead load of a member. Trusses should be erected with enough camber to offset at least the dead-load deflection.

29. Loads and Stresses Various formulas for estimating the weight of bridge trusses have been developed, but designs vary so widely that formulas are not usually dependable. The dead load may be approximated by comparison with similar bridges, or by estimate. Primary stresses may be computed by the methods of joints or of sections. A tabular solution, based on moment increments, is useful for dead-load stresses in parallel-

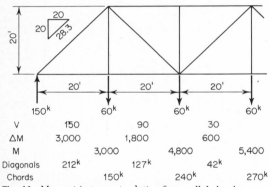

Fig. 44 Moment-increment solution for parallel-chord truss.

chord trusses (Fig. 44). Influence lines are useful for live loads, particularly axle loads, and are easily constructed (Fig. 45).

Secondary Stresses. Trusses having fixed joints generate bending in the members. The resulting bending stresses are called secondary stresses. Such stresses may also result from eccentric connections, frozen pins, or transverse loading on the members. Secondary stresses can be minimized by avoiding stiff members. Members whose width is not more than one-tenth their length will usually not have secondary stresses larger than 25 percent of the primary stresses. Both AASHTO and AREA require that secondary stresses in excess of 4000 psi in tension members and 3000 psi in compression members be treated as primary stresses.

30. Truss Members So far as is possible, all members of a truss should have the same depth so that they can be connected at the joints with a minimum of filler plates. For the shorter spans, W sections make suitable, clean-looking members. For longer spans welded built-up I's or box sections with perforated cover plates are used.

Truss members should be designed so that all surfaces can be reached and painted, and with adequate clearances for riveting or bolting. Pockets which may trap water or debris should be avoided.

Combinations of Steels.   Several grades of steel can be used to advantage in large bridge trusses to meet varying stress conditions, with the stress grade for each member

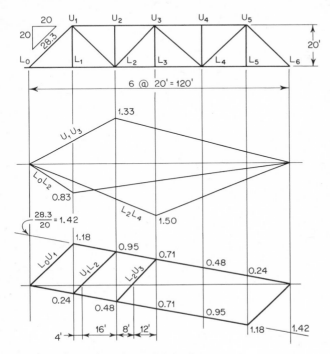

Fig. 45   Influence lines for parallel-chord truss.

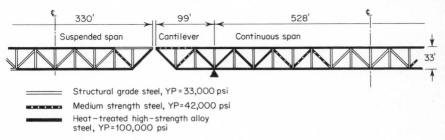

Fig. 46   Benicia-Martinez Bridge.

based on maintaining, in general, a minimum thickness of plate. Thus, lower stressed members will be made of structural-grade steel, with the next higher stress grade being used for members with larger stress, etc. For example, three grades of steel were used in the Benicia-Martinez Bridge (Fig. 46).

The efficiency of bolted truss members can be increased by butt-welding a section of higher-strength steel on each end. This offsets the loss in area from holes, so that the gross area of the member can be fully effective. In other cases the same result can be achieved by welding on a thicker section of steel of the same grade as the member. This device can also help reduce the size of the joint. These ideas are illustrated in Fig. 47. Note in this detail that the weld is 1 ft out from the first transverse line of bolts in the joint. Where several grades of steel are brought together in a joint, the gusset plates are made of the highest stress grade of steel represented in the joint.

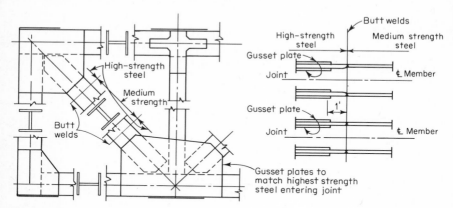

Fig. 47   Benicia-Martinez Bridge.

31. Lateral Forces   The design of truss bridges is usually based on the assumption that wind load on the structure is carried to the supports by the upper and lower lateral systems, where they are transmitted to the bearings by the end sway frames or portal bracing. Wind on the live load is assumed to be transmitted by the deck. The intermediate sway frames are then designed to resist only the overturning force in each panel. Stresses in lateral trusses will usually be so small that minimum slenderness ratios will determine the sizes of the members.

CONCRETE BRIDGES

In the 1977 edition of the AASHTO specifications, the 28-day strength of class A concrete was increased from 3000 to 4000 psi. Some of the examples in this section are based on 3000 psi, some on 4000 psi.

32. Camber, Plastic Flow, and Shrinkage   The determination of the camber to provide in the deck of a bridge is an important consideration. The geometrics of some structures, such as those with skewed bents and those on horizontal curves, or a combination of both, can make this a complex problem.

Initial deflection is a function of dead load. It is also affected by the sequence of construction. The secant modulus $E = 1000f'_c$ can be used to determine the deflection which occurs immediately upon striking the falsework. Structures cambered only for initial deflection will ride smoothly at first, but undulations develop with time. This is due to plastic flow or creep and, to some extent, shrinkage.

Deflection due to plastic flow increases rapidly at first, but the rate of increase decreases with time, so that most of it occurs in a period of about 4 or 5 years following construction. Sufficient camber should be provided to allow for the ultimate deflection, which may amount to three or four times the initial deflection. For calculating this camber, values of E of 250 to $350f'_c$ may be used, depending upon the characteristics of the concrete.

The AASHTO specifications require that shrinkage be considered in the design of concrete structures. The effectiveness of shrinkage in producing stress in a reinforced-concrete member has been questioned by some authorities. Because of the restraining effects of reinforcing steel, shrinkage manifests itself in producing cracks in the concrete (Art. 8). Reinforcement distributes cracking by producing more cracks of smaller magnitude. Shrinkage stresses can be minimized or practically eliminated by prudent planning of pouring sequence, and the designer should always consider this alternative to that of allowing shrinkage stresses to develop so that they must be accounted for in the design calculations.

SLAB BRIDGES

33. Simple Spans A longitudinally reinforced slab is the simplest form of concrete bridge superstructure. It is economical for simply supported spans up to 20 ft or more in length, and may show economy in continuous spans up to 40 ft. It can be supported on piles, column bents, or piers. Typical sections of concrete slab structures are shown in Fig. 48.

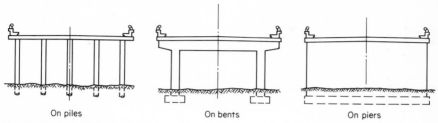

On piles   On bents   On piers

Fig. 48 Typical slab bridges.

Piles generally provide the most economical intermediate support for continuous slab bridges. They are particularly suited to locations where drift is not a problem. Concrete-column bents and solid piers are the next most economical substructure types. On continuous spans of ±26 ft and longer, pile caps can be eliminated, since for such spans the pile spacing is close enough to allow the slab itself to act as a cap within the requirements of the AASHTO specifications. This results in a minimum of formwork and provides a pleasing and simple structure.

Distribution of Live Load.   For design of the slab the AASHTO specifications assume a wheel load to be distributed transversely over the effective width $E = 4 + 0.06S$, where S is the effective span in feet.

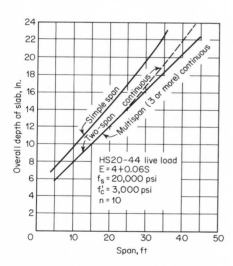

Fig. 49 Approximate depths of slab bridges.

Example Design a simple-span slab bridge for the HS20-44 loading. Span = 20 ft (the span length is defined as the distance center to center of supports, but not to exceed the clear span plus thickness of slab), f_c = 1200 psi, f_s = 20,000 psi, n = 10. Impact allowance for a 20-ft span is 30 percent.

From Fig. 49, which gives slab depths that can be used for preliminary design, the probable depth of the slab is 14½ in., for which the dead load is 180 psf. The effective width $E = 4 + 0.06 \times 20 = 5.2$ ft. Therefore,

$$DL \text{ moment} = 180 \times \frac{20^2}{8} = \quad 9000 \text{ ft-lb}$$

$$LL + I \text{ moment} = \frac{16{,}000 \times 20}{4 \times 5.2} \times 1.30 = 20{,}000$$

$$M = 29{,}000 \text{ ft-lb}$$

$$\text{Effective depth} = 12.5$$

$$\text{Center of steel to bottom of slab} = 2$$

$$\text{Depth of slab} = 14.5 \text{ in.}$$

$$A_s = \frac{12 \times 29{,}000}{20{,}000 \times 0.88 \times 12.5} = 1.58 \text{ in.}^2$$

$$f_c = \frac{2 \times 29{,}000 \times 12}{0.37 \times 0.88 \times 12 \times 12.5^2} = 1140 \text{ psi} \qquad \text{O.K.}$$

For No. 8 bars at 6 in., $A_s = 1.58$ in.² Lengths of these bars are determined from the maximum-moment envelope in Fig. 50. Figure 51 shows typical sections and steel layout.

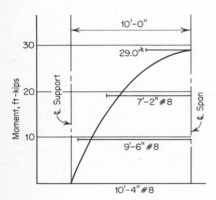

Fig. 50

The specifications require that one-third of the main steel extend beyond the face of the support at least 6 in. The remaining bars must be extended 15 diameters, but not less than one-twentieth of the span, beyond the point at which computations show they are no longer needed to resist stress. For No. 8 bars in a 20-ft span this gives 15 in. These extensions are shown in Fig. 51.

Distribution steel in the bottom of the slab, transverse to the main reinforcement, is determined as the percentage $100/\sqrt{S} = 100/\sqrt{20} = 22.4$ percent of the main reinforcement. The area required is $0.224 \times 1.58 = 0.35$ in.²/ft of slab, which is furnished by No. 6 bars at 15 in. These bars are placed in the center half of the slab. Distribution steel can be reduced by 50 percent in the outer quarters as shown in Fig. 51. Temperature reinforcement shown in the top of the slab is the required minimum. No. 4 bars at 18 in.

Top and bottom reinforcing bars can be uniformly spaced and held securely by the use of bar chairs. These are made from No. 4 bars, bent as shown in the detail in Fig. 54. They should be placed about 4 ft on centers measured along the span and 3 ft on centers transversely.

34. Continuous Spans Where good foundation material is encountered and differential settlement is not a problem, continuous slab bridges are economical. In some locations, they may be economical up to spans of about 40 ft. Above 40 ft, reinforced-concrete T-beams, cored slabs, or box girders should be considered.

For aesthetic and economic reasons the continuous slab bridge should have a uniform thickness. If the spans are equal, this thickness can be held to a minimum by using compressive steel at points of large moment. For example, in a four-span slab of uniform depth, the maximum moment occurs at the first interior bent. To hold the volume of concrete to a minimum, the slab depth for balanced reinforcement can be based on the positive moment in the end span, and the larger negative moment at bent 2 can be provided for with compression steel in the bottom of the slab. The positive steel which must be extended through the support can be used for this purpose.

Continuous slab bridges on the pile bents can be built up to about 150 ft in length without expansion joints, with the end bent constructed monolithically with the slab. For this length the passive resistance of the approach material to pressure caused by a temperature increase in the deck does not produce excessive compressive stresses in the concrete.

Live-Load Positions.   For continuous slabs of spans up to about 40 ft the HS20-44 truck governs for both positive and negative moment. Above 44 ft ± the lane loading governs for

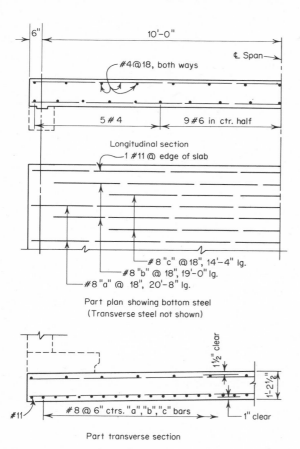

Fig. 51   Details of slab bridge.

negative moment. Effective width of slab for distribution of live load is the same as for simple-span slabs (Art. 33). It is not necessary to investigate shear in slabs designed for bending moment in accordance with this distribution.

Placement of Main Steel.   Required extension of reinforcement at the exterior support is the same as for simple spans. At the interior supports, at least one-fourth of the required positive reinforcement must be carried through the support. In the regions of positive moment it is advisable to use top reinforcement not less than the equivalent of No. 5 bars at 18 in.

Distribution steel is determined as a percentage of the main reinforcement in the bottom of the slab as for simple spans. In skewed spans, distribution steel should be placed parallel to the bents with the spacing measured along the centerline of bridge.

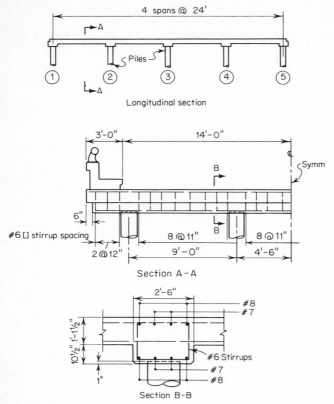

Fig. 52 Pile bent for slab bridge.

35. Design of Bents Where drift is not a problem and stream flow is small, the pile bent can be used to advantage in continuous slab bridges. Figure 52 will be used to illustrate the design procedure.

Dimensions of the cap are shown in section *BB*. The position of the HS20-44 truck to produce maximum live-load reaction is shown in Fig. 53a. The reaction on one bent is

$$
\begin{array}{lr}
DL \text{ slab} & 1.14^* \times 0.169 \times 24 \times 34 = 157 \\
DL \text{ cap, below slab } 0.33 \times 33 & = 11 \\
DL \text{ railing and curb } 1.14 \times 2 \times 0.53 \times 24 & = 29 \\
\hline
& 197 \text{ kips} \\
LL \text{ at 56 kips per lane} = 2 \times 56 = 112 \text{ kips} \\
\hline
DL + LL = 309 \text{ kips}
\end{array}
$$

Required number of piles at 45 tons per pile = 309/90 = 3.4; use 4.

DESIGN CAP FOR MOMENT. The truck positioned for positive moment (Fig. 53b) is a satisfactory criterion for both top and bottom steel. Although the negative moment may be theoretically slightly larger, the transverse steel in the slab will act with the cap steel to give additional resisting moment.

The rear wheel of the truck will act as a concentrated load on the bent, the front wheels (axle 3) and rear wheels (axle 1) as uniform loads (Fig. 53a). The latter two produce a bent reaction of 24 kips per lane, which can be distributed over the 10-ft width of lane.

*The multiplier 1.14 in these calculations is a continuity factor.

The maximum positive moment in the cap is determined from an influence line for a three-span continuous beam of uniform cross section.

DL moment: $0.079 \times 9^2 \times 197/34 = 37.1$
LL moment, including 30 percent impact
    Axle 2 $0.18 \times 16 \times 9 \times 1.3 = 33.8$
    Axles 1 and 3 $0.07 \times 2.4 \times 9^2 \times 1.3 = \underline{17.7}$
                                              $M = 88.6$ ft-kips

$$A_s = \frac{12 \times 88.6}{20 \times 0.88 \times 21.5} = 2.80 \text{ in.}^2$$

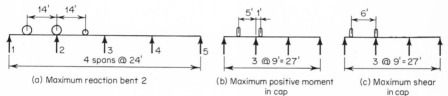

(a) Maximum reaction bent 2   (b) Maximum positive moment in cap   (c) Maximum shear in cap

Fig. 53 Live-load positions for design of bent.

Use two No. 8 bars + two No. 7 bars = 2.78 in.²
Maximum shear in the cap is produced with the truck positioned as shown in Fig. 53c and is determined from an influence line for a three-span continuous beam of uniform cross section.

DL shear $0.553 \times 5.8 \times 9 = 28.1$
LL + impact
    Axle 2 $1.41 \times 16 \times 1.3 = 29.4$
    Axles 1 and 3 $0.517 \times 2.4 \times 9 \times 1.3 = \underline{14.5}$
                                          $V = \overline{72.0}$ kips

$$v = \frac{V}{bd} = \frac{72,000}{30 \times 21.5} = 112 \text{ psi}$$

$$v_c = 0.95 \sqrt{f_c'} = 0.95 \sqrt{4000} = 60 \text{ psi}$$

$$\text{Spacing of No. 6 stirrups} = \frac{f_v A_v}{b(v - v_c)} = \frac{20,000 \times 0.88}{30 \times 52} = 11.3 \text{ in.}$$

AASHTO specifications limit the maximum stirrup spacing to 0.5d, or 0.5 × 21.5 = 10.8 in., say 11 in.

$$A_{v,\text{min}} = \frac{50 b_w s}{f_y} = \frac{50 \times 30 \times 11}{20,000} = 0.81 \text{ in.}^2$$

Therefore, No. 6 U-stirrups at 11 in. are satisfactory. Reinforcement details are shown in Fig. 52.

36. Typical Details, Continuous Slabs   Figure 54 shows typical superstructure details for a continuous slab bridge. Figure 55 shows typical details for an end-wall abutment with wing walls parallel to the roadway. This type of abutment has a massive appearance, yet it is economical to build. When appearance is not an important factor, control of the earth approach can be provided by short wing walls parallel to the abutment.

Figure 56 shows typical details at a bent; a is used for short spans, b for longer spans where the bent reactions are greater and the spacing between piles less. The latter also provides economy through a minimum of formwork and should be considered when structurally feasible.

Figure 57 shows typical superstructure details for a slab span with an expansion hinge located within the span. Hinges of this type have been used extensively and, when properly designed and constructed, have proved to be very durable. Figure 58 shows an expansion detail which can be used at a bent with cap. It is particularly adaptable to short spans where caps are required to extend below the slab. This type of joint is more economical than the steel expansion hinge of Fig. 57. However, it would not be economical to use caps extending below the slab just to use this type of joint, since the cost of formwork for the caps exceeds the additional cost of the fabricated steel hinge.

T-BEAM BRIDGES

37. Economics   In simple spans longer than about 25 ft, and in continuous spans longer than 30 to 35 ft, the economy of reinforced-concrete T-beams becomes apparent. Depending upon span length, and labor and material prices at various geographical locations, T-beams compare in economy with reinforced-concrete slabs, steel stringers (rolled beams), welded plate girders (both composite and noncomposite), reinforced-concrete box girders, and prestressed girders. In spans around 25 ft, T-beams are comparable in economy with longitudinally reinforced slabs, while in longer spans they should be compared with the other types.

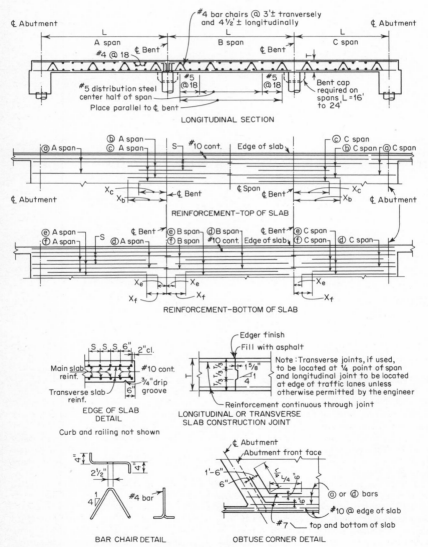

Fig. 54   Superstructure details, continuous slab bridge.

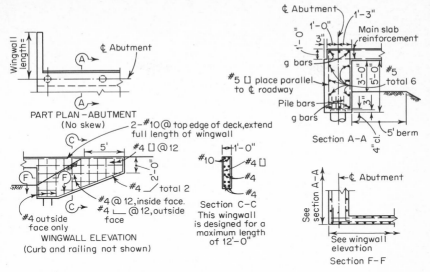

Fig. 55 Abutment details, continuous slab bridge.

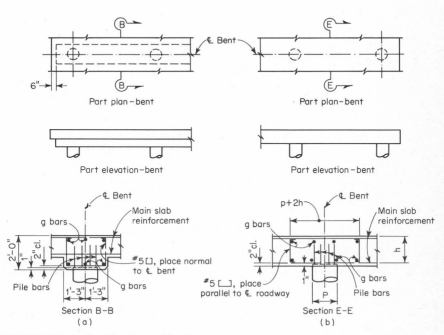

Fig. 56 Bent details: (*a*) with cap; (*b*) without cap.

Depth-Span Ratios. Economical depth-span ratios for simple spans vary from about 0.07 to 0.09, for continuous spans from 0.065 to 0.085. These ratios are for girders having uniform depth. In continuous spans, maximum economy usually results when the depth-span ratio is 0.065 to 0.070. These ratios require a small amount of compressive reinforcement over the supports, which can usually be supplied by the positive reinforcement which is required to be extended beyond the face of the support.

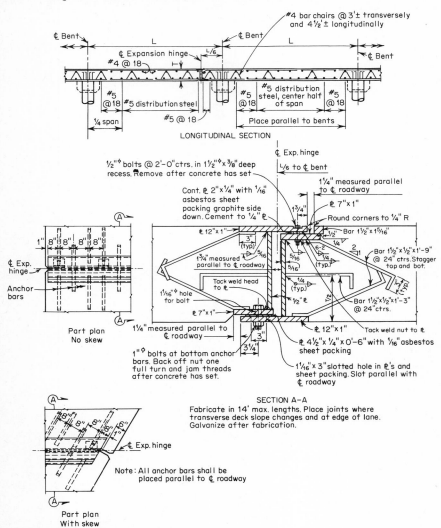

Fig. 57   Expansion hinge for slab bridge.

In some cases economical depth cannot be determined by considering the economy of the structure itself. Situations arise where the girder depth has to be established for overall economy through a consideration of approach fill costs or the necessity for meeting existing street grades.

Girder Spacing. Spacing of girders in T-beam bridges is dictated by economy. The most commonly used spacings are 6 ft and 10 ft, although in long spans it may be as much as 13 or 14 ft. Spacing depends upon the overall width of structure and the necessity for maintaining a slab overhang beyond the exterior girder which, according to the AASHTO specifications, should not exceed half the spacing of girders. When widening of a structure is anticipated, it is preferable to keep the slab overhang to one-third the girder spacing in order to eliminate thickening the slab over the exterior girder as a consequence of

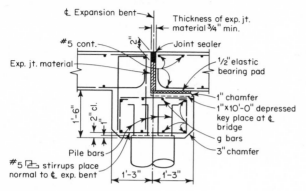

Fig. 58 Expansion detail at bent.

increased negative moments. In structures with three or more girders the spacing should be such as to make the girders as nearly alike as possible.

Girder stem width is determined by the number, spacing, and arrangement of reinforcing bars at the center of the span. The least width which can be used to enclose the positive-moment bars usually provides maximum economy. When required for shear, or for compression over the bents in continuous spans, the stem can be flared as shown in Fig. 59. The increased width of the stem at the support is usually governed by moment,

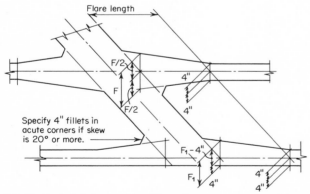

Fig. 59 Typical girder-stem flare. Where the increase in stem width is not more than 2½ in. on a side, stirrups may be detailed to uniform width.

while the length of flare is governed by shear (Art. 38). On continuous structures with spans of 60 ft or less, sufficient compressive reinforcement can be added for negative moment at the piers to keep the stem width uniform. This adds to the economy through simplifcation of formwork. When compressive reinforcement is used, the amount should not exceed 50 percent of the tension reinforcement required at the same section.

Table 6 facilitates the design of beams with compressive reinforcement. It can be used for inverted T-beams (flange in tension) as well as for rectangular beams and slabs.

TABLE 6   Beams Reinforced for Compression

f_s = 20,000 psi, f_c = 1200 psi, n = 10 (tensile reinforcement), n = 20 (compressive reinforcement), M = moment (tensile reinforcement) per ft b, A_s = area tensile reinforcement per ft b, M_s' = moment per in.² compressive reinforcement.

Effective depth d, in.	M, ft-kips per ft b	A_s, in.² per ft b	M_s', ft-kips per in.² A_s' Depth of compressive reinforcement d', in.					
			1½	2	2½	3	4	5
4	3.15	0.538	0					
¼	3.56	0.572	0.29					
½	3.99	0.605	0.60					
¾	4.44	0.639	0.92					
5	4.92	0.673	1.26					
¼	5.43	0.706	1.60					
½	5.96	0.740	1.97	0.16				
¾	6.51	0.773	2.33	0.49				
6	7.09	0.807	2.70	0.80				
¼	7.69	0.841	3.08	1.12				
½	8.32	0.874	3.46	1.45				
¾	8.97	0.908	3.85	1.79	0.09			
7	9.65	0.942	4.24	2.14	0.38			
¼	10.35	0.975	4.64	2.50	0.69			
½	11.07	1.01	5.04	2.86	1.00			
¾	11.82	1.04	5.44	3.23	1.32			
8	12.60	1.08	5.85	3.60	1.65	0.00		
¼	13.40	1.11	6.26	3.98	1.99	0.29		
½	14.22	1.14	6.67	4.36	2.33	0.58		
¾	15.07	1.18	7.08	4.74	2.68	0.89		
9	15.95	1.21	7.50	5.13	3.03	1.20		
¼	16.85	1.24	7.92	5.53	3.39	1.52		
½	17.77	1.28	8.34	5.92	3.76	1.85		
¾	18.72	1.31	8.76	6.32	4.13	2.18		
10	19.69	1.35	9.18	6.72	4.50	2.52		
½	21.71	1.41	10.03	7.53	5.26	3.21		
11	23.82	1.48	10.88	8.34	6.03	3.93	0.38	
½	26.04	1.55	11.74	9.17	6.81	4.65	0.98	
12	28.35	1.61	12.60	10.00	7.60	5.40	1.60	
½	30.76	1.68	13.46	10.84	8.40	6.16	2.24	
13	33.27	1.75	14.33	11.68	9.21	6.92	2.91	
½	35.88	1.82	15.20	12.52	10.02	7.70	3.59	0.19
14	38.59	1.88	16.07	13.37	10.84	8.49	4.29	0.77
½	41.39	1.95	16.94	14.22	11.67	9.28	5.00	1.38
15	44.30	2.02	17.82	15.08	12.50	10.08	5.72	2.00
½	47.30	2.08	18.70	15.94	13.34	10.89	6.46	2.64
16	50.40	2.15	19.28	16.80	14.18	11.70	7.20	3.30
½	53.60	2.22	19.95	17.66	15.02	12.52	7.95	3.97

TABLE 6 Beams Reinforced for Compression (*Continued*)

Effective depth d, in.	M, ft-kips per ft b	A_s, in.² per ft b	M_s', ft-kips per in.² A_s'					
			Depth of compressive reinforcement d', in.					
			1½	2	2½	3	4	5
17	56.90	2.29	20.62	18.53	15.86	13.34	8.72	4.66
½	60.29	2.35	21.28	19.40	16.71	14.17	9.49	5.36
18	63.79	2.42	21.95	20.27	17.57	15.00	10.27	6.07
½	67.38	2.49	22.61	21.14	18.42	.15.83	11.05	6.79
19	71.07	2.56	23.28	22.01	19.28	16.67	11.84	7.52
½	74.86	2.62	23.94	22.88	20.14	17.52	12.64	8.25
20	78.75	2.69	24.61	23.76	21.00	18.36	13.44	9.00
21	86.82	2.82	25.93	25.27	22.73	20.06	15.06	10.51
22	95.29	2.96	27.27	26.60	24.46	21.76	16.69	12.05
23	104.15	3.09	28.60	27.93	26.20	23.48	18.34	13.62
24	113.40	3.23	29.93	29.26	27.95	25.20	20.00	15.20
25	123.05	3.36	31.26	30.59	29.70	26.93	21.67	16.80
26	133.09	3.50	32.58	31.92	31.26	28.66	23.35	18.42
27	143.52	3.63	33.92	33.25	32.58	30.40	25.04	20.04
28	154.35	3.77	35.24	34.58	33.92	32.14	26.74	21.68
29	165.57	3.90	36.58	35.91	35.24	33.89	28.45	23.34
30	177.19	4.04	37.90	37.24	36.58	35.64	30.16	25.00
31	189.20	4.17	39.24	38.57	37.90	37.39	31.88	26.67
32	201.60	4.30	40.56	39.90	39.24	38.57	33.60	28.35
33	214.40	4.44	41.90	41.23	40.56	39.90	35.33	30.04
34	227.59	4.57	43.22	42.56	41.90	41.23	37.06	31.73
35	241.17	4.71	44.56	43.89	43.22	42.56	38.79	33.43
36	255.15	4.84	45.88	45.22	44.56	43.89	40.53	35.13
37	269.52	4.98	47.22	46.55	45.88	45.22	42.28	36.84
38	284.29	5.11	48.54	47.88	47.22	46.55	44.02	38.56
39	299.45	5.25	49.88	49.21	48.54	47.88	45.77	40.28
40	315.00	5.38	51.20	50.54	49.88	49.21	47.52	42.00
41	330.95	5.51	52.54	51.87	51.20	50.54	49.27	43.73
42	347.29	5.65	53.86	53.20	52.54	51.87	50.54	45.46
43	364.02	5.78	55.20	54.53	53.86	53.70	51.87	47.19
44	381.15	5.92	56.52	55.86	55.20	54.53	53.20	48.93
45	398.67	6.05	57.86	57.19	56.52	55.86	54.53	50.67
46	416.59	6.19	59.18	58.52	57.86	57.19	55.86	52.41
47	434.90	6.32	60.52	59.85	59.18	58.52	57.19	54.15
48	453.60	6.46	61.84	61.18	60.52	59.85	58.52	55.90
49	472.70	6.59	63.18	62.51	61.84	61.18	59.85	57.65
50	492.19	6.73	64.50	63.84	63.18	62.51	61.18	59.40
51	512.07	6.86	65.84	65.17	64.50	63.84	62.51	61.15
52	532.35	6.99	67.16	66.50	65.84	65.17	63.84	62.51
53	553.02	7.13	68.50	67.83	67.16	66.50	65.17	63.84
54	574.09	7.26	69.82	69.16	68.50	67.83	66.50	65.17
55	595.55	7.40	71.16	70.49	69.82	69.16	67.83	66.50
56	617.40	7.53	72.48	71.82	71.16	70.49	69.16	67.83
57	639.65	7.67	73.82	73.15	72.48	71.82	70.49	69.16
58	662.29	7.80	75.14	74.48	73.82	73.15	71.82	70.49
59	685.32	7.94	76.48	75.81	75.14	74.48	73.15	71.82

TABLE 6   Beams Reinforced for Compression (*Continued*)

Effective depth d, in.	M, ft-kips per ft b	A_s, in.² per ft b	M_s', ft-kips per in.² A_s' Depth of compressive reinforcement d', in.					
			1½	2	2½	3	4	5
60	708.75	8.07	77.80	77.14	76.48	75.81	74.48	73.15
61	732.57	8.20	79.14	78.47	77.80	77.14	75.81	74.48
62	756.79	8.34	80.46	79.80	79.14	78.47	77.14	75.81
63	781.40	8.47	81.80	81.13	80.46	79.80	78.47	77.14
64	806.40	8.61	83.12	82.46	81.80	81.13	79.80	78.47
65	831.80	8.74	84.46	83.79	83.12	82.46	81.13	79.80
66	857.59	8.88	85.78	85.12	84.46	83.79	82.46	81.13
67	883.77	9.01	87.12	86.45	85.78	85.12	83.79	82.46
68	910.35	9.15	88.44	87.78	87.12	86.45	85.12	83.79
69	937.32	9.28	89.78	89.11	88.46	87.78	86.45	85.12
70	964.69	9.42	91.10	90.44	89.78	89.11	87.78	86.45
71	992.45	9.55	92.44	91.77	91.10	90.44	89.11	87.78
72	1,020.60	9.68	93.76	93.10	92.44	91.77	90.44	89.11
73	1,049.15	9.82	95.10	94.43	93.76	93.10	91.77	90.44
74	1,078.09	9.95	96.42	95.76	95.10	94.43	93.10	91.77
75	1,107.42	10.09	97.76	97.09	96.42	95.76	94.43	93.10
76	1,137.15	10.22	99.08	98.42	97.76	97.09	95.76	94.43
77	1,167.27	10.36	100.42	99.75	99.08	98.42	97.09	95.76
78	1,197.79	10.49	101.74	101.08	100.42	99.75	98.42	97.09
79	1,228.70	10.63	103.08	102.41	101.74	101.08	99.75	98.42
80	1,260.00	10.76	104.40	103.74	103.08	102.41	101.08	99.75
81	1,291.70	10.89	105.74	105.07	104.40	103.74	102.41	101.08
82	1,323.79	11.03	107.06	106.40	105.74	105.07	103.74	102.41
83	1,356.27	11.16	108.40	107.73	107.06	106.40	105.07	103.74
84	1,389.15	11.30	109.72	109.06	108.40	107.73	106.40	105.07
85	1,422.42	11.43	111.06	110.39	109.72	109.06	107.73	106.40
86	1,456.09	11.57	112.38	111.72	111.06	110.39	109.06	107.73
87	1,490.15	11.70	113.72	113.05	112.38	111.72	110.39	109.06
88	1,524.60	11.84	115.04	114.38	113.72	113.05	111.72	110.39
89	1,559.45	11.97	116.38	115.71	115.04	114.38	113.05	111.72
90	1,594.69	12.11	117.70	117.04	116.38	115.71	114.38	113.05
91	1,630.32	12.24	119.04	118.37	117.70	117.04	115.71	114.38
92	1,666.35	12.37	120.36	119.70	119.04	118.37	117.04	115.71
93	1,702.77	12.51	121.70	121.03	120.36	119.70	118.37	117.04
94	1,739.59	12.64	123.02	122.36	121.70	121.03	119.70	118.37
95	1,776.80	12.78	124.36	123.69	123.02	122.36	121.03	119.70
96	1,814.40	12.91	125.68	125.02	124.36	123.69	122.36	121.03
97	1,852.40	13.05	127.02	126.35	125.68	125.02	123.69	122.36
98	1,890.79	13.18	128.34	127.68	127.02	126.35	125.02	123.69
99	1,929.57	13.32	129.68	129.01	128.34	127.68	126.35	125.02
100	1,968.75	13.45	131.00	130.34	129.68	129.01	127.68	126.35

Example   Given $M = 600$ ft-kips, $d = 40$ in., $d' = 2$ in., $b = 18$ in., $f_s = 20,000$ psi, $f_c = 1200$ psi. Determine the reinforcement.

SOLUTION.   From Table 6, $M = 315$ ft-kips for $b = 1$ ft. For $b = 18$ in., $M = 1.5 \times 315 = 473$ ft-kips; so compressive reinforcement must provide $600 - 473 = 127$ ft-kips.

From Table 6, $M_s' = 50.54$ ft-kips sq in. of compressive reinforcement. Therefore, $A_s' = 127/50.54 = 2.51$ in.² Also, $A_s = 5.38$ in.²/ft of b, so that $A_s = 5.38 \times 600/315 = 10.25$ in.²

Example   Given $M = 525$ ft-kips, $d = 40$ in., $d' = 2$ in., $A_s' =$ two No. 11 $= 3.12$ in.², $f_s = 20{,}000$ psi, $f_c = 1200$ psi. Determine b and A_s.

SOLUTION.   From Table 6, $M_s' = 50.54$ ft-kips/sq in. of compressive reinforcement. For $A_s' = 3.12$ in.², $M_s' = 50.54 \times 3.12 = 158$ ft-kips. Moment to be provided $= 525 - 158 = 367$ ft-kips. For $b = 1$ ft, $M = 315$ ft-kips. Therefore, $b = 12 \times 367/315 = 14$ in. and $A_s = 5.38 \times 525/315 = 8.98$ in.².

Bundling of Reinforcement.   Bundled bars are very effective in reducing the size of concrete sections and in lessening the maze of reinforcing steel which frequently hampers the placing of concrete. For example, the stem of a T-beam reinforced with eight No. 11 bars in two layers requires a width of 16 in. By placing these bars in two bundles of three bars each and two bars in the top layer, the width can be reduced to 13 in. In addition, placing of both steel and concrete is simplified. Four bars may be bundled, although it is preferable to limit the number to three. Not more than one bar should be terminated at any one point in a bundle. The development length for each bar must be that for the individual bar increased by 20 percent for a three-bar bundle and 33 percent for a four-bar bundle. Where lap splices are permitted (for bars No. 11 and smaller), the length of laps for bundled bars should be increased by the same percentages.

The table in Fig. 60 shows the preferred minimum spacings for the bundled bar

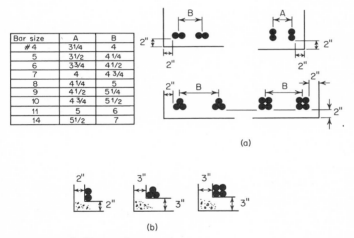

Bar size	A	B
#4	3 1/4	4
5	3 1/2	4 1/4
6	3 3/4	4 1/2
7	4	4 3/4
8	4 1/4	5
9	4 1/2	5 1/4
10	4 3/4	5 1/2
11	5	6
14	5 1/2	7

(a)

(b)

Fig. 60   Bundled bar reinforcement.

combinations in a. Preferred bar coverings for beam and girder sections are shown in a. Figure 60b shows minimum coverings for bundled bars in mass sections such as footings, abutments, piers, caps, and large columns. When bundled bars are used in work exposed to seawater, additional covering should be used.

38. Design of a T-Beam Bridge   Points to be considered in the design of a T-beam bridge will be discussed for the four-span continuous structure with 28-ft roadway shown in Fig. 61. Allowable stresses are $f_s = 20{,}000$ psi, $f_c = 1200$ psi, $v_c = 0.95 \sqrt{f_c'}$, $v_{max} = v_c + 4 \sqrt{f_c'}$.

DESIGN OF SLAB. The design of transverse slab continuous over three or more supports in T-beam structures is similar to that of transverse slabs on steel stringers (Art. 11). The only difference is in the effective span, which for T-beam slabs is the clear distance between faces of support. Where fillets making an angle of 45° or more with the axis of a continuous or restrained slab are built monolithic with the slab and support, the span can be measured from the section where the combined depth of slab and fillet is at least one and one-half times the thickness of the slab.

Figure 62 shows a typical section of the slab, which has been designed in accordance with the AASHTO specifications. The effective span has been taken as the clear distance between girder stems. Longitudinal girder steel in the top of slab is not shown except for the continuous bars at the stirrup hooks and the bars at the bend point of the transverse steel. At the supports, these bars at the bend points are usually main girder bars, while at the centers of spans they are extensions to main bars in the form of temperature steel.

Transverse slab steel can be placed parallel to the bents for skews up to 20°. For larger skews, consideration should be given to placing it normal to the centerline of structure. This keeps the amount of transverse slab steel to a minimum, since the steel spacing is measured along centerline of structure regardless of the amount of skew.

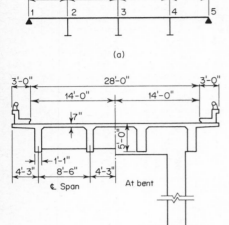

(a)

(b)

Fig. 61 T-beam bridge. (*a*) Layout; (*b*) typical section.

DESIGN OF STEM. The total depth chosen for the structure in Fig. 61 is 0.07 × span = 0.07 × 70 = 5 ft. Economy dictates that the width of stem be kept to a minimum at the center of span, yet be sufficient to enclose the positive-moment reinforcing and provide ample clearance between bars to facilitate concrete placement. In this example, bundled bars are used. In a T-beam of uniform depth, the stem width can be increased near the support to meet bending-moment requirements. The increase should be just sufficient to keep the compressive stress within the allowable when one-fourth of the positive-moment steel is extended through the support. Length of the stem flare is determined by the criterion of uniform spacing of stirrups throughout its length.

At the center of span 2, the *DL* + *LL* + *I* positive moment in an interior girder is 955 ft-kips. To resist this moment

$$A_s = \frac{12 \times 955,000}{20,000 \times 0.92 \times 57} = 10.9 \text{ in.}^2$$

Six No. 11 bars = 6 × 1.56 = 9.36
Two No. 8 bars = 2 × 0.79 = 1.58

10.94 in.²

The No. 11 bars can be arranged in two bundles of three each, with the two No. 8 bars placed above, as shown in section *AA* of Fig. 63. The required width of stem (Fig. 60) is $b' = B$ + two No. 11 bar diameters + 2 × 2 = 6 + 2 × 1.41 + 4 = 12.82 in. Use 13 in.

The compressive stress f_c in the flange can be determined approximately from the formula

$$f_c = \frac{2M}{bt(d - t/2)}$$

where $b = 12t + b'$ but not to exceed the center-to-center spacing of girders
b' = stem width
t = slab thickness
d = effective depth of beam

$$f_c = \frac{2 \times 12 \times 955,000}{98.5 \times 7.13(57 - 3.56)}$$
$$= 610 \text{ psi} < 1200$$

At bent 3, the negative moment for an interior girder is 1260 ft-kips. With three No. 11 bars for compressive reinforcement, $A_s' = 3 \times 1.56 = 4.68$ in.[2] From Table 6, for $d = 57$ in. and $d' = 3$ in., $M_s = 71.82 \times 4.68 = 336$ ft-kips. Moment to be provided by the concrete is $1260 - 336 = 924$ ft-kips. From Table 6, $M = 639.65$ for $b = 1$ ft; therefore, $b = {}^{924}\!/_{640} = 1.45$ ft $= 18$ in. and $A_s = (1260/640) \times 7.67 = 15.1$ in.[2] Use 10 No.11 bars $= 15.6$ in.[2] The resulting arrangement is shown in Fig. 63.

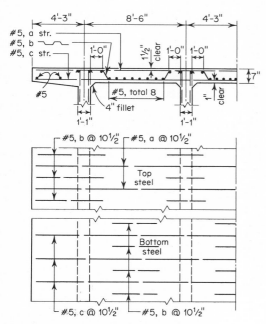

Fig. 62   Typical section of T-beam bridge slab.

SHEAR. Figure 64 shows the $DL + LL + I$ shear diagram for an interior girder of span 2. The maximum shear at bent 2 is 121 kips, so that

$$v = \frac{V}{bd} = \frac{121,000}{18 \times 57} = 118 \text{ psi}$$

$$v_c = 0.95 \sqrt{3000} = 52 \text{ psi}$$

$$v_c + 4\sqrt{3000} = 271 > 118 \text{ psi}$$

Spacing of No. 5 stirrups =

$$\frac{f_v A_v}{b(v - v_c)} = \frac{20,000 \times 0.62}{18 \times 66} = 10.4 \text{ in.}$$

Use 10 in. The length of flare of the stem is determined by locating the point where the shear requirement can be met by stirrups at 10 in. in the 13-in. stem:

$$v - v_c = \frac{20,000 \times 0.62}{13 \times 10} = 95.4 \text{ psi}$$
$$v = 52 + 95.4 = 147.4 \text{ psi}$$
$$V = \frac{147.4 \times 13 \times 57}{1000} = 109 \text{ kips}$$

From the shear diagram, this is the shear at a point 8 ft from bent 2. Stirrup spacings at other points are shown in Fig. 63.

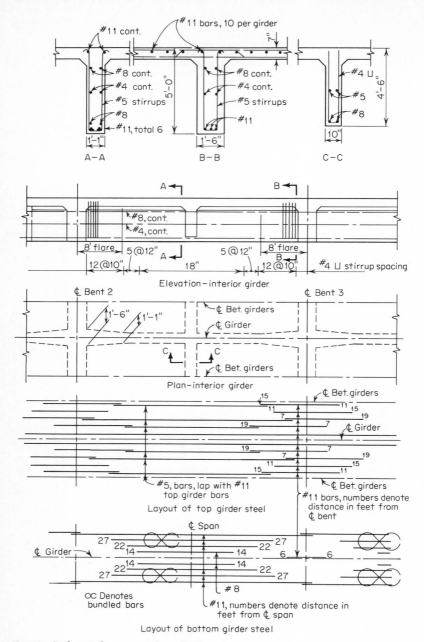

Fig. 63 Girder reinforcement, T-beam bridge.

Diaphragms.   Intermediate diaphragms (Fig. 63, section *CC*) act in conjunction with the slab as transverse distributing elements for the live load. For the ordinary T-beam bridge it is not necessary to proportion them or determine the reinforcement by actual analysis (stresses could be determined by an analysis in which the diaphragms constitute a beam continuous over elastic supports). Width of diaphragms should permit ease in

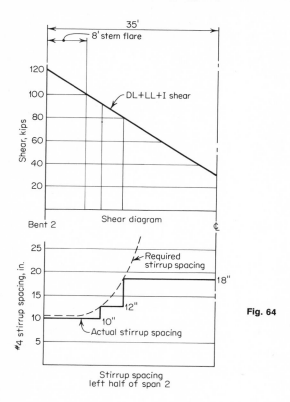

Fig. 64

placement of concrete. While an 8-in. wall will accommodate two rows of steel, 10 in. is better for placing of reinforcement. Transverse slab steel resists the negative moments in the diaphragms. The bottom steel can be nominal but should not be less than two No. 6 bars. Two longitudinal bars should be placed at the midheight of diaphragm to act as temperature steel and as tie bars for the stirrups. For girder spacings of 10 ft or less, stirrups can be No. 4 bars at 18 in. Usually the unit shear is less than allowable, and the stirrups act more nearly in the capacity of temperature steel than as diagonal tension reinforcement.

39. Design of Substructure   Intermediate supports for T-beam bridges may be
  1. Reinforced-concrete piers.
  2. Reinforced-concrete pier walls constructed integrally with the superstructure.
  3. Reinforced-concrete bents consisting of two or more columns; these may be integral with the superstructure.
  4. Reinforced-concrete single-column bents; these may be integral with the superstructure.
  5. Reinforced-concrete piles.
  6. Steel piles.
Choice of type depends upon skew, foundation conditions, streamflow, drift, ice, and the manner in which it will function in the structure. In T-beam bridges across streams,

piers or pier walls are commonly used since they are less affected by drift, streamflow, and ice. Where ice conditions do not exist, multicolumn bents, single-column bents, or piles can be used. Single-column and multicolumn bents are particularly suited to overcrossings where grades for two or more roadways are separated. The single-column bent can be used to advantage on skewed crossings because bent caps can be placed normal to the bridge centerline, which results in simpler superstructure framing.

The following discussion covers some of the more important points to be considered in the design of a two-column bent (Fig. 65) for the T-beam bridge of Art. 38.

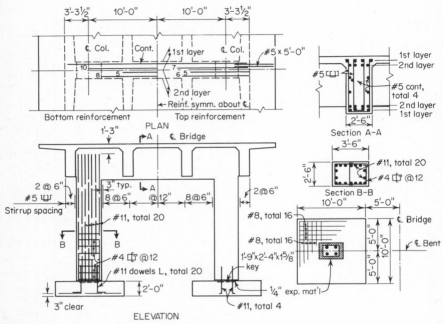

Fig. 65 Two-column bent for T-beam bridge.

Loads. The various loads for which the bent should be investigated are dead load, live load, impact, wind load on structure, wind load on live load, longitudinal force from live load, shrinkage, temperature, and earthquake. AASHTO 1.2.22 specifies the allowable unit stresses which may be used with various combinations of these loads. The bent is designed as a rigid frame with the columns hinged at the base.

Design of the cap for moment is determined by $DL + LL + I$ at basic unit stresses. The cap can be considered as a T-beam with an effective width equal to the width of stem plus twelve times the slab thickness. Girder dead loads are applied as concentrations, with the weight of the cap itself a uniform load.

Live-load positive moment is determined by placing two trucks adjacent to and symmetrical about the centerline of bridge, with the rear axles over the centerline of bent. Continuity of superstructure should be taken into account in determining the reaction on the bent of the truck's front axle and the trailer axle. This reaction can be applied to the bent uniformly over the 10-ft width of lane. For the spacing of the columns in this bent, and considering the AASHTO limitations on load positions in the design lanes, live-load negative moments are determined for the same load position as for positive moment. In bents with wider column spacings, the loads should be positioned separately for maximum positive moment and maximum negative moment.

Double stirrups (Fig. 65, Section A-A) are usually required in caps of this type to provide spacing large enough for easy placing of concrete. For the determination of maximum shearing stresses, live load can be positioned in the same manner as for moment.

The main steel in the cap should be located so as not to interfere with the column steel or the deck steel. In structures skewed more than 20 or 25° some designers place the deck steel normal to the centerline of bridge. This necessitates placing all the top bars in the cap below the steel in the slab.

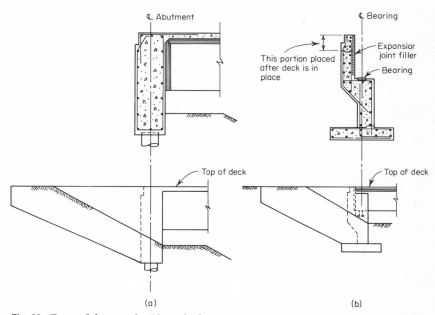

Fig. 66 Types of abutment for T-beam bridges.

Design of Columns. The columns for this structure are cast integrally with the cap and superstructure to form a rigid frame both longitudinally and laterally. A hinge is provided at the top of the footing. Their design involves an analysis for direct stress, and bending in two directions.* All the loading groups should be investigated to determine maximum column stresses. When investigating bending in two directions, it is important to remember that live-load stresses that are combined must be compatible; i.e., if the column is being investigated for a particular position of the live load, this position should be used in determining all the stresses that are to be combined.

In a continuous structure without intermediate hinges the deck is extremely stiff laterally, so that wind and earthquake loads are transmitted to the abutments rather than to the bents. In such cases, the bent need not be analyzed for such loads.

Design of Footings. The footing for the bent in Fig. 65 is designed for a column reaction of 380 kips. Weight of the footing itself is 30 kips. Maximum allowable soil pressure is 2 tons/sq ft.

The AASHTO specifications require that the critical section for bending be taken at the face of the column. For this footing the actual bearing area is provided by the 1 ft 9 in. × 2 ft 4 in. key. Taking moments at the 2 ft 4 in. edge,

*Lu-Shien Hu, Eccentric Bending in Two Directions of Rectangular Concrete Columns, ACIJ., May 1955. See also Sec. 11, Art. 19.

$$M = \frac{3.8 \times 10 \times 4.13^2}{2} = 325 \text{ ft-kips}$$

$$A_s = \frac{12 \times 325,000}{20,000 \times 0.88 \times 20.5} = 10.8 \text{ in.}^2$$

The AASHTO specifications allow this steel to be distributed over an effective width equal to the width of the column (in this case, the key) plus twice the effective depth of

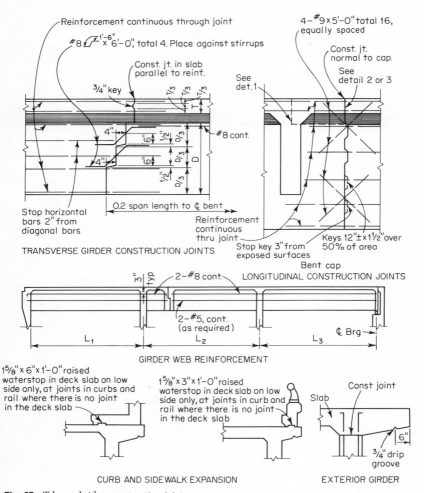

Fig. 67 T-beam bridge construction joints.

the footing, plus half the remaining width of footing, in this case 7 ft 10½ in. Spacing of reinforcement outside this width may be double the spacing within the effective width. However, it is desirable to maintain uniform bar size and spacing. Therefore, 16 No. 8 uniformly spaced bars are used, which gives the area required within the effective width.

Shear must be investigated at two sections: (1) for one-way action on a section across the full width of the footing at a distance d from the face of the reaction area, and (2) for two-way action on a concentric vertical section $d/2$ from the faces of the reaction area. The respective values of v_c are $0.95 \sqrt{f_c}$ and $1.8 \sqrt{f_c}$, which give 52 and 99 psi respectively, for $f_c' = 3000$ psi. The reaction area is the $1'9'' \times 2'4''$ key (Fig. 65).

For one-way action,

$$V = 3.8 \times 2.42 \times 10 = 92 \text{ kips}$$
$$v = \frac{92,000}{120 \times 20.5} = 37 \text{ psi} < 52$$

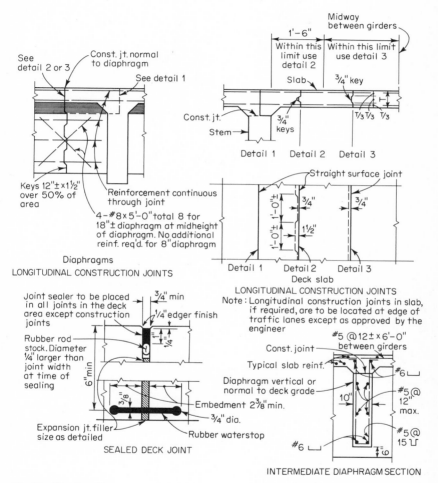

Fig. 67 (*Continued*)

For two-way action,

$$V = 3.8 (10 \times 10 - 4.21 \times 5.21) = 297 \text{ kips}$$
$$b = 2 (3.46 + 4.04) = 15.0 \text{ ft.}$$
$$v = \frac{297,000}{15 \times 12 \times 20.5} = 80.5 \text{ psi} < 99$$

Abutments.   Figure 66 shows a typical section and elevation for each of two types of abutments which can be used for T-beam bridges. The abutment in *a* is cast integrally with the deck and rests directly on the pile substructure. Expansion in the deck must overcome the passive resistance of the earth adjacent to the abutment wall. Therefore, to avoid excessive compression in the deck, this type of abutment should be used only when the total length of bridge does not exceed 150 ft. However, it can be used for longer bridges if a sufficient number of expansion hinges are provided. Wing walls can be placed normal to the structure or, as in this case, parallel to it.

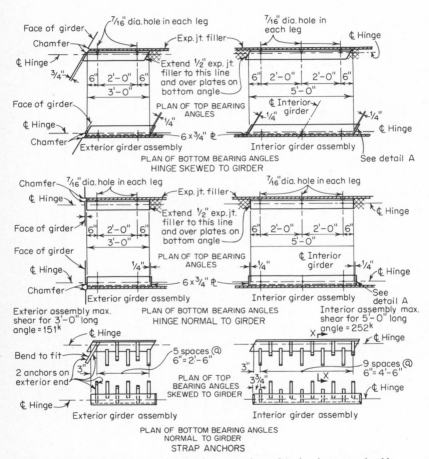

Fig. 68 T-beam bridge expansion details. Abutting surfaces of steel and concrete should be separated with expansion filler to provide free movement. Calk and tape joints between steel angles and all remaining openings to prevent seepage of grout or concrete between abutting surfaces of hinge.

The abutment in *b* provides for expansion and is a type which can be used for bridges of any length. The back wall may rest directly on the footing or, as shown in the figure, on a wall (which may be 35 ft or more in height). Active earth pressure is taken by the wall. The abutment can be founded on either a spread footing or a footing on piles, depending upon conditions.

40. Typical Details Typical details for construction joints and intermediate expansion hinges for T-beam bridges are shown in Figs. 67 and 68, respectively. Fixed and expansion bearing assemblies are shown in Fig. 69 and concrete placing diagrams in Fig. 70.

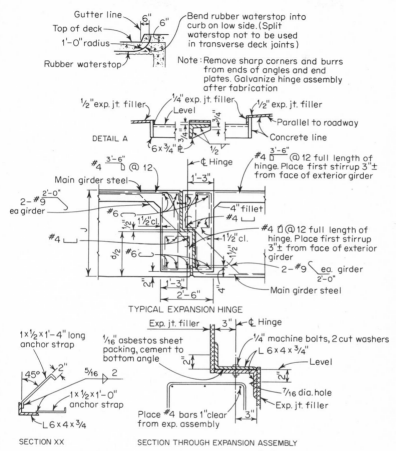

Fig. 68 (*Continued*)

BOX-GIRDER BRIDGES

Many states use box girders as a solution to some of the bridge problems encountered in the development of the Interstate Highway system. The relatively shallow depth of the box-girder bridge is an advantage where, as is often the case in cities, headroom is limited. In addition, they provide ideal space for utilities. The smooth soffit is pleasing. Because of their torsional stiffness they can be built on rather sharp horizontal curves; this is a desirable characteristic where curved alignment has a bearing on the overall cost of a highway interchange.

41. Economics Box-girder bridges are seldom justified for short spans on the basis of cost alone and are rarely used on spans of less than 50 ft; if they are, it is usually for the sake of uniformity in a continuous structure of varying spans. In the 50- to 100-ft range, box girders cost about one-third more than T-beams. However, the experience of the contractor is a big factor, and contractors in some areas claim that they can build box girders about as economically as T-beams. One reason is in the method of forming the webs, where, instead of finishing the tops of the forms to exact grade, as in T-beam construction, they are extended above grade and the webs are poured to finishing strips.

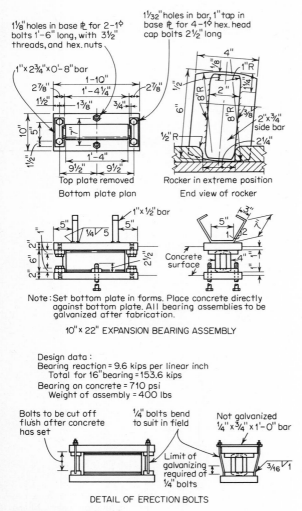

Top plate removed
Bottom plate plan

Rocker in extreme position
End view of rocker

Note: Set bottom plate in forms. Place concrete directly against bottom plate. All bearing assemblies to be galvanized after fabrication.

10" x 22" EXPANSION BEARING ASSEMBLY

Design data:
Bearing reaction = 9.6 kips per linear inch
   Total for 16" bearing = 153.6 kips
Bearing on concrete = 710 psi
   Weight of assembly = 400 lbs

Bolts to be cut off flush after concrete has set

¼" bolts bend to suit in field

Not galvanized
¼" x ¾" x 1'-0" bar

Limit of galvanizing required of ¼" bolts

DETAIL OF ERECTION BOLTS

Fig. 69   Bearing assemblies.

Another saving results from reduced costs of finishing, since only exterior girder faces and the bottom slab soffit require a high-quality finish.

42. Proportions   Depth-span ratios of 0.06 can be used for simple spans, but 0.07 is the generally accepted minimum for economy. The recognized minimum for continuous spans of uniform depth is 0.05. Ratios of 0.06 to 0.07 are common for continuous spans, but the best appearance is achieved in the range 0.05 to 0.06. In continuous structures with

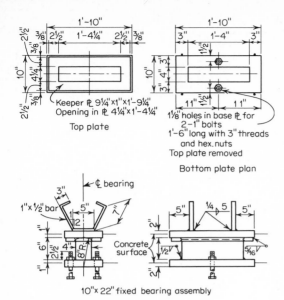

**1⅛" holes in base ℞ for
2–1" bolts
1'–6" long with 3" threads
and hex. nuts
Top plate removed**

Bottom plate plan

Top plate

Keeper ℞ 9¼"x1"x1'–9¼"
Opening in ℞ 4¼"x1'–4¼"

10"x 22" fixed bearing assembly

Fig. 69 (*Continued*).

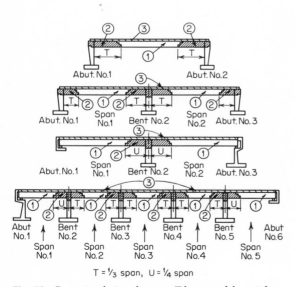

T = ⅓ span, U = ¼ span

Fig. 70 Concrete placing diagram, T-beam and box-girder
bridge superstructures. Numbers 1 and 2 indicate sequence of
placing girder-stem concrete; may be placed simultaneously
provided sections 1 are placed in adjoining spans. Top slab
concrete to be placed separately in box-girder bridges and may
be placed separately in T-beam bridges; may be placed continuously or in parts as approved by the engineer.

parabolic soffits, midspan depth-span ratios can be even smaller. The Portland Cement Association's publications "Continuous Hollow-Girder Concrete Bridges" and "Continuous Concrete Bridges" give information on parabolic-soffit structures.

Webs.   In spans of less than 100 ft where minimum depth-span ratios are used, webs can be spaced 7 to 8 ft on centers. The optimum spacing is likely to be between 7 and 9 ft for spans to 150 ft. In spans longer than 150 ft, spacing up to 12 ft may be found economical. Since the optimum spacing depends on depth-span ratio, comparative studies are advisable for each case.

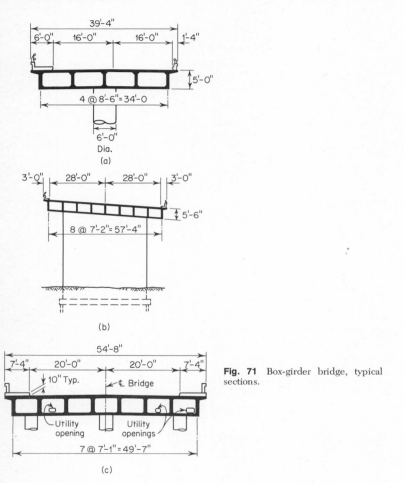

Fig. 71 Box-girder bridge, typical sections.

The minimum practicable width of web is 8 in. to provide for double-leg stirrups and easy placement of concrete. Webs have been built 7 in. thick and, in rare instances, as thin as 6 in. The 8-in. web can be used for depths to about 8 ft, above which 9 or 10 in. is recommended. Webs can be flared to meet shear requirements.

Typical sections are shown in Fig. 71. The substructure for *a* consists of single columns 6 ft in diameter. In *b* solid piers 2 ft 6 in. × 42 ft are used. Three-column bents, each column 3 ft 6 in. in diameter, are used in *c*.

Drainage holes should be provided in the bottom slab for release of curing water, leakage through the top deck, and condensation. Occasionally there is leakage from utilities. Serious overload can result from accumulation of water because of inadequate drainage. There should be a 4-in. hole in each cell adjacent to piers and abutments, and each transverse diaphragm should have a 6-in. opening.

It is comparatively simple to provide for utilities such as gas and water lines, power and telephone ducts, and storm drains and sewers in the cells of the box-girder bridge. Spacing of girder webs can be adjusted to meet specific locations of these utilities, if necessary. Examples of reinforcement at utility openings in abutments, bents, and diaphragms are shown in Figs. 72 and 73. If utilities pass through an abutment where there is

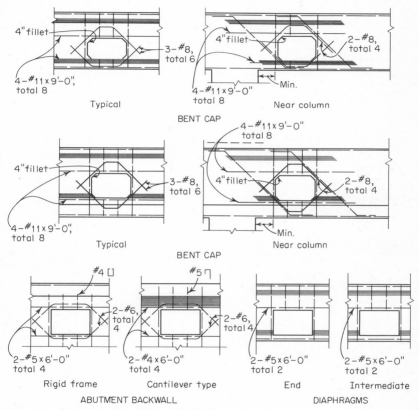

Fig. 72 Utility opening details, box-girder bridge.

provision for expansion in the superstructure, it is usually advisable to provide a passage between the end diaphragm and the back wall for access to the cells for inspection and maintenance (Fig. 74). When such access cannot be provided, the sidewalk manhole is an alternative. Manholes can be placed in the roadway deck if no other location is feasible, but this is a potential for accidents to maintenance personnel. Access between adjacent cells can be through openings in the webs. They should be located near the center of the span, where shear is not significant.

43. Design Some engineers prefer to design the cross section as a unit. As a general rule, this can be done when the ratio of span to width exceeds 2. The unit girder (cell) method is recommended for smaller ratios. In fact, unless and until further research proves otherwise, the latter is suggested for all box girders, since with present allowable live-load distributions it is conservative.

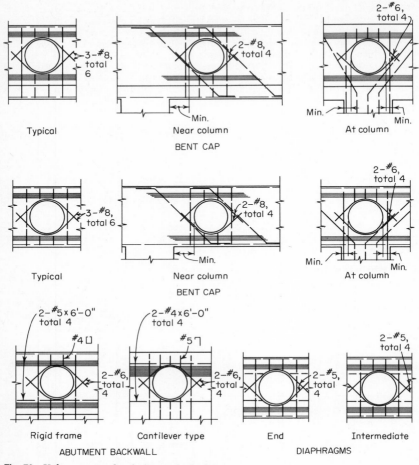

Fig. 73 Utility opening details, box-girder bridge.

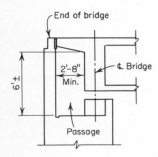

Fig. 74 Utility access at abutment, box-girder bridge.

Moments of inertia of the concrete gross cross section for one interior cell are given in Table 7. These values include the effect of 4-in. fillets (Fig. 77), but since they assume 8-in. stems, they do not account for web flare in regions of large shear. Straight-line interpolation can be used for intermediate values. If the moment of inertia of an entire superstructure is required, the moment of inertia of the exterior girder can be approximated at 72 to 78 percent of that of an interior girder. Values in this table can be used to compute deflections of spans of uniform depth and to determine fixed-end moments, distribution and carryover factors, etc., for variable-depth girders where the variation in moment of inertia approximates that assumed in the preparation of various charts and tables, for example, those of the Portland Cement Association's "Continuous Concrete Bridges." Weights of these cross sections are given in Table 8.

Top slab design is similar to that for the steel-stringer bridge (Art. 11) and the T-beam bridge (Art. 38), the effective span being determined as for the latter.

Distribution of Loads. The fraction of a wheel load carried by each interior girder is $S/7$, provided $S < 16$ ft, where S is the spacing of girders (AASHTO 1.3.1). For the exterior girder S is replaced by W_e, the width of the exterior girder, which is the distance from the midpoint between girders to the outside edge of the slab. In the case of a bridge with sidewalks this distribution of vehicular load assures an exterior box adequate for future widening. Furthermore, sidewalk live load can be disregarded. If railing, curb, and sidewalk are to be placed after the slab is cured sufficiently, their weight can be assumed to be distributed equally among all the girders of the span.

TABLE 7   Moment of Inertia of Interior Box Girders, 8-in. Stem, 4-in. Fillets

Top slab, in...	6	6	6	6⅛	6¼	6¼	6¼	6⅜	6½	6⅝	6¾	7
Bottom slab, in.....	5½	5½	5½	5½	5½	5½	5½	5½	5½	5½	5¾	5⅞
Girder center to center, ft-in........	5-9	6-0	6-3	6-6	6-9	7-0	7-3	7-6	7-9	8-0	8-3	8-6
Depth, ft-in.	Moment of inertia, ft⁴											
3-6	13.9	14.4	15.0	15.6	16.2	16.8	17.4	18.0	18.7	19.3	20.3	21.2
3-9	16.4	17.0	17.7	18.4	19.2	19.8	20.5	21.3	22.0	22.8	23.9	25.0
4-0	19.2	19.9	20.7	21.5	22.4	23.2	23.9	24.8	25.7	26.6	27.9	29.2
4-3	22.2	23.0	23.9	24.9	25.9	26.8	27.6	28.7	29.7	30.7	32.3	33.8
4-6	25.4	26.4	27.4	28.5	29.7	30.7	31.6	32.8	34.0	35.2	36.9	38.7
4-9	28.9	30.0	31.1	32.4	33.7	34.8	35.9	37.3	38.6	40.0	42.0	43.9
5-0	32.7	33.9	35.2	36.6	38.1	39.3	40.6	42.1	43.6	45.1	47.3	49.5
5-3	36.7	38.1	39.5	41.1	42.7	44.1	45.5	47.2	48.8	50.5	53.1	55.5
5-6	41.0	42.6	44.1	45.9	47.7	49.2	50.7	52.6	54.5	56.3	59.2	61.9
5-9	45.6	47.3	48.9	50.9	52.9	54.6	56.3	58.4	60.4	62.5	65.6	68.7
6-0	50.5	52.3	54.1	56.3	58.5	60.4	62.2	64.5	66.7	69.0	72.4	75.8
6-3	55.6	57.6	59.6	62.0	64.4	66.4	68.4	70.9	73.4	75.9	79.7	83.4
6-6	61.0	63.2	65.4	68.0	70.6	72.8	75.0	77.7	80.4	83.1	87.3	91.3
6-9	66.7	69.1	71.5	74.3	77.1	79.5	81.9	84.8	87.8	90.8	95.2	99.6
7-0	72.8	75.3	77.9	80.9	84.0	86.6	89.2	92.4	95.5	98.8	103.6	108.4
7-3	79.1	81.8	84.6	87.9	91.2	94.0	96.8	100.2	103.7	107.1	112.4	117.6
7-6	85.7	88.7	91.6	95.2	98.8	101.8	104.8	108.5	112.2	115.9	121.6	127.1
7-9	92.7	95.8	99.0	102.8	106.7	109.9	113.1	117.1	121.1	125.1	131.2	137.2
8-0	99.9	103.3	106.7	110.8	114.9	118.4	121.8	126.1	130.3	134.6	141.2	147.6
8-3	107.5	111.2	114.8	119.1	123.5	127.2	130.9	135.4	140.0	144.6	151.6	158.5
8-6	115.5	119.3	123.2	127.8	132.5	136.5	140.4	145.2	150.1	155.0	162.4	169.8
8-9	123.7	127.8	131.9	136.9	141.9	146.1	150.2	155.4	160.6	165.8	173.7	181.6
9-0	132.3	136.7	141.0	146.3	151.6	156.0	160.5	165.9	171.4	177.0	185.4	193.8
9-3	141.3	145.9	150.5	156.1	161.7	166.4	171.1	176.9	182.7	188.6	197.6	206.5
9-6	150.6	155.5	160.3	166.2	172.2	177.2	182.1	188.3	194.5	200.7	210.2	219.6

Width of web at the supports is governed by shear. The web can be tapered from this width to the minimum practicable width, the length of flare being determined by the allowable shear for the minimum width. An example is shown in Fig. 75, where the effective depth $d = 45$ in. With the AASHTO allowable shear stress $v = v_c + 0.4\sqrt{f_c}$, where $v_c = 0.95\sqrt{f_c}$, the 8-in. web is just sufficient at 16 ft from the support, as shown in the following table.

Spacing of No. 5 U Stirrups (Fig. 75)

Distance, ft	Web, in.	V, kips	v, psi	$v - v_c$	Spacing, in.
0	12	140	260	208	$8^1/_2$
4	11	129	261	209	9
8	10	117	260	208	10
12	9	106	262	210	11
16	8	97	269	217	12
20	8	89	247	195	$13^1/_2$
24	8	80	222	170	$15^1/_2$

If the increase in width of stem does not exceed 2½ in. on each side, stirrups may be of uniform width.

Bottom Slab. The AASHTO specifications require a bottom flange thickness not less than 5½ in. or one-sixteenth the clear distance between webs, whichever is larger. However, in the regions of negative moment in shallow, continuous girders, flange thickness will be governed by stress.

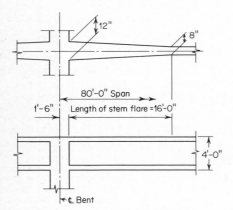

Fig. 75   Flaring box-girder stems.

Reinforcement not less than 0.4 percent of flange area is required in the longitudinal direction in the bottom flange. A single layer may be centered in the slab. Bar spacing must not exceed 18 in. These requirements can be met by distributing the girder tensile reinforcement throughout the slab or by adding steel if the girder reinforcement is concentrated near the web. Minimum transverse reinforcement in the bottom slab is 0.5 percent of flange area, distributed over both surfaces. The bars must be bent up into the exterior girder stems and anchored by a standard hook.

Diaphragms. The box-girder bridge is less dependent on diaphragms for transverse rigidity than is the case for the T-beam bridge. However, diaphragms do improve torsional resistance of curved boxes. Diaphragms need only nominal reinforcement at the top and the bottom for support of stirrups, as the transverse steel in top and bottom slabs provides for their continuity at the girders. Nominal shear reinforcement, i.e., No. 4 at 12 in., is ample except where utility openings are required. Both intermediate and end diaphragms having such openings should be proportioned for load, usually a truck wheel midway between girders.

Horizontally Curved Girders. The average length of span can be used for determining the frame constants for continuous, horizontally curved box girders. In designing the

TABLE 8   Weight of Interior Box Girders, 8-in. Stem, 4-in. Fillets

Top slab, in...	6	6	6	6⅛	6¼	6¼	6¼	6⅜	6½	6⅝	6¾	7
Bottom slab, in.....	5½	5½	5½	5½	5½	5½	5½	5½	5½	5½	5¾	5⅞
Girder center to center ft-in........	5-9	6-0	6-3	6-6	6-9	7-0	7-3	7-6	7-9	8-0	8-3	8-6

Depth, ft-in.	Weight, kips per ft											
3-6	1.11	1.15	1.19	1.23	1.28	1.31	1.35	1.40	1.45	1.49	1.57	1.64
3-9	1.14	1.17	1.21	1.26	1.30	1.34	1.38	1.42	1.47	1.52	1.59	1.67
4-0	1.16	1.20	1.24	1.28	1.33	1.36	1.40	1.48	1.50	1.54	1.62	1.69
4-3	1.19	1.22	1.26	1.31	1.35	1.39	1.43	1.47	1.52	1.57	1.64	1.72
4-6	1.21	1.25	1.29	1.33	1.38	1.41	1.45	1.50	1.55	1.59	1.67	1.74
4-9	1.24	1.27	1.31	1.36	1.40	1.44	1.48	1.52	1.57	1.62	1.69	1.77
5-0	1.26	1.30	1.34	1.38	1.43	1.46	1.50	1.55	1.60	1.64	1.72	1.79
5-3	1.29	1.32	1.36	1.41	1.45	1.49	1.53	1.57	1.62	1.67	1.74	1.82
5-6	1.31	1.35	1.39	1.43	1.48	1.51	1.55	1.60	1.65	1.69	1.77	1.84
5-9	1.34	1.37	1.41	1.46	1.50	1.54	1.58	1.62	1.67	1.72	1.79	1.87
6-0	1.36	1.40	1.44	1.48	1.53	1.56	1.60	1.65	1.70	1.74	1.82	1.89
6-3	1.39	1.42	1.46	1.51	1.55	1.59	1.63	1.67	1.72	1.77	1.84	1.92
6-6	1.41	1.45	1.49	1.53	1.58	1.61	1.65	1.70	1.75	1.79	1.87	1.94
6-9	1.44	1.47	1.51	1.56	1.60	1.64	1.68	1.72	1.77	1.82	1.89	1.97
7-0	1.46	1.50	1.54	1.58	1.63	1.66	1.70	1.75	1.80	1.84	1.92	1.99
7-3	1.49	1.52	1.56	1.61	1.65	1.69	1.73	1.77	1.82	1.87	1.94	2.02
7-6	1.51	1.55	1.59	1.63	1.68	1.71	1.75	1.80	1.85	1.89	1.97	2.04
7-9	1.54	1.58	1.61	1.66	1.70	1.74	1.78	1.82	1.87	1.92	1.99	2.07
8-0	1.56	1.60	1.64	1.68	1.73	1.76	1.80	1.85	1.90	1.94	2.02	2.09
8-3	1.59	1.63	1.66	1.71	1.75	1.79	1.83	1.87	1.92	1.97	2.04	2.12
8-6	1.61	1.65	1.69	1.73	1.78	1.81	1.85	1.90	1.95	1.99	2.07	2.14
8-9	1.64	1.68	1.71	1.76	1.80	1.84	1.88	1.92	1.97	2.02	2.09	2.17
9-0	1.66	1.70	1.74	1.78	1.83	1.86	1.90	1.95	2.00	2.04	2.12	2.19
9-3	1.69	1.73	1.76	1.81	1.85	1.89	1.93	1.97	2.02	2.07	2.14	2.22
9-6	1.71	1.75	1.79	1.83	1.88	1.91	1.95	2.00	2.05	2.09	2.17	2.24

girders, two girders in a given span (for example, the longest and the shortest interior girders) can be designed and the others proportioned from them by interpolation.

Dead load and live load both produce torsion in horizontally curved girders. Where the width of the box exceeds the height, as is usually the case in bridges, torsion can be neglected in continuous spans if the central angle between bents is less than 30°. Various papers on the subject are available where torsion must be considered in the analysis.*

*Osterblom, I.: Bending and Torsion in Horizontally Curved Beams, *ACI J.*, November 1932. Schulz, Martin, and Mauricio Chedrau: Tables for Circularly Curved Horizontal Beams with Symmetrical Uniform Loads, *ACI J.*, May 1957. Curved Steel Box-Girder Bridges: State of the Art, ASCE-AASHTO Task Committee on Curved Box Girders, *J. St. Div. ASCE,* November 1978.

Deflection. Dead-load deflection must be determined so that forms can be cambered. Because of creep, ultimate deflection usually develops in 4 or 5 years (Art. 32). The moment of inertia of the gross concrete cross section may be used in computing deflection; the transformed section is an unnecessary refinement in view of the many uncertainties. Methods of computing deflection are discussed in the Portland Cement Association ST Chart 70. A camber diagram for a three-span continuous box girder is shown in Fig. 76. Immediately after removal of falsework, deflections will be about one-fourth the given values.

Suggested concrete placing sequences are shown in Fig. 70. Midspan portions of the girders are cast first to avoid possible cracking at the bents, which could occur if these portions were placed last. Alternative procedures are suggested in the notes accompanying the figure.

A typical layout of a box-girder bridge superstructure is shown in Fig. 76. A typical, partial cross section is shown in Fig. 77, where only distribution and temperature reinforcement and those girder bars which are continuous are shown. The remaining girder main bars in both top and bottom slabs are shown in a girder-reinforcement layout such as the one for a two-span continuous structure in Fig. 78. The continuous No. 8 bars in the stems just under the top-slab fillets in Fig. 77 are primarily for negative moments from possible falsework settlement before placement of the top slab.

Construction joints and miscellaneous details are shown in Fig. 79, and access openings in Fig. 80. End-bearing assemblies for T-beam bridges can also be used for box-girder bridges (Fig. 69).

44. Substructure   The same types of substructure can be used for box-girder bridges as for T-beam bridges. Both top and bottom of the column must be fixed in single-column bents for continuous structures. In multicolumn bents either end, or both, may be fixed, depending upon requirements for stability, stiffness, and provisions for changes in temperature. Structures of three or four spans can be planned so that all expansion is taken at the abutments, with longitudinal stability provided by the bents. Columns in bents farthest from the center of movement may be hinged in the longitudinal direction at the top of the footing to minimize temperature and live-load moments.

Caps in single- or multicolumn bents can be flush with the box-girder soffit. This simplifies forming and promotes economy. Opinions differ as to the aesthetics of this procedure, and some engineers prefer deeper caps. In any case, caps in single-column bents for wide bridges must be deeper than the superstructure to satisfy stress requirements.

Torsion is not usually significant in single- or multicolumn-bent caps which are cast integrally with the superstructure of a box-girder bridge. For example, the critical loading for the cap of a single-column bent results with live load positioned for maximum bending moment in the cap at the face of the column; this produces little torsion in the cap. On the other hand, torsion is produced with live load positioned for maximum bending moment in the column in the longitudinal direction of the bridge. However, the resulting combination of bending and torsion in the cap is less critical than the maximum-moment condition.

Example Design the cap for the single-column bent of Fig. 81. Lane live load is positioned for a reaction which will maximize the moment about the column. The 100-kip exterior-girder dead load shown in the figure includes the weight of curb and railing. The moments are:

$$
\begin{aligned}
LL + I &= 9 \times 100 = 900 \\
DL \text{ girder} &= 7.5 \times 120 = 900 \\
DL \text{ girder} &= 15 \times 100 = 1500 \\
DL \text{ cap} &= 7.5 \times 65 = \underline{490} \\
& 3790 \text{ ft-kips}
\end{aligned}
$$

For a rectangular section, with $f_s = 20,000$ psi and $f_c = 1200$ psi, and with tension reinforcement 6 in. below the top of the slab, the area of steel required is

$$
A_s = \frac{M}{f_s jd} = \frac{12 \times 3790}{20 \times 0.88 \times 54} = 47.8 \text{ in.}^2
$$

Use 12 No. 18 bars, $A_s = 48$ in.^{2}. Bar cutoffs can be determined from a moment diagram.

The allowable moment for the rectangular cross section with no reinforcement for compression is

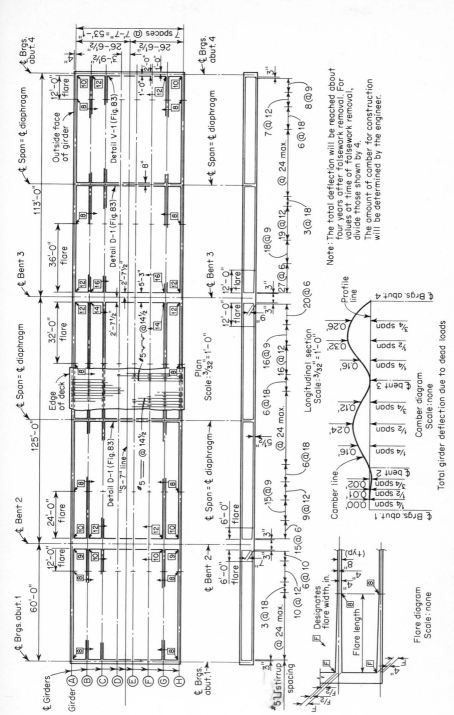

Fig. 76 Typical layout, box-girder bridge.

$$M = \frac{f_c k j b d^2}{2 \times 12} = \frac{1.2 \times 0.375 \times 0.875 \times 72 \times 54^2}{2 \times 12} = 3440 \text{ ft-kips}$$

Compression steel must provide $3790 - 3440 = 350$ ft-kips. From Table 6, for $d = 54$ in. and $d' = 3$ in., the allowable moment is 67.8 ft-kips/sq in. of compression steel. Therefore, $A'_s = 350/67.8 = 5.17$ in.^{2}. Use four No. 11 bars, $A'_s = 6.24$ in.^{2}.

Since the lane-loading concentrated load is 8 kips more for shear than for moment, the maximum shear is

$$
\begin{aligned}
LL + I &= 108 \\
DL \text{ girder} &= 120 \\
DL \text{ girder} &= 100 \\
DL \text{ cap} &= \underline{65} \\
&\ 393 \text{ kips}
\end{aligned}
$$

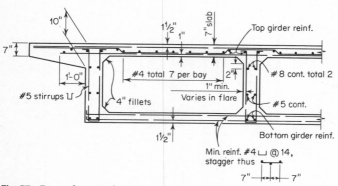

Fig. 77 Box-girder typical section.

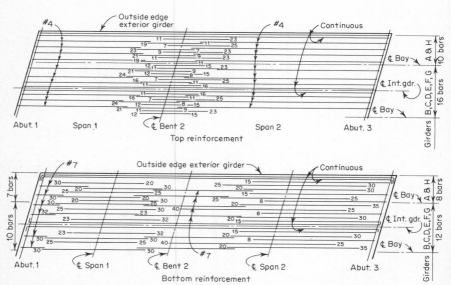

Note: All reinforcement #11 unless otherwise noted. Numbers at ends of bars indicate distance in feet from ₵ bent for top reinforcement or from ₵ span for bottom reinforcement

Fig. 78 Typical reinforcement, box-girder bridge.

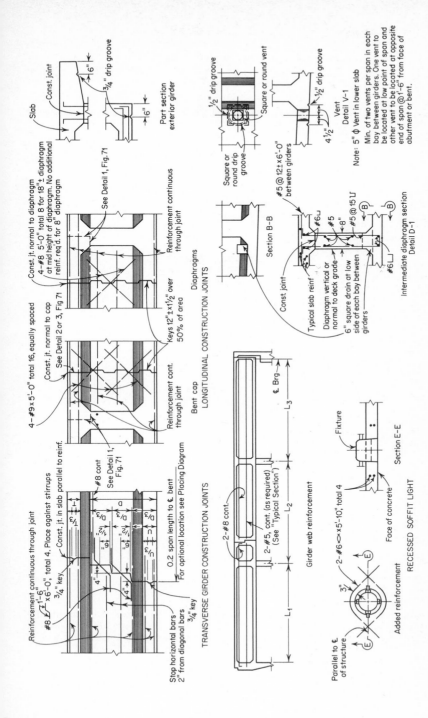

Fig. 79 Typical details, box-girder bridge.

18-85

$$v = \frac{V}{bd} = \frac{393,000}{72 \times 54} = 101 \text{ psi}$$

Assume No. 6 double stirrups as consistent with the size of the cap.

$$A_v f_v = 4 \times 0.44 \times 20,000 = 35,200 \text{ lb}$$
$$s = \frac{A_v f_v}{b(v - v_c)} = \frac{35,200}{72(101 - 52)} = 9.98 \text{ in.}$$

Uniform spacing of 10 in. throughout the cap is suggested. Details of the reinforcement are shown in section *AA* of Fig. 81.

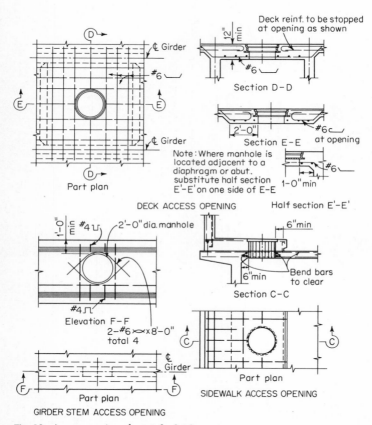

Fig. 80 Access openings, box-girder bridge.

Columns. Although tied columns have a lower allowable stress than do spirally reinforced columns, they are suitable for most bridges. Of course, where higher allowable stresses are required, the spirally reinforced columns may be used. The choice of type is not necessarily related to shape. Thus, the circular column may be designed as a tied column, using either individual hoops or a spiral, while the square column may have a spiral meeting the requirements for spirally reinforced columns.

All the various combinations of loading specified in AASHTO 1.2.22 should be investigated. Both transverse and longitudinal moments must be combined with the appropriate direct load. For purposes of design, the length of a column supported on a spread footing may be measured from the neutral axis of the box girder to the point where the column extends into the footing one-third of the depth of the footing. For a footing founded on piles, the length is usually measured to a point below the footing whose location depends upon the lateral resistance of the soil.

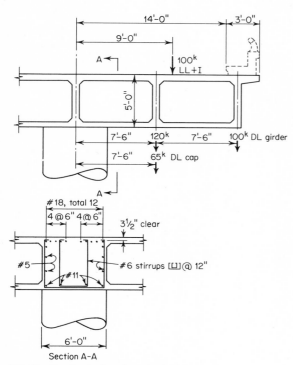

Fig. 81   Cap for single-column bent.

Where foundation conditions permit, round columns piles 3 to 6 ft or more in diameter can be used economically. These consist of piles cast in place in drilled holes, with a reinforced-concrete column to the superstructure. An example of this type of construction is shown in Fig. 82. The piles must extend a minimum of five diameters below the surface of the ground (but not less than 20 ft) and may reach 60 ft or more below ground. Tests show that the section of maximum moment is usually within two diameters of the surface. The point of effective fixity for determining column stiffness is usually one or two pile diameters below the surface. It can be approximated by

$$d_f = \frac{\sqrt{Ph}}{Q_a D}$$

where d_f = distance below ground to point of effective fixity, ft
P = lateral load on column, kips
h = height of P above ground, ft
Q_a = allowable bearing value of the soil, tons
D = diameter of pile, ft

The following design loads can be used for pile columns in bridges:

Diam, In.	Load, Kips
36	300– 600
48	500–1,000
60	700–1,400
72	900–1,800

The various load combinations (AASHTO 1.2.22) to be investigated for the design of the column of Fig. 82 are shown in Fig. 83. Load combinations IV, V, and VI are not tabulated because this column is at the center bent of a four-span structure, for which temperature moments are practically zero. Group III loading, at 125 percent basic allowable stress, governs. Assuming a tied column, analysis shows that 44 No. 11 bars are required at the

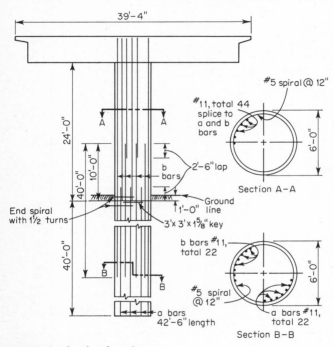

Fig. 82 Single pile-column bent.

lower end. Since the column is also fixed at the upper end, the difference in required reinforcement is small, so that all 44 bars are carried to the top. The bars can be cut off below ground as moment decreases. In this example, 22 bars are carried to the bottom of the pile.

PRESTRESSED-CONCRETE BRIDGES
(See also Sec. 12)

45. Standard Sections Standardization of prestressed-concrete bridge members has been established by several joint committees of the American Association of State Highway and Transportation Officials and the Prestressed Concrete Institute. Standards were developed for the following types: slabs for spans to 55 ft (Fig. 84, Table 9), beams for spans 30 to 100 ft (Fig. 85, Table 10), and box beams for spans up to 103 ft (Fig. 86, Table 11). These sections may be either pretensioned or posttensioned, and any acceptable type of prestressing may be used, i.e., straight or deflected strands, combination of pretensioning and posttensioning, etc. The span limits shown in the tables are approximate. They are based on 28-day compressive strength f'_c = 5000 psi, compressive strength at transfer of prestress f'_c = 4000 psi, and weight of concrete 150 pcf. Over 90 percent of the highway bridges built in the United States have spans within the range of these standards.

Slab Sections. Span limits for the standard slab sections are based on a 28-ft roadway and an allowance of 30 psf for surfacing. Straight tendons and AASHTO allowable stresses

By _____A.P.S._____ Job _____Ohio Street Overcrossing_____

Date _____4-15-63_____ Bent _____No. 3_____

Loading Group		Longitudinal		Transverse		Axial	
		Top of col	Bot of col	Top of col	Bot of col	Top of col	Bot of col
I DL		O	O	975	975	1,048	1,197
2 LL+I max. R.		O	O	1,640	1,640	205	205
	ΣM	O	O	2,615	2,615	1,253	1,402
II @ 125%	DL	O	O	975	975	1,048	1,197
Wind + overturning		O	O	652	2,001		
	ΣM	O	O	1,627	2,976	1,048	1,197
III @ 125% Group I			O		2,615		1,402
Overturning					213		
	30%W				387		
	WL				325		
	ΣM				3,540		1,402
IV @ 125% Group I							
	T						
	ΣM						
V @ 140% Group II							
	T						
	ΣM						
VI @ 140% Group III							
	T						
	ΣM						
VII @ 133⅓% DL		O		975		1,197	
	EQ		O		2,620		1,197
	ΣM		O		3,595		1,197

Fig. 83

are assumed. The 15-in.(minimum) end block requires mild-steel reinforcement sufficient to resist tensile forces resulting from the concentrated prestress forces. Diaphragms are recommended at midspan for spans to 40 ft and at the third points for spans exceeding 40 ft. Lateral ties equivalent to a 1¼-in. mild-steel bar tensioned to 30,000 lb, or the

TABLE 9   Properties of AASHTO-PCI Slab Sections (Fig. 84)

Section	Area, in.^{2}	Moment of inertia, in.^{4}	Section modulus, in.^{3}	Range of span, ft	
				HS20-44	H20-44
SI-36.............	432	5,180	864	20–28	20–29
SII-36.............	439	9,720	1,296	27–35	28–38
SIII-36.............	491	16,510	1,835	34–42	37–46
SIV-36.............	530	25,750	2,452	41–49	45–54
SI-48.............	576	6,910	1,152	20–28	20–29
SII-48.............	569	12,900	1,720	27–35	28–38
SIII-48.............	628	21,850	2,428	34–42	37–46
SIV-48.............	703	34,520	3,287	41–50	45–55

equivalent in high-strength tendons, are required at each diaphragm. The shear keys are filled with high-strength, nonshrinking mortar after the ties have been tightened.

Beam Sections.   Span limits for the standard beam sections are based on a cast-in-place deck 6 to 8 in. thick, with $f'_c = 3000$ psi. Dead load is assumed to be carried by the beams, and live load by composite action of beams and slab. With straight prestressing, spacing of these beams at the lower limits of span will be 8 to 9 ft (four beams for a 28-ft roadway) for HS20-44 loading. At the upper limits of span, with a sufficient number of tendons draped to compensate for dead load of the beams and to maintain allowable stresses at the ends,

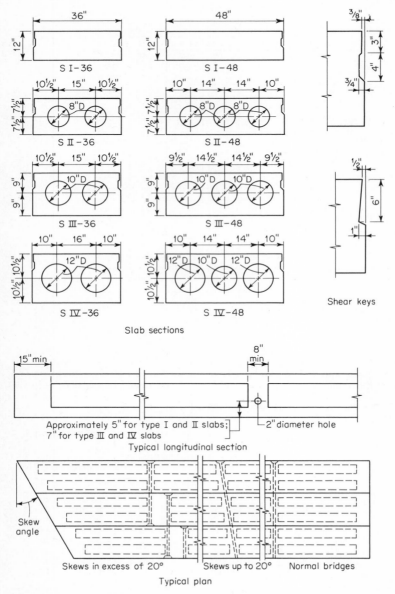

Fig. 84   AASHTO-PCI prestressed concrete slab sections. (See Table 9.)

spacing is about 4 ft 8 in. for beams I and II (seven beams for a 28-ft roadway) and 5 ft 6 in. (six beams for a 28-ft roadway) for beams III and IV.

End blocks are required for posttensioning of the standard beam sections. They should be as wide as the narrower flange of the beam, with a length equal to three-fourths the depth of the beam but not less than 24 in. A closely spaced grid of vertical and horizontal

TABLE 10   Properties of AASHTO-PCI Beam Sections (Fig. 85)

Section	Area, in.²	Moment of inertia, in.⁴	c_b, in.*	Range of span, ft
I	276	22,750	12.59	30–45
II	369	50,980	15.83	40–60
III	560	125,390	20.27	55–80
IV	789	260,730	24.73	70–100

* From center of gravity to bottom surface.

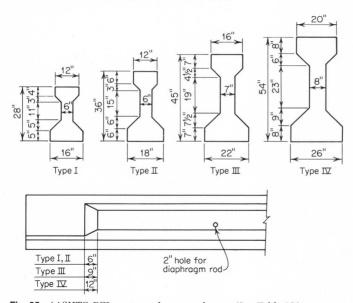

Fig. 85   AASHTO-PCI prestressed concrete beams. (See Table 10.)

bars is required at the end face of the end block, and throughout its length, for posttensioned beams. End blocks are not required if all tendons are pretensioned wires or seven-wire strand. Vertical stirrups at 20,000 psi, sufficient to resist at least 4 percent of the total prestressing force, are required in the length $d/4$ at each end of the beam for pretensioned beams.

Either precast or cast-in-place diaphragms, with either prestressed or nonstressed reinforcement, are recommended at the ends of the span, at midspan for spans to 60 ft, and at the third points for spans longer than 60 ft.

Box Beams.   Span limits for the standard box beams are based on a 28-ft roadway with an allowance of 30 psf for wearing surface. The 18-in. end blocks require mild-steel reinforcement sufficient to resist tensile forces resulting from the concentrated prestress forces. Diaphragms within the box are recommended at midspan for spans to 50 ft, at the third points for spans to 75 ft, and at the quarter points for spans longer than 75 ft. Lateral ties equivalent to a 1¼-in. mild-steel bar tensioned to 30,000 lb, or the equivalent in high-strength tendons, are required at each diaphragm. If adjacent units of the 39- and 42-in.

TABLE 11   Properties of AASHTO-PCI Box Sections (Fig. 86)

Section	Area, in.^{2}	Moment of inertia, in.^{4}	c_b, in.*	Span limit, ft	
				Draped strand	Straight strand
BI-36..........	560	50,330	13.35	74	62
BI-48..........	692	65,940	13.37	73	63
BII-36..........	620	85,150	16.29	86	73
BII-48..........	752	110,500	16.33	86	74
BIII-36.........	680	131,140	19.25	97	83
BIII-48.........	812	168,370	19.29	96	83
BIV-36.........	710	158,640	20.73	103	87
BIV-48.........	842	203,090	20.78	103	88

* From center of gravity to bottom surface.

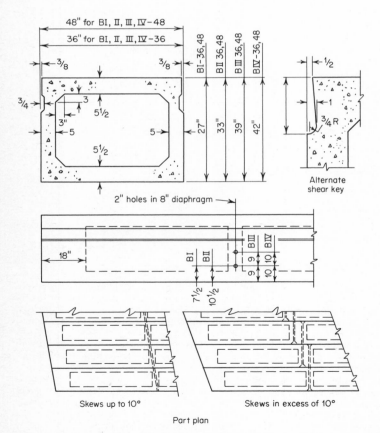

Fig. 86   AASHTO-PCI prestressed concrete box beams. (See Table 11.)

sections are tied in pairs, one tie, centered between the bottom of the beam and the bottom of the key, is permitted. The shear keys are filled with high-strength, nonshrinking mortar after the ties are tightened.

Combining Methods of Prestress.   Pretensioning and posttensioning can be used to advantage in combination in some cases. For example, a beam may be pretensioned for dead load only, and lifted from the bed and subsequently posttensioned to complete the operation. This procedure allows a more continuous use of prestressing beds, which may result in economy.

46. Stresses   Concrete for prestressed bridges should be preferably not less than 5000 psi. The AASHTO specifications permit strengths up to 6000 psi, which, in exceptional cases, may be increased. While economy may be realized with higher-strength concretes, limitations imposed by availability of suitable materials must be considered. The allowable stresses of the AASHTO specifications are the following (see Sec. 12 for notation):

Prestressing steel
1. Temporary stress before losses due to creep and shrinkage................. $0.70f_s'$
   Overstressing to $0.80f_s'$ for short periods of time may be permitted provided the stress after seating of the anchorage does not exceed $0.70f_s'$.
2. Stress at design load, after losses, $0.60f_s'$ or $0.80f_{sy}$, whichever is smaller.

Concrete
1. Temporary losses before losses due to creep and shrinkage:
   Compression
   Pretensioned members....................................... $0.60f_{ci}'$
   Posttensioned members...................................... $0.55f_{ci}'$
   Tension
   Members without nonprestressed reinforcement:
   Single element... $3\sqrt{f_{ci}'}$
   Segmental element.. Zero
   Members with nonprestressed reinforcement sufficient to resist tensile forces in the concrete without cracking when computed on the basis of an uncracked section:
   Single element... $6\sqrt{f_{ci}'}$
   Segmental element (within the element itself)............... $3\sqrt{f_{ci}'}$
2. Stress at design load after losses have occurred:
   Compression.. $0.40f_c'$
   Tension (in precompressed tensile zone)
   Posttensioned members...................................... Zero
   Pretensioned members (but not to exceed 250 psi)............. $3\sqrt{f_c'}$
3. Cracking stress:
   Modulus of rupture from tests or if tests not available............ $7.5\sqrt{f_c'}$
4. Anchorage bearing stress:
   Posttensioned anchorage (but not to exceed f_{ci}').................. $0.6f_{ci}'(A_c/A_b)^{1/2}$

Loss of prestress from all causes except friction may be assumed at 25,000 and 35,000 psi for posttensioned and pretensioned members, respectively. These losses consist approximately of the following:

Friction losses are discussed in Art. 49.

Item	Posttension	Pretension
Concrete shrinkage............	4,000	6,000
Concrete creep................	11,000	11,000
Steel creep...................	5,000	6,000
Elastic shortening............		12,000
Sequence stressing............	5,000	
Total, psi................	25,000	35,000

While these specifications allow prestressing steel stresses of $0.60f'_s$ or $0.08f_{sy}$, whichever is smaller, prestress losses will not always permit these values to be attained. For example, for wire with an ultimate strength $f'_s = 250,000$ psi and yield strength $f_{sy} = 200,000$ psi, the allowable stress is 150,000 psi. Deducting the prescribed prestress loss (excluding losses from friction) of 35,000 psi from the allowable initial stress of $0.7 \times 250,000 = 175,000$ psi leaves only 140,000 psi.

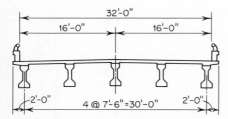

Fig. 87 Typical section of 70-ft composite prestressed girder bridge.

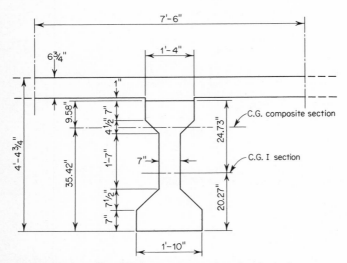

Fig. 88 Interior girder of 70-ft composite prestressed girder bridge.

47. Example Design the prestressed composite concrete girder for the structure shown in Fig. 87. Live load HS20-44, design span 70 ft. Use a Type III AASHTO PCI beam (Fig. 85). Dimensions of the composite girder are shown in Fig. 88.

$$LL \text{ per girder} = \frac{7.5}{5.5} = 1.36 \text{ wheel lines}$$

$$I = \frac{50}{L + 125} = \frac{50}{195} = 25.6 \text{ percent}$$

DECK SLAB. Design procedure for the slab is the same as that for the slab of Art. 11.
Girder:

DL girder (Table 10) 560 × 150/144	= 585 plf
DL slab (0.563 × 7.5 + 0.083 × 1.33)150	= 655 plf

DL curb and rail   = 100 plf
Girder DL moment = $0.585 \times 70^2/8$ = 360 ft-kips
Slab DL moment = $0.655 \times 70^2/8$ = 400 ft-kips
Curb and rail DL moment = $0.100 \times 70^2/8$ = 61 ft-kips
$LL + I$ moment (Table 1) = $1.36 \times 1.26 \times 986/2$ = 845 ft-kips

Properties of I section (Table 10):

$A = 560$ in.^{2} $y_b = 20.27$ in. $y_t = 24.73$ in.

$I = 125,390$ in.^{4} $\dfrac{I}{y_b} = 6185$ in.^{3} $\dfrac{I}{y_t} = 5070$ in.^{3}

$r^2 = 224$ in.^{2} $\dfrac{ey_b}{r^2} = 0.0905e$ $\dfrac{ey_t}{r^2} = 0.1104e$

Properties of composite T (Fig. 88):

Effective flange = $12 \times 6\frac{3}{4} + 16 = 97$ in.
                 = 90 in. center-to-center girders
Flange area = $90 \times 6.75 = 608$ in.^{2}
Flange moment of inertia = $90 \times \dfrac{6.75^3}{12} = 2310$ in.^{4}

	A	y	Ay	I
I.................	560	20.27	11,350	125,390
Flange...........	608	49.38	30,023	2,310
	1,168		41,373	127,700

$$y_b = \frac{41,373}{1168} = 35.42 \text{ in.}$$

I: $Ad^2 = 560 \times 15.15^2 = 128,530$
Flange: $Ad^2 = 608 \times 13.96^2 = \underline{118,490}$
$247,020 + 127,700 = 374,720$
$y_t = 45 - 35.4 = 9.6$ in.

$$\frac{I}{y_t} = \frac{374,720}{9.6} = 39,030 \text{ in.}^3$$

$$\frac{I}{y_b} = \frac{374,720}{35.4} = 10,585 \text{ in.}^3$$

STRESSES IN I

	Top fiber		Bottom fiber	
Girder DL...........	$12 \times \dfrac{360,000}{5,070} =$	850	$12 \times \dfrac{360,000}{6,185} =$	700
Slab DL.............	$12 \times \dfrac{400,000}{5,070} =$	950	$12 \times \dfrac{400,000}{6,185} =$	775
$LL + I$..............	$12 \times \dfrac{845,000}{39,030} =$	260	$12 \times \dfrac{845,000}{10,585} =$	960
Curb and rail........	$12 \times \dfrac{61,000}{39,030} =$	20	$12 \times \dfrac{61,000}{10,585} =$	70
		$f_t = \overline{2,080}$		$f_b = \overline{2,505}$

EFFECTIVE PRESTRESSING FORCE AFTER LOSSES. Estimated value of $e = 20.27 - 3.5 = 16.77$.

$$\frac{ey_t}{r^2} = 0.1104 \times 16.77 = 1.85$$

$$\frac{ey_b}{r^2} = 0.0905 \times 16.77 = 1.52$$

$$P_f = \frac{f_b A}{ey_b/r^2 + 1} = \frac{2505 \times 560}{2.52} = 556,700 \text{ lb}$$

AREA OF STEEL. Assume bar $f'_s = 145,000$ psi.

$$\begin{aligned}
\text{Allowable design-load stress} &= 0.6 \times 145,000 = 87,000 \text{ psi} \\
\text{Allowable initial stress} &= 0.7 \times 145,000 = 101,500 \text{ psi} \\
\text{Less losses} &= \underline{25,000} \\
\text{Allowable stress} &= 76,500
\end{aligned}$$

$$A_s = \frac{556,700}{76,500} = 7.28 \text{ in.}^2$$

INITIAL PRESTRESSING FORCE

$$P_i = 101,500 \times 7.28 = 739,000 \text{ lb}$$
$$\frac{P_i}{A} = \frac{739,000}{560} = 1320 \text{ psi}$$
$$\frac{P_f}{A} = \frac{556,700}{560} = 990 \text{ psi}$$

STRESSES IN CONCRETE

$$f_b = \frac{P}{A}\left(1 + \frac{ey_b}{r^2}\right) = 2.52\,\frac{P}{A} \qquad f_t = \frac{P}{A}\left(1 - \frac{ey_t}{r^2}\right) = 0.85\,\frac{P}{A}$$

At time of prestressing (+ = compression)

$$f_b = 1320 \times 2.52 - 700 = +2630 \text{ psi} \qquad f_t = -1320 \times 0.85 + 850 = -270 \text{ psi}$$

At time slab is placed

$$f_b = 990 \times 2.52 - 1475 = +1020 \text{ psi} \qquad f_t = -990 \times 0.85 + 1800 = +960 \text{ psi}$$

At design load

$$f_b = 990 \times 2.52 - 2.505 = -10 \text{ psi} \qquad f_t = -990 \times 0.85 + 2080 = +1240 \text{ psi}$$

Maximum allowable temporary compression for posttensioning $= 0.55 f'_{ci}$

$$\text{Required } f'_{ci} = \frac{2630}{0.55} = 4780 \text{ psi}$$

Maximum allowable design compression $= 0.40 f'_{ci}$

$$\text{Required 28-day compressive strength} = \frac{1240}{0.40} = 3100 \text{ psi}$$

Maximum allowable tension in concrete $= 3\sqrt{f'_{ci}} = 3\sqrt{4780} = 210 \text{ psi}$

At time of prestressing the tension is 270 psi, so that mild-steel reinforcing at 20,000 psi should be provided to resist the total tension (Fig. 89).

$$T = 270 \times 4.4 \times \text{\small16}\!\tfrac{1}{2} = 9500 \text{ lb}$$
$$A_s = \frac{9500}{20,000} = 0.48 \text{ in.}^2$$

Use two No. 5 bars.

48. Path of Prestressing Force In simple-span girders it is desirable for the tendons to follow a parabolic path. Minimum prestressing steel results when it is placed as near as possible to the bottom of the member. The effective depths that can be achieved with various prestressing systems vary. If a system is not specified, the prestressing force required for several different eccentricities can be shown on the plans—usually, 3, 4, and 5 in. above the bottom of the girder will cover all systems. At the ends of the span the tendons should be at or near the center of gravity of the beam or, at least, within the kern, to eliminate tensile stresses at the beam ends.

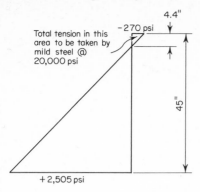

Total tension in this area to be taken by mild steel @ 20,000 psi

−270 psi

4.4"

45"

Fig. 89 Concrete tension at time of prestressing.

+2,505 psi

49. Friction Losses Frictions losses in posttensioned members occur with angle changes in draped cables and from wobble of the ducts. According to AASHTO these losses can be accounted for by the formula

$$T_0 = T_x e^{(KL + \mu \alpha)}$$

or, if $KL + \mu \alpha \lesssim 0.3$, the following may be used

$$T_0 = T_x (1 + KL + \mu \alpha)$$

where μ = friction curvature coefficient
α = angle between T_0 and T_1, radians
K = wobble coefficient per ft
T_0 = steel stress at jacking end
T_x = steel stress at any point x

Values of K and μ recommended by AASHTO are given in Table 12. A nomograph for the

TABLE 12   Values of K and μ for Friction Losses*

Type of duct	Wire or ungalvanized strain		High-strength bars	
	K	μ	K	μ
Bright-metal sheathing	0.0020	0.30	0.0003	0.20
Galvanized-metal sheathing	0.0015	0.25	0.0002	0.15
Galvanized rigid	0.0002	0.25		
Greased or asphalt-coated and wrapped	0.0020	0.30		

*From Ref. 3. See also Sec. 12, Table 10.

solution of the first of these equations is given in Fig. 90. For parabolic tendons, the angle α between the tangent at the vertex and the tangent at any point x from the vertex can be found from $\alpha = \tan \alpha = 2y/x$, where y = rise of the tendon in the distance x.

Example   Determine the jacking force required for the girder of Art. 47. Assume that jacking will be done from one end. Average values of K and μ for high-strength bars in bright-metal sheathing are 0.0003 and 0.20, respectively (Table 12). With the tendon 3½ in. from the bottom at midspan and passing through the center of gravity of the I at each end, the rise is $20.27 - 3.50 = 16.77$ in (Fig. 88). The angle α between tangents at each end is

$$\alpha = 2 \times \frac{2y}{x} = 2 \times 2 \times \frac{16.77}{35} \times 12 = 0.16 \text{ radian}$$
$$\mu \alpha + KL = 0.20 \times 0.16 + 0.0003 \times 70 = 0.053$$
$$e^{\mu \alpha + KL} = 1.054$$

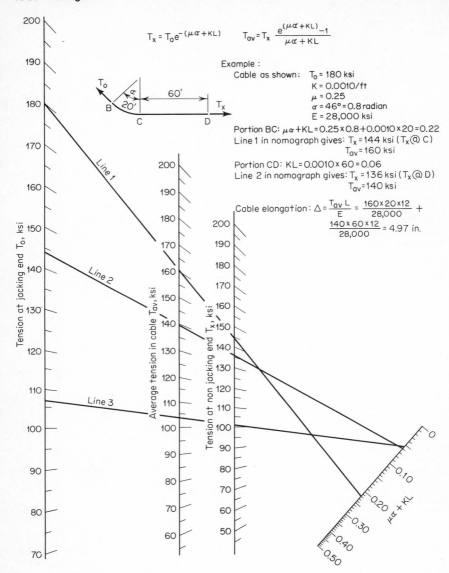

Fig. 90 Nomograph for frictional loss in prestress.

The allowable initial tension is $T_i = 101.5$ ksi (Art. 47).

$$T_0 = 101.5 \times 1.054 = 107 \text{ ksi}$$
$$\text{Allowable jacking tension} = 0.75 \times 145 = 109 \text{ ksi}$$
$$T_{av} = 101.5 \times \frac{0.054}{0.053} = 103.7 \text{ ksi}$$
$$\text{Bar elongation} = \frac{T_{av}L}{E} = \frac{103.7 \times 70 \times 12}{24{,}000} = 3.6 \text{ in.}$$

Line 3 in Fig. 90 shows the nomographic determination of T_0 and T_{av}.

The bar length was assumed 70 ft in calculating elongation. The draped length can be approximated by $L[1 + \frac{2}{3}(2d/L)^2]$ where d is rise of the tendon.

50. Ultimate Load The AASHTO specifications (1.6.6) require that prestressed beams on simple spans of moderate length have an ultimate load capacity not less than $1.3[DL + 1.67(LL + I)]$. The ultimate resisting moment M_u of underreinforced beams can be closely approximated by $M_u = 0.9A_s f_s' d$. For the beam of Art. 47,

$$M_u = 0.9 \times 7.28 \times 145 \times \frac{49.25}{12} = 3900 \text{ ft-kips}$$

The required ultimate moment is

$$1.3(760 + 1.67 \times 845) = 2822 < 3900 \text{ ft-kips}$$

51. Web Reinforcement The AASHTO specifications (1.6.13) recommend that, for beams carrying moving loads, the shear be investigated only in the middle half of the span, with the web reinforcement required at the quarter points used throughout the outer quarters. The area A_v of web reinforcement is determined by

$$A_v = \frac{(V_u - V_c)s}{2f_{sy}jd} \text{ but not less than } \frac{100b's}{f_{sy}}, \text{ with } f_{sy} \lesssim 60{,}000 \text{ psi}$$

For the beam of Art. 47,

DEAD-LOAD SHEAR

Weight of girder, slab, curb, and rail = 1.34 klf
$V = 0.5 \times 1.34 \times 35 = 23.4$ kips at quarter point
At the quarter point $y = 16.77(\frac{1}{2})^2 = 4.19$ in.
$\alpha = 2 \times \dfrac{4.19}{12 \times 17.5} = 0.04$
Vertical compoment of prestress force = $557 \times 0.04 = 22.2$ kips
Net dead-load shear = $23.4 - 22.2 = 1.2$ kips

LIVE-LOAD SHEAR

$$\left(52.5 + 38.5 + \frac{24.5}{4}\right) 16 \times \frac{1.36}{70} = 30.2$$

$$\frac{50}{52.5 + 125} = 0.28; 0.28 \times 30.2 = 8.5$$

$$LL + I = \overline{38.7} \text{ kips}$$
$$V_u = 1.5 \times 1.2 + 2.5 \times 38.7 = 98.6 \text{ kips}$$

STIRRUPS. The shear taken by the concrete is limited by AASHTO to $V_c = 0.06f_c'b'jd$, but not to exceed $180b'jd$.

$$V_c = 0.180 \times 7 \times 0.9 \times 49.25 = 55.8 \text{ kips}$$
$$A_v = \frac{(98.6 - 55.8)s}{2 \times 40 \times 0.9 \times 49.25} = 0.012s$$

Use No. 4 stirrups, $A_v = 0.40$, $s = 0.40/0.012 = 33$ in.

$$\text{Maximum spacing} = \frac{f_{sy}A_v}{100b'} = \frac{40{,}000 \times 0.40}{100 \times 7} = 22.9 \text{ in.}$$
$$\text{Maximum spacing} = \frac{1}{2}d = \frac{1}{2} \times 49.25 = 24.6 \text{ in.}$$

Although the allowable spacing is 23 in., it is good practice to provide a transition in spacing from the end block toward the center; the end block itself requires closely spaced vertical reinforcement throughout to control the formation of horizontal cracks which may develop because of a nonuniform distribution of the prestressing force on the end block. Details of the girder reinforcement are shown in Fig. 91.

To complete the investigation for shear, principal tensile stresses should be computed at the bottom of the top fillets at the end of the end block and at the quarter point (Sec. 12, Art. 25). These principal tensions must not exceed $0.03\sqrt{f_c'}$ at design load or $0.08\sqrt{f_c'}$ at ultimate load.

52. Uplift Uplift due to prestressing can be computed by the formula (Fig. 92)

$$\Delta_P = \frac{5Pe_2L^2}{48EI} + \frac{Pe_1L^2}{8EI}$$

where e_1 is positive as shown in the figure. For the beam of Art. 47, $e_1 = 0$ and the upward deflection is

$$\Delta_P = \frac{5 \times 739{,}000 \times 16.75 \times (12 \times 70)^2}{48 \times 3{,}800{,}000 \times 127{,}700} = 1.87 \text{ in.}$$

The deflection due to the dead load of the girder is

$$\Delta = \frac{5ML^2}{48EI} = \frac{5 \times 360{,}000 \times 70^2 \times 12^3}{48 \times 3{,}800{,}000 \times 127{,}700} = 0.65 \text{ in.}$$

Net uplift at time of prestressing is $1.87 - 0.65 = 1.22$ in.

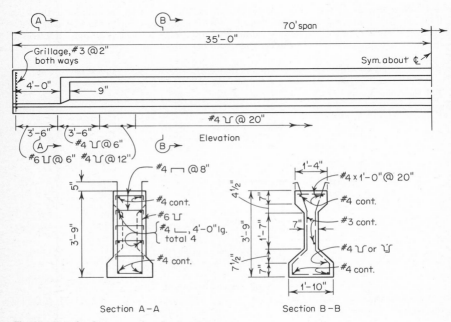

Fig. 91   Details of prestressed girder for 70-ft span.

Assume that the prestress force has reduced from 739,000 to 600,000 lb at the time of placing the slab. Assume the coefficient of creep = 1.5. Then

$$\Delta = 1.5 \, (1.87 \times 600/739 - 0.65) = 1.31 \text{ in.}$$

Upon placing of the slab the girder will deflect

$$\Delta = 0.65 \times 655/585 = 0.73 \text{ in.}$$

Net deflection of the girder is $1.31 - 0.73 = 0.58$ in.

When a large number of similar beams are to be used in a project, periodic deflection measurements should be made on the first girders to check the assumed modulus.

53. Live-Load Deflection   For the purpose of computing live-load deflections, the AASHTO specifications allow the live load to be distributed equally among the girders if the span has adequate diaphragms to ensure lateral distribution. Present information indicates that if the ratio of span to width is 2 or more, this assumption is safe. There is usually no difficulty in satisfying allowable live-load deflection requirements with composite prestressed girders whose depth span ratio is 0.05 or more.

Example Compute the $LL + I$ deflection for the girder of Art. 47. Assuming the live load distributed equally to all girders, the number of lanes per girder is $\frac{2}{5} = 0.4$. Then, per girder,

$$LL + I = 0.40 \times 1.26 \times 986 = 496 \text{ ft-kips}$$

Computing the deflection as outlined in Art. 23, the correction factor for this moment is 1.14.

$$M = 1.14 \times 496 = 565 \text{ ft-kips}$$

Equivalent concentrated load $P = 4 \times 565/70 = 32.3 \text{ kips}$

$$\Delta = \frac{PL^3}{48EI} = \frac{32.3 \times (12 \times 70)^3}{48 \times 3800 \times 374{,}720} = 0.28 \text{ in.}$$

Allowable deflection $\Delta = 12 \times 70/800 = 1.05 \text{ in.}$

In the calculations above, the modulus of elasticity was assumed to be the same as that for computing dead-load deflection. If the predicted live-load deflection is excessive, consideration should be given to the probable increase in modulus of elasticity which will have developed by the time live load is applied.

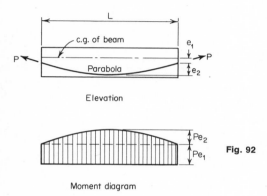

Fig. 92

BRIDGE RAILINGS

Highway railings are built to either deflect, decelerate and contain a vehicle, or withstand impact with little deflection. Since bridge railings must protect structural members as well as prevent vehicles from going over the side, they must withstand impact without appreciable deflection. As a result of a number of full-scale crash tests conducted on a wide variety of railing types, AASHTO established in 1964 static-load requirements for bridge railings which appear to be adequate to contain most present-day vehicles. Four different types of railings and loading conditions are illustrated in Fig. 93.

54. Railing Design Railings may be made of any material strong enough to withstand the load and ductile enough not to shatter or explode under load but to yield in a shock-absorbing, bending failure. To this end, it is customary to require that the metal portions of railings exhibit 10 percent ductility at failure. Rails, posts, and concrete barriers should reamin intact and not break away, possibly to fall on a roadway below.

It is desirable to have some continuity in the railing, but from the standpoint of maintenance it should be possible to remove damaged sections easily without dismantling the entire railing. One method of accomplishing this is to fasten one end of each horizontal rail to a post and make the connection to the next post with a sliding joint. This gives good continuity and yet enables individual sections to be replaced by merely loosening one adjacent section. This semicontinuous condition is recognized by specifying that the horizontal members be designed for a moment of $PL/6$, where L = spacing of posts. In the AASHTO specifications, $P = 10{,}000$ lb and $w = 50$ plf (Fig. 93).

Parapet-type railings, being uniformly supported along the deck, are easiest to anchor. It is extremely difficult to anchor a post to the deck adequately when all the loads for a span of 10 ft or so must be carried by the anchoring device. The deck slab is often too thin to provide adequate anchorage for embedded bolts, and there is often insufficient edge distance between the post and the edge of the slab to develop the required resistance. Even with the continuous concrete-barrier type the designer should make certain that anchorage is adequate.

It is desirable to provide a smooth contact face so that vehicles will rub along the rail and not be caught or pocketed by the posts. Pocketing the rail exerts extreme deceleration

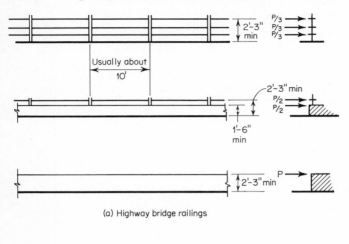

Usually about
10'

(a) Highway bridge railings

(b) Sidewalk railing
(Where there is no barrier at edge of roadway)

Fig. 93 Bridge railings.

forces on the vehicle and causes severe damage to the post. Therfore, the post should not project in front of the rubbing face of the horizontal railing. The most satisfactory bridge railing developed to date is shown in Fig. 94a. Because of the sloping front face, a vehicle approaching at a flat angle will be lifted and tilted away from the rail so the fenders usually will not rub. At angles under 15° or so the driver will usually not lose control and the vehicle will be directed back into a course parallel to the railing and may escape with no damage whatsoever. This railing is widely used in the United States. The barrier shown in Fig. 94b is used for narrow medians.

55. Curbs and Sidewalks The effectiveness of curbs or rubbing strips on the face of the railing varies somewhat with the design of vehicles. The AASHTO specifications require that the face of a curb be at least 9 in. from the face of the railing. Tests made with vehicles approaching at flat angles show that a curb projecting 5 in. or more from the face of the railing will act as a springboard, raising the car so that it strikes the railing higher or even goes over it completely. Curbing has been largely discredited as a means of keeping a high-speed vehicle in the roadway. It is safer for present-day vehicles to contact a strong barrier at least 27 in. above the roadway surface. Therefore, wherever a curb is used, the railing should be at least 27 in. above it. On structures where the face of the bridge railing is in line with the outside edge of the roadway shoulder, curbs can be omitted.

56. Pedestrian Railings A pedestrian railing used in connection with a highway structure should be impossible to crawl through or under and as difficult as possible to climb. It should be narrow and rounded on top so as to be nearly impossible to walk upon. In many cases it has been found necessary to place a high chain-link fence along the railing, curving it back over the sidewalk to prevent things from being thrown onto the roadway below. Where the structure is a pedestrian overcrossing, a cage over the entire walkway is desirable. In the case of a full cage, it should be made round or pointed on top to prevent children from using it as a trampoline. A heavy, pointed, steel picket railing curving back over the sidewalk has been found to be effective on high bridges which invite suicides.

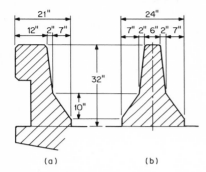

Fig. 94 (*a*) Bridge rail; (*b*) median barrier.

A pedestrian railing should be not less than 36 in. high (42 in. is preferred). It should have balusters spaced with openings not to exceed 5 in. Horizontal rails facilitate climbing and are to be avoided on pedestrian railings. An additional rail placed above an adequately designed traffic railing at least 27 in. high is classed as an ornamental rail and is designed for sidewalk loading.

Part 2. Steel-Plate-Deck Bridges

ROMAN WOLCHUK

Partner, Wolchuk and Mayrbaurl, Consulting Engineers, New York, N.Y.

57. Applications The use of the steel deck as an integral part of the bridge superstructure has gained general acceptance in long-span bridge construction. In this type of design the steel-deck plate, the stiffening ribs, and the transverse floor beams are integrated into one deck structure, with the deck plate as the common top flange. The deck structure, in turn, acts as a part of the main bridge system, in most cases as the top flange of the main girders or trusses (Fig. 95).

The behavior of a steel deck stiffened by closely spaced parallel ribs is comparable with that of a plate having different rigidities in the two perpendicular directions, and in the design of decks with closed ribs it is advantageous to idealize the deck system as such a plate. For these reasons the terms "orthotropic steel plate deck" (from *ortho*gonal-aniso*tropic*) or "orthotropic deck" have often been used to designate structures of this type. However, it is important to distinguish clearly between the mathematical model used in the analysis and the actual structure.

Steel decks are most often used in conjuction with box-girder bridges. An example of the long spans that have been built is the 984-ft span of the Rio-Niteroi Bridge, Brazil (Fig. 96). Longer spans can be attained by the use of cable-stiffened systems. In truss bridges steel decks have been used as top chords of deck trusses or (less often) bottom chords of through structures. The Port Mann Bridge near Vancouver, Canada, is an example of steel-deck design in arch bridges.

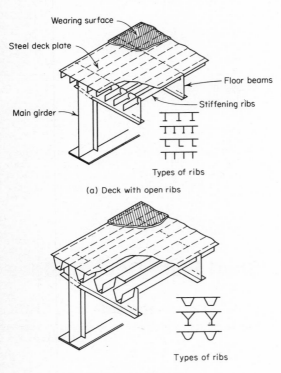

Fig. 95   Elements of steel-deck bridges.

In suspension bridges the steel deck may act as the upper flange of stiffening girders of conventional type (Cologne-Muelheim Bridge) or may form a part of a box section, which provides great flexural and torsional rigidity (Severn River Bridge).

Railroad bridges with steel decks may be designed with standard ballast and timber ties, or with the rails mounted directly on the deck. Significant saving in weight is achieved in both cases because the deck acts as a fully integrated member of the bridge. Other advantages are low structural depth and quick erection.

In the design of movable bridges (lift or bascule type) steel decks are valuable for the dead-weight saving and the rigidity of the superstructure. Steel-deck movable trestles, consisting of prefabricated units, have been found useful for temporary rerouting of traffic during roadway and bridge construction work in congested urban areas.

Examples of steel-deck structures are given in Table 13.

58. Economic Considerations Economy of steel-deck bridge construction is due to savings in total dead weight and in weight of steel. In the three cases shown in Table 14

TABLE 13 Some Major Steel-Deck Bridges

Bridge	Country	Year	Type	Spans, ft	Type of rib
1. Cologne-Muelheim	Germany	1951	Suspension (plate girder)	279-1033-279	Open
2. Save, Belgrade	Yugoslavia	1956	Plate girder	246-856-246	Open
3. Cologne-Severin	Germany	1959	Cable-stiffened box girder	161-292-157-990-492-172	Open
4. Firth of Forth Road	Great Britain	1964	Suspension (truss)	1340-3300-1340	Closed
5. Port Mann, Vancouver	Canada	1965	Arch	360-1200-360	Closed
6. Concordia, Montreal	Canada	1965	Box girder	340-525-525-525-340	Closed
7. Severn River	Great Britain	1966	Suspension (box girder)	1000-3240-1000	Closed
8. Poplar St., St. Louis	U.S.A.	1967	Box girder	300-500-600-500-265	Closed
9. San Mateo-Hayward	U.S.A.	1967	Box girder	375-750-375*	Open
10. Duisburg	Germany	1970	Cable-stiffened box girder	613†-1148-789‡	Open
11. Winningen	Germany	1973	Box-girder	516-716-559-479-439-356	Closed
12. Bosporus	Turkey	1973	Suspension (box girder)	758§-3523-837†	Closed
13. Rio-Niteroi	Brazil	1974	Box girder	656-984-656	Closed
14. St. Nazaire	France	1975	Cable-stiffened box girder	518-1325-518	Closed
15. Humber River	Great Britain	Under construction	Suspension (box girder)	920-4626-1395	Closed
16. Luling	U.S.A.	Under construction	Cable-stiffened box girder	260-495-1235-495-260	Closed

*Channel spans only.
†Total length of four spans.
‡Total length of three spans.
§Total length of five spans.

conventional steel bridges destroyed during World War II were replaced by new structures using steel decks.

In continuous-girder bridge design in the center-span range of 300 to 600 ft the use of a steel deck may result in a steel weight saving of the order of 15 to 30 percent, compared with conventionally designed girder bridges with concrete decks. Steel weight saving in

TABLE 14   Weight of Steel and Total Dead Weight of Old and New Designs

Structure	Span, ft	Year	Type	Total weight steel, tons	Total dead weight of super-structure, tons
Cologne-Muelheim (Germany)	279-1,033-279	1929	Suspension	14,150*	20,000
		1951	Suspension	6,400	9,500
Duesseldorf-Neuss (Germany)	338-676-338	1930	Truss	9,300	22,000
		1951	Box girder	6,970	8,400
Save River, Belgrade (Yugoslavia)	246-856-246	1934	Suspension	7,500*	13,000
		1956	Girder	4,200	5,200

* Includes weight of towers and cables.

girder and other types of bridges increases considerably as the span lengths increase, since the dead weight becomes a governing factor in the design.

A further cost saving in steel-deck construction results from the cost differential between a concrete deck and a wearing surface on steel deck. The cost of the latter is likely to be less than one-half the cost of a concrete deck.

Additional factors that contribute to overall economy are possible reduction of structural depth, as compared with conventional structures, and increased erection speed due to prefabrication of large deck units in the shop. Construction efficiency is further enhanced by the fact that concrete construction is confined to one operation—the erection of the substructure.

Substructure cost saving may be due either to smaller loads on the individual footings because of a substantially reduced dead load of the superstructure, resulting in reduced footing dimensions or number of piles, or to a reduction of the number of supports and their placing at convenient locations as longer superstructure spans become economically feasible.

For shorter-span bridges the economy of steel-deck construction is limited by the minimum steel weight of about 40 to 45 psf of deck, due to the required minimum sizes of deck plate, ribs, and supporting members. This and the higher unit cost of fabrication make this system unlikely to be economical for spans shorter than 300 to 400 ft. However, special conditions requiring light weight, low depth, and erection efficiency may make the use of steel-deck construction advantageous even for shorter spans.

59. Structural Behavior   Although the functions and the resulting stresses of the component parts of a steel-deck bridge are closely interrelated, it is necessary for design purposes to treat separately the three basic structural systems, as follows:

System I. The main bridge system, with the steel deck acting as a part of the main carrying members of the bridge. In the computation of stresses in this system the effective cross-sectional area of the deck (including longitudinal ribs) is considered as flange. Design criteria for box girders as main members are discussed in Arts. 66 to 72.

System II. The stiffened steel-plate deck (acting as bridge floor between the main members) consisting of the ribs, the floor beams, and the deck plate as their common upper flange. The stresses in this system in the ribs and floor beams due to wheel loads are computed by methods discussed in Arts. 62 to 65.

System III. The deck plate, acting in local flexure between the ribs, transmitting the wheel loads to the ribs. The local stresses in the deck plate act mainly in the direction perpendicular to the supporting ribs and floor beams and do not add directly to their other stresses.

The governing stresses in the design of the deck are obtained by superposition of the effects of Systems I and II (Fig. 97).

An important characteristic of steel decks is their capacity for carrying concentrated loads exceeding many times the values computed on the basis of a linear load-stress

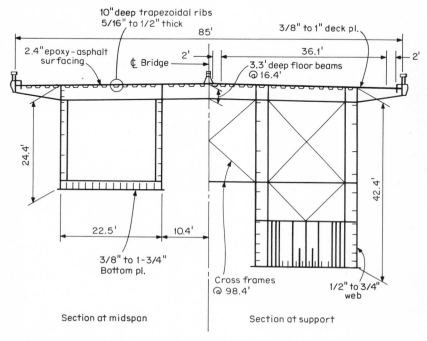

Fig. 96   Cross section of Rio-Niteroi Bridge, Rio de Janeiro, Brazil.

relationship. Tests made on decks with open and with closed ribs have shown that stresses and deflections of the ribs under working loads agree fairly well with the values computed on a linear basis. However, as the loads are increased, the relationship becomes nonlinear and the stresses increase at a smaller rate than the loads because of the partial membrane action of the deck, the latter being still in the elastic range. Furthermore, reaching yield stress at a critical location under load does not result in a marked flattening of the load deflection curve; this is because of the redistribution of the stresses to the adjoining areas of the deck. The ratio of the actual strength of the rib-stiffened deck to the computed allowable-load capacity is of the order of 10 to 40 in the tests.

While practical utilization of the high static strength of the deck is limited by undesirable deformations, it is obvious that stresses in the deck at design loads computed by elastic analyses are not an index of its actual safety. Therefore, the usual values of allowable stresses of existing design codes, based on a prescribed factor of safety against yield of simple tension specimens, are not appropriate in the design of steel-deck members. It also follows that undue refinements of the design computations by the currently available methods of analysis, all of which are based on the ordinary flexural theory, are not warranted.

60. Deck Plate   A deflection criterion limiting the live-load deflection of the deck plate to 1/300 of the rib spacing has often been used in determining deck-plate thickness. Based

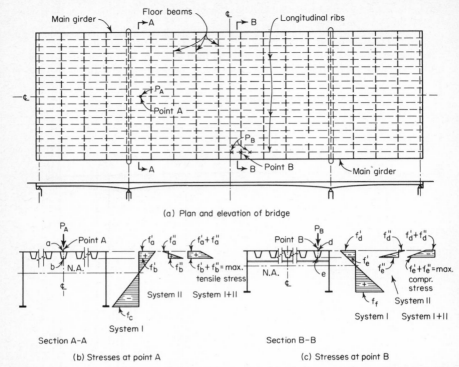

(a) Plan and elevation of bridge

Section A-A

(b) Stresses at point A

Section B-B

(c) Stresses at point B

Fig. 97 Superposition of the overall stresses (System 1) and local rib stresses (System II) in the bridge deck.

on this criterion, the minimum thickness t_p may be determined by the semiempirical formula by Kloeppel as

$$t_P \geqslant 0.07a \sqrt[3]{p} \qquad (1)$$

where a = spacing of deck-plate supports, in.

p = pressure under design wheel, psi

For the usual spacing of supports of about 12 in. this formula results in a plate thickness of $\frac{3}{8}$ in., which is the minimum practical thickness because of welding distortions. The AASHTO[3] design provisions for steel deck plate, based on fatigue considerations, require that the maximum local transverse bending stress caused in the deck due to the specified wheel load not exceed 30 ksi. This criterion results in deck thickness about the same as that by Eq. (1). A somewhat heavier thickness may be desired for satisfactory performance of the wearing surface and for lowering the stress in the fatigue-sensitive webs of the closed ribs at their juncture with the deck plate. The possibility of greater economy by increasing both the deck-plate thickness and the rib spacing to reduce fabrication work may also be considered.

In the design of the deck plate and the ribs composite action of the wearing surface with the steel plate is usually disregarded, although this effect considerably reduces the local stresses and the deflections of the deck plate and the ribs.[5,9]

Further criteria for deck-plate thickness are provided by the required cross-sectional area of the deck acting as the flange of the main bridge members, and by fabrication and erection requirements. Deck-plate corrosion does not seem to be a critical consideration.[7]

61. Rib Criteria   Within the conventional allowable-stress approach the characteristic structural behavior of the ribs has been considered in design either by applying a reduction factor to the stresses of System II prior to adding them to those of System I, and requiring that the sum not exceed the customary allowed values, or by using higher allowable stresses for the combination of System I and System II effects. The second method is used in the AASHTO specifications,[3] which permit a 25 percent increase of the allowable stresses for ribs under combined System I and System II effects.

The high peak values of stresses under working loads obtained by superposition of the theoretical maximum values of the overall and local stresses in the deck would be objectionable only if they affected the fatigue strength of the deck members. However, the maximum values of the System I stresses in conjunction with the peak values of System II stresses may seldom be reached, and the actual maximum stresses in the ribs are, as a rule, considerably below the fatigue level.

In recognition of their good corrosion resistance AASHTO permits the use of $^3\!/_{16}$-in. plate material for closed ribs.

In the design of the deck plate and the ribs in accordance with AASHTO either the 12-kip wheel load or two 8-kip wheels spaced 4 ft apart may be used, whichever produces the larger stress. An impact factor of 30 percent is specified for the deck plate. In the design of the floor beams and in computations of the effects of floor-beam flexibility the 8-kip and the 32-kip axle loads are used.

Wheel-load dimensions, $2g \times 2c$ (Fig. 98), at the top surface of the steel plate, are

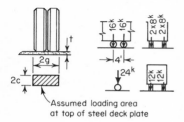

Assumed loading area
at top of steel deck plate

Fig. 98 Design load recommended for steel deck and ribs.

specified as follows: $(20 + 2t) \times (8 + 2t)$ in. for the 8-kip and 12-kip wheels and $(24 + 2t) \times (8 + 2t)$ in. for the 16-kip wheels, where t is the thickness of the wearing surface in inches.

OPEN-RIB DECKS

62. Design of Ribs Open stiffening ribs are usually flat bars. Inverted tees, angles, and bulb sections have also been used (Fig. 99). Open ribs are simple to fabricate and are easily spliced in the field. The disadvantage of this type is the large surface area of deck and ribs exposed to corrosion and the difficulty of maintaining a good paint thickness, especially at the lower edges of flat bar ribs and the hidden surfaces of the tee and angle ribs.[9] The economic span range of open ribs is 5 to 10 ft.

The longitudinal rigidity of a deck stiffened with open ribs is much greater than the rigidity of the deck plate alone in the direction perpendicular to the ribs, so that lateral distribution of concentrated wheel loads is very small. Therefore, the load is carried almost entirely by the directly loaded ribs. Because of the deflection of the floor beams, each rib may be regarded as a continuous beam on elastic supports.

In the design procedure developed by Pelikan and Esslinger[6] bending moments in the ribs are computed in two steps: (1) the ribs are treated as continuous beams on rigid supports and the bending moments computed accordingly; (2) bending-moment corrections due to the elastic flexibility of the floor beams are determined.

Ribs on Rigid Supports.   The load on one rib due to a wheel load directly over it is determined as the reaction of the deck plate assumed continuous over rigid ribs. This assumption is conservative.

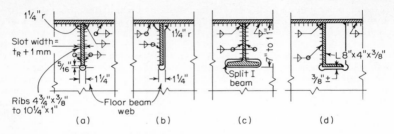

Open ribs : (a) Save River Bridge. (b) Arrangement minimizing residual welding stresses (c) Duesseldorf–Neuss Bridge. (d) Duesseldorf North Bridge.

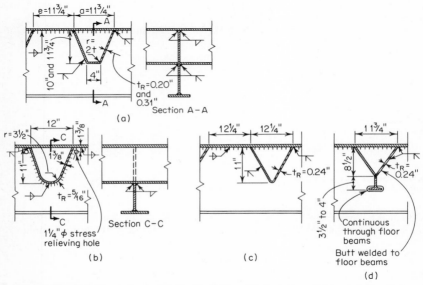

Closed ribs : (a) Mannheim–Ludwigshafen Bridge. (b) Port Mann Bridge. (c) Haseltal Bridge. (d) Fulda River Bridge.

Fig. 99 Details of open and closed ribs.

The positive and negative bending moments in the ribs may be determined by evaluation of influence lines for a continuous beam on rigid supports or may be obtained directly from charts given in Ref. 5.

Effects of Floor-Beam Flexibility.   In a system with rigid floor beams only the loaded ribs are subject to flexure. However, if the floor beams deflect, all ribs in the deck are affected, and even ribs over which there is no load are stressed. This effect is greatest near the middle of the deck, where the floor-beam deflection is maximum, and decreases gradually toward the main girders, where the floor-beam deflection is zero.

The magnitude of the effect of floor-beam flexibility on the rib moments depends on the relative rigidity of the ribs and the floor beams, which is expressed by a coefficient

$$\gamma = \frac{l^4 I_R}{a s^3 \pi^4 I_F} \tag{2}$$

where I_R = moment of inertia of one rib, including effective width of deck plate
  I_F = moment of inertia of one floor beam, including effective width of deck plate
  a = rib spacing
  s = floor-beam spacing
  l = span of floor beam

In the case of continuous floor beams an equivalent simple span length l may be used as an approximation.

Equation (2) shows that decks with closely spaced, relatively slender floor beams have large values of the rigidity coefficient γ. In such cases the effects of floor-beam flexibility are considerable. On the other hand, if the floor beams are relatively stiff and are spaced far apart (as may be the case in decks with closed ribs), the value of γ and the effects of floor-beam flexibility are small. Generally, if γ is less than about 0.01, floor-beam flexibility may be disregarded.

Treatment of the ribs as beams on elastic supports is possible only if the deflection of the floor beam under each rib is proportional to the rib reaction. For the usual case of simply supported floor beams this condition is satisfied only if the load on the deck is sinusoidal over the entire width of the deck. Therefore, the actual loading must be represented by a Fourier series. Formulas for such analysis for various actual loading configurations may be found in Ref. 5.

For practical design purposes it is sufficient to consider only the first sinusoidal component of the actual loading. This substitute loading is then used to compute the desired bending-moment corrections in the ribs ΔM_R due to the elastic flexibility of the floor beams. Formulas for ΔM_R and charts for their evaluation are given in Ref. 5. The values of the moment corrections depend on the loading configuration wih respect to the floor beams and the relative rigidity of the floor beams and the ribs [Eq. (2)].

Generally, floor-beam deflections cause an increase of the positive bending moments in the ribs at their midspans and a decrease of their negative moments over the floor beams. The maximum values of the moment corrections result from the maximum deflections of the adjoining floor beams due to loading of all bridge lanes between the main girders. For the purposes of superposition, this loading must be consistent with the loading used on the critical rib in step 1 of the computation.

The total design moments in the ribs are obtained by adding the bending-moment corrections ΔM_R to the bending moments obtained in step 1 of the computation.

63. Design of Floor Beams Bending moments and shears in the floor beams are determined in two steps. First, the floor beams are assumed to act as rigid supports of the continuous deck, and their loads are obtained accordingly by evaluation of the influence lines for a continuous beam on rigid supports. Multiple-lane loading, in accordance with the AASHTO specifications, should be used to obtain the maximum floor-beam loads.

In the second step the effect of floor-beam flexibility is evaluated, which tends to distribute the load on a directly loaded floor beam to the adjoining floor beams. The magnitude of this effect is a function of the relative rigidities.

The bending-moment and shear corrections in the floor beams may be computed by formulas given in Ref. 5, which include the case of a deck with nonuniform rigidities of the floor beams (heavy cross frames and light intermediate floor beams). They may also be obtained for the AASHTO loadings from charts given in Ref. 5.

64. Stresses in Ribs and Floor Beams Section properties of the ribs and the floor beams are determined with appropriate effective widths of the deck plate acting as the upper flange (Fig. 100). Effective width of plate acting with a stiffener (rib or floor beam) depends on the span of the stiffener and the load distribution, and is not a function of the plate thickness or stiffener rigidity. The effective width is larger (with an upper limit of about one-third the rib span) when only one rib is loaded, and smaller if all ribs are uniformly loaded.

For open ribs in the usual span range and a spacing a of about 12 in. the effective width of plate a_0 varies between $1.1a$ and $1.4a$ and may be determined by charts in Ref. 5. An accurate determination of the effective width is not essential, since the stresses at the lower fibers of the unsymmetrical rib or floor-beam cross sections are not very sensitive to variations of the effective width.

With section moduli of the ribs and the floor beams determined, stresses are computed due to the maximum values of the bending moments. In the design of the ribs the overall stresses (System I) in the deck are superimposed on the flexural stresses in the ribs.

Elastic stability of slender open ribs may have to be checked in the zones of overall compression. Erection conditions and resulting stresses in the ribs and the floor beams should also be considered.

CLOSED-RIB DECKS

65. Design of Closed Ribs   The principal types of closed rib are shown in Fig. 99. Closed ribs have a considerable torsional rigidity and a greater flexural rigidity than open ribs, for the same amount of material. Although the amount of welding required is also smaller, bevel groove welds are required for the rib-to-deck connections. The lateral load-distributing capacity of a deck with closed ribs is large, which reduces the design load per rib.

The usual spacing of ribs is about 24 in. A rib span of 15 ft has been used most often in American practice, but an even larger spacing of floor beams is possible and may be more economical.[8]

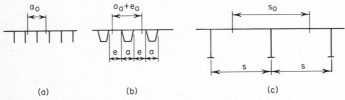

(a)   (b)   (c)

Fig. 100   Effective widths of deck plate acting with ribs and floor beams.

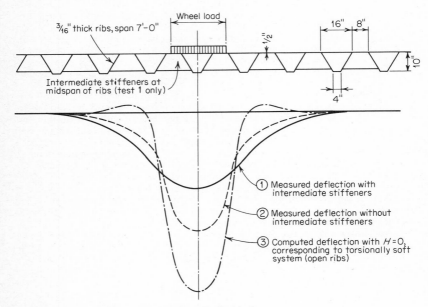

Fig. 101   Stiffening diaphragms in decks with closed ribs.

Transverse load distribution may be increased by welding intermediate diaphragms between the closed rib sections at points between the floor beams (Fig. 101). Such diaphragms limit the relative deflections of the individual ribs and the deformations of the deck plate between the ribs. However, the decrease in load per rib is limited, since a large lateral load-distribution capacity results in significant overlap of the effects of adjoining wheels, placed in accordance with the multiple-lane loading of AASHTO. Thus, advantages of diaphragms between ribs have to be weighed against the increased cost of fabrication.

A steel deck stiffened with closed ribs is a highly statically indeterminate system. Practical design is facilitated by idealizing the deck structure as an equivalent orthogonal-anisotropic plate continuous over the floor beams. For this purpose the flexural rigidities in both directions and the torsional rigidity of the deck and the ribs are assumed to be continuously distributed.

For the usual cases of decks without transverse diaphragm stiffeners between the floor beams, further simplification of the analysis is possible.[5] In this approach, only the longitudinal flexural rigidity and the torsional rigidity of the ribs are considered. The transverse flexural rigidity of the deck plate between the ribs is used only indirectly in determining the properties of the substitute orthotropic plate representing the actual system. This leads to an abbreviated form of Huber's differential equation of an orthotropic plate,

$$D_y \frac{\partial^4 w}{\partial y^4} + 2H \frac{\partial^4 w}{\partial x^2 \partial y^2} = p(x,y) \tag{3}$$

where w = vertical deflection
D_y = flexural rigidity of the substitute orthotropic plate in the direction of the ribs
H = equivalent torsional rigidity of the substitute orthotropic plate
The effect of the flexural rigidity D_x of the deck in the direction perpendicular to the ribs is included in the computation of the value of H.

Flexural and Torsional Rigidity. The flexural rigidity D_y of the substitute system in the longitudinal direction is obtained as the rigidity of one rib divided by the rib spacing,

$$D_y = \frac{EI_R}{a + e} \tag{4}$$

where I_R = moment of inertia of a rib cross section, computed with the effective width of deck plate $a_0 + e_0$ (Fig. 100b). Charts for determination of the value of $a_0 + e_0$ are given in Ref. 5.

The effective torsional rigidity H of the substitute orthotropic plate representing a steel deck with closed ribs must be computed with consideration of the local flexural deformations of the deck plate and the rib walls. Figure 102 shows that, because of these

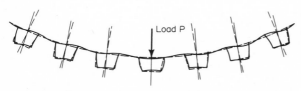

Fig. 102 Deformations of the loaded deck with closed ribs.

deformations, torsional rotations of the ribs in the loaded deck are smaller than they would be if the deck did not deform locally. Thus, the torsional rigidity of the ribs is not fully developed and the effective rigidity of the system is reduced.

The effective torsional rigidity is expressed by

$$H = \mu GK/2(a + e) \tag{5}$$

where G = modulus of elasticity of steel in shear
K = cross-section property characterizing torsional resistance of one rib
μ = reduction coefficient, depending on flexibility of the deck plate, shape of the ribs, and the rib span
Generally, values of μ can be only approximately determined. For the usual cases of decks with closed ribs unstiffened by additional diaphragms inside or between the ribs, formulas for μ for the various shapes may be found in Ref. 5. These formulas are based on the assumption that the deck plate and the rib walls deform locally, as shown in Fig. 102, but that the ribs do not change their shape (that is, there is no diagonal distortion of the

cross section). Although this assumption seems to be acceptable for practical design purposes, it becomes less accurate as the rib spans increase. Only triangular ribs are not subject to diagonal distortion, but they are less efficient than trapezoidal ribs.

The values of D_y and H are combined in the relative rigidity coefficient H/D_y, which is a parameter characterizing the load-distributing capacity of the deck in the direction perpendicular to the ribs. For any given rib size, spacing, and deck-plate thickness, H does not remain constant but increases with the rib span. Therefore, the parameter H/D_y is also a function of span, and the transverse load distribution of the deck structure improves as the span of the ribs is increased.

Bending Moments in the Substitute System. With the relative rigidity coefficient H/D_y determined, bending moments in the substitute orthotropic plate due to the given loading can be computed by formulas derived from the solution of Eq. (3). The solution can only be given as an infinite series. Therefore, values of the influence ordinates, bending moments, etc., must be expressed as sums of the component values for each term n of the series. In order to evaluate these expressions, the applied loads must also be represented as a series of component loads Q_{nx}.

The values needed for the design of the ribs are the bending moments in the orthotropic plate over the floor beams and at the midspan between the floor beams. These moments are obtained by multiplying the values of the component loads Q_{nx} by corresponding influence-ordinate components η_n:

$$M = \Sigma Q_{nx}\eta_n \qquad (6)$$

This is illustrated in Fig. 103.

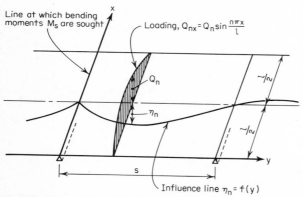

Fig. 103 Computation of the bending moments in orthotropic plate by Eq. (6). The first component of the substitute sinusoidal load and the influence line are shown.

Formulas for Q_{nx} and η_n for the various cases are given in Ref. 5. Numerical evaluation of these expressions may be avoided in cases of AASHTO truck loadings, for which design charts have been prepared.[5]

Moments in the Deck Structure. The bending moments M_y in the direction of the ribs are obtained in the substitute orthotropic plate system in units of in.-kips per inch of width of the deck. Usually only the maximum moment ordinate M_{ymax} at the center of the loaded rib is computed and the moment acting on one rib is then obtained, conservatively, as

$$M_R = M_{ymax}(a + e) \qquad (7)$$

If the effects of adjoining wheels overlap significantly because of a large load-distributing capacity in the transverse direction of the deck, the design moment per rib is obtained by superposition. Such cases may occur in decks with large floor-beam spacing.

Another problem encountered in the design of steel decks with long ribs is that of local stresses in the deck plate near the main girder webs, which act as rigid supports for the plate, while the adjoining ribs deflect elastically under wheel loads. In such cases appropriate design measures should be taken to prevent excessive local stresses and deformations of the deck plate.[8]

Rib moments computed under the assumption of rigid floor beams must be corrected for the elastic flexibility of the floor beams. Computation procedures and formulas are similar to those for decks with the open ribs.

The floor-beam loads and moments are computed as for the open-rib system.

BOX GIRDERS

66. Analysis of Box Girders Bending moments, shears, and support reactions of box girders under symmetrical loading are obtained by the usual methods, but in the general case of unsymmetrical load producing torsion the elementary approach is not sufficient.

An eccentric load P may be represented as a sum of symmetrical loading, torsional loading, and distortional loading (Fig. 104). Symmetrical loading is resisted by longitudi-

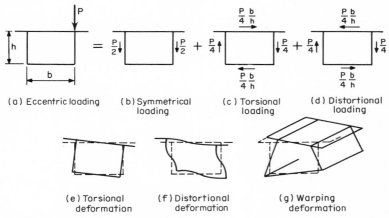

(a) Eccentric loading   (b) Symmetrical loading   (c) Torsional loading   (d) Distortional loading

(e) Torsional deformation   (f) Distortional deformation   (g) Warping deformation

Fig. 104 Box girder under eccentric load.

nal flexural stresses and shear stresses, but the distribution of the flexural stresses in wide flanges is nonuniform because of shear lag. Therefore, in calculations in accordance with the simple theory of bending, an *effective width* of flange must be used. The effective width is a function of the ratio of the span L to the width B of the box, the cross-sectional area of the stress-carrying stiffeners, and the type and position of loading. For continuous girders the effective widths are obtained separately for the individual equivalent simple spans between the points of inflection. Values of effective widths for the case of uniform loading based on a study by Moffatt and Dowling[13] are given in Fig. 105.

Effective-width considerations are applicable for stress calculations in the elastic range. For ultimate-strength calculations a uniform stress across the entire width of the flange is usually assumed. Therefore, higher allowable peak stresses at the flange/web junctures are appropriate, provided that the factor of safety against local collapse of the panel is satisfactory (Art. 69).

The *torsional component* of the antisymmetrical loading produces a twist (Fig. 104e). Although the shape of the box is preserved, it warps. If warping is prevented (by adjoining sections or rigid supports), longitudinal stresses and associated shear stresses result. These stresses, which are concentrated near the points of restraint, are usually small but may contribute locally to fatigue or brittle fracture. Some cross sections twist without warping (e.g., square or circular tubes), but generally twist is resisted by a combination of St. Venant's torsion and warping torsion.

The *distortional component* of loading, when applied to a box section without diaphragms or rigid cross frames, results in distortion of the section (Fig. 104f) and causes transverse bending moments and shears in the webs and the flange, and associated longitudinal warping stresses. Stiff diaphragms or cross frames at points of load application resist the distortions of the box, and are subject to flexural stresses. Therefore, cross frames or diaphragms must be designed for full distortional loading. However, if load is applied between diaphragms, distortional stresses in the box must be considered.

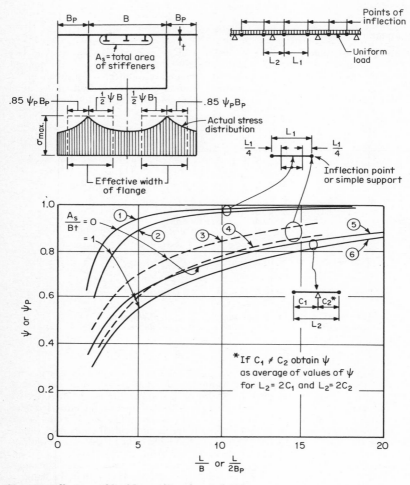

Fig. 105 Effective width of flange. *(Based on Ref. 13.)*

The available analytical methods considering all three loading components (flexural, torsional, and distortional) are based on folded-plate theory, thin-walled-beam theory, or finite-element analysis.[11] Calculations based on these methods require the use of the computer.

67. Design in Accordance with Linear Elastic Theory In the design of webs the critical loading is the theoretical buckling load of an initially flat, unstressed web or web panel. Webs may be unstiffened, or may be subdivided into panels by transverse (vertical) stiffeners in the case of predominating shear stress, or a combination of transverse and horizontal stiffeners, where flexural compression in the web governs. In such cases the transverse and the horizontal stiffeners are often placed at opposite faces of the web in order to avoid undesirable discontinuities and costly intersections. Spacing of transverse stiffeners is governed by the critical shear stress in the web. Location of the longitudinal stiffeners, which may be from one to several in number, is governed by the bend-buckling strength of the web panels. Plate-buckling formulas for uniaxial compression and combined uniaxial or biaxial compression with shear may be found in Refs. 14 and 15.

For slender panels ($b/t > 70$ for structural-grade steel) the critical stress is assumed to be the elastic (Euler) buckling stress. Stocky panels ($b/t < 30$) are expected to yield prior to buckling. For panels of intermediate slenderness the design buckling stress is given by a transition curve which is intended to account for the destabilizing effect of residual

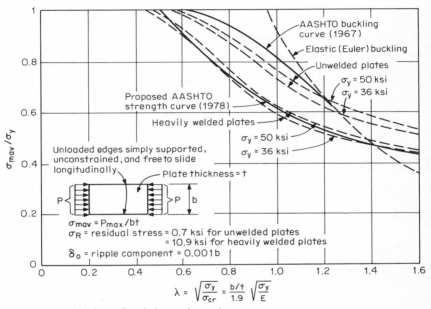

Fig. 106 Strength of unstiffened plate under axial compression.

stresses. It should be noted, however, that transition curves for plates used in design in accordance with elastic theory were based on the assumed similarity of plate and column buckling and not on considerations of residual stresses in plates or on actual plate tests. Such a transition curve, introduced in the 1967 AASHTO specifications, is shown in Fig. 106.

The design of box-girder compression flanges is similar to the web design, except that axial compression, rather than shear, predominates. Near supports the interaction of flexural and torsional shear with compression must be considered. Box-girder top flanges designed for local traffic loads have ample rigidity, and the overall stability problem in top-flange compression zones usually does not arise.

The required rigidities of web and flange stiffeners are given by theoretical and semiempirical formulas in design codes.

Web and flange panels in tensile zones of box girders should be checked for yielding under combined stresses. The Mises yield criterion, Eq. (8), may be used for this investigation.

$$\sigma_y^2 = \sigma_1^2 + \sigma_2^2 - \sigma_1\sigma_2 + 3\tau^2 \tag{8}$$

where σ_1 and σ_2 are axial stresses in two perpendicular directions and τ is the shear stress.

Fatigue may be an important consideration in the design of local deck members (flexure in the deck plate and rib walls, System III). For deck ribs governed by the combination of maximum local and overall stresses, and for box-girder main members where the maximum stresses are due to simultaneous loading of several lanes or tracks, fatigue criteria and load cycles stipulated for stringer-type bridges[18] will be conservative because of the relatively infrequent occurrence of the peak design stresses in such members. Therefore, fatigue criteria and the number of design stress cycles should be determined realistically lest the structure's economy be affected by unduly low allowable stresses. For members subject to stress cycles of various intensities (as in the case of local and overall stresses in the deck ribs) the equivalent number of stress cycles for design may be determined by Miner's rule.[17,18]

It is also important to avoid details that lower the fatigue resistance. Allowable stresses for various types of structural details, based on their susceptibility to fatigue, are given in the AASHTO specifications and other design codes.

NONLINEAR ANALYSIS OF BOX GIRDERS

68. Effect of Imperfections on Behavior of Steel Plating   *Initial Deformation.*   Because of shrinkage of welds between the plate and the stiffeners a fabricated thin steel-plate panel tends to dish in the direction of the stiffeners (Fig. 107). The plate stiffeners are also

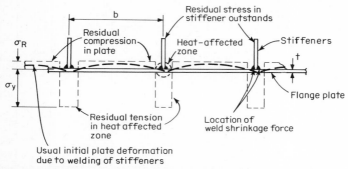

Fig. 107   Schematic representation of residual stresses in plating due to welding of stiffeners.

subject to deviation from straightness due to fabrication and erection inaccuracies. These effects may be noticeable in large stiffened thin-plate panels but are usually insignificant in stocky panels.

Generally, initial deformation of a plate panel in compression causes departure from linear behavior (Fig. 108). Under increasing load the initial bulge tends to magnify and the stress distribution in the plate becomes nonuniform, which decreases the overall compressive rigidity of the panel. In addition to the axial stresses shown in Fig. 108 the plate is also subject to local flexural stresses corresponding to changes under load of the local out-of-plane curvature of the plate. Superposition of the axial and the local flexural stresses results in maximum local surface stresses which can reach yield even under relatively low loading level. Thus, in a slender plate, elastoplastic behavior may be initiated long before reaching the ultimate strength of the panel.

The effects of the initial deformations on strength and behavior of plate panels depend primarily on the shape of the deformation and not on the maximum dishing Δ of the panel. Uniform cylindrical dishing of a panel with a large aspect ratio a/b does not necessarily decrease the strength of the panel, and may even increase it. The important factor is the affinity of the deformation to the theoretical buckling mode of the plate. Therefore, in the derivation of formulas and curves for plate strength only the deformation within the half wavelength of the buckling mode is considered, and this sinusoidal "ripple component" is assumed to govern the behavior of the otherwise randomly deformed panel. Numerical values of the amplitude of the ripple component for analytical purposes are usually expressed as a fraction of the plate width and thickness, or may be given as a function of the specified plate out-of-flatness fabrication tolerance Δ. Since the probability of initial deformations sympathetic to the plate-buckling mode is low, plate-strength predictions based on this assumption tend to be conservative.

Initial bowing of the stiffeners decreases their compressive strength. Secondary reactions due to geometric imperfections of the stiffeners at the cross frames should also be considered in the design of the cross frames.

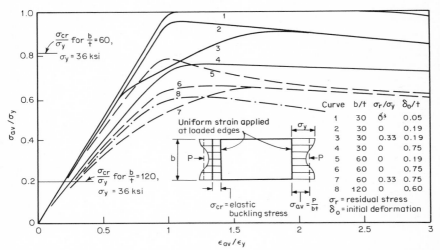

Fig. 108   Typical average stress-average strain curves for axial compression of unstiffened plate panels with initial deformation. Unloaded edges simply supported and free to slide longitudinally, but constrained to remain straight. (*Refs. 20 and 21.*)

Residual Stresses.   Residual stresses caused by welding affect the behavior of plate elements. Shrinkage of the weld and the heat-affected surrounding material causes tension yield stress in the weld region, which is balanced by residual compression σ_R in the plate between the welds (Fig. 107). The magnitude of the residual stresses cannot be estimated reliably, since the weld-shrinkage force varies widely, depending on the type of weld, the number of welds in a joint, the number of welding passes, the method and sequence of welding, etc. For a single-pass fillet weld the shrinkage force has been estimated to be about 1500 kips times the cross-sectional area of the weld in square inches. Thus a 1/4-in. single-pass fillet weld may produce a tensile force of about 45 kips. This force is smaller if the weld is made in several passes.[16] A rough value of the resulting residual compressive stress in the plate may be obtained by dividing the weld-shrinkage force by the gross cross-sectional area of the plating, including the stiffeners. Determination of the magnitude and distribution of the residual stresses in the stiffeners is less reliable; even the sign of the stress at the tip of the stiffener cannot be reliably predicted.[23]

In addition to welding, residual stresses in steel plating are also induced by flame cutting, beveling, cold bending for fit-up of plates and stiffeners, and straightening of dished plates by heating. Thus the residual stresses and the plate deformations are interrelated, and elimination of initial deformations to satisfy specified flatness, or for aesthetic reasons, can be obtained only at the expense of increased residual stresses.

The effect of residual stresses, distributed as shown in Fig. 107, on the behavior of a plate panel in the low and intermediate range of slenderness b/t, and with a moderate amount of initial deformation, is generally a small reduction of the ultimate strength. However, the plate compressive rigidity (slope of the stress-strain curve) decreases considerably at the point at which the combined applied and residual stress in the middle portion of the panel reaches the yield stress (Fig. 108).

Residual compression at the tips of stiffeners causes the onset of yielding at an earlier stage of loading and thus reduces the strength of stiffeners acting as compressive struts.

Direct quantified treatment of geometric imperfections and residual stresses has been attempted in design specifications,[12] but incorporating appropriate allowances in the design curves, depending on the classification of the structural member, appears to be more practical.

Both residual stresses and initial deformations can be minimized by avoiding thin plates with closely spaced stiffeners, limiting the panel slenderness, and keeping the weld sizes to the minimum required in the design.

69. Unstiffened Plate Panel Under Axial Compression   In the theoretical treatment of unstiffened plate panels the longitudinal (unloaded) edges of the panel are assumed to be simply supported, free to slide longitudinally, and unrestrained transversely in the plane of the plate. Therefore this category applies to unstiffened bottom flanges of girders, where these conditions may be assumed to be satisfied.

Individual plate panels between stiffeners are restrained against longitudinal motion by the stiffeners and must be treated only in conjunction with the stiffeners, and not as independent structural elements (Art. 70).

The stress-strain curves of unstiffened panels (Fig. 108) shows that only very stocky panels with small imperfections behave linearly up to the value of the maximum load. Very slender plate panels may exhibit no characteristic strength peak at all, and may reach a load many times greater than the theoretical elastic-buckling (Euler) load. In such cases developing the maximum strength may require considerable axial shortening.

Figure 106 shows typical plate-strength curves proposed for design. The curves for "unwelded" and "heavily welded" plating, developed by Dwight and Little[19], have been obtained semiempirically. The assumed magnitudes of the geometric imperfections (the initial ripple component δ) and the residual stresses are given in the figure. The residual stresses level is assumed to be independent of steel quality; therefore, the curves for higher-grade steels show somewhat greater strength than steel with lower yield stress for the same slenderness ratio. The effect of shear acting simultaneously with axial stress is considered by using curves for a lower equivalent yield stress.

Design curves based on elasto-plastic computer analysis have been proposed by Frieze, Dowling, and Hobbs[20] for two levels of residual stress, for longitudinal edges restrained as well as unrestrained. Strength reduction for the case of simultaneous shear is obtained by an interaction diagram.

Results by the two methods discussed above are similar. Plate strengths drop substantially below the ideal elasto-plastic strength in the range of $25 < b/t < 75$ for $\sigma_y = 36$ ksi but are significantly above the elastic-buckling strength for plates with greater slenderness.

Figure 106 also shows the plate-strength curve proposed (1978) for the revised AASHTO specifications for box girders.[31]

70. Stiffened Plate Panel under Axial Compression   Bottom flange panels of box girders stiffened by longitudinal stiffeners can fail by buckling out of the plane of the plate either towards the interior of the box (in which case the compressive strength of the flange plate governs), or in the opposite direction (in which case the buckling strength of the stiffeners governs). Proportions of the stiffeners must be such as to preclude their local torsional buckling failure; thus torsional buckling stress of the stiffeners greater than the yield stress is usually stipulated in the design.

The strength of the stiffened compression flange depends on the slenderness b/t of the plate between the stiffeners, the slenderness L/r of the stiffeners acting with a portion of the flange plate as compressive struts between the cross frames, the magnitude of the geometric imperfections, and the residual stresses. The strength is generally different for an initial bow Δ of the stiffeners towards the interior of the box (positive Δ) and one away from the box (negative Δ). In a wide flange panel with several stiffeners their imperfections and even the directions of the initial bows may differ. Because the individual stiffeners in a panel are interdependent in their collapse behavior, the resulting failure of the panel is a complex process not easily amenable to theoretical predictions for design purposes. In practice the positive Δ prevails because of the shrinkage of the welds between the plate and the stiffeners, however, the effects of negative Δ are important in consideration of the continuity of the flange over the cross frames.

In practical design the strength of a flange panel is obtained as a sum of the strengths of the individual struts consisting of stiffeners each acting with a portion of the plate, assuming that initial bows and other imperfections of all struts are identical. Imperfections are accounted for by assuming an initial bow and an appropriate reduction of the strength of the flange plate.

Horne and Narayanan[22] use an effective width of plate, determined as a function of the panel slenderness and the assumed plate imperfections. The critical stress is obtained by the Perry strut formula, using the value of α appropriate to the strut out-of-straightness.

In the design method proposed by Dwight and Little[19,24] the concept of "effective yield stress" is used instead of the effective width to account for the plate-stiffener interaction. The effective yield stress is the average strength of the plate-stiffener combination, assuming a very low L/r ratio, with the flange plate strength obtained from the curves discussed in Art. 69. Imperfection coefficients α suggested for use in the Perry formula reflect four practical levels of strut out-of-straightness and correspond to the values used for the proposed European column curves. Flange plate failure by buckling is considered to be critical in this method. All calculations are based on the full cross section of the plate-stiffener combination. Perry curves for $\sigma_y = 50$ ksi for the selected values of α are shown in Fig. 109a. Dwight curves for determining the strength of the strut with a known effective yield value are shown in Fig. 109b for $\alpha = 0.0035$.

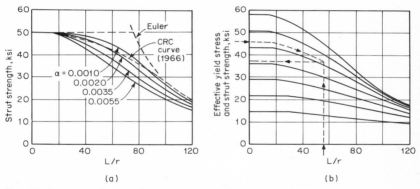

Fig. 109   Strut strength curves for (a) $\sigma_y = 50$ ksi; (b) $\alpha = 0.0035$.

The design method proposed for the AASHTO specifications[31] uses an interaction diagram giving directly the ultimate strength of the stiffener strut as a function of the column slenderness, λ_{col}, of the strut and the plate slenderness, λ_{pl}, of the flange plate between the stiffeners (Fig. 110). The diagram is based on the values obtained by the numerical second-order iterative method by Little[25] underlying the Dwight-Little approach discussed above. The following imperfections are assumed: residual stress in plate $\sigma_R = 10.5$ ksi and stiffener out-of-straightness $\Delta = L/500$ (equal to fabrication tolerance). Both positive and negative values of Δ have been considered. Lehigh column curve 2 has been taken as the lower limit of strength of struts with low plate slenderness.

In flanges with nonuniform stress distribution due to shear lag, developing the full capacity of the flange presupposes transverse load redistribution by plastic deformation of the exterior flange panels. Tests on stiffened box-girder flanges appear to confirm such load redistribution capability,[23] but some allowance for the effect of shear lag may be appropriate for very wide flanges.

The effective length of strut L for wide flanges and relatively close spacing of the cross frames may be assumed to be equal to the cross-frame spacing. For long and narrow flange panels, however, this approach is too conservative. A formula for an effective length for this case has been suggested by Rogers.[24,31]

71. Plate Panel in Shear   Web panels in shear, with or without coincident flexural axial stress, behave nonlinearly. The initial buckling of the panel, which usually develops as gradual bulging of the initial out-of-flatness, causes a stress redistribution in the form of a diagonal tension field. The ultimate shear strength of the panel is the sum of the buckling strength (beam action of the web) and the postbuckling strength (tension field action). Additional shear strength is provided by the frame action of the girder flanges; however, this contribution is disregarded in the design of webs of box girders. If longitudinal stiffeners are used to enhance the web carrying capacity, they must be sufficiently rigid to enforce and maintain the nodal lines of the web subpanels after buckling.

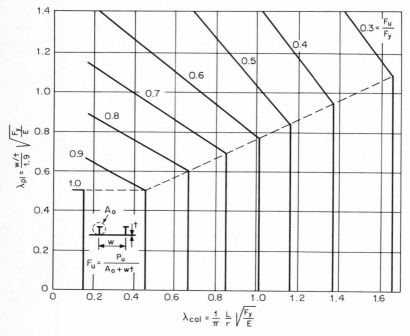

Fig. 110 Interaction diagram for stiffened plates. F_u = ultimate strength of stiffener strut consisting of one longitudinal stiffener and a portion of flange plate of width w equal to spacing of stiffeners; F_y = yield strength of steel; L = spacing of cross frames supporting the longitudinal stiffeners; r = radius of gyration of stiffener strut.

Since the web is considered incapable of carrying additional flexural stresses in the tension-field stage, additional bending moment must be resisted by the flanges alone. This results in forces in the flanges in excess of those computed according to elastic theory.

Webs acting as tension fields are incapable of carrying direct local, vertical compressive stresses. Vertical web stresses due to standard AASHTO wheel loads on bridge decks are low and may be neglected, but webs of box girders to be erected by launching[28] should be checked for the effects of concentrated reactions during erection. A method for such investigation has been proposed by Bergfelt.[26]

Slender webs under live loads may be subject to fluctuating local tensile stresses at the plate surface because of combined effects of the overall web action and the local flexural stresses caused by the web bulging, which may cause fatigue problems. Therefore, web slenderness is subject to limitations.

Proposed (1978) AASHTO provisions for box-girder webs are based on tension-field theory, with the assumption that the stiffness of the flanges is negligible.[29,31]

72. Load-Bearing Diaphragms Load-bearing diaphragms at the supports of box girders transmit the vertical shears from the webs and the torque from the box into the bearings. Stress conditions in the diaphragm are generally complex and require careful consideration. In addition to high local compressive stresses above the bearings the diaphragm is subject to flexure in its plane, since the bearings are usually placed at some distance from the box webs. Transverse compressive stresses will be introduced into the bottom part of the diaphragm and the bottom flange if the webs of the box are sloped. The shear distribution along the web-diaphragm intersection is nonuniform and tends to concentrate near the bottom of the diaphragm.

Because of the large torsional rigidity of the box girder, bearing reactions and diaphragm stresses are sensitive to vertical and angular misalignment of the bearings. Therefore, bearing-alignment tolerances must be small and the bearing position should be adjusted in the field to fit the fabricated shape of the box.

Longitudinal eccentricity of the bearing reactions due to thermal movements may cause out-of-plane flexure of the diaphragm and must be considered.

As a general rule, eccentricities should be avoided, and simple, stocky diaphragms should be preferred to thin diaphragms with multiple stiffeners. Diaphragm-design problems are discussed in Ref. 30.

CONSTRUCTION DETAILS

73. Fabrication and Erection of Decks Typical details of open and closed ribs are shown in Fig. 99. For economy, the ribs are usually made of low-alloy structural steel. However, in cases where deck rigidity governs, carbon-steel ribs may be preferable.

The high residual stresses in the steel deck due to rib welding, the various stress raisers at the rib and floor-beam junctures with the deck plate, and the rapid stress rate of the deck under wheel loads make the deck susceptible to brittle fracture. Therefore, deck material should have adequate notch toughness and stress-raising details should be avoided, especially in structures in cold climates.

The thickness of the flat-bar stiffeners is governed by the acceptable depth-thickness ratio. For the inherently stable closed ribs, plate thicknesses of ¼ in., or even ³⁄₁₆ in., have been used in European installations. Fabrication of the closed-rib sections from plate thicknesses up to ⁵⁄₁₆ in. on bending presses does not present unusual difficulties, but bending heavier plates of low-alloy steel may be uneconomical. Rib lengths up to 25 ft and more can be fabricated with equipment currently available.

Experience has shown that closed ribs welded airtight to the deck need no protection against inside corrosion.

Open ribs run continuously through slots in the floor beams. The closed ribs of the original European installations were usually cut and butt-welded at the floor beams, but later designs have used continuous ribs, which are much to be preferred.

Open ribs are free from transverse local flexural stresses, and may be connected to the deck by fillet welds. Closed ribs are subject to transverse local flexural stresses due to direct wheel loads, and since the maximum values of these stresses occur at the junction of the rib web and the deck, fillet welds are not suitable for this joint because of their low fatigue resistance; bevel groove welds must be used. Full-penetration groove welds require well-fitted backup plates, which is costly. Partial-penetration groove welds are less expensive but have somewhat lower fatigue life.[9,10]

Floor beams are inverted T-sections using the deck plate as the upper flange. If the main girder spacing is large, it may be advantageous to use welded I-sections as lower parts of the floor beams to serve as erection supports for the deck-plating panels, which consist of the deck plate, the ribs, and the floor-beam web stubs and which are joined with the lower floor-beam sections in the field (Fig. 111).

In some cases longitudinal ribs have been placed on top of the floor beams to avoid costly rib-floor beam intersections.

At the juncture of the deck plate, the rib wall, and the floor-beam web, rounded cutouts are usually provided in the floor-beam web to avoid intersections of the welds and to relieve possible stress concentrations. In decks with open ribs it may be advisable to weld the ribs to the floor-beam webs on one side only (Fig. 99) in order to minimize the stresses due to weld shrinkage, which may lead to cracks in the welds.

Generally, in order to prevent large residual welding stresses and distortions of the relatively thin deck plate and the rib, it is important to keep weld sizes to the minimum indicated by design and to avoid overwelding. Submerged-arc welding induces less heat than manual welding and is to be preferred. Welding sequences should be planned carefully.

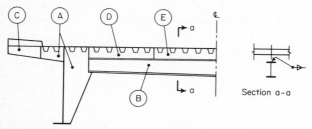

Fig. 111   Erection scheme of a steel-deck bridge with split floor beams.

The usual shop-fabrication procedure of the deck panels is first to join the ribs to the deck plate by continuous submerged-arc welds and then to fit and manually weld the cutout floor beams to the deck and the ribs in reversible cradle jigs. The overall dimensions of the shop-fabricated panels are governed by transportation and erection facilities. If the width of the shop units is limited for transportation by rail or truck platforms, the deck panels are fabricated with their long sides in the direction of the ribs, in order to minimize the number of field rib splices. If transportation by water is possible, the full width of the deck between the main girders may be fabricated as one weldment.

Sidewalks have been made of concrete or steel plate stiffened with ribs. The advantage of the latter is in its lighter weight and the additional contribution to the top flange area of the main members.

Field splices of steel decks, longitudinal ribs, and floor beams may be made by high-strength bolting, welding, or combinations of welding (deck plate) and bolting (ribs and

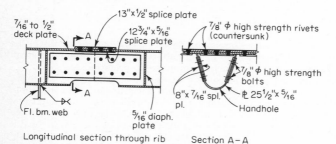

Fig. 112   Bolted rib splice—Port Mann Bridge.

floor beams). The advantage of bolted field connections is in easier fitting of the deck plating units that may be preassembled in the shop. The disadvantages are in the loss of cross section of the members at splices and a greater susceptibility to corrosion. Protruding splice plates and bolt heads make it difficult to place the wearing surface on the deck in a satisfactory manner and preclude the use of thin, lightweight surfacing. Bolted splices of decks with closed ribs require handholes in the bottoms of the ribs (Fig. 112).

Welded field splices of steel decks require a high degree of precision in shop fabrication and erection to obtain a satisfactory fit. Because the effects of weld shrinkage cannot be estimated accurately, and small fabrication errors are unavoidable, some cutting and edge preparation in the field are usually necessary.

An example of a welded splice of a deck with open ribs is shown in Fig. 113a. The transverse splice of the deck plate was welded first from both sides of the plate. Then the 20-in. pieces of the longitudinal ribs below the deck splice were fitted and welded.

Closed ribs may be butt-welded indirectly, through a vertical diaphragm plate (Fig. 113b) or directly (Fig. 113c, d).

In order to minimize the stresses due to weld shrinkage, it is always necessary to allow

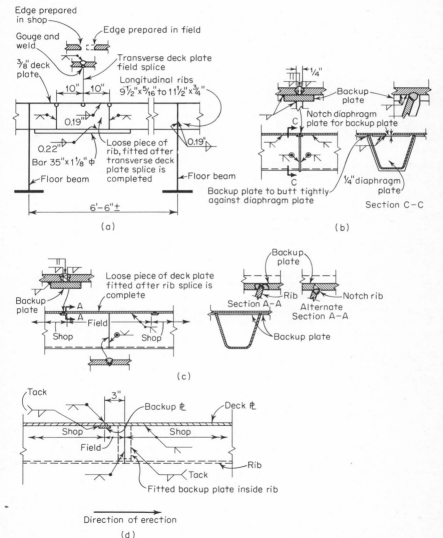

Fig. 113 Welded rib splices. (a) Open ribs (Cologne-Severin Bridge). (b) Indirect splice of closed ribs. (c) Direct splice of closed ribs, welded from above. (d) Direct splice of closed ribs welded from below (Poplar Street Bridge).

an unrestrained motion of the deck panel being connected in the direction perpendicular to the weld made.

74. Erection of Box Girders   Where site conditions preclude the subassembly and transportation of complete box-girder sections, individual elements (webs, bottom flanges, decks) are erected successively and joined at the longitudinal and transverse field splices. This method has been used for the erection of many overland viaducts in Europe. Segmental cantilever erection presents fewer stability problems and is especially advantageous where water transportation of the completed segments is possible. Box segments, with watertight diaphragms at each end, can be floated and towed to the erection site without the use of barges. The Kansas City Southern Railway Bridge over the Arkansas River was erected by preassembling the box girders on the bridge-approach embankments and pushing them out over the piers.[28] Lifting of entire spans preassembled near the bridge site has been successfully accomplished in erection over waterways. Box girder sections weighing over 600 tons have been lifted by floating derricks in the United States and in Japan. The 958-ft twin box-girder sections of the side spans of the Rio-Niteroi bridge, weighing 5275 tons, were lifted into position by means of jacks mounted on the piers (Fig. 96).

Since the web and flange panels of box girders are slender elements susceptible to buckling, their erection requires care. Successive erection-stress conditions must be established with due consideration of such factors as the weight of the erection equipment and material stored on the girder, effects of temperature, and wind. Factors of safety for erection conditions should be sufficient to cover the effects of unintended eccentricities at splices, residual stresses, and other contingencies that may affect the elastic stability.

WEARING SURFACES

Wearing surfacings are placed on steel decks to provide a safe riding surface with good traction characteristics, and to protect the steel-deck plate. It should be noted, however, that some battledeck-type bridges were built with unprotected checkered steel-plate decks that have withstood exposure and wear quite well.

Since the prime aim of steel-deck construction is saving of bridge dead weight, light weight of the surfacing is an essential requirement. Therefore, the total thickness generally does not exceed $2\frac{1}{2}$ in. for bituminous surfacings, and may be only $\frac{1}{8}$ to $\frac{1}{4}$ in. for resin coatings.

Surfacings consist generally of a seal and tack coating topped with a suitable surface course.

75. Seal and Tack Coating   The purpose of the seal and tack coating is to prevent access of air and moisture to the steel-deck surface and to provide a good, permanent bond between the deck and the overlying surfacing. In order to achieve a satisfactory bond the steel surface must be cleaned by sandblasting immediately before the prime coat is placed.

The following materials and their combinations have been used:

1. Bituminous or resinous varnishes and paints, with or without rubber powder, red lead, and other additives. This group also includes various proprietary bituminous adhesive compounds. In some early installations dimpled aluminum or copper foil membranes were used in conjunction with these materials.

2. Inorganic zinc paints.

3. Zinc metallizing, about 1 mil thick, with or without an additional adhesive coating on top.

4. Asphalt mastic coatings (a dense, voidless mixture of bitumen, mineral dust, and fine sand) $\frac{1}{8}$ to $\frac{3}{8}$ in. thick. Mastic coating is considered to be effective in partly absorbing the thermal shear stresses between the steel deck and the bituminous surfacing.

5. Epoxy resin and sand mixtures, with embedded grit, to provide bond with the overlaying surfacing.

All the above materials have given satisfactory service as prime coating when properly installed.

76. Surface Courses   For the surface courses bituminous mixes are predominantly used. Bituminous-mix pavements on steel bridge decks are subject to more severe service conditions than similar highway pavements. The surfacing, bonded to the deck, acts compositely with the steel-deck plate and should be able to withstand stresses due to the different temperature-expansion coefficients of the steel and the asphalt, and the stresses due to the deck-plate flexure. The paving mix should be dense and durable, stable at high temperatures, and yet not brittle when cooled.

Because of their smaller heat-absorption capacity, lightweight steel decks are more sensitive to weather changes than the heavier concrete decks and are more susceptible to sudden freezing. Therefore, a good traction value of the surfacing is essential. These requirements may be reasonably satisfied by appropriate selection of the physical properties of the bitumen in the mix and its content, aggregate quality and gradation, suitable admixtures, and surface treatment.

Bituminous mixes used on steel decks are of two basic types: asphalt concretes and asphalt-mastic systems.

Asphalt-concrete pavements on steel decks have a total thickness ranging from 1¼ to 2½ in. and are usually placed in two courses, with the lower (binder) course somewhat softer than the upper (surface) course. Mixes used on German steel-deck bridges have a relatively high content of bitumen (7 to 10 percent) of original penetration grade of about 25 to 40 for the surface course and 40 to 60 for the binder course. The crushed-stone aggregate, ⅜ in. maximum size, is dense-graded and contains about 30 percent of mineral filler (passing the No. 140 sieve). Because of the fluidity of the mix at placing temperatures of about 375°F, it is usually placed by hand.

The performance of these surfacings has been generally satisfactory. In some cases defects have developed, of which cracking attributed to temperature and flexural stresses has been the more common. However, cracks in surfacing have not been of serious consequence.

Machine-laid double-course asphalt-concrete surfacings 2 to 2½ in. thick have been placed on the Port Mann, Concordia, and Poplar Street bridges.

Asphalt-mastic surfacings consist of a layer of mastic into which crushed-stone aggregate ½ to 1 in. size is embedded. The total thickness of the surfacing may range from 1½ to 2 in. The mastic matrix contains 12 to 17 percent bitumen of a penetration grade varying between 15 to 70, mixed with sand and mineral filler. Mastic surfacings have been used with flat anchor bars about 1 × ¼ in. welded to the deck plate, spaced from 3 to 6 in. apart, or without such anchorage. The bar-anchored type has the advantage of providing an additional stability of the mix and stiffening the deck plate but is expensive to install. Both types have performed well.

Epoxy-asphalt surfacings are mixes of epoxy binder, asphalt, and mineral aggregates. The usual thickness of this surfacing, which can be applied by conventional asphalt-paving equipment, is 2 in. Epoxy-asphalt surfacings have shown remarkable endurance and fatigue resistance and very good traction characteristics.[9,27] They have been used on the San Mateo, San Diego, Queens Way (Long Beach), Fremont Street (Portland), and Rio-Niteroi bridges.

Thin Coatings. Various thin coatings are used. The epoxy-resin and polyester-resin surfacings are made by applying a thin layer of liquid-resin binder on the cleaned steel deck and broadcasting abrasive grit over the surface. Two or more resin coatings are usually applied, resulting in a total thickness of surfacing of ³⁄₁₆ to ⅜ in. Such surfacings have been used on some American steel-deck bridges.[5]

Application of very thin surfacings requires a high degree of flatness of the steel deck, and welding of all deck splices since splice plates and bolt heads cannot be accommodated.

Because of the severe service conditions, surfacings must be designed and placed more carefully than ordinary highway pavements. However, surfacing performance criteria and the associated steel-deck rigidity requirements should not be overemphasized, since this may lead to uneconomical solutions.

RAILROAD BRIDGES

In railroad bridges a fully participating steel deck results generally in reduced weight of structure, lower depth, and greater rigidity. As a rule, welded box-girder sections are used with smooth exterior surfaces which decrease susceptibility to corrosion, facilitate maintenance, and look good. If necessary, such sections can be designed with a very shallow depth. Shorter-span structures can be completely shop-assembled and installed in place in a short time, with a minimum of field work.

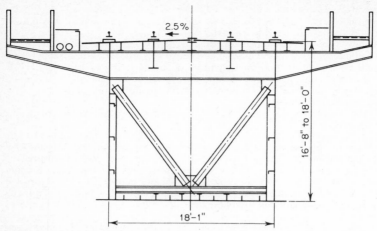

Fig. 114   Railroad bridge over the Rhine at Koblenz, Germany, two spans at 371 ft.

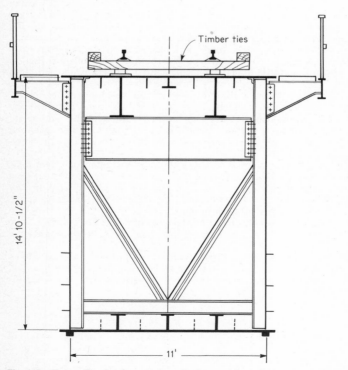

Fig. 115   Kansas City Southern Railway Bridge over the Arkansas River at Redland, Oklahoma, spans 200-230-3@250-330-250-2@175 ft.

If standard ballasted track is desired, the deck is shaped to retain the ballast. In such cases the deck plate must be drained and must receive appropriate protective coating.

In long-span design, steel decks may be used in conjunction with box girders or deck trusses. Aside from the structural and economic advantages of the participating deck, this construction is preferable to the open timber deck by offering protection of the underlying highways or waterways from railroad spillage. Such structures may be designed with the rails mounted directly on the deck or with standard timber ties fastened to the steel deck. The former method is preferable if good rail alignment can be achieved without too much expense. The available fasteners usually consist of a beveled baseplant, rail-positioning plates, and rail grip plates and bolts, with interposed elastoeric plates for damping and insulation. Such devices may be welded, bolted, or glued to the steel deck. An example of this type of bridge is shown in Fig. 114.

A bridge with timber ties is shown in Fig. 115. This 2110-ft continuous structure required only 8 piers, against 20 required in the conventional alternative consisting of a 330-ft main through truss flanked byplate-girder spans. This resulted in considerable cost saving. In addition to the cost advantage, the structure constitutes a considerably smaller obstruction of the waterway and has a superior appearance. It also has a better overload capacity owing to both statical indeterminacy of the structural system and the structural characteristics of the steel deck.[28]

REFERENCES

1. Bridge Department, California Department of Transportation: "Manual of Bridge Design Practice," 3d ed., 1971.
2. ACI Committee 443: "Analysis and Design of Reinforced Concrete Bridge Structures," 1974.
3. American Association of State Highway and Transportation Officials: Specifications for Highway Bridges, 1977.
4. Bureau of Public Roads: "Strength and Serviceability Criteria Reinforced Concrete Bridge Members, Ultimate Design," 1966.
5. "Design Manual for Orthotropic Steel Plate Deck Bridges," American Institute of Steel Construction, New York, 1963.
6. Pelikan, W., and M. Esslinger: Die Stahlfahrbahn, Berechnung und Konstruktion, *Forschungsh.* 7, Maschinenfabrik Augsburg-Nuernberg, 1957.
7. Wolchuk, R.: Old Bridges Give Clues to Steel Deck Performance, *Eng. J. AISC*, no. 4, 1964.
8. Wolchuk, R.: Steel Deck Bridges with Long Rib Spans, *Civil Eng.*, February 1964.
9. Seim, C.: Practical Considerations in Design of Orthotropic Decks, *ASCE Natl. Struct. Eng. Convention*, Apr. 14–18, 1975, New Orleans, La., Meeting Preprint 2480.
10. Seim, C., and R. Ferwerda: Fatigue Study of Orthotropic Bridge Deck Welds, *Mater. Res. Dept. Res. Rep.* CA-HWY-MR666473 (1) 72-22, California Division of Highways, April 1972.
11. Progress Report on Steel Box Girder Bridges, by the Subcommittee on Box Girders of the ASCE-AASHTO Task Committee on Flexural Members, *J. Struct. Div. ASCE.*, April 1971.
12. "Inquiry into the Basis of Design and Method of Erection of Steel Box Girder Bridges. Interim Design and Workmanship Rules," H. M. Stationery Office, London, 1973.
13. Moffatt, K. R., and P. J. Dowling: Shear Lag in Steel Box Girder Bridges, *Struct. Eng.*, October 1975, discussion August 1976, London.
14. Bleich, F.: "Buckling Strength of Metal Structures," McGraw-Hill Book Company, New York, 1952.
15. "Stabilitätsfälle: Berechnungsgrundlagen Vorschriften," DIN 4114, Deutscher Normenausschuss, Beuth-Vertrieb, Berlin, West Germany, 1952.
16. Elliott, P., B. W. Young, and G. Bowers: Residual Stress Measurements and Tolerances, *Proc. Int. Conf. Steel Box Girder Bridges*, The Institution of Civil Engineers, London, 1973.
17. Fisher, J. W.: Guide to 1974 AASHTO Fatigue Specifications, American Institute of Steel Construction, New York, 1974.
18. Wolchuk, R., and R. M. Mayrbaurl: Stress Cycles for Fatigue Design of Railroad Bridges, *J. Struct. Div. ASCE*, January 1976.
19. Dwight, J. B., and G. H. Little: Stiffened Steel Compression Panels—A Design Approach, Tech. Rep. CUED/C—Struct./TR 38, Cambridge University, 1974.
20. Frieze, P. A., P. J. Dowling, and R. E. Hobbs: Ultimate Load Behaviour of Plates in Compression. Paper 2 in "Steel Plated Structures," Crosby Lockwood Staples, London, 1977.
21. Harding, J. E., R. E. Hobbs, and B. G. Neal: Ultimate Load Behaviour of Plates under Combined Direct and Shear In-Plane Loading, Paper 15 in "Steel Plated Structures," Crosby Lockwood Staples, London, 1977.

22. Horne, M. R., and R. Narayanan: An Approximate Method for the Design of Stiffened Steel Compression Panels, *Proc. Inst. Civ. Eng.*, September 1975, London.
23. Dowling, P. J.: Strength of Steel Box-Girder Bridges, *J. Struct. Div. ASCE*, September 1975.
24. Dwight, J. B., and G. H. Little: Stiffened Steel Compression Flanges—A Simpler Approach, *The Structural Engineer*, December 1976, London.
25. Little, G. H.: Stiffened Steel Compression Panels—Theoretical Failure Analysis, *The Structural Engineer*, December 1976, London, and unpublished additional reports on investigation of compression panels with positive and negative out-of-straightness.
26. Bergfelt, A.: The Behaviour and Design of Slender Webs Under Partial Edge Loading, Paper 20 in "Steel Plated Structures," Crosby Lockwood Staples, London, 1977.
27. Balala, B.: Studies Leading to Choice of Epoxy Asphalt for Pavement on Steel Orthotropic Bridge Deck of San Mateo-Hayward Bridge, *High. Res. Rec.*, no. 28, p. 12, Highway Research Board, Washington, D.C.
28. Wolchuk, R., and J. C. Bridgefarmer: Long-Span Steel Deck Railroad Bridge, *Civil Eng.*, April 1970.
29. Rockey, K. C., and D. M. Porter: A Design Method for Predicting the Collapse Behavior of Plate Girders, *Proc. Inst. Civ. Eng.*, March 1978, London.
30. Wood, J. G. M., and A. R. Flint: The Design of Box Girder Diaphragms. Paper 18 in "Steel Plated Structures," Crosby Lockwood Staples, London, 1977.
31. Wolchuk and Mayrbaurl: Reports to ASCE Review Committee for FHWA Steel Box Girder Bridge Specifications Contract, 1978.

Section 19

Buildings

Part 1. General Design Considerations

STEPHEN J. Y. TANG
Professor of Architecture, University of Oregon, Eugene

S. G. HAIDER
Professor of Architecture and Technology, Carleton University

PLANNING BUILDING STRUCTURES

The design of the structural system for a building may be visualized as consisting of the following steps:

1. Programming, or definition of the problem. Analysis of the functional requirements dictated by the building use, location, character, etc.

2. Planning, or preliminary design. Determination of the type, material, layout, and scale of the system

3. Evaluation of the loads. Determination of the type, magnitude, and possible combinations of the loads which act on the building

4. Investigation of the structural performance or behavior under load. Both qualitative and quantitative determination of the external and internal forces generated by the loads

5. Definitive design of the system. Selection of the size, shape, quality, etc., of all the parts of the structure

6. Integration of the structure in the building. Relation of the structure to architectural details, mechanical equipment, building-occupancy requirements, etc.

7. Construction detailing. Study of the methods, details, and sequence of fabrication and assembly of the structure

The planning, or preliminary design, must be done with an intelligent anticipation of the problems which will arise in the later stages of design and construction. In many respects, this makes it the most difficult as well as the most important step. A poor job at this stage may void all the following work and make a completely new start necessary. Since the preliminary design must be based largely on experience and judgment, some amount of reworking must be anticipated where the designer lacks experience or the job is of a complex or unique nature. In fact, except for the very simple and ordinary structures, even the experienced designer will usually study several alternate systems, materials, and layouts before the final scheme is set.

1. Selection of Structural Scheme The selection of the structural system for a particular building involves the comparison of the unique requirements of the problem with the general characteristics of the various systems available. In most designs, the relative merits of several systems will need to be carefully weighed in order to make the best possible selection. The feasibility of various systems must first be established with respect to fire resistance, exposure, appearance, size of elements, spans, loads, adaptability to openings and special loads, attachment of nonstructural elements, etc. They must be compared with respect to the total cost of the finished structure, including the cost of materials, labor for fabrication and erection, transportation to the site, temporary forms or bracing required, and any maintenance required during the life of the building. The extent to which a particular system reduces or increases the cost of other elements of the construction must be considered. Thus, the ability of the structure to develop the required fire resistance with or without applied protection, to facilitate the inclusion or attachment of nonstructural elements, etc., must be considered in any comparative study of systems. The relative depth of a floor system may be critical as it affects the required height of the building in multistory construction.

Complete and fair economic comparisons of structurally equivalent systems are very difficult to make. In addition to considerations of total cost of the structure in place and cost of the complete building, other constantly changing relationships influence the comparative cost. Some of these are: local availability of materials; competition of suppliers; development of new materials which make established ones less competitive; changing building code criteria which make better—and usually more economical—designs possible in certain materials; familiarity of local builders with new systems, which reflects itself in fewer contingencies in bids; etc.

Intelligent selection of the structural scheme can only be made after the criteria for the particular problem have been thoroughly itemized. This entails considerable judgment and the cooperation of all those involved in the planning. Because of the unique or complex nature of a problem, or the lack of experience of any or all of the members of the design team, it may be highly unlikely that the best solution can be arrived at in a single attempt. In fact, except for repeated similar projects, the typical design effort usually involves partial development of alternate schemes, in order to establish the optimal choice.

2. Spatial Requirements The architectural design will also determine the clear spans and clear heights required. Structural considerations may influence these decisions but usually only by way of establishing practical limitations. The determination of the modular system of the building, if any, may require some coordination of architectural and structural requirements.

Limitations on the size of the structural members may be imposed by the architectural design. Reduction of column sizes may be critical in conserving space, especially in the lower floors of multistory buildings. Limitations on the depth of horizontal elements of the structural frame may be important in reducing the building height and volume, resulting in savings on exterior walls, stairs, piping, ducts, wiring, etc. In some cases structural efficiency may be sacrificed if it results in savings in other aspects of the building construction.

Special building uses may dictate the shape and layout of the structure and in turn influence the choice of the type of structural system. Thus, in the design of an auditorium the requirements for an efficient acoustical form and good sight lines may set the shape of the building and indicate the use of a particular system. For a gymnasium or large assembly arena, requiring a long clear span and high open space, an efficient long-span system, such as a truss, arch, dome, or suspension system, will be most feasible.

3. Wind Systems Wind pressure on exterior vertical surfaces of the building is transmitted to the roof and floors and through them to the shear walls, rigid frames, vertical X-bracing, wind towers, or whatever combination of these elements is provided. Transfer of the horizontal wind force through unbraced roofs and floors may involve considerable judgment with regard to the diaphragm capabilities of the deck and the adequacy of the assembly details. Monolithic concrete structures will usually be more than adequate in this respect, although attention must be given to the effect of openings and expansion joints.

Floors and roofs may be stiffened to increase their resistance to in-plane distortion. Intersecting beams may be designed for Vierendeel truss action, horizontal X-bracing may be provided, and single panel strips of the floor system may be stiffened so that loads are transferred to them by other beams in column action. Masonry walls, cast-in-place or precast concrete walls, and stiff sandwich panels of various materials may act as shear panels.

Provision must be made for both overturning and sliding. Codes specify that the resisting dead-load moment be 1.5 to 2 times the overturning moment, or else anchorage must be provided. Horizontal resistance is provided by frictional resistance of the footings and slabs on ground, and by passive earth pressure on foundation walls. The latter resistance may be increased by specifying a highly compacted, granular backfill.

4. Deflection Deformation is an important consideration in design. Major considerations have to do with deflection, vibration, and with the somewhat nebulous but very real quality of solidity. Deflections may be objectionable for various reasons. Excessive deflections may result in curvatures or misalignments perceptible to the eye. Lesser deflections may result in fracture of architectural elements, such as plaster or masonry. Where neither of these is critical, deflection may still be objectionable if it results in the transfer of load to nonstructural elements, such as window frames or interior partitions. The latter situation must often be handled by detailing connections which allow for limited movements. This requires close coordination of the architectural and structural detailing.

Vibration of building structures may result from various sources and may be objectionable because it is perceptible to the occupants, because the use of the building involves delicate work or sensitive machinery, or because it causes actual physical damage, such as fracture of window glass.

Limits and thresholds of tolerable physiological and psychological response to movements are difficult to determine. It is generally agreed that vertical deflection of $\frac{1}{180}$ to $\frac{1}{360}$ of the span and lateral deflection of $\frac{1}{300}$ to $\frac{1}{600}$ of the building height are reasonable and acceptable. However, rate of movement and frequency of vibration play an important part. Table 1 can be used as a guide.

TABLE 1   Comfort Criteria for Building Occupants

Acceleration, % g	Motion
Under 0.5	Not perceptible
0.5–1.5	Threshold of perceptibility
1.5–5	Annoying
5–15	Very annoying
Over 15	Unbearable

5. Structural Materials Considerations to be made in the selection and use of structural materials are:

Structural properties—strength, stress-strain characteristics, time-dependent behavior, temperature effects on strength and stiffness, etc.

Physical properties—form, weight, ductility, hardness, durability, fire resistance, coefficient of expansion, resistance to corrosion, rot, insect attack, etc.

General properties—cost, availability, reliability of quality control, forming limits, general workability, etc.

Structural properties of various materials are given in the following sections: structural steel, Secs. 4 and 6; steel for cold-formed members, Sec. 9; prestressing tendons, Sec. 12; cables for suspension roofs, Sec. 21; aluminum, Sec. 10; masonry, Sec. 15; and timber, Sec. 16.

6. Fire Resistance Thermal characteristics of structural materials are given in Table 2. Fire-resistant ratings required for the various parts of the building, including the members of the structural system, will be established by the building location and use and will usually be specified in detail by the local building code.

Fire-resistive ratings are expressed in hours and refer to the length of time the member can be subjected to a standard fire test without failing. Failure may consist of any of the following:

1. Transmission failure. Rise of temperature of, or passage of hot gases to, the side

opposite the fire. This type of failure may be due to conduction or to formation of holes or cracks in the member.

2. Structural failure at high temperature. The element must retain sufficient strength to support the building and its contents during the fire. Failure may be one of actual collapse of the element or of excessive deflection resulting in failure of other parts of the structure.

TABLE 2   Thermal Characteristics of Structural Materials

Material description	Reaction to high temperature Temperatures given in Fahrenheit (Celsius)
Wood: structural softwoods, pine, fir, spruce	Combustion at 550+° (260°) Considerable different rates of expansion parallel and perpendicular to grain No appreciable change in strength or stiffness with temperature rise Large-dimension members burn slowly, because of the insulating effect of the charred exterior
Concrete: ordinary structural, stone aggregate	Noncombustible. Spalls or pops when exposed to high temperatures for long duration No appreciable change in strength or stiffness due to temperature rise Cover on reinforcing critical for insulation. Thin elements crack, permitting transmission of flame and gas More insulating with lightweight aggregate
Steel: structural grade, rolled sections, bars, plates	Noncombustible, melts at 2400 to 2750° (1300 to 1500°) High conductivity permits rapid heating Strength and stiffness vary with temperature Generally unserviceable above 1000° (540°)
Aluminum: structural alloy, wrought	Noncombustible, melts at 900 to 1200° (480 to 650°) High conductivity and rate of expansion Strength and stiffness variation:

	−320 (−196)	−110 (−79)	−20 (−29)	75 (24)	212 (100)	300 (150)	400 (200)	500 (260)	600 (320)	700 (370)
Temperature										
% of tensile strength at room temp	133	110	105	100	91	71	41	15	8	6
% of tensile modulus at room temp	112	105	102	100	98	95	90	80		

Material description	Reaction to high temperature
Plastic	Most have low fire resistance. Behavior varies from high flammability (celluloid) to slow-burning or self-extinguishing (nylon) Thermosetting types (epoxy, melamine, mylar, phenolics) usable to 300° ±, retaining strength and stiffness Thermoplastic types (cellophane, nylon, polyethylene, vinyl) melt when heated, then return to solid form when cooled. Usable to 150° ±

3. Failure when subjected to a stream of water. The structure should retain its strength when subjected to a hose stream during or after the fire.

It should be noted that standard fire ratings refer only to the action of a member during a fire and in no way certify the usefulness of the structure after the fire. They are established, in most cases, by fire tests on full-scale elements. Fire-resistive ratings of typical construction elements are shown in Table 3.

Since there are many materials and building systems, coupled with numerous design approaches such as liquid-filled steel columns and beams, flame-shielded exterior surfaces of exposed spandrel girders, and fire-protection paints and coatings, etc., many building codes permit architects and engineers to certify the fire-resistance quality and rating of the submitted material and system by procuring a Fire Underwriters' label.

TABLE 3   Hourly Fire-Resistive Ratings of Typical Construction Elements

Elements	Ratings and criteria

Beams

	Reinforced concrete, no protection, cast in place or precast	4 hr if minimum of 1½-in. cover on reinforcing
	Steel section, poured concrete cover, stone concrete	1 to 4 hr, depending on thickness of cover, type of tie or wire wrap around beam, specific type of aggregate
	Steel section, lath and plaster cover	1 to 4 hr, depending on type of plaster, thickness of plaster, type of lath, space between lath and beam
	Steel section, sprayed fiber protection	3 to 4 hr, depending on type of fiber, thickness of covering
	Steel section, enclosed between floor and ceiling	1 to 4 hr, depending on type of ceiling, space between ceiling and beam
	Wood, solid beam glue-laminated	1 to 2 hr, depending on deck and ceiling thickness
	Steel section, fireproof paint cover	1 to 2 hr, depending on thickness and fire test
	Steel section, water and antifreeze filled, auxiliary mechanical pump system	4 hr with constant circulating water
	Steel section	Rating to be determined by fire test, 1 to 4 hr
	Steel section, masonry cover	2 to 4 hr, depending on thickness and material
	Precast concrete	1 to 4 hr, depending on type of aggregate, amount of cover, and fire test

Columns

	Reinforced concrete, tied, spiral, or composite	3 to 4 hr, depending on type of aggregate, size of column, cover on reinforcing
	Steel section, poured concrete cover	1 to 4 hr, depending on thickness of cover, type of aggregate, and size of column

19-5

TABLE 3   Hourly Fire-Resistive Ratings of Typical Construction Elements *(Continued)*

Elements	Ratings and criteria

Columns

	Element	Rating
	Steel section, lath and plaster cover	1 to 4 hr, depending on thickness of plaster, type of plaster, type of lath, space between lath and column, size of steel section
	Steel section, sprayed fiber protection	1 to 4 hr, depending on type of fiber, thickness of cover, details of the application process
	Steel section masonry cover	1 to 4 hr, depending on thickness of cover
	Steel section, fireproof paint cover (paint expands 20 or 30 times original thickness)	1 to 2 hr, depending on thickness and fire test
	Steel section with oversized flange and web so that actual required section temp. <1000°F (537.7°C)	1 to 4 hr, depending on fire test
	Steel section, water and antifreeze filled, auxiliary mechanical pump system	4 hr, with constant circulating water
	Precast	2 to 4 hr, depending on type of aggregate size and cover thickness

Floor and Roof Decks

	Element	Rating
	Steel deck, concrete fill, hung ceiling of lath and plaster	3 to 4 hr, depending on thickness and type of concrete fill, space between deck and ceiling, type and thickness of plaster
	Steel deck, concrete fill, sprayed fiber on underside of deck	2 to 4 hr, depending on thickness and type of concrete, thickness and type of fiber cover
	Concrete slab on steel deck, composite action, negative reinforcing in concrete, no protection on underside	1 to 3 hr, depending on overall thickness of slab, type of aggregate, cover on negative reinforcing
	Reinforced-concrete slab, no protection	1 to 4 hr, depending on overall thickness of slab, cover on reinforcing, type of aggregate
	Steel open-web joists, enclosed between deck and ceiling	1 to 4 hr, depending on type of deck, type of ceiling, space (if any) between ceilings and joists
	Wood joists enclosed between deck and ceiling	1 to 2 hr, depending on thickness of wood floor and ceiling
Topping 1½ to 2″ Precast cellular concrete slab		1 to 3 hr, with concrete topping depending on type of concrete fill and size of cellular core and coverage of reinforcing steel

Fire separation and compartmentation to minimize the spread of fire may require special types of walls and certain structural materials and systems. Another concept of fire containment is to introduce positive pressure mechanically in the corridors, stairways, vertical mechanical and elevator shafts, and certain designated areas.

A variety of automatic fire-prevention and -detection devices and systems, such as smoke and fire sensor systems, alarm and communication systems, and wet and dry sprinkler systems, are used in conjunction with fire-resistant structural systems to provide the required rating for the building.

Design for specific fire ratings will usually require careful coordination of the architectural and structural detailing. Where plastering or other covering of the structure is provided, it may, with proper detailing, be utilized as fire insulation for the bare structure. This is especially important with framed structures of metal.

7. Deterioration Building structures must resist deterioration due to exposure to various corrosive or wearing actions (Table 4). Metal structures may be painted, galvanized, or otherwise protected. However, certain metals, such as aluminum alloys and some highly corrosion-resistant steels, may be left unfinished, as they form their own protective finish because of initial oxidation which inhibits further deterioration. Nonaccessible thin metal elements should be protected from electrolysis, rusting, and fumes.

Concrete structures may be treated with sealers or have their surfaces artificially hardened with chemicals. Exposed concrete structures should use air-entrained cement to minimize the freezing effects. If the exposed surface has a sandblast or bushhammer finish, the concrete should be stronger and the concrete cover of reinforcing steel should be increased.

Wood structures may be painted or treated with preservatives.

The necessity for any of these treatments may be reduced or eliminated where architectural details provide protection for the structure. Coordination of the architectural and structural detailing is required in order to determine the extent of protection required.

8. Provision for Environmental-Control Systems Integrated planning of the environmental-control system and the structural and architectural systems helps to minimize cost. One may elect to use larger voids in structural elements, and deeper beams or joists, at additional structural cost, to accommodate a mechanical circulation pattern. On the other hand, one may use a more expensive environmental-control system, such as a high-velocity or high-pressure system with smaller ducts, in order to maintain the floor-construction depth. In high-rise buildings a mechanical floor is sometimes located at midheight to reduce the required floor area for vertical circulation. This may save enough floor area, depending on the number of floors, to compensate for the extra floor. This floor also can be used as a fire-resistance compartment in helping to solve fire-resistance problems.

9. Limitations of Various Systems For each particular type of structural system, executed in various materials, practical ranges may be established for the size of elements, spans, load capacities, modular units, etc. In addition, certain functional limitations may be inherent in the system, such as the necessity for multiple bays in concrete flat-slab construction, or practicable rise-span ratios in arch construction. Each system lends itself to certain geometric configurations which limit its applicability to particular architectural solutions. Systems vary in the degree to which they adapt to variations of shape or dissymmetry and to the accommodation of openings, special loads, etc. Some systems may be varied or blended with other systems to facilitate special conditions. Thus, a concrete flat slab may be changed to a framed system locally to accommodate openings or concentrated loads; an arch system may use suspended tendons or struts to support a horizontal surface; a linear, rectangular frame may utilize trussing or local rigid-frame action to achieve spatial stability or increased stiffness; a suspension membrane may be stiffened for reverse action as a vault or dome to reduce flutter; etc.

Each structural system has unique fabrication and erection problems. Some of these are:

1. Availability of materials, skilled labor, special equipment, etc. This is especially critical with respect to new systems, special materials, and highly complex or unique systems.

2. Speed of erection. This factor may influence the rejection of certain systems. The time rate of progress of other aspects of the construction relative to that of the structure should be considered. Prefabrication may speed field erection.

3. Maximum size of elements for handling, transporting, and erecting. Several factors

TABLE 4 Various Types of Deterioration of Structural Materials and Their Prevention

Material	Type of deterioration	Prevention or remedy
Wood	Dry rot, fungus	Chemical coating or impregnation Ventilate structure Avoid wetting or contact with ground
	Insect, rodent attack	Avoid contact with ground Initial coating or impregnation with repellent Periodic treatment with repellant
	Wear	Paint or other surface finish or applied finish elements Overdesign for some loss
Concrete	Freezing, thawing cycles	Avoid crack development as much as possible Use air-entraining cement or additive Paint or plastic coating to avoid moisture penetration
	Corrosive gases, liquids	Use corrosion-resistant cement Coat or artificially harden surface
	Wear	Coat, cover, or artificially harden surface Overdesign for some loss Use metal nosings and edgings
	Cracking, shrinking	Use mesh, reinforcing steel to minimize cracking Use expandable-type concrete to minimize shrinking Use densifier and other admixtures to minimize cracking in poured-in-place concrete Use denser and low-slump concrete for precast-concrete members For prestressed, precast members use initial prestressing force to minimize cracking
Metals	Oxidation	Paint, galvanize, etc., to protect surface Use nonprogressive oxidizing alloys Use specified minimum thicknesses for safety against dimensional loss Certain high-strength steels possess self-protective rust coating that prevents further rusting (used as architectural exposed finish) Aluminum has self-protecting film that prevents further oxidation
	Corrosive gases, liquids	As for oxidation. Plastic coatings effectively used
	Electrolysis	Avoid contact of dissimilar metals, especially where water is present
	Wear	Coat, cover, or harden surface
Plastic	Decomposition, chemical change	Avoid contact with noncompatible materials, especially adhesives, joint-sealing compounds, cleansing agents. Avoid conditions of exposure not suitable to material, e.g., temperature, sun, corrosive atmosphere
	Thermal extremes	Select material with tested adequacy for extremes of heat or cold anticipated
	Wear	Use surface hardening or coating

3. Maximum size of elements for handling, transporting, and erecting. Several factors must be considered, such as the degree to which the structure is prefabricated or shop-assembled, the distance from the shop to the site, the equipment available for handling, transporting, and erection, and the accessibility of the site.

4. Location of erection joints and sequence of erection. Construction joints should be located where they interfere least with the structural action, or where they can be designed for the least or simplest stresses.

5. Erection loads and stresses. Special conditions usually exist during the construction owing to the location of construction joints, handling of elements, loads caused by the weight of construction joints, handling of elements, loads caused by the weight of construction equipment or materials, etc. In some cases the structure itself may be designed for these loads, usually at an increase in allowable stresses due to the temporary nature of the condition. In other situations, temporary supports, bracing, or shoring may be required. For some systems these considerations may be an important aspect of the structural design.

LOADS

10. Dead Load Dead load includes the weight of the structure plus any permanently attached elements of the construction. Table 5 gives the weights of various building materials. Table 6 gives the average weights of various typical elements.

Evaluations of the dead load on any individual structural element are sometimes difficult to make. Approximation must be made for openings in walls, and the weight of

TABLE 5   Weights of Building Materials

Material	Weight, pcf	Material	Weight, pcf
Steel, rolled......................	490	Glass, common.................	156
Aluminum alloy, 6061-T6,		Glass, plate......................	161
structural wrought............	170	Plaster, cement and sand........	100
Iron, cast......................	450	Plaster, gypsum and fiber........	30
Bronze........................	509	Earth, clay, dry.................	65
Copper, cast or rolled............	556	Earth, clay, wet.................	110
Lead..........................	710	Earth, sand and gravel, dry, loose .	100
Wood, eastern spruce............	28	Earth, sand and gravel, wet, loose .	125
Wood, redwood.................	28	Earth, sand and gravel, dry,	
Wood, Douglas fir, coastal........	34	packed......................	115
Wood, southern longleaf pine.....	41	Earth, sand and gravel, wet,	
Wood, white oak................	47	packed......................	135
Wood, red cedar................	23	Earth, glacial till, mixed grain	
Wood, hard maple...............	42	size, dry.....................	130
Concrete, sand and gravel aggre-		Earth, glacial till, mixed grain	
gate, structural grade: 5-bag mix	145	size, wet.....................	145
Concrete, expanded shale aggre-		Earth, loam, loose, dry..........	75
gate, structural grade: 7-bag mix	105	Earth, loam, moist, packed......	95
Concrete, perlite aggregate, insu-		Plastic, soft (vinyl).............	85
lating fill, 5:1 mix..............	31	Plastic, hard, acrylate (Lucite)...	90
Brick, soft, common.............	100	Fiberboard, hard-pressed	
Brick, hard-burned, common.....	120	(Masonite)....................	45
Brick, pressed, finish............	140	Fiberboard, soft, structural, deck	
Limestone.....................	165	(Tectum).....................	25
Marble........................	170	Fiberboard, soft, structural, wall	
Granite.......................	165	(Celotex)....................	20

items such as doors, windows, hardware, trim, electrical fixtures and wiring, ductwork, plumbing fixtures and piping, and railings.

Although the structure is usually designed for the dead load of the completed building, for certain types of structures it may be necessary to investigate several stages of loading during the construction with respect to the changing effect on the stress distribution and stability of the structure. Temporary bracing or reinforcing, or even modifications in the design, may be necessary to provide for the sequence of loadings which occur during construction.

11. Live Load Live loads are usually specified by local or national building codes. Table 7 lists minimum live loads for building occupancies.

TABLE 6   Weights of Elements of Building Construction
All weights given are in psf and are average

Walls:		Floor and roof construction:	
4-in. brick, soft common	35	Wood joists, 2 × 8, 16 in. o.c., wood subfloor	6
4-in. brick, hard common	40	Wood joists, 2 × 12, 16 in. o.c., wood subfloor	8
4-in. brick, pressed finish	45	Metal deck, 22 gage, 1½ in. deep, 1 in. insulation board	3
4-in. hollow concrete block, stone aggregate (heavy)	30	Metal deck, 18 gage, 1½ in. deep, 1 in. insulation board	4
6-in. hollow concrete block, stone aggregate (heavy)	40	Metal deck, 20 gage, 1½ in. deep, stone concrete fill, 3 in. thick overall	38
8-in. hollow concrete block, stone aggregate (heavy)	55	Metal deck, 20 gage, 1½ in. deep perlite concrete fill, 2½ in. thick overall	10
10-in. hollow concrete block, stone aggregate (heavy)	68	Wood deck, solid, per inch	2½
12-in. hollow concrete block, stone aggregate (heavy)	80	Reinforced-concrete slab, stone aggregate, per inch	12½
4-in. hollow concrete block, lightweight aggregate	20	Reinforced-concrete slab, shale or slag aggregate, per inch	9½
6-in. hollow concrete block, lightweight aggregate	30	Gypsum Pyrofill concrete on formboard, bulb tee frame, 3-in. thick	10–12
8-in. hollow concrete block, lightweight aggregate	38	Wood W-trusses, 20–30 ft span, 24 ft o.c., ½-in. plywood sheathing	8–10
10-in. hollow concrete block, lightweight aggregate	45	Finish materials:	
12-in. hollow concrete block, lightweight aggregate	55	Floor finishes—see Table 16	
3-in. gypsum block	10	Hung ceiling, metal lath, sand plaster (portland cement)	10–25
4-in. gypsum block	14	Hung ceiling, metal lath, gypsum and fiber plaster	8
2-in. solid plaster on metal lath and studs	20	Hung ceiling, metal runners with fiber or metal tile	5–8
4-in. glass block	20	Siding, wood	1–2
2 × 4 studs, plaster board and plaster, 2 sides	15	Siding, asbestos shingles	1½–2
2 × 4 studs, ½-in. gypsum dry wall, 2 sides	8	Siding, 20-gage corrugated metal	2
Glass wall, small single glazed panes, light mullions	5–10	Roofing, built-up, tar and felt, gravel topping	5–8
Glass wall, large plate, heavy mullions	10–15	Roofing, shingles, asphalt or wood	2–3
Movable metal partitions	5–10	Roofing, slate shingles	10–18
Plaster 1 side of wall	4–5	Roofing, tile, clay	10–20
Miscellaneous elements:		Roofing, copper or monel metal	1½–2½
Skylight, ⅜-in. wire glass, light frame	8		
Insulation, loose or rigid, per inch	½–1		
Stairs, steel risers, treads, stringers, terrazzo fill	30–35		
Stairs, steel treads and stringers, open risers, no fill	15		
Stairs, wood	8–10		

Some types of live loads may be practically permanent in nature, although subject to removal or relocation. Movable partitions, hung ceilings, and building equipment are in this category. They should be considered as dead load when evaluating settlement of clay foundations, creep of concrete, sag of timber, etc.

For structural elements that support large areas of floor, live loads are often reduced based on the probability of full loading on the area (Table 7). Reductions are usually not permitted in the case of roof loads or of floor loads in warehouses or other buildings with storage-type occupancy. Caution and judgment should be exercised in reducing specified live loads for high live-load to dead-load ratios and when deflections are critical.

TABLE 7   Minimum Loads for Floors

Use or occupancy	Uniform load[a]	Concentrated load
Armories	150	0
Assembly areas,[b] auditoriums, and balconies therewith		
Fixed seating areas	50	0
Movable seating and other areas	100	0
Stage areas and enclosed platforms	125	0
Cornices, marquees, and residential balconies	60	0
Exit facilities, public (corridors, exterior exit balconies, stairways, fire escapes, and similar uses)	100	0
Garages		
General storage and/or repair	100	c
Private pleasure car storage	50	c
Hospital wards and rooms	40	1000[d]
Libraries		
Reading rooms	60	1000[d]
Stack rooms	125	1500[d]
Manufacturing		
Light	75	2000[d]
Heavy	125	3000[d]
Offices	50	2000[d]
Printing plants		
Press rooms	150	2500[d]
Composing and linotype rooms	100	2000[d]
Residential (includes private dwellings, apartments and hotel guest rooms)	40	0
Rest rooms[e]		
Reviewing stands, grand stands, and bleachers	100	0
School classrooms	40	1000[d]
Sidewalks and driveways for public access	250	c
Storage		
Light	125	
Heavy	250	
Stores		
Retail	75	2000[d]
Wholesale	100	3000[d]

[a]Except for places of public assembly, and except for live loads greater than 100 psf, the design live loads for columns, piers, walls, foundations, trusses, beams, and flat slabs supporting more than 150 sq ft may be reduced at the rate of 0.08% per sq ft of supported area, but not more than 40% for horizontal members or for vertical members receiving load from one level only, or 60% for other vertical members, or 23.1 $(1 + D/L)$ where D = dead load, psf, and L = live load, psf, on member.

[b]Includes dance halls, drill rooms, gymnasiums, playgrounds, plazas, terraces, and similar occupancies generally accessible to the public.

[c]Garages for storage of private pleasure cars shall have the floor system designed for a concentrated wheel load of not less than 2000 lb without uniform live load.

[d]Provision shall be made for this load acting on any space 2½ ft square wherever on an otherwise unloaded floor it produces stresses greater than those caused by the specified uniform load.

[e]Not less than the load for the associated occupancy, but need not exceed 50 psf.

Reproduced from the 1976 edition of the Uniform Building Code, copyright 1976, with permission of the publishers, the International Conference of Building officials.

12. Snow Load   Figure 1 gives the ground snow load for a 50-year mean recurrence interval. It is recommended that design snow loads for permanent structures be based on these ground loads. However, if there is an unusually high degree of hazard to life and property, a 100-year recurrence interval should be used, and where risk to human life is negligible a 25-year interval may be used.[4] Maps for these intervals are given in Ref. 4.

Snow load on a roof is obtained by multiplying the ground values by a coefficient which depends on the exposure and configuration of the roof. Reference 4 gives values of this coefficient, denoted by C_s, for sheltered shed roofs as follows:

$$\text{Roof slope } \alpha \text{ from 0 to 30°} \quad C_s = 0.8$$

$$\text{Roof slope } \alpha \text{ from 30 to 70°} \quad C_s = 0.8 - \frac{\alpha - 30}{50}$$

$$\text{Roof slope } \alpha \gtrsim 70° \quad C_s = 0$$

These values also apply to gable and hip roofs. However, if $\alpha > 20°$, such roofs must also be checked for snow loads 25 percent larger on only one side of the gable.

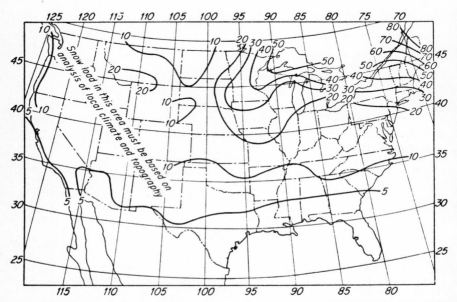

Fig. 1   Ground snow load, psf, 50-year recurrence interval (*United States Weather Bureau Map 12158, January 1969.*)

TABLE 8   Minimum Live Loads* for Roofs

Roof slope		Tributary area for member, ft²		
Shed or gable	Arch or dome	0–200	201–600	Over 600
$0 \leq \text{slope} < 4/12$	$h/l < 1/8$	20	16	12
$4/12 \leq \text{slope} < 12/12$	$1/8 \gtrsim h/l < 3/8$	16	14	12
$12/12 \leq \text{slope}$	$3/8 \gtrsim h/l$	12	12	12

*On horizontal projection, psf.
SOURCE: ANSI A58.1-1972 Building Code Requirements for Minimum Design Loads in Buildings and Other Structures.[4]

In areas where wind is strong enough to remove snow, values of C_s may be reduced 25 percent for roofs exposed on all sides.[4]

Provision should be made for large accumulations due to drifting on irregularly shaped roofs and flat roofs at their intersection with vertical surfaces. Values of C_s as large as 3 are given in Ref. 4 for various cases.

If the minimum roof load from Table 8 exceeds the snow load, design should be based on the value from the table.

13. Wind Loads Wind pressures on a structure depend on the wind velocity and the shape, size, and exposure of the structure. The pressure p is given by

$$p = C_p q \tag{1}$$

where C_p = pressure coefficient (also called shape factor)
q = velocity pressure
The velocity pressure, psf, is given by

$$q = 0.00256 V^2 \tag{2}$$

where V = wind velocity, mph
  Wind velocity varies with height and with the terrain, and the velocity pressure q_z at a height z above ground is given by

$$q_z = 0.00256 K_z V_{30}^2 \tag{3}$$

where K_z = coefficient from Eq. (4)
V_{30} = basic wind speed, mph, 30 ft above ground
The velocity at 30 ft above ground is used because it is the standard elevation at which wind speeds are measured. The coefficient K_z is given by

$$K_z = 2.64 \left(\frac{z}{z_g}\right)^{2/\alpha} \tag{4}$$

where z_g = gradient height = height above which wind speed is assumed to be constant
α = coefficient
The following values of α and z_g for the exposures A, B, and C defined in Table 9 are recommended in Ref. 3:

Exposure	α	z_g, ft
A	3	1500
B	4.5	1200
C	7	900

The *effective pressure* q_{eff} on a structure depends on the dynamic response of the structure to wind, and can be determined by

$$q_{eff} = Gq = 0.00256 G K_z V_{30}^2 \tag{5}$$

where G = gust factor.
  Table 9 gives values of q_{eff} for the exposures A, B, and C for V = 100 mph and including the effects of K and G in Eq. (5). ($G = 1$ for internal pressures.) Values for other velocities are found by multiplying the tabular values by $(V/100)^2$. Values from part (b) of the table should be used for windows, girts, etc., and for tributary areas less than 200 ft.^{2} Values from part (a) should be used for tributary areas of 1000 ft^2 or more. Values for intermediate areas may be interpolated.
  The effective velocity pressures in Table 9 do not account for the effects of vortex shedding or instability due to galloping or flutter. A procedure for dynamic analysis of these effects is given in Ref. 4.
  Wind velocities with a mean recurrence interval of 50 years should ordinarily be used (Fig. 2). However, if there is an unusually high degree of hazard to life and property in case of failure, a 100-year interval should be used, and where there is negligible risk to human life, a 25-year interval may be used.
  Walls of Buildings. Pressure coefficients are given in Tables 10 and 11. The net pressure on the wall of the building is the algebraic sum of the external and internal pressures. Since the internal pressure is assumed to be the same on all walls at a given height, the resultant pressure on the building is the sum of the external pressure on the windward wall and the external suction on the leeward wall. Thus the resultant wind pressure on a building for which the height-width ratio and height-length ratio both exceed 2.5 is obtained by multiplying the effective velocity pressure from Table 9 by C_p = 0.8 − (−0.6) = 1.4.

The Uniform Building Code[1] (UBC) prescribes the wind pressures given in Table 12. These pressures are based on wind velocities at the 30-ft. level, multiplied by 1.3 to account for gusts. This corresponds to a gust factor $G = 1.3^2 = 1.69$ in Eq. (5). A shape factor of 1.3 was then applied to obtain the design pressures. Thus they differ from the ANSI code values in that they are resultant pressures (sum of windward and leeward wall

TABLE 9 Effective Velocity Pressure q_{eff} for $V = 100$ mph*

| | External pressure | | | | | | Internal pressure on walls and roofs of buildings | | |
| | Ordinary buildings and structures | | | Parts and portions of buildings and structures | | | | | |
Height, ft	A†	B‡	C§	A†	B‡	C§	A†	B‡	C§
Less than 30	9	16	26	13	24	38	5	13	26
30	11	20	33	13	24	38	5	13	26
50	13	24	38	16	27	42	7	16	30
100	18	30	44	21	34	49	11	22	36
200	26	38	51	29	42	57	18	30	44
300	31	43	56	35	48	61	23	36	49
400	36	48	59	41	53	66	28	40	54
500	40	51	62	45	57	69	33	45	57
600	44	55	65	49	61	72	37	48	60
700	48	58	67	53	64	74	41	52	63
800	51	61	70	57	67	76	45	55	65

*Values for velocities ranging from 50 to 130 mph are given in ANSI A58.1. Pressures in this table may be multiplied by $(V/100)^2$ to obtain pressures for velocity V. This gives results in agreement with values from ANSI A58.1 except for slight differences for heights of 30 ft or less.
†Large cities and very rough, hilly terrain.
‡Suburban areas, towns, city outskirts, wooded areas, and rolling terrain.
§Flat, open country, open flat coastal belts, and grassland.
SOURCE: ANSI A58.1-1972 Building Code Requirements for Minimum Design Loads in Buildings and Other Structures.[4]

pressures) instead of effective velocity pressures which must be multiplied by a net pressure coefficient as noted above.

Local Pressures on Walls.   Wind pressures at building corners may be considerably higher than the average pressure on a wall. Suction at the corners is important in respect to fastening of elements such as exterior finish panels. Reference 4 prescribes a pressure coefficient of -2 for wind pressures at corners, with the suction assumed to act on a vertical strip $0.1w$ wide, where w is the least width of the building. This pressure is not to be combined with the overall net external pressure.

Roofs.   For flat roofs of buildings for which the ratio of height to width is less than 2.5, $C_p = -0.7$. If the height-width ratio exceeds 2.5, $C_p = -0.8$.

For gabled roofs $C_p = -0.7$ for the leeward slope. Table 13 gives values for the windward slope. These coefficients also apply for shed roofs.

Local Pressures on Roofs.   Table 14 gives net pressure coefficients for ridges, eaves, cornices, and 90° corners of roofs. The pressure is assumed to act on strips $0.1w$ wide, where w is the least width of the building normal to the ridge. This pressure is not to be combined with the overall net external pressure.

Tall Buildings.   Wind-resistant design of tall buildings is discussed in Part 3 of this section.

Tanks, Chimneys, and Solid Towers.   Net pressure coefficients for these structures are given in Table 15. They are to be used with the effective velocity pressures from (a) of Table 9. They should not be used with Table 12, since the values in that table already include a pressure coefficient of 1.3. UBC gives factors of 1, 0.8, and 0.6 for square or rectangular, hexagonal or octagonal, and round or elliptical shapes, respectively, by which values in Table 12 are to be multiplied to obtain pressures for any ratio h/d.

Resultant pressures are determined by multiplying unit pressures by the projected area normal to the wind direction.

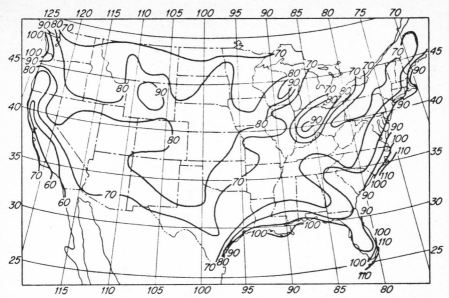

Fig. 2 Annual extreme fastest-mile speed 30 feet above ground, 50-year mean recurrence interval. (*Ref. 5.*)

TABLE 10   External-Pressure Coefficient C_p for Walls*

Location	C_p
Windward	0.8
Leeward:	
h/w and $h/l \geq 2.5\dagger$	−0.6
Other	−0.5
Sidewalls	−0.7

*Use values of q_{eff} for ordinary buildings and structures from Table 9. Negative values denote suction.

$\dagger h, w, l$ = height, width, length of building, respectively.

SOURCE: ANSI A58.1-1972 Building Code Requirements for Minimum Design Loads in Building and Other Structures.[4]

TABLE 11   Internal-Pressure Coefficients C_{pi} for Buildings*

	Openings uniformly distributed	Openings mainly in	
		Windward wall	Leeward and sidewall
0–0.3	±0.3	+0.3 + 1.67n†	−0.3 − n†
>0.3	±0.3	0.8	−0.6

*Use values of q_{eff} for internal pressure from Table 9. Negative values denote suction.

$\dagger n$ = ratio of open area to solid area of wall with majority of openings.

SOURCE: ANSI A58.1-1972 Buiding Code Requirements for Minimum Design Loads in Buildings and Other Structures.[4]

TABLE 12 Horizontal Wind Pressure on Vertical Surfaces, psf

Height zone, ft	Wind-pressure-map area, psf, Fig. 3						
	20	25	30	35	40	45	50
Less than 30	15	20	25	25	30	35	40
30–49	20	25	30	35	40	45	50
50–99	25	30	40	45	50	55	60
100–499	30	40	45	55	60	70	75
500–1,199	35	45	55	60	70	80	90
1,200 and over	40	50	60	70	80	90	100

Reproduced from the 1976 edition of the Uniform Building Code, copyright 1976, with permission of the publishers, the International Conference of Building Officials.[1]

TABLE 13 External-Pressure Coefficients C_p for Windward Slope of Gabled Roofs*

h/w†	Roof slope θ								
	10–15°	20°	25°	30°	35°	40°	45°	50°	≥60°
≤0.3	−1.0‡	0.2	0.25	0.3	0.35	0.4	0.45	0.5	0.01θ
0.5	−1.0	−0.75	−0.5	−0.2	0.05	0.3	0.45	0.5	0.01θ
1.0	−1.0	−1.0	−0.8	−0.55	−0.3	−0.05	0.2	0.45	0.01θ
≥1.5	−1.0	−1.0	−1.0	−0.9	−0.6	−0.35	−0.1	0.2	0.01θ

*Use values of q_{eff} for ordinary buildings and structures from Table 9. Negative values denote suction.
† h = wall height at eave, w = least width of building normal to ridge.
‡ Use $C_p = 0.01\theta$ for this case if $h/w = 0$.
SOURCE: ANSI A58.1-1972 Building Code Requirements for Minimum Design Loads in Buildings and Other Structures.[4]

TABLE 14 Local External-Pressure Coefficients C_p for Roofs

Roof slope θ, deg*	Ridges and eaves	Corners
0–30	−2.4	0.1θ − 5
>30	−1.7	−2

*For arched roofs θ = angle between horizontal and tangent to roof at springing.
SOURCE: ANSI A58.1-1972 Building Code Requirements for Minimum Design Loads in Buildings and Other Structures.[4]

TABLE 15 Net Pressure Coefficients for Tanks, Chimneys, and Solid Towers*

Shape	Type of surface	h/d†		
		1	7	25
Square (wind normal to face)	Smooth or rough	1.3	1.4	2.0
Square (wind along diagonal)	Smooth or rough	1.0	1.1	1.5
Hexagonal or octagonal ($d\sqrt{q} > 2.5$)	Smooth or rough	1.0	1.2	1.4
Round ($d\sqrt{q} > 2.5$)	Moderately smooth‡	0.5	0.6	0.7
	Rough§ ($d'/d \sim 0.02$)	0.7	0.8	0.9
	Very rough§ ($d'/d \sim 0.08$)	0.8	1.0	1.2

*Use values of q_{eff} for ordinary buildings and structures from Table 9.
† h = height, ft; d = diameter or least horizontal dimension, ft.
‡ Concrete, metal, timber.
§ d' = depth of protruding elements such as ribs and spoilers, ft.
SOURCE: ANSI A58.1-1972 Building Code Requirements for Minimum Design Loads in Buildings and Other Structures.[4]

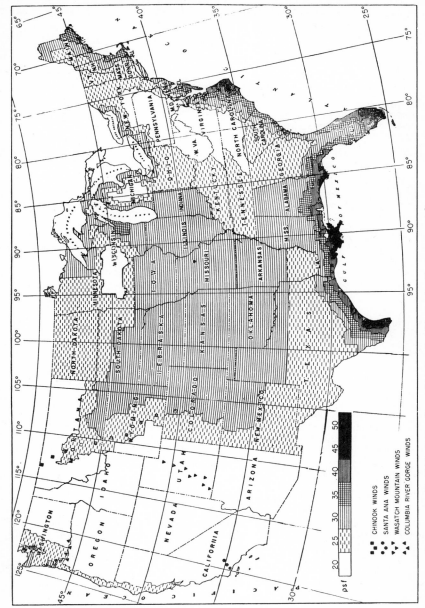

Fig. 3 Wind-pressure map for Table 12. Resultant of windward and leeward wall presures on ordinary square buildings at 30 ft above the ground (Ref. 1).

For slender structures such as flagpoles Ref. 4 prescribes $C_p = 1.2$ minimum if $d\sqrt{q} < 2.5$.

Trussed Towers.   Coefficients for these structures, to be used with the wind pressures of Table 12, are given in Table 16. Pressures are to be computed on the projected area, normal to the wind, of all the members of one face. Pressure coefficients to be used with Table 9 are given in Ref. 4.

TABLE 16   Shape Factors for Radio Towers and Trussed Towers

Type of exposure	Factor*
Wind normal to one face of tower:	
Four-cornered, flat or angular sections, steel or wood	2.20
Three-cornered, flat or angular sections, steel or wood	2.00
Wind on corner, four-cornered tower, flat or angular sections	2.40
Wind parallel to one face of three-cornered tower, flat or angular sections	1.50
Factors for towers with cylindrical elements are approximately ⅔ of those for	
similar towers with flat or angular sections	
Wind on individual members:	
Cylindrical members:	
2 in. or less in diameter	1.00
Over 2 in. in diameter	0.80
Flat or angular sections	1.30

*To be used with wind pressures of Table 12.

Reproduced from the 1976 edition of the Uniform Building Code, copyright 1976, with permission of the publishers, the International Conference of Building Officials.[1]

Outdoor Signs.   Table 17 gives net pressure coefficients to be used with Table 9.

14. Seismic Loads*   Figure 4 shows seismic-risk zones of the United States as defined by the Uniform Building Code. This code recommends that the minimum total lateral seismic force V, assumed to act nonconcurrently in the direction of each of the main axes of a building, be determined by

$$V = ZIKCSW \tag{6}$$

where Z = seismic-zone coefficient = ³⁄₁₆, ³⁄₈, ³⁄₄, and 1 for zones 1, 2, 3, and 4, respectively (Fig. 4)

I = occupancy importance factor (Table 18)

K = coefficient from Table 19

C = coefficient from Eq. (7)

S = site-structure resonance coefficient from Eqs. (9)

W = total dead load (for storage and warehouse occupancies total dead load plus 25 percent of floor live load). Snow load to be included if it exceeds 30 psf but may be reduced if justified by duration

The coefficient C is given by

$$C = \frac{1}{15\sqrt{T}} \tag{7}$$

but need not exceed 0.12. The product CS need not exceed 0.14. T is the fundamental period of vibration, in seconds, in the direction considered. In the absence of properly substantiated technical data for the contemplated structure, T may be determined from

$$T = \frac{0.05h_n}{\sqrt{D}} \tag{8a}$$

where h_n = height above base of uppermost level in main portion of structure, ft

D = dimension of structure parallel to applied forces, ft

In buildings in which the lateral resisting system consists of a moment-resisting space

*See also Sec. 3, Earthquake-Resistant Design.

TABLE 17a   Net Pressure Coefficients for Wind Normal to Surface of Solid Signs[a,b]

At ground level[c]							
H^d	≤ 3	5	8	10	20	30	≤ 40
C_p^e	1.2	1.3	1.42	1.52	1.75	1.84	2.0

Aboveground level[c]							
a/b^f	≤ 6	10	16	20	40	60	≤ 80
C_p^e	1.2	1.3	1.42	1.52	1.75	1.84	2.0

[a]Signs with openings less than 30% of the gross area are classified as solid. Signs with openings greater than 30% of the gross area are classified as open (Table 17b).

[b]To allow for wind oblique to the surface of solid signs, the net pressure normal to the surface may be assumed to vary linearly from a maximum value of $1.6KC_p$ at the windward edge to a minimum of $0.4KC_p$ at the leeward edge, where $K = 1$ for rectangular signs having the shorter edge upwind and 1.15 for square signs and for rectangular signs having the longer edge upwind.

[c]Solid signs are classified as being at the ground if $g/h < 0.25$; where g = distance between bottom of sign and ground and h = vertical dimension of sign.

[d]H = ratio of height to width of surface.

[e]Use of values of q_{eff} for ordinary buildings and structures from Table 9.

[f]a = larger dimension, b = smaller dimension.

SOURCE: ANSI A58.1 Building Code Requirements for Minimum Design Loads in Buildings and Other Structures.[4]

TABLE 17b   Net Pressure Coefficients for Latticed Frameworks*

		Rounded members	
ϕ†	Flat-sided members	$d/\sqrt{q} < 2.5$	$d/\sqrt{q} > 2.5$
<0.1	2.0	1.2	0.8
0.1–0.3	1.8	1.3	0.9
0.3–0.7	1.6	1.5	1.1

*Use values of q_{eff} for ordinary buildings and structures from Table 9.

†Within ± 0.05 of the common boundary of ranges of ϕ, mean values of pressure coefficients for the two ranges may be used.

SOURCE: ANSI A58.1 Building Code Requirements for Minimum Design Loads in Buildings and Other Structures.[4]

TABLE 18   Occupancy Importance Factor

Occupancy	I
Essential facilities*	1.5
Buildings where primary occupancy is assembly of more than 300 persons in one room	1.25
All other	1

*Examples: hospitals and other medical facilities having surgery or emergency-treatment areas, fire and police stations, vital municipal-government disaster-operation and communication centers.

Reproduced from the 1976 edition of the Uniform Building Code, copyright 1976, with permission of the publishers, the International Conference of Building Officials.[1]

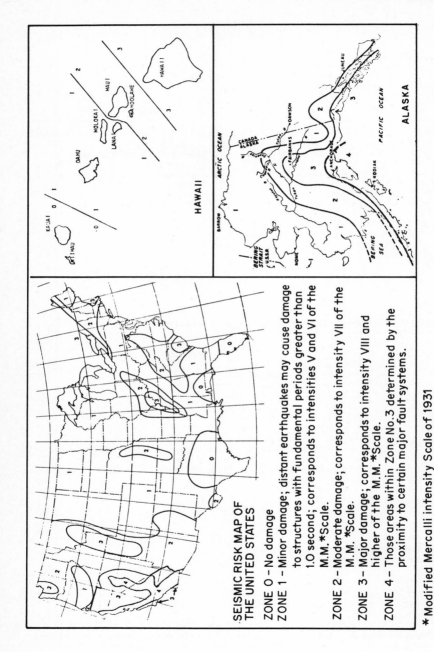

SEISMIC RISK MAP OF
THE UNITED STATES

ZONE 0 – No damage
ZONE 1 – Minor damage; distant earthquakes may cause damage
to structures with fundamental periods greater than
1.0 second; corresponds to intensities V and VI of the
M.M. *Scale.
ZONE 2 – Moderate damage; corresponds to intensity VII of the
M.M. *Scale.
ZONE 3 – Major damage; corresponds to intensity VIII and
higher of the M.M. *Scale.
ZONE 4 – Those areas within Zone No.3 determined by the
proximity to certain major fault systems.

* Modified Mercalli Intensity Scale of 1931

Fig. 4   Seismic zone map of the United States. (Ref. 1.)

frame which resists 100 percent of the lateral forces, and provided that the frame is not enclosed by or adjoined by more rigid elements which tend to prevent it from resisting lateral forces, T may be determined from

$$T = 0.10N \tag{8b}$$

where N = number of stories above base to uppermost level in main portion of structure.

TABLE 19 Horizontal Force Factor K for Buildings or Other Structures

Type or arrangement of resisting elements	K
All building framing systems except as hereinafter classified	1.00
Buildings with a box system*	1.33
Buildings with a dual bracing system consisting of a ductile moment-resisting space frame and shear walls, for which (a) the frame and the shear walls are designed to share the lateral force according to their relative rigidities, (b) the shear walls, acting independently, are designed to resist the entire lateral force, and (c) the space frame is designed to resist not less than 25% of the lateral force	0.80
Buildings with a ductile moment-resisting space frame designed to resist the entire lateral force	0.67
Elevated tanks plus contents, on four or more cross-braced legs, not supported by a building	2.50†
Structures other than buildings and other than those covered by Eq. (6)	2.00

*Defined as a structural system without a complete vertical load-carrying space frame, the lateral forces being resisted by shear walls or braced frames.

†KC in Eq. (6) shall be at least 0.12 but need not exceed 0.25. The tower shall be designed for a torsion of 5 percent. For tanks not supported in the manner described and for tanks supported by buildings use Eq. (11) with $C_p = 0.2$.

Reproduced from the 1976 edition of the Uniform Building Code, copyright 1976, with permission of the publishers, the International Conference of Building Officials.[1]

The coefficient S is given by

$$S = 1 + \frac{T}{T_s} - 0.5 \left(\frac{T}{T_s} \right)^2 \geq 1 \qquad \text{if } \frac{T}{T_s} \leq 1 \tag{9a}$$

$$S = 1.2 + 0.6 \frac{T}{T_s} - 0.3 \left(\frac{T}{T_s} \right)^2 \geq 1 \qquad \text{if } \frac{T}{T_s} \geq 1 \tag{9b}$$

where T_s = characteristic site period.

T in Eqs (9) must be taken not less than 0.3 sec. If the value determined by properly substantiated analysis exceeds 2.5 sec, $T = 2.5$ may be used.

A range of values of T_s may be established by properly substantiated geotechnical data, but T_s is not to be taken less than 0.5 sec or more than 2.5 sec. The value to be used in Eqs. (9) is that value, within the range of site periods, nearest to T. If T_s is not properly established, S must be taken equal to 1.5.

The lateral force V is to be distributed over the height of the structure in the following manner:

A force F_t at the uppermost level n given by

$$F_t = 0.07TV \tag{10a}$$

This force need not exceed $0.25V$ and may be taken zero if $T \leq 0.7$ sec.

A force F_x at each level, including the uppermost level n, given by

$$F_x = (V - F_t) \frac{w_x h_x}{\sum\limits_{i=1}^{n} w_i h_i} \tag{10b}$$

where w_x, w_i = portion of W located at or assigned to level x, i

h_x, h_i = height of level x, i above base, ft

Parts of Structures (see also Sec. 3, Art. 21). The lateral force F_p on parts of structures, according to the UBC, is

$$F_p = ZIC_p SW_p \tag{11}$$

where W_p = weight of part. Values of C_p are given in Table 20. If $C_p = 1$ or more, the values of I and S need not exceed 1.

Overturning. See Sec. 3, Art. 14.

FLOOR AND ROOF CONSTRUCTION

15. Floor and Roof Systems   Table 21 illustrates various types of deck elements and their weight, fire resistance, and load capacity for various beam spacings. Selection of the proper deck must be based on:

   1. Span and load, with consideration being given to type (simple-span or continuous), deflection limitations, etc.

   2. Acoustical and thermal insulating properties, if critical.

   3. Fire resistance.

   4. Framing details: attachment to supports, support of hung ceilings, ducts, light fixtures, piping, etc.

   5. Diaphragm action required for transfer of lateral loads.

   6. Cost of material, installation, finishing, etc. Supports for the deck may be rolled

TABLE 20   Horizontal-Force Factor for Elements of Structures

Part or portion of building	Direction of force	C_p
Exterior bearing and nonbearing walls, interior bearing walls and partitions, interior nonbearing walls and partitions. Masonry or concrete fences	Normal to flat surface	0.20
Cantilever parapet	Normal to flat surface	1
Exterior and interior ornamentations and appendages	Any	1
When connected to, part of, or housed within a building:		
1. Towers, tanks, towers and tanks plus contents, chimneys, smokestacks, and penthouses	Any	0.2^a
2. Storage racks with upper storage level higher than 8 ft, plus contents	Any	$0.2^{a,b}$
3. Equipment or machinery not required for life safety systems or for continued operation of essential facilities	Any	$0.2^{a,c}$
4. Equipment or machinery required for life safety systems or for continued operation of essential facilities	Any	$0.5^{c,d}$
When resting on the ground, tank plus effective mass of contents	Any	0.12
Suspended ceiling (zones 2, 3, and 4 only)	Any	0.20^e
Floors and roofs acting as diaphragms	Any	0.12^f
Connections for exterior panels or for precast, nonbearing, nonshear wall panels or similar elements attached to or enclosing the exterior	Any	2
Connections for prefabricated structural elements other than walls, with force applied at center of gravity of assembly	Any	0.3^g

[a]When located in the upper portion of any building where $h_n/D \geq 5$, use $C_p = 0.3$.

[b]W_p = weight of racks plus contents. For racks over two storage support levels high, use $C_p = 0.16$ for levels below the top two.

[c]For flexible and flexibly mounted equipment and machinery C_p shall be determined with consideration given to the dynamic properties of the equipment or machinery and of the building or structure in which it is placed, but shall not be less than the values listed.

[d]Design and detailing of equipment which must remain in place and be functional following a major earthquake shall consider effects induced by structure drifts not less than $2/K$ times the story drift caused by seismic forces or less than the story drift due to wind. The product IS need not exceed 1.5.

[e]Ceiling weight shall include light fixtures and other equipment laterally supported by the ceiling. A ceiling weight of not less than 4 psf must be used to determine the lateral force.

[f]$W_p = w_x$ in Eq. (11), but if F_x from Eq. (10b) exceeds W_p, it shall be used instead.

[g]W_p shall include 25 percent of the floor live load in storage and warehouse occupancies.

Reproduced from the 1976 edition of the Uniform Building Code, copyright 1976, with permission of the publishers, the International Conference of Building Officials.[1]

TABLE 21 Comparison of Structural Decks

Deck	Weight, psf	Max fire resistance, hr Un-pro-tected‡	Max fire resistance, hr Pro-tected	Use	Total superimposed load, psf 25	50	75	100	150
½-in. plywood, 2 or more spans	1.5	X	1	Roof	2.67	2.50	2.0	2.0	2.0
				Floor		1.33	1.33	1.33	1.33
¾-in. plywood, 2 or more spans	2.2	X	1	Roof	4.0	3.33	2.75	2.5	2.25
				Floor		2.0	2.0	2.0	2.0
1⅝-in. T & G plank, 2 or more spans	3.5	X	1	Roof	11	10	8	7	6
				Floor	6*	5.5*	5*	4.5*	4*
2⅝-in. T & G plank, red cedar, 2 or more spans	5	X	1	Roof		15.5	13	11.5	9.5
				Floor	9.5*	9*	8*	7.5*	6.5*
3⅝-in. T & G plank, red cedar, 2 or more spans	7	X	1	Roof		21	18.5	16	13
				Floor	14.5*	12.5*	11.0*	10*	9*
22-gage steel deck, narrow rib, 1½ in. deep, 3 or more spans	1.8	X	3	Roof	7.0	5.5	4.5	4	3.5
18-gage steel deck, wide rib, 1½ in. deep, 3 or more spans	3	X	3	Roof	10.5	8.33	6.7	6.0	5.0
18-gage steel deck, 4½ in. deep, simple span	5.0	X	3	Roof		18	15.5	13.0	11.0
Steel cellular units, 16-gage plate + 18-gage channel, 4½ in. deep, simple span	6.5	X	3	Roof			16†	16	14†
Steel cellular units, 16-gage corrugated + 16-gage corrugated, 3 in. thick, 2½ in. concrete fill. Noncomposite	40	X	4	Floor			13*	12*	10*
18-gage steel deck, wide rib, 1½ in. deep, 2½ in. concrete fill	36	X	4	Floor		7.5	7	6	5
Reinforced-concrete slab on 22-gage, 1⁵⁄₁₆-in.-deep steel deck, 4 in. total thickness, negative reinforcement	48	3	4	Floor		9.5	8.5	7.5	6.5
Composite concrete slab on 22-gage, 1½-in. steel deck, 4½ in. total thickness. Stone concrete	50	1½	4	Floor		13	11.5	10	9
Composite concrete slab on 20-gage, 3-in. steel deck, 5½ in. total thickness. Stone concrete	56	2	4	Floor		14	13	11	9.5
Perlite concrete fill 1:5 on 28-gage, ⁹⁄₁₆-in. deck, 2½ in. total thickness, 3 or more spans	8	X	3	Roof	4.0†	3.0†	3.0†		
Precast concrete plank, slag aggregate, 2 in. deep	14	X	3	Roof		7	6	5	3
Precast concrete cellular unit, 8 in. deep:				Roof					
No topping	55	2	4	and	26	24	22	19	16
2-in. topping	80	3	4	floor	26	25	23	21	18

TABLE 21   Comparison of Structural Decks *(Continued)*

| | | Max fire resistance, hr | | | Load-span limits, ft | | | | |
| | | | | | Total superimposed load, psf | | | | |
Deck	Weight, psf	Un-pro-tected‡	Pro-tected	Use	25	50	75	100	150
Reinforced-concrete slab, stone aggregate, 150 pcf, 4½ in. deep, $\rho = 0.5\,\rho_b$ simple span									
$f_c = 3000$	56	2	4	Roof	13*	12*	11*	10*	9*
$f_y = 50{,}000$ 4000				and	14.3*	13.3*	12.3*	11.5*	10.5*
5000				floor	14.5*	13.5*	12.5*	12*	11*
Reinforced-concrete slab, stone aggregate, 150 pcf, 6 in. deep, $\rho = 0.5\,\rho_b$, simple span									
$f_c = 3000$	75	3	4	Roof	16.5*	15.5*	14.5*	14*	13*
$f_y = 50{,}000$ 4000				and	18*	16.5*	15.5*	15*	13.5*
5000				floor	18.5*	17*	16*	15.5	14
Reinforced-concrete slab, lightweight aggregate, 110 pcf, 4½ in. deep, $\rho = 0.5\,\rho_b$, simple span,									
$f_c = 3000$	42	4	4	Roof	11.5*	11*	10*	9.5*	8.5*
$f_y = 50{,}000$ 4000				and	13*	12*	11*	10*	9*
5000				floor	13.5*	12.5*	11.5*	10.5	9.5*
Reinforced-concrete slab, lightweight aggregate, 110 pcf, 6 in. deep, $\rho = 0.5\,\rho_b$, simple span									
$f_c = 3000$	55	4	4	Roof	15.5	14.5	13.5	12.5	11.5
$f_y = 50{,}000$ 4000				and	16	15	14	13	12
5000				floor	17	15.5	14.5	13.5	12.5
Wood fiberboard, 2 in. thick, plank form	6	X	2	Roof	3	3	2.5		
2 in. gypsum concrete slab on 1 in. insulation, bulb tees 32⅝ in. c. to c., tees No. 158 at 1.33 lb/ft	11	X	3	Roof	6	4.5	3.75	3	
2 in. gypsum concrete slab on 1 in. insulation, bulb tees 32⅝ in. c. to c., tees No. 258 at 4.67 lb/ft	12	X	3	Roof	12	11	9	8	

Flexure stress governs except as noted:
*Deflection $L/360$ governs.
†Deflection $L/240$ governs.
‡X = no rating.

steel sections, solid or laminated wood beams, poured or precast concrete beams, trusses, etc. Selection of the supports will be based on:

   1. Span of the deck.
   2. Span of the supports, with consideration being given to the type (simple span or continuous).
   3. Deflection limitations. Consideration must be given to the additive deflection of the deck, the supports for the deck, and any additional supporting beams or girders.

4. Framing details: attachment of the deck to the supports and attachment of the supports to the columns, walls, or girders.

5. Lateral support provided by the deck.

6. Fire resistance.

7. Cost.

Table 22 compares typical assemblies of roof construction. Table 23 presents a similar comparison of typical floor constructions.

16. Floor Finish   In some cases the surface of the structure may serve as the finished floor. More often, a finish of some kind is applied and consideration must be made for its weight, thickness (for depression of the structure), and fireproofing quality. Table 27 lists various floor-finishing materials and their properties.

Some floors may require the addition of a fill material, usually concrete. This may be required for leveling, as in the case of corrugated metal decks or precast concrete elements which do not align perfectly. More often, however, the fill is required for covering underfloor wiring, raceways, or conduits. The fill may also function as insulation for fireproofing or acoustical separation. A minimum of 2½ in. will usually be required to cover underfloor wiring. This fill may occasionally be utilized structurally, as when it is combined with metal deck or precast concrete elements in composite action.

17. Roofing   In most climates provision must be made for roof drainage. On nominally flat roofs a slight pitch of the surface is usually desirable to facilitate drainage. This pitch may be as slight as ⅛ in./ft, or 1 percent. Construction of drains and downspouts may require special detailing and framing. Where the structure itself cannot be conveniently sloped, a fill may be required which may double as insulation and fireproofing.

The roof surface must be protected against penetration of water by some type of roofing. On flat roofs this will usually be the membrane type. Commonest of these is that consisting of multiple layers of felt mopped with tar and topped with pea gravel for protection during hot weather. Plastic membranes consisting of multiple layers of materials such as neoprene and Hypalon are especially useful for complex formed roofs and surfaces exposed to view. These plastic membranes may also be used on pitched and vertical surfaces. Sheets of tin, copper, lead, aluminum alloy, stainless steel, and galvanized or enameled steel may be used as roofing on both flat and sloping surfaces. On slopes of 20 percent or more various types of lapped shingles, such as wood, asphalt, clay tile, slate, or metal, may be used.

A major problem in roofing is the carrying of a continuous seal across discontinuities such as ridges, valleys, and intersections of horizontal and vertical surfaces. These often require elaborate and expensive flashing with copper or some other ductile and noncorrosive metal. In some cases, plastic membrane roofing may eliminate the need for flashing. Where thermal movements of the structure are not the same as those of the roofing, rupture may occur if the membrane is not sufficiently resilient to absorb the differential movement. Expansion joints are required in extremely long buildings.

The roof must usually provide some insulation against heat transmission. Ceiling, dead-air space, structural deck, and roofing act together to retard this heat transfer, but some additional material with a purely insulating function is often required. A few structural decks such as timber, gypsum or perlite concrete, or fiberboard may be adequate without added insulation.

Most roofs are designed for a nominal amount of traffic, such as that occurring during construction and that due to maintenance and inspection after construction. Where more than the usual amount of traffic is anticipated, walkways or duckboards may be advisable. Roof areas to be used for promenades, terraces, sun decks, etc., will usually require a membrane protected by paving.

A unique, though not uncommon, situation is that of a flat roof used as a reservoir for water for the air-conditioning system or as a heat moderator in warm weather. Detailing of roofing and flashing usually requires special consideration.

Most contemporary buildings have a collection of miscellaneous items on their roofs, such as ventilators, cooling towers, antennas, signs, skylights, and hatches, which may require special framing. Occasionally these items are collected in a penthouse structure in one section of the roof.

18. Types of Walls   Walls may be classified with respect to their function and degree of permanence as follows:

TABLE 22 Comparison of Typical Roof Construction

Items of comparison / Elements	Weight of structure, psf	Fire resistance, NBFU rating, hr	Insulating value of bare structure, U factor, Btu per hr per sq ft per °F per in.	Min overall thickness or depth, in.	Limiting span, ft	Diaphragm action for distribution of lateral loads	Variations	Special considerations
Solid wood decks	2½–3t	Combustible, but slow burning; no rating without ceiling. Pressurized or painted fire-retardant treatments can provide rating	Fair to good 1⅝:0.30 2⅝:0.20 3⅝:0.16	1⅝	t, in. / Live load, psf — 20 30 40 — 1⅝: 7 6 6 — 2⅝: 11 10 9 — 3⅝: 15 14 13 — All limited by deflection	Fair to good	Thick units may be glue-laminated for economy and better surface finish control	Large openings difficult to frame without subframe
Wood joists at 12, 16, or 24 in. o.c. with ⅜- to ⅝-in. plywood deck	4–7	Poor. Combustible. No rating without ceiling	Nil without ceiling. No rating	8 with 2 × 8 joist	Joists at 16 in. c to c / Live load, psf — 20 30 40 — 2 × 8: 14 12½ 11½ — 2 × 10: 17 16 14½ — 2 × 12: 21 19 17½ — All limited by deflection	Poor unless plywood is nailed at all edges—then fair to good	¾-in. TG boards instead of plywood. Joist spacings of 12, 18, 24 in., etc.	Blocking or bridging of joists required for long spans
Wood beams 3 to 5 ft o.c. with 1⅝-in. wood deck	5–10	Combustible but slow burning. None without ceiling	Fair. 0.30 with 1⅝-in. deck	7 with 6-in. beam	Beams at 5 ft 0 in. / Live load, psf — 20 30 40 — 3 × 6: 9 8 7½ — 3 × 8: 12 11 10 — 4 × 6: 10 9 8 — 4 × 8: 13 12 11 — 4 × 12: 20 18 17 — All limited by deflection	Fair	Thicker deck for wider beam spacing. Glu-lam for larger beams	Rigid specification of materials if quality finish is desired

Wood nailed trusses and plywood sheathing trusses 24 in. o.c.	5–10	Poor. Combustible	Nil without ceiling	Min pitch 2½:12	Usual spans 20–30	Fair	Closer truss spaces for long spans	Creates virtually useless attic space
Wood trusses with wood joists and sheathing. Trusses 15 to 20 ft o.c.	15–25	Poor. Combustible	Nil without ceiling	Min pitch 3 or 4:12	Usual spans 40–150	Fair	Flat, pitched, bowstring, etc.	Some trusses stock by manufacturers
Glue-laminated beam with 1⅝-in. deck	6–12	Combustible. No rating without ceiling or retardant treatment	Nil without ceiling	7½ with 6½-in. beam	Beams at 5 ft 0 in. — Live load, psf: 20 / 30 / 40 3¼ × 6¼: 10.5 / 10 / 9 3¼ × 8⅛: 13 / 12 / 11 4¼ × 6½: 11.5 / 10.5 / 10 4¼ × 8⅛: 14.5 / 13 / 12 4¼ × 13: 23 / 21 / 19 All limited by deflection	Fair	Thicker deck for wider beam spacing. Shaped or tapered beams can be used	Strict specification of materials, glue, moisture content if quality is desired
Plywood box beams with 2⅝-in. deck 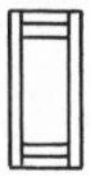	9–15	Combustible. No rating without ceiling	Fair	22½ in. with 20-in. beam	Beams at 10 ft 0 in. — Live load, psf: 20 / 30 / 40 20 in. deep, ⅝ in. web, two 2 × 6 each flange: 24 / 22 / 20 24 in deep, ¾ in. web five 2 × 4 each flange: 32 / 28 / 24 36 in. deep 1 in. web five 2 × 8 each flange: 70 / 60 / 52 All limited by shear	Good	3 to 6 webs can be used to increase capacity	Careful fabrication required. Shear transfer by glue and/ or nails critical

TABLE 22 Comparison of Typical Roof Construction (*Continued*)

Items of comparison Elements	Weight of structure, psf	Fire resistance, NBFU rating, hr	Insulating value of bare structure, U factor, Btu per hr per sq ft per °F per in.	Min overall thickness or depth, in.	Limiting span, ft	Diaphragm action for distribution of lateral loads	Variations	Special considerations
Stressed-skin wood panel	4–10	Combustible. No rating without ceiling	Fair	4½ in.	Usual span 8–30 ft	Good	Usually 4 ft 0 in. module. Other widths can be used	Top plywood usually thicker than bottom. Joints must be connected by pressured glue or/and nails
Concrete: slab and beam system	Stone aggregate 55 up	Excellent. Up to 4 hr	Poor	3½-in. slab plus beam but not < $L/20$	Dead weight becomes impractical for spans over 30–35	Excellent	Prestress for long spans	Accommodates openings easily. DL/LL high
Concrete: one-way pan-formed joists	35–90	1–2 hr. Non-combustible.	Poor	6-in. pan plus 2-in. slab but not < $L/24$	See Table 24. Practical range 20–40	Good	Forms of pressed wood, cardboard, or fiber glass	Modular coordination of pan layout. DL/LL high
Precast concrete: Slab units with voids	40–80 without topping	¾ to 2 hr depending on thickness of unit, unit weight of concrete, aggregate, voids, and topping	Fair. 0.5	4 in. min. $t > L/40$	(see table below)	Fair to good	Various depths and widths of units and void configurations	Alignment of adjacent members with different camber or use concrete topping

Unit	Live load, psf		
	20	30	40
6 in.	24	23	22
8 in.	27	25	24
10 in.	31	27	27
12 in.	40	34	30

All limited by deflection

System	Weight, psf	Fire rating	Acoustic	Depth	Span	Economy/Forms	Variations	Remarks
Concrete: two-way pan-formed joists (waffle system)	70–120 up to 160	1–2 hr depending on slab thickness and reinforcing steel coverage	Poor	Joist width 4 in. pan plus 2½	See Table 25. Practical range 25–50	Excellent	Forms of cardboard or fiber glass can be almost any size to satisfy modulation	Modular coordination of pan layout. Openings somewhat restrictive. Solid panel or fingering out to take care of shear
Concrete: flat-slab system	12.5t plus 10% for drop panels and column caps	Excellent. Up to 4 hr	Poor	4	Practical range 18–40. See Table 26	Excellent	Flat plate (without drops and caps) for simpler forming	Requires the least story height. Minimize openings in the intersection of column strips
Precast prestressed concrete, single and double tees	35–80. With topping 60–105	¾–2 hr depending on thickness of concrete flange, tendon coverage, and concrete unit weight	Not good. Better with topping. 1.0–0.75	t > L/35 to L/40	(see Unit table below)	Fair to good	Width 4 ft 0 in. to 10 ft 0 in. Depth 12–48 in. Shape: tee, double tee, F-section, channel	Check camber for roof loading and long span
Steel formed deck units with rigid insulation	2–7 plus insulation board	Poor. Noncombustible	Fair to good with insulation	1½	(see Gage table below)	Fair to good	Deeper ribs for longer spans. Core units and channel units. Poured insulating concrete fill instead of board	Detail so deck serves as vapor barrier

Precast prestressed concrete — span units:

Unit	Live load, psf		
	20	30	40
12 in. deep double tee	38	30	28
18 in. deep	50	44	40
24 in. deep	72	66	60
36 in. single tee	100	90	80

Steel formed deck units — span gages:

Gage	Live load, psf		
	20	30	40
22	10	9	8
18	10	10	10
All limited by deflection			

TABLE 22 Comparison of Typical Roof Construction (*Continued*)

Items of comparison / Elements	Weight of structure, psf	Fire resistance, NBFU rating, hr	Insulating value of bare structure, U factor, Btu per hr per sq ft per °F per in.	Min overall thickness or depth, in.	Limiting span, ft	Diaphragm action for distribution of lateral loads	Variations	Special considerations
Steel formed deck on steel beams	8–20	Poor. Noncombustible. No rating without ceiling. 1–3 hr with sprayed-on fireproofing	Poor. Good with insulation	2½ plus beam. Usually 14½–20½	Varies with beam size and spacing. For beams 8 ft o.c. Beam — Live load, psf (20, 30, 40) W12 × 16 — 25 23 21 W14 × 17.2 — 28 26 24 W16 × 26 — 34 32 29 W18 × 46 — 40 39 38	Fair to good	Various decks for different beam spacings	Highly flexible for openings, cantilever, special loads
Steel formed deck on open-web steel joists	4–8	Poor. Noncombustible. None without ceiling	Poor. Good with insulation	2½ plus joist. Usually 14½–26½	All limited by deflection Varies with joist size. Lightest joist 6-ft spacing — Live load, psf (20, 30, 40) 12 — 20 17 15 16 — 28 24 22 20 — 35 31 28 24 — 40 35 31	Fair to good	Various decks for different joist spacings	Bridging, anchoring, and other accessories required
Poured gypsum deck on bulb tees	10–12	Poor. Noncombustible. None without ceiling	Fair to good. 0.15–0.40	2½	Function of bulb tee size: 4–12	Fair to good	Different form boards for insulation, acoustics, appearance. Various tees for different spans	
Poured gypsum deck and bulb tees on open-web steel joists	12–18	Poor. Noncombustible. None without ceiling	Fair to good. 0.15–0.40	2½ plus joist. 14½–26½	Varies with joist. Practically same as steel formed deck on open-web steel joists	Fair to good	Joist spacing	Bridging, anchoring, and other accessories required

TABLE 23 Typical Floor Construction

Items of comparison — Elements	Weight of structure, psf	Live-load to dead-load ratio, LL = 100 psf	Fire resistance, NBFU rating, hr	Acoustical properties	Min overall thickness, in.	Span limit, ft; * indicates deflection limitation	Diaphragm action for lateral loads	Deflection characteristics	Impact resistance	Variations	Special considerations
Solid wood deck	2½–3t	t 1⅝ 33 2⅝ 20 3⅝ 14	Combustible, but slow burning. None without ceiling. Pressurized or painted fire-retardant treatments can provide rating	Fair	1⅝	Live load, psf 50 100 1⅝ 5½* 5* 2⅝ 9* 7* 3⅝ 13* 10*	Fair to good	Fair for two-span units. 1⅝ thick bouncy on spans over 4 ft	Fair to good. Material dents and splinters easily	Thick units laminated for economy and control of surface	Large opening difficult to frame without subframe
Wood joists 12, 16, or 24 in. o.c. with ⅜ to ⅝ in. plywood deck	4–7	Joists at 16 in. 2 × 8 20 2 × 10 17 2 × 12 14	Poor. None without ceiling. Pressurized or painted fire-retardant treatments can provide rating	Fair with ceiling	Nominal depth of joist	Joists 16 in. c. to c. Live load, psf 50 100 2 × 8 11* 9* 2 × 10 14* 11* 2 × 12 17* 14*	Poor. Fair to good if plywood nailed at edges	Plywood bouncy for joist spacing greater than 16 in.; check deflection	Deck may puncture	¾ in. T & G boards. Double joists for concentrated loads	Blocking or bridging required for deep beams and long spans
Wood joists 3 to 5 ft o.c. with 1⅝-in. deck	5–10	Joists at 4 ft 3 × 6 20 4 × 8 17 6 × 12 13	Combustible, but slow burning. None without ceiling. Pressurized or painted fire-retardant treatments can provide rating	Fair	Nominal depth of joist	Joists 4 ft to c. Live load, psf 50 100 3 × 6 6* 5* 4 × 8 9* 7* 6 × 12 16* 13*	Fair	See solid wood deck	Fair	Thicker deck for wider joist spacing. Glulam for larger beams	Rigid specification for quality finish

TABLE 23 Typical Floor Construction *(Continued)*

Elements \ Items of comparison	Weight of structure, psf	Live-load to dead-load ratio, LL = 100 psf	Fire resistance, NBFU rating, hr	Acoustical properties	Min overall thickness, in.	Span limit, ft;* indicates deflection limitation	Diaphragm action for lateral loads	Deflection characteristics	Impact resistance	Variations	Special considerations
Glue-laminated beams with 2⅝-in. deck	9–15	12–7	Combustible. No rating without ceiling or retardant treatment	Nil without ceiling	Actual depth of beam	Beams 6 ft o.c. — Live load, psf: 50 / 100 3¼ × 8⅛: 11* / 9* 4¼ × 11⅜: 15* / 12* 4¼ × 14⅝: 19* / 15* 5 × 16¼: 23* / 18* 7 × 24⅜: 38* / 30* All limited by deflection	Fair	Selection of higher E values of wood to increase rigidity	Fair	Thicker deck for wider beam spacing. Shaped or tapered beams can be used	Rigid specification of materials, glue, moisture content if quality is required
Plywood box beams with 3⅝-in. deck 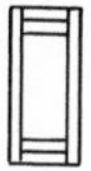	12–20	8–5	Combustible. No rating without ceiling	Fair	Nominal beam depth	Beams 8 ft o.c. — Live load, psf: 50 / 100 20 in. deep ⅝-in. web two 2 × 6 each flange: 20 / 10 24 in. deep ¾-in. web five 2 × 4 each flange: 28 / 14 36 in. deep 1-in. web five 2 × 8 each flange: 52 / 28 All limited by shear	Good	Deflection is not critical	Fair	3 to 6 webs can be used to increase capacity of bending and shear	Fabrication important. Critical requirement is shear, transfer by glue and/or nail
Stressed-skin wood panel 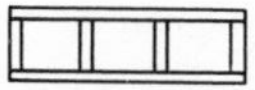	6–12	16–8	Combustible. No rating without ceiling	Fair	Min 6½–12½ in.	Usual span 8–30 ft	Good	Deflection usually governs	Good	Usually 4 ft 0 in. module. Other width available	Top plywood usually is thicker than the bottom panel. Joints must be connected by pressure glue and/or nails

System										
One-way concrete slab	12.5t (stone), 9t (lightweight)	Stone aggregate t 4 2 6 1.33 8 1 10 0.8	Good. Up to 4 hr depending on t, cover, aggregate	Good. Transmits surface noise	3¾-4	See Table 21	Excellent	Creep deflection may be 1.5 to 2 times elastic deflection	Good	Lighten thick slabs by inserting fiber tubes	Construction and expansion joints required
Two-way reinforced-concrete slab on edge beams	Same as one-way + 30% for beams	Stone aggregate t 4 1.54 6 1.02 8 0.77 10 0.62	Same as one-way slab	Good. Transmits surface noise	4	min $t = L/42$ (average)	Excellent	See one-way. Two-way action reduces deflection	Good	Waffle slab (two-way joists) for lighter structures and/or longer spans. Rectangular bays (up to 1:1.5)	Openings require careful planning. Multiple bays most effective
Reinforced-concrete one-way slab-and-beam system	Same as one-way + 50% for beams	t 4 1.43 6 0.95 8 0.72 10 0.57	Same as one-way slab	Good. Transmits surface noise	3¾ plus beam. Not less than $L/20$ preferred	Depends on beam. Practical limit 50-60	Excellent	See one-way concrete slab	Good	Multiplicity	Slab thickness affected by live load, fire rating, beam spacing
Reinforced-concrete one-way pan-formed joists	110	2.9-1.1	1-2 hr, depending on t, cover, aggregate	Good. Transmits surface noise	Pan plus slab. Not less than $L/24$ preferred	See Table 24, Practical range 20-40	Good	See one-way concrete slab	Good	Joists formed with pressed wood fiber, cardboard, or fiberglass forms	Usually requires some coordination with modular layout
Reinforced-concrete two-way pan-formed joists (waffle system)	70-120	2.14-1.25	1-2 hr, depending on t, cover, aggregate	Good. Transmits surface noise	Pan plus slab. Not less than $L/22$ preferred	See Table 25. Practical range 30-55 ft	Good	See one-way concrete slab Deflection is not a problem	Good	Forms of cardboard or fiber glass can be almost any size to satisfy modulation	Modular coordination of pan layout. Openings somewhat restrictive. Solid panel or fingering out to increase shear capacity

TABLE 23 Typical Floor Construction *(Continued)*

Elements	Weight of structure, psf	Live-load to dead-load ratio, LL = 100 psf	Fire resistance, NBFU rating, hr	Acoustical properties	Min overall thickness, in.	Span limit, ft;* indicates deflection limitation	Diaphragm action for lateral loads	Deflection characteristics	Impact resistance	Variations	Special considerations
Reinforced concrete flat slab	Same as one-way plus 10% for drop panels and column caps	t 4 1.82 6 1.21 8 0.91 10 0.73	Same as one-way slab	Good Transmits surface noise	5	Square bay, live load, psf t 50 100 200 5 16 15 13 6 19 18 16 8 25 24 23 10 33 32 29	Excellent	See one-way concrete slab	Good	Flat plate (without drops and caps) for simpler forming. Waffle slab (2-way joists) for lighter structures and/or longer spans	Requires multiple bays
Precast prestressed concrete slab units with voids	40–50 without topping	(With topping) 6×12 8×16 1.54 1.25	Fair to good. None by NBFU. With topping 3 hr by others	Good. Transmits surface noise	6-in. unit + 2-in. topping = 8	Unit Live load, psf 50 100 6 × 12 20 16 8 × 16 26 21	Good	Good. Prestressing reduces deflection	Fair to good	Unit width, thickness, void configurations	Monolithic action with topping. Variation in camber of units requires special treatment to align adjacent units
Precast prestressed rectangular L-shaped inverted tee, I-shaped beams with slab	200–1000 plf	20–3	1–4 hr depending on flange and web thickness, tendon coverage aggregate	Good with ceiling	Slab plus beam	Beam depth varies from 16 to 60 in. span range 16–60 ft	Good with slab	Deflection not a problem	Good	Slab thickness can vary, wide range of depth	Composite action with cast-in-place concrete slab
Precast concrete slab units on steel beams	60–100	1.67–1	1–3 hr, depending on thickness of units, aggregate, concrete topping, spray-on fireproofing	Good, excellent with ceiling	Slab unit plus beam probably 20	Varies with beam spacing and size, and with thickness and strength of concrete slab units. Practical range 18–45 ft	Fair, good with concrete topping		Good	Unit width, thickness, void configuration. Top of unit may line up with top of steel beam to save story	Requires more story height. Care in alignment of units

Precast prestressed concrete, single and double tees	35–80, 60–105 with topping	2.84–0.63	¾–2 hr, depending on thickness of flange and tendon coverage, conc. wt, aggregate, concrete topping	Good, transmits surface noise	$> \frac{L}{35}$ to $\frac{L}{40}$	Unit 4–10 ft — Live load, psf: 50 / 100; 14-in. TT 28–36 / 20–28; 18-in. TT 32–42 / 26–34; 24-in. TT 40–50 / 32–42; 36-in. T 60–72 / 44–62; 48-in. T 78–94 / 62–80	Fair to good	Good. Check camber	Fair to good	Width 4–10 ft, depth 12–48 in. Tee, double tee, channel, F-section, I-section	Adjustment for adjacent members with different camber
Single-unit steel deck: 18-gage 1½-in. deck plus 2½-in. concrete fill	35	2.50–1.50	Noncombustible. With ceiling, insulation sprayed on underside, or composite deck, 1–2 hr	Fair. Excellent with ceiling or insulation	4	Depends on gage of deck, effectiveness of composite action, and number of continuous spans — Live load, psf: 50 / 100; Composite 9* / 8*; Noncomposite 7* / 6	Fair to good	May be bouncy on long spans. Continuous spans better	Fair. Recommend structural-grade topping to minimize puncture	Deck gage and thickness of topping. Deeper decks (3, 4½, and 7½ in.) for longer spans	If composite check construction loads
Cellular-unit steel deck: 18-gage 1½ in. upper bonded to 16-gage 4½-in. lower, 2½-in. concrete fill	40–45	2.2–1.25	Noncombustible. With ceiling, or insulation sprayed on underside, or composite deck, 1–2 hr	Fair to good. Excellent with ceiling or insulation	4½ + 1½ deck + 2½ fill = 8½	Function of deck depth and gages, effectiveness of composite action, and number of continuous spans — Live load, psf: 50 / 100; Composite 23 / 19; Noncomposite 21* / 17*	Good	Stiff. No sag. Composite action of topping may reduce bounciness	Good. Recommend structural-grade topping to minimize puncture	Size and gage of deck	Header ducts required to use cores for wiring. Check construction loads
18-gage 1½-in. deck with 2½-in. concrete fill, on steel beams	45–50	2.22–2.00	Noncombustible. Insulation on beams and underside of deck, or composite deck, 1–2 hr	Fair. Excellent with ceiling or insulation	4 + beam. Probably 14–20 total	Varies with size of beam. Practical range 18–40, 15–40% longer for composite beams	Fair to good	Deck deflection and beam deflection additive	Good	Deck, beam size and spacing	

TABLE 23 Typical Floor Construction (*Continued*)

Items of comparison / Elements	Weight of structure, psf	Live-load to dead-load ratio, LL = 100 psf	Fire resistance, NBFU rating, hr	Acoustical properties	Min overall thickness, in.	Span limit, ft;* indicates deflection limitation	Diaphragm action for lateral loads	Deflection characteristics	Impact resistance	Variations	Special considerations
Concrete deck on corrugated steel centering with open-web steel joists	30–35	2.9–3.3	Noncombustible. None without ceiling. 1–2 hr with fireproof ceiling	Fair. Good with ceiling	Deck plus joists, probably 12–24 total	Varies with joist size and spacing. Practical range 18–60	Fair to good	May be bouncy if design load is small		Forming techniques for deck	Composite joist system may be used to decrease height (about 2½ in.), save weight, and minimize deflection and bounciness
Concrete slab with rolled steel beams. Composite action	55–100	1.82–1.00	Rating depends on slab thickness; 1–4 hr if beams have sprayed-on insulation or encased in concrete	Good. Excellent with ceiling	Slab plus beam. Probably 14–25	Varies with beam spacing and size, and with thickness and strength of concrete. Practical range 18–50	Good	Good	Good. Excellent if beam encased	Thickness and strength of concrete, size and spacing of beams. Metal deck as form and positive steel	Design of shear developers

Structural walls: walls essential to the structure as supporting or bracing elements.

Curtain walls: exterior walls serving only as enclosure elements.

Fixed partitions: nonstructural walls likely to remain in place, such as walls around elevator shafts and toilets.

Movable partitions: nonstructural walls subject to removal or relocation due to changing occupancy requirements.

TABLE 24   Span Limit, ft, Reinforced-Concrete One-Way Pan-Formed Joists

$f_c = 4000$ $f_y = 50,000$ $\rho = 0.33\rho_b$ $\Delta = \dfrac{L}{240}$(creep considered)

Slab thickness t, in.	Pan dimensions, in.		Width joist, in.	Live load, psf					
				50			100		
	H	W		Span* (capacity)	Δ	Max recommended span	Span (capacity)	Δ	Max recommended span
2½	6	20	4	18	13	17	14	12	17
2½	10	20	5	29	19	25	24	18	25
2½	14	20	5	38	25	33	31	23	33
3	6	20	4	19	14	18	16	12	18
3	10	20	5	30	23	26	25	18	26
3	14	20	5	38	25	34	32	23	34
2½	6	30	4	16	12	17	13	10	17
2½	10	30	5	26	20	25	21	16	25
2½	14	30	6	36	24	33	30	22	33
2½	14	30	8	39	26	33	33	24	33
3	6	30	4	17	12	18	13	11	18
3	10	30	5	26	18	26	22	16	26
3	14	30	6	36	24	34	30	22	34
3	14	30	8	39	26	34	34	23	34

*Tapered ends may be required.

Mobile partitions: sliding, folding, or rolling walls, subject to frequent relocation.

Table 28 shows various typical wall assemblies and some characteristics and properties to be considered in their selection and use.

19. Nonbearing Walls Even though they make little or no contribution to the basic structure of the building, some consideration must be given to the structural character of nonbearing walls. Important considerations are:

Stability and stiffness to resist wind, impact, and construction handling, and to be capable of receiving attached elements.

Anchorage to the structural frame, especially against lateral forces. Exterior walls must be designed for both inward and outward wind forces, which most codes specify to be equal.

Isolation from the structural frame to prevent loading of the wall due to deflection or other movement of the frame. This may necessitate the use of expansion material or flexible or sliding joints between the frame and the walls.

20. Bearing Walls Important considerations in the design of bearing walls are:

Load capacity with respect to both strength and stability. Concentrated or eccentric loads may require distributing elements or reinforcing for the wall.

Openings may require lintels or other reinforcing.

Stability against lateral forces. Intersecting walls may provide mutual support, if adequately attached. Fixity at the base or rigid attachment to the element it supports may be sufficient to stabilize a wall. Bracing may be provided in the form of pilasters, wing walls, or buttresses.

21. Windows Wind force is the chief factor in determining glass thickness. Factors to be considered are total area of panes, proportions of the sides of the panes, and the degree of freedom of the edges of the panes in the frame. Sizes of individual panes are also limited by the difficulty of transporting and handling large flexible sheets of glass. This is especially critical with long, narrow strips. It is wise to make inquiry with local suppliers whenever large panes are to be used.

Anchorage of window heads is a difficult problem, since provision must be made for deflection of the spandrel beam. Lateral bending and deflection of mullions may also be critical. Tall mullions should have sufficient stiffness not to display visible deflection due

TABLE 25   Span Limitations, ft, Reinforced-Concrete Two-Way Joists (Waffle)*

Slab thickness t, in.	Pan dimensions		Width joist, in.	Live load, psf	
	H	W		100	150
3	6	19	5	12–18	12–17
3	8	19	5	14–22	14–20
3	10	19	5	16–26	15–24
3	12	19	5	22–31	20–30
3	14	19	5	25–36	24–34
4	6	19	5	15–20	14–20
4	8	19	5	16–26	15–24
4	10	19	5	20–30	18–28
4	12	19	5	24–34	22–32
4	14	19	5	26–38	24–36
3	8	30	6	14–24	13–22
3	10	30	6	16–30	15–28
3	12	30	6	18–32	16–30
3	14	30	6	22–36	20–34
3	16	30	6	26–42	24–40
3	20	30	6	30–48	28–46
4	8	30	6	16–26	14–24
4	10	30	6	18–30	18–28
4	12	30	6	20–34	20–32
4	14	30	6	24–38	24–36
4	16	30	6	27–44	25–42
4	20	30	6	32–50	30–48

*Upper limits on span are approximately 24 times depth of joist.

TABLE 26   Span Limitations, ft, Reinforced-Concrete Flat Slab*

Slab thickness t, in.	With drop panel		Without drop panel	
	Live load, psf		Live load, psf	
	100	150	100	150
6	16–18	15–17	13–15	11–13
7	18–21	16–20	14–19	13–18
8	20–24	18–23	17–22	14–20
9	22–26	20–25	19–24	16–22
10	23–29	22–28	20–26	18–25
11	25–32	23–31	22–30	20–27
12	27–35	26–35	23–32	22–30

*Upper limits on span are 36 to 40 times depth of slab.

to reversal of the wind pressure on the window. While wood has a low modulus of elasticity, wood mullions are usually solid and may thus actually be stiffer than light-gage, hollow, metal mullions, of the same outside dimensions. If proper separation is made to inhibit electrolysis, interior steel elements may be used to stiffen light mullions of aluminum and bronze.

STAIRS

22. Planning Figure 5 illustrates layout criteria for stairs, indicating the basic considerations of rise and run, clear width, and overhead clearances. The riser may be omitted. For purposes of toe room it should be recessed.

TABLE 27 Properties of Floor Finishes

Type of finish	Weight, psf	Usual thickness, in.	Fireproofing quality	Sound-deadening value*
Resilient tile: vinyl, cork, asphalt, rubber	½–1	1/16–¼	None	Nil, except ¼ in. cork fair
25/32-in. hardwood, block or strip	4	⅞ with felt	None	Nil
Hardwood on sleepers, over concrete slab	5	1½–2½	None	Fair
Ceramic tile, glue on	3–5	⅜–⅝	Nil	Nil
Ceramic tile set in mortar	15–25	1½–2½	Good	Fair
Terrazzo, thin	5–8	⅝	Nil	Nil
Terrazzo, usual mortar bed	20–30	1½–2½	Good	Fair
Asphalt mastic paving, over structural slab	15–20	1½–2	Fair	Fair
Cut stone, on mortar bed	30–40	3–4	Good	Fair
Cement, troweled topping	12 per in.	1 min	Good	Nil
Carpet with underlay	1–2	½–1	None	Good

* Transmission of surface impact noises.

For ease and safety a relatively narrow range of stair angle is desired. Indoor stairs for public use should be between 30 and 40°. Lower angles may be used for outdoor stairs or for grand staircases, but for angles below 20° a ramp is preferable. Angles of 45° or more become very steep in sensation to the user, and at some point—say above 50°—the stair becomes more like a ladder and requires very narrow treads and open risers.

The height of an individual riser and width of an individual tread have certain optimal ranges. The simplest relationships between tread and riser are: $17 < R + T < 18$, and $70 < R \times T < 80$. For optimal stair angles between 30 and 40°, acceptable risers range from 6½ to 8 in. The usually accepted ideal riser is approximately 7¼ in. The actual riser dimension may be an uneven fraction and can merely be specified as the total rise divided by the number of risers. The tread dimension may be similarly expressed, although it is usually specified.

23. Types The simplest form of stairs is the single flight, straight plan. When prescribed by code, or necessary in the designer's judgment, an intermediate landing may be inserted. Additional landings may be added.

The stair may also be curved in plan. It may radiate tightly from a central pole or may be a graceful helix on a large diameter. These may cantilever from their landings without intermediate support or hang by a series of delicate rods or cables. Curved stairs must be carefully laid out so that the varying tread width functions in some acceptable range with the constant riser.

24. Framing Stair framing depends largely on the materials and on the nature of the enclosing or supporting structure. In stairs of wood or steel, the treads and/or risers usually span in the direction across the stair and are supported at the edge by the walls or by stringers which span from landing to landing. The stringer may be combined with the landing to form an angled bent. Concrete stairs are often designed as tilted slabs—in single flights or combined with either or both landings.

While the simplest forms of stair framing are either the continuous edge support or the top and bottom support (as with a single rafter), several other possibilities exist, especially for the free-standing, or open, stair. The flights may be supported on cantilevered landings

TABLE 28 Typical Wall Construction

Items of comparison Elements	Weight, psf of wall surface	Fire resistance	Acoustical properties	Insulating properties	Size limits: thickness t, in., length L, height H	Use	Bearing capacity, psi of gross area	Use as diaphragm stiffener for structure	Attachment of elements	Special considerations
Reinforced concrete	(Stone) 12.5t	Excellent	Good barrier. Surface noises readily transmitted	Poor	$t = H$ of pour, ft $= 6$ min if bearing $= 8$ min if foundation $L = 100–150$ ft between expansion joints $H = 25t$ if bearing	Fixed, bearing stiffening, foundation	0.15 to 0.25f_c', depends on H/t	Excellent	Difficult. Attach before placing concrete	Sandblast, bushhammer, or plain exposed surfaces
Precast concrete panels	12.5t, 8–10t ltwt., plus 1–2 for insulated panels	Good	Good barrier. Surface noises readily transmitted	Good through insulation of sandwich panels. Poor at edges and joints of units and for solid panels	$t = 4–8$ L and H depend on fabricator, and transporting and erecting facilities	Fixed, curtain, bearing	Depends on thickness and height	Good composite structure	Difficult. Cast in elements	Transporting. Erecting heavy units without damage. Exposed aggregate sandblast and bushhammer finishes
Solid brick	8–12t	Excellent	Good barrier	Fair	$t = 8$ up. Recommend control joint at 50 to 100 ft. H or L between bracing 20t max if bearing, 25t for reinforced grouted masonry	Fixed, bearing, stiffening	Depends on mortar and brick strength. 50–400	Excellent if bonded to structure	Difficult. Embed elements in mortar joints or drill in anchors	Slow construction. Weather effects. Construction sequence of bearing walls and reinforced walls

Brick plus block	Brick plus block	Excellent	Good barrier	Fair	$t = 8$ up. Control joints at 20–35 ft. H or L between bracing 18t max if bearing	Fixed, bearing, stiffening	Depends on mortar. 70–85. Block governs	Excellent if bonded to structure	Difficult. Embed elements in mortar joints or drill in anchors	Bonding of elements. Bearing limited on block
Concrete block	5 + 4t (lightweight concrete), 10 + 5t (stone concrete)	Good to excellent	Good barrier. Cavity in thick wall deadens some surface impacts	Fair	$t = 4, 6, 8, 10, 12$ nominal. $L = 20$–35 ft between control joints. H or $L = 18t$ max between bracing, 25t if reinforced	Fixed, bearing, stiffening, foundations, partitions	Depends on mortar. 70–85	Good if bonded to structure. Limited shear capacity.	Difficult. Heavy objects should have separate or built-in frame	Solid block or filled block required where elements bear on wall. Reinforcing required in horizontal courses. Mortarless construction
Brick veneer on wood frame	45–55	Poor	Good barrier	Good	$t = 8$ up. No length limit, control joints in brick advisable. Height depends on frame; 9–10 ft with 2 × 4 at 16	Fixed, limited bearing, and stiffening	Stud wall with 2 × 4 at 16, 9 ft high: 35–40	Good	See brick. Wood frame simple for light objects. Need reinforcing for heavy ones	Differential movements of brick and wood frame
Wood siding on wood frame	8–10	Poor	Fair	Fair to good	$t = 5\frac{1}{2}+$	Exterior bearing wall. Curtain wall	On 2 × 4 at 16: 35–40	Fair to good	Easy. Reinforce for heavy objects	Rapid erection
Corrugated sheet steel on steel frame	Sheet: 3–4 Frame: 5–10	Nil. Non-combustible	Nil	Nil	$t = \frac{1}{2}$ plus framing	Curtain wall, partition	0	Poor	To framing only. Relatively easy	
Metal sandwich panel	3–4	Nil. Non-combustible	Poor to fair	Fair	$t = 1\frac{1}{2}$ up	Curtain wall, partition	0. Some for panel with vertical members	Poor	To framing only	

TABLE 28 Typical Wall Construction *(Continued)*

Items of comparison Elements	Weight, psf of wall surface	Fire resistance	Acoustical properties	Insulating properties	Size limits: thickness t, in., length L, height H	Use	Bearing capacity, psi of gross area	Use as diaphragm stiffener for structure	Attachment of elements	Special considerations
Solid plaster on metal lath	$10t$	Fair to good. Best with gypsum-perlite plaster	Poor, depending on texture	Fair to poor	$t = 2$ up. Height or length between bracing $36t$ max	Partitions	0	Fair	Difficult. Reinforce for heavy objects	
Gypsum block. Plaster both sides	With plaster 3 in. 4 in. 5 in. 20 21 25	Fair to excellent	Fair to good	Fair to good	$t = 3$ up. Height or length between bracing $36t$ max	Partitions	0	Fair	Difficult. Reinforce for heavy objects	Not recommended for damp conditions
Steel lath and studs. Plaster both sides	25	Fair to good. Best with gypsum-perlite plaster	Fair	Fair	$t = 3$ up	Partitions, Limited bearing	35–40	Fair	Difficult. Reinforce for heavy objects	
Wood studs. Gypsum wallboard (drywall) both sides	8	Poor. Increase by filling voids or using double layers of wallboard	Fair	Fair	$t = 4$ up	Partitions, Limited bearing	35–40	Fair	Difficult. Reinforce for heavy objects	
Structural clay tile	$4t$ av.	Fair to excellent	Fair to good	Fair to good	$t = 2$ up. H or L between bracing $18t$, $36t$ for nonbearing	Partitions, bearing	70–85	Good if bonded to structure	Difficult	

or the landings may be supported from cantilevered flights. Landings, flights, or even individual treads may be suspended by rods or cables which can also function as enclosing screens for safety.

It is best to use adequate depth-span ratios (1:14 to 1:16) for treads and stringers and support framing in order to avoid bouncing. Care should also be exercised in the use of high-strength steel suspenders whose elongation may add to the movement of the stair.

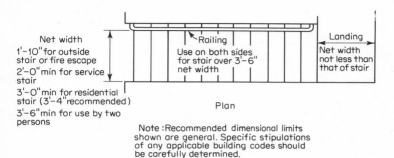

Net width
1'-10" for outside stair or fire escape
2'-0" min for service stair
3'-0" min for residential stair (3'-4" recommended)
3'-6" min for use by two persons

Railing
Use on both sides for stair over 3'-6" net width

Landing
Net width not less than that of stair

Plan

Note: Recommended dimensional limits shown are general. Specific stipulations of any applicable building codes should be carefully determined.

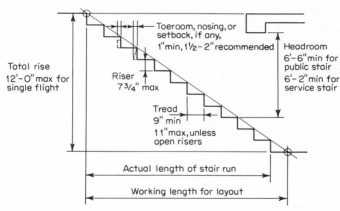

Total rise
12'-0" max for single flight

Toeroom, nosing, or setback, if any,
1" min, 1½-2" recommended

Headroom
6'-6" min for public stair
6'-2" min for service stair

Riser
7¾" max

Tread
9" min
11" max, unless open risers

Actual length of stair run

Working length for layout

Fig. 5 Layout criteria for stairs.

In addition to the framing of the stair itself, some consideration must be given to the relation of the stair to the framing of the building. The two major factors are the nature of the loads imposed on the frame and the necessity for openings and clearances. With regard to the latter, it should be remembered that net stair widths are usually determined by deducting the projection of the handrails into the stair opening. Also, the inside finish of the stair walls may need to be carried past the floor openings, requiring an additional net framed opening size for the thickness of the plaster, paneling, masonry, etc. The difference in traditional referencing techniques in architectural and structural detailing makes this a frequent source of difficulty in building design.

Stair construction should be detailed so that the stair itself may be built some time after the surrounding structure is completed. This is usually preferable for other than service stairs in order to prevent construction workers from using (and abusing) the stairs.

Stair construction must be carefully studied in order to determine exactly where the stair begins and the major frame stops. Landings at floor levels may be part of the stair or part of the major framed floor. This decision may influence the stair location and the location and/or size of major framing members, so that it should not be arbitrarily changed at a later time in the design development. Framing around the stair must be designed for the load of the stair, including intermediate landings, and the weight of the usual surrounding walls. Fire restrictions will often require that these walls be relatively heavy.

25. Steel Stairs   A large variety of stock steel stairs, consisting of tread and riser combinations framed into light steel stringers, is available. Treads may be checkered surface steel plate or open steel grids, or may be depressed to receive a finish of cut stone, wood, precast concrete, poured concrete, or terrazzo. Landings, railings, and even subframing systems may be all part of the same component package. Actual detailing of these systems is usually done by the fabricator from layouts and simple schematic suggested details furnished by the designer.

26. Concrete Stairs   Poured-in-place stairs of reinforced concrete may take various forms. Stair and landing may be one continuous folded slab or may be designed as a monolithic slab-and-beam system. Construction jointing should be carefully worked out, since it is usually not possible to form and pour a series of flights and landings without a construction joint. This becomes more critical when flight and landing are continuous in span, as in a Z-shaped folded slab.

Precast-concrete stairs may consist of single cast treads and risers or tread-riser combinations, supported on subframes of precast concrete or steel. Entire flights may also be precast. The higher quality of finish and detail possible with precast concrete makes this a desirable material for stairs which are architectural features.

27. Escalators   Escalators are designed by the manufacturers, who provide all necessary information with respect to the loads and framing details. It is wise to provide flexibility in the building framing layout to allow for alternative models. The usual low angle plus the overrun space makes a long horizontal run for the typical escalator flight. When the floor-to-floor distance is great, as it is in many commercial buildings, it is often advisable to make the flight straddle a column line, that is, to have the upper floor opening in one bay and the lower in the adjacent bay.

Because of the dynamic character of the heavy moving parts and the vibration of the machinery, it is advisable to have some redundancy in the support structure. Beams supporting the escalator should have high depth-span ratios, probably a minimum of 1:10.

MISCELLANEOUS CONSIDERATIONS

28. Openings and Voids   Some structural elements contain natural voids. Trusses facilitate passage of elements of a size limited only by the scale of the truss. The Vierendeel truss is most accommodating in this respect. Hollow units, such as cored metal deck, voided precast-concrete slab units, and plywood or metal sandwich panels, may receive wiring, piping, or even circulating air in their voids.

Certain items may be embedded in poured concrete elements. Roof drains may thus be piped through columns, conduit and piping may be placed in thick walls and slabs, etc. Piping for hot water, steam, and corrosive liquids should not be so embedded. The ACI Code limits the size of embedded pipe or conduit to one-third of the slab thickness.

Walls of masonry or poured concrete accept openings up to about 1 ft diameter without difficulty. Larger openings may require special reinforcing, such as lintels in masonry and extra reinforcing in concrete. Very large openings in walls must be framed.

Suitable locations of openings in floors depend on the framing system (Fig. 6).

Beams and girders accept small openings in their webs without seriously impairing their strength. Location, as well as size, is important. Shear considerations usually dictate web openings near midspan, while for bending the optimal location is near the neutral axis of the cross section. Flanges of steel beams may be punctured or notched in zones of low moment. Round openings, or square openings with rounded corners, minimize stress concentrations. Reinforcing should be provided around large web openings in steel or concrete beams.

It is always best to anticipate the need for openings in steel members and detail them for shop forming (Fig. 7). Holes in concrete structures should be designed for and formed before the member is poured. Indiscriminate cutting of structural members in the field by other trades is a practice which should not be allowed. Where errors or late changes make cutting unavoidable, specifications should be given as to where and how cutting may be done, and careful field supervision should be provided.

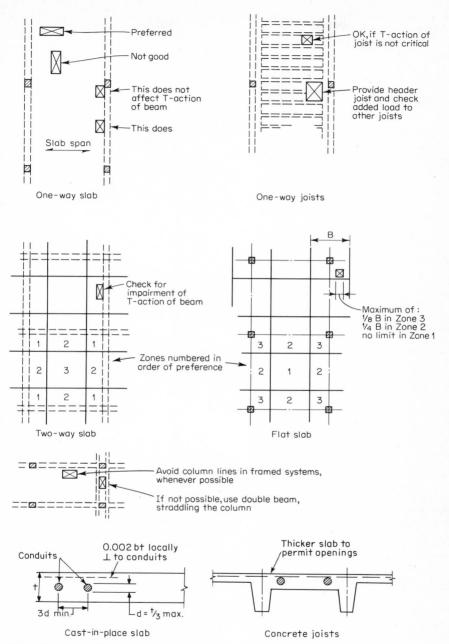

Fig. 6 Openings in floors.

29. Thermal and Seismic Movement Exterior elements and surfaces of buildings are subjected to a range of temperature which in northern climates can be as much as 140°F. The resulting expansion and contraction presents a number of problems.

1. Movement of the entire building structure. Since the substructure is kept at a relatively constant temperature by the ground, there is a buildup of differential length

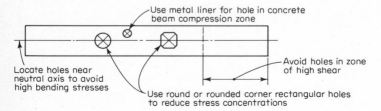

Openings in steel beam webs: hole should not exceed one-half beam depth

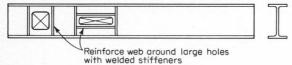

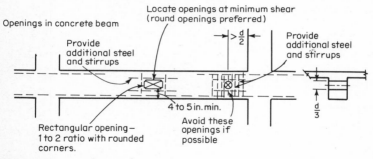

Fig. 7 Openings in beam webs.

between the lower and upper portions of the building. This can cause critical conditions of shear in end walls or sidesway deflection of end columns. Where separate wings of the building are joined, as in L- or H-shaped plans, one wing may pull away from or push against the other. The solution to this problem is usually to provide expansion joints (Fig. 8).

An effective expansion joint is usually a clumsy, elaborate, and expensive construction detail. The expansion continuity must be broken without losing the weather seal and structural integrity. This often calls for elaborate sliding or sealing elements and duplication of structural elements. Details of some typical expansion joints are shown in Fig. 9.

Seismic shocks may produce large-scale sawing motions at the joints and result in the working out or rupture of the joint filler or sealing material. If possible, such joints should be carefully detailed so that these materials can be easily replaced.

2. Differential movement between internal and external structure. Where any portion of the structure is exposed, there is the possibility of critical differential movement due to the expansion and contraction of the exposed portion while the interior is maintained at a relatively constant temperature. Some of these conditions may be remedied by inserting heating devices in the exposed structure, since the largest differentials occur during cold weather. It may be questionable, however, to have the safety of the structure depend on the reliability of heating devices. Since the exposed portions are likely to be restrained to some degree by the rest of the structure, there will be some reduction of their actual movement. Even though these movements may be accommodated structurally by

joints or reinforcing, it may be difficult to allow for them in the architectural features, such as the skin walls and interior partitions.

3. Differential movement between structure and skin. Where the structure is protected by a considerable thickness of surface materials, a critical temperature differential

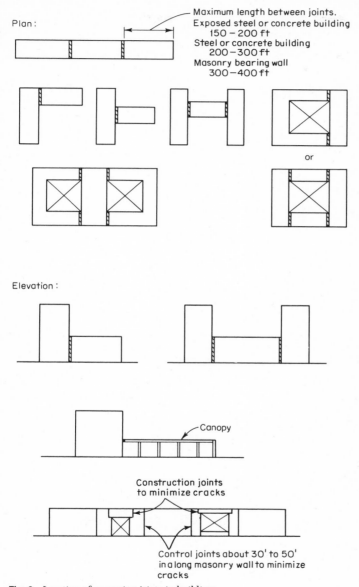

Fig. 8 Location of expansion joints in buildings.

may develop between the surface materials and the structure to which they are attached. In this case, expansion jointing should be located only in the surface elements. This may be easily done if the elements are in separate units and can be joined to allow the accumulation of movement in one unit. Where the surfacing elements are continuous—as

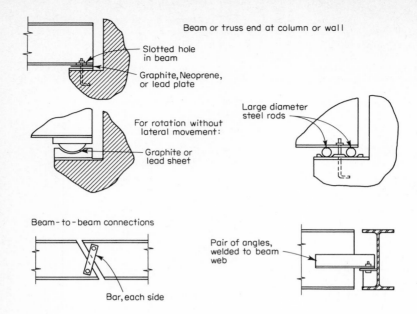

Beam or truss end at column or wall

Slotted hole in beam

Graphite, Neoprene, or lead plate

For rotation without lateral movement:

Graphite or lead sheet

Large diameter steel rods

Beam-to-beam connections

Pair of angles, welded to beam web

Bar, each side

Roof and floor deck connections:

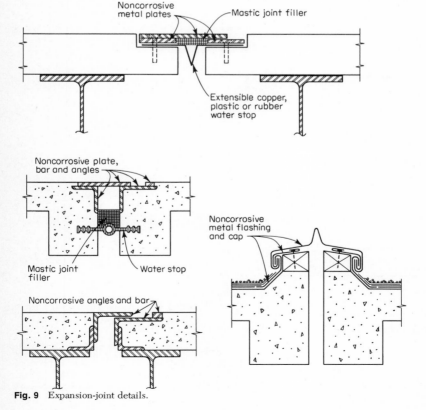

Noncorrosive metal plates

Mastic joint filler

Extensible copper, plastic or rubber water stop

Noncorrosive plate, bar and angles

Noncorrosive metal flashing and cap

Mastic joint filler

Water stop

Noncorrosive angles and bar

Fig. 9 Expansion-joint details.

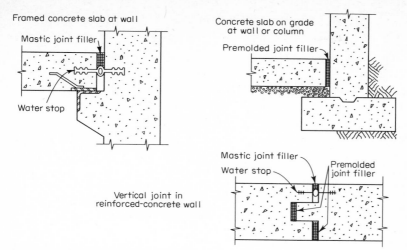

Framed concrete slab at wall

Mastic joint filler

Water stop

Concrete slab on grade
at wall or column

Premolded joint filler

Vertical joint in
reinforced-concrete wall

Mastic joint filler

Water stop

Premolded
joint filler

Fig. 9 Expansion-joint details *(Continued)*

TABLE 29　Coefficients of Linear Expansion—Typical Building Materials

Material	Coefficient of linear expansion per unit of length, per °F, value given $\times 10^{-7}$	Approximate total length change in 10 ft for thermal change of 100°F, in.
Metals:		
Aluminum.....................	128	0.154
Bronze.......................	101	0.121
Copper.......................	98	0.118
Iron, cast, gray................	60	0.072
Iron, wrought.................	67	0.080
Lead.........................	159	0.191
Steel, mild (structural).........	65	0.078
Steel, stainless, 18-8...........	96	0.115
Zinc, rolled...................	173	0.208
Stone and masonry:		
Ashlar masonry................	35	0.042
Brick masonry.................	35–50	0.042–0.060
Cement mortar, portland........	70	0.084
Concrete, stone................	55–70	0.066–0.084
Concrete, perlite...............	43–61	0.052–0.073
Limestone.....................	40	0.048
Marble.......................	45–55	0.054–0.066
Plaster, cement and sand........	90	0.108
Plaster, gypsum and fiber.......	85	0.102
Sandstone....................	55	0.066
Slate.........................	45	0.054
Timber:		
Fir, parallel to grain...........	20–30	0.024–0.036
Fir, perpendicular to grain......	200–300	0.240–0.360
Miscellaneous:		
Glass.........................	45	0.054
Plastic, lucite, plexiglas.........	450–500	0.540–0.600
Plastic, nylon.................	2,000	2.40
Plastic, polyethylene...........	1,000	1.200
Plastic, styrene................	330–450	0.396–0.540
Plastic, foam (styrofoam).......	400	0.480

with masonry or stucco—control or expansion joints must be provided at relatively short intervals.

4. Differential movement between dissimilar materials. Where materials with considerably different expansion rates are attached, critical distortions may develop over relatively short distances. Copper, bronze, aluminum, and stainless steel all have expansion rates higher than structural steel, and facings of these materials should be allowed to expand free of any supporting structure of steel. Similarly, reinforcing elements of steel encased by light mullions of aluminum or bronze should be allowed to slip longitudinally. Long elements of steel supporting masonry or plaster may accumulate critical differential expansion, unless adequate expansion jointing is provided in either, or both, elements.

Table 29 gives the coefficients of linear expansion for most common building-construction materials. Some materials, such as wood and paper, have different rates of expansion in different directions. Some expand differently in different directions because of their configuration, e.g., corrugated metal deck, which tends to accumulate movement in a direction parallel to the corrugations but absorbs motion in the other direction by slight flexing of the corrugations.

Part 2. Industrial Buildings

E. ALFRED PICARDI
Consulting Engineer, Richmond, Va.

30. Design Philosophy The modern industrial building may be defined as any building designed and constructed to support and house a manufacturing process, or to store the raw materials or products of a manufacturing process. Such a definition in its broadest sense would include manufacturing plants and warehouses of all kinds, ranging from a simple roof structure on an open frame affording some protection of an area from the elements, to a highly sophisticated building housing a manufacturing process in spaces requiring specific and accurately controlled environmental conditions and elaborate appurtenant services.

In the proper design of an industrial building, *function,* more than any other factor, will dictate the degree of sophistication. Toward this end, the designer should have an intimate knowledge of the industrial process or purpose for which the building is intended. Often this can be attained only after considerable study of the client's methods. In many instances such study will result in buildings or structures which become an integral part of the manufacturing process equipment, even to the extent of the development of new and better methods through mutual education of client and designer resulting from their working together.

Economy is almost always a basic factor in the design of industrial buildings. Initial cost, maintenance, and operating cost of an industrial building in a real sense become an item in the cost of the manufactured goods. The designer, therefore, should understand each industrial building project from the standpoint of achieving the optimum balance between function and economy.

Architecture is the third important element. The industrial building is seldom or for long isolated from a community, and the building itself, even in the most automated type of manufacturing process, will always have some human population. Therefore, internal architecture of the building, insofar as it provides for the development of the design to satisfy the needs of the occupants, becomes one of the primary factors. External architecture is a primary factor to the extent that the building must be properly related aesthetically to the community and its natural environment.

31. Planning One of the first considerations in the layout of an industrial building is the determination of total area and volume requirements. The first determination of these parameters should be based on function alone. They may then be varied during the development of the design for reasons of economy and architecture, but the designer should always be conscious of function as being of paramount importance.

Having determined total area and volume requirements, the designer must then develop exterior dimensions of the structure surrounding the process. On large structures having multiple bays in both directions, substantial economy can often be realized if the overall structure plan is a square or nearly square area. Such a layout affords economy in exterior wall length. Figure 10 demonstrates the increase in wall perimeter as building length-width ratio varies from 1 to 8. This function is independent of building area and demonstrates one of the economies afforded by the square plan.

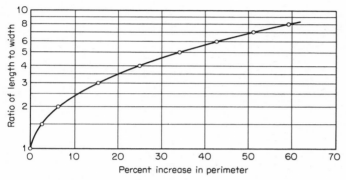

Fig. 10 Variation in perimeter of rectangle with aspect ratio.

Bay Size.   The designer must also develop unit area-volume requirements, i.e., a bay size within the total area volume. In some industrial buildings, span of individual bays in one or both directions is often dictated by the manufacturing process. In other buildings, perhaps a minimum area and clear height will be the only functional requirement for the individual bays. This minimum bay would be of the smallest dimensions capable of providing machinery and/or storage space and aisle requirements between columns and from floor to underside of structure. This is often the case in storage warehouses or plants where machinery consists of small individual units. Most often, however, the industrial building is complex in its functions, requiring both process areas and storage areas. Thus, the unit area volumes or bay sizes may have to be developed to suit all conditions, particularly where flexibility of use is a design requirement.

The unit cost of individual bays will vary substantially with bay area, volume, and ratio of length-to-width dimensions. A study of a series of hypothetical building frames is presented to demonstrate the effect of variation in bay areas and ratios of length to width of bays. Figure 11 shows the weight of steel superstructure vs. bay area for three ratios of bay length to width for a common type of one-story steel industrial frame. These curves are based on designs for typical interior bays consisting of continuous steel girders framed over columns in the direction of the short dimension of the bay, and steel bar joists or long-span joists supported on the girders and framed in the direction of the long dimension of the bay. The columns were assumed to be 20 ft long. The design assumes a roof live load of 25 psf and a dead load of 25 psf consisting of built-up roofing, insulation, metal deck, and electrical and mechanical service equipment. A comparison of the curves for length-width ratios of 1, 2, and 3 shows the advantages of square bays. The penalty in increased weight of steel that results from a departure from the square bay can be estimated.

The single-story industrial building, consisting of floor slabs on grade and a flat-roof superstructure, is by far the most popular design. This solution is the most economical from the standpoint of both first cost and maintenance. In most industrial manufacturing processes and warehousing, it is also the most economical from the standpoint of operating cost. Decentralization of industrial processes has led to new construction outside the

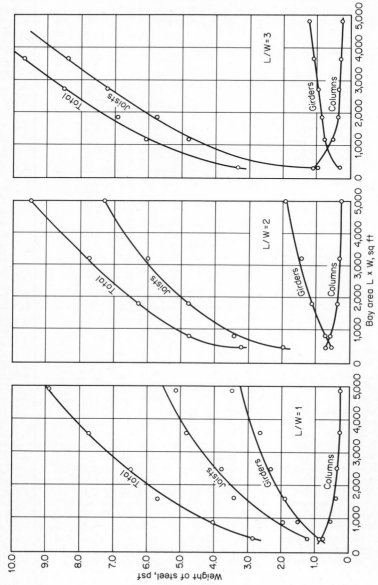

Fig. 11 Variation in weight of steel with bay area.

urban areas where large and relatively inexpensive tracts of land can be procured, which permits construction of large-floor-area, single-story structures most economically.

Multistory industrial buildings are built where one or more of the following conditions exist:

1. An urban location on high-cost land is needed.
2. Sufficient area is lacking on a site which is otherwise desirable.
3. The industrial process lends itself best to a vertical alignment and flow.
4. The industrial process requires a gravity flow.

The degree of sophistication of an industrial building will be dictated by the process requirements. The designer should study the process space arrangement and special requirements of the structure prior to selection of the framing system of the basic shelter and its skin. Special requirements associated with the process, such as heating, cooling, refrigeration, insulation, provision for roof-hung equipment, craneways, monorails, conveyors, mezzanines, penthouses, monitors, high bays, and provisions for additions, should be considered in development of the basic structure requirements. An early and quantitative evaluation of all special requirements will aid greatly in rapid development of a proper and economical framing and skin system for the building.

Concurrently with evaluation of the building requirements, the designer should procure data on subsoil conditions, roads, railroads, and utilities which may be necessary for proper preparation of preliminary studies and designs. Knowledge of pertinent local codes and statutes should also be acquired as early as possible.

32. Framing Systems No material can be singled out as the ideal material for the structural framing of the industrial building. Only after careful study of the particular project can the designer select a framing system and conclude that it is the best solution. While the advantages and disadvantages of steel, concrete, and wood frames will continually be debated and analyzed, valid conclusions can be made only as these materials are applied to a specific project. Even then, the validity of the conclusions will be dependent upon the designer's skill and ingenuity and the effort at exploring the numerous possible framing schemes using the various materials available.

Light-Gage Metal. A large variety of steel framed structures are used in industrial building construction. Light-gage metal-arch construction employing corrugated curved sheets to form a structural load-carrying skin is often among the least expensive. These are popular where the requirements are for low-cost, basic shelters. Primary disadvantages are the variable clear height, inability of the structure to support heavy hung concentrated loads, and possible objections to its appearance.

Prefabricated Buildings. Standard-design rigid-frame structures are presently marketed as prefabricated buildings. Like the metal-arch structures, these may also prove to be adequate solutions to many industrial building problems. They offer the advantages of low cost and speed of erection. These structures should be checked for compliance with local building codes and for adequacy in supporting any concentrated loads associated with the manufacturing process. Many standard rigid-frame buildings can be procured with bents capable of carrying substantial hung loads at specific locations. However, as variation from the standard heights and bent capacities becomes necessary, the cost often increases, and it is advisable to compare these structures carefully with custom designs.

Custom-Design Buildings. Economical, custom-designed, single-story industrial buildings are currently designed in steel, concrete, and wood framing systems.

Steel frames are primarily of two basic types. The first type, most popular for short-span construction, consists of a frame composed of continuous or double-cantilevered girders over the columns in one direction with simple-span bar joists or long-span joists in the other direction. In short-span buildings and/or where heavy roof-hung loads are encountered, rolled sections are often substituted for the bar joists. Simple-span rolled sections may also be substituted for continuous or double-cantilevered girders where spans are short, and the additional steel weight is often found to be offset by simplicity of fabrication and erection. The second type consists of trusses framed in the long dimension of the bay and purlins framing between trusses in the short dimension. This system is often employed where a long span, i.e., in excess of 70 ft, is required in one dimension of the bay and much shorter spans, i.e., in the order of 20 to 30 ft, can be tolerated for the other.

In structures of single-bay width these trusses may be pitched for roof drainage and knee-braced for lateral load stability, or parallel-chord trusses may be used where flat roofs or slightly pitched roofs are dictated by the design considerations. Figure 12 shows types of steel roof-truss configurations most often used. Trusses should be considered where spans exceed 70 ft. Truss systems are advisable where roof-hung ductwork, piping, and mechanical equipment must be placed within the volume of the framing system and clear of craneways or other equipment below. Heavy concentrated loads can be supported at panel points if the industrial process requires them.

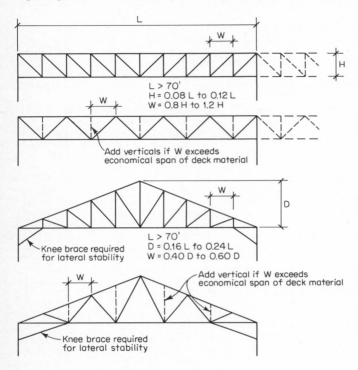

Where spans must exceed 70 ft in both dimensions of a bay, the designer should investigate the use of primary trusses in one direction with secondary trusses framing between them. The secondary trusses should be spaced on centers which afford an economical purlin span, i.e., 20 to 30 ft.

On long spans continuous trusses in one or both directions may be used as an alternate to this system, affording economy in steel weight, and minimizing deflections, especially where heavy equipment is supported by the roof structure. If continuity is considered, the designer should determine both stress and strain effects on the structure due to temperature changes both during and after construction. Strain effects from temperature (expansion and contraction) may often limit the extent of continuity to two or three spans between expansion joints. Additional fabrication cost per ton for the more sophisticated connections necessary in continuous structures should be discussed with a reliable fabricator and erector to determine if saving in weight of steel will offset the additional cost of details and erection time required for the continuous design. An alternate to this system would be a two-way grid system of trusses, or a space-frame solution.

Typical connections and details often employed in steel-frame industrial buildings are shown in Figs. 13 and 14.

Concrete building frames are primarily used where short-span heavy-load multistory fireproof structures are required. The most popular and economical system is the flat slab with drop panels and column capitals. Where possible, capitals are omitted or decreased to minimum dimensions to provide the maximum clear height around the columns. Beam-and-slab construction is often used where the industrial process requires numerous large openings in the floor structure or where unusually heavy concentrated loads must be supported. Frames of this type are usually cast-in-place conventionally reinforced concrete.

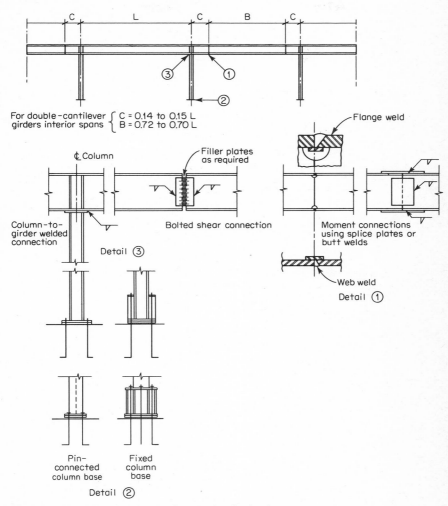

For double-cantilever $\left\{\begin{array}{l} C = 0.14 \text{ to } 0.15\ L \\ B = 0.72 \text{ to } 0.70\ L \end{array}\right.$
girders interior spans

Fig. 13   Continuous and double-cantilever roof-girder details.

Extensive use is made of precast, prestressed members to form single-story industrial building frames. Economical plant-produced single and double T-slabs and hollow-core slabs are available for roof spans from 50 to 100 ft. These units are often employed for roof structures spanning between bearing walls or framing into precast girders. The girders may be prestressed simple-span plant-produced units or cast-in-place posttensioned continuous-span elements. Where girder spans are less than 25 ft, the column-and-girder system is often cast in place or precast with conventionally reinforced concrete. Figure 15 shows typical details, connections, and framing of these systems.

Wood-framed structures are particularly popular in those sections of the country where lumber is procured at low cost. They should be considered where fire hazard is small, where at least one bay dimension can be kept to a minimum, i.e., approximately 12 to 20 ft, and where they afford an economy not available in other materials.

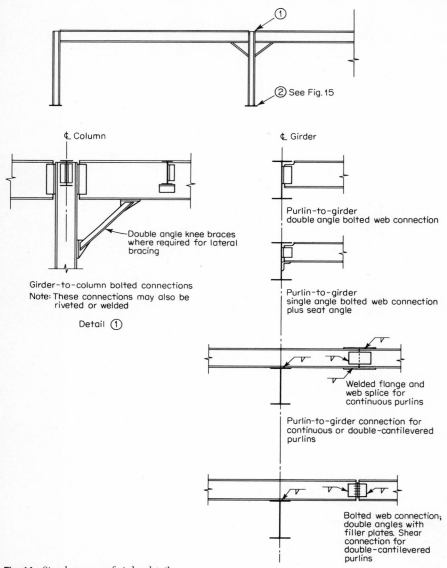

Fig. 14 Simple-span roof-girder details.

The simplest wood-framed structure is the pole-type building in which graded wood poles are set in holes augered in the ground to form the columns. Primary members between poles may be wood timbers on short spans (up to 20 ft), wood trusses, or laminated or built-up wood girders on long spans. Wood joists are framed between the primary members to complete the frame. Where floor-to-roof height is in the order of 12 to

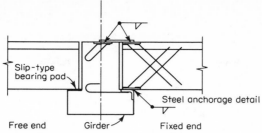

Free end   Girder   Fixed end

Prestressed T or TT slab-to-girder connections

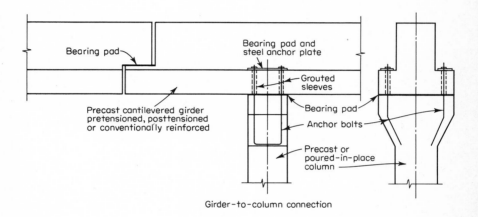

Girder-to-column connection

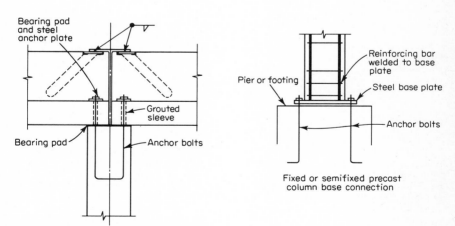

Simple span girder-to-column connection

Fig. 15 Details of concrete frames.

18 ft, the poles can often be designed to act as cantilevered columns to take wind and other lateral loads without additional knee braces. Details of this type of framing are shown in Fig. 16.

The exterior walls of these structures are generally metal skins attached to a wood horizontal girt system. The poles may be more closely spaced around the perimeter to permit an economical girt system.

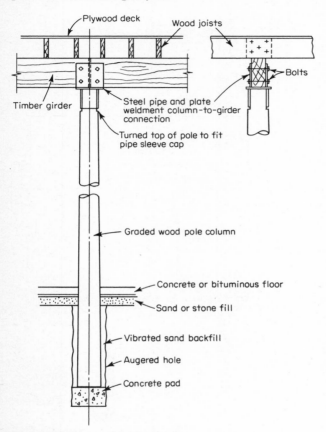

Fig. 16 Pole-type wood-framing details.

33. Wall Systems Exterior wall systems for industrial buildings may be either load-bearing or non-load-bearing. Generally, small one-story structures with a single clear span between exterior walls using steel joists, trusses, or prestressed-concrete roof framing can be developed with load-bearing masonry walls. Large-area single-story structures having multiple spans in both directions, with roof framing in either steel or concrete, will generally be most economically developed with a complete free-standing frame and exterior curtain walls of masonry, precast-concrete panels, or metal panels attached to a steel-girt or steel-stud system. The sophistication of any particular system will be determined by the functional, economic, and architectural requirements of the structure.

Perimeter wall details for load-bearing masonry walls and panel curtain walls attached to pitched or flat-roof steel frames are shown in Fig. 17.

34. Bracing Systems for Lateral Loads All structures subject to wind and other lateral loads must be designed to resist these loads. Unless the primary frame alone is designed to resist these loads, some secondary bracing members must be added.

Steel-Frame Bracing. In simple steel frames consisting of continuous girders over columns, with rolled section purlins or bar joists framed between girders, bracing generally consists of bridging between the bar joists to make the roof diaphragm rigid, and knee braces or moment connections on the girder-to-column connections. In the exterior walls, diagonal bracing or X-bracing may be substituted for the knee braces or moment connec-

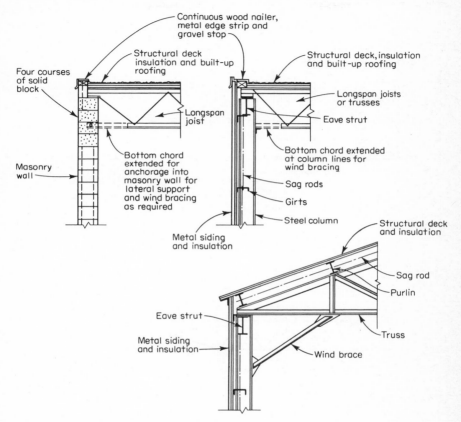

Fig. 17 Perimeter wall details.

tions. Where trusses are used, a horizontal bracing system in the plane of the top chord is generally necessary together with a lateral bracing system connecting top and bottom chords. The lateral bracing system not only serves to carry wind load but also reduces the unsupported length of the bottom chord of the trusses.

Additional wall bracing is required to transmit lateral loads from the roof diaphragm to the foundations. These loads consist of a portion of the wind load on the walls, which is carried to the roof, horizontal components of forces from moving loads attached to the roof structure, and a friction force from wind over the plane of the roof. The latter force is generally neglected in the design of flat-roof structures, but the remaining loads can be estimated with reasonable accuracy and should be used in the design of the X-bracing in the walls. Thus, the number and spacing of X-braced bents in the walls will depend on the magnitude of total horizontal load to be transmitted to the foundations. In general, there should be a minimum of one X-braced bent between each expansion joint in the frame and additional braced bents if the horizontal loads are so large that a single X-bracing would be architecturally objectionable or structurally inadequate. Usually the X-braced bent is located in the same bays as the horizontal bracing system in the plane of the top chord of trussed structures or in the bays adjacent to expansion joints and end walls.

The designer should not overbrace a structure, as light fabricated bracing is expensive and often interferes with functional use of the building, location of wall openings, and

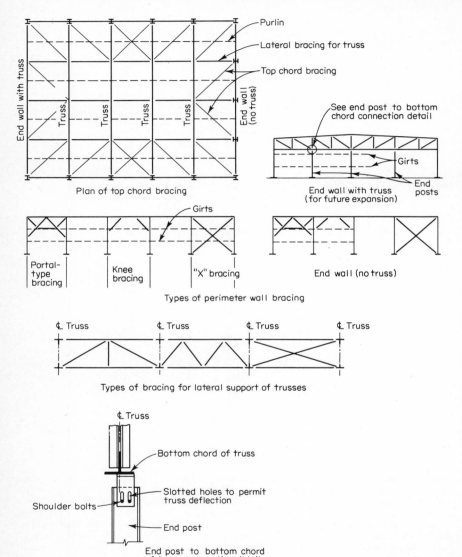

Fig. 18 Bracing for steel frames.

architectural appearance. A good rule to follow is that the geometry must be such that the structure will be stable under lateral loads and all lateral loads can be transmitted through the structure to the foundation system via the combined geometry of the primary structure and the bracing system. Figure 18 shows various bracing schemes for steel-frame structures.

35. Materials Handling   Materials-handling equipment will vary considerably depending upon the function of the plant; usually many different systems will be required.
Piping systems, duct systems, cranes, monorails, and conveyors often are included in the
design contract; if not, provision for their installation must be made.

Piping systems and ductwork for handling liquids, gases, and dry powders are generally
part of the mechanical contract except for supports and hangers, which are usually
designed by the structural engineer using load data and recommendations established by
the mechanical engineers and equipment manuafacturers.

Cranes, monorails, and conveyors of various types are used in industrial plants. Jib
cranes are used to position and handle loads within a limited area. These consist of a hoist

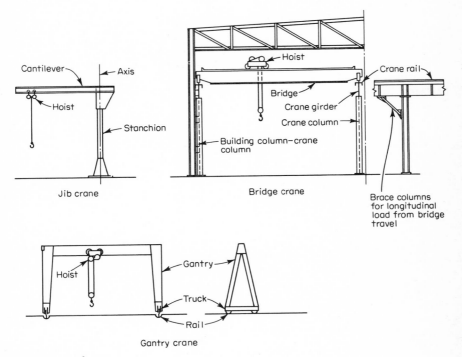

Fig. 19   Typical cranes.

mounted on a rail which is pivoted at one end and cantilevered from a vertical stanchion
or one of the building columns. Bridge cranes are used for handling loads over an entire
floor area of one or more bays between any two lines of building columns. Gantry cranes,
which are mounted on rails in the floor, are used in the same manner as bridge cranes but
where structural support for the craneway is not practical. Hoists are used for simple
vertical positioning of loads or on monorails for continuous movement of loads to specific
destinations. Conveyors are used for continuous high-volume flow of materials or loads
over a fixed route. This handling equipment is shown in Fig. 19. Such equipment is
usually purchased from manufacturers who specialize in its design and construction, and
the structural engineer should consult the various manufacturers early in the design phase
of a project to obtain loads, dimensions, clearances, or other pertinent data to incorporate
the equipment in the structure properly.

The design of bridge-crane supporting girders and columns is usually done by the
building designer. These structures are proportioned from wheel loads and wheel spacing
provided by the manufacturers. Craneway girders may be simple-span or continuous,
either rolled sections or built-up plate girders, depending upon the loads and spans.

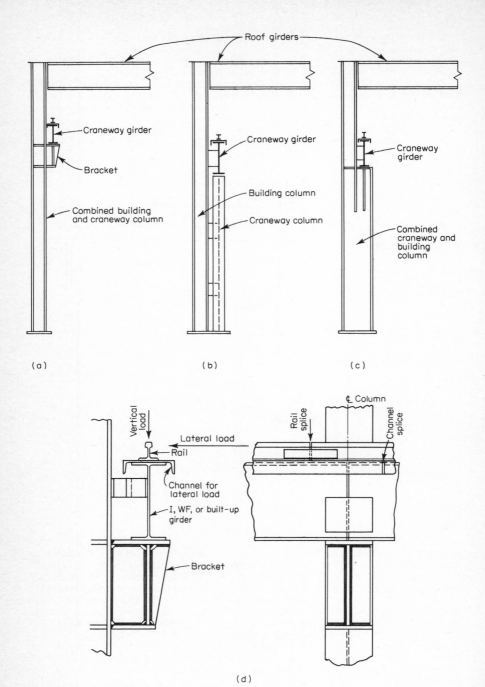

Fig. 20 Crane girder columns.

Continuity is generally desirable to decrease deflection, and for economy in steel weight or to decrease girder depth where headroom under the girder may be critical.

Craneway columns and girders are generally of three types as shown in Fig. 20. The bracket support, type *a*, is generally used on light craneways; *b* and *c* on heavy craneways. Choice of column type should be made from trial designs for the particular structure column and craneway loading. Type *b* for heavy craneways is quite popular as it affords the economy of rolled sections, a minimum of fabrication work, and good stiffness in the combined columns both laterally and longitudinally. Type *c* offers more stiffness on the lateral axis of the craneway, which may be desirable under certain conditions of hoist and bridge travel speeds or building geometry.

Loading Docks. Most industrial plants and warehouses require loading and unloading facilities from railroad cars or trucks or both. In general, both types of docks should be depressed so that loading or unloading may be accomplished with truck bed or rail-car floor at the same level as the plant floor. Slight variations between car level and plant floor are usually compensated for with movable ramps which bridge the gap between the dock and car bed or with automatic levelers in truck docks. These are standard manufactured items generally selected by the designer and built into the docks. Truck docks should be provided with fenders to prevent damage to both truck and structure from careless operation of the truck. Both railroad and truck docks are usually of reinforced-concrete construction and may be inside the building under cover or flush with exterior walls. Sidings in railroad docks should be terminated with a suitable car bumper at the end of the siding track and should be level for the length of the unloading distance. Clearances and railroad dock heights may be obtained from various graphic standards but should be checked by the servicing railroad for their specific requirements. In all cases, the walls of depressed docks should be designed for surcharge of the loaded platform and the retained earth.

Part 3. Tall Buildings

MORTON H. ELIGATOR AND ANTHONY F. NASSETTA

**Partners, Weiskopf & Pickworth, Consulting Engineers,
New York, N.Y.**

FRAMING

36. Bay Sizes Multistory buildings fall mainly into two categories: office and residential. Room and apartment layouts in residential structures are determined by size and efficiency of arrangement and therefore usually have no relation to a grid of any sort, resulting in an irregular column layout. Columns are fitted in where they are least disturbing to the architecture, but at spacings close enough to allow a minimum depth of floor. In both single- and multitenant office buildings a regular column grid usually can be established, resulting in repetitive bays in one or both directions. Regularity of bays is important, since duplication leads to economies.

Architecture and aesthetics control exterior column spacing, and therefore bay sizes, to a great extent. A panel wall design with columns completely inside the wall allows maximum freedom of bay selection, whereas a panel wall in which columns are integrated into the modular design results in the least flexibility. Modular exterior wall design, on the whole, will generally produce a modular structural design.

Bay sizes ordinarily should be selected to produce a minimum story height. A reduction of only 3 in. per floor of a 20-story building will save 5 ft of exterior wall, partitioning, columns, risers, etc. On the other hand, columns cannot be spaced so closely as to detract from the usefulness of the spaces they pass through. Selection of bay sizes is always a compromise between these two considerations.

There is more than "structure" in the depth allowed for floor construction (finished ceiling to finished floor). Mechanical and other services, such as ductwork for heating, ventilating, and air conditioning, and some plumbing and electrical work, may also occupy this space. An economic study is justified to determine whether the best framing solution in steel is shallow (heavier) beams with services below, or deeper (lighter) beams with services running through openings in the steelwork. Large openings in beam webs usually require reinforcing, but for long spans (large column spacing), deep members are necessary for wind loads, if not for gravity loads, and penetrating the beams for duct runs is a common solution.

Figure 21a illustrates typical bay framing and floor construction of a steel-framed building with moderate-sized bays. By holding girder depths to 18 in. (using the W18 × 96 instead of a lighter W24 × 84), a reduction in height of 6 in. per floor is effected. The added cost for the extra weight of floor steel (about 1 psf) is far less than that for 6 in. of extra story height. This is the framing for the upper 15 stories of the 1120 Avenue of the Americas Building in New York City.

In a concrete frame the problem is less acute for moderate spans, which usually become flat slabs (solid or ribbed). Figure 21b shows typical bay framing and floor construction for a concrete building with moderate-sized bays. This is the framing for the lower six stories of the 1120 Avenue of the Americas Building. This building was built in stages, on the site of the old Hippodrome Building, between 1950 and 1961. The lower stories are framed in concrete because steel was in short supply during the Korean War.

Figure 22a illustrates typical bay framing and floor construction of a steel-framed building with very large bays. Here deep members are required for gravity and wind loads, and for wind deflection. By piercing the webs of the wind girders for passage of ductwork, a reduction in height of 12 in. per floor is effected. The extra cost of the extensive web reinforcing required at these large openings is more than offset by the reduction in area of exterior walls, amount of column steel, partitioning, and mechanical risers. Section B-B in Fig. 22a shows another height-saving detail. Here the girders are 2 in. higher than the beams; the metal deck floor, spanning parallel to the girders, is discontinuous between bays.

Figure 22b shows the solution to the problem in a concrete structure with long spans. A formed and reinforced web hole is expensive, but a substitute for a 24 × 36 in. beam, shallow enough to get a duct under, in the same story height, would be perhaps 44 × 24 in. Not only is more material required in the latter, but the depth is not commensurate with the span and would result in excessive elastic and creep deflection.

Type of foundation has an effect on selection of bay sizes. An attempt should be made to minimize the number of individual supports where foundation work is difficult and costly. Fewer columns (larger bays) will reduce total foundation costs where deep piers or caissons are required. For pile foundations, fewer and larger pile groups will normally produce a more efficient utilization of piles, whereas a plethora of columns on many small pile groups is uneconomical.

37. Columns Once bay sizes and column locations have been established in a steel structure, the orientation of the columns is determined. Columns are set with flanges parallel to the long axis of the structure, since the transverse wind condition is the more severe, in order to achieve optimum utilization of column properties. For heavier loads, the required area in compression can be obtained by adding flange plates to rolled sections, which increases the bending properties for wind moments as well. Although web plates can be added for additional compressive area, they add little to column stiffness and should be used alone (without flange plates) only where wind stresses are not a design feature, or where architectural requirements so dictate. Welded built-up columns, usually in the form of an H or a box, are also used for the heavier loads.

In the modern high-rise building with unusually large bays, cover-plated rolled shapes may be inadequate. In such cases, special built-up columns must be designed to suit the various loading and geometric requirements. The design of the heavy column for the lower parts of high-rise structures requires study in each case, since the problem is not simply one of obtaining a cross section of the required area; the wind-bracing scheme is as much a governing consideration as is the load in the proportioning of such columns. If the building is a long, narrow one, wind may be a major problem in one direction only. If the plan is that of an approximately square tower, moment connections may be needed at all faces of a column, and the magnitudes of the maximum moments will require details that

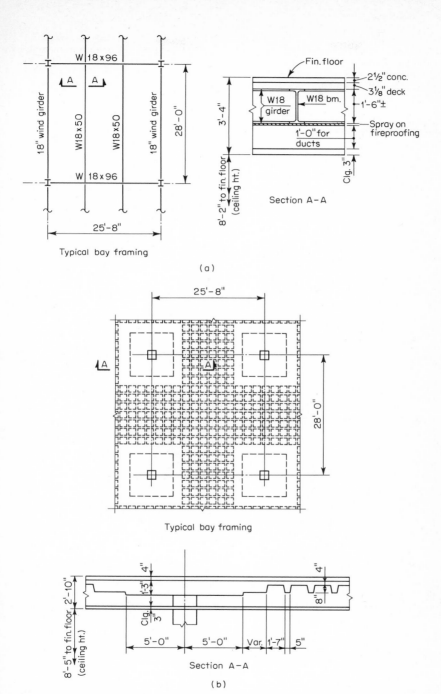

Fig. 21 Typical framing for bays of moderate size; 1120 Avenue of the Americas, New York City.

lend to a grading or modification up through the frame without abrupt change of type and without a shifting of centerlines in either direction. Furthermore, strength is not the only requirement; stiffness must be obtained so that occupants are not conscious of sway in slender towers.

Figure 23 illustrates a column section used at the base of a long, narrow steel-framed

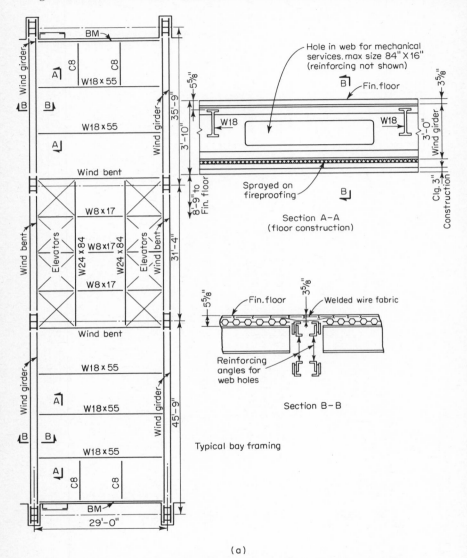

(a)

Fig. 22a   Typical framing for large bays: Chase Manhattan Bank Building, New York City.

tower. Since the major wind problem is in the transverse direction with this rectangular plan, the major axis of the column is oriented in the longitudinal direction. To satisfy the gravity loads, 348 sq in. of material is required. Proper development of a column with utmost section modulus about the desired axis results in economy, since bending stresses due to wind can then usually be absorbed in the increased unit stress allowed for combined loads by most building codes and specifications, with the result that the cross-sectional area need be no more than that required for gravity load.

Figure 24 illustrates a column section near the base of a relatively square tower in which wind stresses in both directions are of similar magnitude. With moment connec-

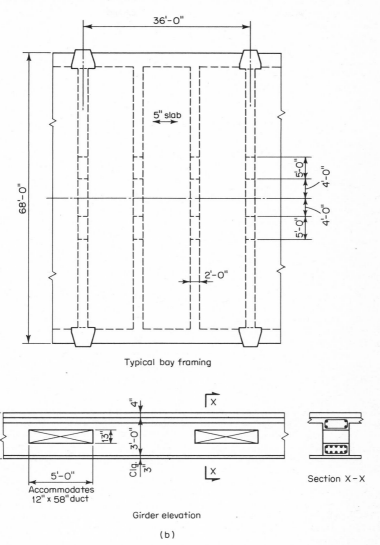

Typical bay framing

Girder elevation

(b)

Fig. 22b Typical framing for large bays: American Bible Society Building, New York City.

tions occurring on all four faces, this optimum column is shaped and detailed to have similar properties about both axes.

With concrete columns, the importance of orientation is normally not so critical as with steel columns. Because of the increase in size of concrete columns from gravity loads alone, every attempt is made to resist wind forces by other means, such as shear walls (Art. 42). Thus, more of the wind is resisted by the shear walls and less by the columns, and orientation of columns becomes a detail depending on framing conditions and space requirements.

When space limitations exist, it is good practice to specify higher-strength concretes and reinforcing steels to control column sizes. For example, a 20-story structure for which the floor framing can be held economically to the use of 3000-psi concrete may use the same

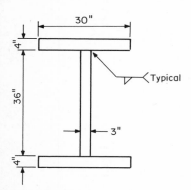

Fig. 23 Exterior column for 54-story office building, One Liberty Plaza, New York City.

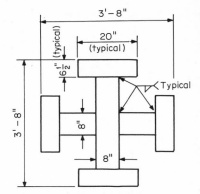

Fig. 24 Interior column for 57-story office building, Commerce Court Building, Toronto, Canada.

grade of concrete in the upper eight floors of columns, and 4000- to 6000-psi concrete in the lower columns, with 90,000- or 100,000-psi (ultimate) reinforcing steel in the four or five lowest floors for additional size control.

Splices. Steel columns are usually erected in two-story lengths, with splices occurring about 1 ft 6 in. above the finished floor line. End surfaces are milled for proper bearing. Splice plates are shop-connected to the upper end of the lower shaft and field-connected to the lower end of the upper shaft. The AISC "Detailers Manual" illustrates various standard column splices. Splices do not transfer load; they are used primarily for stability and alignment. In general, there is no tension on a wind column since the compressive stress from gravity loads normally exceeds the tensile stress in bending from wind loads. In unusual cases, where the base dimension of a frame is small compared with the structure's height, net uplift may result. In such cases, splice plates must be designed for tension and column bases must be anchored to the foundations. Foundations are anchored to rock, or to tension piles, or are made massive enough for their weight to overcome the tension. The safety factor against overturning should be at least 1⅓.

Every attempt should be made to frame to a column on both axes at each floor for proper bracing. Unless this is done, the unsupported length is increased, and the column section may be heavier (and more costly) because of the increased slenderness ratio.

Concrete columns are normally spliced by lapping the reinforcing bars, but with the heavy No. 14 and No. 18 bars a lap splice is both costly and cumbersome. Butt welding or mechanical splicing of bars is an accepted practice. Butt welding can be accomplished either by conventional methods or by the thermite process. Mechanical splicing methods include a mechanism which locks in place the smooth-cut ends of the bars, and a sleeve device fusion-locked to the deformations of the upper and lower bars. Only the latter type develops tension.

38. Elevator Shafts   The vertical transportation system is one of the most important elements in a multistory building. A large amount of floor area is devoted to elevator-shaft space. In addition, basement pits and penthouse machine rooms are required. Consequently, banks of elevators are generally spaced very close together to hold unusable building volume to a minimum.

The four basic parts of an elevator system that affect the structural design are (1) the machine room, (2) the shaftway, (3) the buffers and guide rails, and (4) the pit. Figure 25 illustrates a typical elevator system for a medium-rise steel-framed building, showing the

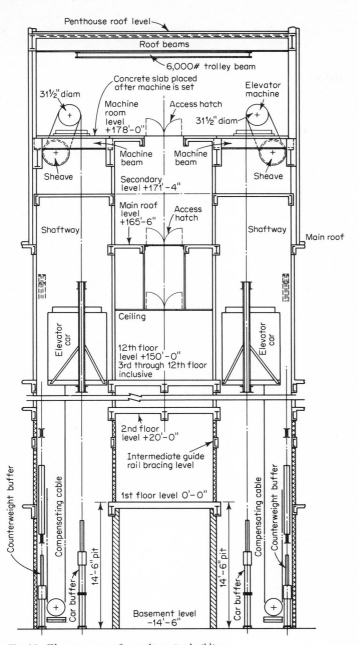

Fig. 25 Elevator system for medium-rise building.

arrangement of a machine room, shaftway, and pit. The elevator system for a medium-rise concrete-framed building would be identical.

Machine Room.   The elevator machinery which supports the car and counterweights is generally located at the top of the shaft in multistory buildings. Normally, a special machine room or penthouse is required, since the machinery must always be placed at least one story above the last story served by the elevator car. Access to elevator machine rooms is generally by stairs, and equipment is lifted to the machine-room level by hoists suspended from overhead trolley beams. Speed governors are usually located on secondary levels, within the shaft, a few feet below the elevator machinery. The machine-room floor adjacent to the area over the shaft is used to support motor generators, controls, relays, and other miscellaneous equipment.

The structural members that support the equipment in the elevator machine rooms are the overhead machine beams and sheave beams, and their immediate supporting members. All these members must satisfy special allowances for impact, factor of safety, and deflection. The following requirements are typical:

1. Machine and sheaves shall be supported by steel beams and shall be held so effectually to prevent any part from becoming displaced.
2. Loads on overhead beams and their immediate building supports shall be computed as follows:
    a. The total load on overhead beams shall be assumed to be equal to the weight of all apparatus resting on such beams plus twice the maximum load suspended from such beams (impact).
    b. The load resting on such beams shall include the complete weights of machine, sheaves, controller, and similar equipment. The suspended load shall include the sum of the tensions of all cables suspended from such beams.
3. The unit stresses for all machinery and sheave beams and floors and their supports shall not exceed 80 percent of those permitted for static loads by the following codes:
    a. Structural steel. AISC Specification for the Design, Fabrication and Erection of Structural Steel for Buildings.
    b. Reinforced concrete. Building Code Requirements for Reinforced Concrete, ACI 318.
4. The deflections of machinery and sheave beams and their immediate supports under static load shall not exceed 1/1666 of the span.

Figures 26a and 27a illustrate the structural framing for the elevator machine room and the secondary level, respectively, of the system shown in Fig. 25. Figures 26b and 27b illustrate the structural framing for the elevator machine room and the secondary level in a similar medium-rise concrete-framed building.

Shaftway.   The shaftway or hatch is the fireproof enclosure for the elevator car, counterweight, cables, guide rails, etc. The floor framing surrounding the shaftway supports the enclosure walls and the floor construction around the shaftway. In addition, this floor framing serves the extremely important function of bracing the guide rails at each floor. In a steel structure, the framing rarely can be positioned on column centerlines, and narrow-flanged, deep, built-up girders frequently become necessary. In a reinforced-concrete structure, the difficulties of beam and column framing can be overcome by using reinforced-concrete bearing walls. The bearing wall combines the function of the enclosure wall, floor framing, and columns, and often provides space economy as well. Between elevators, in shafts enclosing a bank of elevators, steel divider beams are required to brace guide rails and compression flanges of main framing members.

Figure 28a illustrates the structural framing around the shaftway at a floor level of the system shown in Fig. 25. Figure 28b illustrates the structural framing around the shaftway in a similar medium-rise concrete-framed building.

Buffers and Guide Rails.   Buffers of the spring, oil, or equivalent type are installed symmetrically under the elevator car and counterweight at the bottom of the elevator shaft or pit. Steel guide rails are installed symmetrically at each side of the elevator car and counterweight for the full height of the shaft. During normal operation of the elevator, the guide rails serve only to maintain truly vertical motion without lateral shifting during ascent or descent. In the event of cable failure or accidental overspeed, car and counterweight safeties provide a retarding force capable of stopping and sustaining the car or counterweight load. The retarding force is generally developed by clamping or gripping

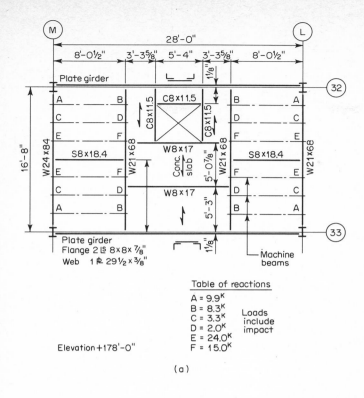

Table of reactions

A = 9.9^K
B = 8.3^K Loads
C = 3.3^K include
D = 2.0^K impact
E = 24.0^K
F = 15.0^K

Elevation +178'-0"

(a)

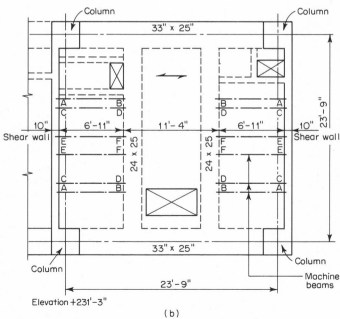

Elevation +231'-3"

(b)

Fig. 26 Machine-room framing. (*a*) Steel frame; (*b*) concrete frame.

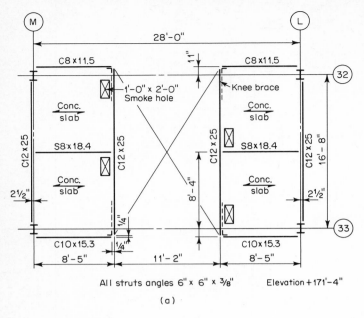

All struts angles 6" x 6" x 3/8"   Elevation +171'-4"

(a)

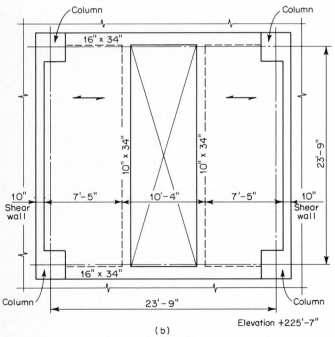

(b)

Fig. 27   Secondary-level framing. (*a*) Steel frame; (*b*) concrete frame.

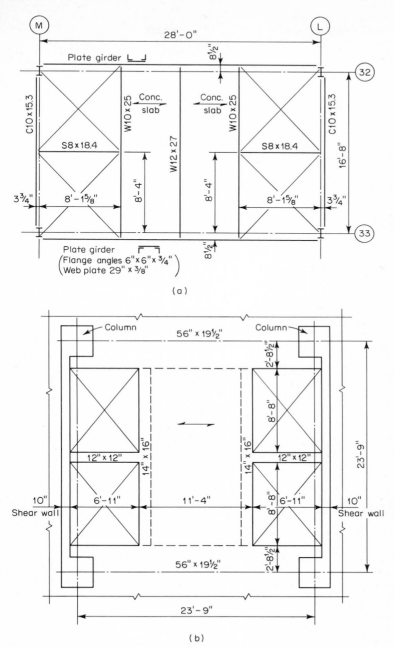

Fig. 28 Typical floor framing plan around shaftway, third to twelfth floors. (*a*) Steel frame; (*b*) concrete frame.

the guide rails with a gradually increasing frictional force. If the car or counterweight is close to the bottom of the shaft, the frictional retarding force on the guide rails may not have enough time to develop. The buffers then provide the additional force necessary to bring the car or counterweight to rest. It is obvious that guide rails and buffers must be capable of supporting the weight of car or counterweight, including impact, in case of emergency.

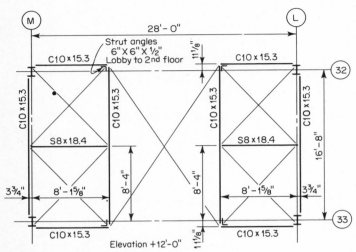

Fig. 29 Guide-rail bracing at intermediate level.

The guide rails are, in effect, long slender columns braced at each floor level to reduce their slenderness ratios. Guide rails can be subject to moment as well as direct stress should the retarding force on one guide rail become greater than that on the other because of nonsynchronous gripping of the guide rails by the safeties. This moment also produces a lateral force on divider beams at floor framing levels.

Figure 29 illustrates guide-rail bracing at an intermediate level for the 20-ft first floor of the system shown in Fig. 25. Figure 30 illustrates the nature of the forces transmitted to the guide rails and divider beams under retarding-force conditions.

Pits.   The pit is the portion of the elevator shaft which extends below the level of the bottom landing to provide for bottom overtravel and clearance, and for parts which require space below the bottom limit of car travel. The floor of the pit supports the buffers and guide rails. Normally the pit is immediately below the lowest floor level served by the elevator car. If this level is the basement, the elevator-pit floor and walls are soil-bearing and do not affect the building framework. If this level is at a higher floor, the elevator pit must be specially supported by the building framework. Particular attention should be paid to occupied spaces below the pit floor when the pit is supported in this manner.

39. Moving Stairs   The vertical transportation system is sometimes a combination of elevators and moving stairs in a multistory building. Load concentrations from the moving-stair truss stringers and machinery must be included in the design of floor framing surrounding the stairway. Figure 31 illustrates a typical moving-stairway framing system.

40. Stairwells   In the event of fire, or of elevator-equipment failure, the interior stairways and fire tower afford additional vertical transportation and means of egress. The stairways in multistory buildings sometimes require as much floor space as elevator shafts. Consequently, stairways, fire towers, elevator shafts, and mechanical shafts are generally grouped very close together to hold unusable building volume to a minimum. In a steel structure, floor beams around this combined grouping of shafts rarely remain on column centerlines and often require deep, narrow-flanged girders. In a reinforced-concrete structure, combinations of bearing walls and beam and column framing can be developed to satisfy all requirements. Wind systems can also become complicated by such shaft

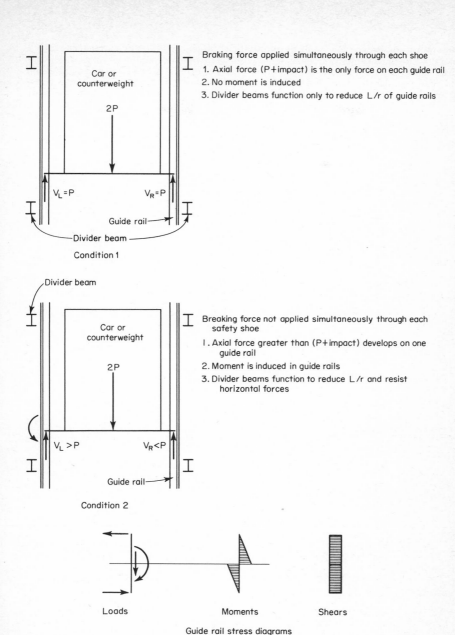

Braking force applied simultaneously through each shoe
1. Axial force (P+impact) is the only force on each guide rail
2. No moment is induced
3. Divider beams function only to reduce L/r of guide rails

Car or counterweight

2P

$V_L = P$ $V_R = P$

Guide rail

Divider beam

Condition 1

Divider beam

Car or counterweight

2P

Breaking force not applied simultaneously through each safety shoe
1. Axial force greater than (P+impact) develops on one guide rail
2. Moment is induced in guide rails
3. Divider beams function to reduce L/r and resist horizontal forces

$V_L > P$ $V_R < P$

Guide rail

Condition 2

Loads   Moments   Shears

Guide rail stress diagrams

Fig. 30 Retarding forces on guide rails.

framing arrangements. Figure 32 illustrates the structural-steel framing around a stairwell at a floor level. Each pair of beams framing the stairwell is designed to support full stair load, since the exact location of stair struts or hangers is determined later, and depends on the final stair construction selected. Figure 33 illustrates the structural framing around a stairwell at a floor level in a concrete building.

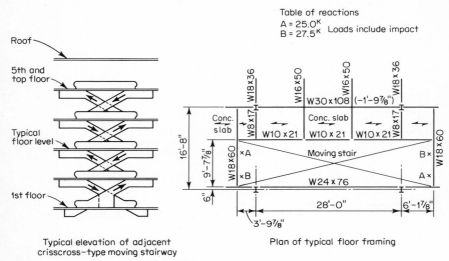

Typical elevation of adjacent crisscross-type moving stairway

Plan of typical floor framing

Fig. 31 Typical moving-stair framing system.

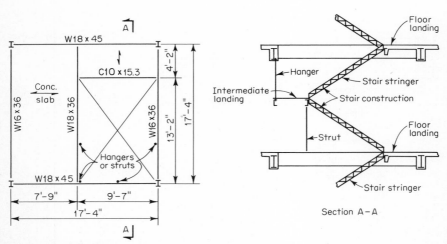

Section A–A

Fig. 32 Steel framing at stairwell.

41. Transfer Girders and Trusses The transfer girder or truss is frequently employed in multistory buildings to interrupt columns at any desired level in the building. If a column-free lobby or auditorium is required in a lower story, huge column transfer loads on long spans will result. Very often the magnitude of the loads and spans will approach those commonly found in heavy bridge work. Reinforced-concrete transfer girders are seldom used when large loads on long spans must be accommodated; greater depth, larger elastic and creep deflection, and added girder weights prove to be severe limitations. Consequently, even in all-concrete structures, structural-steel transfer girders are usually employed to handle major load transfers.

Depth.   When depth of floor construction is great, or when a full story height can be used for structural framing, trusses can efficiently provide the load transfer system

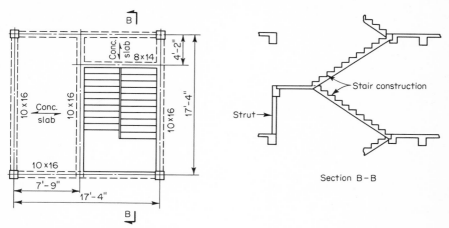

Fig. 33   Concrete framing at stairwell.

necessary. However, these conditions seldom obtain. Consequently, the designer must develop the necessary girder strength in minimum depth so as to reduce the waste space between ceiling and floor, and hence the waste height of the building.

Deflection and Camber.   It is customary and desirable to eliminate dead-load deflection in trusses and girders by cambering. This is particularly necessary in floors or roofs where fills and finishes are omitted, or if many stories are supported by the transfer girder or truss. If a floor deflects under dead load even as little as ½ in. in a building where floor and ceiling finishes are attached directly to the structural members, unsightly finished appearance as well as difficulty in achieving level floors results. Roof dead-load deflection, where roof fills are omitted, will result in undrainable roof areas. When many stories are to be carried by a truss or girder, difficulty in alignment and erection of the upper stories of the supported framework can result if dead-load deflections are not properly considered. In structural-steel framing, cambering is accomplished by cold bending in girders, or by draw at panel points of trusses. In concrete, cambering is accomplished by setting forms at the desired elevation prior to pouring concrete. Particular attention should be paid to modulus of elasticity and creep in concrete-deflection calculations.

Effect on Wind System.   Transfer girders generally interrupt an orderly wind-system arrangement since the elimination of a main column or columns obviously reduces the number of wind connections in the lower portions of the framework where wind moments and shears are the greatest. Consequently, transfer-girder floor levels also become transfer floors for horizontal wind shear. Horizontal wind shear can be readily transferred if the floor is a stiff diaphragm, such as a concrete slab or a metal deck. If numerous floor openings or light roof construction also occur at the transfer floor level, additional horizontal diagonal bracing should be used between column bays. Figure 34 illustrates the effect on the wind system at a column-transfer floor level.

WIND BRACING

42. Medium-Rise Buildings (20 to 60 Stories)—Braced Bents, Rigid Frames, and Shear Walls   In a steel structure, the most efficient system of wind bracing (optimum use of materials) is one in which bending is kept to a minimum. The K-brace system, in which the horizontal element (floor beam) is supported at midspan between columns, is more efficient than the X-brace system for several reasons. In the X-brace system, the floor beam spans full length between columns, resulting in larger floor-beam bending moments, and the total length of bracing material is greater because of the redundant member. The K-brace also offers greater freedom in the use of aisle space, since it is possible to fit doors beneath its apex. If still more usable space is desired, the braces are spread apart and

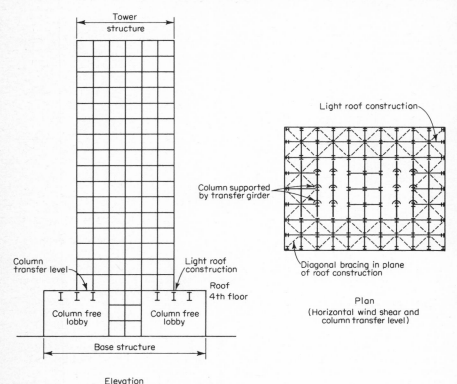

Fig. 34   Union Carbide Building, New York City.

separated from a common junction at the top, resulting in a full-story knee-braced bent. More bending is induced in the floor beam, reducing slightly the efficiency of this system.

The use of these systems is limited to building core areas, as between elevator or duct shafts and in particular adjacent to corridors or passageways. Where full usable aisle space is required, beams rigidly connected to columns must substitute for bracing. The system then becomes one of maximum bending, and hence the least efficient.

Examples 1, 2, 3, and 4 illustrate the analysis and relative efficiency of each of the four systems of Fig. 35. For the 29-story tower shown in Fig. 36, using a bay width of 30 ft 0 in. and a wind pressure of 20 psf, the wind force at the point of contraflexure midway between the ninth and tenth floors is $0.020(246 \times 30) = 147.6$ kips. Assuming wind design by the portal method, each of the three bents resists a wind shear of $147.6/3 = 49.2$ kips. Gravity load on interior columns between ninth and tenth floors is 2300 kips. Uniform load on the horizontal member of the bent is 1.5 kips/ft.

Whereas in a steel structure various components must be jointed in such a manner as to develop rigidity, a poured-in-place concrete building has inherent potential joint rigidity. A preliminary evaluation of any multistory concrete framework must be made to determine its rigidity and what, if any, additional rigidity is needed.

A structure developed as a beam-free framework, with many closely spaced columns

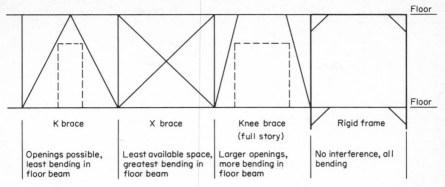

Fig. 35 Wind bracing systems.

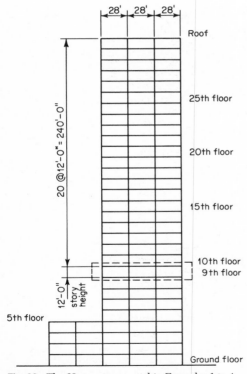

Fig. 36 The 29-story tower used in Examples 1 to 4.

(e.g., an apartment house) and proportioned for gravity loads only, may be adequate for the specified wind moments and shears. If not, reinforcing can be added, and slab thickness or column sizes increased. It is not uncommon to design such structures 20 stories in height without need of any other form of wind bracing. Figure 37 shows a part plan of such a structure built for the New York City Housing Authority.

Example 1: Design example illustrating K-brace bent design

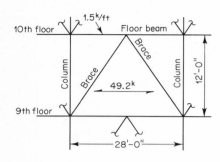

Floor beam

The floor beam is analyzed as a two-span beam continuous over simple supports. Therefore

$$(-) M = \tfrac{1}{8} \times 1.5 \times 14^2 = 36.8^{'k}$$

$$S = \frac{36.8 \times 12}{24} = 18.9 \text{ in.}^3 \text{ use W10} \times 22$$

$$R = (@ \text{ columns}) = \tfrac{3}{8} \times 1.5 \times 14 = 7.9^k$$

$$R = (@ \text{ center support}) = 2 \times \tfrac{5}{8} \times 1.5 \times 14 = 26.2^k$$

Wind direct stress in beam is neglected in this computation, since its effect on total stress is extremely small, especially considering the increase in allowable stresses permitted for combined loadings.

Brace

C–C length of brace: $(12^2 + 14^2)^{1/2} = 18.4'$
Wind force in each brace member:
$\qquad (49.2/2) \times (18.4/14.0) = \pm 32.4^k$
Compression in brace from gravity load:
$\qquad (26.2/2) \times (18.4/12.0) = -20.1^k$
Total compression $= 32.4 + 20.1 = 52.5^k$
Unsupported length of brace $= 18.4 - 1.0 = 17.4'$
For $l/r = 200$ max., $r_{req.} = (17.4 \times 12)/200 = 1.04$
Use L8 × 6 × $\tfrac{7}{16}$
$\qquad f_a = 52.5/5.93 = 8.9$ ksi
$\qquad F_a = 7.34 \times 1.33 = 9.7$ ksi

Column (interior)

Interior columns receive no bending, since all wind stresses are axial, and no direct wind loads, since vertical components of braces cancel each other. Thus, the column is designed for gravity load only.
$\qquad P = 2300^k \qquad l_u = (12.0 - 1.5) \times 12 = 126''$
$\qquad$ Try W14 × 398 $\qquad r = 4.31 \qquad l/r = 126/4.31 = 29$
$\qquad F_a = 20.01$
$\qquad A_{req} = \dfrac{2300}{20.01} = 115.0$ sq in.
$\qquad$ Use W14 × 398 (A = 117 sq in.)

Weight of material

1 column	398#	x	12.0'	=	4,780
1 floor beam	21#	x	26.4'	=	555
2 braces	2(20.2#)	x	17.4'	=	705
					6,040#

Where columns are few and far between, wind forces can no longer be handled by the floor slab alone. It is seldom practical to introduce concrete wind girders in the type of floor system generally employed in multistory buildings; moreover, large moments increase column sizes rapidly, and should be avoided. The wind-bracing systems most often used in this case consist of a series of shear walls placed in exterior walls or interior core areas, with the floors acting as stiff horizontal diaphragms to distribute the story shears to each shear wall. Central circular or rectangular arrangements of shear walls are sometimes necessary where large wind shears occur in two directions.

It is seldom possible to introduce shear walls completely free of openings or penetra-

Example 2: Design example illustrating full story knee brace bent design

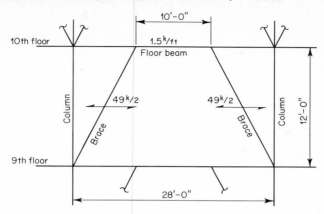

C.–C. length of brace = $(9^2 + 12^2)^{1/2} = 15.0'$
Wind force in each brace member = $(49.2/2) \times (15.0/9.0) = \pm 41.2^k$
Vertical component of brace force:
$(49.2/2) \times (12.0/9.0) = \pm 32.9^k$

Floor beam

For <u>wind applied</u> forces, $R_L = R_R = (32.9 \times 10)/28 = 11.7^k$
(Neglect horizontal component of wind forces in this calculation)

For <u>gravity applied</u> forces, $R_L = R_R \cong 0.1 \times 1.5 \times 28 = 4.2^k$
$R_I \cong 0.4 \times 1.5 \times 28 = 16.8^k$

For <u>combined</u> forces

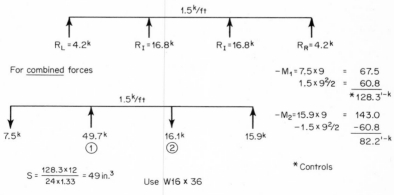

$$-M_1 = 7.5 \times 9 \quad = \quad 67.5$$
$$1.5 \times 9^2/2 \quad = \quad \underline{60.8}$$
$$*128.3'^{-k}$$

$$-M_2 = 15.9 \times 9 \quad = \quad 143.0$$
$$-1.5 \times 9^2/2 \quad \quad \underline{-60.8}$$
$$82.2'^{-k}$$

* Controls

$$S = \frac{128.3 \times 12}{24 \times 1.33} = 49 \text{ in.}^3 \qquad \text{Use W16} \times 36$$

Brace

Compression $= 49.7 \times (15/12) = 62.4^k$
$l_u = (15.0 - 1.5) \times 12 = 162''$
For $L/r = 200$ max., $r_{req} = 162/200 = 0.81$
Try L6×6× $\frac{1}{2}$
$f_a = 62.4/5.75 = 10.8$ ksi
$F_a = 8.70 \times 1.33 = 11.6$ ksi

Example 2:   Design example illustrating full story knee brace design (Continued)

Column (interior)

The column is W14 x 398, as in Example 1, with no bending or wind direct load

Weight of material

1 column	398#	x	12.0'	=	4,780	
1 floor beam	36#	x	26.4'	=	950	
2 braces	2x(19.6#)	x	13.5'	=	530	
					6,260#	

Example 3: Design example illustrating X brace bent design

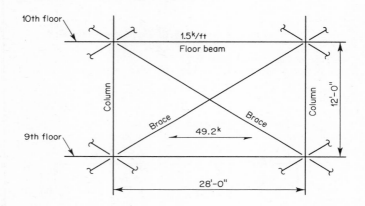

Brace

C/C length of brace $= (12^2+28^2)^{1/2} = 30.5'$

Since l/r must be limited to 200 for compression members, but may be 300 for tension members, bracing will be designed as a tension system. The wind force in either diagonal is

$$49.2 \times 30.5/28 = 53.2^k$$

With the diagonals connected at their intersection, the unsupported length is

$$l_u = (30.5-2)/2 = 14.25'$$

where 2ft is deducted for the approximate depth of connection.

$$r_{req} = 14.25 \times 12/300 = 0.53.\ \text{Use L4} \times 3\tfrac{1}{2} \times \tfrac{5}{16}$$

With deduction in area for one rivet hole

$$f_a = 53.2/(2.25-0.35) = 27.4$$
$$F_a = 22 \times 1.33 = 29.2$$

Floor beam

$$M = \tfrac{1}{8} \times 1.5 \times 28^2 = 147.0'^{-k}\ \text{(simple span)}$$

$$S = \frac{147.0 \times 12}{24} = 73.5\ \text{in.}^3$$

Use W18 x 46

Column (interior)

The column is W14 x 398, as in Example 1, with no bending or wind direct load.

Weight of material

1 column	398#	x	12.0'	=	4,780	
1 floor beam	45#	x	26.4'	=	1,185	
2 braces	2 x (7.7#)	x	28.5'	=	440	
					6,405#	

Example 4: Design example illustrating rigid frame bent design

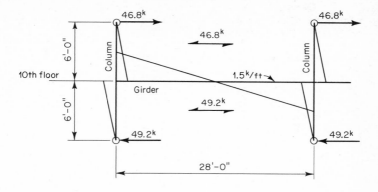

Interior joint

$M = (46.8 + 49.2)6 = 576^{\text{'-k}}$

Column moment corrected to face of beam: $576/2 \times 5/6 = 239^{\text{'-k}}$

Beam moment corrected to face of column:

$$\frac{576}{2} \times \frac{13.2}{14.0} = 271^{\text{'-k}}$$

Girder design

$-M_g = 1/12 \times 1.50 \times 28^2 = 81^{\text{'-k}}$

$-M_w \text{ (see above)} = \underline{271}$

$\qquad\qquad\qquad M_T = 352^{\text{'-k}}$

$$S = \frac{352 \times 12}{24 \times 1.33} = 132 \text{ in.}^3$$

Use W27 x 94

Column design

There are no wind direct loads on interior columns, since girder wind shears at connections are equal and in opposite directions.

Column gravity load = $2,300^k$

Column moment $\quad = \quad 239^{\text{'-k}}$

1) No wind, use W14 x 398 (Example 1)

2) With wind, try W14 x 398

$\qquad L_u = 12'0''$

$\qquad r_{x-x} = 7.16 \qquad S_{x-x} = 656 \text{ in.}^3$

$\qquad L_u / r_{x-x} = 120/7.16 = 16.75$

$\qquad F_A = 20.79; \quad F_B = 24.0$

$$f_a = \frac{2,300}{117} = 19.7 \qquad f_b = \frac{239 \times 12}{656} = 4.37$$

$$\frac{f_a}{0.6F_y} + \frac{f_b}{F_B} = \frac{19.7}{0.6 \times 36} + \frac{4.37}{24} = 1.10 < 1.33 \text{ (OK)}$$

Weight of material

1 column	$398^\#$	x	$12.0'$	=	4,776
1 wind girder	$68^\#$	x	$26.4'$	=	1,795
					6,571 $^\#$

tions for mechanical services. Generally, openings and penetrations can be accommodated, by careful positioning in the shear wall, without seriously affecting stress distribution or stiffness. Example 5 illustrates the analysis of a core of the shear wall system shown in Fig. 38 for wind in the critical directions. A similar analysis must be undertaken to check the stresses produced by wind in the directions normal to the one shown.

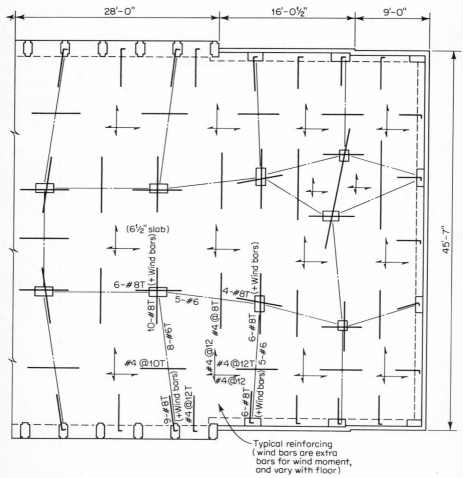

Fig. 37   Part framing plan—flat-plate design—New York City Housing Authority project.

43. High-Rise Buildings (above 60 stories)—Tubular Frames, Tube within a Tube, and Combinations   The conventional systems consisting of two-dimensional wind frames or shear walls in the transverse and longitudinal directions to resist wind effects from either direction are inefficient in very tall buildings. Tubular frames which engage all exterior or interior columns and girders in both transverse and longitudinal directions, creating a vertical tubular cantilever beam to resist wind effects from both directions, are generally more efficient. In steel structures, the most efficient tubular system of wind bracing is one in which exterior columns and girders can be utilized to keep bending to a minimum.

Exterior tubular framing systems are either Vierendeel systems composed of closely spaced building columns and spandrel girders (Fig. 39) or cross-braced systems composed

Example 5: Design example illustrating shear wall design

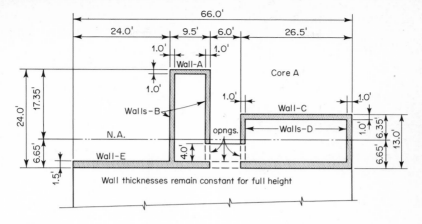

Properties of core

1. Neutral axis (N.A.)

1.5	x	66.0	=	99.0	x	0.75	=	74.3
1.0	x	22.5	=	22.5	x	12.75	=	287.0
1.0	x	18.5	=	18.5	x	14.75	=	273.0
1.0	x	7.5	=	7.5	x	9.25	=	69.4
1.0	x	11.5	=	11.5	x	7.25	=	83.4
1.0	x	7.5	=	7.5	x	23.50	=	176.2
1.0	x	24.5	=	24.5	x	12.50	=	306.0
				191.0ft^2				$1,269.3 \text{ft}^3$

$$\bar{y} = \frac{1,269.3}{191.0} = 6.65 \text{ft}$$

2. Moment of inertia of core concrete (I_C)

(1) $\dfrac{66 \times 1.5^3}{12}$ = 18.6ft^4

(2) 99×5.9^2 = $3,450.0$

(3) $\dfrac{1.0 \times 22.5^3}{12}$ = 949.0

(4) 22.5×6.10^2 = 833.0

(5) $\dfrac{1.0 \times 18.5^3}{12}$ = 527.0

(6) 18.5×8.10^2 = $1,215.0$

(7) $\dfrac{1.0 \times 7.5^3}{12}$ = 35.0

(8) 7.5×2.60^2 = 51.0

(9) $\dfrac{1.0 \times 11.5^3}{12}$ = 127.0

(10) 11.5×0.60^2 = 4.2

(11) 7.5×16.85^2 = $2,130.0$

(12) 24.5×5.85^2 = 840.0

I_C = $10,179.8 \text{ft}^4$

3. Moment of inertia of core reinforcing (I_R)

Assume $\frac{1}{2}\%$ reinforcing in walls

$I_R = (0.005)(n-1)I_C$

$0.005 \times (8-1)10,179.8 = 356 \text{ft}^4$

4. Total core moment of inertia (I_T) and section modulus (S)

$I_T = 10,179.8 + 356 = 10,536 \, ft^4$

$S_{max} = \dfrac{10,536}{6.65} = 1584 \, ft^3 \qquad S_{min} = \dfrac{10,536}{17.35} = 607 \, ft^3$

Wind analysis

1. Wind forces (using applicable code)

$F_1 = 20 \times 66 \times 200' = 264,000^{\#}$

$F_2 = 4 \times 66 \times \dfrac{150'}{2} = \quad 19,800^{\#}$

$\qquad\qquad\qquad\qquad \overline{283,800^{\#} = F_T}$

2. Wind moment at base

$M = (264 \times 150) + (19.8 \times 200) = 43,560^{'-k}$

$\qquad\qquad\qquad\qquad\qquad$ (say 44,000)

Assume each core takes half,
M per core $= \pm 22,000^{'-k}$

3. Maximum stress at base (Short wall)

$f_c = \dfrac{\pm 22,000}{607} = \pm 36.3 \, ksf$

Check for tensile stresses in wall at end of core:
$1.0 \times 1.0 \times 0.150 \times 250 = 37.5 \, ksf$ (O.K.)

Design stress at base of wall:
$\dfrac{36,300 + 37,500}{144} \times \dfrac{3}{4} = 385 \, psi$

4. Minimum stress at base (Long wall)

$f_c = \dfrac{\pm 22,000}{1584} = \pm 13.8 \, ksf$ (wind)

$f_c = \dfrac{{}^{*}208^{k/ft}}{1.5} = +139 \, ksf$ (gravity)

$\qquad\qquad$ *208 represents total
$\qquad\qquad$ uniform load at base of
$\qquad\qquad$ 66' long wall

Design stress at base of wall:
$\dfrac{139,000 + 13,800}{144} \times \dfrac{3}{4} = 798 \, psi$ (with wind)

or $\quad \dfrac{139,000}{144} = 970 \, psi$ (no wind) controls

5. Wind shear at base

Total area of walls resisting shear:
$(1.0 \times 24) + (1.0 \times 20) + (1.0 \times 9) + (1.0 \times 13)$
$= 66 \, ft^2$

$F_T = 283.8/2 = 141.9^k$

$f_v = \dfrac{141,900}{66 \times 144} = 15 \, psi$ (OK)

6. Wall design

Use 4,000 psi concrete $(f_c = 0.25 \times 4,000 = 1,000 \, psi)$
with $\frac{1}{2}\%$ reinforcing steel.

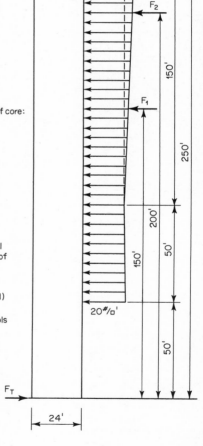

7. Check for stability (overturning)

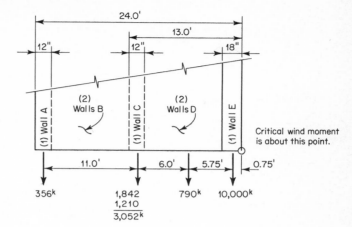

Using dead loads only:

Wall A = 9.5 × 1.0 × 0.150 × 250 = 356   × 1 = 356^k

B = 22.17 × 1.0 × 0.150 × 250 (wall) = 833

    22.17 × 7.5/2 × 0.050 × 21 (floor) = 88

                        921   × 2 = 1,842

C = 26.5 × 1.0 × 0.150 × 250 (wall) = 990

    26.5 × 10.5/2 × 0.075 × 21 (floor) = 220

                      1,210   × 1 = 1,210

D = 10.5 × 1.0 × 0.150 × 250 = 395   × 2 = 790

E = 66.0 × 1.5 × 0.150 × 250 (wall) = 3,700

    66.0 × 72/2 × 0.125 × 21 (floor) = 6,300

                     10,000   × 1 = 10,000

Overturning moment (wind) = 22,000$^{\prime-k}$

Stabilizing moment (gravity)

10,000 ×	0.75	=	7,500
790 ×	6.50	=	5,100
3,052 ×	12.50	=	38,200
356 ×	23.50	=	8,370
			59,170

Factor of safety = $\dfrac{59,170}{22,000}$ = 2.69 (OK – min. = 1.33)

of multistory diagonals, building columns, and spandrel girders (Fig. 40). Of course, architectural considerations must be integrated with structural requirements in any exterior tubular system.

Tube-within-a-tube framing systems are basically multiple-tube frameworks. Various arrangements are possible when interior columns and girders are also engaged to create vertical tubular cantilever beams. Combinations of exterior tubular framework with interior shear walls or braced bents are generally more practical. Interior space arrangements in very tall buildings usually require grouped elevator shafts, mechanical shafts, and fire stairs, which can readily accommodate shear walls and braced bents but not tubular frameworks.

The principles of the behavior of a tubular framework can be compared with a stack or chimney, which is a very simple type of tubular cantilever beam. In the case of a building, however, the analysis of the tube is very complex because of the numerous component parts, which are statically indeterminate, and the three-dimensional nature of the problem.

Figure 41 illustrates the pattern of the stresses that must be determined. Note the shear-lag effect due to deformations of the girders. To properly analyze this system a computer program with three-dimensional capabilities must be used.

44. Fixed and Partially Fixed Joints in Steel Structures   The design of wind connections in rigid-frame bents must take into consideration the degree of fixity of the joint. If the joint is

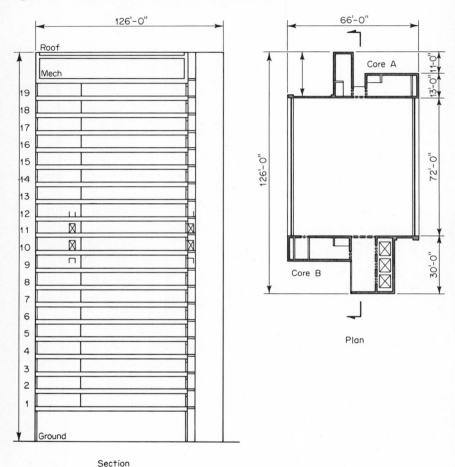

Section

Fig. 38   American Life Insurance Building, Wilmington, Del.

fully fixed, gravity moments must be added to the wind moments in designing the connection, girder, and column. If the joint is partially fixed, the gravity moment is generally neglected.

The degree of joint fixity is established in accordance with the type of fastener and connection material. Welding or high-strength bolting, together with T-flange or gusset-plate connection material, usually can be considered to fix a joint fully. Lesser combinations of connection material and fasteners are commonly considered to fix a joint partially.

45. Wind-Load Determination   Wind loads on buildings are established in all building codes. Depending on the experience of the design engineer and on the building location, height, shape, and exposure, and in certain cases on its mass and stiffness (damping), wind-tunnel testing to determine the adequacy of code wind loads may be necessary. If testing is needed to arrive at design loads, the following procedure is suggested:

1. Determine the mass and stiffness, using applicable building-code wind loads. This is a preliminary design based on approximate methods of analysis, either manual or computer-assisted.

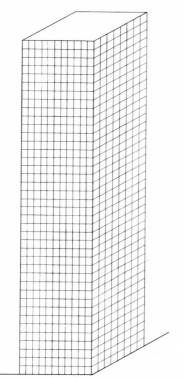

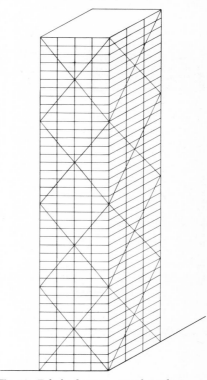

Fig. 39   Tubular framing—Vierendeel system.

Fig. 40   Tubular framing—cross-braced.

2. Build an aerolastic model of the building, with at least three degrees of freedom, capable of reproducing the dynamic responses of the building.

3. Construct a model of the building environment within an adequate radius (300 to 500 ft).

4. Make preliminary wind-tunnel tests to check for poor basic dynamic responses such as resonance, coupling, and general instability. Modify the design if required.

5. Run tests using gradient wind speeds of at least four levels, i.e., 75, 100, 125, and 150 mph, from all wind directions.

6. Measure tip deflections and accelerations, and the static and fluctuating pressure coefficients on all facades.

7. Calculate base moments and shears using the measured data.

8. Select a design gradient wind speed from a study of meteorological data.

9. Determine an effective static wind-load diagram for design purposes.

10. Rerun the wind-tunnel test on a modified model if the final design results in significant changes in mass and stiffness.

There are no universally accepted procedures and standards for wind-tunnel testing of buildings. Each study is unique and requires individual attention and programming.

46. Wind Deflection   The horizontal displacement of a floor relative to the floor below is called the deflection or drift per story. This deflection is the result of column shortening, and joint translation and rotation. The column-shortening component is a function of the chord wind stress only and is independent of the type, size, and arrangement of the web system, but the component due to joint translation and rotation involves deformations of all elements of the system.

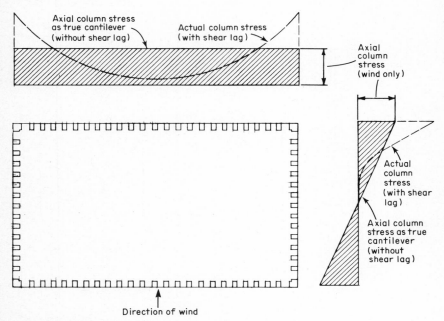

Fig. 41   Wind-load stresses in columns of tubular frame.

A tall, slender framework must be stiff enough to limit drift. If a tower is too limber, partitions and finishes will crack, and groaning noises and swaying during windstorms will produce uncomfortable psychological sensations. A measure of the deflection characteristics of a building framework is called the deflection, or drift, index. This index is the ratio of the deflection per story to the story height, Δ/h, where Δ = deflection and h = height.

The selection of a proper index for a tall tower building is as important as the selection of the proper wind force. Although building codes establish the minimum wind force that must be used in the design, seldom is a deflection index specified, since safety is not involved. If too low a deflection index is selected, the wind system may become unrealistically large and costly. Factors which should be considered in establishing a deflection index are (1) type of building and type of occupancy, (2) stiffening effects of interior and exterior walls and floors, (3) shielding against wind, and (4) magnitude of wind forces established by code.

Engineering judgment must recognize economic values involved in a building constructed as a speculative or purely commercial venture; the usual result would be a building designed primarily for strength and stability. A single-occupancy corporate or prestige building, however, can and must afford the luxury of added stiffness to minimize unnerving movement, unsightly plaster cracks, and eerie noises. Hotels, residences, and hospitals require special consideration also, since movement and noise are apt to be more disturbing.

The stiffening effects of floors and interior and exterior walls are highly indeterminate and are neglected in deflection calculations. Masonry exterior walls and masonry core and stair-tower walls add considerable stiffness by shear wall action; glass and metal curtain walls do not. Concrete fireproofing of beams and columns adds stiffness; sprayed-on fireproofing or protective ceilings do not.

Buildings located in the center of major cities, and buildings of moderate height, are less likely to be subjected to the full design wind pressures required by code. Deflection criteria are less critical in such cases.

A building code establishes standards of safety. Unrealistic calculated deflections will result for buildings that must conform to codes which prescribe high standards with respect to strength and stability. The selection of a deflection index should be tempered by this factor, other data, experience, and judgment.

In New York City, many high-rise tower office buildings have been designed with a deflection index of 0.0020 to 0.0030 for masonry buildings and 0.0015 to 0.0025 for curtain wall buildings, calculated on the basis of 20 psf above the 100-ft level. A change in the New York City building code in 1968 that mandates higher wind loads would justify larger deflection indexes. Table 30 may be used as a guide.

TABLE 30   Weiskopf and Pickworth Deflection-Index Guide

Occupancy	Type of construction		Exposure	Wind requirements	Deflection index
	Walls	Framing system			
Commercial	Curtain wall	Steel, no concrete fireproofing encasement	Maximum to minimum	Minimum to extreme	0.0015–0.0035
		Concrete, or steel with concrete encasement of beams and columns	Minimum	Moderate	0.0030
	Masonry	Steel, no concrete fireproofing encasement	Maximum	Minimum	0.0020
		Concrete, or steel with concrete encasement of beams and columns	Average	Moderate	0.0030
Residential	Curtain wall	Steel, no concrete fireproofing encasement	Maximum	Minimum	0.0010
		Concrete, or steel with concrete encasement of beams and columns	Maximum	Minimum	0.0015
	Masonry	Steel, no concrete fireproofing encasement	Maximum	Minimum	0.0015
		Concrete, or steel with concrete encasement of beams and columns	Minimum	Extreme	0.0025

Example 6a illustrates a method of manually calculating the deflection index for a steel-framed building. Example 6b illustrates a method of manually calculating the deflection index for the concrete-framed structure shown in Fig. 38 in which the wind forces are resisted by shear walls. These procedures are helpful in preliminary design, but when they begin to yield acceptable deflection indexes after adjustment of the framework, the final design can usually be completed more efficiently by computer methods.

The usual method of designing a wind system is based on an assumed wind-shear distribution which may or may not be in accordance with the relative stiffness of the

Example 6a: Design example illustrating Weiskopf and Pickworth method of calculating deflection index
(Lever House, New York City)

The deflection is assumed to be entirely due to drift; column shortening is neglected because of relatively low building height.

10th story shear = 100^k

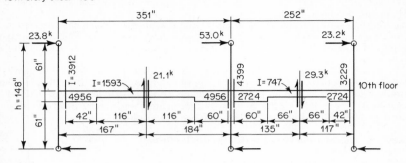

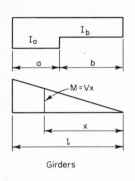

Girders

External work = internal work

$\frac{1}{2}$ x total story shear x Δ = Σ columns + Σ girders

Work done by girders (girders + end stubs)

$$W = \int \frac{M^2 dx}{2EI} = \frac{1}{2E} \left[\int_b^l \frac{M^2}{I_a} dx + \int_0^b \frac{M^2}{I_b} dx \right]$$

$$= \frac{V^2}{2E} \left[\int_b^l \frac{x^2}{I_a} dx + \int_0^b \frac{x^2}{I_b} dx \right]$$

$$= \frac{V^2}{6E} \left[\frac{l^3 - b^3}{I_a} + \frac{b^3}{I_b} \right]$$

Work done by columns ($I_a = I_b = I$)

$$W = \frac{V^2 l^3}{6EI}$$

Work done by girders, including stubs

$$W_G = \frac{21.1^2}{6E} \left[\frac{158^3 - 116^3}{4,956} + \frac{2 \times 116^3}{1,593} + \frac{176^3 - 116^3}{4,956} \right] + \frac{29.3^2}{6E} \left[\frac{126^3 - 66^3}{2,724} + \frac{2 \times 66^3}{797} + \frac{108^3 - 66^3}{2,724} \right]$$

$$= \frac{21.1^2}{6E} \left[\frac{6,274}{4,956} + \frac{3,122}{1,593} \right] \times 10^3 + \frac{29.3^2}{6E} \left[\frac{2,686}{2,724} + \frac{574}{797} \right] \times 10^3$$

$$= \frac{485 \times 10^3}{E}$$

Work done by columns

$$W_C = 2 \times \frac{23.8^2 \times 61^3}{6E \times 3,912} + 2 \times \frac{53.0^2 \times 61^3}{6E \times 4,399} + 2 \times \frac{23.2^2 \times 61^3}{6E \times 3,229}$$

$$= \frac{11 \times 10^3}{E} + \frac{48.4 \times 10^3}{E} + \frac{12.6 \times 10^3}{E} = \frac{72 \times 10^3}{E}$$

Total work $= \frac{485 \times 10^3}{E} + \frac{72 \times 10^3}{E} = \frac{557 \times 10^3}{E}$

$\frac{1}{2} \times 100 \times \Delta = \frac{557 \times 10^3}{E}$

$$\Delta = \frac{557 \times 10^3 \times 2}{30 \times 10^3 \times 100} = 0.371"$$

Deflection index = 0.371/148 = 0.0025

Example 6b: Design example illustrating method of calculating the deflection index (concrete shear wall design)

The deflection is assumed to be entirely due to shortening of the core walls; drift due to openings is neglected.

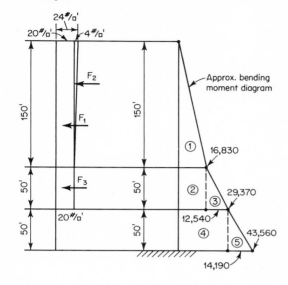

Wind load diagram

$F_1 = 20 \times 66 \times 150 = 198^k$

$F_2 = \frac{1}{2} \times 4 \times 66 \times 150 = 19.8^k$

$F_3 = 20 \times 66 \times 50 = 66^k$

Moment diagram

M = 150' from top

$F_1 \times 75 = 198 \times 75 = 14,850$

$F_2 \times 100 = 19.8 \times 100 = \underline{1,980}$

$16,830^{\prime-k}$

M = 200' from top

$F_1 \times 125 = 198 \times 125 = 24,750$

$F_2 \times 150 = 19.8 \times 150 = 2,970$

$F_3 \times 25 = 66 \times 25 = \underline{1,650}$

$29,370^{\prime-k}$

M @ base

$F_1 \times 175 = 198 \times 175 = 34,650$

$F_2 \times 200 = 19.8 \times 200 = 3,960$

$F_3 \times 75 = 66 \times 75 = \underline{4,950}$

$43,560^{\prime k}$

Deflection at top Δ by area-moment method

$\Delta = \Sigma \dfrac{\text{moment of area}}{EI}$ about top of building

① $16,830 \times \dfrac{150}{2} \times 100 \times \dfrac{1}{EI} = 126,200 \times \dfrac{10^3}{EI}$

② $16,830 \times 50 \times 175 \times \dfrac{1}{EI} = 147,300 \times \dfrac{10^3}{EI}$

③ $12,540 \times 50 \times \dfrac{1}{2} \times 183 \times \dfrac{1}{EI} = 57,400 \times \dfrac{10^3}{EI}$

④ $29,370 \times 50 \times 225 \times \dfrac{1}{EI} = 330,400 \times \dfrac{10^3}{EI}$

⑤ $14,190 \times 50 \times \dfrac{1}{2} \times 233 \times \dfrac{1}{EI} = 82,700 \times \dfrac{10^3}{EI}$

$$744,000 \times \dfrac{10^3}{EI} \text{ ft}^3\text{-k}$$

$\Delta = \dfrac{744,000 \times 10^3}{3.64 \times 10^3 \times 10,536 \times ② \times 12} = 0.8''$ ⌐ 2 cores

Deflection index $= \dfrac{\Delta}{h} = \dfrac{0.8}{12 \times 250} = 0.00027$

various wind bents. The floor system acts as a stiff plate or horizontal diaphragm at each floor and tends to distribute shears to each bent in accordance with relative stiffness, regardless of design assumption. Twisting or excessive deflection in parts of a framework can result if the various members are not tuned up or adjusted to deflect equally. Tuning of portals to the same deflection index in all wind bents not only controls deflection of the framework to any desired amount but also adjusts the stiffnesses to agree with whatever shear distribution was assumed. Example 7 illustrates a method of tuning portals in a steel-tower structure by manual calculation. When computer methods are used, resulting member stresses are examined and members resized until the stresses in each portal are optimized and deformations made acceptable.

Example 7: Design example illustrating Weiskopf and Pickworth method of tuning portals to the desired deflection index in a tower structure (Chase Manhattan Bank Building – New York City)

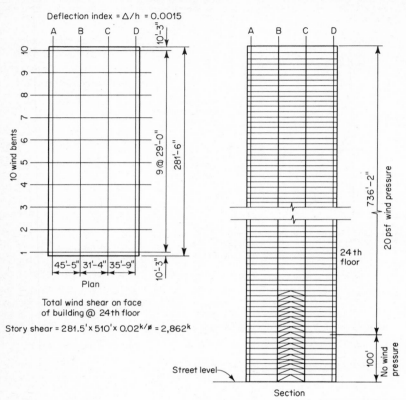

Deflection index = Δ/h = 0.0015

10 wind bents

A B C D

$10'-3"$

$9 @ 29'-0"$

$281'-6"$

$10'-3"$

45'-5" 31'-4" 35'-9"

Plan

Total wind shear on face
of building @ 24th floor

Story shear = $281.5' \times 510' \times 0.02^{k/\#} = 2,862^k$

A B C D

$736'-2"$

20 psf wind pressure

24th
floor

Street level

100'

No wind
pressure

Section

Showing typical wind bent with wind girders in all
portals except for the diagonals in lower stories
in middle portal

(A) Deflection from column shortening

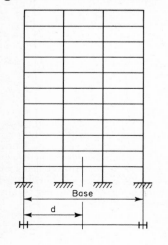

Base

d

$*\Delta = wl^4/6EI_0$

Where w = wind pressure

1 = height of building

$I_0 = 2ad^2$ = moment of inertia of exterior columns

a = approx. column area

d = base/2

Then w = $20 \times 29/12 = 48.3^\#$/in.

1 = $736.2 \times 12 = 8,880$ in.

E = 30×10^6

$I_0 = 500^{\square"} \times 675^2 \times 2 = 455 \times 10^6$

$\Delta = \dfrac{48.3 \times 8,880^4}{6 \times 30 \times 10^6 \times 455 \times 10^6} = 3.80"$

$\Delta/h = 3.80/8,880 = 0.00043$ say 0.0005

Leaving 0.0015 – 0.0005 or 0.0010
available for wind drift.

* See "Wind Bracing" – Spurr, page 127.

19-94

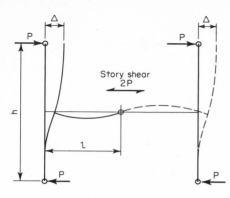

External work = internal work

½ P × Δ = Σ columns + Σ girders

For simplicity ½ of portal is analyzed

Work done by columns = $\dfrac{M_C^2 l_C}{6EI_C}$

Work done by girders = $\dfrac{M_G^2 l_G}{6EI_G}$

$\therefore$ ½ P × Δ = $\dfrac{M_G^2 l_G}{6EI_G} + 2\dfrac{M_C^2 l_C}{6EI_C}$

$\Delta = \dfrac{2M_G^2 l_G}{6EI_G P} + \dfrac{4M_C^2 l_C}{6EI_C P}$

When I_C/l_C is greater than 8 to 10 times I_G/l_G, strain energy of column may be neglected

$\therefore \Delta = \dfrac{2M_G^2 l_G}{6EI_G P}$ where $M_G = P \times h$

or $I_G = \dfrac{2M_G^2 l_G}{6EP\Delta}$

The above formula can be applied to determine the moment of inertia of each wind girder in each portal for stiffness at any floor level

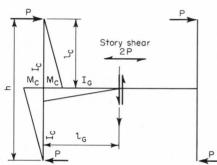

At the 24th floor the moment of inertia of each wind girder required to limit the story deflection due to drift to Δ = 0.0010 × h = 0.151" is determined as follows :

24th story shear = $2{,}862 \div 10 = 286.2^k$/bent

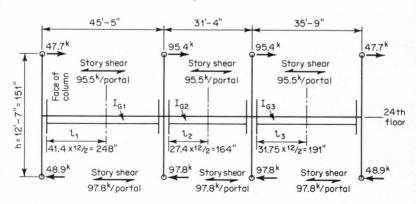

23rd story shear = $2{,}934 \div 10 = 293.4^k$/bent

Δ (each portal) = 0.0010 × 151" = 0.151"   E = 30 × 10⁶

P = column wind shear   M = P × h × correction factor

Example 7: Design example illustrating Weiskopf and Pickworth method of tuning portals to the desired deflection index in a tower structure (Chase Manhattan Bank Building-New York City) (Continued)

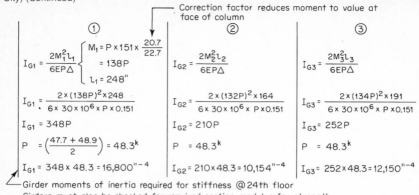

Correction factor reduces moment to value at face of column

①

$$I_{G1} = \frac{2M_1^2 \ell_1}{6EP\Delta}$$

$$\begin{cases} M_1 = P \times 151 \times \dfrac{20.7}{22.7} \\ \quad = 138P \\ \ell_1 = 248'' \end{cases}$$

$$I_{G1} = \frac{2 \times (138P)^2 \times 248}{6 \times 30 \times 10^6 \times P \times 0.151}$$

$$I_{G1} = 348P$$

$$P = \left(\frac{47.7 + 48.9}{2}\right) = 48.3^k$$

$$I_{G1} = 348 \times 48.3 = 16,800''^{-4}$$

②

$$I_{G2} = \frac{2M_2^2 \ell_2}{6EP\Delta}$$

$$I_{G2} = \frac{2 \times (132P)^2 \times 164}{6 \times 30 \times 10^6 \times P \times 0.151}$$

$$I_{G2} = 210P$$

$$P = 48.3^k$$

$$I_{G2} = 210 \times 48.3 = 10,154''^{-4}$$

③

$$I_{G3} = \frac{2M_3^2 \ell_3}{6EP\Delta}$$

$$I_{G3} = \frac{2 \times (134P)^2 \times 191}{6 \times 30 \times 10^6 \times P \times 0.151}$$

$$I_{G3} = 252P$$

$$P = 48.3^k$$

$$I_{G3} = 252 \times 48.3 = 12,150''^{-4}$$

Girder moments of inertia required for stiffness @ 24th floor
Girders must also be checked for required section modulus for strength

Ⓒ Deflection from wind drift – wind bent with girders and diagonals

*Δ of portals with wind girders = $\dfrac{2M_G^2 \ell_G}{6EI_G P} + \dfrac{4M_C^2 \ell_C}{6EI_C P}$

(*See above for derivation)

Δ of portals with diagonals

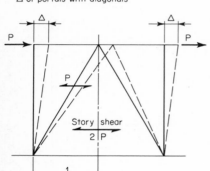

Story shear
2 P

External work = internal work
$\frac{1}{2} \times P \times \Delta = \Sigma$ diagonals
For simplicity $\frac{1}{2}$ of portal is analyzed
Work done by diagonals = $\dfrac{F^2 L}{2AE}$

where F = wind stress in diagonal
L = length of diagonal
A = area of diagonal

$\therefore \frac{1}{2} \times P \times \Delta = \dfrac{F^2 L}{2AE}$

$\Delta = \dfrac{F^2 L}{AEP}$

$\dfrac{P}{\ell} = \dfrac{F}{L}$

$F = \dfrac{P}{\ell} \times L$

At a lower floor level the required moment of inertia of each wind girder and area of each diagonal to limit the story deflection due to drift to Δ = 0.0010 × h is determined as follows:

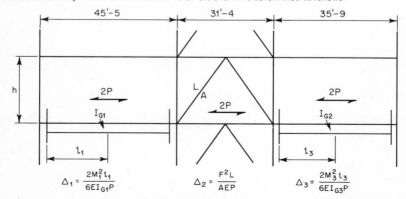

45'-5   31'-4   35'-9

h

2P

I_{G1}

ℓ_1

L / A

2P

2P

I_{G2}

ℓ_3

$\Delta_1 = \dfrac{2M_1^2 \ell_1}{6EI_{G1}P}$

$\Delta_2 = \dfrac{F^2 L}{AEP}$

$\Delta_3 = \dfrac{2M_3^2 \ell_3}{6EI_{G3}P}$

Design for required area of diagonal for direct wind stress. Compute Δ_2 using above formula and correct area so that $\Delta_2 = \Delta_1 = \Delta_3 = 0.0010 \times h$

Portal frameworks are generally not feasible, economically or architecturally, in a concrete multistory building. However, when portal or a combination of portal and shear-wall frameworks are used the procedure of Example 7 is applicable.

47. Wind-Shear Dissipation   The purpose of the wind-bracing system in a multistory framework is to dissipate the wind shear throughout the framework and ultimately into the foundations. The usual framework consists of a series of wind bents placed in a more or less symmetrical pattern, with floors acting as stiff horizontal girders to distribute the story shears to each bent. The wind bents may be rigid frame bents, braced bents, or a combination of both. Fig. 42 illustrates a relatively symmetrical wind-bracing system combining rigid frame and braced bents.

There is no preferable form of wind bracing. Each framework must be evaluated with respect to the particular architectural, mechanical, and structural problems to be resolved. The elimination of columns to permit large, column-free lobbies or auditoriums, unsymmetrical positioning of wind bents to permit special column arrangements, confining wind bents to core areas and exterior walls, and shifting from one system of wind bracing to another at a lower story are a few of the considerations which can influence the design of the wind-bracing system. Figure 43 illustrates a system of braced bents confined to core areas, thus eliminating deep wind girders in floor areas and permitting maximum utilization of ceiling space for mechanical work.

In the case of a high-rise tubular frame the floor system must be capable of distributing wind shears to the tube frames. Figures 48 and 49 illustrate a tube-frame wind system combined with interior shear walls.

Torsional moments due to eccentricity of the wind-load resultant with respect to the center of rigidity of the structure often become significant in concrete buildings employing shear-wall systems. Figure 44 illustrates a long rectangular concrete building with unsymmetrically located shear walls.

High-rise concrete buildings with column-free spaces are entirely dependent upon interior core walls and exterior shear-bearing walls for wind-shear dissipation. Figure 45 illustrates a relatively square tower with no interior columns.

48. Approximate Methods of Analysis   There are various approximate methods of computing wind stresses in rigid-frame wind bents of multistory buildings. Wind bents made up of more than one aisle or portal of equal or unequal widths are generally analyzed as statically determinate frameworks based on simplifying assumptions with respect to points of contraflexure, story shear distribution, etc. Uncalculated stiffness which is added to a structure by its floors, walls, and partitions, and conservative wind-force assumptions in buildings of moderate height, are usually significant enough to offset the errors in the usual approximate methods. However, when floors, walls, and partitions afford little

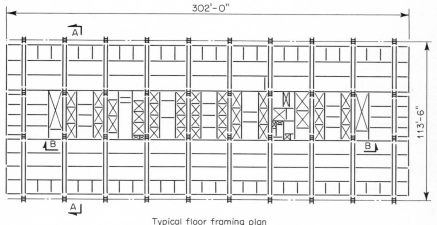

Typical floor framing plan
(6th through 10th floor)

Fig. 42a   Chase Manhattan Bank Building, New York City: floor framing.

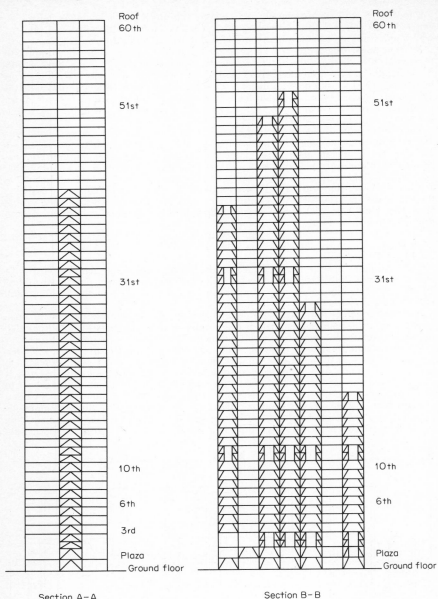

Fig. 42b   Chase Manhattan Bank Building, New York City: wind bracing.

stiffness, or when building height or location dictates accurate wind-stress analysis, the suitability of any approximate method of analysis requires careful consideration. It should also be noted that approximate methods of analysis are applicable only to wind bents with all rigid-frame portals. If one of the portals is a truss knee-braced portal or a shear wall, and the others are rigid frames, approximate methods applied to the rigid-frame portals without proper consideration of relative portal stiffnesses could lead to serious error.

Treatment of wind stresses in building frameworks depends upon a reasonably accurate approximate method of analysis to first determine the moments of inertia of columns and girders. Since accurate member-size determination is seldom possible in the early design stages because of architectural and mechanical requirements, it is not practical to attempt a final analysis immediately. Instead, a reasonable first approximation, subsequent refinement, and then final analysis and correction are indicated. Accordingly, a procedure of analysis of wind bents in any multistory building (steel or concrete) of any height-width ratio should be as follows:
1. First approximation of member sizes
2. Tuning the portals of each wind bent (Example 7)
3. Final analysis
4. Correction if necessary

It can be shown by comparison with exact methods that differences among various simplified methods of analysis are usually not of major significance, provided the following conditions are met:

1. The story shears can be accurately determined. Irregularity of plan and elevation require careful treatment.

2. All the wind bents have the same base-width dimensions and are symmetrically located.

3. The stiffnesses of the wind bents are proportional to the assumed distribution of story shears.

There are two approximate methods of analysis that can be used if these conditions are met. The *portal method* treats a wind bent as a series of consecutive independent portals or aisles in the determination of the direct stress in the columns due to overturning effect.

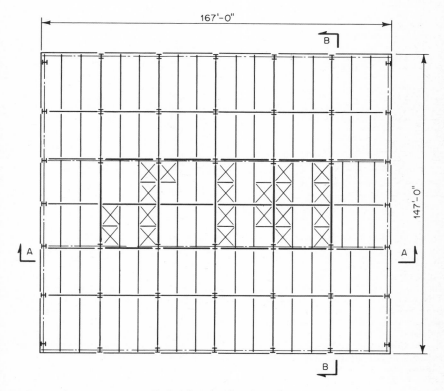

Typical floor framing plan

Fig. 43a   Equitable Life Assurance Building, New York City: floor framing.

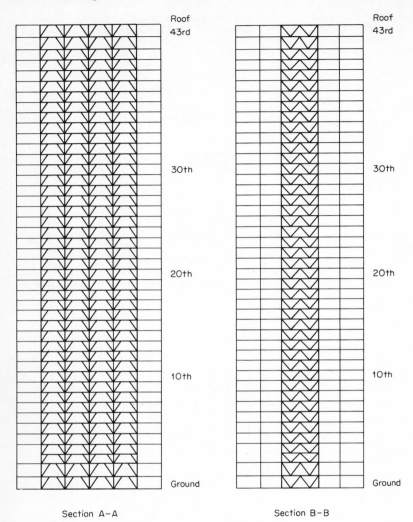

Section A–A   Section B–B

Fig. 43b   Equitable Life Assurance Building, New York City: wind bracing.

Interior columns each serve two such portals, and the direct compression arising from the overturning effect on one portal is offset by the direct tension arising from the overturning effect on the adjacent portal. If the widths of the different portals are unequal, the portion of the transverse load resisted by each portal may be assumed proportional to its width to keep the interior columns free from direct stress. On the other hand, it is common practice to disregard unequal aisle widths as affecting transverse load distribution and obtain calculated direct stresses in the interior columns. The portal method assumes that points of inflection of the columns are at midheight and points of inflection of the girders at midspan. Figure 46 illustrates a wind-stress analysis by the portal method.

In addition to the simplicity of calculation, the portal method results in economical duplication of wind-girder moments and connection details. The outstanding weakness of the method appears when it is used for building over 300 ft high. The axial interior column deformations, essentially overlooked in this method in which most of the direct

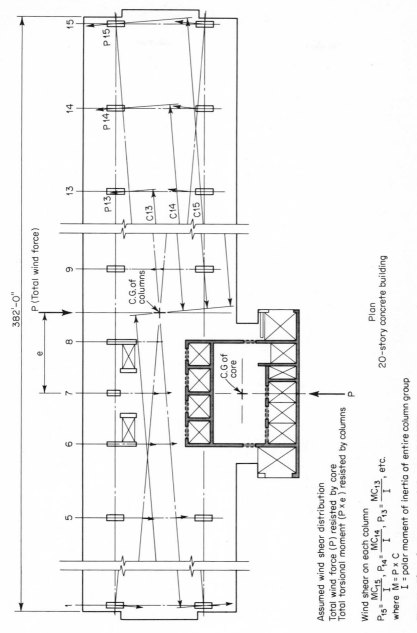

Assumed wind shear distribution
Total wind force (P) resisted by core
Total torsional moment (P x e) resisted by columns

Wind shear on each column

$$P_{15} = \frac{MC_{15}}{I}, \quad P_{14} = \frac{MC_{14}}{I}, \quad P_{13} = \frac{MC_{13}}{I}, \quad \text{etc.}$$

where $M = P \times C$
       $I =$ polar moment of inertia of entire column group

Fig. 44 Plan of unsymmetrical shear-wall building.

Plan
20-story concrete building

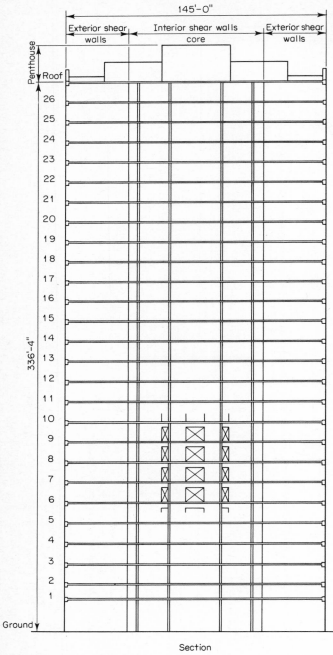

Fig. 45a Symmetrical shear-wall building.

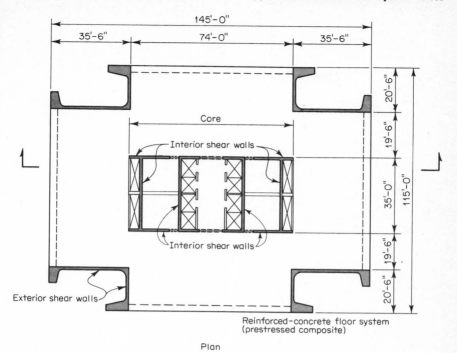

Plan

Fig. 45b Symmetrical shear-wall building *(Continued).*

wind load is assumed to be resisted by the exterior columns, become significant and must be accounted for because large errors in calculated moments will result, particularly in the upper stories of the frame. If wind stresses control the design of members and connections, the portal method is unsuitable for these upper stories. On the other hand, if gravity stresses control, errors in wind stress of even 100 percent may be insignificant. The latter is the usual case, especially in the upper floors, because of the increased allowable stress for wind.

Table 31 compares moments at the ends of the 45-ft 5-in. wind girder of the structure shown in Example 7, obtained by the portal method and by a computer analysis using slope-deflection equations.

The *cantilever method* assumes that the bent acts as a cantilever beam and that, in resisting the bending moments produced by the wind forces, it develops in its columns axial stresses distributed linearly about the neutral axis of the bent. Points of inflection are assumed at midheight of columns and at midspan of girders. Figure 47 illustrates a wind-stress analysis by the cantilever method.

The axial interior-column deformation is more fully considered in the cantilever method, since the moment of inertia of the bent, which is a function of the column areas and distances between columns, is the basis for the determination of the axial stresses. For buildings up to 500 ft high and with height-width ratios less than 5, this refinement generally suffices in overcoming inaccuracies in calculated moments, particularly in the upper stories of the framework.

Since the relative stiffnesses of columns and girders is disregarded in both the portal and the cantilever methods, the latter may or may not yield more accurate results. Many refinements have been developed to account for relative stiffnesses. The Witmer method of K percentages, the Spurr, Goldberg, and Grinter methods, and the tuning-portal method are examples. No approximate method has been found to be entirely satisfactory

Portal Method Wind Stress Analysis

Assumptions

1. Exterior column shear (H) = $S/6$
   Interior column shear (2H) = $S/3$
2. Point of inflection of columns at midheight = $h/2$
3. Point of inflection of girders at midspan = $1/2$

Shears and moments

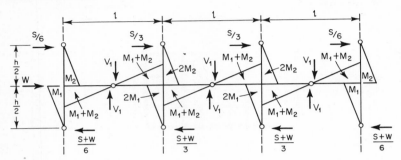

S	=	story shear
M_1	=	lower column bending moment – exterior column
$2M_1$	=	lower column bending moment – interior column
M_2	=	upper column bending moment – exterior column
$2M_2$	=	upper column bending moment – interior column
M_1+M_2	=	girder bending moment at each end
V_1	=	girder shear $\left(2 \times \dfrac{M_1+M_2}{1}\right)$

Column direct stresses

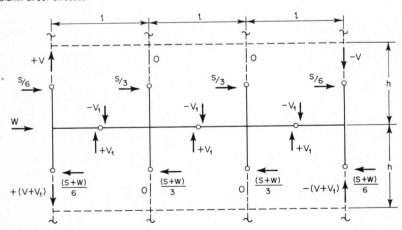

V = column direct stress from above
V_1 = column direct stress increments and girder shear
S = total story shear from above
W = increment of wind force at this level

If adjacent portal widths are unequal, adjacent girder shears are unequal and the direct column stress increment in their common column is $\pm\left[V_1 \text{ (left)} - V_1 \text{ (right)}\right]$.

Fig. 46

19-104

Numerical example
(Direct) stresses, loads and shears in kips, bending stresses in foot-kips

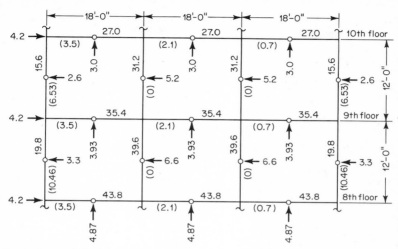

Fig. 46 *(Continued).*

for highly irregular frames. However, they are useful in such cases in obtaining trial sections for a more rigorous solution by methods capable of electronic computation.

49. Computer Methods Many computer programs are available for exact solutions for both static and dynamic analysis, and there is virtually no limit to the size and complexity of structures that can be programmed and solved in reasonable machine time. A computer solution for a static analysis of a representative structure is outlined below.

1. Building geometry—70-story, irregular plan (teardrop) shape (Fig. 48).
2. Wind system—combination interior shear walls and exterior framed tube (tube-in-tube) (Fig. 49).
3. Perform preliminary design to establish wall thicknesses and tube-member sizes based on gravity and wind loads using approximate methods of manual analysis. Drift limitation considered. For this building 0.00175 was used.
4. Verify assumed shear distribution with a computer program, using member sizes and wall thicknesses derived from preliminary design.
5. Modify preliminary-design member sizes, using a computer program, based on more accurate wind-shear distribution (tuning).
6. Investigate dynamic effects of wind in wind tunnel to arrive at an effective static wind load (Art. 45).
7. Use the effective static load as computer input in a program with larger capability to determine wind shears, moments, and deflection (possible final computation).
8. Review all member sizes and wall thicknesses and adjust if necessary, followed by a final computer run.

The state of the art is progressing toward computer methods employing direct dynamic-wind-load input instead of an equivalent static loading. Where height of building, frame flexibility, or sensitivity to motion require a more rigorous design, dynamic methods of analysis should be used. A direct dynamic analysis of a building frame can be made by utilizing computer programs such as ICES DYNAL. This program permits load input in various forms. A time-load history may be used whereby a given magnitude of load is applied at a specific time. The magnitude of load over the next time interval may be different, and the load history describes a dynamic-load condition. Another form of dynamic-load input may be a base acceleration such as that produced by an earthquake. Shock loadings are also possible. In each case, the computer solution produces a time-related response to the dynamic-load input. Accelerations, displacements, moments, shears, and axial forces are calculated and printed for each time interval requested. The maximum response is sometimes also recorded for ease in selecting maximum stresses for design.

Since a dynamic approach is basically a consideration of a time-dependent variable load

TABLE 31   Comparison of Wind Analyses

Floor	Wind moment		Floor	Wind moment	
	Portal method	Exact analysis		Portal method	Exact analysis
Roof	94	151	30	491	484
60	73	127	29	498	508
59	80	132	28	513	514
58	95	140	27	526	602
57	109	149	26	540	600
56	123	158	25	554	594
55	136	166	24	567	570
54	150	175	23	581	570
53	230	237	22	596	596
52			21	610	602
51	261	275	20	623	606
50	208	199	19	638	628
49	219	216	18	651	650
48	233	229	17	666	650
47	246	238	16	680	662
46	261	247	15	694	638
45	275	256	14	710	616
44	289	264	13	1,000	875
43	302	271	12		
42	316	278	11	1,060	970
41	330	301	10	775	676
40	344	314	9	780	692
39	358	346	8	793	728
38	373	358	7	805	730
37	378	380	6	811	724
36	400	383	5	811	714
35	415	418	4	811	692
34	429	420	3	1,550	1,308
33	613	625	2		
32			1	1,710	1,570
31	651	630			

which produces a time-related response, the load input is very significant. For wind analysis, it has been determined by wind-tunnel tests as well as wind-measuring devices that wind loads can be generally considered static at a given level of intensity accompanied by a time-related variation above and below this static, or mean, pressure. The variations are produced by the effects of wind gusts (changes in velocity), turbulence produced by surrounding terrain, or the effects of the building shape.

Cantilever Method Wind Stress Analysis
Assumptions
1. Column axial stresses distributed linearly about the neutral axis
2. Points of inflection of column at midheight $= \frac{h}{2}$
3. Points of inflection of girder at midspan $= \frac{l}{2}$

Column direct stresses

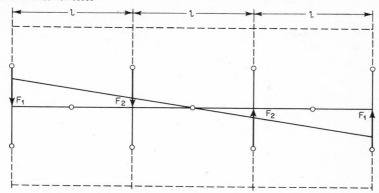

Story moment = Moment of total wind force above this level computed at this level

Story moment $= F_2 l + 3 F_1 l$

$F_1 = 3 F_2$

Story moment $= 10 F_2 l$

$F_2 = \dfrac{\text{Story moment}}{10 l}$

Girder shears

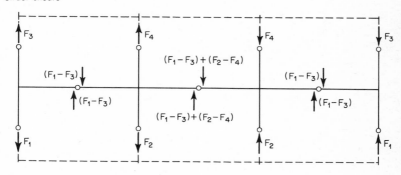

Column and girder moments

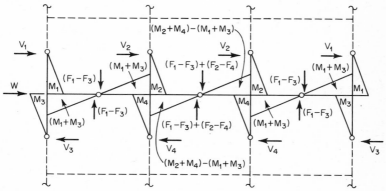

Fig. 47

M_1 and M_3 are moments in exterior columns computed in proportion to the upper- and lower-story shears

M_2 and M_4 are moments in interior columns computed in proportion to the upper- and lower-story shears

V_1, V_2, V_3 and V_4 are column shears computed from the column moments

Numerical example
   (Direct) stresses, loads and shears in kips, bending stresses in foot-kips.

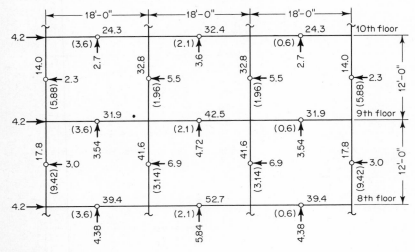

Fig. 47 *(Continued).*

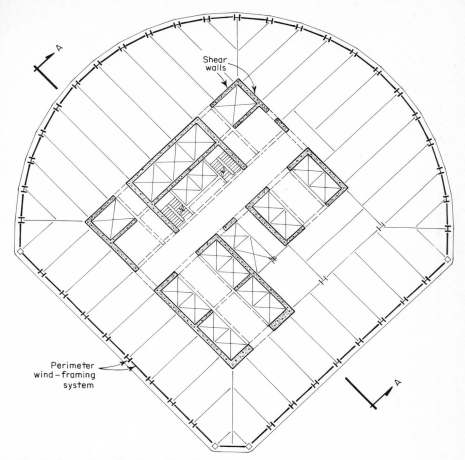

Fig. 48   Lower floor of building of Fig. 49.

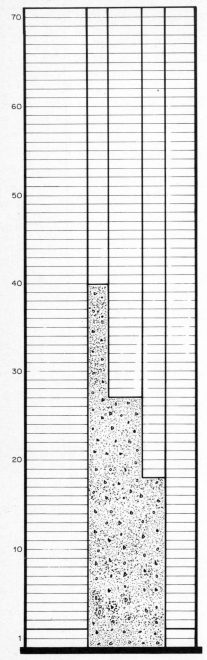

Fig. 49 Section A-A of building of Fig. 48.

Loading conditions commonly assumed for dynamic analysis of building frames are illustrated in Figs. 50 and 51. Figure 50 illustrates a basic static wind load with initial buildup to mean (code) level over a given period of time and periodic reduction with gradual buildup again. Figure 51 illustrates a wind load which fluctuates about a mean which is normally the code wind load. The degree of fluctuation can be determined by

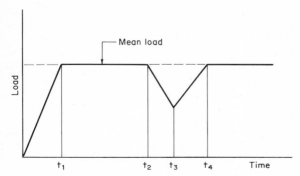

Fig. 50

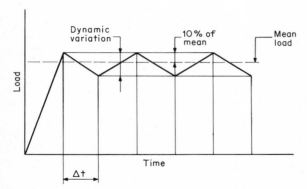

Fig. 51

wind-tunnel testing or taken as a variation of 10 percent of the mean. A third variation consists of the sawtooth load shown in Fig. 51, with the time interval being equal to the natural period of the building frame. This produces a resonant loading condition, which is the most conservative approach. This form of dynamic analysis yields a time history of building-frame response to the variable load to which the building is subjected. In addition to displacements and member loads, accelerations are also computed with respect to time to measure occupant comfort. Results of research on human sensitivity to accelerations permit design judgments to be made from dynamic-analysis output.

REFERENCES

1. Uniform Building Code, International Conference of Building Officials, Whittier, Calif., 1976.
2. National Building Code, American Insurance Association, New York, 1976.
3. Davenport, A. G.: A Rationale for the Determination of Design Wind Loads, *J. Struct. Div. ASCE,* May 1960.
4. American National Standard A58.1, Building Code Requirements for Minimum Design Loads in Buildings and Other Structures, 1972. Copies of the Standard may be purchased from the American National Standards Institute, 1430 Broadway, New York, N.Y., 10018.
5. Thom, H. C. S.: New Distributions of Extreme Winds in the United States, *J. Struct. Div. ASCE,* July 1968.

Section **20**

Thin-Shell Concrete Structures

DAVID P. BILLINGTON
Professor of Civil Engineering, Princeton University

Thin shells are defined as curved or folded slabs whose thicknesses are small compared with their other dimensions. They may be classified in a number of ways, two of which are type of curvature and method of generation.

Singly curved shells are developable surfaces (Fig. 1a). They are usually cylindrical or conical. Shells of positive double curvature (two curvatures in the same direction) usually have a domelike appearance (Fig. 1b). They are nondevelopable and, therefore, are stiffer than singly curved shells if they are properly supported. Shells of negative double curvature (two curvatures in opposite directions) usually have the form of a saddle or warped plate (Fig. 1c), and are also nondevelopable.

Thin-shell surfaces may be generated by rotation or by translation of a plane curve. Surfaces of rotation are formed by revolving a plane curve about an axis in its plane. Such surfaces may all be analyzed in a similar manner, but this does not necessarily imply similarity of behavior.

Surfaces of translation are formed by moving a plane curve along some other plane curve. Again, there is no necessary similarity of behavior within this class. For example, elliptic paraboloids and spherical domes can be built of nearly the same shape, except at the edges, and their behavior is nearly the same although their methods of generation are quite different. On the other hand, elliptic paraboloids and hyperbolic paraboloids covering the same areas can behave quite differently even though their methods of generation are the same.

Many shell structures which have been built do not fit easily into the categories defined above. Some examples are *wave-form translation shells* (Fig. 2), which have alternating positive and negative curvature, and *free-form shells*. The overall behavior pattern of such shells may not be too difficult to establish, but a detailed analysis, particularly for bending moments, may require substantial idealization of the structure which, in the end, may best be confirmed by physical-model analyses.

1. Thin-Shell Concrete Roofs Thin-shell concrete structures have been used for roofs because of their possibilities for wide spans, for impressive appearance, and for economical construction. In some cases all three factors are important, but more commonly one or two predominate.

Thin shells for wide spans are limited by high cost of the extensive centering or scaffolding required. Where the roof can be cast in a number of similar and self-supporting parts, scaffolding can be reused and its cost per square foot of covered area substantially

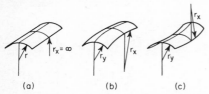

Fig. 1 (a) Singly curved shell; (b) shell of positive double curvature; (c) shell of negative double curvature.

Fig. 2 Waveform translation shell.

reduced. A further limitation is the lack of engineering experience, particularly with regard to analysis for creep, large deflections, and buckling. Thin-shell structures which have performed well on short spans may exhibit difficulties when scaled up to much larger dimensions, because commonly accepted analyses and tolerances in construction may fail to account for some critical factor on very large spans.

The architect's desire for roofs of impressive appearance may lead to complex shapes which, even for small spans where scaffolding costs are not excessive, tend to increase analytic and construction difficulties. Precasting techniques, such as those of Nervi, can be exploited to produce attractive structures with a reduction in field-built formwork.

Architectural exuberance may lead to another difficulty, that of enforcing a shape and a type of support which are not well suited to each other. For example, a roof composed of intersecting circular semicylinders requires diagonal arches of elliptical profile, which is a relatively inefficient arch shape, whereas the doubly curved, hyperbolic-paraboloid cross vault leads to diagonal arches of parabolic profile. Although the parabola is a better arch shape, form costs for the doubly curved thin shell may be greater.

The use of thin shells for reasons of construction economy is limited by the complexity of the shape and the form reuse possible. It is generally true that the carrying action of doubly curved shells is more efficient than that of singly curved ones, but the formwork is more complex. In addition, the unit cost of materials in place will often be higher for the thinner sections of the more complex structure. Furthermore, since minimum concrete thickness is often controlled by construction practice or by buckling, and minimum reinforcement by temperature and shrinkage requirements, increases in the complexity of shape often cannot be offset by materials saving. This is particularly true in structures of smaller span.

2. Behavior of Roof Structures A roof structure is given curvature so that the loading is carried predominantly by in-plane forces. Ideal moment-free behavior cannot be attained in practice, but in well-designed thin shells such behavior should predominate. Ideal

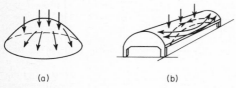

Fig. 3

behavior can be roughly explained in terms either of arch action, where normal loads are carried directly to supports by axial forces (Fig. 3a), or of beam action, where loads are carried by compression, tension, and shear, all in the plane of the middle surface (Fig. 3b). In many thin shells both actions are present.

In addition to in-plane forces (Fig. 4a), there will be moments and radial shears (Fig. 4b) whose importance generally depends upon conditions at the boundaries where the shell is supported, joined to an edge member, or connected to another shell. Generally speaking, if displacements of the edges are not excessive, the bending forces can be kept small and the behavior explained by considering only the in-plane forces.

Most thin-shell structures can resist a variety of loading patterns by in-plane forces and, therefore, usually need be analyzed only for uniformly distributed loadings. Except for the supporting system, it is usually not necessary to consider partial live loadings, e.g., snow load on one side of a roof.

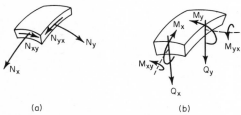

(a)   (b)

Fig. 4   (a) In-plane forces; (b) bending forces.

These comments refer to thin shells under distributed loads. Where heavy concentrated loads occur, large local bending moments may arise even where the boundaries are carefully designed. However, the local failure which may result does not necessarily cause collapse of the roof; there are many examples of locally damaged thin shells which adjusted to their defects.[1] If heavy load concentrations are anticipated, local stiffening of the shell should be provided.

3. Thin-Shell Concrete Walls   Thin shells are frequently used for ground-storage reservoirs, elevated water tanks, and natural-draft cooling towers because they are relatively easy to analyze, inexpensive to build, and have served well over long periods. In such structures the dead load causes predominantly in-plane compression and the live load (internal pressure) causes uniform horizontal in-plane tension which can be carried by reinforcement or, for vertical walls, by prestressing. Where the predominant live load is wind, as for high cooling towers, the resulting in-plane tension (uplift forces) on the windward side of the tower is partly balanced by dead-load compressions and partly taken by reinforcement. Even though the maximum tension is normally below the cracking strength of the concrete, reinforcement should be designed to take all of the uplift tension.

STRUCTURAL ANALYSIS

4. Thin-Shell Theory   A summary of thin-shell theory is given in Ref. 2. For the shells considered here, two simplifications are often admissible: (1) the membrane theory and (2) a simplified bending theory. The principal guide for the choice of a theory is the successful long-term structural behavior of the type of thin shell and its scale. Very simplified theories have proved successful for dome roofs under gravity loading and circular cylindrical walls under internal pressure.

Analysis is commonly based on the following assumptions:

1. The material is homogeneous, isotropic, and linearly elastic. Although none of these assumptions is correct for concrete, tests have indicated that, under working loads, the concrete thin shell behaves very nearly as if they were.[3]

2. The system behaves according to the *small-deflection theory*, which essentially requires that deflections under load be small enough so that changes in geometry do not alter the static equilibrium. The usual measure of validity for this theory is that the radial displacements of the shell be small compared with its thickness.

3. The thickness of a thin shell is denoted by h and is always considered small in comparison with its radii of curvature r. Thin-shell theory has been used where r/h is as low as 30.[4] The surface bisecting the thickness is called the *middle surface*, and by specifying its form, and h at every point, the shell is defined geometrically.

4. The in-plane forces (Fig. 4a) are distributed uniformly over the thickness. They are often expressed as *stress resultants*, defined as forces per unit length on the middle surface. A stress resultant divided by h yields a stress. Membrane stress resultants are denoted by N'_x, N'_y, and $N'_{xy} = N'_{yx}$ and can be obtained solely from equations of equilibrium. Unprimed resultants signify the more accurate values, which include the effects of bending. An extensive compilation of membrane-theory formulas is given in Ref. 5.

The membrane theory is based upon the assumption of no bending or transverse shear in the shell; only in-plane forces are considered. In many thin shells it provides a reasonable basis for design except at the boundaries where the shell is supported or stiffened. This is because local restraints usually exist at boundaries, or because the reactions required by the membrane theory cannot be supplied by the edge members. The substantial bending that can occur at the boundaries is usually evaluated by an approximate bending theory in which the effects of edge loads and edge displacements on both stress resultants and bending moments are considered.

Bending theory in general implies a formulation which includes both in-plane and bending forces. For most thin shells this theory is complicated and difficult to use as a basis for design. For surfaces of rotation a simplified theory works well for many practical cases; this is given in Art. 11. Where this simplification is used, the following general method of analysis is recommended:

1. The loading is assumed to be resisted solely by the membrane stress resultants. Thus the shell is made statically determinate by releasing its internal moment resistance and its external edge restraints. This corresponds to the primary system (reduced structure) in the analysis of a statically indeterminate structure.

2. The forces and displacements at the boundaries of the primary system will, in general, be incompatible with the actual boundary conditions. These discrepancies are errors which must be corrected.

3. Corrections corresponding to unit edge effects are determined.

4. Compatibility is obtained by determining the magnitudes of the corrections which will eliminate the errors.

This approach is illustrated for domes in Art. 10 and cylindrical tanks in Art. 11. For some other types of thin shells the membrane theory provides a useful approximate analysis, but for structures of large scale or for shell types without a record of successful long-term behavior it is essential to use a more complete analysis procedure, usually either a numerical solution or an empirical solution on a well-scaled model. Reference 6 gives a summary of general numerical programs applied to thin-shell concrete structures.

5. Stability *Shell Roofs.*   Buckling can be initiated by relatively high permanent-load compressive stresses, which with time can lead to creep and eventually large displacements. Displacements of a shell normal to its middle surface imply a change in curvature, so that thin shells, since they rely on their curvature for resistance to load, are particularly susceptible to large displacements and thus to buckling. Although much has been written on the stability of thin shells,[7,8,9] there is little experimental work on thin-shell concrete structures of the shape and boundary conditions usually found in roofs.

The buckling pressure q_{cr} on a spherical thin shell[8] is given by

$$q_{cr} = CE \left(\frac{h}{a}\right)^2 \qquad (1)$$

where $C = 2/\sqrt{3(1 - \nu^2)}$

ν = Poisson's ratio

h = thickness of shell

a = radius of shell

Experiments show that C is much smaller than $2/\sqrt{3(1 - \nu^2)}$. Schmidt[10] shows values as low as 0.06 with none above 0.32. For translational shells, he suggests that Eq. (1) be written in the form

$$q_{cr} = CE \frac{h^2}{r_x r_y} \qquad (2)$$

where r_x and r_y are the two principal radii of curvature.

These equations show that instability becomes a problem for shells which are very thin, flat, or of low-modulus concrete. Concrete creep can contribute to large deflections; it may be reduced by providing reinforcing steel in both faces of the shell. The effect of creep can be estimated by assuming a reduced value of E or, if the principal membrane stresses at any point are known, by determining the tangent modulus of elasticity and dividing it by a factor for long-term deflections. The factor should not be less than 2.

Shell Walls.   Permanent load does not play as important a role as transient loadings such as wind in the buckling of shell walls. For a cylinder, fixed at its base, free at its top, and under wind pressure, a bifurcation analysis gives the critical pressure as

$$q_{cr} = KE \left(\frac{h}{r}\right)^3 \tag{3}$$

Values of K for a cylinder with $r/h = 100$ can be found from Fig. 5 for the typical wind-pressure distribution shown in the figure.[11] Comparisons between these results and wind-tunnel test results show the former to be about 36 percent high.[12] Critical pressures with K from Fig. 5 and using $c = 1$ are 1860 and 430 psf for $H/r = 1$ and 6, respectively. Since r/h

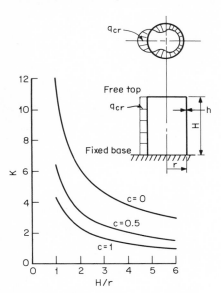

Fig. 5   Values of K in Eq. (3) for $r/h = 100$.

Note: c = internal suction coefficient
as a multiple of q_cr.

seldom exceeds 100 for thin concrete cylindrical shells, this shows that such shells will not buckle under wind pressure unless there is a high thermal gradient, as in cooling-tower shells, and insufficient circumferential steel to control cracking.

For hyperboloids under wind loading, simplified bifurcation results are from two to three times higher than wind-tunnel results.[13] The latter lead to an estimate of buckling pressure as[14]

$$q_{cr} = CE \left(\frac{h}{r}\right)^\alpha \tag{4}$$

where q_{cr} = buckling pressure, psi, along windward meridian
   C = empirical coefficient = 0.052
   h = thickness at throat, i.e., at point on meridian where horizontal radius is minimum
   r = radius of shell parallel circle at throat
   α = empirical coefficient = 2.3

The value of α has been derived from theoretical considerations as ⅔ (Ref. 15). The coefficient C shows wide scatter[14] and in Europe has been taken as 0.077 ± 0.009 (Ref.

16). A restudy of the results reported in Ref. 14 shows 0.052 to be a more reasonable lower-bound estimate.[13] The value of q_{cr} by Eq. (4) should be compared with the design wind pressure at the top of the tower to ensure an adequate factor of safety against buckling.

For cooling towers, q_{cr} should be at least twice q_z, as defined in Art. 11, when dead load is included and cracking considered. This is because thermal gradients across cooling-tower shells, and possibly other types, are large enough to cause vertical cracks and hence to reduce substantially the circumferential stiffness of the shell.[17] Since this bending stiffness plays a major role in the buckling capacity, its reduction can reduce the buckling pressure drastically. The solution is to provide enough circumferential reinforcement to control the cracking and to use a thickness sufficient to keep $q_{cr} > 2q$.

Imperfections in the geometry of a completed shell can lead to overstressing or to buckling or both. In the failure of the Ardeer cooling tower in 1973 imperfections were recorded of over 18 in. in a 6-in.-thick shell, and these directly contributed to the failure.[18] Nevertheless, field tolerances must be permitted on very large structures, but always subject to specification by an experienced designer of thin-shell concrete structures.

6. Dynamic Behavior All the analyses presented in this section are for static behavior. For concrete-shell roofs wind is not usually a dynamic problem because of the weight of the concrete; a more usual problem is the tearing off of roofing because of the wind suction that arises over much of a shell roof. Seismic loading on shell roofs can be a difficult problem if the roof is supported at a few isolated points, in which case the horizontal shears from seismic motions can be large and need to be considered.

For shell walls the principal design loading can frequently be either seismic or wind. Quasistatic approximations, such as those in Ref. 42, to either loading have normally led to successful structures. If there appears to be a possibility that the natural frequencies of a shell are close to important wind-gust frequencies, special dynamic studies may be warranted. Reference 18 gives a discussion of dynamic wind loading on cooling towers.

DOME ROOFS

7. Behavior of Domes The behavior of domelike shells of revolution consists ideally of meridional (latitudinal) arch action and circumferential (longitudinal) hoop action. In the hemispherical dome (Fig. 6a), uniformly distributed surface loads are transmitted by the arch action of meridional forces directly to the continuous support. The hoop action is compressive near the crown and tensile near the edge. If the dome is a partial hemisphere supported by uniformly distributed forces tangent to the edge, the hoop forces can be restricted to compression only (Fig. 6b).

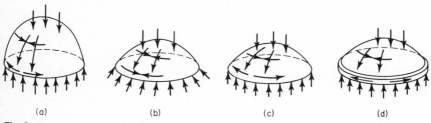

(a)   (b)   (c)   (d)

Fig. 6

In the dome of Fig. 6c, since the support provides only vertical reactions, the edge moves out and appreciable meridional bending is developed. This can be reduced by an edge ring (Fig. 6d). Dome rings are often prestressed so that tension is eliminated and meridional bending reduced. A monolithically connected cylindrical wall support acts somewhat like an edge ring, but tensile forces are distributed over a wider edge region and meridional bending can be larger than with the ring.

Bending can be greatly reduced by using a meridional curve such as a cycloid or an ellipse, which permits shallow-rise domes having a free edge with vertical tangent. The difficult formwork and casting problems tend to make such domes impractical.

The behavior of a surface of revolution is strongly affected by boundary conditions. For example, where only column supports are provided along a circular base, that part of the shell edge region between columns acts as a deep beam to carry the meridional forces to the columns. Figure 7 shows a dome built on a triangular plan with the boundaries formed by passing planes or surfaces through the sphere. Here the behavior departs radically

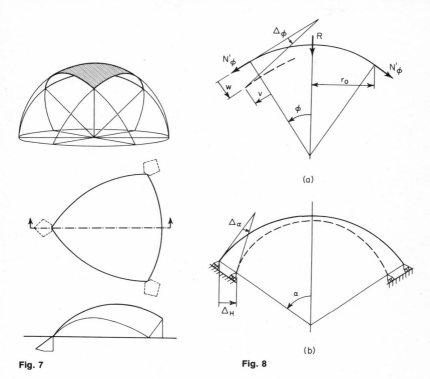

(a)

(b)

Fig. 7   Fig. 8

from the ideal behavior of arch action with hoop restraints, because there is support under only a few meridians and only the hoops near the crown are continuous. The portion at the crown tends to act like a dome, but the arch forces must be carried to the corners by the edge regions. The vertical reactions can be provided by slender columns and the horizontal reactions by prestressing within the shell edge region.

8. Membrane Theory *Surfaces of Revolution.* The equation for vertical equilibrium (Fig. 8a) gives directly an expression for the meridional stress resultant:

$$N'_\phi = - \frac{R}{2\pi r_0 \sin \phi} \tag{5}$$

where R = resultant of the vertical components of all loads above the angle ϕ.
A second equilibrium equation for the forces acting on a differential element (Fig. 9) in the direction normal to the surface gives

$$\frac{N'_\phi}{r_1} + \frac{N'_\theta}{r_2} + p_z = 0 \tag{6}$$

where r_1 = radius of curvature of meridian
$\quad\quad r_2$ = length of shell normal from surface to shell axis
These are the principal radii of curvature of the shell at the point under consideration.

Using Eq. (5) and the relation $r_0 = r_2 \sin \phi$, Eq. (6) gives

$$N'_\theta = \frac{R}{2\pi r_1 \sin^2 \phi} - p_z \frac{r_0}{\sin \phi} \tag{7}$$

The displacements v and w (Fig. 8a) are given by

$$v = \sin \phi \left[\int \frac{f(\phi)}{\sin \phi} \, d\phi + C \right] \tag{8a}$$

$$w = v \cot \phi - \frac{r_2}{Eh} (N'_\theta - \nu N'_\phi) \tag{8b}$$

where ν = Poisson's ratio
C = constant determined by support conditions

$$f(\phi) = \frac{1}{Eh} [N'_\phi (r_1 + \nu r_2) - N'_\theta (r_2 + \nu r_1)] \tag{8c}$$

The rotation Δ_ϕ of the tangent to the meridian at any point is

$$\Delta_\phi = \frac{v}{r_1} + \frac{dw}{r_1 d\phi} \tag{8d}$$

   Values of the displacements are usually needed only at the supports. For supports as shown in Fig. 8b, $v_\alpha = 0$, and the horizontal displacement is given by

$$\Delta_H = - \frac{r_2 \sin \alpha}{Eh} (N'_\theta - \nu N'_\phi)_{\phi = \alpha} \tag{9a}$$

The rotation of the meridional tangent at the edge is

$$\Delta_\alpha = \frac{\cot \alpha}{r_1} f(\alpha) + \frac{1}{r_1} \frac{d}{d\alpha} \frac{\Delta_H}{\sin \alpha} \tag{9b}$$

where f is the function defined in Eq. (8c). The positive directions of Δ_H and Δ_α are shown in Fig. 8b.

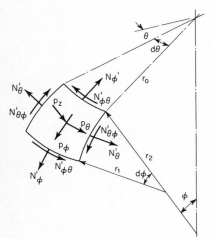

Fig. 9

   Spherical Dome of Constant Thickness.   With q = weight per sq ft of dome surface, p_z = $q \cos \phi$ (Fig. 9). Then, with $r_1 = r_2 = a$, Eqs. (5) and (7) give

$$N'_\phi = - \frac{aq}{1 + \cos \phi} \tag{10a}$$

$$N_\theta' = aq \left(\frac{1}{1 + \cos \phi} - \cos \phi \right) \tag{10b}$$

The distribution of these resultants is shown in Fig. 10. If the dome is a spherical segment, there is no hoop tension at the edge if $\phi \lessgtr 51°50'$ and there is continuous support by forces tangent to the edge.

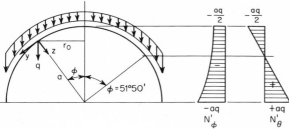

Fig. 10

With load p per sq ft of horizontal projection of the surface, $p_z = p \cos^2 \phi$, and Eqs. (5) and (7) give

$$N_\phi' = -\frac{ap}{2} \tag{11a}$$

$$N_\theta' = -\frac{ap}{2} \cos 2\phi \tag{11b}$$

Stress resultants for other loadings are found similarly. Results for several cases are given in Table 1.

With N_ϕ' and N_θ' known, edge displacements for domes supported as in Fig. 8b are found from Eq. (9). Corresponding to Eq. (10),

$$\Delta_H = -\frac{a^2 q}{Eh} \left(\frac{1 + \nu}{1 + \cos \alpha} - \cos \alpha \right) \sin \alpha \tag{12a}$$

$$\Delta_\alpha = -\frac{aq}{Eh} (2 + \nu) \sin \alpha \tag{12b}$$

and, corresponding to Eq. (11),

$$\Delta_H = -\frac{a^2 p}{2Eh} (- \cos 2\alpha + \nu) \sin \alpha \tag{13a}$$

$$\Delta_\alpha = -\frac{ap}{2Eh} (3 + \nu) \sin 2\alpha \tag{13b}$$

Spherical Dome of Variable Thickness. A good approximation of the edge forces and displacements in a spherical dome of variable thickness is obtained by assuming a uniform thickness equal to that at the arc distance S_a from the edge,[19]

$$S_a = 0.5 \sqrt{ah_a} \tag{14a}$$

where a = radius of dome
h_a = average thickness of thickened portion
This approximation involves the assumption that the thickened portion extends the distance S from the edge,

$$S = 2\sqrt{ah_a} \tag{14b}$$

Parabolic Dome of Constant Thickness. The membrane stress resultants for two cases of loading are given in Table 2. Displacements Δ_H and Δ_α can be determined by substituting these values in Eqs. (9).

TABLE 1 Membrane Forces in Spherical Domes

Loading	N_ϕ'	N_θ'	$N_{\phi\theta}'$
q per ft² of surface	$-aq\dfrac{\cos\phi_0 - \cos\phi}{\sin^2\phi}$	$-N_\phi' - aq\cos\phi$	0
	For $\phi_0 = 0$ (no opening)		
	$-aq\dfrac{1}{1+\cos\phi}$	$-N_\phi' - aq\cos\phi$	0
p per ft² projection	$-\dfrac{ap}{2}\left(1 - \dfrac{\sin^2\phi_0}{\sin^2\phi}\right)$	$-N_\phi' - ap\cos^2\phi$	0
	For $\phi_0 = 0$ (no opening)		
	$-\dfrac{ap}{2}$	$-\dfrac{ap}{2}\cos 2\phi$	0
p_L per ft of edge	$-p_L\dfrac{\sin\phi_0}{\sin^2\phi}$	$-N_\phi'$	0
	For $\phi_0 = 0$ (load P_L at vertex)		
	$-P_L\dfrac{1}{2\pi a\sin^2\phi}$	$-N_\phi'$	0
p_w per ft² projection	$-p_w\dfrac{a\cos\phi\cos\theta}{\sin^3\phi}[3(\cos\phi_0 - \cos\phi) - (\cos^3\phi_0 - \cos^3\phi)]$	$-N_\phi' - ap_w\sin\phi\cos\theta$	$N_\phi'\dfrac{\tan\theta}{\cos\phi}$
	For $\phi_0 = 0$ (no opening)		
	$-p_w\dfrac{a\cos\phi\cos\theta}{\sin^3\phi}(2 - 3\cos\phi + \cos^3\phi)$	$-N_\phi' - ap_w\sin\phi\cos\theta$	$N_\phi'\dfrac{\tan\theta}{\cos\phi}$

9. Bending Theory   For the bending near the edges of domes it is usually sufficient to use the Geckeler approximation, which can be interpreted physically as the substitution of an equivalent cylinder for the dome. The radius of curvature $r_2 = r_0/\sin\phi$ (Fig. 9) at the edge of the dome is taken as the radius of the equivalent cylinder, and the solution for bending in a full cylinder is used for the dome edge. Table 3 gives values for domes

TABLE 2   Membrane Forces in Parabolic Domes

Loading	N_ϕ'	N_θ'	$N_{\phi\theta}'$
q per ft^2 of surface Normal 	$-\dfrac{qr_0}{3}\dfrac{1-\cos^3\phi}{\sin^2\phi\cos^2\phi}$	$-\dfrac{qr_0}{3}\dfrac{2-3\cos^2\phi+\cos^3\phi}{\sin^2\phi}$	0
p per ft^2 of projection 	$-\dfrac{pr_0}{2\cos\phi}$	$-\dfrac{pr_0}{2}\cos\phi$	0

NOTE: r_0 = radius of curvature at vertex.

TABLE 3   Forces and Displacements in Domes of Revolution Loaded by Edge Forces Uniform around a Parallel Circle

N_ϕ	$C_1 H \sin\alpha \cot(\alpha-\psi)$	$-2C_3\beta M_\alpha \cot(\alpha-\psi)$
N_θ	$2C_2\beta a H \sin\alpha$	$2C_1\beta^2 a M_\alpha$
M_ϕ	$\dfrac{C_3}{\beta} H \sin\alpha$	$C_4 M_\alpha$
Δ_H	$2\beta\dfrac{a^2}{Eh} H \sin^2\alpha$	$\dfrac{2\beta^2 a^2 \sin\alpha}{Eh} M_\alpha$
Δ_α	$2\beta^2\dfrac{a^2}{Eh} H \sin\alpha$	$\dfrac{4\beta^3 a^2}{Eh} M_\alpha$

NOTE: $\beta^4 = 3(1-\nu^2)/a^2 h^2$.   See Table 4 for C.

loaded by edge forces and moments. Values of the coefficient C are given in Table 4. Values of N_ϕ, N_θ, and M_ϕ are given for any point in the shell defined by the angle ψ. The edge displacements Δ_H and Δ_α are positive in the positive directions of H and M_α, respectively.

TABLE 4   Coefficients C for Tables 3 and 8*

βs or βx†	C_1	C_2	C_3	C_4
0	1.0000	1.0000	0	1.0000
0.1	0.8100	0.9003	0.0903	0.9907
0.2	0.6398	0.8024	0.1627	0.9651
0.3	0.4888	0.7077	0.2189	0.9267
0.4	0.3564	0.6174	0.2610	0.8784
0.5	0.2415	0.5323	0.2908	0.8231
0.6	0.1431	0.4530	0.3099	0.7628
0.7	0.0599	0.3798	0.3199	0.6997
0.8	−0.0093	0.3131	0.3223	0.6354
0.9	−0.0657	0.2527	0.3185	0.5712
1.0	−0.1108	0.1988	0.3096	0.5083
1.1	−0.1457	0.1510	0.2967	0.4476
1.2	−0.1716	0.1091	0.2807	0.3899
1.3	−0.1897	0.0729	0.2626	0.3355
1.4	−0.2011	0.0419	0.2430	0.2849
1.5	−0.2068	0.0158	0.2226	0.2384
1.6	−0.2077	−0.0059	0.2018	0.1959
1.7	−0.2047	−0.0235	0.1812	0.1576
1.8	−0.1985	−0.0376	0.1610	0.1234
1.9	−0.1899	−0.0484	0.1415	0.0932
2.0	−0.1794	−0.0563	0.1230	0.0667
2.1	−0.1675	−0.0618	0.1057	0.0439
2.2	−0.1548	−0.0652	0.0895	0.0244
2.3	−0.1416	−0.0668	0.0748	0.0080
2.4	−0.1282	−0.0669	0.0613	−0.0056
2.5	−0.1149	−0.0658	0.0492	−0.0166
2.6	−0.1019	−0.0636	0.0383	−0.0254
2.7	−0.0895	−0.0608	0.0287	−0.0320
2.8	−0.0777	−0.0573	0.0204	−0.0369
2.9	−0.0666	−0.0534	0.0132	−0.0403
3.0	−0.0563	−0.0493	0.0071	−0.0423
3.1	−0.0469	−0.0450	0.0019	−0.0431
3.2	−0.0383	−0.0407	−0.0024	−0.0431
3.3	−0.0306	−0.0364	−0.0058	−0.0422
3.4	−0.0237	−0.0323	−0.0085	−0.0408
3.5	−0.0177	−0.0283	−0.0106	−0.0389
3.6	−0.0124	−0.0245	−0.0121	−0.0366
3.7	−0.0079	−0.0210	−0.0131	−0.0341
3.8	−0.0040	−0.0177	−0.0137	−0.0314
3.9	−0.0008	−0.0147	−0.0140	−0.0286
4.0	0.0019	−0.0120	−0.0139	−0.0258
4.1	0.0040	−0.0095	−0.0136	−0.0231
4.2	0.0057	−0.0074	−0.0131	−0.0204
4.3	0.0070	−0.0054	−0.0125	−0.0179
4.4	0.0079	−0.0038	−0.0117	−0.0155

*From Ref. 4.

†βs for domes (Table 3), βx for circular cylinders (Table 8).

The formulas in Table 3 derive from a bending-theory differential equation of the fourth order. In order to simplify the solution, the radius r_2 is taken constant, which is correct only for the spherical dome. It is because edge effects damp out so rapidly that the formulas can be used for most shells of revolution. The formulas can be relied on only if the opening angle α is greater than 25 or 30°.

Typical dimensions of spherical domes are given in Table 5.

TABLE 5   Typical Dimensions for Spherical Domes*

Section	D, ft	h, in.†	ϕ, deg	R, ft	a, ft
	100	3	30	13.4	100
			45	20.7	70.7
	125	3	30	16.8	125
			45	25.9	88.4
	150	3.5	30	20.1	150
		(3)	45	31.0	106.0
	175	4	30	23.5	175
		(3.5)	45	36.2	123.7
	200	4.5	30	26.8	200
		(4)	45	41.4	141.4

* From Portland Cement Association.
† Thickness usually increased by 50 to 75 percent at periphery.

10. Examples. Rigidly Supported Spherical Dome   The dome of Fig. 11 is 2½ in. thick, except that it is thickened at the edge to 6 in. with a uniform taper over a length of 8 ft. The average dead load for the shell is taken at 40 psf. Roofing plus live load is 50 psf. Edge displacements will be calculated for a uniform thickness of 4 in. (Fig. 12).

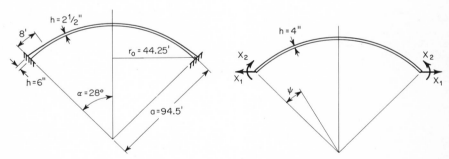

Fig. 11   **Fig. 12**

PRIMARY SYSTEM. The stress resultants are obtained from Eqs. (10):

$$N'_\phi = -\frac{94.5q}{1 + \cos \phi} \qquad N'_\theta = 94.5q \left(\frac{1}{1 + \cos \phi} - \cos \phi \right)$$

At the edge

$$N'_\alpha = -\frac{94.5q}{1 + 0.883} = -50.1q$$

and the horizontal and vertical components are

$$N'_{\alpha H} = N'_\alpha \cos \alpha = -50.1q \times 0.883 = -44.5q$$
$$N'_{\alpha V} = N'_\alpha \sin \alpha = -50.1q \times 0.469 = -23.8q$$

ERRORS. Edge displacements are considered positive in the directions of the redundant forces X_1 and X_2 (Fig. 12). Then from Eqs. (12)

$$\delta_{10} = \frac{94.5^2 q}{0.33E} \left(\frac{1.167}{1.883} - 0.883 \right) 0.469 = -3340 \frac{q}{E}$$

$$\delta_{20} = \frac{94.5 q}{0.33E} \times 2.167 \times 0.469 = 291 \frac{q}{E}$$

CORRECTIONS. The displacements of the edge resulting from unit values of the edge redundant forces X_1 and X_2 are found from the appropriate formulas in Table 3:

$$\beta^4 = \frac{3(1 - 0.167^2)}{(94.5 \times 0.33)^2} = 0.00300$$

$$\beta = 0.234 \qquad \beta^2 = 0.0548 \qquad \beta^3 = 0.0128$$

$$\delta_{11} = \frac{2 \times 0.234 \times 94.5^2 \times 0.469^2}{0.33E} = \frac{2780}{E}$$

$$\delta_{12} = \delta_{21} = \frac{2 \times 0.0548 \times 94.5^2 \times 0.469}{0.33E} = \frac{1390}{E}$$

$$\delta_{22} = \frac{4 \times 0.0128 \times 94.5^2}{0.33E} = \frac{1380}{E}$$

COMPATIBILITY. The equations of compatibility are

$$\delta_{11}X_1 + \delta_{12}X_2 + \delta_{10} = 0$$
$$\delta_{12}X_1 + \delta_{22}X_2 + \delta_{20} = 0$$

so that

$$2780X_1 + 1390X_2 = 3340q$$
$$1390X_1 + 1380X_2 = -291q$$

from which

$$X_1 = 2.63q = 0.237 \text{ kip/ft}$$
$$X_2 = -2.86q = -0.260 \text{ ft-kip/ft}$$

With $H = X_1$ and $M_\alpha = X_2$ known, N_ϕ, N_θ, and M_ϕ can be determined from the formulas in Table 3. The final stress resultants are obtained by adding these values to the membrane resultants N'_ϕ and N'_θ (see following example).

TEMPERATURE CHANGE. No forces result from expansion or contraction of the primary (membrane) structure. The only error in geometry is the lateral expansion caused by a temperature rise T, which is

$$\Delta_H = \delta_{10} = r_0 T \epsilon = 44.25 \times 6 \times 10^{-6} T = 265 \times 10^{-6} T$$

The equations of compatibility are

$$2780X_1 + 1390X_2 = -265 \times 10^{-6} TE$$
$$1390X_1 + 1380X_2 = 0$$

Example—Ring-Stiffened Spherical Dome The dome of this example is the same as that of the preceding example except that the edge is stiffened with a ring which is monolithic with the dome and which is free to slide and rotate on an immovable support (Fig. 13a).

PRIMARY SYSTEM. This consists of the shell and the ring as separate structures, subjected to the membrane stress resultant N'_α, which is assumed to act at the idealized junction of the shell and ring (Fig. 13b). Displacements are taken positive in the directions of the redundants X_1 and X_2.

The stress resultants and their horizontal and vertical components were determined in the preceding example. For the dead load $q = 40$ psf

$$N'_\phi = -\frac{3.78}{1 + \cos \phi} \quad \text{kips/ft}$$

$$N'_\theta = 3.78 \left(\frac{1}{1 + \cos \phi} - \cos \phi \right) \quad \text{kips/ft}$$

$$N'_{\alpha H} = -1.78 \text{ kips/ft} \qquad N'_{\alpha V} = -0.95 \text{ kip/ft}$$

ERRORS. These consist of displacements at the junction of the shell and the ring. For the shell, using the results from the preceding example:

$$\delta_{10}^S = -3340 \times \frac{0.040}{E} = \frac{-133}{E} \quad \text{ft}$$

$$\delta_{20}^S = 291 \times \frac{0.040}{E} = \frac{11.6}{E} \quad \text{rad}$$

Figure 14a shows a portion of the ring under a horizontal force H and a moment M_α, each per unit length of arc. Neglecting the distance $b/2$, which is small compared with r, the horizontal displacement due to H is

$$\Delta_H = \frac{r^2 H}{EA} \tag{15}$$

where A is the area of the ring.

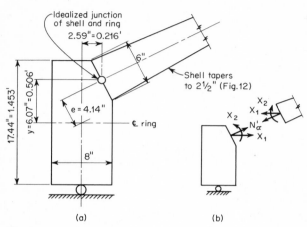

(a)   (b)

Fig. 13

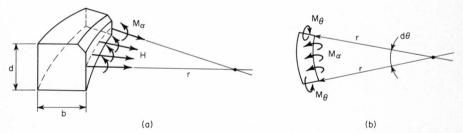

(a)   (b)

Fig. 14

From Fig. 14b, $M_\theta = rM_\alpha$. Assuming a linear stress distribution $f = M_\theta y/I$, the rotation Δ_α of the ring due to M_α is given by

$$\Delta_\alpha = \frac{\epsilon r}{y} = \frac{fr}{Ey} = \frac{M_\theta r}{EI} = \frac{r^2 M_\alpha}{EI} \tag{16}$$

The forces acting on the ring are

$$H = N'_{\alpha H} = -1.78 \text{ kips/ft}$$
$$M_\alpha = -0.506 N'_{\alpha H} + 0.216 N'_{\alpha V}$$
$$= -0.506(-1.78) + 0.216(-0.95) = 0.695 \text{ ft-kip/ft}$$

Then, from Eq. (16)

$$\delta_{20}^R = \frac{44.25^2 \times 0.695}{0.667(1.453^3/12)E} = \frac{7980}{E} \quad \text{rad}$$

The horizontal displacement δ_{10} is, from Eq. (15),

$$\delta_{10}^R = \frac{44.25^2(-1.78)}{0.667 \times 1.453E} - 0.506\delta_{20}^R$$

$$= -\frac{3596}{E} - 0.506\frac{7980}{E} = -\frac{7634}{E} \quad \text{ft}$$

The last term in this equation is the horizontal displacement at the junction due to rotation of the ring. Thus, the δ_0 displacements are

$$\delta_{10} = \delta_{10}^S + \delta_{10}^R = -\frac{7767}{E}$$

$$\delta_{20} = \delta_{20}^S + \delta_{20}^R = \frac{7992}{E}$$

CORRECTIONS. With $X_2 = 1$ acting on the ring, Eq. (16) gives

$$\delta_{22}^R = \frac{44.25^2}{(0.667 \times 1.453^3/12)E} = \frac{11,490}{E} \quad \text{rad}$$

This rotation produces a displacement at the junction of the shell with the ring:

$$\delta_{12}^R = \delta_{21}^R = -0.506\frac{11,490}{E} = -\frac{5810}{E} \quad \text{ft}$$

With $X_1 = 1$ acting on the ring, the contribution to δ_{11} of the moment -1×0.506 resulting from the eccentricity of X_1 must be added to the value given by Eq. (15). This moment gives the rotation $-0.506 \times 11,490/E = -5810/E$. The resulting displacement at the junction is $0.506 \times 5810/E = 2940/E$. Thus, the displacement due to X_1 is

$$\delta_{11} = \frac{44.25^2}{0.667 \times 1.453E} + \frac{2940}{E} = \frac{4960}{E}$$

The displacements of the shell edge due to unit values of the redundants were determined in the preceding example: $\delta_{11}^S = 2780/E$, $\delta_{12}^S = \delta_{21}^S = 1390/E$, and $\delta_{22}^S = 1380/E$. Then, from $\delta = \delta^R + \delta^S$

$$\delta_{11} = \frac{7740}{E} \qquad \delta_{12} = \delta_{21} = -\frac{4420}{E} \qquad \delta_{22} = \frac{12,870}{E}$$

Compatibility is obtained by satisfying the equations

$$7740X_1 - 4420X_2 - 7767 = 0$$
$$-4420X_1 + 12,870X_2 + 7992 = 0$$

for which $X_1 = 0.81$ kip/ft and $X_2 = -0.34$ ft-kip/ft.

The stress resultants and moments throughout the dome can now be obtained by using X_1 and X_2 in the equations of Table 3 and combining the results with the membrane stress resultants. Table 6 gives these values for dead load plus live load (90 psf).

TABLE 6.   Stress Resultants in Dome of Fig. 11

(Surface Load q = 90 psf)

ψ, deg	Dome with ring			Ring prestress = 3.24 kips		
	N_ϕ, kips/ft	N_θ, kips/ft	M_ϕ, ft-kips/ft	N_ϕ, kips/ft	N_θ, kips/ft	M_ϕ, ft-kips/ft
0	-2.92	26.64	-0.77	-0.94	-25.65	-0.37
1	-3.69	17.65	0.25	-0.28	-15.25	-0.85
2	-4.22	9.25	0.68	0.10	-7.35	-0.93
3	-4.52	2.83	0.74	0.28	-2.21	-0.77
5	-4.69	-3.82	0.44	0.28	1.80	-0.37
10	-4.38	-4.15	-0.03	0.00	0.37	+0.04
28	-4.25	-4.25	0.00	0.00	0.00	0.00

PRESTRESSED RING. The effect of prestressing the ring is easily determined. The radial pressure H_F is given by $H_F = F/r$, where F is the prestressing force. If F is not eccentric with respect to the center of gravity of the ring, Eqs. (15) and (16) give

$$\delta_{10} = \frac{44.25^2 H_F}{0.667 \times 1.453E} = \frac{2020 H_F}{E} \text{ ft} \qquad \delta_{20} = 0$$

The equations of compatibility are

$$7740 X_1 - 4420 X_2 + 2020 H_F = 0$$
$$-4420 X_1 + 12,870 X_2 = 0$$

from which $X_1 = -0.328 H_F$ kip/ft and $X_2 = -0.114 H_F$ ft-kip/ft.

The value of F required to counteract ring tension due to the dome dead load plus live load is obtained by combining the two solutions. For the dome load

$$T = r_0(N'_{\alpha H} + X_1) = 44.25(-1.78 + 0.81)90/40 = 96.5 \text{ kips}$$

and for the prestressing

$$C = r_0(H_F - 0.328 H_F) = 29.7 H_F \text{ kips}$$

With $C = T$, $H_F = 3.24$ kips/ft.

The stress resultants and moments caused by prestressing can be determined by using $H = X_1 = -0.328 \times 3.24 = -1.06$ kips/ft and $M_\alpha = X_2 = -0.114 \times 3.24 = -0.374$ ft-kip/ft in the equations of Table 3. The results are given in Table 6.

REINFORCEMENT

1. Dome hoop reinforcement is provided to take all the hoop tension. For the segment $\psi = 0°$ to $\psi = 1°$ (an arc length of 1.65 ft) the tensile force (Table 6) is

$$T_a = \frac{(26.64 - 25.65) + (17.65 - 15.25)}{2} \times 1.65 = 2.8 \text{ kips}$$

for which

$$A_s = \frac{2.8}{20} = 0.14 \text{ in.}^2$$
$$= \frac{0.14}{1.65} = 0.085 \text{ in.}^2/\text{ft, No. 3 at 15 in.}$$

Similarly, for the regions:

1 to 2°, $A_s = 0.18$ in.$^2 = 0.11$ in.^{2}/ft, No. 3 at 12 in.
2 to 3°, $A_s = 0.10$ in.$^2 = 0.06$ in.^{2}/ft, No. 3 at 22 in.

Beyond 3° there is no appreciable tension owing to edge effects.

2. Minimum dome reinforcement will be supplied throughout the shell equal to at least $0.0018bh$ for welded-wire fabric, as specified in ACI 318-77, Sec. 7.12.2, for slabs. For the 2½-in. shell, about 0.054 in.^{2}/ft in each direction is required. A welded-wire fabric $6 \times 6 - 6/6$ provides 0.058 in.^{2}/ft in each of two directions.

3. Meridional bending reinforcement will be provided to resist the combined effects of N_ϕ and M_ϕ. At the dome edge ($\psi = 0$) from Table 6, for total surface load plus prestressing:

$$M_\phi = -0.77 - 0.37 = -1.14 \text{ ft-kips/ft}$$
$$N_\phi = -2.92 - 0.94 = -3.86 \text{ kips/ft}$$

If the small compressive stress ($3860/72 = 54$ psi) is neglected, the resulting simplification is conservative, and with $d = 6 - 1.5 = 4.5$ in.,

$$A_s = \frac{M}{f_s j d} = \frac{1.14 \times 12}{20 \times \frac{7}{8} \times 4.5} = 0.16 \text{ in.}^2/\text{ft}$$

At $\psi = 1°$, $M_\phi = +0.25 - 0.85 = -0.60$ ft-kip/ft, from which, with $d = 6 - 3.5 \times 1.65/8 - 1.5 = 3.8$ in.,

$$A_s = \frac{0.60 \times 12}{20 \times \frac{7}{8} \times 3.8} = 0.11 \text{ in.}^2/\text{ft}$$

Since the moments from prestressing and surface loads are of opposite sign at this location, it is necessary to check for initial prestressing. Assuming initial prestress at a 25 percent increase and only dome dead load acting,

$$M_\phi = +0.25 \times \tfrac{40}{90} - 1.25 \times 0.85 = -0.95 \text{ ft-kip/ft}$$
$$A_s = 0.11 \times \tfrac{95}{65} = 0.16 \text{ in.}^2/\text{ft}$$

This moment is temporary and probably will not increase; so that $f_s = 20,000$ psi is conservative. It is more logical to use an ultimate-load analysis, where

$$M_u = -0.95 \times 1.5 = -1.42 \text{ ft-kips/ft}$$

From ACI 318, with $f'_c = 4000$ psi and $f_y = 40,000$ psi,

$$M_u = \phi A_s f_y \left(d - \frac{a}{2} \right) = 1.42$$
$$A_s f_y = 0.85 f'_c ab$$

from which $a = 0.12$ in. and $A_s = 0.13$ in.^{2}/ft.

At $\psi = 2°$, $M_\phi = +0.68 - 0.93 = -0.25$ ft-kip/ft. With $d = 6 - 3.5 \times 3.30/8 - 1.5 = 3.0$ in.,

$$A_s = \frac{0.25 \times 0.12}{20 \times \text{⅞} \times 3.0} = 0.06 \text{ in.}^2/\text{ft}$$

Initially,

$$M_\phi = +0.68 \times {}^{40}\!/_{90} - 1.25 \times 0.93 = -0.86 \text{ ft-kip/ft}$$
$$M_u = \phi A_s f_y \left(d - \frac{a}{2} \right) = 1.5 \times 0.86 = 1.29 \text{ ft-kips/ft}$$

from which

$$A_s = 0.15 \text{ in.}^2/\text{ft}$$

Similar calculations at $\psi = 3°$, with $h = 3.8$ in., give $A_s = 0.14$ in.^{2}/ft and at $\psi = 5°$, with $h = 2½$ in., $A_s = 0.11$ in.^{2}/ft.

These computations show the need, largely due to initial prestressing, for top radial reinforcement of No. 3 at 8 in. ($A_s = 0.17$ in.^{2}/ft) extending 10 ft from the edge (just beyond $\psi = 5°$).

The layout of reinforcement is shown in Fig. 15.

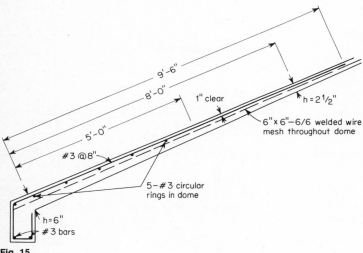

9'-6"

8'-0"

1" clear

$h = 2½"$

6"x 6"—6/6 welded wire mesh throughout dome

5'-0"

#3 @8"

5–#3 circular rings in dome

$h = 6"$
#3 bars

Fig. 15

RING PRESTRESSING. The final tension force required is

$$T = rH_F = 44.25 \times 3.24 = 143 \text{ kips}$$

Prestress will be furnished by circular rings of tensioned steel at an assumed final stress of 120,000 psi:

$$A_s = {}^{143}\!/_{120} = 1.2 \text{ in.}^2$$

This area may be supplied by 25 wires each 0.25 in. in diameter. An initial stress of 150,000 psi is required to compensate for losses assumed to be 20 percent.

Dome on Wall.   If the dome is supported by a cylindrical wall, as in a tank, the analysis is the same as that for the ring-stiffened dome except that the displacements of the ring are replaced by the displacements of the wall (Art. 11). If the dome and wall are joined monolithically through a ring, four redundant forces are involved: two between the dome and the ring and two between the ring and the wall. An example is given in Ref. 20.

Where the dome is built integrally with a wall, or where the ring of a dome-ring structure is restrained either through friction or continuity with a wall, temperature effects will be important (as in the edge-fixed dome of Art. 10) and should be investigated.

SHELL WALLS

11. Cylindrical Tanks   The membrane theory for cylindrical shells results in the following general equations (Fig. 16):

$$N'_\phi = -p_z r \tag{17a}$$

$$N'_{\phi x} = - \int \left(p_\phi + \frac{1}{r} \frac{\partial N'_\phi}{\partial \phi} \right) dx + f_1(\phi) \tag{17b}$$

$$N'_x = - \int \left(p_x + \frac{1}{r} \frac{\partial N'_{x\phi}}{\partial \phi} \right) dx + f_2(\phi) \tag{17c}$$

where f_1 and f_2 are determined from support conditions.

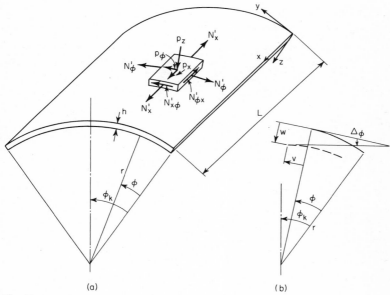

(a)   (b)

Fig. 16

The displacements u, v, w in the directions of the axes x, y, z are

$$u = \frac{1}{Eh} \int (N'_x - \nu N'_\phi) dx + f_3(\phi) \tag{17d}$$

$$v = - \int \frac{1}{r} \frac{\partial u}{\partial \phi} dx + \frac{2(1+\nu)}{Eh} \int N'_{\phi x} dx + f_4(\phi) \tag{17e}$$

$$w = \frac{\partial v}{\partial \phi} - \frac{r}{Eh} (N'_\phi - \nu N'_x) \tag{17f}$$

where f_3 and f_4 are determined from support conditions.

The rotation of the tangent to the circle at any point, positive as shown in Fig. 16b, is given by

$$\Delta_\phi = \frac{v}{r} + \frac{\partial w}{r \partial \phi} \tag{17g}$$

Internal Liquid Pressure (Fig. 17a).   With γ = density of liquid, $p_\phi = p_x = 0$, and $p_z = -y(H - x)$, Eqs. (17) give (Fig. 18)

$$N'_\phi = \gamma(H - x)r \qquad (18a)$$
$$N'_x = N'_{x\phi} = 0 \qquad (18b)$$

The resulting displacements are

$$w = -\frac{\gamma r^2}{Eh}(H - x) \qquad (19a)$$

$$\frac{dw}{dx} = \phi_x = \frac{\gamma r^2}{Eh} \qquad (19b)$$

Variable Thickness.   If the thickness of the tank wall varies linearly (Fig. 17c) the stress resultants for liquid pressure are given by Eq. (18). The displacements are

$$w = -\frac{\gamma r^2}{Eh_x}(H - x) \qquad (20a)$$

$$\frac{dw}{dx} = \phi_x = \frac{\gamma r^2}{Eh_x}\frac{h_{\text{top}}}{h_x} \qquad (20b)$$

where h_x = thickness at elevation x

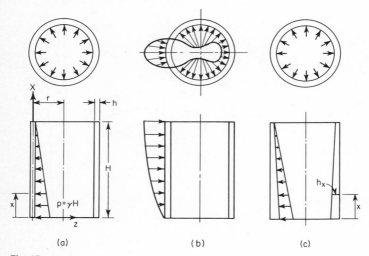

(a)   (b)   (c)

Fig. 17

Seismic Loading.   If the horizontal acceleration can be considered as a percentage α of the dead weight q, so that $q_s = aq$, the loading is $p_\phi = q_s \sin\theta$, $p_x = 0$, $p_z = q_s \cos\theta$, and Eqs. (17) give, for $N'_{\phi x} = N'_x = 0$ at $x = H$,

$$N'_\phi = -q_s r \cos\phi \qquad (21a)$$
$$N'_{\phi x} = 2q_s(H - x)\sin\phi \qquad (21b)$$

$$N'_x = \frac{q_s}{r}(H - x)^2 \cos\phi \qquad (21c)$$

Wind Load.   The wind pressure p_z on shell walls of rotation can be defined by (Fig. 17b).

$$p_z = K_z H_\phi p_{30} \qquad (22)$$

where

$$K_z = 2.64\left(\frac{z}{z_g}\right)^{2/\alpha} \qquad (23)$$

H_ϕ in Eq. (22) is a horizontal distribution factor. Values of α are given in Sec. 19, Art. 13. Coefficients H_ϕ from Ref. 21 for smooth cylinders with several ratios of L/D are given in Table 7a. Results of tests in Germany[22] on a full-scale hyperbolic cooling tower with small vertical ribs are also shown in this table. Measurements in the United States have given similar results.[23]

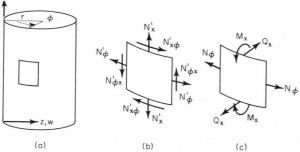

(a)   (b)   (c)

Fig. 18

TABLE 7a   Values of H_ϕ in Eq. (22)

ϕ, deg	L/D 1	7	25	Ref. 22
0	1.0	1.0	1.0	1.0
15	0.8	0.8	0.8	0.8
30	0.1	0.1	0.1	0.2
45	−0.7	−0.8	−0.9	−0.5
60	−1.2	−1.7	−1.9	−1.2
75	−1.6	−2.2	−2.5	−1.3
90	−1.7	−2.2	−2.6	−0.9
105	−1.2	−1.7	−1.9	−0.4
120	−0.7	−0.8	−0.9	−0.4
135	−0.5	−0.6	−0.7	−0.4
150	−0.4	−0.5	−0.6	−0.4
165	−0.4	−0.5	−0.6	−0.4
180	−0.4	−0.5	−0.6	−0.4

TABLE 7b   Fourier Coefficients for H_ϕ in Table 7a*

n	A_n $L/D = 1$	Ref. 22
0	−0.6000	−0.3923
1	0.2979	0.2602
2	0.9184	0.6024
3	0.3966	0.5046
4	−0.0588	0.1064
5	0.0131	−0.0948
6	0.0609	−0.0186
7	−0.0179	0.0468

*From Ref. 24.

The circumferential distribution of wind pressure may be represented by the Fourier series

$$H_\phi = \Sigma A_n \cos n\phi \qquad (24)$$

The first eight harmonics are generally sufficient.[24] Values of A_n corresponding to the test values of H_ϕ in Table 7a are given in Table 7b, along with values for the cylinder with $L/D = 1$.

For each term in the series Eqs. (17) gives

$$N'_\phi = p_z \, rA_n \cos n\phi \tag{25a}$$
$$N'_{\phi x} = p_z(H - x)A_n n \sin n\phi \tag{25b}$$

$$N'_x = \frac{p_z}{2r}(H - x)^2 A_n n^2 \cos n\phi \tag{25c}$$

Bending Theory—Axisymmetrical Loading. Edge effects are confined to a narrow zone. Therefore, except for the unusual case of a cylinder whose length is of the same order of magnitude as the affected zones, edge effects at one end can be treated independently of those at the other. Forces and displacements are given in Table 8.

TABLE 8   Forces (Fig. 18c) and Displacements in Circular Cylinders Loaded by Edge Forces Uniform around a Parallel Circle

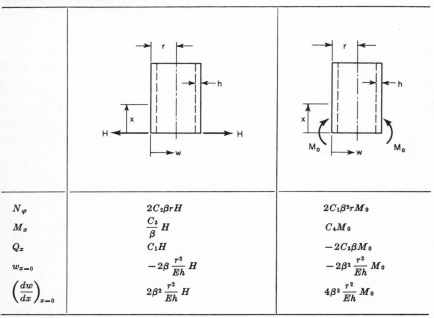

N_φ	$2C_2\beta rH$	$2C_1\beta^2 rM_0$
M_x	$\dfrac{C_3}{\beta} H$	$C_4 M_0$
Q_x	$C_1 H$	$-2C_3\beta M_0$
$w_{x=0}$	$-2\beta \dfrac{r^2}{Eh} H$	$-2\beta^2 \dfrac{r^2}{Eh} M_0$
$\left(\dfrac{dw}{dx}\right)_{x=0}$	$2\beta^2 \dfrac{r^2}{Eh} H$	$4\beta^3 \dfrac{r^2}{Eh} M_0$

NOTE: $\beta^4 = 3(1 - \nu^2)/r^2 h^2$.   See Table 4 for C.

Analysis   For a tank with a flat roof which is continuous with the wall, edge effects can be calculated using Tables 8 and 9.

Primary System.   The roof slab is assumed to be freely supported on the wall.

Errors.   Displacements are considered positive in the direction of the redundant forces (Fig. 19). Conditions at the base of the wall (i.e., whether it is fixed, hinged, free to slide, etc.) are usually immaterial in this analysis. This is because of the localized nature of edge forces on the wall. For the slab under uniform load, radial displacements are zero. Rotation of the edge is given in Table 9. Thus,

$$\delta_{10}^S = 0 \qquad \delta_{20}^S = -\frac{dw}{dR} = \frac{3qr^3}{28D} \qquad \text{for } \nu = \tfrac{1}{6}$$

For the tank filled with liquid, from Eq. (19),

$$\delta_{10}^W = 0 \qquad \delta_{20}^W = -\frac{dw}{dx} = -\frac{\gamma r^2}{Eh}$$

TABLE 9 Symmetrical Bending of Circular Plates

	M_r uniform on circumference	Uniform q over surface	Uniform q over surface
w	$\dfrac{M_r(r^2 - R^2)}{2D(1+\nu)}$	$\dfrac{q}{64D}(r^2 - R^2)\left(\dfrac{5+\nu}{1+\nu}r^2 - R^2\right)$	$\dfrac{q}{64D}(r^2 - R^2)^2$
$\dfrac{dw}{dR}$	$-\dfrac{M_r R}{D(1+\nu)}$	$-\dfrac{qR}{16D}\left(\dfrac{3+\nu}{1+\nu}r^2 - R^2\right)$	$-\dfrac{qR}{16D}(r^2 - R^2)$
$\dfrac{d^2w}{dR^2}$	$-\dfrac{M_r}{D(1+\nu)}$	$-\dfrac{q}{16D}\left(\dfrac{3+\nu}{1+\nu}r^2 - 3R^2\right)$	$-\dfrac{q}{16D}(r^2 - 3R^2)$
M_R	M_r	$\dfrac{q}{16}(3+\nu)(r^2 - R^2)$	$\dfrac{q}{16}[r^2(1+\nu) - R^2(3+\nu)]$
M_T	M_r	$\dfrac{q}{16}[r^2(3+\nu) - R^2(1+3\nu)]$	$\dfrac{q}{16}[r^2(1+\nu) - R^2(1+3\nu)]$
M_r	M_r	0	$-\dfrac{qr^2}{8}$
$M_{\mathbb{C}}$	M_r	$\dfrac{qr^2}{16}(3+\nu)$	$\dfrac{qr^2}{16}(1+\nu)$

$D = Eh^3/12(1 - \nu^2)$, h = thickness, M_R = radial moment, M_T = tangential moment.

20-23

Corrections.   For the wall, using Table 8 with $H = -X_1 = -1$ and $M_0 = X_2 = 1$,

$$\delta_{11}^W = \frac{2\beta r^2}{Eh} \qquad \delta_{12}^W = -\frac{2\beta^2 r^2}{Eh} \qquad \delta_{22}^W = \frac{4\beta^3 r^2}{Eh}$$

The radial displacement of the circumference of a circular slab subjected to unit radial forces at the perimeter is given by

$$\delta_{11}^S = \frac{r}{Eh}(1 - \nu)$$

and the edge rotation is zero, so that

$$\delta_{12}^S = 0$$

The edge rotation resulting from X_2 is found from Table 9. With $M_r = X_2 = 1$,

$$\delta_{22}^S = -\frac{dw}{dR} = \frac{r}{D(1 + \nu)}$$

Compatibility is established by satisfying the simultaneous equations in X_1 and X_2, as in Art. 10, where $\delta_{11} = \delta_{11}^W + \delta_{11}^S$, etc.

With X_1 and X_2 known, the internal forces to be provided for in the tank are evaluated using $H = -X_1$ and $M_0 = X_2$ in the formulas of Table 8, to which the membrane stress resultants [Eq. (18)] must be added. The internal forces in the roof are obtained similarly from Table 9.

The analyses for a tank roofed with a cylindrical dome and for the interaction between wall and floor (if they are continuous) are identical. However, the loads acting on the floor, and the corresponding errors and corrections, may be more difficult to determine.[20]

12. Hyperboloids   The type of shell wall shown in Fig. 20 has been used frequently for natural-draft cooling towers, for which the two principal loads are dead weight and wind. The membrane stress resultants for dead weight are computed from Eqs. (5) and (7) by numerical means as follows:

$$N'_{\phi j} = \frac{\sum\limits_{i=1}^{j} r_{oi} q_i \Delta s_i}{r_{oj} \sin \phi j} \tag{26a}$$

$$N'_{\theta j} = -\frac{r_{oj}}{\sin \phi j}\left(\frac{N'_{\phi j}}{r_{1j}} + q_j \cos \phi j\right) \tag{26b}$$

where $N'_{\phi j}$ is the sum of the weights of j rings above, each of average unit weight q_i, average horizontal radius r_{oi}, and meridional length Δs_i.

Wind Load.   Membrane stress resultants are not easily computed since the load is not axisymmetrical. The meridional stress resultant is critical for design. Comparisons of meridional stress resultants computed for wind distributed as described in Art. 11 with results from a bending-theory solution show very little difference except in the values of N'_θ near the base.

Because several cooling towers have collapsed during wind storms, sufficient reinforcement must be provided in both the meridional and the circumferential directions. Recommendations are given in Ref. 24.

BARREL SHELLS

Segments of cylinders, often called barrel shells, transfer load by a combination of longitudinal beam action and transverse arch action. In short barrels, the loads are carried essentially by arch action to the longitudinal edges, where they are transferred to the transverse supports by the edge sections of the shell acting as deep beams. In long barrels, the shell behaves primarily as a beam of thin, curved cross section, although there is still some arch action near the crown.

Short barrels are used for aircraft hangars and auditoriums. They have been built with transverse spans of about 150 to 330 ft, and with longitudinal spans of 20 to 50 ft between stiffening ribs. Long barrels are more commonly used for warehouses and factories, where longitudinal spans of about 50 to 150 ft are required, with transverse spans of 20 to 40 ft.

Other simply curved shells, such as truncated segments of cones, are sometimes used. They behave similarly to cylindrical segments, provided they are not too radically tapered.

Membrane stress resultants for the cylindrical barrel are given in Table 10; the boundary conditions are $N'_x = 0$ at $x = 0$ and $x = L$ (Fig. 16).

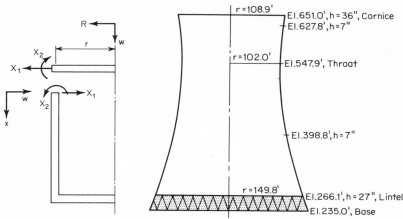

Fig. 19

Fig. 20   Martins Creek cooling tower on Delaware River at Easton.

13. Long Barrels   Barrels for which the ratio r/L is less than 0.6 may be considered to be long.[25] In this case, the in-plane stresses are approximated well by

$$f_x = \frac{N_x}{h} = \frac{M_x y}{I} \tag{27}$$

$$v = \frac{N_{x\phi}}{h} = \frac{VQ}{Ib} \tag{28}$$

where M_x = bending moment about centroidal axis
I = moment of inertia of shell cross section
V = total shear at cross section
b = total cross-sectional thickness of concrete measured horizontally

Transverse moments M_ϕ (arch action) may be approximated by considering the slice from the barrel shown in Fig. 21, where the vertical load q is held in equilibrium by the vertical component of the in-plane shearing forces. The arch moments and thrusts may be determined by any of the methods for arch analysis.

Implicit in the beam-arch analysis are the assumptions that all points on a transverse cross section deflect equally in the vertical direction and not at all horizontally, and that the radial shears Q_x, the longitudinal bending moments M_x, the twisting moments $M_{x\phi}$, and the strains from in-plane shearing forces can be neglected.

Figure 22 compares the stresses from Eq. (27) with those given by the shallow-shell theory. The correspondence is good for $r/L = 0.2$. But for barrels of intermediate and short length the maximum tensile stress is considerably larger than that given by Eq. (27).

Analysis by Eq. (27) is simple, but the arch analysis tends to be lengthy.[26] Table 11, based on the beam-arch equations, gives numerical values which may be used for interior shells of multibarrel systems and for single barrels without edge beams or with relatively flexible edge beams. These values may be corrected for horizontal edge deflection by Tables 12 and 13, which are given only for uniform load p because the error for dead load q is negligible. The corrections are made by adding to the values from Table 11 the product of the horizontal displacement from Table 12 by the corresponding coefficient in Table 13. Table 14 gives values of I for calculation of deflections. Extensive tables based on the shallow-shell theory are given in Ref. 27.

TABLE 10 Membrane Forces in Cylindrical Shells (Fig. 16)

Loading	N_ϕ'	N_x'	$N_{\phi x}'$
q per ft² of surface	$-qr\cos(\phi_k - \phi)$	$-q\dfrac{x}{r}(L - x)\cos(\phi_k - \phi)$	$-q(L - 2x)\sin(\phi_k - \phi)$
p per ft² projection	$-pr\cos^2(\phi_k - \phi)$	$\tfrac{3}{2}p\,\dfrac{x}{r}(L - x)[1 - 2\sin^2(\phi_k - \phi)]$	$-\tfrac{3}{2}p(L - 2x)\sin(\phi_k - \phi)\cos(\phi_k - \phi)$
p_w per ft² projection	$-p_w r\sin(\phi_k - \phi)$	$-p_w\dfrac{x}{2r}(L - x)\sin(\phi_k - \phi)$	$p_w\left(\dfrac{L}{2} - x\right)\cos(\phi_k - \phi)$

Typical dimensions of barrel shells are given in Table 15.

An elliptical cross section substantially improves the behavior by reducing moments and longitudinal stresses. The effect of the nearly vertical edge is somewhat the same as the addition of an edge member, except that no edge moments are created. The elliptical cross section was often chosen when barrel shells were first used but was soon given up because of construction difficulties.

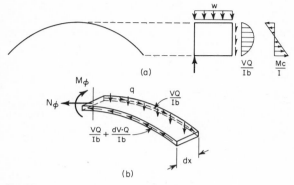

Fig. 21

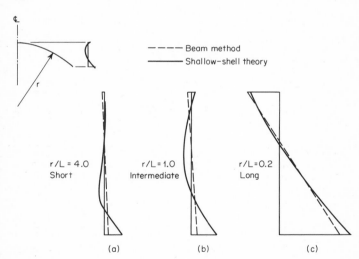

Fig. 22

Prestressing can be conveniently used in the edge members of long-span barrels, primarily to counteract longitudinal tension forces and transverse bending moments and to control deflection and cracking. Ultimate-load behavior must not be overlooked, however, because design moments increase directly with overloads whereas prestressing moments do not.

Transverse frames projecting below the roof complicate construction. Frames projecting above the roof complicate roofing and insulation. Thus, long-barrel roofs have sometimes been designed as ribless shells, in which a considerable portion of the shell is thickened and thus acts as a flat arch.[29] These arches should be investigated for partial loadings as well as for buckling and are thus used only on relatively small spans.

TABLE 11 Symmetrically Loaded Interior Circular Cylindrical Shells*

$\phi_k,$ deg	$\dfrac{\phi\dagger}{\phi_k}$	Uniform transverse load $p\ddagger$				Dead-weight load $q\ddagger$			
		$\dfrac{N_x}{pL^2/r}$	$\dfrac{N_\phi}{pr}$	$-\dfrac{N_{x\phi}}{pL}$	$\dfrac{M_\phi}{pr^2}$	$\dfrac{N_x}{qL^2/r}$	$\dfrac{N_\phi}{qr}$	$-\dfrac{N_{x\phi}}{qL}$	$\dfrac{M_\phi}{qr^2}$
22.5	1	-6.010	-1.411	0.000	-0.00292	-6.167	-1.433	0.000	-0.00309
	0.75	-4.875	-1.189	2.211	-0.00112	-5.003	-1.205	2.269	-0.00118
	0.50	-1.482	-0.614	3.533	0.00232	-1.521	-0.615	3.626	0.00245
	0.25	4.137	0.049	3.084	0.00235	4.245	0.065	3.165	0.00249
	0	11.927	0.361	0.000	-0.00662	12.239	0.384	0.000	-0.00702
25.0	1	-4.855	-1.402	0.000	-0.00353	-5.012	-1.430	0.000	-0.00378
	0.75	-3.937	-1.182	1.985	-0.00135	-4.064	-1.202	2.049	-0.00145
	0.50	-1.193	-0.612	3.170	0.00280	-1.232	-0.613	3.273	0.00300
	0.25	3.342	0.044	2.765	0.00282	3.451	0.064	2.855	0.00304
	0	9.617	0.347	0.000	-0.00797	9.929	0.374	0.000	-0.00857
27.5	1	-4.000	-1.393	0.000	-0.00417	-4.158	-1.426	0.000	-0.00453
	0.75	-3.242	-1.175	1.799	-0.00159	-3.370	-1.199	1.869	-0.00173
	0.50	-0.980	-0.609	2.871	0.00331	-1.018	-0.610	2.985	0.00360
	0.25	2.755	0.038	2.503	0.00332	2.863	0.063	2.602	0.00363
	0	7.908	0.331	0.000	-0.00938	8.220	0.363	0.000	-0.01025
30.0	1	-3.350	-1.383	0.000	-0.00482	-3.508	-1.422	0.000	-0.00533
	0.75	-2.714	-1.166	1.643	-0.00183	-2.842	-1.195	1.720	-0.00203
	0.50	-0.817	-0.606	2.622	0.00384	-0.856	-0.607	2.746	0.00424
	0.25	2.308	0.032	2.284	0.00383	2.417	0.061	2.392	0.00426
	0	6.608	0.314	0.000	-0.01082	6.920	0.352	0.000	-0.01204
32.5	1	-2.844	-1.372	0.000	-0.00548	-3.002	-1.418	0.000	-0.00618
	0.75	-2.303	-1.158	1.511	-0.00207	-2.431	-1.191	1.595	-0.00235
	0.50	-0.691	-0.603	2.410	0.00438	-0.729	-0.603	2.544	0.00492
	0.25	1.960	0.026	2.098	0.00434	2.069	0.060	2.215	0.00492
	0	5.596	0.297	0.000	-0.01227	5.908	0.339	0.000	-0.01393
35.0	1	-2.442	-1.361	0.000	-0.00615	-2.601	-1.414	0.000	-0.00707
	0.75	-1.977	-1.148	1.397	-0.00232	-2.105	-1.186	1.488	-0.00268
	0.50	-0.591	-0.599	2.227	0.00491	-0.629	-0.600	2.372	0.00565
	0.25	1.684	0.019	1.938	0.00484	1.793	0.058	2.064	0.00561
	0	4.794	0.278	0.000	-0.01370	5.105	0.326	0.000	-0.01591
37.5	1	-2.118	-1.349	0.000	-0.00679	-2.278	-1.409	0.000	-0.00800
	0.75	-1.174	-1.138	1.298	-0.00255	-1.842	-1.181	1.396	-0.00302
	0.50	-0.510	-0.596	2.069	0.00544	-0.548	-0.596	2.224	0.00640
	0.25	1.461	0.012	1.798	0.00532	1.571	0.057	1.933	0.00632
	0	4.146	0.260	0.000	-0.01509	4.458	0.312	0.000	-0.01796
40.0	1	-1.853	-1.335	0.000	-0.00742	-2.013	-1.404	0.000	-0.00897
	0.75	-1.498	-1.127	1.211	-0.00277	-1.627	-1.176	1.315	-0.00337
	0.50	-0.443	-0.592	1.929	0.00595	-0.482	-0.592	2.095	0.00719
	0.25	1.279	0.005	1.675	0.00578	1.389	0.055	1.819	0.00705
	0	3.616	0.241	0.000	-0.01641	3.928	0.297	0.000	-0.02006
45.0	1	-1.449	-1.307	0.000	-0.00853	-1.610	-1.393	0.000	-0.01096
	0.75	-1.170	-1.104	1.065	-0.00316	-1.299	-1.165	1.183	-0.00408
	0.50	-0.343	-0.585	1.694	0.00688	-0.381	-0.583	1.882	0.00883
	0.25	1.001	-0.011	1.468	0.00657	1.112	0.052	1.630	0.00854
	0	2.809	0.202	0.000	0.01872	3.120	0.266	0.000	-0.02437

TABLE 11 Symmetrically Loaded Interior Circular Cylindrical Shells* *(Continued)*

ϕ_k, deg	$\dfrac{\phi\dagger}{\phi_k}$	Uniform transverse load $p\ddagger$					Dead-weight load $q\ddagger$			
		$\dfrac{N_x}{pL^2/r}$	$\dfrac{N_\phi}{pr}$	$-\dfrac{N_{x\phi}}{pL}$	$\dfrac{M_\phi}{pr^2}$		$\dfrac{N_x}{qL^2/r}$	$\dfrac{N_\phi}{qr}$	$-\dfrac{N_{x\phi}}{qL}$	$\dfrac{M_\phi}{qr^2}$
50.0	1	-1.160	-1.276	0.000	-0.00939		-1.322	-1.380	0.000	-0.01301
	0.75	-0.935	-1.079	0.947	-0.00344		-1.065	-1.152	1.079	-0.00480
	0.50	-0.271	-0.578	1.504	0.00762		-0.308	-0.574	1.713	0.01054
	0.25	0.802	-0.029	1.300	0.00714		0.914	0.049	1.481	0.01002
	0	2.232	0.164	0.000	-0.02052		2.543	0.234	0.000	-0.02871
55.0	1	-0.946	-1.242	0.000	-0.00989		-1.109	-1.367	0.000	-0.01506
	0.75	-0.761	-1.053	0.849	-0.00358		-0.892	-1.139	0.995	-0.00549
	0.50	-0.217	-0.572	1.347	0.00807		-0.255	-0.563	1.578	0.01227
	0.25	0.655	-0.048	1.161	0.00742		0.767	0.045	1.360	0.01144
	0	1.805	0.128	0.000	-0.02130		2.115	0.201	0.000	-0.03293
60.0	1	-0.783	-1.205	0.000	-0.00992		-0.947	-1.352	0.000	-0.01705
	0.75	-0.629	-1.025	0.766	-0.00355		-0.761	-1.124	0.927	-0.00613
	0.50	-0.177	-0.566	1.213	0.00815		-0.214	-0.552	1.467	0.01398
	0.25	0.543	-0.068	1.043	0.00734		0.656	0.042	1.261	0.01275
	0	1.481	0.095	0.000	-0.02118		1.790	0.167	0.000	-0.03688

*From Ref. 28.
†ϕ measured from edge (Fig. 16).
‡See Fig. 16 for N_x, N_ϕ, $N_{x\phi}$; Fig. 21 for M_ϕ.

TABLE 12 Horizontal Edge Displacement Δ of Circular Cylindrical Shells under Uniform Load p*

$$\Delta = -\frac{pr^2}{Eh} \times \text{(constant from table)}$$

ϕ_k, deg	r/L						
	0.100	0.125	0.150	0.175	0.200	0.225	0.250
22.5	91.83	38.37	18.88	10.37	6.16	3.87	2.52
25.0	123.80	51.79	25.54	14.09	8.42	5.33	3.52
27.5	161.58	67.67	33.43	18.49	11.09	7.07	4.71
30.0	205.26	86.02	42.55	23.59	14.20	9.08	6.09
32.5	254.82	106.85	52.91	29.38	17.72	11.38	7.66
35.0	310.10	130.69	64.47	35.84	21.66	13.94	9.42
37.5	370.84	155.63	77.17	42.94	25.99	16.77	11.37
40.0	436.65	183.31	90.94	50.65	30.70	19.83	13.47
45.0	581.48	244.21	121.26	67.61	41.05	26.59	18.13
50.0	739.50	310.68	154.34	86.14	52.37	33.98	23.22
55.0	904.36	380.03	188.87	105.47	64.18	41.70	28.54
60.0	1,068.79	449.20	223.31	124.77	75.97	49.41	33.85

*From Ref. 28.

TABLE 13. Effect of Unit Horizontal Displacement of Circular Cylindrical Shells under Uniform Load p^*

ϕ_k, deg	$\dfrac{\phi\dagger}{\phi_k}$	N_ϕ	M_ϕ	ϕ_k, deg	$\dfrac{\phi\dagger}{\phi_k}$	N_ϕ	M_ϕ
22.5	1	2,463.0	62.82	37.5	1	199.2	13.92
	0.75	2,451.1	50.96		0.75	196.5	11.26
	0.50	2,415.6	15.49		0.50	188.6	3.35
	0.25	2,356.9	−43.34		0.25	175.7	−9.60
	0	2,275.5	−124.66		0	158.0	−27.24
25.0	1	1,461.9	45.95	40.0	1	145.5	11.53
	0.75	1,453.2	37.26		0.75	143.3	9.32
	0.50	1,427.2	11.30		0.50	136.7	2.76
	0.25	1,384.3	−31.63		0.25	126.0	−7.96
	0	1,324.9	−91.02		0	111.4	−22.50
27.5	1	912.9	34.65	45.0	1	82.2	8.20
	0.75	906.4	28.09		0.75	80.7	6.62
	0.50	886.8	8.49		0.50	76.0	1.94
	0.25	854.4	−23.86		0.25	68.4	−5.66
	0	869.8	−68.50		0	58.2	−15.89
30.0	1	594.6	26.80	50.0	1	49.6	6.06
	0.75	589.5	21.71		0.75	48.4	4.88
	0.50	574.3	6.54		0.50	44.9	1.41
	0.25	549.3	−18.46		0.25	39.3	−4.19
	0	514.9	−52.86		0	31.9	−11.65
32.5	1	401.2	21.17	55.0	1	31.5	4.62
	0.75	397.2	17.14		0.75	30.6	3.72
	0.50	385.2	5.14		0.50	27.9	1.66
	0.25	365.4	−14.59		0.25	23.7	−3.20
	0	338.4	−41.66		0	18.1	−8.81
35.0	1	279.0	17.03	60.0	1	20.9	3.62
	0.75	275.8	13.79		0.75	20.2	2.91
	0.50	266.1	4.12		0.50	18.1	0.82
	0.25	250.3	−11.74		0.25	14.8	−2.51
	0	228.6	−33.43		0	10.5	−6.84

*From Ref. 28.
†ϕ measured from edge (Fig. 16).

TABLE 14 Moment of Inertia of Circular Cylindrical Shells*

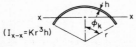

$(I_{x-x} = Kr^3h)$

ϕ_k, deg	K	ϕ_k, deg	K
22.5	0.00041	37.5	0.00502
25.0	0.00068	40.0	0.00687
27.5	0.00110	45.0	0.01216
30.0	0.00168	50.0	0.02017
32.5	0.00249	55.0	0.03174
35.0	0.00358	60.0	0.04782

*From Ref. 28.

Continuous Shells. The behavior of beamlike thin shells continuous over three or more transverse supports is similar to that of continuous beams with regard to in-plane stresses, but transverse bending moments are affected less by longitudinal continuity.

14. Short Barrels Barrels for which $r/L > 0.6$ may be considered to be short. Such shells carry load essentially by arch action and are usually shaped to the arch pressure line. Because of the large value of r relative to L, the shell thickness is often controlled by buckling rather than by strength (Art. 5). Intermediate stiffeners have been used to prevent buckling. Here the cost of extra formwork must be weighed against the saving in cost of materials.

TABLE 15 Typical Dimensions for Barrel Shells*

Section†	Span, ft	Bay width, ft	R, ft	r, ft	h, in.	Rein-forc-ing‡
	80	30	8	25	3	3.5
	100	30	10	30	3	4.0
	120	35	12	30	3	4.5
	140	40	14	35	3	5.0
	160	45	16	35	3.5	6.5

* From Portland Cement Association.
† For long-span multiple barrels, the usual depth-span ratio varies from 1:10 to 1:15.
‡ Pounds per square foot of projected area.

Since most of the roof load is carried to the supporting transverse frames by beam action only near the longitudinal edges, the supporting arch is subjected to outward thrusts near the springing lines and hence may have only a small compression, or even some tension, near the crown. But the adjacent shell is under compression; so that there is incompatibility of strain or, more properly, the ideal behavior is modified and some load is transferred into the arch by bending forces. The effect is to increase the compression in the supporting arch and reduce the compression in the shell adjacent to the arch or, in other words, to force part of the shell to act as a flange for the arch in T-beam behavior.

15. General Procedure for Shallow Shells The theory of shallow shells is based on the following assumptions: (1) the slope of the shell is small compared with some plane of reference (usually the horizontal plane for roofs); (2) the curvature of the surface is small; (3) the shell boundaries are such that the loads are carried primarily by the in-plane stress resultants N_x, N_ϕ, and $N_{x\phi}$; and (4) changes in curvature of the surface are small. The problem can be reduced to the solution of an eighth-order partial differential equation in one unknown.

In the following analysis the shell is assumed to be supported at $x = 0$ and $x = L$ by transverse frames which are rigid in their vertical planes, so that displacements v and w are zero at each end, and completely flexible in respect to out-of-plane displacements, so that N_x and M_x are zero at each end.

Complementary Function. With the displacement w (Fig. 16b) as the unknown, the complementary function has the form

$$w = \sum_{n=1,3\ldots} A_n e^{Mn\phi} \sin kx$$

where $k = n\pi/L$. Table 16 gives the complementary functions for the displacements and resultants for a single, circular, cylindrical shell with symmetrical geometry and loading. Displacements u, v, and w are positive in the positive directions of x, y, and z, respectively (Fig. 16a), while θ corresponds to Δ_ϕ in Fig. 16b. Note that ϕ is measured from the crown, rather than from the edge as in Fig. 16.

Except for Q'_ϕ and Q'_x, the stress resultants and couples in this table can be identified in Fig. 4a and b by substituting ϕ for y (see also Fig. 16a). Q'_ϕ and Q'_x are combinations analogous to those arising in the theory of plates:

$$Q'_\phi = Q_\phi + \frac{\partial M_{\phi x}}{\partial x}$$

$$Q'_x = Q_x + \frac{\partial M_{\phi x}}{R\,\partial \phi}$$

Thus, although there are five stress resultants at the longitudinal edge (M_ϕ, N_ϕ, Q_ϕ, $N_{\phi x}$, and $M_{\phi x}$) with only four boundary conditions to be satisfied, Q'_ϕ reduces the number to four.

The shell constants Q and γ in Table 16 are defined as follows:

$$Q^8 = 3(kr)^4 \left(\frac{r}{h}\right)^2 \qquad \gamma = \left(\frac{kr}{Q}\right)^2$$

Values of the coefficients of ϕ are

$$\alpha_1 = Q\left(\frac{\sqrt{(1+\gamma)^2 + 1} + (1+\gamma)}{2}\right)^{1/2} = Qm_1 \tag{29a}$$

$$\beta_1 = Q\left(\frac{\sqrt{(1+\gamma)^2 + 1} - (1+\gamma)}{2}\right)^{1/2} = Qn_1 \tag{29b}$$

$$\alpha'_1 = Q\left(\frac{\sqrt{(1-\gamma)^2 + 1} - (1-\gamma)}{2}\right)^{1/2} = Qm_2 \tag{29c}$$

$$\beta'_1 = Q\left(\frac{\sqrt{(1-\gamma)^2 + 1} + (1-\gamma)}{2}\right)^{1/2} = Qn_2 \tag{29d}$$

Particular Integral. To determine the particular integral, the load is expressed in a Fourier series

$$p' = \frac{4}{\pi}q \sum_{n=1,3\ldots} \frac{1}{n}\sin kx \qquad k = \frac{n\pi}{L}$$

where q = uniformly distributed load per square foot of shell surface. The first term of this

TABLE 16   Shell Coefficients (ϕ Measured from the Crown)*

$$F = 2\bar{R}\left[\begin{array}{l}(aB_1 - bB_2)\cos\beta_1\phi\cosh\alpha_1\phi - (aB_2 + bB_1)\sin\beta_1\phi\sinh\alpha_1\phi \\ (cB_3 - dB_4)\cos\beta_1'\phi\cosh\alpha_1'\phi - (cB_4 + dB_3)\sin\beta_1'\phi\sinh\alpha_1'\phi\end{array}\right]$$

F	$\bar{R}$	B_1	B_2	B_3	B_4
M_ϕ	$-\dfrac{2D}{r^2}\sin kx$	$Q^2(1+\gamma)$	Q^2	$Q^2(\gamma-1)$	Q^2
M_x	$2Dk^2\sin kx$	1	0	1	0
Q_x	$-\dfrac{2Dk^3}{\gamma}\cos kx$	1	1	-1	1
N_ϕ	$\dfrac{4Drk^4}{\gamma^2}\sin kx$	0	1	0	-1
N_x	$-\dfrac{4Drk^4}{\gamma^3}\sin kx$	-1	$1+\gamma$	1	$1-\gamma$
Q_x'	$-\dfrac{2Dk^3}{\gamma}\cos kx$	$\gamma+2$	2	$\gamma-2$	2
u	$\dfrac{4Drk^3}{hE\gamma^3}\cos kx$	-1	$1+\gamma$	1	$1-\gamma$
w	$2\sin kx$	1	0	1	0

TABLE 16 Shell Coefficients (ϕ Measured from the Crown)* (Continued)

$$F = 2\bar{R}\left[\begin{array}{l}(aB_1 - bB_2)\cos\beta_1\phi\sinh\alpha_1\phi - (aB_2 + bB_1)\sin\beta_1\phi\cosh\alpha_1\phi \\ (cB_3 - dB_4)\cos\beta_1'\phi\sinh\alpha_1'\phi - (cB_4 + dB_3)\sin\beta_1'\phi\cosh\alpha_1'\phi\end{array}\right]$$

F	$\bar{R}$	B_1	B_2	B_3	B_4
Q_ϕ	$-\dfrac{2Dk^3}{(\sqrt{\gamma})^3}\sin kx$	$m_1 - n_1$	$m_1 + n_1$	$-(m_2 + n_2)$	$m_2 - n_2$
Q_ϕ'	$-\dfrac{2Dk^3}{(\sqrt{\gamma})^3}\sin kx$	$m_1(1-\gamma) - n_1$	$m_1 + n_1(1-\gamma)$	$-m_2(1+\gamma) - n_2$	$m_2 - n_2(1+\gamma)$
$N_{x\phi}$	$\dfrac{4Drk^4}{(\sqrt{\gamma})^5}\cos kx$	$-n_1$	m_1	n_2	$-m_2$
v	$\dfrac{4Drk^3}{Eh(\sqrt{\gamma})^7}\sin kx$	$m_1 + n_1(1-\gamma)$	$n_1 - m_1(1-\gamma)$	$-m_2 + n_2(1+\gamma)$	$-n_2 - m_2(1+\gamma)$
$\theta\dagger$	$\sin kx$	$\dfrac{2\alpha_1}{r} + \dfrac{(\bar{R}B_1)_v}{r}$	$\dfrac{2\beta_1}{r} + \dfrac{(\bar{R}B_2)_v}{r}$	$\dfrac{2\alpha_1'}{r} + \dfrac{(\bar{R}B_3)_v}{r}$	$\dfrac{2\beta_1'}{r} + \dfrac{(\bar{R}B_4)_v}{r}$
$M_{x\phi}$	$\dfrac{2Dk}{r}\cos kx$	α_1	β_1	α_1'	β_1'

*After Ref. 30 with $\nu = 0$, so that $D = Eh^3/12$.

†Observe that one part of coefficients B_1, etc., for θ is obtained from v, that is, $(\bar{R}B_1)_v/r = [4Dk^3/Eh(\sqrt{\gamma})](\sqrt{\gamma})[m_1 - n_1(1 - \gamma)]$.

series is sufficiently dominant that good results are obtained (except possibly for short shells) by retaining only it:

$$p' = \frac{4}{\pi} q \sin \frac{\pi x}{L}$$

The particular integral is approximated quite closely by the membrane stress resultants. Thus, in Fig. 16a,

$$p_x = 0 \qquad p_\phi = -p' \sin (\phi_k - \phi) \qquad p_z = p' \cos (\phi_k - \phi)$$

and Eqs. (17) give

$$N'_\phi = -\frac{4q}{\pi} r \cos (\phi_k - \phi) \sin kx \tag{30a}$$

$$N'_{\phi x} = -\frac{4q}{\pi} \frac{2}{k} \sin (\phi_k - \phi) \cos kx \tag{30b}$$

$$N'_x = -\frac{4q}{\pi} \frac{2}{rk^2} \cos (\phi_k - \phi) \sin kx \tag{30c}$$

where $k = \pi/L$. Both $f_1(\phi)$ and $f_2(\phi)$ vanish, the former because $N'_{x\phi} = 0$ at $L/2$, the latter because $N'_x = 0$ at $s = 0$ and $x = L$.

The corresponding displacements are found by Eqs. (17):

$$u = \frac{1}{Eh} \frac{4q}{\pi} \frac{2}{rk^3} \left(1 - \frac{\nu r^2 k^2}{2}\right) \cos (\phi_k - \phi) \cos kx \tag{31a}$$

$$v = -\frac{1}{Eh} \frac{4q}{\pi} \frac{2}{r^2 k^4} \left[1 + \left(2 + \frac{3\nu}{2}\right) r^2 k^2\right] \sin (\phi_k - \phi) \sin kx \tag{31b}$$

$$w = \frac{2}{Eh} \frac{4q}{\pi} \frac{1}{r^2 k^4} \left[1 + \left(2 + \frac{\nu}{2}\right) r^2 k^2 + \frac{r^4 k^4}{2}\right] \cos (\phi_k - \phi) \sin kx \tag{31c}$$

$$\theta = \frac{1}{Eh} \frac{4q}{\pi} r \left(1 - \frac{2\nu}{r^2 k^2}\right) \sin (\phi_k - \phi) \sin kx \tag{31d}$$

Both $f_3(\phi)$ and $f_4(\phi)$ vanish, the former because it represents a lengthwise rigid-body translation of the shell, the latter because v is zero at each end.

Similarly, for uniform load p on the horizontal projection of the surface,

$$p' = \frac{4}{\pi} p \sin \frac{\pi x}{L}$$

$$p_x = 0 \qquad p_\phi = -p' \sin (\phi - \phi_k) \cos (\phi - \phi_k) \qquad p_z = p' \cos^2 (\phi_k - \phi)$$

$$N'_\phi = -\frac{4p}{\pi} r \cos^2 (\phi_k - \phi) \sin kx \tag{32a}$$

$$N'_{\phi x} = -\frac{4p}{\pi} \frac{3}{k} \sin (\phi_k - \phi) \cos (\phi_k - \phi) \cos kx \tag{32b}$$

$$N'_x = -\frac{4p}{\pi} \frac{3}{rk^2} [1 - 2 \sin^2 (\phi_k - \phi)] \sin kx \tag{32c}$$

and the displacements

$$u = \frac{1}{Eh} \frac{4p}{\pi} \frac{3}{rk^3} \left[1 - 2 \sin^2 (\phi_k - \phi) - \frac{\nu r^2 k^2}{3} \cos^2 (\phi_k - \phi)\right] \cos kx \tag{33a}$$

$$v = -\frac{1}{Eh} \frac{4p}{\pi} \frac{12}{r^2 k^4} \left[1 + \left(\frac{1}{2} + \frac{\nu}{3}\right) r^2 k^2\right] \sin (\phi_k - \phi) \cos (\phi_k - \phi) \sin kx \tag{33b}$$

$$w = \frac{1}{Eh} \frac{4p}{\pi} \frac{12}{r^4 k^4} \left\{ \left[1 + \left(\frac{1}{2} + \frac{\nu}{12}\right) r^2 k^2\right] [1 - 2 \sin^2 (\phi_k - \phi)] \right.$$

$$\left. + \frac{r^4 k^4}{12} \cos^2 (\phi_k - \phi) \right\} \sin kx \tag{33c}$$

$$\theta = \frac{1}{Eh} \frac{4p}{\pi} \frac{36}{r^3 k^4} \left(1 + \frac{r^2 k^2}{2} + \frac{r^4 k^4}{18}\right) \sin (\phi_k - \phi) \cos (\phi_k - \phi) \sin kx \tag{33d}$$

For concrete ν is about $\frac{1}{6}$ and appears to have little influence on deformations; so that Eqs. (31) and (33) are usually simplified by taking $\nu = 0$.

Solution. The complete solution is found by adding the complementary function and the particular integral as given by the membrane stress resultants. The four arbitrary constants a, b, c, and d in the function F of Table 16 are found by solving four simultaneous equations given by the boundary conditions. The boundary conditions are obtained by an appropriate choice of four among the stress resultants M_ϕ, N_ϕ, Q'_ϕ, and $N_{\phi x}$ and the displacements u, v, w, and θ. These conditions may range from those for the unsupported edge

$$M_\phi = N_\phi = Q'_\phi = N_{\phi x} = 0$$

to those for the fixed edge

$$u = v = w = \theta = 0$$

A computer program (MULEL) for the solution of these equations is available.[43] More general (finite-element) programs that can be used for the barrel shell are NASTRAN and EASE. Extensive tabulated results have been published.[27]

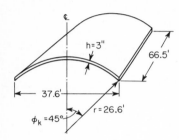

Fig. 23

16. Example The loads for the shell of Fig. 23 are:

Shell dead load	40
Roofing and mechanical equipment	10
Snow	30
	$q = 80$ psf

The first two are distributed essentially uniformly over the surface of the shell. Distribution of snow load is not usually specified in codes. Live load is often assumed to be distributed uniformly over the horizontal projection of the surface, but it is assumed here to be distributed in the same manner as the dead load.

The longitudinal edges are assumed to be free. Three stress resultants N_ϕ, N_x, and $N_{\phi x}$ and one stress couple M_ϕ are important in the proportioning of the shell. In the case of the shell with free edge, Q'_ϕ also is needed to establish one of the boundary conditions.

Only the first term of the Fourier series for w and p' is used, i.e., $n = 1$, so that $k = \pi/L$. Equations (30) give the membrane stress resultants

$$N'_\phi = -2709 \cos(\phi_k - \phi) \sin kx$$
$$N'_{\phi x} = -4313 \sin(\phi_k - \phi) \cos kx$$
$$N'_x = -3430 \cos(\phi_k - \phi) \sin kx$$

For the complementary functions,

$$kr = \frac{\pi r}{L} = \frac{26.6\pi}{66.5} = 1.2566$$

$$Q^8 = 3 \times 1.2566^4 \times \left(\frac{26.6}{0.25}\right)^2$$

$$Q = 4.1303$$

$$\gamma = \left(\frac{1.2566}{4.1303}\right)^2 = 0.09257$$

Substituting the values of Q and γ into Eqs. (29), we get

$$\alpha_1 = 4.6854 \qquad \beta_1 = 1.8205 \qquad \alpha'_1 = 1.9437 \qquad \beta'_1 = 4.3884$$

The required coefficients B_1, B_2, B_3, and B_4 are, from Table 16,

	B_1	B_2	B_3	B_4
M_ϕ	18.6384	17.0593	−15.4801	17.0593
N_ϕ	0	1	0	−1
N_x	−1	1.0926	1	0.9074
$N_{\phi x}$	−0.4408	1.1344	1.0625	−0.4706
Q_ϕ'	0.5886	1.5344	−1.5766	−0.6903

Because ϕ is measured from the crown in Table 16, while in Eqs. (30) to (33) it is measured from the edge, it must be taken negative in evaluating the functions in Table 16. It will be noted that this results in no changes in F for M_ϕ to w, inclusive, but changes the sign of F for Q_ϕ to $M_{x\phi}$, inclusive.

Values of $\cos \beta_1\phi \cosh \alpha_1\phi$, etc., must be determined for the values of ϕ at which the internal forces are to be investigated. Only those at the edge, where $\phi = \phi_k = -45°$, are given here:

$$\cos \beta_1\phi_k \cosh \alpha_1\phi_k = 2.7870 \qquad \cos \beta_1\phi_k \sinh \alpha_1\phi_k = -2.7835$$
$$\sin \beta_1\phi_k \sinh \alpha_1\phi_k = 19.6115 \qquad \sin \beta_1\phi_k \cosh \alpha_1\phi_k = -19.6365$$
$$\cos \beta_1'\phi_k \cosh \alpha_1'\phi_k = -2.2986 \qquad \cos \beta_1'\phi_k \sinh \alpha_1'\phi_k = 2.0913$$
$$\sin \beta_1'\phi_k \sinh \alpha_1'\phi_k = -0.6585 \qquad \sin \beta_1'\phi_k \cosh \alpha_1'\phi_k = 0.7238$$

With these values and the values of B, the required edge stress resultants are determined from Table 16, to which must be added the membrane stress resultants (evaluated at the edge). With the boundary conditions

$$M_{\phi k} = N_{\phi k} = Q'_{\phi k} = N_{\phi k x} = 0$$

we get

$$M_{\phi k} = -2D(-0.7988a - 1.1676b + 0.1323c + 0.08202d) \sin kx = 0$$
$$N_{\phi k} = -2D(1.2130a + 0.1724b + 0.04073c + 0.1422d) \sin kx - 1915 \sin kx = 0$$
$$N_{\phi k x} = -2D(-4.7777a + 1.1176b - 0.5209c - 0.04374d) \cos kx - 3050 \cos kx = 0$$
$$Q'_{\phi k} = 2D(-0.2133a - 0.1185b + 0.02095c - 0.01935d) \sin kx = 0$$

from which

$$a = \frac{1177}{D} \qquad b = \frac{-2989}{D} \qquad c = \frac{-13,504}{D} \qquad d = \frac{-9292}{D}$$

COMPUTATION OF FORCES. With a, b, c, and d known, values of M_ϕ, N_ϕ, N_x, and $N_{\phi x}$ can now be determined throughout the shell. Thus, from Table 16,

$$M_\phi = -\frac{4}{26.6^2}(72,927 \cos \beta_1\phi \cosh \alpha_1\phi + 35,631 \sin \beta_1\phi \sinh \alpha_1\phi$$
$$+ 367,560 \cos \beta_1'\phi \cosh \alpha_1'\phi + 86,528 \sin \beta_1'\phi \sinh \alpha_1'\phi) \sin kx$$

N_ϕ, N_x, and $N_{\phi x}$ are found similarly. The results are given in Table 17, where M_ϕ, N_ϕ, and N_x are given for $x = L/2$ and $N_{\phi x}$ for $x = 0$. Values for any other value of x are obtained by multiplying by $\sin kx$ or $\cos kx$, as the case may be.

TABLE 17   Stress Resultants and Couples for Barrel of Fig. 23

ϕ, deg from edge	M_ϕ, ft-kips/ft, $x = L/2$	N_ϕ, kips/ft, $x = L/2$	N_x, kips/ft, $x = L/2$	$N_{\phi x}$, kips/ft, $x = 0$
45	−2.49	−3.49	+0.57	0
40	−2.44	−3.50	−1.47	−0.01
30	−2.02	−3.45	−14.63	−1.63
20	−1.15	−2.80	−24.33	−6.21
10	−0.16	−1.18	+1.09	−9.77
0	0	0	+105.25	0

17. Shell with Edge Beams Vertical edge beams (Fig. 24) are usually used for long shells, where the principal structural action is longitudinal bending. Horizontal beams are commonly used with short shells, where the principal action is transverse arching.

If the vertical edge beam is slender, it is reasonable to assume that it offers negligible resistance to rotation and horizontal displacement, so that $M_{\phi k} = 0$ and $H_b = 0$ become two boundary conditions. The third boundary condition is found by equating the vertical displacement of the shell edge to the corresponding vertical deflection of the edge beam. The fourth condition is obtained from compatibility of edge displacements u_k of the shell and the corresponding displacements of the edge beam or, what amounts to the same thing, equality of edge stress in the shell and the corresponding stress in the beam.

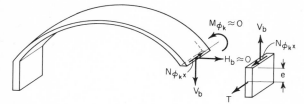

Fig. 24

The vertical deflection of the shell edge (positive upward) is given by

$$\Delta_{vs} = v_k \sin \phi_k - w_k \cos \phi_k$$

where v_k and w_k are found by adding the membrane displacements [Eq. (31) or (33), depending upon the distribution of load] to those of Table 16.

The reaction components V_b and H_b are given by

$$V_b = N_{\phi k} \sin \phi_k - Q'_{\phi k} \cos \phi_k \qquad (34)$$
$$H_b = N_{\phi k} \cos \phi_k + Q'_{\phi k} \sin \phi_k \qquad (35)$$

Note that $N_{\phi k}$ and $Q'_{\phi k}$, and therefore V_b and H_b, vary sinusoidally from $x = 0$ to $x = L$.

The axial tension T (Fig. 24) at any section of the beam is

$$T = -\int_0^x N_{\phi k x} dx \qquad (36)$$

Because $N_{\phi k x}$ varies as $\cos kx$, T varies as $\sin kx$.

The bending moments and deflections of the beam due to V_b and T are found by successive integrations of $EI d^4 y/dx^4 = w$. The combined effect is

$$M = \frac{V_b}{k^2} + Te \qquad (37)$$

$$\Delta_{vb} = \frac{1}{k^2 EI} \left(\frac{V_b}{k^2} + Te \right) \qquad (38)$$

where e is the eccentricity of T. The bending moment M is considered positive with the top fiber in tension and Δ is positive upward.

The stress at the top fiber of the beam is

$$f = \frac{T}{A} + \frac{V_b/k^2 + Te}{Z} \qquad (39)$$

Using the above relations, the four boundary conditions become

$$M_{\phi k} = 0 \qquad (40a)$$
$$H_b = 0 \qquad (40b)$$
$$v_k \sin \phi_k - w_k \cos \phi_k = \frac{1}{k^3 EI} \left(\frac{V_b}{k} - eN_{\phi k x} \right) \qquad (40c)$$
$$\frac{N_{xk}}{h} = \frac{1}{kZ} \left(\frac{V_b}{k} - eN_{\phi k x} \right) - \frac{N_{\phi k x}}{kA} \qquad 40d)$$

It should be noted that the factor $\sin kx$ is common to every term in Eq. (40). Therefore, when v, w, and the various resultants are taken from Table 16 and Eq. (30), (31), and/or (32), (33), the trigonometric term should be omitted.

Following the solution of Eq. (40) for the constants a, b, c, and d, computation of forces and moments throughout the shell is carried out as in Art. 16.

The computer program MULEL can be used to solve the equations for the shell with edge beams.[43] Also, an analysis that requires the solution of only two simultaneous equations (involving V and H) can be made with the use of tables.[20,27]

Prestressed Edge Beam.   The bending moment and deflection caused by a prestressing force F in a tendon draped to a parabolic curve passing through the centroids at each end of the edge beam can be computed by considering the equivalent uniform load

$$w = \frac{8Fe_c}{L^2}$$

where e_c = eccentricity of tendon at midspan. The prestressing force must be represented by a Fourier series

$$F_x = \frac{4}{\pi} F \sum_{n=1,3\ldots} \frac{1}{n} \sin kx$$

and the moment and deflection determined by successive integration of $EId^4y/dx^4 = w$.

With the effects of F added to Eq. (40), the four boundary conditions for the prestressed edge beam are

$$M_{\phi k} = 0 \tag{41a}$$
$$H_b = 0 \tag{41b}$$

$$v_k \sin \phi_k - w_k \cos \phi_k = \frac{1}{k^3 EI}\left(\frac{V_b}{k} - eN_{\phi kx} + \frac{4}{\pi}\frac{8Fe_c}{kL^2}\right) \tag{41c}$$

$$\frac{N_{xk}}{h} = \frac{1}{kZ}\left(\frac{V_b}{k} - eN_{\phi kx} + \frac{4}{\pi}\frac{8Fe_c}{kL^2}\right) - \frac{N_{\phi kx}}{kA} - \frac{4}{\pi}\frac{F}{A} \tag{41d}$$

The factor $\sin kx$ is omitted in the terms involving F because it is common to all other terms, as was noted in connection with Eq. (40).

18. Transverse Frames   Figure 25 shows an arch rib loaded by the in-plane forces $N_{x\phi}$ and the radial shears Q_x. The latter are so small that they may be neglected, so that the rib may be analyzed as an arch for the forces $N_{x\phi}$ alone. Tables[31] or standard computer programs simplify the analysis.

Fig. 25

19. Barrel-Shell Reinforcement   Figure 26 shows a layout of reinforcement for the barrel shell of Fig. 27. Reinforcement is provided to take all the principal tensile stresses, and may be placed either in the direction of the stress trajectories or in two directions, usually orthogonal. Principal stresses are computed from the values of the stress resultants N_ϕ and N_x. Stress trajectories for the barrel of Fig. 27 are shown in Fig. 28.

At the corners, where the shear is a maximum, reinforcement is generally placed at 45°. Transverse (hoop) reinforcement (No. 3 bars in Fig. 26) is based on M_ϕ, usually neglecting N_ϕ. If the torsional and lateral stiffnesses of the edge beam are neglected in the analyses, $M_\phi = 0$ at the edge. However, it is advisable to provide reinforcement for some positive moment caused by edge-beam stiffness. This can be done by placing the principal-tension reinforcement in this region near the underside of the shell (the No. 4 bars at 11 in. in Fig. 26).

Longitudinal reinforcement at the juncture of the shell and the transverse rib is usually based on the assumption that the rib permits no radial movement and no rotation of the shell. This produces a moment $M_x = -0.29hN_\phi$. At $x = 0$, $N_\phi = 0$ because of the

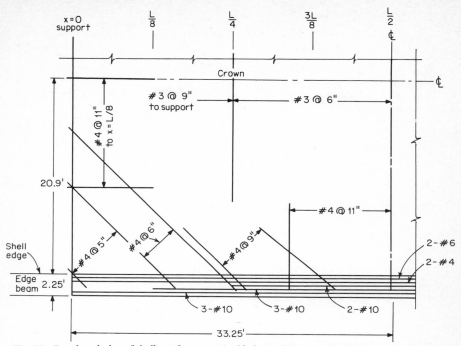

Fig. 26 Developed plan of shell reinforcement (welded wire fabric not shown).

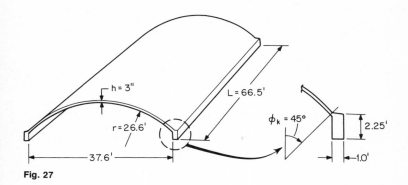

Fig. 27

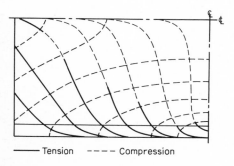

—— Tension – – – – Compression

Fig. 28 Stress trajectories for shell and edge beams.

assumption of sinusoidal loading, but since the actual loading is uniform, N_ϕ must be constant along the span. Therefore, the value of N_ϕ at $x = L/2$ should be used to determine M_x at $x = 0$. An allowance for additional moment resulting from arch deflection of the rib can be made.[20] The No. 4 bars at 11 in. in Fig. 26 are proportioned on this basis.

Minimum reinforcement of 0.35 percent in each of two directions in tensile zones and 0.18 percent in other zones, spaced no farther apart than five times the shell thickness, is recommended. This can be supplied in the form of fabric.

Edge-beam reinforcement is usually sized for the tensile force in the beam. This can be determined by using Simpson's rule to evaluate Eq. (36), or by computing the stresses at the top and bottom of the beam. Tests show that it is advisable to place most of the reinforcement near the bottom (Fig. 29).

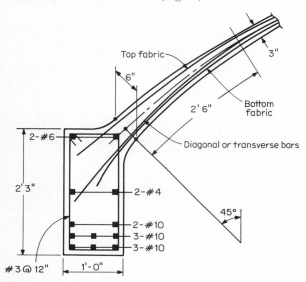

Fig. 29 Section at edge beam.

FOLDED PLATES

Figure 30 shows typical folded-plate cross sections. Typical dimensions for folded-plate roof structures are given in Table 18. These structures have no curvature but can be considered as approximations of shell geometry; i.e., the shape of Fig. 30a approximates a barrel shell with vertical edge beams and the shape in Fig. 30c a multiple-barrel shell. An

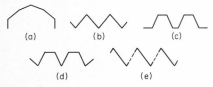

Fig. 30

important difference between folded plates and barrel shells is the difference in the transverse bending moments M_ϕ (Fig. 31), which results from the plates acting as one-way slabs to carry roof loads transversely to the ridges by bending. The entire cross section then carries the loads longitudinally as a deep beam, with high in-plane tension stresses at the bottom and in-plane compression stresses near the top. Because of the longitudinal bending, the ridges deflect. Therefore, the one-way slabs rest on yielding supports and

Table 18 Typical Dimensions for Folded Plates*

Span, ft	Bay† width, ft	D, ft-in. Max	D, ft-in. Min	h, in.‡	Reinforcing§
Two-segment plate, 25° < φ < 45°					
40	15	4-0	2-9	4	1.2-1.6
60	20	6-0	4-0	4-6	1.9-2.7
75	25	7-6	5-0	4-6	2.6-3.7
100	30	10-0	6-9	5-6	4.0-5.2
Four-segment plate, 30° < φ < 45°					
40	20	5-0	2-6	3	1.5-2.0
60	25	6-0	4-0	3-3.5	2.0-3.0
75	30	7-7	5-0	3-4	2.5-4.0
100	40	10-0	6-6	4-5	4.0-6.0

Two-segment plate, 25° < φ < 45°

Four-segment plate 30° < φ < 45°

* From Portland Cement Association.
† Varies with design.
‡ Average thickness.
§ In psf of projected area.
Usual ratio of depth to span varies from 1:10 to 1:15.   Cantilevers can help counterbalance the span.

their bending moments depart radically from those that would be obtained by assuming the ridges unyielding as in Fig. 31.

20. Analysis of Folded Plates* The classical method of analysis of folded plates is based upon a simplification similar to the beam-arch method for long barrels.[32] A number of computer programs, such as MULEL, MULTPL, MUPDI (direct stiffness harmonic analyses), and FINPLA (a finite-element analysis) are available.[43]

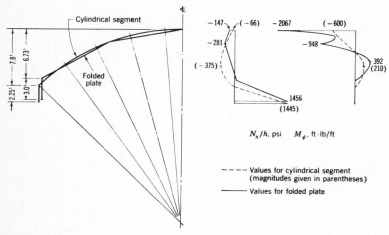

Fig. 31

The system is considered in two parts, (1) a continuous one-way slab spanning transversely between joints, and (2) a series of simple beams spanning longitudinally between end diaphragms, ribs, or trusses, etc.

1. *The primary system* consists of the plates, considered hinged at the joints, loaded with the reactions R from the continuous one-way slab analysis (Fig. 32). The resulting inplane plate loads are given by

$$P_n = -R_{n-1} \frac{\cos \phi_{n-1}}{\sin \alpha_{n-1}} + R_n \frac{\cos \phi_{n+1}}{\sin \alpha_n} \qquad (42)$$

R_n is positive downward, P_n is positive in the direction from joint n to joint $n-1$, and ϕ is positive counterclockwise from the horizontal at joint $n-1$. Furthermore,

$$\alpha_n = \phi_n - \phi_{n+1} \qquad (43)$$

The uniformly distributed loads P_n produce stresses in the beam of

$$f = \pm \frac{M_n}{Z_n} = \pm \frac{3P_n L^2}{4 h_n d_n^2} \qquad (44)$$

if the edges are free to slide longitudinally at the joints. Tension is positive. These stresses ordinarily will not be equal in the plates common to a given joint. Since they must be equal if the plates are monolithic, adjustments are required. The differences can be removed by a method of stress distribution in which, for plates of constant thickness,

$$\text{Stress stiffness factor} = \frac{4}{A_n} = \frac{4}{h_n d_n}$$
$$\text{Carryover factor} = -\tfrac{1}{2}$$

The free-edge stresses correspond to the fixed-end moments of moment distribution.

---

*A review of various methods of analysis and a comprehensive bibliography are given in Ref. 32.

2. *The errors* arise from relative rotation of the plates common to a joint, because of the in-plane deflections of the plates. With the edge stresses due to the loading P_n, the plate beam deflections y (Fig. 32c) are given by

$$y_n = (f_{n-1,n} - f_{n,n-1}) \frac{5L^2}{48d_n E} \tag{45}$$

where y_n is positive in the direction of positive P_n, and $f_{n-1,n}$ denotes the stress at joint $n - 1$ in the plate bounded by joints $n - 1$ and n (plate n). The rotation of each plate is given by

$$\delta_n^B = -\frac{1}{d_n}\left[\frac{y_{n-1}}{\sin \alpha_{n-1}} - y_n(\cot \alpha_{n-1} + \cot \alpha_n) + \frac{y_{n+1}}{\sin \alpha_n}\right] \tag{46}$$

where a positive rotation is taken as clockwise. The errors at any joint will then be

$$\delta_{n0} = \delta_n^B - \delta_{n+1}^B \tag{47}$$

The rotations given by Eqs. (46) and (47) occur at midspan and reduce to zero at the supports.

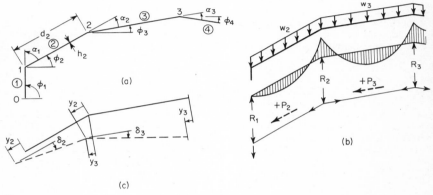

Fig. 32

3. *The corrections* consist of transverse joint moments, which must have the same longitudinal variation as the errors. Such a variation is closely approximated by a sine wave. A correction moment $X_n = 1$ at joint n produces plate rotations which consist of the sum of the rotation δ^S due to the action of the plate as a slab and the relative rotations δ^B of the plates due to their in-plane beam deflections. The rotations δ^S are given by

$$\delta_{n,n}^S = \frac{4d_n}{Eh_n^3} + \frac{4d_{n+1}}{Eh_{n+1}^3} \tag{48a}$$

$$\delta_{n-1,n}^S = \frac{2d_n}{Eh_n^3} \tag{48b}$$

$$\delta_{n+1,n}^S = \frac{2d_{n+1}}{Eh_{n+1}^3} \tag{48c}$$

The rotations δ^B are found by computing the joint reactions

$$R_n = +\frac{1}{d_n \cos \phi_n} + \frac{1}{d_{n+1} \cos \phi_{n+1}} \tag{49a}$$

$$R_{n-1} = -\frac{1}{d_n \cos \phi_n} \tag{49b}$$

$$R_{n+1} = -\frac{1}{d_{n+1} \cos \phi_{n+1}} \tag{49c}$$

from which the corresponding plate loads are found from Eq. (42). The stresses f are given by

$$f = \pm \frac{6P_n L^2}{\pi^2 h_n d_n^2} \tag{50}$$

As is the case in the primary system, these stresses ordinarily will not be equal in the plates common to a given joint. Corrections by the stress-distribution procedure mentioned previously are required.

The deflections y are found from the corrected stresses f by

$$y_n = (f_{n-1,n} - f_{n,n-1}) \frac{L^2}{\pi^2 d_n E} \tag{51}$$

Equations (50) and (51) differ from the corresponding equations (44) and (45) only because of the assumed sinusoidal variation of the correction moment.

With y known, the rotations δ^B are found from Eq. (46). The total correction at joint n is then

$$\delta_{nn} = \delta_{nn}^S + \delta_{nn}^B \tag{52}$$

in which

$$\delta_{nn}^B = \delta_n^B - \delta_{n+1}^B \tag{53}$$

4. *Compatibility of displacements* is obtained in the usual way for statically indeterminate structures. Thus, for symmetrical loading of the structure of Fig. 32a, and assuming joint 1 simple, the equations of compatibility are

$$\delta_{22}X_2 + \delta_{23}X_3 + \delta_{20} = 0$$
$$\delta_{32}X_2 + \delta_{33}X_3 + \delta_{30} = 0$$

where X_2 and X_3 are the statically indeterminate correction moments at joints 2 and 3.

5. *Final values* of the transverse moments are obtained by adding the correction moments X to those of the continuous slab analysis. Final values of the longitudinal stresses are found by adding the stresses f resulting from the moments X to those of the primary system.

6. The significant results of the analysis are the transverse bending moments and the longitudinal principal stresses. The latter are obtained by combining the stresses f with the shearing stresses v given by

$$v = \frac{4N}{h_n L} \left(1 - \frac{2x}{L}\right) \tag{54}$$

where, at the joints,

$$N_n = N_{n-1} - \frac{h_n d_n}{2} (f_{n-1,n} + f_{n,n-1}) \tag{55}$$

and, midway between joints,

$$N_{d_n/2} = \frac{N_{n-1} + N_n}{2} - \frac{h_n d_n}{8} (f_{n-1,n} - f_{n,n-1}) \tag{56}$$

Equation (54) is based on uniformly distributed loading. An equation corresponding to the assumed sine distribution of the correction loads can be derived, but since the correction loads are usually small relative to the loads on the primary system, it is sufficient for practical purposes to use Eq. (54) for both.

21. Example[*] The folded-plate system of Fig. 33 will be analyzed for a uniformly distributed load consisting of

Live load	20
Roofing, insulation, etc.	10
Plate at 4 in.	50
	80 psf

Table 19 gives the results of the one-way slab analysis, from which $R_1 = 535$, $R_2 = 908$, $R_3 = 764$ plf.

[*]From Ref. 20.

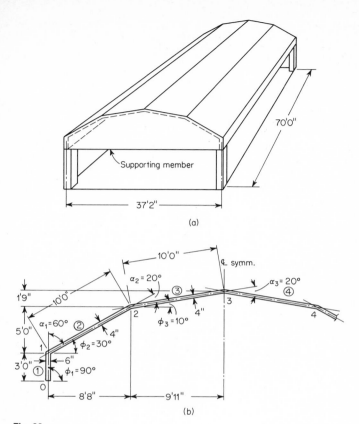

Fig. 33

TABLE 19 Elementary Analysis of One-Way Slab at Midspan

0	1	2	3	Joint
1	2	3		Plate

	1	0.428	0.572	0.500	Distribution factor
		867	−656	656	Fixed-end moment, ft-lb/ft
0	0	−90	−121		Distribute
				−60	Carryover
0	0	777	−777	596	Final moment, ft-lb/ft
0	−90	90	18	−18	$M/(d_n \cos \phi_n)$, lb/ft
225*	400	400	400	400	$wd_n/2$, lb/ft
225	310	490	418	382	Total shear, lb/ft

535	908	764 Joint reaction, lb/ft

* Weight of edge beam = wd_1.

1. PRIMARY SYSTEM. The plate loads are obtained by substituting the joint reactions R into Eq. (42), which gives $P_1 = 535$, $P_2 = 2615$, $P_3 = -97$ plf. Because of symmetry, only half the system need be considered. The resulting free-edge stresses are, from Eq. (44),

$$f_{01} = -f_{10} = +3034 \text{ psi}$$
$$f_{12} = -f_{21} = +2003$$
$$f_{23} = -f_{32} = -74$$

Distribution to balance these stresses is shown in Table 20.

TABLE 20   Stress Distribution for Elementary Analysis

0		1		2		3 Joint	
	1		2		3		Plate
0	0.69	0.31	0.5	0.5	0		Distribution factor
	−0.5	−0.5	−0.5	−0.5			Carryover factor
+3,034	−3,034	+2,003	−2,003	−74	+74		Free-edge stress*
	+3,475	−1,562	+964	−965			Distribution
−1,738		−482	+781		+482		Carryover
	−333	+149	−390	+391			Distribution
+166		+195	−75		−195		Carryover
	+135	−60	+38	−37			Distribution
−67		−19	+30		+19		Carryover
	−14	+5	−15	+15			Distribution
+7		+8	−2		−8		Carryover
	+5	−3	+1	−1			Distribution
+1,402	+234	+234	−671	−671	+372		Final stress*

* Stress in psi, tension +, compression −.

2. ERRORS. Substituting the final stresses from Table 20 into Eqs. (45), with $E = 2 \times 10^6$ psi, gives $y_{10} = +1.193$, $y_{20} = +0.277$, $y_{30} = -0.319$, $y_{40} = +0.319$ in., which, when substituted into Eq. (46), gives

$$\delta_2^B = -\frac{1}{120}\left[\frac{1.193}{\sin 60°} - 0.277(\cot 60° + \cot 20°) + \frac{-0.319}{\sin 20°}\right]$$
$$= 0.398 \times 10^{-2}$$
$$\delta_3^B = -\frac{1}{120}\left[\frac{0.277}{\sin 20°} - (-0.319)(\cot 20° + \cot 20°) + \frac{0.319}{\sin 20°}\right]$$
$$= -2.914 \times 10^{-2}$$

Then, from Eq. (47),

$$\delta_{20} = +0.398 \times 10^{-2} - (-2.914 \times 10^{-2}) = +3.312 \times 10^{-2}$$
$$\delta_{30} = -2.914 \times 10^{-2}$$

Note that, because of symmetry, joint 3 does not rotate, so that the error is only the rotation due to plate 3.

3. The *corrections* are first obtained for a unit moment at joint 2. From Eq. (48),

$$\delta_{22}^S = 2\frac{4 \times 10}{2 \times 10^6 \times 144 \times (\frac{1}{3})^3} = +0.75 \times 10^{-2}$$
$$\delta_{32}^S = \frac{2 \times 10}{2 \times 10^6 \times 144 \times (\frac{1}{3})^3} = +0.1875 \times 10^{-2}$$

The joint reactions from Eq. (49) are $R_1 = -115$, $R_2 = +217$, $R_3 = -203$ plf, and from Eq. (42), the plate loads are $P_1 = -115$, $P_2 = +625$, $P_3 = -1138$ plf. The free-edge stresses, from Eq. (50), are

$$f_{01} = -f_{10} = -530 \text{ psi}$$
$$f_{12} = -f_{21} = +388$$
$$f_{23} = -f_{32} = -704$$

Table 21 shows the stress distribution. The plate deflections are obtained by substituting the final

Example   20-47

stresses from this table into Eq. (51) to give $y_{11} = -0.985$, $y_{21} = +0.308$, $y_{31} = -0.351$ in. The plate rotations are found from Eq. (46):

$$\delta_2^B = -\frac{1}{120}\left[\frac{-0.985}{\sin 60°} - 0.308(\cot 60° + \cot 20°) + \frac{-0.351}{\sin 20°}\right]$$
$$= 2.65 \times 10^{-2}$$
$$\delta_3^B = -\frac{1}{120}\left[\frac{0.308}{\sin 20°} - (-0.351)(\cot 20° + \cot 20°) + \frac{0.351}{\sin 20°}\right]$$
$$= -3.212 \times 10^{-2}$$

The joint rotations are, from Eq. (53),

$$\delta_{22}^B = +2.65 \times 10^{-2} - (-3.212 \times 10^{-2}) = 5.862 \times 10^{-2}$$
$$\delta_{32}^B = -3.212 \times 10^{-2}$$

The total joint rotations are, from Eq. (52),

$$\delta_{22} = +0.75 \times 10^{-2} + 5.862 \times 10^{-2} = +6.612 \times 10^{-2}$$
$$\delta_{32} = +0.188 \times 10^{-2} - 3.212 \times 10^{-2} = -3.024 \times 10^{-2}$$

The same order of computations for a unit moment at joint 3 yields

$$\delta_{33}^S = +0.375 \times 10^{-2} \qquad \delta_{23}^S = +0.1875 \times 10^{-2}$$
$$R_1 = 0 \qquad R_2 = -102 \qquad R_3 = +203 \text{ plf}$$
$$P_1 = 0 \qquad P_2 = -292 \qquad P_3 = +842 \text{ plf}$$
$$f_{01} = -f_{10} = 0 \qquad f_{12} = -f_{21} = -182 \qquad f_{23} = -f_{32} = +522 \text{ psi}$$

The distribution to balance these stresses is shown in Table 22. From the final stresses in this table $y_{12} = +0.285$, $y_{22} = -0.155$, $y_{32} = +0.225$ in. The plate rotations are

$$\delta_2^B = -1.25 \times 10^{-2} \qquad \delta_3^B = +1.957 \times 10^{-2}$$

and the joint rotations are

$$\delta_{33}^B = 1.957 \times 10^{-2} \qquad \delta_{23}^B = -3.207 \times 10^{-2}$$

so that the total joint rotations are

$$\delta_{33} = 2.332 \times 10^{-2} \qquad \delta_{23} = -3.019 \times 10^{-2}$$

4. The equations of compatibility are

$$6.612X_2 - 3.02X_3 + 3.312 = 0$$
$$-3.02X_2 + 2.332X_3 - 2.914 = 0$$

from which $X_2 = 0.172$, $X_3 = 1.472$ ft-kips/ft.

TABLE 21   Stress Distribution for Correction Moment $X_2 = 1$

0		1		2		3 Joint
	1		2		3	Plate
0	0.69	0.31	0.5	0.5	0	Distribution factor
	-0.5	-0.5	-0.5	-0.5		Carryover factor
-530.4	+530.4	+387.8	-387.7	-704.0	+704.0	Free-edge stress*
	-98.5	+44.2	-158.1	+158.2		Distribution
+49.2		+79.0	-22.1		-79.1	Carryover
	+54.5	-24.5	+11.1	-11.0		Distribution
-27.2		-5.5	+12.2		+5.6	Carryover
	-3.8	+1.7	-6.1	+6.1		Distribution
+1.9		+3.0	-0.9		-3.0	Carryover
	+2.1	-0.9	+0.4	-0.5		Distribution
-1.0		-0.2	+0.4		+0.2	Carryover
	-0.2		-0.2	+0.2		Distribution
-507.5	+484.5	+484.5	-551.0	-551.0	+627.7	Final stress*

* Stress in psi, tension +, compression −.

These corrections are entered in Table 23 to determine the final values of the transverse bending moments. Final values of the longitudinal stresses are computed in Table 24. The uncorrected stresses are from Table 20, while the corrections are stresses from Tables 21 and 22 proportional to the respective correction moments. The final stresses in this table enable the longitudinal shears to be determined from Eqs. (55) and (56). These are entered in Table 25, together with the corresponding stresses from Eq. (54).

Principal stresses throughout the system are given in Table 26. The stress trajectories are plotted in Fig. 34.

TABLE 22   Stress Distribution for Correction Moment $X_3 = 1$

0	1		2		3	Joint
	1		2		3	Plate
0	0.69	0.31	0.5	0.5	0	Distribution factor
	−0.5	−0.5	−0.5	−0.5		Carryover factor
		−181.5	+181.5	+521.5	−521.5	Free-edge stress*
	−125.2	+56.3	+170.0	+170.0		Distribution
+62.6		−85.0	−28.2		+85.0	Carryover
	−58.6	+26.4	+14.1	−14.1		Distribution
+29.3		−7.0	−13.2		+7.0	Carryover
	−4.8	+2.2	+6.6	−6.6		Distribution
+2.4		−3.3	−1.1		+3.3	Carryover
	−2.3	+1.0	+0.5	−0.6		Distribution
+1.2		−0.3	−0.5		+0.3	Carryover
	−0.2	+0.1	+0.3	−0.2		Distribution
+95.5	−191.1	−191.1	+330.0	+330.0	−425.9	Final stress*

* Stresses in psi, tension +, compression −.

TABLE 23   Transverse Bending Moments at Midspan, Ft-lb/Ft

	Joint 1	Plate 2	Joint 2	Plate 3	Joint 3
Elementary analysis.........	0	+478	−777	+299	−596
Correction.................	0	−86	−172	−822	−1,472
Final value................	0	+392	−949	−523	−2,068

TABLE 24   Longitudinal Stresses at Midspan*

Joints	0	1	2	3
Elementary analysis........	0	0	0	0
Uncorrected stresses........	+1,402	+234	−671	+372
Corrections:				
$X_2 = 0.172$ ft-kip........	−87	+83	−95	+108
$X_3 = 1.472$ ft-kips.......	+141	−281	+485	−627
Final stress...............	+1,456	+36	−281	−147

* Stresses in psi, tension +, compression −.

Reinforcement   The pattern of reinforcement is shown in Fig. 35. The requirements are based on the principal stresses of Table 26. Along the support, starting at joint 1 in plate 1 and proceeding to the longitudinal centerline, the required reinforcement at 45° is

Plate 1, Joint 1:

$$A_s = \frac{129 \times 6 \times 12}{20,000} = 0.463 \text{ in.}^2/\text{ft}$$

Plate 2:

Joint 1: $A_s = \dfrac{193 \times 4 \times 12}{20,000} = 0.463$ in.^{2}/ft, No. 4 at 5 in.

Middle: $A_s = \dfrac{179 \times 48}{20,000} = 0.43$ in.^{2}/ft, No. 4 at 5 in.

Joint 2: $A_s = \dfrac{123 \times 48}{20,000} = 0.295$ in.^{2}/ft, No. 4 at 8 in.

Plate 3:

Middle: $A_s = \dfrac{51 \times 48}{20,000} = 0.122$ in.^{2}/ft, No. 3 at 10 in.

Joint 3: $A_s = 0$

Next, the steel required along joint 1 is computed. The angle of the line of principal stress is always $45° \pm 15°$ to very near midspan, so that we may place the steel at $45°$.

At support: $A_s = 0.463$ in.^{2}/ft, No. 4 at 5 in.

At $L/8$: $A_s = \dfrac{153 \times 48}{20,000} = 0.368$ in.^{2}/ft, No. 4 at 7 in.

At $L/4$: $A_s = \dfrac{111 \times 48}{20,000} = 0.267$ in.^{2}/ft, No. 4 at 9 in.

At $3L/8$: $A_s = \dfrac{68 \times 48}{20,000} = 0.164$ in.^{2}/ft, No. 4 at 15 in.

TABLE 25 Longitudinal Shearing Stresses at Midspan

Location	0	01*	1		12	2	23	3
Final shear force, kips......	0	−119	−161		−150	−102	−43	0
Final shear stress, psi.......	0	−95	−128†	−192‡	−179	−121	−51	0

* Denotes point midway between joints 0 and 1.
† Stress in plate 1 ($h = 6$ in.).
‡ Stress in plate 2 ($h = 4$ in.).

TABLE 26 Principal Stresses, psi, Tension +, Compression −

Plate	Location	$x = 0$	$L/8$	$L/4$	$3L/8$	$L/2$
3	Joint 3	0	0	0	0	0
		0	−64	−110	−138	−147
	Middepth	+51	+13	+4	+1	0
		−51	−107	−164	−199	−214
	Joint 2	+123	+45	+17	+4	0
		−123	−167	−227	−268	−281
2	Joint 2	+123	+45	+17	+4	0
		−123	−167	−227	−268	−281
	Middepth	+179	+110	+56	+16	0
		−179	−164	−146	−130	−122
	Joint 1	+193	+153	+111	+68	36
		−193	−137	−83	−34	0
1	Joint 1	+129	+105	+79	+53	+36
		−129	−89	−51	−19	0
	Middepth	+95	+341	+564	+700	+745
		−95	−15	−4	0	0
	Joint 0	0	+635	+1,090	+1,362	+1,456
		0	0	0	0	0

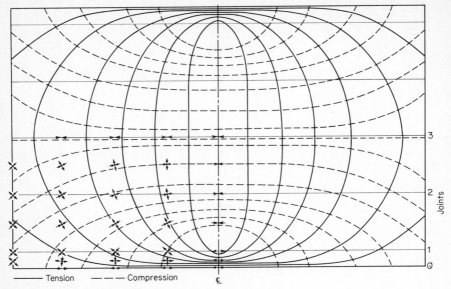

Tension ———— Compression   ¢

Fig. 34   Stress trajectories in developed folded plate of Fig. 33.

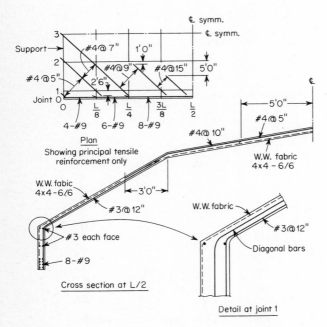

Fig. 35

From $7L/16$ to $L/2$, the minimum reinforcement required by ACI 318 is sufficient. For simplicity, these bar spacings are continued through to the support, even though somewhat less reinforcement would suffice.

In plate 2 at $L/2$ there is a longitudinal tension of

$$T = \frac{f_{12}^2}{f_{12} - f_{21}} \frac{h_2 d_2}{2} = \frac{36^2}{36 + 281} \frac{4 \times 120}{2} = 985 \text{ lb}$$

for which a minimum welded-wire-fabric reinforcement, as required for slabs in ACI 318, Sec. 7.12.2, of $0.0018 \times 12h = 0.087$ in.²/ft each way, is sufficient. The fabric 4×4–$6/6$ is provided.

In plate 1 at $L/2$ the total tensile force is

$$T = \frac{f_{01} + f_{10}}{2} h_1 d_1 = \frac{1456 + 36}{2} \times 6 \times 36 = 161 \text{ kips}$$

and $A_s = {}^{161}\!/\!20 = 8.05$ in.², for which eight No. 9 bars are provided in the bottom 16 in. A minimum tension reinforcement of $0.0035 \times 12h = 0.126$ in.²/ft is provided in the remaining portion by two No. 3 bars in each face.

The tension forces at other sections in plate 1 are

$$T_x = 4 \frac{x}{L} \left(1 - \frac{x}{L}\right) T_{L/2}$$

from which $T_x = 151$ kips at $3L/8$, 121 kips at $L/4$, and 71 kips at $L/8$. These are all nearly horizontal. The forces will be taken by eight No. 9, six No. 9, and four No. 9 bars, respectively. The four No. 3 and four No. 9 bars are carried into the support.

Reinforcement for Transverse Moments. At the middle of plate 2, $M = +392$ ft-lb/ft (Table 23) and $h = 4$ in. With a cover of ½ in., $d = 3$ in. Then, with $f_s = 20,000$ psi and $f'_c = 4000$ psi,

$$A_s = \frac{392 \times 12}{20,000 \times \frac{7}{8} \times 3} = 0.09 \text{ in.}^2/\text{ft}$$

for which No. 3 at 12 in. is provided at the bottom of the plate. At joints 2 and 3, respectively,

$$A_s = \frac{949 \times 12}{20,000 \times \frac{7}{8} \times 3} = 0.217 \text{ in.}^2/\text{ft}$$

$$A_s = \frac{2,068 \times 12}{20,000 \times \frac{7}{8} \times 3} = 0.473 \text{ in.}^2/\text{ft}$$

for which No. 4 at 10 in. and No. 4 at 5 in. are provided. At the middle of plate 3 the moment is 55 percent of that at joint 2, so that half the No. 4 bars coming from joint 3 are cut off at that point.

The longitudinal distribution of these moments is not well defined in the analysis. For cylindrical-segment thin shells, it appears that the moment drops off toward the supports somewhat more slowly than a sine curve. In this case we shall provide the full amount of steel in the center half of the span and reduce it to roughly 75 percent in the outer quarter span lengths. The reinforcement is shown in Fig. 35.

Influence of Span. Figure 36 shows how the longitudinal stresses from in-plane plate bending and transverse moments from one-way slab bending are influenced by the span L. The dashed line in Fig. 36c shows values of M for unyielding ridge lines, which is the limiting case as $L \to 0$. When $L = 35$ ft, the moments have approximately the same values as for $L = 0$, but for $L = 105$ ft the maximum moments are from 2.5 to 6.5 times the values for $L = 0$. On the other hand, the in-plane stresses (Fig. 36b) exhibit the same general behavior whether the span is 35 or 105 ft, with relatively high tension at the bottom and much lower compression near the top.

Significance of Joint Rotations. The effects of joint rotations ordinarily cannot be neglected in monolithic concrete folded plates, as is clearly shown in the example. They should be neglected only where previous analysis by more rigorous theory has demonstrated that it is safe to do so.[32] The relative magnitude of the corrections depends on the longitudinal and transverse rigidities of the component plates as well as on the shape of the structure.

Figure 36 compares the uncorrected slab moments M and beam stresses f for the structure of Fig. 33 with the values corrected for plate rotation. Values for the 70-ft span are from Tables 23 and 24. It is noted that transverse behavior approximates the ideal behavior fairly well for the 35-ft span but differs drastically as the span increases. This is because the increased flexibility of the edges permits a large relative deflection between

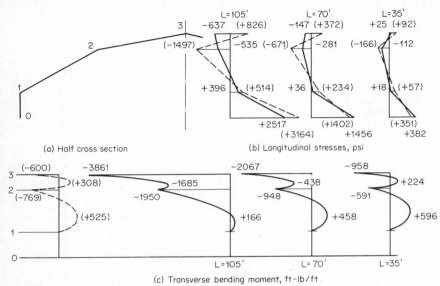

(a) Half cross section   (b) Longitudinal stresses, psi

(c) Transverse bending moment, ft-lb/ft

Fig. 36   Longitudinal stresses and transverse bending moments for folded plate of Fig. 33. Dashed lines (values in parentheses) denote plate rotation ignored; final values shown in solid lines.

the exterior and interior joints, which gives rise to large negative moments at the crown. Longitudinal beam action is concentrated in the region close to the edge for the 35-ft span but tends to spread over the entire cross section as the span increases.

In Fig. 30b, each plate tends to act in an ideal manner except for the edge plates, where there tends to be larger relative deflection between exterior and interior joints, with a resulting increase in negative moment at the first interior ridge and some deviation from longitudinal beam stresses in the exterior plates. However, ideal behavior is approached if an edge beam is used.

Where joints are not monolithic, as is usually the case in folded-plate structures in metal and wood, the analysis can be simplified to that of a one-way slab under surface loads and a series of inclined beams acted upon by the loads P_n.

Transverse Frames.   The T-beam behavior of the transverse frame is similar to that of the transverse arch for cylindrical shells (Art. 18). The behavior of the longitudinally continuous folded-plate structure is also similar to that of the continuous cylindrical shell.

Other Configurations.   Folded plates of other cross sections are common, e.g., Fig. 30b and c. For such cases, where there are many folds, the analysis presented in this article would lead to a large number of equations of compatibility. However, the interior plates will normally behave nearly as beams, which enables the analysis to be simplified. The analysis can be completed after one moment distribution, as in Table 19, and, with the resulting values of R and P, one computation from Eq. (44) for the beam stresses. Only the exterior plates need be analyzed by stress distribution and compatibility of plate rotations. For this simplification in analysis, it is necessary to establish a joint interior to which the beam method is applicable. For trapezoidal cross sections with four plates per repeated unit, it is suggested in Ref. 44 that the beam method be used for units interior to the sixth joint.

22. Continuous Folded Plates When folded plates are continuous over transverse supports, design practice consists of determining stresses by using the ratios of continuous beam moments to simple beam moments. For example, for a folded plate with the cross section shown in Fig. 36a, continuous for two spans of 70 ft each, the midspan tension of 1456 psi at the lower edge (Fig. 36b) would be reduced to 728 psi (ratio of $wL^2/16$ to $wL^2/8$). The lower-edge compression at the interior support would be 1456 psi (the same as the midspan tension for a simple span of 70 ft) because the negative moment at the interior support of a fully loaded two-span continuous beam is $wL^2/8$. Results of more accurate analyses confirm this practice as a reasonable basis for design.[33]

23. Prestressed Folded Plates If the longitudinal tensile stress in a folded-plate system is large, prestressing can be used to advantage (Fig. 37). The analysis proceeds as in Art. 21. The equivalent uniform load for a prestressing force F in a parabolic profile at an eccentricity e is $p = 8Fe/L^2$. The resulting free-edge plate bending stresses are given by Eq. (44). They must be brought into balance by stress distribution as in Table 20. The axial compressive stresses $f = F/A = F/hd$ must be balanced separately by a second distribution. This is because they are constant throughout the span, while the bending stresses vary parabolically.

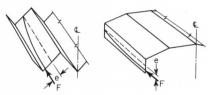

Fig. 37

With the stresses in balance, deflections y are computed, using Eq. (45) for the bending stresses and the following equation for the uniform compression:

$$y_n = (f_{n-1,n} - f_{n,n-1}) \frac{L^2}{8d_nE}$$

The plate rotations for each set of stresses are determined next from Eq. (46) and the errors at the joints from Eq. (47).

Results of prestressing the edge beam of the folded plate of Art. 21 with a force of 146 kips are given in Table 27. They show the prestressing essentially eliminates the in-plane

TABLE 27

	DL and LL	Prestress	Combined
f_3	-147	-83	-230
f_2	-281	$+85$	-196
f_1	34	-31	3
f_0	1456	-1456	0
M_3	-2067	$+954$	-1113
M_{32}	-523	$+888$	$+365$
M_2	-984	$+822$	126
M_{21}	$+392$	$+411$	$+803$
M_1	0	0	0

tension and reduces substantially the maximum negative transverse moments, but increases the positive moment in plate 2.

TRANSLATION SHELLS OF DOUBLE CURVATURE

24. Membrane Theory Figure 38 shows the stress resultants on a differential element. The relations between them and their projections on a horizontal plane are given by

$$N'_x = \overline{N}_x \frac{\cos\theta}{\cos\phi} \qquad N'_y = \overline{N}_y \frac{\cos\phi}{\cos\theta} \qquad N'_{xy} = \overline{N}_{xy} \tag{57}$$

where $\tan\phi = \partial z/\partial x$ and $\tan\theta = \partial z/\partial y$.

The horizontal projections are given by

$$\overline{N}_x = \frac{\partial^2 F}{\partial y^2} - \int \overline{p}_x dx \qquad \overline{N}_y = \frac{\partial^2 F}{\partial x^2} - \int \overline{p}_y dy \qquad \overline{N}_{xy} = -\frac{\partial^2 F}{\partial x \partial y} \tag{58a}$$

where F is a stress function which satisfies the equation

$$\frac{\partial^2 F}{\partial x^2}\frac{\partial^2 z}{\partial y^2} - 2\frac{\partial^2 F}{\partial x \partial y}\frac{\partial^2 z}{\partial x \partial y} + \frac{\partial^2 F}{\partial y^2}\frac{\partial^2 z}{\partial x^2} = q \tag{58b}$$

in which

$$q = -\overline{p}_z + \overline{p}_x \frac{\partial z}{\partial x} + \overline{p}_y \frac{\partial z}{\partial y} + \frac{\partial^2 z}{\partial x^2}\int \overline{p}_x dx + \frac{\partial^2 z}{\partial y^2}\int \overline{p}_y dy \tag{58c}$$

In Eq. (58), $\overline{p}_x$, $\overline{p}_y$, and $\overline{p}_z$ are loads per unit of area of the horizontal projection of the surface. The relation between these components and the load p_x, p_y, and p_z per unit of area of the shell surface is

$$\frac{\overline{p}_x}{p_x} = \frac{\overline{p}_y}{p_y} = \frac{\overline{p}_z}{p_z} = \frac{\sqrt{1 - \sin^2\phi \sin^2\theta}}{\cos\phi\cos\theta} = \sqrt{1 + \tan^2\phi + \tan^2\theta} \tag{59}$$

Where only uniform vertical loads need be considered, $\overline{p}_x = \overline{p}_y = 0$, so that $q = -\overline{p}_z$ and Eq. (58) are greatly simplified. Furthermore, the vertical load can be assumed to be uniform over the horizontal projection for fairly flat shells, so that Eq. (59) is eliminated, i.e., $\overline{p}_z = p_z$.

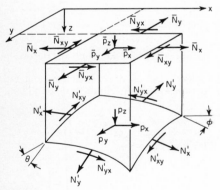

Fig. 38

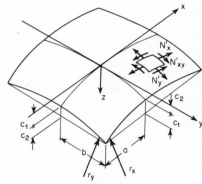

Fig. 39   Elliptical paraboloid shell.

25. Elliptic Paraboloids   The elliptic paraboloid (Fig. 39) is defined by

$$z = c_1 \frac{x^2}{a^2} + c_2 \frac{y^2}{b^2} \tag{60}$$

This surface intersects a horizontal plane in an ellipse. Sections cut by vertical planes parallel to the xz and yz planes are parabolas.

Substitution of Eq. (60) into Eq. (58b) yields

$$\frac{2c_2}{b^2}\frac{\partial^2 F}{\partial x^2} + \frac{2c_1}{a^2}\frac{\partial^2 F}{\partial y^2} = q$$

Once a stress function F which satisfies this equation is found, the membrane stress resultants can be determined by Eqs. (57) and (58a). Various solutions are possible, depending upon the assumptions relative to the boundaries.[20] Most important is the case where all four edges are supported so that they are subjected only to in-plane shears N'_{xy} (Fig. 39). In this case, the shell needs only four point supports, with edge members to carry the shear. Derivation of the stress resultants is given in Ref. 34:

$$N'_x = -\frac{\bar{p}_z a^2 k}{c_1} \times \text{coefficient} \tag{61a}$$

$$N'_y = -\frac{\bar{p}_z b^2}{k c_2} \times \text{coefficient} \tag{61b}$$

$$N'_{xy} = -\frac{\bar{p}_z ab}{\sqrt{c_1 c_2}} \times \text{coefficient} \tag{61c}$$

where $k^2 = \dfrac{1 + (2c_1/a)^2 (x/a)^2}{1 + (2c_2/b)^2 (y/b)^2}$

$\bar{p}_z$ = uniform load per unit area of horizontal projection

For shallow shells, k approaches unity and usually may be taken as 1 in practical applications since N'_x and N'_y are not large.

Values of the coefficients in these equations are given in Table 28. Table 29 gives values of the coefficients to determine the edge shears at closer intervals. It will be noted that these shears tend to become infinitely large at the corners. This means that transverse shearing forces and bending moments develop so that edge shears can remain finite. It is suggested in Ref. 34 that N'_{xy} can be considered to be maximum at

$$x = a - 0.4 \sqrt{r_x h} \qquad y = b - 0.4 \sqrt{r_y h} \tag{62}$$

where r_x and r_y are the corner radii (Fig. 39). This assumption is based on the fact that edge disturbances damp out rapidly.

The Geckeler approximation (Art. 9) may be used for an approximate evaluation of the bending near the edges. If the edge arch is much stiffer than the shell, the shell edge can be assumed fixed. Thus, for the edges $x = \pm a$,

$$w = -2\beta \frac{r^2}{Eh} Q_{xa} - 2\beta^2 \frac{r^2}{Eh} M_{xa} + \frac{\bar{p}_z r^2}{Eh} = 0$$

$$\frac{dw}{dx} = 2\beta^2 \frac{r^2}{Eh} Q_{xa} + 4\beta^3 \frac{r^2}{Eh} M_{xa} = 0$$

In these equations, the terms involving the edge shear Q_{xa} and the edge moment M_{xa} are obtained from Table 8. The last term in the first equation is the radial (hoop) displacement due to $\bar{p}_z$. The solution is

$$Q_{xa} = \frac{\bar{p}_z}{\beta} \qquad M_{xa} = -\frac{\bar{p}_z}{2\beta^2} \tag{63a}$$

With the edge shear and moment known, moments at points interior to the boundary can be determined from Table 8. Thus

$$M_x = \frac{C_3}{\beta} Q_{xa} + C_4 M_{xa} = \frac{\bar{p}_z}{\beta^2} \left(C_3 - \frac{C_4}{2} \right) \tag{63b}$$

If the torsional stiffness of the supporting arch is small, the shell edge can be assumed to be simply supported, in which case $w = M_{xa} = 0$, so that

$$Q_{xa} = \frac{\bar{p}_z}{2\beta} \tag{64a}$$

TABLE 28 Coefficients for Computing Stress Resultants in Elliptical Paraboloids [Eqs. (61)]*

Value of y/b

x/a	Stress resultants	(a) $c_1/c_2 = 1.0$					(d) $c_1/c_2 = 0.8$				
		0	0.25	0.50	0.75	1.0	0	0.25	0.50	0.75	1.0
0.00	N_y	0.250	0.233	0.182	0.101	0	0.289	0.270	0.213	0.119	0
	N_x	0.250	0.267	0.318	0.399	0.500	0.211	0.230	0.287	0.381	0.500
	N_{xy}	0	0	0	0	0	0	0	0	0	0
0.25	N_y	0.267	0.250	0.199	0.111	0	0.304	0.285	0.228	0.130	0
	N_x	0.233	0.250	0.301	0.389	0.500	0.196	0.215	0.272	0.370	0.500
	N_{xy}	0	0.029	0.068	0.096	0.108	0	0.034	0.069	0.100	0.114
0.50	N_y	0.318	0.301	0.250	0.150	0	0.347	0.331	0.277	0.169	0
	N_x	0.182	0.199	0.250	0.350	0.500	0.153	0.169	0.223	0.331	0.500
	N_{xy}	0	0.068	0.140	0.210	0.244	0	0.065	0.139	0.215	0.255
0.75	N_y	0.399	0.389	0.350	0.250	0	0.416	0.406	0.369	0.270	0
	N_x	0.101	0.111	0.150	0.250	0.500	0.084	0.094	0.131	0.230	0.500
	N_{xy}	0	0.096	0.210	0.356	0.465	0	0.091	0.201	0.353	0.480
1.0	N_y	0.500	0.500	0.500	0.500	0	0.500	0.500	0.500	0.500	0
	N_x	0	0	0	0	0	0	0	0	0	0
	N_{xy}	0	0.108	0.243	0.465	∞	0	0.101	0.229	0.443	∞

x/a	Stress resultants	(b) $c_1/c_2 = 0.6$					(e) $c_1/c_2 = 0.4$				
		0	0.25	0.50	0.75	1.0	0	0.25	0.50	0.75	1.0
0.00	N_y	0.336	0.316	0.252	0.143	0	0.395	0.374	0.307	0.180	0
	N_x	0.164	0.184	0.248	0.357	0.500	0.105	0.126	0.193	0.320	0.500
	N_{xy}	0	0	0	0	0	0	0	0	0	0
0.25	N_y	0.348	0.329	0.267	0.155	0	0.403	0.383	0.319	0.192	0
	N_x	0.152	0.171	0.233	0.345	0.500	0.097	0.117	0.181	0.308	0.500
	N_{xy}	0	0.031	0.067	0.103	0.120	0	0.026	0.060	0.101	0.125
0.50	N_y	0.383	0.367	0.312	0.197	0	0.425	0.410	0.357	0.235	0
	N_x	0.117	0.133	0.188	0.304	0.500	0.075	0.090	0.143	0.265	0.500
	N_{xy}	0	0.060	0.132	0.216	0.265	0	0.049	0.115	0.208	0.274
0.75	N_y	0.436	0.426	0.392	0.296	0	0.459	0.451	0.419	0.331	0
	N_x	0.064	0.074	0.108	0.204	0.500	0.041	0.049	0.081	0.169	0.500
	N_{xy}	0	0.081	0.185	0.342	0.494	0	0.065	0.156	0.316	0.506
1.00	N_y	0.500	0.500	0.500	0.500	0	0.500	0.500	0.500	0.500	0
	N_x	0	0	0	0	0	0	0	0	0	0
	N_{xy}	0	0.089	0.208	0.413	∞	0	0.070	0.173	0.363	∞

x/a	Stress resultants	(c) $c_1/c_2 = 0.2$				
		0	0.25	0.50	0.75	1.0
0.00	N_y	0.462	0.446	0.388	0.248	0
	N_x	0.038	0.054	0.112	0.252	0.500
	N_{xy}	0	0	0	0	0
0.25	N_y	0.465	0.451	0.396	0.261	0
	N_x	0.035	0.049	0.104	0.239	0.500
	N_{xy}	0	0.014	0.040	0.088	0.128
0.50	N_y	0.473	0.462	0.414	0.303	0
	N_x	0.027	0.038	0.086	0.197	0.500
	N_{xy}	0	0.027	0.074	0.174	0.280
0.75	N_y	0.485	0.480	0.456	0.383	0
	N_x	0.015	0.020	0.044	0.117	0.500
	N_{xy}	0	0.034	0.098	0.246	0.510
1.00	N_y	0.500	0.500	0.500	0.500	0
	N_x	0	0	0	0	0
	N_{xy}	0	0.038	0.108	0.262	∞

*From Ref. 34.

TABLE 29   Shear along Edges of Elliptical Paraboloids [Eq. (61c)]*

y/b	c_1/c_2				
	1.0	0.8	0.6	0.4	0.2
			At $x = \pm a$		
0.0	0.0000	0.0000	0.0000	0.0000	0.0000
0.1	0.0419	0.0389	0.0342	0.0307	0.0137
0.2	0.0854	0.0793	0.0701	0.0550	0.0286
0.3	0.1319	0.1231	0.1096	0.0872	0.0481
0.4	0.1836	0.1721	0.1546	0.1254	0.0731
0.5	0.2432	0.2294	0.2081	0.1728	0.1075
0.6	0.3204	0.3066	0.2859	0.2493	0.1818
0.7	0.4071	0.3897	0.3627	0.3173	0.2296
0.8	0.5363	0.5178	0.4887	0.4400	0.3443
0.85	0.6279	0.6090	0.5791	0.5292	0.4306
0.9	0.7570	0.7378	0.7074	0.6667	0.5659
0.95	0.9777	0.9582	0.9276	0.8763	0.7741
1.0	∞	∞	∞	∞	∞

x/a			At $y = \pm b$		
0.0	0.0000	0.0000	0.0000	0.0000	0.0000
0.1	0.0419	0.0444	0.0468	0.0488	0.0500
0.2	0.0854	0.0903	0.0950	0.0990	0.1014
0.3	0.1319	0.1391	0.1460	0.1519	0.1553
0.4	0.1836	0.1930	0.2019	0.2095	0.2140
0.5	0.2432	0.2545	0.2652	0.2743	0.2798
0.6	0.3204	0.3317	0.3425	0.3516	0.3571
0.7	0.4071	0.4213	0.4348	0.4463	0.4532
0.8	0.5363	0.5515	0.5659	0.5782	0.5855
0.85	0.6279	0.6434	0.6582	0.6707	0.6782
0.9	0.7570	0.7728	0.7878	0.8005	0.8081
0.95	0.9777	0.9935	1.0087	1.0215	1.0290
1.0	∞	∞	∞	∞	∞

*From Ref. 34.

With the edge shear known, moments at interior points are determined from Table 8. Thus,

$$M_x = C_3 \frac{\overline{p_z}}{2\beta^2} \tag{64b}$$

Example Given $a = 40$ ft, $b = 50$ ft, $c_1 = 8$ ft, $c_2 = 10$ ft, $h = 3$ in., $p_z = 60$ psf. From Eqs. (61), assuming $k = 1$,

$$N'_x = -\frac{60 \times 40^2}{8} = -12{,}000 \times \text{coefficient}$$

$$N'_y = -\frac{60 \times 50^2}{10} = -15{,}000 \times \text{coefficient}$$

$$N'_{xy} = -\frac{60 \times 50 \times 40}{\sqrt{80}} = -13{,}400 \times \text{coefficient}$$

Stress resultants along the edge $x = a = 40$ ft will be calculated. From Table 28,

$$N'_x = 0 \qquad N'_y = -0.5 \times 15,000 = -7500 \text{ lb/ft}$$

The edge shears N'_{xy} are determined from the coefficients in Table 29, following which principal tensions are calculated from

$$N' = \frac{N'_y}{2} \pm \sqrt{\left(\frac{N'_y}{2}\right)^2 + N^2_{xy}}$$

The results are shown in Table 30.

From Eq. (60)

$$\frac{\partial z}{\partial y} = \frac{2c_2 y}{b^2} \qquad \frac{\partial^2 z}{\partial y^2} = \frac{2c_2}{b^2}$$

from which the radius of curvature r_y is

$$r_y = \frac{[1 + (\partial z/\partial y)^2]^{3/2}}{\partial^2 z/\partial y^2} = 125 \left[1 + \left(\frac{y}{125}\right)^2\right]^{3/2}$$

At the corner, $r_y = 125 \times 1.16^{3/2} = {}^\cdot 56$ ft, and from Eq. (62) the edge shear can be considered maximum at

$$y = 50 - 0.4\sqrt{156 \times 0.25} = 47.5 \text{ ft} = 0.95b$$

Thus, according to Table 30, the largest principal tension along the edge is 9.64 kips/ft.

Reinforcement for the tension at the corners is usually placed diagonally. The controlling tension is usually that at the edge, but principal stresses at several interior points should also be computed to determine the extent of the area to be reinforced. Minimum reinforcement of at least $0.0018bh$ for welded-wire fabric (ACI 318, Sec. 7.12.2) should be supplied throughout.

The transverse shear and moment along the edge $x = a = 40$ ft can be estimated by Eq. (63) or (64), depending upon the assumption as to relative stiffnesses of the shell and its supporting arch. The corner radius r_x is found to be 125 ft, from which $\beta = 0.24$. Equation (63a) gives $Q_{xa} = 250$ lb/ft and $M_{xa} = -520$ ft-lb/ft. However, if the edge is assumed to be simply supported, $Q_{xa} = 125$ lb/ft [Eq. (64a)] and the corresponding maximum moment is, from Eq. (64b)

$$M_x = 0.32 \frac{60}{2 \times 0.24^2} = 167 \text{ ft-lb/ft}$$

The maximum value 0.32 of C_3 is from Table 8, and corresponds to $\beta s = 0.8$. Therefore, this maximum moment is located about $0.8/0.24 = 3.3$ ft from the edge.

The supporting arches should be designed for the edge shears N'_{xy}, as in the case of the barrel shell (Art. 18).

26. Hyperbolic Paraboloids These shells have a simplicity of geometry and a potential for shape that is economical and attractive. The simplicities, however, may mislead the designer into geometries or details that violate good practice in reinforced concrete. Specifically, if the rise is too slight, compressions are high and moments due to creep can cause instabilities. Also, simplified procedures for determining stresses often provided no criterion for reinforcement except for the minimum dictated by good practice.

Many hypars designed on the basis of membrane theory have performed well. However, numerical analyses of bending have shown that membrane-theory stresses can be misleading at times, especially with regard to edge beams. For example, Schnobrich has shown that in a 3-in. hypar roof shaped as in Fig. 42d, spanning 80 ft on a side and with a crown rise of 8 ft, the axial force in the horizontal ridge beams at midspan is between 5 and 35 percent of that computed by the membrane theory, depending upon the beam size.[36] (See also Sec. 1, Fig. 101.)

TABLE 30   Shear and Principal Tension, Kips per Ft, on Edge $x = a$

y/b	0	0.1	0.2	0.3	0.4	0.5	0.6	0.7	0.8	0.85	0.90	0.95
N'_{xy}	0	-0.52	-1.06	-1.65	-2.31	-3.08	-4.11	-5.23	-6.95	-8.17	-9.90	-12.85
N'	0	$+0.04$	$+0.15$	$+0.35$	$+0.65$	$+1.10$	$+1.82$	$+2.68$	$+4.14$	$+5.24$	$+6.84$	$+9.64$

The design of hypars must be based on a study of successful structures as well as those that have experienced difficulties.[37,38,39]

The equation of the hyperbolic paraboloid is (Fig. 40a)

$$z = c_2 \frac{y^2}{b^2} - c_1 \frac{x^2}{a^2}$$

(65)

For x (or y) constant, z describes a parabola in a plane parallel to yz (or xz), so that the surface can be generated by translating the parabola EOF along the parabola GOH.

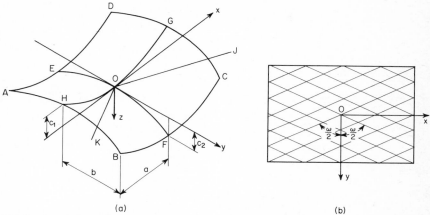

(a)   (b)

Fig. 40   Hyperbolic paraboloid shell.

The surface contains two sets of straight generators; OJ in Fig. 40a represents one set, OK the other. The projections of these generators on the xy plane are shown in Fig. 40b, where ω is given by

$$\tan^2 \frac{\omega}{2} = \frac{c_2}{c_1} \frac{a^2}{b^2}$$

(66)

Alternatively, the surface can be referred to axes x,y parallel to the projections of the straight generators (Fig. 41a). The equation is

$$z = \frac{c_0}{a_0 b_0} xy$$

(67)

In this position, the surface can be considered to be generated by allowing the line OJ to rotate around OK as it moves from OJ to KL, always remaining parallel to the xz plane.

Orthogonal Straight Boundaries.   Where the boundaries of the shell are straight generators, and the latter are orthogonal, Eq. (67) can be substituted into Eq. (58b) of Art. 24, which gives

$$\frac{\partial^2 F}{\partial x \partial y} = \frac{q a_0 b_0}{2 c_0}$$

Once a stress function F which satisfies this equation is found, the membrane stress resultants can be determined by Eqs. (57) and (58a). Various solutions are possible, depending upon the assumptions for boundary conditions.[20] When the load is $\bar{p}_z$ uniform over the horizontal projection of the surface, a solution of practical interest is (Fig. 41b)

$$\overline{N}_{xy} = -\frac{\bar{p}_z a_0 b_0}{2 c_0} \qquad \overline{N}_x = \overline{N}_y = 0$$

(68a)

The edges are subjected only to shearing-stress resultants, which must be resisted by edge members. The projected principal-stress resultants are inclined 45° with the edges:

$$\overline{N}_1 = -\overline{N}_2 = \frac{\overline{p}_z a_0 b_0}{2c_0} \tag{68b}$$

Various combinations of the surface of Fig. 41b can be made. Continuing it into the other three quadrants gives the saddle surface in Fig. 42a. Figure 42b shows four

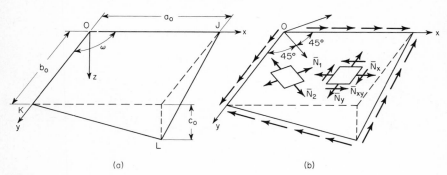

(a)   (b)

Fig. 41

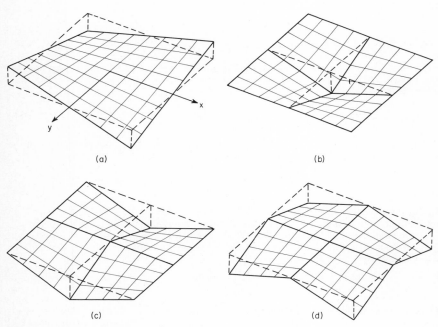

(a)   (b)

(c)   (d)

Fig. 42 Hyperbolic paraboloid roof surfaces.

quadrants combined to produce a surface which is usually called an inverted umbrella. Other combinations are shown in c and d.

Skewed Surfaces. If the surface is bounded by straight generators and $\omega \neq 90°$ (Fig. 41a), a load $\overline{p}_z$ uniform over the horizontal projection can be carried by

$$\overline{N}_{xy} = \overline{N}_{yz} = -\frac{\overline{p}_z a_0 b_0}{2c_0} \sin \omega \qquad (69a)$$

$$\overline{N}_x = \overline{N}_y = 0 \qquad (69b)$$

where $\overline{N}_{xy} = \overline{N}_{yz}$ are shearing stress resultants on an element bounded by the (nonorthogonal) projections on the plane xy of the straight generators. $\overline{N}_x$ and $\overline{N}_y$, being parallel to x and y, respectively, are nonorthogonal.

The corresponding principal stress resultants in the surface lie approximately in vertical, orthogonal planes, one of which bisects the quadrant xy,

$$N_1' = -N_2' = \frac{\overline{p}_z a_0 b_0}{2c_0} \sin \omega \qquad (69c)$$

Unsymmetrical Load. If some quadrants of the surfaces in Fig. 42 are unloaded, stress resultants for the loaded quadrants can be determined by Eq. (68). However, this results in unbalanced forces in certain edge members. If only two quadrants of the surface in Fig. 42d are loaded (Fig. 43), the edge-member forces F must be reacted. Unless there is external restraint at these points, F must be equilibrated in each case by the other half of the edge member. This subjects the unloaded panels to shearing forces along their edges. This problem is discussed in Ref. 35.

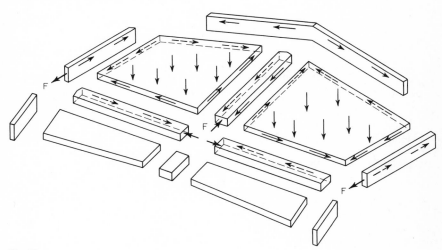

Fig. 43

Example* For the inverted umbrella shown in Fig. 44, $a_0 = b_0 = 20$ ft and $c_0 = 5.5$ ft. The weight of the 3-in. shell is 37.5 psf, to which is added 5 psf to account for the weight of the edge beams. The live load is 30 psf. Then, from Eqs. (68)

$$-\overline{N}_{xy} = \overline{N}_1 = -\overline{N}_2 = \frac{72.5 \times 20 \times 20}{2 \times 5.5} = 2640 \text{ lb/ft}$$

The required reinforcement in the direction of $\overline{N}_1$ is

$$A_s = \frac{2640}{20,000} = 0.132 \text{ in.}^2/\text{ft}$$

Minimum reinforcement in the orthogonal direction for temperature and shrinkage is $0.002 \times 36 = 0.072$ in.²/ft (ACI 313-7.12.2).

*From Ref. 35.

For easier placement, steel is often placed along the straight-line generators. Assuming $A_{sx} = A_{sy}$, Eq. (72) gives, since $\theta = 45°$,

$$A_{sx} = A_{sy} = \frac{N_p}{f_s} = \frac{2640}{20,000} = 0.132 \text{ in.}^2/\text{ft}$$

Thus, where it follows the straight-line generators, the steel required is $2 \times 0.132 = 0.264$ in.²/ft. On the other hand, only $0.132 + 0.072 = 0.204$ in.²/ft is required if it is placed along the parabolas.

The compressive stress is only

$$f_c = \frac{2640}{3 \times 12} = 74 \text{ psi}$$

The tension T at midlength of the edge members is

$$T = 2640 \times 20 = 52,800 \text{ lb}$$

for which

$$A_s = \frac{52,800}{20,000} = 2.64 \text{ in.}^2$$

The shearing forces on both sides of the sloped edge members contribute to its axial force. The compression C is

$$C = 2 \times 52,800 \times \frac{\sqrt{20^2 + 5.5^2}}{20} = 109,560 \text{ lb}$$

For a tied column with $p_g = 0.01$,

$$A_g = \frac{C}{0.8(0.225f_c' + f_s p_g)} = \frac{109,560}{540 + 160} = 157 \text{ in.}^2$$

For the triangular cross section shown in section AA of Fig. 44 the area furnished is

$$A_g = \frac{20d^2}{5.5} = 157 \text{ in.}^2$$

from which $d = 7$ in. This depth is made 9 in. in Ref. 35 to provide bending strength for unsymmetrical loading.

Nonuniform Load.   Although dead load is not uniform over the horizontal projection of sloping surfaces, it is a good approximation for shells of uniform thickness, provided the surface is not too steep. Equations (68) must be modified for steep surfaces.

The true load on the horizontal projection is given by Eq. (59) (see Fig. 38):

$$\overline{p}_z = p_z\sqrt{1 + \tan^2 \phi + \tan^2 \theta}$$

For the hyperbolic paraboloid with orthogonal straight boundaries on axes x, y [Eq. (67)],

$$\overline{N}_{xy} = -\frac{p_z a_0 b_0}{2c_0}\sqrt{1 + k^2 x^2 + k^2 y^2} \qquad (70a)$$

$$\overline{N}_x = \frac{p_z y}{2} \log \frac{kx + \sqrt{1 + k^2 x^2 + k^2 y^2}}{\sqrt{1 + k^2 y^2}} + f_1(y) \qquad (70b)$$

$$\overline{N}_y = \frac{p_z x}{2} \log \frac{ky + \sqrt{1 + k^2 x^2 + k^2 y^2}}{\sqrt{1 + k^2 x^2}} + f_2(x) \qquad (70c)$$

where $k = c_0/a_0 b_0$ and $f_1(y)$ and $f_2(x)$ are constants of integration. With $\overline{N}_{xy}$, $\overline{N}_x$, and $\overline{N}_y$ known, the stress resultants in the surface are determined from Eq. (57).

Equations (70) show that edge members are no longer subjected only to shear. However, by assigning appropriate values to f_1 and f_2, two adjoining edges can be freed of normal stress. The other two edges will be subjected to normal forces, which must be carried by the edge members or by bending in the shell.

Equations (70) cannot be used for the skewed surface, for which the solution is more difficult.[40]

Parabolic Boundaries. Possible systems of stress resultants for a surface bounded as in Fig. 40a, and with x,y directed as in that figure, are

$$\overline{N}_x = \frac{\overline{p}_z a^2}{2c_1} \qquad \overline{N}_y = \overline{N}_{xy} = 0 \tag{71a}$$

$$\overline{N}_y = -\frac{\overline{p}_z b^2}{2c_2} \qquad \overline{N}_x = \overline{N}_{xy} = 0 \tag{71b}$$

$$\overline{N}_x = \frac{\overline{p}_z a^2}{4c_1} \qquad \overline{N}_y = -\frac{\overline{p}_z b^2}{4c_2} \qquad \overline{N}_{xy} = 0 \tag{71c}$$

The first system requires anchorage of the surface at $x = \pm a$, the second at $y = \pm b$, and the third at both $x = \pm a$ and $y = \pm b$. The resultants $\overline{N}_x$ and $\overline{N}_y$ normal to the vertical planes *AHB, AED*, etc., constitute heavy load which it will usually be impracticable to support.

Combinations of the surface can be made. An example is the groined vault of Fig. 45. Reference 35 contains tables of coefficients for this structure.

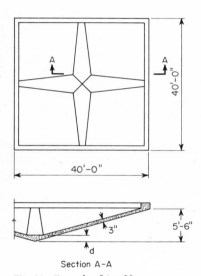

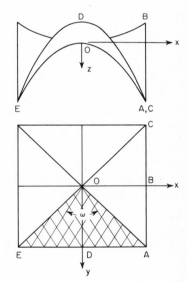

Fig. 44 Example of Art. 26.   **Fig. 45** Groined vault.

DIMENSIONING

The following comments on proportioning of concrete and reinforcement in thin shells are based on Ref. 25.

Thickness is rarely based on allowable stress, but usually on construction requirements or stability.

Reinforcement should be provided to resist completely the principal tensile stresses, assumed to act at the middle surface. It may be placed either in the general direction of the lines of principal stress or in two or three directions. In regions of high tension it is advisable to place it in the general direction of the principal stress. Whenever possible, such reinforcement may run along lines practical for construction, such as straight lines.

Reinforcement which does not deviate in direction more than 15° from the direction of the principal stress may be considered to be parallel to it. A slightly greater deviation can be tolerated in areas where the stress in the reinforcement is two-thirds, or less, of the allowable.

For reinforcement placed in two directions at right angles, as along x and y axes, and where A_{sy} forms the angle θ with the direction of the principal stress resultant N_p, reinforcement is required such that

$$N_p = f_{sy}(A_{sy} \cos^2 \theta + A_{sx} \sin^2 \theta \tan \theta) \qquad (72a)$$

$$\tan 2\theta = \frac{2N_{xy}}{N_x - N_y} \qquad (72b)$$

For positive values of $\tan 2\theta$, θ is measured counterclockwise from the face on which N_p acts.

Minimum reinforcement should be provided as required in ACI 318 even where not required by analysis.

In areas where the computed tensile stress in the concrete exceeds 300 psi, at least one layer of reinforcement must be parallel to the principal tensile stress, unless it can be proved that a deviation is permissible because of the geometrical characteristics of the shell and because, for reasons of geometry, only insignificant and local cracking could develop.

The allowable stress for reinforcement may be used at any point in the shell independently of the magnitude of the stress in the concrete at that point.

Additional reinforcement to resist bending moments must be proportioned and provided in the conventional manner. Generally, where moments are significant in thin shells, the effect of direct compression forces may be neglected. Either working-load or ultimate-load analysis may be used.

Where the computed principal tensile stress (psi) exceeds $2\sqrt{f_c'}$ (f_c' in psi), the spacing of reinforcement should not exceed three times the thickness of the shell. Otherwise, reinforcement should be spaced not more than five times the thickness of the shell, or more than 18 in.

Splices in principal tensile reinforcement should be kept to a practical minimum. Splices should be staggered, with not more than one-third of the bars spliced at any one cross section. Bars should be lapped only within the same layer. The minimum lap for draped reinforcing bars should be 30 diameters, with a minimum of 18 in. unless more is required by ACI 318, except that the minimum may be 12 in. for reinforcement not required by analysis. The minimum lap for welded-wire fabric should be 8 in. or one mesh, whichever is greater, except that ACI 318 governs where wire fabric at a splice must carry the full allowable stress.

Concrete cover should be at least ½ in. for bars, ⅜ in. where precast and for welded-wire fabric, and 1 in. for prestressing tendons, provided the concrete surfaces are protected from weather and are not in contact with the ground. In no case should the cover be less than the diameter of the bar or tendon. If greater cover is required for fire protection, it need apply only to principal tensile and moment reinforcement whose yielding would cause failure.

CONSTRUCTION

A thin shell should be as thin as is practicable in order to induce in-plane stress as opposed to bending. However, the high cost of labor in the United States usually makes it more economical to use a thicker shell than to enforce the careful casting techniques essential to construction of very thin shells. Three layers of steel are usually used, and to place concrete properly a 3-in. thickness is about minimum.

Slope of the shell should be less than 45° to avoid top forms, which increase the difficulty of eliminating honeycombing. Where the slope is 30°, concrete with a slump of 1 to 3 in. can be cast to reasonable tolerances without a top form. Where the slope exceeds 45° it may be possible to cast without top forms if the slump is low, but the cost of placing is increased. In some cases concrete has been shot onto vertical surfaces successfully.

Location of ribs can be important. Application of insulation and roofing is relatively simple when the surface is free of ribs. However, ribs projecting below the soffit complicate movement of forms. Form movement may constitute a large item of cost in continuous, cylindrical-segment roofs, in which case ribs are built above the roof. These projections must be carefully flashed. Ribless shells, i.e., shells with wide, flat ribs, have been successfully built for short spans where buckling was not a major factor in design.[29]

Maximum aggregate size should not exceed one-half the shell thickness, or the clear distance between bars, or 1½ times the cover. Where top forms are required, maximum size of aggregate should not exceed one-fifth the minimum clear distance between forms, or the cover over the reinforcement.

Forms should be carefully built. The designer should consider the structural effect of small deviations from the plans and should set tolerances. The stability of the shell depends upon its radii of curvature, and relatively small radial deviations in surface dimensions can cause large variations in radius. For example, a variation of ½ in. radially in an arc length of 10 ft in the dome of Art. 10 causes a 45 percent change in radius if the surface remains spherical.

Concrete should be cast in a symmetrical pattern to avoid bracing the scaffolding for the effect of unbalanced loads. It is recommended that concreting commence at the low point, or points, and proceed upward. Concrete should be deposited as nearly as possible in its final position. Vibration of thin sections is difficult. It has been done by vibrating the reinforcement or the forms, but these systems must be rigid to withstand such treatment. Vibrating screeds have been successfully used.

Thin shells are susceptible to shrinkage cracking if curing of the concrete is faulty. In hot weather, the use of retarders, preliminary fog-spray curing, and wet burlap or water curing is advisable. In cold weather, accelerators and special precautions against freezing are usually required. In moderate weather (40 to 70°), ordinary methods such as membrane-curing compounds are usually satisfactory, although wet curing may produce better results.

The method of form removal (decentering) is usually specified by the designer, in order particularly to avoid any unwanted temporary supports or concentrated reactions on the shell. It is best to begin decentering at points of maximum deflections and progress toward points of minimum deflection, with decentering of edge members proceeding simultaneously with that of the adjoining shell. It is important to control deflections at the time of decentering, and it is common to specify a modulus of elasticity that must be obtained before permission to decenter is granted. Small, lightly reinforced beams tested in flexure have been used successfully to determine E.[41]

Thin shells of dramatic shape and lightness, which can be successfully built in warm, dry climates, cannot always be duplicated in harsher regions. Furthermore, it is not easy to predict whether a particular construction scheme will be economical.

REFERENCES

1. Kalinka, J. E.: Monolithic Concrete Construction for Hangars, *Mil. Eng.*, January–February 1940.
2. Koiter, W. T.: A Consistent First Approximation in the General Theory of Thin Elastic Shells, *Proc. IUTAM Symp. Theory of Thin Elastic Shells* (Delft), North-Holland Publishing Company, Amsterdam, 1959.
3. Bouma, A. L., A. C. Van Riel, H. Van Koten, and W. J. Beranek: Investigations of Models of Eleven Cylindrical Shells Made of Reinforced and Prestressed Concrete, *Proc. Symp. Shell Research* (Delft), North-Holland Publishing Company, Amsterdam, 1961.
4. Timoshenko, S. P., and S. Woinowsky-Krieger: "Theory of Plates and Shells," 2d ed., McGraw-Hill Book Company, New York, 1959.
5. Pflüger, Alf: "Elementary Statics of Shells," 2d ed. (English translation by Ervin Galantay), McGraw-Hill Book Company, New York, 1961.
6. Medwadowski, S., W. C. Schnobrich, and A. C. Scordelis (eds.): "Concrete Thin Shells," ACI Publication SP-2B, Detroit, 1971.
7. Flugge, W.: "Stresses in Shells," Springer-Verlag OHG, Berlin, 1960.
8. Timoshenko, S. P., and J. M. Gere: "Theory of Elastic Stability," 2d ed., McGraw-Hill Book Company, New York, 1961.
9. Collected Papers on Instability of Shell Structures, 1962 NASA TN-D1510, 1962.
10. Schmidt, H.: Ergebnisse von Beulversuchen mit doppelt gekrummten Schalen-modellen aus Aluminium, *Proc. Symp. Shell Research* (Delft), North-Holland Publishing Company, Amsterdam, 1961.
11. Wang, Y-S., and D. P. Billington: Buckling of Cylindrical Shells by Wind Pressure, *J. Eng. Mech. Div. ASCE*, October 1974.
12. Kundurpi, P. S., et al.: Stability of Cantilever Shells under Wind Loads, *J. Eng. Mech. Div. ASCE* October 1975.
13. Cole, P. C., et al.: Buckling of Cooling-Tower Shells: State of the Art, *J. Struct. Div. ASCE*, June 1975.

14. Der, T. J., and R. Fidler: A Model Study of the Buckling Behavior of Hyperbolic Shells, *Proc. Inst. Civil Eng.*, vol. 41, London, September 1968.
15. Ewing, D. J. F.: The Buckling and Vibration of Cooling Tower Shells, Part II: Calculations, *Lab. Rept.* RD/L/R 1764, Central Electricity Research Laboratories, Leatherhead, England, November 1971.
16. Krätzig, W., Statische und dynamische Stabilitat der Kühlturmschale, *Konstr. Ingenieurbau Ber.*, vol. 1, Vortage der Tagung Naturzug-Kühlturme, Haus der Technik, Essen, Germany, Apr. 19, 1968.
17. Cole, P. P., J. F. Abel, and D. P. Billington: Buckling of Cooling-Tower Shells: Bifurcation Results, *J. Struct. Div. ASCE*, June 1975.
18. Billington, D. P., and J. F. Abel: Design of Cooling Towers for Wind, *Proc. Specialty Conf. ASCE*, Madison, Wis., Aug. 22–25, 1976.
19. Hanna, M. M.: Thin Spherical Shells under Rim Loading, Fifth Congress IABSE, Lisbon, 1956.
20. Billington, D. P.: "Thin Shell Concrete Structures," McGraw-Hill Book Company, New York, 1965.
21. Wind Forces on Structures, *Trans. ASCE*, vol. 126, part II, 1961.
22. Niemann, H. J.: "Zur stationären Windbelastung rotationssymmetrischer Bauwerke im Bereich Transkritischer Reynoldszahlen," Technischwissenschaftliche Mitteilung, no. 71-2, des Instituts für Konstruktiven Ingenieurbau der Ruhr-Universität, Bochum, 1971.
23. Sollenberger, N. J., and Robert H. Scanlan: Pressure Differences across the Shell of a Hyperbolic Natural Draft Cooling Tower, *Proc. Int. Conf. Full Scale Testing of Wind Effects*, London, Ontario, June 1974.
24. Reinforced Concrete Cooling Tower Shells—Practice and Commentary, Report by ACI-ASCE Committee 334, *J. ACI*, January 1977.
25. Concrete Shell Structures, Practice and Commentary, *J. ACI*, September 1964.
26. Chinn, J.: Cylindrical Shell Analysis Simplified by Beam Method, *J. ACI*, May 1959.
27. Design of Cylindrical Concrete Shell Roofs, "ASCE Manual of Engineering Practice 31," New York, 1952.
28. Parme, A. L., and H. W. Conner: Discussion of Ref. 26, *J. ACI*, December 1959.
29. Tedesko, A.: Multiple Ribless Shells, *J. Struct. Div. ASCE*, October 1961.
30. Gibson, J. E.: "The Design of Cylindrical Shell Roofs," 2d ed., D. Van Nostrand Company, Inc., Princeton, N.J., 1961.
31. Design Constants for Circular Arch Bents, *Adv. Eng. Bull. 7*, Portland Cement Association, Chicago, 1963.
32. Phase I Report of the Task Committee on Folded Plate Construction, *J. Struct. Div. ASCE*, December 1963.
33. Pultar, M., et al.: Folded Plates Continuous over Flexible Supports, *J. Struct. Div. ASCE*, October 1967.
34. Parme, A. L.: Shells of Double Curvature, *Trans. ASCE*, vol. 123, 1958.
35. Elementary Analysis of Hyperbolic Paraboloid Shells, Portland Cement Association, Chicago, 1960.
36. Schnobrich, W. C.: Analysis of Hipped Roof Hyperbolic Paraboloid Structures, *J. Struc. Div. ASCE*, July 1972.
37. Students Clear Gym Moments Before Roof Fails, *Eng. News-Rec.*, Sept. 24, 1970.
38. 15-Year-Old H.P. Roof Fails, Injuring 18, *Eng. News-Rec.*, July 10, 1975.
39. Tedesko, A.: Shell at Denver—Hyperbolic Paraboloid Structure of Wide Span, *J. ACI*, October 1960.
40. Candela, F.: General Formulas for Membrane Stresses in Hyperbolic Paraboloid Shells, *J. ACI*, October 1960.
41. Tedesko, A.: Construction Aspects of Thin Shell Structures, *J. ACI*, February 1953.
42. Building Code Requirements for Minimum Design Loads in Buildings and Other Structures, American National Standards Institute, Inc. New York, 1972.
43. "Concrete Thin Shells," ACI Publication SP-28, Detroit, 1971.
44. Whitney, C. S., B. G. Anderson, and H. Birnbaum: Reinforced Concrete Folded Plate Construction, *J. Struct. Div. ASCE*, October 1959.

Suspension Roofs

LEV ZETLIN
Zetlin-Argo Corporation, New York, N.Y.

I. PAUL LEW
Associate, Lev Zetlin Associates, Inc., New York, N.Y.

Suspension roofs are tension structural systems which have negligible flexural stiffness and depend solely on adjusting their geometry and tensile resistance to carry loads. They are of two types: cable structures and membrane structures. A cable is a linear structural element that is always under tension distributed uniformly over its cross section. A suspension membrane is a planar structural layer that can only resist tensile forces within its own plane.

The high-strength cable is approximately six times as strong as structural steel, which results in less weight of structural material but costs only about twice as much per pound. Furthermore, it is as easy to string a 400-ft cable as a 50-ft one. These factors are conducive to economy of construction. Suspension structures also offer new forms for architectural design of buildings.

In a suspension roof the cables themselves and their erection usually represent the smaller portion of its cost; the larger portion is in the fittings, their connections, and the anchorage members. Because of this, suspension roofs are not likely to be economical for spans less than 150 ft unless the design and method of erection eliminate fittings and anchorage at the end of each cable, as when a continuous cable is used to traverse a smaller span several times. With large spans, however, cables become very economical, since the cost of fittings per unit of area covered is reduced. For example, a 1000-ft span might require the same number of fittings as a 150-ft span and cost no more to erect. This emphasizes an important potential of long-span suspension roofs: their cost per unit of covered area tends to decrease with increasing span. On the other hand, the cost of conventional framing increases with increasing span.

Tensile membranes can be reinforced with cables to span large distances. The advantage and economy of the membrane tensile structure lie in its high degree of transportability and its ability to provide inherently waterproof roof systems. The tensile membrane also has controllable translucency.

1. Examples of Suspension Structures *Cable Systems.* The Utica Municipal Auditorium, Utica, N.Y. (Fig. 1), a circular building of 240-ft clear span, employs a system similar to Fig. 11a. The cables are anchored into a concrete ring. Prestressing of the cables was achieved by jacking apart the two central steel rings. Optimum curvature of the cables required to eliminate flutter and minimize deflections due to eccentric loads was obtained by the vertical struts between layers of cables. The entire roof was prefabricated and erected in 2½ weeks, using only one tower as scaffolding. Mechanical and air-conditioning equipment was placed between the layers of cables.

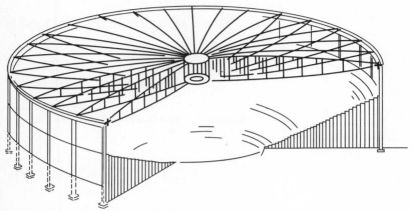

Fig. 1   Utica Auditorium. (Gehron & Seltzer, Architects. Lev Zetlin Associates, Inc., Structural Engineers.)

The roof of the New York State Pavilion at the 1964–1965 New York World's Fair (Fig. 2) is oval in plan with a major axis of 350 ft and a minor axis of 260 ft. The system is similar to Fig. 11b. The cables are anchored into a steel ring. The entire roof floats on lubrite plates supported on steel brackets which protrude from 16 exterior towers. The ring and cable suspended system were all erected on the ground. The entire suspension system was assembled and prestressed within a few days. The entire roof was lifted into its final position.

One of the largest prestressed cable suspension roofs in the world is the Salt Lake Civic County Auditorium (Fig. 3). It has a diameter of 360 ft and utilizes a system similar to the one used on the New York State Pavilion, but incorporates an integral double compression ring to anchor the top and bottom cables.

The Travelers Insurance Company Pavilion at the 1964–1965 New York World's Fair (Fig. 4) is an example of efficient utilization of cable roofs. In this case the tension roof reduces stresses in the steel supporting structure below the roof. The building is 160 ft in diameter. The system consists of 24 boomerang-shaped prefabricated steel ribs. The ribs were assembled in place with a single temporary support at the outer edge of each boomerang. The tops of the boomerangs were connected by pair-sets of cables of the type shown in Fig. 11a. The tension required to keep the cables dynamically stable was adequate to lift the ribs off their temporary supports. To effect further economies and stiffness, equatorial cables were wrapped around the ribs. Spreading the cables away from the ribs at the equator induced a horizontal force in the ribs, which reduced the bending moments in the ribs. Each rib ties into a secondary compression ring. Each cable is socketed to the top of one rib, runs through a saddle, is connected to a tension plate, and returns to an adjoining rib. This eliminates a large number of costly sockets and fittings.

Membrane Systems.   The Valley Curtain (Fig. 5) is the largest membrane structure yet constructed. The curtain was supported by a set of four 3-in. cables that span between two cliffs almost a quarter of a mile apart. The structural nylon fabric for the curtain was lifted up to the main cables by a top boundary cable. The fabric was unfurled and attached to the ground over 250 ft below via earth anchors and a bottom boundary cable along the curtain.

DESIGN OF SUSPENSION SYSTEMS

The design of suspension structures differs from design of conventional structural components mainly in two respects: anchorage forces and dynamic behavior.

2. Anchorage Forces Tension in cables can be resisted by (1) anchoring the ends into foundation abutments, (2) anchoring the ends into a continuous structural member, such

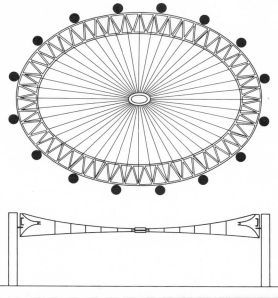

Fig. 2 New York State Pavilion, 1964–1965 New York World's Fair. (Philip Johnson Associates, Architects. Lev Zetlin Associates, Inc., Structural Engineers.)

as a ring or similar closed structural shape which can resist anchorage loads within itself, and (3) choosing the geometry of the structure so that the structure itself offers the resistance to the tension anchorage forces; i.e., stability of the structure requires external forces which coincide in magnitude and direction with the anchorage tension forces

Fig. 3 Salt Lake County Civic Auditorium. (Bonneville, Architects. Hoffman C. Hughes, Engineer. Lev Zetlin Associates, Inc., Structural Engineers.)

created by the cables (Fig. 4). The first method is the most expensive, while the third is not only the most economical but, in many cases, can reduce the construction cost of a structure below that of a conventional frame.

3. Dynamic Behavior Suspension structures require consideration of their dynamic behavior. Individual suspended cables and grids of cables are observed from time to time

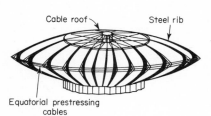

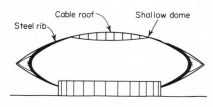

Fig. 4 Travelers Insurance Company Pavilion, 1964–1965 New York World's Fair. (Kahn and Jacobs, Architects. Lev Zetlin Associates, Inc., Structural Engineers.)

to exhibit a motion usually called "flutter." This can result from exterior dynamic forces such as wind and mobile loads and, in extreme cases, from sound waves or from vibrations in the ground set up by vehicular traffic.

4. Single Cable under Uniformly Distributed Load A curved cable is a basic component of any suspension structure. Since the configuration corresponding to a uniformly distributed load occurs quite often in suspension structures, an understanding of its behavior under static as well as dynamic loads is basic.

The solid line in Fig. 6 represents the initial configuration of an unloaded hanging cable attached to immovable supports at its ends. Since the cable will assume a parabolic configuration under uniform load, the initial curve could also be assumed to be a parabola. The assumed initial configuration has no effect on further analysis. The problem is to find the deflected (dotted) configuration, the force in the cable, and the forces on the supports. It is assumed that the uniformly distributed load acts vertically and that the supports at each end of the cable are on the same horizontal line. Notation is given in Art. 8.

Static Behavior. The initial length L of the cable can be expressed accurately enough for design purposes by

$$L = l \left[1 + \frac{8}{3} \left(\frac{f}{l} \right)^2 \right] \tag{1}$$

The tension T is given by

$$T = \frac{ql^2}{8f} \sqrt{1 + 16 \left(\frac{f}{l} \right)^2} \tag{2}$$

The approximate elastic elongation of the cable is

$$\Delta L = \frac{TL}{EA} \tag{3}$$

The increase in sag is

$$\Delta f = \frac{\Delta L}{^{16}\!/_{15}(f/l)[5 - 24(f/L)^2]} \tag{4}$$

The angle α is given by

$$\tan \alpha = \frac{4(f + \Delta f)}{l} \tag{5}$$

The vertical and horizontal reactions are

$$V = T \sin \alpha \tag{6}$$
$$H = T \cos \alpha \tag{7}$$

Dynamic Behavior.   The frequency W_n of vibration of a suspended cable is given by

$$W_n = \frac{n\pi}{l} \sqrt{\frac{Tg}{q}} \tag{8}$$

where $n = 1, 2, 3, \ldots, \infty$. The fundamental mode is given by $n = 1$. The frequency of the fundamental mode is the *natural frequency*.

Suppose a cable at rest is subjected to an externally applied pulsating force having a frequency W_e. If one plots the ratio W_e/W_n along the horizontal in Fig. 7, the increase in amplitude of the modes of vibration will be as shown by the ordinates of the two-branched curve. Of significant concern is the range between points a and b, where the amplitudes increase rapidly. When $W_e/W_n = 1$, the amplitude is infinite (resonance). This means that a cable of any strength would break. Within the range of values between a and b, serious destructive effects could be experienced. If the ratio W_e/W_n is to the left of a or to the right of b, the cable is not in danger and can be considered stable.

Theoretical treatment of flutter of an entire roof and of its dynamic instability is extremely complex. Since the problem is akin to resonance in vibrations, the complete solution requires exact knowledge of the externally applied dynamic loads as well as the internal dynamic properties, such as natural frequencies, of the entire system. Nevertheless, simplifications are possible.

One approach in designing suspension structures is to stay outside the range ab of Fig. 7 by increasing the mass q/g [Eq. (8)]. This is why some suspension structures have been built with concrete placed on top of the cables, while in others guy wires have been strung from the suspension roof into the foundation. However, the designer has no control over the frequencies of the externally applied loads (in most cases designers cannot even predict them). Therefore, since the value of W_e/W_n cannot be predicted, there can be no assurance that the resulting structure will fall outside the range ab of Fig. 7.

If the cable is damped, the increase in amplitude is represented by the dashed line of Fig. 7. It is seen that the amplitudes are always controlled; that is, there is no possibility of resonance or prolonged oscillations. In general, a sufficiently damped cable, when displaced, will return directly to its initial configuration without oscillations. Two interconnected curved cables, each with a predetermined natural frequency, are equivalent to a damped system of two cables (Fig. 10). With such a system, the designer does not have to predict or be concerned with the frequency or nature of the external dynamic loads, or the natural frequency of the entire suspension roof.

5. Configuration and Shapes of Suspension Structures   Many forms of suspension roofs are possible, and there are a variety of configurations of cables that can be used. If adequately treated in the conceptual design stage, structural suspension systems offer numerous architectural forms, not only for roofs but for the entire building. The most common types are:

Catenary.   The most elementary structural suspension system is a catenary similar to that used in suspension bridges. This system requires end towers and abutments to resist

Fig. 5   Valley Curtain. (Christo, Artist. Lev Zetlin Associates, Inc., Structural Engineers.)

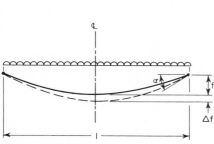

Fig. 6   Cable profile.

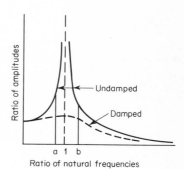

Ratio of natural frequencies

Fig. 7   Damping graph.

the tension in the catenary; it also requires a stiff structure (e.g., the suspended roof) to eliminate flutter in the cables.

Preloaded Catenary.   The preloaded catenary uses precast or cast-in-place concrete to weigh down and reduce flutter for radial cable systems. The system consists of a central tension ring and a perimeter compression ring tied together by radial cables (Fig. 8). The system does not always act as a tension system, since the cables normally are jacked to prestress the concrete roofing. This prestressing creates a rigid inverted dome rather than a flexible suspension system.

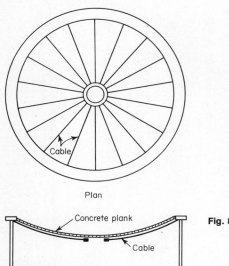

Plan

Fig. 8   Preloaded cable.

Section

Cable Grids.   To avoid flutter without imposing heavy dead weight on the structure, grids of interlacing cables are used. The surfaces created by these grids have reverse curves created by intersecting cables along the directions of principal opposite curvatures. Typically, the convex cables have an initial tension and press against the concave cables as illustrated in Fig. 9. These opposite curvature surfaces are known as hyperbolic surfaces.

Membranes.   Membranes have a structural behavior similar to that of cable grids. Membranes have two-way tensile resistance and additional in-plane tear resistance. The structural membranes consist of high-strength fabrics and films. Certain tents, where configuration stability is not required, do not have doubly curved surfaces. However, for the membrane structure to be stable, double curvature is necessary.

DOUBLE LAYER OF PRESTRESSED CABLES*

The method presented here consists of an approach at damping the suspension system. The system consists of a double layer of interconnected prestressed cables. Examples are shown in Figs. 1, 2, and 3. A number of such systems which have been built demonstrate a complete absence of flutter and a high degree of rigidity. Although they are much lighter

---

*A structural system consisting of a double layer of prestressed cables, initially conceived by the senior author in 1956 and used by him and Tyge Hermansen for the Utica Municipal Auditorium, constructed in 1958. The design method presented here was subsequently developed by the senior author.

in weight, their rigidity is comparable with, or higher than, that of conventional steel trusses or girders.

6. Damped Suspension System   The fundamental features of the damped suspension system are (1) two interconnected cables, (2) initial tension in the cables, and (3) control of natural frequencies of individual cables. The possible configurations of two intercon-

Fig. 9   Cable net.

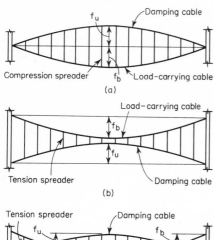

Fig. 10   Damping systems.

nected curved cables (called a pair-set) are shown in Fig. 10. The following must be considered:

1. The concave cable may be considered as the load-carrying cable and the convex cable as the damping cable. Load, such as roofing, can be applied to either cable or to both.

2. Equation (8) shows that the natural frequency of a cable depends, among other factors, upon its tension. However, the tension depends upon both the initial tension at

erection and the superimposed loads, which, in turn, can be applied in stages and are therefore variable (e.g., erection load, dead load, live load).

3. The fundamental frequencies (first mode) of the cables of a pair-set must always be different from each other. Therefore, if a designer starts with a set of two cables which have slightly different frequencies under initial tension, he should make sure that the difference increases as the tension changes because of application of superimposed loads.

4. It is seen from Fig. 10 that, when a downward load is applied to any pair-set of any configuration *a*, *b*, or *c*, the load-carrying cable will be subjected to tension and the damping cable to compression. Therefore, a pair-set should always be erected with initial tension sufficient to assure a residual tension in the damping cable at all times under all superimposed loads.

5. The initial tension helps control the natural frequencies required. The choice of initial tension depends on the magnitude of the superimposed design loads and, for economy, should be the minimum necessary to satisfy item 3.

6. Since the spreaders are installed simultaneously with the cables, they are either in compression or in tension (Fig. 10). The only load acting on the pair-set is its own weight.

7. Structural Relationships In establishing the relationships for one pair-set of cables, the following is assumed:

1. The anchorage points at each end are on the same horizontal line.

2. The pair-set is subjected to a vertical uniformly distributed design load, which implies that each cable is a parabola under all loading conditions.

3. The spreaders exert a vertical uniformly distributed load on each cable, which means that the spreaders from end to end of a pair-set represent a continuous diaphragm.

The procedure for a pair-set involves two basic stages: (1) as erected and (2) superimposed loads.

With these assumptions, the structural relationships for a pair-set of cables are established by the coupling of the load-carrying and the damping cables via the compression or tension spreaders. The coupling requires that the two cables deflect equally, which gives the basic compatibility condition. The deflection relationship for a given span is then a function of the sag and the area of the cables.

A superimposed uniformly distributed load is supported by both the load-carrying cable and the damping cable. The portion supported by the load-carrying cable increases the load in that cable while the portion supported by the damping cable decreases its load.

8. Notation The following notation is used in the interaction equations for the double-layer cable. The term "diaphragm load" refers to the load per unit length of cable induced by the spreaders. In the equations to follow, subscripts *b* and *u* refer to the load-carrying cable and damping cable, respectively.

A = area of cable
E = modulus of elasticity of cable
f = sag of cable
F_s = allowable stress in cable
g = acceleration of gravity
H = horizontal component of T at support
L = developed length of cable
l = horizontal distance between supports
p = uniformly distributed superimposed load per unit length
q = weight of cable per unit length
q_d = final diaphragm load
q_{id} = initial diaphragm load
q_w = weight of cables and spreaders per unit length
S = stress in cable
S_i = initial stress in cable
T = tension in cable at support
T_i = initial tension (erection force) in cable
V = vertical component of T at support
W_n = frequency of vibration
α = angle between horizontal and tangent to cable at support
Δf = change in sag of cable
Δq_d = change in diaphragm load per unit length
ΔL = increase in length of cable due to T

ΔS_i = change in initial stress in cable
ΔT_i = change in initial tension in cable

9. Preliminary Design of Double-Layer Cable System   A condition of loading must be assumed for preliminary sizing of the cables. The condition of full load is best suited for this purpose. Under this condition all the external loads act on the load-carrying cable. In addition, there must still be a prestress in the system to assure that the damping cable does not go slack. The prestress load on the damping cable is counterbalanced by the same load being imposed on the load-carrying cable by the spreaders. It is recommended that the final prestress diaphragm load equal at least the live load.

To eliminate flutter under all conditions, the natural frequency of the load-carrying cable must be greater than that of the damping cable. For a given span, this is achieved if the stress in the load-carrying cable is always higher than the damping-cable stress. Thus, for erection conditions

$$\frac{T_{iu}}{A_u} < \frac{T_{ib}}{A_b} \tag{9}$$

This criterion is satisfied if the erection stress of the damping cable is between 90 and 96 percent of the erection stress of the load-carrying cable. This criterion must also be satisfied for any uplift condition. The stress in the erection state in the load-carrying cable can be related to the known final load state to obtain Eq. (14). The net change Δq_d in the diaphragm load is given by Eq. (12). These two equations with Eq. (2) for cable tension enable the cable system to be sized once a geometry has been chosen. The procedure is as follows:

1. Select a geometry (l, f_b, f_u).
2. Determine the loads and assume a final prestress diaphragm load q_d.
3. Compute T_b under full load:

$$T_b = \frac{(q_w + p + q_d) \, l^2}{8 f_b} \sqrt{1 + 16 \left(\frac{f_b}{l}\right)^2} \tag{10}$$

4. Determine the required area of the load-carrying cable:

$$A_b = \frac{T_b}{F_s} \tag{11}$$

5. Solve the following three equations for A_u:

$$\Delta_{qd} = \frac{f_u^2 A_u}{f_u^2 A_u + f_b^2 A_b} \, p \tag{12}$$

$$T_u = \frac{(\Delta q_d + qd) \, l^2}{8 f_u} \sqrt{1 + 16 \left(\frac{f_u}{l}\right)^2} \tag{13}$$

$$T_u = \frac{0.96 A_u (q_w + q_d + \Delta q_d) \, T_b}{A_b} \frac{}{q_w + q_d + p} \tag{14}$$

Combining these equations after substituting the known data gives a quadratic equation in A_u. If this equation does not give a real solution, the final diaphragm force is not large enough.

With values of A_b and A_u determined, the more exact analysis which follows can be accomplished.

10. Analysis of Double-Layer Cable System   The analysis consists of two stages: (I) as erected and (II) full load.

Analysis of Stage I (as Erected).   At this stage the external load consists of only the weight of the cable system. This can be determined from the preliminary sizes and an initial diaphragm load q_{id}.

The procedure for stage I analysis is as follows:

1. Determine $q_{id} = q_d + \Delta q_d$ \hfill (15)

2. Compute

$$T_{iu} = \frac{q_{id}l^2}{8f_u} \sqrt{1 + 16 \left(\frac{f_u}{l}\right)^2} \tag{16}$$

$$T_{ib} = \frac{(q_{id} + q_w)l^2}{8f_b} \sqrt{1 + 16 \left(\frac{f_b}{l}\right)^2} \tag{17}$$

3. Determine the natural frequencies of each cable: For the load-carrying cable,

$$W_{nb} = \frac{n\pi}{l} \sqrt{\frac{T_{ib}g}{q_b}} \tag{18}$$

For the damping cable,

$$W_{nu} = \frac{n\pi}{l} \sqrt{\frac{T_{iu}g}{q_u}} \tag{19}$$

Analysis for Stage II (Design Loads).   At this stage, tension in the load-carrying cable is increased while that in the damping cable is decreased, and the diaphragm force of the spreaders changes. Design for superimposed loads may have to be performed several times, e.g., for dead load (or a number of times for partial dead loads) and then for live loads (or a number of times for partial live loads). This may be necessary to determine the natural frequencies or to ascertain forces and/or deflections at various loading conditions. Cables are not linear structural systems and do not follow the laws of superposition; so the designer must consider the sequence of loading. Load p_2 applied after load p_1 will produce different forces and deflections in the pair-set than if it is applied before p_1.

Consider the cable configuration shown in Fig. 11a.

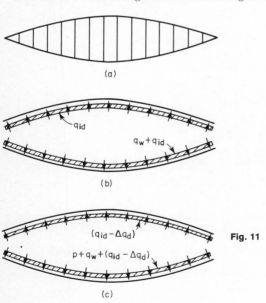

(a)

(b)

(c)   **Fig. 11**

1. In the as-erected condition the damping cable is subjected to an upward uniformly distributed load q_{id}, while the load-carrying cable is subjected to a downward uniformly distributed load $q_w + q_{id}$ (Fig. 11b).

2. When a uniformly distributed superimposed load p is applied to the load-carrying cable, the diaphragm load q_{id} becomes $q_{id} - \Delta q_d$ (Fig. 11c). Thus the damping cable is

subjected to a uniformly distributed load $q_{id} - \Delta q_d$ and the load-carrying cable to a uniformly distributed load $p + q_w + (q_{id} - \Delta q_d)$.

3. If the load p is applied to the damping cable, the damping cable is subjected to a uniformly distributed load $(q_{id} - \Delta q_d) - p$ and the load-carrying cable to a uniformly distributed load $q_w + (q_{id} - \Delta q_d)$.

The procedure for the stage II analysis is as follows:

1. Assuming the superimposed load is applied to the load-carrying cable:

$$\Delta q_d = p \frac{f_u^2 A_u}{f_b^2 A_b + f_u^2 A_u} \tag{20}$$

2. Compute tension in load-carrying cable:

$$T_b = \frac{p + q_w + (q_{id} - \Delta q_d)}{q_{id} + q_w} T_{ib} \tag{21}$$

3. Compute tension in damping cable:

$$T_u = \frac{q_{id} - \Delta q_d}{q_{id}} T_{iu} \tag{22}$$

4. Deflection of a pair-set is given by

$$\Delta f_u = \Delta f_b = \frac{(T_b - T_{ib})(L_b/EA_b)}{{}^{16}\!/_{15} (f_b/l)[5 - 24 (f_b/l)^2]} \tag{23}$$

5. Determine the natural frequencies of each cable:
   Load-carrying cable,

$$W_{nb} = \frac{n\pi}{l} \sqrt{\frac{T_b g}{q_b}} \tag{24}$$

   Damping cable,

$$W_{nu} = \frac{n\pi}{l} \sqrt{\frac{T_u g}{q_u}} \tag{25}$$

6. The angle of the cable tangent with the horizontal at the supports is given by

$$\alpha = \tan^{-1} \frac{4f}{l} \tag{26}$$

from which the components H and V of the tension in each cable are determined.

Equations (21) and (22) are based on the initial values of f_b and f_w respectively. This could be accurate enough for practical purposes. Actually, f_b and f_u change, which affects the magnitudes of T_b and T_w. If more accurate values are desired, T_b and T_u can be recomputed, taking into consideration the change in sag given by Eq. (23).

If load is applied to the damping cable, the numerators of Eqs. (20), (21), and (22) are changed to $f_b^2 A_b$, $q_w + (q_{id} - \Delta q_d)$, and $(q_{id} - \Delta q_d) - p$, respectively.

11. Behavior of Pair-Set of Cables   1. The load-carrying cable and the damping cable deflect equal amounts [Eq. (23)]. This is true even if f_b is not equal to f_w.

2. When the assembly deflects, the gain ΔT_{ib} of the load-carrying cable is not generally equal to the loss in tension ΔT_{iu} of the damping cable. In general, a pair-set of cables cannot be considered as a simply supported beam in which the bending moment due to a superimposed load is resisted by a couple consisting of changes in cable tensions. In other words, the relationship $M = \Delta T_{ib}(f_b + f_u) = \Delta T_{iu}(f_b + f_u)$ does not hold generally.

3. If $f_b = f_w$ the value of Δq_d under a uniformly distributed load p is given by

$$\Delta q_d = p \frac{A_u}{A_u + A_b}$$

4. The length L of as-erected cable having sag f_b or f_u is given by Eq. (1). As-erected cable, however, has tensions due to $q_w + q_i$; therefore, the as-erected length is longer than the fabricated length of the cable prior to erection.

5. When both cables in a pair-set are under tension, with spreaders (either tension or compression) between them as shown in cross section in Fig. 12, the two cables tend to remain in the vertical plane. In other words, if one of the cables is displaced laterally by an external force, it tends to spring back into the vertical plane. Therefore, a series of pair-sets of cables, which together form a suspension roof, do not require lateral bracing for stability. This suggests that a suspension roof consisting of a double-layer system of cables as described here can be designed and built of individual pair-sets without need of crisscrossing or interconnection in each layer. If, for example, a suspension roof of this type has identical parallel pair-sets of cables, all with uniformly distributed load, it is sufficient to design only one pair-set. On the other hand, if the roof is circular or elliptical (Figs. 1 and 2), it may be necessary to design individually each radial pair-set, since either their length or the load per pair-set is different.

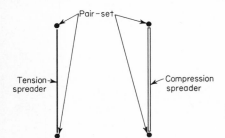

Fig. 12   Spreaders.

6. When the loads on various pair-sets are not equal, the elongations and deflections of each will be different. This is equivalent to load redistribution in conventional continuous structural frames. Basically, however, this type of suspension roof requires only design considerations for each individual pair-set.

7. If the superimposed load is not uniform, or if the force exerted by the spreaders is to be considered as a series of concentrated forces, the design calculations are somewhat more lengthy, but they are amenable to electronic-computer programming. However, if the superimposed load does not vary greatly from a uniform distribution, or if the distance between the spreaders is not too large, so that the segmental shape of the cables closely approximates a parabola, the assumptions of uniform distribution of superimposed and diaphragm loads can be used in designing many suspension roofs.

8. There must be enough residual tension in the damping cable, under the most critical superimposed load, to keep its sag between adjacent spreaders to a permissible maximum, depending on the type of roof deck used.

Example 1   Determine the cable sizes for a double-layer cable system to span 300 ft and support a live load of 120 lb/ft and a dead load of 80 lb/ft.

Additional data: $f_u = f_b = 20$ ft, system weight load $q_w = 25$ lb/ft, $F_s = 75$ ksi.

From Eq. (1),

$$L = 300 \left[1 + \frac{8}{3} \left(\frac{20}{300} \right)^2 \right] = 303.55 \text{ ft}$$

Assume the final prestress diaphragm load q_d to be 120 lb/ft. Then the tension in the load-carrying cable from Eq. (10) is

$$T_b = \frac{(25 + 200 + 120)300^2}{8 \times 20} \sqrt{1 + 16 \left(\frac{20}{300} \right)^2} = 201 \text{ kips}$$

$$A_b = \frac{201}{75} = 2.68 \text{ in.}^2$$

Use $2\frac{1}{8}$-in. steel-strand cable, $A_b = 2.71$ in.^{2}, $q_b = 9.49$ lb/ft (Table 2).

The area of the damping cable is determined from Eqs. (12), (13), and (14). With the loads in kips, and with $f_b = f_u$ and $p = 120 + 80 = 200$ lb/ft,

$$\Delta q_d = \frac{A_u \times 0.200}{A_u + 2.71}$$

$$T_u = \frac{(\Delta q_d + 0.120)300^2}{8 \times 20} \sqrt{1 + 16 \left(\frac{20}{300}\right)^2} = 582.2\Delta q_d + 69.9$$

$$T_u = \frac{0.96 A_u}{2.71} \frac{(0.025 + 0.120 + \Delta q_d)201}{0.025 + 0.120 + 0.200} = (29.9 + 206.4\Delta q_d)A_u$$

Solving these three equations for A_u gives $A_u = 2.52$ in.^{2}

Use $2\tfrac{1}{16}$-in. cable, $A_u = 2.55$ in.^{2}, $q_u = 8.94$ lb/ft. Substituting A_u in the above equation for Δq_d gives $\Delta q_d = 97$ lb/ft.

Stage I: Erection [Eqs. (15) to (19) inclusive]

$$q_{id} = 120 + 97 = 217 \text{ lb/ft}$$

$$T_{iu} = \frac{0.217 \times 300^2}{8 \times 20} \sqrt{1 + 16 \left(\frac{20}{300}\right)^2} = 126.3 \text{ kips}$$

$$T_{ib} = \frac{(0.217 + 0.025)(300)^2}{8 \times 20} \sqrt{1 + 16 \left(\frac{20}{300}\right)^2} = 140.9 \text{ kips}$$

The frequencies of the fundamental modes ($n = 1$) are:

$$W_{1b} = \frac{\pi}{300} \sqrt{\frac{140,900 \times 32.2}{9.49}} = 7.25 \text{ Hz}$$

$$W_{1u} = \frac{\pi}{300} \sqrt{\frac{126,300 \times 32.2}{8.94}} = 7.06 \text{ Hz}$$

Stage II: Full load [Eqs. (20), (21), and (22)]

$$\Delta q_d = \frac{200 \times 2.55}{2.71 + 2.55} = 97.0 \text{ lb/ft}$$

$$T_b = \frac{200 + 25 + 217 - 97}{217 + 25} \times 140.9 = 201 \text{ kips}$$

$$T_u = \frac{217 - 97}{217} \times 126.3 = 69.8 \text{ kips}$$

From Eqs. (24) and (25)

$$W_{1b} = \frac{\pi}{300} \sqrt{\frac{201,000 \times 32.2}{9.49}} = 8.65 \text{ Hz}$$

$$W_{1u} = \frac{\pi}{300} \sqrt{\frac{69,800 \times 32.2}{8.94}} = 5.25 \text{ Hz}$$

Comparing these values with those for Stage I shows that the frequences diverge. Therefore, flutter is prevented.

12. Application to Preliminary Design of Cable Grids and Membranes Cable grids and membranes must have opposing curvatures of their surface to remain stable. An important class of these surfaces is known as translation surfaces. One such surface is the hyperbolic paraboloid. The equation for this surface [Sec. 20, Eq. (65)] shows that the generators satisfy the assumptions for the analysis of the double-layer cable system; namely, the cable profiles are parabolic and are interconnected and of opposite curvature.

If the boundary is a square with its center at the origin, i.e., $c_2/b^2 = c_1/a^2$, and the sides are parallel to the x and y axes, then under uniform load all cables in the grid or unit strip in the membrane in each direction are equally loaded. Also, any boundary that is symmetric with respect to the origin will be equally loaded in each direction. Therefore, it is possible to use the procedure for preliminary sizing of the double-layer cable system to size the grid or membrane by selecting for analysis any two cables or strips of opposite curvature and equal span on the hyperbolic-paraboloid surface.

The analysis of cable grids and membranes is valid only in those areas where the deflection of the system is small with respect to the sag of the system. In those regions of the surface where the sag-span ratio is less than 1:25, deflections must be considered in the analysis. Localized loadings can result in reversal of curvature for the flat regions of the surface, and the potential for localized flutter conditions exists. For this as well as other reasons, the boundaries of the cable net are usually chosen so as to exclude the flat regions of the surface.

Example 2 A hyperbolic-paraboloid roof 100 ft on each side has a height of 30 ft. One diagonal has a rise of 15 ft, the other a sag of 15 ft. The diagonal span is $100\sqrt{2}$ ft (Fig. 13). The dead load is 100 lb/ft and the live load 200 lb/ft.

$l = 141.4$ ft $f_b = 15$ ft $f_u = 15$ ft
$p = 100 + 200 = 300$ lb/ft
Assume $q_w = 20$ lb/ft
From Eq. (1), $L = 145.64$ ft
From Eq. (10), $T_b = 94$ kips. Then with $F_s = 80$ ksi $A_b = 1.18$ in.² Use 1⁷⁄₁₆ in. diameter, $A_b = 1.24$ in.²
Equations (12), (13), and (14) give, with $f_b = f_w$

$$\Delta q_d = \frac{A_u}{3.33A_u + 4.13}$$
$$T_u = 181\Delta q_d + 36.2$$
$$T_u = 30.8A_u + 140\Delta q_d A_u$$

The solution of these three equations gives $A_u = 1.22$ in.². Use 1⁷⁄₁₆ in. diameter, $A_u = 1.24$ in.² Then $\Delta q_d = 150$ lb/ft.

With this preliminary design a more exact analysis can be made as in Example 1.

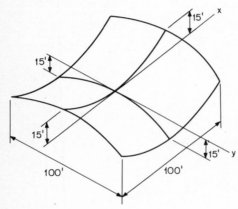

Fig. 13 Example 2. Hyperbolic paraboloid.

13. Load Combinations for Selection of Cables According to Ref. 1, the effective design breaking strength of cables should not be less than the largest value for the following cable-tension conditions: 2.2T1, 1.6T1 + 2.7T2, 2.2T3, 2.0T4, 2.0T5, 2.0T6,

where $T1$ = net tension in cable due to dead load and prestress
     $T2$ = change in cable tension due to application of live load
     $T3$ = net tension in cable due to dead load, prestress, and live load
     $T4$ = net tension in cable due to dead load, prestress, live load, and wind or earthquake forces
     $T5$ = net tension in cable during erection of structure
     $T6$ = net tension in cable due to deal load, prestress, and wind load
These safety factors are for interior benign environments. Higher safety factors should be considered for exterior and/or corrosive environments.

14. Types of Cables Bridge wire is used in two types of cables. *Strand* is an arrangement of wires helically laid around a center wire to produce a symmetrical section (Fig. 14). *Wire rope* consists of a plurality of strands laid helically around a core of either fiber rope, wire rope, or another steel strand. Strand is more compact than wire rope and therefore requires smaller fittings. Wire rope, on the other hand, is more flexible. For most building uses, the flexibility of strand is sufficient.

The requirements for tensile strength, yield strength, and elongation of galvanized bridge strand are given in Tables 1 and 2. Bridge strand is prestressed under a tension of not more than 55 percent of the tabulated breaking strength. For class A coating, the minimum modulus of elasticity is 24,000 ksi for strands of nominal diameters from ½ to

2⁹⁄₁₆ in. and 23,000 ksi for diameters 2⅝ in. and larger. These values are reduced by about 1000 ksi for heavier coatings.

15. Fittings In general, fittings should be capable of developing at least the breaking strength of the cables to which they are attached. Fittings which are commonly used are those which have been developed for bridges, towers, and aerial tramways. Saddles are

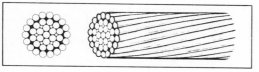

Strand

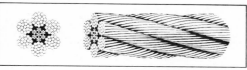

Wire rope

Fig. 14 Strand and wire rope. *(From Ref. 1.)*

TABLE 1   Properties of Galvanized Bridge Wire

Coating class	Diam, in.	Min tensile strength, ksi	Min yield strength* at 0.7 % extension under load, ksi	Min total elongation in 10 in., %
A	0.041 and over	220	160	4
B	All	210	150	4
C	All	200	140	4

* For actual cross section including zinc coating.

used where changes of curvature of the cables are encountered. They are made of cast steel and are designed expressly for each installation.

Figures 15, 16, and 17 each show the connection of a spreader to a cable. The detail in Fig. 17 accommodates the angle change between the compression spreader and the cable to which the roofing is attached. The detail shown in Fig. 18 can be used to allow a continuous cable to traverse a small span several times.

A number of different types of standard end fittings used with bridge strand and rope are shown in Table 3. Alternatively, cables can be anchored by an end rope thimble plus cable clips. This approach is normally used only for relatively small cables because the number of clips required to anchor the cable soon becomes impractical. For instance, a 2-in.-diameter cable requires eight clamps for anchorage.

16. Membranes Membrane materials can be classified as fabrics and films. A *fabric* is a material of woven threads. It has a warp direction, the direction of roll of the fabric, which is normally the stronger direction. The fabric also has a fill direction across the roll which is the more elastic direction of the fabric. Fabric strength can be measured by the threads per inch. A *film* is a thin sheet of homogeneous material. Films are temperature-dependent. At high temperatures they are weaker and more elastic and at low temperatures stronger but more brittle. Test procedures to determine the strength and fire

TABLE 2 Minimum Breaking Strength of Bridge Strand in Tons (2000 lb)

Nominal diam, in.	Class A coating throughout	Class A coating inner wires, class B coating outer wires	Class A coating inner wires, class C coating outer wires	Approx metallic area, in.^{2}	Approx weight per ft, lb
1/2	15.0	14.5	14.2	0.150	0.52
9/16	19.0	18.4	18.0	0.190	0.66
5/8	24.0	23.3	22.8	0.234	0.82
11/16	29.0	28.1	27.5	0.284	0.99
3/4	34.0	33.0	32.3	0.338	1.18
13/16	40.0	38.8	38.0	0.396	1.39
7/8	46.0	44.6	43.7	0.459	1.61
15/16	54.0	52.4	51.3	0.527	1.85
1	61.0	59.2	57.9	0.600	2.10
1 1/16	69.0	66.9	65.5	0.677	2.37
1 1/8	78.0	75.7	74.1	0.759	2.66
1 3/16	86.0	83.4	81.7	0.846	2.96
1 1/4	96.0	94.1	92.2	0.938	3.28
1 5/16	106.0	104.0	102.0	1.03	3.62
1 3/8	116.0	114.0	111.0	1.13	3.97
1 7/16	126.0	123.0	121.0	1.24	4.34
1 1/2	138.0	135.0	132.0	1.35	4.73
1 9/16	150.0	147.0	144.0	1.47	5.13
1 5/8	162.0	159.0	155.0	1.59	5.55
1 11/16	176.0	172.0	169.0	1.71	5.98
1 3/4	188.0	184.0	180.0	1.84	6.43
1 13/16	202.0	198.0	194.0	1.97	6.90
1 7/8	216.0	212.0	207.0	2.11	7.39
1 15/16	230.0	226.0	221.0	2.25	7.89
2	245.0	241.0	238.0	2.40	8.40
2 1/16	261.0	257.0	253.0	2.55	8.94
2 1/8	277.0	273.0	269.0	2.71	9.49
2 3/16	293.0	289.0	284.0	2.87	10.05
2 1/4	310.0	305.0	301.0	3.04	10.64
2 5/16	327.0	322.0	317.0	3.21	11.24
2 3/8	344.0	339.0	334.0	3.38	11.85
2 7/16	360.0	355.0	349.0	3.57	12.48
2 1/2	376.0	370.0	365.0	3.75	13.13
2 9/16	392.0	386.0	380.0	3.94	13.80
2 5/8	417.0	411.0	404.0	4.13	14.47
2 11/16	432.0	425.0	419.0	4.33	15.16
2 3/4	452.0	445.0	438.0	4.54	15.88
2 7/8	494.0	486.0	479.0	4.96	17.36
3	538.0	530.0	522.0	5.40	18.90
3 1/8	584.0	575.0	566.0	5.86	20.51
3 1/4	625.0	616.0	606.0	6.34	22.18
3 3/8	673.0	663.0	653.0	6.83	23.92
3 1/2	724.0	714.0	702.0	7.35	25.73
3 5/8	768.0	757.0	745.0	7.88	27.60
3 3/4	822.0	810.0	797.0	8.44	29.53
3 7/8	878.0	865.0	852.0	9.01	31.53
4	925.0	911.0	897.0	9.60	33.60

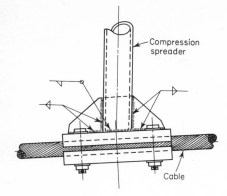

Fig. 15 Connection for compression spreader.

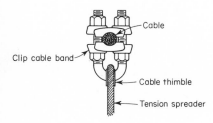

Fig. 16 Connection for tension spreader.

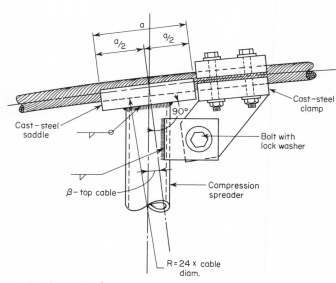

Fig. 17 Connection for compression spreader.

TABLE 3 Standard End Fittings*

Name	Description	Attach-ment	Sizes available diam. in.	
			Strand	Rope
Open socket		Poured zinc	$1/2-4$	$3/8-4$
Closed socket		Poured zinc	$1/2-4$	$3/8-4$
Open bridge socket		Poured zinc	$1/2-4$	$3/8-4$
Closed bridge socket		Poured zinc	$1/2-4$	$3/8-4$
Button socket	Bearing surface	Poured zinc	$1/2-4$	$3/8-4$
Bearing sockets	(a) / Bearing surface (b)	Poured zinc	$1/2-4$	$3/8-4$
Threaded socket	Bearing surface / Internal threads, optional	Poured zinc	$1/2-4$	$3/8-4$
Threaded stud socket	Bearing surface	Poured zinc	$1/2-4$	$3/8-4$
Bridge socket bowl	Anchor rods by others	Poured zinc	$1/2-4$	$3/8-4$
Closed swaged socket		Swaged	$1/2-1\tfrac{3}{8}$	$3/8-2$
Open swaged socket		Swaged	$1/2-1\tfrac{3}{8}$	$3/8-2$
Threaded swaged socket		Swaged	$1/2-1\tfrac{3}{8}$	$3/8-2$

*From Ref. 1. These fittings develop the breaking strength of the cable.

TABLE 4 Fabric and Film Evaluation Grid

Factors			Materials				
	Designation	Description	Nonsupported vinyl film	Vinyl-coated nylon fabric	Vinyl-coated polyester fabric	Vinyl-coated glass fabric	Teflon-coated glass fabric
I	Time effects						
	A	Life, years	3*	5*	8*	10*	20*
	B	Aging characteristic	Turns yellow with time noticeably at end of life	Turns brown with time unless tan initially	Turns brown with time unless tan initially	Turns brown with time but not as much as other fabrics unless tan initially	Bleachs white from initial tan color
	C	Cleanliness	Picks up dirt but as smooth film easier to wash off than fabrics from rain	Without special in-place coating, dirt catcher	Without special in-place coating, dirt catcher	Without special in-place coating, dirt catcher	Teflon coating does not pick up dirt and washes off easily
	D	Temperature 1. Hot 2. Cold	Loses strength Becomes brittle but can be corrected	None Minimal	None Minimal	None Minimal	None Minimal
II	Structural characteristic						
	A	Tensile strength	Very low at room temp, 20-mil film 16 lb/in.	Low, 200–400 lb/in.	Medium, 300–700 lb/in.	300–800 lb/in.	Highest, 300–1000 lb/in.
	B	Tear strength	High at high temperature, low during winter	Medium	High	Medium	Medium
	C	Dimensional stability	Minimal	Low	High	Very high	Very high

TABLE 4 Fabric and Film Evaluation Grid (Continued)

Factors			Materials				
Designation		Description	Nonsupported vinyl film	Vinyl-coated nylon fabric	Vinyl-coated polyester fabric	Vinyl-coated glass fabric	Teflon-coated glass fabric
III							
Construction and cost	D	Special characteristic	Essentially nonstructural			Requires special handling because more brittle	Requires special handling because more brittle
	A	Ease of fabrication	Very good	Good	Very good	Good	Requires more time
	B	Ease of handling	Good	Good	Good	Needs care because of brittleness	Needs care because of brittleness
	C	Cost, $/sq ft, 2 layers	Lowest	Low	Medium	Medium	High
IV							
Light transmission	A	Translucency per layer	Excellent, 30–40% transmission	Poor, general light for seeing, 8% transmission	Acceptable light for seeing, 12% transmission	Good light for seeing, 13% transmission	Good, 20–25% transmission
	B	Foster plant-growth support	Possible	No	Only in combination with artificial lights	Only in combination with artificial lights	Marginally, shade plants
V							
Fire resistance	A	Level of fire protection	By additives that decrease translucency, meets code	Requires more additives in vinyl coating to meet code	Requires additives in vinyl coating	Need less fire protection in vinyl coating than other no-glass fabrics to meet codes	Fire-rated, fire-resistant
	B	Toxic-gas emission	Not dangerous but noticeable	Less emission	Less emission	Minimal	None
	C	Hot drippings	Less than other	Yes	Yes	None	None

*± 10 percent.

resistance of films and fabrics have been established by ASTM and the Underwriters' Laboratories.

The five commonly used membrane materials for structures are nonsupported vinyl film, vinyl-coated nylon fabric, vinyl-coated polyester fabric, vinyl-coated fiberglass fabric, and Teflon-coated fiberglass fabric. The characteristics of each are evaluated in Table 4.

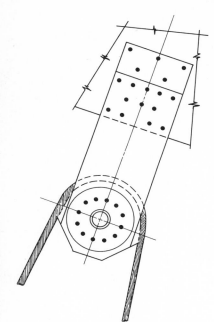

Fig. 18

The purpose of the coatings on fabrics is to protect the strength of the fabric under exposure to sunlight and the elements and to improve its fire resistance.

REFERENCES

1. "Manual for Structural Applications of Steel Cables for Buildings," American Iron and Steel Institute, Washington, D.C., 1973.
2. Krishna, Prem: "Cable-Suspended Roofs," McGraw-Hill Book Company, New York, 1978.

Section **22**

Reinforced-Concrete Bunkers and Silos

GERMAN GURFINKEL
Professor of Civil Engineering, University of Illinois, Urbana

1. Introduction A bin is an upright container for the storage of bulk granular materials. Shallow bins are usually called bunkers; tall bins are called silos. Typical bunkers are shown in Fig. 1. Vertical cross sections of silos are shown in Fig. 2. While bins may be erected individually, they are commonly grouped for increased efficiency in operation. Typical configurations of grouped silos are shown in Fig. 3.

Janssen[1] in 1895 was the first to develop a theory to predict pressures from stored material at rest. Design using his formulas produced acceptable structures for many years. Theimer[2] notes, however, that early designers used low allowable stresses that rendered a factor of safety of about 2.5. With the use of higher-strength steel and concrete, together with higher allowable stresses, the safety margin is reduced considerably. Thus old silos can withstand pressures substantially greater than those predicted by the Janssen formulas, while some newer ones may not.

A number of investigations throughout Europe and the United States have reported pressures generated during material emptying higher than those predicted by Janssen's formulas for the material at rest.[3-11] Provisions for the calculation of these effects have appeared for many years in the German Specification[12] DIN 1055, Sheet 6. In the United States, ACI Committee 313 published a standard where provisions for the calculation of overpressures are given.[13]

Pieper[14] explains emptying phenomena by identifying three zones within the cell (Fig. 4). In zone 3 the stored material is likely to move as a mass, although with different velocities across the section. In zone 2, funnel flow develops. This contrasts with mass flow in that only the central portion of the mass moves toward the outlet, within a contracting channel formed within a stagnant mass. While the latter has poorly defined boundaries in zone 2, it is sharply defined in zone 1, where the material underlying the funnel is at a complete standstill. The relative heights of the zones may be quite different, and any of the three may vanish.

The largest lateral pressures occur in region A, Fig. 4, because of the abrupt change in material cross section. In region B development of domes may inhibit material flow. These domes are short-lived. Following the collapse of a dome the entire material above falls rapidly and creates a shock. Periodic repetition of this phenomenon creates a hammering effect. In region C the firm walls of the funnel offer the necessary support for development of a large dome which may halt material flow.

Material flow is influenced by the inclination α of the hopper walls with the horizontal. Hoppers with steep inclinations ($\alpha > 70°$) promote major mass flow; flat bottoms and hoppers with gentle inclinations generate conditions for funnel flow of the material.

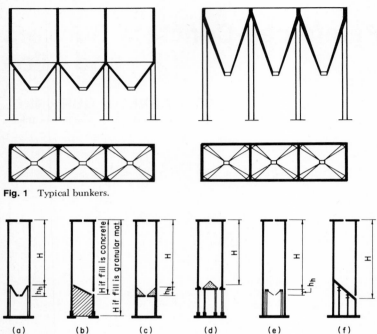

Fig. 1 Typical bunkers.

Fig. 2 Typical silos: (a) On raft foundation, independent hopper on pilasters; (b) with wall footings and independent bottom slab on fill; (c) with hopper-forming fill and bottom slab on thickened lower walls; (d) with multiple discharge openings and hopper-forming fill on bottom slab, all supported by columns; raft foundation has stiffening ribs; (e) on raft foundation with hopper on ring beam and columns; (f) on raft foundation with hopper on concrete walls and steel frames.

2. Bin Pressures Formulas by Janssen[1] and Reimbert[6] for computing bin pressures are given in Table 1. Notation is defined in the table and in Fig. 5. These formulas give pressures for the stored material at rest (static pressures). They give substantially the same results, and are commonly used in the United States.

The Janssen formulas are used in DIN 1055 (Table 1), but with two sets of values of μ' and k, one for filling pressures and one for emptying pressures, and with a coefficient c_1 by which the Janssen values must be multiplied to obtain emptying pressures. Values of c_1 are given by

$$c_1 = 1 + 0.2 \left(c_2 + \frac{eL}{1.5A} \right) \qquad (1)$$

where $c_2 = 1$ for organic materials
$c_2 = 0$ for inorganic materials
e = eccentricity of outlet, measured from centroid of cross section
L = perimeter of silo
A = cross-sectional area of silo

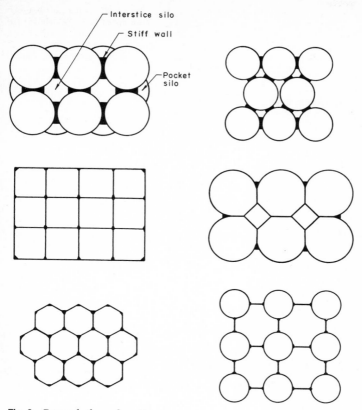

Fig. 3 Grouped silo configurations.

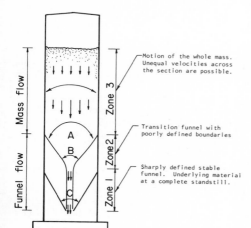

Fig. 4 Flow patterns during emptying. *(From Ref. 14.)*

22-3

Equation (1) is not to be used for sugar; instead, use $c_1 = 1$. Also, values of c_1 by Eq. (1) must be multiplied by 1.3 for corn. This is because European experience shows that corn may contain substantial amounts of dust and broken kernels, which probably results from the number of loadings and unloadings during shipment from foreign sources. This situation would not be expected with corn shipped from the field to storage, but might exist to some extent in dockside elevators.

TABLE 1   Material Pressures in Bins

Value at depth Y for	Janssen (static values only)	Reimbert (static values only)	DIN 1055, Sheet 6 (dynamic effects included) Filling	DIN 1055, Sheet 6 (dynamic effects included) Emptying
Vertical pressure q	$\gamma Y_0(1 - e^{-Y/Y_0})$	$\gamma\left[Y\left(\dfrac{Y}{C}+1\right)^{-1} + \dfrac{h_s}{3}\right]$	$\gamma Y_{0f}(1 - e^{-Y/Y_{0f}})$	$\gamma Y_{0e}(1 - e^{-Y/Y_{0e}})$
Lateral pressure p	qk	$p_{max}\left[1 - \left(\dfrac{Y}{C}+1\right)^{-2}\right]$	$q_f k_f$	$c_1 q_e k_e$
Vertical frictional force per unit width of wall V	$(\gamma Y - q)R^*$	$(\gamma Y - q)R$	$(\gamma Y - q_f)R$	$c_1(\gamma Y - q_e)R$

*Reference 13 gives this as $(\gamma Y - 0.8q)R$. The coefficient 0.8 was introduced so that the formula would give about the same values as the Reimbert formulas.

Notation:

γ = weight per unit volume of stored material (Tables 4 and 5, also Sec. 23, Table 2)

Y = depth defined in Fig. 5

$Y_0 = R/\mu'k$

$Y_{0f} = R/\mu'_f k_f$, filling

$Y_{0e} = R/\mu'_e k_e$, emptying

R = hydraulic radius of horizontal cross section of storage space = A/L (Table 6)

A = cross-sectional area of silo (Table 6)

L = perimeter of silo (Table 6)

μ' = coefficient of friction between stored material and wall (Table 5, also Sec. 23, Table 2)

μ'_f = coefficient of friction between stored material and wall during filling (Table 2)

μ'_e = coefficient of friction between stored material and wall during emptying (Table 2)

c = roughness factor (Table 3)

c_1 = coefficient given by Eq. (1)

$k = p/q = (1 - \sin \rho)/(1 + \sin \rho)$, Janssen and Reimbert

$k_f = p_f/q_f$ (Table 2)

$k_e = p_e/q_e$ (Table 2)

ρ = angle of internal friction or (approximately) angle of repose ϕ (Tables 4 and 5, also Sec. 23, Table 2)

C = characteristic abscissa for Reimbert's formula (Table 8)

h_s = height of sloping top surface of stored material (Fig. 5)

p_{max} = Reimbert's maximum static pressure (Table 8)

To account for the load-reducing effect of the bin bottom, DIN allows the emptying pressure to be reduced over a height $1.2D$, but not to exceed $0.75h$, above the outlet. This is accomplished by prescribing a linear variation from the filling pressure at the outlet to the emptying pressure at the top of the reduced-pressure zone.

Pieper[14] also suggests two sets of values of μ' and k for use in the Janssen formula, with a roughness factor c to determine μ' (Table 2). Five categories of roughness are defined (Table 3). However, it has been observed that rough walls of grain silos have been made smoother by the development with time of a skin of wax and fat-like material. Because of the resulting reduced friction, less weight of grain is transferred to the walls, so that more acts on the hopper. As a result, hoppers have sometimes failed unexpectedly after a number of years of operation.[15] Where this effect is to be expected, a reduction in the value of c in Table 3 is recommended (footnote a).

Bunkers.   Although the formulas in Table 1 can be used for shallow bins, Rankine's active pressure (Sec. 5, Art. 40) is sometimes used for bunkers. In this case $q = \gamma Y$ and $p = kq$. Coulomb's formula for pressure on retaining walls (Sec. 5, Art 42) has also been used for bunkers.

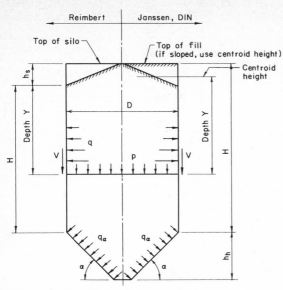

Fig. 5 Silo dimensions and pressures.

TABLE 2   Values of k and μ': DIN 1055[12] and Pieper[14]

| | DIN 1055 | | |
	Grain material, avg diam > 0.2 mm	Powder material, avg diam < 0.06 mm	Pieper
k_f	0.5	0.5	$1 - \sin \rho_f$*
k_e	1	1	$1.5\, k_f$
μ'_f	$\tan 0.75\, \rho_f$	$\tan \rho_f$	$\tan \rho_f c$
μ'_e	$\tan 0.60\, \rho_f$	$\tan \rho_f$	$\tan 0.8\, \rho_f c$

See Table 1 for notation, Table 3 for values of c, Table 4 for values of p_f.
*Use $1 - \sin 1.5\, \rho_f$ for flour.

TABLE 3   Values of Roughness Factor c (From Ref. 14)

| | Roughness category[a] | | | | |
Stored material	1[b]	2[c]	3[d]	4[e]	5[f]
Dustlike[g]	0.85	0.90	0.95	0.95	
Fine-grained[h]	0.75	0.80	0.85	0.90	0.95
Coarse-grained[i]		0.75	0.80	0.85	0.95

[a]For stored material that makes silo walls smoother, the value of c for the calculation of q and p should be reduced by 0.05, except for roughness category 1. When designing against buckling of silo walls, increase c by 0.05.
[b]Glass and enamel with smooth joints, few in number, welded aluminum, coatings of synthetic materials.
[c]Plywood with few joints, riveted or bolted (round heads) aluminum, welded steel, finished concrete, wood (planed boards with vertical joints).
[d]Riveted or bolted steel, perforated steel sheets, wood (unplaned boards with horizontal joints), unfinished concrete.
[e]Corrugated steel sheets, fine-mesh wire screen.
[f]Coarse-mesh wire screen.
[g]Grain diameter < 0.1 mm. Grain flour, finely ground rock, cement.
[h]Maximum grain diameter < 1 cm. Sand, grain, soybeans, granulated artificial materials.
[i]Gravel, cement clinker, coke, ore.

Angle of Internal Friction.   This depends on whether the material is at rest or in motion. The value ρ_f for material at rest can be determined experimentally, approximately as the angle of repose ϕ or exactly by using a shear box. In the latter method, ρ_f is the slope of the line obtained by plotting pairs of values of shear τ and normal stress σ at which the material fails by shear, that is, $\rho_f = \tan^{-1} \tau/\sigma$.

The angle of internal friction ρ_e during emptying of model silos has been measured at approximately $0.8\,\rho_f$.

The static value of ρ_f is used in computing pressures by the Reimbert formulas and, except for DIN and Pieper, Janssen's formulas.

Values of ρ_f for various materials are given in Table 4. Values of ϕ are given in Table 5 and in Sec. 23, Table 2.

TABLE 4   Physical Properties of Storage Materials[a]

	Unit weight, lb/ft³					$k_f{}^d$	
	γ_{min}	γ_{max}	$\gamma_1{}^b$	$\gamma_2{}^b$			
Grain	50	56	54	52	31	0.485	0.727
Grain flour	37	50	46	42	29	0.312	0.467
Rice	50	56	54	52	33	0.455	0.683
Corn	46	52	50	48	28	0.530	0.796
Quartz sand	91	104	100	96	34	0.441	0.661
Cement	75	112	100	87	28	0.530	0.796
Cement clinker	100	128	119	109	36	0.412	0.618
Limestone powder	66	82	77	71	27	0.546	0.819
Gravel	87	125	112	100	32	0.470	0.705
Sugar, refined	52	62	59	56	34	0.441	0.661

[a]Adapted from Ref. 14.
[b]Use is determined by h; see Eqs. (2).
[c]For the angle of internal friction during emptying, take $p_e = 0.8\,p_f$.
[d]$k_f = 1 - \sin p_f$, except for grain flour where $k_f = 1 - \sin 1.5\,p_f$.
[e]$k_e = 1.5\,k_f$.

Weight of Material.   The unit weight γ of the material stored depends not only on its specific weight but also on its compressibility. The latter is influenced by storage height and time, and by vibrations to which the material may be subjected. Pieper defines lower and upper values γ_{min} and γ_{max}, which are determined experimentally. The design value is determined by the height h of the silo (Fig. 5) as follows:

$$h > 10 \text{ m}(32.8 \text{ ft}) \qquad \gamma_1 = \frac{\gamma_{min} + 2\gamma_{max}}{3} \qquad (2a)$$

$$h < 5 \text{ m}(16.4 \text{ ft}) \qquad \gamma_2 = \frac{2\gamma_{min} + \gamma_{max}}{3} \qquad (2b)$$

Values of γ for $5 < h < 10$ m are determined by linear interpolation.

Values of γ_{min}, γ_{max}, γ_1, and γ_2 for various materials are given in Table 4. Values of γ are also given in Table 5 and in Sec. 23, Table 2.

Coefficient of Friction.   Values of μ' for various materials are given in Table 5 and in Sec. 23, Table 2. These values are not to be used in the DIN and Pieper procedures, for which μ' depends on the internal angle of friction ρ_f (Table 2).

Values of k.   Except for the DIN and Pieper procedures, values of k are computed from $k = (1 - \sin\rho)/(1 + \sin\rho)$ (Table 1). DIN and Pieper values are given in Table 2. Values of k for various materials according to Pieper's formulas are given in Table 4.

Geometric Properties.   Formulas for the cross-sectional area A, perimeter L, and hydraulic radius R for various cross sections are given in Table 6.

Janssen Formulas (Table 1).   As Y approaches infinity (practically, and within 1 percent error, when $Y/Y_0 = 4.6$), the maximum pressures are

$$q_{max} = \gamma Y_0 = \frac{\gamma R}{\mu' k} \tag{3a}$$

$$p_{max} = k q_{max} = \frac{\gamma R}{\mu'} \tag{3b}$$

Table 7 gives values of $1 - e^{-Y/Y_0}$ by which q_{max} and p_{max} are multiplied for the region $0 < Y/Y_0 < 4.9$. Linear interpolation is acceptable.

Reimbert Formulas (Table 1). Equations (3) also give the maximum pressures by the Reimbert formulas. Values of p_{max} and C for various cross sections are given in Table 8.

TABLE 5 Physical Properties of Granular Materials*

	γ, lb/ft^3	ϕ, deg	μ' Against concrete	Against steel
Cement, clinker	88	33	0.6	0.3
Cement, portland	84–100	24–30	0.36–0.45	0.30
Clay	106–138	15–40	0.2–0.5	0.36–0.7
Coal, bituminous	50–65	32–44	0.50–0.60	0.30
Coal, anthracite	60–70	24–30	0.45–0.50	0.30
Coke	38	40	0.80	0.50
Flour	38	40	0.30	0.30
Gravel	100–125	25–35	0.40–0.45	
Grains, small†	44–62	23–37	0.29–0.47	0.26–0.42
Gypsum in lumps, limestone	100	40	0.5	0.3
Iron ore	165	40	0.50	0.36
Lime, burned (pebbles)	50–60	35–55	0.50–0.60	0.30
Lime, burned, fine	57	35	0.5	0.3
Lime, burned, coarse	75	35	0.5	0.3
Lime, powder	44	35	0.50	0.30
Manganese ore	125	40		
Sand	100–125	25–40	0.40–0.70	0.35–0.50
Sugar, granular	63	35	0.43	

*From Table 4A of Ref. 13. The properties listed here are illustrative of values which might be determined from physical testing. Ranges of values show the variability of some materials. Design parameters should preferably be determined by test on samples of the actual materials to be stored.

†Wheat, corn, soybeans, barley, peas, beans (navy, kidney), oats, rice, rye.

3. Emptying Pressures in Funnel-Flow Silos. For the purpose of computing emptying pressures, Pieper[14] classifies silos as shallow, intermediate in height, and tall.

Shallow Silos (Fig. 6a). If $h < h_{F1}$, where $h_{F1} = \frac{1}{2}(D - d) \tan \alpha_1$, with $\alpha_1 = 29(\sqrt[4]{\rho_f c})$, emptying does not increase the pressures that were generated during filling. Therefore, only filling pressures need be considered in design.

Tall Silos (Fig. 6b). If $h \geqslant 2h_{F2}$, where $h_{F2} = \frac{1}{2}(D - d) \tan \alpha_2$, with $\alpha_2 = 32 \, (\sqrt[4]{\rho_f c})$, design is governed by emptying pressures. Between the level of the outlet and the height h_{F2} above it, emptying pressures reduce linearly from the value at height h_{F2} to the value of the filling pressure at the outlet.

Silos of Intermediate Height. If $h_{F1} < h < 2h_{F2}$, the height of the reduced pressure zone is determined by linear interpolation between h_{F1} and h_{F2} by

$$h_F = h_{F1} + \frac{h_{F2} - h_{F1}}{2h_{F2} - h_{F1}} (h - h_{F1}) \tag{4}$$

There is a zone immediately above the reduced pressure zone in tall and intermediately tall silos in which emptying pressures exceed p_e. Figure 7c shows a simplified vertical distribution of the pressure increase for tall silos. The height h_{F2} for a central opening (Fig. 7b) also applies for eccentric outlets (Fig. 7a). Lateral distributions of the pressure increase for a central outlet, an outlet at one side, and outlets at opposite sides are shown in Fig. 7d.

TABLE 6 Geometry of Silo Cross Sections

Cross section		Area A	Perimeter L	Hydraulic radius R
Circle		$\frac{\pi}{4}D^2$	πD	$\frac{D}{4}$
Circular sector	α, Radians	$\frac{\alpha D^2}{8}$	$(\frac{\alpha}{2}+1)D$	$\frac{D/4}{1+2/\alpha}$
	α, Degrees	$0.00218\,\alpha D^2$	$(0.0087\alpha+1)D$	$\frac{D/4}{1+114.6/\alpha}$
Regular polygon		$\frac{ND^2}{8}\sin\frac{2\pi}{N}$	$ND\sin\frac{\pi}{N}$	$\frac{D}{4}\cos\frac{\pi}{N}$
		$\frac{ND^2}{4}\cot\frac{\pi}{N}$	Na	$\frac{a}{4}\cot\frac{\pi}{N}$
Rectangle		ab	$2(a+b)$	$\frac{ab}{2(a+b)}$, Long Side$\quad$ $\frac{a}{4}$, Short Side
Long rectangle		—	—	$\frac{a}{2}$
Ring		$\frac{\pi}{4}(D_e^2-D_i^2)$	$\pi(D_e+D_i)$	$\frac{(D_e-D_i)}{4}$
Interstice		$(1-\frac{\pi}{4})D^2$	πD	$\frac{a}{4}=0.104D$

TABLE 7 Values of $1-e^{Y/Y_0}$

Y/Y_0	0	0.1	0.2	0.3	0.4	0.5	0.6	0.7	0.8	0.9
0	0	0.095	0.181	0.259	0.333	0.393	0.451	0.503	0.551	0.593
1	0.632	0.667	0.699	0.727	0.753	0.777	0.798	0.817	0.835	0.850
2	0.865	0.878	0.889	0.900	0.909	0.918	0.926	0.933	0.939	0.945
3	0.950	0.955	0.959	0.963	0.967	0.970	0.973	0.975	0.978	0.980
4	0.982	0.983	0.985	0.986	0.988	0.989	0.990	0.991	0.992	0.993

TABLE 8   Values of p_{max} and C in the Reimbert Formulas

Silo	p_{max}	C		Remarks
Circular	$\dfrac{\gamma D}{4\mu'}$	$\dfrac{D}{4\mu'k}$	$\dfrac{h_s}{3}$	
Polygonal, of more than four sides	$\dfrac{\gamma R}{\mu'}$	$\dfrac{L}{\pi}\dfrac{1}{4\mu'k}$	$\dfrac{h_s}{3}$	R from Table 6
Rectangular, on shorter wall a	$\dfrac{\gamma a}{4\mu'}$	$\dfrac{a}{\pi\mu'k}$	$\dfrac{h_s}{3}$	
Rectangular, on longer wall b	$\dfrac{\gamma a'}{4\mu'}$	$\dfrac{a'}{\pi\mu'k}$	$\dfrac{h_s}{3}$	$a' = \dfrac{2ab}{a+b}$

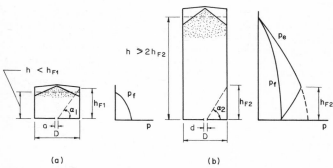

Fig. 6   (*a*) Shallow silo; (*b*) tall silo.

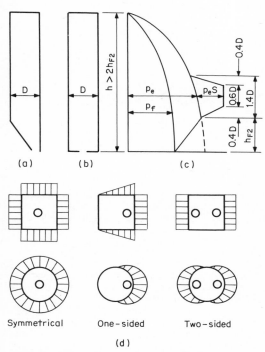

Fig. 7   Emptying pressure increase in tall silos.

Values of the factor S in Fig. 7c have been determined for quartz sand and wheat in square silos with a variety of outlet locations. Values ranged from 0.15 for a central outlet to 0.25 for edge and corner outlets. A value 0.4 was reached for a bin with a long slot outlet on each of the four edges. The pressure increase appears to be material-dependent, since an experiment with rice in a silo with a slot outlet on one side gave $S = 0.90$.

Values of S and the height over which the corresponding pressure increase is distributed in intermediately tall silos can be determined by linear interpolation between zero for $h = h_{F1}$ and the values for $h = h_{F2}$.

The increased pressures are assumed to be distributed over the entire width of the wall and the height shown, and pressures at some points in the region of increased pressure may be several times larger than the values shown. However, the average values can be used to dimension reinforced-concrete silos because their walls are stiff enough to distribute load concentrations.

Effect of Eccentric Outlets on Pressures below h_{F2}. Figure 8a shows the flow funnel in a circular silo with an eccentric outlet.[14] The emptying pressure p_e in the funnel is less than the filling pressure in the silo, and varies from p_f at the top of the funnel to zero at the level of the outlet (Fig. 8b). The differential pressure $\Delta p = p_f - p_e$ at the conical surface of the funnel is assumed to be transferred to the silo walls by arching, which causes a concentrated increase in pressure on the wall at the intersection of the funnel with the wall (Fig. 8c). For design, Δp is assumed to be most unfavorable at the center of the height $h_F/2$ from the top of the funnel, the corresponding value of p_e being given by

$$p_e = \frac{r_F}{D} p_f$$

where r_F is the radius of the funnel from point A on the silo wall.

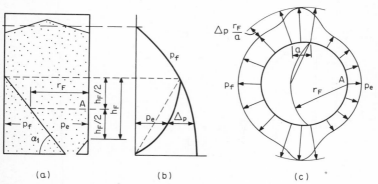

Fig. 8   Emptying pressures in silo with eccentric outlet.

The circumferential distribution of pressure is shown in Fig. 8c. At each intersection of the funnel boundary with the silo wall there is a zone where the pressure exceeds p_f. The width a of this zone is given by $D/10 \gtrless a \gtrless r_F/2$ and the average pressure over the zone by $r_F \Delta p/a$.

The angle α in Fig. 8a should be taken equal to α_1 except that the resulting intercept on the silo wall should not be taken larger than h_{F2}. Also, in low silos the funnel may intersect the surface of the stored material instead of the silo wall.

Pressures due to eccentric outlets in rectangular silos may be determined similarly.

4. Emptying Pressures in Funnel-Flow Silos—ACI 313. Emptying pressure for silos with central outlets are obtained by multiplying filling pressures by *overpressure factors C_d* (Table 9) or *impact factors C_i* (Table 10). The larger values are to be used for design. Values of C_d for the Janssen filling pressures are from the Soviet Silo Code,[16] with some slight modifications. Safarian[17] computed the values for the Reimbert formulas so as to give the same pressures as the Janssen formula using the Soviet Code values.

TABLE 9 Recommended Minimum Overpressure Factor C_d*

Portion of silo	$\frac{H}{D} < 2$		$2 = \frac{H}{D} < 3$		$3 = \frac{H}{D} < 4$		$4 = \frac{H}{D} < 5$		$\frac{H}{D} \geq 5$	
	J	R	J	R	J	R	J	R	J	R
Upper zone, H_1 high	1.35	1.10	1.45	1.20	1.50	1.25	1.60	1.30	1.65	1.35
Next zone, $(H-H_1)/4$ high	1.45	1.20	1.55	1.30	1.60	1.35	1.70	1.40	1.75	1.50
Next zone, $(H-H_1)/4$ high	1.55	1.45	1.65	1.55	1.75	1.60	1.80	1.70	1.90	1.75
Next zone, $(H-H_1)/4$ high	1.65	1.65	1.75	1.75	1.85	1.85	1.90	1.90	2.00	2.00
Lower zone, $(H-H_1)/4$ high	1.65	1.65	1.75	1.75	1.85	1.85	1.90	1.90	2.00	2.00
Hopper†										
Bottom:										
Concrete	1.35	1.50	1.35	1.50	1.35	1.50	1.35	1.50	1.35	1.50
Steel	1.50	1.75	1.50	1.75	1.50	1.75	1.50	1.75	1.50	1.75

*Adapted from Ref. 13.

†For hoppers use pressure for zone above, uniform throughout l_h, or reduce pressure in accordance with R. Pressures for hopper-forming fill may be reduced linearly from top of fill to top of flat slab.

J = Janssen, R = Reimbert.
H = height defined in Fig. 2
H_1 = D tan ρ for circular silos
H_1 = a tan ρ and b tan ρ for rectangular silos
C_d applies to the lateral pressure at bottom of each height zone.
Bottom pressures need not be considered larger than the pressure from the weight of silo contents.
Values in this table are too small for mass flow.
In the region of a flow-correcting insert (e.g., Buhler Nase) lateral pressures may be much larger than static pressures, and values in this table are too small.

Eccentric Outlets. It is suggested in the commentary that the increase in lateral pressure due to eccentric discharge be taken to be at least 25 percent of the static (filling) pressure at the bottom of the silo if the outlet is adjacent to the wall. For outlets with smaller eccentricities this value may be reduced linearly to zero for a central outlet. The increase may also be reduced linearly to zero at the top of the silo, and need not be multiplied by the overpressure factor C_d. Thus the design pressure at depth Y is

$$p = C_d p_f + 0.25 p_f \frac{e}{r} \frac{Y}{H} \tag{5}$$

where r = radius of bin.

TABLE 10   Recommended Minimum Impact Factor C_i*

Ratio of volume dumped in one load to silo capacity	1:2	1:3	1:4	1:5	1:6 and less
Concrete bottom	1.4	1.3	1.2	1.1	1.0
Steel bottom	1.75	1.6	1.5	1.35	1.25

*Adapted from Ref. 9.

For circular silos it is conservative to compute the increased hoop forces by assuming the increased pressure to be distributed uniformly around the circumference. This will not enable the bending moments and shears caused by the actual distribution to be evaluated, but if the walls are reinforced at both faces the additional steel required by the increased hoop force should take care of the moments. However, for circular silos with large eccentricities of discharge, and for rectangular silos, it is advisable to consider more realistic pressure distributions such as those of Fig. 8.

5. Shock Effects from Collapse of Domes   During emptying of a silo, domes of the stored material may form. They eventually collapse as emptying continues and their spans become too long, although for cohesive materials it is sometimes necessary to destroy them by external means to restore flow. Collapse of a dome is followed by free falling of the material above, and shocks are created when it strikes the material below. These shocks increase the emptying lateral pressure p_e and the vertical filling pressure q_f. The increased pressures p_d and q_d can be computed from[14]

$$p_d = p_e \left(1 + \frac{1.8 S_d}{\sqrt{a}} \right) \qquad a \geqslant 3 \tag{6a}$$

$$q_d = q_f (1 + 2 S_d) \tag{6b}$$

where S_d = coefficient which depends on material
a = horizontal distance, ft, between center of outlet and corresponding point on wall

The following values of S_d are from Ref. 14:

Material	S_d
Quartz gravel	0.10
Grain	0.15
Corn	0.30
Cement clinker	0.30–0.40

Shock effects may be disregarded if $1.8 S_d / \sqrt{a}$ is less than 0.05.

Maximum shocks occur in silos with height-width ratios equal to or greater than 4. Experience has shown that shock effects are not significant if this ratio is less than 2. Linear interpolation can be used for intermediate height-width ratios.

Shock effects increase with decrease in the distance a from outlet to wall. In addition, the unevenly distributed shock pressure from eccentric outlets induces bending moments and shears in the silo walls. Therefore, it is recommended that only central outlets be used in silos for materials that are known to generate domes.

6. Pressures Induced by Dustlike Materials Dustlike materials such as cement, wheat flour, and limestone powder (talc) with an average particle diameter under 0.06 mm (0.0024 in.) become mixed with air during rapid filling of a silo. The mixture has been shown[14] to possess hydrostatic characteristics and is able to develop a maximum lateral pressure

$$p_{max} = 1.6 \, \gamma_{min} \, vt_s k_f \tag{7}$$

where γ_{min} = minimum unit weight, Table 4
$\quad\quad v$ = filling velocity measured as vertical rise of surface of material per hour
$\quad\quad t_s$ = setting time of material, hr. Use 0.24 hr for talc, 0.19 hr for cement, and 0.14 hr for wheat flour
$\quad\quad k_f = p_f/q_f$ (Table 2)

The lateral pressure increases linearly from zero at the surface of the material to the maximum pressure, which occurs at a depth vt_s from the surface. The maximum pressure remains constant down to the bottom of the silo.

After the dustlike material has settled, and during emptying, pressures may be determined as for grainlike materials (Table 1).

It is necessary for design purposes to consider not only the effects caused by the rapid filling, but also the conventional pressures after the material has settled, and during emptying. Design of silos with small hydraulic radius may be controlled by p_{max} generated during rapid filling (Fig. 9a). Design of silos with larger hydraulic radius should be based on the envelope of maximum lateral pressures. This means that design will be governed at the top portion of the silo by lateral pressures generated during rapid filling and at the lower portion by lateral pressures generated during emptying (Fig. 9b).

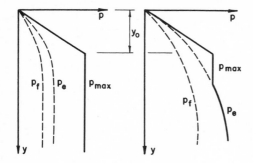

a) Silos with small R   b) Silos with large R

Fig. 9 Wall pressures from dust-like materials.

Pneumatic Emptying. Pneumatic devices to assist emptying (Fig. 10a) prevent a buildup of dustlike material and subsequent closing of the outlet. If the imposed air pressure p_u exceeds the emptying pressures p_e and q_e, it must be used for design of the lower portion of the silo (Fig. 10b). The range of height Δh above which this pressure must be accounted for is given by Pieper[14] as

$$\Delta h = 1.6 \, \frac{p_u}{\gamma_{min}} \tag{8}$$

The pressure variation for design is shown in Fig. 10b.

Homogenizing Silos. In these silos dustlike materials are thoroughly mixed with air. Design must satisfy the conventional pressures and loads of Table 1 as well as the hydrostatic pressures generated during mixing. For the latter

$$p = q = C\gamma_1 Y \tag{9}$$

where γ_1 is unit weight from Table 4. The coefficient C is given as 0.6 and 0.55 by ACI 313[13] and Pieper,[14] respectively.

7. Earthquake Forces The following minimum requirements have been proposed for design.[13] The total lateral seismic force H_e for shear at the base is given by

$$H_e = ZC_p(W_g + W_{\text{eff}}) \tag{10}$$

where Z = earthquake-zone factor = $\frac{3}{16}$, $\frac{3}{8}$, $\frac{3}{4}$, and 1 for zones 1, 2, 3, and 4, respectively (Sec. 19, Fig. 4)

$\quad W_g$ = weight of structure

$\quad W_{\text{eff}}$ = 80 percent of weight of stored material, applied at centroid of volume

$\quad C_p$ = 0.2 for silos with material stored on bottoms above ground and 0.1 when silo walls extend to ground and stored material rests directly on ground. For intermediate cases C_p may be obtained by linear interpolation

If the bin bottom-supporting system is independent of the walls, W_{eff} may be distributed between the two independent structures according to their relative stiffnesses.

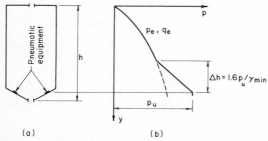

Fig. 10 Additional pressure generated by pneumatic equipment used during emptying of dust-like materials.

A dynamic analysis, using a design earthquake spectrum compatible with the seismic zone and with local foundation conditions, may be used instead of Eq. (10).

WALL FORCES

8. Circular Silos The hoop force F per unit height of wall due to the radial pressure p of the stored material is given by

$$F = \frac{pD}{2} \tag{11}$$

The increase ΔD in the diameter can be determined from

$$\Delta D \approx \frac{pD^2}{4tE_c} \tag{12}$$

where t = wall thickness.

Walls are also subjected to vertical compression from the roof, from their weight, and from wall friction of the stored material.

Eccentric Outlets. Forces due to the additional pressures caused by eccentric outlets (Fig. 7) are given in Figs. 11 and 12.

Grouped Circular Silos. Some groups frequently used are shown in Fig. 3. Initial design of individual circular cells of these groups is done as if they were isolated. The effects of interstices are determined by considering the adjacent circular cells to be empty (Fig. 13). Formulas for the case of an empty central circular cell with adjacent full circular cells are given in Fig. 14.

Analysis of other cross-section configurations and empty-full cases is possible by treating the whole horizontal cross section as a rigid frame. The curved members and variable wall thicknesses at wall intersections may make an "exact" analysis quite complex and laborious. Approximate methods based on local analysis of elements fixed at the sections where silo walls intersect are acceptable and may be the only practical thing to do in many cases.

Attention to local effects at wall intersections is important. Good design should provide strength for proper transfer of bending moments, shear, and direct axial force at these locations.

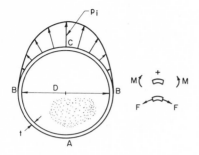

Bending moments

$M_A = -0.0183\, p_i\,(D+t)^2$

$M_B = 0.0208\, p_i\,(D+t)^2$

$M_C = 0.0234\, p_i\,(D+t)^2$

Hoop forces

$F_A = 0.1238\, p_i\,(D+t)$

$F_B = 0.1667\, p_i\,(D+t)$

$F_C = -0.2905\, p_i\,(D+t)$

Fig. 11 One-sided additional pressure on circular silo.

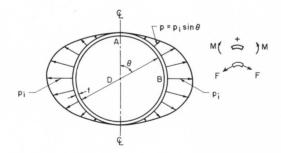

Bending moments

$M = p_i\dfrac{(D+t)^2}{4}\left[\left(\dfrac{\pi}{4}-\dfrac{\theta}{2}\right)\cos\theta + \dfrac{\sin\theta}{2}-\dfrac{2}{\pi}\right]$

$M_A = 0.0372\, p_i\,(D+t)^2$

$M_B = 0.0342\, p_i\,(D+t)^2$

Maximum Shear $V = 0.140\, p_i\,D$, at $\theta = 40.7°$

Hoop forces

$F = p_i\dfrac{(D+t)}{8}\left[(\pi-20)\cos\theta + 2\sin\theta\right]$

$F_A = 0.393\, p_i\,(D+t)$

$F_B = 0.250\, p_i\,(D+t)$

Fig. 12 Two-sided additional pressure on circular silo. *(After G. Ruzicka.)*

9. Rectangular and Polygonal Silos Walls in the pressure zones of square, rectangular, and polygonal silos are subjected to bending moment, horizontal shear, and horizontal tension due to the lateral pressure, and to vertical compression from the roof, from their weight, and from wall friction of the stored material.

Walls whose height is more than twice the width may be analyzed for one-way bending in the horizontal direction. Since adjoining walls are continuous at their junctures, moments may be determined as for a frame. Formulas for M, F, and V for a horizontal strip are given for rectangular walls in Fig. 15 and regular polygonal walls of N sides in Fig. 16.

Rectangular walls whose height is less than half the width may be analyzed for one-way bending in the vertical direction. The lower edge can usually be assumed fixed. The upper edge may be assumed fixed or simply supported, depending on the attached construction, or free if there is none.

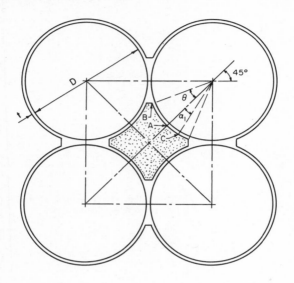

Bending moments

$$M_A = \frac{p}{4}(D+2t)(D+t)\sin\theta\left(1-\frac{\sin\theta}{\theta}\right)$$

$$M_B = \frac{p}{4}(D+2t)(D+t)\sin\theta\left(\cos\theta-\frac{\sin\theta}{\theta}\right)$$

$$M_C = \frac{p}{4}(D+2t)(D+t)\sin\theta\left(\cos\alpha_1-\frac{\sin\theta}{\theta}\right)$$

Compression forces

$$F_A = \frac{p}{2}(D+2t)(1-\sin\theta)$$

$$F_B = \frac{p}{2}(D+2t)(1-\sin\theta\cos\theta)$$

$$F_C = \frac{p}{2}(D+2t)(1-\sin\theta\cos\alpha_1)$$

Fig. 13 Full interstice and empty adjacent cells. *(Adapted from Ref. 18.)*

Moments in walls whose height is more than half the width but less than twice the width should be determined as for a plate supported on four edges or, if the upper edge is free, as a plate supported on three edges. Tables of moment coefficients for various cases are given in Refs. 8, 19, and 20.

Walls that are supported on columns are subjected to in-plane bending due to the load from an attached bottom. Analysis depends on the height of the wall relative to the spacing of the columns (Art. 17).

10. Thermal Effects A differential temperature ΔT between the interior and exterior faces of a silo creates a strain gradient $\alpha_t \Delta T/t$ in the wall section, where t is the thickness of the wall and α_t is the thermal coefficient of expansion of concrete. Because of the closed nature of silos, rotational restraint is imposed by continuity, and thus bending moments are generated in the presence of thermally induced strain gradients. For a section subject to bending moment M and axial force P caused by loading unrelated to thermal effects, Gurfinkel[21] has shown that the additional thermal moment M_t for a given strain gradient depends on the existing strain distribution, which in turn depends on M and P; iteration is required to determine M_t. ACI 313[13] calculates the thermal bending moments M_{xt} per unit of wall height and M_{yt} per unit of wall width as if generated in an uncracked section of wall subjected to a state of plane strain. Thus,

$$M_t = \frac{E_c t^2 \alpha_t \, \Delta T}{1 - \nu} \tag{13}$$

where E_c and ν are the concrete modulus of elasticity and Poisson's ratio, respectively.

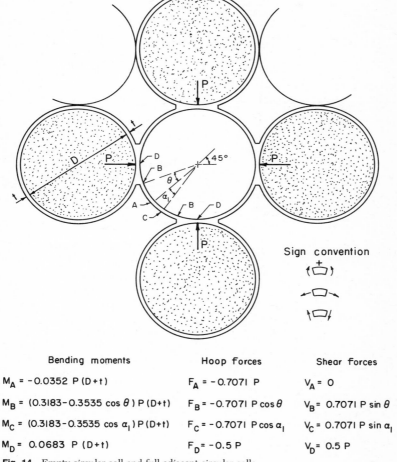

Bending moments	Hoop forces	Shear forces
$M_A = -0.0352\ P\,(D+t)$	$F_A = -0.7071\ P$	$V_A = 0$
$M_B = (0.3183 - 0.3535\cos\theta)\,P\,(D+t)$	$F_B = -0.7071\ P\cos\theta$	$V_B = 0.7071\ P\sin\theta$
$M_C = (0.3183 - 0.3535\cos\alpha_1)\,P\,(D+t)$	$F_C = -0.7071\ P\cos\alpha_1$	$V_C = 0.7071\ P\sin\alpha_1$
$M_D = 0.0683\ P\,(D+t)$	$F_D = -0.5\ P$	$V_D = 0.5\ P$

Fig. 14   Empty circular cell and full adjacent circular cells.

For normal-weight concrete $E_c = 57{,}000\sqrt{f_c'}$, $\alpha_t = 5.5 \times 10^6/°F$, and $\nu = 0.3$. This formula gives conservative values for the thermal moments, since any cracking of the wall section would reduce its stiffness and result in lower values.

Temperature Differential.   The temperature differential between external and internal faces of a concrete silo wall containing hot stored material can be calculated from

$$\Delta T = (T_{i,\mathrm{des}} - T_0)K_T \tag{14}$$

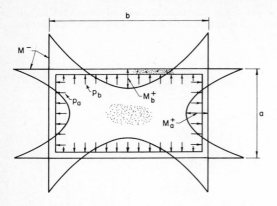

$$M^- = \frac{1}{12(1+n)}\left(p_a a^2 + np_b b^2\right) \text{ where } n = \frac{b}{a}\left(\frac{t_a}{t_b}\right)^3$$

$$M_a^+ = \frac{p_a a^2}{8} - M^-$$

$$M_b^+ = \frac{p_b b^2}{8} - M^-$$

$$F_b = p\frac{a}{2} \qquad V_b = p\frac{b}{2}$$

$$F_a = \frac{pb}{2} \qquad V_a = \frac{pa}{2}$$

Fig. 15 Lateral pressure on rectangular silo.

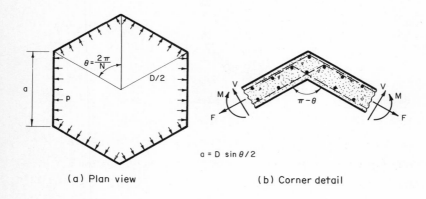

$$a = D\sin\theta/2$$

(a) Plan view   (b) Corner detail

	At corner	At side midspan
F	$\dfrac{pD}{2}\cos\dfrac{\theta}{2}$	$\dfrac{pD}{2}\cos\dfrac{\theta}{2}$
V	$\dfrac{pD}{2}\sin\dfrac{\theta}{2}$	0
M	$-\dfrac{pD^2}{12}\sin^2\dfrac{\theta}{2}$	$\dfrac{pD^2}{24}\sin^2\dfrac{\theta}{2}$

Fig. 16 Lateral pressure on polygonal silo.

where $T_{i,\text{des}} = T_i - 80$, T_i = temperature of stored material, T_0 = design winter dry-bulb temperature, and K_T = ratio of thermal resistance of wall alone to that of the wall plus an outside surface film of air plus a thickness t_m of stored material acting as insulating material. If R_m and R_c = 0.08 represent the thermal resistances per unit thickness (resistivity*) of the stored material and the concrete wall, respectively, and R_a = 0.17 is the thermal resistance of the outer surface film of air, then

$$K_T = \frac{0.08t}{t_m R_m + 0.08t + 0.17} \tag{15a}$$

For silos storing hot cement ACI 313 suggests t_m = 8 in. and $t_m R_m$ = 3.92. This reduces K_t to

$$K_t = \frac{0.08t}{4.09 + 0.08t} \tag{15b}$$

The analysis is illustrated in Example 6.

DESIGN OF WALLS

Except as noted, design formulas in the following articles are in terms of strength design. Subscripts u denote ultimate values obtained by multiplying service-load forces by load factors. ACI 313 prescribes load factors of 1.7 for live load and 1.4 for dead load. ϕ is the capacity-reduction factor.

11. Minimum Thickness of Circular Walls To allow for noncalculable moments due to transient nonuniform pressure on the walls of circular silos, the following minimum thickness is recommended:[13]

$$t_{\min} = p\,\frac{D}{2}\,\frac{mE_s + f_s - nf_{ct}}{f_s f_{ct}} \tag{16}$$

in which m = shrinkage coefficient (may be taken as 0.0003), f_s = allowable steel stress (between 0.4 and $0.5f_y$), and f_{ct} = concrete stress in uncracked section under static lateral pressure (may be taken as $0.1\,f'_c$).

Final wall thickness is governed by practical considerations and by load requirements and permissible crack width. A minimum thickness of 6 in. should be used for cast-in-place silo walls.

12. Maximum Crack Width An important design consideration is the maximum crack width that may be tolerated. A limit of 0.008 in. is suggested[13] for the case of grain or cement storage silos and for other silos exposed to the weather. This helps to avoid penetration of water that causes corrosion of the reinforcement, and spoils the silo contents by inducing germination of the grain and hydration of the cement.

Lipnitski's method[9] allows a simple determination of crack width for walls which are subjected mainly to hoop tensions, and includes silo walls and hoppers and other membrane-type structures. The width w_{cr} of a vertical crack caused by the simultaneous action of short- and long-term loadings may be determined by

$$w_{cr} = w_1 - w_2 + w_3 \tag{17}$$

Values of w in Eq. (17) are given by

$$w_1 = \psi_1\,\frac{A\beta}{\Sigma o}\,\frac{f_{s,\text{tot}}}{E_s} \qquad \text{where } \psi_1 = 1 - 0.7\,\frac{0.8Af'_t}{F_{\text{tot}}} \gtrless 0.3 \tag{18a}$$

---

*The unit of conductivity (Btu/ft²/hr/°F/in.) is the amount of heat in Btu that will flow in 1 hr through 1 ft² of a layer 1 in. thick of a homogeneous material per 1°F temperature difference between surfaces of the layer. Resistivity, which measures the insulating value of a material, is the reciprocal of conductivity.

$$w_2 = \psi_2 \frac{A\beta}{\Sigma o} \frac{f_{s,st}}{E_s} \qquad \text{where } \psi_2 = 1 - 0.7 \frac{0.8Af'_t}{F_{st}} \geq 0.3 \qquad (18b)$$

$$w_3 = \psi_3 \frac{A\beta}{\Sigma o} \frac{f_{s,st}}{E_s} \qquad \text{where } \psi_3 = 1 - 0.35 \frac{0.8Af'_t}{F_{st}} \geq 0.65 \qquad (18c)$$

where $\beta = 0.7$ and 1.0 for deformed and plain bars, respectively
   A = cross-sectional area of wall per unit height
   st = subscript indicating static load
   tot = subscript indicating static load plus overpressure
   $f'_t = 4.5\sqrt{f'_c}$ = tensile strength of concrete
   f_s = steel stress
   Σo = sum of perimeters of horizontal reinforcing bars per unit height of wall
Crack-width evaluation is illustrated in Example 2.

13. Walls in Tension   The required reinforcement A_s per unit height of wall is given by

$$A_s = \frac{F_u}{\phi f_y} \qquad (19)$$

ϕ in this equation may be taken as 0.9.

14. Walls in Tension and Flexure   The following formulas are from Ref. 13. There are two cases.
   Case I, $e = M_u/F_u \gtrless t/2 - d''$ (Fig. 17a)
On the side nearer to F_u

$$A_s = \frac{F_u e'}{\phi f_y (d - d')} \qquad (20)$$

and on the opposite side

$$A'_s = A_s \frac{e''}{e'} \qquad (21)$$

   Case II, $e = M_u/F_u = t/2 - d''$ (Fig. 17b)

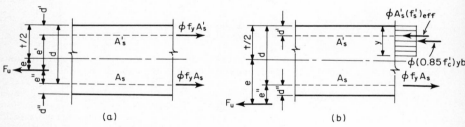

Fig. 17 Wall under tension and flexure.

1. Determine the depth y_L of a rectangular-section compression-stress block with 75 percent of balanced reinforcement (Sec. 11, Art. 7) from

$$\frac{y_L}{d} = 0.75\beta_1 \frac{87}{87 + f_y, \text{ksi}} \qquad (22)$$

$$\beta_1 = 0.85 - 0.05 \frac{f'_c - 4000}{1000} \geq 0.85 \qquad (23)$$

Values of y_L/d for common values of f'_c and f_y are given in the following table.

	f_y, ksi		
f'_c, ksi	40	50	60
To 4	0.436	0.405	0.378
5	0.411	0.381	0.355
6	0.386	0.357	0.333

2. Determine the effective compressive-steel stress

$$(f_s')_{\text{eff}} = 87\,\frac{y_L - \beta_1 d'}{y_L} - 0.85\,f_c' \lessgtr f_y - 0.85\,f_c' \qquad \text{ksi} \tag{24}$$

3. If $(f_s')_{\text{eff}}$ in step 2 is positive,

$$A_s' = \frac{F_u e''/\phi - 0.85 f_c' b y_L (d - y_L/2)}{(d - d')(f_s')_{\text{eff}}} \tag{25}$$

4. If A_s' in step 3 is positive,

$$A_s = \frac{F_u/\phi + 0.85\,f_c' b y_L + A_s'(f_s')_{\text{eff}}}{f_y} \tag{26}$$

5. If A_s' in step 3 is negative, no compressive steel is needed. In this case, whether or not steel is provided on the compression side, the wall is designed as singly reinforced according to

$$A_s = \frac{F_u/\phi + 0.85 f_c' b y}{f_y} \tag{27}$$

where $y \approx d - \sqrt{d^2 - \dfrac{2 F_u e''}{0.85 \phi f_c' b}}$

If $(f_s')_{\text{eff}}$ in step 2 is negative, compression steel will be ineffective, and if a singly reinforced member would not be acceptable, either d must be increased or d' decreased.

Shear.   The shear stress v_u is given by

$$v_u = \frac{V_u}{bd} \tag{28}$$

where v_u should not exceed v_c given by

$$v_c = 2\phi \left(1 - 0.002\,\frac{F_u}{bt} \right) \sqrt{f_c}, \text{psi} \tag{29}$$

15. Walls in Compression   ACI 318[22] allows walls for which the compressive force falls within the middle third to be considered as concentrically loaded. If buckling is not involved, the permissible compression is

$$f_c = 0.55\phi f_c' \tag{30}$$

where $\phi = 0.70$.

For rectangular walls, where slenderness may influence strength, buckling should be accounted for by using[13]

$$f_c = 0.55\phi f_c' \left[1 - \left(\frac{H_0}{40t} \right)^2 \right] \tag{31a}$$

where $\phi = 0.70$ and H_0 = clear vertical distance between supports. If $H_0 > l_0$, where l_0 = clear horizontal distance between supports, use l_0 in place of H_0.

Circular walls in the pressure zone may be designed for the allowable stress of Eq. (30) if there are no openings. If there are unreinforced openings, Eq. (31a) should be used, with H_0 = height of opening.

Circular walls below the pressure zone, continuous throughout, should be designed for[13]

$$f_c = 0.55\phi f_c' \left[1 - \left(\frac{D}{120t} \right)^3 \right] \tag{31b}$$

If there are unreinforced openings, use

$$f_c = 0.55\phi f_c' \left[1 - \left(\frac{H_0}{40t} \right)^2 - \left(\frac{D}{120t} \right)^3 \right] \tag{31c}$$

16. Walls in Compression and Flexure   These may be designed using the provisions of ACI 318, Chap. 10.[22] Combined compression and flexure is also discussed in Sec. 11, Art.

17. The interaction diagrams in that section for rectangular columns with reinforcement on opposite faces can be used for walls.

17. In-Plane Bending of Walls   The in-plane bending behavior of a wall supported on columns depends on the height of the wall relative to the spacing of the columns.

Low Walls.   The stiffness of a wall of a low bunker is of the same order as that of the hopper wall, and the two can be assumed to act together in transferring vertical load to the columns. For bin walls with $H/a \gtrless 0.5$ Ciesielski et al.[23] recommend that the wall and that portion of the hopper wall whose vertical projection is $0.4a$, where a is the length of the wall, be analyzed as a beam (Fig. 18a). The resulting bending stresses (Fig. 18b) are computed from $f = M/S$, where M is the moment due to the vertical loads. The stress in the part of the hopper wall not considered to be part of the beam is assumed to decrease linearly from the value of the bottom-fiber beam stress to zero at the vertex of the hopper wall. These stresses must be considered in combination with the moments and axial tensions due to lateral pressure (Fig. 15) in determining the wall reinforcement.

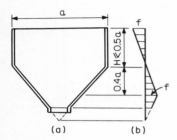

Fig. 18

(a)   (b)

A folded-plate analysis of the joint action of a low bunker wall and an adjoining hopper wall can also be made.[8]

High Walls.   Experimental studies by a number of investigators of the beam behavior of reinforced-concrete walls are discussed in Ref. 24. Walls for which $H/l \gtrless 1$, where $H =$ height and $l =$ length, can be designed by the usual procedure for reinforced-concrete beams. For a single-panel wall with $H/l > 1$, simply supported on columns spaced l center-to-center and carrying a uniformly distributed load w, the tension T that must be furnished by tensile reinforcement is given by

$$T = \frac{0.14wl}{\sqrt{H/l}} \qquad 1 < \frac{H}{l} \gtrless 2 \tag{32}$$

The value of T for $H/l = 2$ is to be used for walls with $H/l > 2$. The reinforcement is to be distributed over a depth $0.1l$.

If one-half to two-thirds of the tensile reinforcement is bent up, the shear strength V of the panel is given by

$$V \gtrless 0.54 f'_c t^2 \sqrt{H/t} \tag{33}$$

If the shear $w_u l/2$ exceeds V by Eq. (33), or if no bars are bent up, web reinforcement must be provided to resist the tension T_s given by

$$T_s = \frac{w_u l/2}{\sqrt{2H/l}} \qquad 1 \gtrless \frac{H}{l} \gtrless 2 \tag{34}$$

It is assumed in this formula that the necessary web reinforcement is inclined about 60° to the horizontal. Of course, equivalent reinforcement in the form of stirrups can be used.

Equation (32) is based on tests in which the load was applied to the top of the panel. However, tests on panels loaded along the bottom edge showed that the formula can also be used for this case.

18. Walls Subjected to Thermal Stresses   The required additional vertical and horizontal reinforcement per unit width or height is given by

$$A_{s,t} = \frac{1.4M_t}{f_y(d - d'')} \tag{35}$$

where M_t is given by Eq. (13). This steel should be placed near the cooler (usually outer) face of the wall. In singly reinforced walls it should be added to the main hoop steel, which should be near the outer face. In doubly reinforced walls it should be added to the outer layer, but for simplicity an equal amount is often added to the inner layer to avoid having bar sizes or spacings differ from one layer to the other.

Vertical tensile thermal stress is usually offset by vertical dead-load compressive stress so that additional temperature steel is often not needed.

19. Vertical Reinforcement   This is required not only in outside walls of silo groups but also on all inside walls. Vertical steel distributes lateral overpressures to adjacent horizontal reinforcement. Gurfinkel reported a silo where failure was averted when vertical steel redistributed lateral pressures that could not be resisted by hoop reinforcement that was in an advanced stage of corrosion.[25] Vertical steel also resists tension caused by bending moments due to restraint against circular elongation, eccentric loads from hopper edges or attached auxiliary structures, and temperature differentials between inside and outside wall surfaces or between silos.

20. Details and Placement of Reinforcement   Table 11 summarizes the requirements of ACI 313.[13] Bar splices, both horizontal and vertical, are staggered. Adjacent hoop-reinforcing splices in the pressure zone are staggered horizontally by not less than one lap length in 3 ft and do not coincide in vertical array more frequently than every third bar.

TABLE 11   Minimum Reinforcement Requirements

Region	Horizontal steel	Vertical steel (No. 4 or larger)
Pressure zone	As required by calculations	Exterior walls: $0.0020t$; max spacing $4t$ or 18 in. Interior walls: $0.0015t$; max spacing $4t$ or 24 in.
Below pressure zone	Continue A_s from above for a distance equal to six times wall thickness. Below this provide $0.0025t$ per unit height of wall	$0.0020t$
Bottom of walls and columns		Dowels as needed to prevent uplift and shifting by earthquake or wind loading
Wall intersections subjected to moment	Provide as required	Provide as required
Adjoining silos	Provide as required to prevent separation	
Circular walls, single-reinforced	Place nearest to the outer faces	

Slipforming should not be considered an excuse for not tying reinforcement together. Haeger considers the normal tying of the ends of hoop reinforcement, with additional ties every 4 to 5 ft between, to be acceptable.[26] Vertical steel should not be omitted to provide access for concrete buggies in slipform construction; instead it may be spaced farther apart at specified access locations. The total amount of vertical steel is unchanged; only the spacing is affected. The conventional practice of leaving the slipforming jack rods embedded in the concrete is fine, but widely spaced jack rods should not be construed as the equivalent of vertical reinforcement.

Typical reinforcing patterns at wall intersections are shown in Fig. 19.

Wall Openings.   Table 12 summarizes the requirements of ACI 313.[13] Figure 20 shows a typical detail of the reinforcement of a narrow silo wall between openings.

DESIGN OF BOTTOMS

21. Bottom Pressure   Static unit pressure q_α normal to a surface inclined at an angle α to the horizontal is given by

$$q_\alpha = p \sin^2 \alpha + q \cos^2 \alpha \qquad (36)$$

Silo bottoms are designed to resist q_α. In seismic zone 4, q in Eq. (36) should be computed for the effective weight (80 percent) of the stored material because of the loss of friction against the silo walls due to seismically induced lateral vibrations, that is, $q = 0.8\gamma Y$. In other seismic zones q should be increased by the following fractions of the increased pressure $(0.8\gamma Y - q)$ for zone 4: ¾ for zone 3, ⅜ for zone 2, and ³⁄₁₆ for zone 1. ACI 318 load factors are suggested for ultimate-strength design under seismic load.

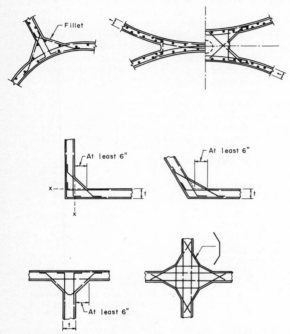

Fig. 19 Reinforcement at intersecting walls. (*Adapted from Ref. 13.*)

TABLE 12   Reinforcement at Wall Openings

Openings	Horizontal steel	Vertical steel
In pressure zone	Add at least 1.2 times area of interrupted reinforcement, ½ above the opening and ½ below	Provide by assuming narrow strip of wall, $3t$ in width on each side of opening, to act as column within opening height subjected to its own vertical load plus that from ½ wall span above opening. Add steel at least equal to that eliminated by opening
Outside pressure zone	Add no less than the normal reinforcement interrupted by opening, distributed as above	As above
Closely spaced		See Fig. 20

22. Plane Bottoms   Design loads for horizontal slab bottoms are dead load, vertical pressure q, and thermal load from the stored material. For inclined slab bottoms q_α should be used. Allowance for earthquake forces should be made as described in Art. 21.

Formulas for bending moments and deflections of circular slabs, with and without a central hole, are given in Ref. 19. Tables of coefficients are given in Ref. 8. Moments and deflections in rectangular slabs can be computed by the ACI 318 procedure for two-way slabs.

23. Conical Hoppers Walls of these structures are subjected to meridional and circumferential tensile membrane forces F_m and F_t (Fig. 21). Values of F_m and F_t per unit width at any horizontal section are given by

$$F_m = \frac{qD}{4 \sin \alpha} + \frac{W}{\pi D \sin \alpha} \qquad (37)$$

$$F_t = \frac{q_\alpha D}{2 \sin \alpha} \qquad (38)$$

where q, q_α = pressures computed at the section
D = diameter at the section
W = weight of that portion of hopper and hopper contents below the section

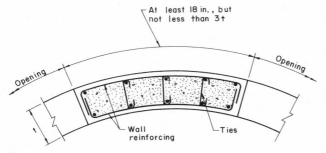

Fig. 20 Reinforcement for narrow silo wall between openings. *(From Ref. 13.)*

F_{mu} and F_{tu} for strength design are obtained by multiplying q and q_α by the load factor 1.7 and computing W by

$$W = 1.4W_h + 1.7W_m \qquad (39)$$

where W_h = weight of hopper below the section and W_m = weight of hopper contents below the section. These weights are given by

$$W_h = \frac{\pi(D^2 - d^2)\gamma_h}{4 \cos \alpha} \qquad (40)$$

$$W_m = \frac{\pi(D^3 - d^3)\gamma_m \tan \alpha}{24} \qquad (41)$$

where γ_h = weight of hopper wall per unit area and γ_m = weight of material per unit volume.

The required reinforcement is given by Eq. (19). A minimum wall thickness of 5 in. is recommended, and the crack width should not exceed an acceptable value.

Design of a conical hopper is illustrated in Example 4.

24. Pyramidal Hoppers Walls of these structures are subjected to tensile membrane forces F_m and F_t (Fig. 22) and plate-type bending. There will also be in-plane bending if the hopper is not supported continuously along its upper edge.

If the vertical components of the meridional forces are assumed to be distributed uniformly on the perimeter of a horizontal section of a symmetrical hopper of rectangular cross section, F_m is given by

$$F_m = \frac{W + (q_a + q_b)\, ab/2}{2(a + b) \sin \alpha} \qquad (42)$$

where a is the length and b the width of the section, q_a and q_b are the vertical pressures corresponding to sides a and b, respectively, W is the weight of that portion of the hopper and hopper contents below the section, and α is the angle with the horizontal of the hopper wall.

If $a > b$, $q_a < q_b$ (because the hydraulic radius is smaller for the longer side). This suggests that the vertical component of the meridional force on wall b may be larger than

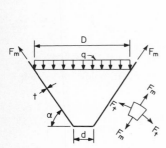

Fig. 21   Forces in conical hopper.

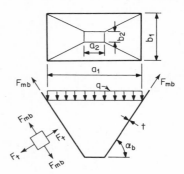

Fig. 22   Forces in pyramidal hopper.

on wall a. Assuming $W/4$ and the resultant vertical pressure on the triangular area adjacent to each wall to be carried by that wall, the following formulas result:

$$F_{ma} = \left(\frac{W}{a} + q_a b\right) \frac{1}{4 \sin \alpha_a} \tag{43a}$$

$$F_{mb} = \left(\frac{W}{b} + q_b a\right) \frac{1}{4 \sin \alpha_b} \tag{43b}$$

The horizontal membrane force F_t is given by

$$F_{ta} = \frac{1}{2}(q_{ab} + \gamma_h \cos \alpha_b) \, b \, \sin \alpha_a \tag{44a}$$

$$F_{tb} = \frac{1}{2}(q_{aa} + \gamma_h \cos \alpha_a) \, a \, \sin \alpha_b \tag{44b}$$

where γ_h = weight of hopper wall per unit of area.

F_{mu} and F_{tu} for strength design are obtained by multiplying q by the load factor 1.7 and computing W by Eq. (39). The weights W_h and W_m are given by

$$W_h = h_h \gamma_h \left(\frac{a_1 + a_2}{\sin \alpha_a} + \frac{b_1 + b_2}{\sin \alpha_b}\right) \tag{45}$$

$$W_m = \frac{h_h \gamma_m}{6}\left[(2a_1 + a_2) \, b_1 + (2a_2 + a_1)b_2\right] \tag{46}$$

where a_1 is the length and b_1 the width of the section at which F is being computed, and a_2 and b_2 are the corresponding dimensions of the hopper opening.

Plate Bending under Normal Pressures.   Bending moments in triangular walls may be approximated by the bending moments in the equivalent rectangular plate shown in Fig. 23a.[8] Tables of coefficients for the analysis of triangular walls with various types of edge support are available.[8]

Bending moments in trapezoidal walls for which $a_2/a_1 \geqq 4$ can be approximated by the moments in the triangular wall formed by extending the sloping sides to their intersection[8] (Fig. 23b). Therefore, trapezoidal walls of these proportions can also be solved by using the equivalent rectangle of Fig. 23a.

Bending moments in trapezoidal walls for which $a_2/a_1 < 4$ can be approximated by the moments in an equivalent rectangular wall (Fig. 23c) with the dimensions[8]

$$a_{eq} = \frac{2a_2(2a_1 + a_2)}{3(a_1 + a_2)} \qquad (47a)$$

$$b_{eq} = h - \frac{a_2(a_2 - a_1)}{6(a_1 + a_2)} \qquad (47b)$$

Tables of coefficients for the analysis of trapezoidal plates with $a_1 = \frac{3}{8}a_2$ and $a_1 = \frac{1}{2}a_2$ with various types of edge support are available.[8]

Edge conditions (fixed, simply supported, etc.) of hopper walls depend on the adjoining construction. They should be considered fixed at their junctures with adjoining walls. In adjoining walls of unequal lengths, the average of the unequal end moments may be used, or they may be distributed in the ratios given by the negative-moment formula in Fig. 15; moments in the central regions of the plates should be adjusted to correspond. Upper edges which are continuous with silo or bunker walls, with or without an intervening edge beam, may be assumed to be fixed. On the other hand, the upper edge of a pyramidal bunker which has no roof must be considered to be free.

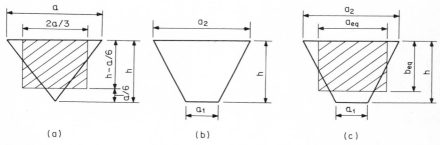

Fig. 23

Bending in Plane of Wall. In-plane bending of a hopper wall acting in conjunction with a low vertical wall is discussed in Art. 17 (Fig. 18). A similar analysis can be made for the bunker without vertical walls, using an effective beam depth of $0.5a$ at midspan[8] (Fig. 24). The stress in the portion below the effective-beam depth is assumed to vary linearly to zero at the vertex, as shown. The bending moment can be computed from $M = F_m a^2/8$, where F_m is the meridional tension computed by Eq. (42) or Eqs. (43). These bending stresses must be combined with the horizontal membrane forces to determine wall reinforcement.

Concentrated Forces at Pyramidal Bunker Supports. Pyramidal bunkers and pyramidal hoppers supported independently on columns at the four corners are subjected to concentrated forces at the supports (Fig. 25). The tensile force T along the edge of adjoining walls is given by

$$T = \frac{P}{\sin \alpha} \qquad (48)$$

where α is the angle of the edge with the horizontal. The compressive forces are given by

$$C_a = P \cos \alpha \cos \beta_a \qquad (49a)$$
$$C_b = P \cos \alpha \cos \beta_b \qquad (49b)$$

where β_a and β_b are the angles between the edges a and b and the diagonal of the horizontal cross section of the hopper.

Since these forces are localized, they need be provided for only in the vicinity of the column.

25. Hopper-Supporting Beams Concrete hoppers are usually supported by edge beams cast integrally with the hopper wall (Fig. 26a,b). A conical steel hopper supported by a ring beam is shown in Fig. 26c. The beams may be supported continuously by a wall, or by pilasters or columns.

If the hopper wall does not intersect the centroid of the supporting-beam cross section, a twisting moment acts on the beam. In ring beams this produces a bending moment which is uniform around the circumference. Both moments can be neglected if the beam is supported on a wall, because the deformation that would be generated is prevented by the silo wall and the bearing wall. Therefore, such a ring beam can be designed for only the horizontal component of the hopper meridional tension, which produces a uniform compressive force, equal to $(F_m \cos \alpha)(D/2)$, in the ring (Example 4).

Ring beams supported on columns or pilasters, as in Fig. 26a, must be designed for the moments, shears, and torsion due to F_m, as well as the ring compression. A design procedure for this case is discussed by Safarian.[27]

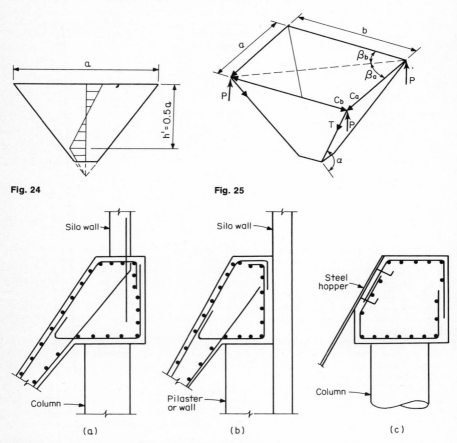

Fig. 24

Fig. 25

Fig. 26 Typical hopper-supporting beams. *(From Ref. 13.)*

No bending moments are produced by twisting of an eccentric edge beam of a rectangular hopper. Torsion can be neglected in edge beams on bearing walls. Edge beams on columns should be doweled to the silo walls to resist torsion.

Where supporting beams are not used and the hopper is keyed to the silo wall, reinforcement must be provided for the hopper-wall bending moments at the juncture.

26. Columns Columns supporting silo shells, and particularly silo bottoms, will be subjected to a live load due to the stored material that is substantially larger than the dead load of the structure. Long sustained periods of material storage cause reinforced-concrete columns to creep. As a result, concrete stresses decrease and the load carried by the steel reinforcement increases. Subsequent emptying of the silo may place the concrete in tension as the reinforcement recovers elastically. Once the tensile stress in the concrete exceeds its tensile strength, the column develops severe transverse cracking. The situation can be dangerous if transverse cracking is accompanied by longitudinal cracking, as could occur with high bond stresses during unloading. To prevent this condition, it is recommended that the reinforcement ratio ρ not exceed 0.02 and that the total amount of reinforcement not exceed L/f_y, where L is the live load on the column. If lateral loads must be resisted, larger columns should be used to keep the steel ratio low. All other provisions of ACI 318, Chap. 10 for the design of columns should be followed.

The analysis described here is illustrated in Example 5.

27. Roofs   Designers are divided on the subject of attachment of concrete roof slabs to silo walls. Some believe that the slab should be supported only vertically at the walls (on elastomeric material or heavy tar paper) so as to be free to contract or expand with temperature changes and to move slightly during earthquakes. To prevent total freedom of horizontal displacement, the slab may be attached at one central location, usually the elevator tower. On the other hand, attaching the roof slab to the walls stiffens tall silos against wind and earthquake loads and reduces lateral deformations. Continuity with the silo walls also makes the roof slab stiffer and reduces vertical deflections under live load.

In long installations expansion joints are provided to reduce cracking of the slab in winter and undue longitudinal forces on the silos in summer. Good design calls for an expansion joint to cut across the silo group by extending down to the foundation, especially if the roof slab is attached to the silo walls.

The steel beams which support the wood platform during slipforming of silo walls are used later to support the roof slab, thus reducing its span considerably. Ample bearing on the concrete wall should be provided at the ends of the steel beams, and the concrete below and to each side of the beam should be reinforced to prevent undue cracking or even a concrete fallout after some years of service.

28. Failures   Three major reasons for failures of reinforced-concrete silos are foundation failure, incorrect determination of loads, and improper detailing and faulty workmanship.

Concrete silos are supported on pile foundations, or on extended rafts in the case of a stiff subsoil. Theimer[2] cites various reasons for failures: (1) the weight is usually great and may shift considerably with unbalanced filling and emptying of the numerous cells, causing major overstress in the foundation, (2) dredging in an adjoining river may weaken the pile foundation, (3) batter piles may fail after having been damaged by a ship collision, (4) piles which have been eroded by aggressive groundwater may buckle, (5) underlying soft soil may shift, causing tilting of the raft and elevator. Prevention of foundation failure requires thorough investigation of subsoil conditions including a number of test-pile loadings. Thorough control and inspection of pile-driving operations and cast-in-place piles (especially those without steel shells to prevent mud intrusions) are necessary. Batter piles should always be used to support high silos against wind action and seismic motions. Raft foundations should be designed as continuously reinforced concrete mats with two grids to resist bending moments caused by unbalanced loading of the silos.

Incorrect determination of silo pressures has resulted invariably in underreinforced structures. Application of the conventional Janssen theory without any allowance for overpressures generated during emptying and by eccentric outlets, combined with the use of higher allowable stresses for concrete and steel, has caused unacceptable major cracking and even total failure in numerous instances. ACI 313,[13] DIN 1055,[12] or Pieper's recommendations[14] will give a more accurate evaluation of load effects.

Major damage and even collapse of silos have resulted from improper detailing and faulty construction. Irregular and excessive spacing of hoop reinforcement, particularly in slipformed silos, may seriously reduce strength. In addition, radial displacement of hoop bars is frequent, and by reducing concrete cover, the capacity of the lap splices is limited. If hoop bars are placed without tying, circumferential shifts may occur, thus making some lap lengths shorter and some longer than the specified length. Absence of vertical steel in combination with excessive spacing of hoop reinforcement may leave large portions of concrete unreinforced and cracked, a situation which may eventually result in concrete fallouts.[28] Vertical reinforcement of walls and columns should be anchored to the foundation by dowels to prevent uplift and shifting under earthquake or wind loading. Insufficient cover for the hoop reinforcement causes it to corrode in a rather short period of time. Sloppy workmanship may cause all splices of hoop reinforcement to be at the same locations without staggering, thereby increasing the possibility of bond failure (with splitting) and thus weakening the wall. Improper detailing at wall intersections of interstices and pocket bins may omit wall fillets and double layers of reinforcement that are necessary to rest local bending moments and shears. Proper attention to detail, followed by inspection at the jobsite and proper control and organization of construction are required to prevent mistakes that may lead to collapse.

29. Dust Explosions in Grain Elevators and Flour Mills   Major destruction of these installations occurs when dust from grain products ignites and releases great amounts of energy. An extremely rapid pressure rise, of the order of 2000 tons/ft²/sec, originates a pressure wave of such high intensity that normal vents for the release of explosion pressures are insufficient to prevent the installation from blowing up. Theimer[29] cites three principal causes for these explosions: a dust cloud, a source of ignition, and the presence of oxygen.

It is necessary to have a minimum concentration of 0.02 oz/ft³ of grain or flour dust (resembling a dense fog) before it can become ignited. If the concentration is greater than 2 oz/ft³, incomplete combustion of the particles retards ignition and prevents the explosion. Dry dust that accumulates on floors, walls, ledges of doors and windows, steel beams, overhead ducts, etc., is highly oxygenated and quite dangerous. Good housekeeping calls for constant removal of such dust. Suction is the principal method to control dust clouds by inducing air currents supplied through dust-collecting systems. Venting is also recommended for bins, heads of bucket elevators, and scale hoppers.

Ignition temperatures vary between 750 and 930°F when the air relative humidity is between 30 and 90 percent. Sources of heat that can ignite dust clouds are: (1) open flames (lights, matches, burning cigarettes); (2) heat generated on pulleys of bucket elevators due to belt slip; (3) hot surfaces of radiators, bearings, light bulbs; (4) sparks caused by metal parts in rotating machinery, electric equipment, and friction; (5) static electricity; and (6) welding, cutting, and soldering. In addition, Theimer[29] cites a case of spontaneous ignition due to constant increase in the temperature of the material caused by inadequate heat dispersion. Obviously, all prevention measures should be taken to avoid ignition heat sources.

Designers should consider the layout of various buildings in an installation to decrease its vulnerability to explosions as a whole. Theimer suggests leaving as large a space as possible between the various buildings. Explosion reliefs such as vents, light brick walls, and light roof construction should be provided.

EXAMPLES

Example 1   Determine the hoop and meridional reinforcement for a 30-ft-diameter silo 100 ft high with a 6-in. wall, and a conical hopper 20 ft deep, to contain wheat (Fig. 2a). Use ACI 313 (Art. 4) with f'_c = 4000 psi and A615 Grade 60 deformed bars.

From Table 5, γ = 50 lb/ft³, ρ = 28°, and μ' = 0.44. The basic pressures and forces are obtained from Janssen's formulas (Table 1) except that $V = (\gamma Y - 0.8q) R$ as required by ACI 313.

$$R = \frac{D}{4} = \frac{30}{4} = 7.5 \text{ ft} \qquad k = \frac{1 - \sin\rho}{1 + \sin\rho} = \frac{1 - \sin 28°}{1 + \sin 28°} = 0.361$$

$$Y_0 = \frac{R}{\mu' k} = \frac{7.5}{0.44 \times 0.361} = 47.22 \text{ ft}$$

$$q = \gamma Y_0 (1 - e^{-Y/Y_0}) \qquad p = qk$$

Basic pressures are multiplied by the overpressure factors C_d from Table 9. The results are given in Table 13.

Hoop reinforcement is determined in Table 14, using the design pressures of Table 13. The hoop force $T = p_{des}D/2$ and $F_u = 1.7F$. The required steel area is given by $A_s = T_u/(0.9 \times 60)$. Spacing of No. 5 bars can be computed from $s = 0.31 \times 12/A_s$.

Spacing is given by zones starting from a minimum No. 5 at 12 in. for the top 40 ft of wall and ending with No. 5 at 5½ in. for the lower 20 ft above the hopper level. The latter spacing is continued for a distance of six times the wall thickness below the pressure zone (Table 11). With t = 6 in. this calls for No. 5 at 5½ in. to be continued another 3 ft below hopper level. From there down to the foundation, Table 11 requires a minimum A_s = 0.0025t. Thus A_s = 0.0025 × 6 = 0.015 in.²/in. = 0.18 in.²/ft, i.e., No. 5 at 20.7 in.; use No. 5 at 12 in. (0.31 in.²/ft).

Meridional reinforcement is usually determined to satisfy minimum requirements (Table 11). For this purpose A_s = 0.0020t = 0.0020 × 6 = 0.012 in.²/in. = 0.144 in.²/ft, i.e., No. 4 at 16.3 in.; use No. 4 at 16 in. (0.147 in.²/ft).

WALL COMPRESSION. The maximum meridional force on the walls above the hopper level is at $Y = 100$ ft. The factored loads are

$$
\begin{aligned}
\text{Weight of wall} &= 1.4 \times 100 \times 0.15 \times {}^{6}\!/_{12} & = 10.5 \\
\text{Weight of 6-in. roof} &= 1.4 \times ({}^{30}\!/_{4}) \times 0.15 \times {}^{6}\!/_{12} &= 0.8 \\
1.7\, C_d V \text{ from Table 13} &= 1.7 \times 47.56 & = 80.9 \\
& & \overline{92.2 \text{ kips/ft}}
\end{aligned}
$$

$$
f_c = \frac{92.2}{12 \times 6} = 1.28 \text{ ksi}
$$

The allowable value from Eq. (30) is $f_c = 0.7 \times 0.55 \times 4 = 1.54$ ksi.

TABLE 13   Values of q, p, and V for Silo of Example 1

			Basic pressures and forces				Design pressures and forces		
Y, ft	Y/Y_0	$1 - e^{-Y/Y_0}$	q, lb/ft²	p, lb/ft²	V, kips/ft	C_d	$C_d q$, lb/ft²	$C_d p$, lb/ft²	$C_d V$, kips/ft
0	0	0	0	0	0	0	0	0	0
10	0.212	0.191	451	163	1.04	1.6	722	261	1.66
20	0.424	0.346	817	295	2.60	1.7	1389	502	4.42
30	0.635	0.470	1110	401	4.59	1.7	1887	682	7.80
40	0.847	0.571	1348	487	6.91	1.8	2426	877	12.44
50	1.059	0.653	1542	557	9.50	1.8	2776	1003	17.10
60	1.271	0.719	1698	613	12.31	1.9	3226	1165	23.39
70	1.483	0.773	1825	659	15.30	1.9	3468	1252	29.07
80	1.694	0.816	1927	695	18.44	1.9	3661	1321	35.04
90	1.906	0.851	2009	725	21.70	1.9	3817	1378	41.23
100	2.118	0.880	2078	750	25.03	1.9	3948*	1425*	47.56
110	2.330	0.903	2132	770			3061	1105	
120	2.541	0.921	2174	785			2174	785	

*Design pressures are reduced linearly from this level to basic pressures at bottom of hopper.

TABLE 14   Hoop Reinforcement for Silo of Example 1

Depth Y, ft	Design pressure p_{des}, lb/ft²	Hoop forces F, kips/ft	F_u, kips/ft	Steel required A_s, in.²/ft	Steel provided A_s, in.²/ft	Spacing
0	0	0	0	0	0.31	No. 5 at 12 in.
10	261	3.92	6.66	0.12		
20	502	7.53	12.80	0.24		
30	682	10.23	17.39	0.32		
40	877	13.16	22.37	0.41	0.47	No. 5 at 8 in.
50	1003	15.05	25.59	0.47		
60	1165	17.48	29.72	0.55	0.62	No. 5 at 6 in.
70	1252	18.78	31.93	0.59		
80	1321	19.82	33.69	0.62	0.68	No. 5 at 5½ in.
90	1378	20.67	35.14	0.65		
100	1425	21.38	36.35	0.67		

ECCENTRIC DISCHARGE. If the silo of this example is provided with an inclined hopper slab (Fig. 2f) instead of a central conical hopper, the design pressures should be determined by Eq. (5). Since the outlet is at the wall, $e = r$ and $p_{des} = C_d p + 0.25 p Y/H$, where p and $C_d p$ are given in Table 13. These pressures are compared in Table 15 with the values from Table 13 for a central outlet.

Example 2   Determine the maximum crack width in the walls of the silo of Example 1. Maximum effects can be expected just above the top of the hopper ($Y = 100$ ft). From the data of Table 13 the hoop tensions for Eqs. (18) are

$$
F_{st} = pD/2 = 750 \times 30/2 = 11,250 \text{ lb/ft}
$$
$$
F_{tot} = p_{des} D/2 = 1425 \times 30/2 = 21,375 \text{ lb/ft}
$$

From Table 14, $A_s = 0.68$ in.^{2}/ft. Then

$$f_{s,st} = 11,250/0.68 = 16,640 \text{ psi}$$
$$f_{s,tot} = 21,375/0.68 = 31,620 \text{ psi}$$

From Eqs. (18), with $f_t' = 4.5\sqrt{f_c'} = 4.5\sqrt{4000} = 285$ lb/in., $\beta = 0.7$ for deformed bars, and $A = 6 \times 12 = 72$ in.^{2}

$$\psi_1 = 1 - \frac{0.7 \times 0.8 \times 72 \times 285}{21,375} = 0.462$$

$$\psi_2 = 1 - \frac{0.7 \times 0.8 \times 72 \times 285}{11,250} = -0.02 < 0.3; \text{ use } 0.3.$$

$$\psi_3 = 1 - \frac{0.35 \times 0.8 \times 72 \times 285}{11,250} = 0.489 < 0.65; \text{ use } 0.65.$$

For No. 5 bars at 5½ in. $\Sigma o = 1.96 \times 12/5.5 = 4.28$ in./ft

$$w_1 = \frac{0.462 \times 72 \times 0.7 \times 31,620}{4.28 \times 29 \times 10^6} = 0.00593 \text{ in.}$$

$$w_2 = \frac{0.3 \times 72 \times 0.7 \times 16,640}{4.28 \times 29 \times 10^6} = 0.00203 \text{ in.}$$

$$w_3 = \frac{0.65 \times 72 \times 0.7 \times 16,640}{4.28 \times 29 \times 10^6} = 0.00439 \text{ in.}$$

Crack width at $Y = 100$ ft is given by Eq. (17)

$$w_{cr} = w_1 - w_2 + w_3 = 0.00829 \text{ in.}$$

This crack width is 3.6 percent larger than the suggested limit 0.008 in., but may be acceptable. Reducing the spacing to 5 in. would reduce the crack width to $0.00829 \times 5/5.5 = 0.0075$ in.

TABLE 15   Pressures with Concentric Outlet and Eccentric Outlet for Silo of Example 1

Y, ft	p_{des}	
	Concentric	Eccentric
0	0	0
10	261	265
20	502	517
30	682	712
40	877	926
50	1003	1073
60	1165	1257
70	1252	1367
80	1321	1460
90	1378	1541
100	1425	1613

Example 3   Determine the pressures and forces for a flat-bottomed silo 30 ft in diameter and 120 ft high to contain wheat, using Pieper's recommendations (Art. 3).

Assume the outlet diameter $d = 2$ ft. From Table 3, $c = 0.85$ for unfinished concrete walls. From Fig. 6,

$$\alpha_2 = 32(\sqrt[4]{\rho_f c}) = 32\sqrt[4]{31 \times 0.85} = 72.5°$$
$$h_{F2} = \frac{D - d}{2} \tan \alpha_2 = \frac{30 - 2}{2} \tan 72.5° = 44.4 \text{ ft}$$

Since $H = 120 > 2h_{F2}$, the silo is classified as tall. Therefore, the design is governed by emptying pressures (Art. 3).

The data and formulas are shown in Fig. 27 and the calculated pressures and forces in Table 16. Comparison of values in this table shows that the largest lateral pressures are generated during emptying and the largest vertical pressures during filling.

Over the height $h_{F2} = 44.4$ ft above the outlet, emptying pressures vary linearly from the value of p_e at 44.4 ft ($Y = 75.6$ ft) to p_f at the level of the outlet, while increases occur in the region $1.4D = 1.4 \times 30$

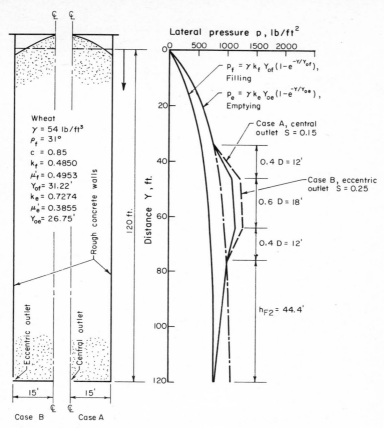

Fig. 27 Silos of Example 3.

TABLE 16 Pressures and Forces for Silo of Example 3

		Filling				Emptying			
Y, ft	Y/Y_{0f}	$1 - e^{-Y/Y_{0f}}$	q_f, lb/ft²	p_f, lb/ft²	V, kips/ft	Y/Y_{0e}	$1 - e^{-Y/Y_{0e}}$	q_e, lb/ft²	p_e, lb/ft²
0	0	0	0	0	0	0	0	0	0
10	0.320	0.274	463	224	0.58	0.374	0.312	451	328
20	0.641	0.473	797	387	2.12	0.748	0.527	761	553
30	0.961	0.618	1041	505	4.34	1.122	0.674	974	708
40	1.281	0.722	1218	591	7.07	1.495	0.776	1121	815
50	1.602	0.798	1346	653	10.16	1.869	0.846	1222	889
60	1.922	0.854	1439	698	13.51	2.243	0.894	1291	939
70	2.422	0.894	1507	731	17.05	2.617	0.927	1339	974
80	2.563	0.923	1556	755	20.73	2.991	0.950	1372	998
90	2.883	0.944	1592	772	24.51	3.365	0.965	1395	1014
100	3.203	0.959	1618	785	28.37	3.738	0.976	1410	1026
110	3.523	0.971	1636	794	32.28	4.112	0.984	1421	1034
120	3.844	0.979	1650	800	36.23	4.486	0.989	1428	1039

= 42 ft above the 44.4-ft level (Fig. 7c). From Art. 3, values of S are taken at 0.15 for a central outlet (Case A of Fig. 27) and 0.25 for an outlet at the wall (Case B). Final lateral design pressures are given in Table 17. The horizontal distribution of these additional pressures is shown in Fig. 7d. The pressures in Case A produce additional hoop forces, while Case B produces bending moments as well as additional hoop forces (Fig. 11).

TABLE 17   Final Lateral Design Pressure, Example 3

Depth Y, ft	Basic pressure p_e, lb/ft^2	Central outlet p_e, lb/ft^2	Eccentric outlet p_e, lb/ft^2
0	0	0	0
10	328	328	328
20	553	553	553
30	708	708	708
33.6	751	751	751
45.6	860	989	1075
63.6	953	1096	1191
75.6	988	988	988
80	998	969	969
90	1014	927	927
100	1026	885	885
110	1034	842	842
120	1039	800	800

Example 4   Design a conical hopper for the silo of Example 1 according to ACI 313 (Art. 4). The hopper is 8 in. thick, 20 ft high, and has top and bottom diameters of 30 and 3 ft, respectively.

The tangential and meridional forces are given by Eqs. (37), (38), and (39), with W_h and W_m by Eqs. (40) and (41). Values of q_α are computed by Eq. (36).

Results are given in Table 18. Areas of reinforcement per unit length required for strength are given in the last two columns. Some adjustment upward was necessary to satisfy a crack width $w < 0.008$ in. This was easily accomplished by increasing the length of the meridional bars and reducing the spacing of the hoop bars.

TABLE 18   Design of Conical Hopper for Silo of Example 1

Depth Y, ft	Basic pressures q, lb/ft^2	p, lb/ft^2	Design pressures q_{des}, lb/ft^2	p_{des}, lb/ft^2	$q_{a, des}$, lb/ft^2	Tangential force F_{tu}, lb/ft	Meridional force F_{mu}, lb/ft	Tangential steel $A_{s,t}$, in.^{2}/ft	Meridional steel $A_{s,m}$, in.^{2}/ft
100	2078	750	3117	1125	1725	52,600	55,400	0.97	1.03
105	2105	760	3158	1140	1748	41,300	42,500	0.76	0.79
110	2132	770	3198	1155	1771	29,700	29,800	0.55	0.55
115	2154	778	3231	1167	1789	17,700	17,400	0.33	0.32
120	2174	785	3261	1178	1806	5,500	5,000	0.10	0.09

RING BEAM (Art. 25). From Table 18, $F_{mu} = 55.4$ kips at the top of the hopper. From Fig. 28, the slope of the hopper wall is 56°.7. Therefore, the ring compression is

$$P = (F_{mu} \cos \alpha)(D/2) = (55.4 \cos 56°.7)(30/2) = 456 \text{ kips}$$

The ring shown in Fig. 28 is 15×20 in. with 8 No. 6 bars. Then

$$P_u = \phi[0.85f'_c(A_g - A_s) + f_y A_s]$$
$$= 0.7[0.85 \times 4(300 - 3.52) + 60 \times 3.52] = 853 \text{ kips}$$

Although P_u is considerably larger than P, use of a smaller ring is questionable. The 15-in. width gives projections to facilitate forming, and the depth gives bending strength to bridge openings that might later be cut into the bearing wall.

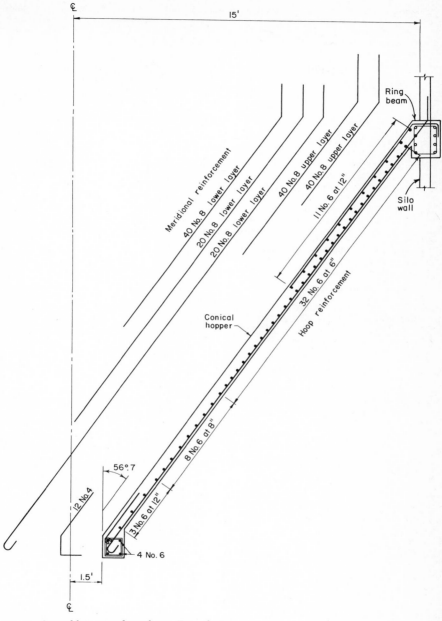

Fig. 28 Conical hopper and ring beam, Example 4.

Example 5   A 20-in.-square reinforced-concrete column with 4 No. 18 bars and No. 4 ties at 12 in. is one of a group supporting a silo hopper as shown in Fig. 2e. The column loads are D = 150 kips and L = 600 kips. f'_c = 3 ksi, f_y = 60 ksi, E_c = 3000 ksi, and E_s = 29,000 ksi. Unloading of the silo may occur a long time after filling. Check the suitability of the design (Art. 26).

The initial strain ϵ_i in the concrete is

$$\epsilon_i = \frac{D + L}{[A_g + (E_s/E_c - 1)A_s]\,E_c}$$

$$= \frac{150 + 600}{[20 \times 20 + (29{,}000/3000 - 1)16]3000} = 0.000464$$

Assume the strain has trebled because of creep. Thus ϵ_t = 3 × 0.000464 = 0.0014. The stress in the steel is $f_s = \epsilon_t E_s$ = 0.0014 × 29,000 = 40.6 ksi and can support a load $P_s = f_s A_s$ = 40.6 × 16 = 649.6 kips. The load supported by the concrete is $P_c = D + L - P_s$ = 150 + 600 − 649.6 = 100.4 kips. The stress in the concrete is $P_c /(A_g - A_s)$ = 100.4/(20 × 20 − 16) = 0.261 ksi.

Upon removal of the live load, elastic unloading occurs. The strain $\Delta\epsilon$ which is recovered is given by

$$\Delta\epsilon = \epsilon_i \frac{L}{D + L} = 0.000464 \times \frac{600}{150 + 600} = 0.000371$$

The stress in the concrete is reduced from 0.261 in compression to 0.261 − 0.000371 × 3000 = 0.852 ksi in tension. This exceeds the cracking strength of the concrete, estimated at $4.5\sqrt{f'_c} = 4.5\sqrt{3000} = 246$ psi. As a result, concrete will crack during unloading of the silo. The stress in the steel after unloading is given by $f_s = D/A_s$ = 150/16 = 9.38 ksi.

To avoid cracking, the amount of steel in the column should be limited to the smaller of $0.02A_c$ or L/f_y. This gives 0.02 × 20 × 20 = 8 in.² and 600/60 = 10 in.². Thus A_s for this column should be 8 in.² instead of 16 in.² This reduction requires f'_c = 4000 psi instead of 3000 psi.

Example 6 (See Art 10)   Determine the thermal reinforcement required for a concrete silo in a region where T_0 = −20°F. The silo has doubly reinforced walls 8 in. thick and stores cement for which T_i = 400°F. f'_c = 4000 psi, f_y = 60,000 psi.

Using Eqs. (15b), (14), and (13) gives

$$K_t = \frac{0.08 \times 8}{4.09 + 0.08 \times 8} = 0.135$$
$$\Delta T = [400 - 80 - (-20)]0.135 = 46°$$
$$M_t = \frac{57{,}000\sqrt{4000} \times 8^2 \times 5.5 \times 10^{-6} \times 46}{1 - 0.3} = 83{,}200 \text{ ft-lb}$$

Assume net covers of 1.5 and 0.75 in., respectively, for the exterior and interior layers of hoop reinforcement. If No. 5 bars are used, d = 8 − 1.5 − 0.62/2 = 6.2 in. and d' = 0.75 + 0.62/2 = 1.1 in. Then from Eq. (35)

$$A_{s,t} = \frac{1.4 \times 83{,}200}{0.9 \times 60{,}000(6.2 - 1.1)} = 0.42 \text{ in.}^2/\text{ft}$$

Place this amount in the outer layer and, for the sake of equal spacing, the same amount in the inner layer.

REFERENCES

1. Janssen, H. A.: Versuch über Getreide-druck in Silozellen, *VDI Z.* (Düsseldorf), vol. 39, Aug. 31, 1895.
2. Theimer, O. F.: Failures of Reinforced Concrete Grain Silos, *ASME Publ.* 68-MH-36, New York, 1968.
3. Kovtun, A. P., and P. N. Platonov: The Pressure of Grain on Silo Walls, *Mukomol'no Elevat. Promst.*, Moscow, USSR, vol. 25, no. 12, December 1959.
4. Pieper, K., G. Mittelman, and F. Wenzel: Messungen des Horizontalen Getreidedruckes in einer 65 m hohen Silozelles, *Beton und Stahlbetonbau*, Berlin, vol. 59, no. 1, November 1964.
5. Pieper, K.: Investigation of Silo Loads on Measuring Models, *J. Eng. Ind. Trans. ASME*, May 1969.
6. Reimbert, Marcel, and André Reimbert: "Silos—Traité Theoretique et Practique," Editions Eyrolles, Paris, 1961.
7. Jenike, A. W., and J. R. Johanson: Bin Loads, *J. Struct. Div. ASCE*, April 1968.
8. Fischer, W.: "Silos und Bunker in Stahlbeton," VEB Verlag für Bauwesen Berlin, DDR, 1966.
9. Lipnitski, M. E., and S. P. Abramovitsch: "Reinforced Concrete Bunkers and Silos" (in Russian), Izdatelstvo Literaturi Po Stroitelstvu, Leningrad, 1967.
10. Turitzin, A. M.: Dynamic Pressure of Granular Material in Deep Bins, *J. Struct. Div. ASCE*, April 1963.

11. Homes, A. G.: Lateral Pressures of Granular Materials in Silos, *ASME Publ.* 72-MH-30, New York, 1972.
12. "Lastannahmen für Bauten. Lasten in Silozellen," DIN 1055 Sheet 6, November 1964. Also supplementary provisions, May 1977.
13. ACI 313–77: Recommended Practice for Design and Construction of Concrete Bins, Silos, and Bunkers for Storing Granular Materials, *J. ACI*, October 1975.
14. Pieper, K., Technische Universität Braunschweig, unpublished work in private communication to E. H. Gaylord, 1977.
15. Gurfinkel, G.: Collapse Investigation of Inclined Hopper of Reinforced Concrete Silo in Shellburn, Indiana, Report for Sullivan County Farm Bureau Co-op, Sullivan, Ind., March 1976.
16. Ukazania P_0 Proectirovaniu Silosov Dlia Siputschich Materialov (Instructions for Design of Silos for Granular Materials), Soviet Code CH-302-65, Gosstroy, USSR, Moscow, 1965.
17. Safarian, S. S.: Design Pressures of Granular Material in Silos, *J. ACI*, August 1969.
18. Kellner, M.: Silos à Cellules de Grande Profondeur, *Travaux*, October 1960.
19. Timoshenko, S., and S. Woinowski-Krieger: "Theory of Plates and Shells," 2d ed., McGraw-Hill Book Company, New York, 1959.
20. Rectangular Concrete Tanks, Portland Cement Association, IS003.020, Chicago, 1969.
21. Gurfinkel, G.: Thermal Effects in Walls of Nuclear Containments, Elastic and Inelastic Behavior, *Proc. 1st Int. Conf. Structural Mechanics in Reactor Technology*, vol. 5, part J, Berlin, Germany, September 1972.
22. Building Code Requirements for Reinforced Concrete, ACI 318-77, American Concrete Institute, Detroit.
23. Ciesielski, R., et al.: Behalter, Bunker, Silos, Schornsteine, Fernsehturme und Freileitungsmaste, Wilhelm Ernst & Sohn KG, Berlin, 1970.
24. Schütt, H.: Über das Tragvermögen wandartiger Stahlbetonon und Stahlbetonbau, Beton and Stahlbetonbau, October 1956.
25. Gurfinkel, G.: Investigation of Silos at Seneca, Illinois, Report for Continental Grain Co., Regional Office in Chicago, Ill., October 1974.
26. Discussion of ACI 313–77, *J. ACI*, June 1976.
27. Safarian, S. S.: Design of a Circular Concrete Ring-Beam and Column System Supporting a Silo Hopper, *J. ACI*, February 1969.
28. Gurfinkel, G.: Structural Adequacy of Reinforced Concrete Silo Complex in Leverett, Illinois, Report for Thomasboro Grain Co., Thomasboro, Ill., December 1976.
29. Theimer, O. F.: Cause and Prevention of Dust Explosions in Grain Elevators and Flour Mills, *ASME Publ.* 72-MH-25, New York, 1972.

Section **23**

Steel Tanks

ROBERT S. WOZNIAK
Senior Engineer, Chicago Bridge & Iron Company, Oak Brook, Ill.

Water-storage tanks may be reservoirs, standpipes, or elevated tanks. *Reservoirs* are large-diameter, vertical, cylindrical, ground-supported, flat-bottom tanks, relatively low in height (20 to 50 ft). On flat terrain the water from these reservoirs must be pumped into a water-distribution system to provide the required pressure and flow. On hilly terrain the tank can be located on high ground to provide pressure and flow. *Standpipes* also are vertical, cylindrical, ground-supported, flat-bottom tanks, but of such height that the upper portion of the storage is available at some minimum desired pressure in the distribution system. The water in the lower portion is reserve storage requiring pumping to maintain minimum pressure. *Elevated tanks* are storage vessels on structural towers, the entire volume of which is available at required minimum pressure. Reservoirs and standpipes use identical procedures of design.

The American Water Works Association (AWWA) has established standards for water-storage containers, and the American Petroleum Institute (API) for tanks for the petroleum industry.

RESERVOIRS

1. Capacity The capacity Q of a reservoir is determined by local requirements. The maximum permissible height of the tank is determined by the supporting power of the soil on which the tank will be built. Steel reservoirs 330 ft in diameter, with capacities as large as 28 million gallons, have been built. Shell plates up to 2 in. thick have been used. Several trial calculations are needed to determine the most economical dimensions. The dimensions may be determined by the equations

$$H = \frac{P}{62.5g} \tag{1}$$

$$D = 0.4126 \sqrt{\frac{Q}{H}} \tag{2}$$

where H = maximum permissible height, ft

   D = tank diameter, ft

   Q = required capacity, gal

   P = allowable soil pressure, psf

   g = specific gravity

Since tanks must be tested with water, a specific gravity g less than 1 is not recommended for design purposes.

   2. Shell Design   The shell plate is made up of one or more horizontal plate courses of width w (usually about 8 ft). Several plates may be required to make up each course. The vertical seams are staggered relative to the vertical seams in adjacent plate courses. Tank shells are cylindrical membranes designed to resist hoop tension (Fig. 1). Plate thickness is calculated at the bottom edge of each course.

$$T_h = 2.60hDg \tag{3}$$

$$t_h = \frac{T_h}{fE} = \frac{2.60hDg}{fE} \tag{4}$$

where T_h = shell tension, lb per in., at depth h

   h = depth from top of tank, ft

   E = joint efficiency factor

   f = allowable unit stress, psi

   t_h = shell plate thickness, in.

For welded construction, E varies from 0.35 to 1.0 depending on the type of joint used and the required welding-inspection procedure. Lap-welded joints, where used, are based on fillet welds the full thickness of the plates joined. Lap welds in tank bottoms and roof plates not in contact with water are single-lap-welded from the top side only. Lap welds in roofs and shell plates in contact with water are double-lap-welded.

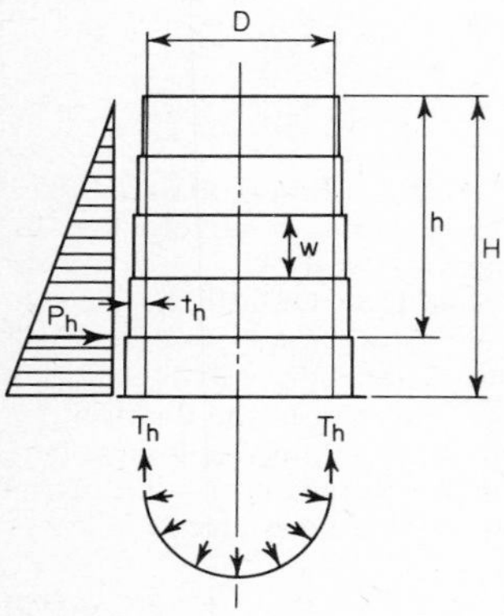

   The joint efficiencies specified in Table 1 are based on nominal inspection of the welding. The owner may specify full magnafluxing in addition to nominal inspection. Only full-penetration butt-welded joints are permitted at joint efficiencies greater than 0.75, and the joint efficiency may exceed 0.85 only when rigid inspection (extended radiographic examination) is provided. Partial-penetration butt welds are permissible only in tank shell joints subjected to secondary stress, such as horizontal seams, for which the direct vertical stress is usually negligible. However, when heavy vertical loads are supported by the shell, the vertical stress is considered to be a principal stress and the horizontal seams require full-penetration butt welding.

Fig. 1

   3. Bottom Plates   The flat bottoms in reservoirs are usually grade-supported, the liquid load on the top side being resisted by an equal upward foundation soil pressure. Steel bottoms usually have a minimum thickness of ¼ in. The plates are lap-welded to each other top side only for liquid tightness (Fig. 2). Figure 3 shows the simple detail required for connecting the bottom to the tank shell.

   Support of relatively light shells requires no special consideration. When the allowable soil pressure directly under the shell is exceeded, increasing the projection a of the bottom plate may suffice to reduce it to an acceptable value. The projection a is given by

$$a = \frac{W'}{2P - 62.5Hg} \tag{5}$$

where W' = shell load, lb per ft of circumference

   P = allowable soil pressure, psf

   H = shell height, ft

The minimum projection should be 1 in.; the practical limit is about 6 in.

The large shell loads caused by earthquakes would require impractical projections. Experience shows that the shell can be allowed to settle during seismic loading with no detrimental effect except that it must be releveled. However, the possibility of a slope failure of the foundation must be investigated, since this could result in loss of support over a large area of the bottom and trigger a bottom failure, with a consequent loss of contents and possible complete failure of the tank.

TABLE 1   Efficiencies of Welded Joints

Type of welded joint	E	Remarks
Single lap.................................	0.35	Continuous welds
Double lap.................................	0.75	Continuous welds
Partial-penetration butt ($\frac{3}{8}$ min)........	0.66	Tension joints
Partial-penetration butt ($\frac{3}{8}$ min)........	1.00	Compression joints
Full-penetration butt....................	0.85	Tension joints
Full-penetration butt....................	1.00	Compression joints

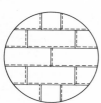

Fig. 2   Lap-welded bottom.   **Fig. 3**   Connection of bottom to shell.

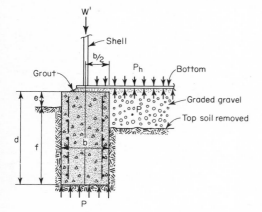

Fig. 4   Foundation ringwall.

4. Concrete Ringwall   If the allowable soil pressure cannot be reduced sufficiently by projecting the bottom, a concrete ringwall must be provided. The ringwall is so proportioned that soil pressure under the ringwall equals the soil pressure under the bottom at the center of the tank (Fig. 4):

$$b = \frac{W'}{31.25Hg - 44d} \qquad (6)$$

where d = ringwall height, ft
b = ringwall thickness, ft

The height e of ringwall above grade should be at least 0.5 ft, and the bottom should be at least 0.5 ft below frost line. A map of extreme frost penetration in the United States is given in Fig. 1, Sec. 5.

Circumferential reinforcing steel hoops must be provided in the concrete ringwall to develop the hoop stress produced by lateral soil pressure within the ringwall. The required area A_s, in.^{2}, of such circumferential steel, based on an allowable stress of 20,000 psi, is determined by

$$A_s = 0.00052HDdg \tag{7}$$

Vertical steel which meets minimum temperature-steel requirements must be installed to provide supports and spacers for the hoop steel.

5. Roofs Tank roofs may be flat, conical, or spheroidal. They can be structurally supported or self-supported. Structurally supported roofs are usually conical with a minimum slope of ¾ in. in 12 in. to ensure drainage. Greater slope is preferable. Columns and beams are arranged as shown in Fig. 5 and positioned so all columns carry approximately equal loads. In freezing climates, the roof framing must be located above the high-water line to avoid damage by ice loading.

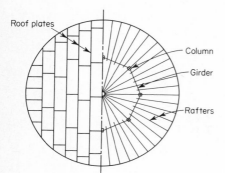

Fig. 5 Roof framing.

Rafters are arranged radially with a maximum spacing of 2π ft. The roof plate is laid directly on the rafters without attachment to them. Roof plates are lap-welded to each other, top side only, for tightness. The rafters are clip-attached to the shell and to the girder (Fig. 6). They may be welded or bolted to the clips. If they are bolted, ASTM A307 bolts should be used. Bolt size and number are determined by the connection loading. The girders, which form a polygon, are bolted to the columns to avoid induced bending in the girders due to limited differential settlement of the columns.

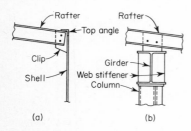

Fig. 6 Rafter connections.

Roof framing is designed for dead load plus 25 psf live load in areas subject to high wind or normal snow load. In climates with very little snowfall a minimum live load of 15 psf may be used, but in areas subject to extremely heavy snowfall a larger load must be specified.

Roof framing and columns are designed in accordance with AISC specifications. The columns are designed as secondary members with a maximum l/r of 180. Compression flanges of rafters are considered laterally supported by the friction of the roof plates, and compression flanges of the girders are laterally supported by the rafters. Columns are supported by structural-steel base spreaders resting directly on the bottom plate. Size of the base spreader is determined by the column load and the allowable soil pressure. Two types are used: (1) circular steel plate and (2) H-shaped structural frames made of channels (Fig. 7). When soil strength is not known the spreader is designed for 1500 lb/ft of channel base, and the circular baseplate for 1500 psf over an area 1 ft larger in radius than the plate.

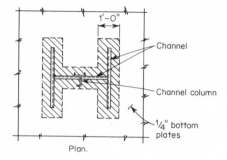

1'-0"

Channel

Channel column

¼" bottom plates

Plan.

Fig. 7   H-shaped column base.

Tanks 100 ft or less in diameter may have rafter-formed roofs supported entirely by the tank shell. The roof framing is steeply sloped to provide economical design. Such rafters are designed as beams loaded with end thrust and vertical loads. The reactions are a horizontal thrust H at the inner end and an equal horizontal thrust and a vertical reaction V at the outer end (Fig. 8a). Because of possible ice loading in freezing climates, the roof framing must be located above the high-water line.

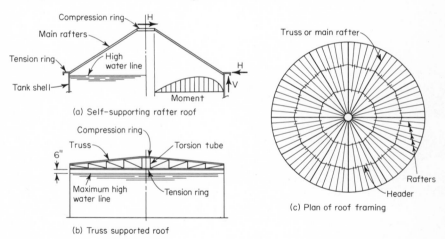

Compression ring

Main rafters

Tension ring

High water line

Tank shell

H

H

V

Moment

(a) Self-supporting rafter roof

Compression ring

Truss

6"

Torsion tube

Maximum high water line

Tension ring

(b) Truss supported roof

Truss or main rafter

Rafters

Header

(c) Plan of roof framing

Fig. 8   Roof framing.

On large-diameter tanks where it is desirable or necessary to avoid columns, the tank roof may be supported by trusses and rafters carried by the shell. Such rafters and trusses are arranged in a radial pattern (Fig. 8c). They are supported by clips welded to the top shell ring and are connected to rings at the center of the tank. The top chords of the trusses are connected to a compression ring, the bottom chords to a tension ring. A torsion tube is provided at the center for truss stability. Eccentricities caused by radial misalignment of truss chord members generate torques tending to rotate the center rings. Interconnecting the top-chord and bottom-chord center rings with a torsion tube provides rotation stability; thus, torque induced by the top-chord compression forces is counteracted with torques generated by bottom-chord tension forces (Fig. 8b).

Tension and Compression Rings on Self-Supported Roofs. The ring is loaded with N equal radial loads W, considered positive outward (Fig. 9). This loading system produces bending and direct stress in the ring with the maximum stresses occurring at the load points and at midpoints between loads.

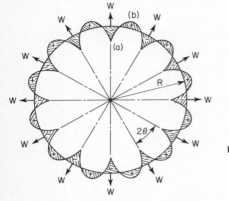

Fig. 9

The moments M, thrusts T, and shears V are determined by the following equations: At load point,

$$M_a = \frac{WR}{2}\left(\cot\frac{180°}{N} - \frac{N}{\pi}\right) \tag{8}$$

$$T_a = \frac{W}{2}\cot\frac{180°}{N} \tag{9}$$

$$V_a = \frac{W}{2} \tag{10}$$

At midpoint between loads,

$$M_b = \frac{WR}{2}\left(\csc\frac{180°}{N} - \frac{N}{\pi}\right) \tag{11}$$

$$T_b = \frac{W}{2}\csc\frac{180°}{N} \tag{12}$$

$$V_b = 0 \tag{13}$$

where R = ring radius, ft
      N = number of radial loads W
Positive thrust produces tension in the ring and positive moment tension at the inside face. The stress in the ring is

$$f = +\frac{T}{A} \pm \frac{Mc}{I} \tag{14}$$

Details of a ring made of a channel are shown in Fig. 10.

STANDPIPES

6. Design The height of a standpipe is usually determined by the required usable gravity-head range and its diameter by the required capacity. Standpipes as tall as 150 ft with shell plates 2 in. thick have been built.

The design of shells and bottoms for standpipes is the same as that for reservoirs. Self-supported roofs are used for economy and simplicity of design and construction. Cone roofs without rafters are limited to tank diameters less than 30 ft and must have a slope of 30° or greater. Standpipes whose diameters are less than 50 ft can be provided with dished, spheroidal, plate roofs without rafters. Such roofs comprise a segment of a sphere

(Fig. 11) whose radius R may range from 0.8 to 1.2 times the tank diameter. The thickness of such a dome can be determined by

$$t = 0.00071CD\sqrt{W} \qquad 0.8 \lessgtr C \lessgtr 1.2 \tag{15}$$

where $C = \dfrac{\text{roof radius, ft}}{\text{tank diameter, ft}}$

D = tank diameter, ft
W = live load + dead load, psf on roof surface
t = plate thickness, in.

Single-lap-welded joints may be used on plates up to ⅜ in. in thickness. Thicker plates must be butt-welded.

7. Anchorage   Standpipes must be checked for overturning by wind or seismic forces. Wind governs when the tank is empty and seismic overturning when the tank is full. Overturning anchorage consists of anchor bolts embedded in the ringwall or concrete mat and connected to the lower part of the tank shell by anchor-bolt chairs (Fig. 12).

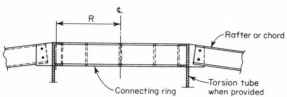

Fig. 10   Details of ring.

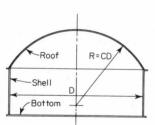

Fig. 11   Spheroidal roof tank.

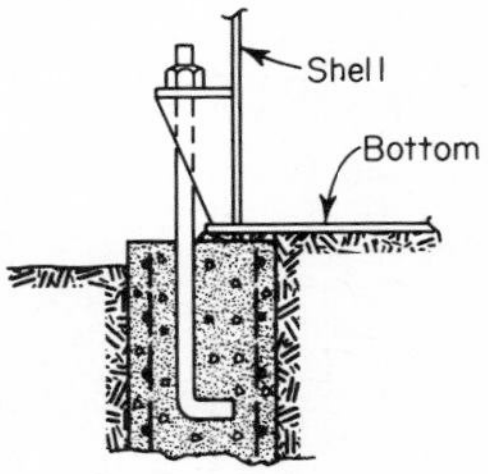

Fig. 12   Shell anchor.

Wind Loading.   An approximate, but practical, solution for anchor-bolt wind loading is

$$T_w = -\frac{W'}{N} + \frac{4M_w}{ND} \tag{16}$$

where T_w = anchor-bolt tension, lb
M_w = overturning moment about base of tank, ft-lb
D = tank diameter, ft
N = number of anchor bolts
W' = total weight of shell and that portion of the roof carried by the shell

The overturning moment is determined by

$$M_w = 0.0015KDH^2V^2 \tag{17}$$

where $K = 0.6$ for cylinders
V = wind velocity, mph
H = shell height, ft

Seismic Loading.   For seismic anchor-bolt loading

$$T_s = -\frac{W'}{N} + \frac{4M_s}{ND} \tag{18}$$

$$M_s = (W_s + W_w)\frac{H}{2}\,S \qquad (19)$$

$$W_w = 49.0HD^2 \qquad (20)$$

where T_s = seismic bolt tension, lb
M_s = seismic overturning moment, ft-lb
W_s = total weight of tank shell and roof
S = seismic factor

No vertical liquid load is included in W' because the anchor bolts develop full tensile load before any liquid on the thin flexible bottom plates can develop effective downward ballast.

Spacing of anchor bolts should not be less than 2 ft or greater than 10 ft; that is,

$$0.31D < N < 1.57D \qquad (21)$$

In any case, N must not be less than 4.

Unanchored Tanks.   Flat-bottomed tanks may survive seismic load without anchorage. In this case, a portion of the liquid contents near the shell resists the overturning forces. The overturning forces produce vertical compression in the bottom shell ring which may cause it to buckle. Tanks with a large height-diameter ratio are more susceptible to buckling and overturning. On the other hand, tanks with a small height-diameter ratio may experience overstress in hoop tension in the middle and upper courses.[15]

Figure 13 can be used to check the seismic resistance of API and AWWA standard unanchored tanks. Thus, according to Fig. 13, an 80-ft API tank 40 ft high has a lateral seismic factor of 0.18 without hoop overstress, but an AWWA tank of the same dimensions would require an increase in the shell thickness for the same seismic factor. In using the charts, the design earthquake should be multiplied by the design factor of safety.

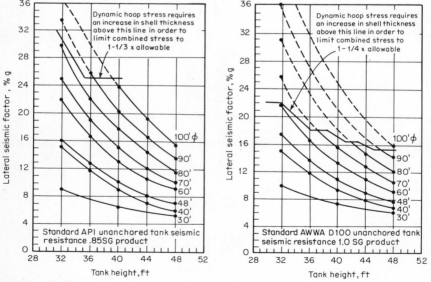

Fig. 13   Seismic resistance of unanchored flat-bottomed tanks.

8. Foundations   Tall, ground-supported standpipes require strong supporting soils. Ringwalls may be required to reduce soil bearing pressure to acceptable limits under the tank shell. The design of ringwalls for standpipes is the same as for reservoirs except that the ringwall must provide adequate ballast to resist maximum anchor-bolt load.

When the supporting soil strength is less than $P = 62.5H$, spread foundations are required (Fig. 14). The required base diameter Z is approximately

$$Z = 1.128 \sqrt{\frac{W_m + W_w + W_e + W_c}{P}} \qquad (22)$$

where W_m = total weight of tank metal, lb
W_w = weight of contents, lb
W_c = weight of concrete, lb
W_e = weight of earth on foundation, lb

Tensile Reinforcement.   The bending moment M about a centerline (Fig. 15) is

$$M = \tfrac{1}{2}(VX_p - W_mX_m - W_wX_w - W_cX_c - W_eX_e) \qquad (23)$$

where V = total foundation reaction, lb
X = distance from the centerline to the centroid of each respective load on one-half the foundation, ft

Reinforcing steel should be distributed uniformly across the diameter, with an equal amount in the transverse direction to form a rectangular grid. If the required area of steel is excessive, greater foundation depth may be required. The entire foundation should be poured as a monolith to eliminate construction-joint difficulties and shear-key problems.

When soil conditions require deep foundations (over 10 ft) it may be more economical to use a slab and ringwall design (Fig. 16a) or a pile-supported foundation (Fig. 16b). The ringwall in Fig. 16a provides an extension of foundation depth and should not be considered a moment-resisting element. It is reinforced for hoop tension produced by internal soil pressure and must be anchored to the base slab to resist overturning moments due to wind or seismicity. When piling foundations are used, the pile capacity should be such that all the units will be driven within the area covered by the tank; otherwise a thick slab will be required to resist the bending moment in the foundations.

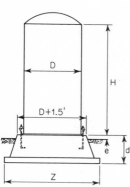

Fig. 14 Spread foundation.

ELEVATED TANKS

Modern elevated tanks are designed as membrane structures having their supporting columns attached directly to the tank shell. A standard tank consists of the roof, shell, and suspended bottom. To enhance its appearance the entire tank may be made of double-curvature plates to form the desired shape.

The elevated tank may have a conical or an elliptical roof, a vertical cylindrical shell to which the columns are attached, and a suspended bottom which may be elliptical, hemispherical, or conical. Such tanks have been constructed with capacities varying from 5000 to 4,000,000 gal. Maximum head range determines the proportions. Tank capacity is measured between the lip of the overflow (top capacity level) and the bottom capacity level set by the specified head range (Fig. 17).

9. Roofs When conical roofs are used, the top capacity level is always maintained below the junction of the roof and the shell. The rules for designing conical roofs are the same as for roofs for reservoirs and standpipes. Ellipsoidal roofs permit establishing the top capacity level within the roof. Water within the roof develops biaxial membrane stresses in the roof plates, which can be calculated when the membrane alone can be completely cut by a transverse plane. One of the biaxial stresses can be determined by statics. The upward hydrostatic pressure on a horizontal plane at the distance h below the top capacity level will be balanced by a system of meridional membrane stresses T_2 (Fig. 18). Thus

$$T_2 = \frac{\gamma h \pi D'^2/4 - W_w}{\pi D' \cos \theta} \qquad (24)$$

where D' = diameter of membrane at cut section
γ = density of product stored

The weight of the metal should be included in the determination of T_2.

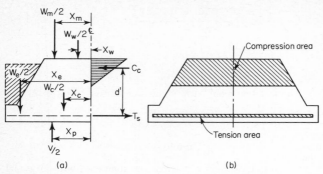

Fig. 15

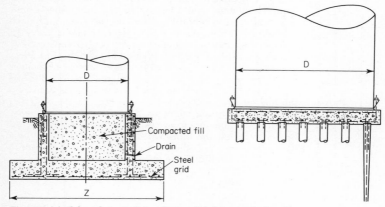

Fig. 16 (a) Slab and ringwall foundation; (b) slab-on-pile foundation.

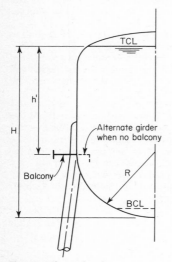

Fig. 17 Elevated tank.

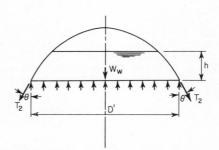

Fig. 18 Membrane forces on roof.

To determine the other (latitudinal) membrane stress T_1, equilibrium of T_1 and T_2 with the hydrostatic pressure p normal to the surface at the depth h gives

$$T_1 = R_1 \left(p - \frac{T_2}{R_2} \right) \tag{25}$$

where R_1 = latitudinal radius
$\quad\quad R_2$ = meridional radius

Several points should be checked to determine maximum compressive stress. The computed maximum stress will determine the required roof-plate thickness. The curvature in the meridional direction determines the compressive buckling strength in the latitudinal direction, whereas the latitudinal curvature determines the compressive buckling strength in the meridional direction.

Having computed T_1 and T_2 the next step is to determine the required plate thickness. Allowable tensile unit stresses are governed by the tensile properties of the material. Allowable compressive unit stresses are governed by buckling strength, which can be determined by the Boardman formula for mild steels,

$$f_a = 2{,}000{,}000 \frac{t}{R} \left(1 - \frac{100}{3} \frac{t}{R} \right) \tag{26}$$

where t = plate thickness, in.
$\quad\quad R$ = curvature normal to direction of stress, in.
$\quad\quad f_a$ = allowable compression stress, psi* (limited to 15,000 psi in current AWWA standard)

10. Suspended Bottoms Design of suspended bottoms formed by a surface of revolution is similar to that of the roof. The bottom is sectioned by a transverse plane, the stresses T_1 and T_2 determined, and the required plate thickness calculated. In hemispherical bottoms, where $R_1 = R_2 = R$, the maximum tensile stress, which occurs at the very bottom (Fig. 17), is

$$T_1 = T_2 = \frac{\gamma H R}{2} \tag{27}$$

The stresses at the spring line are

$$T_1 = \gamma R \left(\frac{h'}{2} - \frac{R}{3} \right) \tag{28}$$

$$T_2 = \gamma R \left(\frac{h'}{2} + \frac{R}{3} \right) \tag{29}$$

where R = spherical radius of bottom
$\quad\quad H$ = distance from top capacity level to bottom of tank
$\quad\quad h'$ = distance from top capacity level to spring line of bottom

The stresses T_1 and T_2 in suspended conical bottoms are determined independently (Fig. 19). When the cylindrical portion is filled to a depth X above the cone-to-cylinder junction, the stresses in the cone at any point h_c below the spring line are

$$T_2 = \frac{\gamma}{2 \cos \theta} \left(\frac{D}{2} - h_c \tan \theta \right) \left(X + \frac{2h_c}{3} + \frac{D}{6} \cot \theta \right) \tag{30}$$

$$T_1 = \frac{\gamma}{\cos \theta} \left(\frac{D}{2} - h_c \tan \theta \right) (X + h_c) \tag{31}$$

At the spring line the stresses are

$$T_2 = \frac{\gamma}{2 \cos \theta} \frac{D}{2} \left(X + \frac{D}{6} \cot \theta \right) \tag{32}$$

$$T_1 = \frac{\gamma D X}{2 \cos \theta} \tag{33}$$

where θ = apex angle
At the apex $T_2 = T_1 = 0$.

*Factor of safety = 2.

Compression stresses must also be determined at the cone-to-cylinder junction, where a compression girder is required to resist the inward pull of the cone bottom (Fig. 31a). The compression force C in the girder is

$$C = \frac{\gamma}{8}\left(X + \frac{D}{6}\cot\theta\right)D^2 \tan\theta \qquad (34)$$

Portions of cone and shell act with the girder. The effective width of each strip is assumed to be $0.78\sqrt{Rt}$ but not to exceed $16t$. Therefore, the effective area is the smaller of

$$A_{\text{eff}} = 0.78(t_c\sqrt{R_c t_c} + t_1\sqrt{R_1 t_1}) \qquad (35a)$$
$$A_{\text{eff(max)}} = 16(t_c^2 + t_1^2) \qquad (35b)$$

where R_c, R_1 = radius of shell, cone
 t_c, t_1 = thickness of shell, cone

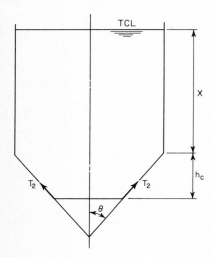

Fig. 19 Membrane forces in suspended bottom.

11. Balcony or Ring Girder   The shell of a column-supported tank is considered to be a circular girder uniformly loaded over its periphery and supported by columns, equally spaced on the shell circumference, attached directly to the tank shell. The supporting tower generates concentrated radial and tangential forces on the tank structure. These forces may be caused by sloping columns and/or the diagonal bracing system in the tower, and a ring girder must be provided to resist them. The ring girder is located at the intersection of the column's neutral axis with the tank shell and is usually positioned at the spring line of the suspended bottom. It also functions as the top strut line of the tower. A tank balcony can serve as a ring girder. Balconies should be wide enough to permit walking upright around the tank and should provide easy passage at the columns. If a balcony is not used, a continuous ring girder must be provided.

The force system on the balcony or ring girder of a tank having vertical columns consists of the shears q in the shell resulting from the horizontal load H at strut line or balcony due to wind or seismic forces, and the resisting forces in the bracing system (Fig. 20). The beam formula VQ/I gives

$$q = \frac{H}{\pi R'}\sin\beta \qquad (36)$$

where R' = horizontal radius of column circle at balcony or strut line
 β = angle from line of action of H to any point on the shell

The resisting force T_{nH} in a bracing rod is

$$T_{nH} = \frac{2H \sin \alpha_n}{N} \qquad (37)$$

where α_n = angle between line of action of H and normal to horizontal projection of bracing rod being considered
$\qquad N$ = number of columns or active rods in a tower panel
$\qquad$ The force system on the balcony or ring girder of a tank having sloping legs consists of

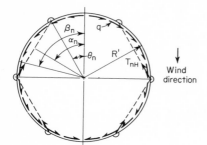

Wind direction

Fig. 20 Forces on ring girder of tank with vertical columns.

the shears q given by Eq. (36), radial thrusts produced by dead and vertical live load and by wind or seismic loads, together with the resisting forces of the bracing system (Fig. 21).
$\qquad$ The radial thrusts P_r from vertical loads are

$$P_r = \frac{SV}{N} \qquad (38)$$

where S = slope of column
$\qquad V$ = sum of vertical load at balcony or strut line
$\qquad$ The radial thrusts P_{nR} due to wind or seismic loads are

$$P_{nR} = \frac{2HaS \cos \theta_n}{R'N} \qquad (39)$$

where a = distance from balcony to center of gravity of horizontal loads
$\qquad \theta_n$ = angle from line of action of H to radial line from center to column n
$\qquad$ The resisting force T_{nsH} in a bracing rod is

$$T_{nsH} = \frac{2H}{N} \left(1 - \frac{aS}{R'}\right) \sin \alpha_n \qquad (40)$$

12. Columns Tower columns may be rolled structural shapes or tubular sections. Tubular columns permit use of longer unbraced lengths and are easier to maintain. Design of the supporting tower follows conventional procedures. The diameter of the tank and height of the tower influence the choice between sloped and vertical columns. Appearance of the structure may also be a deciding factor. The vertical component P_n of the column loads in each panel is

$$P_n = -\frac{V}{N} + \frac{2M}{R'N} \cos \theta_n \qquad (41)$$

where M = overturning moment of wind or seismic forces about the strut line
Columns must also be checked for uplift.
$\qquad$ The panel points on tubular columns must be adequately stiffened with internal, transverse diaphragms to prevent local buckling by surface- or skin-applied loads. This requirement applies particularly to members acting normal to the axis of the columns. When rigid frame connections are used, full internal transverse diaphragms, or the equivalent, must be provided in the column at both the tension and compression flanges of the connecting members (Fig. 22).

The length of contact of the column with the tank is chosen to keep shear buckling stresses in the shell within acceptable limits. Because the column-to-tank connection defies exact analysis, designs have been based on rationalization, experience, and tests (see Art. 17).

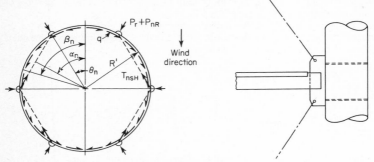

Fig. 21 Forces on ring girder of tank with sloping columns.

Fig. 22 Connection to tubular column.

Tower Rods. The bracing-rod load T_{nR} in each panel of a tower with vertical columns is

$$T_{nR} = \frac{2H \sin \alpha_n}{N \cos \phi_n} \tag{42}$$

For towers with sloped columns the stresses are (Fig. 23)

$$T_{nR} = \frac{(P_B - P_A) \sin \alpha_n}{2 \sin \phi_n \sin (180°/N)} \tag{43}$$

where P_A = leeward column load at strut line A
P_B = leeward column load at strut line B
ϕ_n = angle of bracing rod with horizontal

Tower rods are made of either square or round steel bars. They must be readily weldable. The ends are provided with double clevis plates and fitted pins for attachment to gusset plates welded to the columns. Turnbuckles provide for adjustment of the rods and lining up of the tower. Turnbuckles for the rods of large tanks may exceed practical size limits and be difficult to adjust by manual methods. In such cases the rods may be welded directly to the gusset plates and adjustment made by heating and upset shrinking of the rods. Turnbuckles should be located in the lower end of each panel just above the strut line for convenient access during adjustment.

Tower Struts. The strut load T_{ns} is the horizontal component of the tower rod load:

$$T_{ns} = \frac{2H \sin \alpha_n}{N} \tag{44}$$

Struts may be rolled structural shapes or built-up members. Figure 23a shows a common type. Struts are frequently used to support vertical loads, in which case they must be designed for combined compression and bending. When the thrust is small a maximum l/r = 175 is permissible. Struts may be pin-connected or rigidly connected to the gusset plates. No allowance should be made for fixed-end conditions when calculating l/r. Strut sizes are sometimes determined by erection-loading conditions, in which case they may appear oversized. It is good practice to design the struts so that their ultimate strength is sufficient to develop the yield strength of the rods, since combined pretension and tension due to lateral forces have on occasion stretched rods, especially under seismic loading.

13. Foundations Prior to design and start of construction, every proposed tank site should be explored by a competent soils engineer to determine its suitability. Such investigations avert costly delays and assist in selecting proper foundation construction. The investigation should determine soil strength, magnitude of expected general settlement, and nature and magnitude of differential settlements. It must be remembered that water tanks are continuous, not articulated, and that they generally impose sustained design load during the life of the structure. Only small differential settlements can be tolerated, and repeated releveling or underpinning is expensive. It is prudent to test a site prior to acquisition.

Foundations for elevated tanks are designed so that the column load passes through the centroids of the top and bottom of the pier in order to obtain uniform soil pressure on the base. The pier must be founded 6 in. below the greatest frost penetration. This consideration can be waived when the piers are founded on ledge rock. The gross weight of the pier (which includes concrete and earth directly above the base) must be equal to or greater than the maximum calculated uplift at a column baseplate. Maximum uplift due to wind occurs when the tank is empty, while that due to seismic forces occurs when the tank is full. Figure 24 shows a typical pier and footing. For durability, the 28-day strength of the concrete should be at least 3000 psi. Design should comply with the ACI Building Code. However, because of the reduced load factor for dead load, water load should be considered as live load even if it is defined as dead load in the code being used.

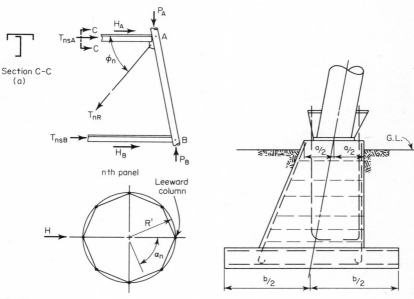

Fig. 23

Fig. 24   Tank-column pier.

The center pier for an elevated tank will vary with piping and center-riser requirements. The pier should be designed to carry the water and metal loads and be proportioned to have uniform pressure under the base. Large valve vaults and control pits are more economically designed as separate units not attached to the center pier. If possible, valve vaults should be located outside the tank-foundation area.

ACCESSORIES

Overflows must be provided for every water tank. They should be sized to handle the maximum pumping rate in the system. Siphon overflows (Fig. 25) are simplest to build. Internal weir-type overflows are subject to ice damage.

The tank should have an external ladder extending to the roof. Internal ladders are subject to ice damage and should be avoided. However, pedestal-type tanks (golf ball on a tee) have internal ladders for aesthetic reasons. Such ladders are acceptable, since they are protected by internal risers and are not exposed to water or ice damage. A manway should be provided in the roof to permit entrance into the tank and to provide ventilation during painting and maintenance of the structure (Fig. 25). Additional manways should be provided in the shell of reservoirs and standpipes located a sufficient distance above the bottom to permit reinforcement of the shell cutout (Fig. 26).

Adequate air vents must be provided in tank roofs to prevent internal pressure or vacuum buildup. The free area for air flow through the vent should equal the area of the

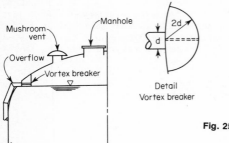

Fig. 25 Roof accessories.

outlet pipe. This will permit the maximum water discharge velocity to reach 40 fps before vacuum on the tank shell will approach ½ in. water pressure. Screening on tank vents should not be smaller than 4 mesh; finer screening clogs up with frost and ice and makes the vent inoperative. Clogged vents can cause collapse of the tank roof during withdrawal and can also cause internal overpressure that may rupture the tank. However, many states require screens finer than 4 mesh. Fail-safe vents which lift a pressure-relief pallet should the screens become clogged are used in these cases. Such vents require maintenance to ensure trouble-free operation.

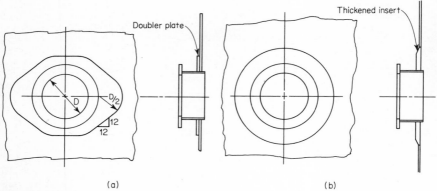

Fig. 26 Manways.

Float-type water-level gages are practical only in warm climates. If gages are desired they should be of the mercury-manometer or bourdon-tube dial-gage type and located in heated areas.

Heaters.   In areas where the low 1-day mean temperature may be +5°F, frost protection must be given consideration. If water does not flow either into or out of the tank at all times, some means should be provided to heat the water. The best method is by means of a heat exchanger that takes cold water out of the base of the riser and discharges heated water into the tank. Such heaters are most commonly needed in fire-protection tanks. Reservoirs and standpipes do not require them as long as the connecting piping is protected from extreme cold. Piping in reservoirs and standpipes should enter the tank, at least 3 ft inside the shell, from pipe vaults below the bottom (Fig. 27). A map of low 1-day mean isothermals for the United States is shown in Fig. 28.

During cold spells in extremely cold climates it may be advisable to waste water during periods of low usage, to draw down the tank, and refill with warmer well water. If the water source is a lake or stream, heating of water may be required to prevent freeze-up of inlet and discharge piping.

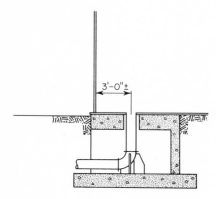

3'-0"±

Fig. 27 Pipe vault.

BINS

Steel storage tanks for granular materials can be classified as (1) structural-type bins made up of structural members and flat plates, and (2) thin, steel, membrane-type bins similar to water-storage tanks. Only the membrane-type bin is discussed here.

Design of the bottom to prevent arching, ratholing, and flow stoppage is discussed in the literature.[2-5,7,8]

14. Forces Factors affecting the behavior of granular materials are temperature, humidity, angle of repose, coefficient of friction between bin walls and fill, depth of fill, duration of storage, and withdrawal-opening location. For purposes of analysis, bins are classified as deep or shallow. A deep bin is defined as one in which the plane of rupture intersects the bin wall rather than the surface of the fill (Fig. 29).

Janssen's formula (see also Sec. 22, Art. 2) is widely used to determine granular pressure and wall drag in deep bins,

$$P = \frac{wR}{f} (1 - e^{-Kfh/R}) \tag{45}$$

$$V = \frac{wR}{Kf} (1 - e^{-Kfh/R}) \tag{46}$$

where P = horizontal pressure at depth h from top, psf
V = vertical pressure at depth h from top, psf
w = weight of fill, pcf
R = hydraulic radius = A/p
A = horizontal cross-sectional area, ft²
p = perimeter of cross section, ft
e = base natural log
f = coefficient of friction of fill on bin wall
$\tan^2 (45° - \phi/2) = \dfrac{1 - \sin \phi}{1 + \sin \phi}$
ϕ = angle of repose of fill
D = bin diameter, ft

Table 2 contains published values of average properties of some common granular materials (see also Sec. 22, Tables 4 and 5). It is advisable to determine by laboratory tests the actual properties of a particular material for which data are not available.

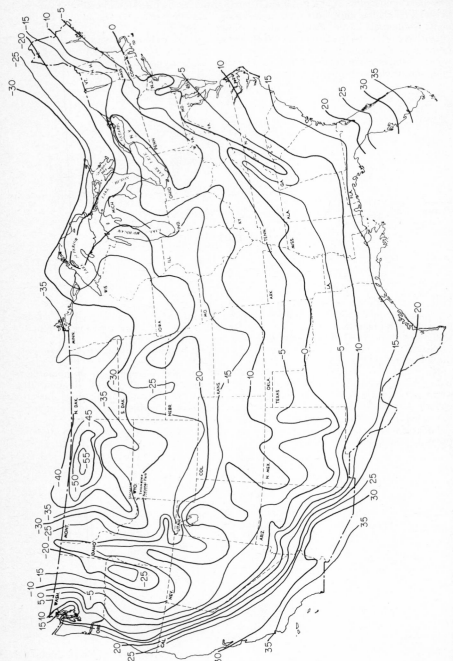

Fig. 28 Isothermal lines, lowest 1-day mean temperatures.

15. Circular Bins   The hoop tension and vertical compression can be derived from Eqs. (45) and (46),

$$T_1 = \frac{wD^2}{96f}\left(1 - e^{-4Kfh/D}\right) \tag{47}$$

$$T_2 = -\frac{wDh}{48} + \frac{T_1}{2K} \tag{48}$$

where T_1 = hoop force, lb/in.
T_2 = meridional force, lb/in.

For thin circular steel bins, T_2 will govern. That is, the shell will fail by buckling from vertical drag rather than bursting due to hoop tension. The allowable vertical compressive

TABLE 2   Properties of Some Common Materials

Material	Weight, pcf	ϕ, deg	K	f
Barley.....................	39	27	0.38	0.38
Beans.....................	46	32	0.31	0.37
Corn......................	44	27	0.38	0.38
Flaxseed..................	41	24	0.42	0.34
Oats......................	28	28	0.36	0.41
Peas......................	50	25	0.40	0.26
Bauxite ore..............	85	35	0.27	0.70
Cement...................	84–92	39	0.22	0.55
Cinders...................	40–45	39	0.22	0.70
Coal, anthracite..........	52–56	27	0.38	0.40
Coal, bituminous.........	45–55	45	0.17	0.70
Gravel, bank.............	120	30–39	0.22–0.33	0.58
Sand, screened...........	90–110	32	0.31	0.55
Stone, crushed...........	100–110	32–39	0.22–0.31	0.80

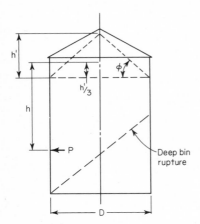

Fig. 29

stress is governed by Eq. (26); an increase of 50 percent is considered reasonable for granular products with properties which are well known and dependable.

Suspended Bottoms.   The suspended bottom is designed to carry the full weight of the fill. From Fig. 30,

$$T_2 = \frac{R_1}{2}\left(q + \frac{W}{\pi R_1^2 \cos^2 \theta}\right) \tag{49}$$

where R_1 = radius at section A-A, in.
$\quad\quad\theta$ = angle of T_2 with vertical
$\quad\quad W$ = weight of fill below section A-A, lb
$\quad\quad q$ = pressure at section A-A, psi
For conical bottoms (Fig. 19), Eq. (49) reduces to

$$T_2 = \frac{w}{24 \cos \theta} \left(\frac{D}{2} - h_c \tan \theta \right) \left(X + \frac{2}{3} h_c + \frac{D}{6} \cot \theta \right) \tag{50}$$

where h_c = depth of cone to point considered, ft
$\quad\quad X$ = depth of fill above cone-cylinder intersection, ft
$\quad\quad w$ = weight of fill, pcf
The hoop force T_1 is

$$T_1 = q_n R_1 \tag{51}$$

where $q_n = q(K \cos^2 \theta + \sin^2 \theta)$ = normal pressure on wall, psi
For a double curved bottom, equilibrium will determine the force T_2, the hoop force T_1 being given by

$$T_1 = q_n R_1 - T_2 \frac{R_1}{R_2} \tag{52}$$

where R_2 is defined in Fig. 31b.

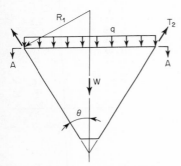

Fig. 30   Forces on suspended bottom.

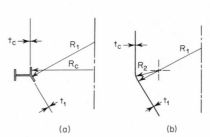

(a)   (b)

Fig. 31   Cylinder-to-cone transitions.

Compression Junction. The transition from cylinder to cone can be made abruptly (Fig. 31a), or with a knuckle as shown in Fig. 31b. If the abrupt change is used, a continuous compression ring must be provided to resist the horizontal pull from the cone. Steel rings should be designed for an allowable stress of 10,000 psi. The relatively low value is used to minimize deflection, and hence the secondary bending stress, which may cause fatigue. For structures subjected to few or infrequent cycling, stresses on the order of 15,000 to 18,000 psi may be permissible. The compression ring must also be checked for buckling. Using a factor of safety of 3 in Levy's formula for buckling of a ring under uniform pressure,

$$I = \frac{T_H R_R^3}{E} \tag{53}$$

where T_H = horizontal component of T_2, lb/in.
$\quad\quad R_R$ = centroidal radius of ring, in.
$\quad\quad E$ = Young's modulus, psi
$\quad\quad I$ = minimum moment of inertia, in.^{4}
The effective area of shell acting with the compression ring is given by Eqs. (35).

The stress T_1 for a knuckle transition is given by Eq. (52). This stress should be computed at two points: $0.78 \sqrt{R_1 t_1}$ from the junction of the knuckle with the shell and $0.78 \sqrt{R_c t_c}$ from its junction with the cone. Allowable values are given by Eq. (26).

Bin-Support Details.   Bins may be supported on the ground, on columns, or on skirts, as shown in Fig. 32*a*, *b*, and *c*. The final choice may depend on the relative cost, on service requirements, on aesthetics, and on customer preference. For relatively low structures, the skirt support may be the most economical and may offer the added advantage of providing a housing for equipment. For high, elevated structures, a tower is preferable.

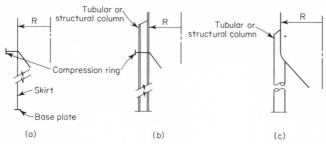

Fig. 32   Bin supports.

The handling process (filling and emptying) requires careful analysis. The filling method should be such that the surface of the granular material is at a uniform height at its periphery. The same rule applies to withdrawal. Extreme unsymmetrical loading will result in distorted and buckled shells.

Difficulty has been experienced in membrane-type bins having side withdrawal. The unsymmetrical loading on the bin, coupled with increased drag on the shell in the region of the withdrawal opening, has caused buckling in the shell directly above the opening.

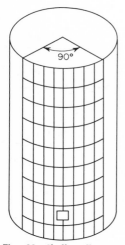

Fig. 33   Shell stiffening at withdrawal opening.

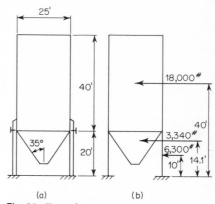

Fig. 34   Example.

Side withdrawal also causes a loss of radial pressure, resulting in a flattening of the shell in the area of the opening. The disturbance extends the full height of filling. The area affected extends the full height of the tank and 30° on either side of the opening. This situation can be remedied by stiffening the shell vertically and circumferentially over a 90° arc as shown in Fig. 33. Material removal by fluidizing can create the same effects if such removal systems are extended too close to the shell.

16. Shallow Bins Hoop tension will control design of the shell for shallow bins. Two different formulas have been used,

$$T_1 = \frac{whD}{24}\left(\frac{1}{\sqrt{\phi(\phi + f)} + \sqrt{1 + \phi^2}}\right)^2 \qquad (54)$$

$$T_1 = \frac{KwhD}{24} \qquad (55)$$

These equations give similar results over a large range of values.

17. Example Crushed-stone bin (Fig. 34a).

$w = 100$ pcf	$D = 25$ ft
$\phi = 37°$	$R_c = 150$ in.
$K = 0.25$	$H = 40$ ft
$f = 0.80$*	$\theta = 35°$

SHELL. From Eq. (47),

$$T_1 = \frac{100 \times 25^2}{96 \times 0.80}(1 - e^{-4\times0.25\times0.80h/D})$$
$$= 813(1 - e^{-0.80h/D})$$

Try $t = 0.3125$ in.

$$\frac{h}{D} = \frac{40}{25} = 1.60 \qquad T_1 = 587 \text{ lb/in.}$$

$$\sigma_1 = \frac{587}{t} = 1880 \text{ psi}$$

From Eq. (48),

$$T_2 = -\frac{100 \times 25 \times 40}{48} + \frac{587}{0.50} = -909 \text{ lb/in.}$$

$$\sigma_2 = -\frac{909}{t} = -2909 \text{ psi}$$

From Eq. (26),

$$\frac{t}{R_c} = \frac{0.3125}{150} = 0.00208$$
$$\sigma_{\text{all}} = -3870 \times 1.50 = -5800 \text{ psi (50\% increase)}$$
$$\sigma_{\text{all}} > \sigma_2 \qquad t = 0.3125 \text{ O.K.}$$

h	T_1	T_2	t	σ_1	σ_2	σ_{all}
40	587	−909	0.31	1,880	−2,909	−5,800
32	520	−627	0.25 (min)	2,080	−2,507	−4,720
24	436	−378	0.25 (min)	1,742	−1,512	−4,720
16	...	.....	0.25 (min)			
8	...	.....	0.25 (min)			

CONE

$$q = 100 \times {}^{40}\!/_{144} = 27.8 \text{ psi}$$
$$q_n = 27.8(0.25 \times 0.818^2 + 0.573^2) = 13.77 \text{ psi}$$

From Eq. (49),

$$W = 2920 \times 100 = 292,000 \text{ lb}$$

$$R_1 = \frac{150}{\cos 35°} = 183 \text{ in.}$$

$$T_2 = \frac{R_1}{2}\left(27.8 + \frac{292,000}{\pi R_1^2 \cos^2 \theta}\right) = 2920 \text{ lb/in. (governs)}$$

*$f > \tan \phi$; hence material may slide on itself. The designer may wish to check this condition.

Example 23-23

From Eq. (51),

$$T_1 = 13.77 \times 183 = 2520 \text{ lb/in.}$$

For 85 percent joint efficiency (Table 1),

$$t = \frac{2920}{18,000 \times 0.85} = 0.191 \qquad \text{Use } \tfrac{1}{4} \text{ in. (min)}$$

COMPRESSION RING

$$T_H = 2920 \sin 35° = 1675 \text{ lb/in.}$$
$$F_H = T_H R_c = 1675 \times 150 = 251,500 \text{ lb}$$
$$\text{Allowable stress} = 10,000 \text{ psi (Art. 15)}$$
$$A_{\text{reqd}} = \frac{251,500}{10,000} = 25.2 \text{ in.}^2$$

From Eq. (35a),

$$A_{\text{eff}} = 0.78(0.3125\sqrt{0.3125 \times 150} + 0.25\sqrt{0.25 \times 183}) = 2.99 \text{ in.}^2$$

From Eq. (35b),

$$A_{\text{eff}} = 16(0.3125^2 + 0.25^2) = 2.56 \text{ in.}^2 \qquad \text{(use)}$$

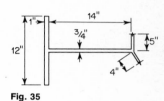

Additional required ring area $= 25.2 - 2.56 = 22.64 \text{ in.}^2$
Use web $14 \times 0.75 = 10.5$
Flange $12 \times 1.0 = \underline{12.0}$
$ 22.5 \text{ in.}^2 \qquad \text{(Fig. 35)}$

Fig. 35

Required moment of inertia, Eq. (53),

$$I = \frac{T_H R_R^3}{E} = \frac{1675 \times 160.1^3}{29,000,000} = 237 \text{ in.}^4$$
$$I = 791 \text{ in.}^4 \text{ provided O.K.}$$

TOWER AND SHELL SUPPORT

$$\text{Total product load} = 2,260,000 \text{ lb}$$
$$\text{Metal} = 50,000 \text{ lb (approximately)}$$
$$\text{Wind} = 30 \text{ psf } (\times \text{ shape factor})$$

From Fig. 34b,

$$25 \times 40 \times 30 \times 0.60 = 18,000 \times 40 \quad = 720,000$$
$$25 \times 17.8 \times 0.5 \times 30 \times 0.50 = 3,340 \times 14.1 = 47,100$$
$$6 \times 20 \times 1.167 \times 30 \times 1.5 = \underline{6,300} \times 10 = \underline{63,000}$$
$$ 27,640 \text{ lb} 830,100 \text{ lb-ft}$$

Assume six columns: $N = 6$, $R' = 12.5$ ft.
From Eq. (41),

Tank empty + wind:

$$P_{WE} = -\frac{50,000}{6} + \frac{2 \times 830,100}{6 \times 12.5} = 13,900 \text{ lb} \uparrow \text{(uplift)}$$

Tank full + wind:

$$P_{WF} = -\frac{2,260,000}{6} - \frac{2 \times 830,100}{6 \times 12.5} = -398,200 \text{ lb} \downarrow$$

Tank full, no wind:

$$P_F = -\frac{2,260,000}{6} = -376,000 \text{ lb} \downarrow$$
$$\frac{398,200}{376,000} = 1.06 < 1.25$$

(25 percent increase in allowable stress for wind; no-wind case governs.)

COLUMN ATTACHMENT* (Fig. 36)

From Fig. 36*b*,

$$q_1 = \frac{376,000}{2 \times 72} = 2610 \text{ lb/in.}$$

From AISC 1.10.5.2,

$$a = 25 \times \frac{12\pi}{6} = 157 \text{ in.} \qquad h = 72 \text{ in.} \qquad \frac{a}{h} = 2.18$$

Try $t = \frac{7}{16}$ in., $h/t = 72/0.4375 = 165$, $v_{\text{all}} = 6600$ psi.

$$v = \frac{2610}{0.4375} = 5970 < 6600 \text{ psi} \qquad \text{O.K.}$$

Therefore, use $\frac{7}{16}$ plate for first course (8 ft).

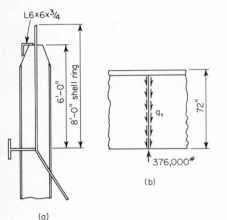

(a)

(b)

Fig. 36

Girder stresses:

$$M = -\frac{WR}{N}\left(\frac{1}{\alpha} - \frac{1}{2}\cot\frac{\alpha}{2}\right) \qquad \text{where } \alpha = \frac{2\pi}{N}$$

$$N = 6 \qquad \alpha = \frac{2\pi}{6} = 1.047 \qquad M = 418,000 \text{ lb-ft}$$

Girder consists of: Top flange $6 \times 6 \times \frac{3}{4}$ angle
Bottom blange 12×1 plate $+ 14 \times \frac{3}{4}$ plate
Web $72 \times \frac{7}{16}$

Top,
$$\sigma = 418,000 \times \frac{12}{1090} = 4600 \text{ psi}$$

Bottom,
$$\sigma = 418,000 \times \frac{12}{1750} = -2870 \text{ psi}$$

COLUMNS. Try W14 $\times$ 90, $A = 26.5$ in.², $r_x = 6.14$ in., $r_y = 3.70$ in.

$$\frac{KL}{r_y} = 0.85 \times \frac{240}{3.70} = 55$$
$$\frac{KL}{r_x} = 1 \times \frac{240}{6.14} = 39$$

---

*As noted in Art. 12, this connection is difficult to analyze and, in large part, is designed on the basis of empirical relationships and experience. The approach here is conservative. There are a number of alternatives to the girder, such as insert plates or pad plates. These more sophisticated designs are better left to experienced designers of tanks.

By AWWA,

$$F_A = 15,000 \text{ psi for } \frac{L}{r} < 60$$

$$f_a = \frac{376,000}{26.5} = 14,190 < 15,000 \qquad \text{O.K.}$$

RODS [Eq. (42)]

$$\cos \phi_n = \frac{25}{\sqrt{20^2 + 25^2}} = 0.78$$

$$T_{nR} = \frac{2 \times 27,640}{6 \times 0.78} = 11,800 \text{ lb} \qquad (\text{max when } \alpha_n = 90°)$$

With 25 percent increase in allowable stress for wind,

$$A = \frac{11,800}{15,000 \times 1.25} = 0.63 \text{ in.}^2$$

Use 1-in. upset rod, $A = 0.79$ in.^{2}

FOUNDATION (Fig. 37). Assume allowable soil pressue = 4000 psf. Estimated weight of foundation 80,000 lb.

$$\frac{376,000 + 80,000}{4000} = 114 \text{ ft}^2$$

Try 11×11-ft base = 121 ft^2.

$$
\begin{aligned}
\text{Weight: Concrete } 3 \times 3 \times 5.5 &= 49.5 \\
11 \times 11 \times 1 &= \underline{121.0} \\
170.5 \times 144 &= 24,500 \\
\text{Earth } (11^2 - 3^2) \times 5.0 = 560 \times 100 &= \underline{56,000} \\
&\,\, 80,500 \text{ O.K.}
\end{aligned}
$$

Since $80,500 > 13,900$, foundation weight will resist uplift.

Since these footings overlap, trapezoidal footings or a ring footing must be used.

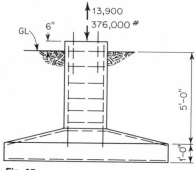

Fig. 37

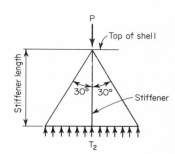

Fig. 38

18. Miscellaneous Details Concentrated loads supported by the bin shell must be distributed by means of stiffeners. The load is assumed to spread on a 30° angle on each side of the stiffener (Fig. 38). Stiffener length is extended to a point where the stress T_2 is within the allowable value given by Eq. (26).

When multiple bins are connected by conveyors or walkways, sliding support connections must be provided to avoid structural damage caused by movements due to differential settlement or tipping of the bins.

Adequate bottom connections must be provided between bins and foundations. Bottom connections are especially vulnerable when side-withdrawal openings are used, because of the tendency of the shell to go out of round under uneven pressure. Anchor bolts more than 2 in. in diameter have been sheared off by this distortion. A steel bottom with continuous ¼-in. lap weld is equivalent to 1½-in. anchor bolts 3½ ft on centers in its capacity to maintain roundness. When concrete bottoms are used, adequate shell anchorage must be provided.

Reinforcement at Openings. Openings for access or equipment in the side of the shell must be reinforced to carry the biaxial-stress system. Reinforcement may consist of a thickened insert plate or structural framing. The area of reinforcement must equal the area of the cutout material, and the openings must have rounded corners so as to avoid stress concentrations (Fig. 39).

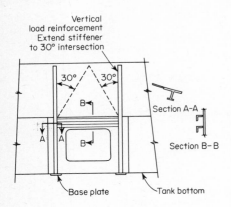

Fig. 39   Reinforcement at opening.

MATERIALS

Plates. The plate material most commonly used for construction of storage tanks and granular storage bins is ASTM A283 Grades A, B, C, and D. A36 and A131 Grades A, B, and C and A573 Grade 58 are also commonly used, the latter two in particular for more severe service conditions. Shell plates of A131 Grade C and A36 should be limited to 2 in. and normalized in thicknesses over 1.5 in. A36 plates over 1 in. through 2 in. should be ordered to A36 supplementary requirement S2 (silicon-killed fine-grain practice).

Recent trends have limited A283 material used in tension to thicknesses of 1 in. or less. Maximum permissible shell-plate thickness for storage tanks is 1.5 to 2 in., depending on the applicable code. Where shell plates in excess of maximum permissible thicknesses are required, the designer should refer to the applicable high-strength design basis in AWWA Appendix C or the API Code.

Base plates. A283 Grade C and A36 may be used for base plates regardless of thickness or ambient temperature. Material ordered to ASTM A36, Sec 4.2 is not acceptable.

Structural shapes are generally A36 steel. Forgings for turnbuckles and clevises are generally A668 Class D. Pins may be made from A36, A307 Grade A, or A108 Grade 1018 or 1025.

Welding Electrodes. Manual arc-welding electrodes should conform to the requirements of the AWS-ASTM Specifications for Mild Steel Arc-Welding Electrodes (AWS A5.1, ASTM A233) and should be any E6015, E6016, E6018 or E7015, E7016, E7018 Low Hydrogen classification suitable for the electric-current characteristics, the position of welding, and other conditions of intended use.

Allowable stresses for tanks vary considerably with the governing code. The following are typical:

1. Tensile stresses in tank shells: AWWA 15,000 psi (elevated and ground-supported tanks), API 21,000 psi (ground-supported tanks)
2. Tensile stresses in bin shells: 18,000 psi
3. Compressive stresses in tank and bin shells: Eq. (26)
4. Compressive stresses in tubular columns (AWWA):

$$\frac{P}{A} = XY$$

where $X = \dfrac{18,000}{1 + L^2/18,000r^2}$ (maximum 15,000 psi)

$Y = \frac{2}{3} \times 100t/R(2 - \frac{2}{3} \times 100t/R)$ for $t/R < 0.015$

$Y = 1$ for $t/R > 0.015$

R = radius to exterior surface of tubular member, in.

t = thickness of tubular member, in.; minimum $\frac{1}{4}$ in.

The slenderness ratio L/r must not exceed 120 for compression members carrying weight or pressure of tank contents and 175 for those carrying wind and/or seismic load.

REFERENCES

1. Ketchum, M. S.: "Design of Walls Bins and Grain Elevators," 3d ed., McGraw-Hill Book Company, New York, 1919.
2. Cooper, F. D., and J. R. Garvey: "Flow of Coal in Bins," ASME Paper 57-FU-2 presented at ASME-AIME Joint Fuels Conference, Quebec, Canada, Oct. 10–12, 1957.
3. Gardner, G. C.: The "Best" Hopper Profile for Cohesive Material, *Chem. Eng. Sci.*, vol. 18, pp. 35–39, 1963.
4. Lee, Yee: Hyperbolic Hopper Outlet Means, U.S. Patent 3071297, 1963.
5. Lee, Chesman A.: Hopper Design Up to Date, *Chem. Eng.*, April 1963.
6. Reimbert, Marcel: Excess Pressure during the Emptying of Grain Silo, *Acier Stahl Steel*, vol. 20, pp. 359–365, 1955.
7. Jenike, A. W.: Why Bins Don't Flow, *Mech. Eng.*, May 1964, pp. 40–43.
8. Johanson, J. R.: New Design Criteria for Hoppers and Bins, Applied Research Laboratory, *U.S. Steel Tech. Rept.*, April 1964.
9. AWWA Standard for Steel Tanks, Standpipes, Reservoirs and Elevated Tanks, for Water Storage, AWWA D-100-59.
10. API Recommended Rules for the Design and Construction of Large, Welded, Low-pressure Storage Tanks, API Standard 620.
11. Welded Steel Tanks for Oil Storage, API Standard 650.
12. Wilson, W. M., and N. M. Newmark: The Strength of Thin Cylindrical Shells as Columns, *Univ. Ill. Bull.* 255, 1933.
13. Wilson, W. M.: Tests of Steel Columns; Thin Cylindrical Shells; Laced Channels; Angles, *Univ. Ill. Bull.* 292, 1937.
14. Nuclear Reactors and Earthquakes, U.S. Atomic Energy Commission, TID 7024, August 1963.
15. Wozniak, Robert S.: Lateral Seismic Loads on Flat Bottomed Tanks, Chicago Bridge & Iron *Water Tower*, November 1971.

Towers and Transmission Pole Structures

MAX ZAR
Partner and Manager of Structural Department,
Sargent & Lundy, Engineers, Chicago, Ill.

JOSEPH R. ARENA
Transmission Line Consultant, Chicago, Ill.

1. Types of Towers All towers are space structures, single-plane structures, or single-pole structures. Towers with two or more legs are designed to support electrical transmission lines, tanks, radio and television antennas, radar and microwave equipment, floodlights, bridges, etc.

The tall towers used in the electronic industry are usually guyed at intermediate levels. Transmission towers can be designed as self-supported cantilevers or as guyed structures. The members may be lattice trussed or cold-formed from steel plate welded to form a pole. Each transmission-pole supplier has a preferred shape depending upon his fabricating facilities. The shapes vary from circular to multisided polygonal. Welded-plate poles are considered by many to be more attractive.

2. Materials The most commonly used materials for towers of all types are ASTM A36 steel and A572 of Grade 50 or higher. A36 and A572 Grade 50 are usually used for lattice towers. A36 and A572 Grades 50, 55, 60, and 65 are usually used in steel-pole towers. It is economical to use steels with minimum yield strengths of 100 ksi in very tall television towers.

All lattice steel towers are composed of hot-rolled angles. When either A36 or A572 steel is used, tower members are hot-dipped galvanized to obtain maximum resistance to corrosion. The thickness of the galvanizing specified is 0.0034 in., which is equivalent to 2 oz/sq ft. Galvanizing, if done properly, gives the best protective coating. Because of the high silicon content of A572 steel, its pickling time should be shorter than for A36 steel to avoid a loss of material thickness. Furthermore, galvanized A572 steel will not hold its spangled luster as long as galvanized A36, because of the high silicon content, but this does not diminish its corrosion resistance.

Ungalvanized A242 steel has been used for towers in environments which promote the development of a protective iron-oxide coating, and theoretically no other surface treatment is necessary. However, the coating has sometimes caused "packing" in the joints, which traps water, and in some cases members have buckled. Special joint details have been developed to solve this problem.[1]

Aluminum transmission towers are usually fabricated with Type 6061-T6 structural aluminum. Bare aluminum has good corrosion resistance in all environmental exposures. The weight of an aluminum tower is usually about half that of a similar steel tower designed for the same loading conditions, but the cost per pound is greater.

Where transmission lines cross relatively inaccessible terrain, aluminum offers a construction advantage because a tower can be flown in by helicopter and bolted to the foundation.

Practically all lattice towers, whether steel or aluminum, are of bolted construction. Steel-tower bolts usually conform to ASTM A394. Bolts for A242 steel towers should be made of corrosion-resistant material. Aluminum towers are usually erected with 2024-T4 bolts.

3. Height Limitations   All towers above 150 ft in height must be approved by the Federal Aviation Agency (FAA), and if the proposed structure is to support radio, radar, television, microwave, or other electronic antenna or equipment, a construction permit is required from the Federal Communications Commission (FCC). When a tower is located near an airport or on an aeronautical flight path, the tower height may be restricted by these agencies. Tall towers are required to have aeronautical-obstruction lighting and painting, in accordance with Sub-Part 17-C of the FAA Rules and Regulations.

In determining the configuration of electrical transmission towers the minimum distances between a conductor and the ground and between any tower member and any conductor are a function of line voltage.

4. Loads   The loads used in the design of transmission towers are discussed in Art. 17. On other towers the basic loads are those due to dead weight, live load, wind, earthquake, and, in northern climates, ice or snow.

Since the forces due to wind are critical in tall structures, careful thought should be given to the wind pressures to be provided for. Minimum Design Loads in Buildings and Other Structures of the American National Standards Institute (ANSI) recommends wind loading for all parts of the country and suggested increases at various heights above the ground.

In a four-legged tower the maximum wind loading on the columns occurs with a diagonal wind direction. Column loads are maximum on a three-legged tower with the wind normal to one side. Maximum stresses in the diagonals result when the wind is parallel to the tower face.

Ice and snow loads are related to climatic conditions; in the northern part of the country it may be advisable to provide for an ice load corresponding to ½ in. of ice around all members, including guys. Occasionally, specifications require that towers be designed for full wind and ice loads acting together. However, most codes do not require this, since ice is unlikely to build up under maximum wind.

Earthquake forces are significant in the design of some towers, especially those supporting water tanks.

In the design of guyed towers the stresses resulting from temperature changes in the guys must be considered.

5. Candelabra   It is often desirable to support more than one antenna on a single tower. The preferred arrangement is to mount them one atop the other. When height restrictions require multiple antennas at one level, the structure is called a candelabra.

During erection, or as a result of the structural failure of an antenna in service, candelabra may be subjected to eccentric and/or torsional load. This must be taken into consideration, in addition to the usual loads for single-antenna towers.

FREE-STANDING TOWERS

6. Stresses   *Towers with Constant Batter.*   Towers whose legs are straight can be analyzed by assuming that each of the three or more tower faces is a plane truss. Loads can be resolved into components that lie in the planes of the tower faces at the joints. Each planar truss can be analyzed for the force components lying in its plane. The leg stresses are the sums of the stresses in the trusses of which the leg is a common chord.

When the faces of a tower are identical in configuration, it is advantageous to compute the stresses in each member of a face resulting from applying separately at each joint in the face a unit horizontal load and a unit load parallel to the tower leg. Stresses from any combination of external loads can then be readily determined.

The diagonal and horizontal members in each face should be arranged so that they are subjected only to concentric axial load. A typical arrangement is shown in Fig. 1. The diagonals between A and B resist compression as well as tension. The diagonals between B and C also carry both compression and tension but form two bracing systems each of which may be analyzed separately by assuming alternate panel points loaded. All diagonals between C and D are assumed to carry tension only. The lower panel illustrates the use of redundants (a and b) which carry no calculated stress but serve to reduce the unsupported lengths of the legs and the long diagonals.

Towers without Constant Batter.   When the legs are not straight (Fig. 2) the tower must be analyzed as a three-dimensional structure using the six equations of static equilibrium.

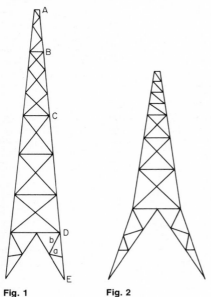

Fig. 1   **Fig. 2**

When a system of horizontal bracing is added, as at the level where the leg batter changes, the tower becomes statically indeterminate. The stresses in these horizontal diagonals may be assumed to be zero (unless forces due to torsion are present), which makes the structure determinate.

7. Foundations   The legs of free-standing towers must be anchored to resist uplift. Pile and caisson footings will be adequate to resist uplift provided that sufficient side surface area exists to develop the required skin friction. One-half the value used for designing the pile under downward loads may be used as its resistance to uplift forces. Spread footings should be covered with enough backfill so that the concrete and the overburden weigh about 1½ times the uplift forces. Proper reduction for the submerged weight of concrete and soil below the high-water line should be made. Although very little bearing area is required when founding on rock, extreme care must be taken to provide tight bond for the rock anchors.

GUYED TOWERS

For economy, tall antenna structures are guyed. Part I of Standard RS-222 of the Electronic Industries Association is the accepted guide for the structural requirements for steel guyed-tower design. This standard presents data on materials, wind loads, allowable unit stresses, fabrication, and foundations. Maximum allowable movements in microwave towers, a necessary requirement to avoid excessive misalignment in the transmitted beam, are also given.

It is economical to use solid round bars for members of tall, highly stressed masts, since their narrow projection and circular shape reduce the wind load. Similarly, high-strength bridge rope for guys reduces the dead weight and the exposed surface for wind and ice. The cable and fittings should be galvanized for weather protection. To avoid elastic stretch during construction and to ensure a stable modulus of elasticity, the cable is usually prestretched.

It is possible to have the cables, together with their hardware end attachments, proof-loaded by the cable manufacturer. This serves a dual purpose: it is a load test for the cable and fittings and at the same time the cable is prestretched. If proof loading and prestretching are done in one operation at the manufacturing plant, overall guy lengths including fittings can be provided within a tolerance of 1 in. in 100 ft. Final lengths can be corrected by threaded takeup rods.

8. Wind Wind loads for tower design in any county within the continental United States are described in EIA Standard RS-222; gust factors are included. Wind on members with circular cross section is reduced by one-third. On four-legged trussed towers area exposed to wind is assumed to be 1.75 times the area of one face. On towers of triangular cross section the exposed area is taken to be 1.5 times the area of one face. Wind pressures are given for three height zones: 0 to 300 ft, 301 to 650 ft, and over 650 ft. For towers over 1200 ft in height, consideration should be given to loads greater than those for the last height zone. It is customary to use a uniform average wind pressure on the mast between supports. Since gusts can occur at any elevation, 25 percent of the load may be considered equivalent to a moving load and positioned for maximum moments and shears.

Masts of guyed towers are usually triangular in cross section and are assumed to resist lateral wind loads as continuous beams with the guyed points as supports. The portion which cantilevers above the top guy support is designed as a free-standing tower.

Example 1   The solution by moment distribution of the wind moments and shears in a 340-ft mast is shown in Fig. 3. For simplicity it is assumed that moments of inertia and wind loading are uniform throughout the height. Deflections of the tower resulting from stretching and slacking of the guys are ignored. This requires that the sizes and initial tensions of the guys to be adjusted so that the wind deflections of the mast at the guy levels are proportional to their distances above the ground. Thus, if the base is hinged, a straight line can be drawn from the mast support through each deflected guy support; this avoids the more involved solution based on elastic supports.

9. Design of Guys The direction of the wind for maximum stress in guy X is shown in Fig. 4. For equally spaced guys, and neglecting initial tension, the horizontal component is

$$H_x = 1.155P \tag{1}$$

The value of P in this equation is the force due to wind on the mast, which is given by the shear diagram (Fig. 3), plus the wind force on the guy. The corresponding horizontal components of the stresses in guys Y and Z are

$$H_y = 0.577P_1 \qquad H_z = 0 \tag{2}$$

The relation between the horizontal component H of the guy tension and sag s of the guy is given by

$$s = \frac{wa^2}{8H} \tag{3}$$

where a = horizontal distance from mast to guy anchor
  w = average weight of cable per horizontal foot
The length L of cable is given approximately by

$$L = a \left(\frac{1}{\cos \alpha} + \frac{8s^2 \cos^3 \alpha}{3a^2} \right) \tag{4}$$

where α is defined in Fig. 4. If $\alpha \geqq 45°$, L will not be significantly longer than the straight chord S.

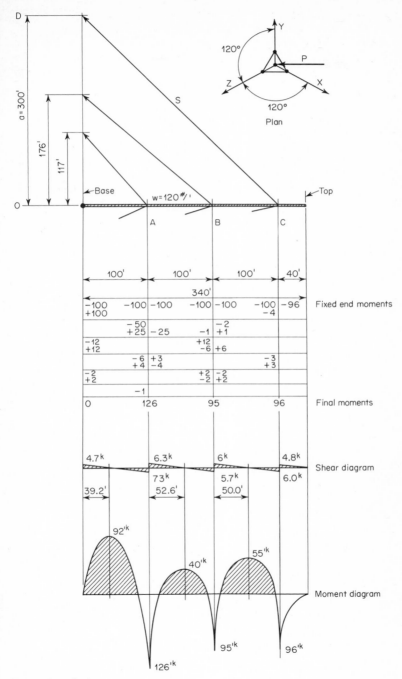

Fig. 3 Example 1.

To design a guy, the size of the cable is assumed and the horizontal component of guy tension determined from Eq. (1). The effect of the initial tension is determined next. Deflection of the mast at the upper connection of the guys must be taken into account in calculating this effect. The deflection d is given by

$$d = \frac{5PS}{3AE} \tag{5}$$

where S = chord dimension of cable, ft
A = cross-sectional area of cable, in.^{2}
E = modulus of elasticity of cable, psi

The reduction in the tension of the slack guy must be assumed to start the solution. This enables the sag s in the slack cable to be computed:

$$s^2 = \frac{3(a - d)}{8 \cos^4 \alpha} [L \cos \alpha - (a - d)] \tag{6}$$

The change in L resulting from change in stress in the slack cable should be taken into account in using this equation. With s known, H can be computed from Eq. (3) and

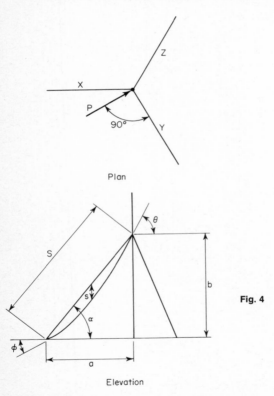

Plan

Elevation

Fig. 4

compared with the assumed value. If the correspondence is satisfactory, the value of the horizontal component of tension in guy X is corrected for the effect of initial tension in guy Z, and the angle θ (Fig. 4) is determined:

$$\tan \theta = \frac{wa}{2H} + \frac{b}{a} \tag{7}$$

The true stress in guy X can now be found.

Example 2   The procedure for design of guys will be illustrated for the tower of Fig. 3. Top guys should be designed first.

From the figure $S = 300\sqrt{2} = 424.26$ ft. Assume $\frac{3}{4}$-in. bridge strand Class A coating, for which $A = 0.338$ in.^{2} weight $= 1.18$ lb/ft, breaking strength $= 68,000$ lb, and $E = 24 \times 10^{6}$ psf (Table 2, Sec. 21). For a wind load of 30 psf, the load p per foot of cable is

$$p = \tfrac{2}{3} \times 30 \times \tfrac{1}{12} \times \tfrac{3}{4} = 1.25$$

From Fig. 3,

$$\text{Wind on the mast at } C = 10,800$$
$$\text{From wind on two guys: } 2 \times 424.26 \times 1.25/2 = \underline{530}$$
$$11,330 \text{ lb}$$

From Eq. (1),

$$H_x = 11,330 \times 1.155 = 13,100 \text{ lb}$$

The breaking strength of $\frac{3}{4}$-in. bridge strand is 68,000 lb. The initial tension is usually one-eighth the breaking strength. Therefore, the initial tension is $68,000/8 = 8,500$ lb. Assume the initial tension in guy Z to be reduced 7100 lb when tension in guy X is maximum. This leaves 1400 lb of tension in Z.

From Eq. (5),

$$d = \frac{5 \times 11,330 \times 424.6}{3 \times 0.338 \times 24 \times 10^6} = 1.017 \text{ ft}$$

Guy Z will shorten by the amount

$$\frac{PL}{AE} = \frac{7100 \times 424.26}{0.338 \times 24 \times 10^6} = 0.382 \text{ ft}$$

and its corrected length $L = 424.26 - 0.382 = 423.88$ ft. The horizontal distance $a - d = 300 - 1.017 = 298.98$ ft. Then, from Eq. (6),

$$s^2 = \frac{3 \times 298.98}{8 \times 0.707^4}(423.88 \times 0.707 - 298.98) = 334.11$$
$$s = 18.28 \text{ ft}$$

The corresponding stress in guy Z is, from Eq. (3),

$$Z = \frac{1.61 \times 298.98^2}{8 \times 18.28 \times 0.707} = 1396 \text{ lb}$$

This compares favorably with the assumed 1400 lb.

The horizontal component of the stress in guy X is

$$H_x = 13,100 + 1396 \times 0.707 = 14,087 \text{ lb}$$

From Eq. (7),

$$\tan \theta = \frac{1.61 \times 300}{2 \times 14,087} + \frac{300}{300} = 1.017$$
$$\theta = 45°29'$$

from which $X = 14,087/\cos 45°29' = 14,087/0.7011 = 20,092$ lb.

TEMPERATURE. Since a drop in temperature increases the tension in the guys, this condition should be investigated. A temperature drop of about 80°F is a reasonable allowance when combined with maximum wind loading. The increase in unit stress f_t is given by

$$f_t = TCE \cos^2 \alpha \qquad (8)$$

where T = temperature change, °F
C = coefficient of linear expansion
α = angle with horizontal (Fig. 4)

Example 3   For the top guys of the tower of Fig. 3,

$$X_t = 80 \times 65 \times 10^{-7} \times 24 \times 10^{6} \times 0.707^{2} = 6240 \text{ lb}$$

which is to be added to the wind-load stress. Thus,

$$X = 20,092 + 6240 = 26,332 \text{ lb}$$

The factor of safety is $68,000/26,332 = 2.58$.

Guys at lower levels must be designed so that the deflections at their upper ends are proportional to their heights above ground (Example 1). The required chord length for a cable to maintain a given deflection d is given by

$$S = \sqrt{\left(\frac{R}{2p}\right)^2 + \frac{3AEd}{5p}} - \frac{R}{2p} \tag{9}$$

where R = wind reaction at connection to mast

Example 4   For the guys at level B of the tower of Fig. 3, the required deflection at B is

$$d = 1.017 \times \tfrac{2}{3} = 0.678 \text{ ft}$$

From the shear diagram (Fig. 3) the wind reaction at B is $R = 11,700$ lb. For ¾-in. cable,

$$S = \sqrt{\left(\frac{11,700}{2 \times 1.25}\right)^2 + \frac{3 \times 0.338 \times 24 \times 10^6 \times 0.678}{5 \times 1.25}} - \frac{11,700}{2 \times 1.25}$$
$$= 266.4 \text{ ft}$$

and $a = \sqrt{266.4^2 - 200^2} = 175.9$ ft, which locates the lower connection of the guy (Fig. 3).

$$\begin{aligned}
\text{Wind on mast at } B &= 11,700 \\
\text{Wind on two guys} = 2 \times 266.4 \times 1.25/2 &= \underline{333} \\
& 12,033 \text{ lb}
\end{aligned}$$

$$H_x = 12,033 \times 1.155 = 13,900 \text{ lb}$$

Assuming a 7500-lb reduction in initial tension in the slack guy under full wind, the reduced length of guy Z is found to be 266.15 ft. The weight of the cable is $1.14 \times {}^{266}\!/_{176} = 1.73$ lb per horizontal ft. The sag s_z is found from Eq. (6) to be 12.29 ft, and from Eq. (3) the cable tension is $Z = 821$ lb compared with 1000 lb assumed. Then,

$$H_x = 13,900 + 821 \times 0.66 = 14,440 \text{ lb}$$

From Eq. (7), $\tan\theta = 1.1490$. The resulting cable tension is 22,000 lb. Adding the temperature effect [Eq. (8)],

$$X = 22,000 + 5440 = 27,440 \text{ lb}$$

The factor of safety is $68,000/27,440 = 2.48$.

LOWER GUYS. The proportional deflection of the mast at the lower guy connection A is

$$d = 1.017 \times \tfrac{1}{3} = 0.339 \text{ ft}$$

The wind reaction R at level A is 13,600 lb (Fig. 3). Assuming a ¾-in. cable, Eq. (9) gives $S = 120.0$ ft. This requires the distance a from the lower connection to the tower base to be $\sqrt{120^2 - 100^2} = 66.4$ ft. By comparison with the wind reaction at B and C, it appears that the resulting guy angle will be too steep for the reaction at A, so that the ¾-in. cable will be overstressed.

For a ⅞-in. cable, $S = 154$ ft and $a = 117$ ft.

10. Ice Loading   Stresses in the guys due to ice load can be calculated as follows. Assuming ½ in. radial ice on the guys, the ice load for the upper guys of the tower of Fig. 3 is about 1.2 lb per horizontal ft. The dead-load sag corresponding to an initial tension of 8500 lb is found from Eq. (3) to be

$$s = \frac{1.18\sqrt{2} \times 300^2}{8 \times 8500 \times 0.707} = 3.01 \text{ ft}$$

Using this value of s the horizontal component of the guy tension due to ice load is calculated from Eq. (3):

$$H = \frac{1.2 \times 300^2}{8 \times 3.01} = 4500 \text{ lb}$$

$$T = 4500\sqrt{2} = 6360 \text{ lb}$$

The increase in length of the cable is

$$\frac{6360 \times 300\sqrt{2}}{0.338 \times 24 \times 10^6} = 0.342 \text{ ft}$$

and the cable length is $300\sqrt{2} + 0.342 = 424.60$.

From Eq. (6) with $d = 0$,

$$s^2 = \frac{3 \times 300}{8 \times 0.707^4}(424.60 \times 0.707 - 300) = 86.4$$
$$s = 9.31 \text{ ft}$$

The true value of s lies between 9.31 and 3.01 ft. Assume it to be 5.5 ft. The corresponding increase in the length of the cable is

$$0.342 \times \frac{3.01}{5.5} = 0.187$$

and the cable length is $300\sqrt{2} + 0.187 = 424.45$.

The revised value of s from Eq. (6) is $s = 6.17$. Assume the true sag to be 5.8 ft. Then

$$H = 4500 \times \frac{5.8}{3.01} = 2590 \text{ lb}$$

and from Eq. (7),

$$\tan \theta = \frac{1.2 \times 300}{2 \times 2590} + \frac{300}{300} = 1.0695$$
$$\theta = 46°55'$$

from which $X = 2590/0.683 = 3800$ lb. This is small relative to the safe load for a ¾-in. cable.

Most codes do not require that ice load be combined with wind load (Art. 4).

11. Guy Tensioning It is customary to install the guys with an initial tension of one-eighth the breaking strength of the cable. This will be sufficient to assure that the leeward cable will be under tension when the tower is subjected to full wind load.

Cables which are prestretched in the factory are simpler to set for a prescribed initial tension in the field. The guys should be tightened by means of hydraulic jacks, calibrated to permit determination of cable tension. It is essential to maintain the jacks in accurate calibration throughout the course of the work. Initial tension can be measured with mechanical tensiometers and sag-determining devices. Allowance for temperature should be made.

12. Guy Vibration Aeolian vibration is quite common on tower guys. Cables which are undamped and free to vibrate have been known to gallop as much as 30 ft. This can be disastrous for the entire structure. Mechanical dampers, such as the Stockbridge damper which is often installed near the bottom end of each guy, can reduce vibration to an insignificant amplitude.

13. Design of Mast Details of a 20-ft section of the mast of Fig. 3 are shown in Fig. 5. The following example illustrates the design of a section just below support C.

Example 5 The maximum stress in a leg of the tower occurs when the wind pressure P_u is perpendicular to the opposite side (Fig. 6b). From Fig. 3, the moment at C is 96 ft-kips. The resulting compression in leg a is found by taking moments about bc. To this must be added the weight of the 40-ft section above C, at 36 plf, and the vertical component of the guy tension. The guy attached to leg a will be slack; its tension is assumed to be the same as that calculated for the wind direction corresponding to maximum tension in guy X, namely, 1400 lb (Example 2). Therefore, the load on leg a is

$$\frac{96 \times 12}{0.866 \times 32} = 41.5 \text{ kips from moment}$$
$$\frac{36 \times 40}{3} = 0.5 \text{ kip weight above } C$$
$$1400 \times 0.707 = \underline{1.0} \text{ kip guy tension}$$
$$P = 43.0 \text{ kips}$$

Try a leg section 1⅝ in. in diameter, A36 steel:

$$r = \frac{D}{4} = \frac{1.62}{4} = 0.40 \text{ in.}$$
$$\frac{L}{r} = \frac{30}{0.40} = 75 < 120$$

Using the AISC allowable stress,

$$F_c = 15.90 \times 1.33 = 21.2 \text{ ksi}$$
$$P = 21.2 \times 2.07 = 43.8 \text{ kips}$$

TOWER SECTION BC. The horizontal component of the slack-guy tension is

$$0.707 \times 1.4 = 1.0 \text{ kip}$$

With the wind force $P_u = 11.3$ kips normal to bc (Fig. 6b) the total horizontal force to be resisted is 12.3 kips. This force is shared equally by the guys at b and c, and since the guys are equally spaced, the

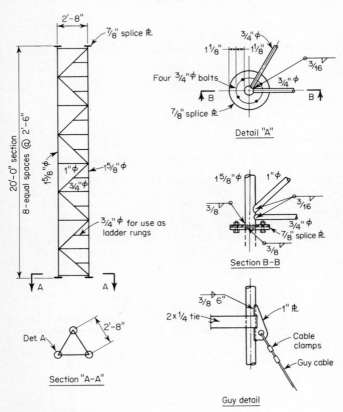

Fig. 5 Example 5.

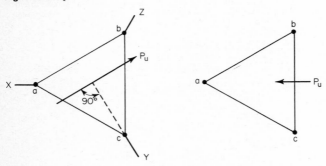

Fig. 6

force polygon is equilateral. Therefore, the horizontal component of each guy tension is 12.3 kips, which is also the vertical component, and the vertical load on the 100-ft section is

Guys at b and c: $12.3 \times 2 = 24.6$
Slack guy $\qquad\qquad = 1.0$
Dead load: $\qquad 0.5 \times 3 = 1.5$
$\overline{\qquad\qquad\qquad 27.1 \text{ kips}}$

The resulting direct stress is

$$f_a = \frac{P}{A} = \frac{27.1}{3 \times 2.07} = 4.4 \text{ ksi}$$

The bending stress for section BC is the direct stress in leg a calculated previously,

$$f_b = \frac{43.0}{2.07} = 20.6 \text{ ksi}$$

The least radius of gyration of the section is given by

$$I = 2A \left(\frac{h}{2}\right)^2$$

$$r = \sqrt{\frac{I}{3A}} = \frac{h}{\sqrt{6}} = 0.408h$$

where h = width of side. Then

$$r = 0.408 \times 32 = 13 \text{ in.}$$
$$\frac{L}{r} = 100 \times \frac{12}{13} = 93 < 120$$
$$F_c = 1.33 \times 13.84 = 18.5 \text{ ksi}$$
$$F_b = 1.33 \times 22 = 29.3 \text{ ksi}$$
$$\frac{f_a}{F_a} + \frac{f_b}{F_b} = \frac{4.4}{18.5} + \frac{20.6}{29.3} = 0.235 + 0.705 = 0.940 < 1$$

WELD TO SPLICE PLATE. The leg is welded at the top and bottom of the ⅞-in. splice plate (B-B of Fig. 5). The circumference of the leg is

$$c = 1.625\pi = 5.1 \text{ in.}$$

so that

$$\frac{43}{2 \times 5.1} = 4.2 \text{ kips/in.}$$

For a ⅜-in. fillet weld,

$$3.6 \times 1.33 = 4.8 \text{ kips/in.}$$

SPLICE-PLATE BOLTS. Maximum tension in leg a results when the wind is in the direction opposite to that shown in Fig. 6b. Therefore, it is numerically the same as the maximum compression (41.5 kips) found previously, reduced by the vertical component of tension in guy a and by dead load. Thus,

$$41.5 - 12.3 - 1.0 = 28.2 \text{ kips net tension}$$
$$\frac{28.2}{4} = 7.1 \text{ kips/bolt}$$

For A307 ¾-in. bolts,

$$14 \times 0.44 \times 1.33 = 8.2 \text{ kips}$$

SPLICE PLATE. The plate is shown in detail A of Fig. 5. Bolt tension produces a moment at the circumference of the leg. Assuming a circumferential line of inflection midway between the bolt line and leg face,

$$M = 4 \times 7.1 \times \tfrac{9}{16} = 16 \text{ in.-kips}$$
$$S = \frac{ct^2}{6} = \frac{5.1 \times 0.87^2}{6} = 0.65 \text{ in.}^3$$
$$f_b = \frac{16}{0.65} = 24.5 \text{ ksi} < 22 \times 1.33$$

LACING. Maximum stress in the lacing results when the wind is parallel to one of the faces (Fig. 6a). For the conditions shown in the figure, face ab will resist two thirds the shear produced by P_w. From Fig. 3, the maximum shear in section BC is 6 kips. Therefore,

$$P = 6 \times \tfrac{2}{3} = 4 \, \text{kips}$$
$$\frac{L}{r} = \frac{44 - 2}{0.25} = 168 < 200$$
$$F_c = 5.29 \times 1.33 = 7.0 \, \text{ksi}$$

For 1-in. round,

$$P = 7.0 \times 0.78 = 5.5 \, \text{kips}$$
$$\frac{P}{c} = \frac{5.5}{\pi} = 1.8 \, \text{kips/in.}$$

Use ³⁄₁₆-in. fillet weld: $1.8 \times 1.33 = 2.4$ kips/in.

HORIZONTAL MEMBERS. These members are redundant, for which the maximum $L/r = 250$.

$$r = \frac{32 - 1.5}{250} = 0.122 \, \text{in.}$$
$$D = 4r = 4 \times 0.122 = 0.488 \, \text{in.}$$

Use ¾-in. bars to serve as ladder rungs.

GUY DETAIL (Fig. 5). The stress caused by wind and temperature in the ¾-in. guys is 26,332 lb (Example 3). The allowable bearing stress is $27 \times 1.33 = 36$ ksi, and the required plate thickness is

$$t = \frac{26,332}{36,000 \times 0.75} = 0.98 \, \text{in.}$$

Use 1-in. plate.

The stress in the ties is

$$F = \frac{26,332 \times 0.707}{2 \times 0.866} = 10,500 \, \text{lb}$$
$$A = \frac{10.5}{22 \times 1.33} = 0.36 \, \text{in.}^2$$

Use $2 \times$ ¼-in. ties.

14. Foundations The topography of the tower site may affect the foundation layout. Available land may dictate the maximum permissible distance from the guy-anchor foundation to the mast. All guys at one support level should make the same angle with the mast, even though, where the ground is irregular, this may result in guys of slightly different lengths and foundations which are not equidistant from the mast.

The guy foundations resist uplift and horizontal shears. It is common practice to increase the design values by 100 percent when sizing the foundation. If the guy foundations can resist large loads, it is sometimes economical to anchor guys from more than one level to a common foundation. This must be evaluated against possible increases in cable costs.

The foundation for the tower base must be designed for full vertical loading and horizontal shear. In addition, where the base is fixed instead of hinged, the overturning moment must be provided for.

15. Erection Bolts are the most suitable form of fastener for tall towers. It is quite common to specify lock nuts and washers.

Although most guyed towers have hinge-supported masts, this requires temporary guys as soon as erection begins. A fixed-base tower can be erected for some height before temporary guying is needed. Very tall towers often are equipped with personnel elevators for maintenance, and a tapered, hinge-supported mast base makes it impossible for the bottom elevator landing to be close to the ground.

The amount of ground preassembly is dependent upon the erection method and the available equipment. Crawler cranes can be used for the lower 200 ft; after that, a gin pole, raised as the tower grows, is used. As soon as a set of permanent guys is installed and tensioned, the temporary guys below are removed.

TRANSMISSION TOWERS

16. Types Most transmission towers are designed to support one or two circuits, although some have been designed to support three or four. Each circuit consists of three phases. In all areas except those that have no incidence of lightning the lines must be shielded by ground wires (sometimes called static wires). Typical towers are shown in Figs. 7 and 9.

Transmission towers are of three general types:

1. Tangent towers are used where the line is straight or has an angle not exceeding 3°. They support vertical loads; transverse and longitudinal wind loads; a transverse load from the angular pull of the wires; and a longitudinal load due to unequal spans, forces resulting from the wire-stringing operation, or a broken wire.

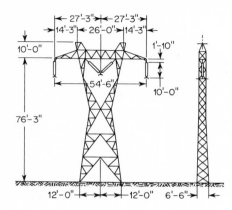

Fig. 7 Elevation of one-circuit 345-kV transmission tower.

2. Angle towers are used where the line changes direction by more than 3°. They support the same kinds of load as the tangent tower. The number of groups and the range of line angle in each group depend on the layout of the line.

3. Dead-end towers (also called anchor or strain towers) must take the dead-end pulls from all the wires on one side, in addition to the vertical and transverse loads.

Because of the large duplication of tangent towers (70 to 80 percent of all the towers in a long line) the designer should focus his attention primarily on this type. Three items should be considered: cost of material, cost of erection, and cost of foundations. The cost of the material is directly related to the weight of the tower. The cost of erection is directly related to the number of pieces and number of bolts to be installed. The lightest tower may not be the most economical if it contains many more pieces per ton than a slightly heavier one, because more pieces must be handled in fabrication, shipment, storage, and erection. The cost of foundations is directly related to the spread of the tower legs. A tower with closely spaced legs will have shorter lacing members, but the higher stresses in the legs will increase foundation costs.

17. Loads The National Electrical Safety Code[3] (NESC) is mandatory by law in some states and advisory in others. When it is specified, towers must meet the *minimum* requirements of Grade B construction. The code divides the United States into three loading zones (Fig. 8). In the heavy zones, towers are designed for a 4-psf wind acting on a surface consisting of the wire plus ½ in. of radial ice and 6.5 psf on the tower steel with no ice, both at 0°F. This is the equivalent of a 40-mph wind.

It is extremely important that, in addition to the minimum wind requirements of the NESC, the tower be designed for a wind speed from any direction that has a recurrence interval commensurate with the expected loss should failure occur.

Figure 2 in Sec. 19 shows isotachs for winds having recurrence intervals of 50 years. Wind is the primary enemy of tangent towers; it is usually these that fail during a heavy windstorm, and design for a 100-year wind should be considered.

All wind speeds in the United States are recorded at a height of 30 ft above ground. The design speed should be increased according to tower height, using[4]

$$V_z = V_{30} \left(\frac{z}{30} \right)^{1/n} \qquad (10)$$

where z = height above ground

V_z = speed at height z

V_{30} = wind speed at 30 ft

n = constant (Sec. 19, Art. 13)

Wind pressure should be calculated for an area twice the exposed area of one face of a lattice tower to allow for pressure on the opposite face, but it should not exceed the loading on a solid structure of the same outside dimensions.

There is no NESC requirement for extreme summertime wind. This is left to the discretion of the designer, as is also the overload factor. Common overload factors are 1.0 or 1.1.

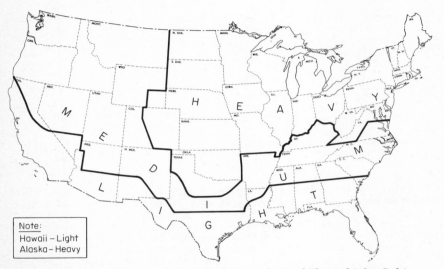

Fig. 8   United States loading zones for transmission towers (National Electrical Safety Code).

If wind records are not available for 50- or 100-year periods but are available for shorter periods, the highest wind velocity that may be expected can be extrapolated for longer periods by the following formulas, which are based on Type I distribution (also known as the Gumbel distribution).

$$V_n = V_2(0.93 + 0.24 \log_{10} n)$$
$$V_n = V_5(0.86 + 0.22 \log_{10} n)$$
$$V_n = V_{10}(0.80 + 0.20 \log_{10} n) \tag{11}$$
$$V_n = V_{15}(0.77 + 0.19 \log_{10} n)$$
$$V_n = V_{20}(0.75 + 0.19 \log_{10} n)$$

where $V_2, \ldots, V_{20}$ = highest velocity in 2, ..., 20 years and V_n = predicted highest velocity in n years.

Loads are multiplied by overload factors, and the tower members are designed to yield or buckle under the resulting member forces. Most towers are designed in accordance with overload factors specified by NESC.

The incidence of conductor breakage has become so rare on modern transmission lines, because of the use of stronger conductors, that many utilities have deleted the broken-wire condition. However, it should be recognized in the design that longitudinal wind, unequal spans, unequal tower heights, ice drop-off, and stringing loads can cause longitudinal loads on the tower. If the longitudinal load is less than the transverse load, the tower can be designed with the four legs forming a rectangle instead of a square, with the long dimension transverse to the direction of the line. This results in a substantial savings of materials. If possible, such towers should be spotted for almost equal spans with no great difference in elevation of adjacent towers.

18. Vibration   There are two types of wind-induced vibration of conductors which affect tower loading and clearances:

1. Galloping of conductors occurs when ice or sleet forms on a circular conductor to create a streamlined profile. When the conductor is set in motion, wind forces can keep it galloping for hours. This motion, which fortunately is quite rare, can cause buckling of members and, in some cases, complete collapse of structures, especially if ice is vibrated off in alternate spans.

2. Aeolian vibration, which almost always exists, is a vibration of the conductor in a plane normal to the wind. In addition to damage to the electrical equipment and cables, aeolian vibration can cause fatigue failures of redundant members and may loosen bolts if appropriate locking devices are not installed, or if bolts are not properly tightened.

19. Stress Analysis   In analyzing the stresses in the various tower members, it is customary to make the following assumptions:

1. All transverse forces are resisted by the transverse faces of the tower.
2. All longitudinal forces are resisted by the longitudinal faces of the tower.
3. All vertical loads are carried by the arms and legs.
4. In panels composed of crossed tension members, all shears are resisted by tension members in the panel.
5. In panels composed of members designed to take compression or tension, the shear is divided equally between them.
6. Torsional moments resulting from a broken-wire condition or a wire-stringing force are resisted by the horizontal bracing at the level where the moment is applied.

Most transmission towers were analyzed graphically until the advent of the computer. The graphical solution of a 230-kV two-circuit tower is shown in Fig. 9. This method is still in use where computer programs are not available and is sometimes used to check computer output.

Computer programs[5] for tower analysis incorporate features such as:

1. Simplicity of input
2. Tower-configuration capability
3. Convergence to a final design within a reasonable number of cycles regardless of initial member size. This relieves the designer of the burden of estimating member size accurately.
4. Tension-only members. These are usually omitted in analysis by manual methods if the loading produces compression in them. Computer programs can be written to allow such members to carry load up to their capacity in compression and distribute the rest to other members.
5. Simplicity of design. A number of tower configurations can be designed to optimize the tower weight.

Where transverse and longitudinal loads are equal or almost equal, it is economical to use a square cage and a square base. With this arrangement the torsion and shear forces are determined as follows (Fig. 10):

$$T = \frac{Pl}{2a} \tag{12}$$

$$F = \frac{P}{2} \tag{13}$$

Classical solutions for determining the torsional shears in towers with rectangular cages or bases are time-consuming, and the computer is an ideal aid. An approximate analysis is shown in Fig. 11.

20. Steel Tension Members   The allowable stress on the net section of concentrically loaded members is taken to be the yield stress F_y. However, eccentricities between bolt gage lines and the centroidal axis of an angle are usually neglected, and according to Ref. 6 the unconnected leg of an angle connected by one leg may be considered fully effective unless it is wider than the connected leg, in which case it should be assumed to have the same width as the connected leg.

The net section of members with a zigzag line of holes is determined by the $s^2/4g$ rule.

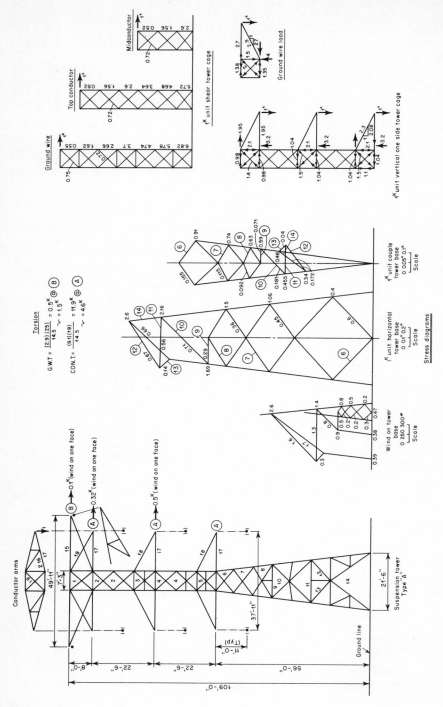

Ground wire

Top conductor

Midconductor

1ᴷ unit shear tower cage

Ground wire load

1ᴷ unit vertical one side tower cage

Torsion

$GWT = \dfrac{(2.9)(25)}{14.5} \sim = 0.5^K$ @ Ⓑ

$CON.T = \dfrac{(9.1)(19)}{14.5} \sim = 1.5^K$ $\sim = 11.9^K$ @ Ⓐ $\sim = 4.6^K$ @ Ⓐ

1ᴷ unit couple tower base
0 0.05" 0.1"ᴷ
Scale

1ᴷ unit horizontal tower base
0 0.1" 0.2"
Scale

Stress diagrams

Wind on tower base
0 250 300#
Scale

Conductor arms

0.1ᴷ (wind on one face) Ⓑ

0.32ᴷ (wind on one face) Ⓐ

0.5ᴷ (wind on one face) Ⓐ

Suspension tower
Type "A"

Ground line

49'-1"

37'-11"

21'-6"

7'-3"

8'-0"

22'-6"

22'-6"

56'-0"

11'-0" (Typ)

109'-0"

24-16

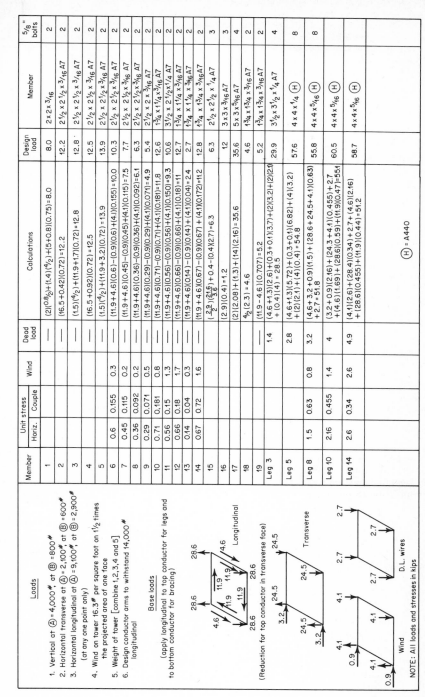

Loads

1. Vertical at Ⓐ =4,000# at Ⓑ =800#
2. Horizontal transverse at Ⓐ = 2,100#, at Ⓑ =600#
3. Horizontal longitudinal at Ⓐ =9,100#, at Ⓑ =2,900# (at any one point only)
4. Wind on tower 16.3# per square foot on 1½ times the projected area of one face
5. Weight of tower [combine 1,2,3,4 and 5]
6. Design conductor arms to withstand 14,000# longitudinal

(apply longitudinal to top conductor for legs and to bottom conductor for bracing)

Base loads

(Reduction for top conductor in transverse face)

Ⓗ = A440

NOTE: All loads and stresses in kips

Member	Unit stress Horiz.	Couple	Wind	Dead load	Calculations	Design load	Member	$^5/_8$" bolts
1				—	$(2)(0.8_2)+(1.4)(^4/_2)+(5+0.8)(0.75)=8.0$	8.0	$2\times2\times^3/_{16}$	2
2				—	$(16.5+0.42)(0.72)=12.2$	12.2	$2\tfrac12\times2\tfrac12\times^3/_{16}$ A7	2
3			0.3	—	$(1.5)(^4/_2)+(11.9+1.7)(0.72)=12.8$	12.8	$2\tfrac12\times2\tfrac12\times^3/_{16}$ A7	2
4				—	$(16.5+0.92)(0.72)=12.5$	12.5	$2\tfrac12\times2\tfrac12\times^3/_{16}$ A7	2
5				—	$(1.5)(^4/_2)+(11.9+3.2)(0.72)=13.9$	13.9	$2\tfrac12\times2\tfrac12\times^3/_{16}$ A7	2
6	0.6	0.155	0.3	—	$(11.9+4.6)(0.6)-(0.9)(0.6)+(4.1)(0.155)=10.0$	10.3	$2\tfrac12\times2\tfrac12\times^3/_{16}$ A7	2
7	0.45	0.115	0.2	—	$(11.9+4.6)(0.45)-(0.9)(0.45)+(4.1)(0.115)=7.5$	7.7	$2\tfrac12\times2\tfrac12\times^3/_{16}$ A7	2
8	0.36	0.092	0.2	—	$(11.9+4.6)(0.36)-(0.9)(0.36)+(4.1)(0.092)=6.1$	6.3	$2\tfrac12\times2\tfrac12\times^3/_{16}$ A7	2
9	0.29	0.071	0.5	—	$(11.9+4.6)(0.29)-(0.9)(0.29)+(4.1)(0.071)=4.9$	5.4	$2\tfrac12\times2\times^3/_{16}$ A7	2
10	0.71	0.181	0.8	—	$(11.9+4.6)(0.71)-(0.9)(0.71)+(4.1)(0.181)=11.8$	12.6	$1\tfrac34\times1^1/_4\times^3/_{16}$ A7	2
11	0.56	0.15	1.3	—	$(11.9+4.6)(0.56)-(0.9)(0.56)+(4.1)(0.150)=9.3$	10.6	$3\tfrac12\times2\tfrac12\times^1/_4$ A7	2
12	0.66	0.18	1.7	—	$(11.9+4.6)(0.66)-(0.9)(0.66)+(4.1)(0.18)=11$	12.7	$1\tfrac34\times1^1/_4\times^3/_{16}$ A7	2
13	0.14	0.04	0.3	—	$(11.9+4.6)(0.14)-(0.9)(0.14)+(4.1)(0.04)=2.4$	2.7	$1\tfrac34\times1^1/_4\times^3/_{16}$ A7	2
14	0.67	0.72	1.6	—	$(11.9+4.6)(0.67)-(0.9)(0.67)+(4.1)(0.172)=12$	12.8	$1\tfrac34\times1\tfrac34\times^3/_{16}$A7	2
15					$(\tfrac{2.9}{2})(\tfrac{21.6}{3.6})+0.4-(0.4)(2.7)=6.3$	6.3	$2\tfrac12\times2\tfrac12\times^1/_4$ A7	3
16					$(2.9)(0.4)=1.2$	1.2	$3\times3\times^3/_{16}$ A7	3
17					$(2)(2.08)+(1.3)+(14)(2.16)=35.6$	35.6	$5\times5\times^5/_{16}$ A7	3
18					$^4/_2(2.3)=4.6$	4.6	$1\tfrac34\times1\tfrac34\times^3/_{16}$ A7	2
19				1.4	$(11.9-4.6)(0.707)=5.2$	5.2	$1\tfrac34\times1\tfrac34\times^3/_{16}$A7	2
Leg 3				2.8	$(4.6+1.3)(2.6)+(0.3+0.1)(3.7)+(2)(3.2)+(2)(2.1)$ $+(0.4)(4)=28.5$	29.9	$3\tfrac12\times3\tfrac12\times^1/_4$ A7	4
Leg 5					$(4.6+1.3)(5.72)+(0.3+0.1)(6.82)+(4)(3.2)$ $+(2)(2.1)+(4)(0.4)=54.8$	57.6	$4\times4\times^1/_4$ Ⓗ	8
Leg 8	1.5	0.63	0.8	3.2	$(4.6+3.2+0.9)(1.5)+(28.6+24.5+4.1)(0.63)$ $+2.7=51.8$	55.8	$4\times4\times^5/_{16}$ Ⓗ	8
Leg 10	2.16	0.455	1.4	4	$(3.2+0.9)(2.16)+(24.3+4.1)(0.455)+2.7$ $+(4.6)(1.69)+(28.6)(0.59)+(11.9)(0.47)=55.4$	60.5	$4\times4\times^5/_{16}$ Ⓗ	
Leg 14	2.6	0.34	2.6	4.9	$(4.1)(2.6)+(28.4)(0.34)+2.7+(4.6)(2.16)$ $+(28.6)(0.455)+(11.9)(0.44)=51.2$	58.7	$4\times4\times^5/_{16}$ Ⓗ	

Fig. 9 Graphical solution for 230-kV transmission tower.

24-17

Allowable loads for single angles, assuming deduction for a single hole in the connected leg and the unconnected leg to be fully effective, are given in Table 1.

21. Aluminum Tension Members The allowable stress on the net section of concentrically loaded members is taken to be the yield stress (35 ksi for 6061-T6). It is recommended in Ref. 7 that connection eccentricity be taken into account by using a net area A_n given by the following equations:

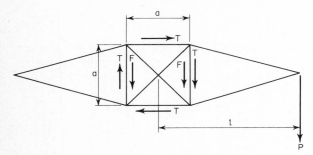

Fig. 10 Plan of crossarm showing torsion due to broken wire or stringing load.

Single angles with bolt line at center of one leg,

$$A_n = A_1 + \frac{A_2}{3} \tag{14}$$

Double angles bolted to one face of gusset plate, channels bolted through web, tees bolted through flange,

$$A_n = A_1 + \frac{A_2}{2} \tag{15}$$

Double angles on opposite faces of gusset plate, bolted at center of leg,

$$A_n = A_1 + \frac{3A_2}{4} \tag{16}$$

In these equations A_1 = net area of connected leg and A_2 = net area of unconnected leg, each determined by the $s^2/4g$ rule.

22. Steel Compression Members The allowable stress F_a on the gross section of angles in compression recommended in Ref. 6 is given by

$$\frac{F_a}{F_y} = 1 - \frac{1}{2}\left(\frac{KL/r}{C_c}\right)^2 \qquad \frac{KL}{r} \lessgtr C_c \tag{17}$$

$$F_a = \frac{286{,}000}{(KL/r)^2} \quad \text{ksi} \qquad \frac{KL}{r} \gtrless C_c \tag{18}$$

provided the largest width-thickness ratio b/t is not greater than

$$\left(\frac{b}{t}\right)_{\text{lim}} = \frac{2500}{\sqrt{F_{y,\text{psi}}}} \tag{19}$$

In these equations

$\quad C_c = \pi \sqrt{2E/F_y}$
$\quad E$ = modulus of elasticity = 29,000 ksi
$\quad F_y$ = guaranteed minimum yield strength
KL/r = largest effective slenderness ratio of any unbraced segment of member
$\quad b$ = distance from edge of fillet to extreme fiber
$\quad t$ = thickness

If b/t exceeds $(b/t)_{\lim}$, F_y in the above equations is replaced by the value of F_{cr} given by

$$\frac{F_{cr}}{F_y} = 1.8 - \frac{0.8 b/t}{(b/t)_{\lim}} \qquad \left(\frac{b}{t}\right) \gtrless \frac{b}{t} \lesssim \frac{3750}{\sqrt{F_{y,\text{psi}}}} \tag{20}$$

$$F_{cr} = \frac{8400}{(b/t)^2} \quad \text{ksi} \qquad \frac{b}{t} \gtrsim \frac{3750}{\sqrt{F_{y,\text{psi}}}} \tag{21}$$

Values of F_a for F_y = 36 and 50 ksi are given in Tables 2, 3, and 4.

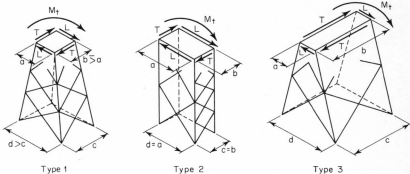

Type 1   Type 2   Type 3

Table of structure properties
From indeterminate solutions of actual structures

Type	Height (ft)	a	b	c	d	Point	K	C
		(in feet)						
1	85.0	5.0	5.0	10.0	25.0	1	2.19	2.00
1	33.0	5.0	5.0	6.9	12.8	2	1.67	1.50
1	62.0	5.0	5.0	10.0	15.0	3	1.43	1.33
1	30.0	5.0	5.0	7.4	9.8	4	1.35	1.19
1	65.0	4.0	4.0	4.0	18.5	5	3.50	2.82
1	34.9	4.0	4.0	4.0	11.8	6	2.60	1.98
1	90.0	4.0	4.0	4.0	18.4	7	3.40	2.80
1	39.2	4.0	4.0	4.0	10.3	8	2.10	1.79
*2	—	6.0	5.5	5.5	6.0	9	1.26	1.09
3	75.0	4.5	23.0	23.0	23.0	10	1.54	1.67
3	20.0	2.0	7.8	7.8	7.8	11	1.84	1.60
3	31.5	4.0	16.0	16.0	16.0	12	1.70	1.60
3	51.5	3.5	15.0	15.0	15.0	13	1.56	1.62
3	75.0	6.0	33.0	33.0	33.0	14	2.37	1.69
3	45.0	6.0	33.0	33.0	33.0	15	3.21	2.37
1	40.0	8.6	6.8	10.0	15.0	16	1.52	1.40

*From beam solution

Applicability of types:
Type 1 Rigid frame columns or individual towers
Type 2 Rectangular columns, beams, or towers
Type 3 Wedge type towers

Formulas:

(1) $T = \dfrac{M_t}{Kb+a}$

(2) $L = TK$

in which:
M_t = horizontal torsional moment, ft–lb
K = constant from chart
L and T = horizontal shears in faces, lb
a, b, c, d = dimensions, ft

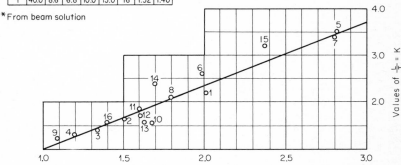

Values of $C = \dfrac{a+d}{b+c}$ or $\dfrac{b+c}{c+d}$, whichever is greater than unity

Note: Points shown are solutions in table.

Fig. 11 Distribution of torsional shears for columns, towers, and beams.

Values of KL/r in Eqs. (17) and (18) are to be determined as follows:
1. Leg sections with $0 < L/r \leqslant 150$, $KL/r = L/r$
2. All other members with $0 < L/r \leqslant 120$:
    a. Concentric loading at both ends, $KL/r = L/r$
    b. Concentric loading at one end, normal framing eccentricities at the other, $KL/r = 30 + 0.75\ L/r$
    c. Normal framing eccentricities at both ends, $KL/r = 60 + 0.5L/r$

TABLE 1   Allowable Load in Kips for Single-Angle Tension Members Based on Allowable Stresses of Art. 20

Angle	Max length, ft		P, kips				Weight, lb/ft
			⅝-in bolts*		¾-in. bolts*		
	$L/r_x = 500$	$L/r_z = 500$	$F_y = 36$	$F_y = 50$	$F_y = 36$	$F_y = 50$	
1¾ × 1¼ × ³⁄₁₆	23.0	11.2	14.4	20.1			1.80
2 × 1¼ × ³⁄₁₆	26.5	11.3	15.9	22.1	15.1	20.9	1.96
1¾ × 1¾ × ³⁄₁₆	22.4	14.3	17.7	24.5			2.12
2 × 1½ × ³⁄₁₆	26.3	13.4	17.7	24.5	16.8	23.4	2.12
1¾ × 1¼ × ¼	22.7	11.1	18.6	25.9	17.6	24.4	2.34
2 × 2 × ³⁄₁₆	25.7	16.4	20.9	29.1	20.1	27.9	2.44
2½ × 1½ × ³⁄₁₆	33.5	13.6	20.9	29.1	20.2	28.0	2.44
2 × 1¼ × ¼	26.2	11.2	20.8	28.9	19.7	27.4	2.55
2½ × 2 × ³⁄₁₆	33.0	17.8	24.6	34.1	23.7	32.9	2.75
2½ × 2½ × ³⁄₁₆	32.4	20.6	27.7	38.6	26.9	37.4	3.07
3 × 2 × ³⁄₁₆	40.3	18.3	27.7	38.6	26.9	37.4	3.07
2 × 2 × ¼	25.4	16.3	27.6	38.4	26.5	36.9	3.19
2½ × 2 × ¼	32.7	17.7	32.0	44.4	30.9	42.9	3.62
3 × 3 × ³⁄₁₆	39.1	24.8	34.6	48.1	33.8	46.9	3.71
3 × 2 × ¼	39.9	18.1	36.7	50.9	35.6	49.4	4.10
2½ × 2½ × ¼	32.0	20.5	36.7	50.9	35.6	49.4	4.10
3 × 2½ × ¼	39.4	22.0	41.0	56.9	39.8	55.4	4.50
3 × 3 × ¼	38.8	24.7	45.6	63.4	44.5	61.9	4.90
3½ × 2½ × ¼	46.7	22.7	45.6	63.4	44.5	61.9	4.90
3½ × 3 × ¼	46.3	30.3	50.0	69.4	48.9	67.9	5.40
3 × 2½ × ⁵⁄₁₆	39.0	21.9	50.6	70.3	49.2	68.3	5.60
4 × 3 × ¼	53.3	27.1	54.7	75.9	53.6	74.4	5.80
3½ × 3½ × ¼	45.4	28.9	54.7	75.9	53.6	74.4	5.80
3½ × 2½ × ⁵⁄₁₆	46.3	22.5	56.4	78.3	54.9	76.3	6.10
4 × 3½ × ¼	52.9	30.6	59.0	81.9	57.8	80.4	6.20
4 × 4 × ¼	52.1	33.1	63.6	88.4	62.4	86.9	6.60
4 × 3 × ⁵⁄₁₆	52.9	27.0	67.4	93.8	66.2	91.8	7.20
3½ × 3½ × ⁵⁄₁₆	45.0	28.8	67.4	93.8	66.2	91.8	7.20
4 × 3½ × ⁵⁄₁₆	52.5	30.4	73.4	101.8	71.9	99.8	7.70
5 × 3 × ⁵⁄₁₆	67.1	27.4	78.7	109.3	77.2	107.3	8.20
4 × 4 × ⁵⁄₁₆	51.7	33.0	78.7	109.3	77.2	107.3	8.20
5 × 3½ × ⁵⁄₁₆	67.1	31.9	84.4	117.3	83.1	115.3	8.70
4 × 3½ × ⅜	52.1	30.3			85.1	118.2	9.10
4 × 4 × ⅜	51.3	32.8			91.9	127.8	9.80
5 × 3 × ⅜	67.1	27.3			91.9	127.8	9.80
5 × 5 × ⁵⁄₁₆	65.4	41.4	101.2	140.8	100.0	138.8	10.30
5 × 3½ × ⅜	66.7	31.8			100.5	139.8	10.40
6 × 3½ × ⅜	80.8	32.0			112.0	155.8	11.70
5 × 5 × ⅜	65.0	41.2			119.2	165.3	12.30
6 × 4 × ⅜	80.4	36.5			119.2	165.3	
5 × 5 × ⁷⁄₁₆	64.6	41.1			137.8	191.2	
6 × 6 × ⅜	78.3	49.6			145.8	202.8	

*Diameter of hole = diameter of bolt plus ¹⁄₁₆ in.

3. All other members with $L/r > 120$:
   a. No rotational restraint at either end, $KL/r = L/r$ for L/r to 200
   b. Partial restraint at one end, $KL/r = 28.6 + 0.762L/r$ for L/r to 225
   c. Partial restraint at both ends, $KL/r = 46.2 + 0.615L/r$ for L/r to 250

Suggestions for determining the member length L for various tower configurations are given in Ref. 6.

TABLE 2   Allowable Compression F_a in ksi for $F_y = 36$ ksi (Art. 22)

KL/r \ b/t	≈13.2	14	15	16	17	18	19	20
5	36.0	34.2	32.0	29.8	27.6	25.4	23.3	21.0
10	35.9	34.1	31.9	29.8	27.6	25.4	23.2	21.0
15	35.7	34.0	31.8	29.7	27.5	25.3	23.2	20.9
20	35.5	33.8	31.7	29.5	27.4	25.2	23.1	20.8
25	35.3	33.6	31.5	29.3	27.2	25.1	23.0	20.8
30	35.0	33.3	31.2	29.1	27.0	24.9	22.8	20.7
35	34.6	32.9	30.9	28.9	26.8	24.8	22.7	20.5
40	34.2	32.6	30.6	28.6	26.6	24.6	22.5	20.4
45	33.7	32.1	30.2	28.3	26.3	24.3	22.3	20.2
50	33.2	31.6	29.8	27.9	26.0	24.0	22.1	20.0
55	32.6	31.1	29.3	27.5	25.6	23.7	21.8	19.8
60	31.9	30.5	28.8	27.0	25.2	23.4	21.6	19.6
65	31.2	29.9	28.2	26.5	24.8	23.1	21.3	19.4
70	30.5	29.2	27.6	26.0	24.4	22.7	21.0	19.1
75	29.6	28.4	27.0	25.5	23.9	22.3	20.6	18.8
80	28.8	27.7	26.3	24.9	23.4	21.8	20.2	18.5
85	27.8	26.8	25.5	24.2	22.8	21.4	19.9	18.2
90	26.8	25.9	24.8	23.5	22.2	20.9	19.4	17.9
95	25.8	25.0	23.9	22.8	21.6	20.3	19.0	17.5
100	24.7	24.0	23.1	22.1	21.0	19.8	18.5	17.1
105	23.5	22.9	22.1	21.3	20.3	19.2	18.1	16.8
110	22.3	21.8	21.2	20.4	19.6	18.6	17.5	16.3
115	21.0	20.7	20.2	19.6	18.8	18.0	17.0	15.9
120	19.7	19.5	19.1	18.6	18.0	17.3	16.5	15.5
125	18.3	18.2	18.0	17.7	17.2	16.6	15.9	15.0
130	16.9	16.9	16.9	16.7	16.4	15.9	15.3	14.5
135	15.7	15.7	15.7	15.7	15.5	15.1	14.7	14.0
140	14.6	14.6	14.6	14.6	14.6	14.4	14.0	13.5
145	13.6	13.6	13.6	13.6	13.6	13.6	13.3	12.9
150	12.7	12.7	12.7	12.7	12.7	12.7	12.6	12.3
155	11.9	11.9	11.9	11.9	11.9	11.9	11.9	11.7
160	11.2	11.2	11.2	11.2	11.2	11.2	11.2	11.1

See Table 4 for $KL/r > 160$.

A single-bolt connection is not considered to be rotationally restrained. A multiple-bolt connection detailed to minimize eccentricity is considered to be rotationally restrained if the connection is to a member capable of resisting rotation of the joint. Points of intermediate support are considered to provide rotational restraint if they satisfy this criterion.

If specific details are known by test experience to provide greater restraint than is described above, KL/r values may be modified accordingly.

23. Aluminum Compression Members   According to Ref. 7 the allowable stress F_a on the gross section of concentrically loaded members of alloy 6061-T6 is given by

$$F_a = 35 \text{ ksi} \qquad\qquad 0 < KL/r \lessapprox 15.4 \qquad\qquad (22)$$

$$F_a = 39 - 0.26 \frac{KL}{r} \text{ ksi} \qquad 15.4 \lessapprox KL/r \lessapprox 63 \qquad\qquad (23)$$

$$F_a = \frac{90,000}{(KL/r)^2} \text{ ksi} \qquad\qquad 63 \lesssim KL/r \qquad\qquad (24)$$

Formulas for other alloys are given in Ref. 7. These equations are applicable for single- or double-angle members and members of tee cross section only if $5\ b/t \lesssim L/r$. Premature local (torsional) buckling may occur if this value of b/t is exceeded. Furthermore, for cross sections with only one axis of symmetry the slenderness ratio about the axis of symmetry should not exceed 0.8 times the ratio for the other principal axis (equal-legged single-angle members satisfy this criterion).

Single-angle leg members are considered to be concentrically loaded.

TABLE 3   Allowable Compression F_a in ksi for F_y = 50 ksi (Art. 22)

b/t $\diagdown$ KL/r	$\lesssim$11.2	12	13	14	15	16	17	18
5	49.9	47.0	43.4	39.9	36.3	32.7	29.0	25.9
10	49.8	46.9	43.3	39.8	36.2	32.7	29.0	25.9
15	49.5	46.6	43.1	39.6	36.1	32.6	28.9	25.8
20	49.1	46.3	42.8	39.4	35.9	32.4	28.8	25.7
25	48.6	45.9	42.5	39.0	35.6	32.2	28.6	25.6
30	48.0	45.3	42.0	38.7	35.3	31.9	28.4	25.4
35	47.3	44.7	41.5	38.2	34.9	31.6	28.2	25.2
40	46.5	44.0	40.8	37.7	34.5	31.3	27.9	25.0
45	45.6	43.1	40.1	37.1	34.0	30.9	27.6	24.7
50	44.5	42.2	39.4	36.4	33.5	30.4	27.2	24.5
55	43.4	41.2	38.5	35.7	32.8	29.9	26.8	24.2
60	42.1	40.1	37.5	34.9	32.2	29.4	26.4	23.8
65	40.8	38.9	36.5	34.0	31.5	28.8	25.9	23.4
70	39.3	37.6	35.4	33.1	30.7	28.2	25.4	23.0
75	37.7	36.2	34.2	32.1	29.9	27.5	24.9	22.6
80	36.0	34.7	32.9	31.0	29.0	26.8	24.3	22.2
85	34.2	33.1	31.6	29.9	28.0	26.0	23.7	21.7
90	32.3	31.4	30.1	28.6	27.0	25.2	23.1	21.2
95	30.3	29.6	28.6	27.4	25.9	24.3	22.4	20.6
100	28.2	27.7	27.0	26.0	24.8	23.4	21.7	20.1
105	25.9	25.7	25.3	24.6	23.6	22.4	20.9	19.5
110	23.6	23.7	23.5	23.1	22.4	21.4	20.1	18.8
115	21.6	21.6	21.6	21.5	21.1	20.4	19.3	18.2
120	19.9	19.9	19.9	19.9	19.7	19.3	18.4	17.5
125	18.3	18.3	18.3	18.3	18.3	18.1	17.5	16.8
130	16.9	16.9	16.9	16.9	16.9	16.9	16.6	16.0
135	15.7	15.7	15.7	15.7	15.7	15.7	15.6	15.2
140	14.6	14.6	14.6	14.6	14.6	14.6	14.6	14.6
145	13.6	13.6	13.6	13.6	13.6	13.6	13.6	13.6
150	12.7	12.7	12.7	12.7	12.7	12.7	12.7	12.7
155	11.9	11.9	11.9	11.9	11.9	11.9	11.9	11.9
160	11.2	11.2	11.2	11.2	11.2	11.2	11.2	11.2

See Table 4 for $KL/r > 160$.

TABLE 4   Allowable Compression F_a for F_y = 36 to 50 ksi, $b/t \leqslant 20$ (Art. 22)

KL/r	165	170	175	180	185	190	195	200	205
F_a, ksi	10.6	9.90	9.34	8.83	8.36	7.92	7.52	7.15	6.81

KL/r	210	215	220	225	230	235	240	245	250
F_a, ksi	6.48	6.19	5.91	5.65	5.41	5.18	4.97	4.77	4.58

See Tables 2 and 3 for $KL/r < 165$.

Eccentricity of load in single angles fastened by one leg is accounted for by using an equivalent slenderness ratio $(KL/r)_{eq}$ given by

$$\left(\frac{KL}{r}\right)_{eq} = \sqrt{\left(\frac{K_zL}{r_z}\right)^2 + 25\left(\frac{b}{t}\right)^2} \tag{25}$$

where r_z is the least radius of gyration, and F_a is limited to $0.5F_y$ for single-bolt connections and $0.67F_y$ for two bolts.

Values of K in these equations are given for various framing configurations in Ref. 7.

24. Limiting Slenderness Ratios The following slenderness-ratio limits have been used for many years, and are recommended in Ref. 6 for steel members and Ref. 7 for aluminum members.

	Steel	Aluminum
Compression members:		
Legs and crossarms	150	120
All others carrying calculated load	200	150
Tension members	500	200
Members carrying no calculated load*	250	200

*These members are usually called "redundant" members in tower terminology.

Tension members are required by Ref. 6 to be fabricated short so as to be in tension (draw) when erected. According to Ref. 7, the limit 200 for tension members can be increased to $200\sqrt{1+f/F_E}$ if draw is specified, where f = initial tension and F_E = Euler stress.

25. Bolts The following allowable stresses are suggested for bolts:

	A394	2024-T4
Shear on shank area, ksi	30	24
Bearing, ksi	60	45

Values for A394 bolts are the same as in Ref. 6. In Ref. 7, however, the allowable shear for 2024-T4 bolts is 37 ksi on the gross or root area, as applicable.

26. Foundations Most tower foundations can be classified as steel grillage, concrete spread footing, concrete caisson, augered hole, piles, and rock anchors. The commonly used types are steel grillages (Fig. 12) and augered concrete caissons (Fig. 14). The augered concrete caisson is the most reliable and economical footing for resisting uplift. The belled caisson can be used in clay soils and is very reliable because there is no backfill. The bottom of the bell will push up against undisturbed clay, which has a cohesion value of half the unconfined compressive strength.

It is extremely difficult to bell the caisson in granular soils. A straight shaft, whose ability to resist uplift is directly related to the granular friction along the vertical surfaces of the caisson, can be used instead.

The steel grillage is likely to be less economical than a caisson. It is also more difficult to design against uplift, especially when clay is used for backfill.

Grillages are purchased with the towers before soil-boring information is available, because the right of way has not yet been acquired. When this information becomes available, some of the grillages may prove to be inadequate and must be encased in concrete. This is uneconomical.

Figure 13 is an example of a spread footing with a cross member on the leg stub encased in concrete. This design can be adapted to various soils by altering the dimensions to develop the required bearing and uplift. Control of the quality of the backfill material and its compaction are essential for this footing and the galvanized-steel grillage to resist uplift forces adequately.

A typical pile foundation is shown in Fig. 15. Footings of this type are used only where water or unusually poor soil are encountered.

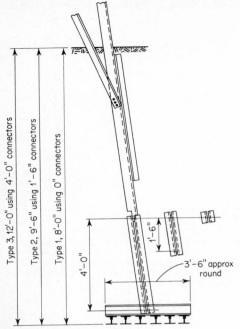

Type 3, 12'-0" using 4'-0" connectors

Type 2, 9'-6" using 1'-6" connectors

Type 1, 8'-0" using 0" connectors

1'-6"

4'-0"

3'-6" approx round

Fig. 12 Steel grillage footing with legs adjustable for depth.

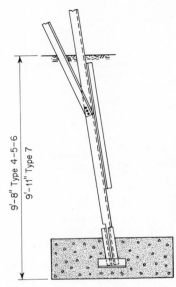

9'-8" Type 4-5-6

9'-11" Type 7

Fig. 13 Typical concrete foundation.

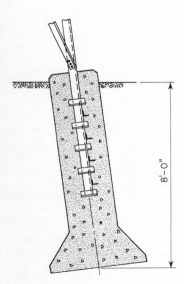

8'-0"

Fig. 14 Tower leg in cylindrical concrete caisson.

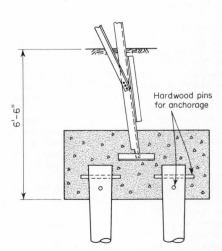

Hardwood pins for anchorage

6'-6"

Fig. 15 Detail of pile foundation.

Rock anchors are required where rock is too close to the surface to obtain enough backfill to overcome the anticipated uplift loads. A typical rock-foundation detail is shown in Fig. 16.

A sufficient number of soil borings should be made to determine the soil conditions. The groundwater level must be established in order to evaluate the reduced weight of soil when it is buoyant, since this affects the uplift resistance. Uplift forces are always less than those for bearing.

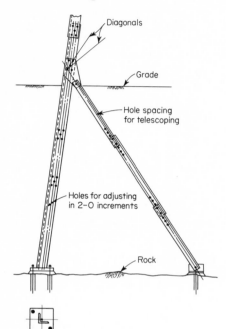

Fig. 16 Adjustable rock anchor.

27. Tower Tests Failure of a tower is defined as any permanent deformation that exists after all loads are removed. The purpose of a full-scale test is to confirm, under accurately simulated loading conditions, the adequacy of a tower to support the specified design loads and to establish the accuracy of the design, detailing, and fabrication. The material used for the test tower should be, as far as possible, a random selection representative of the material properties in a production run.

The result of a full-scale test of one tower is not necessarily a prediction of the behavior of all towers in the line. However, because of cost and the effect on the fabrication schedule, usually only one tower is tested. The tallest tower of a series should be tested to obtain the maximum benefit from testing a single tower. For economic reasons, most full-scale tests are performed on a tangent tower.

The tower is tested on a rigid base because it is designed as a cantilever fixed at the base. If it is tested on a rigid base any anomaly in its behavior cannot be charged to the foundation or soil conditions and the results are related only to materials, design, and fabrication.

Tower-testing facilities usually include dead weights, hand and electric winches, and single-control hydraulic rams. Loads are measured by dynamometers or load cells. In some cases it is not possible to install them at the points of attachment of the conductors and shield wires, in which case they are often installed at locations somewhat removed from the correct points of load application. As a result, the load on the tower is slightly reduced because of frictional losses between the cables and the pulleys in the load-application system.

Two techniques are used to assess structural adequacy:
  1. Visual inspection to see if members show signs of impending failure.
  2. Measurement by transit of the movement of a graduated scale attached to the tower. An increasing rate of movement under steady load shows that a member or connection is failing, while a decreasing rate shows that the tower is redistributing and holding the applied load.

The loads are commonly applied in increments of 25, 50, 75, 87½, and 100 percent of the total load and are maintained for 5 minutes each. The 5-minute period permits the tower to adjust to load and the deflection readings to be recorded.

If the tower successfully passes all tests, it is customary to continue the test to destruction by increasing the transverse loads in small increments. This is done to determine whether the tower was designed economically. At the completion of the destruction test, the tower should be dismantled and inspected. Critical connections should be examined for signs of distress. Specimens for chemical and physical analysis should be selected from tower members to correlate the quality of steel in the test tower with the specified material.

In assessing the results of a tower test, it is important to evaluate design concepts rather than merely review the results.

POLE STRUCTURES

28. Design   Suggested procedures for the design of steel transmission-pole structures given in Ref. 8 are discussed in this article. A typical pole is shown in Fig. 17.

Loadings and overload factors should be the same as for latticed towers. Loadings should not be less than those for NESC Grade B construction.

Stress calculations are based on elastic analysis. The effects of bracing, interpole ties, etc., in multipole structures should be accounted for, and the analysis should include the effects of deflection unless they are known to be negligible. The self-supported pole is designed as a cantilever.

Axial Compression.   Equations (17) and (18) are recommended for axially loaded compression members of uniform cross section. However, poles are usually tapered. For this case, the following equivalent slenderness ratio can be used:

$$\frac{KL}{r_{eq}} = \frac{1}{P^*} \frac{KL}{r_0} \tag{26}$$

where r_0 = radius of gyration of small end
   P^* = coefficient from charts in Ref. 9

Axial compression and bending in members not subject to lateral-torsional buckling must satisfy

$$f_a + f_b \gtrless F_y \tag{27}$$

$$\frac{f_a}{F_a} + \frac{f_b}{F_y} \frac{C_m}{1 - f_a/F_E} \gtrless 1 \tag{28}$$

Equation (27) applies at points of support in the plane of bending, with f_b calculated for the larger end moment. Equation (28) applies at a point between supports, with f_b calculated at the point of maximum moment for members with transverse loads and for the larger end moment for members without transverse load. In these equations

   f_a = axial stress
   f_b = bending stress
   F_a = allowable axial stress by Eqs. (17) and (18)
   F_E = Euler load = $\pi^2 E/(KL/r)^2$

The coefficient C_m may be taken equal to unity, but for members acted upon by end moments M_1 and M_2, with no intermediate loads, the value given by

$$C_m = 0.6 + 0.4 \frac{M_1}{M_2} \geqq 0.4 \tag{29}$$

may be used, where $M_1 \gtrless M_2$ and M_1/M_2 is positive for single-curvature bending.

Poles are usually analyzed for the combination of axial and transverse loads by computer, taking into account the secondary moment of P due to the pole deflection[8] and satisfying Eq. (27). If P/P_E is small, as is usually the case, the secondary moment can be accounted for quite accurately[10] by multiplying the moment due to the transverse loads by the amplification factor $1/(1 - P/P_E)$, where P_E is determined for the equivalent L/r by Eq. (26).

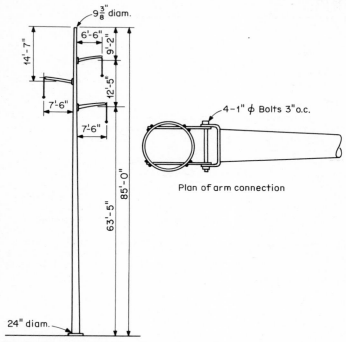

Fig. 17 One-circuit 138-kV steel transmission pole.

Local Buckling. The allowable local-buckling stress for circular and elliptical cross sections in axial compression is

$$F_a = F_y \qquad\qquad \frac{D}{t} \lesssim \frac{3800}{F_y} \tag{30}$$

$$F_a = 0.75F_y + \frac{950}{D/t} \qquad \frac{3800}{F_y} \lesssim \frac{D}{t} \lesssim \frac{12{,}000}{F_y} \tag{31}$$

$$F_b = F_y \qquad\qquad \frac{D}{t} \lesssim \frac{6000}{F_y} \tag{32}$$

$$F_b = 0.7F_y + \frac{1800}{D/t} \qquad \frac{6000}{F_y} \lesssim \frac{D}{t} \lesssim \frac{12{,}000}{F_y} \tag{33}$$

where F_y = yield strength, ksi
$\quad\ D$ = outside diameter, in.
$\quad\ t$ = wall thickness, in.

For circular and elliptical cross sections in combined bending and axial compression,

$$\frac{f_a}{F_a} + \frac{f_b}{F_b} \lesssim 1 \tag{34}$$

Formed regular polygonal tubular members for which the compressive stress $P/A + Mc/I$ on the extreme fiber equals the yield stress should be proportioned so that

$$\frac{w}{t} \lesssim \frac{240}{F_{y,\text{ksi}}} \tag{35}$$

where w is the flat width of a side and t its thickness. The inside bend radius of the pole should be used to determine w, but if the radius exceeds $4t$ it should be taken equal to $4t$. These values of w/t may be exceeded if the compressive stress does not exceed the value F_a given by

$$F_a = 1.45F_y\left(1 - 0.00129\sqrt{F_y}\,\frac{w}{t}\right) \qquad \frac{240}{\sqrt{F_y}} \lesssim \frac{w}{t} \lesssim \frac{385}{\sqrt{F_y}} \tag{36}$$

$$F_a = \frac{108,000}{(w/t)^2} \qquad \frac{385}{\sqrt{F_y}} \lesssim \frac{w}{t} \tag{37}$$

F_y in Eqs. (36) and (37) is in units of ksi.

29. Material   Selection of material is an important factor. A36 and A572 steels are usually used. Since maximum loadings may occur at low temperatures, the possibilities of brittle fracture must be evaluated (Sec. 4). Brittle behavior tends to be more likely in thick material. Triaxial tensile stresses from welding are more likely to occur in thick sections, and it is more difficult to be assured of metallurgical uniformity in thick sections. Many purchasers specify Charpy V-notch values for protection against brittle fracture.[8]

Material for anchor bolts should be a fine-grained fully killed heat-treated steel meeting a minimum Charpy V-notch requirement of 15 ft-lb at $-20°F$ when tested in the longitudinal direction. Certification of weldability should be obtained if welding is required.

Fabrication.   Practically all poles are cold-formed from flat plates on a press brake. The effects of cold-forming on properties of steel are discussed in Sec. 9, Art. 15.

Welds.   Complete-penetration groove welds should be designed for the same stresses as the base metal, provided a suitable electrode is specified. Partial-penetration groove welds in compression normal to the throat should be designed for the corresponding stresses for the base metal, provided a suitable electrode is specified.

Partial-penetration groove welds in shear or in tension normal to the throat should be designed for the same stresses as for fillet welds. The allowable stress on the throat of a fillet weld should be 0.50 times the tensile strength of the weld metal.

In the design of vang or stiffener connections to the pole, care should be taken to distribute the load sufficiently to protect the wall of the pole against local buckling.

Field welding should be avoided whenever possible.

30. Pole Splices   *Slip Splices.*   Sections joined by friction (telescoping) splices should be detailed for a minimum lap of 1.5 times the largest inside diameter of the female section, with allowances for fabrication tolerances. Tests have shown that this amount of lap is sufficient to develop the yield strength of the member in bending, but supplemental locking devices may be needed if relative movement at the joint is critical. Complete-penetration welds should be used in the splice area of the female section. Some manufacturers also use complete-penetration welds in the male section.

Circumferential Welded Splices.   Complete-penetration welds should be used for sections joined by circumferential welds. Longitudinal welds within 6 in. of circumferential welds should be complete-penetration welds.

Base Connection.   Welded pole-to-base plate connections should be complete-penetration welds.

Flange Connections.   Flange connections may be made with fillet welds or complete-penetration groove welds.

31. Foundations   The augered concrete caisson is a reliable footing for steel poles. Its resistance to overturning depends on the passive resistance of the surrounding soil. There are several theories and methods for the design of concrete caissons subject to moment and horizontal shear. Generally, the caisson is considered to be a rigid member restrained laterally by springs. The length-diameter ratio of the caisson determines whether it can be classified as a rigid body. The greater the ratio the less sound is the assumption of rigidity, and if the critical ratio is exceeded, the caisson should be designed as a flexible shaft. This analysis is discussed in Ref. 11.

The pole is usually anchored to its foundation by a cluster of anchor bolts formed from high-strength reinforcing bars. These come from the manufacturer prepositioned and preassembled so that they serve as their own template. High-strength bars should not be welded because they are susceptible to cracking or breaking.

Some users have anchored poles by direct embedment without a concrete foundation, using the excavated soil, crushed stone or sand, or both, for backfill. This may be considered for light loads, provided the design is carefully done, recognizing, among other things, that the pole might oval below ground and suffer a reduction of certain section properties. The pole may deflect laterally in the soil, and its effect on the pole deflection under load should be evaluated. It may be difficult to realign an embedded pole if it has a large permanent deflection after heavy loading conditions occur.

REFERENCES

1. Brockenbrough, R. L., and R. J. Schmitt: Considerations in the Performance of Bare High-Strength Low-Alloy Steel Transmission Towers, Paper C75 041-9, IEEE Winter Conference, New York, January 1975.
2. Building Code Requirements for Minimum Design Loads in Buildings and Other Structures, American National Standards Institute, Inc., New York.
3. National Electrical Safety Code, American National Standards Institute, Inc., New York.
4. Davenport, A. G.: Rationale for Determining Design Wind Velocities, *J. Struct. Div. ASCE*, May 1960.
5. Lo, D. L. C., A. Marcos, and S. K. Goel: Use of Computers in Transmission Tower Design, *J. Struct. Div. ASCE*, July 1975.
6. "Guide for Design of Steel Transmission Towers," ASCE Manuals and Reports on Engineering Practice, no. 52, New York, 1971.
7. Guide for the Design of Aluminum Transmission Towers, *J. Struct. Div. ASCE*, December 1972.
8. Design of Steel Pole Transmission Structures, *J. Struct. Div. ASCE*, December 1974.
9. Gere, J. M., and W. O. Carter, Critical Buckling Loads for Tapered Members, *J. Struct. Div. ASCE*, February 1962.
10. Design of Steel Transmission Pole Structures (Closure), *J. Struct. Div. ASCE*, May 1976
11. DiGioia, A. M., T. D. Donovan, and F. J. Cortese: Laterally Loaded Caisson Foundations for Single Pole Transmission Structures, Pennsylvania Electrical Association, Engineers Section, Structures and Hydraulic Committee, January 1972.
12. Arena, J. R.: How Safe Are Your Poles? *J. Struct. Div. ASCE*, December 1974.

Section **25**

Buried Conduits

RAYMOND J. KRIZEK
Professor of Civil Engineering, Northwestern University

1. Types of Conduits Buried conduits are constructed of many types of materials, such as concrete, clay, steel, aluminum, cast iron, asbestos, plastic and wood, and in a variety of shapes, such as circular, elliptical, rectangular, square, arched, and various combinations thereof. Installations involve a variety of techniques, such as riveting, bolting, welding, slipforming, and gluing. Other variables which affect performance include the type of installation (ditch or embankment), bedding, backpacking, degree of compaction, and construction sequence.

Helical Corrugated-Metal Pipe. A major benefit of helically corrugated pipe is its improved hydraulic characteristics relative to circumferentially corrugated pipe. The main disadvantages are that the helical corrugations have slightly less resistance to diametral deflections than their annular counterparts, and tight connections between pipe sections are difficult to obtain.

The structural design is the same as that for pipe with annular corrugations.

Smooth-Lined Pipe. Smooth-lined pipe has a double wall with corrugated outer wall and a smooth inner wall. Its primary advantages are a low friction factor and high abrasion resistance. In some cases, the improved hydraulic characteristics may permit the use of a smaller pipe with possible cost savings. Smooth-wall pipe contains approximately the same total metal area per unit length as standard corrugated pipe of the same nominal diameter, and its strength is approximately the same as standard pipe.

Cast-in-Place Concrete Pipe. Cast-in-place concrete pipe involves the application of a slipform technique to cast a concrete pipe in a trench with a rounded bottom and a width equal to the outside diameter of the pipe; the bottom and sides of the trench are used as stationary forms, and the inside diameter is shaped with temporary forms that are removed when the concrete sets. The conduit walls are not normally reinforced; structural strength is derived from lateral confinement by the soil. A variation of the slipformed cast-in-place pipe is a precast reinforced-concrete core around which concrete is poured. Reinforcement of the precast core conforms to requirements for standard reinforced-concrete pipe and provides a section with carefully controlled properties and dimensions.

Prestressed Concrete Pipe. Prestressed pipe is used extensively for pressure pipe, and is also practical for large-diameter conduits. Circumferential prestressing reduces the tensile bending stresses induced by high fills and may provide more economical use of materials. Claimed benefits include high strength, equal strength around the circumference, service-load moment resistance without cracking, elastic behavior under overloads, high shear resistance, and proof loading during prestressing.

Concrete-Polymer Composites. Concrete-polymer composites are portland-cement concretes in which a monomer and a resin are added to the mix and subsequently polymerized. The resulting material has strengths that are greater than normal concrete by a factor of 3 or more and can reduce the weight and increase the strength of concrete pipe.

Asbestos-Cement Pipe. This is an alternate to some of the more commonly used pipes. Potential advantages over concrete pipe are higher flow capacity and lower installation and excavation costs. Asbestos-cement pipe is generally limited to diameters less than 36 in.

Plastic Pipe. There are at least three different concepts for plastic pipe; one consists of a fiberglass-reinforced wall with helically wound fibers; another uses helical ribs for added bending stiffness; and a third is made from a thermosetting compound that is heated and formed in place by rolling pipe flat on a preformed bed and pulling a heated mandrel through to rigidize the walls. Types for transportation drainage systems include PVC sewer pipe, corrugated PE tubing, and ABS composite sewer pipe. Other types that merit consideration are PE sewer pipe, RPM sewer pipe, corrugated PE tubing (large-diameter), and ABS smooth-wall pipe.[5] Plastic pipe has been used primarily in sewage or corrosive chemical environments, but it has features (such as relatively light weight, corrosion resistance, and smooth inner walls) that may enhance its use for culverts and underdrains. However, it has a lower strength and greater susceptibility to creep than metal and concrete pipes, and tends to become brittle at low temperatures and weak at high temperatures. Many plastics are also potentially flammable. One composite pipe (truss pipe) consists of concentric plastic skins that are held apart by longitudinal plastic stiffeners; the approximately triangular spaces between these thin plastic elements are filled with foamed cement to achieve stiffness and strength.

Table 1 lists types of pipe and various agency specifications for each.

2. Analysis and Design The analysis and design of a buried conduit is essentially a problem of soil-structure interaction, and full cognizance must be given to this fundamental coupling phenomenon. Historically, buried conduits have been divided into two general categories—flexible and rigid—and independent analysis and design procedures have been developed for each. However, despite the individual treatments given these extreme situations, the load acting on the conduit in each case is determined in basically the same manner; this, of course, is inconsistent with the effects of the interaction between the soil and the conduit.

Design procedures developed by Marston, Spangler, and their coworkers over 50 years ago,[16] and refined by others as additional data became available, are used to design most buried conduits. However, these procedures become increasingly inadequate as conduit sizes or cover heights increase and as conduits of intermediate stiffness are encountered, and more sophisticated methods should be considered for these cases.

Although several failure modes are possible, the quantification of "failure" of a buried conduit is often quite arbitrary, and the specification of a safety factor for any given set of circumstances is ambiguous and subject to considerable misunderstanding. This becomes especially difficult when durability is considered.

The ratio of a given design criterion to the corresponding performance, either predicted or measured, is termed either the *safety factor* or the *performance factor*, both of which constitute convenient measures for evaluating a particular design. If the selected design criterion is tantamount to failure, the ratio is termed a safety factor; hence safety factors should be substantially greater than unity for safe design. If the selected design criterion is a measure of some allowable response level, the resulting ratio is called a performance factor, and it can be equal to or even less than unity for safe design.

The inherent difficulty in applying either of these concepts is the lack of factual data to evaluate either the actual failure condition or the actual state of the conduit in a particular situation. Both these shortcomings are due to the high cost of obtaining reliable performance data for in-place conduits and the rarity of a failure (especially due to design considerations). Many aspects must be considered in providing appropriate safety margins; among these are (1) the degree of certainty of the values for the material properties, (2) the degree of certainty of the load values, (3) the accuracy of the formulations used to determine the system response, (4) the cost of variations in the structure dimensions or material property values, and (5) the cost (including intangibles) of a failure.

LOADS ON CONDUITS

Loads on buried conduits consist of earth loads and traffic loads, both of which depend on the relative stiffness of the soil and conduit. In the case of pressure conduits, the combined effect of these loads and the internal pressure must be considered when

TABLE 1   Specifications for Drainage Pipe

Pipe material	Specification			
	AASHTO	ASTM	Federal	Other
Steel:				
Galvanized corrugated steel	M36		WW-P-405	
Corrugated steel structural plate	M167			
Precoated, galvanized steel	M245			
Aluminum:				
Corrugated aluminum alloy	M196		WW-P-402	
Aluminum-alloy structural plate	M219		WW-P-402	
Concrete:				
Reinforced	M170	C76		
Reinforced, box sections	M259	C789		
Reinforced, elliptical	M207	C507		
Nonreinforced	M86	C14		
Cast-in-place, nonreinforced				ACI 346
Reinforced arch	M206	C506		
Asbestos-cement	M217	C428	SS-P-331	
		C663		
Cast iron	M64	A142	WW-P-421	
Clay	M65	C700	SS-P-361	
Clay liner plates		C479		
Plastic:				
Polyethylene (PE)	M252	F405		
Polyvinyl chloride (PVC)		D3033		
		D3034		
Acrylonitrile-butadiene-styrene		D2680		
(ABS)		D2751		
Fiberglass-reinforced (FRP)		D2996		
		D2997		
Reinforced plastic mortar (RPM)		D3262		
Stainless steel, culvert grade				AISI Type 409

determining the stresses in the conduit wall. According to Marston-Spangler theory,[16] conduits are classified as (1) ditch, (2) positive projecting, (3) negative projecting, (4) imperfect ditch, and (5) tunnel.

3. Loads on Ditch Conduits   A ditch conduit is defined as one that is installed in a relatively narrow ditch in undisturbed soil and then covered with backfill to the ground surface. For flexible conduits and/or poor soil compaction the backfill tends to settle in relation to the adjacent undisturbed soil, and upward shear forces are generated on the sides of the backfill. The load W per unit length for this type of installation is given by

$$W = C_d \gamma_t B_d^2 \tag{1}$$

where C_d = load coefficient, γ_t = unit weight of soil backfill, and B_d = width of ditch at elevation of top of conduit. The load coefficient is a function of H/B_d, where H is the height of fill above the top of the conduit, and the frictional characteristics of the backfill soil and the sides of the trench. Values of C_d are given in Fig. 1. Most designers base load calculations on the curve for ordinary clay. The unit weight of the backfill soil is taken to be the wet weight of the materials, and should never be assumed less than 120 pcf.

The load calculated by Eq. (1) is the total load on the horizontal plane through the top of the conduit between the sides of the ditch. If the conduit is relatively rigid, such as a concrete, clay, or heavy-walled cast-iron pipe, practically all this load will be carried by the conduit itself, because the side columns of the soil are relatively compressible and cannot support more than a small portion of the total load on the horizontal plane. If the

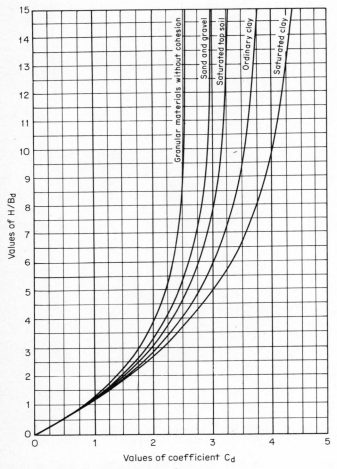

Fig. 1 Load coefficients for ditch conduits.

conduit is relatively flexible, such as a thin corrugated metal, smooth steel, or ductile iron pipe, the stiffness of the sidefill soil may approach that of the conduit itself, and the sides of the ditch (undisturbed soil) will carry their proportionate share of the total load on the plane; in this case Eq. (1) may be modified by multiplying the right-hand side by B_c/B_d, where B_c is the outside width of the conduit, which gives

$$W_c = C_d \gamma_t B_c B_d \tag{2}$$

Interpolation may be used if the situation lies between the conditions described.

The load on a ditch conduit can be minimized by keeping the width of the ditch as small as possible, consistent with providing sufficient working space to place and joint the pipes properly and to compact the sidefill, particularly under the haunches of the pipe. It is especially important to ascertain by competent and effective inspection that the width of the ditch used in the design calculations is not exceeded during construction; if caving causes an excessive ditch width, it may be necessary to use stronger pipe than originally specified. Since the ditch width that controls the load on a pipe is the width at the elevation of the top of the pipe, it may be advantageous to use a ditch with sloping sides above the top of the conduit (Fig. 2).

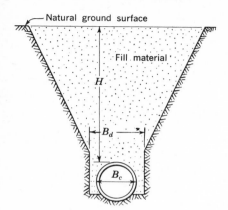

Fig. 2   Subditch.

When conduits are installed in tunnels instead of by open-cut methods, the load on the conduit may by estimated by modifying Eq. (1) to take into account the cohesion c of the undisturbed soil above the tunnel; such an expression is

$$W = C_d B_t (\gamma_t B_t - 2c) \tag{3}$$

where B_t is the maximum width of the tunnel excavation. When determining C_d in this case, H is taken as the distance from the top of the tunnel to the ground surface. Equation (3) is applicable only when the tunnel is excavated in relatively homogeneous soils which are not subjected to unusual internal stresses. When the tunnel is excavated through materials which tend to squeeze or swell, such as some types of clay or shale, or through blocky or seamy rock, the vertical load cannot be determined solely from consideration of the factors upon which Eq. (3) is based.

4. Loads on Projecting Conduits   A *positive projecting conduit* is one which is installed in shallow bedding, with its top projecting some distance above the surface of the adjacent natural ground, and then covered with an embankment. A *negative projecting conduit* is one which is installed in a relatively narrow and shallow ditch, with its top at some elevation below the natural ground surface, and then covered with an embankment. In general, the load produced by a given height of fill is less for a negative projecting conduit than for a positive projecting conduit. Best results from the standpoint of load reduction are obtained if the ditch between the top of the conduit and the natural ground surface is refilled with loose, uncompacted soil. A transition or intermediate class between positive and negative projecting conduits is the *zero projecting conduit;* this is the case in which the conduit is installed in a narrow shallow ditch with its top approximately at the level of the natural ground surface.

The load acting on a positive projecting conduit is governed by the magnitude and direction of the relative movements of the soil overlying, underlying, and adjacent to the conduit, as well as the deflection of the conduit. These factors (Fig. 3) have been combined into an abstract parameter, called the *settlement ratio* r_{sd} defined as

$$r_{sd} = \frac{(s_m + s_g) - (s_f + d_c)}{s_m} \tag{4}$$

where s_m = compression strain in side columns of soil of height pB_c
s_g = settlement of natural ground surface adjacent to conduit
s_f = settlement of conduit into its foundation
d_c = vertical shortening of conduit

The height pB_c is the distance between the natural ground surface and the top of the conduit; p is called the *projection ratio.*

The load on a positive projecting conduit is determined from

$$W = C_c \gamma_t B_c^2 \tag{5}$$

Values of the coefficient C_c are given in Fig. 4.

Although the settlement ratio r_{sd} is a rational quantity in the development of the load formula, it is not feasible to determine it rationally in advance of construction. Instead, it

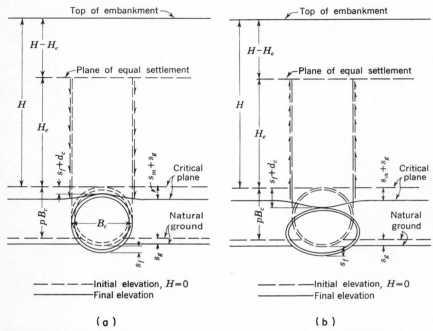

(a)   (b)

Fig. 3 Positive projecting conduit: (*a*) positive settlement ratio; (*b*) negative settlement ratio.

must be considered as an empirical factor whose values are determined from observations of the performance of conduits under embankments. Recommended values are given in Table 2.

5. Loads on Conduits in Wide Ditches According to Eq. (1), the load on a ditch conduit increases with the square of B_d. However, there is a limit above which this does not apply. The width of ditch which yields a load on the conduit equal to that to which it would be subjected if installed as a positive projecting conduit is defined as the *transition width.* When the ditch width at the top of the conduit is less than the transition width, the actual width should be substituted in Eq. (1), but when it is equal to or greater than the transition width the latter should be used instead of the actual width. A diagram for estimating the transition width is given in Fig. 5.

6. Loads on Negative Projecting and Imperfect-Ditch Conduits The analysis of loads on negative projecting conduits is based on the concept that the soil prism of width B_d overlying the conduit settles downward in relation to the adjacent soil and is therefore subjected to upward shear forces along the sides. In this case the negative projection ratio p' is defined as the distance between the natural ground surface and the top of the conduit to the ditch width B_d, and the settlement ratio is given by

$$r_{sd} = \frac{s_g - (s_d + s_f + d_c)}{s_d} \tag{6}$$

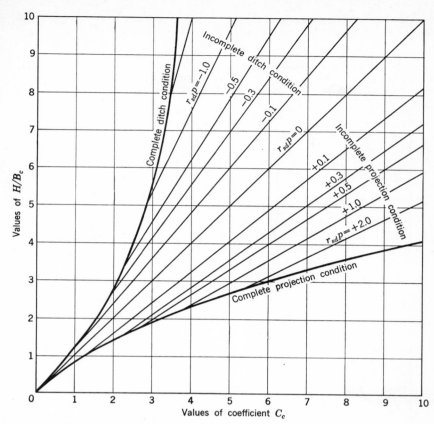

Fig. 4 Load coefficients for projecting conduits.

TABLE 2 Design Values for Settlement Ratio (Ref. 2)

	Settlement ratio r_{sd}	
Installation and foundation condition	Usual range	Design value*
Positive projecting:	0 to +1.0	
Rock or unyielding soil	+1.0	+1.0
Ordinary soil	+0.5 to +0.8	+0.7
Yielding soil	0 to +0.5	+0.3
Zero projecting		0
Negative projecting:	−1.0 to 0	
$p = 0.5$		−0.1
$p = 1.0$		−0.3
$p = 1.5$		−0.5
$p = 2.0$		−1.0
Imperfect trench:	−2.0 to 0	
$p = 0.5$		−0.5
$p = 1.0$		−0.7
$p = 1.5$		−1.0
$p = 2.0$		−2.0

*Flexible conduits are usually designed for a settlement ratio of zero.

where the notation is shown in Fig. 6. Suggested design values are given in Table 2. The load per unit length on such a conduit is

$$W = C_n \gamma_t B_d^2 \tag{7}$$

where values of the load coefficient C_n are given in Fig. 7. Factual data relative to values of the settlement ratio for this type of installation are meager; the few measurements which have been made indicate that values in the range of -0.3 to -0.5 are appropriate.

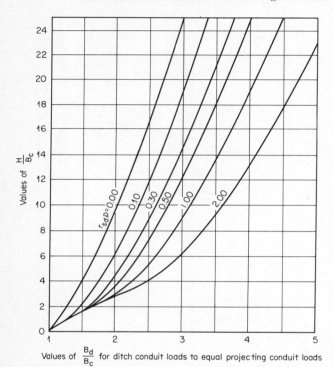

Values of $\dfrac{B_d}{B_c}$ for ditch conduit loads to equal projecting conduit loads

Fig. 5 Transition-width ratios.

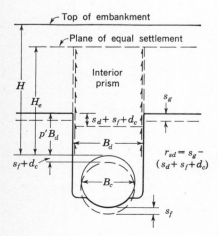

Fig. 6 Negative-projecting and imperfect-ditch conduits.

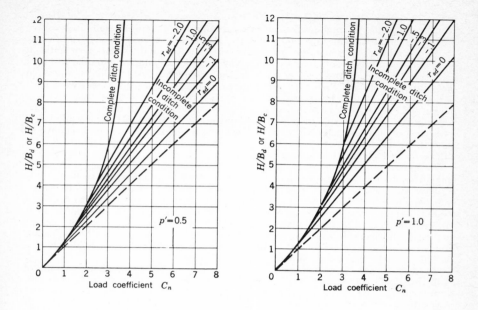

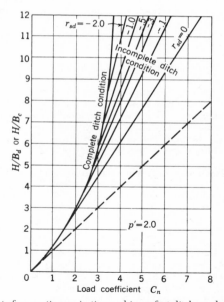

Fig. 7 Load coefficients for negative-projecting and imperfect-ditch conduits.

The *imperfect-ditch method* of construction was devised by Marston more than 50 years ago to reduce the high loads that usually result when pipes are installed as positive projecting conduits under high embankments. This goal is accomplished by artificially creating conditions in which the soil prism overlying a conduit is caused to settle in relation to the adjacent soil, thereby mobilizing shear forces and reducing the resultant load on the conduit. In this method of construction the pipe is first installed as a positive projecting conduit, and then surrounded by soil which is thoroughly compacted for a distance equal to at least two pipe diameters on each side and up to some distance above the top of the pipe. Next, a trench having a width equal to the outside diameter of the pipe is dug in the compacted soil directly above the structure. This trench is then refilled with loose compressible soil, after which the embankment is completed in a normal manner. In general, the deeper the trench and the more compressible the trench backfill, the greater will be the reduction of load on the conduit.

The imperfect-ditch concept is used to minimize costs of installations under high embankments, and there are a number of installations on record in which such materials as straw, hay, sawdust, tree leaves, and pine straw have been used successfully. If the embankment is sufficiently high (and this method is not economical except in cases of medium to high embankments) no differential settlement will occur at the embankment surface. Equation (7) may be used to estimate loads on imperfect-ditch conduits, except that the width should be taken as B_c and the ratio H/B_c should be used when selecting values of C_n from Fig. 7. Also, p' is the depth of the imperfect trench divided by B_c. Design values for the settlement ratio are given in Table 2.

7. Surface Loads When a conduit is placed under a relatively shallow covering of earth, live loads due to highway, railway, or airplane traffic, as well as construction equipment, may be of major importance. Experiments show that a concentrated static surface load is transmitted through the soil to a conduit substantially in accordance with the solution for stress distribution in a semi-infinite linear elastic medium, and have also indicated the magnitude of the impact produced by moving wheel loads. As a result of these observations, the live load W_t per unit length on an underground conduit may be determined from

$$W_t = \frac{I_c C_t P}{L} \tag{8}$$

where L = length of conduit section, I_c = impact factor, C_t = load coefficient, and P = concentrated surface load.

For a precast, segmental section of pipe which is 3 ft or less in length, L is the actual length, but for continuous conduits or those constructed of segmental sections longer than 3 ft L is the effective length, which is defined as the length of pipe over which the average live load produces the same effect on stress or deflection as does the actual load which is of varying intensity along the pipe. No information is available concerning the effective length of continuous pipelines, and an effective length of 3 ft is suggested for the design of longer pipes.

The impact factor I_c is equal to unity when the surface load is static. For moving loads it depends on the speed of the vehicle, its vibratory action, the effect of wing uplift, and most importantly the roughness characteristics of the roadway surface. It is independent of the depth of cover. Values range from 1.5 to 2.0.

The load coefficient C_t in Eq. (8) represents the fractional part of the concentrated load that is transmitted through the soil to the conduit. It depends on the length and width of the conduit section, the depth below the surface, and the point of application of the concentrated load. Figure 8 gives the coefficient for the load on a rectangular area, one corner of which lies directly below the wheel load. These coefficients are a function of m = X/H and $n = Y/H$, where X and Y are the length and width, respectively, of the area and H is the height of fill above the conduit. Equation (8) and Fig. 8 may be applied to a wide variety of cases with reference to the lateral position of the load relative to the projected area of the conduit section, as illustrated in Fig. 8. In Case I the load is directly above a corner of the area, for which $X = L$, the effective length of the conduit, and $Y = B_c$, the outside width, so that C_t is determined directly from Fig. 8. When the load is directly above the center of the area ($X = L/2$ and $Y = B_c/2$), as in Case II, C_t is obtained by multiplying by 4 the coefficient from Fig. 8 for one quadrant of the area. Other cases

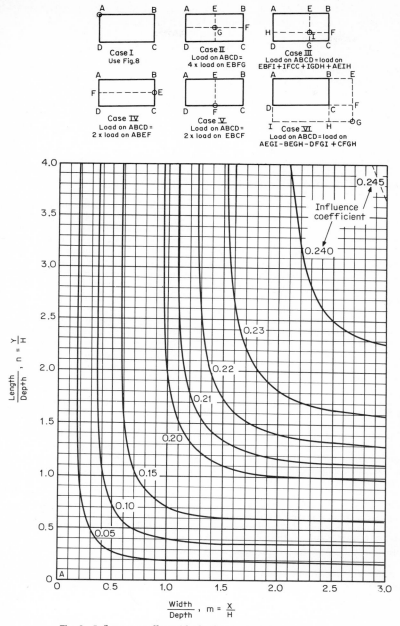

Fig. 8 Influence coefficient for load on rectangular area.

shown in Fig. 8 are handled similarly by combining coefficients for rectangles, each of which has one corner directly under the point of load application.

RIGID CONDUITS

8. Supporting Strength   The standard method to determine the strength of rigid pipe is the three-edge bearing test (ASTM C497 for concrete pipe or tile, C301 for vitrified-clay pipe, A142 for cast-iron culvert pipe). Under this method of loading the pipe is subjected to concentrated line loads at the crown and invert, and the load is increased until the ultimate strength of the pipe has been reached, or for concrete pipe until a 0.01-in. crack has occurred throughout a length of 1 ft or more. The bending moment in the pipe wall at the springline in this test is $M = 0.182Pr$.

Another means of expressing the strength of reinforced-concrete pipe is in terms of the D-load, which is the three-edge bearing strength (pounds per linear foot) per foot of inside diameter of the pipe. The D-load concept enables strength classification of pipe independent of pipe diameter. ASTM C76-66T describes five strength classes based on D-load at 0.01-in. crack and/or ultimate load.

The three-edge bearing test is probably the most severe loading to which any pipe will be subjected. To relate the three-edge bearing strength to the in-place supporting strength, load factors have been developed. The load factor L_f is defined as the ratio of the supporting strength of a pipe under any stated condition of loading in the field to the supporting strength of a similar pipe as determined in a three-edge bearing test. The two primary conditions that influence the load factor are the distribution of the bottom reaction on the pipe and the magnitude and distribution of any lateral stress acting on the sides of the pipe. Load factors for various trench and embankment installations have been determined either experimentally or on the basis of assumed stress distributions on the pipe.

Factor of Safety.   The safe supporting strength of a pipe is equal to its in-place supporting strength divided by a factor of safety. However, because the in-place supporting strength depends to a large degree on installation conditions and local quality control, the value of the factor of safety must be based on judgment. Based on the minimum strengths of a representative number of pipe specimens, values from 1.2 to 1.5 are suggested for unreinforced-concrete pipe. The suggested factor of safety for reinforced pipe is 1, based on the minimum 0.01-in. crack strength or 80 percent of the minimum ultimate strength, whichever is smaller. In the latter case the residual strength of the pipe between the 0.01-in. crack strength and the ultimate strength provides some margin of safety.

Actual factors of safety are larger than the values suggested above because long before the ultimate strength is reached in a field installation the flexibility of the pipe allows the horizontal diameter to increase, thereby mobilizing some resistance from the surrounding soil, which greatly increases the ultimate strength of the pipe. This concept is supported by the fact that, although many reinforced-concrete pipes have developed longitudinal cracks wider than 0.01 in., apparently none have collapsed under earth load.

9. Bedding Classes for Trench Conduits   Four general classes of bedding for the installation of circular pipe in a trench are defined in Fig. 9. The left-hand illustration for each bedding class represents conditions that were achieved a number of years ago when hand labor was used extensively, and the associated load factors were determined experimentally. In general, these load factors do not take into account any lateral stresses on the pipe, because such installations are usually performed under conditions where effective compaction of the soil under and around the pipe cannot be achieved with any degree of confidence. The right-hand illustration for each bedding class is representative of conditions achieved by use of modern installation techniques with selected granular materials and much less hand labor.

10. Bedding Classes for Embankment Installation   Four general classes of bedding for circular pipe in an embankment installation are defined in Fig. 10. The associated load factors based on assumed conditions and relationships proposed by Spangler are given in Table 3. The trench load factors shown in Fig. 9 should be used for negative projecting embankment installations. Table 4 gives the ranges of load factors for imperfect-ditch installations. The jacking method of construction usually approximates an ideal bedding

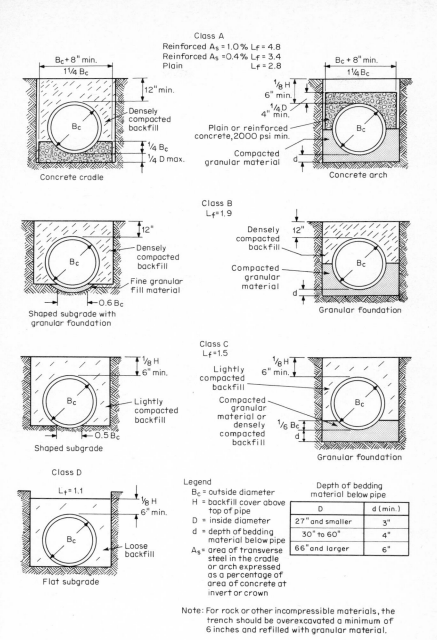

Class A
Reinforced $A_s = 1.0\%$ $L_f = 4.8$
Reinforced $A_s = 0.4\%$ $L_f = 3.4$
Plain $L_f = 2.8$

$B_c + 8"$ min.
$1\frac{1}{4} B_c$

12" min.

Densely compacted backfill

B_c

$\frac{1}{4} B_c$
$\frac{1}{4} D$ max.

Concrete cradle

$\frac{1}{8} H$
6" min.
$\frac{1}{4} D$
4" min.

Plain or reinforced concrete, 2000 psi min.

Compacted granular material

$B_c + 8"$ min.
$1\frac{1}{4} B_c$

B_c

d

Concrete arch

Class B
$L_f = 1.9$

12"

Densely compacted backfill

B_c

Fine granular fill material

$\leftarrow 0.6 B_c$

Shaped subgrade with granular foundation

Densely compacted backfill 12"

Compacted granular material

B_c

d

Granular foundation

Class C
$L_f = 1.5$

$\frac{1}{8} H$
6" min.

Lightly compacted backfill

B_c

$\leftarrow 0.5 B_c$

Shaped subgrade

$\frac{1}{8} H$
6" min.

Lightly compacted backfill

Compacted granular material or densely compacted backfill

$\frac{1}{6} B_c$
d

B_c

Granular foundation

Class D
$L_f = 1.1$

$\frac{1}{8} H$
6" min.

B_c

Loose backfill

Flat subgrade

Legend
B_c = outside diameter
H = backfill cover above top of pipe
D = inside diameter
d = depth of bedding material below pipe
A_s = area of transverse steel in the cradle or arch expressed as a percentage of area of concrete at invert or crown

Depth of bedding material below pipe

D	d (min.)
27" and smaller	3"
30" to 60"	4"
66" and larger	6"

Note: For rock or other incompressible materials, the trench should be overexcavated a minimum of 6 inches and refilled with granular material.

Fig. 9 Trench beddings for circular concrete pipe. (*From Ref. 2.*)

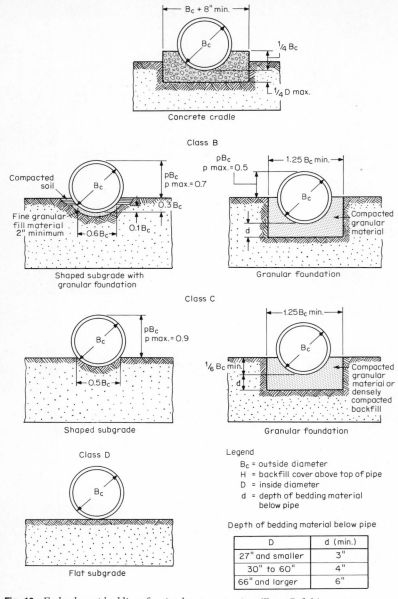

Fig. 10 Embankment beddings for circular concrete pipe. *(From Ref. 2.)*

ROCK OR OTHER NONCOMPRESSIBLE FOUNDATION—Where ledge rock, rocky or gravelly soil, hard pan or other unyielding foundation material is encountered, the hard unyielding material should be excavated before the elevation of the concrete cradle (Class A) or the bottom of the pipe or pipe bell (Class B & C Beddings) for a depth of at least 6 inches or $\frac{1}{2}$ inch for each foot of fill over the top of the pipe, whichever is greater, but not more than $\frac{3}{4}$ the nominal diameter of the pipe. For Class D Bedding, the depth should be 6 inches. The width of the excavation should be one foot greater than the outside diameter of the pipe. The excavation should be refilled with selected fine compressible material, such as silty clay or loam, lightly compacted and shaped as required for the specified class of bedding.

condition by providing positive contact around the lower exterior surface of the pipe and the surrounding soil. If the bore is overexcavated, the space between the pipe and the soil can be filled with sand, grout, concrete, or other suitable material; for this type of installation a load factor of 3 is recommended. If the bore is slightly overexcavated and the space between the pipe and the soil is not filled, a minimum load factor of 1.9 is

TABLE 3   Load Factors for Circular Conduits in Positive Projecting Embankment Installations (Ref. 2)

Projection ratio p	Bedding class			
	A	B	C	D
0.9	3.74–11.26	Not recommended	1.98–3.01	1.20–1.51
0.7	3.65–7.52	2.33–3.00	1.89–2.35	1.17–1.33
0.5	3.40–4.84	2.17–2.37	1.79–1.94	1.13–1.19
0.3	3.10–3.49	2.07–2.11	1.73–1.76	1.10–1.12
0	2.83	2.02	1.70	1.10

Table includes values for H/B_c ranging from 0.5 to 1.5 and $r_{sd}p$ ranging from 0 to 1; within the individual ranges given, the higher load factors are associated with lower values for H/B_c and $r_{sd}p$, and the load factor decreases as either parameter increases.

TABLE 4   Load Factors for Circular Conduits in Imperfect-Trench Installations (Ref. 2)

Projection ratio p	Bedding class		
	A	B	C
0.9	7.11–47.57	Not recommended	2.62–3.75
0.7	5.91–22.05	2.78–3.72	2.21–2.76
0.5	4.45–8.20	2.32–2.63	1.91–2.12
0.3	3.43–4.29	2.10–2.18	1.76–1.81
0	2.83	2.02	1.70

Table includes values for H/B_c ranging from 1 to 25 and r_{sd} ranging from -0.1 to -2.0; within the individual ranges given, the higher load factors are associated with the higher values for H/B_c and the lower values for r_{sd} (where $-2n$ is lower than $-n$).

recommended. A maximum live-load factor of 1.5 is recommended for highway loads on an unsurfaced roadway and aircraft loads on flexible pavements.

Too much emphasis cannot be placed on the need to bed a conduit in such a manner that the bottom vertical reaction is well distributed over the width of bedding. Failure to achieve this favorable distribution is one of the major causes of structural distress in rigid pipes and may contribute to excessive deflection of flexible pipes. Contractors frequently lay pipe on a flat bed of soil and then attempt to compact backfill material into the wedge-shaped spaces under the lower haunches of the pipe. Unless the backfill material is a very favorable type, such efforts will usually not be successful, especially since compaction in a vertical direction causes little or no compaction of soil in a lateral direction.

11. Monolithic Conduits   Monolithic reinforced-concrete conduits (arch and box culverts and sewers) may be satisfactorily designed as rigid-frame structures. Numerous measurements of the vertical earth load on both rectangular and circular conduits have shown that it is essentially uniformly distributed over the width of the conduit. Such structures may or may not be subjected to active lateral soil pressure on all or part of their sidewall areas, depending on the local situation. The sidewalls of arch-shaped conduits of masonry or concrete, especially those under deep overburden, should be designed to follow the pressure line obtained by construction of an equilibrium polygon of the vertical and horizontal forces. This procedure results in a horseshoe cross-sectional shape where the crown and sidewalls are subjected only to compression.

Example 1 A 60-in. reinforced-concrete culvert pipe having 6-in. sidewalls is to be installed as a projecting conduit in a Class C bedding on a foundation of ordinary soil and covered with an embankment 24 ft high. The bottom of the trench is 2 ft below existing ground level, and the unit weight of the soil cover is 120 pcf. Determine the required strength of pipe which will not develop a crack wider than 0.01 in.

SOLUTION. Settement ratio $r_{sd} = 0.7$ (Table 2)

Projection ratio $p = 4/6 = 0.67$

From Fig. 4, for $r_{sd}p = 0.7 \times 0.67 = 0.47$ and $H/B_c = 24/6 = 4$, $C_c = 6.4$.

From Eq. (5), $W = C_c \gamma_t B_c^2 = 6.4 \times 120 \times 6^2 = 27,650$ plf

From Table 3, for Class C bedding, $p = 0.67$ and $H/B_c = 4$, the estimated load factor $= 2$.

Required three-edge bearing strength $= 27,650/2 = 13,825$ plf or $2765D$. Use standard ASTM (C76-66T) Class V pipe rated at 15,000 plf or $3000D$.

FLEXIBLE CONDUITS

The predominant source of supporting strength for a flexible conduit is the lateral pressure exerted by the soil at the sides of the pipe. The pipe itself has relatively little bending strength, and a large part of its ability to support vertical loads is obtained from the membrane action that develops in the wall as the conduit deforms to redistribute any imposed nonuniform loads. Two design procedures are commonly used for circular conduits; one is concerned with limiting conduit deformations and the other with limiting the compressive load in the conduit wall. In general, design manuals recommend selection of the wall thickness in accordance with the latter and a deflection check by the former. Local buckling is not usually considered except for conduits of very large diameter. This appears justifiable because no report of a buckling failure of an in-service circular conduit has been found unless the failure was preceded by excessive deflection. However, buckling considerations may be significant with smooth-wall plastic pipe and large-diameter corrugated-metal pipe.

Most of the early design criteria are empirical in nature and were established by observational studies aimed toward correlating pipe performance with deflection. However, studies show that a conduit will carry load in ring compression up to its wall strength in crushing or buckling, without deflection distress, provided the sidefill soil is compacted to at least 85 percent of the standard Proctor density.

12. Ring Compression According to the ring-compression theory, the compressive stress σ_c in the conduit wall can be calculated by assuming the conduit to be loaded hydrostatically. Thus

$$\sigma_c = \frac{pD}{2A} \tag{9}$$

where p = design pressure per unit length of conduit, D = conduit diameter, and A = wall area per unit length of conduit.

The following American Iron and Steel Institute (AISI) formulas[3] for the wall strength of corrugated pipes of steel with a yield strength of 33,000 psi, with backfill compacted to 85 percent of AASHTO Standard T99, are based on tests:

$$\sigma = \sigma_y = 33,000 \text{ psi} \qquad \frac{D}{r} < 294 \tag{10a}$$

$$\sigma = 40,000 - 0.081 \left(\frac{D}{r}\right)^2 \qquad 294 < \frac{D}{r} < 500 \tag{10b}$$

$$\sigma = \frac{4.93 \times 10^9}{(D/r)^2} \qquad 500 < \frac{D}{r} \tag{10c}$$

where D = diameter and r = radius of gyration of conduit wall.

Values of r in Eq. (10) can be determined from Table 5 $(r = \sqrt{I/A})$. However, they are

TABLE 5 Moment of Inertia and Cross-Sectional Area of Corrugated-Steel Sheets and Plates for Underground Conduits (Ref. 3)

Corrugation pitch × depth, in.	Specified thickness, in.											
	0.034	0.040	0.052	0.064	0.079	0.109	0.138	0.168	0.188	0.218	0.249	0.280
	Moment of inertia I, in.⁴/ft width											
1½ × ¼	0.0025	0.0030	0.0041	0.0053	0.0068	0.0103	0.0145	0.0196				
2 × ½	0.0118	0.0137	0.0184	0.0233	0.0295	0.0425	0.0566	0.0719				
2⅔ × ½	0.0112	0.0135	0.0180	0.0227	0.0287	0.0411	0.0544	0.0687				
3 × 1	0.0514	0.0618	0.0827	0.1039	0.1306	0.1855	0.2421	0.3010				
6 × 2						0.725	0.938	1.154	1.296	1.523	1.754	1.990
	Cross-sectional wall area, in.²/ft width											
1½ × ¼	0.380	0.456	0.608	0.761	0.950	1.331	1.712	2.093				
2 × ½	0.409	0.489	0.652	0.815	1.019	1.428	1.838	2.249				
2⅔ × ½	0.387	0.465	0.619	0.775	0.968	1.356	1.744	2.133				
3 × 1	0.444	0.534	0.711	0.890	1.113	1.560	2.008	2.458				
6 × 2						1.556	2.003	2.449	2.739	3.199	3.658	4.119

nearly independent of wall thickness in the larger corrugations, and the following are values for the commonly used corrugations:

Corrugation	r, in.
2⅔ × ½	0.172
3 × 1	0.344
6 × 2	0.688

Equations (10) are plotted in Fig. 11. Values of σ can be read directly in terms of various pipe diameters for the three standard corrugations.

The AISI recommended factor of safety is 2. Using this value and equating σ_c in Eq. (9) to $\sigma/2$ gives

$$A = \frac{pD}{\sigma} \tag{11}$$

Areas of corrugated-steel sheets and plates are given in Table 5.

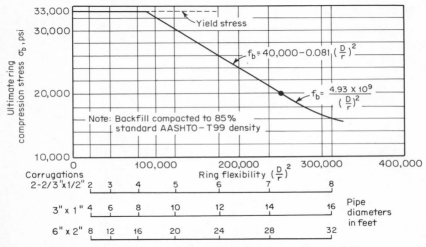

Fig. 11 Ultimate ring-compression stress for corrugated steel pipe. (*From Ref. 3.*)

Design Pressure.   Since the load-carrying capacity of a flexible conduit depends on the external lateral pressure on the walls as well as on the strength of the pipe wall, and since the external lateral pressure depends on the density of the soil, Eqs. (10) give the apparent strength of the wall only for pipes in soil at the 85 percent standard Proctor density for which the equations were derived. Installations in soils of other densities could be handled by developing apparent-strength formulas for pipes in soils of various densities, or by using Eqs. (10) for all densities but with an adjusted or apparent load on the pipe. The latter method is used in the AISI procedure. The total load (dead plus live) above the horizontal plane at the top of the pipe must be multiplied by the load factor from Fig. 12 to compute the apparent load on the pipe. If the height of cover is less than one pipe diameter, the total load should be assumed to act on the pipe.

The design value chosen for the backfill soil density should be based on the importance and size of the structure and the quality that can be reasonably expected. The recommended value for ordinary installations is 85 percent of the AASHTO standard (most specifications call for 90 percent). Higher values and higher-quality backfill should be required for important structures in high-fill situations.

Seams.   Standard factory-made seams are satisfactory for pipe designed for the recommended maximum allowable stress of 16,500 psi, but shop- or field-bolted seams should be evaluated by tests of straight unsupported columns.[3] The test strength should be not less than twice the design value of the ring compression. Allowable values of bolted joints, based on short-column tests, are given in Table 6. The wall stress corresponding to the allowable joint strength is also shown.

13. Deflection   Although flexible pipes do not usually fail by deflection collapse if the soil density is 85 percent or more of the AASHTO standard, vertical deflection may need to be considered in situations where the fill is high. In this case it is close enough to assume that the vertical deflection is equal to the vertical deformation of an adjacent free-field layer of soil. Such a determination requires a knowledge of the stress-strain relationship (constrained modulus) of the soil.

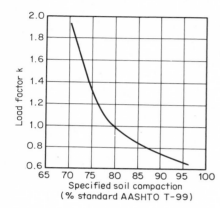

Fig. 12   Load factors for corrugated steel pipe. (*From Ref. 3.*)

TABLE 6   Bolted-Seam Design Data (Ref. 3)

Thickness, in.	Structural plate pipe 6- × 2-in. corrugation (four ¾-in. bolts per ft)		Corrugated-steel pipe 3- × 1-in. corrugation (eight ½-in. bolts per ft)	
	Allowable strength (½ ultimate), lb/ft	Corresponding wall stress, psi	Allowable strength (½ ultimate), lb/ft	Corresponding wall stress, psi
0.064			14,400	16,200
0.079			17,900	15,800
0.109	21,000	13,500	26,500	17,000
0.138	31,000	15,500	31,900	15,900
0.168	40,500	16,500	35,400	14,400
0.188	46,500	17,000		
0.218	56,000	17,500		
0.249	66,000	18,100		
0.280	72,000	17,500		

For low to moderate fill heights Spangler assumed that the horizontal deflection of the conduit is inversely proportional to the modulus of passive resistance of the soil, and showed that the change Δx in the horizontal diameter of a conduit can be calculated from

$$\Delta x = \frac{D_1 K W r^3}{EI + 0.061 E' r^3} \tag{12}$$

where D_1 = deflection lag factor, usually assumed to range from 1.25 to 1.5
K = bedding constant; values range from 0.110 for a bedding angle* of 0° to 0.083 for a bedding angle of 90°; may be interpolated linearly
r = mean radius of pipe

*The bedding angle is one-half the angle subtended by that portion of the pipe which is in contact with the pipe bedding and over which the bottom reaction is distributed.

E = modulus of elasticity of pipe material
I = moment of inertia of pipe wall per unit length of pipe (Table 5)
E' = modulus of soil reaction

The major shortcoming of Eq. (12) is that little is known about the nature of E'. Although experimental data from actual installations indicate that it varies from about 200 to 8000 psi, a value of 700 psi is usually recommended for design with backfill of 85 to 90 percent of standard Proctor density and 1400 psi with backfill of 95 percent Proctor density.

Based on inspections of numerous pipe installations, the average deflection before failure was determined to be about 20 percent of the pipe diameter. The use of a so-called safety factor of 4 established the commonly accepted design deflection of 5 percent.

Example 2 A 60-in. 10-gage ($t = 0.138$ in.) corrugated-metal pipe (standard corrugations, ½ in. deep at 2⅔-in. centers) is to be installed as a projecting conduit with 60° bedding (bedding angle = 30°) and covered with an embankment 20 ft high. The bottom of the trench is 18 in. below the existing ground level, and the unit weight of the soil cover is 120 pcf. Determine the long-time deflection of the pipe and the compressive stress in the pipe wall.

SOLUTION. Assume modulus of soil reaction $E' = 700$ psi (Art. 13). Assume settlement ratio $r_{sd} = 0$ (Table 2). Projection ratio = $p = 3.5/5 = 0.7$. From Fig. 4, for $r_{sd}p = 0$ and $H/B_c = 20/5 = 4$, $C_c = 4$. From Eq. 5,

$$W = C_c \gamma_t B_c^2 = 4 \times 120 \times 5^2 = 12{,}000 \text{ plf} = 1000 \text{ pli}$$

From Table 5, $I = 0.0544$ in.^{4}/ft $= 0.0045$ in.^{4}/in. $K = 0.1$ for bedding angle of 30°. From Eq. (12), with $D_1 = 1.25$

$$\Delta x = \frac{D_1 K W r^3}{EI + 0.061 E' r^3} = \frac{1.25 \times 0.1 \times 1000 \times 30^3}{30 \times 10^6 \times 0.0045 + 0.061 \times 700 \times 30^3} = 2.63 \text{ in.}$$

$$\frac{\Delta x}{D} = \frac{2.63}{60} = 0.044 < 5 \text{ percent} \qquad \text{O.K.}$$

From Table 5, $A = 1.744$ in.^{2}/ft $= 0.145$ in.^{2}/in.
Overburden stress = $p = 20 \times 120 = 2400$ psf = 16.7 psi

$$\text{From Eq. (9), } \sigma_c = \frac{pD}{2A} = \frac{16.7 \times 60}{2 \times 0.145} = 3455 \text{ psi}$$

From Fig. 11, $\sigma = 30{,}500$ psi, which with the recommended safety factor of 2 gives $\sigma_{\text{allowable}} = 15{,}250$ psi.

14. Pipe Arches The pressure at the corners of a pipe arch is greater than the pressure in the fill, and is the practical limiting design factor. If the bending strength of the conduit wall is neglected, the ring compression is constant around the arch, and since $C = pR$, the pressure normal to the wall is inversely proportional to the wall radius (Fig. 13). The corner pressure p_c is given by

$$p_c = \frac{pR_c}{R_t} \tag{13}$$

where p = design pressure, R_c = corner radius, and R_t = top radius. Therefore, proper installation of pipe arches is essential because the limiting design pressure is established by the allowable soil pressure at the corners. Special backfill material, such as crushed stone or soil cement, at the corners can extend this limit. A maximum value of 3 tons/sq ft is suggested for routine use.

Ignoring the wall bending strength yields results which would preclude the use of standard sizes of corrugated pipe for pipe arches carrying heavy loads such as Cooper E80. However, the longitudinal distribution of the live load throughout the pipe crown together with the bending strength of the small-radius corners gives acceptable performance. The basis for such designs is empirical tables.[3]

15. Arches on Rigid Foundations Because of the restraint at the base of an arch on an unyielding foundation the ring cannot move into the backfill at these points, so that inward local buckling of the wall is possible. The compressive strength of such an arch that is semicircular, or less, in shape (Fig. 14a) is significantly smaller than for the equivalent round pipe. Such arches are usually designed for an allowable wall stress of one-half the allowable value for an equivalent round pipe (factor of safety of 4).

Horseshoe-shaped arches whose reentrant angle θ at the base is 20° or more to the vertical (Fig. 14b) can seat into the soil at the base and prevent local buckling. An allowable stress of 75 percent of that for an equivalent round pipe can be used for reentrant angles of 20 to 30°, while the full allowable can be used if the reentrant angle is greater than 30°.

PRESSURE CONDUITS

When a conduit is subjected to both external load and internal pressure, the hoop stress due to the internal pressure combines with the bending stress due to the external load; thus a pressure conduit will not carry as much external load as one of the same strength which is not subjected to internal pressure.

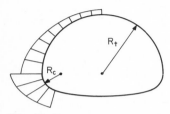

Fig. 13 Pressures on pipe arch. (*From Ref. 3.*)

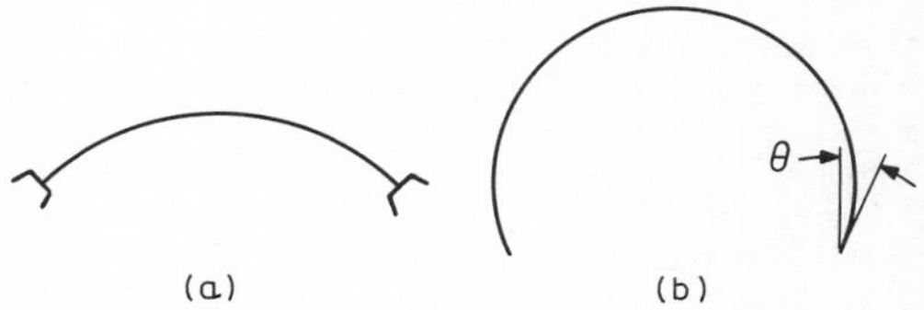

Fig. 14 Typical arch conduits

(a)   (b)

16. Flexible Pressure Conduits Conduits are usually laid and backfilled before being pressurized, and the final system of loads on a flexible pressure conduit will impose a shape that is intermediate between a circle and the deflected shape under external load alone. If the bending stresses in the wall and the hoop stress due to internal pressure are assumed to be additive, the maximum stress s under the combined loading may be approximated by

$$s = \frac{p(D - 2t)}{2t} + 1.4 \frac{C_d \gamma_t B_d^2 E t r}{E t^3 + 2.6 p r^3} \tag{14}$$

where p = internal pressure
  D, t, r = outside diameter, wall thickness, mean radius of conduit
  E = modulus of elasticity of conduit material
  B_d = trench width at top of pipe
  γ_t = unit weight of backfill
  C_d = load coefficient for ditch conduit (Fig. 1)

Example 3 A 30- × ⅜-in. steel conduit carrying natural gas at a maximum operating pressure of 600 psi is installed under 5 ft of ordinary soil cover in a trench 3½ ft wide. The unit weight of the soil is 120 pcf. Determine the combined stress in the conduit wall due to the external load and internal pressure.
  SOLUTION. $H/B_d = 5/3.5 = 1.43$. From Fig. 1, $C_d = 1.15$. From Eq. (14), with $r = (30 - 0.375)/2 = 14.8$ in.

$$s = \frac{600(30 - 0.75)}{0.75} + 1.4 \frac{1.15(120/12^3)(3.5 \times 12)^2 \times 30 \times 10^6 \times 0.375 \times 14.8}{30 \times 10^6 \times 0.375^3 + 2.6 \times 600 \times 14.8^3}$$
$$= 23{,}400 + 4950 = 28{,}350 \text{ psi}$$

17. Rigid Pressure Conduits The three-edge bearing load s under combined internal pressure and external loading may be determined from

$$\frac{s}{S} = \left(\frac{T - t}{T}\right)^{1/2} \tag{15}$$

where S = three-edge bearing strength (Art. 8)
  T = internal pressure at failure with no external load
  t = internal pressure under combined loading

Example 4 A 20-in. cast-iron conduit with a bursting strength of 620 psi and a three-edge bearing strength of 4200 plf is to be used as a water main at an operating pressure of 200 psi, which includes an allowance for water hammer. The pipe is to be laid in a shaped (Class C bedding) ditch 3 ft wide and covered with ordinary clay backfill that has a unit weight of 120 pcf. What depth of lightly compacted soil cover can be placed over the conduit with a safety factor of 2?

SOLUTION. From Eq. (15), $s = \left(\dfrac{620 - 200}{620}\right)^{1/2} \times 4200 = 3460$ plf

From Fig. 10, $L_f = 1.5$

Field supporting strength $= 1.5 \times \dfrac{3460}{2} = 2590$ plf

From Eq. (1), $C_d = \dfrac{W}{\gamma_t B_d^2} = \dfrac{2590}{120 \times 3^2} = 2.4$

From Fig. 1, $H/B_d = 3.8$

Therefore, $H = 3.8 \times 3 = 11.4$ ft

MODERN DESIGN METHODOLOGY

With the advent of plastic, bituminous fiber, and asbestos-cement pipe there is a need for a design method that can handle pipes of intermediate stiffness and, if possible, provide better results for rigid and flexible conduits. In addition, large spans and fill heights dictate the need for a more fundamental approach. Modern numerical techniques treat the conduit and the surrounding soil as a system. In general, however, these methodologies should be used with caution until a greater background of experience is accumulated.

18. Elasticity Solution   Burns and Richard[4] have developed a closed-form plane-strain solution for a thin circular conduit of a linearly elastic material buried deep in a weight-less, homogeneous, isotropic linearly elastic soil. Stresses and deformations are determined for two limiting cases: full slip (zero shear stress between the soil and the conduit wall) and no slip. The load is a surface overpressure. For the full-slip case and a conduit with infinite resistance to circumferential stress, but no bending resistance, the radial pressure on the conduit wall is hydrostatic and equal in magnitude to the weight of the overburden, as in the ring-compression theory (Art. 12). The solution provides an assessment of soil-structure interaction and a basis for the development of design methods. Comparisons with Eq. (12) and other limiting cases indicate that its application to some of the cases covered by traditional design methods produces results of similar magnitude.

The assumption of elastic behavior of the soil is open to question. At points of high stress the shear strength of a soil may be exceeded and plastic flow may occur; also, creep may occur at stresses well below the shear strength. However, high-density compacted fills can probably be approximated more closely by elastic behavior than many natural soils.

Whether a compacted fill responds in a manner approximating an isotropic medium is uncertain; in general, there is likely to be some difference between the respective moduli in the vertical and horizontal directions. The homogeneity of the fill may also be questioned. For the compacted fill itself, field-control requirements probably ensure that, relative to the geometrical scale of the conduit, the required uniformity is reasonable. However, the underlying soil may have a higher or lower modulus than the fill. In general, the conduit bedding should be designed to avoid any large difference in soil stiffness; a softer foundation may lead to difficulties in controlling the conduit elevation, whereas a harder foundation will lead to undesirable nonuniform deformations and excessive loads in the conduit. In many cases some overexcavation and replacement of the foundation soils will be required to obtain uniform bedding conditions. Alternatively, if the bedding is not appropriate, some adjustment must be made in the design values obtained.

The vertical overpressure is assumed to act beyond the zone of influence of the conduit, which requires that special consideration be given to shallow fills and nonlinearity of the embankment stress-strain relationship. Since conduit behavior during the fill from spring line to crown is largely controlled by construction practices, only the response to fill above the crown is considered in the solution. For this reason, accurate results cannot be expected for shallow cover (height of cover less than one pipe diameter).

Because the stress-strain relationship of the soil is nonlinear, the total load cannot be considered to be applied instantaneously, and a stepwise calculation is desirable to simulate the progress of loading in the field (Example 6).

Displacements, thrusts, and bending moments at the spring and crown lines can be determined from the curves of Fig. 15. Extensional flexibility was assumed to be negligi-

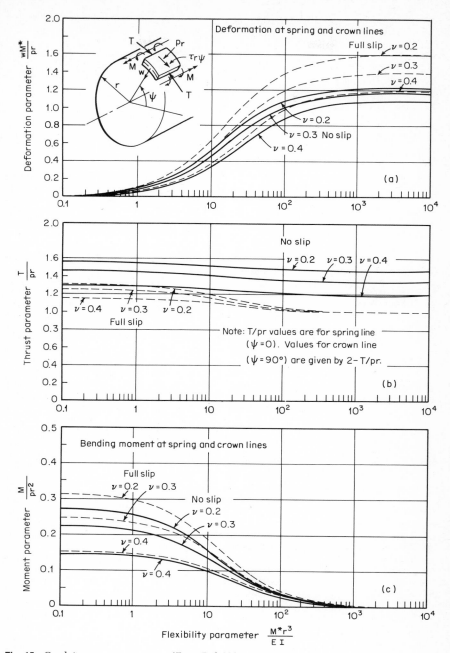

Fig. 15 Conduit response parameters (*From Ref. 10.*)

ble in computing these values. In addition to the notation shown on the figure, M^* is the constrained soil modulus, EI is the circumferential bending stiffness of the conduit per unit length, and ν is Poisson's ratio for the soil.

Based on the results of many uniaxial strain tests on a variety of compacted and undisturbed soils,[6,10,14] Fig. 16 has been developed to estimate the constrained modulus M^* for most commonly encountered soils. If possible, however, M^* should be determined directly from test data on the soils being used.

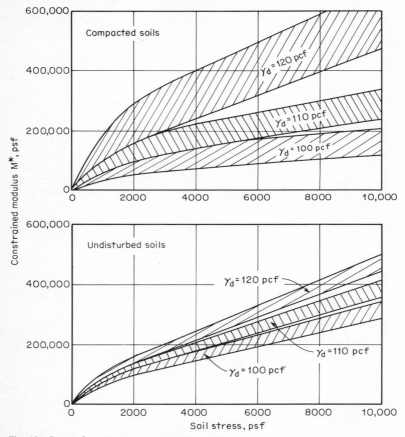

Fig. 16   Curves for estimating constrained modulus.

Kay and Abel[9] have developed correction factors which can be applied to results determined from Fig. 15 to yield improved estimates for normal field installations, as opposed to the idealized installation assumed by Burns and Richard.[4] Although the correction factor required to account for the finite-cover height over most conventional conduits is small, consideration of finite-cover height becomes more important as the conduit stiffness increases.

Example 5   Same data as in Example 1. The constrained modulus M^* of the soil at its in-place dry density and approximate state of stress must be determined by test or estimated from correlation with other soil parameters, as from Fig. 16. Assume the dry density to be 100 pcf. At midheight of fill $p = 12 \times 120 = 1440$ psf. Then from Fig. 16 $M^* = 80,000$ psf $= 555$ psi and

$$\frac{M^* r^3}{EI} = \frac{555 \times 30^3}{3 \times 10^6 \times 1 \times 6^3/12} = 0.278$$

Assume $\nu = 0.3$. From Fig. 15 for the full-slip case:

$$\frac{M}{pr^2} = 0.250 \qquad M = 0.25 \times 24 \times 120 \left(\frac{30}{12}\right)^2 = 4500 \text{ ft-lb/ft}$$

Since $M = 0.182\ Pr$ in the three-edge bearing test (Art. 8), the three-edge bearing load required to produce the calculated bending moment is

$$P = \frac{M}{0.182r} = \frac{4500 \times 12}{0.182 \times 30} = 9890 \text{ plf} = 1980D$$

A standard ASTM C76-66T Class IV pipe, which is rated at 10,000 plf or 2000 D could be used. The Class V pipe determined in Example 1 would certainly be safe.

If the no-slip case is assumed, Fig. 15 gives $M/pr^2 = 0.22$, from which $M = 3960$ ft-lb/ft and $P = 8700$ plf or $1740D$.

Example 6 Same data as in Example 2. Divide the 20-ft cover into four 5-ft layers, each of which contributes a vertical stress of $5 \times 120 = 600$ psf. The average vertical pressures at the springline are 600, 1200, 1800, and 2400 psf. Assume the dry density to be 104 pcf. From Fig. 16, $M^* = 40,000, 65,000,$ 87,000, and 110,000 psf. From Table 5, $I = 0.0544$ in.^{4}/ft $= 0.0045$ in.^{4}/in. For $M^* = 40,000$ psf $= 280$ psi,

$$\frac{M^* r^3}{EI} = \frac{280 \times 30^3}{30 \times 10^6 \times 0.0045} = 56$$

Values for $M^* = 65,000, 87,000,$ and 110,000 psf are 90, 121, and 153. Assume $\nu = 0.3$. From Fig. 15 for the full-slip case the deformation parameters wM^*/pr are 1.07, 1.18, 1.22, and 1.25.

The conduit deformation caused by the top layer is $w = 1.07 \times 600 \times 30/40,000 = 0.48$ in. Adding this to the values of w for the other three layers and doubling to obtain diametrical deformation gives $w = 2.52$ in. Calculations for the no-slip case gives $w = 2.16$ in. These values are about 4.2 and 3.6 percent, respectively, of the conduit diameter. The actual deformation would be expected to lie within this range.

The average vertical pressure at the midpoint of the soil mass above the springline of the conduit is $(20 + 2.5) \times 120/2 = 1350$ psf. From Fig. 16, $M^* = 77,000$ psf $= 535$ psi. Then

$$\frac{M^* r^3}{EI} = \frac{535 \times 30^3}{30 \times 10^6 \times 0.0045} = 107$$

With $\nu = 0.3$, Fig. 15 gives $T/pr = 1.03$, from which

$$T = 1.03 \times 22.5 \times 120 \times 2.5 = 6950 \text{ plf}$$

From Table 5, $A = 1.744$ in.^{2}/ft. The compressive stress σ in the conduit wall is $6950/1.744 = 3985$ psi.

For the no-slip case $T = 9110$ plf and $\sigma = 5225$ psi. With a factor of safety of 2, the allowable stress from Fig. 11 is $30,500/2 = 15,250$ psi.

To check for buckling, Eq. (16) gives

$$p_{cr} = 4.5 \sqrt{\frac{M^* EI}{D^3}} = 4.5 \sqrt{\frac{535 \times 30 \times 10^6 \times 0.0045}{60^3}} = 82 \text{ psi}$$

The stress at the springline (assumed to be hydrostatic) is $22.5 \times 120/144 = 19$ psi. Hence the safety factor against buckling is $82/19 = 4.3$.

19. Finite-Element Solution The only numerical technique that offers a reasonable potential for achieving adequate simulation of soil-conduit systems is the finite-element method. Three important advantages of this technique are (1) the ease with which irregular boundaries can be handled; (2) the ability to assign different mechanical properties to any region of the fill, the underlying soils, or the conduit; and (3) the capability to model incremental construction and nonlinear behavior. Although the method is uniquely capable of handling the complicated analyses usually associated with specialized high-cost projects, its greatest value and usefulness probably lie in the development of design charts and graphs (as opposed to automated design) to handle a wide range of conditions. In either case, studies can be made to ascertain how changes in bedding, soil placement, or conduit type influence the response of the system.

Although a number of general-purpose finite-element programs (NONSAP, SAP IV, STRUDL, ANSYS, NASTRAN, MARC, and ASKA) could be applied, all are oriented toward structural mechanics.[12] Among programs specifically designed for buried conduits are CANDE,[8] NUPIPE,[18] and FINLIN.[11,13]

20. CANDE (Culvert ANalysis and DEsign) This program is reasonably comprehensive in its treatment of virtually all types of pipe in the usual installations, and does not require a high level of expertise and experience to define the problem and interpret the results.

The structure of CANDE involves three main areas: the main control, the pipe library, and the solution library. Three basic selections (execution mode, solution level, and pipe type) in the main control characterize the problem to be considered; any pipe type from the pipe library can be matched with any solution level from the solution library, and the pair can be run in either a design or an analysis mode. The analytical modeling techniques include incremental construction, nonlinear soil models, nonlinear interface models, and nonlinear pipe models ranging from ductile yielding to brittle cracking.

Execution Mode. The execution mode provides the decision between design or analysis. Analysis output consists of the structural response (displacements, stresses, and strains) and an evaluation of the pipe performance in terms of safety factors against potential modes of failure. In the design mode the geometrical section properties of the conduit wall are unknown; desired safety factors are input and a design is achieved by a direct search; that is, a series of analyses are performed wherein an initial trial section is successively modified until the desired safety factors are achieved. Design output includes the required wall properties (e.g., corrugation and gage size for metal pipes and wall thickness and steel area for reinforced-concrete pipe), safety factors, and structural response.

Solution Level. There is a choice of three solution levels corresponding to successively increased levels of analytical capability. The successive increase is accompanied by an increase in the input preparation and computer cost. Level 1 is based on an enhanced closed-form elasticity solution, while Levels 2 and 3 are based on the finite-element method. Level 2 provides completely automated finite-element meshes suitable for the vast majority of conduit installations, whereas Level 3 requires a user-defined mesh for special installations. All solution levels are cast in incremental form to accommodate nonlinear processes, and all assume small-strain theory (the treatment of buckling is external and is based on simplifying assumptions) and plane strain (out-of-plane effects, such as longitudinal bending, must be investigated externally). All solution levels also neglect time-dependent response (however, long-time material moduli can be used to assess long-term effects indirectly), durability (supplemental material thickness must be provided to satisfy durability considerations), and dynamic loadings.

The three solution levels allow the engineer to select a degree of rigor and cost commensurate with the confidence of input parameters and relative worth of the project. The best guideline is to use the level that allows a description of everything that is known about the soil-conduit system. Level 1 should not be used for shallow (less than one pipe radius) depths of cover; Level 2 is intended to handle most routine design problems, but Level 3 will probably be required for box, arch, or open conduits.

Pipe Library. This library contains subroutines for corrugated steel, corrugated aluminum, plastic, and reinforced concrete; in addition, a subroutine called BASIC allows for the analysis of nonstandard or built-up properties. The pipe subroutines are the key control areas of CANDE and monitor the design process. Nonlinear stress-strain laws for pipe materials are accommodated by an interaction loop with the design loop.

Flexible Pipe. Table 7 summarizes the design criteria for corrugated steel, corrugated aluminum, and a class of plastic pipe that is linear up to brittle rupture. The traditional concepts for displacement and buckling are adopted for each pipe material, and only the

TABLE 7   CANDE Design Criteria for Flexible Pipe (Ref. 8)

	Design criteria			
Flexible pipe	Thrust stress σ_N	Outer-fiber strain ϵ	Relative pipe displacement ΔX	Buckling pressure σ_a
Corrugated steel	$\leqslant \sigma_y/SF$, $SF = 2.0\text{--}3.0$		$\leqslant 0.2D/SF$, $SF = 3.5\text{--}4.0$	$\leqslant p_{cr}/SF$, $SF = 2.0\text{--}3.0$
Corrugated aluminum	$\leqslant \sigma_y/SF$, $SF = 2.0\text{--}3.0$	$\leqslant \epsilon_u/SF$, $SF = 2.0\text{--}3.0$	$\leqslant 0.2D/SF$, $SF = 3.0\text{--}4.0$	$\leqslant p_{cr}/SF$, $SF = 2.0\text{--}3.0$
Smooth plastic		$\leqslant \epsilon_u/SF$, $SF = 2.5\text{--}3.5$	$\leqslant 0.2D/SF$, $SF = 3.0\text{--}4.0$	$\leqslant p_{cr}/SF$, $SF = 2.5\text{--}3.5$

SF = safety factor, σ_y = initial yield stress, ϵ_u = strain at rupture, p_{cr} = buckling pressure.

values of the suggested safety factors differ slightly. The criterion for steel is to limit the thrust to the yield stress divided by a suitable safety factor; this is identical to the traditional concept of ring compression. Brittle types of plastic pipe rupture under excessive outer fiber strain; accordingly, the maximum strain (bending plus thrust) must be limited to the ultimate strain of the material. Both thrust and outer-fiber strain criteria are employed for aluminum.

Rigid Pipe. The 0.01-in. crack criterion (Art. 8) is used in CANDE but can be excluded. Because it represents allowable cracking, it defines a performance factor rather than a safety factor. Table 8 shows the crack criterion with a suggested performance factor of unity. The program does not determine the load producing complete collapse of the pipe, but uses instead the design criteria, with suggested safety factors, given in Table 8. Another proposed criterion limits the diametrical displacement to $D^2/1200h$, where D is the pipe diameter and h is the wall thickness. Although seldom used in design, this criterion is a useful performance factor.

TABLE 8   CANDE Design Criteria for Reinforced-Concrete Pipe (Ref. 8)

Parameter	Design criteria	Relationship*
Concrete crushing	Maximum compressive stress σ_c	$\sigma_c \leq f'_c/SF,$ $SF = 1.5–2.0$
Diagonal cracking	Maximum shear stress τ_c	$\tau_c \leq f'_t/SF,$ $SF = 2.0–3.0$
Steel yielding	Maximum steel stress f_s	$f_s \leq f_y/SF,$ $SF = 1.5–2.0$
Crack width	Maximum crack width C_w	$C_w \leq 0.01$ in./$PF,$ $PF = 1.0$
Bowstringing	Maximum radial stress along steel-concrete bond f_b	$f_b \leq f'_t/PF,$ $PF = 1.0$
Displacement	Maximum diametrical displacement ΔX	$\Delta X \leq d_L/PF,$ $PF = 1.0$

*f'_c = concrete compressive strength, f'_t = concrete tensile strength, f_y = steel yield stress, $d_L = D^2/1200h$ = allowable deflection, SF = safety factor, PF = performance factor.

Buckling. For long cylinders deeply embedded in a soil, the hydrostatic buckling pressure can be approximated by[1]

$$p_{cr} = 4.5 \sqrt{\frac{M^*EI}{D^3}} \tag{16}$$

where M^* is the constrained modulus of the soil, EI is the in-plane bending stiffness of the pipe, and D is the pipe diameter. The inherent limitations of Eq. (16) are (1) the soil is linear elastic with no surface influence, (2) the pipe is circular and linearly elastic, and (3) the loading is hydrostatic. Some of these limitations can be mitigated by using average elastic soil properties representative of the current state of stress and continually adjusting the pipe stiffness to represent the current stiffness. For most corrugated pipe and all thick-wall pipe, elastic buckling is seldom a controlling design factor; deflection predictions will usually overshadow any potential buckling problem.

ADDITIONAL DESIGN CONSIDERATIONS

21. Handling Criteria   These are intended to assure conduits sufficiently sturdy to withstand all loads and shocks that may be imposed prior to the backfilling operation. The traditional criterion for flexible conduits is stated in terms of a flexibility factor $F_f = D^2/EI$. Recommended minimum values are 0.0433 for corrugated-steel pipe (except for 6- × 2-in. pipe, for which 0.02 is recommended) and 0.06 to 0.09 for corrugated-aluminum pipe. Most design procedures for plastic pipe do not consider handling, and those that do show little uniformity. The wall of a reinforced-concrete pipe is usually assumed to be stiff enough to withstand normal handling loads.

Based on the handling criterion for corrugated-steel pipe, handling criteria can be established for other pipe types by utilizing the fact that F_f is proportional to the diameter change of a circular ring with diametrically opposite concentrated loads. Hence, for two conduits of equal diameter but different materials, and restricted to the same allowable deflection, F_f is inversely proportional to EI. To account for the fact that the allowable deflection K should not be the same for all pipe materials, F_f should be adjusted according to the ratio of the allowable deflections. This yields the relationship[8]

$$\frac{(F_f)_n}{(F_f)_s} = \frac{(EI)_s}{(EI)_n} \frac{K_n}{K_s} \tag{17}$$

where the subscripts s and n refer to steel and another material, respectively. Based on the assumptions that (1) the characteristic moments of inertia for flexible pipes are equivalent, (2) K_n/K_s is proportional to the ratio of the respective yield strengths, (3) the yield value for plastic pipe can be taken as one-half its rupture stress but not to exceed 12,500 psi, and (4) the ratio of the characteristic moments of inertia of the corrugated-steel pipe and the cracked reinforced-concrete pipe is on the order of 0.1, proposed flexibility factors are given in Table 9. These criteria should be viewed as design aids rather than absolute requirements. In some cases they can provide relationships for starting the design process.

TABLE 9   Flexibility Factors for Handling Criteria (Ref. 8)

Type of pipe	$\dfrac{(EI)_s}{(EI)_n}$	$\dfrac{K_n}{K_s}$	$F_f = \dfrac{D^2}{EI}$
Corrugated steel	1.0	1.00	0.043
			0.02*
Corrugated aluminum	3.0	0.73	0.09
			0.042*
Plastic (fiberglass)	18.5	0.38	0.30
Reinforced concrete	1.0	0.167	0.0072

*For structural-plate corrugations.

22. Durability   Because durability greatly influences the service life of a conduit, it often forms the basis for choosing a particular material, as well as the thickness of the material or the protective coating that should be applied. Obviously, the design life of the structure plays a major role in this decision, and durability considerations rather than structural considerations sometimes govern the final design.[10]

Concrete Pipe.   Under most normal conditions concrete pipe performs satisfactorily. However, under certain conditions (for example, when the flow contains chlorides associated with deicing salts, or sulfates originating from mine waters) durability problems may arise. Air-entrained concrete, sulfate-resistant cement, richer mixes, and/or thicker walls may provide solutions. Expansive reactions between certain types of aggregates and high-alkali cements may cause random cracking and disintegration. Some rock types that are known to have such reactive properties are opaline silica, siliceous limestone, chalcedony, some cherts, andesites, ryolite, dacite, and certain phyllites. If aggregate with reactive components must be used, a cement having an alkali content below some critical value (a value of 0.6 percent is sometimes specified) should be used and/or a compound that will react with the harmful components in such a way as to prevent further reaction after the concrete has hardened should be added.

Metal Pipe.   The corrosion problem for metal culverts is extremely complex and dependent on a variety of factors, such as (1) hydrogen-ion concentration (acidity or alkalinity), (2) presence of various other ions (sulfides, sulfates, chlorides, nitrates, ammonia, ferrous iron, etc.), (3) water hardness, (4) electrical resistivity, (5) flow velocity, (6) temperature, (7) oxygen concentration, and (8) sulfate-reducing bacteria. Research directed toward relating metal loss with various physical and chemical properties of the soil at a given site has provided limited (and sometimes misleading) results, but experience has shown that metal pipes perform very well in most localities. Some of the strongest evidence suggests that metal loss in culvert applications can be predicted on the basis of pH and water velocity, but no reliable correlation has been found for the other factors mentioned above.

Design Suggestions. The following design suggestions are based on a statistical evaluation of past performance. The anticipated velocity at peak design flow, the pH of the soil and water at the site under normal climatic conditions, the desired conduit life, and the degree of importance (based on economic and other considerations) of having the conduit reach its desired life span should be evaluated. Pipes in the area should be examined. Design should take into account the following considerations:

1. Where the peak flow velocity is greater than about 8 fps and the water contains significant amounts of sediment, allowance should be made for abrasion. Design measures include using bituminous paving at the pipe invert, to provide some protection for abrasion-prone materials, designing the conduit entrance to reduce the approach velocity and sediment flow, increasing the thickness of the pipe wall, and installing an oversize pipe to reduce flow velocities.

2. Where the normal pH of the water is less than about 4.5, the use of either metal or concrete conduits is questionable, and an acid-resistant plastic conduit should be considered.

3. For pH in the range of 4.5 to 9, relatively short-term test results appear to indicate that aluminum conduits are highly resistant to corrosion.

4. Where the water pH exceeds 4.5, steel conduits (generally galvanized, asbestos-bonded, or bituminous-coated) may be used, but it is advisable to provide some additional thickness of metal to supplement the structural requirement. The additional thickness can be estimated from Fig. 17.

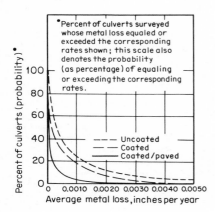

Fig. 17 Average metal loss for galvanized steel conduits. (*From Ref. 7.*)

23. Camber The weight of a highway embankment or other surface load causes consolidation of underlying layers of compressible soil. Because of the trapezoidal cross-sectional shape of most highway embankments, the settlement is greatest beneath the central portion of the fill and decreases appreciably toward the toes. Differential settlements may cause longitudinal stresses in the conduit walls. These are at right angles to the stresses from ring action, and the combination of the two complicates the design procedure. The importance of beam action depends on the rigidity of the conduit. Watkins[17] discusses the implication of the resultant stresses and their relative effect on flexible conduits. For design, it is suggested that beam stresses and ring stresses be considered separately.

Standard methods of predicting soil settlement may lead to unjustifiable expense and delay when applied to conduit problems. An approximate scheme for predicting settlement is presented in Ref. 10. After the anticipated settlement profile is determined, one or more of the following steps may be taken:

1. Install the conduit with a camber so that the settlement will eventually lower the conduit to approximately the desired grade.

2. Replace some or all of the compressible soils with well-compacted soils.

3. Preload the area to induce the major portion of the settlement before the conduit is installed; tunneling may be considered under these conditions.

4. Select a conduit composed of short articulated sections.

5. Use flexible joints.

24. Wrappings and Coatings A variety of wrappings and coatings may be used on the pipe exterior in corrosive or otherwise severe environments. Such materials modify interface conditions and thereby alter the soil-conduit interaction. A comparison of elasticity solutions for the cases of no slip and full slip, within the practical range of conduit diameters and soil-to-conduit stiffnesses, indicates that no-slip displacements, thrusts, and moments are, respectively, about 15 percent less, 25 percent greater, and 15 percent less than those for corresponding full-slip conditions. Thus, if thrust controls the design, wrappings and coatings that promote slippage can be structurally beneficial.

CONSTRUCTION CONSIDERATIONS

25. Site Preparation In most cases there is little or no choice in selecting the site for a conduit installation, and particular attention must be directed toward adequate site preparation. One important aspect is concerned with the minimization of settlements. Conduits (especially highway culverts) are often located in stream beds, where there is frequently an accumulation of soft compressible sediments and concentrations of vegetation. Complete removal of vegetation is necessary before construction begins, and the removal of compressible stream-bed material is highly desirable. Every effort should be made to obtain as high a degree of homogeneity as possible between the conduit foundation soil and the adjacent compacted embankment soil. In areas where large settlements are expected, it is usually desirable to maximize the pipe gradient so that a reverse gradient is avoided as settlement occurs.

Differential settlements are minimized when the compressibility of the foundation soil is reasonably uniform throughout the length of the conduit. It is highly undesirable to have one portion of a conduit founded on stream sediments while another rests on the less compressible soils of the adjacent stream bank. Total and differential settlements may also be reduced to some extent by using the conduit bedding to spread the load.

26. Bedding The main objective in providing adequate pipe bedding is to produce as nearly as possible a uniform distribution of loading (or reaction to loading) over the area of the pipe surface that cannot be reached during the fill-compaction process. Experience has shown that inadequate bedding is one of the major factors that contribute to unacceptable performance. Commonly found undesirable situations are (1) a hard (rock) surface in close proximity to the underside of the conduit (this is likely to produce loading conditions approaching those of the three-edge bearing test while for a flexible pipe it may lead to excessive deformations) and (2) air voids or soft pockets in the vicinity of the pipe wall (this is a common occurrence where a pipe is placed on a relatively flat soil surface and an attempt is made to obtain the bed by shoveling bedding material into the haunch areas adjacent to the pipe).

Although a material of uniform compressibility preformed to the shape of the pipe is generally ideal, there is some difference in the requirements for rigid and flexible conduits. For concrete pipes, for which structural deformations are small, a bedding material with a uniformly low compressibility is ideal. Concrete provides excellent bedding for a rigid pipe if uniform contact is achieved, but significant stress concentrations can develop with improper seating; this is one reason why it is not generally used. On the other hand, for flexible conduits a bedding material with extremely low compressibility may not deform in accordance with the pipe deformations produced by the compacted fill, and as a result, load concentrations or shape discontinuities may occur at the upper level of the bedding. Therefore, the bedding for flexible conduits should ideally be of a compressibility similar to the compacted backfill surrounding the upper portions of the pipe.

27. Fill Construction Concrete pipe normally has sufficient strength to withstand loads from compaction equipment used in the adjacent fill, but considerable care is required in constructing the fill adjacent to a corrugated-metal pipe. Possible excessive distortion of flexible pipe may prevent the use of heavy equipment close to the pipe wall. This normally necessitates the use of hand-compaction methods close to the pipe wall. Because the performance of flexible conduits depends largely on the passive resistance of the fill within a distance of one pipe diameter, good compaction in this area is absolutely essential, particularly if the fill is high or the borrow material is poor.

Close control of the diameter of large flexible conduits as the fill increases is recommended. Should distortion exceed a few percent of the diameter, horizontal strutting may be necessary. However, under no conditions should proper compaction be sacrificed. Special precautions must be taken to compact the fill reasonably symmetrically on both sides of the conduit; this is especially necessary in the case of flexible pipes. Also, care must be taken to ascertain that the pipe does not rise as the fill is being compacted below the spring line.

28. Compaction Procedures An end-product type of specification is desirable for the soil compaction within approximately one diameter of the conduit wall. There are substantial differences in the effectiveness of commercially available compaction equipment, and it is very difficult to write a suitable method-type specification for the hand-operated equipment to correspond with that for the larger motorized equipment.

29. Strutting Considerable reduction in conduit deformation may be obtained either by elongating the conduit in the vertical direction or by maintaining its shape during construction and releasing it after the fill has reached a certain height. The latter objective is commonly achieved by installing closely spaced vertical wooden struts throughout the length of the pipe; normally a small amount of deformation is permitted by using compressible softwood blocks at the top of the struts. Provision should be made to spread the strut load both longitudinally and laterally on the pipe wall. Because considerable structural support for a flexible conduit is obtained by allowing it to increase its horizontal dimension, struts must be removed at an appropriate time. Since the soil modulus manifests its largest rate of increase as the stress increases from 0 to 2000 or 3000 psf, it seems logical to remove the struts after the fill has reached 20 to 30 ft above the crown of the conduit.

30. Joints The performance of transverse joints is vitally important to the successful service of a conduit. In the case of rigid pipes they must permit limited longitudinal and rotational movements to account for such factors as shrinkage and differential settlement. If a pipe fractures or if there is a separation at the joint, the backfill at the periphery of the pipe may gradually pass through the opening and eventually cause a failure of the pipe or excessive settlement of the surface above the pipe, particularly if the soil is a fine sand or silt. Corrugated-metal pipes are usually sufficiently flexible in the longitudinal direction to prevent this problem.

31. Backpacking This term denotes the introduction of a low-modulus material, such as uncompacted soil, polyurethane foam, or straw, into the confining medium near a conduit or liner to reduce or redistribute the interface stresses. When properly used, it reduces interface stresses by a factor of 2 or 3 for high fills or other large loadings, but improperly used it can cause serious imbalances in interface stresses, gross distortions of shape, or premature transitional buckling. Care must be taken to assure that the backpacking is not initially crushed when the fill is compacted around the conduit; crushing should not occur until the maximum height of the fill is approached.

LONG-SPAN CORRUGATED-METAL CONDUITS

Depending on the shape of the structure, long-span conduits are usually defined as having spans greater than 15 to 25 ft, and changes in the design criteria and construction procedures relative to those for ordinary conduits must be made to ensure satisfactory performance.[15]

The fundamental idea of the long-span corrugated-metal structure, as with all relatively flexible structures, is to use the soil as the principal load-bearing component; the basic difference is that the principle is exploited to a greater extent in long-span structures. Typical construction problems are compacting the soil against the side of the structure, preventing excessive peaking of the crown and distortion of the shape during backfilling, buckling due to live loads imposed by construction equipment, and excess flattening as fill is placed above the crown. In general, special design features such as thrust beams, compaction wings, fin and rib plates, relieving slabs, and temporary tension strutting have been developed by various fabricators to aid in construction.

In general, individual fabricators have developed specific design approaches with varying degrees of success. Most of the special features associated with each design approach are patented; these include thrust beams, compaction wings, soil bins, relieving slab, etc., which are intended to transfer most of the applied load to the soil. A common requirement of all designs is that good control be maintained during construction. In most cases design is handled in the offices of the fabricator, and several fabricators insist on having at the site during construction a representative with the authority to stop work that is not being performed properly.

Excellent control of soil placement and compaction must be maintained during construction; of particular importance is control of the shape of the structure. Regular checks must be performed during the backfilling operation to assure that distortions more than 2 percent of the span or rise, whichever is greater, or more than 5 in., are not occurring. For arches with top-to-side radius ratios of 3 or more the deviation should not exceed 1 percent of the span. Stable side slopes must be provided, but to avoid deterioration prior to backfilling they should not be cut back too soon. Every effort must be made to proceed more or less symmetrically with the backfill operation, never getting one side more than about 2 ft higher than the other. Experience has shown that very little further change in structural shape will occur after the fill has reached a level of 2 to 4 ft above the crown of the structure; so care must be taken to guarantee that the proper shape prevails when the crown is covered.

REFERENCES

1. Allgood, J. R., and S. K. Takahashi: Balanced Design and Finite Element Analysis of Culverts, *Highw. Res. Rec.*, no. 413, 1972.
2. "Concrete Pipe Design Manual," American Concrete Pipe Association, Arlington, Va., 1974.
3. "Handbook of Steel Drainage and Highway Construction Products," 2d ed., American Iron and Steel Institute, New York, 1971.
4. Burns, J. Q., and R. M. Richard: Attenuation of Stresses for Buried Conduits, *Proc. Symp. Soil-Structure Interaction*, University of Arizona, Tucson, 1964.
5. Chambers, R. E.: "Design Manual for Buried Plastic Pipe for Drainage of Transportation Facilities," National Cooperative Highway Research Program, Washington, D.C., 1977.
6. Corotis, R. B., A. S. Azzouz, and R. J. Krizek: Statistical Evaluation of Soil Index Properties and Constrained Modulus, *Proc. 2d Int. Conf. Applications of Statistics and Probability to Soil and Structural Engineering*, Aachen, West Germany, 1975.
7. Haviland, J. E., P. J. Bellair, and V. D. Morrell: Durability of Corrugated Metal Culverts, *Rept.*, Department of Transportation, State of New York, 1967.
8. Katona, M. G., J. M. Smith, R. J. Odello, and J. R. Allgood: CANDE: A Modern Approach for the Structural Design and Analysis of Buried Culverts, Civil Engineering Laboratory, Naval Construction Battalion Center, Port Hueneme, Calif., 1976.
9. Kay, J. N., and J. F. Abel: A Design Approach for Circular Buried Conduits, *Transp. Res. Rec.* 616, 1977.
10. Krizek, R. J., R. A. Parmelee, J. N. Kay, and H. A. Elnaggar: Structural Analysis and Design of Pipe Culverts, *Rept.* 116, National Cooperative Highway Research Program, Washington, D.C., 1971.
11. Leonards, G. A., and M. B. Roy: Predicting Performance of Pipe Culverts Buried in Soil, *Rept.* JHRP-76-15, Purdue University, West Lafayette, Ind., 1976.
12. Pilkey, W. D., K. Saczalski, and H. G. Schaeffer: "Structural Mechanics Computer Programs: Surveys, Assessments, and Availability," University Press of Virginia, Charlottesville, 1974.
13. Roy, M. B.: FINLIN User's Manual, *Rept.* JHRP-76-16, Purdue University, West Lafayette, Ind.
14. Salazar Espinosa, J. H., R. J. Krizek, and R. B. Corotis: Statistical Analysis of Constrained Soil Modulus, *Transp. Res. Rec.* 537, 1975.
15. Selig, E. T., and J. F. Abel: Review of the Design and Construction of Long-Span, Corrugated-Metal, Buried Conduits, *Tech. Rept.* (sponsored by Federal Highway Administration) to Civil Engineering Laboratory, Naval Construction Battalion Center, Port Hueneme, Calif., 1977.
16. Spangler, M. G., and R. L. Handy: "Soil Engineering," 3d ed., Intext Educational Publishers, New York, 1971.
17. Watkins, R. K.: Failure Conditions of Flexible Culverts Embedded in Soil, *Proc. Highw. Res. Bd.*, vol. 39, 1960.
18. Wenzel, T. H., and R. A. Parmelee: Computer Aided Structural Analysis and Design of Concrete Pipe, *ASTM Spec. Tech. Publ.* 630, 1977.
19. White, H. L., and J. P. Layer: The Corrugated Metal Conduit as a Compression Ring, *Proc. Highw. Res. Bd.*, vol. 39, 1960.

Section 26

Chimneys

MAX ZAR
Partner and Manager of Structural Department
Sargent & Lundy, Engineers, Chicago, Ill.

SHIH-LUNG CHU
Associate and Head, Structural Analytical Division
Sargent & Lundy, Engineers, Chicago, Ill.

1. Materials Almost all chimneys in the United States are built of reinforced concrete or welded steel. A few chimneys for commercial installations are constructed of radial brick, but because of the labor cost they are used only for short chimneys. The decision to build a chimney of concrete or steel (the latter are frequently called stacks) is based on cost. The economic comparison must include the cost of linings, foundations, painting, aeronautical-obstruction lighting, access platforms, ladders, and in some cases, elevators. Steel stacks superimposed on building roofs impose a cost penalty on the building structural steel and foundations.

2. Diameter and Height Chimneys are constructed for flue-gas emission; the height and top diameter depend on the gas temperatures, volumes, and desired exit velocities. The chimney configuration should be selected to minimize construction cost.

DESIGN LOADS

The chimney and its supporting system should be designed to resist stresses resulting from dead load, wind or earthquake loads (whichever are greater), stack draft, and temperature gradients. Furthermore, the resonant vibration due to dynamic wind and the ovaling due to wind pressure should be considered. However, the design need not be limited to these loadings.

3. Dead Loads Dead loads should include the weight of all permanent construction and fittings, insulation, fly ash, clinging ash, and other loads.

4. Wind Loads Wind pressure acting in the direction of the wind can be taken as the resultant horizontal pressure on the projected area. Wind-loading provisions are given in ACI 307 (Ref. 1) and ANSI Standard A58.1 (Ref. 2).

ACI 307-69 specifies the resultant horizontal pressures on the projected area as bands of uniform load in accordance with the height zones and wind-pressure areas in Table 1.

These values are generally the same as the UBC wind pressures (Sec. 19, Table 12) multiplied by the UBC-recommended factor 0.6 for round chimneys.

ANSI A58.1-72, which contains meteorological data and provides a reasonable evaluation of gust factors, is discussed in Sec. 19, Art. 13. It gives design wind loads lower than the ACI. In using this standard, a 50-year recurrence interval may be used to determine the basic wind velocity V_{30}.

Wind Pressures due to Transverse Resonant Vibrations (See also Art. 16). When the chimney is subjected to a steady wind, the periodic shedding of vortices will cause swaying oscillations in a direction transverse to that of the wind (Fig. 1a). If the vortex-

TABLE 1   Wind Pressures on Circular Chimneys (From Ref. 1)

Height zone, ft, above ground	Wind pressure, psf				
	Map area < 35*	35	40	45	50
0–100	23	23	26	29	32
100–500	31	33	36	42	45
500–1000	34	36	42	48	54
1000 and over	36	42	48	54	60

*Wind-pressure map, Sec. 19, Fig. 3.

(a)   (b)

Fig. 1 Oscillation due to vortex shedding: (a) transverse; (b) ovaling.

shedding frequency is resonant with the natural frequency of the chimney, it may result in large vibrations. This dynamic influence may be approximated by assuming an equivalent static force F_L per foot of height, acting in the direction of oscillations, given by

$$F_L = \frac{1}{2\beta} C_L D q_{cr} \tag{1}$$

where β = critical damping ratio of chimney
C_L = Von Karman (lift) coefficient = 0.2 for circular cylinders
D = diameter of chimney, ft, at position under consideration
q_{cr} = dynamic wind pressure at critical wind velocity, psf
The dynamic wind pressure is given by

$$q_{cr} = 0.00119 V_{cr}^2 \tag{2}$$

where V_{cr}, the critical wind velocity for resonant transverse vibration, is given (fps) by

$$V_{cr} = \frac{f_t D}{S} \tag{3}$$

where f_t = natural frequency of transverse vibration of stack, cps (Art. 8)
S = Strouhal number, varying from 0.18 to 0.22
A simplified formula for circular stacks is obtained by substituting q_{cr} from Eq. (2) and $C_L = 0.2$ into Eq. (1)

$$F_L = 1.2 \times 10^{-4} V_{cr}^2 \frac{D}{\beta} \tag{4}$$

In computing the bending moment due to vortex shedding, the equivalent static force obtained from Eq. (4) is applied to the entire length for a constant-diameter chimney. For a tapered chimney there is a limiting height over which the vortex shedding pressures act. This is the height over which the diameter changes by only ±5 percent at the position under consideration.

Critical Wind Velocity for Ovaling Vibrations (See also Art. 16). In addition to transverse swaying oscillations, a steel stack may also be subject to flexural vibration in the circular cross-sectional plane as a result of vortex shedding (Fig. 1*b*). The frequency of the lowest mode of ovaling vibration in a circular shell is computed by

$$f_0 = 0.126 \frac{t\sqrt{E}}{D^2} = 678.5 \frac{t}{D^2} \tag{5}$$

where f_0 = first-mode ovaling frequency, cps
    t = shell-plate thickness, in.
    E = modulus of elasticity, psi
    D = shell diameter at position under consideration, ft

A resonant condition occurs when the first-mode ovaling frequency is nearly twice the vortex-shedding frequency. Thus, from Eq. (3) the critical wind velocity V_0 for ovaling vibration of the stack is (fps)

$$V_0 = \frac{f_0 D}{2S} \tag{6a}$$

Then with f_0 from Eq. (5) and $S = 0.2$,

$$V_0 = 1696 \frac{t}{D} \tag{6b}$$

Circumferential Wind Moment. The variation in wind pressure along the circumference of a chimney shell (Fig. 2) produces circumferential moments which should be accounted for. The maximum moments are

$$M_{max} = 0.314qR^2 \text{ (tension on inside)} \tag{7a}$$
$$M_{max} = 0.272qR^2 \text{ (tension on outside)} \tag{7b}$$

where q = wind pressure at level considered, psf
    R = mean radius of chimney at same level, ft

Values of q can be obtained by dividing values from Table 1 by 0.6, or the ANSI or UBC values given in Tables 9 and 12, respectively, of Sec. 19, may be used.

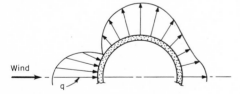

Wind

q

Fig. 2   Wind pressure on circular section.

5. Earthquake Forces   Chimneys in earthquake areas should be designed for seismic resistance. The total lateral force (base shear) can be determined by the UBC formula[3]

$$V = ZIKCSW \tag{8}$$

This formula and other UBC provisions are discussed in Sec. 19, Art. 14. A suggested value of the importance factor I is 1.3 (Ref. 4). The horizontal-force factor $K = 2$ should be used. The site-structure resonance factor S may be determined by the UBC formula [Sec. 19, Eq. (9)] or taken from Table 2. The distribution of V and the resulting moments should be according to UBC.

If the chimney has a lining built as an integral part of the shell, the weight W should include the weight of the lining.

For cases in which the chimney has an independent lining, dynamic seismic analysis by the response-spectrum method discussed in Sec. 3 may be used to evaluate the interaction effect. The peak horizontal ground acceleration at the site should be selected based on the results of a seismic-risk analysis. In lieu of such an analysis, seismic-risk maps[6] may be used as a guide. Average elastic response spectra for a peak horizontal ground acceleration of 1.0g are shown in Fig. 3. Design response spectra are obtained by scaling these spectra

TABLE 2   Site-Structure Resonance Factors S (From Ref. 5)

Soil type	S
Profile Type A	1
This profile is one with:	
• Rock of any characteristic, either shalelike or crystalline in nature; such material may be characterized by a shear-wave velocity greater than 2500 fps, or	
• Stiff soil conditions where the soil depth is less than 150 ft and the soil types overlying rock are stable deposits of sand, gravels, or stiff clays	
Profile Type B	1.2
This profile is one with deep cohesionless or stiff clay conditions, including sites where the soil depth exceeds 200 ft and the soil types overlying rock are stable deposits of sands, gravels, or stiff clays	
Profile Type C	1.5
This profile is one with soft to medium-stiff clays and sands, characterized by 30 ft or more of soft to medium-stiff clay with or without intervening layers of sand or other cohesionless soils	

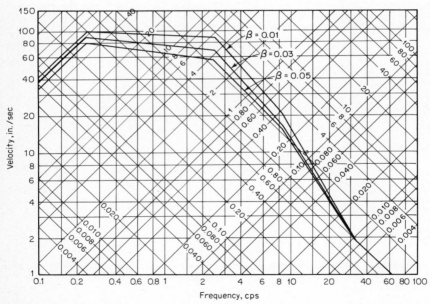

Fig. 3   Elastic average horizontal response spectra for 1-g maximum ground acceleration. (*From Ref. 7.*)

by the ratio of the peak horizontal ground acceleration for the site to the reference ground acceleration, 1.0g, of the figure. The following damping ratios are suggested: 5 percent for concrete chimneys, 3 to 5 percent for lined steel stacks, and 1 to 2 percent for unlined steel stacks.

In general, the dynamic interaction effect on the chimney shell is negligible for reinforced-concrete chimneys with independent steel lining. However, it is important to determine the seismic force exerted on the steel lining as a result of this interaction.

6. Pressure Differentials   When a chimney conveys hot flue gas of a specific weight less than the surrounding atmosphere the pressure on the inside of the liner is less than on the outside. This negative pressure at the flue-gas entrance, often called stack draft, can be determined from

$$D_s = 0.52HP\left(\frac{1}{T_a} - \frac{1}{T_g}\right) \tag{9}$$

where D_s = stack draft, in. of water
  H = stack height above gas entrance, ft
  P = atmospheric pressure at plant level, psig
  T_a, T_g = temperatures of atmosphere and gas, respectively, °F absolute (°F + 460°)

The curves in Fig. 4 can be used to determine the negative pressure at the flue-gas entrance for most chimneys without appreciable error. Negative pressure varies linearly with height and reaches zero at the flue-gas exit.

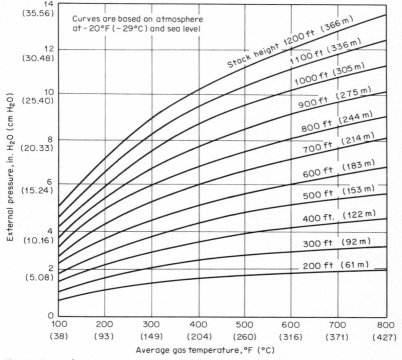

Fig. 4   External pressure curves. (*From Ref. 7.*)

For cases where an induced-draft fan forces gas through the chimney the negative pressure is reduced and the chimney may operate under a net positive pressure. This pressure may also be assumed to vary linearly with height.

The pressure differential is resisted by circumferential stresses in the liner itself if the liner is independent, and by composite action if it is built integrally with the shell.

7. Temperature Differentials   All chimneys should be investigated for stresses resulting from temperature differentials. A procedure to calculate the temperature gradient across the chimney shell, air space (if any), insulation, and liner is discussed in Ref. 1.

A chimney with two or more inlet breeching openings will develop an uneven temperature distribution around the circumference of the liner. These differentials produce an unequal expansion which causes lateral deflection of the liner. Significant longitudinal stresses will develop when the liner is restrained. This maximum differential occurs at the level of the breeching opening and exponentially decays along the chimney height. A method to estimate the maximum temperature differential along the circumference and its vertical profile is discussed in Ref. 7.

8. Natural Frequency of Vibration   The natural frequency of vibration of a stack or chimney with a constant diameter, uniform thickness, and fixed base is given by

$$f_t = \frac{1}{T} = \frac{3.52}{4\pi} \frac{D}{H^2} \sqrt{\frac{Eg}{2w_s}} = \frac{3.9D}{H^2} \sqrt{\frac{E}{w_s}} \tag{10}$$

where f_t = fundamental frequency, cps
  T = fundamental period, sec
  w_s = unit weight of shell, lb/in.^{3}
  g = gravity acceleration = 386 in./sec^2
  E = modulus of elasticity, psi
  H = height, in.
  D = diameter, in.
  t = thickness, in.

Reinforced-concrete chimneys are usually built with a uniform taper and with the thickness varying from a minimum at the top to the thickness required for strength at the

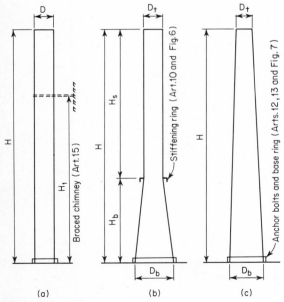

(a)   (b)   (c)

Fig. 5   Steel stacks.

bottom. Steel chimneys are built with or without taper, or with a tapered lower section, and with several thicknesses of plates (Fig. 5). Several formulas for an effective diameter D_e and effective height H_e have been developed to enable the natural frequencies of these types to be determined by Eq. (10).[1,8] The following formulas are suggested in Ref. 8:

Chimneys with straight taper or straight top and flared bottom,

$$D_e = D_t \left(\frac{t_b}{t_t}\right)^{0.27}$$

(11)

Chimneys with uniform taper,

$$H_e = H \left(\frac{2D_t}{D_t + D_b}\right)^{0.9}$$

(12)

Chimneys with straight top and flared bottom,

$$H_e = H \left[\frac{H_s}{H} + \frac{H_b}{H}\left(\frac{2D_t}{D_t + D_b}\right)^2\right]^{0.9}$$

(13)

In these equations D_t, D_b = mean diameter at top, bottom; t_t, t_b = thickness at top, bottom; H_s = height of straight segment; H_b = height of bottom (tapered) segment (Fig. 5). Frequencies calculated by Eq. (10) using these effective dimensions are in good agreement with computer determination of the frequencies of chimneys with a wide range of proportions.

If a chimney is lined and the weight of the lining is carried by the shell, there is an increased mass to be set in vibration. Since there is no significant increase in moment of inertia, such a chimney will have a lower natural frequency than an unlined chimney of the same dimensions. An approximate natural frequency can be determined from Eq. (10) by using for w_s the weight of the shell plus lining, divided by the thickness of the plate. The weight and thickness should be taken at about one-fourth the height above the base.

Base flexibility may need to be considered in determining frequency. For example, when stacks are supported on a roof structure or a steel frame, the translational and rotational spring constants of the support must be taken into account.

STEEL STACKS

There are many types of steel stacks, including self-supported, guyed, and braced. The choice of a particular type should be based on the evaluation of its comparative costs and the site conditions. The three profile types commonly used are shown in Fig. 5. Short stacks, less than 100 ft high, may be straight cylinders. For taller stacks a bell base may be used to reduce plate thickness and anchor-bolt size. The bell height is usually between one-fourth and one-third of the total stack height. The diameter of the flared base D_b is usually about 1⅓ to 1¾ times the cylinder diameter D_t.

Most steel stacks are built from plate conforming to ASTM A36. In some cases, A242, A588, A131B or CS, A283C, or other grades of steel have been used as dictated by individual experience and specific requirements. Stiffeners are normally A36. Steel for stacks in cold climates should have a low transition temperature (Sec. 4).

The steel stack and its anchorage should be designed for the loads discussed in preceding articles. When subjected to wind and/or earthquake loads, the stack may be treated as a beam column and analyzed by the conventional beam theory. Except in guyed and braced stacks, nonuniform temperature differentials will not induce bending moments. Particular attention is required in the design of reinforcing at the cone-to-cylinder junction and at the breeching opening.

To reduce heat loss, insulation of the exterior surface of unlined steel stacks, including the projecting flanges of all attachments, is recommended. This is common practice in England to reduce soot fallout. Proper lining should be applied to the interior surface of the shell to protect the bare steel from high temperature, abrasion, and corrosion from the flue gases. The weight of insulation and lining should be taken into account in the frequency calculation for resonance under the lined condition. Unless an integral shotcrete or brick lining is used, no credit should be given to the lining in calculating stack stiffness.

9. Allowable Stresses   The allowable longitudinal compressive stresses due to vertical load and bending moment can be determined by[7]

$$F = XY$$

(14)

where

$$
X = \begin{cases} 0.0625Et/R & \text{for } 0 \leqslant \dfrac{t}{R} \leqslant \dfrac{8F_p}{E} \\[2mm] 0.5[F_y - k_s(F_y - F_p)] & \text{for } \dfrac{8F_p}{E} \leqslant \dfrac{t}{R} \leqslant \dfrac{20F_y}{E} \\[2mm] 0.5F_y & \text{for } \dfrac{t}{R} > \dfrac{20F_y}{E} \end{cases}
$$

$$
k_s = \left(\frac{F_y - 0.05Et/R}{F_y - 0.4F_p} \right)^2
$$

$$
Y = \begin{cases} 1 & \text{for } \dfrac{L}{r} \leqslant 60 \text{ and } F_y \leqslant 50 \text{ ksi} \\[2mm] \dfrac{21,600}{18,000 + (L/r)^2} & \text{for } \dfrac{L}{r} > 60 \text{ and } F_y \leqslant 50 \text{ ksi} \end{cases}
$$

F_y = yield strength at mean shell temperature, ksi
F_p = proportional limit at mean shell temperature, ksi; may be taken as $0.7F_y$
E = modulus of elasticity at mean shell temperature, ksi
t = shell-plate thickness, in., at the section under consideration
R = radius of shell, in.
L = length of stack between points of lateral support. For a self-supporting stack, L should be taken as the effective length, i.e., $L = 2 \times$ stack height, in.
r = radius of gyration = $0.707R$, in.

The factor Y in Eq. (14) is intended to account for a possible interaction of cylindrical shell buckling, which depends on t/R, and column buckling, which depends on L/r.

The allowable stress given by Eq. (14) is based on a factor of safety of 2, and is suggested for load combinations which include either wind or earthquake forces.

Because of possible corrosion the computed required thickness should be increased. The allowance may vary from $\frac{1}{16}$ to $\frac{1}{8}$ in., depending on the properties of flue gases, the types of insulation and lining provided, and the operating gas temperature. Including the corrosion allowance, it is recommended that the shell thickness be not less than $\frac{5}{16}$ in. for unlined stacks and $\frac{1}{4}$ in. for lined stacks. Outstanding elements of rolled shapes and built-up members should have a minimum thickness of $\frac{1}{4}$ in.

10. Cone-to-Cylinder Junction A stiffening ring is required at the junction of the cone and the straight cylinder sections of stacks. It is normally designed to resist the circumferential compression that results from the vertical loads and bending moments at the junction. Where external pressure due to stack draft is significant, the resulting additional circumferential forces should also be considered.

The maximum vertical force N_x per unit length of circumference in the cylinder at the junction is

$$
N_x = \frac{W}{2\pi R} + \frac{M}{\pi R^2} \tag{15}
$$

where W = axial load at junction
M = wind or other moment at junction

The total circumferential compression Q in the ring is

$$
Q = R[N_x \tan \theta + 0.78P_d(\sqrt{Rt_1} + \sec \theta \sqrt{Rt_2 \sec \theta})] \tag{16}
$$

where θ = acute angle between cone wall and cylinder
P_d = external pressure per unit area, psi
t_1 = thickness of cylinder wall, in.
t_2 = thickness of cone wall, in.

The required area A_s and moment of inertia I_s of the ring are

$$
A_s = \frac{Q}{F_a} \tag{17}
$$

$$
I_s = \frac{QR^2}{E} \tag{18}
$$

The allowable ring compression F_a in Eq. (17) is usually limited to 8000 psi to minimize the secondary vertical bending stresses. For stacks of diameters greater than 15 ft, or where higher values of F_a are used, it would be advisable to evaluate the secondary stresses. In addition, bending stresses due to the circumferential variation in wind pressure should be checked by Eqs. (7).

In determining the section properties of the stiffening ring the area of a portion of the shell (Fig. 6) can be included, but the area so included should not exceed the area of the ring itself to ensure a nominal-size stiffener. The maximum permissible longitudinal compressive stresses in the cone may be determined by Eq. (14) with the horizontal radius R replaced by the cone radius $R \sec \theta$.

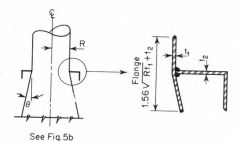

Fig. 6 Stiffening ring at cone-cylinder junction.

See Fig. 5b

11. Circumferential Stiffeners   In addition to the stiffener at the cone-cylinder junction a stiffener is required at the top of the stack. Intermediate ring stiffeners may also be required. The purpose of such stiffening is to prevent excessive deformations of the stack shell under wind pressure and to provide adequate structural resistance to negative draft.

Intermediate stiffener spacing L_s can be determined by

$$L_s = 60 \sqrt{\frac{Dt}{P_w}} \tag{19}$$

where P_w = static wind pressure (Table 1), t = thickness of thinnest plate, in., in section under consideration, and D = stack diameter, in. To ensure a nominal size of intermediate stiffener, the spacing should be kept within 1.5 times the chimney diameter. Intermediate stiffeners should meet the following minimum requirements:

$$S = \frac{P_w L_s D^2}{1100 F_b} \tag{20a}$$

$$I_s \geq \frac{P_d L_s D^3}{8E} \tag{20b}$$

$$A_s \geq \frac{P_d L_s D}{2F_a} \tag{20c}$$

where S = section modulus, in.^{3}, I_s = moment of inertia, in.^{4}, and A_s = cross-sectional area, in.^{2}, of stiffeners. Values of L_s and D should be in inches. The allowable bending stress F_b should be taken as $0.6F_y$. The allowable axial compressive stress F_a may be taken as 12,000 psi. In the calculation of stiffener section properties, an area of the shell equal to $1.56t\sqrt{Rt}$ or the area of the stiffener, whichever is smaller, may be included.

12. Anchor Bolts   These should be designed to resist the net tension resulting from the dead load W_b and the wind or earthquake moment M_b at the stack base. The bolt tension F can be determined from

$$F = \frac{4M_b}{ND'} - \frac{W_b}{N} \tag{21}$$

where D' = diameter of bolt circle, in., and N = number of anchor bolts. Since the elongation of the anchor bolts contributes to the lateral deflection of the stack, it is advisable to limit the allowable bolt tension at the root of the thread to 15,000 psi even if high-strength bolts are used. Also, the bolt spacing should not exceed 5½ ft and at least eight bolts should be used.

13. Base Ring for Anchor Bolts Anchor-bolt tension is eccentric with respect to the chimney shell, and an unstiffened base angle is normally insufficient to take the bending. Deflection of the base angle will increase lateral movements of the stack and will cause large vertical secondary bending stresses in the chimney shell, and in many cases it is necessary to provide a continuous stiffened base ring at the location shown in Fig. 7. Alternately, a separate chair for each anchor bolt may be used.

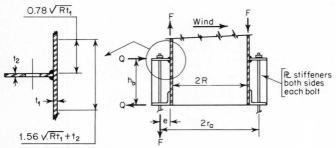

Fig. 7 Base ring for anchor bolts.

Circumferential compressive stresses in the continuous base ring may be calculated as though it were loaded with equally spaced concentrated loads $Q = Fe/h_b$ in the plane of the ring (Fig. 7). The maximum circumferential bending moment, which occurs at the windward anchor bolt and produces compression in the outside of the ring, is given by

$$M_a = CQr_a \tag{22}$$

where C = coefficient from Table 3
$\quad\quad r_a$ = radius of bolt circle, in.
The portion of the chimney shell within $0.78\sqrt{Rt}$ on either side of the attached ring plate may be counted as part of the ring. If the calculated size of the ring is excessive, its height should be increased, or the eccentricity e should be reduced if possible. Vertical stiffeners should be placed on both sides of each anchor bolt.

TABLE 3

No. of bolts N	C	No. of bolts N	C
8	0.191	36	0.510
12	0.217	40	0.563
16	0.258	44	0.616
20	0.305	48	0.670
24	0.355	52	0.724
28	0.406	56	0.778
32	0.457	60	0.832

Suggested designs of anchor-bolt chairs are given in Ref. 9.
For a stack with a bell base the base angle must be designed for the ring tension caused by the horizontal component of forces in the base cone, in addition to the bending stresses due to the bearing load. The allowable tension in the base angle should be limited to 10 ksi or less to keep secondary bending stresses in the cone within safe limits.

Example 1 Design a ground-supported A36 steel stack 160 ft high with the profile of Fig. 5b. Given $H_b = 40$ ft, $H_s = 120$ ft, $D_t = 10$ ft, $D_b = 15$ ft, operating gas temperature = 350°F, 2-in. shotcrete lining, weight of shotcrete 110 pcf, site in UBC wind-map zone of 35 psf and in earthquake zone 0.

PLATE THICKNESS. Trial section 60 ft below top

$$\text{Assume } \tfrac{1}{4}\text{-in. steel plate} = 10.2 \text{ psf}$$
$$\text{Lining } 110 \text{ pcf} \times \tfrac{2}{12} = \underline{18.3 \text{ psf}}$$
$$28.5 \text{ psf}$$

Weight per foot of height = $28.5 \times 10\pi = 894$ lb/ft. From Table 1,

$$P_w = 33 \text{ psf. Then}$$
$$M = 10 \times 60 \times 33 \times 30 = 594{,}000 \text{ ft-lb}$$

Deducting $\tfrac{1}{16}$ in. from the plate thickness for corrosion allowance,

$$\frac{t}{R} = \frac{3/16}{60} = 0.003125$$
$$A = 2\pi \times 60 \times 0.1875 = 70.7 \text{ in.}^2$$
$$S = \pi R^2 t = \pi \times 60^2 \times 0.1875 = 2120 \text{ in.}^3$$
$$f = \frac{W}{A} + \frac{M}{S} = \frac{894 \times 60}{70.7} + \frac{594{,}000 \times 12}{2120} = 4120 \text{ psi}$$

Determine the allowable stress from Eq. (14), using the equivalent height from Eq. (13):

$$H_e = 160 \left[\frac{120}{160} + \frac{40}{160} \left(\frac{2 \times 10}{10 + 15} \right)^2 \right]^{0.9} = 147 \text{ ft}$$
$$L = 2 \times 147 = 294 \text{ ft} \qquad \frac{L}{r} = \frac{294}{0.707 \times 5} = 83$$
$$Y = \frac{21{,}600}{18{,}000 + 83^2} = 0.868$$
$$\frac{8F_p}{E} = \frac{8 \times 0.7 \times 36}{28{,}000} = 0.0072 > \frac{t}{R}$$
$$X = 0.0625 \times 28 \times 10^6 \times 0.003125 = 5470 \text{ psi}$$
$$F = 0.868 \times 5470 = 4750 \text{ psi} > 4120 \qquad \text{O.K.}$$

Trial section 90 ft below top. Assume $t = \tfrac{5}{16}$ in., including $\tfrac{1}{16}$ in. corrosion allowance.
Weight per ft of height $= (12.76 + 18.3) \times 10\pi = 974$ lb/ft
$M = 10 \times 60 \times 33 \times 60 + 10 \times 30 \times 23 \times 15 = 1{,}291{,}000$ ft-lb

$$\frac{t}{R} = \frac{0.25}{60} = 0.00417$$
$$A = 2\pi \times 60 \times 0.25 = 94.2 \text{ in.}^2$$
$$S = \pi \times 60^2 \times 0.25 = 2827 \text{ in.}^3$$
$$f = \frac{894 \times 60 + 974 \times 30}{94.2} + \frac{1{,}291{,}000 \times 12}{2827} = 6370 \text{ psi}$$
$$X = 0.0625 \times 28{,}000{,}000 \times 0.00417 = 7290 \text{ psi}$$
$$F = 0.868 \times 7290 = 6330 \text{ psi} \qquad \text{O.K.}$$

The required shell thickness 120 ft below the top is found to be $\tfrac{7}{16}$ in. Also, the base cone must be $\tfrac{7}{16}$ in. thick to resist the wind moment and weight at the base:

$$M_b = 33 \times 10 \times 60 \times 130 + 23 \times 10 \times 60 \times 70 + 23 \times \frac{2 \times 10 + 15}{3} \times 40 \times 20 = 3{,}750{,}000 \text{ ft-lb}$$

$W_b = 894 \times 60 + 974 \times 30 + 1051 \times 30 + 1313 \times 40 = 167{,}000$ lb
RING AT CONE-TO-CYLINDER JUNCTION (Art. 10)

$$W = 894 \times 60 + 974 \times 30 + 1051 \times 30 = 114{,}400 \text{ lb}$$
$$M = 33 \times 10 \times 60 \times 130 + 23 \times 10 \times 60 \times 70 = 3{,}540{,}000 \text{ ft-lb}$$
$$N_x = \frac{114{,}400}{2\pi \times 60} + \frac{3{,}540{,}000 \times 12}{\pi \times 60^2} = 4300 \text{ lb/in.}$$

Assume external pressure = 2 in. water

$$P_d = 2 \times 62.4/12^3 = 0.0722 \text{ psi}$$
$$\tan\theta = \frac{2.5}{40} = 0.0625 \qquad \theta = 3.58° \qquad \sec\theta = 1.002$$

From Eq. (16), with $t_1 = t_2 = \tfrac{7}{16}$ in.

$$Q = 60[4300 \times 0.0625 + 0.78 \times 0.0722(\sqrt{60 \times 0.4375} + 1.002\sqrt{60 \times 0.4375 \times 1.002})] = 16{,}450 \text{ lb}$$

From Eqs. (17) and (18),

$$A_s = \frac{16,450}{8000} = 2.05 \text{ in.}^2$$

$$I = \frac{16,450 \times 60^2}{28,000,000} = 2.12 \text{ in.}^4$$

Try $3 \times 3 \times \frac{1}{4}$ angle: $A = 1.44$ in.^{2} $I = 1.24$ in.^{4} $x = 0.842$ in.

Effective flange (Fig. 6) $= 0.4375(1.56\sqrt{60 \times 0.4375} + 0.25) = 3.60$ in.^{2}

Ring area $A = 3.60 + 1.44 = 5.04$ in.^{2}

Ring c.g. $= \dfrac{1.44 \times 0.842 + 1.80 \times 3.22 + 1.80(3.22 - 2 \tan 3.58°)}{5.04} = 2.50$ in.

Ring $I = 1.24 + 1.44(2.50 - 0.842)^2 + 1.80(3.22 - 2.50)^2 + 1.80(3.22 - 2 \tan 3.58° - 2.50)^2 = 6.77$ in.^{4}

$3 \times 3 \times \frac{1}{4}$ angle O K.

INTERMEDIATE STIFFENERS IN UPPER 60 FT. From Eq. (19),

$$L_s = 60 \sqrt{\frac{10 \times 0.25}{33}} = 16.5 \text{ ft}$$

Maximum allowable spacing $= 1.5D = 15$ ft. Try $3 \times 3 \times \frac{1}{4}$ angles spaced 10 ft. From Eq. (20a),

$$S = \frac{33 \times 10 \times 10^2}{1100 \times 21,600} = 2.4 \text{ in.}^3$$

The section modulus provided is 2.92 in.^{3}. Since the draft is negligible near the top, there is no need to compute I_s and A_s.

ANCHOR BOLTS. Assume radius of bolt circle = 93 in. For 20 anchor bolts the required net area per bolt is, from Eq. (21),

$$F = \frac{4 \times 3,750,000 \times 12}{20 \times 93 \times 2} - \frac{167,000}{20} = 40,000 \text{ lb}$$

$$A = \frac{40,000}{15,000} = 2.67 \text{ in.}^2$$

Use $2\frac{1}{4}$-in. bolts, for which the area at the root of the thread is 3.02 in.^{2}.

14. Guyed Stacks For most plant sites the space required for guys and their anchors makes the use of guyed stacks (Fig. 8) undesirable except where the guys can be anchored to adequately braced plant structures so as to provide ample ground clearance. Generally, one set of guys spaced 120° apart around the stack circumference with an angle β of 45 to 50° between the guy and the vertical axis of the stack is satisfactory. The vertical components of the guy tensions must be taken into account when computing stresses in the shell.

The maximum horizontal component P_h of the force in a single guy due to wind or earthquake can be computed from

$$P_h = \frac{M - M_b}{H_1} \tag{23}$$

where M = moment of wind or earthquake force about base of stack, ft-lb

M_b = moment at base due to restraint of foundation

H_1 = height of guy ring above base of stack, ft

In determining moments for the design of the shell it should be noted that the stack is not held rigidly at the guy ring but moves laterally owing to the decrease under lateral forces in the initial sag of the windward guy. With adequate initial tension, this movement is small. The height of the guy ring can be chosen so that the moment of the cantilever section at the guy ring is approximately the same as the moment at the base. The latter is based on an estimated partial fixity of the base. It is not good practice to use a thinner shell between the guy ring and the base, even where the computed stresses might permit a decrease.

An evaluation of the true maximum tension in the windward guy depends on the wind force, the components in line with the windward guy of the residual initial tension in the two leeward guys, the wind force on, and the weight of, the guys themselves, and other factors. The approximations given below are believed to be sufficiently accurate to determine the size of guys required for a normal guyed stack with a height less than 200 ft.[10]

The breaking strength B.S. of the guy cables should be based on a factor of safety of about 2.5. An initial tension of 0.2 × B.S. in all guys is recommended. With this initial tension, the corresponding live-load capacity of the windward guy can be taken as 0.3 × B.S. and the slack force in the leeward guys as 0.1 × B.S. The required breaking strength can be taken as B.S. = $P_H/0.3 \sin \beta$ and a cable selected which has a corresponding minimum breaking strength.

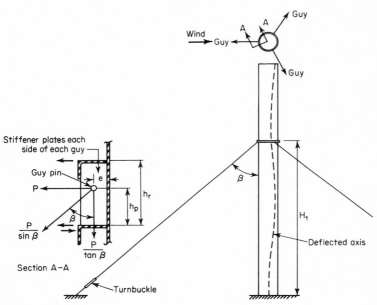

Fig. 8 Guyed stack.

The moments in the plane of the guy ring can be taken as

$$M_{rt} = 0.25R\left(\frac{h_p}{h_r} P + \frac{e}{h_r} \frac{P}{\tan \beta}\right) \tag{24a}$$

$$M_{rb} = 0.25RP - M_{rt} \tag{24b}$$

where M_{rt} = moment in top flange of guy ring, in.-lb
M_{rb} = moment in bottom flange of guy ring, in.-lb
P = maximum horizontal force due to wind or earthquake, lb
Other notation is shown in Fig. 8.

The height of the shell acting as a flange of the ring can be determined as for stiffening rings (Fig. 6). Vertical stiffeners should be provided on each side of each guy.

Example 2 Design a guyed, unlined A36 steel stack 6 × 130 ft for a map wind zone of less than 35 psf and an earthquake zone of 0. Locate the guy ring 30 ft below the top, $\beta = 45°$. Since the foundation of a guyed stack is not massive, assume
Base moment:

$$M = \frac{6 \times 23 \times 100^2}{16} = 86,300 \text{ ft-lb}$$

Cantilever moment:

$$M = \frac{6 \times 31 \times 30^2}{2} = 83,700 \text{ ft-lb}$$

Horizontal wind at guy ring [Eq. (23)]:

$$P = \frac{6 \times 31 \times 30 \times 115 + 6 \times 23 \times 100 \times 50 - 86,300}{100} = 12,480 \text{ lb}$$

$$\text{Required cable B.S.} = \frac{12,480}{0.3 \times 0.707} = 58,800 \text{ lb}$$

Use ¾ × 19 wire extra-high-strength grade with minimum breaking strength = 58,300 lb.

Use initial tension = 0.2 × 58,300 = 11,660 lb
Leeward slack guy tension = 0.1 × 58,300 = 5830 lb
Horizontal component in line with slack guy = 5830 sin 45° = 4120 lb
Horizontal component in line with windward guy = 4120 cos 60° = 2060 lb

Horizontal wind force at windward guy = 12,480
Horizontal slack guy force at windward guy 2 × 2060 =   4,120
Total horizontal force resisted by windward guy = 16,600 lb

$$\text{Tension in windward guy, full wind} = \frac{16,600}{\sin 45°} = 23,450 \text{ lb}$$

$$\text{Factor of safety} = \frac{58,300}{23,450} = 2.48$$

Chimney shell at guy ring: Minimum plate thickness ¼ in. Deduct $\frac{1}{16}$ in. corrosion allowance and compute stresses for $\frac{3}{16}$-in. plate.

$$A = 2\pi \times 36 \times 0.1875 = 42.4 \text{ in.}^2 \qquad \frac{d}{t} = \frac{72}{0.1875} = 384$$

$$\frac{I}{c} = \pi \times 36^2 \times 0.1875 = 762 \text{ in.}^3$$

Stresses at windward guy:

$$\text{Dead load} = 2\pi \times 3 \times 7.6 \times \frac{30}{42.4} = -101 \text{ psi}$$

$$\text{Wind moment} = \frac{Mc}{I} = \frac{83,700 \times 12}{762} = +1320 \text{ psi}$$

Vertical component of guy tension, assumed distributed by guy ring over 15 in. of shell $= \frac{16,600}{0.1875 \times 15}$
$= -5900$ psi.

$$\text{Net stress} = -101 + 1320 - 5900 = -4681 \text{ psi}$$

$$\text{Stress at base} = \frac{86,300 \times 12}{762} = 1360 \text{ psi}$$

$$\text{Allowable } F = XY = 0.0625E \frac{t}{R} \times 1.0$$

$$= 0.0625 \times 28 \times 10^6 \times \frac{0.1875}{3} \times 1.0$$

$$= 9114 \text{ psi}$$

Use ¼-in. plate for entire height of chimney.

15. Braced Stacks   Where a steel stack is adjacent to a properly braced building or structure, it is economical to provide a brace from the stack to the structure at a substantial distance above the base (Fig. 5a). The design of the shell is similar to that of a guyed stack except that the stack is assumed to be rigidly held at the brace. For two struts from the ring of the stack to the adjacent structure (Fig. 9), the maximum reaction at each strut can be computed from Eq. (23).

The moment M_r in the brace ring can be determined from

$$M_r = 0.25RP \tag{25}$$

16. Resonant Vibrations   Steel stacks are more susceptible to wind-induced vibrations than other types of chimneys. There are numerous records of steel stacks, conservatively designed for maximum probable wind or earthquake forces, which have experienced serious vibrations from steady-state winds of velocities of the order of 10 to 40 mph. As a consequence, it is generally necessary to investigate them for dynamic wind action. Most of the serious low-wind vibration problems have been with unlined stacks.

The wind velocity at which the natural frequency of vortex shedding equals the natural frequency of the stack is given by Eq. (3). With $S = 0.2$, this formula gives

$$V_{cr} = \frac{Df_t}{0.2} \times \frac{3600}{5280} = 3.41 Df_t \qquad (26)$$

with V_{cr} in miles per hour and D in feet. The value of D for stacks of the profile in Fig. 5c may be taken as the diameter at one-eighth the height below the top.

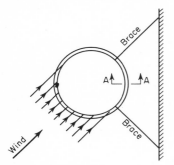

Fig. 9 Stack braces.

Experience indicates that periodic vortex shedding is inhibited by the natural turbulence of the airstream at the higher wind velocities, and that it is not likely to occur at velocities greater than 60 mph. However, it is not always necessary to proportion the stack so that the critical velocity exceeds 60 mph, or the highest sustained velocity that is likely to occur. If the critical velocity is low enough (about 15 to 40 mph), the stresses due to dynamic deflection, which may be determined from the equivalent static force of Eq. (4), may be within design limits. In that case, the stack is structurally adequate if noticeable movement is not objectionable. If the stresses are excessive, a mechanical damping device, which may consist of wind spoilers or a refractory or similar lining, is recommended. If this does not reduce the stresses to safe limits, the only solution is to change the stack diameter.

If the critical wind velocity of a proposed stack falls between 40 and 60 mph, a change of configuration to raise the frequency is recommended.

Stacks which are to be lined should be investigated for vibration in the unlined condition. They may need to be guyed or braced temporarily during these periods. Also, to prevent ovaling of the lining during erection, it is advisable to install temporary spiders. Circumferential stiffening rings should be provided to protect shotcrete lining during curing.

Unlined stacks are also subject to ovaling vibrations. To guard against this, the thickness of an unlined stack should not be less than $\frac{1}{250}$ of the diameter. If this limitation is not met, circumferential stiffening rings are required to raise the resonant velocity above 60 mph. The ovaling frequency of a stiffened stack is given by

$$f_0 = \frac{3}{7}\sqrt{\frac{EIg}{wR^4}} \qquad (27)$$

where w = weight per unit length of circumference of ring and stack between adjacent stiffeners. An effective width of shell equal to $1.56\sqrt{Rt}$ should be included in computing the moment of inertia of the ring.

Example 3 Investigate the stack of Example 1 for resonant wind vibration.

TRANSVERSE VIBRATION. The equivalent height H_e was determined in Example 1 to be 147 ft and the shell thicknesses $\frac{1}{4}$ in. at the top and $\frac{7}{16}$ in. at the bottom. From Eq. (11),

$$D_e = 10 \left(\frac{0.4375}{0.25}\right)^{0.27} = 11.6 \text{ ft}$$

The equivalent unit weight of the shell, including the shotcrete lining at 18.3 psf, and computed at one-fourth the height above the base (Art. 8) is

Shell $= 0.4375 \times 0.283 = 0.124$
Lining $= 18.3/144 \qquad = 0.127$
$$\overline{0.251/0.4375 = 0.574 \text{ lb/in.}^3}$$

Then, from Eq. (10)

$$f_t = \frac{3.9 \times 11.6 \times 12}{(160 \times 12)^2} \sqrt{\frac{28 \times 10^6}{0.574}} = 1.03 \text{ cps}$$

The critical wind velocity by Eq. (26) is

$$V_{cr} = 3.41 \times 10 \times 1.03 = 35 \text{ mph} = 51.3 \text{ fps}$$

Using Eq. (4) and assuming $\beta = 0.03$ the equivalent dynamic wind force is

$$F_L = 1.2 \times 10^{-4} \times 51.3^2 \times \frac{10}{0.03} = 105 \text{ lb/ft}$$

The calculated dynamic force is less than the design wind forces, which are $33 \times 10 = 330$ lb/ft on the upper 60 ft, $23 \times 10 = 230$ lb/ft on the next 60 ft, and an average of $23 \times 12.5 = 288$ lb/ft on the 40-ft base. Consequently, the design is considered to be satisfactory as to resonant transverse vibration.

OVALING VIBRATION. The critical velocity is computed by Eq. (6),

$$V_0 = 1696 \frac{t}{D} = 1696 \frac{t}{10} = 42.4 \text{ fps for } t = \tfrac{1}{4} \text{ in.}$$
$$= 53.0 \text{ fps for } t = \tfrac{5}{16} \text{ in.}$$
$$= 74.2 \text{ fps for } t = \tfrac{7}{16} \text{ in.}$$

Since these velocities are less than 60 mph, the shell may be subject to ovaling; so intermediate stiffeners spaced 10 ft on centers should be provided. From Eq. (27) the ovaling frequency of the stack in the unlined condition, stiffened with the $3 \times 3 \times \tfrac{1}{4}$ angles determined in Example 1, is

$$w = 0.283(1.44 + 10 \times 12 \times 0.25) = 8.90 \text{ lb/in.}$$

$$f_0 = \frac{3}{7} \sqrt{\frac{28,000,000 \times 6.77 \times 386}{8.90 \times 60^4}} = 10.8 \text{ cps}$$

With $S = 0.2$, the corresponding value of V_0 by Eq. (6a) is

$$V_0 = \frac{10.8 \times 10}{0.4} = 270 \text{ fps} = 184 \text{ mph}$$

Hence the stiffeners provided are adequate to guard against wind-induced ovaling vibration.

Example 4 Compute the resonant wind velocity for an A36 lined stack of the profile in Fig. 5c. Given $H = 224$ ft, $D_t = 13$ ft, $D_b = 28$ ft, $t = \tfrac{1}{4}, \tfrac{5}{16}, \tfrac{3}{8}, \tfrac{7}{16}$, and $\tfrac{1}{2}$ in. for top 48 ft, next 48 ft, next 44 ft, next 44 ft, and bottom 40 ft, respectively. Shotcrete lining 2 in. thick weighs 110 pcf.

From Eqs. (11) and (12),

$$D_e = 13 \left(\frac{0.50}{0.25}\right)^{0.27} = 15.7 \text{ ft}$$

$$H_e = 224 \left(\frac{2 \times 13}{13 + 28}\right)^{0.9} = 149 \text{ ft}$$

The equivalent unit weight of the shell, including the shotcrete lining at 18.3 psf, and computed at one-fourth the height above the base (Art. 8) is

Shell $= 0.4375 \times 0.283 = 0.124$
Lining $= 18.3/144 \qquad = 0.127$
$$\overline{0.251/0.4375 = 0.574 \text{ lb/in.}^3}$$

Then from Eq. (10),

$$f_t = \frac{3.9 \times 15.7 \times 12}{(149 \times 12)^2} \sqrt{\frac{28 \times 10^6}{0.574}} = 1.61 \text{ cps}$$

From Eq. (26) with $D = 15$ ft, the value at $H/8$ below the top,

$$V_{cr} = 3.41 \times 15 \times 1.61 = 82 \text{ mph}$$

Since $V_{cr} > 60$ mph, the design is satisfactory for a site subject to steady-state winds.

REINFORCED-CONCRETE CHIMNEYS

Construction costs of reinforced-concrete chimneys and ground-supported steel stacks are generally comparable up to a height of about 150 ft. Superimposed steel stacks may compete with ground-supported reinforced-concrete chimneys up to 200 ft above the building roof. At greater heights, reinforced-concrete chimneys are usually more economical. Reinforced-concrete chimneys also cost less to maintain than steel chimneys because they are free from atmospheric corrosion.

It is generally economical to taper tall concrete chimneys, with the top diameter determined by gas exit requirements and liner construction details. The base diameter results from optimizing the costs of concrete, forms, and reinforcing steel for various shapes. It is common to use a variable rather than a straight taper, for aesthetic reasons. This gives the silhouette a slight concave appearance.

17. ACI Standard   ACI 307-69 (Ref. 1) gives material, construction, and design requirements for reinforced-concrete chimneys. It sets forth recommended loadings and methods for determining the resulting stresses. Because the formulas for calculating the stresses are complex, charts are included to aid in the solutions.

18. Vibration due to Wind   Reinforced-concrete chimneys are unlikely to exhibit ovaling resonant vibrations because of the relatively large shell thickness. However, there have been cases of resonant vibration transverse to the wind at wind speeds considerably below the design velocity. Measurements of vibrations are rare, and are for concrete chimneys over 300 ft high. Since many concrete chimneys have not experienced this problem, the authors believe that tapered concrete chimneys of normal diameters and up to 300 ft in height are not likely to experience transverse resonant vibration. For shorter chimneys of unusually slender proportions the procedure discussed in Art. 4 may be used to investigate this effect.

LININGS

A chimney lining has many functions. It provides better chimney draft by maintaining the flue-gas temperature, thus reducing fan installation and operating costs. If it is insulated it reduces the temperature of the chimney shell. It also protects the shell from the acid attack of flue gases. Depending on the material, linings may be self-supporting or may be placed against the chimney shell.

Independent, self-supporting steel linings (Fig. 10a) are used most frequently in reinforced-concrete chimneys. They are fully insulated and can be supported at the bottom, top, or some intermediate level of the shell. Steel liners are essentially corrosion-free where entering gas temperatures are above the acid dew point, provided that the exterior surfaces are properly insulated. A coating for the interior surface should be considered if entering gas temperatures below 250°F are expected and/or the gas is saturated. Most steel liners have been built of A36 steel; A242 steel has been used occasionally because of its improved atmospheric corrosion resistance. Since the A242 steels are not acid-resisting, their added expense may not be justified. Detailed discussions of the design and construction of steel chimney liners are given in Ref. 7.

Independent, self-supporting brick linings (Fig. 10b) have been used for concrete chimneys where earthquake requirements are not severe. To reduce the circumferential stress due to the temperature gradient across the brick thickness, the lining should have enclosing steel bands at regular intervals in its height and, in particular, immediately below each change in lining thickness, where the weight of the lining above is eccentric to the section below the offset.

Partial brick linings supported on corbels (Fig. 10c), which for economic reasons were quite common in the past, have been the largest source of trouble for concrete chimneys, owing to gas leakage and acid attack resulting from differential vertical expansion at the lining support.

Shotcrete linings applied over mesh reinforcement attached by welded studs have been used extensively for steel stacks. Experience indicates that calcium-aluminate cement gives better protection against corrosion than portland cement. The sand should have a high silica content, and the lining should be thoroughly sprayed with a membrane curing compound immediately after shotcreting, since proper curing is essential. Shotcrete linings have given satisfactory service on concrete chimneys in moderate climates.

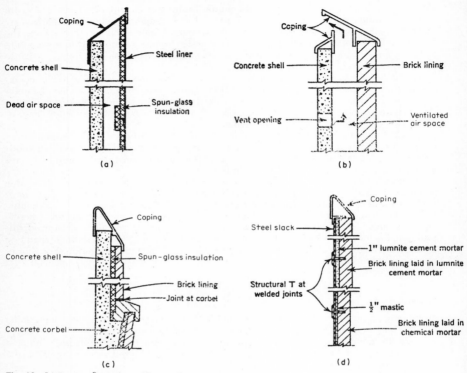

Fig. 10 Lining configurations.

Brick Linings for Steel Stacks.   If the gases are highly corrosive, a more effective, but more expensive, lining consisting of brick laid in calcium-aluminate mortar with the same mortar placed between the brick and the steel has been used (upper portion of Fig. 10*d*). The quality of the mortar can be improved by using high-silica sand and by curing the joints with a sprayed membrane. The most satisfactory brick for this lining is one conforming to ASTM Specification C279. A more costly lining that appears to be more maintenance-free where high-sulfur fuels are burned consists of brick in sodium silicate mortar with a mastic between the brick and the steel (lower portion of Fig. 10*d*).

FOUNDATIONS

Foundations of chimneys and stacks may be subjected to large overturning moments as a result of wind or earthquake loadings, and it is important to investigate the supporting soil or rock because a small rotation or unequal settlement of the foundation magnifies any movement of the top of the chimney.

Large concrete foundations for chimneys are usually circular in shape, with reinforcing steel placed in the circumferential and radial directions except in the center portion, where a grid of straight bars eliminates bending circumferential steel and avoids the

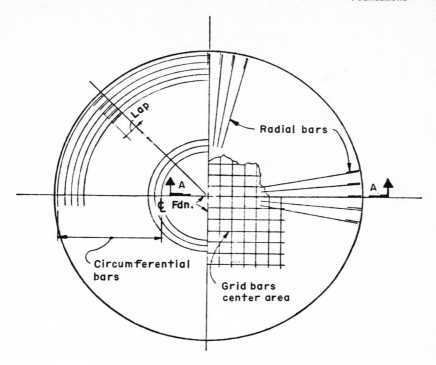

Top and bottom reinforcing plan

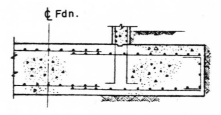

Section A–A

Fig. 11 Circular foundation.

congestion of bars in the radial direction (Fig. 11). Where chimneys are founded on competent rock or on caissons or piles it is economical to support them on ring-shaped foundations (Fig. 12). Medium and small chimneys are commonly supported on octagonal foundations with reinforcing bars placed in four directions (Fig. 13).

The foundation may be of uniform thickness or may be stepped or sloped. Stresses resulting from diagonal tension should be checked. In designing an octagonal foundation the maximum soil pressure is usually determined by considering the foundation as a circle whose radius is the mean of the inscribed and circumscribed circle of the octagon.

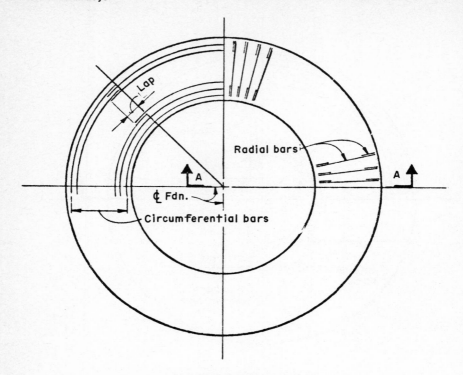

Top and bottom reinforcing plan

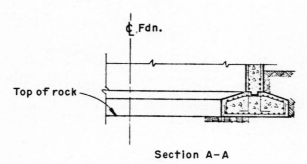

Section A–A

Fig. 12 Ring foundation.

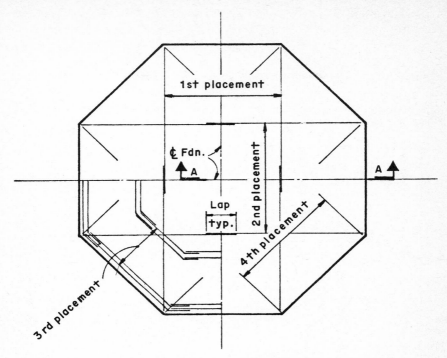

Top and bottom reinforcing plan

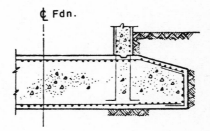

Section A—A

Fig. 13 Octagonal foundation.

The soil pressure p at the leeward toe of a circular foundation, or of a ring-shaped foundation with inside radius R_0, is given by

$$p = \frac{W}{A} k_1 \qquad (28)$$

where $A = \pi(R^2 - R_0^2)$. The distance z to the line of zero pressure is given by

$$z = Rk_2$$

Values of k_1 and k_2 are given in Fig. 14a and b. In these figures the eccentricity $e = M/W$, where $M =$ overturning moment and $W =$ total load.

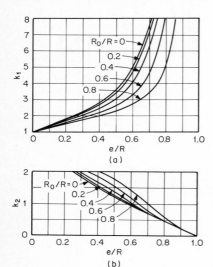

Fig. 14 Soil pressure constants for circular and ring foundations.

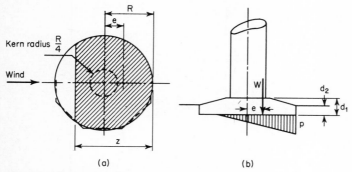

Fig. 15 Soil pressure under foundation.

Dead load used in calculations should include the weight of the wall, liner, foundation, backfill, and the expected dust accumulation inside the chimney, unless the dust load reduces the effect of horizontal forces. Checks should be made to ensure the minimum factors of safety against overturning and sliding as indicated in Sec. 5, Part 3. For ultimate-strength design in accordance with ACI 318-77, and in the absence of live loads, the total design load should be the larger of $0.9D + 1.3W$ and $0.9D + 1.4E$, but not less than $1.4D$.

Example 5 Design a tapered octagonal foundation for the steel stack of Example 1. Allowable soil bearing pressure 3500 psf, angle of internal friction $\phi = 30°$. Try a mean diameter of 30 ft with a 16-ft top. Assume $d_1 = 3.5$ ft, $d_2 = 1.5$ ft (Fig. 15).

Weight of chimney and lining = 167,000 lb. Trial weight of foundation

$$\pi \times 15^2 = 706 \text{ ft}^2$$
$$\underline{\pi \times 8^2 = 202} \text{ ft}^2 \text{ at } 3.5 \times 150 \text{ psf} = 106,000$$
$$504 \text{ ft}^2 \text{ at } 2.5 \times 150 \text{ psf} = \underline{189,000}$$

Weight of chimney, lining, and foundation = 462,000 lb

Horizontal shear at top of foundation = $33 \times 10 \times 60 + 23 \times 10 \times 60 + 23 \times 12.5 \times 40 = 45,100$ lb

Overturning moment at top of foundation = 3,770,000 ft-lb
$$45,100 \times 3.5 = \underline{158,000}$$

Moment at bottom of foundation = 3,928,000 ft-lb

$$e = \frac{3,928,000}{462,000} = 8.50 \text{ ft} \qquad \frac{e}{R} = \frac{8.50}{15} = 0.567$$

From Fig. 14a and b, $k_1 = 4.46$ $k_2 = 1.05$
From Eqs. (30 and (31),

$$p = \frac{4.46 \times 462,000}{706} = 2919 \text{ psf}$$
$$z = 15 \times 1.05 = 15.75 \text{ ft}$$

Shear resistance available at base = $462,000 \tan 30° = 267,960$ lb

Factor of safety against sliding = $\dfrac{267,960}{45,100} = 5.94$

Factor of safety against overturning = $\dfrac{462,200 \times 15}{3,928,000} = 1.76$

Use octagon based on 30-ft mean-diameter circle, check assumed depths, and determine required reinforcement.

REFERENCES

1. Standard Specification for Design and Construction of Reinforced Concrete Chimneys, ACI 307-1969, American Concrete Institute, Detroit, Mich., 1969.
2. Building Code Requirements for Minimum Design Loads in Building and Other Structures, ANSI A58.1-1972, American National Standards Institute, New York, 1972.
3. Uniform Building Code, International Conference of Building Officials, Whittier, Calif., 1976.
4. Commentaries on Part 4 of the National Building Code of Canada, 1975, National Research Council of Canada, Ottawa, Canada, 1975.
5. Recommended Comprehensive Seismic Design Provisions for Buildings, Applied Technology Council, San Francisco, Calif., January 1977 Draft.
6. Algermissen, S. T., and D. M. Perkins: A Probabilistic Estimate of Maximum Acceleration in Rock in the Contiguous United States, U.S. Geological Survey Open File *Rept.* 76-416, 1976.
7. "Design and Construction of Steel Chimney Liners," American Society of Civil Engineers, New York, 1975.
8. Chu, Kuang-Han, and J. Park: Approximate Fundamental Frequencies for Chimneys, *J. Power Div. ASCE*, November 1973.
9. "Steel Plate Engineering Data," vol. 2, American Iron and Steel Institute, Washington, D.C.
10. Rowe, R. S.: Amplified Stress and Displacement in Guyed Towers, *Trans. ASCE*, vol. 125, p. 199, 1960.
11. Zar, M.: Problems with High Chimneys, *Civil Eng.*, February 1964.

Appendix

TABLE A1   Torsional Properties of Solid Cross Sections*

Cross section	Torsional stiffness J	Shear stress
circle, $2r$	$\frac{1}{2}\pi r^4$	$\frac{2T}{\pi r^3}$
ellipse, $2a$, $2b$	$\frac{\pi a^3 b^3}{a^2 + b^2}$	$\frac{2T}{\pi a b^2}$ at ends of minor axis
square, a, a	$0.141a^4$	$\frac{T}{0.208a^3}$ at midpoint each side
rectangle, $t<b$, b	$\frac{bt^3}{3}\left[1 - 0.63\frac{t}{b} + 0.052\left(\frac{t}{b}\right)^2\right]$	$\frac{3T}{bt^2}\left(1 + 0.6\frac{t}{b}\right)$ at midpoint each long side
triangle, a, a, a	$\frac{a^4\sqrt{3}}{80}$	$\frac{20T}{a^3}$ at midpoint each side

* $T = GJ\theta$, where T = torque, G = shearing modulus of elasticity, J = torsional stiffness, θ = angle of twist, radians per unit length.

TABLE A2   Torsional Properties of Closed Thin-walled Sections*

Cross section	Torsional stiffness J	Shear stress
	$\dfrac{4A^2}{\int ds/t_s}$	$\dfrac{T}{2At}$
	$2\pi r^3 t$	$\dfrac{T}{2\pi r^2 t}$
	$b^3 t$	$\dfrac{T}{2b^2 t}$
	$\dfrac{2a^2 b^2}{\dfrac{a}{t_1} + \dfrac{b}{t_2}}$	$\dfrac{T}{2abt}$

* $T = GJ\theta$, where T = torque, G = shearing modulus of elasticity, J = torsional stiffness, θ = angle of twist, radians per unit length.

A = area bounded by midline of wall.

NOTE: Warping constant C is usually negligible for closed thin-walled cross sections.

TABLE A3 Torsional Properties of Open Cross Sections

Cross section	Location e of shear center S	Warping constant C
		$\dfrac{d^2 I_y}{4}$
	$\dfrac{c_1 I_1 - c_2 I_2}{I_y}$	$\dfrac{d^2 I_1 I_2}{I_y}$
	$\dfrac{\bar{x}}{4}\left(\dfrac{d}{r_x}\right)^2$	$\dfrac{d^2 I_y}{4}\left[1 - \dfrac{\bar{x}(e-\bar{x})}{r_y^2}\right]$
		$(b_1{}^3 + b_2{}^3)\dfrac{t^3}{36}$
		$\dfrac{t_1{}^3 b^3}{144} + \dfrac{t_2{}^3 d^3}{36}$
		$\dfrac{d^2}{4} I_a$

NOTE: The torsional stiffness J for cross sections in this table can be determined closely enough for most applications by $J = \Sigma b t^3/3$.   The warping constant C is usually negligible for the angle and the T.

TABLE A4 Effective Length Coefficients for Columns

Case	Ends	K
	er–fr	$\left(\frac{10}{\beta_1}+4\right)^{1/2}$
	er–fx	$\frac{4+\beta_1}{2+\beta_1}$
	er–p	$\frac{3+0.7\beta_1}{3+\beta_1}$
	er–er	$\frac{2+0.5\beta_1}{2+\beta_1}$
	er–fx	$\frac{2.1+0.5\beta_1}{3+\beta_1}$
Central elastic support	p–et–p	$\frac{1}{(1+\beta_2)}$ if $\beta_2 \leq 3$
	p–et–p	0.5 if $\beta_2 > 3$
$\leftarrow L \rightarrow\!\mid\! kL \!\mid\!\leftarrow$	p–p–fr	$1+2k$
	fx–p–fr	$0.7+2k$
$\leftarrow L \rightarrow\!\mid\! kL \!\mid\!\leftarrow$ $k<1$	p–p–p	$0.7+0.3k$
	fx–p–fx	$0.5+0.2k$
$\leftarrow L \rightarrow\!\mid\! \leftarrow L \rightarrow$	p–p–p	$0.9+0.1\frac{P_o}{P}$
Central load	p–p	$0.75+\frac{P_o}{4P}$
Intermediate loads $\mid\!\leftarrow k_iL \rightarrow\!\mid$	p–p	$\left(\frac{\Sigma P_i k_i}{\Sigma P_i}\right)^{1/2}$

NOTE: p = pinned, fx = fixed, fr = free, er = elastic rotational restraint, et = elastic translational restraint.

$\beta_1 = \dfrac{\alpha_1 L}{EI}$ where α_1 = moment to produce 1 radian rotation

$\beta_2 = \dfrac{\alpha_2 L^3}{53EI}$ where α_2 = load to produce unit deflection of the central support

* From "The Strength of Aluminum," Aluminum Company of Canada, Ltd., 1965.

TABLE A5   Buckling of Plate under Edge Stress, Four Edges Simply Supported*

Case	Range of application	Buckling coefficient $k^{\dagger}$
1	$\alpha \gtrsim 1$	$5.34 + \dfrac{4}{\alpha^2}$
	$\alpha \lesssim 1$	$4.00 + \dfrac{5.34}{\alpha^2}$
2	$0 \lesssim \psi \lesssim 1$ $\alpha \gtrsim 1$	$\dfrac{8.4}{1.1 + \psi}$
	$\alpha < 1$	$\dfrac{2.1}{1.1 + \psi}\left(\alpha + \dfrac{1}{\alpha}\right)^2$
3	$\alpha \gtrsim \dfrac{2}{3}$	23.9
	$\alpha < \dfrac{2}{3}$	$15.87 + \dfrac{1.87}{\alpha^2} + 8.6\alpha^2$
4	$\psi \gtrsim 1$	Same as Case 3 except use $b = 2b_c$ to compute α and b/t
5	$0 \lesssim \psi \lesssim 1$	$k = (1-\psi)k_2 + \psi k_3 - 10\psi(1-\psi)$ where $k_2 = k$ for Case 2 with $\psi = 0$ $k_3 = k$ for Case 3

* From German Specification DIN4114.

$\dagger$ Critical (buckling) stress σ (or τ) $= \dfrac{k\pi^2 E}{12(1 - \nu^2)(b/t)^2}$.

TABLE A6 Stiffened Beam Webs

Case	λ	Buckling coefficient		Stiffener	
		k	Range	I_s/bt^3	Range
1	$1/2$	35.6	$\alpha > \tfrac{2}{3}$	0.12	
				$0.3\left[3.7\alpha^2\left(1+4\,\dfrac{A_s}{A_w}\right)-\alpha^3\right]$	$\alpha < 1.6$
	$1/4$	101	$\alpha > 0.4$	$0.22\left(1+7.7\,\dfrac{A_s}{A_w}\right)$	$\alpha < 0.5$
				$1.1\left(1+7.7\,\dfrac{A_s}{A_w}\right)(\alpha-0.3)$ but need not exceed $1.47\left(1+12.5\,\dfrac{A_s}{A_w}\right)$	$\alpha > 0.5$
	$1/5$	129		$1.15\left[0.4+\left(1+6\,\dfrac{A_s}{A_w}\right)\alpha^2\right]$	$0.5 < \alpha < 1.5$
				$0.355+0.47\alpha+0.81\alpha^2\left(1+8.8\,\dfrac{A_s}{A_w}\right)$	$0.5 < \alpha < 1.5$
2	$1/2$	Table A6, Case 1		$0.5\alpha^2(-1+2\alpha+2.5\alpha^2-\alpha^3)$	$0.5 < \alpha < 2$
	$1/4$	Table A6, Case 1		$0.66\alpha^2(1-3.3\alpha+3.9\alpha^2-1.1\alpha^3)$	$0.5 < \alpha < 2$
3		Table A6, Case 1		$0.37\left(\dfrac{7}{\alpha}-5\alpha\right)$	

NOTE: t = web thickness; I_s = moment of inertia of stiffener (for one-sided stiffeners usually taken at face of web); A_s = cross-sectional area of stiffener; A_w = area of web = bt.

On the basis of tests, Massonnet suggests that, to keep longitudinal stiffeners practically straight to collapse of girder, the theoretical values of I_s be multiplied by n, where $n = 3$ for $\lambda = \tfrac{1}{2}$, 4 for $\lambda = \tfrac{1}{3}$, 6 for $\lambda = \tfrac{1}{4}$, and 7 for $\lambda = \tfrac{1}{5}$.

Index

Abutments, 5-56, 18-57, 18-68, 18-72
  bank-block, 5-56
  spill-through, 5-56
Aeolian vibration, 24-9
Aluminum alloys:
  alclad, 10-2
  electrical conductivity, 10-2
  expansion coefficient, 10-2
  mechanical properties, 10-3
  modulus of elasticity, 10-2
  physical properties, 10-2
  thermal conductivity, 10-2
  yield strength, 10-4
Aluminum structures:
  beams, 10-14 to 10-20
    holes, 10-15
    lateral buckling, 10-16, 10-18
    local buckling, 10-18
    round tubes, 10-18
    shape factor, 10-14
    shear buckling, 10-19
    ultimate strength, 10-14
    welded, 10-15
    (*See also* plate girders, *below*; Beams; Cold-
      formed steel structures; Steel beams)
  bolted connections, 10-24 to 10-25
    factor of safety, 10-24
    tightening, 10-25
    (*See also* Bolted connections)
  compression members: buckling formulas, 10-5,
    24-21
    effective length, A-4
    factor of safety, 10-6
    lacing, 10-7
    round tubes, 10-10
    torsional buckling, 9-16, 10-6
    welded, 10-12
    (*See also* Cold-formed steel structures; Steel
      compression members)
  corrosion resistance, 10-2
  fabrication, 10-2
  fatigue, 10-27, 10-28
  flat plates: buckling, A-5, A-6
    combined-stress buckling, 10-22
    effective width, 10-9
    postbuckling strength, 10-8
    stiffened, 10-9 to 10-11

Aluminum structures (*Cont.*):
  local buckling, 10-7
    round tubes, 10-11, 10-14
    (*See also* flat plates, *above*)
  plate girders: bearing stiffeners, 10-21
    lateral buckling, 10-20
    longitudinal stiffeners, 10-21, A-6
    vertical stiffeners, 10-21, A-6
    web, 10-20 to 10-22, A-6
    (*See also* beams, *above*; Steel plate girders)
  riveted connections, 10-22 to 10-24
    factor of safety, 10-23
    holes, 10-23 to 10-24
    rivet spacing, 10-24
    shear strength of rivets, 10-23
    (*See also* Riveted connections)
  shapes, 10-1
  specifications, 10-1 to 10-2
  tension members: net section, 10-4
    welded, 10-4
    (*See also* Cold-formed steel structures; Steel
      tension members)
  welded connections: butt, 10-25 to 10-26
    fillet, 10-25 to 10-26
    (*See also* Welded connections)
  welding, heat-affected zone, 10-4, 10-25
Analysis:
  deflections (*see* Deflection)
  degrees of freedom, 1-1
  determinacy, 1-2, 1-9 to 1-12
  direction cosines, 1-13 to 1-15
  displacement methods, 1-2, 1-15, 1-56 to 1-71, 1-
    87 to 1-93
  equilibrium, 1-2, 1-8
  finite-element (*see* Finite-element method)
  force methods, 1-2, 1-15, 1-45 to 1-56, 1-78 to 1-
    84
  graphical, 1-13 to 1-21
  Henneberg's method, 1-13
  inelastic, 1-8
  matrix (*see* Matrix methods of analysis)
  method of joints, 1-13 to 1-16
  method of sections, 1-13 to 1-17, 1-19
  moment distribution (*see* Moment distribution)
  nonlinear, 1-71

Analysis (*Cont.*):
  numerical integration, 1-14, 1-21 to 1-22
  plastic (*see* Plastic design)
  tension coefficients, 1-13, 1-18 to 1-19
  wind stresses (*see* Tall buildings)
  (*See also* Statically indeterminate structures)
Arches:
  analysis, 17-4 to 17-8
  buckling, 17-46 to 17-47
  centering, 13-6, 17-48
  circular, 17-10 to 17-13, 17-46
  classification, 17-1
  column analogy, 17-5, 17-43
  concrete placement, 17-48
  continuous, 17-43
  design: final, 17-29
    intermediate, 17-13
    preliminary, 17-6
  elastic center, 17-6
  elastic theory, 17-4
  elliptical, 17-13
  energy methods, 17-4 to 17-6
  erection, 8-15
  falsework, 17-48
  filled-spandrel, 17-1
  fixed, 17-9 to 17-15
    formulas, 17-4
  formwork, 13-6, 17-48
  foundations, 17-52
  influence coefficients, 17-22 to 17-24, 17-28 to 17-29
  kern points, 17-4
  lateral loads, 17-47
  neutral point, 17-5
  nomenclature, 17-1
  open-spandrel, 17-2
  parabolic, 17-6 to 17-15, 17-46
  pressure line, 17-1, 17-6
  reinforcement, 17-48
  rib shortening, 17-8, 17-16 to 17-23
  Rio Blanco bridge, Mexico, 17-52
  second-order theory, 17-44
  skewed barrel, 17-47
  temperature effect, 17-5, 17-9, 17-11, 17-16, 17-22
  three-hinged, 17-8
  tied, 17-1, 17-9
  timber (*see* Timber arches)
  two-hinged, 17-8 to 17-10
  ultimate design, concrete, 17-30
  unsymmmetrical, 17-30
  (*See also* Reinforced-concrete bridges)
Atterberg limits, 5-5, 5-13, 5-21

Barrel shells (*see* Shells, barrel)
Beams:
  analysis (*see* Analysis)
  deflection (*see* Deflection)
  formulas, 1-23 to 1-26
  maximum moment (*see* Bridges)
  openings, 19-46
  (*See also* Aluminum structures; Cold-formed steel
    structures; Composite beams; Prestressed
    concrete beams; Reinforced-concrete beams;
    Steel beams; Timber beams)
Bend test for brittle fracture, 4-13

Bins:
  arching, 22-12, 23-17
  deep, definition of, 23-17
  dust explosions, 22-30
  earthquake forces, 22-14
  emptying pressure in funnel flow: ACI 313, 22-10
    collapse of dome, 22-12
    DIN 1055, 22-2, 22-4
    dustlike materials, 22-13
    eccentric outlets, 22-10, 22-12, 23-21
    flow patterns, 22-4
    Pieper, 22-4
    pneumatic, 22-13
  filling pressure: DIN 1055, 22-2, 22-4
    Janssen's formula, 22-2, 22-4, 23-16
    Pieper, 22-4
    Reimbert's formulas, 22-2, 22-4
  flow patterns, 22-4
    effect of hopper-wall slope, 22-2
  flow stoppage, 23-17
  funnel flow, 22-1, 22-3
  geometry of cross section, 22-8
  granular material properties, 22-6, 22-7, 23-19
  mass flow, 22-1, 22-3
  ratholing, 23-17
  reinforced-concrete (*see* Reinforced-concrete silos
    and bunkers)
  shallow, definition of, 23-17
  steel (*see* Steel bins)
  wind pressure, 19-13 to 19-17
Boardman formula, 23-11
Bolted connections:
  beam: framed, 6-68
    moment-resistant, 6-70 to 6-75
    seat, 6-65 to 6-68
    T-stub, 6-70, 7-30
  bearing-type (steel), 6-61
  in cold-formed steel structures, 9-27
  column splices, 6-81 to 6-83
  eccentrically loaded: fasteners in shear, 6-62 to 6-64
    fasteners in tension, 6-64 to 6-65
  efficiency, 6-61
  fatigue (*see* Fatigue)
  friction-type (steel), 6-60
  prying action, 6-64, 6-70, 7-31
  (*See also* Aluminum structures; Plastic design)
Bolts:
  aluminum, 10-24
  high-strength steel, 6-55 to 6-59
    inspection, 6-58 to 6-59
    installation, 6-56 to 6-58
      calibrated-wrench method, 6-58
      tension, 6-57
      turn-of-nut method, 6-58
    proof loads, 6-57
    properties, 6-56
    washers, 6-57
  steel: bearing, 6-59
    interference-type, 6-59
    ribbed, 6-59
    turned, 6-59
    unfinished, 6-59
Boussinesq equation:
  influence chart, 5-63
  influence coefficients, 25-11
Bow's notation, 1-18

Brick:
  Building: concrete, 15-4
    grades, 15-1
    physical requirements, 15-2
  facing, 15-2
  walls, 15-10, 19-40, 19-41
  (*See also* Reinforced masonry)
Bridge strand, 21-18
Bridge wire, galvanized, 21-17
Bridges:
  AASHTO loading, 18-1
    continuous spans: negative moments, 18-14
      positive moments, 18-12
      shear, 18-16
    impact, 18-16
    simple span: maximum-moment envelope, 18-3,
      18-39, 18-40
      maximum-moment formulas, 18-3
      maximum-moment position, 18-3
      maximum-moment table, 18-4 to 18-7
      maximum-shear formulas, 18-3
      maximum-shear table, 18-4 to 18-7
  centrifugal force, 18-19
  Cooper's loading, 18-8
    maximum moments, shears, reactions, 18-9 to
      18-11
    moment table, 18-8
    position for maximum moment, 18-8
  curbs, 18-102
  earth pressure, 18-20
  earthquake load, 18-18
  ice pressure, 18-20
  longitudinal force, 18-19
  moving-water pressure, 18-20
  railings, 18-102
    pedestrian, 18-102
  sidewalks, 18-102
  thermal forces, 18-19
  wind load, 18-18
  (*See also* Composite beams; Prestressed-concrete
    bridges; Reinforced-concrete bridges; Steel
    bridges; Steel-plate-deck bridges)
Brittle fracture of structural steel:
  attachments, effect of, 4-19
  cold-work-effect, 4-17, 4-20
  crack arrest, 4-15, 4-19
  designing against, 4-18
  examples, 4-14
  fabrication, quality of, 4-20
  factors affecting, 4-16
  fracture appearance, 4-13
  fully killed steel, 4-13
  geometry effect, 4-17, 4-19
  hydrogen embrittlement, 4-15, 4-17
  initiation, 4-15
  inspection, 4-21
  laboratory tests, 4-13
  nil-ductility transition, 4-19, 4-20
  normalizing, 4-17
  notch toughness, 4-13
  peening, 4-21
  propagation, 4-15
  radiational embrittlement, 4-15
  residual stress effect, 4-17
  rimmed steel, 4-13
  semikilled steel, 4-13
  sharp-crack-propagation concept, 4-19

Brittle fracture of structural steel (*Cont.*):
  size effect, 4-17
  strain-aging effect, 4-17, 4-20
  strain rate, 4-17
  strakes, 4-15
  transition temperature, 4-13, 4-17, 4-19
  triaxial stresses, 4-17
Buildings:
  attachment devices, 13-5, 13-6
  bracing, 19-58
  column-core, 8-15
  comfort criteria, 19-3
  construction (concrete) (*see* Concrete
    construction)
  deflection, 19-3, 19-90
  environmental control, 19-7
  erection (*see* Erection of structural steel)
  expansion joints, 19-46 to 19-49
    coefficients of expansion, 19-49
  fabrication (*see* Fabrication of structural steel)
  fire resistance, 19-3 to 19-7, 19-24 to 19-36
  floors: acoustical properties, 19-32 to 19-36
    deflection, 19-3
    deflection characteristics, 19-30 to 19-36
    diaphragm qualities, 19-32 to 19-36
    finish, 19-25, 19-39
    framing, 19-63 to 19-67
      concrete, 11-57
      (*See also* Reinforced-concrete slabs)
    impact resistance, 19-32 to 19-36
    openings, 11-50, 19-44
    span limits, 19-23, 19-32 to 19-36
    structural decks, 9-5, 19-23
    tile, 15-3
  footings (*see* Footings)
  high-rise (*see* Tall buildings)
  live load, 19-10 to 19-12
  materials (*see* Materials)
  openings in beams, 19-46
  partitions (*see* walls, *below*)
  preliminary planning, 19-1
  roofs:
    diaphragm qualities, 19-26 to 19-30
    insulating values, 19-26 to 19-30
    pitch, 19-25
    ponding, 16-27
    roofing, 19-25
    span limits, 19-26 to 19-30
    structural decks, 9-5, 19-23, 19-24
    trusses, 19-53 to 19-54
    (*See also* Reinforced-concrete slabs; Shells;
      Suspension roofs; Timber roofs)
  seismic loads, 19-18 to 19-22
    (*See also* Earthquake-resistant design)
  snow load, 19-11 to 19-12
  stairs (*see* Stairs)
  vibration, 14-8, 19-3
  walls: acoustical properties, 19-40 to 19-42
    concrete, 11-48, 19-40
    diaphragm qualities, 19-40 to 19-42
    fire resistance, 19-40 to 19-42
    footings, 11-41 to 11-42
      (*See also* Footings)
    insulating values, 19-40 to 19-42
    openings, 11-48, 13-3, 19-45
    reinforced-masonry, 15-8
    shear, 19-80, 19-84 to 19-87

Buildings (*Cont.*):
   weights of materials, 19-9, 19-10, 19-23 to 19-36
   wind bracing, 19-2
   (*See also* Tall buildings)
   wind load, 19-13 to 19-18
   windows, 19-37
   (*See also* Industrial buildings; Tall buildings)
Bulkheads:
   anchorage, 5-58, 5-61
   forces, 5-57
   penetration of piles, 5-58
Bunkers (*see* Bins; Reinforced-concrete silos and
   bunkers; Steel bins)
Buried conduits:
   analysis, 25-2
      CANDE, 25-27
      elasticity solution, 25-22
      finite-element solution, 25-25
   buckling, 25-27
   constrained modulus of soil, 25-24
   construction: backpacking, 25-31
      bedding, 25-30
      compaction, 25-31
      fill, 25-30
      joints, 25-31
      site preparation, 25-30
      strutting, 25-31
   corrugated metal: aluminum, 25-3, 25-26
      steel, 25-3, 25-26
         bolted seams, 25-18, 25-19
         cross-sectional properties, 25-17
         example, 25-20
         long-span, 25-31
         wall strength, 25-16
         (*See also* flexible, *below*)
   design: camber 25-29
      coatings, 25-30
      durability, 25-28
      handling criteria, 25-27
      wrapping, 25-30
      (*See also* analysis, *above*)
   flexible, 25-16
      deflection, 25-19
      design criteria, 25-26
      design pressure, 25-18
      pipe arch: corner pressure, 25-20
         on rigid foundation, 25-20
      ring compression, 25-16
      (*See also* corrugated metal, *above*)
   loads: ditch conduit, 25-3 to 25-5
      imperfect-ditch conduit, 25-6
      negative projecting conduit, 25-6
      projecting conduit, 25-5 to 25-6
      surface, 25-10 to 25-11
      wide-ditch conduit, 25-6
   performance factor, 25-2
   pressure: flexible, 25-21
      rigid, 25-21
   reinforced-concrete (*see* rigid, *below*)
      example, 25-16
   rigid, 25-12
      embankment installation, 25-13
      factor of safety, 25-12
      monolithic, design, 25-15
      supporting strength, 25-12
      three-edge bearing test, 25-12
      trench installation, 25-12

Buried conduits (*Cont.*):
   safety factor, 25-2
   settlement ratio, 25-5 to 25-7
   types, 25-1, 25-3

Cable-supported roofs (*see* Suspension roofs)
Caisson foundations:
   allowable bearing pressure, 5-69
   box, 5-69
   open, 5-69
   pneumatic, 5-69
Casagrande piezometer, 5-39
Castigliano's theorems, 1-7, 1-14, 1-15, 1-45, 1-51
Cement, high-early-strength, 13-5
Charpy test, 4-18
Chimneys:
   dead load, 26-1
   diameter of top, 26-1
   draft intensity, 26-5
   earthquake forces, 26-3
   foundation design, 26-18 to 26-23
   height, 26-1
   linings, 26-17
   materials, 26-1
   pressure differential, 26-5
   reinforced-concrete, 26-17
   steel: allowable stresses, 26-7
      anchor bolts, 26-9
      base ring, 26-10
      braced, 26-14
      circumferential stiffeners, 26-9
      guyed, 26-12 to 26-14
      profile, 26-7
      stiffening ring, 26-8
   temperature differentials, 26-5
   vibration: natural frequency, 26-6
      oviling, 26-15, 26-16
      transverse, 26-14 to 26-17
   wind pressure, 19-16, 26-3
      circumferential moment, 26-3
Cold-formed steel structures:
   allowable stresses: basic, 9-9
      beam, 9-19 to 9-21
      beam web, 9-23
      bending and compression, 9-26
      compression member, 9-16 to 9-21
      lateral buckling, 9-24
      spot weld, 9-27
      unstiffened element, 9-15 to 9-16
   beam-column, 9-26
   beams: allowable stress, 9-21
      crippling, 9-24
      deflection, 9-21
      lateral buckling, 9-24
   bending and compression, 9-26
   cellular panels, 9-5
   channels, bracing, 9-26
   compression element: effective width, 9-12
      multiple-stiffened, 9-12
      stiffened, 9-11, 9-18, 9-21
      stiffeners, 9-11
      unstiffened, 9-11, 9-13, 9-18 to 9-21
   compression member, 9-16
      allowable stress, 9-18
   connections, 9-27
   decks, 19-23, 19-29, 19-30, 19-35

Cold-formed steel structures (*Cont.*):
deflection, 9-12 to 9-17, 9-22
flat-width ratio, 9-8
folded-plate roof, 9-6
framing members, 9-3
hyperbolic paraboloid, 9-8
line-element properties, 9-10
local buckling, 9-18
materials, 9-1 to 9-3
member types, 9-3
nailability, 9-4
roof deck, 9-5
section properties, 9-9
shape factor, 9-18
shapes, 9-3
shear diaphragms, 9-5
sheet thickness, 9-2
steel properties, 9-3
affected by cold forming, 9-28
stiffeners: edge, 9-4, 9-11
intermediate, 9-11
lip, 9-4
multiple, 9-12
stress-strain curve, 9-2
surface members, 9-4
tests, 9-24
torsional buckling, 9-16
wall panels, 9-5
Z's bracing, 9-26
(*See also* Aluminum structures)
Column analogy, 1-15, 1-53 to 1-56
Columns (*see* Aluminum structures; Cold-formed
steel structures; Reinforced-concrete colums;
Steel compression members; Timber columns)
Composite beams:
bridges, 14-32 to 14-39, 18-37
assumptions, 14-32
example, 14-35 to 14-39
flange width, 14-33
plate girder, 14-33
rolled beam, 14-33
with cover plate, 14-33
shear connectors, 14-34
allowable fluctuating loads, 14-34
spacing, 14-34
buildings: assumptions, 14-9
encased beams, 14-9
example, 14-28 to 14-32
flange width, 14-9
rolled beam, 14-10
shear connectors, 14-10
allowable static loads, 14-10
spacing, 14-11
concrete-encased, 14-1, 14-9
creep, 14-8, 14-9, 14-23
deflection, 14-8
effective width of flange, 14-9, 14-33
elastic properties: neutral axis below slab, 14-4
neutral axis in slab, 14-4
rolled beam with tension plate, 14-3
slab in tension, 14-5
unsymmetrical beam, 14-4
erection, 8-8
formed-steel deck, 14-10
modular ratio, 14-10
negative moments, 14-8
notation, 14-1 to 14-3

Composite beams (*Cont.*):
plastic properties: hybrid steel beam, 14-5
neutral axis below slab, 14-6
neutral axis in slab, 14-5
slab in tension, 14-6
section properties, rolled beam, 14-12 to 14-27
shear connectors, 8-8, 18-37
allowable loads, 14-28, 14-34, 14-35
channel, 14-7
cover, 14-34
formulas, 14-10, 14-34
spacing, 14-11, 14-34
stud, 14-7
ultimate strength, 14-34
shores, 14-9
shrinkage, 14-9, 14-33
temperature change, 14-9, 14-32
vibration, 14-8
Compression members:
effective length, A-4
(*See also* Aluminum structures; Cold-formed steel
structures; Reinforced-concrete columns;
Steel compression members; Timber
columns)
Computers:
applications: analysis, 2-15
detailing, 2-16
final documents, 2-16
proportioning, 2-15
basic concepts, 2-1
decision tables, 2-8
advantages, 2-9
flowcharting, 2-8
hardware, 2-10
Hollerith card, 2-12
interaction, 2-11
level of communication, 2-12
media, 2-12
modes, 2-11
organization: data, 2-9
programs, 2-9
programming: analysis, 2-3
development, 2-3
documentation, 2-4
evaluation, 2-4
languages: problem-oriented, 2-7
procedural, 2-7
symbolic, 2-7
maintenance, 2-4
problem definition, 2-3
problem design, 2-3
schematic diagram, 2-2
structured, 2-4
programs: sources, 2-14
types, 2-13
structural design process, 2-1
schematic diagram, 2-2
structured programming, 2-4
Concrete:
bridges (*see* Prestressed-concrete bridges;
Reinforced-concrete bridges)
buildings (*see* Buildings; Industrial buildings; Tall
buildings)
compressive strength, 11-1
construction methods, 13-1 to 13-10
creep, 12-2, 12-10, 12-28
curing, 13-5

Concrete (*Cont.*):
  high-early-strength cement, 13-5
  modulus of elasticity, 12-28
  prestressed (*see* Prestressed concrete)
  shell roof (*see* Shells)
  shrinkage, 12-2, 18-20
  tensile strength, 11-1
  testing, 13-5
  (*See also* Prestressed concrete; Reinforced concrete)
Concrete block:
  dimensions, 15-4
  physical requirements, 15-2
  walls, 19-41
  (*See also* Reinforced masonry)
Concrete construction:
  contract documents: designer's responsibility, 13-9
    drawings, 13-7, 13-9
      errors, 13-9
      revisions, 13-9, 13-10
    intent, 13-9
    preparation, 13-8
    scope of work, 13-9
    specifications, 13-3, 13-4, 13-8
  embedded items, 13-5, 13-6
  formwork: accessories, 13-3
    beams, 13-3
    column, 13-2
    cost, 13-2
    openings, 13-3
    reuse, 13-2
    slabs, 13-3
    stripping, 13-5, 13-6
    tie beams, 13-3
    walls, 13-3
  information for contractor, 13-6 to 13-7
  inspection, 13-8
  material samples, 13-7
  reinforcing steel: columns, 13-3
    concentrations, 13-4
    dowels, 13-4
    long bars, 13-4
    splices, 13-4
  resident engineer, 13-8
  shop drawings, 13-7
  stability, 13-6
  tolerances, 13-7
Conduits (*see* Buried conduits)
Conjugate-beam method, 1-15
Conservative systems, 1-3
Coulomb's theory, 5-48
Crane girders, 6-20, 19-62
Cranes, 19-62
  erection, 8-13
Culverts (*see* Buried conduits)
Cumulative damage, 4-4
  (*See also* Fatigue)

Deflection:
  beams, 1-33 to 1-37
    camber, 19-77
    composite, 14-8
    formulas, 1-26 to 1-29
    prestressed-concrete, 12-23 to 12-26, 18-100
    reinforced-concrete, 11-25, 18-50
    steel, 6-22, 9-20
    timber, 16-27

Deflection (*Cont.*):
  bridges: concrete, 14-8, 18-50, 18-80, 18-100
    steel, 18-37, 18-44
  buildings, 19-3, 19-90 to 19-97
  Castigliano's theorem, 1-14, 1-39
  complementary energy method, 1-39
  composite beams, 14-8
  conjugate-beam method, 1-14
  dummy load method, 1-14
  elastic-weights method, 1-14
  flat roofs, 16-18
  matrix methods, 1-14, 1-71, 1-78
  Maxwell-Mohr method, 1-14, 1-15
  moment-area method, 1-14
  nonlinearly elastic structures, 1-38
  numerical integration, 1-14, 1-28
  pin-jointed frames, 1-33, 1-37
  plastic structures (*see* Plastic design)
  self-straining, 1-39, 1-83, 1-92
  trusses, 1-33, 1-37
    timber, 16-35
  unit-load method, 1-14, 1-30 to 1-33
  virtual-work method, 1-14
  Williot-Mohr method, 1-14, 1-27 to 1-28
Derricks, 8-13
Digital computers (*see* Computers)
Displacement methods of analysis, 1-2, 1-15, 1-56 to 1-71, 1-87 to 1-93
Domes (*see* Shells; Timber domes)
Drop-weight test for brittle fracture, 4-13

Earth pressure:
  active, 5-44
    coefficients, 5-59
  Coulomb's theory, 5-44
  equivalent-fluid method, 5-45, 18-19
  Mohr circle, 5-43
  passive, 5-44
    coefficients, 5-59
  Rankine's theory, 5-44
  retaining walls, 5-47, 5-48
  stresses in earth, 5-43
  surcharge, 5-47
  trial-wedge method, 5-45 to 5-47
Earthquake-resistant design:
  behavior, 3-18
  core walls, 3-22
  cost, 3-26
  curtain-wall buildings, 3-21
  damping, 3-1, 3-20, 3-21
    critical, 3-1
  detailing: reinforced-concrete, 3-25
    steel, 3-25
  El Centro, Calif., earthquake, 3-2 to 3-4, 3-6, 3-8, 3-19
  forced vibrations, 3-1
  foundations, 5-65
  general considerations, 3-18
  lateral force for design, 3-19 to 3-20, 19-18 to 19-22
  Latino-Americana Tower, Mexico City, 3-24
  multi-degree-of-freedom systems, 3-9
    fundamental mode, 3-9, 3-10
  overturning, 3-20
  parts of buildings, 3-22 to 3-23, 19-22

Earthquake-resistant design (*Cont.*):
  period of vibration, 3-10 to 3-18, **19**-18
    fundamental mode, 3-10 to 3-13
    higher modes, 3-13
    modal participation factor, 3-13 to 3-15
    Rayleigh's method, 3-11 to 3-14
    successive approximations, 3-11 to 3-13
  response to earthquake motion, 3-1
  response spectra, 3-3 to 3-9
    amplification factor, 3-5 to 3-7
    arithmetic plots, 3-4
    ductility factor, 3-6, 3-8, 3-19, 3-25
    elastic systems, 3-3
    inelastic systems, 3-6
    logarithmic plot, 3-4
    pseudoacceleration, 3-3
    pseudovelocity, 3-3
  shear beam, 3-9
    spring constant, 3-15
  shear walls, 3-16, 3-22
  soil conditions, 3-23
  spring constant: assemblies, 3-16
    frames, 3-15
    girder flexibility, 3-16
    shear beam, 3-15
    shear walls, 3-16
  torsion, 3-21
  Uniform Building Code, 3-19, 3-20, **19**-18 to **19**-22
Elevated steel tanks:
  air vents, **23**-16
  allowable stresses, **23**-26
  balconies, **23**-12
  Boardman formula, **23**-11
  bottom: conical, **23**-11
    hemispherical, **23**-11
  bracing, **23**-14
  capacity, **23**-9
  columns, **23**-13
  foundations, **23**-14
  heaters, **23**-16
  ladders, **23**-15
  manway, **23**-15
  overflow, **23**-15
  plate materials, **23**-26
  ring girder, **23**-12
  roofs, **23**-9
  tower rods, **23**-14
  tower struts, **23**-14
  welded-joint efficiency, **23**-3
  welding electrodes, **23**-26
  wind pressure, **19**-13, **19**-16
  (*See also* Reservoirs; Shells; Standpipes)
Elevators, **19**-68 to **19**-74
Elliptic paraboloid (*see* Shells)
Energy:
  complementary, 1-4
  strain, 1-4
  theorems: Castigliano, 1-7
    definitions of, 1-3
    least-work principle, 1-7, 1-44
    minimum complementary potential energy, 1-6, 1-15, 1-45, 1-71
    minimum potential energy, 1-6, 1-15, 1-45, 1-71
    unit-displacement, 1-6
    unit-load, 1-6
    virtual displacements, 1-5
    virtual work, 1-4

Equivalent loading, 1-9, 1-57
Erection of structural steel:
  arches, 8-15 to 8-16
  bids, 8-7
  bolting, 8-11
  bridges, 8-15
  buildings, 8-14 to 8-15
  composite construction, 8-8
  construction joints, 8-11
  cost estimates, 8-7
  design for economy, 8-7 to 8-8
  drawings, 8-6
  equipment, 8-13 to 8-14
  erection diagram, 8-2
  expansion joints, 8-11
  falsework, 8-7, 8-16
    allowable stresses, 8-16
    responsibility for, 8-16
  plate girders, 8-9, 8-15
    camber, 8-9
    splices, 8-9
  reaming, 8-11
  schedules, 8-7
  shipping, 8-13
  specifications, 8-6, 8-16
  storage, 8-13
  strain-gage measurements, 8-12
  stress participation, 8-10
  stresses, 8-12
  subpunching, 8-11
  trusses: camber, 8-9, 8-10
    cambered shape, 8-9
    geometric shape, 8-9
    reaming assembled, 8-9, 8-10
    secondary stresses, 8-10
  weighing reactions, 8-11 to 8-12
  welding, 8-11
Escalators, **19**-44, **19**-74
Expansion joints:
  bridges, 8-11, **18**-33, **18**-34
  buildings, 8-5, **19**-46 to **19**-50
Explosion-bulge test for brittle fracture, 4-13

Fabrication of structural steel:
  bending, 8-3
  bolting, 8-4
  bridge assembling, 8-4
  chipping, 8-4
  cutting, 8-3
  drilling, 8-3
  finished parts bill, 8-2
  finishing, 8-4
  fitting, 8-3
  laying out, 8-3
  machining, 8-4
  material bill, 8-2
  material orders, 8-1
  painting, 8-4
  punching, 8-3
  reaming, 8-3
  receiving yard, 8-2
  riveting, 8-4
  shipping, 8-2, 8-5
  shop bill, 8-2
  shop drawings, 8-2
  straightening, 8-3

Fabrication of structural steel (*Cont.*):
  suggestions for designers, 8-5
  template shop, 8-2
  welding, 8-4
Fatigue:
  attachments, effect of, 4-11
  beams and girders, 4-6
  bolted connections: aluminum, 10-27
    structural steel, 4-5
  cumulative damage, 4-4
  designing against, 4-7 to 4-12
  factors affecting strength, 4-3
  fracture surface appearance, 4-1
  Goodman diagram, 4-2
  limit, 4-3
  Miner's hypothesis, 4-4
  residual stress effect, 4-5
  rest periods, 4-5
  riveted connections: aluminum, 10-27
    structural steel, 4-5
  S-N curve, 4-1
  stress-concentration effect, 4-5
  structural steel, 4-1
  temperature effect, 4-5
  welded connections: aluminum, allowable
    stresses, 10-27
    structural steel, 4-6
  Wöhler curve, 4-1
Finite-element method:
  beam bending, 1-103
  coordinate transformation, 1-98
    rotation matrix, 1-108
  direct stress and bending, 1-103
  discretization of structure, 1-96
  displacement model, 1-96
  displacement polynomial, 1-98, 1-106
  element: conforming, 1-100
    definition of, 1-96
    isoparametric, 1-101
    nonconforming, 1-100
    overconforming, 1-100
    plane-strain, 1-100
    plane-stress, 1-100, 1-101
    plate-bending, 1-104
    shell, 1-105
    three-dimensional, 1-105
    types, 1-98
  equilibrium equations, 1-108
  example: hyperbolic paraboloid roof, 1-110
    plate-girder haunch, 1-110
  force model, 1-96
  grid selection, 1-97
    curved boundaries, 1-97
    element shape, 1-97
    mesh layout, 1-97
    numbering system, 1-98
    reducible net, 1-100
    subdivision at discontinuities, 1-97
  hybrid model, 1-96
  incidence table, 1-98, 1-109
  interpolation functiom, 1-98, 1-106
  mixed model, 1-96
  natural coordinates, 1-106, 1-107
  nodal displacement, 1-98
  nodal force vector, 1-103
    condensed, 1-106
  nodal line, 1-96

Finite-element method (*Cont.*):
  nodal plane, 1-96
  node, 1-96
    interior, 1-97
    side, 1-97
  patch test, 1-100
  plane-stress analysis, 1-102
  plate bending, 1-103, 1-104
  reduced integration, 1-102
  shells, 1-103, 1-105
  solution for displacements, 1-109
    for element strains and stresses, 1-109
  stiffness matrix: condensed, 1-108
    element, 1-103
    structure, 1-108
  strain-displacement matrix, 1-106
  stress-strain matrix, 1-106
  three-dimensional problems, 1-103
Fixed-end moment, 1-66, 1-68
Flexibility coefficients, 1-45 to 1-49
Flexibility matrix, 1-71 to 1-74
Folded-plate roof:
  cold-formed steel, 9-6 to 9-8
  concrete, 20-40 to 20-53
    example, 20-44
  prestressed, 20-53
  transverse frame, 20-52
  typical dimensions, 20-41
Footings:
  cantilever, 5-65, 11-42, 11-43, 11-47
  on clay, 5-61
  column, 11-42
  combined, 5-65, 11-44 to 11-46
  construction problems, 5-85
  in earthquake regions, 5-65
  on loess, 5-65
  long, narrow, 5-14
  Newmark influence chart, 5-63
  on sand, 5-64
  settlement: clays, 5-62
    sand, 5-64
  shallow, 5-11 to 5-12
  on silt, 5-65
  spread, 5-61
  square, 5-14
  uplift, 5-83
  wall, 11-41, 11-42
Force methods of analysis, 1-2, 1-15, 1-45 to 1-56, 1-78 to 1-84
Foundations (*see* Footings; Pier foundations; Pile foundations; Raft foundations)
Frames:
  complex, 1-11
  compound, 1-11
  Henneberg's method, 1-13, 1-15
  pin-jointed: deflections, 1-33
    number of members, 1-11
  rigid (*see* Rigid frames)
  simple, 1-11
  (*See also* Plastic design; Tall buildings)
Frost penetration in continental U.S., 5-2
Funicular polygon, 1-18

Geckeler approximation, 20-11, 20-55
Glued-laminated lumber (*see* Timber)
Goodman diagram, 4-2

Guy anchors, 5-81
Guyed towers (*see* Towers, guyed)
Gypsum:
   deck, 19-30
   wall, block, 19-42

Henneberg's method, 1-13, 1-15
Hyperbolic paraboloid:
   cable grid, 21-16
   cold-formed metal, 9-8
   timber, 16-43
   (*See also* Shells)

Industrial buildings:
   bay size, 19-51
   bracing systems, 19-58
   concrete, 19-54 to 19-57
   cranes, 19-61
   framing systems, 19-53
   loading docks, 19-63
   materials handling, 19-61
   planning, 19-50
   pole-type, 19-56
   prefabricated, 19-53
   steel-framed, 19-53 to 19-54
   walls, 19-58
      (*See also* Buildings, walls)
   wood, 19-56
   (*See also* Buildings)
Influence coefficients, 1-40
Influence lines, 1-40
   statically determinate systems, 1-40 to 1-42
   statically indeterminate systems, 1-42 to 1-44

Janssen's formula, 22-4, 23-17

Kern point, 17-4

Levy's formula, 23-20
Light-gage metal (*see* Cold-formed steel structures)
Loading docks, 19-63
Lug angles, 6-5

Masonry:
   admixtures, 15-12
   brick (*see* Brick)
   compressive strength by test, 15-11
   concrete block (*see* Concrete block)
   mortar, 15-3, 15-4
   reinforced (*see* Reinforced masonry)
   structural-clay tile (*see* Structural-clay tile)
   workmanship: bed joints, 15-12
      racking, 15-12
      toothing, 15-12
      wetting units, 15-12
Materials:
   durability of, 19-8
   expansion coefficient, 19-49
   fire resistance, 19-3 to 19-6
   granular, properties, 23-19
   soils (*see* Soil mechanics)

Materials (*Cont.*):
   thermal characteristics, 19-4
   weight, 5-60, 19-9, 19-10
   (*See also specific material*)
Matrix algebra:
   addition, 1-93
   definitions, 1-93
   inversion, 1-95
   multiplication, 1-94
   partitioning, 1-95
   submatrix, 1-95
   transposition, 1-94
Matrix methods of analysis:
   choice of, 1-78
   column vector, 1-71, 1-72
   deflections, 1-14, 1-71 to 1-77
   deformation matrix, 1-73
   displacement method, 1-87 to 1-93
   displacement transformation matrix, 1-87
   flexibility coefficients, 1-72
      addition, 1-72
   flexibility matrix, 1-71 to 1-76
      of unassembled elements, 1-74
   force matrix, 1-73
   force method, 1-78
      choice of redundancy, 1-80
   nonlinear structures, 1-93
   self-straining, 1-83, 1-92
   stiffness coefficients, addition, 1-85
   stiffness matrix, 1-57, 1-84
      beam element, 1-57, 1-87
      condensed, 1-90 to 1-92
      by direct calculation, 1-86
      by inversion, 1-84, 1-95
      local axes, 1-86
      partitioned, 1-90
      structural axes, 1-86
      unassembled elements, 1-87
   unit-load matrix: statically equivalent, 1-74
      true, 1-75, 1-79, 1-88
Maxwell-Mohr method, 1-14, 1-30 to 1-33
Maxwell's theorem, 1-7
Miner's hypothesis, 4-4
Mises yield criterion, 18-117
Mohr circle, 5-6, 5-43 to 5-45
Moment-area method, 1-14
Moment distribution, 1-15, 1-66 to 1-71
   axial force effect, 1-69
   carryover factor, 1-67
   curved members, 17-43 to 17-46
   distribution factor, 1-66
   fixed-end moment, 1-66
   nonprismatic members, 1-69
   sidesway, 1-69
   stiffness coefficient, 1-66
Müller-Breslau principle, 1-6
Multistory buildings (*see* Tall buildings)

Nil-ductility transition, 4-19, 4-20
Nonconservative systems, 1-3
Notch toughness, 4-13
Notched tensile test for brittle fracture, 4-13

Pier foundations:
   allowable bearing pressure: on clay, 5-68
      on sand, 5-68

Pier foundation (*Cont.*):
Chicago method, 5-68
construction problems, 5-85
drilled, 5-68, 5-86
intrusion, 5-68
Gow method, 5-68
open excavation, 5-67
uplift, 5-83
(*See also* Caisson foundations)
Pile drivers:
diesel, 5-72
double-acting hammer, 5-71
single-acting hammer, 5-71
vibratory, 5-72
Pile foundations:
batter piles, 5-80
bearing piles: classification, 5-70
materials, 5-70
in clay, 5-75
construction problems, 5-85
driving: equipment, 5-69 to 5-72
formulas, 5-72 to 5-74
heave, 5-86
lateral load, 5-77 to 5-80
negative skin friction, 5-75
pile load tests, 5-38, 5-74
preexcavation, 5-74
in sand, 5-75
settlement, 5-77
spudding, 5-74
test piles, 5-38, 5-74
tests, 5-74 to 5-75
uplift resistance, 5-84
Pinned connections, 6-75 to 6-77
Plastic design:
analysis: mechanism method, 7-5
moment balancing, 7-12
moment check, 7-7
statical method, 7-5
axial force, 7-24 to 7-25
connections: beam to column, 7-28
corner, 7-27
formulas, 7-28 to 7-31
haunched knee, 7-29
high-strength bolts, 7-31
welded, 7-31
deflection, 1-38, 7-15 to 7-20
frame instability, 7-26, 7-35, to 7-38
instantaneous center, 7-9
lateral bracing, 7-25
local buckling, 7-26
lower-bound theorem, 7-4
mechanism, 7-5
gable, 7-9
plastic hinge, 7-1
plastic moment, 7-1
preliminary design, 7-22
redistribution of moment, 7-3
shape factor, 7-2
shear, 7-26
sidesway, 7-26, 7-35 to 7-38
specifications, 7-20
upper-bound theorem, 7-4
(*See also* Rigid frames)
Plate girders (*see* Aluminum structures; Steel plate girders)

Plates:
aluminum, properties, 10-2, 10-3
buckling: under edge stress, A-5
stiffened webs, A-6
circular: radial displacement, 20-24
symmetrical bending, 20-23
folded (*see* Folded-plate roof)
steel, properties, 6-1 to 6-4, 9-3
(*See also* Aluminum structure; Cold-formed steel structures; Reinforced-concrete slabs; Steel plate girders)
Plywood (*see* Timber)
Pole-type buildings, 5-81, 19-56
Poles:
lateral stability, 5-81
(*See also* Transmission poles)
Ponding on roofs, 16-27
Prestressed concrete:
anchorage, 12-10, 12-32
allowable stress, 12-32
beams (*see* Prestressed-concrete beams)
BBRV system, 12-8
bridges (*see* Prestressed-concrete bridges)
columns, 12-40
concrete strength, 12-2
creep, 12-2, 12-10, 12-28
effective prestress, 12-14
end anchorage, 12-10, 12-32
flexure (*see* Prestressed-concrete beams)
Freyssinet system, 12-8
friction coefficient, 12-14
grouting tendons, 12-5
harping, 12-6
lightweight concrete, 12-3
loss of prestress: creep, 12-10
effective prestress, 12-14
elastic shortening, 12-6, 18-93
friction, 12-13, 18-97 to 18-98
relaxation, 12-12
shrinkage, 12-11
slippage, 12-12
nonprestressed reinforcement, 12-37, 12-38
notation, 12-1
partial prestress, 12-37 to 12-38
posttensioning, 12-6
prestressed and reinforced, 12-13
prestressing bars, 12-9
prestressing systems, 12-7
pretensioning, 12-6
shrinkage, 12-2, 18-20, 18-50
steel: bars, 12-5, 12-9
elongation, 12-14
strands, 12-4
stress-strain curve, 12--3
wire, 12-3
tendons: bonded, 12-5
protection, 12-36
spacing, 12-36
unbonded, 12-5
(*See also* steel, *above*)
tensioning methods, 12-5 to 12-7
VSL system, 12-9
wobble coefficient, 12-13, 18-97
Prestressed-concrete beams:
balanced-load design, 12-25
basic concepts, 12-14 to 12-18
bond, prestress transfer, 12-32

Prestressed-concrete beams (*Cont.*):
  box section, 12-34, 18-92
  cable: concordant, 12-42
    layouts, 12-38, 12-40, 12-41
  camber, 12-28, 12-29, 12-47, 18-99
  cantilever, 12-25, 12-31
  composite section, 12-22, 12-48
  concordant cable, 12-42
  continuous: C line, 12-40 to 12-42
    cable layout, 12-40, 12-41
    concordant cable, 12-42
    load-balancing method, 12-42 to 12-45
    secondary moment, 12-40
    ultimate strength, 12-45
  cracking moment, 12-19, 12-47
  deflection, 12-27 to 12-28
    span-depth ratio, 12-36, 12-37
  elastic design, 12-23
    allowable stresses, 12-24
  end anchorage, 12-10, 12-32
  end block, 12-32
  I, 12-34, 12-35, 18-91
  jacking force, example, 18-97
  nonprestressed reinforcement, 12-38
  partial prestress, 12-37 to 12-38
  preliminary design, 12-22
  principal tension, 12-28
  rectangular, 12-14 to 12-22, 12-25, 12-27
  shear, 12-28
  slab sections, 18-89
  span-depth ratio, 12-36, 12-37
  steel (*see* Prestressed concrete)
  stirrups, 12-30 to 12-32
  T, 12-21, 12-23 to 12-25, 12-45
  typical sections, 12-33 to 12-35, 18-89 to 18-92
  ultimate design, 12-25, 12-45
  ultimate moment, 12-19, 12-48, 18-99
    overreinforced, 12-20
    underreinforced, 12-19
  web reinforcement, 12-30 to 12-32, 12-48
  (*See also* Prestressed concrete: Prestressed-
    concrete bridges)
Prestressed-concrete bridges:
  allowable stresses, 18-93
  beam sections, 18-90 to 18-92
  box sections, 18-92
  deflection, 18-100
  diaphragms, 18-89, 18-91
  end blocks, 12-10, 12-32, 18-91
  example, 18-94
  jacking force, 18-97
  prestressing force: losses, 18-93
    path, 18-96
  slab sections, 18-89
  ultimate load, 18-99
  uplift from prestress, 18-99
  web reinforcement, 18-99
  (*See also* Prestressed concrete; Prestressed
    concrete beams)
Proctor test, 5-19

Raft foundations:
  on clay, 5-67
  on sand, 5-67
Rankine's theory, 5-44
Reciprocal theorem, 1-7

Reimbert's formulas, 22-4
Reinforced concrete:
  codes, 11-2
  creep, 12-2, 12-10, 12-28
  reinforcement (*see* Reinforcing bars)
  shrinkage, 12-2, 18-20, 18-50
  strength design, 11-3
    capacity reduction factors, 11-6
    load factors, 11-3
  working-stress design, 11-3
  (*See also* Concrete; Concrete construction)
Reinforced-concrete beams:
  balanced reinforcement, 11-6
  bars: anchorage, 11-20
    areas of groups, 11-10
    areas in slabs, 11-10
    bend points, 11-24
    cutoff, 11-24
    splices, 11-22
    (*See also* Reinforcing bars)
  continuous, 11-7, 11-11, 11-12
    depth, 11-7
  deflection, 11-25
  diagonal tension, 11-16
  doubly reinforced, 11-2
  load factors, 11-3
  rectangular, 11-6
    flexural strength coefficients, 11-8, 11-9
    width, minimum, 11-10
  shear, 11-16
  special shapes 11-16
  stirrups, 11-16
  tee, 11-13 to 11-16
  (*See also* Reinforced-concrete bridges)
  (*See also* Composite beams; Prestressed-concrete
    beams; Reinforced concrete; Reinforced-
    concrete bridges)
Reinforced-concrete bridges:
  abutments, 5-56, 18-57, 18-69, 18-71, 18-72
  arch: approximate formulas, 17-8 to 17-12, 17-19,
    17-22, 17-29
    Cochrane formulas, 17-12
    deck participation, 17-46, 17-49
    depth, 17-11 to 17-13
    Douglas formula, 17-12
    drainage, 17-49
    expansion joints, 17-49
    hinges, 17-48, 17-50
    influence coefficients, 17-10, 17-12, 17-14, 17-15
    influence lines, 17-16, 17-22 to 17-24
    maximum moments, 17-25 to 17-27
    plastic flow, 17-34
    proportions, 17-11 to 17-13
    shrinkage, 17-16, 17-29, 17-30, 17-33
    stress compensation, 17-50
    Wald formula, 17-12
    Whitney's method, 17-13 to 17-29
    (*See also* Arches)
  bents, 18-54 to 18-55, 18-67 to 18-69, 18-82 to 18-
    88
    column footing, 18-69 to 18-71
    columns, 18-69, 18-86
    pile, 18-54 to 18-55, 18-87
    pile column, 18-87
    single-column, 18-82 to 18-88
      point of fixity, 18-87
    two-column, 18-68 to 18-71

Reinforced-concrete bridges (*Cont.*):
  box-girder, 18-73 to 18-88
    access openings, 18-77, 18-88
    bearings, 18-82
    concrete placing sequence, 18-75
    construction joints, 18-85
    deflection, 18-82
    depth, 18-74
    diaphragms, 18-80
    drainage holes, 18-76
    economical span, 18-73
    horizontally curved, 18-81
    intermediate supports, 18-82 to 18-88
      single-column bent, 18-82 to 18-88
    live-load distribution, 18-79
    manholes, 18-77
    moment of inertia, table, 18-79
    slab: bottom, 18-80
      top, 18-79
    typical details, 18-76 to 18-78, 18-80, 18-83 to
      18-88
    utility openings, 18-77, 18-78
    web, 18-76
      tapered 18-80
    weight, table, 18-81
  camber, 18-50
  composite, 18-37
  (*See also* Composite beams)
  construction joints, 18-70, 18-71, 18-85
  deflection, 14-8, 18-50, 18-82, 18-100
  expansion joints, 18-58, 18-72
  plastic flow, 18-50
  shrinkage, 18-20, 18-50
  slab bridge (continuous) 18-52 to 18-55
    bents, 18-54
      (*See also* bents, *above*)
    economical span, 18-52
    expansion joints, 18-58
    intermediate supports, 18-51
    live load, position, 18-53
    thickness, 18-51, 18-52
    typical details, 18-56 to 18-59
  slab bridge (simple-span), 18-51 to 18-52
    economical span, 18-51
    example, 18-51 to 18-52
    live-load distribution, 18-51
    thickness, 18-51
    typical details, 18-51, 18-53
  T-beam bridge, 18-56 to 18-73
    abutments, 18-69
    bearings, 18-74
    compression reinforcement, 18-60 to 18-62
    concrete-placing sequence, 18-75
    construction joints, 18-70 to 18-71
    depth, 18-58
    diaphragms, 18-67
    economical span, 18-56
    example, 18-63
    expansion details, 18-72, 18-73
    girder spacing, 18-58, 18-59
    intermediate supports, 18-67
      two-column bent, 18-68
      (*See also* bents, *above*)
    slab, 18-63
    stem, 18-59, 18-64
      flared, 18-59
    typical details, 18-68 to 18-74

Reinforced-concrete bridges (*Cont.*):
  temperature range, 18-20
  thermal forces, 18-19
  (*See also* Bridges; Concrete construction;
    Prestressed-concrete bridges; Reinforced
    concrete; Reinforced-concrete beams)
Reinforced-concrete builtlings (*see* Buildings; Tall
  buildings)
Reinforced-concrete chimneys, **26**-17
Reinforced-concrete columns:
  with biaxial bending, 11-26
  combined compression and bending, 11-26
    interaction diagrams, 11-27, 11-29 to 11-34
  effective length, 11-26
  footings, 11-42
  spiral, percent of core volume, 11-40
  splices, 11-22, 11-38
  vertical bars: spirally reinforced, 11-36
    tied, 11-35
  (*See also* Reinforcing bars)
Reinforced-concrete shells (*see* Shells)
Reinforced-concrete silos and bunkers, 22-1, 22-4
  columns, 22-28
  examples, 22-30 to 22-36
  failures, 22-9
  grouped, 22-3, 22-16, 22-17
  hoppers: conical, 22-25
    pyramidal, 22-25
    supporting beams, 22-27
  plane bottoms, 22-24
  pressures (*see* Bins)
  roof, 22-29
  walls, 22-14 to 22-23
    circular silos, 22-14
      eccentric outlet, 22-15
      grouped 22-16, 22-17
      minimum thickness, 22-19
    compression, 22-21
    compression and flexure, 22-21
    crack width, 22-19
    in-plane bending, 22-18, 22-22
    openings, 22-24
    polygonal silos, 22-15, 22-18
    rectangular silos, 22-15, 22-18
    reinforcement, 22-23
    tension, 22-20
    tension and flexure, 22-22
    thermal effects, 22-16
    thermal stresses, 22-22
  (*See also* Bins)
Reinforced-concrete slabs:
  one-way, 19-33
    with pan-formed joists (ribbed), 11-51, 19-28,
      19-33, 19-37
    on rolled beams, 19-36
    slab and beam, 19-28, 19-33
  two-way: flat, 11-50, 11-52, 19-29, 19-34, 19-38
    on edge beams, 11-50, 11-53, 19-33
    flat plate, 11-51, 11-53
    with pan-formed joists (waffle), 11-51, 11-53,
      19-29, 19-33, 19-38
Reinforced-concrete walls, 11-48, 19-40
  compression, 19-40, 22-21
  compression and flexure, 22-21
  footings, 11-41
  tension, 22-20
  tension and flexure, 22-22

Reinforced masonry:
  allowable stresses, 15-6
  beams, 15-5
    bond, 15-6
    deep, 15-8
    shear, 15-6
  columns, 15-10
  compressive strength, 15-11
  detailing, 15-11
  diaphragms, 15-10
  foundations, 15-11
    dowels, 15-11
  grout, 15-5
  joints, 15-12
    effect on net section, 15-5
  lintels, 15-8
  net section, 15-5
  reinforcement, 15-5
    cover, 15-11
    spacing, 15-11
  walls, 15-8 to 15-10
    reinforcement, 15-10
  (See also Brick; Concrete block; Masonry)
Reinforcing bars:
  anchorage, 11-20
  ASTM specification, 11-3
  bar dimensions, 11-2
  beams (see Reinforced-concrete beams)
  bending details, 11-4, 11-5
  bundled, 18-63
  cold-bend test, 11-3
  columns (see Reinforced-concrete columns)
  development length, 11-20, 11-21
  hooks, 11-4
  mechanical properties, 11-3
  splices, 11-22
  welded-wire fabric, 11-1, 11-6
Reservoirs:
  bottom plates, 23-2
  capacity, 23-1
  concrete ringwall, 23-3
  roofs, 23-4
    compression ring, 23-5, 23-6
    framing, 23-4
    rafters, 23-4
    tension ring, 23-6
    torsion tube, 23-5
    truss, 23-5
  shell design, 23-2
  welded-joint efficiency, 23-2
  (See also Shells; Standpipes; Steel tanks)
Retaining walls:
  backfill, 5-49
  base pressure, 5-49
  batter, 5-53
  bearing capacity, 5-52
  cantilever, 5-47
    design, 5-54
    reinforcement, 11-23
  counterfort, 5-47
    design, 5-55
  crib walls, 5-47
    design, 5-53
  drainage, 5-53
  earth pressure (see Earth pressure)
  factors of safety, 5-52

Retaining walls (Cont.):
  gravity, 5-47
    design, 5-53
  joints, 5-55
  overturning, 5-52
  on piles, 5-50
  reinforced earth, 5-48
  semigravity, 5-47
    design, 5-53
  sliding, 5-52
  stability, 5-52
  weep holes, 5-53
Rigid frames:
  arched bent, 17-39, 17-40, 17-43
  column analogy, 1-15, 1-53 to 1-56, 17-44
  concrete: base detail, 17-48
    hinges, 17-48
    placing sequence, 17-49
    proportions, 17-34
  determinacy, 1-12, 1-13
  hinged-base reaction, 17-37 to 17-40
  inflection points, 17-36
  steel: example, 17-36
    haunch detail, 17-42
    knee, 17-52
    proportions, 17-34
  (See also Plastic design)
  (See also Frames)
Rings, 17-2, 17-4 to 17-6
Riveted connections (see Aluminum structures; Bolted connections)
Rivets:
  aluminum, 10-23
  steel, 6-55
Roofs (see Buildings; Folded-plate roof; Shells; Suspension roofs; Timber roofs)

Shells:
  analysis, 20-3 to 20-6
  barrel, 20-24 to 20-40
    buckling, 20-5
    dimensions, 20-31
    edge beam, 20-36 to 20-38
    edge displacements, 20-29, 20-34
    elliptical, 20-27
    example, 20-35
    long, 20-25
    membrane forces, 20-19, 20-26, 20-28, 20-34
    moment of inertia, 20-30
    shallow, 20-31 to 20-37
    short, 20-31
    tapered, 20-25
    timber, 16-37 to 16-43
    transverse frame, 20-27, 20-38
    wind, 20-5
  bending theory, 20-4, 20-11, 20-22
  buckling, 20-4
  circular cylinder (see barrel, above; cylindrical tank, below)
  classification, 20-1
  concentrated loads, 20-3
  concrete roofs, 20-1
    creep, 20-4
  construction, 20-64
  cooling towers, 20-6, 20-24

Shells (*Cont.*):
  cylindrical (*see* barrel, *above*)
  cylindrical tank, 20-19
    bending theory, 20-22
    liquid pressure, 20-20
    loaded by edge forces, 20-22
    seismic loading, 20-20
    variable thickness, 20-20
    wind, 20-5, 20-20
  dimensioning, 20-63
  dome, 20-6 to 20-19
    bending theory, 20-11
    elliptical, 20-6
    examples, 20-13, 20-14
    on wall, 20-18
    (*See also* spherical dome, *below*; Timber domes)
  double curvature, 20-2, 20-53 to 20-63
  dynamic behavior, 20-6
  economy, 20-2
  elliptic paraboloid, 20-54 to 20-57
    edge bending, 20-55
    edge shear, 20-58
    examples, 20-57
    membrane forces, 20-55, 20-56
    reinforcement, 20-57
    stress resultants, 20-56
  elliptical dome, 20-6
  folded plate (*see* Folded-plate roof)
  forms, 20-65
  free-form, 20-1
  Geckeler approximation, 20-11
  hyperbolic paraboloid, 20-57 to 20-63
    cold-formed metal, 9-8
    edge beams, 20-58
    examples, 1-110, 20-61
    parabolic boundaries, 20-63
    reinforcement, 20-61
    ridge beam, 1-113
    saddle, 20-60
    skewed, 20-60
    straight boundaries, 20-59
    timber, 16-37
    umbrella, 20-60 to 20-62
    unsymmetrical load, 20-61
  hyperboloid, wind, 20-5, 20-6
  membrane forces: cylindrical shell, 20-19, 20-26, 20-28, 20-34
    elliptic paraboloid, 20-55, 20-56
    parabolic dome, 20-11
    spherical dome, 20-10
  membrane theory, 20-4
    cylindrical segment, 20-19
    surface of revolution, 20-7
    translation shells of double curvature, 20-53
  parabolic dome, 20-9
  principal stresses, 20-58, 20-64
  reinforcement, 20-63
  single curvature, 20-1, 20-19 to 20-40
  small-deflection theory, 20-3
  spherical dome, 20-8 to 20-19
    design example, 20-13
    dimensions, 20-13
    edge displacements, 20-9
    loaded by edge forces, 20-11

Shells (*Cont.*):
    membrane forces, 20-10
    prestressed ring, 20-17 to 20-18
    reinforcement, 20-17
    ring-stiffened, 20-6, 20-14, 20-17 to 20-18
    temperature effect, 20-14
    variable thickness, 20-9
    (*See also* dome, *above*; Timber domes)
  stability, 20-4
  surfaces of translation, 20-1
  wave-form, 20-1
  (*See also* Suspension roofs)
Signs:
  embedment for support, 5-81
  wind load, 19-17, 19-19
Silos (*see* Bins; Reinforced-concrete silos; Steel bins)
Slope-deflection method, 1-15, 1-64 to 1-66, 1-70
S-N curve, 4-1
Snow load, 19-11 to 19-12
Soil exploration:
  aerial photographs, 5-36
  borehole shear device, 5-34
  borings, 5-28 to 5-33
    auger, 5-28
    borehole camera, 5-32
    dry-sample, 5-28
    number, 5-41
    reports, 5-32
    rock drilling, 5-28
    rotary drilling, 5-29
    wash, 5-28
  core barrels, 5-32
  cost, 5-42
  crystalline rock, 5-27
  depth, 5-42
  earth-resources satellite, 5-36
  fact-finding survey, 5-41
  frozen soils, 5-27
  glacial materials, 5-23
  ground movements, 5-40
  load tests, 5-37
  organic soils, 5-26
  penetration test, 5-4, 5-33
    cone penetrometer, 5-33
    Dutch cone, 5-33
  permeability (*see* Soil mechanics)
  pile tests, 5-38
  piston sampler, 5-31
  plate load test, 5-37
  pressuremeter, 5-34
  residual soils, 5-26
  resistivity surveys, 5-37
  sampling, 5-29
  sedimentary rock: limestone, 5-27
    sandstone, 5-26
    shale, 5-27
  seismic surveys, 5-36
  site inspection, 5-41
  sources of information, 5-23
  split spoon, 5-29
  test pit, 5-28
  tube sampler, 5-30
  vane shear device, 5-34, 5-35
  water-laid materials, 5-25
  wind-laid materials, 5-24
  wireline drilling, 5-32

Soil mechanics:
  angle of internal friction, 5-6, 5-7, 5-60
  Atterberg limits, 5-5, 5-13, 5-21
  bearing pressure, allowable, 5-8
  Boussinesq equation, influence chart, 5-63, 25-11
  Casagrande piezometer, 5-39
  coefficient of permeability, 5-21
  cohesion intercept, 5-7
  compaction, 5-19
    field density test, 5-19
    optimum moisture, 5-19
    Proctor test, 5-19
    typical requirements, 5-20
  consolidated-undrained test, 5-22
  consolidation, 5-12 to 5-16
    coefficient of, 5-16
    test, 5-22
    (See also settlement, below)
  cuts, 5-17
    factor of safety, 5-17, 5-18
  direct-shear test, 5-22
  earth pressure (see Earth pressure)
  embankments, 5-17
  frost movement, 5-1
  frost penetration, 5-2
  grout injection, 5-85
  improving subsoil, 5-84
  laboratory tests, 5-22
  liquid limit, 5-5, 5-21
  Mohr coordinate field, 5-5, 5-6
  optimum moisture content, 5-21
  overburden pressure, 5-6, 5-13
  penetration test, 5-4, 5-33
  permeability, 5-20 to 5-21
    Casagrande piezometer, 5-39
    coefficient of, 5-20, 5-21
    constant-head test, 5-21
    falling-head test, 5-21
    fine-grained soil, 5-39
    granular materials, 5-39
  plastic limit, 5-5, 5-21
  plasticity index, 5-21
  pore-water pressure, 5-7
  porosity, 5-21
  Proctor test, 5-19
  repeated loads, 5-8
  sand drains, 5-84
  saturation, degree of, 5-21
  sensitive soils, definition of, 5-13
  settlement, 5-1 to 5-4
    footing: on clay, 5-62
      on sand, 5-64
    Newmark influence chart, 5-63
    permissible, 5-9 to 5-11
    pile foundations, 5-77
    predicted from plate load test, 5-14
    raft: on clay, 5-67
      on sand, 5-67
    time rate, 5-15, 5-16
    (See also consolidation, above)
  shear failures, 5-11 to 5-12
    factor of safety, 5-12
  shrinkage, 5-2
  shrinkage limit, 5-21
  slope stability, 5-17
  soil deformation, 5-3

Soil mechanics (Cont.):
  soil properties: angle of internal friction, 5-6, 5-7, 5-60
    compression index, 5-13
    compressive strength, 5-4, 5-6
    confining pressure, effect of, 5-5
    elastic-plastic deformation, 5-9
    shearing strength, 5-6
    unconfined compressive strength, 5-5
  subsidence, 5-2
  transient loads, 5-8
  triaxial-compression test, 5-22
  unconfined compression test, 5-22
  vertical cut depth, 5-5
  vibroflotation, 5-85
  void ratio, 5-21
  weight of soil, 5-60
  (See also Footings; Pier foundations; Pile foundations; Raft foundations; Soil exploration)
Stairs:
  angle, 19-39
  concrete, 11-41, 19-44
  escalators, 19-44, 19-74
  framing, 19-74 to 19-77
  layout, 19-43
  moving, 19-44, 19-74
  steel, 19-44
  types, 19-39
Standpipes:
  anchorage, 23-7
  design, 23-6
  foundations, 23-8
  seismic loading, 23-7
  wind loading, 23-7
  (See also Reservoirs; Shells; Steel tanks)
Statically indeterminate structures:
  arches (see Arches)
  Castigliano's theorem, 1-7, 1-14, 1-15, 1-45, 1-51
  choice of method, 1-44 to 1-46, 1-78
  column analogy, 1-15, 1-53 to 1-56, 17-43, 17-44
  displacement methods, 1-45, 1-57 to 1-71, 1-87 to 1-89
  elastic center, 1-15, 17-6
  flexibility coefficients, 1-46 to 1-50
    self-straining, 1-49
  force methods, 1-45 to 1-56, 1-78 to 1-84
  matrix methods, 1-15, 1-78 to 1-93
    (See also Matrix methods of analysis)
  minimum complementary energy, 1-15, 1-45, 1-71
  minimum potential energy, 1-15, 1-45, 1-71
  moment-distribution method, 1-15, 1-66 to 1-71
    (See also Moment distribution)
  nonlinear structures, 1-71, 1-93
  number of redundants, 1-9 to 1-12
  rigid frames (see Rigid frames)
  slope-deflection method, 1-15, 1-64 to 1-66
  stiffness coefficients, 1-56 to 1-57
  three-moment equation, 1-15, 1-52 to 1-53
  unit-displacement method, 1-15, 1-57 to 1-63
  unit-load method, 1-15, 1-50 to 1-51
    (See also Analysis)
Steel:
  fully killed, 4-13
  prestressing (see Prestressed concrete)
  reinforcing (see Reinforcing bars)
  rimmed, 4-13

Steel (*Cont.*):
  semikilled, 4-13
  sheet: corrugated, 19-41
    properties, 9-3
    thicknesses and weights, 9-2
  structural: erection (*see* Erection of structural
      steel)
    fabrication (*see* Fabrication of structural steel)
    maintenance, 8-12 to 8-13
    stress-strain curve, 7-1, 9-2
    types, 6-1 to 6-4
Steel beams:
  allowable stresses, 6-18 to 6-20
  beam-column, 6-23 to 6-25
  bearing plates, 6-78
  bending and compression, 6-23 to 6-25
  biaxial bending, 6-21
  compact section, 6-18
  crane runway, 6-22, 19-61 to 19-63
  deflection, 6-22
  fatigue, 4-6
  government anchor, 6-78
  lateral buckling, 6-18
  plastic design (*see* Plastic design)
  shear, 6-21
  (*See also* Aluminum structures; Beams; Cold-
      formed steel structures; Plastic design; Steel
      plate girders)
Steel bins:
  allowable stresses, 23-26
  Boardman formula, 23-11
  bottom connections, 23-25
  columns, 23-23, 23-26
  compression junction, 23-20, 23-23
  deep: definition of, 23-17
    hoop tension, 23-19
  example, 23-22 to 23-25
  hoop tension, 23-19, 23-22
  Levy's formula, 23-20
  materials, 23-26
  meridional force, 23-19
  openings, 23-26
  pressures (*see* Bins)
  shallow: definition of, 23-17
    hoop tension, 23-22
  stiffeners for concentrated loads, 23-25
  supports, 23-21
    sliding, 23-25
    suspended bottoms, 23-19
  welded joint efficiency, 23-3
  wind pressure, 19-13 to 19-16, 23-23
  (*See also* Bins; Reinforced-concrete silos)
Steel bridges:
  abutments, 5-56
  arch, Rio Blanco, Mexico, 17-53
    (*See also* Arches; Rigid frames)
  beam: composite, 18-37
      (*See also* Composite beams)
    continuous spans, 18-37
    deflection, 18-37, 18-44
    depth, 18-38
    diaphragms, 18-37
    economical span, 18-35
    lateral system, 18-37
    rolled, 18-35
    spacing, 18-37

Steel bridges (*Cont.*):
  bearings: elastomeric pads, 18-31 to 18-32
    pedestal, 18-31, 18-33
    rocker, 18-32
    roller, 18-32
    self-lubricating, 18-32
    shoes, 18-33
    sliding plate, 18-31
  box girder (*see* Steel-plate-deck bridges)
  deck expansion, 18-34
  erection (*see* Erection of structural steel)
  expansion hangers, 18-33
  expansion joints, 18-34
  expansion shoes, 8-12
  fabrication (*see* Fabrication of structural steel)
  fatigue, 18-39
    (*See also* Fatigue)
  floor: battledeck, 18-26
      wearing surface, 18-27, 18-126
    concrete slab, 18-21
    lightweight concrete slab, 18-24
    orthotropic (*see* Steel-plate-deck bridges)
    steel-grid: I-Beam Lok, 18-28
      Irving, 18-28
    structural plate, 18-28
      wearing surfaces, 18-29, 18-126
    transverse slope, 18-21
  floor beams, 18-29
  plate-girder, 18-35 to 18-48
    bearing stiffeners, 18-39, 18-45
    camber, 8-9
    composite, 18-37
      (*See also* Composite beams)
    cross frames, 18-37, 18-47
    deflection, 18-37, 18-44
    depth, 18-38
    economical span, 18-36
    field splice, 18-39, 18-48
    flange splice, 18-38
    lateral system, 18-37, 18-46, 18-47
    longitudinal stiffeners, 18-39, 18-45
    spacing, 18-37
    stiffeners, 18-39, 18-45, 18-46
    transverse stiffeners, 18-39, 18-45
    web splice, 18-38, 18-47
    welded, 18-38 to 18-37
      example, 18-39 to 18-47
    (*See also* Steel plate girders)
  temperature range, 18-20
  thermal forces, 18-19
  thermal movement, 18-31
  truss, 18-48 to 18-50
    Benicia-Martinez Bridge, 18-49, 18-50
    camber, 8-9, 8-10, 18-48
    combining grades of steel, 18-49
    influence lines, 18-48
    lateral forces, 18-50
    members, 18-48 to 18-50
    proportions, 18-48
    secondary stresses, 18-48
    stresses, 18-48
    walkways, 8-12
    (*See also* Bridges; Erection of structural steel;
        Fabrication of structural steel; Steel-plate-
        deck bridges)

Steel buildings (*see* Buildings; Industrial buildings; Tall buildings)
Steel chimneys (*see* Chimneys, steel)
Steel compression members:
  allowable stress, 6-7, 6-8
  amplification factor, 6-9 to 6-11, 6-25 to 6-29
  beam-column, 6-23 to 6-25
  columns: bases, 6-78 to 6-81
    crane-girder, 19-61 to 19-63
    leveling, 6-79
    splices, 6-81 to 6-87, 19-68
  effective length, 6-8, A-4
  lacing, 6-14
  local buckling, 6-12
  perforated cover plates, 6-14 to 6-15
  proportioning, 6-12
  radii of gyration, table, 6-13
  tapered, 6-16
  torsional buckling, 9-16
  (*See also* Aluminum structures; Cold-formed steel structures; Plastic design)
Steel connections:
  pinned, 6-75 to 6-77
  rigid, semirigid, 6-50, 19-88
  (*See also* Bolted connections; Plastic design; Riveted connections; Welded connections)
Steel-plate-deck bridges:
  box girders, 18-115 to 18-123, 18-126
    analysis, 18-115
    erection, 18-126
    fatigue, 18-117
    linearly elastic analysis of plates, 18-116
    load-bearing diaphragms, 18-123
    nonlinear analysis of plates, 18-118 to 18-122
      effect of imperfections, 18-118
      residual stresses, 18-119
      shear, 18-121
      stiffened, in compression, 18-120
      unstiffened, in compression, 18-120
  closed-rib deck, 18-112 to 18-115
    deck-structure moments, 18-114
    effective width of plate, 18-112
    intermediate diaphragms, 18-112
    ribs: criteria, 18-109
      spacing, 18-112
      types, 18-112
    substitute orthotropic plate, 18-113
      bending moments, 18-114
      flexural rigidity, 18-113
      torsional rigidity, 18-113
    typical details, 18-110
  deck plate, 18-107
  economical span, 18-106
  economy, 18-104
  fabrication: deck panels, 18-124
    floor beams, 18-128
    ribs, 18-123
  field splices, 18-124
  movable bridges, 18-104
  open-rib deck, 18-109 to 18-112
    floor-beam design, 18-111
    ribs: criteria, 18-109
      design, 18-109
    typical details, 18-110
  railroad bridges, 18-127
  sidewalks, 18-124

Steel-plate-deck bridges (*Cont.*):
  structural behavior, 18-106
  wearing surfaces, 18-126, 18-127
  (*See also* Bridges; Steel bridges)
Steel plate girders:
  bearing stiffeners, 6-37
  example, 6-34 to 6-37
  fatigue, 4-6, 18-39
  flange, 6-32
  lateral buckling, 6-32
  longitudinal stiffener, 6-32, 6-34, A-6
  shear and bending, combined, 6-34
  tension-field web, 6-30, 6-33
  vertical stiffeners, 6-32, 6-33, 6-35 to 6-36, A-6
  web, 6-30, A-5, A-6
  (*See also* Aluminum structures; Steel beams; Steel bridges)
Steel stacks (*see* Chimneys)
Steel tanks (*see* Elevated steel tanks)
Steel tension members:
  allowable stress, 6-3
  eccentric load, 6-5
  effective area, 6-3
  lug angles, 6-5
  net section, 6-3
  shear lag, 6-3
  slenderness ratio, 6-6
  truss, 6-5
  types, 6-5
Stiff-jointed frames (*see* Rigid frames)
Stiffness coefficients, 1-56, 1-66
Stiffness matrix, 1-57, 1-84
String polygon, 1-18
Structural clay tile:
  ceramic glazed, 15-2
  dimensions, 15-3
  facing, 15-2
  floor, 15-2
  load-bearing, 15-2
  non-load-bearing, 15-2
  physical requirements, 15-3
Suspension roofs:
  cable design: anchorage, 21-3
    bridge strand, strength, 21-18
    dynamic behavior, 21-5, 21-6
      damping, 21-6
      frequency, 21-6
    flutter, 21-5
    sag, 21-6
    static behavior, 21-5
    tension, 21-6
    uniformly distributed load, 21-5
  catenary, 21-7
    preloaded, 21-8
  damped suspension system, 21-9
  double layer of prestressed cables, 21-8 to 21-15
    analysis, 21-11
    behavior, 21-13
    example, 21-14
    natural frequency, 21-13
    preliminary design, 21-11
  economy, 21-1
  fittings, 21-17, 21-19, 21-20
  flutter, 21-5
  galvanized bridge wire, 21-17
  grids, 21-8

Suspension roofs (*Cont.*):
 hyperbolic paraboloid, 21-16
 membrane systems, 21-2, 21-17, 21-21, 21-22
 New York State Pavilion, 1964–1965 World's
  Fair, 21-2, 21-3
 overturning, 19-68
 preloaded catenary, 21-8
 Salt Lake Civic County Auditorium, Utah, 21-2,
  21-4
 Traveler's Insurance Company Pavilion,
  1964–1965 World's Fair, 21-2, 21-5
 Utica Municipal Auditorium, Utica, New York,
  21-1, 21-2
 Valley Curtain, 21-2, 21-7

Tall buildings:
 bay framing, 19-63 to 19-67
  flat-plate, 19-78, 19-84
  moderate-sized bays, 19-64
   1120 Avenue of the Americas, New York
    City, 19-65
  large bays, 19-64
   American Bible Society Building, New York
    City, 19-67
   Chase Manhattan Bank Building, New York
    City, 19-66, 19-94 to 19-98
 bay sizes, 19-63
 columns, 19-64
  Commerce Court Building, Toronto, Canada,
   19-67, 19-68
  One Liberty Plaza, New York City, 19-66, 19-
   68
  splices, 6-81 to 6-87, 10-68
  (*See also* Reinforced-concrete columns; Steel
   compression members)
 earthquake design (*see* Earthquake-resistant
  design)
 elevator system, 19-68 to 19-74
 erection (steel), 8-14 to 8-15
 stairs, 19-74 to 19-76
  (*See also* Stairs)
 transfer girders, 19-76
  camber, 19-77
  Union Carbide Building, New York City, 19-78
 wind bracing, 19-77 to 19-88
  analysis, 19-97 to 19-108
   cantilever method, 19-103, 19-107 to 19-108
   computer methods, 19-105
   dynamic, 19-105
   portal method, 19-99 to 19-106
  Chase Manhattan Bank Building, New York
   City, 19-98
  drift, 19-90
   deflection index, 19-91 to 19-93
  Equitable Life Assurance Building, New York
   City, 19-99, 19-100
  high-rise buildings, 19-84
  joint fixity, 6-50, 6-51, 19-88
  K-brace, 19-77, 19-79, 19-80
  knee brace, 19-79, 19-81
  medium-rise buildings, 19-77
  rigid-frame, 19-79, 19-83
  shear walls, 19-80, 19-85 to 19-87, 19-101 to
   19-103
   American Life Insurance Building,
    Wilmington, Del., 19-88

Tall buildings, wind bracing (*Cont.*):
  tubular frames, 19-84, 19-90
   with interior shear walls, 19-105, 19-109, 19-
    110
  tuning, 19-93 to 19-96
  X-brace, 19-78, 19-79, 19-82
  wind-load determination, 19-88, 19-89
  wind-shear dissipation, 19-97
  torsional moments, 19-97
Tank:
 cylindrical, 20-19 to 20-24
 wind pressure, 19-13 to 19-16, 20-20
 (*See also* Elevated steel tanks; Reinforced-
  concrete silos; Reservoirs; Shells; Standpipes;
  Steel bins)
Tear test for brittle fracture, 4-13
Temperature, lowest 1-day mean in continental
 U.S., 23-18
Tension coefficients, 1-13 to 1-15
Tension members (*see* Aluminum structures; Cold-
 formed steel structures; Steel tension members;
 Timber)
Thermal properties:
 expansion coefficients, 19-49
 fire-resistance requirements, 19-3, 19-7
Thin shells (*see* Shells)
Three-moment equation, 1-15, 1-52 to 1-53
Tile:
 ceramic, 15-2
 clay, 15-2
 floor, 15-2
Timber:
 allowable stresses; glued-laminated softwood, 16-
  14 to 16-15
  plywood, 16-17
  stress-grade lumber, 16-4 to 16-10, 16-12 to 16-
   13
 anisotropic nature, 16-1
 compression at angle to grain, 16-11
 compressive strength, 16-2
 creep, 16-1
 decay, 16-3
 deflection, 16-3
 density, 16-2
 duration of load, 16-3
 elastic constants, 16-1
 endurance limit, 16-3
  of glued-laminated, 16-13
 fasteners (*see* Timber fasteners)
 glued-laminated: allowable stresses, 16-14 to 16-
  15
  endurance limit, 16-13
  moisture content, 16-11
  scarf-joint efficiency, 16-11
  widths, 16-11
 impact strength, 16-13
 knots, 16-2
 modulus of elasticity, 16-1, 16-4 to 16-8, 16-12
 moisture content, effect on strength, 16-2, 16-3
 plywood: allowable stresses, 16-17
  grades, 16-16
  section properties, 16-18 to 16-21
 scarf joints, 16-11
 section properties, American standard, 16-26
 shear strength, 16-2

Timber (*Cont.*):
  spring wood, 16-1
  summer wood, 16-1
  tensile strength, 16-2
  time-load effects, 16-3
  working stresses (*see* allowable stresses *above*)
Timber arches:
  buttressed, 16-37, 16-41
  foundation, 16-37, 16-41
  load combinations, 16-34
  three-hinged Tudor: base section, 16-36
    bracing, 16-37
    deflection, 16-37
    haunch section, 16-38 to 16-40
  tied, 16-37, 16-41
  two-hinged, 16-37
  types, 16-36
    circular, tables, 17-10
Timber barrel vault, 16-43
  (*See also* Shells, barrel)
Timber beam column, 16-33
Timber beams:
  bearing stress, 16-27
  bending stress, 16-4 to 16-12, 16-14, 16-15, 16-17,
    16-25
    form factor, 16-25
  camber in flat roofs, 16-28
  cantilevered systems, 16-28 to 16-30
  continuous spans, 16-28 to 16-30
  curved, radial stresses, 16-13
  deflection, 16-27
  in floors, 19-31, 19-32
  lateral stability, 16-28
    rules for support, 16-28
  notched, 16-27
  pitched, 16-30
  in roofs, 19-26 to 19-27
  section properties, 16-26
  shear stress, 16-4 to 16-12, 16-14, 16-15, 16-17,
    16-26, 16-27, 16-31
    concentrated load, 16-27
    notched beam, 16-27
  span limit, 16-25
  tapered, 16-30
Timber columns:
  box, 16-32
  solid, 16-31
  spaced, 16-32
Timber decks:
  floors, 19-31
  roofs, 19-26
Timber domes:
  radial-rib, 16-37, 16-41 to 16-43
    compression ring, 16-42
    diagonal rods, 16-42
    tension ring, 16-42
  shapes, 16-37
  span limit, 16-37
  (*See also* Shells, dome)
Timber fasteners:
  bolts: allowable shear, 16-22, 16-24
    duration of load, 16-23
    edge distance, 16-23
    moisture effect, 16-22
    spacing, 16-23
    in unseasoned lumber, 16-22

Timber fasteners (*Cont.*):
  shear plate, 16-23
  split ring, 16-23
  truss plate, 16-23
Timber hyperbolic paraboloid, 16-43, 16-44
  (*See also* Shells, hyperbolic paraboloid)
Timber roofs, 19-26, 19-27
  camber, 16-28, 16-35
  load combinations, 16-34
  ponding, 16-27
  trussed joists, 16-36
  (*See also* Timber arches; Timber domes)
Timber trusses:
  bracing, 16-35
  camber, 16-28, 16-35
  deflection, 16-27, 16-35
  depth, 16-33
  design of members, 16-33 to 16-35
  load combinations, 16-34
  spacing, 16-33
  span limit, 16-33
  splices, 16-35
  types, 16-33
Torsional buckling, 9-18, 10-6
Torsional properties:
  solid cross section, A-1
  thin-walled section: closed, A-2
    open, A-3
Towers:
  aeronautical obstruction lighting, 24-2
  aluminum, 24-2
  antenna, 24-3
  candelabra, 24-2
  compression member, allowable stress, 24-21, 24-
    22
  free-standing: analysis, 24-2, 24-3
    foundations, 24-3
  guyed, 24-3 to 24-12
    bridge strand properties, 21-18
    erection, 24-12
    foundation anchorage, 5-81
    foundations, 24-12
    galvanized bridge wire, 21-17
    guy design, 24-4 to 24-8
      ice load, 24-8
    initial tension, 24-9
    mast design, 24-9 to 24-12
    proof-loaded cables, 24-4, 24-9
    temperature forces, 24-7
    tensioning, 24-9
    vibration of guys, 24-9
    vibration dampers, 24-9
    wind loads, 19-13 to 19-19, 24-4
  height limitations, 24-2
  loads, 24-2
  materials, 24-1
  steel, 24-1
  tension member (single-angle) allowable load, 24-
    20
  tests, 24-25
  transmission: allowable stresses: aluminum, 24-18
    steel, 24-15
    analysis, 24-15 to 24-17
    anchor tower, 24-13
    angle tower, 24-13
    dead-end tower, 24-13

Towers, transmission (*Cont.*):
    foundations: concrete caisson, 24-23, 24-24
        concrete footing, 24-23, 24-24
        grillage, 24-23, 24-24
        pile, 24-23, 24-24
        rock anchor, 24-25
    galloping of conductors, 24-15
    loading zones, 24-13, 24-14
    loads, 24-13
    overload factors, 24-14
    slenderness-ratio limits of members, 24-23
    strain tower, 24-13
    tangent tower, 24-13
    types, 24-12
    vibration, 24-15
    types, 24-1
    wind load, 19-13 to 19-19, 24-13
Transition temperature, 4-13
Transmission poles:
    axial compression, 24-26
    bending and compression, 24-26
    design, 24-26 to 24-28
    foundations, 24-28
    local buckling, 24-27
    material, 24-28
    splices, 24-28
    (*See also* Towers)
Transmission towers (*see* Towers, transmission)
Travelers (erection), 8-14
Trusses:
    analysis (*see* Analysis)
    bridge, 18-48 to 18-50
    building, 19-53 to 19-54
    deflection (*see* Deflection)
    timber (*see* Timber trusses)

Unit-displacement method, 1-15, 1-57 to 1-62
Unit-load method, 1-14, 1-15, 1-30 to 1-33, 1-50 to
    1-51

Vibration:
    in building applications, 14-8, 19-3
    of chimneys, 26-6, 26-14 to 26-17
    of guy wires, 24-9
    of transmission-line conductors, 24-15
Virtual displacements principle, 1-5
    (*See also* Deflection)
Virtual work principle, 1-4, 1-44, 7-5
    (*See also* Deflection)

Walls (*see* Buildings; Reinforced-concrete walls;
    Reinforced masonry; Retaining walls)
Welded connections:
    beam: to column, 7-28
        framed, 6-48 to 6-50
        moment-resistant, 6-50 to 6-55
        seat, 6-45 to 6-48
    butt, 6-43, 7-31
    column splices, 6-82, 6-84 to 6-87
    end returns, 6-43
    fatigue (*see* Fatigue)
    fillet, 6-42, 7-31
    groove, 6-43
    haunch: curved, 7-29, 17-50, 17-51
        tapered, 7-29, 17-42, 17-49
    (*See also* Aluminum structures; Plastic design)
Welding:
    electrodes, steel, 6-40
    filler wire, aluminum, 10-26
    fillet, 6-39, 10-25
    GMA, 10-25
    groove, 6-39
    GTA, 10-25
    inspection, 6-41
    Mig, 6-38
    plug, 6-39
    processes, 6-37 to 6-39
    puddle, 9-27
    slot, 6-39
    spot, 9-27
    weld types, 6-39
    weldability, 6-40
Wide-plate test for brittle facture, 4-13
Williot-Mohr method, 1-14, 1-27, 1-28
Wind bracing, buildings (*see* Tall buildings)
Wind pressure:
    on bridges, 18-18
    on buildings, 19-13 to 19-17
    on chimneys, 26-1 to 26-3
    in continental U.S., 19-17
    on roofs, 19-16
    on signs, 19-19
    on tanks, 19-16, 23-7
Wire fabric, welded, 11-1, 11-6
Wire rope:
    bridge: galvanized, 21-17
        strand, 21-18
Wohler curve, 4-1
Wood:
    decks, 19-23, 19-26, 19-31, 19-32
    pole-type buildings, 19-58
    walls, 19-41, 19-42
    (*See also* Timber; Timber arches)
Wood beams (*see* Timber beams)
Wood trusses (*see* Timber trusses)